AI and Data-Driven Advancements in Industry 4.0—Volume I

AI and Data-Driven Advancements in Industry 4.0—Volume I

Topic Editors

Teng Huang
Qiong Wang
Yan Pang

MDPI

Basel • Beijing • Wuhan • Barcelona • Belgrade • Novi Sad • Cluj • Manchester

Topic Editors
Teng Huang
Institute of Artificial
Intelligence and Blockchain
Guangzhou University
Guangzhou
China

Qiong Wang
Shenzhen Institute of
Advanced Technology
Chinese Academy of Sciences
Shenzhen
China

Yan Pang
Shenzhen Institute of
Advanced Technology
Chinese Academy of Sciences
Shenzhen
China

Editorial Office
MDPI AG
Grosspeteranlage 5
4052 Basel, Switzerland

This is a reprint of the Topic, published open access by the journals *Remote Sensing* (ISSN 2072-4292), *Cancers* (ISSN 2079-9292), *Mathematics* (ISSN 2227-7390), *Sensors* (ISSN 1424-8220), and *Electronics* (ISSN 2072-4292) , freely accessible at: https://www.mdpi.com/topics/FLN51J9SH1.

For citation purposes, cite each article independently as indicated on the article page online and as indicated below:

Lastname, A.A.; Lastname, B.B. Article Title. *Journal Name* **Year**, *Volume Number*, Page Range.

Volume I
ISBN 978-3-7258-4165-3 (Hbk)
ISBN 978-3-7258-4166-0 (PDF)
https://doi.org/10.3390/books978-3-7258-4166-0

Set
ISBN 978-3-7258-4223-0 (Hbk)
ISBN 978-3-7258-4224-7 (PDF)

Contents

sensors

MDPI

Editorial

AI and Data-Driven Advancements in Industry 4.0

Yan Pang [1,*], Teng Huang [2] and Qiong Wang [1]

1 Shenzhen Institute of Advanced Technology, Chinese Academy of Sciences, Shenzhen 518055, China; wangqiong@siat.ac.cn
2 School of Artificial Intelligence, Guangzhou University, Guangzhou 510700, China; huangteng1220@gzhu.edu.cn
* Correspondence: yanpang@gzhu.edu.cn

1. AI in Industry

Industrial artificial intelligence is rapidly evolving, driven by an unprecedented explosion of diverse data modalities. In today's smart factories and Industry 4.0 environments, vast streams of data—from machine logs and sensor readings to high-resolution imagery and acoustic signals—are fueling a new wave of data-centric innovation [1]. Organizations now collect heterogeneous datasets—ranging from natural language documents [2] and medical images [3] to sensor readings [4] and satellite imagery—to monitor, optimize, and transform industrial processes on an unprecedented scale.

This surge in data availability has sparked a fundamental shift toward a data-centric AI paradigm, where enhancing data quality and coverage becomes the primary driver of performance Zha et al. [5]. Unlike traditional model-centric approaches, data-centric strategies focus on meticulously curating multi-modal datasets, refining annotations, and ensuring that the data reflect the complexity of real-world industrial environments, thereby bolstering model learning.

Recent breakthroughs in deep learning are further accelerated by the rich, multi-sensor data now accessible in industrial settings—spanning high-resolution RGB images, LiDAR scans, hyperspectral imagery, and audio streams captured from various production sources Paheding et al. [6]. As a result, modern AI systems are beginning to leverage this wealth of multi-modal information to achieve superior generalization and even exhibit emergent capabilities once thought beyond reach. For example, large-scale multimodal models that integrate vision and language are now capable of generating detailed narratives about production processes or solving complex operational tasks without explicit instruction Baltrušaitis et al. [7]. In parallel, advancements in natural language processing (NLP), driven by scaling both models and data in large language models (LLMs), are unlocking novel behaviors, such as few-shot reasoning and chain-of-thought problem solving, which are increasingly relevant to industrial applications Xu et al. [8]. Contribution 1 introduces an innovative intelligent sensor software that combines robust machine learning with interactive heat map visualization to optimize resistance spot welding for steel reinforcement. Contribution 2 presents an artificial neural network (ANN) model that accurately predicts process performance. Contribution 3 introduces an innovative U-shaped network that integrates a Space-To-Depth module and a refined attention mechanism to effectively address class imbalance and achieve precise segmentation of micro and fine scratch defects on metal surfaces. These developments highlight how the diversity and volume of data are transforming AI techniques, establishing a robust foundation for the impactful deployment of industrial artificial intelligence across modern manufacturing and beyond.

The convergence of powerful AI methods with Industry 4.0 is driving transformative changes across a wide range of application domains Rai et al. [9]. Computer vision and NLP algorithms, in particular, are now deeply embedded in sectors such as healthcare Hirsch et al. [10], manufacturing Ahmad and Rahimi [11], and cybersecurity Zolanvari et al. [12]. In healthcare, deep learning models can analyze medical images (e.g., radiographs, MRI scans, pathology slides) with accuracy approaching or even exceeding expert human performance Shen et al. [13], thus aiding in diagnostics and treatment planning. Likewise, NLP techniques are used to mine clinical texts and biomedical literature Juhn and Liu [14], enabling predictive analytics and decision support in clinical workflows. These developments reflect a broader trend in which deep learning has fundamentally reshaped how we tackle tasks in vision, language, and healthcare analytics, among other fields (Paheding et al. [6], Abramson et al. [15]).

1.1. AI in Robotic

In recent years, the integration of artificial intelligence with robotics has led to transformative advancements in robotic perception, planning, decision making, and control Du and Yu [16], Sünderhauf et al. [17]. In medical applications, for instance, the fusion of imaging, localization, and magnetic actuation systems has enabled highly precise in vivo operations with miniature robots Aziz et al. [18]. In industrial and service contexts, visual and multimodal sensor fusion have significantly enhanced the robustness of grasping, navigation, and human–robot collaboration tasks Mao et al. [19]. Moreover, inspired by insights from neuroscience, brain-inspired robotics is exploring novel paradigms that emulate biological cognitive processes, thereby facilitating end-to-end autonomous learning and real-time adaptive control Qiao et al. [20]. These developments collectively represent a shift away from traditional, hand-coded control schemes toward intelligent systems that are capable of operating in complex, unstructured, and dynamic environments Hayes-Roth and Thorndyke [21]. Contribution 4 offers a comprehensive narrative review that elucidates machine learning strategies for the development of remotely monitored central nervous system biomarkers using wearable sensors. Contribution 5 presents an innovative task selection and allocation scheme for mobile crowdsensing that integrates Lyapunov optimization with a Double Deep Q-Network framework to dynamically stabilize task queues and optimize resource allocation under constrained conditions. Contribution 6 exemplifies the cutting-edge integration of multi-view imaging and advanced keypoint detection techniques to enhance robot arm monitoring in intelligent industrial environments.

1.2. AI in Medicine

The integration of artificial intelligence (AI) into healthcare within the framework of Industry 4.0 has revolutionized traditional medical practices, enabling data-driven decision making and personalized patient care. By leveraging advanced technologies such as machine learning and predictive analytics, AI facilitates the interpretation of complex medical data, including imaging diagnostics, where it enhances accuracy in identifying patterns in X-rays and MRI scans. This transformation aligns with the vision of Health 4.0, which emphasizes decentralized, AI-enabled home healthcare services, reducing reliance on conventional hospital-centric models. However, scaling AI applications in medical device manufacturing and clinical workflows remains challenging due to interoperability issues, regulatory hurdles, and the need for robust data governance frameworks. These challenges highlight the critical role of Industry 4.0 technologies in optimizing healthcare systems through digitization, automation, and real-time data exploitation.

The integration of artificial intelligence (AI) into healthcare within the framework of Industry 4.0 has revolutionized traditional medical practices, enabling data-driven decision making and personalized patient care Topol [22]. By leveraging advanced technologies, such as machine learning and predictive analytics, AI facilitates the interpretation of complex medical data, including imaging diagnostics, where it enhances accuracy in identifying patterns in X-rays and MRI scans Esteva et al. [23]. This transformation aligns with the vision of Health 4.0, which emphasizes decentralized, AI-enabled home healthcare services, reducing reliance on conventional hospital-centric models Keesara et al. [24]. However, scaling AI applications in medical device manufacturing and clinical workflows remains challenging due to interoperability issues, regulatory hurdles, and the need for robust data governance frameworks Rajkomar et al. [25]. Contribution 7 exemplifies a rigorous methodological advancement by integrating multi-scale feature extraction and channel selection to significantly enhance EEG-based Parkinson's disease classification performance. Contribution 8 highlights the growing role of machine learning in oncology, with a focus on medical image analysis, treatment planning, and patient prognosis. Contribution 9 leverages a novel cross-scale attention mechanism alongside a progressive edge refinement module to enhance retinal vessel segmentation performance. These challenges highlight the critical role of Industry 4.0 technologies in optimizing healthcare systems through digitization, automation, and real-time data exploitation.

1.3. AI in Blockchain

In parallel, blockchain technology is intersecting with AI to enhance trust, security, and data integrity in Industry 4.0 applications Rahman et al. [26]. Blockchain (distributed ledger) systems provide tamper-resistant record keeping and decentralized data governance, which complement AI's data-driven decision making. The amalgamation of AI and blockchain is recognized as a disruptive force of the Fourth Industrial Revolution, holding tremendous potential to create new business models and ecosystems Kumar et al. [27]. For instance, in industrial supply chains, blockchain can verify the provenance and authenticity of goods while AI monitors quality and demand forecasts Liu et al. [28]; in healthcare, blockchain enables the secure sharing of medical data or model parameters across institutions, while AI algorithms analyze those data for insights. Similarly, in finance and digital asset management, blockchain-based smart contracts can work in tandem with AI models to automate decision logic in a transparent, fraud-resistant manner Bathula et al. [29]. Early studies show integrated AI-blockchain platforms being explored for secure IoT networks, intellectual property protection for AI models and data, and federated learning setups where blockchain logs and verifies each update Issa et al. [30]. Contribution 10 combines Groth16 zero-knowledge proofs, blockchain technology, smart contracts, and IPFS to ensure secure. Contribution 11 introduces a novel deep learning and multimodal decision fusion approach that integrates source code, opcode, and control flow information to achieve high detection accuracy for smart contract vulnerabilities. This convergence of AI with decentralized technologies underscores the wide deployment of next-generation AI methods well beyond conventional settings, reaching into cyber-physical systems, smart cities, and global digital ecosystems that define Industry 4.0.

2. Trustworthy AI

As AI becomes deeply embedded in Industry 4.0 systems, ensuring that models are trustworthy, robust, and secure is of paramount importance. Real-world industrial AI applications (whether in autonomous driving, medical diagnosis, or power grid management) demand high reliability and resilience to adversities. However, recent research has revealed that AI models can be vulnerable to a variety of attacks and failure modes Hu et al. [31]. For example, adversarial attacks involve subtle perturbations to inputs that cause misclassification, data poisoning injects malicious data into training sets to subvert model behavior, and model inversion or extraction attacks can leak sensitive information from a model's learned parameters. Such risks pose serious concerns in safety-critical environments. A malfunctioning vision system in a robot or an altered prediction in a clinical AI tool could have dire consequences. Therefore, the development of secure and robust AI models has become a critical research focus. Methods for adversarial defense, robust training (e.g., data augmentation, ensemble methods), and anomaly detection are being actively investigated to harden AI models against these threats. Equally crucial is safeguarding data privacy: Industry 4.0 AI often operates on sensitive personal or proprietary data. Therefore, techniques such as federated learning, differential privacy, and secure multi-party computation are being employed to protect data while still enabling collaborative learning.

In tandem with the robustness, trust, and transparency of AI systems, these systems are also receiving increasing attention. Stakeholders must be confident that AI-driven decisions or predictions can be understood and justified—especially in domains such as healthcare and manufacturing where accountability is essential Murdoch et al. [32]. Efforts in explainable AI (XAI) aim to highlight the reasoning behind a model's output, whether through interpretable model architectures or post hoc explanation tools. Moreover, governance frameworks for Trustworthy AI have emerged Rudin [33], advocating principles such as fairness, accountability, transparency, and ethics in AI system design. For instance, regulators and industry consortia are beginning to recommend or require that AI systems provide audit trails of how data are used and how decisions are made Gunning et al. [34]. In manufacturing, this might mean that an AI quality inspection system can highlight which product features led to a rejection; in smart healthcare, an AI diagnostic assistant should explain its recommendation in terms a clinician can validate. Incorporating these principles mitigates the "black box" nature of many deep learning models and increases user confidence. In summary, achieving trusted AI in Industry 4.0 involves a multi-faceted approach: technical robustness against attacks and failures, privacy-preserving data practices, and transparency and fairness measures. Addressing these challenges is vital for the long-term sustainability and societal acceptance of AI-driven industrial systems.

3. Conclusions

In summary, this Editorial, titled "AI and Data-Driven Advancements in Industry 4.0", has showcased a diverse array of studies that collectively underscore the pivotal role of AI and multi-modal data in modern industrial applications. Industrial automation, robotics, medical diagnostics, and blockchain integration are among the key areas explored, illustrating how advanced data-centric strategies and deep learning techniques are being harnessed to address complex challenges. These works exemplify how cutting-edge AI approaches not only enhance theoretical insights but also drive practical innovations in the evolving landscape of Industry 4.0.

Conflicts of Interest: The authors declare no conflicts of interest.

List of Contributions:

1. Ferreiro-Cabello, J.; Martinez-de Pison, F.J.; Fraile-Garcia, E.; Pernia-Espinoza, A.; Divasón, J. Intelligent Sensor Software for Robust and Energy-Sustainable Decision-Making in Welding of Steel Reinforcement for Concrete. *Sensors* **2024**, *25*, 28.
2. El-Asfoury, M.S.; Baraya, M.; El Shrief, E.; Abdelgawad, K.; Sultan, M.; Abass, A. AI-Based Prediction of Ultrasonic Vibration-Assisted Milling Performance. *Sensors* **2024**, *24*, 5509.
3. Liu, Y.; Qin, Y.; Lin, Z.; Xia, H.; Wang, C. Detection of Scratch Defects on Metal Surfaces Based on MSDD-UNet. *Electronics* **2024**, *13*, 3241.

4. ZhuParris, A.; de Goede, A.A.; Yocarini, I.E.; Kraaij, W.; Groeneveld, G.J.; Doll, R.J. Machine learning techniques for developing remotely monitored central nervous system biomarkers using wearable sensors: A narrative literature review. *Sensors* **2023**, *23*, 5243.
5. Chang, S.; Wu, Y.; Deng, S.; Ma, W.; Zhou, H. Task-Importance-Oriented Task Selection and Allocation Scheme for Mobile Crowdsensing. *Mathematics* **2024**, *12*, 2471.
6. Zhou, L.; Wang, R.; Zhang, L. Accurate Robot Arm Attitude Estimation Based on Multi-View Images and Super-Resolution Keypoint Detection Networks. *Sensors* **2024**, *24*, 305.
7. Wu, H.; Qi, J.; Purwanto, E.; Zhu, X.; Yang, P.; Chen, J. Multi-scale feature and multi-channel selection toward parkinson's disease diagnosis with eeg. *Sensors* **2024**, *24*, 4634.
8. Varlamova, E.V.; Butakova, M.A.; Semyonova, V.V.; Soldatov, S.A.; Poltavskiy, A.V.; Kit, O.I.; Soldatov, A.V. Machine learning meets cancer. *Cancers* **2024**, *16*, 1100.
9. Yuan, Y.; Zhang, Y.; Zhu, L.; Cai, L.; Qian, Y. Exploiting cross-scale attention transformer and progressive edge refinement for retinal vessel segmentation. *Mathematics* **2024**, *12*, 264.
10. Yang, J.; Zhang, W.; Guo, Z.; Gao, Z. TrustDFL: A blockchain-based verifiable and trusty decentralized federated learning framework. *Electronics* **2023**, *13*, 86.
11. Deng, W.; Wei, H.; Huang, T.; Cao, C.; Peng, Y.; Hu, X. Smart contract vulnerability detection based on deep learning and multimodal decision fusion. *Sensors* **2023**, *23*, 7246.

References

1. LeCun, Y.; Bengio, Y.; Hinton, G. Deep learning. *Nature* **2015**, *521*, 436–444. [CrossRef] [PubMed]
2. Otter, D.W.; Medina, J.R.; Kalita, J.K. A survey of the usages of deep learning for natural language processing. *IEEE Trans. Neural Netw. Learn. Syst.* **2020**, *32*, 604–624.
3. Litjens, G.; Kooi, T.; Bejnordi, B.E.; Setio, A.A.A.; Ciompi, F.; Ghafoorian, M.; Van Der Laak, J.A.; Van Ginneken, B.; Sánchez, C.I. A survey on deep learning in medical image analysis. *Med. Image Anal.* **2017**, *42*, 60–88. [PubMed]
4. Zhu, X.X.; Tuia, D.; Mou, L.; Xia, G.S.; Zhang, L.; Xu, F.; Fraundorfer, F. Deep learning in remote sensing: A comprehensive review and list of resources. *IEEE Geosci. Remote Sens. Mag.* **2017**, *5*, 8–36. [CrossRef]
5. Zha, D.; Bhat, Z.P.; Lai, K.H.; Yang, F.; Jiang, Z.; Zhong, S.; Hu, X. Data-centric artificial intelligence: A survey. *ACM Comput. Surv.* **2025**, *57*, 1–42.
6. Paheding, S.; Saleem, A.; Siddiqui, M.F.H.; Rawashdeh, N.; Essa, A.; Reyes, A.A. Advancing horizons in remote sensing: A comprehensive survey of deep learning models and applications in image classification and beyond. *Neural Comput. Appl.* **2024**, *36*, 16727–16767. [CrossRef]
7. Baltrušaitis, T.; Ahuja, C.; Morency, L.P. Multimodal machine learning: A survey and taxonomy. *IEEE Trans. Pattern Anal. Mach. Intell.* **2018**, *41*, 423–443.
8. Xu, P.; Zhu, X.; Clifton, D.A. Multimodal learning with transformers: A survey. *IEEE Trans. Pattern Anal. Mach. Intell.* **2023**, *45*, 12113–12132.
9. Rai, R.; Tiwari, M.K.; Ivanov, D.; Dolgui, A. Machine learning in manufacturing and industry 4.0 applications. *Int. J. Prod. Res.* **2021**, *59*, 4773–4778.
10. Hirsch, L.; Huang, Y.; Luo, S.; Rossi Saccarelli, C.; Lo Gullo, R.; Daimiel Naranjo, I.; Bitencourt, A.G.; Onishi, N.; Ko, E.S.; Leithner, D.; et al. Radiologist-level performance by using deep learning for segmentation of breast cancers on MRI scans. *Radiol. Artif. Intell.* **2021**, *4*, e200231. [CrossRef]
11. Ahmad, H.M.; Rahimi, A. Deep learning methods for object detection in smart manufacturing: A survey. *J. Manuf. Syst.* **2022**, *64*, 181–196.
12. Zolanvari, M.; Teixeira, M.A.; Gupta, L.; Khan, K.M.; Jain, R. Machine learning-based network vulnerability analysis of industrial Internet of Things. *IEEE Internet Things J.* **2019**, *6*, 6822–6834.
13. Shen, D.; Wu, G.; Suk, H.I. Deep learning in medical image analysis. *Annu. Rev. Biomed. Eng.* **2017**, *19*, 221–248. [PubMed]
14. Juhn, Y.; Liu, H. Artificial intelligence approaches using natural language processing to advance EHR-based clinical research. *J. Allergy Clin. Immunol.* **2020**, *145*, 463–469.
15. Abramson, J.; Adler, J.; Dunger, J.; Evans, R.; Green, T.; Pritzel, A.; Ronneberger, O.; Willmore, L.; Ballard, A.J.; Bambrick, J.; et al. Accurate structure prediction of biomolecular interactions with AlphaFold 3. *Nature* **2024**, *630*, 493–500.
16. Du, X.; Yu, J. Image-integrated magnetic actuation systems for localization and remote actuation of medical miniature robots: A survey. *IEEE Trans. Robot.* **2023**, *39*, 2549–2568.
17. Sünderhauf, N.; Brock, O.; Scheirer, W.; Hadsell, R.; Fox, D.; Leitner, J.; Upcroft, B.; Abbeel, P.; Burgard, W.; Milford, M.; et al. The limits and potentials of deep learning for robotics. *Int. J. Robot. Res.* **2018**, *37*, 405–420.

18. Aziz, A.; Pane, S.; Iacovacci, V.; Koukourakis, N.; Czarske, J.; Menciassi, A.; Medina-Sánchez, M.; Schmidt, O.G. Medical imaging of microrobots: Toward in vivo applications. *ACS Nano* **2020**, *14*, 10865–10893.
19. Mao, Q.; Liao, Z.; Yuan, J.; Zhu, R. Multimodal tactile sensing fused with vision for dexterous robotic housekeeping. *Nat. Commun.* **2024**, *15*, 6871.
20. Qiao, H.; Chen, J.; Huang, X. A survey of brain-inspired intelligent robots: Integration of vision, decision, motion control, and musculoskeletal systems. *IEEE Trans. Cybern.* **2021**, *52*, 11267–11280.
21. Hayes-Roth, B.; Thorndyke, P.W. Paradigms for intelligent systems. *Educ. Psychol.* **1985**, *20*, 231–241.
22. Topol, E.J. High-performance medicine: The convergence of human and artificial intelligence. *Nat. Med.* **2019**, *25*, 44–56. [PubMed]
23. Esteva, A.; Kuprel, B.; Novoa, R.A.; Ko, J.; Swetter, S.M.; Blau, H.M.; Thrun, S. Dermatologist-level classification of skin cancer with deep neural networks. *Nature* **2017**, *542*, 115–118.
24. Keesara, S.; Jonas, A.; Schulman, K. COVID-19 and health care's digital revolution. *N. Engl. J. Med.* **2020**, *382*, e82.
25. Rajkomar, A.; Dean, J.; Kohane, I. Machine learning in medicine. *N. Engl. J. Med.* **2019**, *380*, 1347–1358.
26. Rahman, Z.; Yi, X.; Khalil, I. Blockchain-based AI-enabled industry 4.0 CPS protection against advanced persistent threat. *IEEE Internet Things J.* **2022**, *10*, 6769–6778.
27. Kumar, S.; Lim, W.M.; Sivarajah, U.; Kaur, J. Artificial intelligence and blockchain integration in business: Trends from a bibliometric-content analysis. *Inf. Syst. Front.* **2023**, *25*, 871–896.
28. Liu, X.; Sun, S.X.; Huang, G. Decentralized services computing paradigm for blockchain-based data governance: Programmability, interoperability, and intelligence. *IEEE Trans. Serv. Comput.* **2019**, *13*, 343–355.
29. Bathula, A.; Gupta, S.K.; Merugu, S.; Saba, L.; Khanna, N.N.; Laird, J.R.; Sanagala, S.S.; Singh, R.; Garg, D.; Fouda, M.M.; et al. Blockchain, artificial intelligence, and healthcare: The tripod of future—A narrative review. *Artif. Intell. Rev.* **2024**, *57*, 238.
30. Issa, W.; Moustafa, N.; Turnbull, B.; Sohrabi, N.; Tari, Z. Blockchain-based federated learning for securing internet of things: A comprehensive survey. *ACM Comput. Surv.* **2023**, *55*, 1–43.
31. Hu, Y.; Kuang, W.; Qin, Z.; Li, K.; Zhang, J.; Gao, Y.; Li, W.; Li, K. Artificial intelligence security: Threats and countermeasures. *ACM Comput. Surv. (CSUR)* **2021**, *55*, 1–36. [CrossRef]
32. Murdoch, W.J.; Singh, C.; Kumbier, K.; Abbasi-Asl, R.; Yu, B. Definitions, methods, and applications in interpretable machine learning. *Proc. Natl. Acad. Sci. USA* **2019**, *116*, 22071–22080. [CrossRef] [PubMed]
33. Rudin, C. Stop explaining black box machine learning models for high stakes decisions and use interpretable models instead. *Nat. Mach. Intell.* **2019**, *1*, 206–215. [CrossRef] [PubMed]
34. Gunning, D.; Stefik, M.; Choi, J.; Miller, T.; Stumpf, S.; Yang, G.Z. XAI—Explainable artificial intelligence. *Sci. Robot.* **2019**, *4*, eaay7120. [CrossRef] [PubMed]

sensors

MDPI

Article

Cost-Sensitive YOLOv5 for Detecting Surface Defects of Industrial Products

Ben Liu, Feng Gao and Yan Li *

School of Mechanical and Precision Instrument Engineering, Xi'an University of Technology, Xi'an 710048, China
* Correspondence: liyangf@xaut.edu.cn

Abstract: Owing to the remarkable development of deep learning algorithms, defect detection techniques based on deep neural networks have been extensively applied in industrial production. Most existing surface defect detection models assign equal costs to the classification errors among different defect categories but do not strictly distinguish them. However, various errors can generate a great discrepancy in decision risk or classification costs and then produce a cost-sensitive issue that is crucial to the manufacturing process. To address this engineering challenge, we propose a novel supervised classification cost-sensitive learning method (SCCS) and apply it to improve YOLOv5 as CS-YOLOv5, where the classification loss function of object detection was reconstructed according to a new cost-sensitive learning criterion explained by a label–cost vector selection method. In this way, the classification risk information from a cost matrix is directly introduced into the detection model and fully exploited in training. As a result, the developed approach can make low-risk classification decisions for defect detection. It is applicable for direct cost-sensitive learning based on a cost matrix to implement detection tasks. Using two datasets of a painting surface and a hot-rolled steel strip surface, our CS-YOLOv5 model outperforms the original version with respect to cost under different positive classes, coefficients, and weight ratios, but also maintains effective detection performance measured by mAP and F1 scores.

Keywords: defect detection; cost-sensitive learning; YOLOv5; misclassification risk; intelligent industry

1. Introduction

Industrial and manufacturing object detection has been greatly facilitated by deep neural networks [1–4]. Object detection models are exploited to predict the position of the object in the input visual data and the corresponding class information, which plays a role in accelerating intelligent industry transformation [5–7]. The YOLO family [8,9] is one of the most popular approaches in real applications due to their fast inference speed and outstanding accuracy. Moreover, in the industrial sector, product surface defect detection is a general task [10–12]. To improve productivity, many manufacturing industries utilize machine vision systems, within which object detection is the core algorithm [13].

After all, there is no object detection model that can be completely accurate. The vast majority of existing surface defect detection approaches assume the same cost for all detection errors and focus on realizing high detection accuracy [11,14]. However, previous studies have confirmed that different types of detections and misclassifications lead to distinct costs [15–17], which leads to the cost-sensitive problem of models in practical industrial applications. For example, a defect detection model for automobile parts may judge nondestructive parts as defective parts, resulting in the loss of time and efficiency. Conversely, the misclassification of defective parts as nondestructive components may be detrimental to the safety of vehicle users and cause potential risks. Obviously, the cost (decision risk) of the latter is significantly higher than the former. Detection methods based on deep learning have achieved great success [18,19], but most of them cannot directly deal with the cost-sensitive problem, which is an urgent issue needing further investigation. This work mainly focuses on the cost-sensitive defect detection problem caused by discriminative misclassification errors.

Cost-sensitive learning has attracted much attention in past years [17,20–22]. Existing cost-sensitive models can be roughly divided into two categories: external cost-sensitive and internal cost-sensitive methods [23–25]. The objective of the external cost-sensitive method is to deal with the problem of model discriminant bias related to imbalanced data such as the long-tailed distribution of training datasets [26–29]. Differently, the internal cost-sensitive method is aimed at the decision cost (risk) caused by the classification errors among different categories in specific application scenarios [24,30]. Recently, these methods have been applied to scenarios such as face recognition, intelligent decisions, and intelligent healthcare [16,31–36]. However, most existing deep learning-based object detection work does not consider the cost-sensitive problem, but treats the misclassification cost between different classes equally, which may lead to high decision risk and is not applicable to cost-sensitive surface defect detection.

To address the above internal cost-sensitive issue in defect detection, a supervised cost-sensitive YOLOv5 detection model (CS-YOLOv5) is proposed. Considering the insufficiency of after-processing cost-sensitive methods, a direct-type cost-sensitive principle is conducted after re-examining the training process of object detection. Such a principle requires the output class prediction results of the defect detection model to be directly cost-sensitive, rather than being computed in an additional decision stage. That is, the parameters of the object detection model should preserve the classification cost information. The main contributions are listed as follows:

- Therefore, a classification loss function based on a label–cost vector selection method is designed, which can equip YOLOv5 with cost sensitivity after training. The misclassification cost involved might be the labor cost of defect detection, security cost, etc., which can be specified in practice by defining the cost matrix.
- Compared with the original YOLOv5 model, CS-YOLOv5 can solve the internal cost-sensitive problem, exploiting the classification risks defined by a risk matrix in specific applications.
- Experiments on our newly constructed painting defect dataset as well as a hot-rolled steel strip defect dataset demonstrate the superiority of our approach.

The rest of the paper is organized as follows. In Section 2, a newly developed principle and model are proposed. In Section 3, experiments and results analysis are reported to verify the effectiveness of the new model while discussing the proposed method. Finally, Section 4 concludes this paper.

2. Methodology

Figure 1 illustrates the CS-YOLOv5 model and its classification loss structure proposed in this paper. To exploit defect misclassification risks described by a cost matrix, a direct-type cost-sensitive principle based on a label–cost vector selection was designed. Under this principle, we propose a new cost-sensitive loss function. Specifically, the input defect images in training go through the forward process of the deep neural network and are encoded to class predictions and region predictions. Then, cost vectors are selected from the cost matrix via supervised class labels, which are utilized for the cost-sensitive loss calculation along with the class prediction. Meanwhile, classification loss and region loss are also derived. Finally, an object detection model with cost-sensitive features is obtained by backpropagating the loss gradient and updating the model parameters.

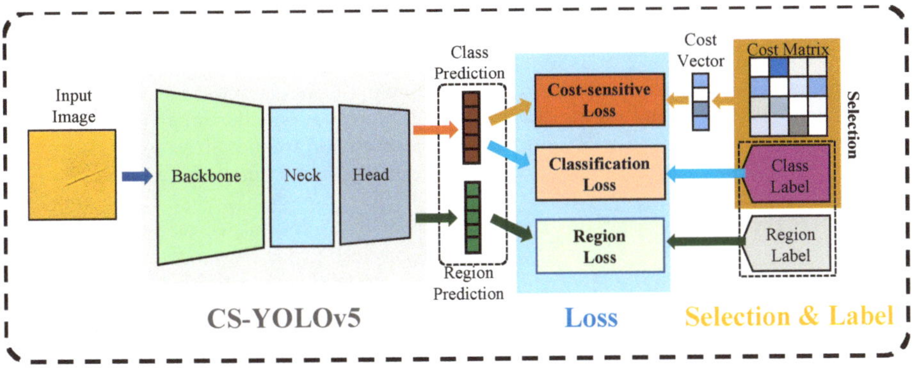

Figure 1. CS-YOLOv5 model and classification loss structure.

2.1. Cost-Sensitive Learning Modeling

Internal cost-sensitive learning attempts to reduce the decision risk defined in specific scenarios in training. This can be formulated as follows [37]:

$$\phi^*(\mathrm{x}) = \arg\min_j \mathrm{loss}(\mathrm{x}, j) \tag{1}$$

where $\mathrm{loss}(\mathrm{x}, j)$ is the expected cost of classifying sample x as j-th class, which is determined by the element C_{ij} in the cost matrix C. Furthermore, the cost matrix C is obtained via analysis from domain experts or data mining, e.g., estimating the cost of leakage and security costs in surface defect detection.

Without loss of generality, in deep learning, the classification loss is expressed as a function of the difference between the supervised class labels and the output prediction distribution. Assuming X and D are the input space and the output space, respectively, then the expectation of classification error can be expressed as:

$$E = \mathbb{E}_{\mathrm{X,D}} l[\varphi(\mathrm{X}), \mathrm{D}] \tag{2}$$

where $\varphi(\mathrm{X})$ is the prediction result of the model for input space X, and $l(\cdot)$ is the measurement method of the classification error. Therefore, the classification error for a single input sample x is described as follows:

$$e_{\mathrm{x},\boldsymbol{d}} = l(\boldsymbol{p}, \boldsymbol{d}) \tag{3}$$

where $p = \varphi(\mathrm{x}) = [p_1, \cdots, p_m]$ is the category probability vector estimated by the model for x, and the element p_j denotes the probability of the input sample x being recognized as the j-th class. Similarly, $\boldsymbol{d}$ is the one-hot label vector of x. For a sampled batch b with size B, Formula (2) can be expressed as:

$$E_b = \frac{1}{B} \Sigma_i^B l\left(\boldsymbol{p}^{(i)}, \boldsymbol{d}^{(i)}\right) \tag{4}$$

In terms of a defect detection model, $p = g(\mathrm{x}, \theta)$, where $g(\cdot, \theta)$ is the deep neural network with a parameter set of θ. The general classification loss function can measure the difference between the input space and the output space. However, there is no specific design for the output space. Classification errors between different classes are treated equally, without cost sensitivity. Generally, the total loss function includes the classification loss and a loss related to the region predictions for targets. Therefore, the general form can be expressed as follows:

$$L = L_{cls} + \beta L_{reg} \tag{5}$$

where L_{cls} is the classification loss designed in a cross-entropy form, L_{reg} is the location region loss, and β is a trade-off parameter.

According to the calculation of the ordinary classification loss described by Equation (4), the following conclusions can be drawn: (1) The classification loss between different categories is not assigned specifically. (2) Moreover, only the difference between class predictions and the correspondence labels is taken into account. In essence, it is still a generic non-cost-sensitive loss function that cannot process misclassification risks in defect detection.

The classification cost matrix $\boldsymbol{C}$ (decision risk matrix) is formulated as:

$$\begin{bmatrix} 0 & C_{12} & \cdots & C_{1N} \\ C_{21} & 0 & \cdots & C_{2N} \\ \vdots & C_{i2} & 0 & \vdots \\ C_{N1} & \cdots & C_{N(N-1)} & 0 \end{bmatrix} \tag{6}$$

The number of rows and columns of $\boldsymbol{C}$ are equal to the classes' quantity N, where $C_{ij} \geqslant 0$ and $C_i \neq 0$. Without loss of generality, $C_{ii} = 0$ means that the classification is correct without cost. Row vector $\boldsymbol{C_i}$ denotes the cost of classifying the samples with the class i into each class [30,38]. In multiclass detection or classification tasks, if all classes are divided into positive classes and negative ones, the misclassification coefficient can be divided into 4 types [16]:

1. False acceptance coefficient λ_{NP}: the risk of misclassifying a target (imposter) that belongs to the negative category into the positive category or positive class (gallery).
2. False rejection coefficient λ_{PN}: the risk of misclassifying a gallery to an imposter.
3. Two types of misidentification coefficients λ_{PP} and λ_{NN}: that is, misclassification between classes of samples with the same nature (the positive class or negative class).

The relationship between the magnitudes of the different coefficients is generally expressed as follows:

$$\lambda_{NP} > \lambda_{PN} > \lambda_{NN} > \lambda_{PP} \tag{7}$$

Moreover, the elements of the cost matrix can be expressed as:

$$C_{ij} = \begin{cases} \lambda_{NP}, & \text{if} & i\epsilon N, j\epsilon P \\ \lambda_{PN}, & \text{if} & i\epsilon P, j\epsilon N \\ \lambda_{NN}, & \text{if} & i\epsilon N, j\epsilon N \\ \lambda_{PP}, & \text{others} & \end{cases} \tag{8}$$

2.2. *Direct-Type Cost-Sensitive Learning*

The goal of cost-sensitive classification is to learn the probability prediction vectors formulated in Equation (4), which not only has high accuracy defined by supervised labels but also low cost with respect to the cost matrix $\boldsymbol{C}$. In this work, we argue that the cost matrix is additional supervised information for object detection. The rationale behind this is that the cost matrix reflects requirements for the model in a specific scenario. Therefore, we describe how to train a cost-sensitive defect detection model using both kinds of supervised information in the following part.

As the label exploitation is already formulated in Equation (4), we construct the cost-sensitive constraint in a label-cost vector selection method. Specifically, the cost vector is selected from the cost matrix $\boldsymbol{C}$ via the label $\boldsymbol{d}$ of the input sample x. This vector also constrains the output of posterior probability $\boldsymbol{p}$, in which the maximum probability in $\boldsymbol{p}$ determines the output class. Obviously, our approach incorporates cost information directly into training, which is different from the post-processing cost-sensitive learning based on Bayes minimum risk methods [31,39].

In the case of supervised learning, the ground-truth class index of sample x with label $\boldsymbol{d}$ is:

$$t = \underset{1 \leqslant i \leqslant N}{\operatorname{argmax}} d_i \tag{9}$$

According to Equation (1), the target of direct-type cost-sensitive learning is formulated as:

$$\begin{cases} \boldsymbol{p}' = \underset{\boldsymbol{p}' \in \mathbb{R}^N}{\operatorname{argmin}} \boldsymbol{p}' \cdot C_t \\ \phi^*(x) = \underset{1 \leqslant i \leqslant N}{\operatorname{argmax}} \boldsymbol{p}' \end{cases} \tag{10}$$

where p' is the output probability vector, and $\boldsymbol{C}_t$ is the aforementioned selected risk (cost) vector. Let $R_{tp}(x)$ represent the misclassification risk of samples of class t under a current probability vector $\boldsymbol{p}$. The risk brought by $\boldsymbol{p}$ can be expressed as:

$$\begin{cases} \boldsymbol{p}' = \underset{p' \in \mathbb{R}^N}{\operatorname{argmin}} R_{tp'}(x) = \underset{p' \in \mathbb{R}^N}{\operatorname{argmin}} \sum_{j=1}^{N} p'_j C_{tj} \\ \phi^*(x) = \underset{1 \leqslant i \leqslant N}{\operatorname{argmax}} (p'_i) \end{cases} \tag{11}$$

Equation (11) indicates that in the training of the direct-type cost-sensitive model: (1) output vector p' is to minimize the overall misclassification cost, i.e., minimizing the risk of classifying the sample belonging to class t into each class. (2) The same as before, the maximum index in the probability vector determines the class output. This is the direct-type cost-sensitive learning principle of the deep learning model with supervised label and cost information. The new principle specifies the calculation form of risk under supervised learning and provides linear constraints for the estimation of probability directly in N-dimensional space.

2.3. CS-YOLOv5 Model and Method Analysis

In order to equip the defect object detection model with cost sensitivity, the modified loss function is proposed and described as follows:

$$\begin{aligned} \text{Loss} &= \lambda_{iou} \textstyle\sum_{i=0}^{S^2} \sum_{j=0}^{B} 1_{ij}^{obj} L_{\text{CIoU}} \\ &- \lambda_{cls} \textstyle\sum_{i=0}^{S^2} \sum_{j=0}^{B} 1_{ij}^{obj} \sum_{c \in classes} \lambda_c [\hat{p}_{ti}(c) \log(p_i(c)) + (1 - \hat{p}_{ti}(c)) \log(1 - p_i(c))] \\ &+ \eta \lambda_{cost} \textstyle\sum_{i=0}^{S^2} \sum_{j=0}^{B} 1_{ij}^{obj} \sum_{c \in \text{llasses}} \lambda_c S(p_i(c)) C_{[t(ij)]c} \\ &+ \lambda_{obj} \textstyle\sum_{i=0}^{S^2} \sum_{j=0}^{B} 1_{ij}^{obj} \lambda_p [\hat{C}_i \log(C_i) + (1 - \hat{C}_i) \log(1 - C_i)] \\ &+ \lambda_{obj} \textstyle\sum_{i=0}^{S^2} \sum_{j=0}^{B} 1_{ij}^{\text{noobj}} [\hat{C}_i \log(C_i) + (1 - \hat{C}_i) \log(1 - C_i)] \end{aligned} \tag{12}$$

where $C_{[t(ij)]c}$ is the cost of that the ij-th sample of the class $t(ij)$ is classified into C class, and is the c-th element of cost vector $C_{[t(ij)]}$ from the risk coefficient matrix $\boldsymbol{C}$. η, λ_{cost}, λ_{cls}, λ_{iou}, and λ_{obj} are trade-off parameters that control the importance of distinct parts of losses. 1_{ij}^{obj} returns the value of 1 if the ij-th region contains an object. $S(y) = (1 + e^{-y})^{-1}$ is the sigmoid logistic regression function. Individually, the new classification loss function is expressed as:

$$\begin{aligned} L_{cls} &= \eta \lambda_{cost} \textstyle\sum_{i=0}^{S^2} \sum_{j=0}^{B} 1_{ij}^{obj} \lambda_c \odot S(p_i) \cdot C_{[t(ij)]} \\ &- \lambda_{cls} \textstyle\sum_{i=0}^{S^2} \sum_{j=0}^{B} 1_{ij}^{obj} \sum_{c \in \text{lasses}} \lambda_c [\hat{p}_{ti}(c) \log(p_i(c)) + (1 - \hat{p}_{ti}(c)) \log(1 - p_i(c))] \end{aligned} \tag{13}$$

where the first term is the cost-sensitive classification loss. When the parameter $\lambda_{cost} = 0$ and $\lambda_{cls} \neq 0$, the loss degenerates to a non-cost-sensitive loss. When $\lambda_{cost} \neq 0$ and $\lambda_{cls} = 0$, the loss degradation calculation is similar to the loss function in [25]. However, the intuitive meaning of improving the loss function in [25] is that a O_2 item is sufficiently close to the label value, which does not follow the cost-sensitive principle in Equation (10). The cost-sensitive principle in Equation (10) is further formulated as:

$$L_{cost} = \sum_{c\text{cclasses}} \lambda_c S(p_i(c)) C_{tc} \tag{14}$$

The use of a sigmoid function and normalization operation for the risk matrix in practical calculation can guarantee the boundary, thereby preventing the vanishing gradient and gradient explosion in training. In Equation (13), the classification loss is constructed as minimizing classification risk under class labels and the cost matrix. In other words, the optimization of the cross-entropy loss ensures the accuracy of the output in the label sense, that is, the estimate of the posterior probability is close enough to the label. The existence of cost-sensitive classification loss ensures that the risk corresponding to the estimated posterior probability is small enough. The loss weights of both realize the balancing effect of classification loss for different parts.

2.3.1. Cost-Sensitive Gradients

During the supervised training of the object detection model, the impact of the modified classification loss function on the implementation of the backpropagation algorithm needs to be taken into consideration [40]. Since the cross-entropy loss in the classification loss function remains unchanged and is superimposed with the cost-sensitive loss function, it will not affect the original backpropagation process. For a sample x of class t, the estimate of its output probability is set as $\boldsymbol{p}$. For the output layer, the gradient with respect to output neuron 0_n is directly associated with the n-th dimension output p_n and can be expressed as:

$$\frac{\partial L_{cost}}{\partial o_n} = \lambda_{cn} C_{tn} \frac{\partial S(p_n)}{\partial p_n} \tag{15}$$

Since $\frac{\partial S(y)}{\partial y} = S(y)(1 - S(y)) = \frac{e^{-y}}{(1+e^{-y})^2}$, then:

$$\frac{\partial L_{cost}}{\partial o_n} = \lambda_{cn} C_{tn} \frac{e^{-p_n}}{(1+e^{-p_n})^2}$$

The gradient of the cost-sensitive loss function directly with respect to the output probability is:

$$\nabla L_{cost} = \sum_{n=1}^{N} \lambda_{cn} C_{tn} \frac{e^{-p_n}}{(1+e^{-p_n})^2} \tag{16}$$

Since $e^{-p_n} > 0$ and $C_i \neq 0$, then $\nabla L_{\cos t} \neq 0$.

2.3.2. Cost-Sensitive YOLOv5 Algorithm

The role of the cost-sensitive classification loss function proposed in this paper is to establish a cost-sensitive parameter optimization space by combining various loss functions. The optimization algorithm is executed on this parameter space, thereby continuously updating the weight parameters of the model. The forward propagation of the established CS-YOLOv5 model has cost sensitivity. The implementation process is expressed in Algorithm 1:

Algorithm 1: Training direct-type cost-sensitive YOLOv5

Input: training space (X, D), validation set (X_v, D_v), minibatch size b, training epochs e, learning rate lr, as well as loss weight $(\lambda_{cost}, \lambda_{cls}, \lambda_{iou}, \lambda_c)$, and model $g(\theta)$
Output: trained model $g(\theta^*)$
1: Initialize parameters randomly of the model $g(\theta_0)$
2: If the current epoch $i \leqslant e$, loop:
3: If the current batch is j, loop:
4: Forward propagation: $O_j = g\left(X_j, D_j \mid \theta_{j-1}\right)$
5: Calculate the loss $L_j = \left(O_j, D_j\right)$ based on Equation (12)
6: Calculate the gradient ∇L_j and backpropagation, execute the SGD algorithm
7: Update parameter $\theta_j \leftarrow \theta_{j-1}$
8: Evaluate the model: $g(X_v, D_v \mid \theta_i)$
9: Return $g(\theta^*)$
End

Through the proposed cost-sensitive loss method, the misclassification cost information is directly involved in the training of the deep neural network model. With the classification loss weight $(\lambda_{cost}, \lambda_{cls})$ set within an appropriate range, we have the following observations:

(Case 1): When the risk coefficient of the cost vector is C_{ii} = 0 or C_{ii} is the smallest element in vector $\boldsymbol{C}_i$, classifying the sample into its ground-truth has the lowest risk. If the risk coefficient C_{in} is corresponding to the large value p_n in the output probability, the classification cost will be large. The model optimization process results in a delineation of the ground-truth and drives away this costly situation.

(Case 2): When the risk coefficient C_{ii} of the cost vector is the element with a larger value in the vector $\boldsymbol{C}_i$ and the risk coefficient corresponding to the larger value p_n in the posterior probability estimation is exactly C_{ii}, the correct classification will cause a large classification cost. After parameter updating, the model optimization process will lead to the division of clear truth value label pointing.

The essence of the classification loss function is to move away or strengthen the decision boundary towards the direction of less risk. Then, the boundary ambiguity is reduced compared with non-cost-sensitive models. From another perspective, the method proposed in this paper uses the prior knowledge of the cost matrix to modify the constraints of the model optimization. Thus, the optimization of model parameters can be regarded as a regularization method. The cost-sensitive learning with the original Bayes minimum risk decision type is to simply concatenate the output of estimating probability through the process of the minimum-risk decision. Instead, our new cost-sensitive learning method proposed in this paper is to obtain the optimal posterior probability vector in the sense of the label and cost matrix in one training process, which makes the model output directly cost-sensitive.

3. Experiments and Results Analysis

3.1. Experimental Dataset

NEU surface defect database: This industrial image dataset of hot-rolled steel strip surface defects is constructed in [41], which contains six types of defects, i.e., rolled-in scale (RS), patches (Pa), crazing (Cr), pitted surface (PS), inclusion (In), and scratches (Sc). There are 300 images in each class for a total of 1800 images. In the experiment, 1500 images were divided into a training set, and the remaining 300 images were divided into a test set. The original data size was 200 × 200 pixels, which was converted to size 192 × 192 in the experiments. This dataset is referred to as the NEU dataset for simplicity in this paper.

Paint surface defect dataset: This is a manufacturing image dataset involving paint spraying surface defects. The author participated in the collection and collation work of the inspection site of construction machinery structural parts. The dataset contains four classes of target defects, including dust, pit, sag, and scratch. There were 1417 images in total, including 1246 training set images and 171 test set images. The original size was 1024×1024, which was set to size 640×640 in the experiments. For simplicity, this dataset is referred to as the Paint dataset in the remainder of this paper.

The class distributions and the size distribution of ground-truth bounding box of these two datasets are counted and shown in Figure 2a,b, respectively. In the bar figures, the values denote the statistics of the sample numbers for each class. For the NEU dataset, the number of sample classes is more uniform than Paint. The Paint dataset has obvious uneven sample distribution, in which sag and scratch are less than the other two classes, and most of the two classes are more than 75% of the total number of samples. This is because the frequency of sag and scratch is so low at the collection site. Thus, it is a long-tail problem, which is very common in application.

(a)

(b)

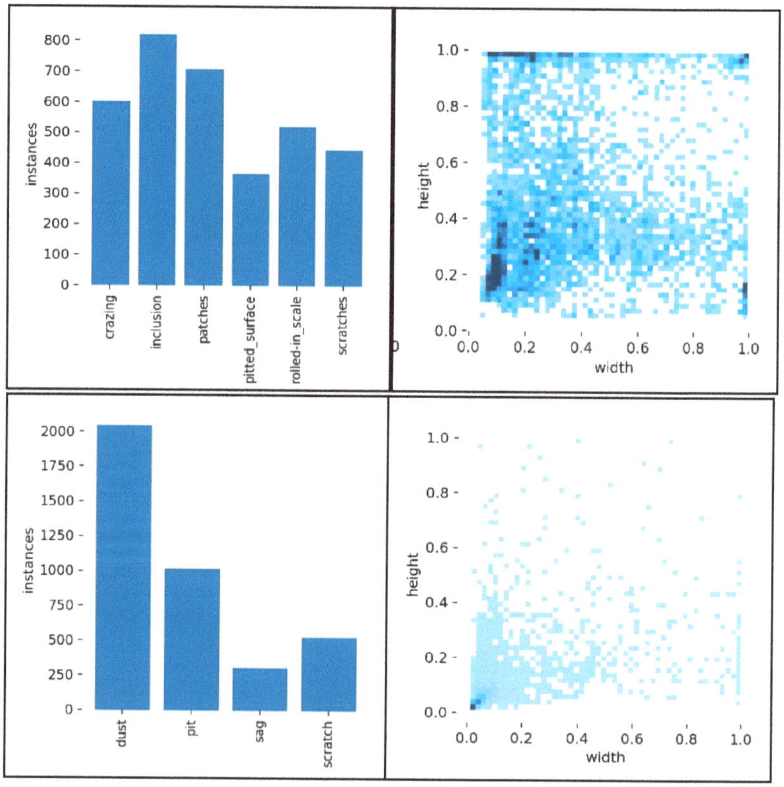

Figure 2. (**a**) NEU dataset sample statistics (**b**) Paint dataset sample statistics.

The scatter figures exhibit the relative size distribution of ground-truth bounding box for the datasets, that is, the distribution of length and width. The darker color indicates more bounding boxes are concentrated here. It can be seen that the bounding box width and height of the NEU dataset are widely distributed. The densest is where the relative width is about 0.2 and the relative height is about 0.1. In the Paint dataset, the bounding box is clearly concentrated in the lower left corner of the scatterplot. This is because the actual bounding boxes of dust samples in the dataset are relatively small, and the corresponding number is much larger than other classes. Based on the above analysis, the Paint dataset has obvious sample imbalance and most sample bounding boxes are too small. Therefore, the object detection problem of the Paint dataset is more difficult than that of the NEU dataset.

3.2. YOLOv5 Model Settings

YOLOv5 was selected as the baseline model in experiments, whose code is provided by [7]. We experimented with several Tesla T4 GPUs and implemented the code with the Pytorch 1.7.1 platform. The initial learning rate was set to 0.002. The cosine annealing strategy was utilized during training. The weight decay coefficient was 0.0005 for the adopted SGD optimizer. The balance parameter β was set to 0.375, and η was set to 1/10,000. For the NEU dataset, the batch size was set to 230, with 90 epochs of training. For the Paint dataset, the batch size was set to 24, with 55 epochs of training. The experiments all used the Mosaic data augmentation method. As the examples show in Figure 3a,b, this augmentation method stitches images together in an overlapping manner to improve the robustness of the models.

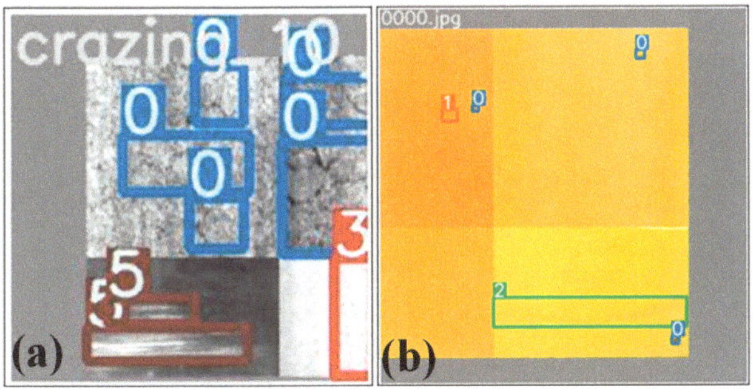

Figure 3. (**a**) NEU Data augmentation (**b**) Paint Data augmentation. Numbers are class indexes.

3.3. Experimental Metrics

3.3.1. Classification Cost Evaluation Metric

The proposed cost-sensitive learning method concentrates on minimizing the decision risk within the acceptable range of accuracy. To quantitatively measure the classification cost, we construct a metric formulated as:

$$\text{Cost} = \frac{1}{F}\sum_{i=1}^{F}\sum_{c\in \text{ cesses}} p_{ic}C_{[t(i)]c} = \frac{1}{F}\sum_{i=1}^{F}\boldsymbol{P}_i \cdot \boldsymbol{C}_{[t(i)]} \tag{17}$$

where $C_{[t(i)]c}$ is the c-th element of the cost vector $\boldsymbol{C}_{[t(i)]}$ from matrix $\boldsymbol{C}$ and denotes the cost of classifying the j-th sample belonging to the class $t(i)$ into the c-th class. F is the number of positive samples used to calculate the loss. Intuitively, such a cost metric is consistent with the cost-sensitive target to be optimized in the loss function.

3.3.2. Comprehensive Metrics

The recall ratio in object detection refers to the proportion of data samples correctly detected by the model, which is one of the basic performance measurements. TP, FP, TN, and FN are true positive, false positive, true negative, and false negative examples, respectively. Then, the recall rate is:

$$\text{Recall} = \frac{\text{TP}}{\text{TP} + \text{FN}} \tag{18}$$

To comprehensively evaluate the trained models, we also adopted the common metric, mean average precision (mAP) and F1-score in the experiments. They are described as follows:

$$\text{mAP} = \int_0^1 \text{Precision}(r)dr \tag{19}$$

where Precison $= \text{TP}/(\text{TP} + \text{FP})$ is the accuracy rate, and r represents Recall.

$$\text{F1} = 2 \times \frac{\text{Precision} \cdot \text{Recall}}{\text{Precision} + \text{Recall}} \tag{20}$$

For the multiclass data (more than two), the calculation method is given in the literature [42].

3.4. Risk Coefficient Experiment

The premise of cost-sensitive learning methods is to construct a classification cost matrix for the current data. Since the cost matrix is directly involved in the calculation of the classification cost of Equation (16), we performed experiments with different classification cost matrices. Different risk coefficient ratios simulate different setting conditions of the cost matrix in real applications to verify the effectiveness of the cost-sensitive learning method.

Each element of the cost matrix is the classification risk coefficient. Moreover, objects in the dataset were divided into a positive class and a negative class. The settings of the four risk coefficients follow Equation (7). In this experiment, we tested the cost-sensitive learning method by adjusting the rate of risk coefficient under fixed positive classes.

3.4.1. Risk Coefficient Setting

For the NEU dataset, positive classes were selected as Pa, PS, In, and Sc, while for Paint, dust and scratch were positive classes. Figure 4a,b shows the examples of positive and negative classes in NEU and Paint dataset, respectively.

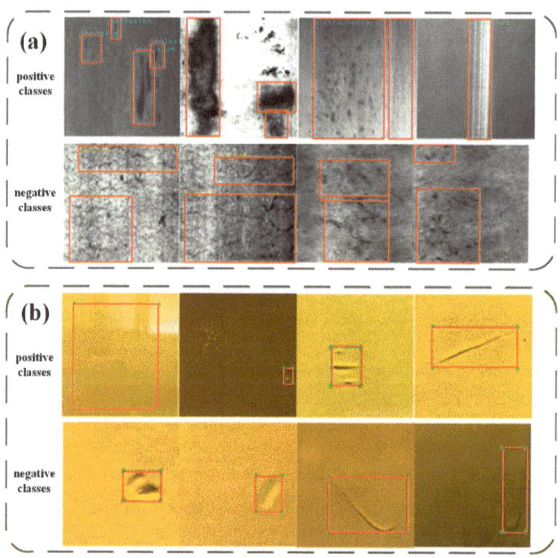

Figure 4. (**a**) Example of NEU division (**b**) Example of Paint division.

According to Equation (7), the base coefficient is selected as $\lambda_{NP} : \lambda_{PN} : \lambda_{NN} : \lambda_{PP} = 200 : 20 : 2 : 1$. The ratios between adjacent coefficients are linearly scaled by equal proportions to obtain four sets of classification risk coefficients. Table 1 shows the risk coefficient grouping.

Table 1. Risk coefficient grouping.

(a): $\lambda_{NP} : \lambda_{PN} : \lambda_{NN} : \lambda_{PP} = 200 : 20 : 2 : 1$	(b): $\lambda_{NP} : \lambda_{PN} : \lambda_{NN} : \lambda_{PP} = 675 : 45 : 3 : 1$
(c): $\lambda_{NP} : \lambda_{PN} : \lambda_{NN} : \lambda_{PP} = 1600 : 80 : 4 : 1$	(d): $\lambda_{NP} : \lambda_{PN} : \lambda_{NN} : \lambda_{PP} = 3125 : 125 : 5 : 1$

3.4.2. Experimental Results and Analysis

The experimental results of classification cost of four groups (a)–(d) with different risk coefficients are shown in Figure 5. For the four groups (a), (b), (c), and (d), the classification costs related to CS-YOLOv5 are 0.000470, 0.000337, 0.000340, and 0.000341, while those for YOLOv5 are 0.00368, 0.00378, 0.00381, and 0.00422. Compared with YOLOv5, the cost of group (a) for CS-YOLOv5 decreases by 87.16%, and the most significant decrease in group (d) is 91.93%. This is because the risk coefficient ratio of different groups directly affects the cost matrix, thus giving CS-YOLOv5 different cost sensitivities during training. According to the cost metric given by Equation (17), different risk matrices also make models have different cost performances. Similar conclusions can be obtained in Figure 5b. According to the results of the NEU and Paint datasets, the classification cost of the cost-sensitive YOLOv5 model is also significantly lower than that of the original YOLOv5. Under different risk coefficients, the output of the cost-sensitive YOLOv5 model itself has lower classification costs, which proves the effectiveness of our method.

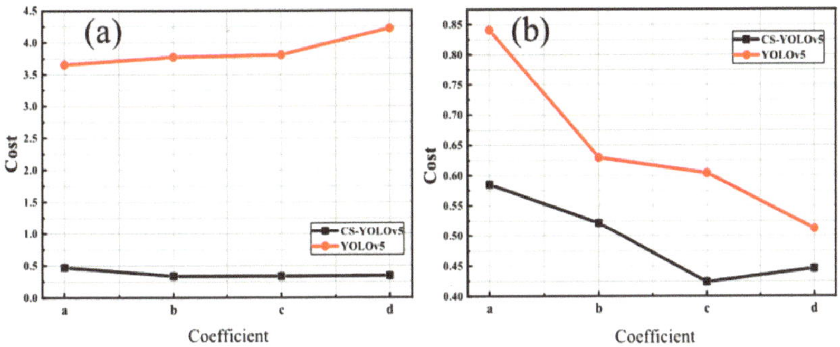

Figure 5. (**a**) Cost ($\times 10^{-3}$) with varying risk coefficients on NEU (**b**) Cost ($\times 10^{-3}$) with varying risk coefficients on Paint.

Moreover, the results of mAP and F1-score are shown in Table 2. The improved model has cost-sensitive capabilities while performing comparably to the original model under two metrics. Even the performance of the proposed approach is better than the original YOLOv5 under some risk coefficients (bold in the table). It can be concluded that when other cost matrix settings such as positive and negative classes are fixed, cost-sensitive YOLOv5 significantly improves the cost-sensitive performance without losing the comprehensive performance.

Table 2. mAP and F1-Score with Different Risk Coefficients.

Model	Group	NEU		Paint	
		mAP	F1	mAP	F1
YOLOv5	a	0.76	0.74	0.61	0.63
	b	0.74	0.72	0.61	0.63
	c	0.76	0.74	0.63	0.63
	d	0.74	0.72	0.62	0.63
CS-YOLOv5	a	0.75	0.74	0.63	0.63
	b	0.76	0.74	0.62	0.63
	c	0.76	0.74	0.62	0.62
	d	0.74	0.72	0.65	0.64

3.5. Experiment of Positive Classes

In real scenarios, positive and negative classes are often determined in accordance with different production requirements. The division of the positive class and negative class plays a crucial role in cost-sensitive learning. According to Section 3.2, the division is directly related to the form of the cost matrix. Therefore, we performed this experiment by setting the number of positive classes as a variable to verify the effectiveness of CS-YOLOv5.

3.5.1. Number Setting of Positive Classes

In this part, the risk coefficient was set to a fixed $\lambda_{NP} : \lambda_{PN} : \lambda_{NN} : \lambda_{PP} = 200 : 20 : 2 : 1$. For the NEU dataset, the number of positive classes was increased from 1 to 4. For the Paint (with a total of 4 classes), the number of positive classes was increased from 1 to 3. Considering the preciseness, the positive classes were randomly selected and tested several times.

3.5.2. Experimental Results and Analysis

For the division, samples belonging to positive classes were classified into the positive set, while samples of other classes are classified into the negative set. The cost matrix conducted based on this division and risk coefficients was exploited in training. The classification cost results under different positive classes are shown in Figure 6.

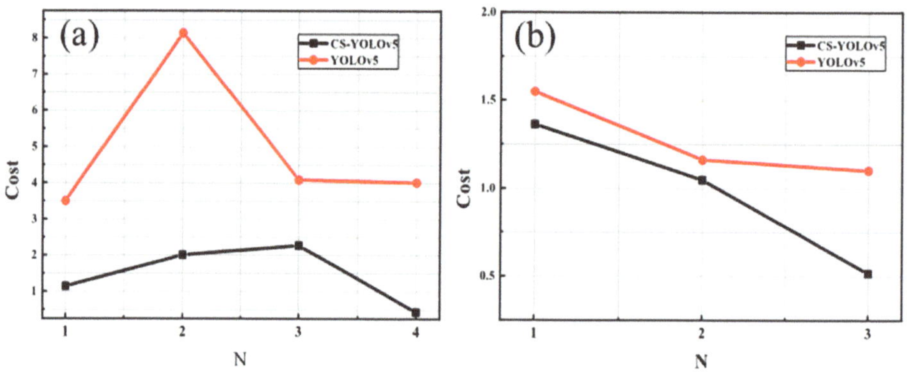

Figure 6. (**a**) Cost ($\times 10^{-3}$) with varying positive classes on NEU (**b**) Cost ($\times 10^{-3}$) with varying positive classes on Paint.

As can be seen from Figure 6, when the number of positive classes in the NEU surface defect database dataset is 1, 2, 3, and 4, the classification cost of CS-YOLOv5 is 67.58%, 75.36%, 44.56%, and 89.27% lower than that of YOLOv5. Different numbers of positive classes will affect the construction of the cost matrix, so the parameters of CS-YOLOv5 have different cost sensitivity. According to the cost metric defined by Formula (17), the change in the cost matrix will lead to different classification cost of YOLOv5. Similarly, the analysis of Figure 6b can also reach a similar conclusion. Therefore, the classification cost of the CS-YOLOv5 model is lower than the original model under various partitions on both datasets, which demonstrates the superiority of the direct-type cost-sensitive method. In addition, the performance under mAP and F1-score is shown in Table 3.

From these results, we have some observations: (1) CS-YOLOv5 has performance which is comparable to or better than the original model under the metrics. (2) When other settings such as the misclassification risk coefficient are fixed, the cost sensitivity of CS-YOLOv5 is significantly improved.

Table 3. mAP and F1-score with different positive classes.

Model	Group	NEU		Paint	
		mAP	F1	mAP	F1
YOLOv5	1	0.73	0.72	0.63	0.61
	2	0.74	0.72	0.63	0.63
	3	0.74	0.73	0.64	0.65
	4	0.77	0.74	—	—
CS-YOLOv5	1	0.75	0.73	0.61	0.63
	2	0.75	0.72	0.64	0.64
	3	0.78	0.75	0.62	0.61
	4	0.74	0.71	—	—

3.6. Weight Ratio Experiment

As stated in Section 3.4, the cost-sensitive classification loss (Equation (14)) conforms to the direct-type cost-sensitive principle proposed for supervised learning. Furthermore, the optimization of the cross-entropy loss in the classification loss function enforces the estimation of the output probability close to the corresponding ground-truth. On the other hand, the role of the cost loss function (Equation (16)) is the total classification risk of the output probability. The proportion and balance of the two targets are determined by the weight ratio $r = \lambda_{cost} / \lambda_{cls}$. Therefore, it is reasonable and necessary to implement experiments on this weight ratio in the classification loss.

3.6.1. Settings for Weight Ratio Experiment

The risk coefficient is set to $\lambda_{NP} : \lambda_{PN} : \lambda_{NN} : \lambda_{PP} = 200 : 20 : 2 : 1$. For the NEU dataset, positive classes are selected as Pa, PS, In, and Sc, and those are dust and scratch for Paint. Moreover, the weight ratio $r = \lambda_{cost} / \lambda_{cls}$ is set to vary from 0 to 1.5 with an increment of 0.25. A total of 7 groups of linearly changing ratios are considered. According to the analysis in Section 2.3, when $r = 0$, the classification loss used in the experiment degenerates into a non-cost-sensitive loss function. In this case, the model is non-cost-sensitive.

3.6.2. Experimental Results and Analysis

The experimental results under different loss weight ratios are developed into coefficient weight–classification cost curves in Figure 7. When $r = 0$, the results denote the original non-cost-sensitive model and the other 6 groups of CS-YOLOv5.

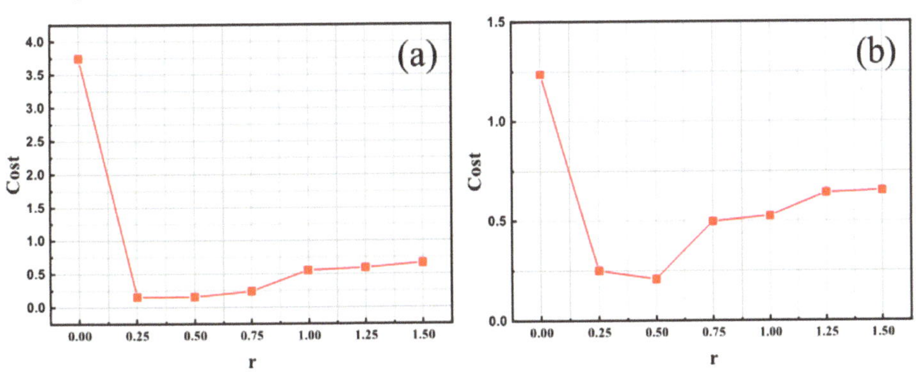

Figure 7. (**a**) Cost ($\times 10^{-3}$) with varying weight ratios on NEU (**b**) Cost ($\times 10^{-3}$) with varying weight ratios on Paint.

As shown in Figure 7a, r varies from 0.25 to 1.5 on NEU, representing 6 groups of CS-YOLOv5 models with different loss weight ratios. $r = 0$ indicates that the model degenerates to the original model YOLOv5. From these results, we can see that the classification cost of 6 groups of CS-YOLOv5 models is lower than that of YOLOv5, which indicates the effectiveness of the cost-sensitive method for model improvement. Too-large r can negatively affect the performance of CS-YOLOv5. Thus, in constructing CS-YOLOv5, an appropriate loss weight ratio can equip the model with cost sensitivity and maintain good performance. Similar conclusions can be drawn from the experiments on the Paint dataset in Figure 7b. The cost of the CS-YOLOv5 model on both datasets is lower than the results of the ordinary YOLOv5 model represented by $r = 0$. Consequently, under the cost-sensitive classification functions of different weights in experiments, the effectiveness of the direct-type cost-sensitive method in reducing the classification cost has been demonstrated. In addition, the comprehensive performance measured by mAP and F1-score is shown in Table 4.

Table 4. mAP and F1-score with different weight ratios.

Weight Ratio	NEU		Paint	
	mAP	**F1**	**mAP**	**F1**
0	0.75	0.73	0.61	0.63
0.25	0.74	0.72	0.61	0.63
0.5	0.72	0.70	0.61	0.64
0.75	0.76	0.73	0.63	0.63
1	0.68	0.67	0.59	0.62
1.25	0.68	0.66	0.60	0.61
1.5	0.68	0.66	0.59	0.59

From the results on the Paint dataset in Table 4, it can be observed that the improved model with appropriate weight ratio performs better than the original model. When the weight ratios are 0.75 and 0.64, the mAP and F1 metrics reach their optimal values, respectively. As the weight of the cross-entropy loss function decreases, the performance of the model degrades. When the weight ratio is set to 1.5, both mAP and F1 drop to the lowest value of 0.59. This is because a small proportion of cross-entropy loss makes the model parameter more inclined to be cost-sensitive, with insufficient accuracy optimization. Similar conclusions can be drawn from the results of the NEU dataset. Therefore, it can be concluded that the two weights λ_{cost} and λ_{cls} in the classification loss need to be properly set, so that YOLOv5 can have cost sensitivity and comprehensive performance.

3.7. Ablations

As clarified in Section 3.1, the resolution of these two datasets is quite different, which is 192 × 192 for NEU and 640 × 640 for Paint. Thus, the batch sizes on them have a gap under same the computing resources. On another, we implemented a smaller batch size on NEU and found that they had few differences, and the results are shown in Table 5.

Table 5. The cost performance on NEU with different batch sizes.

Batch size	64	128	230
cost ($\times 10^{-3}$)	0.48	0.47	0.47

Moreover, since the number of images in these two datasets is not large (1800 and 1417, respectively) and the categories are fewer than 10, many epochs are not necessary. We present the training classification loss and cost curves in Figure 8, and we can observe that the models tend to converge under the adopted settings.

(a)

(b)

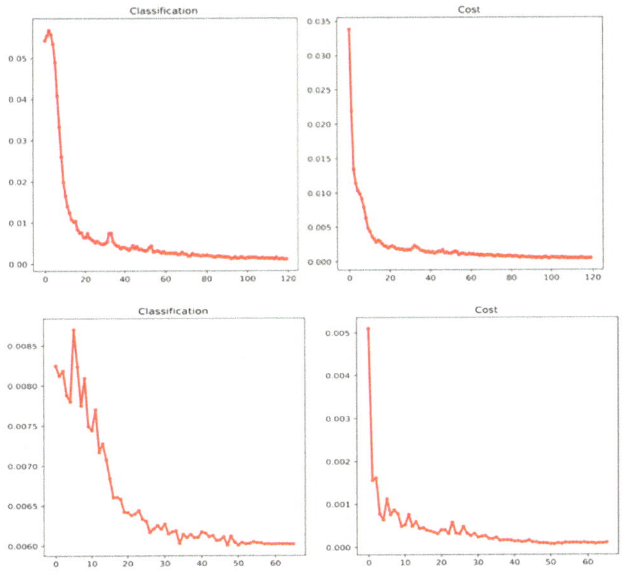

Figure 8. Training curves of (**a**) NEU (**b**) Paint.

3.8. Discussions and Comparisons

In this paper, we design a strategy to address the cost-sensitive issue of defect detection by presenting the CS-YOLOv5. Besides the widely adopted YOLO family in manufacturing, another single-stage SSD method [43] is also exploited in the previous literature [44]. Such methods are comparable with YOLO in detection speed, but the settings of the prior box are sensitive and may lead to instability [45]. On the other hand, the two-stage detector Faster-RCNN [4] is also utilized in defect detection [46], which especially perfroms better in denser detection. However, it is slower than single-stage designs due to their higher complexity. More importantly, both of these vanilla versions of the detector cannot process the internal cost-sensitive problem without a specific design, while our CS-YOLOv5 can not only reduce the misclassification cost but also inherits the speed and accuracy advantages.

4. Conclusions

To tackle the challenge of cost-sensitive defect detection in manufacturing, in this work, we propose a novel general cost-sensitive learning method called the supervised classification cost-sensitive learning method. Specifically, the misclassification risks represented by a cost matrix are directly integrated into the model optimization. Cost vectors are selected with labels as additional supervised information to train. Based on this approach, we enrich the YOLO family with a cost-sensitive YOLOv5 (CS-YOLOv5). It is a method that reduces the misclassification risk while maintaining the original model structure. We also construct a defect detection dataset from an industrial site. Extensive experiments demonstrate the effectiveness of the proposed CS-YOLOv5 for cost-sensitive defect detection. In the future, we will pay attention to some denser detection tasks which are more sensitive to location information. Hence, future research work can focus on feasible principles and methods for location cost in manufacturing.

Author Contributions: B.L. and Y.L. provided the conception and methodology of this study and are responsible for the draft text writing. B.L. and F.G. designed the experiments and completed the data analysis. Y.L. assisted with the validation and guided the writing and revision of the paper. All authors have read and agreed to the published version of the manuscript.

Funding: This work was supported by the Key Industrial Chain Project of Shaanxi Province (No. 2021ZDLGY 10-07) and the Scientific-Technological Innovation Projects in Strategic Emerging Industries of Shandong Province (No. 2017TSCYCX-24).

Institutional Review Board Statement: Not applicable.

Informed Consent Statement: Not applicable.

Data Availability Statement: The datasets used and/or analyzed during the current study are available from the corresponding author on reasonable request.

Acknowledgments: Thanks to the advanced manufacturing lab of Xi'an University of Technology for providing this research with open experiment data and resources.

Conflicts of Interest: The authors declare no conflict of interest.

References

1. Qi, S.; Yang, J.; Zhong, Z. A Review on Industrial Surface Defect Detection Based on Deep Learning Technology. In Proceedings of the 3rd International Conference on Machine Learning and Machine Intelligence, Hangzhou, China, 18–20 September 2020; pp. 24–30. [CrossRef]
2. Mahto, P.; Garg, P.; Seth, P.K.; Panda, J. Refining Yolov4 for Vehicle Detection. *Int. J. Adv. Res. Eng. Technol. (IJARET)* **2020**, *11*, 409–419.
3. He, Y.; He, N.; Zhang, R.; Yan, K.; Yu, H. Multi-scale feature balance enhancement network for pedestrian detection. *Multimed. Syst.* **2022**, *28*, 1135–1145. [CrossRef]
4. Ren, S.; He, K.; Girshick, R.; Sun, J. Faster R-CNN: Towards Real-Time Object Detection with Region Proposal Networks. *IEEE Trans. Pattern Anal. Mach. Intell.* **2017**, *39*, 1137–1149. [CrossRef] [PubMed]
5. Redmon, J.; Farhadi, A. YOLO9000: Better, Faster, Stronger. In Proceedings of the 30th IEEE Conference on Computer Vision and Pattern Recognition, Honolulu, HI, USA, 21–26 July 2017; pp. 6517–6525. [CrossRef]
6. He, K.; Gkioxari, G.; Dollár, P.; Girshick, R. Mask R-CNN. In Proceedings of the 2017 IEEE International Conference on Computer Vision (ICCV), Venice, Italy, 22–29 October 2017; pp. 2961–2969. [CrossRef]
7. Jocher, G.; Stoken, A.; Borovec, J.; NanoCode012; Chaurasia, A.; Liu, C.; Xie, T.; Abhiram, V.; Laughing; Tkianai; et al. Ultralytics/Yolov5: V5.0—YOLOv5-P6 1280 Models, AWS, Supervise.ly and YouTube Integrations. *Zenodo* **2021**, *11*. [CrossRef]
8. Redmon, J.; Divvala, S.; Girshick, R.; Farhadi, A. You Only Look Once: Unified, Real-Time Object Detection. In Proceedings of the 2016 IEEE Conference on Computer Vision and Pattern Recognition (CVPR), Las Vegas, NV, USA, 27–30 June 2016; pp. 779–788. [CrossRef]
9. Yang, G.; Feng, W.; Jin, J.; Lei, Q.; Li, X.; Gui, G.; Wang, W. Face mask recognition system with YOLOV5 based on image recognition. In Proceedings of the 2020 IEEE 6th International Conference on Computer and Communications (ICCC), Chengdu, China, 11–14 December 2020; pp. 1398–1404. [CrossRef]
10. Chen, X.; Lv, J.; Fang, Y.; Du, S. Online Detection of Surface Defects Based on Improved YOLOV3. *Sensors* **2022**, *22*, 817. [CrossRef] [PubMed]
11. Xu, R.; Hao, R.; Huang, B. Efficient surface defect detection using self-supervised learning strategy and segmentation network. *Adv. Eng. Inform.* **2022**, *52*, 101566. [CrossRef]
12. Tabernik, D.; Šela, S.; Skvarč, J.; Skočaj, D. Segmentation-based deep-learning approach for surface-defect detection. *J. Intell. Manuf.* **2019**, *31*, 759–776. [CrossRef]
13. Chen, Y.; Ding, Y.; Zhao, F.; Zhang, E.; Wu, Z.; Shao, L. Surface Defect Detection Methods for Industrial Products: A Review. *Appl. Sci.* **2021**, *11*, 7657. [CrossRef]
14. Da, Y.; Dong, G.; Wang, B.; Liu, D.; Qian, Z. A novel approach to surface defect detection. *Int. J. Eng. Sci.* **2018**, *133*, 181–195. [CrossRef]
15. Elkan, C. The foundations of cost-sensitive learning. In Proceedings of the 17th International Joint Conference on Artificial Intelligence, Seattle, WA, USA, 4–10 August 2001; Morgan Kaufmann Publishers Inc.: Seattle, WA, USA, 2001; Volume 2, pp. 973–978.
16. Zhang, Y.; Zhou, Z.-H. Cost-Sensitive Face Recognition. *IEEE Trans. Pattern Anal. Mach. Intell.* **2010**, *32*, 1758–1769. [CrossRef]
17. Frumosu, F.D.; Khan, A.R.; Schiøler, H.; Kulahci, M.; Zaki, M.; Westermann-Rasmussen, P. Cost-sensitive learning classification strategy for predicting product failures. *Expert Syst. Appl.* **2020**, *161*, 113653. [CrossRef]
18. Tan, M.; Pang, R.; Le, Q.V. EfficientDet: Scalable and Efficient Object Detection. In Proceedings of the IEEE/CVF Conference on Computer Vision and Pattern Recognition, Seattle, WA, USA, 13–19 June 2020. [CrossRef]
19. Wu, X.; Sahoo, D.; Hoi, S.C.H. Recent advances in deep learning for object detection. *Neurocomputing* **2020**, *396*, 39–64. [CrossRef]
20. Fernandez, A.; Garcia, S.; Herrera, F.; Chawla, N.V. SMOTE for Learning from Imbalanced Data: Progress and Challenges, Marking the 15-year Anniversary. *J. Artif. Intell. Res.* **2018**, *61*, 863–905. [CrossRef]
21. Zhu, X.; Yang, J.; Zhang, C.; Zhang, S. Efficient Utilization of Missing Data in Cost-Sensitive Learning. *IEEE Trans. Knowl. Data Eng.* **2021**, *33*, 2425–2436. [CrossRef]
22. Fernández, A.; García, S.; Galar, M.; Prati, R.C.; Krawczyk, B.; Herrera, F. Cost-sensitive learning. In *Learning from Imbalanced Data Sets*; Springer: Berlin/Heidelberg, Germany, 2018; pp. 63–78. [CrossRef]

23. Kukar, M.; Kononenko, I. Cost-Sensitive Learning with Neural Networks. In *European Conference on Artificial Intelligence*; John Wiley & Sons, Ltd.: Hoboken, NJ, USA, 1998; pp. 445–449. [CrossRef]
24. Petrides, G.; Verbeke, W. Cost-sensitive ensemble learning: A unifying framework. *Data Min. Knowl. Discov.* **2022**, *36*, 1–28. [CrossRef]
25. Khan, S.H.; Hayat, M.; Bennamoun, M.; Sohel, F.A.; Togneri, R. Cost-Sensitive Learning of Deep Feature Representations from Imbalanced Data. *IEEE Trans. Neural Netw. Learn. Syst.* **2018**, *29*, 3573–3587. [CrossRef]
26. Zhou, Z.H.; Liu, X.Y. Training cost-sensitive neural networks with methods addressing the class imbalance problem. *IEEE Trans. Knowl. Data Eng.* **2006**, *18*, 63–77. [CrossRef]
27. Li, T.; Cao, P.; Yuan, Y.; Fan, L.; Yang, Y.; Feris, R.; Indyk, P.; Katabi, D. Targeted Supervised Contrastive Learning for Long-Tailed Recognition. In Proceedings of the 2022 IEEE/CVF Conference on Computer Vision and Pattern Recognition (CVPR), New Orleans, LA, USA, 18–24 June 2022; pp. 6908–6918. [CrossRef]
28. Raj, V.; Magg, S.; Wermter, S. Towards Effective Classification of Imbalanced Data with Convolutional Neural Networks. In *Artificial Neural Networks in Pattern Recognition*; Springer: Cham, Switzerland, 2016; pp. 150–162. [CrossRef]
29. Zhou, Z.-H. Cost-sensitive learning. In *Modeling Decision for Artificial Intelligence: 8th International Conference*; Springer: Berlin/Heidelberg, Germany, 2011; pp. 17–18. [CrossRef]
30. Chung, Y.-A.; Lin, H.-T.; Yang, S.-W. Cost-aware pre-training for multiclass cost-sensitive deep learning. In Proceedings of the Twenty-Fifth International Joint Conference on Artificial Intelligence, New York, NY, USA, 9–15 July 2016; AAAI Press: New York, NY, USA, 2016; pp. 1411–1417. [CrossRef]
31. Collell, G.; Prelec, D.; Patil, K.R. A simple plug-in bagging ensemble based on threshold-moving for classifying binary and multiclass imbalanced data. *Neurocomputing* **2018**, *275*, 330–340. [CrossRef] [PubMed]
32. Kuo, W.; Häne, C.; Yuh, E.; Mukherjee, P.; Malik, J. Cost-Sensitive Active Learning for Intracranial Hemorrhage Detection. In *Medical Image Computing and Computer Assisted Intervention—MICCAI*; Springer: Berlin/Heidelberg, Germany, 2018; pp. 715–723. [CrossRef]
33. Zhang, H.; Jiang, L.; Li, C. CS-ResNet: Cost-sensitive residual convolutional neural network for PCB cosmetic defect detection. *Expert Syst. Appl.* **2021**, *185*, 115673. [CrossRef]
34. Liu, M.; Miao, L.; Zhang, D. Two-Stage Cost-Sensitive Learning for Software Defect Prediction. *IEEE Trans. Reliab.* **2014**, *63*, 676–686. [CrossRef]
35. Natarajan, N.; Dhillon, I.S.; Ravikumar, P.; Tewari, A.J.J.M.L.R. Cost-Sensitive Learning with Noisy Labels. *J. Mach. Learn. Res* **2017**, *18*, 5666–5698.
36. Seliya, N.; Khoshgoftaar, T.M. The use of decision trees for cost-sensitive classification: An empirical study in software quality prediction. *WIREs Data Min. Knowl. Discov.* **2011**, *1*, 448–459. [CrossRef]
37. Wan, J.; Yang, M. Survey on Cost-sensitive Learning Method. *J. Softw.* **2019**, 113–136. [CrossRef]
38. Lu, S.; Liu, L.; Lu, Y.; Wang, P. Cost-sensitive neural network classifiers for postcode recognition. *Int. J. Pattern Recognit. Artif. Intell.* **2012**, *26*, 1263001. [CrossRef]
39. Jiang, X.; Mahadevan, S. Bayesian risk-based decision method for model validation under uncertainty. *Reliab. Eng. Syst. Saf.* **2007**, *92*, 707–718. [CrossRef]
40. LeCun, Y.; Bengio, Y.; Hinton, G. Deep learning. *Nature* **2015**, *521*, 436–444. [CrossRef] [PubMed]
41. Song, K.; Yan, Y. A noise robust method based on completed local binary patterns for hot-rolled steel strip surface defects. *Appl. Surf. Sci.* **2013**, *285*, 858–864. [CrossRef]
42. Espíndola, R.P.; Ebecken, N.F.F. On extending f-measure and g-mean metrics to multi-class problems. In *Sixth International Conference on Data Mining, Text Mining and Their Business Applications*; WIT Press: Ashurst Lodge, UK, 2005; Volume 35, pp. 25–34. [CrossRef]
43. Liu, W.; Anguelov, D.; Erhan, D.; Szegedy, C.; Reed, S.; Fu, C.-Y.; Berg, A.C. *SSD: Single Shot Multibox Detector*; Springer International Publishing: Cham, Switzerland, 2016; pp. 21–37. [CrossRef]
44. Chen, J.; Liu, Z.; Wang, H.; Nunez, A.; Han, Z. Automatic Defect Detection of Fasteners on the Catenary Support Device Using Deep Convolutional Neural Network. *IEEE Trans. Instrum. Meas.* **2018**, *67*, 257–269. [CrossRef]
45. Redmon, J.; Farhadi, A. Yolov3: An incremental improvement. *arXiv* **2018**, arXiv:1804.02767. [CrossRef]
46. He, Y.; Song, K.; Meng, Q.; Yan, Y. An End-to-End Steel Surface Defect Detection Approach via Fusing Multiple Hierarchical Features. *IEEE Trans. Instrum. Meas.* **2020**, *69*, 1493–1504. [CrossRef]

sensors

MDPI

Article

An Optimization Method of Production-Distribution in Multi-Value-Chain

Shihao Wang [1,2], Jianxiong Zhang [1,2], Xuefeng Ding [1,2], Dasha Hu [1,2], Baojian Wang [1,2], Bing Guo [1,2], Jun Tang [1,2,3], Ke Du [3], Chao Tang [3] and Yuming Jiang [1,2,*]

[1] College of Computer Science, Sichuan University, Chengdu 610065, China
[2] Big Data Analysis and Fusion Application Technology Engineering Laboratory of Sichuan Province, Chengdu 610065, China
[3] Changhong Central Research Institute, Sichuan Changhong Electronic (Group) Co., Ltd., Mianyang 621000, China
[*] Correspondence: jiangym@scu.edu.cn

Abstract: Value chain collaboration management is an effective means for enterprises to reduce costs and increase efficiency to enhance competitiveness. Vertical and horizontal collaboration have received much attention, but the current collaboration model combining the two is weak in terms of task assignment and node collaboration constraints in the whole production-distribution process. Therefore, in the enterprise dynamic alliance, this paper models the MVC (multi-value-chain) collaboration process for the optimization needs of the MVC collaboration network in production-distribution and other aspects. Then a MVC collaboration network optimization model is constructed with the lowest total production-distribution cost as the optimization objective and with the delivery cycle and task quantity as the constraints. For the high-dimensional characteristics of the decision space in the multi-task, multi-production end, multi-distribution end, and multi-level inventory production-distribution scenario, a genetic algorithm is used to solve the MVC collaboration network optimization model and solve the problem of difficult collaboration of MVC collaboration network nodes by adjusting the constraints among genes. In view of the multi-level characteristics of the production-distribution scenario, two chromosome coding methods are proposed: staged coding and integrated coding. Moreover, an algorithm ERGA (enhanced roulette genetic algorithm) is proposed with enhanced elite retention based on a SGA (simple genetic algorithm). The comparative experiment results of SGA, SEGA (strengthen elitist genetic algorithm), ERGA, and the analysis of the population evolution process show that ERGA is superior to SGA and SEGA in terms of time cost and optimization results through the reasonable combination of coding methods and selection operators. Furthermore, ERGA has higher generality and can be adapted to solve MVC collaboration network optimization models in different production-distribution environments.

Keywords: multi-value-chain collaboration; optimal planning; high-dimensional decision space; production-distribution; genetic algorithm

1. Introduction

As global markets become more efficient, competition no longer occurs between individual firms, but it instead occurs across the entire value chain [1], and more and more firms seek to enter the market as an organization in pursuit of greater efficiency in sourcing, production, and distribution. As value chains become larger and more complex in a globalized world, it is important to work efficiently with partners along the chain [2–4]. Higher levels of business performance will be achieved by companies that can fully integrate value and supply concurrent streams [5]. In this context, collaboration by two or more companies that create competitive advantage and higher profits than acting alone has been studied by scholars. Ref. [6] divided the collaboration approaches proposed by scholars into vertical collaboration, horizontal collaboration, and lateral collaboration. In [7], it is proposed that lateral collaboration includes the ability of horizontal and vertical collaboration overcoming their limitations, which still requires a lot of research. Besides, the effect of inventory strategy on collaboration was studied in the lateral collaboration model in [7]. Ref. [8] developed a framework based on scenarios designed for collaborative supply chains and then suggested novel models of lateral collaboration to optimize 3BL (triple bottom line) sustainability-related objectives. Ref. [9] modeled the MNE (multi-national enterprise) as a complex adaptive system and investigates the effects of lateral collaboration on performance at both the MNE and subsidiary level. It is found that the existing lateral collaboration models are weak in terms of resource and task allocation and node collaboration constraints in the whole production-distribution process.

In MVC collaboration networks, there is both vertical cooperation on the same chain and horizontal cooperation on different chains, so it is necessary to study the MVC collaboration mechanism and construct an optimization model. Actually, when the value chains operate in collaboration, more collaboration constraints arise and higher requirements are imposed on the collaboration mechanism.

In the MVC collaboration network, there are intersecting nodes between multiple value chains. These nodes represent the enterprises that undertake the tasks of multiple value chains. With the influence of enterprise capacity, warehouse inventory capacity, and other uncertain factors during the operation of the MVC collaboration network, there are many constraints on the intersecting nodes. Such constraints include the time to deliver goods to agents cannot exceed the specified time; production tasks assigned to enterprises in the collaboration network cannot exceed the enterprise capacity; and the number of products stored in the warehouse cannot exceed the upper limit of the storage capacity of the transit warehouse. Therefore, it is necessary to construct a collaboration model to effectively coordinate resources; coordinate the division of labor and the cooperation between enterprises of the value chains; meet the constraints of value chain collaboration; and reduce the internal friction caused by the unreasonable assignment of tasks or resources.

The production-distribution decision optimization problem of a MVC collaboration network model is a combinatorial planning problem with constraints, while existing research shows that metaheuristic algorithms can solve it effectively. For example, Ref. [10] proposed a service selection model based on collaboration effects and designed an improved GSA algorithm to solve it. Ref. [11] proposed an enhanced multi-objective gray wolf optimizer to solve the manufacturing service optimization combination problem. Therefore, considering the effectiveness and superiority of heuristic algorithms [12–14], this study uses genetic algorithms to optimize the production-distribution strategy of MVC collaboration networks.

In summary, there are few existing studies on resource management and task assignment in MVC collaboration networks. Therefore, the purpose of this paper is to investigate how to adjust the inter-node constraints in value chains after the partners of a MVC have been identified, and to effectively conduct production-distribution planning with the aim of reducing the total production-distribution cost of the collaboration network. To this end, this paper makes the following contributions and innovations:

(1) Taking the definition of collaboration relationship among value nodes in a MVC as the entry point, the collaborative process of the MVC is modeled in the production-distribution scenario of multi-task, multi-production end, multi-distribution end with multi-level inventory in dynamic alliances of manufacturing enterprises. A MVC collaboration network optimization model is constructed with the lowest total production-distribution cost as the optimization objective and with delivery lead time and task quantity as constraints.

(2) A genetic algorithm is used to solve the optimization model, which maps the collaboration constraints of nodes in the value chain to the constraints among genes and solves the conflict problem of MVC collaboration network nodes. In view of the multi-level characteristics of the production-distribution scenario, two chromosome coding methods are proposed in this paper: staged coding, which is applicable to the high requirements for the

distribution quantity and time of each stage; and integrated coding, which is applicable to the pursuit of lower cost of the whole production-distribution process.

(3) This paper proposes an ERGA with enhanced elite retention based on SGA. The comparative experiment results and population evolution process show that ERGA outperforms SGA and SEGA in terms of time cost and optimization results through the reasonable combination of coding schemes and selection operators. Besides, ERGA has higher generality and can be adapted to the solution of MVC collaboration network optimization models in different production-distribution environments.

2. Related Work

2.1. Research Status of Value Chain Collaboration

Along with the internationalization of markets and the enhancement of communication technologies, scholars have proposed a variety of collaboration models. The first is the proposal of a collaboration framework used to build collaboration and partner selection. Ref. [15] developed a virtual e-chain model that presents a supply chain collaboration framework in a virtual environment for classifying the roles of partners, identifying the key capabilities for building each collaboration, and assessing the readiness of partners for collaboration. Ref. [16] developed a theoretical framework on value chains and outlined a value chain process model. To address the multi-cycle and multi-stage problem, ref. [17] proposed an integrated framework to identify effective collaboration mechanisms in a MVC throughout its life cycle. To investigate the collaboration process, scholars have proposed vertical collaboration, horizontal collaboration, and lateral collaboration models. Vertical collaboration can be defined as collaboration when two or more organizations, such as manufacturers, distributors, operators, and retailers, share their responsibilities, resources, and performance information to serve relatively similar end customers. Ref. [18] focused on the process of vertical collaboration between hotels and their food and beverage suppliers, revealing the impact of trust on transaction costs. Ref. [19] developed three different simulation models based on a simple vertical supply chain with one retailer and one operator to investigate the impact of CTM (collaborative transportation management) on total dealer costs and service levels. Horizontal collaboration refers to the collaboration between two or more companies at the same level in a network to facilitate the achievement of common goals, and it can also be referred to as external collaboration. A horizontal collaboration model is proposed using hierarchical analysis to comprehensively assess the degree of collaboration with a single horizontal collaboration program to check the feasibility of meeting customer needs [20]. Lateral collaboration refers to the combination of vertical and horizontal collaboration to obtain increased flexibility and improved resource sharing [21]. Vertical collaboration in supply chains has been investigated intensively in recent years. However, relatively few studies have focused on the horizontal or lateral modes of collaboration [8]. Among them, even less work has been done on lateral collaboration models in terms of resource and task allocation and node collaboration constraints in the whole production-distribution process. However, in the MVC collaboration network for the manufacturing industry, there are often cross nodes between multiple value chains, and the enterprises represented by these nodes undertake multiple tasks division of multiple value chains and have both horizontal and vertical collaboration with other enterprises, so it is necessary to study the MVC collaboration mechanism and construct an optimization model.

2.2. Research of Genetic Algorithms

The value chain network design problem in the optimization model is a combinatorial planning problem with multiple constraints, and the metaheuristic algorithm has been proven to be an effective tool to solve the combinatorial planning problem with good performance in many scenarios [22–27]. Besides, in recent years, some scholars have combined genetic algorithms into other fields and achieved good results [28–31].

GAs (genetic algorithms) are a special class of bionic methods that mimic Darwin's theory of evolution and natural selection, which are well-known population-based metaheuristic algorithms [32,33]. When implementing GAs, the definition of a chromosome, i.e., a string representing a possible solution to a problem, is a key issue that strongly influences the performance of the algorithm in terms of the quality of the optimal solution and the speed of convergence [34]. Encoding strategies are widely used today, such as real-valued encoding, Prufer encoding, and priority-based encoding [35]. The most important operator in genetic operators is the selection operator, which is the key factor affecting the performance of GAs [36]. For common selection operators, such as the roulette wheel operator and the tournament selection operator [37,38], they cannot guarantee that the best-performing individuals can be left in each evolution, while the advantage of the elite retention strategy is that the best individuals of a certain generation cannot be destroyed by genetic operations during the search process, which can guarantee the convergence of genetic algorithms [39]. Therefore, to improve the convergence of the algorithm, the elite retention strategy has been combined with the selection strategy [40]. Genetic operators, including crossover and mutation, are then applied with some probability to the chromosomes to create new and potentially better solutions. The main purpose of crossover operations is to enable the algorithm to explore the search space more efficiently and to produce better individuals in the next generation [41]. However, for some specific problems, the crossover may produce infeasible solutions, so the crossover operator needs to be selected for specific problems, such as the cryptography problem [42]. In conclusion, genetic algorithms need improvement in specific application scenarios, so this paper investigates the combination of coding and operators in a MVC collaboration optimization model.

3. Optimization Model of MVC Collaboration Network

3.1. Model Mathematical Symbols

The meanings of mathematical symbols used in the paper are shown in the following Table 1.

Table 1. Symbol table.

Symbol	Description	Symbol	Description
VC	Collection of value chains	$MaxQ_{tw_i}$	The maximum storage quantity of transit warehouse tw_i
VC_i	The i-th value chain	T_{a_i}	The maximum lead time of agent a_i
n_i^j	The j-th value node on the i-th value chain	TC	The total cost of the MVC collaboration network
P	Collection of products	TPC	The total production cost of the MVC collaboration network
P_{tol}	Total number of products	TFC	Total fixed cost of the MVC collaboration network
p_i	The products of the i-th category, $p_i \in P$	TTC	Total transportation cost of the MVC collaboration network
E	Collection of enterprises	$C_{e_i}^{p_i}$	Daily capacity of enterprise e_i to produce product p_i
E_{tol}	Total number of enterprises	$T_{e_i}^{p_i}$	Time for enterprise e_i to complete the order of product p_i
e	The e-th enterprise, $e \in E$	$T_{tw_i}^{e_i}$	The time it takes for products to be distributed from enterprise e_i to transit warehouse tw_i
TW	Collection of transit warehouses	$T_{a_i}^{tw_i}$	The time it takes for products to be distributed from transit warehouse tw_i to agent a_i
TW_{tol}	Total number of transit warehouses	F_1	Mapping function between variable cost and product quantity
tw_i	The i-th transit warehouse, $tw \in TW$	F_2	Mapping function of distribution cost, product quantity, and path length
A	Collection of agents	L	Matrix of path length
A_{tol}	Total number of agents	l_i^j	The path length from node i to node j
a_i	The i-th agent, $a_i \in A$	fit	Fitness function
$D_1^{p_i}$	Distribution solution in the first stage of product p_i	$Q_{e_i}^{p_i}$	Number of product p_i produced by enterprise e
$d_{e_i}^{tw_j}$	Total number of the products distributed from enterprise e_i to transit warehouse tw_j	$Q_{tw_i}^{p_i}$	Number of product p_i stored in transit warehouse tw_i
$D_2^{p_i}$	Distribution solution in the second stage of product p_i	$Q_{a_i}^{p_i}$	Quantity of product p_i purchased by agent a_i
$d_{tw_i}^{a_j}$	Total number of the products distributed from transit warehouse tw_j to agent a_j		

3.2. Value Node Collaboration Definition

The essence of MVC collaboration is the collaboration among value nodes, and the collaboration strategy is the central issue of value chain collaboration, including the collaboration of the division of labor of each link in the value chain, such as production collaboration strategy, sales collaboration strategy, logistics collaboration strategy, and so on. To better construct a model of the MVC collaboration network, the following value-node collaborative relationship definitions are given.

Definition 1. *Collaboration between multiple value chains. Define the value chain set* $VC = \{VC_1, VC_2, \ldots, VC_n\}$ *as n value chains that operate collaboratively. Each value chain is composed of multiple value nodes, the value node can be an enterprise, a production line, a workshop, and other units that can complete value-added activities. If* $VC_i = \{n_i^1, n_i^2, \ldots, n_i^m\}$, $VC_i \in VC$ *denotes that the i-th collaborative value chain consists of m nodes.* n_i^1 *and* n_i^m *respectively represent the starting point and ending point of the value chain* VC_i*. If* $VC_i \cap VC_j = \left\{n_i^1, n_i^2, \ldots, n_j^1, n_j^2\right\}$, $VC_i, VC_j \in VC$ *, then value node* $n_i^1, n_i^2, \ldots, n_j^1, n_j^2$ *simultaneously undertake the labor of the value chain* VC_i *and* VC_j*, which are the intersection nodes of the value chains.*

Definition 2. *Collaboration within the value chain. There are many mixed relationships in the product production process. The value chain composition includes parallel, selection, loop, and other mixed link structures, so there are also collaboration relationships on the same value chain.* $CRv(n_l^i, n_l^j) = 1$ *means the i-th and the j-th value node of the l-th value chain have a collaboration relationship. Inversely,* $CRv(n_l^i,\ n_l^j) = 0$ *means they do not have collaboration relationship.*

Definition 3. *Collaboration relationship between nodes of multiple value chains.* n_a^i *and* n_b^j *are value nodes of value chain* VC_a *and* VC_b *, respectively, and the corresponding types of value-added services are* s_a *and* s_b *. When* $s_a = s_b$*, the node* n_a^i *and the node* n_b^j *belong to the same value-added service type, i.e., their capabilities are similar, so they have the basis of collaboration. That is, with certain conditions, they can cooperate. The synthetic relation can be described as* $CRv(n_a^i,\ n_b^j) = 1$*. When there is no collaboration relationship between nodes,* $CRv(n_a^i,\ n_b^j) = 0$*. Extend this relation to the whole network,* $CRv\left\{n_a^i,\ n_b^j, \ldots, n_c^k\right\} = 1$ *means the node* $n_a^i,\ n_b^j,\ \ldots,\ n_c^k$ *have collaboration relationship among them.* $n_a^i,\ n_b^j,\ \ldots,\ n_c^k$ *are different value nodes in different value chains.*

The essence of MVC collaboration lies in the collaboration between value chain nodes to achieve win-win results. Through the above definition, the collaboration relationship between nodes on the same chain and the collaboration relationship between nodes on different chains are described.

3.3. Description of Collaboration Network Scenario

The MVC collaboration network as shown in Figure 1 consists of enterprises, transit warehouses, logistics, and agents. There are three kinds of products in Figure 1, so there are three value chains (VC_1, VC_2, and VC_3) operating collaboratively.

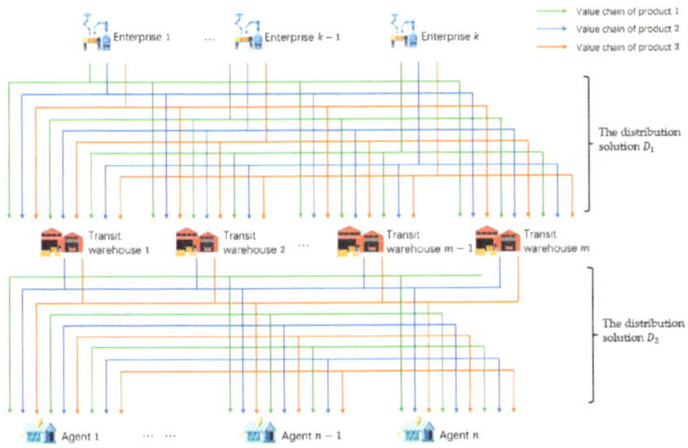

Figure 1. The business scenario of a MVC collaboration network.

Taking VC_1 and VC_2 as an example, the process of obtaining the solution can be divided into three steps. Firstly, the agents order the products, then the task-issuing platform summarizes the order information. Next, according to the order information, the platform chooses $n_1^i, n_1^j, \ldots, n_2^k$ from the value nodes with collaborative basis, i.e., $CRv\left\{n_1^i, n_1^j, \ldots, n_2^k\right\} = 1$, on L_1 and L_2 to complete the orders. Then, based on the information of each node, the solution with the minimum total cost of the collaboration network is formulated. The solution can be divided into two parts, the first part is the production solution, and task issuing platform generates task sets for enterprises according to production capacity. The other part is the distribution solution. Enterprises will produce the specified products within a certain period. If each enterprise directly transfers the products to the nearest transit warehouse, it may not be able to store the products due to the limited storage capacity of the warehouse, which will lead to a conflict between manufacturing nodes and transit nodes. Therefore, it is necessary to find a product distribution solution on the premise of not exceeding the upper limit of each warehouse capacity and meeting the order quantity and time required by agents. The distribution solution includes two stages, that is, the first stage of distribution solution (D_1) from the enterprises to the transit warehouses, and the second stage of distribution solution (D_2) from the transit warehouses to the agents.

The overall distribution route of the product is shown in Figure 1. The products produced by enterprises are distributed to the corresponding transit warehouses for storage according to the distribution solution D_1. The products stored in the transit warehouses are then distributed to the designated agents according to the distribution solution D_2 and complete a transaction.

3.4. Assumptions and Descriptions

The MVC collaboration network model is constructed based on the following assumptions:

(1) The types and quantities of products produced by the enterprise alliance are equal to the types and quantities in the orders of the agents.
(2) The total time of production and distribution shall not exceed the maximum lead time (T_{a_i}) of the order.
(3) The tasks assigned to the enterprise shall not exceed the upper limit of the production capacity of them, and the number of goods distributed by the enterprises to the transit warehouses shall not exceed the upper limit of the storage capacity of the transit warehouses.
(4) The distribution cost and time of a single item from the enterprise to the transit warehouse are known and fixed; the delivery cost and delivery time of a single item from the transit warehouse to the agent are also known and fixed.

(5) There is no secondary processing of manufactured products. The route from the warehouse of enterprises to the transit warehouse and then to the warehouse of agents only has three links: the warehouse of enterprises, the transit warehouse, and the warehouse of agents. Each link only goes through once and there is no loop.
(6) The selected enterprises can meet the product production requirements put forward by agents.
(7) Each value chain of the collaboration network is represented by the value increment process of each product, and there are inter-value-chain and intra-value-chain collaborations in the network.
(8) To meet the customized requirements of agents, the manufacturing of products often involves specific parameter requirements, so the production is not in advance.

3.5. Decision Variables

In the MVC collaboration operation, there is a decision space composed of the production solution and the distribution solution, which has a decisive influence on the production-distribution cost. There are P_{tol} types of products, E_{tol} manufacturers, and C_{tol} agents in the MVC collaboration network model. The production solution of enterprise i is represented as shown in Equation (1).

$$mq_{e_i} = \left(Q_{e_i}^1, Q_{e_i}^2, \ldots, Q_{e_i}^p\right) \tag{1}$$

The overall production solution of the MVC collaboration network is represented as shown in Equation (2).

$$MQ = \begin{bmatrix} Q_{e_1}^1 & \cdots & Q_{e_1}^p \\ \vdots & \ddots & \vdots \\ Q_{em}^1 & \cdots & Q_{em}^p \end{bmatrix} \tag{2}$$

The distribution solution D_1^p of product p in the first stage of the MVC collaboration network is shown in Equation (3).

$$D_1^p = \begin{bmatrix} d_{e_1}^{t_1} & \cdots & d_{e_1}^{t_j} \\ \vdots & \ddots & \vdots \\ d_{e_i}^{t_1} & \cdots & d_{e_i}^{t_j} \end{bmatrix} \tag{3}$$

The distribution solution D_2^p of product p in the second stage of the MVC collaboration network is shown in Equation (4).

$$D_2^p = \begin{bmatrix} d_{t_1}^{c_1} & \cdots & d_{t_1}^{c_j} \\ \vdots & \ddots & \vdots \\ d_{t_i}^{c_1} & \cdots & d_{t_i}^{c_j} \end{bmatrix} \tag{4}$$

3.6. Optimization Objective Function

The MVC collaboration network is optimized mainly from the perspective of cost based on the actual production-distribution situation. Therefore, the minimum total cost TC is select as the optimization objective function, which is composed of the total production cost and total distribution cost.

The total cost in the collaboration model should include the cost of PSs (production sites) and DCs (distribution centers), etc., and the cost of PSs includes the operating and manufacturing cost [8]. But this paper main focus of study is the optimization of production-distribution strategy, so the total production cost (TPC) is divided into total variable cost and total fixed cost, and the formula is shown in Equation (5). The total fixed cost represents the sum of the fixed inputs produced during a period, such as a series of expenses incurred in the operation of a production plant, which does not vary with the number of products produced. The total variable cost will increase with the increased number of products produced by the enterprises. The expression is shown in Equation (6).

$$TPC = TVC + TFC \tag{5}$$

$$TVC = F_1(Q) \tag{6}$$

Total distribution cost (TTC) consists of two parts: one is the distribution cost generated by shipping from the enterprises to the transit warehouses, and the other is the distribution cost generated by shipping from the transit warehouses to the agents. To simplify distribution cost calculation, batch pricing is adopted, and distribution cost is positively correlated with distribution route length and product quantity [35], as shown in Equation (7).

$$TC = F_2(Q, L) \tag{7}$$

The matrix of path length L is shown in Equation (8).

$$L = \begin{bmatrix} l_1^1 & \cdots & l_1^j \\ \vdots & \ddots & \vdots \\ l_i^1 & \cdots & l_i^j \end{bmatrix} \tag{8}$$

In the MVC collaboration network model, the quantity and variety of manufactured products are determined by the orders of agents. Therefore, when the quantity of products is fixed, the distribution route solution is the decisive factor affecting the distribution cost and the production solution is the decisive factor affecting the production cost. Therefore, the optimization objective function of total production cost and total distribution cost can be obtained as Equation (9).

$$\begin{cases} minTC = min(TMC + TTC) \\ TPC = \sum_{p\in P} \sum_{e\in E} F_1\left(Q_e^p\right) + TFC \\ TTC = \sum_{k=1}^{P_{tol}} \sum_{j=1}^{T_{tol}} \sum_{i=1}^{E_{tol}} F_2\left(D_1^{p_k}{}_{ij}, l_{e_i}^{t_j}\right) + \sum_{k=1}^{P_{tol}} \sum_{j=1}^{C_{tol}} \sum_{i=1}^{T_{tol}} F_2(D_2^{p_k}{}_{ij}, l_{t_i}^{c_j}) \end{cases} \tag{9}$$

3.7. Constraint Condition

In the operation of the MVC collaboration network, there are many constraints: the capacity of enterprises; the inventory capacity of transit warehouses; and uncertain factors. According to our assumptions in Section 3.4, the important constraints set in the model are as follows:

1. The enterprises in the MVC collaboration network cooperate to complete all the agents' orders, and the variety and quantity of the products must meet the agents' order requirements.

$$\forall p \in P, \sum_{i=1}^{M_{tol}} Q_{e_i}^p = \sum_{j=1}^{C_{tol}} Q_{c_j}^p \tag{10}$$

2. In Equation (10): $\sum_{i=1}^{M_{tol}} Q_{e_i}^p$ represents the sum of the quantity of product p produced by enterprises; $\sum_{j=1}^{C_{tol}} Q_{c_j}^p$ represents the agents' total need of the quantity of product p.
3. The agent's order has a maximum lead time (T_{a_i}), that is, the time taken from production to delivery to the agents has to less than T_{a_i}. Enterprises are limited by site scale, human resources, production equipment, and other resources, and each enterprise has a limit of daily production capacity. Therefore, the production tasks assigned to enterprises in the collaboration network cannot exceed their production capacity. Production time is positively correlated with daily production capacity, and the distribution time is positively correlated with the distribution route length. Since agent orders cannot be split, products required by agent a_i will be uniformly sent out after

each enterprise finishes production and delivery to the designated transit warehouse, and the following time constraint (Equation (11)) will be imposed on each agent's order:

$$\left(T_e^{p_i} + T_{tw}^e + T_{a_i}^{tw}\right) \leq T_{a_i} \tag{11}$$

4. Limited by the scale and supporting resources, the storage capacity of the transit warehouses is limited, and the number of stored products cannot exceed the upper limit of the storage capacity of the transit warehouses Equation (12).

$$\forall tw \in TW, \sum_{i=1}^{P_{tol}} Q_{tw}^{p_i} \leq MaxQ_{tw} \tag{12}$$

5. The quantity of products sent by the transit warehouse to the agent, i.e., the export of the transit warehouse, depends on the number of products sent by the enterprise to the transit warehouse Equation (13), i.e., the import of the transit warehouse.

$$\sum_{i=1}^{E_{tol}} d_{e_i}^{tw_j} = \sum_{k=1}^{A_{tol}} d_{tw_j}^{a_k} \tag{13}$$

Therefore, the constraints existing in the MVC collaboration network are shown in Equation (14).

$$S.T.\begin{cases} \sum_{i=1}^{E_{tol}} Q_{e_i}^{p_i} = \sum_{j=1}^{A_{tol}} Q_{a_j}^{p_i} \\ \left(T_e^{p_i} + T_{tw}^e + T_{a_i}^{tw}\right) \leq T_{a_i} \\ \forall tw \in TW, \sum_{i=1}^{P_{tol}} Q_{tw}^{p_i} \leq MaxQ_{tw} \\ \sum_{i=1}^{E_{tol}} d_{e_i}^{tw_j} = \sum_{k=1}^{A_{tol}} d_{tw_j}^{a_k} \end{cases} \tag{14}$$

The MVC collaboration network scenario is described in this chapter. Based on the analysis of the collaboration relationship and the production-distribution business scenario of MVC in enterprise dynamic alliance, the optimization model of the MVC collaboration network is constructed.

4. Algorithm for Optimizing the MVC Optimization Model

The above model is a single objective optimization problem with constraints. In the model constructed, the scale of solution space will increase exponentially with the increase of decision variables caused by the increased number of product types, enterprises, warehouses, and agents. It is difficult to solve the problem with conventional solution methods, so the evolutionary algorithm is adopted to solve it. Evolutionary algorithms (EAs) have a long history of successfully solving COPs (constrained optimization problems) [43]. One of the EAs, the GA, is a probability-based random search algorithm [44], which can search for the optimal solution or sub-optimal solution in the solution space, and is an effective method to solve the optimization problem with a high-dimensional solution space [45].

4.1. Algorithm Flow

The process of the genetic algorithm adopted is shown in Figure 2. After setting the node collaboration relationship and constraint conditions of the model, the production-distribution solution that minimizes the total cost of the MVC collaboration network model can be obtained by the algorithm. To select the appropriate encoding scheme and selection operator, the encoding scheme and selection operator will be determined after comparative experiments.

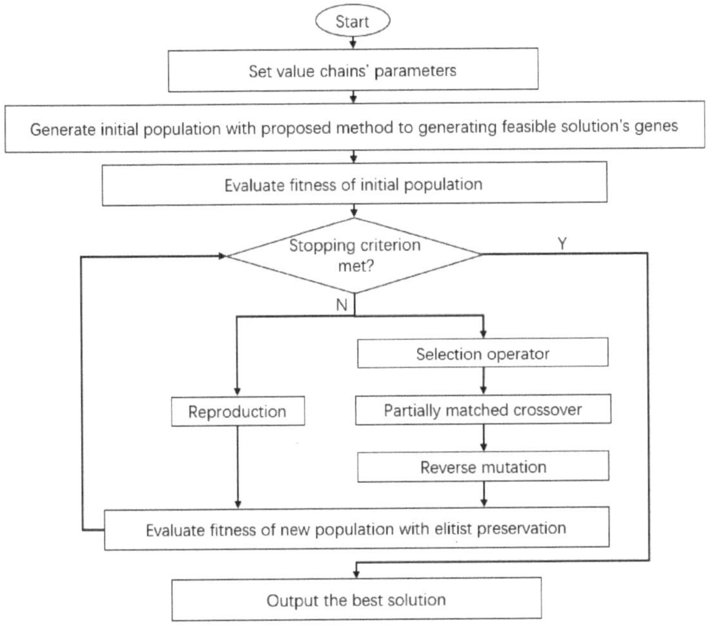

Figure 2. Process of the genetic algorithm.

4.2. Chromosome Encoding

Chromosome encoding is designed to match genes with solutions. There are several chromosome encoding methods, such as binary encoding and real-value encoding. To improve the encoding feasibility and expansibility [33], real-value encoding is adopted.

Considering the characteristics of the product solution and distribution solution, the high-dimensional decision variable matrix composed of production decision variables and distribution decision variables shown in Equations (2)–(4) is converted into a one-dimensional matrix, which is represented by integer encoded chromosomes. Since multiple products on multiple value chains are involved, and the logistics part of this scenario is a two-stage distribution problem, before the encoding of the logistics solution of the two stages, the following two issues need to be taken in account: (1) how to meet the constraints of the two stages; (2) whether the encoding method used has high solving efficiency and high-quality solution. According to these, two encoding methods are proposed: staged encoding and integrated encoding.

Method (1): staged encoding, that is, the genetic algorithm is used to obtain the optimal distribution solution of the first stage firstly, and the solution of the first stage is used as conditions to obtain the solution of the second stage. Three products, three enterprises, four transit warehouses, and three agents were taken as conditions to build an MVC collaboration network model. The possible values of each group of decision variables MQ and D_1 of the first stage were represented by one-dimensional decimal chromosomes, as shown in Figure 3.

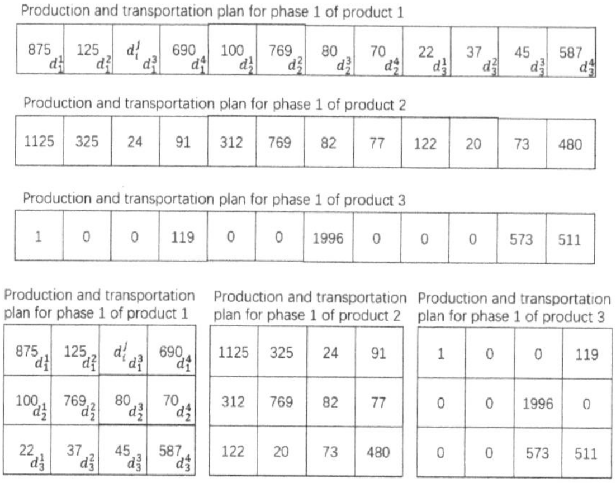

Figure 3. Encoding of solutions in the first stage.

In the MVC collaboration network model, there are 3 enterprises and 4 transit warehouses in the distribution process of the first stage. The genes in the chromosome represent the number of products delivered from the enterprise to the transit warehouse, the subscript represents the serial number of the corresponding enterprise, and the superscript corresponds to the serial number of the transit warehouse.

In Figure 3: d_i^j is part of the distribution solution of the i-th enterprise for illusion, in other words, it is the number of products distributed from the i-th enterprise to the j-th transit warehouse. $\sum_{j=1}^{TW_{tol}} d_i^j$ is the total number of products produced by enterprise i, i.e., the production solution of enterprise i.

The second stage decision variable consists only of D_2, and the possible values for each group of D_2 are represented by the one-dimensional decimal chromosomes, as shown in Figure 4.

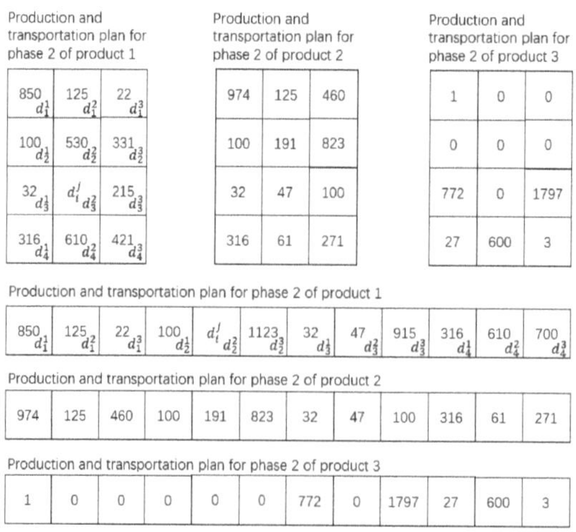

Figure 4. Encoding of solutions in the second stage.

In the MVC collaboration network model, there are 4 transit warehouses and 3 agents in the second stage of the distribution process. The genes in the chromosome represent the number of products transferred from the warehouses to the agents, the subscript represents the serial number of the corresponding transit warehouse, and the superscript corresponds to the serial number of the agent manufacturer.

In Figure 4: d_i^j is the distribution solution of the i-th transit warehouse, in other words, it is the total number of products distributed from the i-th transit warehouse to the j-th agent. $\sum_{j=1}^{TW_{tol}} d_i^j$ is the total number of products distributed by transit warehouse i. $\sum_{i=1}^{A_{tol}} d_i^j$ is the total number of products received by agent j.

Method (2): integrated encoding, that is, the distribution strategies of the two stages are encoded into one chromosome. The production solution of the enterprises, the production-distribution solution of the first stage, and the distribution solution of the second stage, are generated simultaneously according to the constraint conditions when the chromosome is randomly initialized. The relationship between the output and input of transit warehouses is maintained by constraints. The relationship is: the quantity of product p shipped from warehouse i to agents in the second stage is equal to the quantity of product p shipped to warehouse i in the first stage. The MVC collaboration network model is built with the experimental parameters in Section 5.3. The possible values of each group of decision variables (MQ, D_1, and D_2) are represented by one-dimensional decimal chromosomes as shown in Figure 5.

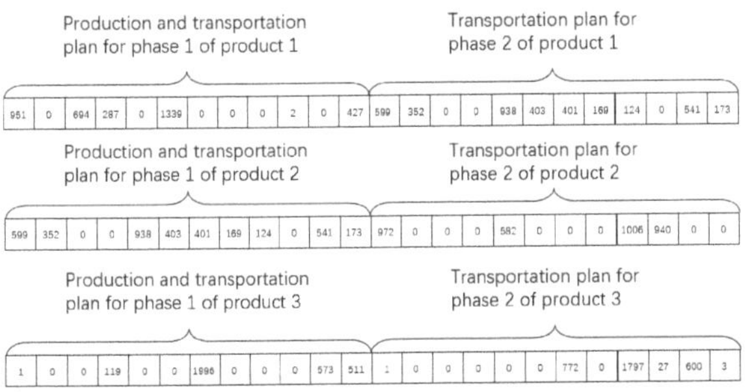

Figure 5. Integrated encoding of decision variables in the whole process.

According to the preliminary analysis of the encoding method, since the staged encoding and solving method minimizes the total cost by stages, there are fewer decision variables and a smaller solution space ranges in each stage, so the optimization effect of the total cost is not as good as the integrated encoding method in theory. However, because of fewer decision variables and more constraints, the solution speed of the staged encoding is faster. In Section 5, a comparative experiment and analysis of these two encoding methods are conducted.

4.3. Fitness Function

The fitness function is the standard used to distinguish the advantages and disadvantages of individuals. Because the research is about the optimization effect, the objective function is the total cost of the MVC collaboration network, that is, the sum of the total production cost and the total distribution cost. The calculation formula is shown in Equation (15).

$$fit = F_1(Q) + F_2(Q, L) \tag{15}$$

4.4. Selection Operator

The selection operator simulates the survival of the fittest in the natural biological evolution process, selecting individuals with high fitness and eliminating individuals with low fitness. The candidate selection operators are the widely used operators: the enhanced elite tournament selection operator and the roulette selection operator.

The roulette operator can make it possible for all individuals to be selected. While valuing the elite, it also leaves little chance for non-elite to survive, therefore the diversity will be better than the elite championship. However, at the same time, it will lead to slow population convergence. In Section 5, a comparative experiment of these two operators is conducted.

4.5. Crossover Operator

The crossover operator is one of the core elements of the genetic algorithm. By randomly selecting the chromosomes of parents to carry out allelic cross-exchange, the excellent genes of the parents can be passed on to the offspring in a certain probability, to produce excellent offspring in the crossover process. Since the proposed encoding methods have few requirements on crossover operator and the research focus is encoding method and selection operator, a simple two-point crossover-based partial matching crossover operator is adopted, randomly generates matching regions on the alleles of the paternal chromosomes, and exchanges genes of two paternal matching regions according to the crossover probability [46]. The partial matching crossover strategy is shown in Figure 6.

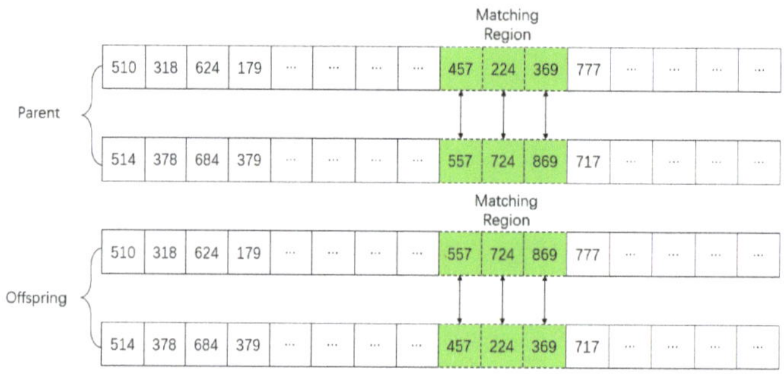

Figure 6. Chromosome crossover operator.

4.6. Mutation Operator

Mutation refers to the mutation of genes on individuals' chromosomes, thus generating new individuals. Mutation operation is one of the important means of genetic algorithms to produce excellent individuals. The reversal mutation operator [47] is used to randomly generate two reversal points on chromosome individuals, and the gene fragments between the two points are arranged in reverse order to generate new individuals, as shown in Figure 7.

Figure 7. Chromosome mutation operator.

According to the characteristics of production-distribution decision space, a genetic algorithm is used to solve the MVC collaboration network optimization problem. Moreover, according to the specific scenario of production-distribution, two chromosome encoding methods are proposed: staged encoding and integrated encoding. The selection operator, crossover operator, mutation operator are used to solve the problem, and specific methods of chromosome encoding are introduced in this chapter.

5. Experiment and Analysis

5.1. Experiment Environment

The experiment environment and device configuration are as follows: operating system: Windows 10; processor: Intel Core (TM) i5-7300HQ CPU @ 2.50 GHz; language: Python 3; integrated development environment: JetBrains PyCharm Community Edition.

5.2. Data of Experiment

Based on the household appliance industry and combined with the production, distribution, and sales data of a dynamic enterprise alliance, a MVC collaboration network of multiple production ends, multiple agents, and multiple products is constructed. There are 3 enterprises, 4 transit warehouses, and 3 agents in the network, and the simulation experiment is carried out for 3 kinds of products. Specific simulation data are shown in Table 2 of enterprises; Table 3 of transit warehouses; and Table 4 of agents.

Table 2. Information of the enterprises.

Enterprise No.	*MC* (¥ Per Piece)	*FC* (Million ¥)	C_e^p, (Daily Production)		
			Product 1	Product 2	Product 3
1	1550	0.5	1000	900	1100
2	1650	0.7	1100	700	1000
3	1600	0.6	1200	1000	900

Table 3. Information of the transit warehouses.

Warehouse No.	$MaxQ_t$ (Pieces)	D_t^e (km)		
		Enterprise 1	Enterprise 2	Enterprise 3
1	4000	100	2500	2300
2	3500	2200	800	2000
3	3000	2000	1400	1800
4	3500	1500	1600	1500

Table 4. Information of the agents.

Agent No.	T_c Per Day	Q_c^p, (Pieces)			D_c^t, (km)			
		Product 1	Product 2	Product 3	Ware-House 1	Ware-House 2	Ware-House 3	Ware-House 4
1	15	1000	2100	800	500	2300	2400	850
2	15	2000	600	600	3000	900	1800	1800
3	15	700	800	1800	3600	1700	700	2870

5.3. Experiment Parameters

5.3.1. Experiment Parameters for Encoding Methods

To ensure the objectivity of the comparison experiment, the algorithm, selection, crossover, and mutation operators used in the experiment are the same except for the encoding method. The algorithm parameters are shown in Table 5.

Table 5. Experiment parameters for the encoding methods.

Encoding Method	Population Size	Algorithm	Evolutionary Generations	Crossover Probability	Mutation Probability
Staged encoding	5000	SEGA	20,000	90%	20%
Integrated encoding	5000	SEGA	20,000	90%	20%

5.3.2. Experiment Parameters of Algorithms

The compared algorithms are SGA with the roulette operator and SEGA with the elite tournament operator. To ensure the objectivity of the comparison experiment, the two algorithms adopted unified parameters. The algorithm parameters are shown in Table 6.

Table 6. Algorithm parameters.

Algorithm	Population Size	Evolutionary Generations	Crossover Probability	Mutation Probability
SGA	5000	20,000	90%	20%
SEGA	5000	20,000	90%	20%

5.4. Analysis of Experiment Results

5.4.1. Analysis of the Results Obtained by Different Encoding Methods

For the staged encoding and integrated encoding, the SEGA algorithm with the same experiment parameters is used to solve the same problem 10 times. The best fitness and solution time of the last generation in ten experiments with different methods were recorded respectively for comparison. The results of different encoding methods are shown in Table 7.

Table 7. Experiment data of the encoding methods.

Best Fitness, f_{min} (¥)			Time Cost (s)	
Staged Encoding	Integrated Encoding	Gap of Best Fitness	Staged Encoding	Integrated Encoding
21,034,179	19,249,951	8.48%	609.305	1033.66
21,072,627	19,969,750	5.23%	610.713	1035.62
21,083,618	20,041,390	4.94%	617.838	1051.23
21,155,280	19,345,940	8.55%	588.161	1088.91
21,987,913	19,460,975	11.49%	592.410	1052.95
21,037,718	19,950,505	5.17%	608.124	1033.63
21,020,710	19,109,735	9.09%	583.551	1039.44
20,958,410	19,771,720	5.66%	572.628	1048.78
21,063,779	20,025,090	4.93%	566.634	1051.49
21,143,800	19,870,990	6.02%	579.786	1037.93

1. Comparative analysis of experiment results

As can be seen from the above results, the optimal fitness of the staged encoding is lower than that of the integrated encoding, because the overall process is divided by the staged encoding into two staged planning problems as shown in Figures 8 and 9.

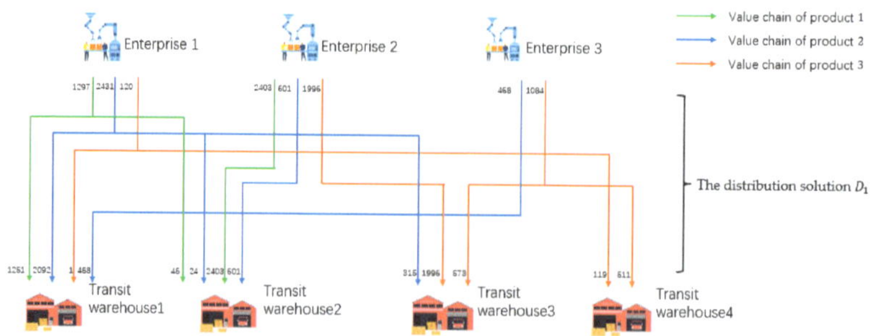

Figure 8. The first stage of production-distribution.

Figure 9. The second stage of production-distribution.

In the scenario of Section 3.3, on the premise that the constraint conditions of Section 3.6 are satisfied, and the experimental data of Section 5.2 are taken as an example, the distribution strategy with the minimum distribution cost in the first stage of the product is set as shown in (16). In (16), the left matrix is the distribution strategy of product 1, the middle matrix is the distribution strategy of product 2, and the right matrix is the distribution strategy of product 3. d_2^3 means d pieces of products shipped from enterprise 2 to transit warehouse 3.

$$\left[\begin{array}{cccc|cccc|cccc} 1251 & 46 & 0 & 0 & 2092 & 24 & 315 & 0 & 1 & 0 & 0 & 119 \\ 0 & 2403 & d_2^3 & 0 & 0 & 601 & 0 & 0 & 0 & 0 & 1996 & 0 \\ 0 & 0 & 0 & 0 & 468 & 0 & 0 & 0 & 0 & 0 & 573 & 511 \end{array}\right] \tag{16}$$

The distribution solution of the second stage is shown in (17). Matrices (16) and (17) are staged solutions with the lowest cost of each stage as the optimization objective function. The integrated method takes the total cost of the entire MVC collaboration network as the optimization objective function. Therefore, there is a solution D in this method in theory: the production and distribution cost of the first stage of D is higher than that of (16), but the total cost of D is lower than the summary cost of (16) and (17).

$$\left[\begin{array}{ccc|ccc|ccc} 626 & 0 & 625 & 2090 & 470 & 0 & 1 & 0 & 0 \\ 374 & 2000 & 75 & 0 & 0 & 625 & 0 & 0 & 0 \\ 0 & 0 & 0 & 10 & 130 & 175 & 772 & 0 & 1797 \\ 0 & 0 & 0 & 0 & 0 & 0 & 27 & 600 & 3 \end{array}\right] \tag{17}$$

2. Comparative analysis of the evolution process

In terms of the speed of the solution, because the optimization is carried out in stages, each stage has more constraints, fewer decision variables, and a smaller solution space, so the speed of the solution is faster. This feature is more prominent in the comparison of the evolution process in the first and second stages (Figure 10).

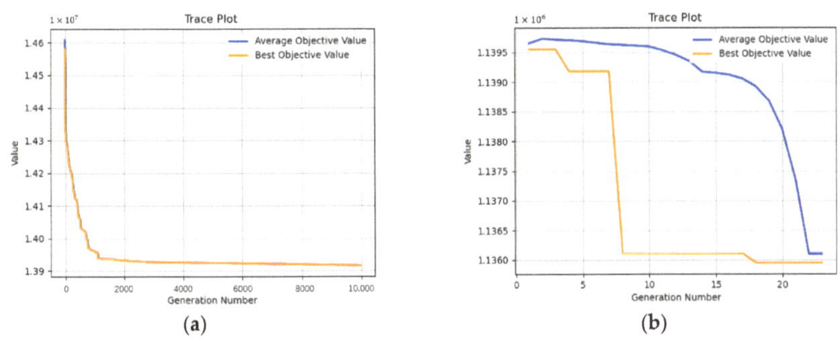

Figure 10. The evolution processes of staged methods: (**a**) the first stage of staged encoding; (**b**) the second stage of staged encoding.

Based on the experiment results and the images of the evolution processes (Figures 10 and 11), it can be concluded that staged encoding is faster than integrated encoding because of more constraints, smaller optimization space, and fewer decision variables, so it is more suitable for the MVC network with multiple products where the time cost is longer. Besides, it is easy to assign different constraints to the MVC network in different stages according to the requirements of nodes, so the staged encoding is suitable for complex scenarios that have many requirements on the distribution quantity and time at each stage. However, in terms of the optimization, the staged encoding is inferior to the integrated encoding, because the integrated encoding is used to optimize the entire production and distribution process, which has more decision variables and larger optimization space. Although integrated encoding needs longer solving time, the cost of the optimization result is 7% lower on average than that of the staged encoding. So, integrated encoding is suitable for scenarios where cost reduction is given priority and there are no strict requirements for each stage of the value chain.

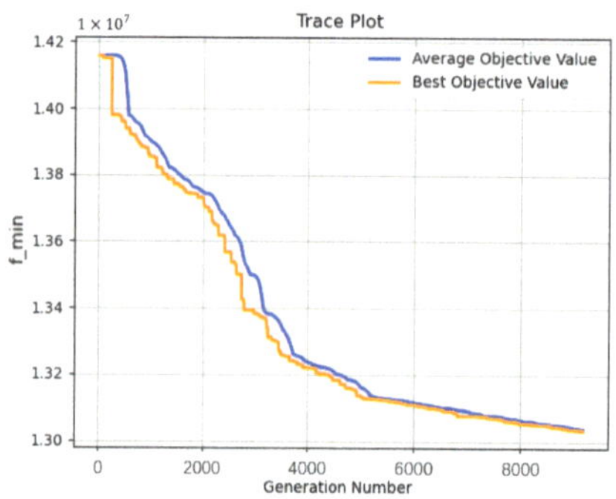

Figure 11. The evolution process of integrated encoding.

5.4.2. Comparative Analysis of Solving Algorithms

1. Comparative analysis of SGA and SEGA

With the above algorithm parameters and the integrated encoding method adopted, the same problem is solved 10 times by both the SEGA algorithm, which adopts the elitist tournament operator, and by the SGA algorithm, which adopts the roulette operator. The worst fitness, optimal fitness, algorithm time cost, and fitness standard deviation of the last generation in ten experiments of different algorithms were recorded respectively for comparison. The results of different algorithms are shown in Table 8 and Figure 12.

Table 8. Experiment results of SGA, SEGA, and ERGA.

Best Fitness, *f_min* (¥)			Worst Fitness, *f_max* (¥)			Time Cost (s)			Standard Deviation of Fitness, *f_std*		
SGA	**SEGA**	**ERGA**	**SGA**	**SEGA**	**ERGA**	**SGA**	**SEGA**	**ERGA**	**SGA**	**SEGA**	**ERGA**
20,030,187	19,849,478	19,651,383	21,285,757	20,086,978	19,888,883	935.75	1065.37	1228.26	134,024.43	23,494.22	21,038.63
19,907,146	19,809,604	19,703,741	21,233,376	20,147,104	19,941,241	928.36	1059.46	1221.01	139,240.78	25,064.57	24,736.33
19,954,510	19,803,118	19,912,110	21,387,070	19,974,908	20,151,510	958.66	1093.96	1244.24	136,504.67	18,943.18	27,164.67
19,992,657	19,892,734	19,760,671	21,328,307	20,130,234	19,998,171	913.71	1098.42	1297.24	137,321.44	24,284.71	24,668.79
19,913,814	19,909,604	19,657,517	21,422,150	20,147,104	19,883,767	899.18	1059.46	1296.35	147,413.45	25,064.57	17,993.85
19,971,522	19,849,478	19,756,053	22,569,942	20,086,978	19,981,053	908.85	1065.37	1287.20	141,977.68	23,494.22	22,493.11
19,992,806	19,882,079	19,868,010	21,844,706	20,119,579	20,093,010	933.69	1044.54	1267.65	150,292.57	23,112.23	19,900.79
19,949,906	19,944,644	19,764,781	21,389,986	20,182,144	20,002,281	875.98	1100.74	1253.51	151,597.91	24,861.79	25,712.40
19,965,013	19,943,682	19,743,073	21,567,413	20,122,282	19,980,573	881.38	1074.60	1270.67	139,943.68	19,186.87	24,335.14
19,954,700	19,955,039	19,887,158	21,304,590	20,177,539	20,124,658	867.82	1073.79	1267.19	141,423.25	20,093.23	20,301.43

Figure 12a shows the images of the population evolution process of SGA and SEGA, with the horizontal coordinate as the number of iterations, and the vertical coordinate as the minimum value of population fitness value, i.e., the minimum value of total cost. Since the minimum value of each generation is close to the mean value, the optimal value curve covers the mean curve. From the solution results, it can be seen that the cost of SEGA is lower than that of SGA in nine out of ten experiments, and the standard deviation of SEGA is lower than that of SGA each time, indicating that the stability of SEGA is better and the diversity of SGA is better; from the comparison of the solution process images, it can be seen that SGA has not reached the convergence state at 20,000 generations, and SEGA has reached the local convergence state at the 12,000th generation. It can be concluded that SEGA has a better convergence speed compared with SGA but is more likely to fall into local convergence. Since SEGA has an elite retention after merging the pre-variant population with the post-variant population, the computation time of each generation of SEGA is longer, but SEGA converges faster. The reason for the transient fluctuations in the optimal target value curve of SGA is that the roulette selection operator unnecessarily selects the optimal individual.

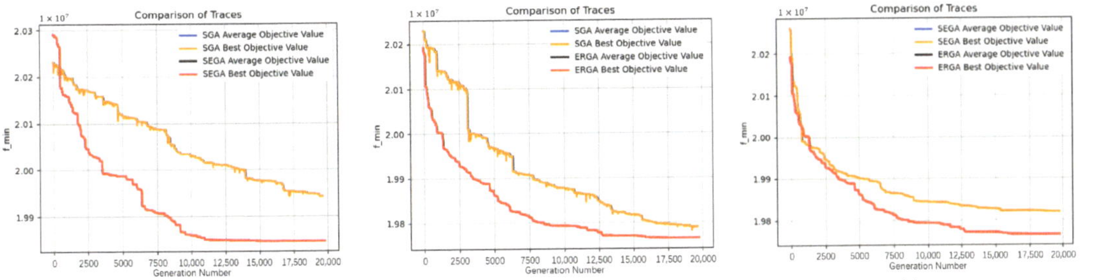

(a) (b) (c)

Figure 12. The evolution processes of algorithms: (**a**) processes of SEGA and SGA; (**b**) processes of ERGA and SGA; (**c**) processes of SEGA and ERGA.

2. Comparative analysis of ERGA and SGA

Since the tournament operator combined with elite retention performs better than the roulette operator, this section will further study the influence of elite retention strategy on the roulette operator based on the scenario above and compare the performance of ERGA and SGA, which is enhanced by elite retention. The experimental results of ERGA and SGA are shown in Table 8 and Figure 12b.

According to the evolution process images in Figure 12 and the experimental results in Table 8, it can be concluded that compared with the original SGA, ERGA reserves more elite individuals, and has a faster convergence rate. However, at the same time, because dominant individuals will be copied and quickly spread to the population, the diversity of SEGA will be reduced, and it is easier to fall into local convergence. Also, because of the extra time needed to join the parent population and sort it, ERGA would be more time-consuming than the original SGA.

3. Comparative analysis of ERGA and SEGA

Finally, ERGA and SEGA, both of which combine the elite retention strategy, are compared and analyzed. The experimental results of ERGA and SEGA are shown in Table 8 and Figure 12c. According to the evolutionary process images in Figure 12c and the experimental results in Table 8, after being enhanced by the elite retention strategy the performance of the SGA algorithm has been significantly improved. The quality of the optimization solution obtained with the same evolutionary generation has changed from a big gap with SEGA to a small gap or even better than SEGA. It can be concluded that the elite retention strategy improves the convergence speed of ERGA and the diversity of the roulette operator is not significantly reduced due to the low percentage of elite retention. Therefore, ERGA explores the solution space more fully and optimizes better. However, ERGA is required to score and rank the fitness of all individuals, while SEGA only needs to rank the individuals selected for tournament play, so ERGA is larger than SEGA in terms of time cost.

5.4.3. Analysis of the Solution

As can be seen from the above comparative analysis, the elite retention strategy can improve the convergence of the algorithm by retaining more elite individuals, but it also reduces the diversity of the algorithm and makes it easier to fall into local convergence. Compared with SEGA, which also combines an elite retention strategy, ERGA performs better in terms of global convergence with little time difference.

In the scenario of Section 3.3, the primary goal of the MVC collaboration optimization network is to reduce the total cost of production and distribution. Therefore, according to the above conclusion, the ERGA algorithm combined with the integrated encoding method should be selected. The production solution is shown in Figure 13.

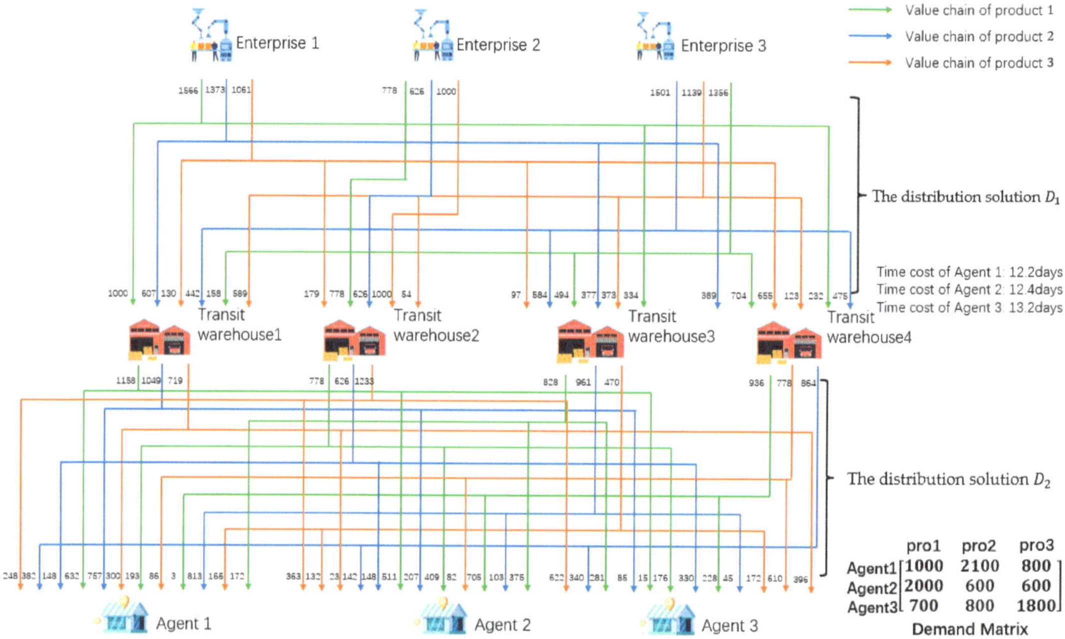

Figure 13. Optimization results of production and distribution solution.

Specifically, the first stage of the solution is shown in Table 9, and the second stage of the solution is shown in Table 10. Taking enterprise 1 as an example; enterprise 1 produced 1566 pieces of product 1, then delivered 1000 pieces to transit warehouse 1, 334 pieces to warehouse 3, and 232 pieces to warehouse 4. Enterprise 1 produced 1373 pieces of product 2, then delivered 607 pieces to transit warehouse 1, 377 pieces to warehouse 3, and 389 pieces to warehouse 4. Enterprise 1 produced 1061 pieces of product 3, then delivered 130 pieces to transit warehouse 1, 179 pieces to warehouse 2, 97 pieces to warehouse 3, and 655 pieces to warehouse 4.

Table 9. The first part of the solution.

	Product 1				Product 2				Product 3			
	Warehouse 1	**Warehouse 2**	**Warehouse 3**	**Warehouse 4**	**Warehouse 1**	**Warehouse 2**	**Warehouse 3**	**Warehouse 4**	**Warehouse 1**	**Warehouse 2**	**Warehouse 3**	**Warehouse 4**
Enterprise 1	1000	0	334	232	607	0	377	389	130	179	97	655
Enterprise 2	0	778	0	0	0	626	0	0	0	1000	0	0
Enterprise 3	158	0	494	704	442	0	584	475	589	54	373	123

Table 10. The second part of the solution.

	Product 1			Product 2			Product 3		
	Agent 1	Agent 2	Agent 3	Agent 1	Agent 2	Agent 3	Agent 1	Agent 2	Agent 3
Warehouse 1	632	511	15	757	207	85	300	23	396
Warehouse 2	193	409	176	148	148	330	248	363	622
Warehouse 3	172	375	281	813	103	45	166	132	172
Warehouse 4	3	705	228	382	142	340	86	82	610

From Figure 13, Tables 9 and 10, it can be concluded that the ERGA algorithm, which combines the integrated coding approach, yields a multi-product production-distribution solution that meets the requirements of the manufacturers, transit warehouses, and agents in each value chain without node conflicts.

In this chapter, based on the background of the household appliance industry and combined with the production, distribution, and sales data of a dynamic enterprise alliance, an MVC collaboration network of multiple products, multiple production ends, and multiple agents is constructed. Due to the multi-stage characteristic of the production-distribution scenario, two chromosome coding methods are proposed in this paper: staged coding and integrated coding. The comparative experimental results and evolutionary process of the coding methods show that staged coding is more suitable for complex scenarios with requirements on the number of distributions and distribution time at each stage. Integrated coding is more suitable for scenarios where cost reduction is a priority and there are no strict requirements for each stage of the value chain.

Since the tournament operator combined with elite retention performs better than the roulette operator, the influence of elite retention strategy on the algorithm has been further studied. A comparative experiment analysis for SGA and SEGA genetic algorithms is conducted and it shows that the elite retention strategy can improve the convergence of the algorithm by retaining more elite individuals, but it also reduces the diversity of the algorithm and makes it more likely to fall into local convergence. Based on this, a roulette algorithm ERGA combined with the elite retention strategy is proposed. The comparative experiment results and the analysis of the population evolution process show that ERGA has the best combined effect in terms of time cost and optimization compared with SGA and SEGA.

6. Conclusions

In the enterprise dynamic alliance, an analysis of the essential characteristics of the MVC collaboration network has been conducted. An MVC collaboration network optimization model aiming at the optimization demand of the MVC collaboration network in production-distribution is proposed. It takes production-distribution solutions as the decision space, the total cost as the optimization objective, and the time limit and capacity as constraints. For the high-dimensional characteristics of the decision space in the multi-task, multi-production end and multi-distribution end scenario, a genetic algorithm is used to solve the MVC collaboration network optimization model. The collaboration constraints of nodes in the chains are mapped as constraints among genes of chromosomes, and the problem of difficult collaboration or even conflicts among nodes in MVC collaboration networks is solved by adjusting the constraints among genes. In view of the multi-level characteristics of the production-distribution scenario, two chromosome coding methods are proposed in this paper: staged coding and integrated coding. The comparative experiments show that the staged coding is more suitable for complex scenarios with high requirements on distribution volume and time at stages; the integrated coding is more suitable for the optimization scenarios pursuing lower costs in the whole production-distribution process. Finally, this paper proposes a roulette algorithm ERGA, which was combined with the elite retention strategy. The comparative experimental results and the analysis of the population evolution process show that ERGA outperforms SGA and SEGA in terms of time cost and optimization, and that ERGA is more generalized. By reasonably combining coding methods and selection operators, the MVC collaboration network optimization model can be made more adaptable to different production-distribution environments and help the dynamic alliance of enterprises to solve the optimal solution under different production-distribution high-dimensional environments.

In future research, the influence of the fullness of the TEU (transmission extension unit) on the transportation cost could be introduced into the optimization model, and the matching modeling analysis of the fullness and the shipping volume could be conducted so that the scenario is more in line with the actual transportation cost, since this paper does not consider the influence of the fullness of the TEU on the transportation cost. Secondly, based on the research scenario of product production-distribution in this paper, which involves product manufacturing, warehousing, logistics, etc., it could be further extended to carry out research on the scenario of the MVC of the whole product process, including raw material supply, product manufacturing, warehousing, logistics, marketing, and sales, etc., involving the whole process of the MVC collaboration planning of suppliers, manufacturers, distributors, and customers.

Author Contributions: Conceptualization, B.W., B.G. and Y.J.; methodology, S.W. and J.Z.; validation, X.D., J.T., K.D. and C.T.; formal analysis, D.H.; writing—original draft preparation, S.W.; writing—review and editing, S.W. and Y.J.; supervision, D.H., B.G. and Y.J. All authors have read and agreed to the published version of the manuscript.

Funding: This work was supported in part by the National Key R&D Program of China under Grant No. 2020YFB1707900 and 2020YFB1711800; the National Natural Science Foundation of China under Grant No. 62262074, 62172061 and U2268204; the Science and Technology Project of Sichuan Province under Grant No. 2022YFG0159, 2022YFG0155, 2022YFG0157, 2021GFW019, 2021YFG0152, 2021YFG0025, 2020YFG0322.

Institutional Review Board Statement: Not applicable.

Informed Consent Statement: Not applicable.

Data Availability Statement: Not applicable.

Conflicts of Interest: The authors declare no conflict of interest.

References

1. Horvath, L. Collaboration: The key to value creation in supply chain management. *Supply Chain Manag.* **2001**, *6*, 205–207. [CrossRef]
2. Chou, S.; Chen, C.W.; Kuo, Y.T. Flexibility, collaboration and relationship quality in the logistics service industry: An empirical study. *Asia Pac. J. Mark. Logist.* **2018**, *30*, 555–570. [CrossRef]
3. Christopher, M.L.; Peck, H.; Towill, D. A Taxonomy for Selecting Global Supply Chain Strategies. *Int. J. Logist. Manag.* **2006**, *17*, 277–287. [CrossRef]
4. Keah, C.T. Supply chain management: Practices, concerns, and performance issues. *J. Supply Chain Manag.* **2002**, *38*, 42–53. [CrossRef]
5. Feller, A.; Shunk, D.; Callarman, T. Value chains versus supply chains. *BP Trends* **2006**, *1*, 1–7.
6. Soosay, C.A.; Hyland, P. A decade of supply chain collaboration and directions for future research. *Supply Chain Manag. Int. J.* **2015**, *20*, 613–630. [CrossRef]
7. Chan, F.T.S.; Prakash, A. Inventory management in a lateral collaborative manufacturing supply chain: A simulation study. *Int. J. Prod. Res.* **2012**, *50*, 4670–4685. [CrossRef]
8. Guo, Y.; Yu, J.; Allaoui, H.; Choudhary, A. Lateral collaboration with cost-sharing in sustainable supply chain optimisation: A combinatorial framework. *Transp. Res. Part E Logist. Transp. Rev.* **2022**, *157*, 102593. [CrossRef]
9. Celo, S.; Lehrer, M. How much lateral collaboration is optimal? Insights from computer simulations of MNEs as complex adaptive systems. *J. World Bus.* **2022**, *57*, 101289. [CrossRef]
10. Ren, M.; Ren, L.; Jain, H. Manufacturing service composition model based on synergy effect: A social network analysis approach. *Appl. Soft Comput.* **2018**, *70*, 288–300. [CrossRef]
11. Yang, Y.; Yang, B.; Wang, S.; Jin, T.; Li, S. An enhanced multi-objective grey wolf optimizer for service composition in cloud manufacturing. *Appl. Soft Comput.* **2020**, *87*, 106003.
12. Wang, Y.; Wang, S.; Gao, S.; Guo, X.; Yang, B. An adaptive multi-objective multi-task service composition approach considering practical constraints in Fog Manufacturing. *IEEE Trans. Ind. Inform.* **2021**, *18*, 6756–6766. [CrossRef]

13. Zhang, F.; Mei, Y.; Nguyen, S.; Tan, K.; Zhang, M. Multitask genetic programming-based generative hyperheuristics: A case study in dynamic scheduling. *IEEE Trans. Cybern.* **2021**, *52*, 10515–10528. [CrossRef] [PubMed]
14. Li, B.; Yang, Y.; Su, J.; Zhang, N.; Wang, S. Two-sided matching model for complex product manufacturing tasks based on dual hesitant fuzzy preference information. *Knowl. Based Syst.* **2019**, *186*, 104989. [CrossRef]
15. Manthou, V.; Vlachopoulou, M.; Folinas, D. Virtual e-Chain (VeC) model for supply chain collaboration. *Int. J. Prod. Econ.* **2004**, *87*, 241–250. [CrossRef]
16. Rietjens, S.; Goedee, J.; Van Sommeren, S.; Soeters, J. Meeting needs: Value chain collaboration in stabilisation and reconstruction operations. *J. Humanit. Logist. Supply Chain Manag.* **2014**, *4*, 43–59. [CrossRef]
17. Wei, X.; Shang, J. Mobile value chain collaboration for product diffusion: Role of the lifecycle. *Expert Syst. Appl.* **2023**, *215*, 119346. [CrossRef]
18. Piboonrungroj, P.; Disney, S.M. Supply chain collaboration in tourism: A transaction cost economics analysis. *Int. J. Supply Chain Manag.* **2015**, *4*, 25–31.
19. Chan, F.T.S.; Zhang, T. The impact of Collaborative Transportation Management on supply chain performance: A simulation approach. *Expert Syst. Appl.* **2011**, *38*, 2319–2329. [CrossRef]
20. Bahinipati, B.K.; Kanda, A.; Deshmukh, S.G. Horizontal collaboration in semiconductor manufacturing industry supply chain: An evaluation of collaboration intensity index. *Comput. Ind. Eng.* **2009**, *57*, 880–895. [CrossRef]
21. Bahinipati, B.K.; Deshmukh, S.G. Lateral Collaboration in Semiconductor Industry Supply Networks: A Procurement Perspective. *Int. J. Inf. Syst. Supply Chain Manag.* **2014**, *7*, 39–79. [CrossRef]
22. Gu, Q.; Luo, J.; Li, X.; Lu, C. An adaptive evolutionary algorithm with coordinated selection strategies for many-objective optimization. *Appl. Intell.* **2022**, *52*, 1–28. [CrossRef]
23. Youping, W.; Liang, L.; Lin, C. An advanced genetic algorithm for traveling salesman problem. In Proceedings of the 2009 Third International Conference on Genetic and Evolutionary Computing, Guilin, China, 14–17 October 2009.
24. Das, S.K.; Pervin, M.; Roy, S.K.; Weber, G.W. Multi-objective solid transportation-location problem with variable carbon emission in inventory management: A hybrid approach. *Ann. Oper. Res.* **2021**, *297*, 1–27. [CrossRef]
25. Mauri, G.R.; Biajoli, F.L.; Rabello, R.L.; Chaves, A.A.; Ribeiro, G.M.; Lorena, L.A.N. Hybrid metaheuristics to solve a multiproduct two-stage capacitated facility location problem. *Int. Trans. Oper. Res.* **2021**, *28*, 3069–3093. [CrossRef]
26. Rezaeipanah, A.; Matoori, S.S.; Ahmadi, G. A hybrid algorithm for the university course timetabling problem using the improved parallel genetic algorithm and local search. *Appl. Intell.* **2021**, *51*, 467–492. [CrossRef]
27. Abdirad, M.; Krishnan, K.; Gupta, D. A two-stage metaheuristic algorithm for the dynamic vehicle routing problem in Industry 4.0 approach. *J. Manag. Anal.* **2021**, *8*, 69–83. [CrossRef]
28. Jiao, J.; Ghoreishi, S.M.; Moradi, Z.; Oslub, K. Coupled particle swarm optimization method with genetic algorithm for the static–dynamic performance of the magneto-electro-elastic nanosystem. *Eng. Comput.* **2022**, *38*, 2499–2513. [CrossRef]
29. Zivkovic, M.; Bacanin, N.; Djordjevic, A.; Antonijevic, M.; Strumberger, I.; Rashid, T.A. Hybrid genetic algorithm and machine learning method for covid-19 cases prediction. In *Proceedings of International Conference on Sustainable Expert Systems*; Lecture Notes in Networks and Systems; Springer: Singapore, 2021; Volume 176, pp. 169–184.
30. Kim, C.; Batra, R.; Chen, L.; Tran, H.; Ramprasad, R. Polymer design using genetic algorithm and machine learning. *Comput. Mater. Sci.* **2021**, *186*, 110067. [CrossRef]
31. Ewees, A.A.; Al-qaness, M.A.A.; Abualigah, L.; Oliva, D.; Algamal, Z.Y.; Anter, A.M.; Ibrahim, R.A.; Ghoniem, R.M.; Abd Elaziz, M. Boosting arithmetic optimization algorithm with genetic algorithm operators for feature selection: Case study on cox proportional hazards model. *Mathematics* **2021**, *9*, 2321. [CrossRef]
32. Mirjalili, S. Genetic algorithm. *Evol. Algorithms Neural Netw.* **2019**, *780*, 43–55.
33. Lambora, A.; Gupta, K.; Chopra, K. Genetic algorithm-A literature review. In Proceedings of the International Conference on Machine Learning, Big Data, Cloud and Parallel Computing (COMITCon), Faridabad, India, 14–16 February 2019.
34. Costa, A.; Celano, G.; Fichera, S.; Trovato, E. A new efficient encoding/deencoding procedure for the design of a supply chain network with genetic algorithms. *Comput. Ind. Eng.* **2010**, *59*, 986–999. [CrossRef]
35. Molla-Alizadeh-Zavardehi, S.; Hajiaghaei-Keshteli, M.; Tavakkoli-Moghaddam, R. Solving a capacitated fixed-charge transportation problem by artificial immune and genetic algorithms with a Prüfer number representation. *Expert Syst. Appl.* **2011**, *38*, 10462–10474. [CrossRef]
36. Katoch, S.; Chauhan, S.S.; Kumar, V. A review on genetic algorithm: Past, present, and future. *Multimed. Tools Appl.* **2021**, *80*, 8091–8126. [CrossRef] [PubMed]
37. Yu, F.; Fu, X.; Li, H.; Dong, G. Improved roulette wheel selection-based genetic algorithm for TSP. In Proceedings of the 2016 International Conference on Network and Information Systems for Computers (ICNISC), Wuhan, China, 15–17 April 2016.
38. Fang, Y.; Li, J. A review of tournament selection in genetic programming. *Int. Symp. Intell. Comput. Appl.* **2010**, *6382*, 181–192.
39. Thierens, D.; Goldberg, D. Elitist recombination: An integrated selection recombination GA. In Proceedings of the First IEEE Conference on Evolutionary Computation, IEEE World Congress on Computational Intelligence, Orlando, FL, USA, 27–29 June 1994.
40. Deb, K.; Agrawal, S.; Pratap, A.; Meyarivan, T. A fast elitist non-dominated sorting genetic algorithm for multi-objective optimization: NSGA-II. In Proceedings of the International Conference on Parallel Problem Solving from Nature, Paris, France, 18–20 September 2000.

41. Koohestani, B. A crossover operator for improving the efficiency of permutation-based genetic algorithms. *Expert Syst. Appl.* **2020**, *151*, 113381. [CrossRef]
42. Manzoni, L.; Mariot, L.; Tuba, E. Balanced crossover operators in genetic algorithms. *Swarm Evol. Comput.* **2020**, *54*, 100646. [CrossRef]
43. Elsayed, S.M.; Sarker, R.A.; Essam, D.L. Multi-operator based evolutionary algorithms for solving constrained optimization problems. *Comput. Oper. Res.* **2011**, *38*, 1877–1896. [CrossRef]
44. Lee, J.E.; Gen, M.; Rhee, K.G. Network model and optimization of reverse logistics by hybrid genetic algorithm. *Comput. Ind. Eng.* **2009**, *56*, 951–964. [CrossRef]
45. Ghosh, S.; Roy, S.K.; Ebrahimnejad, A.; Verdegay, J.L. Multi-objective fully intuitionistic fuzzy fixed-charge solid transportation problem. *Complex Intell. Syst.* **2021**, *7*, 1009–1023. [CrossRef]
46. Khan, I.H. Assessing different crossover operators for traveling salesman problem. *Int. J. Intell. Syst. Appl.* **2015**, *7*, 19–25.
47. Mirjalili, S.; Dong, J.S.; Sadiq, A.S.; Faris, H. Genetic algorithm: Theory, literature review, and application in image reconstruction. *Nat. Inspir. Optim.* **2020**, *811*, 69–85.

electronics
MDPI

Article

Feature Interaction-Based Reinforcement Learning for Tabular Anomaly Detection

Yaoxun Liu, Liangli Ma *, Muyuan Wang and Siyuan Zhang

College of Electronic Engineering, Naval University of Engineering, Wuhan 430033, China; lyx_nue@163.com (Y.L.)
* Correspondence: maliangl@163.com

Abstract: Deep learning-based anomaly detection (DAD) has been a hot topic of research in various domains. Despite being the most common data type, DAD for tabular data remains under-explored. Due to the scarcity of anomalies in real-world scenarios, deep semi-supervised learning methods have come to dominate, which build deep learning models and leverage a limited number of labeled anomalies and large-scale unlabeled data to improve their detection capabilities. However, existing works share two drawbacks. (1) Most of them simply treat the unlabeled samples as normal ones, ignoring the problem of label contamination, which is very common in real-world datasets. (2) Only very few works have designed models specifically for tabular data instead of migrating models from other domains to tabular data. Both of them will limit the model's performance. In this work, we propose a feature interaction-based reinforcement learning for tabular anomaly detection, FIRTAD. FIRTAD incorporates a feature interaction module into a deep reinforcement learning framework; the former can model tabular data by learning a relationship among features, while the latter can effectively exploit available information and fully explore suspicious anomalies from the unlabeled samples. Extensive experiments on three datasets not only demonstrate its superiority over the state-of-art methods but also confirm its robustness to anomaly rarity, label contamination and unknown anomalies.

Keywords: deep reinforcement learning; anomaly detection; semi-supervised learning; feature interaction; tabular data

1. Introduction

Tabular data refers to data that are arranged in the form of a table, in which each row represents a sample, and each column represents a feature. As the most common type of data in real-world applications, tabular data are widely used in many domains, such as network security [1,2], financial transaction [3,4], industrial manufacturing [5,6], marine traffic-cite [7,8], etc. Anomalies (also called outlier or novelty), which exist in almost all domain applications, often indicate malfunctions or malicious behavior and may result in property damage or even casualties. Anomaly detection (AD) for tabular data has been a lasting yet active topic in the last few decades, and dozens of methods have been proposed for different tasks.

Due to the rarity of anomalies, most real-world datasets are severely imbalanced, i.e., negative instances (normal samples) account for the vast majority, while positive instances (anomalous samples) account for only a small minority. Therefore, many researchers consider anomaly detection as an unsupervised learning problem, such as proximity-based methods [9–11], ensemble-based methods [12–14], and neural network-based methods [15–17]. In the last decade, a few studies [18–20] have pointed out the availability of labeled anomalies in some real-world scenarios. A limited number of positive samples that come from the identification of experts or the accumulation of the system are usually available with trivial cost. In recent years, more and more works have considered anomaly detection as a semi-supervised or weakly-supervised learning problem, leveraging a small set of labeled anomalies and a large-scale unlabeled dataset to train AD models. Unfortunately, most existing works [18–22] treat all the samples from the unlabeled dataset as normal for convenience and overlook possible anomalies (also called anomalous contamination). Considering the very large data size, even if the proportion of anomaly samples is extremely low, there would exist quite a few anomalies in the unlabeled dataset. Ignoring these samples may result in a loss of information and then limitations in a model's performance.

Additionally, most existing deep models for tabular data anomaly detection (TAD) [18–20,22,23] try to transfer AD approaches from other domains, such as computer vision (CV) or natural language processing (NLP), instead of modeling the characteristics of tabular data. Studies [24,25] indicate that deep models that excel in CV or NLP cannot achieve the desired performance on tabular data due to its characteristics, including lack of locality, data sparsity and mixed feature of types. The most notable difference between tabular data and other types of data is the associative relationship between columns. Practices [26–29] in Click-Through prediction (CTR) demonstrate that feature interactions, especially high-order feature interactions, are crucial to modeling tabular data. However, to the best of our knowledge, no paper has yet worked on how to apply feature interaction to TAD.

To cope with the problem of anomalous contamination in an unlabeled dataset, we propose a novel method based on deep reinforcement learning (DRL). As can be seen from the name, DRL combines the expression ability of deep learning and the decision-making ability of reinforcement learning (RL). Different from unsupervised learning and semi-supervised learning methods, RL updates parameters by interacting with the environment without requiring data to be given in advance. In this work, we leverage the DRL algorithm to train an anomaly detector that can not only fit the labeled anomalies but also detect possible anomalies from the unlabeled dataset.

To settle the problem that deep models do not model tabular data well, we introduce a tabular data modeling approach named gated adaptive feature interaction network (GAIN) [29]. GAIN exploits multiple parallel interaction units to learn useful high-order feature interactions. The parallel design guarantees a high-efficiency model, as works [30,31] have experimentally demonstrated that parallel architectures can dramatically reduce the processing time of models both on CPUs and on GPUs. GAIN works as a middleware, which can transform raw features into informative representations and can replace the deep module in DRL.

We further instantiate the proposed approach into a model called feature interaction-based reinforcement learning for tabular data anomaly detection (FIRTAD), and the architecture is shown in Figure 1. We choose the soft Actor–Critic (SAC) [32] as the main framework of the FIRTAD. The policy network and the Q-network share the same deep module, which is implemented with GAIN. We create a simulation environment to interact with the SAC, which includes labeled anomalies and an unlabeled dataset. To ensure the exploitation of all samples, we propose a novel sampling strategy that prevents repeated sampling from a densely distributed region. To encourage the agent to explore samples, which can bring more novelty, we extend the reward function with an intrinsic reward. More details are discussed in Section 3.

We summarize the main contributions of this work as follows:

1. We propose a novel DRL-based anomaly detection approach specifically for tabular data and deliberately devise a simulation environment that allows all samples to be fully explored.
2. We introduce a feature interaction module (GAIN) into our approach, which can model the characteristics of tabular data by learning interactions between features. To the best of our knowledge, it is the first effort to apply feature interactions to anomaly detection.
3. We instantiate the proposed approach into a model called FIRTAD and extensively evaluate the model, comparing six baselines on three benchmark datasets. The

experimental results demonstrate that our model performs better than state-of-the-art models and exhibits better robustness to class imbalance, label contamination and unknown anomalies.

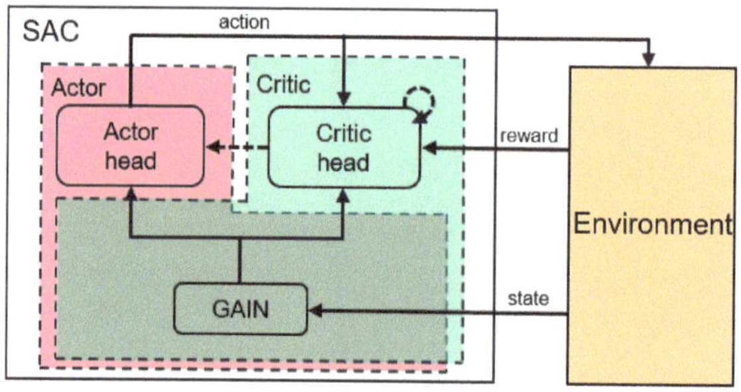

Figure 1. Architecture of FIRTAD. FIRTAD, which is based on a deep reinforcement learning framework, consists of two parts, the agent and the environment. The agent is implemented by an SAC algorithm, in which the Actor and the Critic share the same feature interaction module GAIN as their deep modules. The environment is implemented with a sampling function and a reward function.

2. Related Works

2.1. Anomaly Detection Methods

From the perspective of the availability of supervision information (labels), most existing AD methods can be divided into three categories: unsupervised learning, supervised learning and semi-supervised (weakly supervised) learning methods. Due to the high cost of data-labeling processes in real-world scenarios, supervised learning methods are often impractical, and the other two are much more popular.

2.1.1. Unsupervised Anomaly Detection

Unsupervised AD methods are almost based on the assumption that normal samples have different distributions from anomalous samples. Normal samples are densely distributed, while anomalous samples are sparsely distributed and far from normal ones. Some works [9,10,33] treat samples as data points and detect anomalies by calculating the distance or density; Other works [11–13] judge the degree of abnormality through comprehensive analysis of the distribution of every single dimension. Ref. [34] obtains the anomaly scores of samples by calculating the joint distribution of all dimensions. Conventional machine learning methods fail to work effectively when dealing with high-dimensional samples due to the curse of dimensionality. To tackle this problem, deep learning methods are introduced to anomaly detection; the common practice is to exploit deep neural networks (e.g., multi-layer perception (MLP), autoencoder (AE) or generative adversarial network (GAN)) to project samples into a low-dimensional representation space, and then distinguish anomalies from normal samples [15,35,36]. Due to the lack of supervised information, almost all unsupervised methods detect anomalies by modeling normality. Although the unsupervised AD methods have achieved decent results, their performances are still limited because anomalies that can be easily obtained have not been utilized.

2.1.2. Supervised Anomaly Detection

Although supervised AD methods are not as popular as unsupervised and semi-supervised ones, they still attract the attention of many researchers. Traditional supervised methods treat AD as a binary classification problem, and commonly used methods include Naive Bayesian, Support Vector Machine (SVM) and Gradient Boosting Decision Tree (GBDT). Deep learning-based methods train a classifier to detect anomalies on the basis of representation learning; well-known works include MLP, ResNet and FTTransformer [24]. Both traditional and deep methods have their own drawbacks. Traditional methods do not perform well when processing high-dimensional, heterogeneous or non-independent data. In contrast, the deep methods, although they can handle these problems well, require a large number of labeled samples for training.

2.1.3. Semi-Supervised Anomaly Detection

Semi-supervised AD methods utilize limited supervised information to improve the ability to identify anomalies. Some works leverage labeled anomalies to enhance existing unsupervised AD models. Ref. [18] uses supervised information to push anomalies away from the center of a compact hypersphere, which is built using an unsupervised method [35]. Ref. [22] introduces anomalous samples to a distance-based method (e.g., K-nearest neighbors [10]) and identifies anomalies by calculating the knn-distance between the query sample and a random unlabeled subset. Other works introduce labeled anomalies into supervised AD models, and solve the problem of class imbalance through data augmentation, downsampling, etc. Ref. [37] proposes two strategies to enrich the anomalous samples and distinguishes anomalies using a contrastive learning method. Ref. [19] builds instance pairs to make the proportion of instance pairs containing anomalies reach 50%. Refs. [20,21] train their models on datasets that are equally sampled from both labeled anomalies and unlabeled samples. Ref. [38] leverages anomalies to obtain a prior anomaly score for each sample and uses the score as supervised information to optimize the AD model. Due to the exploitation of labeled anomalies, semi-supervised methods significantly improve the performance of AD models. However, almost all semi-supervised AD models are based on the assumption that the unlabeled subset contains normal samples only or do not consider the impact of possible anomalies in it.

2.2. *Feature Interaction of Tabular Data*

Due to the lack of locality and the complexity of features, tabular data cannot be modeled well with prevailing deep models, such as MLP, convolutional neural network (CNN), recurrent neural network (RNN), etc. Most deep anomaly detection approaches project high-dimensional tabular data into a low-dimensional space using MLP, ResNet, AE or GAN. However, these architectures cannot guarantee the preservation of discriminative information because they overlook feature interaction, which is the most prominent difference between tabular data and other data types.

Feature interaction, especially high-order feature interaction, has been proven to be crucial in improving the model's expressiveness in CTR prediction tasks [26–28,39]. In [40], Rendle argues that a second-order feature interaction can be represented with the inner product of two latent vectors, each of which represents a single feature, and propose factorization machine (FM) to automatically learn all possible second-order feature interactions. Many works [26–28] extend FM to learn higher-order feature interactions. However, these works brutely enumerate all possible feature interactions without differentiating their importance, and the introduction of useless interactions not only increases the computational complexity but also downgrades the model's performance. Xue et al. [41] propose AutoHash to adaptively learn useful high-order interactions. In AutoHash, all features are put into k buckets with randomly initialized probabilities (features can be reused), and every bucket represents a feature interaction. The probabilities are learnable variables that help the buckets preserve useful interactions through training. Liu et al. [42] propose a two-stage algorithm called automatic feature interaction selection (AutoFIS). In the first stage, the model is trained to drop interactions that contribute little to the final prediction; in the second stage, the model is re-trained to learn the importance pf the retained interactions. Chen et al. [43] propose a bayesian higher-order feature interaction selection (BH-FIS). BH-FIS implements the enumeration of all feature interactions by using outer-product and masking techniques and employs spike-and-slab priors to distinguish useful feature interactions from useless ones. Liu et al. [29] propose a gated adaptive feature interaction network (GAIN) that can adaptively learn high-order feature interactions. GAIN consists of a cross-module and a deep module; the former exploits multiple parallel interaction units to explicitly model feature interactions, while the latter leverages an MLP to model feature interactions in an implicit way.

Although the effectiveness of feature interactions has been proven in modeling tabular data, surprisingly, we can hardly find a work that incorporates it into tabular data anomaly detection.

2.3. Deep Reinforcement Learning for Tabular Anomaly Detection

The vanilla DRL algorithm is suitable for time series data due to its dependence on a live environment. Lopez-Martin et al. [44] make a conceptual modification of the vanilla DRL algorithm to make it feasible for tabular data, replacing the environment with a sampling function and designing a reward function based on the detection error. In addition, the authors make a comparison of several DRL algorithms on network intrusion detection datasets. Vimal et al. [3] exploit a DQN to tackle the payment fraud detection problem and use the technology of experience replay to improve the efficiency of sampling.

To solve the AD problem with a small set of labeled anomalies and a large-scale unlabeled dataset, Pang et al. [45] propose an approach called Deep Q-learning with Partially Labeled ANomalies (DPLAN). DPLAN creates an anomaly-biased simulation environment that continuously samples anomalies or suspected anomalies from the whole dataset. Separate sampling functions are designed for the labeled anomaly set and the unlabelled dataset, denoted as g_a and g_u, respectively. g_a selects samples uniformly from the labeled anomaly set, while g_u selects samples that are likely to be anomalous from the unlabeled dataset. g_u is defined as

$$g_u(s_{t+1}|s_t, a_t) = \begin{cases} \underset{s \in S^u}{\operatorname{argmin}}\, d(s_t, s) & if\ a_t = 1 \\ \underset{s \in S^u}{\operatorname{argmax}}\, d(s_t, s) & if\ a_t = 0, \end{cases} \tag{1}$$

where S^u denotes a random subset of the unlabeled dataset, d denotes a function of Euclidean distance and a_t denotes the agent's judgment on s_t. $a_t = 1$ means the agent identifies s_t as an anomaly, the sample nearest to s_t is considered most likely to be anomalous, and is selected as s_{t+1}. In contrast, $a_t = 0$ means s_t is identified as a normal sample, and the sample farthest from s_t is returned to the agent.

DPLAN also designs a combined reward function $r_t = r_t^e + r_t^i$. The external reward r_t^e is defined based on the prediction error. r_t^e will return a positive reward if an anomalous sample is correctly recognized by the agent, no reward will be returned if a normal sample is correctly identified, and a negative reward will be returned if the agent makes a mistake. The external reward function is defined as

$$r_t^e = \begin{cases} 1 & if\ a_t = 1\ and\ s_t \in D^a \\ 0 & if\ a_t = 0\ and\ s_t \in D^u \\ -1 & otherwise, \end{cases} \tag{2}$$

where D^a and D^u denote the labeled set and the unlabeled dataset, respectively.

The intrinsic reward r_t^i is devised to encourage the agent to explore novel anomalies. Hence, samples from lower-density regions receive higher intrinsic rewards as DPLAN assumes that anomalous samples are sparsely distributed and far from normal ones. The iForest [12] algorithm can indicate the degree of abnormality of a sample and is, therefore, used to calculate the intrinsic reward. The intrinsic reward function is defined as $r_t^i = iForest(s_t)$.

Due to the special design of the anomaly-biased sampling function and the combined reward, the agent of DPLAN is encouraged to explore the unlabeled dataset for possible anomalies. However, the DPLAN still has three main drawbacks. First, the discrete action space, i.e., $A = \{0, 1\}$, severely limits the model's performance. The identification ability of the agent is gradually improved through training. Before the model converges, the agent's actions are usually undetermined. However, the discrete action space cannot express the uncertainty accurately, and the step changes between 0 and 1 would hinder the model from learning useful information. Secondly, the external rewards are designed to be too coarse-grained to cover all situations, e.g., when the agent takes an ambiguous action. Thirdly, the anomaly-biased sampling function may result in excessive exploration of anomalies and insufficient exploration of normal samples, which will be sub-optimal.

3. Our Proposed Method

3.1. Problem Definition

Given a training dataset $D = D^a \cup D^u$, where D^a and D^u denote a small labeled anomalous subset and a large-scale unlabeled subset, respectively. The size of D^a is much smaller than that of D^u, and the ratio of their sizes usually does not exceed 10%. The vast majority of the samples in D^u are normal, and only a very small number of samples are anomalous, part of which may come from unknown classes (i.e., classes that have not appeared in D^a). We aim to find out hidden anomalies from D^u by taking full advantage of the whole dataset D. Note that D is a tabular dataset in which all samples are independent of each other, and there is no temporal relationship between samples.

To apply a DRL algorithm to anomaly detection, we formulate the binary classification problem as a sequential decision-making problem. The agent receives a sample s_t from the environment at time t, and takes an action a_t. The environment gives a reward r_t and a new sample s_{t+1} to the agent according to s_t and a_t. The interaction between the agent and the environment can be represented with a Markov Decision Process (MDP), which is defined as below:

- **State space**. The whole dataset D (including unlabeled dataset D^u and labeled anomalies D^a) is defined as the state space. Each $s_t \in D$ denotes the state received from the environment at time t.
- **Action space**. Different from the existing works, we define a continuous action space $A = [0, 1]$. Therefore, the action a_t can also be regarded as the anomaly probability of s_t. The closer a_t is to 1, the more likely s_t is an anomaly and vice versa.
- **State transition**. After the agent takes an action a_t, the environment renders a new state s_{t+1} to the agent. Different from the anomaly-biased strategy used by DPLAN, which is dedicated to sampling anomalous states, we propose a novel sampling strategy that fully explores the entire data space.
- **Reward**. Similar to DPLAN, our proposed method leverages a combined reward function. We design a continuous extrinsic reward function to be compatible with the continuous action space. When s_t comes from D^a, the agent will obtain a large penalty if it fails to recognize s_t. When s_t comes from D^u, identifying s_t as an anomaly should not be given a large penalty as anomalies may be hidden in D^u. In addition, we design a curiosity-driven intrinsic reward function to encourage the agent to explore the entire state space. The reward function is defined as $r_t = r_t^e + \lambda r_t^i$, where λ is a scalar weighting the relevance of the intrinsic reward, and it takes a value from $[0, 1]$.

3.2. Agent

3.2.1. Foundation of the Proposed Approach

The agent of our proposed model is implemented with a SAC, which is a stochastic policy algorithm that can deal with continuous action space. Different from other DRL algorithms that aim to learn a policy to maximize the cumulative rewards, the SAC augments the objective with an entropy regularization term to concurrently maximize the entropy of the agent's action. The introduction of the maximum entropy can not only promote the exploration but also prevent premature convergence. The objective of the SAC is represented as

$$\pi^* = \underset{\pi}{\operatorname{argmax}} \underset{a_t \sim \pi}{\mathrm{E}} \left[\sum_{t=0}^{\infty} \gamma^t \Big(r(s_t, a_t, s_{t+1}) + \alpha H(\pi(\cdot|s_t)) \Big) \right], \tag{3}$$

where π represents the policy network, γ represents the discounting factor, $H(\cdot)$ represents the entropy function, and α is the trade off coefficient.

To improve the utilization of data, the SAC maintains an experience replay buffer $\mathcal{D}$ to store historical transitions (i.e., (s, a, r, s')) so that the minibatch can be sampled from the buffer during training. The SAC incorporates an Actor–Critic architecture, in which the Q-network and the policy network can be updated by temporal difference and policy gradient, respectively. To tackle the problem of overestimation brought by bootstrapping, the SAC adopts the technique of target network. For faster and more stable training, the SAC exploits two Q-networks and chooses the minimum Q-value. The loss function for the Q-networks is represented as

$$L(\phi_i, \mathcal{D}) = \underset{(s,a,r,s') \sim \mathcal{D}}{\mathrm{E}} \left[\left(Q_{\phi_i} - y(r, s') \right)^2 \right], \tag{4}$$

where ϕ represents the parameters of the Q-network. The target y is represented as

$$y(r, s') = r + \gamma \left(\min_{j=1,2} Q_{\phi_{targ,j}}(s', \tilde{a}') - \alpha \log \pi_\theta(\tilde{a}'|s') \right), \qquad \tilde{a}' \sim \pi_\theta(\cdot|s'), \tag{5}$$

where θ represents the parameters of the policy network, and $\tilde{a}'$ represents the next action sampled by updated policy. The loss function of the policy network is represented as

$$L(\theta, \mathcal{D}) = \underset{\substack{s \sim \mathcal{D} \\ \epsilon \sim \mathcal{N}}}{\mathrm{E}} \left[\alpha \log \pi_\theta(\tilde{a}_\theta(s, \epsilon)|s) - \min_{j=1,2} Q_j(s, \tilde{a}_\theta(s, \epsilon)) \right], \tag{6}$$

where ϵ represents a random number sampled from the standard normal distribution. In addition, the target networks of SAC conduct a soft update, i.e., update slowly toward the main networks in each step rather than updating periodically.

3.2.2. Feature Interaction-Based Policy Network and Q-Network

Feature interactions, especially high-order feature interactions, have been proven effective and efficient in modeling tabular data by many studies. In this work, we introduce a feature interaction module to extract expressive vectors from tabular data. Among dozens of feature interaction models that have emerged in recent years, GAIN is chosen by our work due to its effectiveness in learning high-order interactions and computational efficiency.

The structure of a GAIN is shown in Figure 2. GAIN takes raw features of samples as input and outputs a low-dimensional representation vector. GAIN is composed of two main modules: a cross-module and a deep module. The cross-module consists of multiple interaction units, each of which maintains a gate for every feature, and each gate only has two statuses, closed or open. Whether a feature can participate in an interaction is determined by the status of its corresponding gate. Each unit represents a feature interaction, and the interaction order is the number of gates that are open. The statuses of the gates are randomly initialized and dynamically adjusted. Through training, the statuses are gradually stabilizing, and meaningful feature interactions will be preserved. The deep module is implemented with an MLP. The outputs of the two modules concatenate to form the output of the GAIN. In our proposed model, GAIN is used as a middleware learning useful high-order feature interactions. To improve the training efficiency and reduce the parameters, the GAIN is shared by both the policy network and the Q-network.

In addition to learning higher-order feature interactions, the GAIN can also be seen as a transformer that projects the original features into a low-dimensional representation space. The benefits are twofold. On the one hand, the dimensionality reduction can avoid the curse of dimensionality. On the other hand, low-dimensional vectors facilitate the consequent distance calculation of states. In this work, the low-dimensional representation of the state s is called the abstract state, and it is denoted as $\hat{s}$. For the sake of brevity, we will not distinguish the two terms of *state* and *abstract state* in this paper unless necessary.

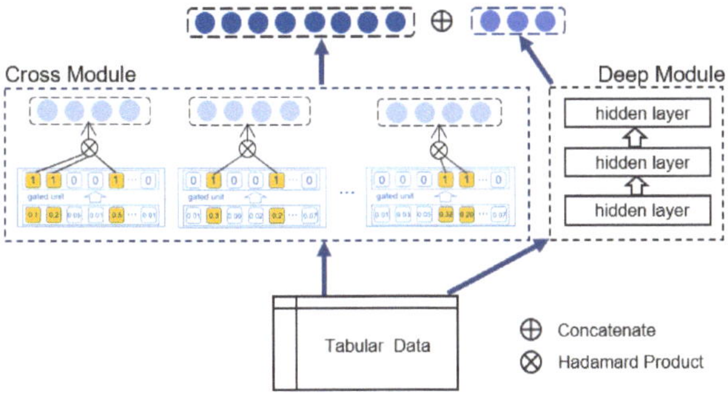

Figure 2. Structure of GAIN. The cross-module is composed of multiple interaction units, each of which can learn a high-order feature interaction using Hadamard product. The output of all units is concatenated with the output of the deep module to obtain the final output of the GAIN.

3.3. Knight Sampling Strategy

With respect to the distribution of samples, AD models typically make the following assumptions. Normal samples are large in number, and most of them are densely distributed, while anomalous samples are relatively scarce and are distributed far away from normal ones.

During the interaction between the environment and the agent, states need to be continuously sampled from the environment. The selection of a state is determined by the sampling strategy, which may affect the training efficiency. A suitable strategy could significantly improve the convergence speed of the model, while an inappropriate strategy would extend the training time or make the model converge to a local optimum. The random strategy adopted by [44] uniformly samples states from the entire state space, in which the probability of a class being selected is proportional to the proportion of states of that class in all states. Considering the severe class imbalance in the state space, normal states have much higher probabilities of being selected, and it will take more steps for the agent to encounter all anomalous states, which might result in much longer training times. To tackle the inefficient sampling of anomalous states, [45] proposes an anomaly-biased strategy, where states that are more likely to be anomalous have higher priorities to be selected. However, repeated exploitation of anomalous states would lead to overfitting of the model. Meanwhile, insufficient exploration of normal samples would also limit the model's performance.

Therefore, a novel sampling strategy needs to be devised so that all regions of the state space can be explored fully and evenly. Inspired by the Knight's Tour problem [46], we propose a novel sampling strategy. A knight-like sampler is devised that can leap to a distant state. Leaping not only make the sampler visit every part of the state space but also help the sampler escape from a cluster. An intuitive idea is to choose the farthest state from the k-nearest neighbors. However, considering the high computational consumption of the KNN algorithm, we take a subset instead. First, a subset $D' \subset D$ is randomly sampled; Secondly, a Euclidean distance is calculated between s_t and each state in D'; Thirdly, the state that is farthest from s_t is selected to be s_{t+1}. Selecting the furthest state from a random subset ensures that the entire state space can be explored and prevents the sampler from being trapped in a certain region. We name the proposed strategy as Knight Sampling Strategy.

3.4. Combined Reward Function

To make a trade-off between exploration and exploitation, we design a combined reward, $r_t = r_t^e + \lambda r_t^i$, where r_t^e and r_t^i represent extrinsic and intrinsic reward, respectively.

The former encourages the agent to exploit known information, while the latter stimulates the agent to explore novelty in the environment. λ is the trade-off coefficient.

3.4.1. Extrinsic Reward Function

Extrinsic reward is an immediate reward given by the simulation environment according to the state-action pair. Since s_t may come from either D^a or D^u, the extrinsic reward function should be designed separately according to the different sources of s_t. If s_t comes from D^a, a large a_t should be given a positive reward. In contrast, a small a_t should be given a negative reward as a penalty. Hence, r_t^e should be proportional to a_t (the anomaly probability) when s_t is sampled from D^u. If s_t comes from D^u, r_t^e is inversely proportional to a_t. Considering the possible anomalies in D^u, to prevent the agent from easily identifying a sample from D^u as normal, we should give a minor reward if $a_t \to 0$. Similarly, if $a_t \to 1$, a minor penalty is more feasible since a large penalty would discourage the agent from detecting hidden anomalies. A coefficient, τ, is utilized to scale the reward, and the value of τ is usually specified as the ratio of the sizes of D_a and D_u. To be compatible with the continuous action space, a continuous extrinsic reward function is required, which is defined below.

$$r_t^e(s_t, y_t, a_t) = \begin{cases} 2a_t - 1 & \text{if } y_t = 1, \\ -\tau \cdot (2a_t - 1) & \text{if } y_t = 0 \end{cases} \tag{7}$$

3.4.2. Intrinsic Reward Function

Inspired by human experiences in playing games that the highest score can only be obtained if the environment is fully explored, we design a curiosity-driven intrinsic reward function to encourage the agent to explore states with high novelty. Intuitively, the novelty of a state will decrease if it is sampled several times. In addition, the same happens with repeated sampling of nearby states. Therefore, the intrinsic reward of a state-action pair is inversely proportional to the visits to the region where the state is located. We exploit a Gaussian kernel function to approximately calculate the number of visits, which is represented as

$$\kappa(x, y) = exp - \frac{\|x - y\|_2}{2\sigma^2}, \tag{8}$$

where $\|\cdot\|$ returns a Euclidean distance, and σ is a hyper-parameter, which is discussed in Section 5.6. If two states are close to each other, κ tends to be 1. In contrast, if two states are far away from each other, κ tends to be 0.

We define an episodic memory $M = [\hat{s}_1, \hat{s}_2, \cdots, \hat{s}_{t-1}]$ to store the states before time step t. Since the kernel function can convert the distance of two states between 0 and 1, we calculate κ of $\hat{s}_t$ and each state in M and take the reduction sum as the approximate counts. The intrinsic reward function is represented as:

$$r_t^i = \frac{1}{\sqrt{\sum_{i=1}^{n-1} \kappa(\hat{s}_t, \hat{s}_i) + 1}}. \tag{9}$$

4. Model Analysis

4.1. Analysis of the Agent

Our proposed model exploits a DRL framework to solve the TAD problem. We make some adaptations to the vanilla SAC by replacing the MLP in the Actor and the Critic with a GAIN that can generate more expressive representations. The output of the Actor can be used to express the uncertainty of the agent's judgment, i.e., $a \to 0$ or $a \to 1$ indicates that the agent is quite confident in its judgment, while $a \to 0.5$ indicates the opposite. In the early stages of training, the model has not learned enough information to make the best decision. Hence the uncertainty should be preserved to avoid premature convergence. The entropy regularization term aims to enhance the exploration of actions, and a coefficient

α is used to make a trade-off between expected rewards and entropy. As training continues, the uncertainty will decrease. To accelerate the convergence of the model, the value of α gradually reduces.

4.2. Explanation of the Knight Sampling Strategy

Because there is no environment that can automatically generate a new state according to the agent's action, we create a simulation environment that returns a state and a reward to the agent at each time step. There exist two basic facts about the anomaly detection task. First, the number of normal states in the training set far exceeds the number of anomalous ones. Secondly, normal states are relatively densely distributed, while anomalous states are sparsely distributed and far from normal ones. Based on the facts above, we propose a knight sampling strategy to sample evenly from both categories of states. The knight sampling strategy, which can be understood as dividing the state space into a chessboard with a grid, will alleviate the problem of severe category imbalance. The distribution of 100 states is shown in Figure 3; normal and anomalous states are indicated by a blue dot and red cross, respectively. The ratio of anomalous states to normal ones is 1/9. After dividing the state space with a grid, the ratio of anomalous squares to the normal ones increases to 9/17. Guided by this strategy, the sampler leaps around the data space like a knight, and each move returns a state from the current square.

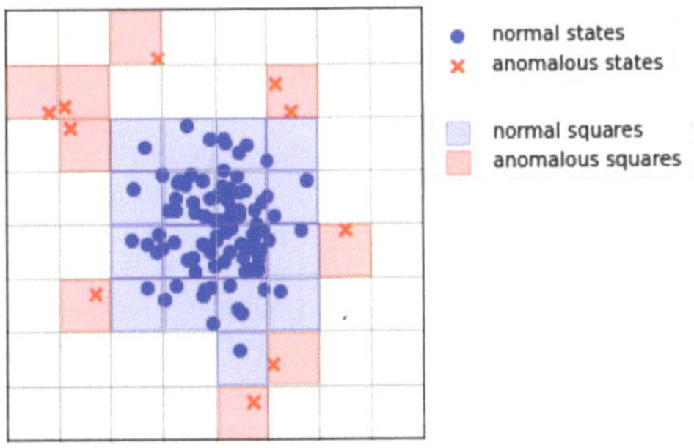

Figure 3. Explanation of Knight Sampling Strategy. Our proposed strategy gives the sampler the leaping capability of a knight and divides the entire data space into grids. The leap capability allows the sampler to fully and uniformly sample from the data space by traversing all the squares.

4.3. Combined Reward

In addition, we deliberately design a combined reward function to make a trade-off between exploitation and exploration of the data space. The extrinsic reward function leverages the continuous reward function to return a reward for each action to settle the sparsity of the rewards. The continuous reward function can be regarded as a generalization of the discrete reward function over the continuous action space. From the perspective of Information Theory, the intrinsic reward aims to encourage the agent to explore the states that are expected to bring high information gains. Although all states contain information, the information gain will decrease when a state is sampled repeatedly. The Gaussian kernel function, which is exploited to approximately count the times of sampling, returns the similarity of the current state and the previous ones. Due to the nature of the exponential function, a few states that are quite similar to the current state will be counted. To cope with this problem, we can adjust the hyper-parameter σ to control the counting radius. The similarities between all points and the center in a square are calculated using Equation (8), and the results are shown in Figure 4. The four subgraphs represent the influence of σ on the distribution of similarities.

We can tell from Figure 4 that if σ is assigned a small value, the similarity decreases sharply with the increase in distance from the center. Hence most neighbors of the center are omitted as they share a similarity close to 0. Therefore, only a few historical states that are very close to the center will be counted. On the contrary, if σ is assigned a larger value, the similarity decreases slowly. Hence more states make contributions to Equation (9), which indicates that more historical states will be counted. Consequently, the hyper-parameter σ can be approximated as the counting radius.

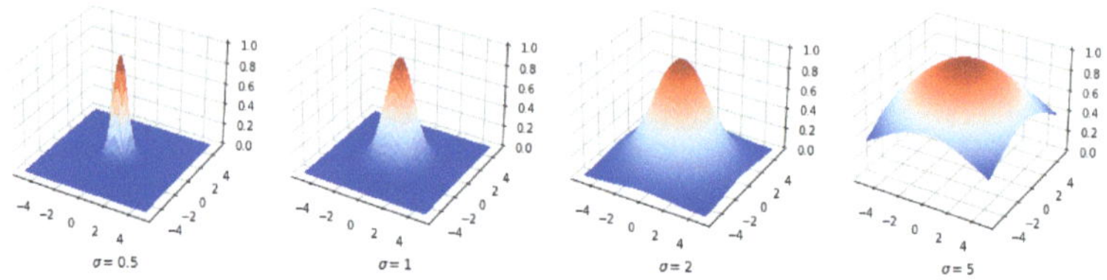

Figure 4. Effect of σ on similarity. The four subplots represent the Gaussian kernel function surfaces of the surrounding points and the center point for different values of the σ. The closer the color is to red, the closer the similarity is to 1. The closer the color is to blue, the closer the similarity is to 0. As the σ increases, the steep surface gradually becomes flatter, and the red area gradually increases, which means that the number of points similar to the center gradually increases.

5. Experiments

5.1. Datasets

To verify the validity of our proposed model, three datasets from different application scenarios are selected for our experiments. NSL-KDD is a dataset from the domain of web security, and each sample in it represents a network traffic record that consists of 40 features. The value of *normal* in feature *attack_type* indicates benign connections, while other values indicate malicious ones. Credit card is a dataset from the domain of finance, which contains credit card transactions conducted by cardholders in Europe over two days in September 2013. Each sample in the dataset represents a transaction record, which consists of 30 features. The value of 1 in *Class* indicates an anomalous transaction, while the value of 0 indicates the opposite. Census is a dataset from the domain of sociology, which contains weighted data extracted from the 1994 and 1995 current population surveys conducted by the U.S. Census Bureau. The dataset contains 40 features, including demographic and employment information, in which the record with *income* "50,000+." are regarded as anomalies. The details of the three datasets are listed in Table 1.

Table 1. Datasets Details. "num." and "cat." denote numerical features and categorical features, respectively.

Dataset	# Samples	# Features	% Anomaly
NLS-KDD	148,517	7 cat. + 33 num.	48.12
Credit Card	284,807	30 num.	0.17
Census	299,285	33 cat. + 7 num.	6.20

The three datasets are very representative for their different composition of features, i.e., NSL-KDD contains 7 categorical features and 33 numerical features, census contains 33 categorical features and 7 numerical features, and credit card only contains numerical features. Recall the assumptions in Section 3.1 that a dataset for AD task consists of a small labeled subset of anomalies and a large-scale unlabeled subset. To make the dataset of NSL-KDD fit our assumption, a downsampling technique is employed to reduce the anomaly ratio. NSL-KDD contains four types of anomalies, each with a large variation in sample size, i.e., *dos*: 45927, *r2l*: 11656, *probe*: 995, *u2r*: 52. Considering the severe intra-imbalance of anomalies, we keep all samples from *probe* and *u2r*, and retain only 11–12% samples from *dos* and *r2l*. The downsampling operation reduces the anomaly ratio from 48.12% to 10.06%.

We select six competing methods to perform a performance comparison with our proposed model. In their official implementations, raw features are not preprocessed in the same manner, e.g., iForest directly removes all categorical features, CatBoost converts categorical features to continuous values, Deep SAD transforms all features to a low-dimensional vector, FT-Transformer transforms all features to embedding vectors with the same length, etc. To reduce the impact of different preprocessing methods on performance, we use the same data preprocessing for all models. First, numerical features are discretized into categorical features; Secondly, all categorical features are embedded into a vector with the same length.

5.2. Competing Methods

Dozens of methods have been proposed to tackle the TAD problems in recent years. According to the difference in the leveraging of supervisory information, those methods can be divided into four categories, such as unsupervised learning methods (UN), supervised learning methods (SU), semi-supervised learning methods (SS) and reinforcement learning methods. In this work, two state-of-the-art methods from each category are selected as the baselines.

- **iForest** [12]. iForest (UN) determines a sample's anomaly degree based on the distribution of each feature value in its feature field. It assumes that the feature values of anomalies are sparsely distributed and, therefore, can be easily distinguished from that of normal samples. A split tree is built for each feature field, and the depth of the feature value represents the anomaly score of a single feature. By combining the anomaly scores of multiple features, the anomaly score of a sample can be obtained.
- **CBLOF** [47]. CBLOF (UN) is a cluster-based anomaly detection method that assumes that anomalies count for a small proportion of the total size and that the samples far away from the large cluster can be considered anomalies. First, a clustering method (e.g., K-Means clustering) is used to cluster samples; Secondly, large clusters are distinguished from small ones; Lastly, distances between samples and large clusters are calculated as anomaly scores.
- **CatBoost** [48]. CatBoost (SU) uses gradient boosting on decision trees. It is an ensemble algorithm that creates a strong learner from an ensemble of multiple weak learners. As its name suggests, CatBoost is capable of handling categorical data. In addition, it solves the problems of gradient bias and prediction shift.
- **FT-Transformer** [24]. FT-Transformer (SU) is a deep model that adapts transformer architecture to tabular data. All features are first transformed to an embedding vector and then passed to a stack of transformer layers to get the prediction.
- **Deep SAD** [18]. Deep SAD (SS) tries to learn a neural network that maps samples to a low-dimensional space in which normal samples cluster in a compact hypersphere while anomalous ones are located outside the hypersphere. A few labeled anomalies can be used in Deep SAD to improve the model's predictive accuracy.
- **DevNet** [20]. DevNet (SS) is an end-to-end semi-supervised method that leverages a limited number of labeled anomalies as prior knowledge to predict anomaly scores. It is based on the assumption that there exists significant statistical deviation between normal samples and anomalous ones.

5.3. Evaluation Metrics

Two popular and complementary metrics, AUC-ROC (Area Under Receiver Operating Characteristic Curve) and AUC-PR (Area Under Precision-Recall Curve), are chosen as the evaluation metrics in this work. AUC-ROC, which summarizes the ROC curve of true positives against false positives, is used to evaluate the classification performance of a model. However, AUC-ROC cannot truly reflect the classification performance when the classes of samples are severely imbalanced. Whereas AUC-PR, which summarizes the ROC curve of precision against the recall and focuses on the performance of anomaly class, is more suitable for this work. A larger AUC-ROC or AUC-PR value reflects better performance.

5.4. Performance Comparison with Competing Models

To verify the effectiveness of our proposed model, comparative experiments are conducted on the three datasets, and the results are presented in Table 2. Our proposed FIR-TAD performs consistently better than its competitors no matter in datasets with more categorical features or in the dataset with more numerical features. In addition, several observations can be obtained from the results. First, unsupervised methods cannot outperform supervised or semi-supervised methods. On the one hand, the absence of labels reduces the information available. On the other hand, unsupervised methods are usually based on the assumption that anomalous samples are significantly different from normal ones in distribution, which may not provide sufficient accuracy. Secondly, semi-supervised methods perform slightly better than supervised methods due to the exploitation of labeled anomalies. Thirdly, the DRL methods show strong competitiveness compared with other deep methods, especially FIR-TAD achieves substantially better performance. We attribute this to the introduction of feature interactions and the exploration ability of SAC.

Table 2. Performance comparison.

Model	NSL-KDD		Credit Card		Census	
	AUC_ROC	AUC_PR	AUC_ROC	AUC_PR	AUC_ROC	AUC_PR
iForest	0.8359	0.8483	0.9469	0.1412	0.5956	0.0781
CBLOF	0.8213	0.5307	0.8772	0.2451	0.5938	0.0742
CatBoost	0.9208	0.9012	0.8446	0.4867	0.8815	0.3510
FT-Transformer	0.9278	0.8530	0.8275	0.4499	0.8376	0.2335
DeepSAD	0.9391	0.9037	0.8902	0.2577	0.7232	0.1855
DevNet	0.9410	0.9278	0.9520	0.5109	0.8354	0.3211
FIRTAD(ours)	0.9457	0.9362	0.9583	0.5870	0.8952	0.3670

5.5. Test on Robustness

Robustness with regard to anomaly ratio. To study the robustness of all models on datasets containing different numbers of anomalies, we create five datasets with different proportions (10%, 5%, 1%, 0.5%, 0.1%) of anomalies based on the NSL-KDD. Comparative experiments are performed on the five datasets, and the results are shown in Figure 5a. We choose AUC-PR as the only evaluation metric in the following because the results of AUC-ROC are often over-optimistic in the case of imbalanced classes. As shown in Figure 5a, the performance of all models degrades with the decrease in the proportion of anomalies, and the unsupervised methods are less affected as they distinguish anomalies based on differences in the distribution of normal samples and are not sensitive to the number of anomalous samples. With the reduction in supervised information, the supervised methods experience significant performance degradation due to their reliance on supervisory information. Although the semi-supervised methods are less affected by the decrease in the ratio of anomalies, we note that the performance of the semi-supervised methods is significantly weaker than that of the unsupervised method when the ratio drops below 0.5%. In contrast, our proposed method shows better robustness in all cases.

Robustness with regard to label contamination. To test the robustness of all models under label-contaminated conditions, which is very common in real-world scenarios, we define five datasets by sampling incremental numbers of labeled anomalies, removing their labels and blending them into the unlabeled samples. The contamination rates of the five datasets are 0%, 2%, 5%, 10% and 20%, respectively. All models are evaluated on these five datasets, and the experimental results in Figure 5b show that the increase in label contamination rate has almost no impact on the unsupervised method. The supervised methods suffer the most due to the decrease in data quality. The semi-supervised methods consistently perform better than the unsupervised methods, proving their good robustness to label contamination. Our proposed model exhibits the best robustness, and we attribute it to the design of the extrinsic reward function, which encourages the exploration of anomalies in normal samples.

Robustness with regard to unknown anomalies.For some online anomaly detection systems, the training and test sets usually have inconsistent sample distributions, i.e., the test set may contain anomalies that never appear in the training set. To compare the robustness of different methods in the face of unknown anomalies, we define a dataset in which the training and test set contains only one type of anomaly and four additional test sets by adding new anomaly types to the test set in turn. We evaluate the trained model on the five test sets and present the results in Figure 5c. The performance of all models decreases with the increase in anomaly types, with the unsupervised methods being the least affected and the supervised methods being the most affected. The semi-supervised methods still exhibit good robustness. Our proposed model performs significantly better than its competitors. The introduction of feature interaction allows our model to learn useful discriminative information, and the exploitation of the DRL algorithm gives our model the ability to explore the unknown. These two factors give our model very good robustness to unknown anomalies.

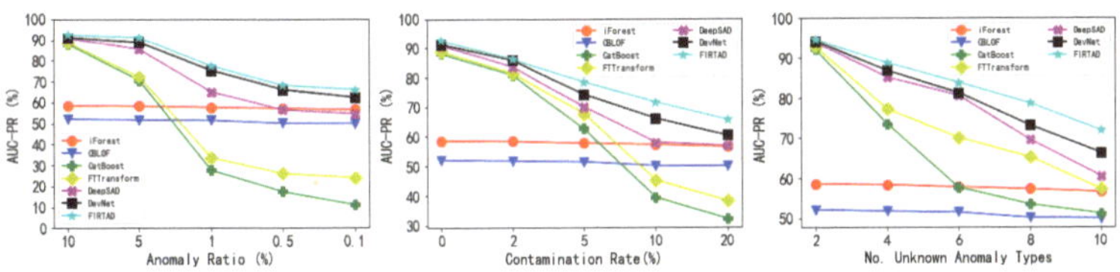

(a) (b) (c)

Figure 5. Test on Robustness. (**a**) Robustness with regard to anomaly ratio; (**b**) Robustness with regard to label contamination; (**c**) Robustness with regard to unknown anomalies. We find that although the unsupervised models exhibit the best robustness, their performance is limited. The supervised models suffer most from the decrease in supervised information and label quality; they perform even worse than unsupervised models. The semi-supervised models show good robustness against the anomaly ratio. However, the performance degrades significantly with respect to label contamination and unknown anomalies. Our proposed FIRTAD shows better robustness and consistently outperforms the other models.

5.6. Impacts of Hyperparameters

In this section, we mainly focus on the impacts of the hyperparameters using the NSL-KDD dataset.

- Impacts of representation dimensionality of each feature (d). In a deep model, the representation vector with a longer length carries more information; hence, it is common to improve the expressiveness of a model by increasing the representation dimensionality of the input. Nevertheless, it is a double-edged sword, as an increase in the dimension of the representation vector would lead to an increase in memory consumption and a decrease in model efficiency. We choose $d = 4, 8, 16, 32$ and plot the AUC-PR in Figure 6a. Apparently, the performance improves as the length of the vector increases. Concretely, when d changes from 8 to 16, the AUC-PR improves from 0.9175 to 0.9437, with an improvement of 2.86%. However, as the length of the vector continues to increase, the improvement in model performance becomes very limited. The AUC-PR improves from 0.9437 to 0.9439 when d increases from 16 to 32. Considering the consequent doubling of memory consumption and computation time, $d = 16$ seems to be a more reasonable option, which creates a balance between effectiveness and efficiency.
- Impacts of coefficient of intrinsic reward (λ). As discussed in Section 3.4, λ is leveraged to provide a tradeoff between exploitation and exploration, which represents the

weight factor corresponding to the intrinsic reward. We choose λ from 0, 0.3, 0.5, 0.8, 1, in which different values correspond to different extensions of exploration, e.g., $\lambda = 0$ represents a deprecation of exploration, while $\lambda = 1$ represents the opposite. The experimental results are plotted in Figure 6b. We can tell from the results that the introduction of intrinsic rewards does improve the model's performance, while a high weight may cause the model to converge prematurely to suboptimal solutions. As shown in Figure 6b, $\lambda = 0.5$ is the most suitable option.

- Impacts of counting radius (σ). To encourage the agent to explore unknown regions in the environment, we designed the intrinsic reward to score the novelty of a region. The number of samplings from a region is approximately counted by a Gaussian kernel function, in which σ can be regarded as the counting radius. For a specific region that has been visited a certain number of times, increasing the radius will lead to an increase in novelty and vice versa. We choose the value of σ from 0.1, 0.5, 1, 2, 5 and plot the results in Figure 6c. Intrinsically, a small radius would result in large intrinsic rewards for new samples, even if similar ones have been sampled many times, which might hinder the agent from exploring unknown regions. Moreover, a large radius would result in small intrinsic rewards for samples from sparse regions, which might result in insufficient exploration of known regions. Both of the above conditions would reduce convergence speed or make the model converge to a suboptimum, and this intuition is verified by the experimental results. As shown in Figure 6c, $\sigma = 1$ is the most reasonable option.

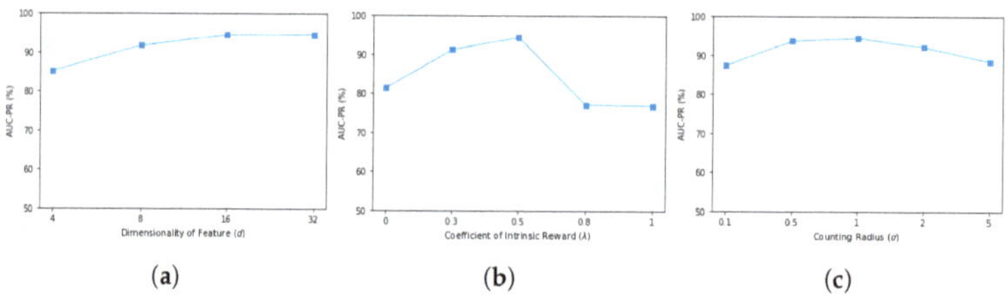

Figure 6. Impacts of Hyperparameters. (**a**) The contribution of representation dimensionality to model performance improvement decreases with the increase in dimension; (**b**) The introduction of intrinsic reward improves the model's performance, but a high weight would be counterproductive; (**c**) A small counting radius will increase the intrinsic reward and thus lower the agent's desire to explore the state space, while a large counting radius will render a small intrinsic reward and hence lead to insufficient exploration of states.

5.7. Ablation Study

To investigate the effects of different components in our proposed model, we propose several variants based on FIRTAD and conduct ablation experiments on all three datasets.

1. MLP-SAC. MLP-SAC replaces the feature interaction module (GAIN) with an MLP. To reduce the number of parameters and accelerate model convergence, the MLP is shared by the Actor and Critic.
2. FIR-DDPG. FIR-DDPG replaces the SAC with the Deep Deterministic Policy Gradient (DDPG), which is an Actor–Critic, model-free algorithm based on the deterministic policy gradient that can operate over continuous action spaces.
3. FIR-AB. FIR-AB replaces the knight sampling strategy with the anomaly-biased strategy used in [45].

Table 3 shows the results of our original model, FIRTAD, and its variants. In the following, we analyze the effects of components in our model. First, compared to the original model, MLP-SAC shows a significant drop in performance on NSL-KDD and census. This implies that the feature interaction module indeed enhances our model's ability for anomaly detection. Further, we note that MLP-SAC achieves comparable results to the original model on credit card, suggesting that GAIN has no advantage in dealing with data consisting entirely of numerical features. Secondly, although both SAC and DDPG are off-policy algorithms for continuous action space, SAC has a stronger exploratory capability than DDPG due to its exploitation of stochastic policy. The comparison result between FIR-DDPG and FIRTAD validates this view and prove the effectiveness of the SAC module. Thirdly, the performance gaps between FIR-AB and FIRTAD in the three datasets indicate that the knight sampling strategy may outperform the anomaly-biased sampling strategy.

Table 3. AUC-PR Performance of FIRTAD and its Three Ablated Variants.

Datasets	FIRTAD	MLP-SAC	FIR-DDPG	FIR-AB
NLS-KDD	0.9362	0.6874	0.9012	0.8829
Credit Card	0.5870	0.5798	0.5022	0.3681
Census	0.3670	0.1453	0.3124	0.2445

6. Conclusions

In this paper, we propose a novel anomaly detection method for tabular data called FIRTAD, which incorporates feature interaction techniques into a deep reinforcement learning framework. The innovative aspects of this article are manifested in the following dimensions: (1) It is an anomaly detection system specifically designed for tabular data; (2) It employs the SAC algorithm to generalize the discrete action space into a continuous space, thereby enhancing the model's expressive power; (3) It creates a simulation environment by devising a novel sampling strategy and a combined reward function; (4) As far as we know, it is the first effort to apply feature interaction to anomaly detection tasks. The experiments demonstrate that our proposed model not only outperforms state-of-the-art models in terms of performance but also exhibits good robustness in situations involving varying anomaly ratios, label contamination and unknown anomalies. Our model is applicable to real-world anomaly detection scenarios, particularly in domains that have accumulated some anomalies. This work serves as an attempt to apply deep reinforcement learning to anomaly detection task and may provide some inspiration to relevant researchers.

Despite the encouraging results, our proposed FIRTAD still has some limitations. First, our model relies heavily on large amounts of data for training, so its performance advantages may not be apparent when only dealing with a small dataset. Second, the anomaly detection capability of our model gradually improves during the interaction between the agent and the environment, which leads to an increased demand for computational resources and training time. Third, the experiments demonstrate that the performance advantages of our model on the balanced dataset are not significant, which limits the applicability of the model.

Future work will consider using multiprocessing for model training and the study of anomaly interpretability.

Author Contributions: Conceptualization, Y.L. and L.M.; methodology and implementation, Y.L. and S.Z.; writing—original draft preparation, Y.L.; writing—review and editing, L.M., M.W. and S.Z. All authors have read and agreed to the published version of the manuscript.

Funding: This research received no external funding.

Institutional Review Board Statement: Not applicable.

Informed Consent Statement: Not applicable.

Data Availability Statement: The NSL-KDD dataset supporting this study was obtained from https://www.unb.ca/cic/datasets/nsl.html (accessed on 13 February 2023). The credit card dataset was obtained from https://www.kaggle.com/mlg-ulb/creditcardfraud (accessed on 13 February 2023). The census dataset was obtained from http://archive.ics.uci.edu/ml/machine-learning-databases/census-income-mld/ (accessed on 13 February 2023).

Acknowledgments: The authors would like to thank the anonymous reviewers for their constructive comments and suggestions.

Conflicts of Interest: The authors declare no conflict of interest.

References

1. Zhang, H.; Yu, X.; Ren, P.; Luo, C.; Min, G. Deep Adversarial Learning in Intrusion Detection: A Data Augmentation Enhanced Framework. *arXiv* **2019**, arXiv:901.07949.
2. Lee, J.; Park, K. GAN-based imbalanced data intrusion detection system. *Pers. Ubiquitous Comput.* **2021**, *25*, 121–128. [CrossRef]
3. Vimal, S.; Kayathwal, K.; Wadhwa, H.; Dhama, G. Application of Deep Reinforcement Learning to Payment Fraud . *arXiv* **2021**, arXiv:2112.04236.
4. Tingfei, H.; Guangquan, C.; Kuihua, H. Using Variational Auto Encoding in Credit Card Fraud Detection. *IEEE Access* **2020**, *8*, 149841–149853. [CrossRef]
5. Mokhtari, S.; Abbaspour, A.; Yen, K.K.; Sargolzaei, A. A Machine Learning Approach for Anomaly Detection in Industrial Control Systems Based on Measurement Data. *Electronics* **2021**, *10*, 407. [CrossRef]
6. Zhou, X.; Hu, Y.; Liang, W.; Ma, J.; Jin, Q. Variational LSTM Enhanced Anomaly Detection for Industrial Big Data. *IEEE Trans. Ind Inform.* **2021**, *17*, 3469–3477. [CrossRef]
7. May Petry, L.; Soares, A.; Bogorny, V.; Brandoli, B.; Matwin, S. Challenges in Vessel Behavior and Anomaly Detection: From Classical Machine Learning to Deep Learning. In *Advances in Artificial Intelligence*; Goutte, C., Zhu, X., Eds.; Springer International Publishing: Cham, Switzerland, 2020; pp. 401–407. ISBN 978-3-030-47358-7.
8. Karata, G.B.; Karagoz, P.; Ayran, O. Trajectory pattern extraction and anomaly detection for maritime vessels. *Internet Things* **2021**, *16*, 100436. [CrossRef]
9. Breunig, M.M.; Kriegel, H.P.; Ng, R.T.; Sander, J. LOF: Identifying Density-Based Local Outliers. In Proceedings of the 2000 ACM SIGMOD International Conference on Management of Data, Dallas, TX, USA, 15–18 May 2000; ACM: New York, NY, USA, 2000 pp. 93–104, ISBN 1581132174 .
10. Ramaswamy, S.; Rastogi, R.; Shim, K. *Efficient Algorithms for Mining Outliers from Large Data Sets*; International Conference on Management of Data; ACM: New York, NY, USA, 2000.
11. Goldstein, M.; Dengel, A. Histogram-based Outlier Score (HBOS): A fast Unsupervised Anomaly Detection Algorithm. In Proceedings of 35th German Conference on Artificial Intelligence, Saarbrücken, Germany, 24–27 September 2012; Springer Berlin/Heidelberg, Germany, 2012; pp. 59–63.
12. Liu, F.T.; Ting, K.M.; Zhou, Z.-H. Isolation Forest. In Proceedings of the 2008 Eighth IEEE International Conference on Data Mining, Pissa, Italy, 15–19 December 2008; IEEE: Piscataway, NJ, USA, 2008; pp. 413–422.
13. Li, Z.; Zhao, Y.; Hu, X.; Botta, N.; Ionescu, C.; Chen, G. ECOD: Unsupervised Outlier Detection Using Empirical Cumulative Distribution Functions. *IEEE Trans. Knowl. Data Eng.* **2022**, *1*.
14. Han, X.; Chen, X.; Liu, L.-P. GAN Ensemble for Anomaly Detection. *Aaai Conf. Artif. Intell.* **2021**, *35*, 4090–4097. [CrossRef]
15. Murase, H.; Fukumizu, K. ALGAN: Anomaly Detection by Generating Pseudo Anomalous Data via Latent Variables. *IEEE Access* **2022**, *10*, 44259–44270. [CrossRef]
16. Fatemifar, S.; Awais, M.; Akbari, A.; Kittler, J. Developing a generic framework for anomaly detection. *Pattern Recognit.* **2022**, *124* 108500. [CrossRef]
17. Zhang, L; Xie, X.; Xiao, K.; Bai, W.; Liu, K.; Dong, P. MANomaly: Mutual adversarial networks for semi-supervised anomaly detection. *Inf. Sci.* **2022**, *611*, 65–80. [CrossRef]
18. Ruff, L.; Vandermeulen, R.A.; Görnitz, N.; Binder, A.; Müller, E.; Müller, A.; Kloft, M. Deep Semi-Supervised Anomaly Detection *arXiv* **2019**, arXiv:1906.02694.
19. Pang, G.; Shen, C.; Jin, H.; van den Hengel, A. Deep weakly-supervised anomaly detection. *arXiv* **2019**, arXiv:1910.13601.
20. Pang, G.; Shen, C.; van den Hengel, A. Deep Anomaly Detection with Deviation Networks. *arXiv* **2019**, arXiv:1911.08623.
21. Zhou, Y.; Song, X.; Zhang, Y.; Liu, F.; Zhu, C.; Liu, L. Feature Encoding with AutoEncoders for Weakly-supervised Anomaly Detection. *IEEE Trans. Neural Netw. Learning Syst.* **2022**, *33*, 2454–2465. [CrossRef]
22. Pang, G.; Cao, L.; Chen, L.; Liu, H. Learning Representations of Ultrahigh-dimensional Data for Random Distance-based Outlier Detection. In Proceedings of the 24th ACM SIGKDD International Conference on Knowledge Discovery & Data Mining, London UK, 19–23 August 2018; Guo, Y., Farooq, F., Eds.; ACM: New York, NY, USA, 2018 ; pp. 2041–2050. ISBN 9781450355520.
23. Qiu, C.; Pfrommer, T.; Kloft, M.; Mandt, S.; Rudolph, M. Neural Transformation Learning for Deep Anomaly Detection beyond Images. *arXiv* **2021**, arXiv:2103.16440.
24. Gorishniy, Y.; Rubachev, I.; Khrulkov, V.; Babenko, A. Revisiting Deep Learning Models for Tabular Data. In *Advances in Neural Information Processing Systems*; Ranzato, M., Beygelzimer, A., Dauphin, Y., Liang, P.S., Vaughan, J.W., Eds.; Curran Associates, Inc New York, NY, USA, 2021; pp. 18932–18943.
25. Shwartz-Ziv, R.; Armon, A. Tabular data: Deep learning is not all you need. *Inf. Fusion* **2022**, *81*, 84–90. [CrossRef]

26. Chen, B.; Wang, Y.; Liu, Z.; Tang, R.; Guo, W.; Zheng, H.; Yao, W.; Zhang, M.; He, X. Enhancing Explicit and Implicit Feature Interactions via Information Sharing for Parallel Deep CTR Models. In Proceedings of the 30th ACM International Conference on Information & Knowledge Management. CIKM '21: The 30th ACM International Conference on Information and Knowledge Management, Virtual Event, Queensland, Australia, 1–5 November 2021; Demartini, G., Zuccon, G., Culpepper, J.S., Huang, Z., Tong, H., Eds.; ACM: New York, NY, USA, 2021; pp. 3757–3766. ISBN 9781450384469.
27. Zhao, Z.; Yang, S.; Liu, G.; Feng, D.; Xu, K. FINT: Field-aware INTeraction Neural Network For CTR Prediction. *arXiv* **2021**, arXiv:2107.01999.
28. Deng, W.; Pan, J.; Zhou, T.; Kong, D.; Flores, A.; Lin, G. DeepLight: Deep Lightweight Feature Interactions for Accelerating CTR Predictions in Ad Serving. In Proceedings of the Fourteenth ACM International Conference on Web Search and Data Mining (WSDM '21), Virtual Event, Jerusalem, Israel, 8–12 March 2021; ACM: New York, NY, USA, 2021. [CrossRef]
29. Liu, Y.; Ma, L.; Wang, M. GAIN: A Gated Adaptive Feature Interaction Network for Click-Through Rate Prediction. *Sensors* **2022**, *22*, 7280. [CrossRef]
30. Klein, S.T.; Wiseman, Y. Parallel Lempel Ziv coding. *Discret. Appl. Math.* **2005**, *146*, 180–191. [CrossRef]
31. Nakasato, N. Implementation of a parallel tree method on a GPU. *J. Comput. Sci.* **2012**, *3*, 132–141. [CrossRef]
32. Haarnoja, T.; Zhou, A.; Hartikainen, K.; Tucker, G.; Ha, S.; Tan, J.; Kumar, V.; Zhu, H.; Gupta, A.; Abbeel, P.; et al. Soft Actor-Critic Algorithms and Applications. *arXiv* **2018**, arXiv:1812.05905.
33. Pang, G.; Ting, K.M.; Albrecht, D. LeSiNN: Detecting Anomalies by Identifying Least Similar Nearest Neighbours. In Proceedings of the 2015 IEEE International Conference on Data Mining Workshop (ICDMW), Atlantic City, NJ, USA, 14–17 November 2015; pp. 623–630.
34. Li, Z.; Zhao, Y.; Botta, N.; Ionescu, C.; Hu, X. COPOD: Copula-Based Outlier Detection. In Proceedings of the 2020 IEEE International Conference on Data Mining (ICDM), Sorrento, Italy, 17–20 November 2020; pp. 1118–1123.
35. Ruff, L.; Vandermeulen, R.A.; Görnitz, N; Deecke, L.; Kloft, M. Deep One-Class Classification. In Proceedings of the International Conference on Machine Learning, PMLR 80, Stockholm, Sweden, 10–15 July 2018.
36. Miele, E.S.; Bonacina, F.; Corsini, A. Deep anomaly detection in horizontal axis wind turbines using Graph Convolutional Autoencoders for Multivariate Time series. *Energy AI* **2022**, *8*, 100145. [CrossRef]
37. Jézéquel, L.; Vu, N.-S.; Beaudet, J.; Histace, A. Semi-Supervised Anomaly Detection with Contrastive Regularization. In Proceedings of the 2022 26th International Conference on Pattern Recognition (ICPR) , Montreal, QC, Canada, 21–25 August 2022; pp. 2664–2671.
38. Gao, F.; Li, J.; Cheng, R.; Zhou, Y.; Ye, Y. ConNet: Deep Semi-Supervised Anomaly Detection Based on Sparse Positive Samples. *IEEE Access* **2021**, *9*, 67249–67258. [CrossRef]
39. Wang, Z.; She, Q.; Zhang, P.; Zhang, J. ContextNet: A Click-Through Rate Prediction Framework Using Contextual information to Refine Feature Embedding. *arXiv* **2021**, arXiv:2107.12025.
40. Rendle, S. Factorization machines. In Proceedings of the 2010 IEEE International Conference on Data Mining, Sydney, Australia, 14–17 December 2010; pp. 995–1000.
41. Xue, N.; Liu, B.; Guo, H.; Tang, R.; Zhou, F.; Zafeiriou, S.P.; Zhang, Y.; Wang, J.; Li, Z. AutoHash: Learning Higher-order Feature Interactions for Deep CTR Prediction. *IEEE Trans. Knowl. Data Eng.* **2020**, 34, 2653–2666. . [CrossRef]
42. Liu, B.; Zhu, C.; Li, G.; Zhang, W.; Lai, J.; Tang, R.; He, X.; Li, Z.; Yu, Y. Autofis: Automatic feature interaction selection in factorization models for click-through rate prediction. In Proceedings of the 26th ACM SIGKDD International Conference on Knowledge Discovery & Data Mining, Virtual Event, Long Beach, CA, USA, 23–27 August 2020; pp. 2636–2645.
43. Chen, Y.; Wang, Y.; Ren, P.; Wang, M.; Rijke, M.d. Bayesian feature interaction selection for factorization machines. *Artif. Intell.* **2022**, *302*, 103589. [CrossRef]
44. Lopez-Martin, M.; Carro, B.; Sanchez-Esguevillas, A. Application of deep reinforcement learning to intrusion detection for supervised problems. *Expert Syst. Appl.* **2020**, *141*, 112963. [CrossRef]
45. Pang, G.; van den Hengel, A.; Shen, C.; Cao, L. Toward Deep Supervised Anomaly Detection. In Proceedings of the 27th ACM SIGKDD Conference on Knowledge Discovery & Data Mining. KDD '21: The 27th ACM SIGKDD Conference on Knowledge Discovery and Data Mining, Virtual Event, Singapore, 14–18 August 2021; Zhu, F., Chin Ooi, B., Miao, C., Wang, H., Skrypnyk, I., Hsu, W., Chawla, S., Eds.; ACM: New York, NY, USA, 2021; pp. 1298–1308. ISBN 9781450383325.
46. Watkins, J.J. *The Knight's Tour Problem: Across the Board: The Mathematics of Chessboard Problems*; Chapter 3; Princeton University Press: Princeton, NJ, USA, 2004.
47. He, Z.; Xu, X.; Deng, S. Discovering cluster-based local outliers. *Pattern Recognit. Lett.* **2003**, *24*, 1641–1650. [CrossRef]
48. Prokhorenkova, L.; Gusev, G.; Vorobev, A.; Dorogush, A.V.; Gulin, A. CatBoost: Unbiased boosting with categorical features. *arXiv* **2017**, arXiv:1706.09516.

sensors

MDPI

Article

Industrial Transfer Learning for Multivariate Time Series Segmentation: A Case Study on Hydraulic Pump Testing Cycles

Stefan Gaugel [1,2,*] and Manfred Reichert [2]

[1] Bosch Rexroth AG, 89081 Ulm, Germany
[2] Institute of Databases and Information Systems, Ulm University, 89081 Ulm, Germany
* Correspondence: stefan.gaugel@boschrexroth.net

Abstract: Industrial data scarcity is one of the largest factors holding back the widespread use of machine learning in manufacturing. To overcome this problem, the concept of transfer learning was developed and has received much attention in recent industrial research. This paper focuses on the problem of time series segmentation and presents the first in-depth research on transfer learning for deep learning-based time series segmentation on the industrial use case of end-of-line pump testing. In particular, we investigate whether the performance of deep learning models can be increased by pretraining the network with data from other domains. Three different scenarios are analyzed: source and target data being closely related, source and target data being distantly related, and source and target data being non-related. The results demonstrate that transfer learning can enhance the performance of time series segmentation models with respect to accuracy and training speed. The benefit can be most clearly seen in scenarios where source and training data are closely related and the number of target training data samples is lowest. However, in the scenario of non-related datasets, cases of negative transfer learning were observed as well. Thus, the research emphasizes the potential, but also the challenges, of industrial transfer learning.

Keywords: time series segmentation; deep learning; multivariate time series; transfer learning; end-of-line testing

1. Introduction

1.1. Problem Statement

The ubiquity of sensor data in modern manufacturing is creating opportunities as well as challenges for manufacturers. Since both computer-aided engineering [1] and time series-based machine learning (ML) have shown significant advancements in recent years, an increasing number of manufacturers rely on ML-based solutions to create value out of sensor data. In particular, fields like predictive maintenance [2], condition monitoring [3], and anomaly detection [4] have achieved immense progress and have become well-known terms in the academic and industrial world.

A lesser-known research area is time series segmentation (TSS), which deals with splitting up a time series into distinct non-overlapping segments. In a manufacturing setting, TSS is usually based on multivariate time series data. TSS can be utilized to detect different operational states of production machines or to recognize human actions via tracking sensors. Multivariate time series data pose special challenges for algorithms. The most successful multivariate TSS approaches proposed in recent years are based on deep learning and use a supervised learning concept (e.g., [5]). In recent work, we analyzed a TSS use case in manufacturing and proposed a novel deep learning architecture [6]. The presented network was successful in splitting up the end-of-line testing cycle data of hydraulic pumps into different testing phases based on nine sensor channels. However, results were based on a sufficiently large training dataset that was available for each pump variant. In this paper, we build on this previous work and revisit the use case with the difference that we face training data scarcity.

The biggest drawback of deep learning models is their heavy dependency on the availability of data. In particular, in many industrial environments, the collection and labeling of required data is expensive and time consuming. In turn, there are many settings where only a limited amount of training data are available. This widespread issue of data scarcity led to the emergence of research fields that aim to find approaches able to cope with a small amount of training data. Examples of corresponding approaches include few-shot learning, data augmentation, and transfer learning (TL). The concept of TL is illustrated by Figure 1. TL denotes the process of utilizing the information created when solving a task in order to solve a similar but distinct task.

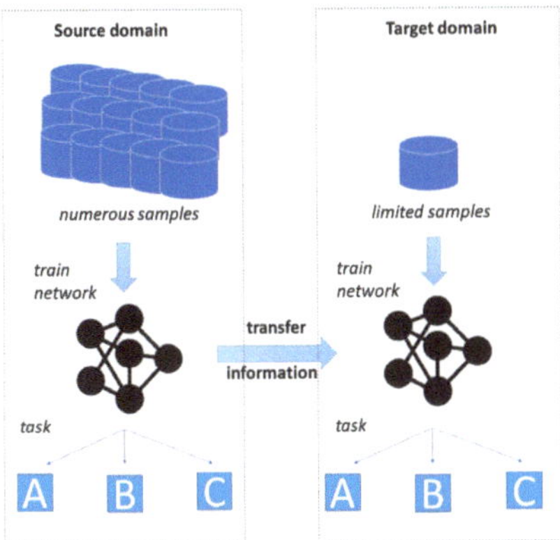

Figure 1. Illustration of the basic concept of transfer learning.

In the context of deep learning, TL describes the pretraining of the layers of a network by a *source dataset* before transferring the learned parameters to a different model, the *target model*. Subsequently, the *target model* is retrained by the *target dataset*, which often comprises a small number of training samples (finetuning). As a basic idea, the information that the model learned by solving the *source problem* might be helpful to boost the training process of the *target problem*. Especially for production plants, TL constitutes a promising approach, as a high number of different product variants and models with widely varying quantities may be manufactured on the same shop floor. In particular, the data from products manufactured in high quantities might provide interesting information for similar goods with less turnover. This idea was recognized by many manufacturers, triggering a high number of publications in recent years about the successful use of TL in industrial settings (see, for example, [7,8]). Particularly in deep learning-based TSS, however, TL has received limited attention compared to its high prevalence in other areas like image recognition or condition monitoring. There is close to no research on applying TL approaches to TSS settings and on analyzing what factors increase or limit the benefits of TL. One of the very few contributions covering the topic showed promising results when applying cross-domain TL to boost medical TSS, even though the scope of the analysis concerning TL is very limited [9]. In general, no authors have ever performed a study focusing on how TL can be used to boost the performance of deep learning TSS models.

1.2. Our Contribution

With the presented problem setting and identified research gap in mind, this paper provides the following contributions:

- We analyze the benefits of using TL for TSS in an industrial setting. The paper provides one of the very first works to even tackle the problem of TL for TSS in general.

- We systematically analyze how pretraining with three different *source datasets* with varying degrees of similarity to the *target dataset* affects the performance of the *target model* after finetuning.
- We analyze to what degree the benefit of TL depends on the amount of available samples in the *target dataset*.
- The use case analyzed in the paper deals with the segmentation of operational states within the end-of-line testing cycle of hydraulic pumps. This is an innovative application of time series-based deep learning for a practical manufacturing problem.

The remainder of the paper is structured as follows: Section 2 provides an overview of related works in the area of industrial transfer learning for time series data. In Section 3, the concept of TL, the datasets used, and the experimental designs are presented. Section 4 describes the obtained experimental results and discusses them, followed by a conclusion and outlook in Section 5.

2. Literature Research

Facilitated by the general breakthrough of deep learning in many domains, the related research field of TL has recently received much attention and has advanced significantly. Different surveys were published in recent years, providing a detailed overview of the state of the art. The most extensive one focuses on the general concept of TL over all domains, including experimental results on text categorization and object detection [10]. More recently, a systematic study was published that had focused on TL in the context of time series data, discovering that the publication frequency has sharply increased since the year 2019 [11]. The study presented in [12] provides a detailed summary of TL concepts, use cases, and trends in the context of industrial automation. Another field that is highly related to TL and recently garnered significant attention is domain adaptation [13,14]. In this paper, however, we deal with a setting that exceeds the scope of domain adaptation (see Section 3.1) and, therefore, we do not cover the various frameworks from the field. Instead, we limit ourselves to reviewing relevant contributions in two areas, namely time series-based TL and deep industrial TL. We clarify that our literature review does not cover recent contributions in the field of domain adaptation.

2.1. Transfer Learning for Time Series

The concept of TL has its origin in the area of image recognition. After showing promising results there, it found its way into time series-based ML. The research field is still relatively young, with most contributions being published after 2018. However, despite the novelty of the research field, time series-based TL has already been applied in various fields like manufacturing [15], finance [16], geoscience [17], mobility [18], and medicine [19]. Successfully solved tasks include time series imaging [20], anomaly detection [21], classification [22], and forecasting [23]. A seminal work applying TL to time series classification analyzed how the pretraining of a model with different *source datasets* affects the classification accuracy of the *target task* [24]. In total, the authors tested more than 85 *source datasets* for pretraining and discovered that the benefit of pretraining significantly varies with the used *source dataset*. While a high similarity between *source dataset* and *target dataset* tended to lead to performance improvement, pretraining with dissimilar datasets did not improve or worsen the classification accuracy. Another interesting contribution demonstrates the successful application of TL to both convolutional neural networks (CNN) and long short-term memory (LSTM) architectures when performing time series classification and time series predictions [25].

Almost none of the previous works in the TL field focused on the task of TSS. While TSS problems might be less prevalent in the everyday life of companies and researchers, they still deserve attention due to their importance for fields like human action recognition [26], sleep staging [5], and operational state detection [6]. One contribution had a first look at the topic [9] and successfully showed that feature extraction process of a TSS–CNN model can be improved when pretraining the model with a large, domain-independent *source dataset*.

However, a broad analysis focusing on TL and a detailed discussion of the implications of the results were not performed.

2.2. Deep Industrial Transfer Learning

Deep industrial transfer learning describes the concept of using formerly gained information and knowledge through deep learning to facilitate the solution of industrial problems [27]. In industrial settings, it is often not feasible (or even not possible) to accumulate a sufficiently large amount of data required for traditional deep learning approaches. The reasons for the data scarcity include the variability and high customization of processes and products [28], data protection regulations, and the high cost of data collection, storage, and labeling [29]. Industrial environments are highly volatile and dynamic. Wear, reconfigurations, and process adaptations [28] often diminish the value of historical data, leading to only a limited amount of short-term data accurately representing the correct relationship between input data and label [30]. Therefore, approaches like TL that attempt to address the issue of data scarcity have gained increasing attention in recent years.

Ref. [15] compared different TL approaches for anomaly detection based on pump vibration data. In turn, ref. [31] used already outdated data for pretraining their network to boost the prediction accuracy compared to a cold-start network. Results were verified on two industrial datasets (solar, furnace). Another approach combined both few-shot learning and TL techniques to overcome the issue of a small training set in rotatory machine fault detection; [7,8] used a huge general-domain dataset to train the feature extractor of a CNN, before finetuning the classificator with the small *target dataset* to diagnose gear faults. Ref. [32] analyzed whether TL can help to identify bearing defects by transferring learned features across data of different working conditions. The authors could show that their proposed TL approach increases the effectiveness of vibration-based fault detection models. Staying in the area of vibration-based bearing fault detection, another contribution provided an analysis of how and to what extent discriminative information found in varying settings (different sensor locations or working loads) can be transferred between networks [33]. In turn, ref. [34] investigated online fault detection of pumps and motor bearings and achieved high accuracy by creating a transfer convolutional neural network compared to classic models. Finally, ref. [35] was able to show that the deep TL network created with a domain adaptation module outperformed other approaches in a piston pump fault detection use case.

To the best of our knowledge, no other works have investigated the application of TL strategies to a TSS problem within an industrial setting. Almost all TL contributions from the time series domain focus on the tasks of forecasting, anomaly detection, and classification. There is a lack of research on TL strategies for TSS in general, and even more so for TSS use cases in industrial settings. The pump testing dataset we use in this paper was published in the context of one of our previous works [6]. In that work, the dataset was used for TSS experiments in which the availability of a sufficient amount of training data was ensured. In this paper, however, we deal with data scarcity and investigate whether TL can help to facilitate the learning of the model when few training samples are available. Therefore, we are the first to provide detailed results about the factors and conditions (sample size, *source dataset*, layer freezing) that determine the benefit of TL in the context of TSS. Potential explanations for why certain conditions favor TL are discussed as well. Our findings extend previous research about time series-based TL and provide insightful implications for practical industrial use cases.

3. Experimental Design and Data

3.1. Transfer Learning Formalization

The idea of TL fundamentally differs from the traditional ML idea. Traditional ML approaches are isolation-based, meaning that they rely on an isolated dataset for the training of a distinct model in order to solve a single task. Usually, the models lack the ability to generalize well across different tasks or data. In turn, transfer learning allows for incorporating previously learned knowledge from other tasks in order to avoid some of the common ML drawbacks. Usually, the transferable knowledge is more valuable if the domains of the different models are related. While TL is a broad field with a number of different approaches and concepts, this paper focuses on a specific TL setting in which both the *source domain* and the *target domain* have labeled data and the transfer consists of pretraining a model with a *source dataset* (see Figure 2).

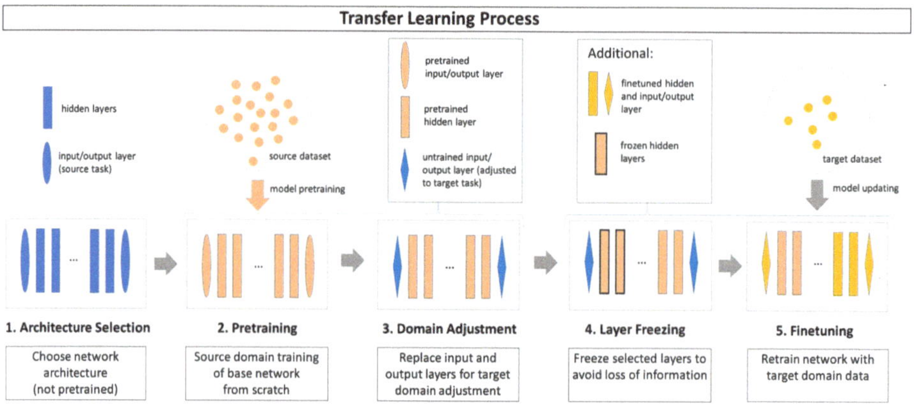

Figure 2. Visualization of 5-step transfer learning approach.

In our work, we reuse the formalization introduced by [36]. In general, an ML problem comprises two components, the domain and the task. A domain is formed by two elements, i.e., the feature space X and the probability distribution $P(X)$. We denote a specific domain D as $D = (X, P(X))$. A task comprises two elements, namely the label space Y and the prediction function $f(x)$ that is used to predict the label of a given instance x. The prediction function is learned by training the model via corresponding pairs of $x_i \in X$ and $y_i \in Y$. Moreover, function $f(x)$ can be also interpreted as conditional probability distribution $P(Y|X)$. Therefore, a specific task is denoted as $T = (Y, P(Y|X))$ or $T = (Y, f(x))$.

With these definitions, transfer learning can be formalized: Assume that we have a *source domain* D_s, a *source task* T_s, a *target domain* D_t, and a *target task* T_t. TL aims to enhance finding the most accurate *target model* $f(x)$ in T_t by incorporating the information in D_s and T_s with either $D_s \neq D_t$, or $T_s \neq T_t$, or both. One subcategory of TL is domain adaptation, which is applicable for scenarios in which only the probability distributions $P(X)$ of D_s and D_t differ, while T_s and T_t are the same [37]. However, the experimental design of our paper does not include any setting in which *source task* T_s and *target task* T_t are equivalent. In turn, domain adaptation frameworks are not considered further.

The TL approach used in this paper is illustrated in Figure 2. It contains 5 steps:

- **Step 1 (Architecture Selection):** An untrained network architecture is selected, whose hidden layer structure is assumed to solve both the *source task* and the *target task*. The input and output layers are chosen to fit the *source task* and *source domain*.
- **Step 2 (Pretraining):** The network is trained with the *source domain dataset* on the *source task*. Usually, the *source dataset* is expected to contain a high number of samples to make the training effective.
- **Step 3 (Domain Adjustment):** The pretrained network is adjusted to the *target domain* and *target task*, whereas the knowledge found in its hidden layers is preserved. To achieve this, only the input and output layers are replaced by untrained layers adapting the network to the *target task* and *target domain*.

- **Step 4 (Layer Freezing):** Depending on the TL strategy, some or all hidden layers can be frozen before training for the *target task*. The parameters of a frozen layer are not updated in future training processes, which ensures that the knowledge learned during *source domain training* is preserved. As a drawback, the adaptation ability of the *target domain data* is limited.
- **Step 5 (Finetuning):** The network is retrained on the *target task* with the *target domain dataset*. Usually, the *target dataset* has only a limited number of instances. The resulting model includes information from both the *source domain* and the *target domain*.

The potential advantages of TL are illustrated by Figure 3. First, we obtain a better starting point for the model training compared to a cold start. This can be observed most often when *source domain* and *target domain* are heavily related and part of the learned knowledge is directly applicable to the *target task*. Second, a higher slope (faster learning process) can be achieved as the result of pretraining the hidden layers. By pretraining, parameters are often set in a generally functional way and just have to be finetuned for the new task. The most important advantage for industrial use cases is the higher asymptote, meaning a higher accuracy of the final model. This can often be observed in cases where the *target dataset* is very small and a part of the missing information in the *target dataset* can be filled by knowledge learned during the *source domain* training. We will observe which of the three benefits can be found in the different experimental settings we describe.

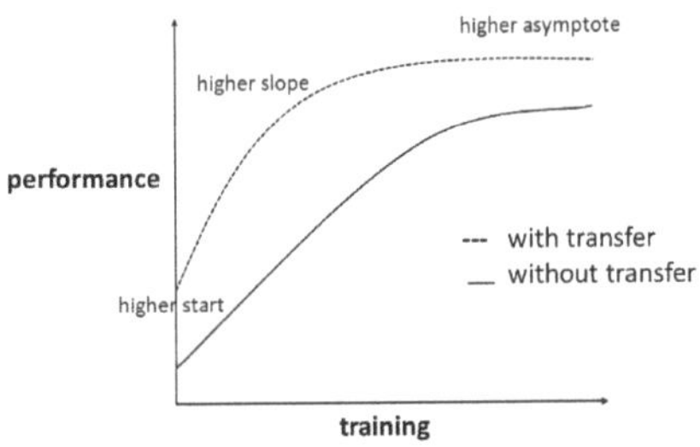

Figure 3. Three potential benefits of transfer learning.

In this paper, we aim to analyze how pretraining with different *source datasets* affects the performance of the *target model*. To enable a high comparability, the selected *target dataset* and *target task* stay constant across all experiments. We define three different settings:

- **Setting 1:** *Target domain* and *source domain* share the same feature space X, and *target task* and *source task* share the same label space Y. However, the domains differ in terms of probability distribution $P(X)$ of the feature space, while the tasks differ in the feature–label relationship (conditional probability $P(Y|X)$).
- **Setting 2:** In addition to non-identical probability distributions $P(X)$ and non-identical conditional probabilities $P(Y|X)$, the label space Y of the *source task* and the *target task* differs as well. Only the feature space X of both domains is identical in this setting.
- **Setting 3:** All four elements (feature space X, label space Y, feature probability distribution $P(X)$, and conditional probability $P(Y|X)$) differ between *source domain* and *target domain* as well as *source task* and *target task*.

3.2. Overview of Used Datasets

In the experiments, two datasets from two different domains were used, namely (1) the Hydraulic Pump End-of-Line dataset (HPEoL) and (2) the Opportunity dataset.

(1) Hydraulic Pump End-of-Line Dataset

The Hydraulic Pump End-of-Line dataset collection was published recently. End-of-line testing is a type of manufacturing quality assurance process that is conducted at the end of the production line. It is used to ensure that the product meets the required specifications and that it is free from defects. In hydraulic pump testing, end-of-line testing usually consists of mounting the pump at a test bench and forcing it through a predefined testing cycle to check its function. The HPEoL dataset contains the end-of-line testing cycle data of around 198 hydraulic pumps measured by hydraulic and mechanical sensors on a hydraulic test bench. The dataset includes the testing cycle data of three different hydraulic pump types:

- Direct control pumps (DC): 120 instances distributed over three versions (V35, V36, V38) differing in size and technical specifications with 40 instances each.
- Speed-based (mechanical) control pumps (SC): 38 instances
- Proportional control pumps (PC): 40 instances

For all control types, each sample corresponds to a multivariate time series with nine sensors representing the testing cycle of one specific pump. All 198 time series are multi-phased, meaning that each sample can be segmented into different segments representing the different testing phases of the testing cycle. While all phases of the testing cycle have to be completed for each pump, variations in the order of the phases and repetitions of phases are common. The data were collected with a frequency of 100 Hz. One testing cycle lasts between 3 and 7 min. Each time stamp has a integer-encoded state label representing the current testing phase of the pump at the time the sensor measures were taken. Supervised time series segmentation aims now to identify the different testing phases within the testing cycle by assigning state labels to all time stamps. Note that the testing cycles of the three different pump control types vary heavily (different label space Y), while the phases of different pump versions within a control type are very similar and only differ in the exact specification of a testing phase (identical label space Y, different conditional probabilities $P(Y|X)$). The number of different state labels for DC-types corresponds to 44, of PC-types to 43, and of SC-types to 44. Figure 4 depicts a hydraulic test bench.

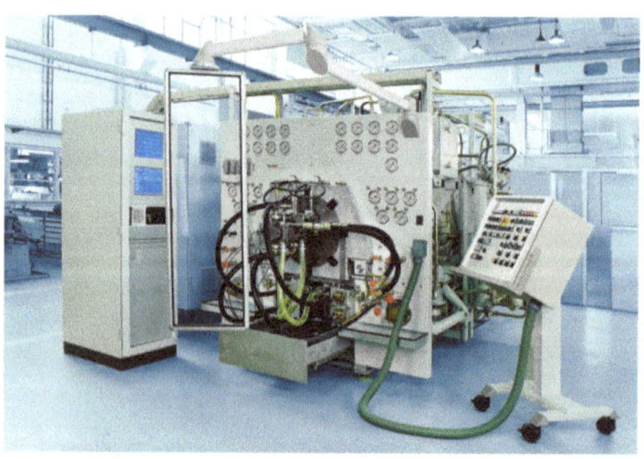

Figure 4. Hydraulic test bench with mounted pump and attached sensors (source: Bosch Rexroth AG).

(2) Opportunity Dataset

The Opportunity dataset was published to examine and compare human activity recognition algorithms, including TSS. It includes sequences created by tracking a number of human subjects via body-worn, object-attached, and ambient sensors (like accelerometers or location sensors). The sequences are annotated by activity labels. It is a popular and widely used dataset in the literature (e.g., [26,38]) that has often been used for benchmarking time series-based approaches. The dataset can be structured into two parts, i.e., the ADL (Activities of Daily Life) section and the Drill section. The ADL section just follows the subjects accomplishing daily life activities (including relaxing and eating) and, in turn, often has longer time periods without a change in the activity label. In the Drill section, however, the subjects were told to perform several short activities (e.g., open doors) without any breaks in between. The activity labels, therefore, are rapidly changing, resulting in a setting that is more similar to the HPEoL dataset, in which many label switches are present within a short period of time. Figure 5 shows the location of different sensors during the data collection process.

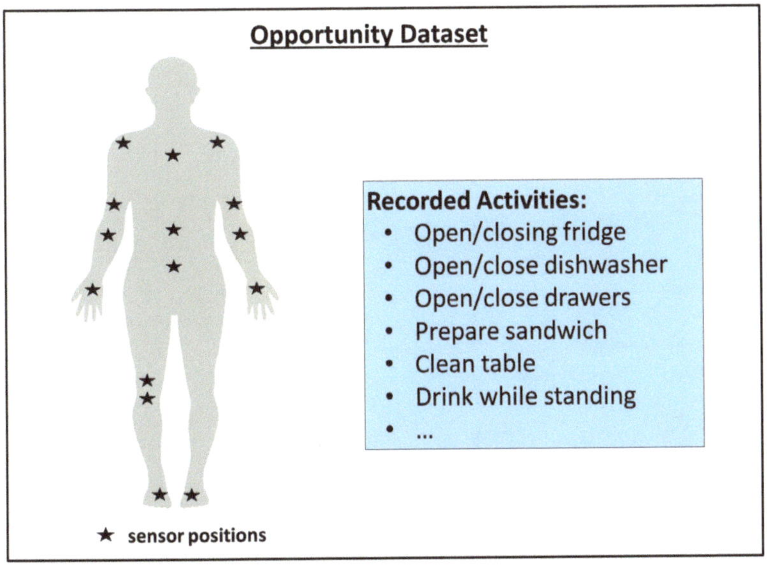

Figure 5. Opportunity dataset: Overview of sensor positions (adapted from [39]).

3.3. Model Architecture

All *source tasks* and *target tasks* investigated in this paper belong to the category of supervised TSS tasks. More precisely, we target to find and train a deep learning model to enable it to predict the state labels for all timestamps of a multivariate time series. This is called dense labeling. The network architecture we use to analyze the impact of TL in this work is called PrecTime. The PrecTime architecture was introduced in 2022 in one of our previous works [6]. It is designed for multivariate TSS and was already successfully applied to the HPEoL dataset and Opportunity dataset in isolation. As the output of the network is a sequence of state label predictions (one per time stamp), it is considered a sequence-to-sequence model. The conceptual backbone of the network is illustrated in Figure 6. PrecTime is a time window-based architecture consisting of three modules, i.e., a CNN-based intra-window feature extractor, an LSTM-based inter-window context detection, and a CNN-based intra-window prediction refinement model. The concrete details about the architecture, implementation, loss functions, and hyperparameters are described in [6]. They were reused in this paper.

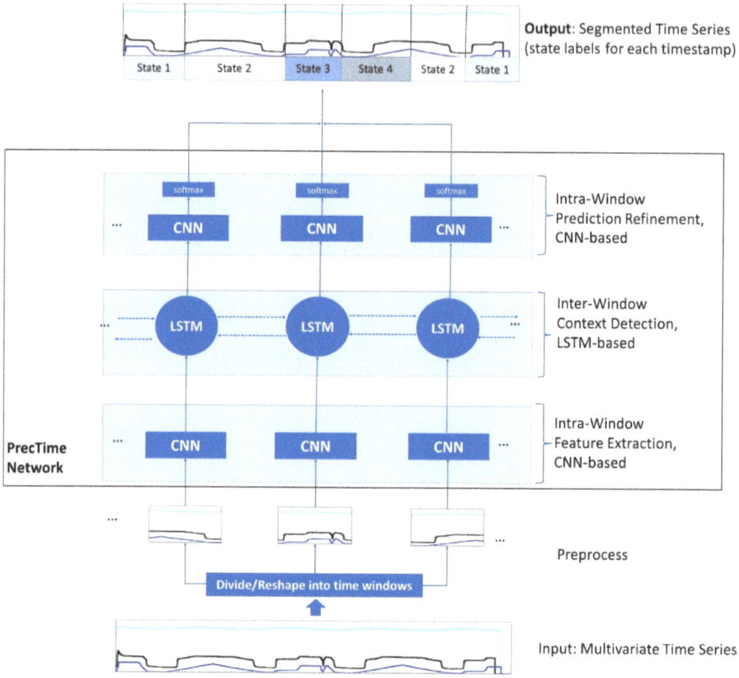

Figure 6. PrecTime architecture backbone with three modules (based on [6]).

3.4. Experimental Setup

The experimental setup consists of three settings with varying *source datasets* (see Figure 7). The *source datasets* were chosen in a way that three varying degrees of similarity between the *source dataset* and the *target dataset* are represented in the settings. The *target domains* and *target tasks* are kept constant over all experiments to enable direct comparability. In each setting, PrecTime is newly pretrained by the *source dataset* before being finetuned by the *target dataset*. As the amount of samples used for training might seem very low for deep learning, it has to be noted that in our TSS use case, each sample consists of several thousand labeled datapoints (one label per time stamp). Thus, a far lower number of samples is required to achieve good results compared to other deep learning tasks. The three settings are explained in detail in the following section.

Setting 1 (Same-Asset Pretraining): Source data closely related to target data

In the first setting, both *source datasets* and *target datasets* are taken from the HPEoL dataset. For the experiment, all three subsets belonging to the DC pump type are selected, with each of the subsets representing the 40 samples of a distinct DC-pump version (V35,V36,V38). The technical design of the three pump versions differs only slightly. In turn, the testing cycles of the three pumps are very similar as well. Only detailed specifications of certain testing phases may vary. Based on the high similarity, we call this setting same-asset pretraining. Using the formalization introduced in Section 3.1, we have a setting where the feature space X and label space Y of *source data* and *target data* are identical, while the probability distribution $P(X)$ and conditional probabilities $P(Y|X)$ slightly differ. In the experiments, each pump version (V35,V36,V38) is used once as the *target dataset*, while the data of the remaining two versions are merged to become the combined *source dataset*. For each of these three scenarios, we perform various experimental runs, in which we limit the sample number in the *target training set* to 1, 3, 5, and 10. This allows us to investigate how the number of samples in the *target training dataset* affects the benefit of pretraining considering training speed and accuracy. Furthermore, two layer freezing approaches are compared. In the first approach, all layers except the output layer were frozen (TL-fr). In the second approach, all layers were set as trainable (TL-tr). A non-pretrained model that was trained only by the *target dataset* is defined as the baseline (BL1). Additionally, as the feature space *X* and label space *Y* are identical, an additional baseline can be created by evaluating the model performance when merging the data of all three pump versions into one big training set (BL2). Note that this additional baseline is only available in Setting 1, as it requires equivalence of feature space and label space.

Setting 2 (Cross-Asset Pretraining): Source data distantly related to target data

As in Setting 1, all *source datasets* and *target datasets* are subsets of the HPEoL dataset. However, in this case, all three pump types (DC, SC, and PC) are considered. Each of the three DC pump versions (V35,V36,V38) is used as the *target data*, while both the SC pump subset and the PC pump subset are distinctly used as *source data*. The testing cycles of the different control types (DC, SC, and PC) differ heavily. However, the used test benches and sensors are identical across all types. Therefore, we have still have an identical feature space *X* across *source data* and *target data* in this setting. We limit the *target domain* training data to three instances to make the impact of TL strongly visible. The TL is performed with all layers unfrozen. A non-pretrained model is set as the baseline (BL1).

Setting 3 (Cross-Domain Pretraining): Source data non-related to target data

In Setting 3, it is explored whether pretraining with a non-related *source dataset* can boost the accuracy of the *target task*. As in Settings 1 and 2, the three DC pump versions are distinctly chosen as *target sets*. In this case, however, the pretraining is not done by HPEoL subsets, but by the Opportunity dataset (Section 3.1). The Opportunity dataset contains sensor location and movement data of human subjects and therefore covers a completely different domain compared to pump testing. As in setting two, the *target domain* training data are limited to three instances, and the TL is performed with all layers unfrozen. Again, the non-pretrained model BL1 is set as the baseline.

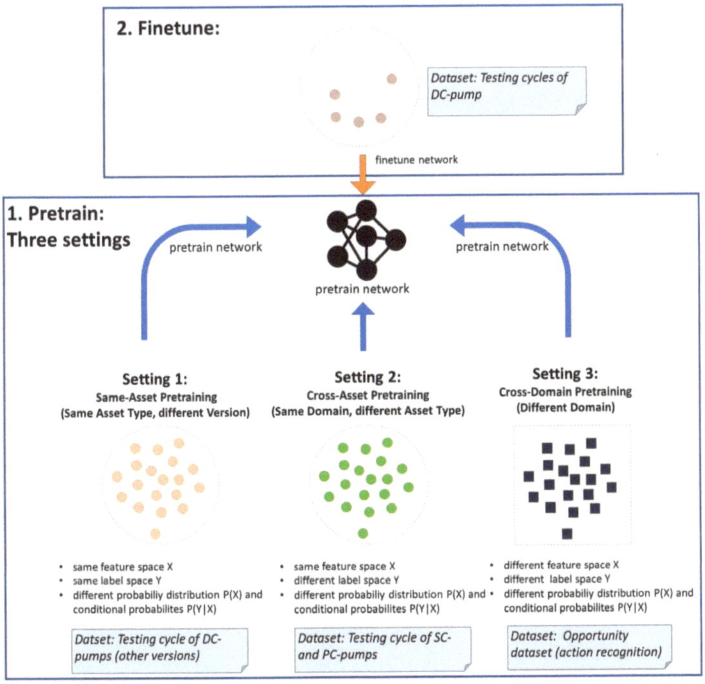

Figure 7. Overview of experimental setup with three settings.

3.5. Implementation Details

The networks were trained on an *NVIDIA Tesla K80* with 6 cores, 56 GB RAM, and 380 GB disk storage in the *Microsoft Azure* cloud. The implementation was done with *Python*, mainly utilizing the *Tensorflow* and *Keras* packages. For the pretraining, all available samples found in the selected *source dataset* were used. The *target dataset* was split into a training set and a validation set. The size of the *target training dataset* was set to a predefined limited size in each Setting (see Section 4.3). The training sample size in Setting 1 was varied in different experimental runs (1, 3, 5, 10), while the training sample size in Settings 2 and 3 was kept constant as 3. To achieve comparability, the *target validation set* was kept constant across all settings. It was defined to include 20 samples from the *target dataset*. The evaluation of the model performance was done by tracking the progress of the validation accuracy for 100 epochs. The accuracy was calculated by dividing the number of time stamps that were correctly labeled by the mode by the total number of time stamps. In pretraining, the batch size was defined as 20, while during finetuning, it was set equal to the number of training samples. Cross-entropy was chosen as the loss function to be optimized during the training process. The window size was defined to be 100 in all experiments. The preprocessing included using zero-padding to bring all instances to a fixed length (45,000 for HPEoL, 60,000 for opportunity). For the Opportunity dataset only the drill-section data were used as samples, and 113 sensors were selected as is the common practice in the literature [26]. Missing sensor values were interpolated.

4. Results and Discussion

The following section gives detailed insight into the experimental results and also discusses them.

4.1. Results for Setting 1: Same-Asset Pretraining

Table 1 provides an overview of the achieved validation accuracies in Setting 1. It can be clearly seen that the TL-fr approach (hidden layers set frozen) showed a very weak performance. Thus, the approach seems to be the wrong one for the use case of supervised TSS. In turn, we focus our analysis on TL-tr (all layers unfrozen), which showed a very strong performance in comparison. Pretraining with the *source dataset* from the same asset type increased the final accuracy of the network in all experiments. A clear negative correlation between the amount of available *target data* and the benefit of TL can be observed, meaning TL shows its greatest potential when facing *target data* scarcity. While the exact numbers differ in detail, the observations are consistent across all three DC pump versions. When having a *target training set* of size one, the accuracy difference between the unfrozen TL-tr approach and the isolated training baseline (BL1) ranged from 10% for the V38 pump to 34% for the V35 pump. On the contrary, in the experimental runs using the highest number of *target training samples* (10), the performance boost decreased to around 2–3%, which is helpful but far less significant. Interestingly, no clear conclusion could be drawn about the benefit of combining *target data* and *source data* into one large training set and training the network with it (BL2). While in some experimental runs, the combined training led to accuracy improvement compared to isolated *target data* training (BL1), in other cases it seemed to impede model performance. However, note that combining *source data* and *training data* into one big dataset (BL2) in no case showed a better performance than the TL-tr approach, emphasizing the superiority of using TL in this setting.

Table 1. Validation accuracies (%) after 100 epochs in Setting 1 for BL1 (no pretraining), BL2 (combine source and target dataset), TL-fr (layers frozen), and TL-tr (layers unfrozen).

Number of Training Samples in Target Data	**V35 as Target Data V36+V38 as Source**				**V36 as Target Data V35+V38 as Source**				**V38 as Target Data V35+V36 as Source**			
	BL1	**BL2**	**TL-fr**	**TL-tr**	**BL1**	**BL2**	**TL-fr**	**TL-tr**	**BL1**	**BL2**	**TL-fr**	**TL-tr**
1	59.0	45.2	53.0	93.8	66.9	45.0	50.3	86.0	77.1	83.9	69.2	88.7
3	92.3	92.5	57.8	95.9	90.9	88.9	56.8	93.5	90.2	91.3	73.0	94.4
5	94.3	92.5	53.5	96.9	92.5	92.5	60.2	95.5	90.4	93.1	70.7	95.8
10	95.7	97.0	58.4	97.4	94.1	95.2	59.3	96.7	93.1	96.3	69.0	96.5

Figure 8a shows the training process of all three pump versions for a *target training set* of size 3. Clearly, we can observe that for unfrozen TL (TL-tr), not only is the final accuracy improved (higher asymptote), but also the training speed is significantly increased. Very high accuracies (>90%) are reached far earlier compared to the two baselines (BL1, BL2). The main reason for the increased training speed seems to be not a steeper slope, but a higher starting point achieved through pretraining. The extent of this observation varies across the three pumps; it is most visible in the V35 training process. The high starting points after pretraining seem to be the result of using *source data* and *training data* of high similarity. For frozen TL, it becomes obvious that freezing all hidden layers almost completely eliminates any training progress in this setting. The models end up with almost the same performance before and after finetuning (TL-fr).

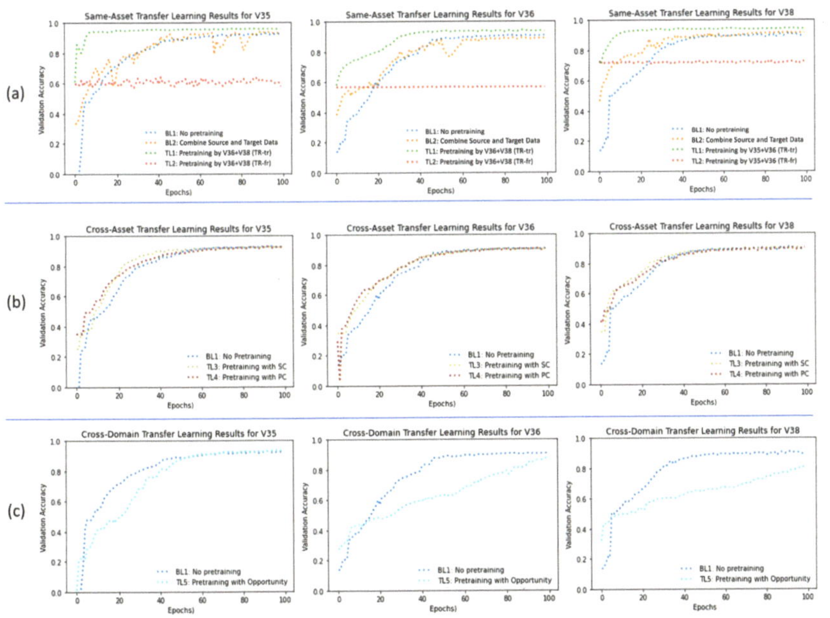

Figure 8. Training process comparison for different settings, all with a target training set of size 3: (**a**) Setting 1: Same-asset transfer learning T1 (unfrozen) and T2 (frozen) compared to two baselines BL1 and BL2, (**b**) Setting 2: Cross-asset transfer learning T3 (SC pretraining) and T4 (PC pretraining) within same domain compared to baseline BL1, (**c**) Setting 3: Cross-domain transfer learning T5 (Opportunity pretraining) compared to baseline BL1.

4.2. Results for Setting 2: Cross-Asset Pretraining

In Setting 2 (pretraining by *source data* from a different asset type), the benefits of TL are diminished clearly compared to Setting 1 (see Table 2). When examining the final model performance, no clear differences (>1%) across the three pretraining scenarios can be observed for any of the three DC pump versions. Therefore, the potential TL benefit of a higher asymptote is not observed in this setting.

Table 2. Validation accuracies (%) after 100 epochs in Setting 2 for BL1, SC-, and PC-pretraining.

Setting	V38 as Target Dataset	V36 as Target Dataset	V35 as Target Dataset
BL1 (no pretraining)	90.2	91.0	92.4
Pretraining by SC pump dataset	90.8	90.5	92.9
Pretraining by PC pump dataset	90.4	90.8	92.4

However, when taking a closer look at the training process in Figure 8b, we can make an insightful observation. Despite all model training processes ending on a similar performance level after 100 epochs, in the first half of the training process, the models with applied TL show a better performance compared to the models without pretraining. This difference disappears more and more in later training stages. This can be explained by the fact that the pretrained models avoid a complete cold start and the pretraining has already set parameters in a certain way that helps the *target data* training process to progress quicker. Despite the non-identical label space of the *source data* and *target data*, the pretrained models seem to benefit from a certain relatedness in the nature of the data (shared feature space) at the start of the training.

4.3. Results for Setting 3: Cross-Domain Pretraining

As opposed to the observed benefits of TL in Settings 1 and 2, we can observe a form of negative TL in Setting 3. As the *source dataset* (Opportunity) in this setting belongs to a completely different domain than the *target dataset* (HPEoL), pretraining does not lead to any benefit, but instead impairs the training speed of the *target data* (see Figure 8c). The different nature of the data prevents a meaningful pretraining that can be exploited later by the *target task*. However, despite the slowed training process, the final asymptote level of the pretrained network (TL5) is similar to or only slightly below that of the non-pretraining scenario (BL1) for all three pump variants. This can be explained by the fact that all layers were set to trainable. Thus, the pretrained models are able to overcome their impaired starting point caused by cross-domain pretraining over the course of the training process and fully adapt to the new domain.

Altogether, the findings provide systematic evidence for the hypothesis that the benefit of TL in TSS is (1) increasing with lack of *target training data* and (2) increasing the similarity between *source data* and *target data*. Table 3 summarizes the observed results. While we assume these results are to some extent transferable to other domains, further research has to be done to confirm the general validity of our finding.

While freezing layers is often recommended in the literature, in this work, it heavily aggravated the model performance in initial experiments (see TL-fr results in Setting 1) and was therefore not investigated deeper. However, checking different layer freezing strategies in a more nuanced way might be an interesting direction to go for future research.

Table 3. Summary table of observed TL effects (target training set of size 3, all layers unfrozen) in the three experimental settings (++: very positive effect, +: positive effect, 0: negligible effect, -: negative effect).

Setting No.	Setting Description	Effect on Asymptote	Effect on Training Start	Effect on Training Slope
1	Same-asset pretraining (closely related source and target data)	+	++	+
2	Cross-asset pretraining (distantly related source and target data)	0	+	+
3	Cross-domain pretraining (non-related source and target data)	0	0	-

5. Conclusions

We performed a systematic investigation of different TL scenarios and strategies for an industrial TSS problem. It was examined whether and how pretraining with close- to non-related *source datasets* improves the network learning performance of PrecTime, a sequence-to-sequence deep learning architecture for TSS. Focus points were the speed of the training process and the achieved validation accuracy after completing the training. The *target task* was specified to be the supervised segmentation of hydraulic pump testing cycles into different testing phases via the prediction of state labels. Three different *source datasets* were selected for pretraining (1) different versions of same pump type, (2) different pump types, and (3) different domains. It was observed that pretraining with a closely-related *source dataset* led to the biggest benefit (Setting 1), while pretraining with a *source dataset* from a different domain even aggravated the training process (Setting 3). Additionally, we could show that TL performed best in comparison to baselines in scenarios in which the *target training dataset* had the smallest sample size.

The main limitation of our study is the focus on a specific use case. While our results provide interesting findings, the domain-independence and general validity of the results still have to be established. Furthermore, the important topic of layer freezing was not investigated in depth. Therefore, future work could look deeper into different layer freezing strategies and aim to provide recommendations as to which layers should be frozen and which layers set to trainable. Additionally, one could check if incrementally unfreezing layers over the course of the training process leads to promising results. Possible research directions could also include performing experiments with different TSS network architectures and datasets to validate and build on the findings of this paper. It could also be examined how pretraining with multiple datasets or a general-domain dataset affects the model performance. Finally, the transfer of knowledge between labeled data and unlabeled data could be investigated in the context of TSS.

Author Contributions: Conceptualization, S.G.; methodology, S.G.; software, S.G.; validation, S.G.; formal analysis, S.G.; investigation, S.G.; resources, S.G.; data curation, S.G.; writing—original draft preparation, S.G.; writing—review and editing, M.R.; visualization, S.G.; supervision, M.R.; project administration, S.G.; funding acquisition, S.G. All authors have read and agreed to the published version of the manuscript.

Funding: This research received no external funding.

Institutional Review Board Statement: Not applicable.

Informed Consent Statement: Not applicable.

Data Availability Statement: The Hydraulic Pump End-of-Line dataset is accessible via https://github.com/boschresearch/Hydraulic-EoL-Testing/ (accessed on 22 February 2023). The Opportunity dataset is accessible via https://archive.ics.uci.edu/ml/datasets/opportunity+activity+recognition (accessed on 22 February 2023).

Conflicts of Interest: The authors declare that the publication of data must be approved by the Bosch Rexroth AG. The authors declare no conflict of interest. The funders had no role in the design of the study; in the collection, analyses, or interpretation of data; in the writing of the manuscript; or in the decision to publish the results.

Abbreviations

The following abbreviations are used in this manuscript:

ADL	activities of daily life
BL1	baseline 1
BL2	baseline 2
CNN	convolutional neural network
DC	direct control
HPEoL	hydraulic pump end-of-line
LSTM	long short-term memory
ML	machine learning
PC	proportional control
SC	speed-based control
TL	transfer learning
TSS	time series segmentation

References

1. Kaveh, A.; Talatahari, S.; Khodadadi, N. Stochastic Paint Optimizer: Theory and application in civil engineering. *Eng. Comput.* **2022**, *38*, 1921–1952. [CrossRef]
2. Hoppenstedt, B.; Pryss, R.; Stelzer, B.; Meyer-Brötz, F.; Kammerer, K.; Treß, A.; Reichert, M. Techniques and Emerging Trends for State of the Art Equipment Maintenance Systems—A Bibliometric Analysis. *Appl. Sci.* **2018**, *8*, 916. [CrossRef]
3. Hoppenstedt, B.; Reichert, M.; Kammerer, K.; Probst, T.; Schlee, W.; Spiliopoulou, M.; Pryss, R. Dimensionality Reduction and Subspace Clustering in Mixed Reality for Condition Monitoring of High-Dimensional Production Data. *Sensors* **2019**, *19*, 3303. [CrossRef]
4. Kammerer, K.; Hoppenstedt, B.; Pryss, R.; Stökler, S.; Allgaier, J.; Reichert, M. Anomaly Detections for Manufacturing Systems Based on Sensor Data–Insights into Two Challenging Real-World Production Settings. *Sensors* **2019**, *19*, 5370. [CrossRef]
5. Phan, H.; Andreotti, F.; Cooray, N.; Chén, O.; de Vos, M. SeqSleepNet: End-to-End Hierarchical Recurrent Neural Network for Sequence-to-Sequence Automatic Sleep Staging. *IEEE Trans. Neural Syst. Rehabil. Eng.* **2018**, *27*, 400–410. [CrossRef] [PubMed]
6. Gaugel, S.; Reichert, M. PrecTime: A Deep Learning Architecture for Precise Time Series Segmentation in Industrial Manufacturing Operations, PrePrint. Available online: www.academia.edu (accessed on 22 February 2023).
7. Lu, N.; Hu, H.; Yin, T.; Lei, Y.; Wang, S. Transfer Relation Network for Fault Diagnosis of Rotating Machinery with Small Data *IEEE Trans. Cybern.* **2021**, *52*, 11927–11941. [CrossRef]
8. Cao, P.; Zhang, S.; Tang, J. Preprocessing-Free Gear Fault Diagnosis Using Small Datasets With Deep Convolutional Neural Network-Based Transfer Learning. *IEEE Access* **2018**, *6*, 26241–26253. [CrossRef]
9. Matias, P.; Folgado, D.; Gamboa, H.; Carreiro, A. Time Series Segmentation Using Neural Networks with Cross-Domain Transfer Learning. *Electronics* **2021**, *10*, 1805. [CrossRef]
10. Zhuang, F.; Qi, Z.; Duan, K.; Xi, D.; Zhu, Y.; Zhu, H.; Xiong, H.; He, Q. A Comprehensive Survey on Transfer Learning. *Proc. IEEE* **2021**, *109*, 43–76. [CrossRef]
11. Weber, M.; Auch, M.; Doblander, C.; Mandl, P.; Jacobsen, H.A. Transfer Learning With Time Series Data: A Systematic Mapping Study. *IEEE Access* **2021**, *9*, 165409–165432. [CrossRef]
12. Maschler, B.; Weyrich, M. Deep Transfer Learning for Industrial Automation: A Review and Discussion of New Techniques for Data-Driven Machine Learning. *IEEE Ind. Electron. Mag.* **2021**, *15*, 65–75. [CrossRef]
13. Li, W.; Gao, H.; Su, Y.; Momanyi, B.M. Unsupervised Domain Adaptation for Remote Sensing Semantic Segmentation with Transformer. *Remote Sens.* **2022**, *14*, 4942. [CrossRef]
14. Liu, X.; Yoo, C.; Xing, F.; Oh, H.; Fakhri, G.; Kang, J.W.; Woo, J. Deep Unsupervised Domain Adaptation: A Review of Recent Advances and Perspectives. *APSIPA Trans. Signal Inf. Process.* **2022**, *11*, e25. [CrossRef]
15. Heistracher, C.; Jalali, A.; Strobl, I.; Suendermann, A.; Meixner, S.; Holly, S.; Schall, D.; Haslhofer, B.; Kemnitz, J. Transfer Learning Strategies for Anomaly Detection in IoT Vibration Data. In Proceedings of the IECON 2021—47th Annual Conference of the IEEE Industrial Electronics Society, IEEE, Toronto, ON, Canada, 13–16 October 2021; pp. 1–6. [CrossRef]
16. He, Q.Q.; Pang, P.C.I.; Si, Y.W. Multi-source Transfer Learning with Ensemble for Financial Time Series Forecasting. In Proceedings of the 2020 IEEE/WIC/ACM International Joint Conference on Web Intelligence and Intelligent Agent Technology (WI-IAT), Melbourne, Australia, 14–17 December 2021; pp. 227–233. [CrossRef]

17. Yan, J.; Wang, L.; He, H.; Liang, D.; Song, W.; Han, W. Large-Area Land-Cover Changes Monitoring With Time-Series Remote Sensing Images Using Transferable Deep Models. *IEEE Trans. Geosci. Remote Sens.* **2022**, *60*, 4409917. [CrossRef]
18. Lian, R.; Tan, H.; Peng, J.; Li, Q.; Wu, Y. Cross-Type Transfer for Deep Reinforcement Learning Based Hybrid Electric Vehicle Energy Management. *IEEE Trans. Veh. Technol.* **2020**, *69*, 8367–8380. [CrossRef]
19. Aldayel, M.S.; Ykhlef, M.; Al-Nafjan, A.N. Electroencephalogram-Based Preference Prediction Using Deep Transfer Learning. *IEEE Access* **2020**, *8*, 176818–176829. [CrossRef]
20. Gross, J.; Buettner, R.; Baumgartl, H. Benchmarking Transfer Learning Strategies in Time-Series Imaging: Recommendations for Analyzing Raw Sensor Data. *IEEE Access* **2022**, *10*, 16977–16991. [CrossRef]
21. Wen, T.; Keyes, R. Time Series Anomaly Detection Using Convolutional Neural Networks and Transfer Learning. *arXiv* **2019**, arXiv:1905.13628. [CrossRef]
22. Gikunda, P.; Jouandeau, N. Homogeneous Transfer Active Learning for Time Series Classification. In Proceedings of the 2021 20th IEEE International Conference on Machine Learning and Applications (ICMLA), Pasadena, CA, USA, 13–16 December 2021; pp. 778–784. [CrossRef]
23. Warushavithana, M.; Mitra, S.; Arabi, M.; Breidt, J.; Pallickara, S.L.; Pallickara, S. A Transfer Learning Scheme for Time Series Forecasting Using Facebook Prophet. In Proceedings of the 2021 IEEE International Conference on Cluster Computing (CLUSTER), Portland, OR, USA, 7–10 September 2021; pp. 809–810. [CrossRef]
24. Fawaz, H.I.; Forestier, G.; Weber, J.; Idoumghar, L.; Muller, P.A. Transfer learning for time series classification. In Proceedings of the 2018 IEEE International Conference on Big Data (Big Data), Seattle, WA, USA, 10–13 December 2018; pp. 1367–1376.
25. Dridi, A.; Afifi, H.; Moungla, H.; Boucetta, C. Transfer Learning for Classification and Prediction of Time Series for Next Generation Networks. In Proceedings of the ICC 2021—IEEE International Conference on Communications, Montreal, QC, Canada, 14–23 June 2021; pp. 1–6. [CrossRef]
26. Yao, R.; Lin, G.; Shi, Q.; Ranasinghe, D. Efficient Dense Labeling of Human Activity Sequences from Wearables using Fully Convolutional Networks. *Pattern Recognit.* **2018**, *78*, 252–266. [CrossRef]
27. Maschler, B.; Vietz, H.; Jazdi, N.; Weyrich, M. Continual Learning of Fault Prediction for Turbofan Engines using Deep Learning with Elastic Weight Consolidation. In Proceedings of the 2020 25th IEEE International Conference on Emerging Technologies and Factory Automation (ETFA), Vienna, Austria, 8–11 September 2020; pp. 959–966. [CrossRef]
28. Kammerer, K.; Pryss, R.; Reichert, M. Context-Aware Querying and Injection of Process Fragments in Process-Aware Information Systems. In Proceedings of the 2020 IEEE 24th International Enterprise Distributed Object Computing Conference (EDOC), Eindhoven, The Netherlands, 5–8 October 2020; pp. 107–114. [CrossRef]
29. Maschler, B.; Knodel, T.; Weyrich, M. Towards Deep Industrial Transfer Learning for Anomaly Detection on Time Series Data. In Proceedings of the 2021 26th IEEE International Conference on Emerging Technologies and Factory Automation (ETFA), Vasteras, Sweden, 7–10 September 2021; pp. 1–8. [CrossRef]
30. Tercan, H.; Guajardo, A.; Meisen, T. Industrial Transfer Learning: Boosting Machine Learning in Production. In Proceedings of the 2019 IEEE 17th International Conference on Industrial Informatics (INDIN), IEEE, Helsinki, Finland, 22–25 July 2019; pp. 274–279. . [CrossRef]
31. Zhou, X.; Zhai, N.; Li, S.; Shi, H. Time Series Prediction Method of Industrial Process with Limited Data Based on Transfer Learning. *IEEE Trans. Ind. Inform.* **2022**, 1–10. [CrossRef]
32. Xu, W.; Wan, Y.; Zuo, T.Y.; Sha, X.M. Transfer Learning Based Data Feature Transfer for Fault Diagnosis. *IEEE Access* **2020**, *8*, 76120–76129. [CrossRef]
33. Wu, S.; Jing, X.Y.; Zhang, Q.; Wu, F.; Zhao, H.; Dong, Y. Prediction Consistency Guided Convolutional Neural Networks for Cross-Domain Bearing Fault Diagnosis. *IEEE Access* **2020**, *8*, 120089–120103. [CrossRef]
34. Xu, G.; Liu, M.; Jiang, Z.; Shen, W.; Huang, C. Online Fault Diagnosis Method Based on Transfer Convolutional Neural Networks. *IEEE Trans. Instrum. Meas.* **2020**, *69*, 509–520. [CrossRef]
35. He, Y.; Tang, H.; Ren, Y. A Multi-channel Transfer Learning Framework for Fault Diagnosis of Axial Piston Pump. In Proceedings of the 2021 Global Reliability and Prognostics and Health Management (PHM-Nanjing), Nanjing, China, 15–17 October 2021; pp. 1–7. [CrossRef]
36. Lin, Y.P.; Jung, T.P. Improving EEG-Based Emotion Classification Using Conditional Transfer Learning. *Front. Hum. Neurosci.* **2017**, *11*, 334. [CrossRef]
37. Sun, S.; Shi, H.; Wu, Y. A survey of multi-source domain adaptation. *Inf. Fusion* **2015**, *24*, 84–92. [CrossRef]
38. Chambers, R.D.; Yoder, N.C. FilterNet: A Many-to-Many Deep Learning Architecture for Time Series Classification. *Sensors* **2020**, *20*, 2498. [CrossRef] [PubMed]
39. Chavarriaga, R.; Sagha, H.; Bayati, H.; Millan, J.d.R.; Roggen, D.; Förster, K.; Calatroni, A.; Tröster, G.; Lukowicz, P.; Bannach, D.; et al. Robust activity recognition for assistive technologies: Benchmarking machine learning techniques. In Proceedings of the Workshop on Machine Learning for Assistive Technologies at the Twenty-Fourth Annual Conference on Neural Information Processing Systems (NIPS-2010), Whistler, BC, Canada, 10 December 2010.

sensors

MDPI

Article

Cluster Validity Index for Uncertain Data Based on a Probabilistic Distance Measure in Feature Space

Changwan Ko [1], Jaeseung Baek [2,3], Behnam Tavakkol [4] and Young-Seon Jeong [1,5,*]

[1] Department of Industrial Engineering, Chonnam National University, Gwangju 61186, Republic of Korea; kcw7536@gmail.com
[2] College of Business, Northern Michigan University, Marquette, MI 49855, USA; jbaek@nmu.edu
[3] Department of Industrial & Systems Engineering, Rutgers University, Piscataway, NJ 08854, USA
[4] School of Business, Stockton University, Galloway, NJ 08205, USA; behnam.tavakkol@stockton.edu
[5] Interdisciplinary Program of Arts and Design Technology, Chonnam National University, Gwangju 61186, Republic of Korea
* Correspondence: young.jeong@jnu.ac.kr; Tel.: +82-62-530-1790

Abstract: Cluster validity indices (CVIs) for evaluating the result of the optimal number of clusters are critical measures in clustering problems. Most CVIs are designed for typical data-type objects called certain data objects. Certain data objects only have a singular value and include no uncertainty, so they are assumed to be information-abundant in the real world. In this study, new CVIs for uncertain data, based on kernel probabilistic distance measures to calculate the distance between two distributions in feature space, are proposed for uncertain clusters with arbitrary shapes, sub-clusters, and noise in objects. By transforming original uncertain data into kernel spaces, the proposed CVI accurately measures the compactness and separability of a cluster for arbitrary cluster shapes and is robust to noise and outliers in a cluster. The proposed CVI was evaluated for diverse types of simulated and real-life uncertain objects, confirming that the proposed validity indexes in feature space outperform the pre-existing ones in the original space.

Keywords: uncertain data; cluster validity index; kernel probabilistic distance; feature space

1. Introduction

The purpose of clustering is to partition objects into groups with criteria such that the similarity within the groups and the dissimilarity among different groups should be maximized [1,2]. Although clustering methods have been widely used in many applications, most clustering algorithms do not provide the optimal number of clusters. Partitional-based clustering algorithms such as K-means clustering [3] must preset the number of clusters [4]. As cluster information is rarely known in the real world, it is crucial to evaluate the clustering results depending on the different numbers of clusters. Although many clustering methods exist for diverse applications, such as pattern recognition [5], semiconductor manufacturing [6], and healthcare [7], they have been developed primarily for only certain data or fixed values. However, the embedded uncertainty of data is essential in many applications. For instance, a patient's blood pressure may not be consistent because of environmental conditions and instrument errors. Furthermore, measurement values are continuously changing because of the positions of instrumentation devices or workers' conditions. Aside from these examples, data randomness, missing data, delayed updates, and worker fatigue are other factors of data uncertainty [8,9].

Uncertain data are assumed to be prevalent information in the real world, e.g., measurement errors and environmental conditions. The uncertainty of uncertain data can be expressed by probability density functions (PDFs). Figure 1 illustrates two uncertain data, each distributed by a PDF. The standard method of converting uncertain data is to transform a summary statistic (e.g., mean or median) into certain data. However, these statistics could lose extra information of uncertainty that is significant to capture the uncertainty information of uncertain objects.

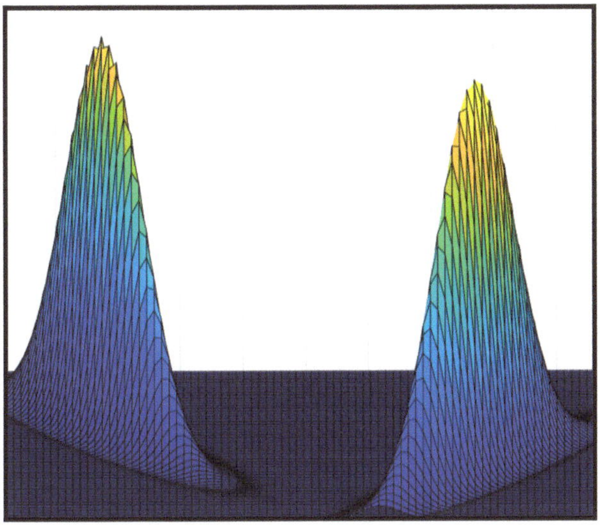

Figure 1. Two uncertain datasets, each expressed by a PDF.

Cluster validity indices (CVIs), which are indicators for validating the quality of clustering algorithms, have been widely used to determine the correct number of clusters for the given data. As the CVIs only use input data information, they must be used according to the characteristics of the data. The two components of a CVI are compactness and separability measures. The former refers to an intra-cluster distance, and the latter represents an inter-cluster distance. Most CVIs indicate that a good partition produces a small compactness value and a high separability value. However, the existing CVIs are vulnerable to validating cluster results when the shapes of the clusters are not spherical clusters [10,11].

For certain data, several CVIs, such as the Dunn [12], Calinski–Harabasz [13], Davies–Bouldin [14], and Xie–Beni [15] indices, have been proposed based on combinations of compactness and separability measures. However, most of the existing CVIs have been developed for certain data. There have been few studies on uncertain data. Moreover, relatively new CVIs are also being designed to incorporate mathematical theories into pre-existing CVIs, such as the K-nearest neighbor algorithm, which is used to compute compactness and separation by taking into account shared/non-shared data pairs [10], and principal component analysis, which is used to capture the geometry of the clusters [16]; or to develop clustering algorithms to cluster more well-separated clusters [1].

To apply uncertain data to the existing CVIs' formulas, they should be changed to calculate distance measures of compactness and separability. In a study of uncertain CVIs, Tavakkol et al. [17] proposed CVIs for uncertain data to calculate the distance between two uncertain objects using probabilistic distance measures in the original space. However, it leads to sensitivity to arbitrary shapes of clusters, sub-clusters, and outliers because of the clusters shape that may cause inaccurate compactness and separability [11].

Consequently, this study proposes new uncertain CVIs for uncertain data objects based on kernel probabilistic distance measures in feature space. The proposed CVIs for uncertain objects are designed to adapt the kernel-based Bhattacharyya probabilistic distance in kernel spaces. In kernel space, the proposed CVIs produce accurate compactness and separability for the arbitrary shapes of clusters by transforming them into elliptical shapes in feature space. Figure 2 illustrates that the ambiguous shape of a dataset in the original space is transformed into a relatively elliptical, circular shape in feature space; thus, the kernel transformation can improve performance in calculating accurate compactness and separability. Furthermore, the proposed approaches could be robust to noise and outliers in a cluster. The superior performance of the proposed CVIs was evaluated through diverse experiments, including simulated and real-life datasets.

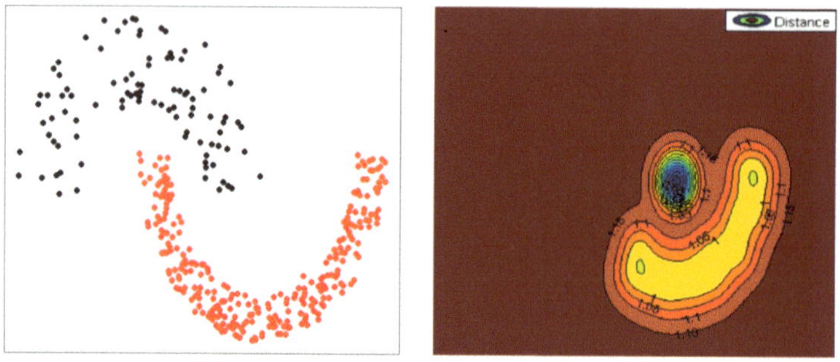

(**a**) (**b**)

Figure 2. Visualization of kernel transformation: (**a**) asymmetry shape in original space; (**b**) transformed shape in feature space.

This paper is organized as follows. Section 2 reviews the previous studies on CVIs. New CVIs for uncertain data based on a kernel probabilistic distance measure are proposed in Section 3. After the extensive experiments are presented in Section 4, the conclusions and future studies are provided in Section 5.

2. Related Work

2.1. CVI for Certain Data

In the past few decades, many CVIs have been developed to determine the optimal number of clusters. Most CVIs focus on calculating compactness and separability measures. The combination of the two measures is composed of a ratio-type or summation-type index. This section presents several popular CVIs that have been evaluated in many applications.

The Dunn (DU) index [12]:

$$DU_K = \frac{min_{i,j=1,\cdots,K,\ i\neq j}\left\{min_{x\in C_i,\ y\in C_j} d(x,y)\right\}}{max_{i=1,\cdots,K}\left\{max_{x,y\in C_i} d(x,y)\right\}}. \tag{1}$$

Compactness and separability are computed using the maximum diameter among all clusters and the minimum pair-wise distance between objects in different clusters. The DU index is integrated by the ratio type of separability to compactness. Thus, the maximum value of the DU index is the optimal number of clusters (max. S/C).

Calinski–Harabasz (CH) index [13]:

$$CH_K = \frac{\sum_{i=1}^{K} n_i \cdot d(z_i \cdot z_{tot})^2}{K-1} \cdot \frac{n-K}{\sum_{1=1}^{K} \sum_{x\in C_i} d(x,z_i)^2} \tag{2}$$

The CH is composed of the ratio type of separability and compactness like the DU index. z_{tot} is the centroid of the entire dataset. Compactness and separability are computed using within- and between-cluster sums of squares. Thus, the maximum value for CH is the optimum partition (max. S/C).

The Davies–Bouldin (DB) index [14]:

$$DB_K = \frac{1}{K} \max_{i=1,\ldots,K,\ i\neq j} \left\{ \left(\sqrt{\frac{1}{n_i} \sum_{x\in C_i} d(x,z_i)^2} + \sqrt{\frac{1}{n_j} \sum_{y\in C_j} d(y,z_j)^2} \right) / d(z_i,z_j) \right\} \tag{3}$$

where z_i and z_j are the centroids of each cluster. Compactness and separability are calculated using the sum of mean squares of individual clusters, unlike the DU index, which considers the compactness and separability of the total cluster. Compactness is the computed sum of the pair-wise distances between different clusters; separability is calculated differently for each cluster. The DB index is comprised of the ratio types of compactness and separability. Therefore, the minimum value of DB is the optimum partition (min. C/S).

The pre-existing CVIs are sensitive to sub-clusters, arbitrary shapes, and noise in clusters for the compactness measure [18]. This study overcomes those drawbacks by conducting a spatial transformation from the original space into feature space using a kernel function that correctly measures cluster compactness and separability.

2.2. CVI for Uncertain Data

Most CVIs have focused on certain data or fixed values [19]. Certain data do not have uncertainty caused by several factors and environments such as sensor measurement error, repeated measurements by workers, or equipment operating environments. Uncertain data objects come in two possible forms: (1) multiple points for each object and (2) a PDF for each object, either given or obtained by fitting the multiple points [20]. Several studies related to clustering uncertain data have been conducted. However, CVIs for uncertain data have rarely been used. The CVIs are crucial criteria for validating the results of clusters [21,22] to find the appropriate number of clusters. Therefore, the study of CVIs for uncertain data is necessary.

In this study, the proposed CVIs use kernel probabilistic distance measures to compute the distance between two uncertain data objects. There are many popular probabilistic distance measures, such as Bhattacharyya distance [23], Wasserstein distance, and Kullback–Leibler divergence [24]. This study uses the Bhattacharyya distance measure. The Bhattacharyya distance measure is one of the widely used probabilistic distance measures and has been generally used in diverse applications.

The Bhattacharyya distance between two probability distributions can be calculated in discrete and continuous cases. Let p and q be the continuous probability distributions over the same space. The definition of the Bhattacharyya distance for a continuous case in original space can be described as follows:

$$PD_{Bhatt}(p,q) = -\ln\left\{ \int_x \sqrt{p(x)q(x)}dx \right\} \tag{4}$$

There are closed-form solutions for many probabilistic distance measures, including the Bhattacharyya distance, for cases where uncertain data objects are modeled with multivariate normal distributions. As probabilistic distance measures can capture the distance between PDFs, they can also be used to capture the distance between uncertain data objects [25]. The Bhattacharyya distance is a special case of Chernoff distance with parameters $\alpha_1 = \alpha_2 = 1/2$, and the closed-from of Bhattacharyya distance for multivariate normal PDFs is defined in Equation (5):

$$PD_{Bhatt}(p,q) = \frac{1}{8}(\boldsymbol{\mu}_p - \boldsymbol{\mu}_q)'(\boldsymbol{\Sigma}_p + \boldsymbol{\Sigma}_q)^{-1}(\boldsymbol{\mu}_p - \boldsymbol{\mu}_q) + \frac{1}{2}\ln\left(\frac{|\boldsymbol{\Sigma}_p + \boldsymbol{\Sigma}_q|}{2(|\boldsymbol{\Sigma}_p| + |\boldsymbol{\Sigma}_q|)^{\frac{1}{2}}}\right) \tag{5}$$

where $\boldsymbol{\mu}_p$ and $\boldsymbol{\mu}_q$ are means, and $\boldsymbol{\Sigma}_p$ and $\boldsymbol{\Sigma}_q$ are covariance matrices of $P \sim MVN(\boldsymbol{\mu}_p, \boldsymbol{\Sigma}_p)$ and $Q \sim MVN(\boldsymbol{\mu}_q, \boldsymbol{\Sigma}_q)$.

This study models the Bhattacharyya distance between two uncertain data objects in kernel space. We can compute the probabilistic distance between two uncertain data objects in feature space using a kernel function.

3. Proposed CVIs for Uncertain Data

3.1. Kernel Probabilistic Distance Measure in Feature Space

Computing the probabilistic distance is a nontrivial problem. We can compute the Bhattacharyya distance in feature space by referring to several steps developed by Zhou and Chellappa [26]. In capturing the probabilistic distance, suppose that $x_1 = \{x_{11}, x_{21}, \ldots, x_{N1}\}$ and $x_2 = \{x_{12}, x_{22}, \ldots, x_{N2}\}$ are the given objects in original space $\mathbb{R}^d$ with a multivariate normal density function:

$$\mathrm{N}(\mathbf{x};\ \mu,\ \boldsymbol{\Sigma}) = \frac{1}{\sqrt{(2\pi)^d |\boldsymbol{\Sigma}|}} \exp\left\{-\frac{1}{2}(\mathbf{x}-\boldsymbol{\mu})^T \boldsymbol{\Sigma}^{-1}(\mathbf{x}-\boldsymbol{\mu})\right\} \tag{6}$$

The radial basis function (RBF) kernel function displayed in Equation (7) can be used to transfer original data into feature space for calculating the distance between uncertain data objects x_1 and x_2. The RBF kernel function is commonly used in various fields and algorithms because it outperforms other kernel functions [27,28].

$$K_{ij} = \exp\left(-\frac{1}{2\sigma^2}\|x_i - x_j\|^2\right),\ i, j = 1, 2 \tag{7}$$

In kernel function $K(x_1,\ x_2)$, where $x_1,\ x_2 \in \mathbb{R}^d$, and the non-linear mapping function ϕ and kernel Gram matrix K are defined as $K = \Phi^T\Phi$, where $\Phi := \Phi_N = [\phi(x_1),\ \phi(x_2),\ \ldots,\ \phi(x_N)] \in \mathbb{R}^f$, and $f \gg d$ represents the data transformed to kernel space. The mean $\boldsymbol{\mu}$ and covariance matrix $\boldsymbol{\Sigma}$ in feature space are estimated as:

$$\boldsymbol{\mu} = \mathrm{N}^{-1} \sum_{\mathrm{n}=1}^{\mathrm{N}} \phi(\mathrm{x}_n) = \boldsymbol{\Phi}, \tag{8}$$

$$\boldsymbol{\Sigma} = N^{-1} \sum_{n=1}^{N} (\phi_n - \mu)(\phi_n - \mu)^{\mathrm{T}} = \boldsymbol{\Phi} \boldsymbol{J} \boldsymbol{J}^T \boldsymbol{\Phi}^{\mathrm{T}}, \tag{9}$$

where $J = \frac{1}{\sqrt{n}}(I_N - s\vec{1})$ with $s_{N\times 1} = \frac{1}{N}\vec{1}^T$ and $\vec{1} = [1, 1,\ \ldots,\ 1]$.

The covariance matrix $\boldsymbol{\Sigma}$ must be converted into approximation form because of its rank-deficient characteristic $f \gg d$. Therefore, we can use the approximation form as follows:

$$C = \boldsymbol{\Phi} \boldsymbol{J} \boldsymbol{J}^T \boldsymbol{\Phi}^{\mathrm{T}} + \rho I_f = \mathbf{W}\mathbf{W}^{\mathrm{T}} + \rho I_f = \boldsymbol{\Phi} \boldsymbol{A} \boldsymbol{\Phi}^T + \rho I_f, \tag{10}$$

where $\mathbf{W} \doteq \boldsymbol{\Phi} \boldsymbol{J}\mathbf{Q}$, $A \doteq \boldsymbol{J}\mathbf{Q}\mathbf{Q}^{\mathrm{T}}\boldsymbol{J}^{\mathrm{T}}$, and ρ is a user parameter that should be pre-specified in advance.

Obtaining the matrix $\mathbf{Q}$ requires computing the top r eigenvalues matrix $\boldsymbol{\Lambda}_r$ and the top r eigenvectors matrix V_r of $\overline{K} = J^T K J$, where top r is a pre-specified parameter; thus, $r = 3$ is used. Q is an $N \times r$ matrix calculated as follows:

$$Q \doteq V_r\left(I_r - \rho\boldsymbol{\Lambda}_r^{-1}\right)^{1/2}. \tag{11}$$

Define matrix $\boldsymbol{P}$ as:

$$\boldsymbol{P}_{(N_1+N_2)\times(r_1+r_2)} = \begin{bmatrix} \sqrt{\alpha_1} J_1 Q_1 & 0 \\ 0 & \sqrt{\alpha_2} J_2 Q_2 \end{bmatrix}. \tag{12}$$

The Bhattacharyya distance is a special case of Chernoff distance; it must be set to $\alpha_1 = \alpha_2 = 1/2$ for all experiments. The $\tau_i, i = 1, \ldots, r_1 + r_2$, are eigenvalues of a $\boldsymbol{L}_{ch}$ matrix, with dimensions of $(r_1 + r_2) \times (r_1 + r_2)$ given by

$$\boldsymbol{L}_{ch} = \mathbf{P}^{\mathrm{T}} \begin{bmatrix} \boldsymbol{\Phi}_1^{\mathrm{T}} \\ \boldsymbol{\Phi}_2^{\mathrm{T}} \end{bmatrix} \begin{bmatrix} \boldsymbol{\Phi}_1^T & \boldsymbol{\Phi}_2^T \end{bmatrix} \boldsymbol{P} = \boldsymbol{P}^T \begin{bmatrix} \boldsymbol{K}_{11} & \boldsymbol{K}_{12} \\ \boldsymbol{K}_{21} & \boldsymbol{K}_{22} \end{bmatrix} \boldsymbol{P}. \tag{13}$$

Scalar values ε_{11}, ε_{12}, ε_{22} are computed by Equation (14).

$$\varepsilon_{ij} = s_i^T \boldsymbol{K}_{ij} s_j - s_i^T [\boldsymbol{K}_{i1} \boldsymbol{K}_{i2}] \boldsymbol{B}_{ch} \begin{bmatrix} \boldsymbol{K}_{1j} \\ \boldsymbol{K}_{2j} \end{bmatrix} s_j \quad (14)$$

where $\boldsymbol{B}_{ch} = \boldsymbol{P}(\rho \boldsymbol{I}_{r_1+r_2} + \boldsymbol{L}_{ch})^{-1} \boldsymbol{P}^T$ with dimensions of $(N_1 + N_2) \times (N_1 + N_2)$.

The kernel-based probabilistic Bhattacharyya distance between two uncertain data objects $\boldsymbol{x}_1$ and $\boldsymbol{x}_2$ in feature space is calculated as follows:

$$KPD_{Bhatt} = 0.5[\alpha_1 \alpha_2 \rho^{-1}(\varepsilon_{11} + \varepsilon_{22} - 2\varepsilon_{12})] + 0.5 \sum_{i=1}^{r_1+r_2} \log \frac{\rho + \tau_i}{\lambda_{i,1}} + \sum_{i=1}^{r_1+r_2} \log \frac{\rho + \tau_i}{\lambda_{i,2}}, \quad (15)$$

where $\lambda_{i,j}$, $i = 1, \ldots, r_j$ are the eigenvalues of C_j:

$$\lambda_{i,j} = \begin{cases} \lambda_{i,j}, & when\ i = 1, \ldots, r_j \\ \rho, & when\ i = r_j + 1, \ldots, r_1 + r_2 \end{cases} \quad (16)$$

3.2. New CVI for Uncertain Data

The uncertain data objects in the cluster are transformed into feature space to compute the compactness and separability in the feature space by applying a kernel function. The mapped uncertain data objects are used to compute the distance between different clusters for calculating compactness and separability, which are used to obtain the values of the proposed CVIs. The calculated value of the indices changes according to the number of clusters K, and the proposed uncertain feature space DU (UFSDU) and uncertain feature space CH (UFSCH) index, are defined in Equations (17) and (18), respectively:

UFSDU index:

$$UFSDU_K = \frac{min_{i,j=1,\cdots,K,\ i \neq j}\left\{min_{x \in C_i,\ y \in C_j} KPD_{Bhatt}(x, y)\right\}}{max_{i=1,\cdots,K}\left\{max_{x,y \in C_i}\ KPD_{Bhatt}(x, y)\right\}} \quad (17)$$

UFSCH index:

$$UFSCH_K = \frac{\sum_{i=1}^{K} n_i \cdot KPD_{Bhatt}(z_i \cdot z_{tot})^2}{K-1} \cdot \frac{n-K}{\sum_{i=1}^{k} \sum_{x \in C_i} KPD_{Bhatt}(x, z_i)^2} \quad (18)$$

These proposed CVI equations are similar to the DU and CH indices, except for the term $KPD_{Bhatt}(x, y)$, which is the computed distance between two uncertain data objects in feature space in Equation (15).

4. Experimental Results

In this study, we propose two CVIs that are calculated probabilistic distances between different uncertain data objects in feature space. The K-medoids clustering algorithm proposed by Jiang et al. [19] was used to compare the performances of the proposed CVIs in feature space. The K-medoids algorithm is one of the most useful algorithms in clustering problems, which uses probabilistic distance measures to capture the similarity between uncertain objects. It differs from the popular K-means clustering algorithm used for clustering data into groups in its robustness to outliers. The K-means method represents each cluster by the mean of all objects in this cluster, whereas the K-medoids method calculates the distance between every pair of all uncertain data objects and the medoid within a cluster [19]. Then, of all calculated distance values, uncertain data with the smallest distances are assigned as a new medoid for the cluster. We proceeded with the experiments by setting the value of K, which is the number of clusters and is used as the probabilistic distance measure. In this study, we varied the number of clusters (K) and the

Bhattacharyya distance measure to compute distances between different uncertain data objects in feature space.

4.1. Experimental Procedure for Uncertain Data

Experiments were performed with artificial and real-world datasets that may have sub-clusters and clusters with asymmetrical, arbitrary, and noisy shapes to evaluate the performances of the proposed CVIs. A normalization process was conducted for each feature of the datasets to reduce the scale gap between different features defined in Equation (19):

$$x_{norm} = \frac{x - x_{min}}{x_{max} - x_{min}}, \tag{19}$$

where x_{min} and x_{max} are the minimum and maximum values of one feature of the dataset. We then simulated uncertain data objects from certain data objects by following the methodology used by [20].

The pre-existent DU and CH indexes were used to compute uncertain data objects in original space—uncertain original space, DU (UOSUD), and uncertain original space, CH (UOSCH)—to confirm the validity of the proposed CVIs. The overall experimental procedure is represented by Algorithm 1. The procedure used to compare the performances of the proposed CVIs with those of the previous CVIs was as follows: The inputs included the number of uncertain data objects N, the number of object features M, and the number of clusters K. We modeled the uncertain data with multivariate normal distributions. The means of the distributions were the original certain data. The covariances were estimated as follows:

$$f\left(S_i^k \middle| \Psi^k, df^k\right) = \frac{\left|\Psi^k\right|^{\frac{df^k}{2}}}{\frac{p\cdot df^k}{2}\Gamma_p\left(\frac{df^k}{2}\right)} \left|S_i^k\right|^{-\frac{df^k+p+1}{2}} e^{-\frac{1}{2}tr\left(\Psi^k\left(S_i^k\right)^{-1}\right)},\ i = 1,\ldots,n_k,\ k = 1,\ldots,K \tag{20}$$

where S_i^k represents the covariance matrices for objects in class k with the inverse Wishart PDF [29], as defined in Equation (20) [20]. Ψ^k is a positive definite scale matrix and df^k is the degree of freedom. p indicates the dimensions of S_i^k, $tr(\cdot)$ is the trace of a matrix, and Γ is the multivariate gamma function.

Algorithm 1: K-medoids for uncertain data using a probabilistic distance measure in feature space.

```
Input: n: The number of objects in cluster k, K: The number of clusters, iter = 0;
Randomly select the cluster medoids C^(0) = {c_1^(0), ..., c_K^(0)} obtained from the initial clusters
Initialize
CVIs = {cvi^(1), ..., cvi^(K)} obtained UOSDU, UOSCH, UFSDU, and UFSCH
Repeat
  for k = 2 to K
     c_k^(old) = c_k^(0); c_k^(new) = 0
     Compute the new medoids:
     while c_k^(old) ≠ c_k^(new)
         p = argmin_{1≤i≤n} Σ_{j=1}^{k} KPD_Bhatt(x_i, c_jk), where j is an index of cluster medoid in c_k
         c_k^(new) = x_p
      end
      Calculate the cvi^(k) using Equations (1), (2), (17), and (18).
  end
  iter = iter + 1
Until (iter = Maxiter)
```

Step 1: Set K initial clusters with uncertain objects randomly for a given dataset. Run a K-medoids clustering algorithm with different values for the K parameter ($2 \leq K \leq 10$).

Step 2: Obtain the medoids of each cluster for which the sum of the probabilistic distance between the objects is the smallest.

Step 3: Calculate CVIs for all the partitions. We calculated the compactness and separability in kernel space using an RBF kernel function with σ (bandwidth in the RBF kernel function). The optimal value was determined through a set of preliminary experiments by taking $[0.1, 0.2, \ldots, 4]$ in σ.

Step 4: We increased the reliability of experimental results by replicating the experiment 100 times for the same dataset with different trial seeds to obtain the initial medoids in Step 1 and used the average value of CVI for each cluster.

Step 5: Finally, we evaluated each CVI and the suggested number of clusters from a CVI; the actual numbers of clusters of a dataset were then compared.

4.2. Experiments with Artificial and Real-World Datasets

Experiments were conducted to evaluate the proposed CVIs in comparison to the pre-existent CVIs. These experiments used 10 datasets with sensitive characteristics containing arbitrariness, sub-clusters, asymmetry, and noise provided by the UCI (https://archive.ics.uci.edu/, accessed on 10 March 2023) [30] and Tomas Barton repositories (https://github.com/deric/clustering-benchmark, accessed on 10 March 2023), which have 122 artificial datasets with arbitrariness, sub-clusters, and asymmetric shapes in two or three features. The datasets from UCI repository, (e.g., D3, D4, D5, and D7) were collected in real environmental conditions; however, the other datasets were artificially created, which can be checked in Tomas Barton repositories.

The summary of datasets used for the experiments is presented in Table 1. Two-dimensional (2D) and three-dimensional (3D) dataset shapes are illustrated in Figure 3. The CVI values were computed by changing the number of clusters (K) in each dataset and then comparing the predicted labels of experiments to the actual labels in the datasets.

Table 1. Summary of datasets.

Dataset Index	Dataset Name	# of Obs.	# of Dim.	# of Clusters	Projection Shape
D1	A.K Jain's Toy	373	2	2	Asymmetry, Arbitrary shape
D2	Flame	240	2	2	Sub-cluster, Noise
D3	Iris	150	4	3	-
D4	Thyroid	215	5	2	-
D5	Wine	178	13	3	-
D6	Wisconsin	683	9	2	-
D7	Harberman	301	3	2	Random shape
D8	Chainlink	1000	3	2	Sub-cluster, Arbitrary shape
D9	Lsun	400	2	3	Asymmetry, Arbitrary shape
D10	Zelnik1	299	2	3	Sub-cluster

4.3. Performance Comparison of the Proposed CVIs

The experimental results are given in Tables 2–11. The actual number of clusters is below the name of the dataset. It is also noted with an asterisk (*) adjacent to the actual number of clusters along the top. Moreover, all the results of the datasets are presented in Table 12, indicating the performance of the proposed CVIs by a quantitative figure. Each cell in Table 12 represents the optimal number of clusters K determined by its CVI criteria.

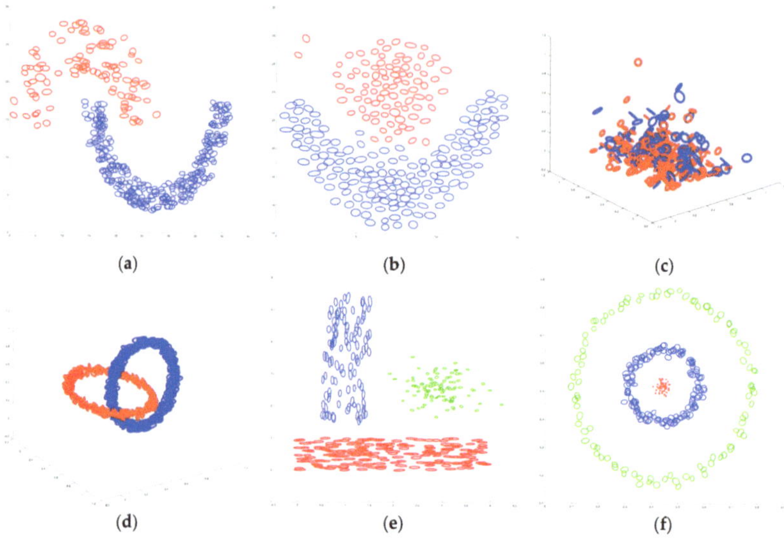

Figure 3. Shapes of 2D and 3D datasets: (**a**) D1 dataset; (**b**) D2 dataset; (**c**) D7 dataset; (**d**) D8 dataset; (**e**) D9 dataset; (**f**) D10 dataset.

Table 2. Performance results for D1.

	CVI \ # of Clusters	2 *	3	4	5	6	7	8	9	10
D1 (2)	UOSDU	**0.00075**	0.00063	0.00049	0.00046	0.00043	0.00047	0.00044	0.00042	0.000410
	UOSCH	554.4796	537.8279	573.5387	**586.5310**	576.5872	562.0666	575.2021	566.6556	567.6008
	UFSDU	**0.011830**	0.00727	0.007410	0.006350	0.006920	0.006390	0.006740	0.00580	0.005630
	UFSCH	**256.0945**	204.767	167.9338	149.5915	138.3076	128.206	122.4676	117.0263	112.4593

Table 3. Performance results for D2.

	CVI \ # of Clusters	2 *	3	4	5	6	7	8	9	10
D2 (2)	UOSDU	0.00578	0.00581	0.00583	**0.00533**	0.00494	0.00494	0.00452	0.00454	0.00448
	UOSCH	**218.9052**	188.6698	201.7685	195.0877	190.2412	190.7961	192.3785	187.7774	186.0032
	UFSDU	**0.01875**	0.01433	0.01619	0.01386	0.01284	0.01261	0.01263	0.0125	0.01271
	UFSCH	**246.7711**	190.3472	184.7522	163.52	150.3938	143.1108	138.9139	131.6189	127.3284

Table 4. Performance results for D3.

	CVI \ # of Clusters	2	3 *	4	5	6	7	8	9	10
D3 (3)	UOSDU	**0.57393**	0.18691	0.06671	0.04599	0.03375	0.03045	0.02475	0.02443	0.02427
	UOSCH	**393.8149**	340.7616	288.9103	257.4766	227.8328	211.7321	193.9894	179.4227	172.1492
	UFSDU	**0.78121**	0.05291	0.0332	0.02818	0.0201	0.02217	0.02033	0.01676	0.01503
	UFSCH	97.24412	**100.9677**	83.47847	74.68629	65.08186	59.80128	55.42499	51.32508	48.54411

Table 5. Performance results for D4.

	CVI \ # of Clusters	2 *	3	4	5	6	7	8	9	10
D4 (2)	UOSDU	**0.01059**	0.00702	0.00447	0.00389	0.00338	0.00285	0.0029	0.00264	0.00254
	UOSCH	**52.44662**	49.27229	45.29772	44.23136	46.29286	43.05835	40.0334	38.99379	36.43862
	UFSDU	**0.09045**	0.02678	0.02097	0.01941	0.0186	0.0166	0.01728	0.01604	0.01577
	UFSCH	**88.16833**	63.62494	54.54528	48.32164	43.33752	38.65073	35.53446	32.89777	30.6346

Table 6. Performance results for D5.

	CVI \ # of Clusters	2	3 *	4	5	6	7	8	9	10
D5 (3)	UOSDU	**0.28546**	0.19218	0.16953	0.13451	0.13042	0.12188	0.1222	0.11775	0.11498
	UOSCH	**46.98845**	41.61822	34.08324	29.45127	26.66111	23.71564	21.97848	20.8878	19.0692
	UFSDU	0.1351	**0.13992**	0.12361	0.11102	0.1058	0.10544	0.10343	0.10402	0.10242
	UFSCH	**166.5115**	94.11775	70.17926	57.15066	48.48219	42.44803	38.19718	34.55733	31.1674

Table 7. Performance results for D6.

	CVI \ # of Clusters	2 *	3	4	5	6	7	8	9	10
D6 (2)	UOSDU	**0.10223**	0.04719	0.02262	0.01209	0.00742	0.00342	0.0014	0.00109	0.00075
	UOSCH	**237.829**	186.8503	145.4631	119.3866	98.36381	89.72379	80.18472	70.83073	66.12163
	UFSDU	**0.22631**	0.10763	0.04928	0.03902	0.01416	0.01228	0.0084	0.00605	0.00391
	UFSCH	**349.3685**	261.4169	205.8692	171.2457	144.4285	124.5582	109.2653	97.50292	88.98401

Table 8. Performance results for D7.

	CVI \ # of Clusters	2 *	3	4	5	6	7	8	9	10
D7 (2)	UOSDU	**0.00198**	0.0014	0.00112	0.00086	0.00078	0.00089	0.00069	0.00079	0.00076
	UOSCH	**128.8359**	117.8517	104.8203	97.56451	95.82686	92.17925	86.85381	84.98897	82.71107
	UFSDU	**0.13021**	0.02577	0.01681	0.01199	0.01108	0.01122	0.01132	0.01028	0.00945
	UFSCH	**319.3255**	171.7169	127.0638	104.5919	90.63319	80.94442	72.86994	67.51974	62.62473

Table 9. Performance results for D8.

	CVI \ # of Clusters	2 *	3	4	5	6	7	8	9	10
D8 (2)	UOSDU	0.00019	0.00017	0.00017	0.00017	0.00017	0.00018	**0.00021**	0.00019	0.00017
	UOSCH	419.8882	371.9768	388.8548	430.2229	426.5956	430.8854	**449.3122**	438.7834	417.3569
	UFSDU	**0.00439**	0.00237	0.00204	0.00114	0.0013	0.00118	0.00149	0.00153	0.0014
	UFSCH	445.5408	**463.2664**	449.4758	439.8262	425.4487	411.5018	422.1565	428.8755	437.9047

Table 10. Performance results for D9.

	CVI \ # of Clusters	2	3 *	4	5	6	7	8	9	10
D9 (3)	UOSDU	**0.01277**	0.00168	0.00087	0.00081	0.00069	0.00062	0.0006	0.00063	0.00054
	UOSCH	316.7407	**406.3877**	395.188	401.578	380.968	363.1193	365.4242	349.9761	351.8199
	UFSDU	0.01439	**0.02006**	0.01697	0.01119	0.00658	0.00574	0.00485	0.00472	0.00416
	UFSCH	190.3465	**205.1745**	189.6315	175.6124	164.8462	154.2108	149.907	141.5702	133.6363

Table 11. Performance results for D10.

	CVI \ # of Clusters	2	3 *	4	5	6	7	8	9	10
D10 (3)	UOSDU	0.030644	**0.049296**	0.048849	0.048798	0.046752	0.044594	0.042478	0.037749	0.041905
	UOSCH	**235.4205**	161.3342	142.117	135.4194	127.012	126.4954	125.6673	123.9964	132.4379
	UFSDU	**0.00368**	0.00123	0.00123	0.00115	0.00103	0.00087	0.00073	0.00077	0.00056
	UFSCH	102.6013	**106.5976**	99.7133	98.79822	97.68495	95.67929	95.82844	96.62246	102.6371

Table 12. Difference between the actual and estimated numbers of clusters in lower-dimensional datasets.

Dataset	Dim	# of Clusters	UOSDU	UOSCH	UFSDU	UFSCH
D1	2	2	⊙	5	⊙	⊙
D2	2	2	4	⊙	⊙	⊙
D3	4	3	2	2	2	⊙
D4	5	2	⊙	⊙	⊙	⊙
D5	13	3	2	2	⊙	2
D6	9	2	⊙	⊙	⊙	⊙
D7	3	2	⊙	⊙	⊙	⊙
D8	3	2	8	8	⊙	3
D9	2	3	2	⊙	⊙	⊙
D10	2	3	⊙	2	2	⊙
# of successes in estimating the optimal number of clusters			5	5	8	8

The bold values with gray-shaded backgrounds indicate the optimal cluster K decided by each CVI. As presented in Table 2, three of the CVIs succeeded in estimating the number of clusters as two in D1. UOSCH failed. The proposed UFSDU and UFSCH also successfully predicted the number of clusters in D2. In contrast, UOSDU failed to estimate the number of clusters in D2.

Although the proposed UFSDU index and the pre-existent CVIs failed to predict the number of clusters in D3, UFSCH was successful. All CVIs correctly predicted the number of clusters for some datasets; see Tables 5, 7 and 8. In contrast, the proposed UFSDU index is the only CVI that correctly predicted the actual number of clusters in D5, as presented in Table 6. Furthermore, the UFSDU index predicted the actual number of clusters of D8. D8's shape (Figure 3) is classified distinctly into two classes when viewed visually. However, it is challenging to calculate the compactness and separability of a cluster in the original space. Nevertheless, the UFSDU index was successful in such predictions; the UFSCH forecasted the number of clusters as three, which is close to the actual number of clusters, two. The kernel transformation facilitates computation to obtain greater compactness and separability in the feature space than the original space, leading to high-performance clustering.

The UOSCH index and the new CVIs predicted the number of clusters to be three in D9, and the UOSDU and UFSCH indexes successfully estimated the number of clusters in D10. Table 12 presents a summary of the results of the 10 datasets above, whereas the symbol of a circled dot (⊙) indicates that the CVI accurately predicted the actual number of clusters. As presented in Table 12, the pre-existent CVIs precisely estimated the number of clusters for five experimental datasets, whereas the newly proposed CVIs accurately predicted the number of clusters for eight datasets—three more than the pre-existent CVIs.

5. Conclusions

In this study, we proposed novel cluster validity indices (CVIs) for uncertain data objects in feature space. Unlike conventional CVIs in original space, the proposed CVIs are used for uncertain data objects with arbitrariness, sub-clusters, and noisy shapes of clusters that are hard to evaluate, by transforming the uncertain data from the original space to the feature space, which is performed by the kernel function. The proposed CVIs measure the compactness and separability of each cluster in kernel space, which transforms the original data into a higher-dimensional space, leading to less sensitivity to the arbitrary shapes of clusters and more robustness to noise and outliers. We compared the performances of the proposed CVIs with those of pre-existent CVIs that only consider for the original space. The Bhattacharyya distance measure, one of the most widely used for calculating distance, was used to perform experiments with several artificial and real-life datasets to capture the distances between probability density functions. Numerical examples, including a real-life case study and artificial datasets, confirmed that our proposed CVIs are robust to arbitrary cluster shapes, especially sub-clusters, and are promising alternatives for evaluating the fitness of clustering results that can find the optimal number of clusters, K. The proposed CVIs outperform the pre-existent CVIs because of the application of kernel functions to uncertain data, transforming them from the original space to the feature space. As for practical significance, the proposed CVIs could be utilized in diverse applications. For example, Kim et al. proposed new a multivariate kernel density estimator for uncertain data classification for mixed defect patterns on DRAM wafer maps [31]. The proposed CVI method could be applied for evaluating the number of defect patterns on wafer maps. However, there are some limitations to the proposed CVIs. The uncertain data are assumed to have multivariate normal distributions in advance to compute the distances between different uncertain data objects. The uncertainty of the uncertain data may have a variety of probability functions (normal distribution, exponential distribution, etc.), and some cannot be strictly modeled by PDFs. This might be overcome through methods for generating random variables and support-measure data description, which is a non-parametric machine learning method that does not require an assumption of a prior distribution to be made in advance.

Future research should consider the compactness measure in kernel space in advanced machine learning algorithms, such as support vector data descriptions or Bayesian frameworks of Bayesian support vector data descriptions. The concepts of our CVIs can also be applied to other clustering algorithms.

Author Contributions: Conceptualization, Y.-S.J.; data curation, C.K.; formal analysis, Y.-S.J.; investigation, B.T. and Y.-S.J.; methodology, C.K. and Y.-S.J.; resources, B.T.; software, B.T.; supervision, Y.-S.J.; validation, J.B.; visualization, J.B.; writing—original draft, C.K.; writing—review and editing, J.B., B.T. and Y.-S.J. All authors have read and agreed to the published version of the manuscript.

Funding: This work was supported by LG Yonam Foundation (of Republic of Korea) and by National Research Foundation of Republic of Korea Grant (No. NRF-2021S1A5A8060639, NRF-2022R1F1A1063174).

Institutional Review Board Statement: Not applicable.

Informed Consent Statement: Not applicable.

Data Availability Statement: The real-world datasets used in this study are available at: https://archive.ics.uci.edu/ml/index.php accessed on 10 March 2023; the artificial datasets that contain data sensitive to shapes are available at: https://github.com/deric/clustering-benchmark/tree/master/ accessed on 10 March 2023.

Conflicts of Interest: The authors declare no conflict of interest.

Abbreviations

The following abbreviations are used in this manuscript:

Bhatt	Bhattacharyya distance measure
C/S	Separability/Compactness
CH	Calinski–Harabasz
CVIs	Cluster validity indices
DB	Davies–Bouldin
DU	Dunn
KPD	Kernel-based probabilistic distance
PD	Probabilistic distance
PDF	Probability density function
RBF	Radial basis function
S/C	Compactness/Separability
UFSCH	Uncertain feature space CH
UFSDU	Uncertain feature space DU
UOSCH	Uncertain feature space CH
UOSDU	Uncertain feature space DU

References

1. Abdalameer, A.K.; Alswaitti, M.; Alsudani, A.A.; Isa, N.A. A new validity clustering index-based on finding new centroid positions using the mean of clustered data to determine the optimum number of clusters. *Expert Syst. Appl.* **2022**, *191*, 116329. [CrossRef]
2. Irani, J.; Pise, N.; Phatak, M. Clustering techniques and the similarity measures used in clustering: A survey. *Int. J. Comput. Appl. Technol.* **2016**, *134*, 9–14. [CrossRef]
3. MacQueen, J.B. Some methods for classification and analysis of multivariate observations. In Proceedings of the Fifth Berkeley Symposium on Mathematical Statistics and Probability, Berkeley, CA, USA, 27 December 1965–7 January 1966; The Regents of the University of California: Santa Barbara, CA, USA, 1967; pp. 281–297.
4. Li, M.J.; Ng, M.K.; Cheung, Y.-m.; Huang, J.Z. Agglomerative fuzzy k-means clustering algorithm with selection of number of clusters. *IEEE Trans. Knowl. Data Eng.* **2008**, *20*, 1519–1534. [CrossRef]
5. Mahesh Kumar, K.; Rama Mohan Reddy, A. A fast DBSCAN clustering algorithm by accelerating neighbor searching using groups method. *Pattern Recognit.* **2016**, *58*, 39–48. [CrossRef]
6. Chien, C.-F.; Wang, W.-C.; Cheng, J.-C. Data mining for yield enhancement in semiconductor manufacturing and an empirical study. *Expert Syst. Appl.* **2007**, *33*, 192–198. [CrossRef]
7. El-shafeiy, E.; Sallam, K.M.; Chakrabortty, R.K.; Abohany, A.A. A clustering based swarm intelligence optimization technique for the internet of medical things. *Expert Syst. Appl.* **2021**, *173*, 114648. [CrossRef]
8. Aggarwal, C.C.; Yu, P.S. A survey of uncertain data algorithms and applications. *IEEE Trans. Knowl. Data Eng.* **2009**, *21*, 609–623. [CrossRef]
9. Shou, L.; Zhang, X.; Chen, G.; Gao, Y.; Chen, K. Mud: Mapping-based query processing for high-dimensional uncertain data. *Inf. Sci.* **2012**, *198*, 147–168. [CrossRef]
10. Duan, X.; Ma, Y.; Zhou, Y.; Huang, H.; Wang, B. A novel cluster validity index based on augmented non-shared nearest neighbors. *Expert Syst. Appl.* **2023**, *223*, 119784. [CrossRef]
11. Lee, S.-H.; Jeong, Y.-S.; Kim, J.-Y.; Jeong, M.K. A new clustering validity index for arbitrary shape of Clusters. *Pattern Recognit. Lett.* **2018**, *112*, 263–269. [CrossRef]
12. Dunn, J.C. Well-separated clusters and optimal fuzzy partitions. *J. Cybern.* **1974**, *4*, 95–104. [CrossRef]
13. Calinski, T.; Harabasz, J. A dendrite method for cluster analysis. *Commun. Stat.-Theory Methods* **1974**, *3*, 1–27. [CrossRef]
14. Davies, D.L.; Bouldin, D.W. A cluster separation measure. *IEEE Trans. Pattern Anal. Mach. Intell.* **1979**, *PAMI-1*, 224–227. [CrossRef]
15. Xie, X.L.; Beni, G. A validity measure for fuzzy clustering. *IEEE Trans. Pattern Anal. Mach. Intell.* **1991**, *13*, 841–847. [CrossRef]
16. Rojas-Thomas, J.C.; Santos, M.; Mora, M. New internal index for clustering validation based on graphs. *Expert Syst. Appl.* **2017**, *86*, 334–349. [CrossRef]

17. Tavakkol, B.; Jeong, M.K.; Albin, S.L. Validity indices for clusters of uncertain data objects. *Ann. Oper. Res.* **2018**, *303*, 321–357. [CrossRef]
18. Wang, J.-S.; Chiang, J.-C. A cluster validity measure with a hybrid parameter search method for the support vector clustering algorithm. *Pattern Recognit.* **2008**, *41*, 506–520. [CrossRef]
19. Jiang, B.; Pei, J.; Tao, Y.; Lin, X. Clustering uncertain data based on probability distribution similarity. *IEEE Trans. Knowl. Data Eng.* **2013**, *25*, 751–763. [CrossRef]
20. Tavakkol, B.; Jeong, M.K.; Albin, S.L. Object-to-group probabilistic distance measure for uncertain data classification. *IEEE Trans. Knowl. Data Eng.* **2017**, *230*, 143–151. [CrossRef]
21. Arbelaitz, O.; Gurrutxaga, I.; Muguerza, J.; Pérez, J.M.; Perona, I. An extensive comparative study of cluster validity indices. *Pattern Recognit.* **2013**, *46*, 243–256. [CrossRef]
22. Rezaee, B. A cluster validity index for Fuzzy Clustering. *Fuzzy Sets Syst.* **2010**, *161*, 3014–3025. [CrossRef]
23. Bhattacharyya, A. On a measure of divergence between two multinomial populations. *Sankhya Indian J. Stat.* **1946**, *7*, 401–406.
24. Kullback, S.; Leibler, R.A. On information and sufficiency. *Ann. Math. Stat.* **1951**, *22*, 79–86. [CrossRef]
25. Tavakkol, B.; Son, Y. Fuzzy kernel K-medoids clustering algorithm for uncertain data objects. *Pattern Anal. Appl.* **2021**, *24*, 1287–1302. [CrossRef]
26. Zhou, S.K.; Chellappa, R. From sample similarity to ensemble similarity: Probabilistic distance measures in reproducing kernel Hilbert space. *IEEE Trans. Pattern Anal. Mach. Intell.* **2006**, *28*, 917–929. [CrossRef] [PubMed]
27. Patle, A.; Chouhan, D.S. SVM kernel functions for classification. In Proceedings of the 2013 International Conference on Advances in Technology and Engineering (ICATE), Mumbai, India, 23–25 January 2013.
28. Tbarki, K.; Ben Said, S.; Ksantini, R.; Lachiri, Z. RBF kernel based SVM Classification for landmine detection and discrimination. In Proceedings of the 2016 International Image Processing, Applications and Systems (IPAS), Sfax, Tunisia, 5–7 November 2016.
29. Nydick, S.W. The wishart and inverse wishart distributions. *Electron. J. Stat.* **2012**, *6*, 1–19.
30. UCI Machine Learning Repository. Available online: https://archive.ics.uci.edu/ (accessed on 28 March 2023).
31. Kim, B.; Jeong, Y.-S.; Jeong, M.K. New multivariate kernel density estimator for uncertain data classification. *Ann. Oper. Res.* **2020**, *303*, 413–431. [CrossRef]

sensors

MDPI

Article

Cloud Based Fault Diagnosis by Convolutional Neural Network as Time–Frequency RGB Image Recognition of Industrial Machine Vibration with Internet of Things Connectivity

Dominik Łuczak *, Stefan Brock and Krzysztof Siembab

Faculty of Automatic Control, Robotics and Electrical Engineering, Poznan University of Technology, 60-965 Poznań, Poland

* Correspondence: dominik.luczak@put.poznan.pl

Abstract: The human-centric and resilient European industry called Industry 5.0 requires a long lifetime of machines to reduce electronic waste. The appropriate way to handle this problem is to apply a diagnostic system capable of remotely detecting, isolating, and identifying faults. The authors present usage of HTTP/1.1 protocol for batch processing as a fault diagnosis server. Data are sent by microcontroller HTTP client in JSON format to the diagnosis server. Moreover, the MQTT protocol was used for stream (micro batch) processing from microcontroller client to two fault diagnosis clients. The first fault diagnosis MQTT client uses only frequency data for evaluation. The authors' enhancement to standard fast Fourier transform (FFT) was their usage of sliding discrete Fourier transform (rSDFT, mSDFT, gSDFT, and oSDFT) which allows recursively updating the spectrum based on a new sample in the time domain and previous results in the frequency domain. This approach allows to reduce the computational cost. The second approach of the MQTT client for fault diagnosis uses short-time Fourier transform (STFT) to transform IMU 6 DOF sensor data into six spectrograms that are combined into an RGB image. All three-axis accelerometer and three-axis gyroscope data are used to obtain a time-frequency RGB image. The diagnosis of the machine is performed by a trained convolutional neural network suitable for RGB image recognition. Prediction result is returned as a JSON object with predicted state and probability of each state. For HTTP, the fault diagnosis result is sent in response, and for MQTT, it is send to prediction topic. Both protocols and both proposed approaches are suitable for fault diagnosis based on the mechanical vibration of the rotary machine and were tested in demonstration.

Keywords: image recognition; HTTP; MQTT; feature extraction; convolutional neural network; time–frequency domain; short-time Fourier transform; sliding discrete Fourier transform

1. Introduction

Modern factories, cites, and households are equipped with an increasing number of electromechanical systems, which consume energy and have limited lifetimes. The appropriate maintenance of these devices can increase their useful life, which is cost-effective and environmentally friendly, reducing the number of electronic waste. In the literature, the term "electronic trash" or "trash" can also be found [1–3], but more often it is "electronic waste", "e-waste" [4,5] or WEEE (Waste Electrical and Electronic Equipment) [6–8]. Maintenance of industrial machines is more demanding due to the complexity of electrical and mechanical parts. Perspective maintenance allows to avoid the unwanted stoppage of the production cycle and prevents unwanted damages of equipment. Deep understanding of available tools and the application of fault diagnosis is more demanding with the number of scientific articles already published in the field. Under keyword "fault diagnosis", Google Scholar lists nearly 1.6 million articles. Limitation of results to "industrial machines" with operator AND gives 1.6 thousand articles. Another aspect of modern factories, cities, and households is increasing capability of exchange data through the global Internet network, which is named IoT (Internet of Things) or more narrowly, IIoT (Industrial Internet of Things) for industrial interconnections. Another term is Industry 4.0 which "describes the organization of production processes based on technology and devices autonomously communicating with each other along the value chain" [9,10]. However, the new term "Industry 5.0" was introduced towards a sustainable, human-centric and resilient European industry [8]. The publication database with the keyword "fault diagnosis" and "IoT" gives almost 12.3 thousand publications in the field. Therefore, both areas of fault diagnosis and IoT are popular and a growing field in the research area. The review of the patent database gives a new view on IoT and fault diagnosis. Since January 2020, the IPC (International Patent Classification) has been extended with the subclass G16Y titled "information and communication technology specially adapted for the internet of things". Additionally, subgroup G16Y40/00 "IoT characterized by the purpose of the information processing" is dedicated to maintenance and management. Further detailed classification symbols are G16Y40/10 for detection; monitoring, G16Y40/20 analytics; diagnosis and G16Y40/40 maintenance of things. The Espacenet patent search gives nearly 2.8 thousand patents in a search by classification symbols G16Y40/10, G16Y40/20, and G16Y40/40. In the following sections, the author makes efforts to present a brief review of fault diagnosis and IoT, however, due to the large number of articles and patents, the limited pages of articles, and the limited time resources of the author, some aspects were omitted.

The manuscript is organized as follows: In chapter 2, general structure of the data collection system with local and Internet connectivity is described, which gives a view of the communication aspects of the fault diagnosis system with IoT for batch and stream data processing; in chapter 3, in detail are described IoT protocols such as HTTP (Hypertext Transfer Protocol) and MQTT (Message Queuing Telemetry Transport) for batch and stream data processing and their usage for data ingestion and data routing in the fault diagnosis system; in chapter 4, the structure of the fault diagnosis system from the signal processing point of view is presented; in chapter 5, different feature extraction methods that can be used for diagnosis in frequency domain such as FFT (fast Fourier transform) algorithms (radix-2 and radix-4), STFT (short-time Fourier transform), and group of algorithms of SDFT (sliding discrete Fourier transform) are explained; in chapter 6, a demonstration rig with MQTT connectivity is presented with dataset description in the time domain, frequency domain, and time–frequency domain. Based on preliminary selected single-axis data (gyroscope Z axis) in frequency domain features calculated by SDFT, a classifier was trained and tested which is an enhancement and reduction of computation requirements in comparison to features calculated by STFT; in Chapter 7, a proposed concept of usage 3-axis acceleration and 3-axis gyroscope sensor to create RGB (red, green, and blue) images from six time–frequency domain features which can be recognized by CNN (convolutional neural network) is presented and verified. The proposed approach proves that CNN can be successfully used in multispectrogram recognition which are organized into RGB images for fault diagnosis without the need for preliminary selection of the vibration axis. The proposed recognition of the RGB six-spectrogram and enhanced usage of SDFT instead of STFT was compared with other available methods and the results are shown in Table 1. CNN is used in vision-based recognition and vision applications [11–14]. Therefore, CNN can be applied to recognize specially prepared time–frequency images, as shown in the proposed method.

Table 1. Comparison of proposed and enhanced methods.

IoT Connectivity	Types of Fault and Labels	Signal or Sensor	Features Extraction Method	Features	Classificator	Article
MQTT and HTTP	demonstration of fan blades' imbalance (normal, fan off, fan with fault)	3-axis accelerometer and 3-axis gyroscope	SDFT or STFT at six axis	RGB image made of six time–frequency domain data	CNN	Proposed

Table 1. *Cont.*

IoT Connectivity	Types of Fault and Labels	Signal or Sensor	Features Extraction Method	Features	Classificator	Article
MQTT and HTTP	demonstration of fan blades' imbalance (normal, fan off, fan with fault)	preliminary selected one axis of 3-axis accelerometer and 3-axis gyroscope	SDFT at one axis enhanced to STFT	frequency domain data	classical classifier	Enhanced
Not specified	bearing (normal, inner ring, outer ring, ball)	vibration one-axis	STFT	color spectrogram of one signal	CNN	[15]
Not specified	bearing normal and four faulty states (ball, inner ring, outer ring, inner + outer)	3-axis accelerometer	frequency transform with weight map	frequency domain for each axis	CNN	[16]
Not specified	blades non-damaged and two fault (5% and 15% broken blades)	from one axis to 3-axis of angular velocity	WPT (wavelet packet transform)—wavelet name not specified	third level of WPT decomposition	LSTM (long and short-term memory)	[17]
Not specified	bearing (normal, outer, ball, inner)	raw data is one-dimensional signal; sensor is not specified	CWT (continuous wavelet transform), STFT	CWT, time domain, and frequency domain features aggregation	MIMTNet (multiple—input, multiple—task CNN)	[18]

2. General Structures of Data Collection and Processing for Fault Diagnosis

The IoT maintenance system is based on measurements delivered in time series to the cloud. The most important part is to prepare data pipelines for fast transmission from sensors to the fault diagnosis server. At low level, analogue sensors are handled by ADC (analogue-to-digital converter). However, modern sensors are equipped with ADC and communication interfaces. Therefore, the sensor is connected to thing (microcontroller or microcomputer) by communication interface like SPI (Serial Peripheral Interface), I^{2}C (Inter-Integrated Circuit), UART (universal asynchronous receiver-transmitter) or another device-specific communication interface. In the first stage of buffer, the classical producer–consumer application is applied between the sensor and the thing (microcontroller) where the sensors are equipped with internal buffers of several samples. The sensor is a producer of data and the thing is a consumer. The next stage is the use of data by microcontroller internal algorithms. The next stage of buffering is a producer–consumer application where the microcontroller is a data producer and the consumer is a cloud (server)-based fault diagnosis system. In each stage of data processing, the producer–consumer uses buffer for data. The general structure of the buffer-connected data pipelines is presented in Figure 1. Modern buffers offer routing to selected destinations which lead to processing in chain by micro services which are easy to modify compared to change in one large service. The collected data can be processed by the fault diagnosis system in two general ways, (a) batch processing or (b) stream processing. Batch processing is performed with data collected in a time window, but not often. Data stream processing is similar to batch processing but is performed repetitively in micro batch and can use the result of the previous micro batch result in the current micro batch processing.

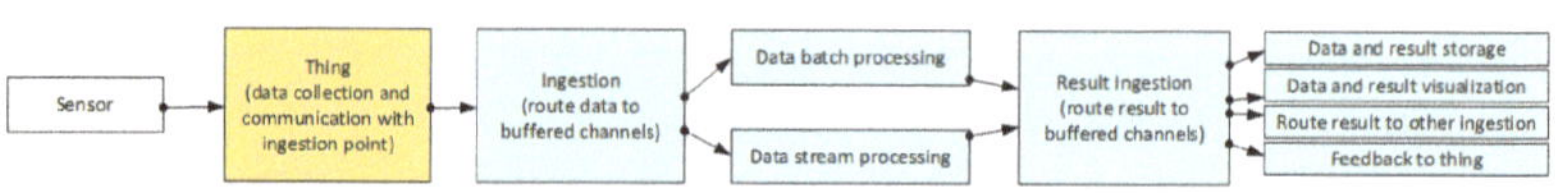

Figure 1. General structure of data pipelines in fault diagnosis.

The general structure of data pipelines presented does not explain which protocol is used by the thing to ingest data to the cloud. Data collected in the thing network called OT

(Operational Technology) are ingested into the cloud in the IT (Information Technology) network by the bridge system (Figure 2). OT part of the network is very wide. In IIoT, fieldbus interfaces can be CAN (Controller Area Network), RS232 (Recommended Standard 232), RS485 (Recommended Standard 485), industrial Ethernet, HART (Highway Addressable Remote Transducer), and other physical layers. The main fieldbus technologies used in industry are presented in the IEC (International Electrotechnical Commission) 61784-1:2019 standard [19]. The 2019 standard groups communication technologies into 19 CPF (Communication Profile Families). The CPFs are divided into 24 profiles: Foundation™ Fieldbus H1 (31.25 kbit/s), HSE (High-speed Ethernet); ControlNet, Ethernet/IP™, DeviceNet®, Profibus DP (Decentralized Peripherals) and PA (Process Automation), P-NET®, WorldFIP® which has 3 profiles, Interbus® which has 3 profiles, CC-Link (Control and Communication Link) which has 3 profiles, HART®, WirelessHART®, SERCOS (acronym for SErial Realtime COmmunications System) I and II, and Mechatrolink-II and M-III. Each CPF profile has described the physical layer, the data link layer, and the application layer of the OSI-ISO (Open System Interconnection—Open Source Initiative) model (Figure 3). The IT network has adopted well the network and transport layer by IPv4 (Internet Protocol version 4) [20], IPv6 (Internet Protocol version 6) [21] and connectionless UDP (User Datagram Protocol) [22], and connection-oriented TCP (Transmission Control Protocol) [23], respectively. Layers above the transport layer are described in the next section. The link layer used by physical and data is Ethernet defined by standard 802.3-2022 [24] approved on 13 May 2022. The physical layer allows for the use of operation over coaxial, twisted pair, or fiber optic cables, or electrical backplanes. The Ethernet standard covers speeds of operation from 1 Mb/s to 400 Gb/s.

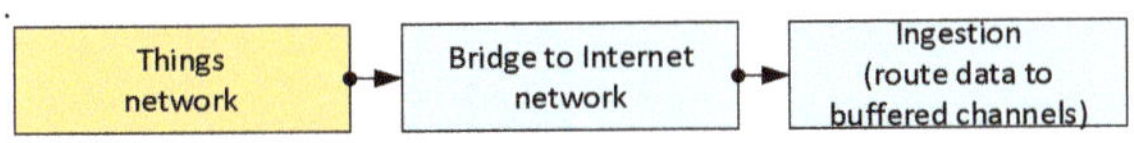

Figure 2. General network structure with network bridge.

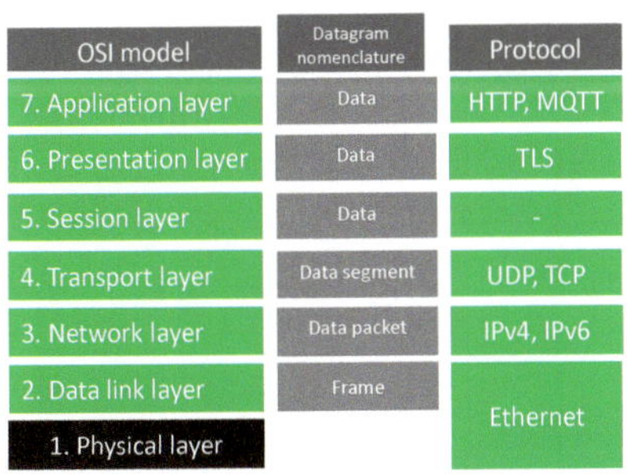

Figure 3. Model OSI-ISO Open System Interconnection—Open Source Initiative.

3. Internet of Things Protocols for Batch and Stream Processing of Fault Diagnosis

The protocols used in the presentation layer in OSI-ISO model (Figure 3) are generally responsible for ensuring the security of transmission in three main aspects: authentication, confidentiality, and integrity of data. The fault diagnosis server side of the channel is always authenticated, and the client (microcontroller) side is optionally authenticated. Confidentiality means that the data sent over the channel after establishment is only visible to the endpoints. Integrity means that data sent over the channel after establishment cannot be modified by attackers without detection. The latest version of TLS (Transport Layer Security) is 1.3 [25] presented in 2018. The layer allows us to use asymmetric cryptography or symmetric cryptography with a pre-shared key. The highest level in the reference model is the application layer, where the HTTP (Hypertext Transfer Protocol) and MQTT (Message Queuing Telemetry Transport) protocols are widely used in IoT applications. The text-based HTTP protocol was widely used in version 1.1 [26] developed in 1997 [27]. HTTP/1.1 can be upgraded to the WebSocket protocol [28], which is useful for established connections to exchange data streams in a bidirectional manner. The next version 2.0 of HTTP was shown in 2015 [29], where the protocol was changed to binary encoding instead of text. The newest version 3.0 of HTTP was introduced in June 2022 [30]. HTTP/3 uses a new transport layer protocol QUIC (Quick UDP Internet Connections) [31] standardized in 2021 instead of TCP. The QUIC transport protocol uses UDP and TLS to provide a stream of binary frames between client and server. Streams can be intended by the client or server as unidirectional or bidirectional. The HTTP/1.1 protocol can be adopted in the batch processing of faults as the request–response principle (Figure 1). The connection is closed by the fault diagnosis server after the response has been sent to the client request. HTTP/1.1 upgraded to WebSocket, HTTP/2, and HTTP/3 are suitable for the stream processing principle (Figure 1) because the connection is not closed after the server sends the response. The HTTP server can be implemented on the microcontroller (thing) or on the remote computer (cloud), see Figure 4.

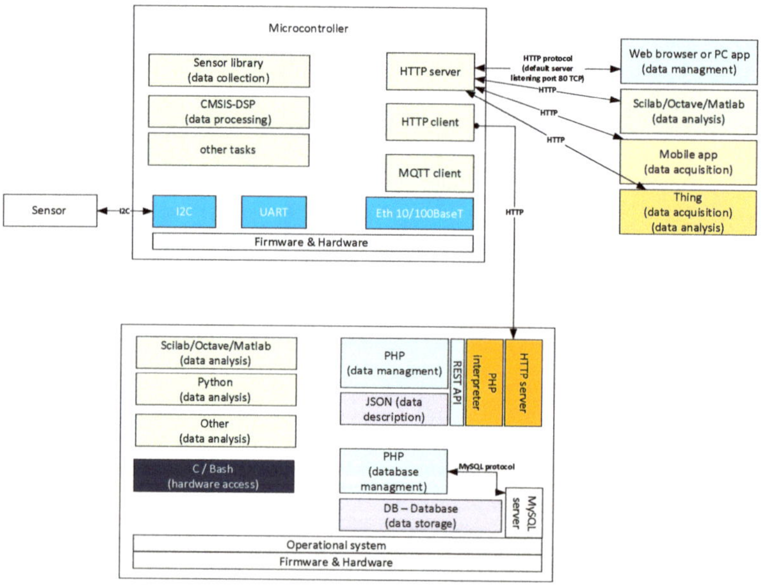

Figure 4. Client–server architecture with HTTP/1.1.

Fault diagnosis as batch processing with HTTP/1.1 request–response header is shown in Figure 5. The microcontroller acts as an HTTP client which sends sensor-collected data in JSON (JavaScript Object Notation) format of 4294 character length by the POST method. The server path is / . . . */predict.php* in the request line. The HTTP server responds *200 OK* in the status line, which means that the data were processed correctly and machine state prediction is given in the response message body in JSON format. Presented response contains predicted class *"prediction":"idle"* and probability *"score"* of each class as array in order: *normal, idle*, and *fault*.

Data from the application layer are encapsulated by lower layer headers. Figure 6 presents the encapsulation of HTTP/1.1 response from server to client. Data from HTTP/1.1 layer are put into the field payload of TCP and the header of TCP needs to be filled. The default TCP port is 80 for the HTTP/1.1 server.

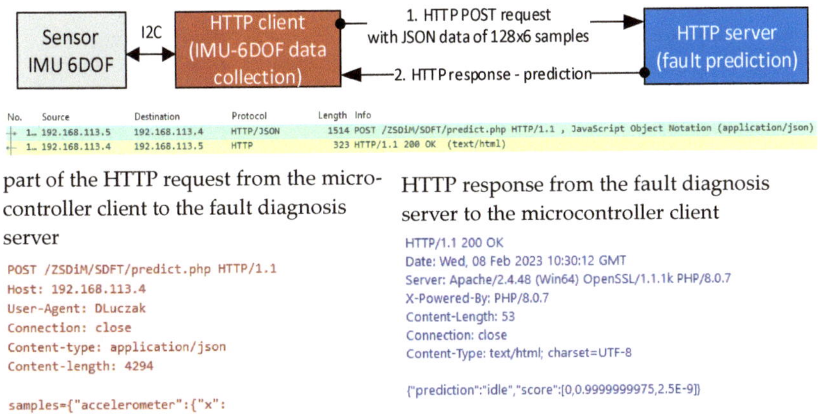

Figure 5. HTTP/1.1 request–response as batch processing of machine state prediction.

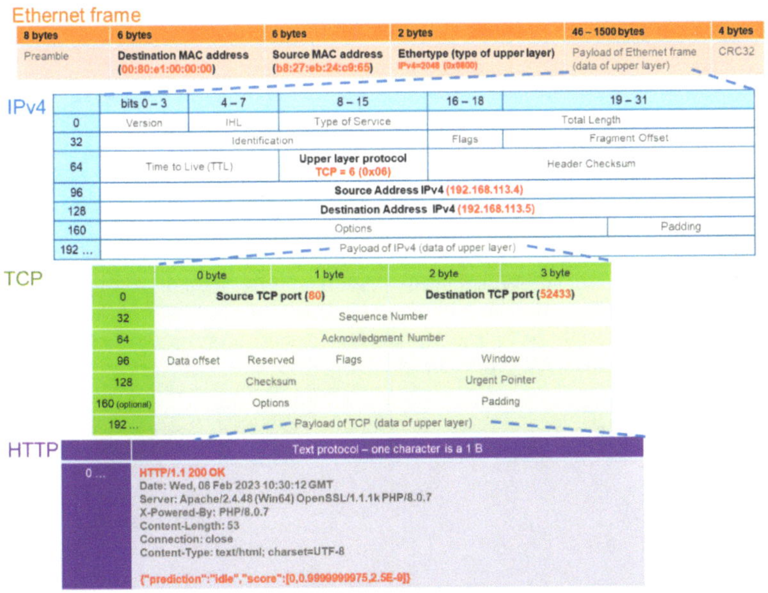

Figure 6. Response from the fault prediction server with HTTP/1.1 encapsulated in the lower layers of the OSI-ISO model.

The data segment is then a payload for IPv4, and the IPv4 header needs to be filled. The IPv4 header field upper-layer protocol is set to 6. The destination address and the source address must be known. The last encapsulation is into an Ethernet frame where the data packet is a payload of the Ethernet frame, and the header of the Ethernet needs to field. The hardware MAC (media access control address) destination address and source address need to be known. If only the IPv4 destination address is known before sending the Ethernet frame, the MAC destination address must be resolved by the ARP (Address Resolution Protocol) [32].

The other application layer protocol adopted for IoT applications is MQTT in version 3.1.1 [33] published in 2015 and version 5.0 [34] updated in 2019. MQTT provides client–server publish/subscribe messaging (Figure 7). The client connects to the server (broker) and subscribes to the topic. Messages published by other clients to the same topic will be pushed to all clients subscribed to this topic by the broker (server). MQTT use TCP at transport layer which establishes a connection as long as the client requires. The default destination TCP port of the broker is 1883.

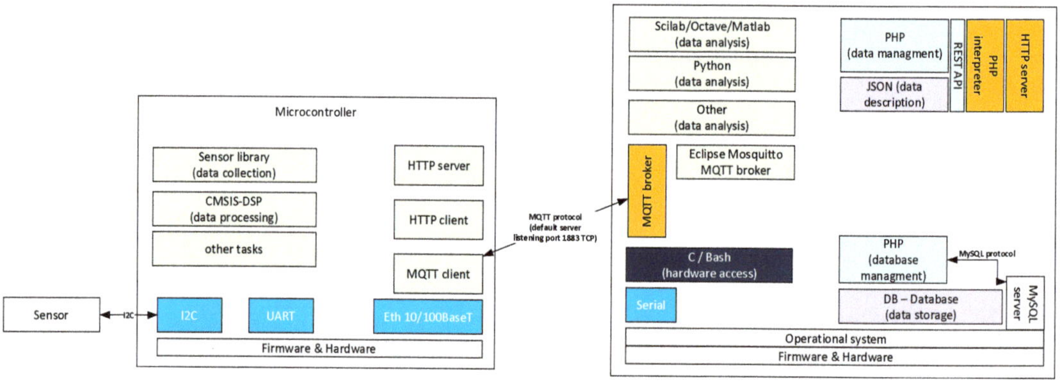

Figure 7. MQTT 3.1 client–server architecture.

The three step TCP handshake from the 52432 client port to the 1883 server port is presented in Figure 8 to establish the connection. After successful subscription, the MQTT microcontroller client send messages to other clients through the broker. The topic can be treated as a buffer name in the producer–consumer architecture. The designed fault diagnosis system consists of two MQTT clients subscribed to topics: *sensor/IMU6DOF/raw* and *sensor/IMU6DOF/predict*. The microcontroller MQTT client writes sensor data to the topic *sensor/IMU6DOF/raw*, and the MATLAB MQTT client retrieves those samples in JSON format. Received samples are transformed into features by digital signal processing methods, which are then evaluated by the classification algorithm. The prediction of classification algorithm is transformed into JSON (e.g., {"prediction":"idle","score":[0,0.9999999955,4.5E-9]}) and sent to *sensor/IMU6DOF/predict* topic.

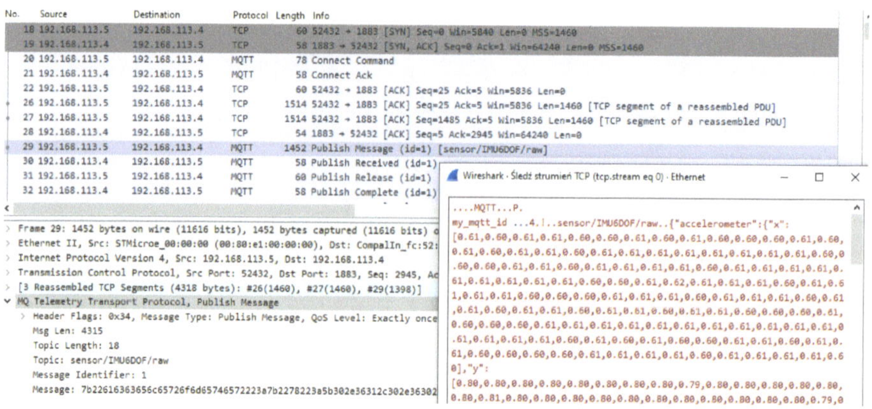

Figure 8. Captured MQTT v3.1 packets captured with TCP handshake for microcontroller MQTT client write sensor data to the topic *sensor/IMU6DOF/raw* as stream processing (micro batch) for machine state prediction.

MQTT can be used in either stream and batch data processing (Figure 1) of a fault diagnosis system. It is important to note that all messages are sent throughout the broker, which leads to central-oriented messaging with routing possibilities. Many clients who subscribed to the same topic received the same data. Therefore, clients can perform different micro services on the same data stream like feature extraction, data storage in a database, or visualization of data stream. This leads to transmission of one-to-many (one-point to multiple points) messages. In contrast to MQTT, WebSocket allows only point-to-point data streaming without the built-in message routing mechanism.

Another protocol used in IoT applications is CoAP (Constrained Application Protocol), defined in 2014 [35]. The protocol is dedicated to 8-bit microcontrollers with limited RAM (random access memory) and ROM (read-only memory). CoAP typical applications include smart energy and building automation. The direct quote from study [35] follows: "The goal of CoAP is not to blindly compress, but rather to realize a subset of REST (representational state transfer) common with HTTP but optimized for M2M (machine-to-machine) applications." Many similarities to HTTP can be found. Similarly to HTTP, the CoAP uses request/response model with GET, PUT, POST, and DELETE methods. In contrast to HTTP, the CoAP uses the UDP protocol in the transport layer (see Figure 3). The use of UDP protocol in transport layer is unreliable which means that it does not handle missing packets, duplicate packets or packets retrieved in a different order caused by a different path of packets in the Internet. Therefore, in 2018 [36], CoAP was designed to use TCP, TLS, and WebSocket transports.

According to the *Content-Type* (see Section 3.1.1.5 in [26]) and *Content-Encoding* (see Section 3.1.2.2 in [26]) part in HTTP 1.1, many different data types are allowed in the message body part. Commonly adopted content types for the IoT application are JSON (JavaScript Object Notation) [37,38], XML (Extensible Markup Language) [39] or raw bytes. The full list of the allowed media types [40] is published by IANA (Internet Assigned Numbers Authority) in [41]. More than 1500 available formats are in the single application section of *Media Types* published on 21 November 2022. All allowed formats are grouped into sections, which are as follows: application, audio, font, example, image, message, model, multipart, text, video. Up-to-date XML media types "application/xml" or "text/xml" were published in 2014 [42] and the up-to-date JSON media type "application/json" was published in 2017 [43].

4. General Structure of Fault Diagnosis and Perspective Maintenance

The fault diagnosis system (Figure 9) detects faults based on changes in features over time. The fault diagnosis system works with the client–server architecture in the IT network [44–50]. The features are extracted from raw measurements and data or pre-processed data. Fault detection means recognition in the change of machine state caused by one or more faults. It can be treated as anomaly state detection, so any state different from normal behavior can be detected. Fault isolation is the recognition of which parts of the machine have faults. Finally, fault identification answers how wide the damages are. The information about the fault can be used by the maintenance team to repair the system. However, if repair is not possible immediately and the system cannot be stopped safely, then FTC (fault tolerant control) is a key component. The system can have hardware redundancy, which will be used if failure occurs, however, it is expensive to duplicate parts and not always technically possible to apply. FTC allows to continue the system usage of the system with performance constrains [51–56].

Fault diagnosis can be made only on data, which is called a data-driven approach. Another option is to prepare a mathematical model of the diagnosed system and analyze the residual data between the model and the system. Today, the investigated system model is called a digital twin [57–64]. Modeling is a very wide area in which static behavior [65–67], dynamic changes and continuous system [68,69], and discrete states [70] of systems can be modeled. Contrary to the analysis of model and real system residuum is the analysis of estimated parameters of system. The fault diagnosis system compares the estimated parameters in a predefined bounded region. The fault is announced if one of the parameters is outside the defined boundaries longer than the chosen amount of time.

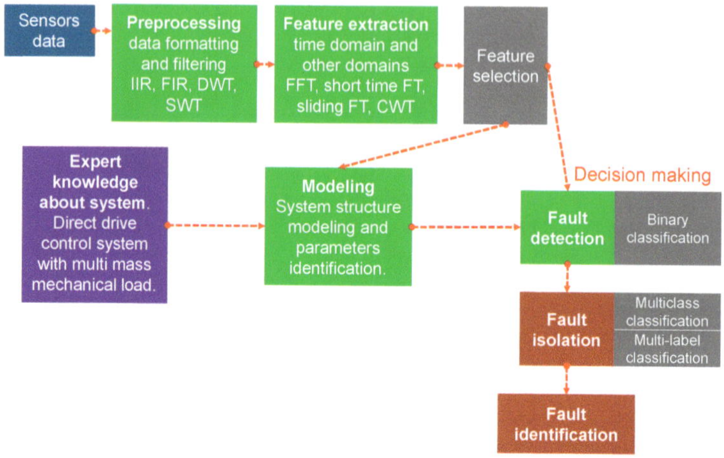

Figure 9. General structure of the fault diagnosis system.

The sensor presented in Figure 9 can be any kind. It can be an additional sensor only for fault diagnosis or a sensor already present in the system used by the control algorithms. The electromechanical machine or power system can be investigated by many different sensors and signals: current [71,72] and voltage [73,74], torque [75,76], angular velocity/position [77,78], linear 3-axis acceleration/speed/position [16,17], Doppler laser vibrometer [79], transmission coefficient and reflexion coefficient of omnidirectional antenna [80], strain/tension [81–84], power consumption [85–88], internal/external temperature at selected points [89,90] or surface temperature by thermal camera [91,92], depending on frequency range: displacement [93], vibrations [15,18,94–96], sound [97–99], sound from several microphones [100] or ultrasound [101,102], vibro-acoustic [103], chemical analysis of lubrication [104,105], chemical analysis by spectral imaging [106–109], camera imaging in human colour spectrum [110–113], and signals to virtual image [114–118].

5. Feature Extraction Methods

The decision made by the fault diagnosis system can be taken based on data in the time domain [119,120] or other domains, for example, the frequency domain [121–124], the time–frequency domain [125–131] or the time-scale domain [132–135].

Analysis with frequency domain is achieved by using one of the FFT (fast Fourier transform) algorithms. These algorithms decompose the analyzed signal into sinusoidal/cosinusoidal components at frequency from 0 Hz up to half of the sampling frequency. The use of the fast Fourier transform algorithm radix-2 [136,137] requires a number of samples equal $N = 2^k, k \in \mathbb{N}$. However, the radix-4 algorithm requires $N = 4^k$ [138,139]. The authors in their publications mostly inform the reader about the use of FFT without mentioning the algorithm used [140–142]. Frequency analysis can be applied to stationary signals in one time window. On the other hand, if knowledge about the sinusoidal/cosinusoidal components and time of their duration in the analyzed signal is required, then time–frequency analysis needs to be applied.

The use of FFT on short time-shifted windows leads to the STFT (short-time Fourier transform) method. STFT changes a one-dimensional time series signal into a two-dimensional time and frequency [143–145]. The STFT parameters are as follows: time window shape (e.g., rectangle, Hamming [146,147], Hanning [148,149]), step size or overlap of next time window that influence time resolution, time window length that influence frequency resolution $f_{res} = f_s/N$, where N—signal length in samples and f_s—sampling frequency. Selection of time window length is a trade-off between good time localization or good frequency localization of the sinusoidal/cosinusoidal components. STFT requires an FFT calculation on each shifted time window. If the step size is equal to 1, it means that FFT will be calculated on $N - 1$ old and 1 new samples, which is not computationally effective. The problem of spectrum update at each new sample was solved using the SDFT method (sliding discrete Fourier transform).

SDFT was presented in 1997 [150] and well described in 2003 (rSDFT, 2003) [151] with an update in 2004 (rSDFT, 2004) [152]. Using study [152], the single spectrum component S_k at frequency k is calculated by $S_k(n) = e^{j2\pi k/N}(S_k(n-1) + x(n) - x(n-N))$, where N—signal length in samples, n—time index, x—signal value. The discrete transfer function of study [152] is $H_{rSDFT,2004}(z) = e^{j2\pi k/N}(1 - z^{-N}) / \left(1 - e^{j2\pi k/N} z^{-1}\right)$, which allows for stability analysis of the algorithm. In 2010, mSDFT (modulated SDFT) was introduced [153]. In study [153], SDFT 2003 with the 2004 update was named rSDFT (recursive SDFT). The mSDFT compared to rSDFT-2004 has reduced the accumulated error and potential instabilities. The stability of algorithms was analyzed and a new guaranteed stable SDFT algorithm (gSDFT) was published in 2015 [154]. The gSDFT algorithm has slightly smaller accumulated errors and better performance compared to mSDFT. Other improvements of mSDFT, called cascade integrator comb (CIC)—SDFT, which allow the embedded usage of the B-spline window function, were presented in 2017 [155]. In the same year, optimal sliding DFT (oSDFT) was presented [156], which reduces the number of multiplications by 73.44%, 64.58%, 82.30%, and 29.17% as compared to FFT, rSDFT, mSDFT, and gSDFT, respectively. The oSDFT accelerates the sliding transform process by 57.02%, 7.51%, 39.77%, and 9.38% compared to the FFT, rSDFT, mSDFT, and gSDFT algorithms, respectively. Misleading may be the name oSDFT (observer-based SDFT) introduced in 2018 [157], which uses the state observer method for digital signal processing purposes.

6. Demonstration of Fault Diagnosis with MQTT Communication

The demonstration shows classification of computer fan work into classes: *idle*, *normal*, or *fault*. The failure is caused by a paper clip added to the fan blade. The prepared demonstration setup (shown in Figure 10) consists of a NUCLEO board with STM32F746ZG microcontroller that handles the IMU-6 DOF (inertial measurement unit with 6 degrees of freedom) MPU6050 sensor. Data are collected synchronously with constant sampling time equal to 5 ms (sampling frequency 200 Hz). The vector of 128 samples from the 3-axis accelerometer and 3-axis gyroscope is then transformed into JSON (JavaScript Object Notation) as the object {"accelerometer":{"x":[],"y":[],"z":[]},"gyroscope":{"x":[],"y":[],"z":[]}}, where the array "[]" contain samples. Data are sent on the topic "sensor/IMU6DOF/raw" by the MQTT client implemented in the microcontroller to the MQTT broker hosted on a personal computer. Eclipse Mosquitto™ was used as an MQTT broker [158]. The data stream was processed in real time by MATLAB R2022b with the MQTT client [159].

The first stage is data collection at predefined class states for several seconds. Data collected in one micro batch in time domain are shown in Figure 11. Data were collected for three states of the fan: (1) *idle* class—fan power is switched off, (2) *normal* class—fan power without paper clip is switched on, and (3) *fault* class—fan power with paper clip is switched on. The time domain signal for the *idle* class has a constant signal with noise, the signal in the *normal* class has small oscillations, and the signal in the *fault* class has significant oscillations compared to previous classes. Data were transformed to the frequency domain by the FFT function in MATLAB, resulting in a complex number vector. The absolute value of the complex numbers is shown in Figure 12. Each class has different frequency features in the range of 0 Hz to 100 Hz. The most significant change in frequency occurs on the Z axis of the gyroscope.

The second stage of preparation was the extraction of features using mSDFT [153]. Each step of mSDFT returns 33 complex numbers at frequencies in the range from 0 Hz to 100 Hz. The algorithm was implemented in MATLAB and the absolute values of the complex number were stored in the files. Each row of the file contains 33 points that were used as a feature vector. The sliding discrete Fourier transform result can be gathered after each new sample. The multiple spectrums calculated at different times are shown in

Figures 13 and 14. The time–frequency analysis showed that for each class, the dominant frequency components are constant and do not vary with time.

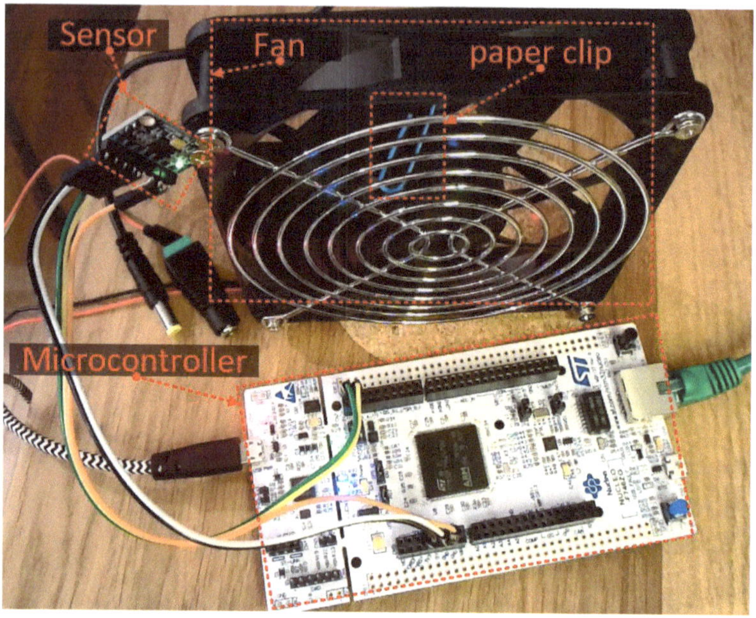

Figure 10. Demonstration system for fan fault diagnosis.

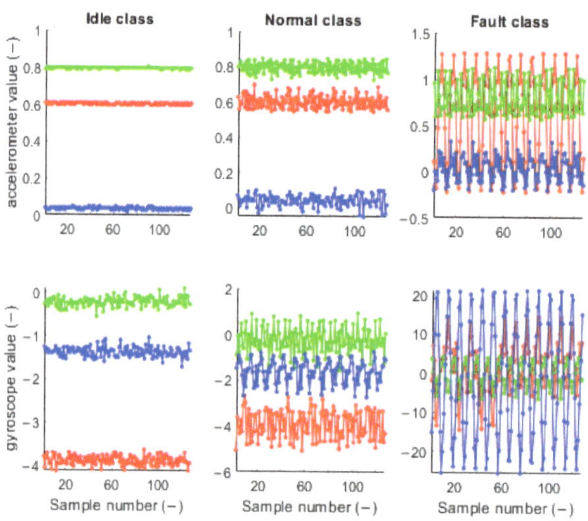

Figure 11. Accelerometer and gyroscope time domain data in 3-axis: red—X, green—Y, and blue—Z.

The third stage of the demonstration requires the preparation of a classification. The features saved in the files for each class were read and loaded into MATLAB Classification Learner [160]. Several classifiers were trained, and the best was used for real-time verification. The dataset has 4200 observation sets randomly divided into 80% training and 20% test dataset. The training process was carried out on 1120 sets of *fault* class, 1120 sets of *idle* class, and 1120 sets of *normal* class. One set consists of 33 frequency features calculated by mSDFT from the gyroscope Z axis signal. The confusion matrix of the trained classifier is shown in Figure 15. The classifier can be used by applying new calculated features to *predictFcn* in MATLAB. The prediction result is sent to the MQTT *sensor/IMU6DOF/predict* topic in the data pipeline (see Figure 1).

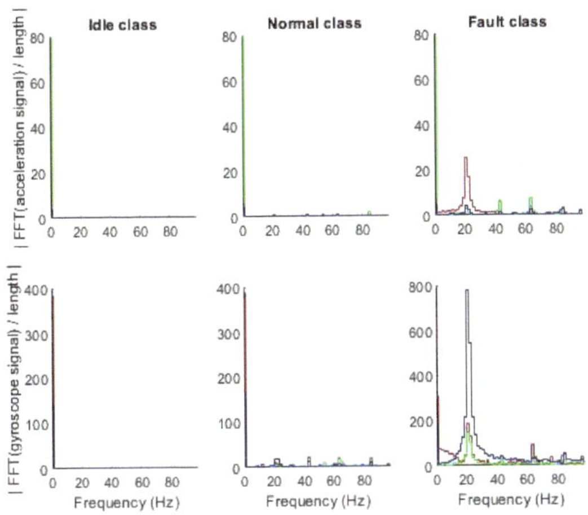

Figure 12. Accelerometer and gyroscope frequency domain data in 3-axis: red—X, green—Y, and blue—Z.

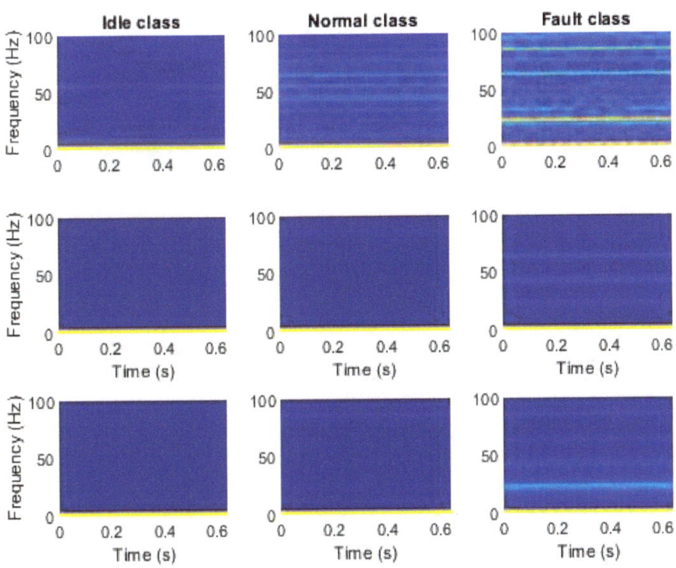

Figure 13. Accelerometer data in the time–frequency domain using mSDFT in the 3-axis: **top** row, Z, **middle** row, Y, **bottom** row—X.

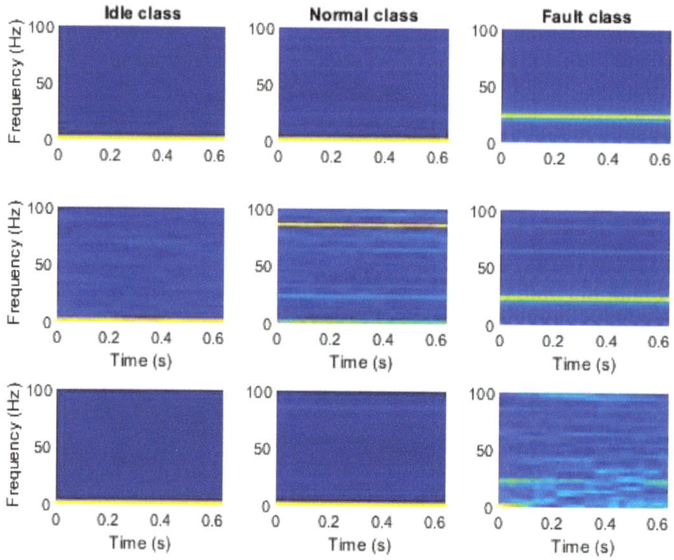

Figure 14. Gyroscope data in the time–frequency domain using mSDFT in 3-axis: **top** row—Z, **middle** row—Y, **bottom** row—X.

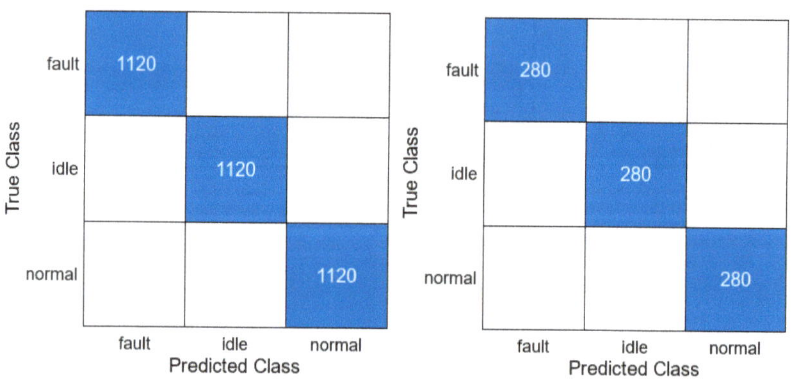

Figure 15. Confusion matrix of the trained classifier (train—**left**; test—**right**).

7. Recognition of a Time–Frequency RGB Image of Vibration

Data from each axis of IMU 6-DOF were transformed into the time–frequency domain by applying STFT (short-time Fourier transform) with 32 samples of window length and 31 samples of overlap (see Figures 16–20). As a result of STFT at 128 time domain samples of single axis, there was a two-dimensional signal of 65 frequencies at 97 time moments. The process was repeated for each axis of accelerometer and gyroscope. Finally, six images for 128 × 6 samples of IMU 6-DOF were obtained in the time–frequency domain for each class: *idle* in Figure 17, *normal* in Figure 18, and *fault* in Figure 19. These six images are combined into one RGB (red, green, and blue) image of size 130 × 97 × 3 in Figure 16; representative RGB images for each class are shown in Figure 20.

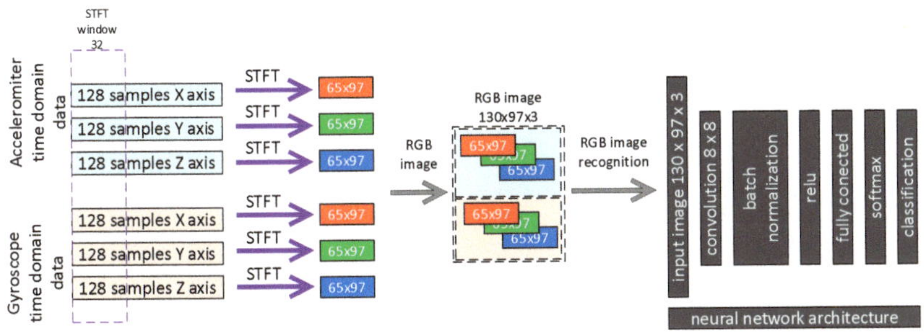

Figure 16. STFT RGB image creation and convolutional neural network architecture.

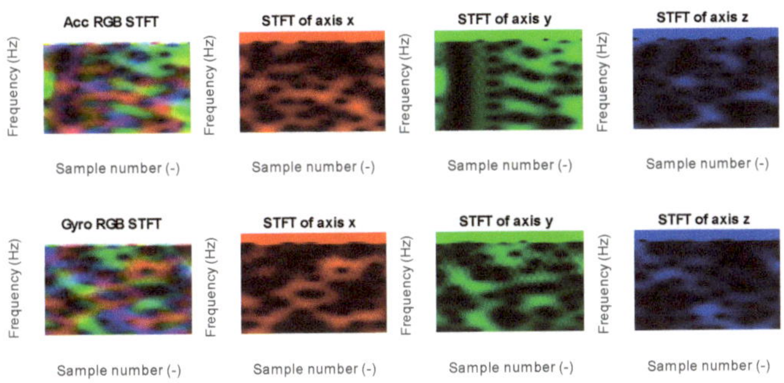

Figure 17. STFT RGB image of idle class for accelerometer x, y, and z axis (**top**) and gyroscope x, y, and z (**bottom**).

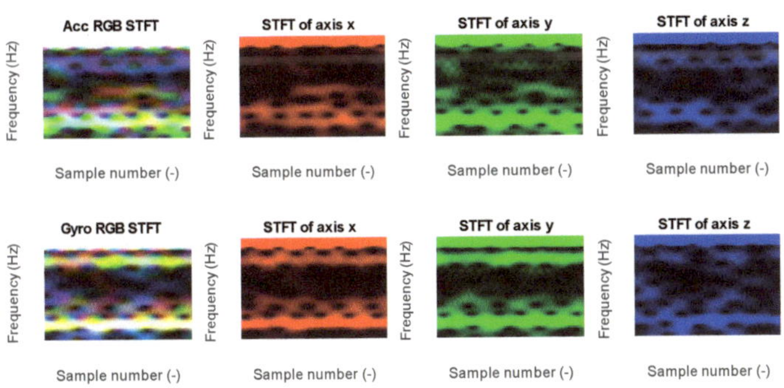

Figure 18. STFT RGB image of normal class for accelerometer x, y, and z axis (**top**) and gyroscope x, y, and z (**bottom**).

Data collected in the time domain for each class were converted into time–frequency RGB images. Together, the dataset has 2670 RGB images divided into classes: *fault* with 890 RGB images, *idle* with 890 RGB images, and *normal* with 890 RGB images. All images, which were divided into training and validation datasets. From the dataset, 80% is the training set (2136 RGB images) and 20% is the testing set (534 RGB images) which were randomly selected. CNN (convolutional neural network) training was performed in MATLAB Deep Learning Toolbox [161] with the support of NVIDIA GPU (graphics processing unit) with CUDA® (Compute Unified Device Architecture). Validation of the trained convolution neural network confirms good classification, shown as a confusion matrix in Figure 21.

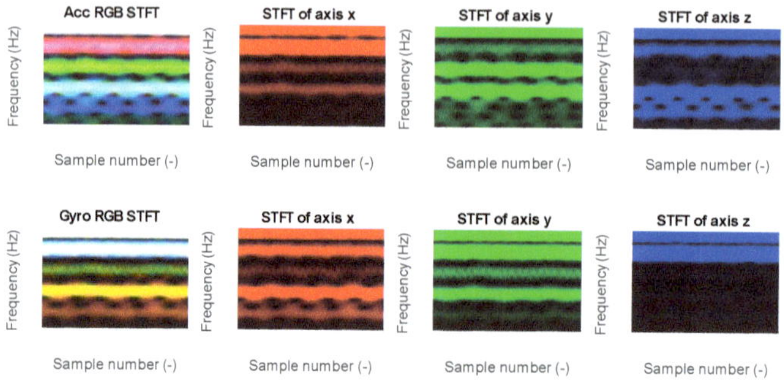

Figure 19. STFT RGB image of the fault class for the accelerometer x, y, and the z axis (**top**) and gyroscope x, y, and z (**bottom**).

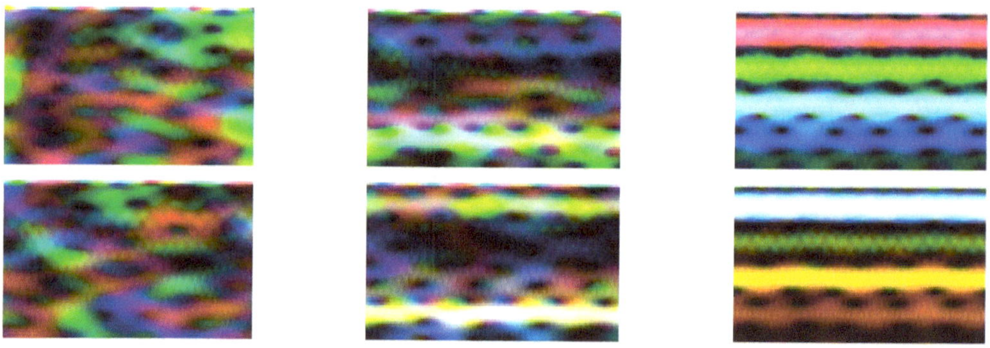

Figure 20. Time–frequency RGB images for class: idle (**left**), normal (**middle**), fault (**right**).

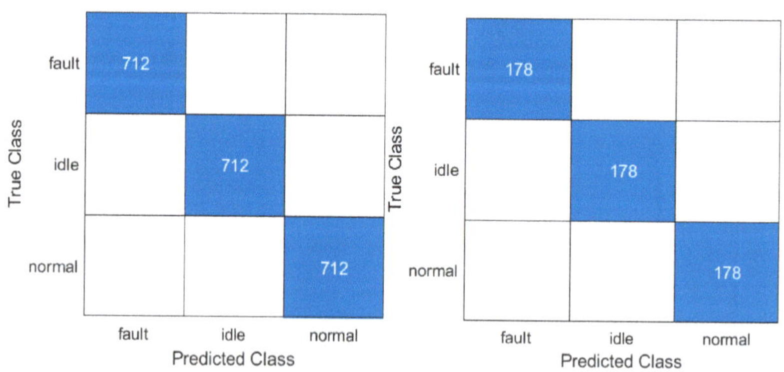

Figure 21. Confusion matrix of the trained (**left**) and tested (**right**) CNN classifier.

8. Discussion

The cloud-based fault diagnosis system is suitable for human-centric and resilient European Industry 5.0 that requires a long lifetime of machines to reduce the amount of e-waste. A large number of industrial protocols used, of which only part are grouped in Communication Profile Families, underline the problem of data collection and ingestion into the cloud system. However, Internet protocols are developed separately without considering the needs of industrial processes. Therefore, the adoption of existing and new Internet protocols for industrial processes is a challenge. Batch and stream processing requires different handling. Batch processing can be applied with the request–response

HTTP/1.1 protocol with a closing connection. However, stream processing requires an established connection for a long time, which can be achieved by WebSocket, HTTP/2.0, HTTP/3.0 or MQTT v3.1 and v5.0. It is worth considering the data exchange problem as a producer–consumer problem with a buffer in the middle. The producer can collect data from many different sensor technologies to investigate the normal and abnormal behavior of an industrial system. An interesting concept is the usage-only energy meter with the energy consumption profile for fault diagnosis. The decision about fault detection is made on the selected signal characteristics. A wide range of diagnostic applications, especially in rotating machines, use frequency analysis or time–frequency analysis. Therefore, short-time Fourier transform (STFT) and sliding discrete Fourier transform algorithms (rSDFT, mSDFT, gSDFT, and oSDFT) were underlined in the article. In batch processing, FFT (radix-2 or radix-4) or STFT can be applied. Real-time stream processing in the frequency domain can be performed using one of the SDFT algorithms.

The manuscript contains a demonstration of the fan fault diagnosis. Data were collected synchronously using a 3-axis accelerometer and a 3-axis gyroscope and microcontroller. The data collected were transformed into a JSON text structure. Those data in JSON are sent to an HTTP server for batch processing and as a result, fault diagnosis server returns a prediction in JSON format. Another investigation was conducted by sending data collection in JSON from the microcontroller client to the MQTT broker in real time as a stream (micro batch) on the *sensor/IMU6DOF/raw* topic for fault prediction. This approach allowed us to use two fault diagnosis predictors (two MQTT clients subscribed to *sensor/IMU6DOF/raw* topic) based on mSDFT and the second based on recognition of the time–frequency RGB image of vibration. For the first predictor, the time series data were transformed into the frequency domain by the mSDFT and evaluated by classification algorithm. The second predictor calculates six STFT and combines the result into an RGB image which is evaluated by the convolutional neural network. The result of both predictors is sent to *sensor/IMU6DOF/predict* topic in JSON format which contains the predicted state and probability for each class. This approach allowed us to classify the state of machine by two different algorithms and can be further extended by other algorithms to ensure voting of prediction from different algorithms to increase software algorithm redundancy.

9. Conclusions

The presented two communication methods allow for rapid prototyping of the fault diagnosis system with cloud connectivity. The HTTP server allows for one-to-one fault diagnosis, where data from one microcontroller are evaluated by one fault diagnosis server endpoint. On the other hand, MQTT allowed for one-to-many fault diagnoses when the same data were evaluated by two fault diagnosis clients with different feature extraction and different classification algorithms. HTTP and MQTT use TCP, however, due to connection closing in HTTP and connection remaining in MQTT, their work differently and both are suitable for fault diagnosis. In the authors' opinion, it is more convenient to use MQTT for rapid prototyping and modularity of a variety of environments of the fault diagnosis system.

The authors have shown two approaches for fault diagnosis. One is based on frequency analysis which was enhanced in comparison to classical ones by usage of sliding Fourier transform (SDFT) instead of fast Fourier transform (FFT). This modification allows us to update the spectrum based on a new sample in time domain and spectrum in the previous step, which work as an IIR (Infinite Impulse Response) filter where filter input is one time domain sample and filter output is one frequency domain sample (complex number). The second proposed approach transforms data from IMU 6-DOF sensor into six spectrograms by STFT, which are used to create an RGB image. This time–frequency RGB image is evaluated by a convolutional neural network for fault diagnosis. The proposed approach proves that CNN can be successfully used in multispectrogram recognition which are organized into RGB images for fault diagnosis without the need for preliminary selection of a vibration axis.

Future research will be focused on increasing the TRL (technology readiness level). Research was carried out at TRL 1 to validate the proof of principles and concept. Further work will be conducted to increase the TRL to higher levels to validate in the laboratory environment the rotary electric machine (electric drive) with fault diagnosis system. At the next TRL, more faults of electric drive will be considered with different natures (electrical and mechanical), e.g., fault in inverter topology, fault in bearing, and shaft cracks.

Author Contributions: Internet of Things, D.Ł.; SDFT and time–frequency RGB image feature extraction, D.Ł.; demonstration of fault diagnosis system, D.Ł.; fan fault selection, D.Ł. and S.B.; electromechanical faults review, D.Ł., S.B. and K.S. All authors have read and agreed to the published version of the manuscript.

Funding: This work has been funded by the National Science Centre, Poland, grant number 2015/17/N/ST7/03796.

Institutional Review Board Statement: Not applicable.

Informed Consent Statement: Not applicable.

Data Availability Statement: Not applicable.

Conflicts of Interest: The authors declare no conflict of interest.

References

1. Kasper, W.C. Power from Trash. *Environ. Sci. Policy Sustain. Dev.* **1974**, *16*, 34–38. [CrossRef]
2. Sullivan, J. Trash or Treasure: Global Trade and the Accumulation of E-Waste in Lagos, Nigeria. *Afr. Today* **2014**, *61*, 89–112. [CrossRef]
3. Kan, Y.; Liu, H.; Yang, Y.; Wei, Y.; Yu, Y.; Qiu, R.; Ouyang, Y. Two Birds with One Stone: The Route from Waste Printed Circuit Board Electronic Trash to Multifunctional Biomimetic Slippery Liquid-Infused Coating. *J. Ind. Eng. Chem.* **2022**, *114*, 233–241. [CrossRef]
4. Ji, X.; Yang, M.; Wan, A.; Yu, S.; Yao, Z. Bioleaching of Typical Electronic Waste—Printed Circuit Boards (WPCBs): A Short Review. *Int. J. Environ. Res. Public Health* **2022**, *19*, 7508. [CrossRef] [PubMed]
5. Rene, E.R.; Sethurajan, M.; Kumar Ponnusamy, V.; Kumar, G.; Bao Dung, T.N.; Brindhadevi, K.; Pugazhendhi, A. Electronic Waste Generation, Recycling and Resource Recovery: Technological Perspectives and Trends. *J. Hazard. Mater.* **2021**, *416*, 125664. [CrossRef] [PubMed]
6. Pan, X.; Wong, C.W.Y.; Li, C. Circular Economy Practices in the Waste Electrical and Electronic Equipment (WEEE) Industry: A Systematic Review and Future Research Agendas. *J. Clean. Prod.* **2022**, *365*, 132671. [CrossRef]
7. Marinello, S.; Gamberini, R. Multi-Criteria Decision Making Approaches Applied to Waste Electrical and Electronic Equipment (WEEE): A Comprehensive Literature Review. *Toxics* **2021**, *9*, 13. [CrossRef]
8. Breque, M.; De Nul, L.; Petridis, A. Directorate-General for Research and Innovation (European Commission). In *Industry 5.0: Towards a Sustainable, Human Centric and Resilient European Industry*; Publications Office of the European Union: Luxembourg, 2021; ISBN 978-92-76-25308-2.
9. Smit, J.; Kreutzer, S.; Moeller, C.; Carlberg, C. *Industry 4.0, European Parliament, Policy Department A: Economic and Scientific Policy*; European Parliament: Strasbourg, France, 2016. [CrossRef]
10. Teixeira, J.E.; Tavares-Lehmann, A.T.C.P. Industry 4.0 in the European Union: Policies and National Strategies. *Technol. Forecast. Soc. Chang.* **2022**, *180*, 121664. [CrossRef]
11. Piechocki, M.; Kraft, M.; Pajchrowski, T.; Aszkowski, P.; Pieczynski, D. Efficient People Counting in Thermal Images: The Benchmark of Resource-Constrained Hardware. *IEEE Access* **2022**, *10*, 124835–124847. [CrossRef]
12. Kraft, M.; Aszkowski, P.; Pieczyński, D.; Fularz, M. Low-Cost Thermal Camera-Based Counting Occupancy Meter Facilitating Energy Saving in Smart Buildings. *Energies* **2021**, *14*, 4542. [CrossRef]
13. Ptak, B.; Pieczyński, D.; Piechocki, M.; Kraft, M. On-Board Crowd Counting and Density Estimation Using Low Altitude Unmanned Aerial Vehicles—Looking beyond Beating the Benchmark. *Remote Sens.* **2022**, *14*, 2288. [CrossRef]
14. Yu, Y.; Samali, B.; Rashidi, M.; Mohammadi, M.; Nguyen, T.N.; Zhang, G. Vision-Based Concrete Crack Detection Using a Hybrid Framework Considering Noise Effect. *J. Build. Eng.* **2022**, *61*, 105246. [CrossRef]
15. Chen, H.-Y.; Lee, C.-H. Vibration Signals Analysis by Explainable Artificial Intelligence (XAI) Approach: Application on Bearing Faults Diagnosis. *IEEE Access* **2020**, *8*, 134246–134256. [CrossRef]
16. Kim, M.S.; Yun, J.P.; Park, P. Deep Learning-Based Explainable Fault Diagnosis Model With an Individually Grouped 1-D Convolution for Three-Axis Vibration Signals. *IEEE Trans. Ind. Inform.* **2022**, *18*, 8807–8817. [CrossRef]
17. Zhang, X.; Zhao, Z.; Wang, Z.; Wang, X. Fault Detection and Identification Method for Quadcopter Based on Airframe Vibration Signals. *Sensors* **2021**, *21*, 581. [CrossRef]

18. Wang, Y.; Yang, M.; Li, Y.; Xu, Z.; Wang, J.; Fang, X. A Multi-Input and Multi-Task Convolutional Neural Network for Fault Diagnosis Based on Bearing Vibration Signal. *IEEE Sens. J.* **2021**, *21*, 10946–10956. [CrossRef]
19. IEC 61784-1:2019 Industrial Communication Networks—Profiles Part 1: Fieldbus Profiles. Available online: https://webstore.iec.ch/publication/59887#additionalinfo (accessed on 28 October 2022).
20. *RFC791*; Internet Protocol. Internet Engineering Task Force: Fremont, CA, USA, 1981.
21. *RFC2460*; Internet Protocol, Version 6 (IPv6) Specification. Internet Engineering Task Force: Fremont, CA, USA, 1998.
22. *RFC768*; User Datagram Protocol. Internet Engineering Task Force: Fremont, CA, USA, 1980.
23. *RFC793*; Transmission Control Protocol. Internet Engineering Task Force: Fremont, CA, USA, 1981.
24. *IEEE Std 802.3-2022 (Revision of IEEE Std 802.3-2018)*; IEEE Standard for Ethernet. IEEE: Piscataway, NJ, USA, 2022. [CrossRef]
25. *RFC8446*; The Transport Layer Security (TLS) Protocol Version 1.3. Internet Engineering Task Force: Fremont, CA, USA, 2018.
26. *RFC2616*; Hypertext Transfer Protocol—HTTP/1.1. Internet Engineering Task Force: Fremont, CA, USA, 1999.
27. *RFC2068*; Hypertext Transfer Protocol—HTTP/1.1. Internet Engineering Task Force: Fremont, CA, USA, 1997.
28. *RFC6455*; The WebSocket Protocol. Internet Engineering Task Force: Fremont, CA, USA, 2011.
29. *RFC7540*; Hypertext Transfer Protocol Version 2 (HTTP/2). Internet Engineering Task Force: Fremont, CA, USA, 2015.
30. *RFC9114*; HTTP/3. Internet Engineering Task Force: Fremont, CA, USA, 2022.
31. *RFC9000*; QUIC: A UDP-Based Multiplexed and Secure Transport. Internet Engineering Task Force: Fremont, CA, USA, 2021.
32. *RFC826*; An Ethernet Address Resolution Protocol: Or Converting Network Protocol Addresses to 48.Bit Ethernet Address for Transmission on Ethernet Hardware. Internet Engineering Task Force: Fremont, CA, USA, 1982.
33. MQTT Version 3.1.1. Available online: http://docs.oasis-open.org/mqtt/mqtt/v3.1.1/errata01/os/mqtt-v3.1.1-errata01-os-complete.html (accessed on 31 October 2022).
34. MQTT Version 5.0. Available online: http://docs.oasis-open.org/mqtt/mqtt/v5.0/mqtt-v5.0.html (accessed on 31 October 2022).
35. *RFC7252*; The Constrained Application Protocol (CoAP). Internet Engineering Task Force: Fremont, CA, USA, 2014.
36. *RFC8323*; CoAP (Constrained Application Protocol) over TCP, TLS, and WebSockets. Internet Engineering Task Force: Fremont, CA, USA, 2018.
37. JSON. Available online: https://www.json.org/json-en.html (accessed on 23 November 2022).
38. ECMA-404; The JSON Data Interchange Syntax; Ecma International, 2nd edition, Geneva, Switzerland, December 2017. Available online: https://www.ecma-international.org/wp-content/uploads/ECMA-404_2nd_edition_december_2017.pdf (accessed on 30 February 2023).
39. Extensible Markup Language (XML). Available online: https://www.w3.org/XML/ (accessed on 23 November 2022).
40. *RFC6838*; Media Type Specifications and Registration Procedures. Internet Engineering Task Force: Fremont, CA, USA, 2013.
41. Media Types. Available online: https://www.iana.org/assignments/media-types/media-types.xhtml (accessed on 23 November 2022).
42. *RFC7303*; XML Media Types. Internet Engineering Task Force: Fremont, CA, USA, 2014.
43. *RFC8259*; The JavaScript Object Notation (JSON) Data Interchange Format. Internet Engineering Task Force: Fremont, CA, USA, 2017.
44. Mendes, M.J.G.C.; Neto, Í.M.M.S.; Calado, J.M.F. Fault Diagnosis System via Internet Applied to a Gantry Robot—A Proposal for Industry 4.0. In Proceedings of the 2018 IEEE International Conference on Autonomous Robot Systems and Competitions (ICARSC), Torres Vedras, Portugal, 25–27 April 2018; pp. 160–166.
45. Jagatheesaperumal, S.K.; Rahouti, M.; Ahmad, K.; Al-Fuqaha, A.; Guizani, M. The Duo of Artificial Intelligence and Big Data for Industry 4.0: Applications, Techniques, Challenges, and Future Research Directions. *IEEE Internet Things J.* **2022**, *9*, 12861–12885. [CrossRef]
46. Natesha, B.V.; Guddeti, R.M.R. Fog-Based Intelligent Machine Malfunction Monitoring System for Industry 4.0. *IEEE Trans. Ind. Inform.* **2021**, *17*, 7923–7932. [CrossRef]
47. Rastogi, V.; Srivastava, S.; Mishra, M.; Thukral, R. Predictive Maintenance for SME in Industry 4.0. In Proceedings of the 2020 Global Smart Industry Conference (GloSIC), Chelyabinsk, Russia, 17–19 November 2020; pp. 382–390.
48. Anil Kumar, T.C.; Bhusal Sharma, N.; Mishra, A.K.; Patil, P.; Sarveswara Reddy, S.; Bhardwaj, R. Fault Identification Model Using IIoT for Industrial Application. *Meas. Sens.* **2022**, *24*, 100526. [CrossRef]
49. Radonjić, M.; Vujnović, S.; Krstić, A.; Zečević, Ž. IoT System for Detecting the Condition of Rotating Machines Based on Acoustic Signals. *Appl. Sci.* **2022**, *12*, 4385. [CrossRef]
50. Behera, S.; Misra, R. Generative Adversarial Networks Based Remaining Useful Life Estimation for IIoT. *Comput. Electr. Eng.* **2021**, *92*, 107195. [CrossRef]
51. Casado-Vara, R.; Vale, Z.; Prieto, J.; Corchado, J.M. Fault-Tolerant Temperature Control Algorithm for IoT Networks in Smart Buildings. *Energies* **2018**, *11*, 3430. [CrossRef]
52. Akay, A.; Lefley, P. Open-Circuit Fault-Tolerant Control of Multi-Phase PM Machines by Compensating the d-q Axes Currents. *Energies* **2021**, *14*, 192. [CrossRef]
53. Patel, H.R.; Shah, V.A. Stable Fault Tolerant Controller Design for Takagi–Sugeno Fuzzy Model-Based Control Systems via Linear Matrix Inequalities: Three Conical Tank Case Study. *Energies* **2019**, *12*, 2221. [CrossRef]
54. Nguyen, N.P.; Hong, S.K. Fault Diagnosis and Fault-Tolerant Control Scheme for Quadcopter UAVs with a Total Loss of Actuator. *Energies* **2019**, *12*, 1139. [CrossRef]

55. Siembab, K.; Zawirski, K. Modified Space Vector Modulation for Fault Tolerant Control of PMSM Drive. In Proceedings of the 2016 IEEE International Power Electronics and Motion Control Conference (PEMC), Varna, Bulgaria, 25–28 September 2016; pp. 1064–1071.
56. Łuczak, D.; Siembab, K. Comparison of Fault Tolerant Control Algorithm Using Space Vector Modulation of PMSM Drive. In Proceedings of the 16th International Conference on Mechatronics—Mechatronika 2014, Brno, Czech Republic, 3–5 December 2014; pp. 24–31.
57. Moghadam, F.K.; Nejad, A.R. Online Condition Monitoring of Floating Wind Turbines Drivetrain by Means of Digital Twin. *Mech. Syst. Signal Process.* **2022**, *162*, 108087. [CrossRef]
58. Leng, J.; Wang, D.; Shen, W.; Li, X.; Liu, Q.; Chen, X. Digital Twins-Based Smart Manufacturing System Design in Industry 4.0: A Review. *J. Manuf. Syst.* **2021**, *60*, 119–137. [CrossRef]
59. Peng, F.; Zheng, L.; Peng, Y.; Fang, C.; Meng, X. Digital Twin for Rolling Bearings: A Review of Current Simulation and PHM Techniques. *Measurement* **2022**, *201*, 111728. [CrossRef]
60. Wang, T.; Cheng, J.; Yang, Y.; Esposito, C.; Snoussi, H.; Tao, F. Adaptive Optimization Method in Digital Twin Conveyor Systems via Range-Inspection Control. *IEEE Trans. Autom. Sci. Eng.* **2022**, *19*, 1296–1304. [CrossRef]
61. Nguyen, T.; Duong, Q.H.; Nguyen, T.V.; Zhu, Y.; Zhou, L. Knowledge Mapping of Digital Twin and Physical Internet in Supply Chain Management: A Systematic Literature Review. *Int. J. Prod. Econ.* **2022**, *244*, 108381. [CrossRef]
62. Teng, S.Y.; Touš, M.; Leong, W.D.; How, B.S.; Lam, H.L.; Máša, V. Recent Advances on Industrial Data-Driven Energy Savings: Digital Twins and Infrastructures. *Renew. Sustain. Energy Rev.* **2021**, *135*, 110208. [CrossRef]
63. Unal, P.; Albayrak, Ö.; Jomâa, M.; Berre, A.J. Data-Driven Artificial Intelligence and Predictive Analytics for the Maintenance of Industrial Machinery with Hybrid and Cognitive Digital Twins. In *Technologies and Applications for Big Data Value*; Curry, E., Auer, S., Berre, A.J., Metzger, A., Perez, M.S., Zillner, S., Eds.; Springer International Publishing: Cham, Switzerland, 2022; pp. 299–319. ISBN 978-3-030-78307-5.
64. Cohen, Y.; Singer, G. A Smart Process Controller Framework for Industry 4.0 Settings. *J. Intell. Manuf.* **2021**, *32*, 1975–1995. [CrossRef]
65. Zhu, G.; Li, Z.; Wu, N. Model-Based Fault Identification of Discrete Event Systems Using Partially Observed Petri Nets. *Automatica* **2018**, *96*, 201–212. [CrossRef]
66. Xu, B.; Yin, X.; Yin, X.; Wang, Y.; Pang, S. Fault Diagnosis of Power Systems Based on Temporal Constrained Fuzzy Petri Nets. *IEEE Access* **2019**, *7*, 101895–101904. [CrossRef]
67. Wang, H.; Wu, Y.; Min, G.; Miao, W. A Graph Neural Network-Based Digital Twin for Network Slicing Management. *IEEE Trans. Ind. Inform.* **2022**, *18*, 1367–1376. [CrossRef]
68. Łuczak, D. Nonlinear Identification with Constraints in Frequency Domain of Electric Direct Drive with Multi-Resonant Mechanical Part. *Energies* **2021**, *14*, 7190. [CrossRef]
69. Luczak, D. Mathematical Model of Multi-Mass Electric Drive System with Flexible Connection. In Proceedings of the 2014 19th International Conference on Methods and Models in Automation and Robotics (MMAR), Miedzyzdroje, Poland, 2–5 September 2014; pp. 590–595.
70. Luczak, D.; Nowopolski, K. Identification of Multi-Mass Mechanical Systems in Electrical Drives. In Proceedings of the 2014 16th International Conference on Mechatronics—Mechatronika (ME), Brno, Czech Republic, 3–5 December 2014; pp. 275–282.
71. Huang, W.; Du, J.; Hua, W.; Lu, W.; Bi, K.; Zhu, Y.; Fan, Q. Current-Based Open-Circuit Fault Diagnosis for PMSM Drives With Model Predictive Control. *IEEE Trans. Power Electron.* **2021**, *36*, 10695–10704. [CrossRef]
72. Wu, C.; Guo, C.; Xie, Z.; Ni, F.; Liu, H. A Signal-Based Fault Detection and Tolerance Control Method of Current Sensor for PMSM Drive. *IEEE Trans. Ind. Electron.* **2018**, *65*, 9646–9657. [CrossRef]
73. Jiang, L.; Deng, Z.; Tang, X.; Hu, L.; Lin, X.; Hu, X. Data-Driven Fault Diagnosis and Thermal Runaway Warning for Battery Packs Using Real-World Vehicle Data. *Energy* **2021**, *234*, 121266. [CrossRef]
74. Chang, C.; Zhou, X.; Jiang, J.; Gao, Y.; Jiang, Y.; Wu, T. Electric Vehicle Battery Pack Micro-Short Circuit Fault Diagnosis Based on Charging Voltage Ranking Evolution. *J. Power Sources* **2022**, *542*, 231733. [CrossRef]
75. Gao, S.; Xu, L.; Zhang, Y.; Pei, Z. Rolling Bearing Fault Diagnosis Based on SSA Optimized Self-Adaptive DBN. *ISA Trans.* **2022**, *128*, 485–502. [CrossRef]
76. Wang, C.-S.; Kao, I.-H.; Perng, J.-W. Fault Diagnosis and Fault Frequency Determination of Permanent Magnet Synchronous Motor Based on Deep Learning. *Sensors* **2021**, *21*, 3608. [CrossRef]
77. Feng, Z.; Gao, A.; Li, K.; Ma, H. Planetary Gearbox Fault Diagnosis via Rotary Encoder Signal Analysis. *Mech. Syst. Signal Process.* **2021**, *149*, 107325. [CrossRef]
78. Ma, J.; Li, C.; Zhang, G. Rolling Bearing Fault Diagnosis Based on Deep Learning and Autoencoder Information Fusion. *Symmetry* **2022**, *14*, 13. [CrossRef]
79. Abbas, S.H.; Jang, J.-K.; Kim, D.-H.; Lee, J.-R. Underwater Vibration Analysis Method for Rotating Propeller Blades Using Laser Doppler Vibrometer. *Opt. Lasers Eng.* **2020**, *132*, 106133. [CrossRef]
80. Dutta, S.; Basu, B.; Talukdar, F.A. Classification of Motor Faults Based on Transmission Coefficient and Reflection Coefficient of Omni-Directional Antenna Using DCNN. *Expert Syst. Appl.* **2022**, *198*, 116832. [CrossRef]
81. Zhang, X.; Niu, H.; Hou, C.; Di, F. An Edge-Filter FBG Interrogation Approach Based on Tunable Fabry-Perot Filter for Strain Measurement of Planetary Gearbox. *Opt. Fiber Technol.* **2020**, *60*, 102379. [CrossRef]

82. Zhang, P.; Lu, D. A Survey of Condition Monitoring and Fault Diagnosis toward Integrated O&M for Wind Turbines. *Energies* **2019**, *12*, 2801. [CrossRef]
83. Wu, J.; Yang, Y.; Wang, P.; Wang, J.; Cheng, J. A Novel Method for Gear Crack Fault Diagnosis Using Improved Analytical-FE and Strain Measurement. *Measurement* **2020**, *163*, 107936. [CrossRef]
84. Fedorko, G.; Molnár, V.; Vasil', M.; Salai, R. Proposal of Digital Twin for Testing and Measuring of Transport Belts for Pipe Conveyors within the Concept Industry 4.0. *Measurement* **2021**, *174*, 108978. [CrossRef]
85. Pu, H.; He, L.; Zhao, C.; Yau, D.K.Y.; Cheng, P.; Chen, J. Fingerprinting Movements of Industrial Robots for Replay Attack Detection. *IEEE Trans. Mob. Comput.* **2022**, *21*, 3629–3643. [CrossRef]
86. Rafati, A.; Shaker, H.R.; Ghahghahzadeh, S. Fault Detection and Efficiency Assessment for HVAC Systems Using Non-Intrusive Load Monitoring: A Review. *Energies* **2022**, *15*, 341. [CrossRef]
87. Sabry, A.H.; Nordin, F.H.; Sabry, A.H.; Abidin Ab Kadir, M.Z. Fault Detection and Diagnosis of Industrial Robot Based on Power Consumption Modeling. *IEEE Trans. Ind. Electron.* **2020**, *67*, 7929–7940. [CrossRef]
88. Sánchez-Sutil, F.; Cano-Ortega, A.; Hernández, J.C. Design and Implementation of a Smart Energy Meter Using a LoRa Network in Real Time. *Electronics* **2021**, *10*, 3152. [CrossRef]
89. Wang, Z.; Tian, B.; Qiao, W.; Qu, L. Real-Time Aging Monitoring for IGBT Modules Using Case Temperature. *IEEE Trans. Ind. Electron.* **2016**, *63*, 1168–1178. [CrossRef]
90. Dhiman, H.S.; Deb, D.; Muyeen, S.M.; Kamwa, I. Wind Turbine Gearbox Anomaly Detection Based on Adaptive Threshold and Twin Support Vector Machines. *IEEE Trans. Energy Convers.* **2021**, *36*, 3462–3469. [CrossRef]
91. Glowacz, A. Fault Diagnosis of Electric Impact Drills Using Thermal Imaging. *Measurement* **2021**, *171*, 108815. [CrossRef]
92. Al-Musawi, A.K.; Anayi, F.; Packianather, M. Three-Phase Induction Motor Fault Detection Based on Thermal Image Segmentation. *Infrared Phys. Technol.* **2020**, *104*, 103140. [CrossRef]
93. Li, Z.; Zhang, Y.; Abu-Siada, A.; Chen, X.; Li, Z.; Xu, Y.; Zhang, L.; Tong, Y. Fault Diagnosis of Transformer Windings Based on Decision Tree and Fully Connected Neural Network. *Energies* **2021**, *14*, 1531. [CrossRef]
94. Rauber, T.W.; da Silva Loca, A.L.; de Assis Boldt, F.; Rodrigues, A.L.; Varejão, F.M. An Experimental Methodology to Evaluate Machine Learning Methods for Fault Diagnosis Based on Vibration Signals. *Expert Syst. Appl.* **2021**, *167*, 114022. [CrossRef]
95. Meyer, A. Vibration Fault Diagnosis in Wind Turbines Based on Automated Feature Learning. *Energies* **2022**, *15*, 1514. [CrossRef]
96. Lee, J.-H.; Pack, J.-H.; Lee, I.-S. Fault Diagnosis of Induction Motor Using Convolutional Neural Network. *Appl. Sci.* **2019**, *9*, 2950. [CrossRef]
97. Cao, Y.; Sun, Y.; Xie, G.; Li, P. A Sound-Based Fault Diagnosis Method for Railway Point Machines Based on Two-Stage Feature Selection Strategy and Ensemble Classifier. *IEEE Trans. Intell. Transp. Syst.* **2022**, *23*, 12074–12083. [CrossRef]
98. Shiri, H.; Wodecki, J.; Ziętek, B.; Zimroz, R. Inspection Robotic UGV Platform and the Procedure for an Acoustic Signal-Based Fault Detection in Belt Conveyor Idler. *Energies* **2021**, *14*, 7646. [CrossRef]
99. Karabacak, Y.E.; Gürsel Özmen, N.; Gümüşel, L. Intelligent Worm Gearbox Fault Diagnosis under Various Working Conditions Using Vibration, Sound and Thermal Features. *Appl. Acoust.* **2022**, *186*, 108463. [CrossRef]
100. Yao, Y.; Wang, H.; Li, S.; Liu, Z.; Gui, G.; Dan, Y.; Hu, J. End-To-End Convolutional Neural Network Model for Gear Fault Diagnosis Based on Sound Signals. *Appl. Sci.* **2018**, *8*, 1584. [CrossRef]
101. Zhang, Z.; Li, J.; Song, Y.; Sun, Y.; Zhang, X.; Hu, Y.; Guo, R.; Han, X. A Novel Ultrasound-Vibration Composite Sensor for Defects Detection of Electrical Equipment. *IEEE Trans. Power Deliv.* **2022**, *37*, 4477–4480. [CrossRef]
102. Wang, W.; Xue, Y.; He, C.; Zhao, Y. Review of the Typical Damage and Damage-Detection Methods of Large Wind Turbine Blades. *Energies* **2022**, *15*, 5672. [CrossRef]
103. Wang, X.; Mao, D.; Li, X. Bearing Fault Diagnosis Based on Vibro-Acoustic Data Fusion and 1D-CNN Network. *Measurement* **2021**, *173*, 108518. [CrossRef]
104. Maruyama, T.; Maeda, M.; Nakano, K. Lubrication Condition Monitoring of Practical Ball Bearings by Electrical Impedance Method. *Tribol. Online* **2019**, *14*, 327–338. [CrossRef]
105. Wakiru, J.M.; Pintelon, L.; Muchiri, P.N.; Chemweno, P.K. A Review on Lubricant Condition Monitoring Information Analysis for Maintenance Decision Support. *Mech. Syst. Signal Process.* **2019**, *118*, 108–132. [CrossRef]
106. Rizk, P.; Younes, R.; Ilinca, A.; Khoder, J. Wind Turbine Ice Detection Using Hyperspectral Imaging. *Remote Sens. Appl. Soc. Environ.* **2022**, *26*, 100711. [CrossRef]
107. Rizk, P.; Younes, R.; Ilinca, A.; Khoder, J. Wind Turbine Blade Defect Detection Using Hyperspectral Imaging. *Remote Sens. Appl. Soc. Environ.* **2021**, *22*, 100522. [CrossRef]
108. Meribout, M. Gas Leak-Detection and Measurement Systems: Prospects and Future Trends. *IEEE Trans. Instrum. Meas.* **2021**, *70*. [CrossRef]
109. Li, Y.; Yu, Q.; Xie, M.; Zhang, Z.; Ma, Z.; Cao, K. Identifying Oil Spill Types Based on Remotely Sensed Reflectance Spectra and Multiple Machine Learning Algorithms. *IEEE J. Sel. Top. Appl. Earth Obs. Remote Sens.* **2021**, *14*, 9071–9078. [CrossRef]
110. Zhou, Q.; Chen, R.; Huang, B.; Liu, C.; Yu, J.; Yu, X. An Automatic Surface Defect Inspection System for Automobiles Using Machine Vision Methods. *Sensors* **2019**, *19*, 644. [CrossRef]
111. Yang, L.; Fan, J.; Liu, Y.; Li, E.; Peng, J.; Liang, Z. A Review on State-of-the-Art Power Line Inspection Techniques. *IEEE Trans. Instrum. Meas.* **2020**, *69*, 9350–9365. [CrossRef]

112. Davari, N.; Akbarizadeh, G.; Mashhour, E. Intelligent Diagnosis of Incipient Fault in Power Distribution Lines Based on Corona Detection in UV-Visible Videos. *IEEE Trans. Power Deliv.* **2021**, *36*, 3640–3648. [CrossRef]
113. Kim, S.; Kim, D.; Jeong, S.; Ham, J.-W.; Lee, J.-K.; Oh, K.-Y. Fault Diagnosis of Power Transmission Lines Using a UAV-Mounted Smart Inspection System. *IEEE Access* **2020**, *8*, 149999–150009. [CrossRef]
114. Ullah, Z.; Lodhi, B.A.; Hur, J. Detection and Identification of Demagnetization and Bearing Faults in PMSM Using Transfer Learning-Based VGG. *Energies* **2020**, *13*, 3834. [CrossRef]
115. Long, H.; Xu, S.; Gu, W. An Abnormal Wind Turbine Data Cleaning Algorithm Based on Color Space Conversion and Image Feature Detection. *Appl. Energy* **2022**, *311*, 118594. [CrossRef]
116. Kreutz, M.; Alla, A.A.; Eisenstadt, A.; Freitag, M.; Thoben, K.-D. Ice Detection on Rotor Blades of Wind Turbines Using RGB Images and Convolutional Neural Networks. *Procedia CIRP* **2020**, *93*, 1292–1297. [CrossRef]
117. Xie, T.; Huang, X.; Choi, S.-K. Intelligent Mechanical Fault Diagnosis Using Multisensor Fusion and Convolution Neural Network. *IEEE Trans. Ind. Inform.* **2022**, *18*, 3213–3223. [CrossRef]
118. Łuczak, D.; Brock, S.; Siembab, K. Fault Detection and Localisation of a Three-Phase Inverter with Permanent Magnet Synchronous Motor Load Using a Convolutional Neural Network. *Actuators* **2023**, *12*, 125. [CrossRef]
119. Liang, M.; Cao, P.; Tang, J. Rolling Bearing Fault Diagnosis Based on Feature Fusion with Parallel Convolutional Neural Network. *Int. J. Adv. Manuf. Technol.* **2021**, *112*, 819–831. [CrossRef]
120. Toma, R.N.; Kim, J.-M. Induction Motor Bearing Fault Diagnosis Using Statistical Time Domain Features and Hypertuning of Classifiers. In *Advances in Computer Science and Ubiquitous Computing*; Park, J.J., Fong, S.J., Pan, Y., Sung, Y., Eds.; Springer: Singapore, 2021; pp. 259–265.
121. Bi, X.; Cao, S.; Zhang, D. Diesel Engine Valve Clearance Fault Diagnosis Based on Improved Variational Mode Decomposition and Bispectrum. *Energies* **2019**, *12*, 661. [CrossRef]
122. Wang, J.; Li, S.; Xin, Y.; An, Z. Gear Fault Intelligent Diagnosis Based on Frequency-Domain Feature Extraction. *J. Vib. Eng. Technol.* **2019**, *7*, 159–166. [CrossRef]
123. Li, H.; Feng, G.; Zhen, D.; Gu, F.; Ball, A.D. A Normalized Frequency-Domain Energy Operator for Broken Rotor Bar Fault Diagnosis. *IEEE Trans. Instrum. Meas.* **2021**, *70*, 3009011. [CrossRef]
124. Luczak, D. Frequency Analysis of Mechanical Resonance in Direct Drive. In Proceedings of the 2012 12th IEEE International Workshop on Advanced Motion Control (AMC), Sarajevo, Bosnia and Herzegovina, 25–27 March 2012; pp. 1–5.
125. Ramteke, S.M.; Chelladurai, H.; Amarnath, M. Diagnosis and Classification of Diesel Engine Components Faults Using Time–Frequency and Machine Learning Approach. *J. Vib. Eng. Technol.* **2022**, *10*, 175–192. [CrossRef]
126. Lim, H.; Kwon, G.-Y.; Shin, Y.-J. Fault Detection and Localization of Shielded Cable via Optimal Detection of Time–Frequency-Domain Reflectometry. *IEEE Trans. Instrum. Meas.* **2021**, *70*, 3092514. [CrossRef]
127. Iglesias-Martínez, M.E.; Antonino-Daviu, J.A.; Fernández de Córdoba, P.; Conejero, J.A. Rotor Fault Detection in Induction Motors Based on Time-Frequency Analysis Using the Bispectrum and the Autocovariance of Stray Flux Signals. *Energies* **2019**, *12*, 597. [CrossRef]
128. Luczak, D. Spectral Analysis of Digital Filter Tuned for Mechanical Resonant Frequency Reduction in Multi-Mass Mechanical Systems in Electrical Direct Drive. *Int. J. Simul. Syst. Sci. Technol.* **2020**, *17*, 11.1–11.8. [CrossRef]
129. Strack, J.L.; Carugati, I.; Orallo, C.M.; Maestri, S.O.; Donato, P.G.; Funes, M.A. Three-Phase Voltage Events Classification Algorithm Based on an Adaptive Threshold. *Electr. Power Syst. Res.* **2019**, *172*, 167–176. [CrossRef]
130. Yoon, Y.; Brahma, A. Air–Fuel Ratio Imbalance Diagnostic of Spark-Ignited Engines With Modulated Sliding Discrete Fourier Transform. *J. Dyn. Syst. Meas. Control* **2020**, *142*, 081003. [CrossRef]
131. Peña-Alzola, R.; Sztykiel, M.; Jones, C.E.; Norman, P.J.; Moore, G.; Pou, J.; Burt, G.M. First-Fault Detection in DC Distribution With IT Grounding Based on Sliding Discrete Fourier-Transform. *IEEE Trans. Power Electron.* **2021**, *36*, 3649–3654. [CrossRef]
132. Li, X.; Bi, F.; Zhang, L.; Yang, X.; Zhang, G. An Engine Fault Detection Method Based on the Deep Echo State Network and Improved Multi-Verse Optimizer. *Energies* **2022**, *15*, 1205. [CrossRef]
133. Gu, J.; Peng, Y.; Lu, H.; Chang, X.; Chen, G. A Novel Fault Diagnosis Method of Rotating Machinery via VMD, CWT and Improved CNN. *Measurement* **2022**, *200*, 111635. [CrossRef]
134. Łuczak, D. Mechanical Resonance Frequensies Identyfication of Direct Drive Using Wavelet Analysis. In Proceedings of the 2012 17th International Conference on Methods & Models in Automation & Robotics (MMAR), Miedzyzdroje, Poland, 27–30 August 2012; pp. 29–32.
135. Łuczak, D. Mechanical Vibrations Analysis in Direct Drive Using CWT with Complex Morlet Wavelet. *Power Electron. Drives* **2023**, *8*, 65–73. [CrossRef]
136. Singleton, R. A Method for Computing the Fast Fourier Transform with Auxiliary Memory and Limited High-Speed Storage. *IEEE Trans. Audio Electroacoust.* **1967**, *15*, 91–98. [CrossRef]
137. Cooley, J.W.; Tukey, J.W. An Algorithm for the Machine Calculation of Complex Fourier Series. *Math. Comput.* **1965**, *19*, 297–301. [CrossRef]
138. Corinthios, M.J.; Smith, K.C.; Yen, J.L. A Parallel Radix-4 Fast Fourier Transform Computer. *IEEE Trans. Comput.* **1975**, *C-24*, 80–92. [CrossRef]
139. Corinthios, M.J. A Fast Fourier Transform for High-Speed Signal Processing. *IEEE Trans. Comput.* **1971**, *C-20*, 843–846. [CrossRef]

140. Soares, M.N.; Mollet, Y.; Kinnaert, M.; Gyselinck, J.; Helsen, J. Multiphysical Time- and Frequency-Domain Fault Detection and Isolation Technique for Power-Electronic Converters in DFIG Wind Turbines. *IEEE Trans. Power Electron.* **2021**, *36*, 3793–3802. [CrossRef]
141. Han, T.; Ding, L.; Qi, D.; Li, C.; Fu, Z.; Chen, W. Compound Faults Diagnosis Method for Wind Turbine Mainshaft Bearing with Teager and Second-Order Stochastic Resonance. *Measurement* **2022**, *202*, 111931. [CrossRef]
142. Li, H.; Fan, B.; Jia, R.; Zhai, F.; Bai, L.; Luo, X. Research on Multi-Domain Fault Diagnosis of Gearbox of Wind Turbine Based on Adaptive Variational Mode Decomposition and Extreme Learning Machine Algorithms. *Energies* **2020**, *13*, 1375. [CrossRef]
143. Liu, D.; Cheng, W.; Wen, W. Rolling Bearing Fault Diagnosis via STFT and Improved Instantaneous Frequency Estimation Method. *Procedia Manuf.* **2020**, *49*, 166–172. [CrossRef]
144. Wang, Z.; Yang, J.; Li, H.; Zhen, D.; Xu, Y.; Gu, F. Fault Identification of Broken Rotor Bars in Induction Motors Using an Improved Cyclic Modulation Spectral Analysis. *Energies* **2019**, *12*, 3279. [CrossRef]
145. Grcić, I.; Pandžić, H.; Novosel, D. Fault Detection in DC Microgrids Using Short-Time Fourier Transform. *Energies* **2021**, *14*, 277. [CrossRef]
146. Fu, X.; Yang, K.; Liu, M.; Xing, T.; Wu, C. LightFD: Real-Time Fault Diagnosis with Edge Intelligence for Power Transformers. *Sensors* **2022**, *22*, 5296. [CrossRef]
147. Liu, J.; Hao, R.; Zhang, T.; Wang, X. Vibration Fault Diagnosis Based on Stochastic Configuration Neural Networks. *Neurocomputing* **2021**, *434*, 98–125. [CrossRef]
148. Miao, Y.; Zhang, B.; Li, C.; Lin, J.; Zhang, D. Feature Mode Decomposition: New Decomposition Theory for Rotating Machinery Fault Diagnosis. *IEEE Trans. Ind. Electron.* **2023**, *70*, 1949–1960. [CrossRef]
149. Shi, H.; Li, Y.; Bai, X.; Zhang, K. Sound-Aided Fault Feature Extraction Method for Rolling Bearings Based on Stochastic Resonance and Time-Domain Index Fusion. *Appl. Acoust.* **2022**, *189*, 108611. [CrossRef]
150. Douglas, S.C.; Soh, J.K. A Numerically-Stable Sliding-Window Estimator and Its Application to Adaptive Filters. In Proceedings of the Conference Record of the Thirty-First Asilomar Conference on Signals, Systems and Computers (Cat. No.97CB36136), Pacific Grove, CA, USA,, 2–5 November 1997; Volume 1, pp. 111–115.
151. Jacobsen, E.; Lyons, R. The Sliding DFT. *IEEE Signal Process. Mag.* **2003**, *20*, 74–80. [CrossRef]
152. Jacobsen, E.; Lyons, R. An Update to the Sliding DFT. *IEEE Signal Process. Mag.* **2004**, *21*, 110–111. [CrossRef]
153. Duda, K. Accurate, Guaranteed Stable, Sliding Discrete Fourier Transform [DSP Tips & Tricks]. *IEEE Signal Process. Mag.* **2010**, *27*, 124–127. [CrossRef]
154. Park, C.-S. Fast, Accurate, and Guaranteed Stable Sliding Discrete Fourier Transform [Sp Tips&Tricks]. *IEEE Signal Process. Mag.* **2015**, *32*, 145–156. [CrossRef]
155. Gudovskiy, D.A.; Chu, L. An Accurate and Stable Sliding DFT Computed by a Modified CIC Filter [Tips & Tricks]. *IEEE Signal Process. Mag.* **2017**, *34*, 89–93. [CrossRef]
156. Park, C.-S. Guaranteed-Stable Sliding DFT Algorithm With Minimal Computational Requirements. *IEEE Trans. Signal Process.* **2017**, *65*, 5281–5288. [CrossRef]
157. Kollar, Z.; Plesznik, F.; Trumpf, S. Observer-Based Recursive Sliding Discrete Fourier Transform [Tips & Tricks]. *IEEE Signal Process. Mag.* **2018**, *35*, 100–106. [CrossRef]
158. Eclipse Mosquitto. Available online: https://mosquitto.org/ (accessed on 22 November 2022).
159. Create MQTT Client Connected to Broker—MATLAB. Available online: https://www.mathworks.com/help/icomm/ug/icomm.mqtt.client.html (accessed on 22 November 2022).
160. Train Models to Classify Data Using Supervised Machine Learning—MATLAB. Available online: https://www.mathworks.com/help/stats/classificationlearner-app.html (accessed on 23 November 2022).
161. Deep Learning Toolbox. Available online: https://www.mathworks.com/products/deep-learning.html (accessed on 7 February 2023).

sensors

Article

SVR-Net: A Sparse Voxelized Recurrent Network for Robust Monocular SLAM with Direct TSDF Mapping

Rongling Lang †, Ya Fan † and Qing Chang *

School of Electronics and Information Engineering, Beihang University, Beijing 100191, China
* Correspondence: changqing@buaa.edu.cn
† These authors contributed equally to this work.

Abstract: Simultaneous localization and mapping (SLAM) plays a fundamental role in downstream tasks including navigation and planning. However, monocular visual SLAM faces challenges in robust pose estimation and map construction. This study proposes a monocular SLAM system based on a sparse voxelized recurrent network, SVR-Net. It extracts voxel features from a pair of frames for correlation and recursively matches them to estimate pose and dense map. The sparse voxelized structure is designed to reduce memory occupation of voxel features. Meanwhile, gated recurrent units are incorporated to iteratively search for optimal matches on correlation maps, thereby enhancing the robustness of the system. Additionally, Gauss–Newton updates are embedded in iterations to impose geometrical constraints, which ensure accurate pose estimation. After end-to-end training on ScanNet, SVR-Net is evaluated on TUM-RGBD and successfully estimates poses on all nine scenes, while traditional ORB-SLAM fails on most of them. Furthermore, absolute trajectory error (ATE) results demonstrate that the tracking accuracy is comparable to that of DeepV2D. Unlike most previous monocular SLAM systems, SVR-Net directly estimates dense TSDF maps suitable for downstream tasks with high efficiency of data exploitation. This study contributes to the development of robust monocular visual SLAM systems and direct TSDF mapping.

Keywords: monocular SLAM; deep learning; TSDF mapping

1. Introduction

A simultaneous localization and mapping (SLAM) system localizes the agent in the environment and reconstructs the environment. It is fundamental to other tasks, such as navigation and planning, etc. In visual SLAM, the agent senses the environment using a monocular, stereo or RGBD camera, and monocular SLAM is the most challenging due to difficulties in robust pose estimation and map construction. Moreover, the demand for dense maps in many applications presents a new set of challenges that must be addressed by existing monocular SLAM methods.

Typically, the pose of the camera is represented by a quantity in the special Euclidean group SE(3), which includes components of translation and rotation. The map in SLAM is often represented by a depth map, which contains the distance between each environmental point corresponding to each pixel in the image and the camera. Traditional SLAM methods are classed as either direct or indirect, according to the way they use information. Direct methods solve for poses and depths through photometric error. Photometric error utilizes abundant information such as lines and intensity variations, but complex patterns of images can cause local minima when optimizing photometric error. The optimization process is also unstable when an image is noisy. Indirect approaches exploit information from feature points. Feature points locate sparse local regions of images and compute descriptors which are similar from different perspectives. Reprojection error is optimized to minimize the distance of projection and matched location of feature points. For example, ORB-SLAM [1] uses a sparse ORB feature for tracking and mapping, which can achieve high performance when feature points are correctly matched. However, in regions with few or repetitive textures, feature points are often erroneously detected or matched, which also leads to a robustness problem.

With the rapid growth of deep learning, neural networks have been successfully applied in image processing due to their powerful feature extraction capability [2,3]. They have also been adopted in SLAM to tackle the robustness problem. Some work has studied replacing submodules of SLAM with networks. In high-level feature extraction, using networks can provide an advantage in computing feature points [4–8] or finding good correspondences [9,10]. Learning-based outlier rejection is explored to produce robust matches [11–13]. Other work has investigated performing end-to-end learned SLAM [14–17]. After training on a wide variety of scenes with ground-truth poses, the networks can utilize prior information to avoid catastrophic errors.

However, previous end-to-end SLAM networks are not trained on applicable dense maps, which limits their availability. Networks such as TartanVo [16] and DROID-SLAM [15] only provide pose results. Other networks including DeepV2D [14] use depth maps or point clouds as map representation, which is not enough for downstream tasks. Post-processing and fusion could directly use depth maps or point clouds to produce maps suitable for downstream tasks, but additional information extracted from the networks would be omitted.

Instead of depth maps or point clouds, denser representation of environment geometry is considered, which is typically required to enable downstream tasks. To reconstruct a dense scene from a set of calibrated views, traditional multi-view stereo techniques rely on local depth map computation and global depth map fusion. The Truncated Signed Distance Function (TSDF) is a fusion format that contains distance information useable for navigation and planning. TSDF assigns each 3D point a signed distance from the nearest environment surface. The sign is decided by whether the point is hidden by the surface, and distances with absolute values greater than 1 are truncated. The result of two-stage reconstruction is prone to being either layered or scattered due to depth inconsistency [18]. Thus, deep neural architectures regressing TSDF for direct 3D reconstruction were designed in recent studies [18–20]. However, these methods require the camera poses as prior information, and they cannot guarantee the quality of maps when pose estimation is not accurate. In monocular SLAM, map construction is more challenging because there are more uncertainties in pose estimation.

In this work, we propose a monocular SLAM system with direct TSDF mapping based on our sparse voxelized recurrent network, SVR-Net. For a sequence of monocular image frames, SVR-Net outputs both poses and a TSDF map. The network takes an end-to-end pipeline that extracts features for correlation and recursively matches them, which is shown to have STOA localization performance in DROID-SLAM. Moreover, to enable TSDF reconstruction in the SLAM pipeline, a novel matching network is designed using sparse 3D convolution on voxels, combining the recursive matching with the TSDF mapping branch. This study makes a contribution towards the advancement of robust monocular visual SLAM systems and direct TSDF mapping.

The main features of the proposed method are summarized as follows:

(1) Robust monocular SLAM. SVR-Net's semantic encoder encodes information such as the scale of scenes to guide matching, which helps monocular pose estimation. SVR-Net utilizes correlation operations to reduce both the size of features and dependence on a specific scene's semantic information, which avoids overfitting. After end-to-end training on ScanNet [21] with ground truth poses and maps, SVR-Net successfully estimates poses for all nine scenes of the challenging TUM-RGBD [22] benchmark, whereas ORB-SLAM fails for most of them.

(2) Accurate pose estimation. Iterative updates are carried out in a recurrent network to search for the optimal match. Gauss–Newton updates are embedded in iterations to impose geometrical constraints, which improves the accuracy of pose estimation.

Experimental results using the TUM-RGBD benchmark show that the pose accuracy of SVR-Net is comparable to that of DeepV2D.

(3) Direct TSDF mapping. SVR-Net directly regresses the TSDF values and occupancy confidences of given voxels. Unlike previous monocular SLAM systems that produce depth maps or sparse 3D points, SVR-Net produces dense TSDF maps suitable for downstream tasks including navigation and planning. Compared with TSDF reconstruction methods that take depth maps as intermediate representation, direct TSDF mapping avoids depth inconsistency. Moreover, SVR-Net's direct TSDF mapping is more data-efficient because both pose and map are estimated using the same features.

2. Related Works

Monocular SLAM uses monocular visual information to locate the carrier and model the environment. Traditional methods are classified as either direct or indirect methods according to the way information is used [23]. Indirect approaches first detect and match key points between two frames, then predict poses and 3D points by minimizing the reprojection error [1,24]. Direct approaches operate directly on pixel intensities, and they estimate poses and depths through photometric error [23,25,26]. However, both direct and indirect methods are vulnerable to outliers.

Deep learning has been proposed to improve the robustness of SLAM systems. Some work has investigated replacing hand-crafted with learned features [4–8], using neural 3D representations [27–33], and combining learned energy terms with classical optimization backends [34,35]. In other work, researchers have tried to produce end-to-end learning for SLAM or VO systems [14–17]. Among these methods, DROID-SLAM consists of end-to-end recurrent iterative updates of camera pose and pixel-wise depth. It achieves large improvements in accuracy and substantially reduces catastrophic failures. Sparse voxel-based networks have been applied for the efficient processing of point clouds and voxels [36–38]. SVR-Net contains an iterative pipeline that is similar to DROID-SLAM, but it estimates TSDFs, instead of depth values, with a sparse voxelized structure for direct mapping.

The implicit representations of surface geometry through SDF are introduced in [39]. The efficacy of TSDFs in fusing high-rate, noisy depth data from consumer-grade depth cameras, which was first demonstrated by Newcombe et al. [40], has led to a rapid rise in their popularity. Furthermore, this representation has led to their recent adoption on robotic platforms [41–43], where the distance field has shown additional utility for optimization-based motion planning [44,45].

Recent studies have shown the advantage of deep networks in TSDF mapping, compared with traditional multi-view methods. NeuralRecon directly reconstructs local surfaces, represented as sparse TSDF volumes for each video fragment sequentially, by a neural network [18]. VoRTX is occlusion-aware, leveraging the transformer architecture to predict an initial, projective scene geometry estimate [20]. VolumeFusion replicates the traditional two-stage framework with deep neural networks, improving both the interpretability and the accuracy of the results [46]. The suggested network is closely related to TSDF mapping networks, but does not need camera poses to be provided. Table 1 shows a comparison of existing learning-based localization and mapping methods.

Table 1. Comparison of existing learning-based approaches.

References	Year	Method	Monocular Input	Pose Estimation	TSDF Mapping
[28]	2020	DeepFactors	Yes	Yes	No
[35]	2020	D3VO	Yes	Yes	No
[14]	2020	DeepV2D	Yes	Yes	No
[32]	2021	IMAP	No	Yes	No
[15]	2021	DROID-SLAM	Yes	Yes	No
[18]	2021	NeuralRecon	Yes	No	Yes
[20]	2021	VoRTX	Yes	No	Yes
[46]	2021	VolumeFusion	Yes	No	Yes
[33]	2022	NICE-SLAM	No	Yes	Yes

3. Method

The SLAM system employs a coarse-to-fine strategy to achieve tracking and global dense mapping, as shown in Figure 1. At the first stage, the raw pose and local map of a pair of input frames are estimated using SVR-Net. The map is represented by sparse voxels with TSDF values. Then, the map is fused with the first-stage global map. At the second stage, the voxels are up-sampled, halving their intervals. Subsequently, the same network is utilized to refine the pose and map. The resulting fine local map is fused with the second-stage global map, and SVR-Net proceeds to track the next pair of frames.

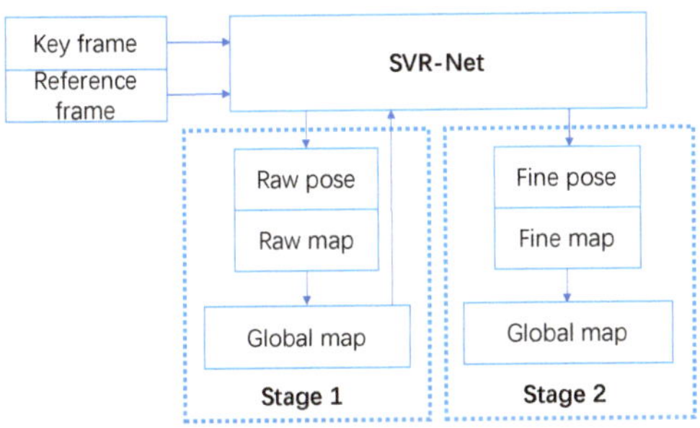

Figure 1. SLAM system.

SVR-Net is an end-to-end network for monocular simultaneous tracking and mapping. Its input consists of a pair of RGB frames with a query set of voxel coordinates. It outputs the relative pose $\boldsymbol{T}_n$ of the frames and TSDF values $\boldsymbol{d}_n$ of the voxels. For each pair of frames, the earlier frame is designated as the keyframe, and the later frame is designated as the reference frame. As shown in Figure 2, SVR-Net first extracts the image features into 2D feature maps. The feature map of the key frame is then transformed into feature voxels and correlated with the features of the reference frame. After sampling based on the current pose estimation, a matching network iteratively matches the features and updates the estimation of pose and map. The iterative pipeline performs a search in a correlation field of the features to find the optimal match, thereby enhancing both accuracy and robustness.

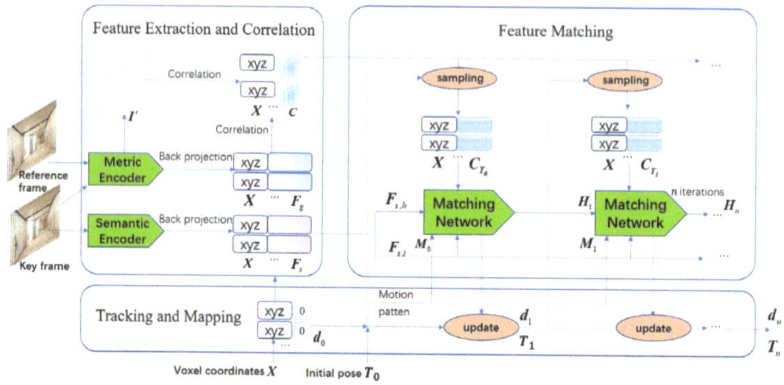

Figure 2. The structure of SVR-Net.

3.1. Voxel Feature Extraction and Correlation

In the feature extraction, deep features of the key frame are computed and back-projected into 3D voxels. Then, correlation maps between features of the voxels and the reference frame are computed to provide similarity information for matching.

3.1.1. Voxel Feature Extraction

2D feature maps are extracted from the input images using 2D convolution networks, namely, a metric encoder and a semantic encoder. Both networks produce feature maps at 1/8 the input image resolution. The metric encoder is a subnet of MnasNet [47] and contains two inverted residual blocks. The metric features of both frames are used to build correlation between voxels and images. The semantic encoder resembles the context network in RAFT, which consists of six residual blocks. The reference frame is only fed into the metric encoder to produce a metric map $\boldsymbol{I}_g^{'}$ for feature correlation, while the semantic features of the key frame provide the initial hidden state and inputs for recurrent feature matching.

Given voxel coordinates, both the semantic and metric feature maps of the key frame are back-projected to produce the 3D feature volume. Assume that there are N input voxels with their coordinates $\boldsymbol{X}_i, 1 \leq i \leq N$. For each voxel coordinate, let the projected coordinate be $\boldsymbol{x}_i$.

$$\boldsymbol{x}_i = \Pi(\boldsymbol{T}_k \boldsymbol{X}_i) \tag{1}$$

Here, Π is the projection function of the camera that maps 3D points onto the image. $\boldsymbol{T}_k$ is the transformation matrix of the key frame. Bilinear interpolation is carried out at $\boldsymbol{x}_i$ on the feature maps of the key frame to produce voxelized features of $\boldsymbol{X}_i$. To exploit sparsity, the coordinates and features are stored in separate matrices $\boldsymbol{X}$, $\boldsymbol{F}_s$ and $\boldsymbol{F}_g$ with N rows. Here, the subscript s and g represent semantic features and metric features, respectively.

3.1.2. Voxel Feature Correlation

After the voxelization of features from the key frame, the feature correlation between the key frame and the reference frame becomes feature correlation between the feature voxels and the reference frame's metric feature map. The correlation maps are computed with dot products within the metric features. For the ith voxel, its correlation map at the first layer is $\boldsymbol{C}_i^1$.

$$\boldsymbol{C}_i^1 \in \mathbb{R}^{h \times w}, C_{ikl}^1 = \sum_{j=1}^{D} F_{g,ij} I_{g,klj}^{'} \tag{2}$$

Here, h, w, and d are the height, width, and feature dimensions of the metric feature map. $F_{g,ij}$ is the jth value of the voxel's metric feature. $I_{g,klj}^{'}$ is the jth value of the reference frame's metric feature at position (k,l). The feature correlation reduces feature dimensions and provides scene-independent information to the downstream networks, which can improve generalization.

To present similarity information at different scales, the C_i^1 is pooled with kernel size 2,4,8. All generated maps are $C, C = \{C_i^m, 1 \leq i \leq N, 1 \leq m \leq 4\}$. C_i^m has a resolution of $\frac{h}{2^{m-1}} \times \frac{w}{2^{m-1}}, m = 1,2,3,4$. The four layers of the correlation maps form a pyramid where the correlation field is sampled at each iteration.

3.2. *Sparse Recurrent Feature Matching*

Feature matching uses an iterative pipeline to estimate the correspondence between the voxels and the reference frame. The voxels are first projected to the reference frame using the estimated pose from the last iteration. Then, correlation values sampled from projection coordinates are fed into the matching network, which predicts both updated TSDF values and corrected projection coordinates.

3.2.1. Sampling

In the sampling operation, correlation vectors are generated for all voxels and their projection coordinates on correlation maps, as demonstrated in Figure 3. For each correlation map of each projection coordinates $\boldsymbol{x}_i$, a grid with radius r is applied. Every point of the grid corresponds to an offset from the coordinate where the correlation value is sampled with bilinear interpolation. For all four correlation maps of the voxel, the values from four grids are concatenated to generate the correlation vector. All correlation vectors are stacked to a matrix $\boldsymbol{C_T} \in \mathbb{R}^{N \times 4(2r+1)^2}$. The subscript $\boldsymbol{T}$, which represents the pose of the reference frame, varies during iteration and results in different correlation values.

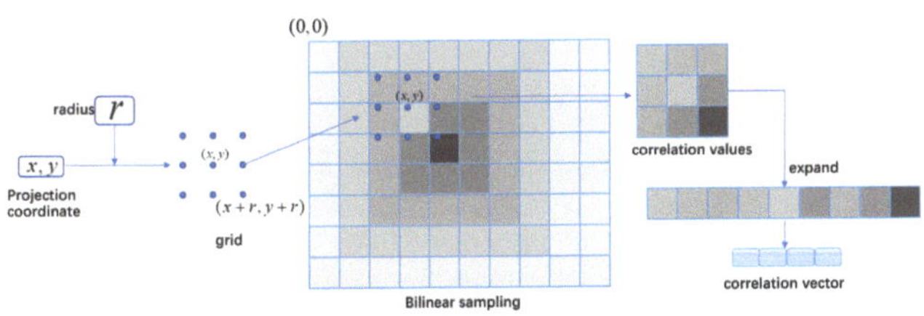

Figure 3. Sampling pipeline.

3.2.2. Matching Network

The coordinates and features of voxels stored in N-row matrices are fed into the matching network. The network produces three outputs: (1) revisions of voxels' TSDF values $\boldsymbol{\Delta d} \in \mathbb{R}^N$; (2) corrections of projection coordinates $\boldsymbol{\Delta x} \in \mathbb{R}^{N \times 2}$; and (3) confidence weights $\boldsymbol{w} \in \mathbb{R}_+^{N \times 3}$. Each row of the outputs corresponds to each input voxel. The first two columns of confidence weights represent the confidence of the projection coordinates. The last column is the confidence of voxel occupancy, whose value is between 0 and 1.

The matching network is implemented with sparse 3D convolution to process voxels with varying spatial distributions and sparsity. A gated recurrent unit (GRU) is also adopted to support iteration. The formulation of a GRU layer is as follows.

$$\begin{aligned} \boldsymbol{Z}_t &= \sigma(SConv_3([\boldsymbol{H}_{t-1}, \boldsymbol{G}_t])) \\ \boldsymbol{R}_t &= \sigma(SConv_3([\boldsymbol{H}_{t-1}, \boldsymbol{G}_t])) \\ \widetilde{\boldsymbol{H}}_t &= \tanh(SConv_3(\boldsymbol{R}_t \odot \boldsymbol{G}_t)) \\ \boldsymbol{H}_t &= (\boldsymbol{1} - \boldsymbol{Z}_t) \odot \boldsymbol{H}_{t-1} + \boldsymbol{Z}_t \odot \widetilde{\boldsymbol{H}}_t \end{aligned} \quad (3)$$

Here, $\odot$ is element-wise multiplication and σ is sigmoid activation. Sparse 3D convolution, $SConv_3$, implicitly utilizes voxel coordinates to fuse features. The hidden state from the last iteration is denoted by $\boldsymbol{H}_{t-1}$. The semantic feature $\boldsymbol{F}_s$ is split into $\boldsymbol{F}_{s,h}$ and $\boldsymbol{F}_{s,i}$, to provide the initial hidden state and input feature, respectively. The input feature

G_t of the GRU is a concatenation of these features: (1) the semantic feature for input $F_{s,i}$; (2) a correlation feature F_c; and (3) a motion feature F_m. The correlation feature is obtained by encoding C_T with two $SConv_3$ layers. The motion feature is obtained by encoding a motion pattern matrix M with a similar network, which is informative for GRU to perceive its progress. The motion pattern consists of flows and errors of projection coordinates, as demonstrated in Section 3.3.

The remaining components of the matching network are three branches to produce outputs at iteration t: Δd_t, Δx_t and w_t. Each branch contains two $SConv_3$ layers and takes H_t as common input, which effectively utilizes information in features.

3.3. Iterative Tracking and Mapping

The outputs of feature matching are utilized to update the estimations of the pose and map in each iteration of SVR-Net. A local map is obtained from the final iteration. Then, map fusion of the SLAM pipeline integrates the local maps from all input frames to enhance global consistency.

3.3.1. Pose Estimation with Gauss–Newton Update

After each iteration of feature matching, the estimated projection coordinates from the last iteration are corrected with Δx. Then, under the constraints of the projection coordinates and the confidence weights, the classical re-projection error E is optimized using Gauss–Newton iteration to produce an updated pose estimation.

$$E = \sum_{i=1}^{N} e_i^T W_i e_i \tag{4}$$

Here, e_i is the re-projection error from the ith voxel. W_i is the corresponding weight matrix based on the confidence weights w.

$$\begin{aligned} e_i &= x_i + \Delta x_i - \Pi(TX_i) \\ W_i &= diag(w_{i1}, w_{i2}) \end{aligned} \tag{5}$$

The Gauss–Newton update on the pose T is performed within Lie algebra. The update is differential with respect to Δx, making it possible to train the entire network end-to-end.

The motion pattern of the updated pose is informative to the matching network because the network can learn to adjust its matching strategy reactively at each iteration. The motion pattern M contains the flows and errors of all voxels. Let the initial pose before the first iteration be T_0 and the estimated pose at iteration t be T_t. Then, the flow of the ith voxel is defined as $\Pi(T_t X_i) - \Pi(T_0 X_i)$, which informs the corresponding optical flow on the image. The error is the re-projection error of T_t.

3.3.2. Map Update and Fusion

For each pair of frames, a local map within the key frame's view frustum is estimated. The map is represented as a set of voxels with TSDF values and occupancy confidences. The coordinates of the voxels are initialized, filling the view frustum with a maximum depth. The distance between neighboring coordinates, i.e., voxel size, controls the resolution of the map. However, when the voxel size is small, the huge number of voxels can cause a heavy memory burden on the GPU. This problem is addressed using the coarse-to-fine strategy.

At the first stage, the TSDF values d_0 in the view frustum are initialized with zero. With iterative running of the matching network, the TSDF values are successively added with revisions Δd. After the final iteration of the network, the first-stage TSDF estimations are obtained with the confidence weights w. Occupancy confidences w_o are the last column of the weights, which are used in map fusion. After the first-stage map fusion, the voxels in the view frustum are up-sampled and fed into SVR-Net again for second-stage estimation, producing a fine-grained map.

As the frames are input in sequence in the SLAM pipeline, all local maps are fused together to produce a global map. The global map is computed using a linear weighting method similar to TSDF integration [40]. The global map is initialized with the local map of the first key frame. For each subsequent local map, the global TSDF values of overlapped voxels are averaged according to global and local occupancy confidences. Then, the global occupancy confidences are updated in a cumulative manner. The data of voxels in a new region are directly attached to the global map.

4. Experiments

SVR-Net is trained on ScanNet(V2), and the full SLAM system is evaluated using TUM-RGBD. The ScanNet dataset includes 1613 indoor scenes with ground-truth camera poses and depth maps. The TUM-RGBD dataset contains nine indoor scenes with ground-truth poses. Without depth maps, the RGB images in TUM-RGBD contain heavy motion blur, which is a challenge for monocular SLAM.

ScanNet is split into a training set and a validation set, with the training set containing 1513 scenes. In order to improve tracking accuracy with various motion patterns, each pair of training frames is sampled using a random selection strategy from a successive subsequence. The length of each subsequence is controlled to satisfy the condition that there are exactly nine frames with mutual distance greater than thresholds. The distance threshold of rotation is 15°, and the threshold of translation is 0.1 m. The ground-truth TSDF global maps are generated by integrating depth frames of the training sequences. For each sequence, there are two scales of ground-truth maps, corresponding to two stages. The voxel sizes of the two scales are 16 cm and 8 cm. For training, the ground-truth map of each image frame is set to the part of the global map within the frame's view frustum. Training supervision includes both pose loss and map loss. The pose loss measures the distance between ground truth and the predicted pose, $L_{pose} = \|Log_{SE3}(\boldsymbol{T}^{-1}\boldsymbol{T}_{gt})\|_2$. The map loss is the mean difference between the TSDF values, $L_{map} = \|\boldsymbol{d} - \boldsymbol{d}_{gt}\|_1$.

The hardware used for training and evaluation includes Intel Xeon Silver 4210R @2.4GHz, 32 GB RAM and Nvidia Quadro RTX6000. The software infrastructure includes CUDA10.1, Python, PyTorch, LieTorch and Open3D.

4.1. Results on Matching

The quality of feature matching influences the performance of both tracking and mapping. To validate the superiority of network-based feature matching, matching results from SVR-Net are compared with results from the traditional method. The results are presented in both 3D and 2D visualizations, and the quality of feature matching is quantified using the end-point-error (EPE) metric.

The visualization of feature matching in 3D form is shown in Figures 4 and 5. The feature matching is performed in a test scene of ScanNet. The blue-colored point clouds represent the central coordinates of voxels. Each voxel is implicitly attached with a feature vector obtained through voxel feature extraction. The black lines are match lines connecting voxels with their matched estimations of projection coordinates on the reference frame. Each pair of panels represents the same match from different perspectives. The left panel shows the pose of the reference frame and the RGB image represented by a point lattice. The right panel shows the result observed from the back of the reference frame, where the voxels in the environment almost overlap with their projection on the image.

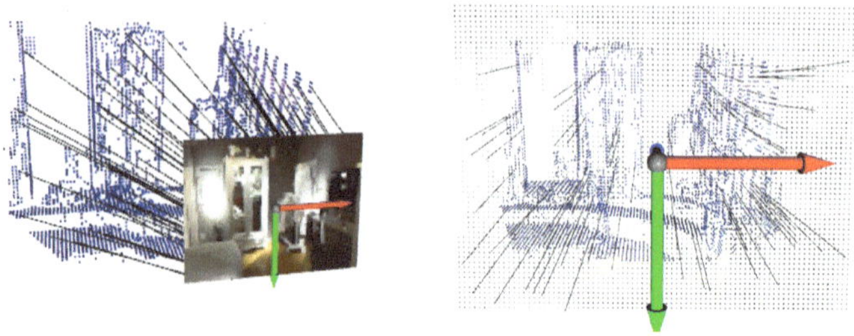

Figure 4. Matching result from SVR-Net(3D).

As illustrated in Figure 4, the match lines of SVR-Net are consistent and conform to the projection rule, where voxels representing an object (such as the wardrobe) are projected onto the corresponding object in the image. The match lines in the right panel have short projection lengths, indicating low end-point-error. This is attributed to the integration capability of the matching network for the features of adjacent voxels.

Figure 5 illustrates match lines obtained through brute-force search, which contain inconsistent parts that indicate outliers. Some of the match lines in the right figure exhibit significant projection length with high end-point-error. The result indicates that estimation from SVR-Net conforms to projection rules, while the estimation of a feature's L2 distance in brute-force search has no constraints from adjacent voxels, and this method is prone to producing outliers. A conventional ratio test can remove outliers, but reduces the available matches at the same time, as shown in the following Figures 6 and 7. All frames in the Figures are included in ScanNet test scenes.

As illustrated in Figure 6, SVR-Net produces dense matches between 2D image frames. To ensure clarity, only the top 200 matches from SVR-Net, possessing the highest confidence weights, are represented. There are no outliers in the matches, and correct matches can be obtained even in low-texture regions. It can be observed that SVR-Net identified the lines in the scene through high-confidence matches. In the matching network, the position information of the lines is propagated to adjacent low-texture regions, enabling the network to match correctly in these areas and improving its robustness.

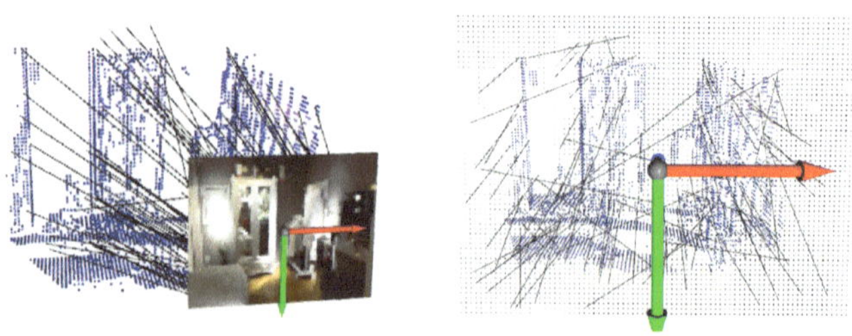

Figure 5. Matching result from brute-force search(3D).

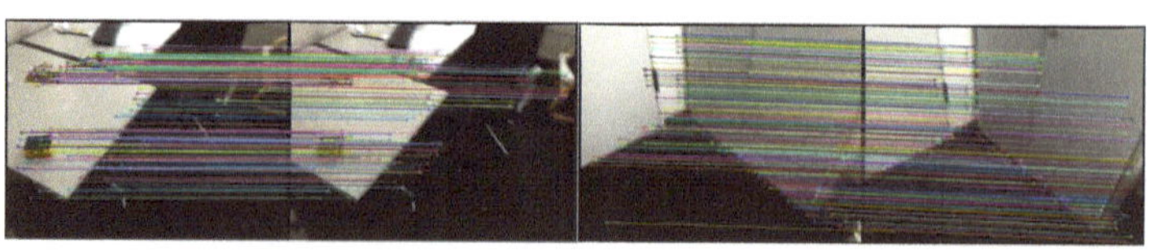

Figure 6. Matching result from SVR-Net(2D).

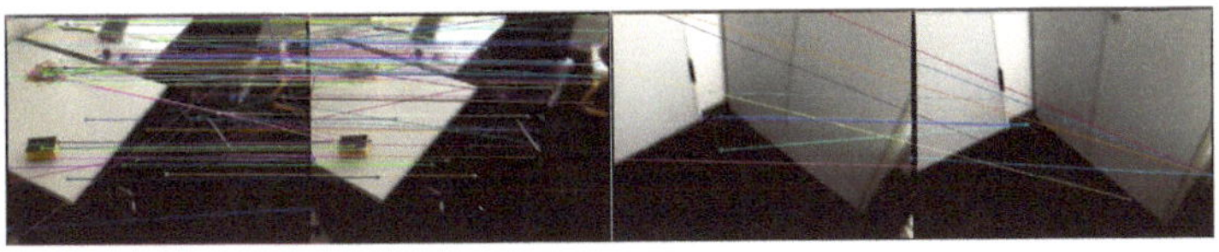

Figure 7. Matching result from SIFT(2D).

For comparison, Figure 7 illustrates the classical approach, which involves the extraction of SIFT feature points and their matches through brute-force search. To eliminate outliers, only matches with a test ratio greater than 0.9 from the classical approach are retained. However, some outliers still remain, due to repetitive features in the low-texture regions, which interfere with pose estimation and reduce robustness. Furthermore, the sparse matches preclude the generation of a dense map.

EPE is computed to evaluate the quality of feature matching. For a set of k feature points in a key frame, we denote their estimated matching coordinates at the corresponding reference frame as $x_1, x_2, \ldots, x_k$, and their true matching coordinates as $\widehat{x}_1, \widehat{x}_2, \ldots, \widehat{x}_k$. Then, the end-point-error is

$$\frac{1}{k}\sum_{i=1}^{k}\|x_i - \widehat{x}_i\|_2. \quad (6)$$

The EPE of a varying number of feature points is plotted in Figure 8. The evaluation involves 304 pairs of frames in two test scenes from the ScanNet dataset. In order to provide a given number of matched features, only matches with the highest confidence weights or test ratios are selected. As the number of matches increases, traditional SIFT-based methods become less effective in distinguishing repeated features, resulting in an increase in EPE. In contrast, SVR-Net experiences a decrease in EPE as the measurement quantity increases. When there are numerous matches, the EPE of SVR-Net is significantly lower than that of traditional methods, indicating that it is more robust for dense matching.

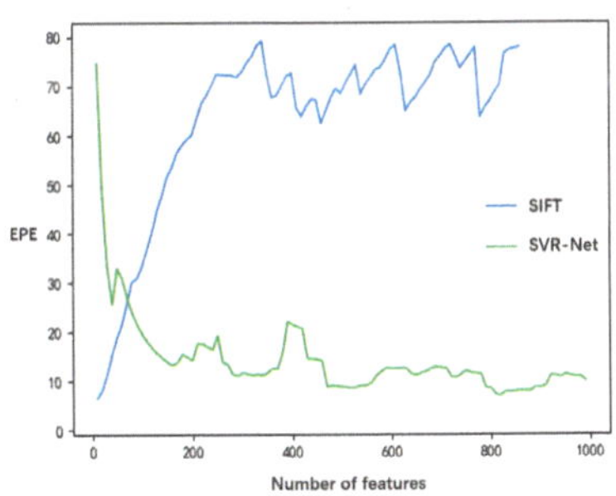

Figure 8. EPE at different numbers of feature points.

4.2. Results On Full System

SVR-Net estimates both raw and fine partial maps for every pair of adjacent frames. The partial dense TSDF maps of a test scene in ScanNet are shown in Figure 9. The maps in (a) and (b) are estimated maps with voxel sizes of 16 cm and 8 cm, and the map in (c) is the ground-truth map with a voxel size of 4 cm. The RGB image in (d) is the corresponding key frame.

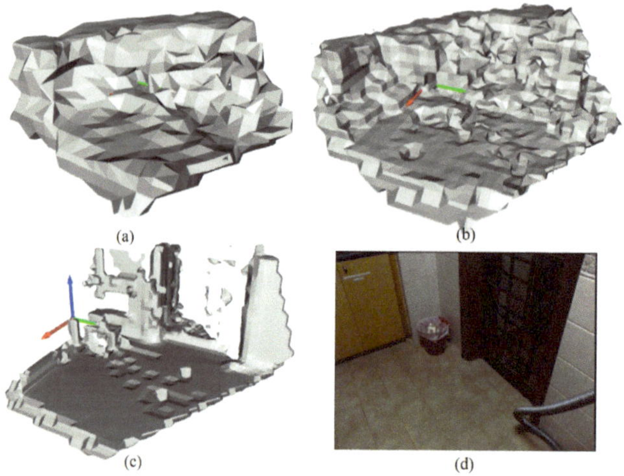

Figure 9. Partial dense TSDF maps and corresponding image. Estimated maps with voxel sizes of 16 cm and 8 cm are shown in (**a**,**b**) respectively. The ground-truth map with a voxel size of 4 cm is shown in (**c**). The RGB image from the corresponding key frame is in (**d**).

The full SLAM results on scene 360 of the TUM-RGBD dataset are shown in Figure 10. Camera trajectory and global maps are processed incrementally in the SLAM system. The estimated trajectory is displayed in blue, while the ground truth is in black. The dense global map is showcased by both the TSDF map and the point cloud of voxel coordinates at a voxel size of 8 cm. It can be observed that the pose estimated by the network almost coincides with the ground truth pose. The TSDF map in the left panel illustrates the floor and wall of the environment. The right panel presents the distribution of environment points through a dense point cloud.

The monocular SLAM system is evaluated on nine indoor scenes from the TUM-RGBD dataset. For each scene, the system runs on sequential RGB frames and outputs the result of tracking and dense mapping. The tracking performance is compared with those of ORB-SLAM and DeepV2D. All trajectories are evaluated up to scale using absolute trajectory error (ATE), which computes the rooted mean square translation error between estimated and ground truth poses. The result is shown in Table 2. All methods are provided mono. In challenging scenarios, the conventional ORB SLAMs fail, whereas deep approaches exhibit greater robustness, effectively computing poses in all scenes. Furthermore, the SVR-Net model outperforms DeepV2D with respect to average ATE. It is noteworthy that, in situations where ORB SLAMs achieve success, learning-based methods are less accurate due to the generalization error of networks. However, in the context of learning-based approaches, SVR-Net, which only estimates the relative pose between adjacent frames, achieves a similar ATE to that of DeepV2D, which utilizes multi-frame joint optimization. This suggests that SVR-Net demonstrates lower generalization error.

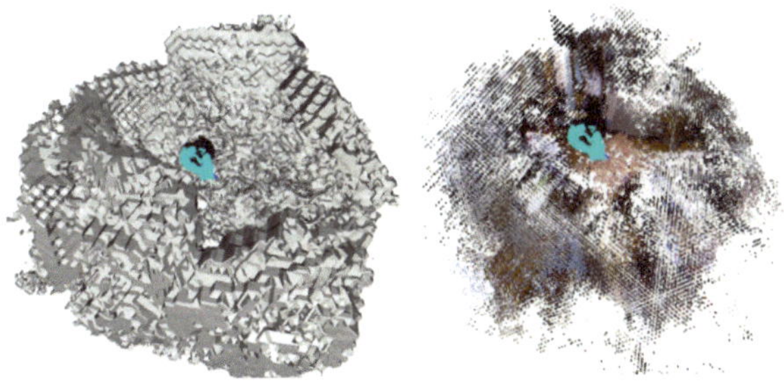

Figure 10. Full SLAM result.

Table 2. ATE on the TUM-RGBD benchmark. Failures are recorded as X.

Methods	360	Desk	Desk2	Floor	Plant	Room	rpy	Teddy	xyz	Average
ORB-SLAM2	X	0.071	X	0.023	X	X	X	X	0.01	-
ORB-SLAM3	X	0.017	0.21	X	0.034	X	X	X	0.009	-
DeepV2D	0.243	0.166	0.379	1.653	0.203	0.246	0.105	0.316	0.064	0.375
Ours	0.205	0.266	0.255	0.433	0.383	0.564	0.521	0.541	0.13	0.366

5. Discussion

Deep learning methods have been applied to the processes of monocular SLAM or dense TSDF mapping in various systems. However, there have been a limited number of studies integrating both of these aspects in a unified network. The suggested system, which is capable of simultaneously estimating pose and TSDF map using an end-to-end network, is more suitable for applications that require both aspects. The same features are exploited to produce two modalities of outputs with accuracy comparable to that of DeepV2D, which indicates highly efficient data utilization.

Real-time performance is an important metric for SLAM. SVR-Net runs at a frame rate of 4Hz on a desktop RTX2080. However, its performance on embedded devices or decentralized systems has not been tested, and this requires further research in the future.

SVR-Net only uses information from adjacent frames, which constrains the accuracy of localization and propagates errors to the map, thereby reducing its precision. A pose graph can significantly reduce pose error using multi-frame joint optimization. Additional research is necessary in the future to incorporate this into the proposed network.

6. Conclusions

This paper proposes SVR-Net, an end-to-end network for monocular TSDF SLAM. SVR-Net produces both poses and global TSDF maps from a sequence of monocular images. A metric encoder is constructed for the purpose of measuring image similarities, while a semantic encoder is devised to encode information, including the scale of scenes, to facilitate matching. In the matching network, sparse 3D convolution is adopted for dense TSDF mapping. A GRU is also employed to enable iterative updates for rectifying matches. Matching results presented in both 3D and 2D visualizations demonstrate that this approach effectively eliminates outliers, producing matches which are dense and more consistent than those produced by the traditional SIFT-based matching method. An experiment using the TUM-RGBD dataset further validates the conclusion that SVR-Net exhibits greater robustness compared with traditional ORB SLAMs. SVR-Net integrates Gauss–Newton updates for accurate pose estimation. Our experiments demonstrate that SVR-Net achieves a level of localization accuracy comparable to that of DeepV2D. Unlike previous monocular SLAM methods, the network directly yields a dense TSDF map during localization, which avoids inconsistency from depth map fusion and is more feasible for downstream tasks that are highly dependent on dense maps. Compared with existing TSDF mapping methods, SVR-Net does not require pre-calibrated poses and can meet real-time requirements.

Author Contributions: Conceptualization, Y.F.; Methodology, Y.F.; Software, Y.F.; Resources, R.L.; Writing—original draft, Y.F.; Writing—review & editing, R.L.; Supervision, Q.C. All authors have read and agreed to the published version of the manuscript.

Funding: This research received no external funding.

Institutional Review Board Statement: Not applicable.

Informed Consent Statement: Not applicable.

Data Availability Statement: Data sharing not applicable.

Conflicts of Interest: The authors declare no conflict of interest.

References

1. Campos, C.; Elvira, R.; Rodríguez, J.J.G.; M. Montiel, J.M.; D. Tardós, J. ORB-SLAM3: An Accurate Open-Source Library for Visual, Visual–Inertial, and Multimap SLAM. *IEEE Trans. Robot.* **2021**, *37*, 1874–1890. [CrossRef]
2. Suryanarayana, G.; Chandran, K.; Khalaf, O.I.; Alotaibi, Y.; Alsufyani, A.; Alghamdi, S.A. Accurate Magnetic Resonance Image Super-Resolution Using Deep Networks and Gaussian Filtering in the Stationary Wavelet Domain. *IEEE Access* **2021**, *9*, 71406–71417. [CrossRef]
3. Yue, Z.; Gao, F.; Xiong, Q.; Wang, J.; Huang, T.; Yang, E.; Zhou, H. A Novel Semi-Supervised Convolutional Neural Network Method for Synthetic Aperture Radar Image Recognition. *Cogn. Comput.* **2021**, *13*, 795–806. [CrossRef]
4. Choy, C.; Gwak, J.; Savarese, S.; Chandraker, M. Universal Correspondence Network. *arXiv* **2016**, arXiv:1606.03558.
5. DeTone, D.; Malisiewicz, T.; Rabinovich, A. SuperPoint: Self-Supervised Interest Point Detection and Description. In Proceedings of the 2018 IEEE/CVF Conference on Computer Vision and Pattern Recognition Workshops (CVPRW), Salt Lake City, UT, USA, 18–22 June 2018; pp. 337–33712. [CrossRef]
6. Luo, Z.; Shen, T.; Zhou, L.; Zhu, S.; Zhang, R.; Yao, Y.; Fang, T.; Quan, L. GeoDesc: Learning Local Descriptors by Integrating Geometry Constraints. In Proceedings of the Computer Vision—ECCV, Munich, Germany, 8–14 September 2018; Lecture Notes in Computer Science; Ferrari, V., Hebert, M., Sminchisescu, C., Weiss, Y., Eds.; Springer International Publishing: Cham, Switzerland, 2018; pp. 170–185. [CrossRef]
7. Mishchuk, A.; Mishkin, D.; Radenović, F.; Matas, J. Working Hard to Know Your Neighbor's Margins: Local Descriptor Learning Loss. *arXiv* **2017**, arXiv:1705.10872.
8. Ono, Y.; Trulls, E.; Fua, P.; Yi, K.M. LF-Net: Learning Local Features from Images. In Proceedings of the Advances in Neural Information Processing Systems, Montréal, QC, Canada, 3–8 December 2018; Curran Associates, Inc.: Red Hook, NY, USA, 2018; Volume 31.
9. Sarlin, P.E.; DeTone, D.; Malisiewicz, T.; Rabinovich, A. SuperGlue: Learning Feature Matching with Graph Neural Networks. *arXiv* **2020**, arXiv:1911.11763.
10. Yi, K.M.; Trulls, E.; Ono, Y.; Lepetit, V.; Salzmann, M.; Fua, P. Learning to Find Good Correspondences. In Proceedings of the IEEE Conference on Computer Vision and Pattern Recognition, Salt Lake City, UT, USA, 18–23 June 2018; pp. 2666–2674.
11. Brachmann, E.; Krull, A.; Nowozin, S.; Shotton, J.; Michel, F.; Gumhold, S.; Rother, C. DSAC — Differentiable RANSAC for Camera Localization. In Proceedings of the IEEE Conference on Computer Vision and Pattern Recognition (CVPR), Honolulu, HI, USA, 21–26 July 2017; pp. 2492–2500. [CrossRef]
12. Brachmann, E.; Rother, C. Neural-Guided RANSAC: Learning Where to Sample Model Hypotheses. In Proceedings of the IEEE/CVF International Conference on Computer Vision, Seoul, Republic of Korea, 27 October–2 November 2019; pp. 4322–4331.
13. Kluger, F.; Brachmann, E.; Ackermann, H.; Rother, C.; Yang, M.Y.; Rosenhahn, B. CONSAC: Robust Multi-Model Fitting by Conditional Sample Consensus. In Proceedings of the 2020 IEEE/CVF Conference on Computer Vision and Pattern Recognition (CVPR), Seattle, WA, USA, 13–19 June 2020; pp. 4633–4642. [CrossRef]
14. Teed, Z.; Deng, J. DeepV2D: Video to Depth with Differentiable Structure from Motion. In Proceedings of the International Conference on Learning Representations, Vancouver, BC, Canada, 30 April–3 May 2018.
15. Teed, Z.; Deng, J. DROID-SLAM: Deep Visual SLAM for Monocular, Stereo, and RGB-D Cameras. In Proceedings of the Advances in Neural Information Processing Systems, Denver, CO, USA, 6–14 December 2021; Curran Associates, Inc.: Red Hook, NY, USA, 2021; Volume 34, pp. 16558–16569.
16. Wang, W.; Hu, Y.; Scherer, S. TartanVO: A Generalizable Learning-based VO. In Proceedings of the 2020 Conference on Robot Learning (PMLR), London, UK, 8–11 November 2020; pp. 1761–1772.
17. Zhou, H.; Ummenhofer, B.; Brox, T. DeepTAM: Deep Tracking and Mapping. In *Proceedings of the Computer Vision–ECCV 2018*; Ferrari, V., Hebert, M., Sminchisescu, C., Weiss, Y., Eds.; Lecture Notes in Computer Science; Springer International Publishing: Cham, Switzerland, 2018; pp. 851–868. [CrossRef]
18. Sun, J.; Xie, Y.; Chen, L.; Zhou, X.; Bao, H. NeuralRecon: Real-Time Coherent 3D Reconstruction from Monocular Video. *arXiv* **2021**, arXiv:2104.00681.
19. Murez, Z.; As, T.; Bartolozzi, J.; Sinha, A.; Badrinarayanan, V.; Rabinovich, A. Atlas: End-to-End 3D Scene Reconstruction from Posed Images. *arXiv* **2020**, arXiv:2003.10432.

20. Stier, N.; Rich, A.; Sen, P.; Höllerer, T. VoRTX: Volumetric 3D Reconstruction with Transformers for Voxelwise View Selection and Fusion. In Proceedings of the 2021 International Conference on 3D Vision (3DV), London, UK, 1–3 December 2021; pp. 320–330. [CrossRef]
21. Dai, A.; Chang, A.X.; Savva, M.; Halber, M.; Funkhouser, T.; Nießner, M. ScanNet: Richly-Annotated 3D Reconstructions of Indoor Scenes. In Proceedings of the 2017 IEEE Conference on Computer Vision and Pattern Recognition (CVPR), Honolulu, HI, USA, 21–26 July 2017; pp. 2432–2443. [CrossRef]
22. Sturm, J.; Engelhard, N.; Endres, F.; Burgard, W.; Cremers, D. A Benchmark for the Evaluation of RGB-D SLAM Systems. In Proceedings of the 2012 IEEE/RSJ International Conference on Intelligent Robots and Systems, Vilamoura-Algarve, Portugal, 7–12 October 2012; pp. 573–580. [CrossRef]
23. Engel, J.; Koltun, V.; Cremers, D. Direct Sparse Odometry. *IEEE Trans. Pattern Anal. Mach. Intell.* **2018**, *40*, 611–625. [CrossRef] [PubMed]
24. Leutenegger, S.; Furgale, P.; Rabaud, V.; Chli, M.; Konolige, K.; Siegwart, R. *Keyframe-Based Visual-Inertial SLAM Using Nonlinear Optimization*; ETH Library: Zurich, Switzerland, 2013. [CrossRef]
25. Engel, J.; Schöps, T.; Cremers, D. LSD-SLAM: Large-Scale Direct Monocular SLAM. In *Proceedings of the Computer Vision–ECCV 2014*; Fleet, D., Pajdla, T., Schiele, B., Tuytelaars, T., Eds.; Lecture Notes in Computer Science; Springer International Publishing: Cham, Switzerland, 2014; pp. 834–849. [CrossRef]
26. Ferrera, M.; Eudes, A.; Moras, J.; Sanfourche, M.; Le Besnerais, G. OV^{2}SLAM: A Fully Online and Versatile Visual SLAM for Real-Time Applications. *IEEE Robot. Autom. Lett.* **2021**, *6*, 1399–1406. [CrossRef]
27. Bloesch, M.; Czarnowski, J.; Clark, R.; Leutenegger, S.; Davison, A.J. CodeSLAM—Learning a Compact, Optimisable Representation for Dense Visual SLAM. In Proceedings of the 2018 IEEE/CVF Conference on Computer Vision and Pattern Recognition, Salt Lake City, UT, USA, 18–23 June 2018; pp. 2560–2568.
28. Czarnowski, J.; Laidlow, T.; Clark, R.; Davison, A.J. DeepFactors: Real-Time Probabilistic Dense Monocular SLAM. *IEEE Robot. Autom. Lett.* **2020**, *5*, 721–728. [CrossRef]
29. Kopf, J.; Rong, X.; Huang, J.B. Robust Consistent Video Depth Estimation. In Proceedings of the IEEE/CVF Conference on Computer Vision and Pattern Recognition, Montreal, QC, Canada, 10–17 October 2021; pp. 1611–1621.
30. Luo, X.; Huang, J.B.; Szeliski, R.; Matzen, K.; Kopf, J. Consistent Video Depth Estimation. *ACM Trans. Graph.* **2020**, *39*, 71:1–71:13. [CrossRef]
31. Sucar, E.; Wada, K.; Davison, A. NodeSLAM: Neural Object Descriptors for Multi-View Shape Reconstruction. In Proceedings of the 2020 International Conference on 3D Vision (3DV), Fukuoka, Japan, 25–28 November 2020; pp. 949–958. [CrossRef]
32. Sucar, E.; Liu, S.; Ortiz, J.; Davison, A.J. iMAP: Implicit Mapping and Positioning in Real-Time. In Proceedings of the IEEE/CVF International Conference on Computer Vision, Montreal, BC, Canada, 11–17 October 2021; pp. 6229–6238.
33. Zhu, Z.; Peng, S.; Larsson, V.; Xu, W.; Bao, H.; Cui, Z.; Oswald, M.R.; Pollefeys, M. NICE-SLAM: Neural Implicit Scalable Encoding for SLAM. In Proceedings of the 2022 IEEE/CVF Conference on Computer Vision and Pattern Recognition (CVPR), New Orleans, LA, USA, 18–24 June 2022; pp. 12776–12786. [CrossRef]
34. Yang, N.; Wang, R.; Stückler, J.; Cremers, D. Deep Virtual Stereo Odometry: Leveraging Deep Depth Prediction for Monocular Direct Sparse Odometry. In Proceedings of the Computer Vision—ECCV 2018, Munich, Germany, 8–14 September 2018; Ferrari, V., Hebert, M., Sminchisescu, C., Weiss, Y., Eds.; Lecture Notes in Computer Science; Springer International Publishing: Cham, Switzerland, 2018; pp. 835–852. [CrossRef]
35. Yang, N.; von Stumberg, L.; Wang, R.; Cremers, D. D3VO: Deep Depth, Deep Pose and Deep Uncertainty for Monocular Visual Odometry. In Proceedings of the 2020 IEEE/CVF Conference on Computer Vision and Pattern Recognition (CVPR), Seattle, WA, USA, 13–19 June 2020; pp. 1278–1289. [CrossRef]
36. Tang, H.; Liu, Z.; Zhao, S.; Lin, Y.; Lin, J.; Wang, H.; Han, S. Searching Efficient 3D Architectures with Sparse Point-Voxel Convolution. In Proceedings of the Computer Vision—ECCV 2020, Glasgow, UK, 23–28 August 2020; Vedaldi, A., Bischof, H., Brox, T., Frahm, J.M., Eds.; Lecture Notes in Computer Science; Springer International Publishing: Cham, Switzerland, 2020; pp. 685–702. [CrossRef]
37. Zhao, L.; Xu, S.; Liu, L.; Ming, D.; Tao, W. SVASeg: Sparse Voxel-Based Attention for 3D LiDAR Point Cloud Semantic Segmentation. *Remote Sens.* **2022**, *14*, 4471. [CrossRef]
38. Liu, L.; Gu, J.; Zaw Lin, K.; Chua, T.S.; Theobalt, C. Neural Sparse Voxel Fields. In Proceedings of the Advances in Neural Information Processing Systems, Vancouver, BC, Canada, 6–12 December 2020; Curran Associates, Inc.: Red Hook, NY, USA, 2020; Volume 33, pp. 15651–15663.
39. Curless, B.; Levoy, M. A Volumetric Method for Building Complex Models from Range Images. In *Proceedings of the SIGGRAPH96: 23rd International Conference on Computer Graphics and Interactive Techniques*; Association for Computing Machinery: New York, NY, USA, 1996. [CrossRef]
40. Newcombe, R.A.; Izadi, S.; Hilliges, O.; Kim, D.; Davison, A.J.; Kohli, P.; Shotton, J.; Hodges, S.; Fitzgibbon, A. KinectFusion: Real-Time Dense Surface Mapping and Tracking. In Proceedings of the 2011 10th IEEE International Symposium on Mixed and Augmented Reality, Basel, Switzerland, 26–29 October 2011.
41. Lin, Y.; Gao, F.; Qin, T.; Gao, W.; Liu, T.; Wu, W.; Yang, Z.; Shen, S. Autonomous Aerial Navigation Using Monocular Visual-Inertial Fusion. *J. Field Robot.* **2018**, *35*, 23–51. [CrossRef]

42. Oleynikova, H.; Taylor, Z.; Fehr, M.; Siegwart, R.; Nieto, J. Voxblox: Incremental 3D Euclidean Signed Distance Fields for on-Board MAV Planning. In Proceedings of the 2017 IEEE/RSJ International Conference on Intelligent Robots and Systems (IROS), Vancouver, BC, Canad, 24–28 September 2017; pp. 1366–1373. [CrossRef]
43. Wagner, R.; Frese, U.; Bäuml, B. Graph SLAM with Signed Distance Function Maps on a Humanoid Robot. In Proceedings of the 2014 IEEE/RSJ International Conference on ntelligent Robots and Systems, Chicago, IL, USA, 14–18 September 2014; pp. 2691–2698. [CrossRef]
44. Oleynikova, H.; Taylor, Z.; Siegwart, R.; Nieto, J. Safe Local Exploration for Replanning in Cluttered Unknown Environments for Microaerial Vehicles. *IEEE Robot. Autom. Lett.* **2018**, *3*, 1474–1481. [CrossRef]
45. Ratliff, N.; Zucker, M.; Bagnell, J.A.; Srinivasa, S. CHOMP: Gradient Optimization Techniques for Efficient Motion Planning. In Proceedings of the 2009 IEEE International Conference on Robotics and Automation, Kobe, Japan, 12–17 May 2009; pp. 489–494. [CrossRef]
46. Choe, J.; Im, S.; Rameau, F.; Kang, M.; Kweon, I.S. VolumeFusion: Deep Depth Fusion for 3D Scene Reconstruction. In Proceedings of the 2021 IEEE/CVF International Conference on Computer Vision (ICCV), Montreal, QC, Canada, 10–17 October 2021; pp. 16066–16075. [CrossRef]
47. Tan, M.; Chen, B.; Pang, R.; Vasudevan, V.; Sandler, M.; Howard, A.; Le, Q.V. MnasNet10-17 October 2021Platform-Aware Neural Architecture Search for Mobile. *arXiv* **2018**, arXiv:1807.11626.

sensors

MDPI

Review

Machine Learning Techniques for Developing Remotely Monitored Central Nervous System Biomarkers Using Wearable Sensors: A Narrative Literature Review

Ahnjili ZhuParris [1,2,3], Annika A. de Goede [1], Iris E. Yocarini [2], Wessel Kraaij [2,4], Geert Jan Groeneveld [1,2] and Robert Jan Doll [1,*]

1 Centre for Human Drug Research (CHDR), Zernikedreef 8, 2333 CL Leiden, The Netherlands; azhuparris@chdr.nl (A.Z.)
2 Leiden Institute of Advanced Computer Science (LIACS), Snellius Gebouw, Niels Bohrweg 1, 2333 CA Leiden, The Netherlands
3 Leiden University Medical Center (LUMC), Albinusdreef 2, 2333 ZA Leiden, The Netherlands
4 The Netherlands Organisation for Applied Scientific Research (TNO), Anna van Buerenplein 1, 2595 DA, Den Haag, The Netherlands
* Correspondence: rjdoll@chdr.nl; Tel.: +31-(71)-5246400

Abstract: Background: Central nervous system (CNS) disorders benefit from ongoing monitoring to assess disease progression and treatment efficacy. Mobile health (mHealth) technologies offer a means for the remote and continuous symptom monitoring of patients. Machine Learning (ML) techniques can process and engineer mHealth data into a precise and multidimensional biomarker of disease activity. Objective: This narrative literature review aims to provide an overview of the current landscape of biomarker development using mHealth technologies and ML. Additionally, it proposes recommendations to ensure the accuracy, reliability, and interpretability of these biomarkers. Methods: This review extracted relevant publications from databases such as PubMed, IEEE, and CTTI. The ML methods employed across the selected publications were then extracted, aggregated, and reviewed. Results: This review synthesized and presented the diverse approaches of 66 publications that address creating mHealth-based biomarkers using ML. The reviewed publications provide a foundation for effective biomarker development and offer recommendations for creating representative, reproducible, and interpretable biomarkers for future clinical trials. Conclusion: mHealth-based and ML-derived biomarkers have great potential for the remote monitoring of CNS disorders. However, further research and standardization of study designs are needed to advance this field. With continued innovation, mHealth-based biomarkers hold promise for improving the monitoring of CNS disorders.

Keywords: machine learning; biomarker; wearables; smartphones; mHealth; remote monitoring; central nervous system; clinical trials

1. Introduction

1.1. Motivation

Disorders that are affected by the Central Nervous System (CNS), such as Parkinson's Disease (PD) and Alzheimer's Disease (AD), have a significant impact on the quality of life of patients. These disorders are often progressive and chronic, making long-term monitoring essential for assessing disease progression and treatment effects. However, the current methods for monitoring disease activity are often limited by accessibility, cost, and patient compliance [1,2]. Limited accessibility to clinics or disease monitoring devices may hinder the regular and consistent monitoring of a patient's condition, especially for patients living in remote areas or for those who have mobility limitations. Clinical trials incur costs related to personnel, infrastructure, and equipment. A qualified healthcare team, including clinical raters, physicians, and nurses, contributes to personnel costs through salaries, training, and administrative support. Trials involving specialized equipment for measuring biomarkers can significantly impact the budget due to costs associated with procurement, maintenance, calibration, and upgrades. Furthermore, infrastructure costs may increase as suitable facilities are required for data collection during patient visits and equipment storage. Patient compliance poses challenges for disease monitoring, as some methods require patients to adhere to strict protocols, collect data at specific time intervals, or perform certain tasks that can be challenging for patients to execute. Low or no compliance can lead to incomplete or unreliable monitoring results, which in turn can hinder the reliability of the assessments. Given these limitations, there is a growing interest in exploring alternative approaches to monitoring CNS disorders that can overcome these challenges. The increasing adoption of smartphones and wearables among patients and researchers offers a promising avenue for remote monitoring.

Patient-generated data from smartphones, wearables, and other remote monitoring devices can potentially complement or supplement clinical visits by providing data during evidence gaps between visits. As the promise of mobile Health (mHealth) technologies is to provide more sensitive, ecologically valid, and frequent measures of disease activity, the data collected may enable the development and validation of novel biomarkers. The development of novel 'digital biomarkers' using data collected from electronic Health (eHealth) and mHealth device sensors (such as accelerometers, GPS, and microphones) offers a scalable opportunity for the continuous collection of data regarding behavioral and physiological activity under free-living conditions. Previous clinical studies have demonstrated the benefits of smartphone and wearable sensors to monitor and estimate symptom severity associated with a wide range of diseases and disorders, including cardiovascular diseases [3], mood disorders [4], and neurodegenerative disorders [5,6]. These sensors can capture a range of physiological and behavioral data, including movement, heart rate, sleep, and cognitive function, providing a wealth of information that can be used to develop biomarkers for CNS disorders in particular. These longitudinal and unobtrusive measurements are highly valuable for clinical research, providing a scalable opportunity for measuring behavioral and physiological activity in real-time. However, these approaches may carry potential pitfalls as the data sourced from these devices can be large, complex, and highly variable in terms of availability, quality, and synchronicity, which can therefore complicate analysis and interpretation [7,8]. Machine Learning (ML) may provide a solution to processing heterogenous and large datasets, identifying meaningful patterns within the datasets, and predicting complex clinical outcomes from the data. However, the complexities involved in developing biomarkers using these new technologies need to be addressed. While these tools can aid the discovery of novel and important digital biomarkers, the lack of standardization, validation, and transparency of the ML pipelines used can pose challenges for clinical, scientific, and regulatory committees.

1.2. What Is Machine Learning

In clinical research, one of the primary objectives is to understand the relationship between a set of observable variables (features) and one or more outcomes. Building a statistical model that captures the relationship between these variables and the corresponding outputs facilitates the attainment of this understanding [9]. Once this model is built, it can be used to predict the value of an output based on the features.

ML is a powerful tool for clinical research as it can be used to build statistical models. A ML model consists of a set of tunable parameters and a ML algorithm that enables the generation of outputs based on given inputs and selected parameters. Although ML algorithms are fundamentally statistical learning algorithms, ML and traditional statistical learning algorithms can differ in their objectives. Traditional statistical learning aims to create a statistical model that represents causal inference from a sample, while ML aims to build generalizable predictive models that can be used to make accurate predictions on previously unseen data [10,11]. However, it is essential to recognize that while ML models can identify relationships between variables and outcomes, they may not necessarily identify a causal link between them. This is because even though these models may achieve good performances, it is crucial to ensure that their predictions are based on relevant features rather than spurious correlations. This enables the researchers to gain meaningful insights from ML models while also being aware of their inherent limitations.

While ML is not a substitute for the clinical evaluation of patients, it can provide valuable insights into a patient's clinical profile. ML can help to identify relevant features that clinicians may not have considered, leading to better diagnosis, treatment, and patient outcomes. Additionally, ML can help to avoid common pitfalls observed in clinical decision making by removing bias, reducing human error, and improving the accuracy of predictions [12–15]. As the volume of data generated for clinical trials and outside clinical settings continues to grow, ML's support in processing data and informing the decision-making process becomes necessary. ML can help to uncover insights from large and complex datasets that would be difficult or impossible to identify manually.

To develop an effective ML model, it is necessary to follow a rigorous and standardized procedure. This is where ML pipelines come in. Table 1 showcases an exemplary ML pipeline, which serves as a systematic framework for automating and standardizing the model generation process. The pipeline encompasses multiple stages, as defined by the authors, to ensure an organized and efficient approach to model development. First, defining the study objective guides the subsequent stages and ensures the final model meets the desired goals. Second, raw data must be preprocessed to remove errors, inconsistencies, missing data, or outliers. Third, feature extraction and selection identifies quantifiable characteristics of the data relevant to the study objective and extracts them for use in the ML model. Fourth, ML algorithms are applied to learn patterns and relationships between features, with optimal configurations identified through iterative processes until desired performance metrics are achieved. Finally, the model is validated against a new dataset that is not used in training to ensure generalizability. Effective reporting and assessment of ML procedures must be established to ensure transparency, reliability, and reproducibility.

Table 1. Representation of a standard machine learning pipeline.

Stage	Objective	Example
1. Study Design	The ML pipeline is provided with a study objective in which the features and corresponding outputs are defined. The ML model aims to identify the associations between the features and outputs.	The study objective is to classify Parkinson's Disease patients and control groups using smartphone-based features.
2. Data Preprocessing	Data preprocessing filters and transforms raw data to guarantee or enhance the ML training process.	To improve the model performance, one may identify and exclude any missing or outlier data.
3. Feature Engineering and Selection	Feature engineering uses raw data to create new features that are not readily available in the dataset. Feature selection selects the most relevant features for the model objective by removing redundant or noisy features. Together, the goal is to simplify and accelerate the computational process while also improving the model process. For deep learning methods, the concept of "feature engineering" is typically embedded within the model architecture and training process, although substantial preprocessing steps may occur prior to that.	An interaction of two or more predictors (such as a ratio or product) or re-representation of a predictor are examples of feature engineering. Removing highly correlated or non-informative features are examples of feature selection. Note: The feature selection step can occur during model training

Table 1. *Cont.*

Stage	Objective	Example
4. Model Training and Validation	During training, the ML model(s) iterates through all the examples in the training dataset and optimizes the parameters of the mathematical function to minimize the prediction error. To evaluate the performance of the trained ML model, the predictions of an unseen test set are compared with a known ground truth label.	Cross-validation can be used to optimize and evaluate model performance. Classification models may be evaluated based on their prediction accuracy, sensitivity, and specificity, while regression models may be evaluated using variance explained (R2) and Mean Absolute Error.

1.3. Objectives

The objective of this narrative literature review is to provide an overview of the ML practices used in studies that use mHealth technologies and ML to develop novel biomarkers for clinical trials. In this review, each component of the ML pipeline has a dedicated section. Based on the results obtained from the review process, each ML component section provides a comprehensive analysis and discussion of the most common and notable practices. These sections delve into the motivations behind these practices, their limitations, and their overall impact on the ML pipeline. This review will not provide precise recommendations for best practices, as much of the research in this area is new and quickly evolving. Rather, the recommendation section discusses the approaches for standardization and validation procedures that are necessary for the development of ML biomarkers to ensure the effectiveness and acceptance of these biomarkers by clinical, scientific, and regulatory committees.

2. Methods

2.1. Information Sources and Search Strategy

Given the wide range of study designs and clinical populations that use smartphones and wearables to collect data, we used the Joanna Briggs Institute (JBI) guidelines to develop a search strategy [16]. Based on an initial limited search of online databases for clinical trials that report using mHealth devices and ML, we developed a custom keyword strategy and performed an in-depth search in PubMed, IEEE Xplore, and CTTI (Table 2). The search terms for the CNS disorder terms were based on the National Library of Medicine's CNS MeSH descriptor data [17]. The relevant papers were selected based on the title and abstract. Finally, other literature review studies that explore the same questions were reviewed; the references cited by these studies were then identified and reviewed if they met our criteria. The date range for the search was between 1 January 2012 and 31 December 2022. The search was conducted on 7 January 2023.

Table 2. An overview of the keyword strategy used for this study.

Domain	Search String
Technology	((“smartphone”[tiab] OR “wearable”[tiab] OR “remote + monitoring”[tiab] OR “home + monitoring”[tiab] OR “mobile + sensors”[tiab] OR “mobile + montoring”[tiab] OR “behavioral + sensing”[tiab] OR “geolocation”[tiab] OR “mHealth”[tiab] OR “passive + monitoring”[tiab] OR “digital + phenotype”[tiab] OR “digital + phenotyping”[tiab] OR “digital + biomarker”[tiab])
Analysis	AND (“machine + learning”[tiab] OR “deep + learning”[tiab] OR “random + forest”[tiab] OR “neural + network”[tiab] OR “time + series”[tiab] OR “regression”[tiab] OR “SVM”[tiab] OR “knn”[tiab] OR “dynamics + model”[tiab] OR “decision + tree”[tiab] OR “discriminant + analysis”[tiab] OR “feature + engineering”[tiab] OR “feature + selection”[tiab] OR “data + mining”[tiab] OR “model”[tiab] OR “classification”[tiab] OR “diagnostic”[tiab] OR “prognostic”[tiab] OR “symptom + severity”[tiab] OR “prediction”[tiab] OR “monitoring”[tiab])

Table 2. *Cont.*

Domain	Search String
Population	AND ("disease"[tiab] OR "disorder"[tiab] OR "diagnosis"[tiab] OR "prognosis" OR "alzheimer"[tiab] OR "parkinson"[tiab] OR "Huntington"[tiab] OR "neurodegenerative"[tiab] OR "degenerative" OR "tremor"[tiab] OR "bipolar"[tiab] OR "depression"[tiab] OR "manic"[tiab] OR "anxiety"[tiab] OR "vocal + biomarker"[tiab] OR "amyotrophic + lateral + sclerosis"[tiab] OR "central + nervous + system"[tiab] OR "symptom"[tiab] OR "psychosis"[tiab] OR "stroke"[tiab] OR "muscular dystrophy"[tiab] OR "Facioscapulohumeral Dystrophy"[tiab] OR "autoimmune"[tiab] OR "seizure"[tiab] OR "multiple + sclerosis"[tiab])
Date	AND ("2012/01/01"[PDAT]:"2022/12/31"[PDAT])
Language	AND (English[lang])
Exclusion Criteria	NOT("animals"[tiab] OR "implant"[tiab] OR "hospital"[tiab] OR "caregiver"[tiab] OR "telemedicine"[tiab] OR "telerehabilitation"[tiab] OR "smartphone + addiction"[tiab] OR "nursing"[tiab] OR"screening"[tiab] OR "recruitment"[tiab] OR "diabetes"[tiab] OR "malaria"[tiab] OR "self-care"[tiab] OR "self-management"[tiab] OR "self-help"[tiab])
Article Type	AND (clinicalstudy[Filter] OR clinicaltrial[Filter] OR clinicaltrialphasei[Filter] OR clinicaltrialphaseii[Filter] OR clinicaltrialphaseiii[Filter] OR clinicaltrialphaseiv[Filter] OR controlledclinicaltrial[Filter] OR meta-analysis[Filter] observationalstudy[Filter] OR randomizedcontrolledtrial[Filter] OR systematicreview[Filter])

2.2. Inclusion Criteria

The authors adopted the Population, Intervention, Comparator, Outcomes, Study type (PICOS) framework to define the inclusion and exclusion criteria [13] (Table 3). The studies included were restricted to those involving participants diagnosed with CNS disorders who were remotely monitored under free-living conditions. The intervention and device criteria were limited to passive data collected from smartphones and other non-invasive remote monitoring sensors, whereas data collected using active engagement from participants, such as disposable blood tests or small scales, were excluded. As we chose to focus on ML pipelines, we selected studies in which a statistical model was used to analyze a dataset and could potentially be used to generate future predictions using an independent dataset. Therefore, traditional statistical models such as linear or logistic regression were included, but statistical models such as ANOVA and correlation analyses were not included. Further, as the focus was on the development and validation of ML models, we did not include studies that did not report on model performance.

Table 3. Table of the inclusion and exclusion criteria used for study selection.

Category	Criteria
Population	The study must be initiated by a research organization and not by the participants. The participants must have a clinical diagnosis that is affected by the CNS. Hence, studies that collected data from participants with no clinically confirmed diagnosis were not considered.
Intervention	The study must include the use of smartphone or non-invasive wearables to remotely monitor and quantify passive biomarkers under free-living conditions.
Comparator	A ground truth comparator for digital phenotyping such as clinical assessment, medical records, or self-reported outcomes.
Outcomes	A ML model that is used to classify a clinical label (such as a diagnosis, or clinical event), estimate symptom severity, or to detect treatment effects.
Study Type	The paper must be about a human-centered observational study (cohort or longitudinal) where the data were collected outside the clinic, lab, or hospital (free-living conditions). Hence, studies that use smartphones or wearables as a form of intervention or as screening tools are not of interest. The study must show if the ML models had ecological validity by validating the models using free-living data. The study has to have been written or translated into English and published within the last 10 years (2012 onwards).

2.3. Data Extraction

Two authors conducted the data extraction following the inclusion criteria, and the results were reviewed by the remaining authors. Data relating to the database source, title, DOI, publication year, trial setting or scenario, objective, devices used, data collection period, number of participants, inclusion of healthy controls, data processing steps, feature engineering, feature selection, machine learning models used, hyperparameters and hyperparameter optimization, model performance, and validation procedure were extracted. The comprehensive data extraction and review conducted by the authors encompassed various essential aspects of the studies, ensuring a thorough analysis of the database source, trial details, data processing steps, machine learning models, and validation procedures.

3. Results

3.1. Study Selection

Our initial keyword search revealed a total of 2310 articles that utilized digital phenotyping devices, such as smartphones and wearables, in a clinical study and applied ML techniques. After screening the titles and abstracts based on our predefined criteria, we narrowed down the articles to 66 studies, which were used for our analysis. Figure 1 provides an overview of the complete selection process.

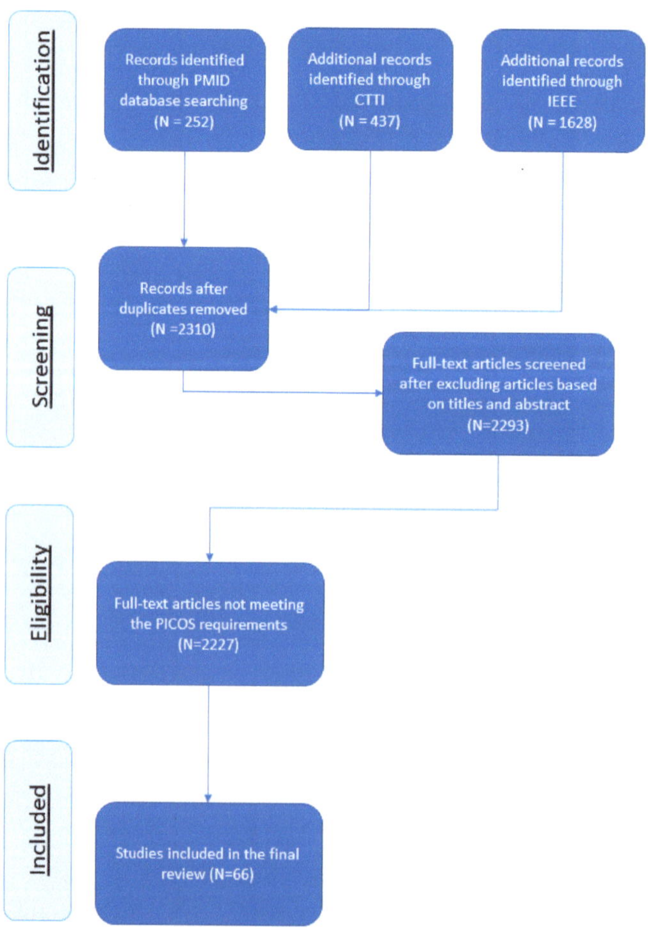

Figure 1. Flow diagram illustrating the paper selection process for this review.

3.2. Study Characteristics

For each of the 66 studies, we extracted information about the clinical population and the ML pipeline that was used to develop the digital biomarkers. We found that only half of the studies included healthy controls (N = 34). As seen in Figure 2, Parkinson's disease (PD) (N = 27) was the most prevalent disorder identified in our search, followed by Bipolar Disorder (BD) (N = 11), and Unipolar Depression or Major Depressive Disorder (MDD) (N = 9). The sample size of the selected studies was heterogenous, ranging from 7 to 6221 participants (Figure 3). Overall, our review provides a comprehensive overview of the characteristics of studies that have utilized mHealth devices and ML techniques, which can help inform future research in this field. In the following sections, we addressed how the selected studies approached the construction of their ML pipelines.

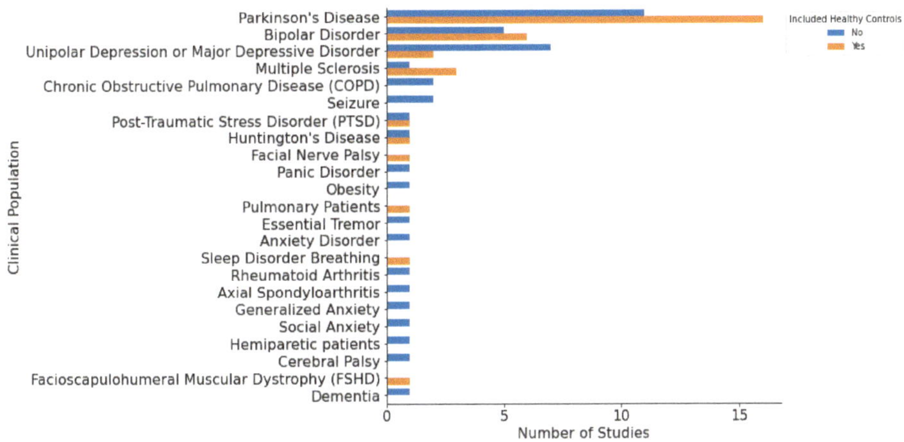

Figure 2. Clinical populations and the use of healthy controls in the selected studies.

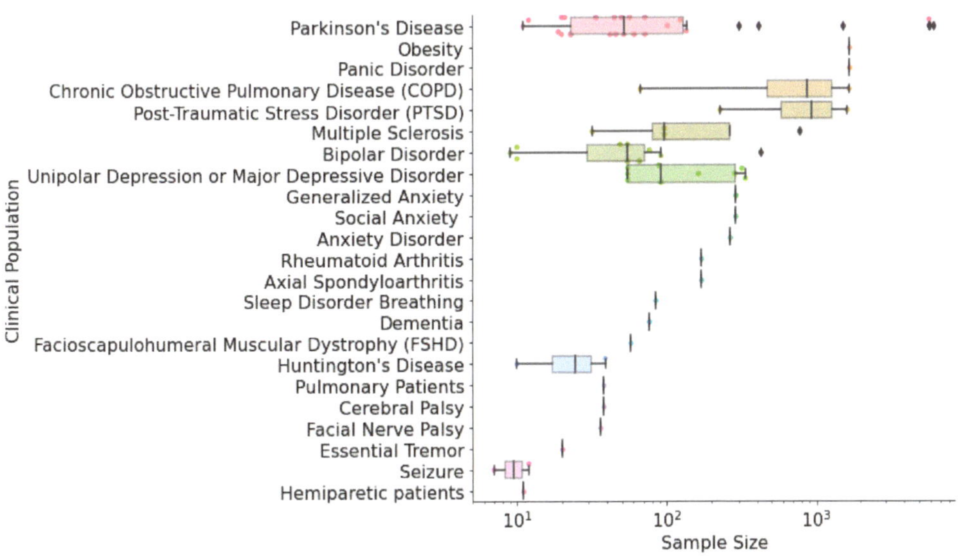

Figure 3. Sample sizes of clinical populations included in selected studies, with x-axis (sample size) presented on a logarithmic scale.

4. Missing and Outlier Data

Missing and outlier data are commonly encountered problems for remote sensing clinical trials. Missing data can be the result of device charging frequency, device robustness, and participant compliance [18]. Outliers can be the result of sensor or device dysfunction or malfunction, incorrect data entry, and incorrect classifications [19]. Data preprocessing, which refers to the dropping or manipulation of data, is required for identifying and removing redundant or irrelevant data and for cleaning the data prior to analysis. Without preprocessing, learning from an imperfect dataset can influence the prediction accuracy of the models [20]. In this section, we address how the selected studies preprocessed their raw data by treating their missing data and outliers, and the limitations of doing so.

4.1. Handling of Missing Data

Missing data can be Missing Completely at Random (MCAR), Missing at Random (MAR), and Missing Not at Random (MNAR) [21]. MCAR assumes that each observation has the same probability of being included or being missed; therefore, there is no difference in the characteristics between participants or observations without missing data and those with missing data. For example, data may be missed due to the battery of the smartphone running out. MAR assumes that missing data may have systematic differences between the missing and non-missing data; however, the cause of the missing data can be explained by the non-missing data. For example, a smartphone may have more missing values when the smartphone battery is low. If the battery percentage is known during the data acquisition, researchers can verify the probability of acquiring missing data depending on the battery percentage. MNAR assumes that missing data are caused by unknown reasons. For example, smartphone sensors may be gradually worn down, which therefore creates more missing data over time. The type of missing data present in the dataset influences whether a researcher should ignore, exclude, or impute the missing data.

Among the selected studies, we found that only 21 of the studies reported the quantity of missing data acquired. Only 29 studies reported how they handled their missing data. We found that complete-case analysis and imputation were the most popular. We identified 14 studies that report using complete-case analysis [22–36]. Complete-case analysis (otherwise known as listwise deletion) is the deletion of an observation involving one or multiple elements of missing data [26,37,38]. While complete-case analysis is the simplest approach to handle missing data, it does reduce the sample size and statistical power of the analysis [39] and can potentially lead to bias if the data are not MCAR [40]. Imputation is the statistical process of replacing missing data with substituted inferred values [41]. We identified studies that imputed their missing data using linear interpolation [29,42,43], forward filling [44], -1 [45], zeros, median, means, and the most frequent value in the column [24,46]. The advantage of imputation is that it enables researchers to use all observations in the dataset. However, the inclusion of imputed values can lead to a false impression of the number of complete cases and reduce the variance in the dataset [47–49].

4.2. Identification of Outliers

Aggarwal's *Data mining: the textbook* states that it is the subjective definition of the researcher that defines an outlier [50]. In cases where the outlier data were discussed in the selected studies, we found that researchers customized their definition of outliers by either defining a range of acceptable values [32] or by defining a threshold based on the mean and standard deviation [51–53]. Visual inspection by the researchers or the optimization of different threshold mechanisms can both be used to define the boundaries of normal or outlier data [54,55]. Maleki et al. defined outliers as observations that were most likely the result of measurement errors [36]. In terms of the handling of outliers, we only identified six studies that explicitly stated that outliers were excluded [26,30,51–53,56].

5. Feature Engineering

5.1. Feature Scaling

Feature scaling is used to normalize the ranges of the features in a dataset [57]. Several feature engineering techniques and ML models (such as Principal Component Analysis and Linear Regression) calculate the distances between two observations. If one feature has a broader range of values compared to the other features, the calculated distances will be heavily influenced by this feature [58]. Therefore, the ranges of all the features should be normalized or standardized so that each feature is appropriately and proportionally considered with respect to the estimated distances [57]. Feature normalization is a common scaling method for rescaling the features into a bounding range using the minimum and maximum values, for example, between 0 and 1. Normalization is an ideal approach when the distribution of the data is not Gaussian, as normalization preserves the original distribution of the data. However, normalization uses minimum and maximum values to define ranges. This makes the method sensitive to outliers [57,59]. Alternatively, feature standardization, also known as z-score normalization, is a method for rescaling the data to fit a standard normalized distribution by using the mean and standard deviation and does not define a bounding range. Consequently, the standardization approach is not sensitive to outliers as it has no bounding range [57,59]. Normalization, log-transformation, and standardization have been reported in a small selection of the selected studies [26,27,36,60,61].

5.2. Expert Feature Engineering

Feature engineering is the process of constructing (new) features from the raw data or existing features while maintaining the original patterns and information in the data [62]. The newly engineered features can be added to or replace features in the original dataset. Engineering of the features can speed up the model performance, improve learning accuracy, and ease the interpretability of the model. The latter is particularly important for clinical trials [63]. Features can be engineered manually by relying on domain-knowledge or automatically by using statistical models, such as Principal Component Analysis (PCA) and Deep Learning (DL) [62–64]. All features aim to increase the separability between the classes or signals, which in turn reduces noise in the dataset. While expert engineered features are easy to interpret and explain and have been widely used in the development of digital biomarkers, these features are typically task- or population-dependent. Due to intra-class variability, some clinically relevant characteristics may be exhibited differently by different individuals (such as different symptom profiles among patients with the same diagnosis). Furthermore, expert engineered features may not be sufficient for representing the most important characteristics of complex patterns and can be time-consuming to acquire, especially when handling large-scale datasets [65,66]. As clinical data has expanded in terms of diversity, availability, and complexity, the aforementioned techniques may be insufficient for developing generic features. In the following sections, we address the notable and generic procedures used to perform feature engineering.

5.3. Signal Processing

To monitor changes in the physical activity of study participants using time series data collected from wearable sensors, signal processing is necessary to detect, clean, and analyze the components of interest. The feature extraction technique used is influenced by the sensor type, study objectives, and signal quality. Typically, signal features are extracted from the frequency, time, or cepstrum domain [67]. Frequency domain features show the prominence of a signal within a given frequency, whereas time-domain features show the changes in the signal of time. Cepstrum domain features represent the rate of change in the different frequency bands. The analysis of the frequency, time, or cepstrum domain features is not mutually exclusive. We identified studies that use both time- and frequency-based features for the estimation of gait speed [68], speech-tasks [69], seizure detection [70], tremor detection [71], and FOG detection [72]. In particular, Tougui et al. built 138 voice-related features extracted from the cepstral, frequency, and time domains [24]. In sum, time series data collected from wearable sensors can be used to monitor the physical activity of study participants, but signal processing is necessary to extract meaningful features. Different feature extraction techniques can be used depending on the sensor type, signal quality, and study objectives. The analysis of these features is not mutually exclusive, and studies that use multiple domains for different clinical applications have been identified.

5.4. Principal Component Analysis

A common linear dimensionality reduction technique for feature engineering and selection is Principal Component Analysis (PCA) [28,73,74]. PCA is used to sufficiently explain a high-dimensional dataset through a few principal components and, therefore, to reduce a high-dimensional dataset to one of fewer dimensions [75]. To this purpose, PCA converts a set of correlated features into a set of uncorrelated features by utilizing orthogonal transformation [75]. The principal components enable a reduction in the feature space by creating a linear combination of the original features, which consequently reduces the storage space and reduces the learning time. Therefore, the periodic components within a concurrent time series dataset can be isolated using PCA, which can subsequently be used to identify any underlying patterns within the dataset. It is important to note that PCA assumes that the data are normally distributed and is sensitive to feature variance [75,76]. Consequently, features with larger ranges will dominate features with smaller ranges. To make the variables comparable, transformation of the data prior to PCA is required [75,76]. Of the studies selected, PCA was used to engineer and select features from times series data sourced from waist-worn triaxial accelerometers and wearable activity trackers [28,73,74]. However, the limitations of PCA are its sensitivity to missing data and outliers and the limited interpretation of the original features. Hence, this observation highlights the need for thorough data preprocessing prior to using PCA.

5.5. Clustering

A clustering algorithm is a common feature engineering method that assigns similar observations to a single cluster and assigns dissimilar observations to another [77]. While PCA compresses the features into principal components, clustering compresses the individual observations into clusters. The grouping of similar observations can improve the model's ability to discriminate between classes [78]. Clustering algorithms, more specifically DBSCAN and K-means clustering, have been deployed in smartphone GPS systems and Wi-Fi-network sensors to extract meaningful location features such as frequented location clusters [79], location patterns [80], and mobility patterns [81]. These studies demonstrate that clustering algorithms are a powerful method for reducing the number of observations into a smaller number of artificial variables that account for the variance within the dataset.

5.6. Deep Learning

The performance of ML models can be limited by the development of manual and arbitrary features, and this potential obstacle can be overcome by DL algorithms. DL algorithms eliminate the need for manual feature engineering, as the DL layers can translate the data into more compact and intermediate abstractions of the data, which in turn can be used as features to predict the final output [82]. While DL can reduce the need for manual data preprocessing and feature extraction, which can potentially improve the generalizability and robustness of a model, the interpretation of the DL model is difficult, as the abstracted features may not be explainable by clinicians. However, it is important to note that the discriminative power of the DL-derived abstractions is strongly influenced by the architecture of the DL algorithm, which is also dependent on the trial-and-error process [59]. Due to DL's representation learning, DL is data-hungry, and therefore requires more data than other ML algorithms [83,84]. For clinical trial data, because of technological limitations and small sample sizes, there may not be enough data to train a sufficiently representative DL model [76,83].

Four studies used DL to engineer features using time series data [23,85–87]. These models were used to extract gait features from accelerometer data [85,87] and tremor characteristics from IMU data [23,86]. However, it should be noted that the DL models do not always outperform the 'shallow learning' models, as shown in a study by Juen et al., in which smartphone accelerometers were used to predict natural walking speed and distance during a six-minute walk test [85].

6. Feature Selection

In recent decades, high-dimensional clinical datasets have relied on feature selection [88]. Feature selection is the process of selecting a subset of the most informative features that will be processed by the ML algorithm [89]. Reducing the features for analysis has both computational and practical benefits. Selecting features can limit storage requirements, increase the algorithm processing speed, increase the interpretability of a model, and improve model performance.

6.1. Overfitting and Underfitting

Overfitting and underfitting are common pitfalls for ML models. Overfitting refers to when a ML model fits too well to its training dataset and is unable to generalize its patterns to unseen data. This problem can occur when the training dataset is small and not representative of the overall potential data distribution. Additionally, if the training dataset contains many outliers, the ML model may also fit the outlier data. Underfitting occurs when the trained ML model is too simple; therefore, it cannot identify the relationship between the features and the outputs. Underfitted models will perform poorly for both the train and validation datasets. To address overfitting, reducing the number of features considered by the model or updating the model architecture to include fewer features can be effective [90]. Underfitting can be improved by adding more features considered by the model or by updating the model architecture to increase the complexity of the feature space [90].

Feature selection identifies the most important features in the dataset and eliminates the irrelevant ones, which thereby reduces noise. However, it is important to strike a balance, as strict feature selection may remove important signals from the data. Therefore, selecting the optimal set of features is important for preventing over- and underfitting. In the following sections, we will elaborate on the three general methods of feature selection that are suitable for ML models [75].

6.2. Filter Methods

Filter methods are used during preprocessing prior to training the ML model. Filtering involves removing features based on domain knowledge, missing data, low variance, or correlation [89,91,92]. As filter methods are independent of any model that is to be used in later steps, they are typically faster to implement and reduce the need for repeating feature selection for different ML models. In our selected studies, we found five studies that used Analysis of Variance (ANOVA), Pearson's Correlation, or Spearman's Correlation to identify features that were statistically significant predictors of the outcomes [24,93–96]. p-value based feature selection, while commonly used in clinical studies, is not always suitable for training a ML model. The use of p-values to identify statistically significant features was a popular approach that relied on the belief that insignificant features were not informative. However, important features can be missed when sample sizes are small. Furthermore, p-values can be biased towards low values due to the increased risk of type 1 errors during multiple comparisons, which in turn increases the probability of random variables being included into the final statistical model [97,98]. Additionally, p-value based feature selection methods may be based on certain assumptions that may not be applicable to ML models, such as assuming that the distribution of scores for the groups among the independent variables are the same [99].

We wanted to highlight one filtering method identified in our selected studies: Relief [100]. Relief is a feature selection technique that also ranks features and selects only the top-scoring features; however, it is notably sensitive to feature interactions [101,102]. Yaman et al., first obtained 177 speech-related features and used Relief to select 66 most predictive vocal biomarkers for the classification of PD [103]. Rodriguez-Molinero used Relief to select frequency features that were subsequently used to predict gait disturbances among PD patients [104]. Overall, Relief has demonstrated its effectiveness in selecting relevant features in various studies related to the prediction of PD using high-dimensional clinical datasets.

6.3. Embedded Methods

The embedded method is a feature selection technique integrated into the ML algorithm itself and is commonly seen in penalized regression [105]. Penalized regression algorithms aim to learn the optimal coefficients for each feature by minimizing its loss function. Regularization (also known as penalization) limits the learning process of the model by increasing the penalty of the loss function [106]. The two common penalized regression methods, identified in the selected studies, are LASSO (also known as L1 penalization) (N = 9) [22,24,29,33,42,95,100,101,107,108] and Ridge (L2 penalization) (N = 2) [109,110]. An advantage of LASSO is that it eliminates non-informative features by reducing their coefficients to zero. The first limitation of LASSO is that, if the number of features f is greater than the number of observations o, LASSO will select a maximum of o predictors as non-zeros, regardless of the relevance of other features. The second limitation is that LASSO also suffers from collinearity; hence, if two or more variables are highly correlated, then LASSO will randomly select one feature and penalize the other correlated features. A disadvantage of Ridge is that it only reduces the weights of the non-informative features by reducing their coefficients towards zero, but it never reduces the number of variables. Therefore, all predictors are included in the final model. However, because of this approach, Ridge protects ML models from overfitting [111].

6.4. Wrapper Methods

Wrapper methods rely on a stand-alone model to select features, but the performance of the selected features is reflected in the performance of the trained model [112]. The wrapper method algorithms tend to be greedy search algorithms that aim to select the optimal feature subset by iteratively selecting the features based on ML performance. As the wrapper method is an iterative process and the model must be evaluated on each feature subset combination, this method is computationally expensive. Wrapper-based feature selection can be completed by ranking the features in terms of relative importance using a ML model (such as decision trees or random forests) [88,101,113]. We identified a handful of feature ranking methods that include two stepwise regression techniques: Forward Selection and Backwards Elimination [29,36,52,114–116], as well as Recursive Feature Selection (RFE) [30,117]. Forward selection starts the modelling process with zero features and adds a new feature to the model incrementally, each time testing for statistical significance. Backwards elimination starts the modelling process with all features and incrementally removes each feature to evaluate its relative importance in predicting the model output [97,118]. RFE fits a model, ranks the features, and removes the least informative features and continues to remove features until a predefined number of features is met [64,119,120]. Senturk et al., illustrated that RFE-based feature selection increased the prediction accuracy of ANN, CART, and SVM when using vocal data to classify a PD diagnosis [121].

7. Machine Learning Algorithms

ML algorithms build a statistical model based on a training dataset, which can subsequently be used to make predictions about a new, unseen dataset. ML algorithms have been used in a wide variety of clinical trial applications, such as the classification of a diagnoses, classification of physical or mental state (such as a seizure or mood), and the estimation of symptom severity. Within the realm of clinical research, ML algorithms can be broadly divided into two learning paradigms: supervised and unsupervised learning [122]. In this section, we will discuss the model objectives of supervised and unsupervised learning and the specific ML models used to achieve these model objectives.

Supervised ML algorithms use labeled data to map the patterns within a dataset to a known label, while unsupervised ML algorithms do not [123]. Rather, the unsupervised ML algorithms learn the structure present within a dataset without relying on annotations. Supervised learning can be used to automate the labelling process, detect disease cases, or predict clinical outcomes (such as treatment outcomes). There are scenarios when experts or participants can provide labelled data; however, it can become labor-intensive or time-consuming to label every observation. For example, a supervised learning algorithm trained to classify human sounds can be used to automatically annotate and quantify hours of coughs [124] and instances of crying [125]. These algorithms can also be used to differentiate between clinical populations and control participants [95] to identify known clinical population subtypes [23] or classify a clinical event (such as a seizure or tremor) [126]. The majority of our selected studies (N = 38) used a clinician to provide the label data. Some studies (N = 22) used a combination of a clinician and self-reported label data, and six studies solely relied on self-reported assessments. Unsupervised ML algorithms can be used to investigate the similarities and differences within a dataset without human intervention. This makes it the ideal solution for exploratory data analysis, subgroup phenotype identification, and anomaly detection. Among digital phenotyping studies, unsupervised learning has been used to identify location patterns [81] and classify sleep disturbance subtypes using wrist-worn accelerometer data [127].

It is important to recognize that unsupervised and supervised methods are not mutually exclusive, and they can be effectively combined. For instance, unsupervised methods can be employed to extract a meaningful latent representation of the input data. Subsequently, these latent vectors, along with the original inputs, can be used as inputs for a supervised model. This type of approach is commonly observed when applying techniques such as PCA, clustering, or other dimensionality reduction methods [29,73,74,128]. By combining unsupervised and supervised methods, valuable information can be extracted from the data and used to enhance the performance and interpretability of the overall model.

In clinical research, supervised ML algorithms have been used to classify class labels or estimate scores. Classification algorithms learn to map a new observation to a predefined class label. These algorithms can be used to classify patient populations and patient population subtypes and identify clinical events. Regression algorithms learn to map an observation to a continuous output. These algorithms are commonly used to estimate symptom severity [129], quantify physical activity, and forecast future events [130]. Among the selected papers that were focused on the classification of a diagnosis or state, the four most common algorithms were Random Forest, Support Vector Machine, Logistic Regression, and k-Nearest Neighbors (Figure 4). Some additional classification algorithm families identified were Naïve Bayes, Ensemble-based methods (including Decision Trees, Bagging, and Gradient Boosting), and Neural Networks (such as Convolutional, Artificial, and Recurring Neural Networks). The three most common algorithms for the regression-focused papers were Linear Regression (including linear mixed effects models), Support Vector Machine, and k-Nearest Neighbors (Figure 4). We found that most studies only considered or reported a single ML algorithm (N = 32). Additionally, 29 of the studies considered or reported two to five ML algorithms, and the remaining 5 studies considered six or more. The following section provides an overview of the most widely used machine learning models, their properties, advantages, and disadvantages. In addition, we discuss some notable off-the-shelf ML approaches and some custom-built ML methods such as transfer learning, multi-task learning, and generalized and personalized models.

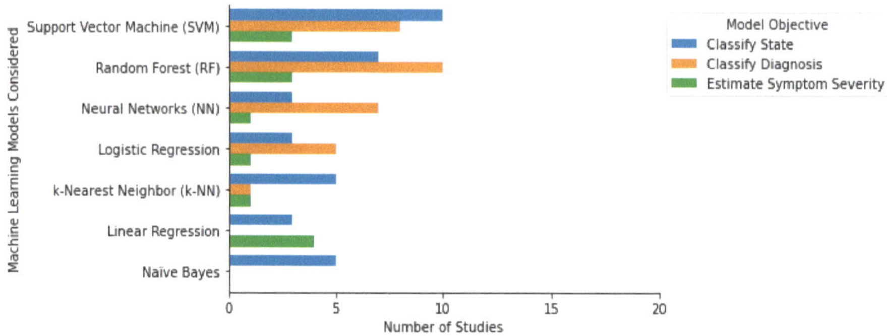

Figure 4. Machine learning algorithms and their respective objectives in the selected studies.

7.1. *Tree-Based Models*

A Decision Tree (DT) is a supervised non-parametric algorithm that is used for both classification and regression. A DT algorithm has a hierarchical structure in which each node represents a test of a feature, each branch represents the result of that test, and each leaf represents the class label or class distribution [131,132]. A Random Forest (RF) algorithm is a supervised ensemble learning algorithm consisting of multiple DTs that aims to predict a class or value [133]. Ensemble learning algorithms use multiple ML algorithms to obtain a prediction [134]. Tree-based models have several benefits. As each tree is only based on a subset of features and data and because they make no assumptions about the relationship between the features and distribution, they are not sensitive to collinearity between features, can ignore missing data, and are less susceptible to overfitting (for multiple trees), making the model more generalizable [135]. Another advantage of RF and DT models is that they can support linear and nonlinear relationships between the dependent and independent variables [136]. Further, as the design of the RF models can be interpreted in terms of feature importance and proximity plots, the interpretability of the RF model is feasible. However, a limitation of using tree-based models is that small changes in the data can lead to drastically different models. Additionally, the more complicated a tree-based model becomes, the less explainable a model becomes. However, pruning the trees can help to reduce the complexity of the model.

According to the selected studies, RF is a versatile and powerful model used for classification and regression tasks across multiple datatypes and populations. RF models have been used for the classification of diagnoses among PD patients [107,110], Multiple Sclerosis [34,118], and BD and unipolar depressed patients [45,61]. It is also a popular classification model for the classification of states or episodes, such as the detection of flares among Rheumatoid Arthritis or Axial Spondylarthritis patients [32] and tremor detection among PD patients [137], to quantify physical activity among cerebral palsy patients [138] and detect the moods of BD patients [69,139]. RF regression algorithms have also been used to predict anxiety deterioration among patients who suffer with anxiety [140].

7.2. *Support Vector Machines*

A Support Vector Machine (SVM) is a supervised algorithm that is used for classification and regression tasks. The objective of a SVM is to identify the optimal hyperplane based on the individual observations, also known as the support vectors. For SVM regression, the optimal hyperplane represents the minimal distance between the hyperplane and the support vectors. Whereas for SVM classification, the objective is to find the hyperplane that represents the maximum distance between two classes [141]. The hyperplanes can separate the classes in either a linear or non-linear fashion [136]. Given that SVMs are influenced by the support vectors closest to the hyperplanes, SVMs are less influenced by outliers, making them more suitable for extreme case binary classification. The performance of a SVM can be relatively poor when the classes are overlapping or do not have clear decision boundaries. This makes SVMs less appealing for classification tasks as inter class similarity is low. SVMs are computationally demanding models as they compute the distance between each support vector; hence, SVMs do not scale well for large datasets [142].

SVM classifiers have been used to classify clinical populations (e.g., facial nerve palsy and their control participants) [143]. SVM classifiers have also been used to classify events or states, such as detecting gait among PD patients [104] and classifying seizures among epileptic children [144]. We identified studies that used SVM regression to estimate motor fluctuations and gait speed among PD and Multiple Sclerosis patients, respectively [74,145].

7.3. k-Nearest Neighbors

A k-Nearest Neighbor (k-NN) algorithm is a non-parametric supervised learning approach that can be used for multi-class classification and regression tasks. Classification k-NN algorithms determine class membership by the plurality vote of its nearest neighbors. They can estimate the continuous value of an output by calculating the average value of its nearest neighbors [136]. Given this, the quality of predictions is not only dependent on the amount of data but also on the density of the data (the number of points per unit). K-NN is simple to implement, intuitive to understand, and robust to noisy training data. However, the disadvantage is that k-NN is computationally slow when it is faced with large multi-dimensional datasets. Further, k-NN does not work well with imbalanced datasets, as under- or over-represented datapoints will influence the classification [146].

The most popular application for k-NN algorithms is for wearable-based time series data. K-NN classification models have been used to classify PD and healthy controls [24], classify tremor severity [147], predict acute exacerbations of chronic obstructive pulmonary disease (AECOPD) [44], and identify mood stability among BD and MDD patients [33,69,148]. Using wearable data, k-NN regression models have been used to predict the deterioration of symptoms associated with anxiety disorder [140].

7.4. Naïve Bayes

A Naïve Bayes (NB) classifier is a supervised multi-class classification algorithm. NB classifiers calculate the class conditional probability—the probability that a datapoint belongs to a given class in the data [141,149]. NB classifiers are computational efficient algorithms; thus, they are suitable for real-time predictions, scale well for larger datasets, and can handle missing values. A limitation of NB is that it assumes that all features are conditionally independent; hence, it is recommended that collinear features are removed in advance. Another limitation is that when new feature-observation pairs do not resemble the data in the training data, the NB assigns a probability of zero to that observation. This approach is particularly harsh, especially when dealing with a smaller dataset. Hence, the training data should represent the entire population.

As NB classifiers help form classification models, we found that NB classifiers have been used for the classification of tremors or for freezing gait among PD patients [52], as well as to classify flares among Rheumatoid Arthritis and Axial Spondylarthritis patients [32] and classify bipolar episodes and mood stability among BD and MDD patients [33,69,148].

7.5. Linear and Logistic Regression

A Linear Regression model is a supervised regression model that predicts a continuous output. It finds the optimal hyperplane that minimizes the sum of squared difference between the true data points and the hyperplane. A Logistic Regression model is a supervised classification model that can be used for binomial, multinominal, and ordinal classification tasks. Logistic Regression classifies observations by examining the outcome variables on the extreme ends and determines a logistic line that divides two or more classes [136]. Linear and Logistic Regression are popular in algorithms as they are easy to implement, efficient to train, and easy to interpret. However, a limitation of both models is that they make multiple assumptions, e.g., that a solution is linear, the input residuals are normally distributed, and that all features are mutually independent [150]. Multicollinearity, the correlation between multiple features, and outliers will inflate the standard error of the model and may undermine the significance of significant features [151]. Further, outliers that deviate from the expected range of the data can skew the extreme bounds of the probability, making both algorithms sensitive to outliers in the dataset [150].

Linear Regression has been used to quantify tremors among Essential Tremor (ET) patients [116] and to estimate motor-related symptom severity among PD patients [31,93]. It has also been used to forecast convergence between body sides for Hemiparetic patients [130]. Logistic Regression was a popular approach for classifying PD diagnosis [107,110], Post-Traumatic Stress Disorder [109], and distinguishing fallers and non-fallers [152]. Logistic Regression has been used to classify drug effects, such as predicting the pre- and post-medication states among PD patients [22].

7.6. Neural Networks

Neural Networks (NN), also known as Artificial Neural Networks (ANN), can be used for unsupervised and supervised classification and regression tasks [153]. NN consists of a collection of artificial neurons (or nodes). Each artificial neuron receives, processes, and sends the signal to the artificial neuron connected to it. The neurons are aggregated into multiple layers, and each layer performs different transformations on the signal. The signal first travels from the input layer into the output layer while possibly traversing multiple hidden layers in between. NN offer several advantages, such as the ability to detect complex non-linear relationships between features and outcomes and work with missing data, while it also requires less preprocessing of the data and offers the availability of multiple training algorithms. However, the disadvantages of NN include increased computational burden, reduced explainability and interpretability (as NN are 'black box' in nature), and the fact that NN are prone to overfitting [154]. However, it is important to highlight the growing number of studies that specifically explore explainable deep learning approaches for biomarker discovery and development. Studies utilizing methodologies such as LIME (Lime Tabular Explainer), SHAP (Shapley Additive exPlanations), and other visual inspections of feature distribution and importance have aided clinicians in understanding the model mechanisms. These approaches also provide patient-specific insights by describing the importance of each feature, which may, in turn, facilitate personalized treatment opportunities [90,155–157].

The most popular applications for neural networks were for the classification of a diagnosis or classification of a state or event. The most popular application is the detection of tremors among PD patients [23,52,86,137,158]. NN have been used to classify unipolar and bipolar depressed patients based on motor activity [45,159], estimate depression severity [159], forecast seizures [160], and classify a treatment response using keyboard patterns among PD patients [161].

7.7. Transfer Learning

Transfer learning (also known as domain adaption) refers to the act of deriving the representations of a previously trained ML model to extract meaningful features from another dataset for an inter-related task [162]. One applicable scenario is the training of a supervised ML model on data collected in a controlled setting (such as in a lab or clinic). The performance of the model may suffer when applied to a dataset collected under free-living conditions. Rather than developing a new model trained solely on a free-living condition dataset, transfer learning can use patterns learned from the controlled setting dataset to improve the learning of the patterns from the free-living conditions dataset.

Transfer learning can also be a valuable technique for enhancing the utilization of limited or rare data [163]. One practical application is to employ pretraining on abundant control data and subsequently finetune the model on the specific population of interest to improve the model's performance [163–165]. This approach not only optimizes the efficiency of utilizing scarce data but also facilitates model personalization. By adapting a pretrained model to individual characteristics or preferences, it becomes possible to create personalized models that better cater to unique needs or circumstances. Transfer learning thus offers a powerful means to leverage existing knowledge and make the most of available data resources, enhancing both the efficiency and personalization of biomarkers.

Given its application, transfer learning reduces the amount of labeled data and computational resources required to train new ML models [162], thus making this method advantageous when the sensor modalities, sensor placements, and populations differ between studies. While we only identified two studies that applied transfer learning to estimate PD disease severity using movement sensor data [166,167], we predict that the application of transfer learning will enable future researchers to overcome the challenges of a limited dataset and develop more sensitive and effective ML models.

7.8. Multi-Task Learning

Multi-task learning (MTL) enables the learning of multiple tasks simultaneously [168]. Learning the commonalities and differences between multiple tasks can improve both the learning efficiency and the prediction accuracy of the ML models [168]. A traditional single-task ML model can have a performance ceiling effect, given the limitations of the dataset size and the model's ability to learn meaningful representations. MTL uses all available data across multiple datasets and can learn to develop generalized models that are applicable to multiple tasks. To use MTL, there should be some degree of information shared between or across all tasks. The correlation allows MTL to exploit the underlying shared information or principles within tasks. Sometimes MTL models can perform worse than single-task models because of 'negative transfers'. This occurs when different tasks share no mutual information or if the information of tasks are contradictory [169]. MTL models have been used to simultaneously model data sourced from two separate sources or to model multiple outcomes [170,171]. For example, Lu et al. explored the use of MTL to jointly model data collected from two different smartphone platforms (iPhone and Android) to jointly predict two different types of depression assessments (QIDS and a DSM-5 survey) [79]. They illustrated that the classification accuracy of the MTL approach outperformed the single-task learning approach by 48%; thus, the classification model benefited from learning from observations sourced from multiple devices.

7.9. Generalized vs. Personalized

ML algorithms can be trained on population data or individual subject data. Generalized models, which are trained on population data, are fed data from all participants for the purpose of general knowledge learning. Conversely, personalized models are trained on an individual's data and take into consideration individual factors such as biological or lifestyle-related variations [172]. We have adopted these terms from Kahdemi et al.'s study, in which they developed generalized and personalized models for sleep-wake prediction [173]. The heterogenous nature of each population or individual can be a potential hinderance for generalizable models. A single individual's deviation from the 'norm' may be viewed as a source of 'noise' in a generalized model. For example, patients with mood disorders such as MDD and BD have large inter-individual symptom variability. Abdullah et al., reliably predicted the social rhythms of BD patients with personalized models using smartphone activity data [30]. Cho et al. compared the mood prediction accuracy of personalized and generalized models based on the circadian rhythms of MDD and BD participants [38]. Their studies illustrated that their personalized model predictions were, on average, 24% more accurate than the generalized models. These studies lay the groundwork for developing personalized models that are more sensitive to individual differences.

8. Model Hyperparameters

The process of building an effective ML model consists of two main steps: selecting the appropriate ML algorithm and optimizing the model performance by tuning its parameters.

Each model consists of two types of parameters: (i) the parameters that are initialized and continuously updated throughout the learning process (e.g., the weights of neurons of a neural networks) and (ii) the hyperparameters that must be set prior to the learning process as they define the model architecture (e.g., the regularization parameters of a Linear Regression model, and the learning rates of a neural network) [174]. Every combination of the selected hyperparameters will have a direct influence on the performance of the learned model. For example, as the number of trees in a RF increases, the more features tend to be selected by the model, which may not always be relevant for the development of biomarkers [175]. Similarly, the number of layers, number of neurons per layer, activation functions, and the regularization techniques used for NN can each influence the model performance [176]. While most ML algorithms come with default values for the hyperparameters, these may not be optimal for the dataset at hand, and even tuned hypermeters are at risk of being non-optimal for a different dataset. The process of selecting the optimal hyperparameter configurations is known as hyperparameter tuning [177].

To identify the optimal hyperparameters for a model, researchers must define the hyperparameter space and the hyperparameter search strategy. When defining the hyperparameter space, the distribution of the hyperparameter ranges can be either uniform or logarithmic. The uniform distribution assigns equal probability to all hyperparameter values within a manually defined range. The log-uniform distribution samples hyperparameter values uniformly between the logarithmic transformations of the lower and upper thresholds. We argue that log-uniform distribution is particularly useful when exploring values that vary over several orders of magnitude. Consider the example of tuning a linear regression model with the hyperparameter alpha, which determines the strength of regularization. To efficiently explore a wide range of alpha values, such as between 0.001 and 10, the log-uniform distribution allows for an evenly distributed search space over different orders of magnitude. Log-uniform distribution can be used for the initial exploration of a large range of hyperparameter values. The range can then be narrowed down to explore with a uniform-distribution to determine the optimal hyperparameters for the respective models.

The manual tuning of hyperparameters is impractical due to the large number of available hyperparameters, hyperparameter configurations, and time-consuming model evaluations. Automated tuning approaches are preferred, and there are a wide variety of approaches available, including GridSearch, RandomSearch, and Bayesian Optimization [177]. GridSearch uses brute force to test a finite combination of hyperparameters to identify the optimal hyperparameter configuration [178]. This approach can suffer from the effects of dimensionality, as more potential hyperparameter configurations can be time-consuming and computationally expensive. An alternative to GridSearch is RandomSearch. RandomSearch only samples a subset of all possible hyperparameter configurations within a specific time or computational budget [179]. While RandomSearch only relies on a subsample of configurations, it has been shown to outperform the GridSearch method [179]. As GridSearch and RandomSearch do not consider previous performance evaluations for their hyperparameter optimization strategy, they are inefficient in exploring the hyperparameter search space. Bayesian Optimization, which uses Bayes Theorem, is a powerful approach. It considers previous hyperparameter evaluations to choose which hyperparameters to evaluate next and disregards potential hyperparameter combinations that are deemed irrelevant [178]. This approach reduces the time and computations required for hyperparameter tuning. The benefit of using these more automated approaches to hyperparameter tuning is three-fold. First, it reduces the time effort required to optimize a ML model. Next, the performance of the ML models is improved as the hyperparameters explore different optimal model configurations for different datasets. Finally, when the hyperparameters and their ranges (together also referred to as the hyperparameter space) and the hyperparameter tuning methods are reported, the models and the findings become reproducible [180]. When similar hyperparameter tuning processes can be used for different ML algorithms for different datasets, researchers can then identify the optimal ML model.

Among the selected studies, 25 discussed which hyperparameters were considered for their models [23,24,34,43,44,46,53,69,73,86,87,94,95,107–110,114,138,158,159,181–184], of which one stated they used the default hyperparameters of the models [69]. Only nine studies discussed how they selected or optimized their hyperparameters. We identified four studies that stated GridSearch was used for the hyperparameter tuning [36,46,95,110]. We did not identify any studies that used RandomSearch or Bayesian Optimization. The limited reporting of hyperparameters and the hyperparameter tuning process poses a problem for the transparency, reproducibility, and comparison of ML models.

9. Model Evaluation

Assessing a ML model's performance is an essential component for determining the usability and reliability of the model. Depending on the objective of the research, it is often necessary to try to compare the performance of multiple ML models to identify the optimal model [185,186]. In ML, the terms metric and measure are often used interchangeably, but they do have slightly different meanings. A metric is a function used to evaluate the performance of a model, while a measure is a numerical summary of the performance of a model obtained using one or more metrics. It is best practice to use multiple metrics and model performance visualizations for the model evaluation, as a model may perform well for one evaluation metric and poorly for another. Using multiple evaluation metrics ensures that the model is operating optimally and correctly. The following sections provide more details about the performance metrics used for classification and regression models. Table 4 provides an overview of the most common performance metrics used in the selected studies, their respective calculations, and their clinical interpretations.

Table 4. Clinical interpretations of common ML performance metrics (True Positive = TP, True Negative = TN, False Positives = FP, False Negatives = FN, Sum of Squares of Residuals = RSS, Total Sum of Squares = TSS, Number of Observations = N).

Term	Title 2	Title 3
Accuracy	$\frac{TP + TN}{All\ Observations}$	Out of all the predictions, how many predictions were correctly identified as positive or negative?
Precision	$\frac{TP}{TP + FP}$	How many predictions were correctly labeled as patients out of all correctly classified patients and misclassified healthy controls?
Specificity	$\frac{TN}{TN + FP}$	How many predictions were correctly labeled as healthy controls out of all healthy controls? In other words, of all healthy controls, who were correctly identified as such?
Recall/Sensitivity	$\frac{TP}{TP + FN}$	Of all the patients, who were correctly classified/identified as such?
F1-score	$\frac{2x(Precision \times Recall)}{Precision + Recall}$	How many predictions were correctly labeled as patients (recall) and what was the accuracy with regards to correctly predicted patients (precision)?
Mean Square Error	$\frac{1}{N}\sum actual - predicted$	What is the absolute difference between the true scores and the predicted scores?
Root Mean Square Error	$\sqrt{\frac{\sum_{i=1}^{N}(Predicted_i - Actual_i)^2}{N}}$	What is the average difference between the true and the predicted scores (in the same unit of the true scores)?
R2	$1 - \frac{RSS}{TSS}$	What fraction of the variance in the data is captured by the model?

9.1. Classification Measures

Classification models have discrete outcomes; thus, a metric must reflect how often an observation belongs to the correct label or class [187]. There are three categories of classification measures: Threshold Metrics, Ranking Metrics, and Error Metrics. Threshold Metrics (such as accuracy and F1 score) quantify the prediction errors of the classification model as a ratio or rate. Ranking Metrics (such as the Receiver Operating Characteristics (ROC) and Area Under the Curve (AUC)) focus on evaluating classification models based on how effective they can discern separate classes. Error Metrics (such as Root Mean Square Error) quantify the uncertainty of the classification model's predictions. While the Threshold and Ranking Metrics are focused on correct and incorrect predictions, the Error Metrics quantify the proportion of classification errors.

As ML models are increasingly being used to perform high-impact tasks pertaining to clinical assessments, an evaluation metric must be selected based on what the stakeholders find to be important regarding the model prediction, which can make the selection of the model metrics challenging. As seen in Table 4, accuracy, sensitivity, specificity, and precision are calculated based on four test results. The True Positive (TP) and True Negative (TN) indicate the presence or absence of a diagnostic or characteristic. The False Positive (FP) and False Negative (FN) indicate the opposite of the true condition.

Binary classification models typically involve a decision threshold hyperparameter that determines how the model assigns labels based on the predicted probabilities. The default threshold is typically 0.5, meaning that if the predicted probability is greater than 0.5, the positive label is assigned, and vice versa. However, it is important to note that this threshold can be adjusted to accommodate specific needs or domain considerations. To evaluate the performance of binary classification models across different decision thresholds, the ROC curve is commonly used. The ROC curve provides an overview of the model's performance by illustrating the trade-off between TP and FP rates at various threshold values. ROC can aid the assessment of the model's performance across a range of decision thresholds and enable the selection of the threshold that aligns with a specific objective.

It is worth noting that many classification metrics, including accuracy, precision, recall, and F1 score, assume binary labels. However, when dealing with multiclass classification problems, another approach is to use one-vs-rest or one-vs-one strategies, wherein the problem is decomposed into multiple binary classification tasks. The performance of the model on each task can then be evaluated using the binary classification metrics, and the results can be aggregated or averaged to provide an overall assessment of the model's performance on the multiclass problem.

Class imbalance can be an obstacle for assessing model performance. In particular, accuracy, AUC, ROC, may be sensitive to such imbalances [188]. Hence, when facing class imbalance, there are two approaches to consider: one can choose a metric that accounts for class imbalance or one can choose to balance the classes. Metrics such as balanced accuracy, F1-score, or Matthews Correlation Coefficient (MCC) are common metrics for handling class imbalance, as identified by 15 studies [23,24,29,36,44,60,61,107,108,110,114,140,159,161,189]. Balanced accuracy represents the mean of the sensitivity and specificity, while the F1-score represents the mean of the precision and recall [190]. The MCC measures the correlation coefficient of the binary and even multiclass classes. Therefore, the MCC score is high only if the classification model correctly predicts both the positive and negative predictions [190,191].

The other approach to handling class imbalances is adjusting the class distribution using oversampling or undersampling. We identified eight studies that used random over/under sampling or SMOTE [29,44–46,61,95,109,192]. Oversampling techniques duplicate the samples of the minority class, while undersampling removes samples of the majority class. However, these techniques also have their disadvantages, as the duplication of multiple samples can lead to overfitting of a model, while undersampling reduces the diverse representation of the majority class. Thus, we would specifically recommend using the Synthetic Minority Oversampling Technique (SMOTE) with Tomek Links or Edited

Nearest Neighbor (ENN)—two undersampling techniques [193,194]. SMOTE is first applied to create an artificial minority class to minimize the class imbalance. Next, Tomek Links or ENN can be used to remove samples that are close to the boundaries between the classes, which would further separate the classes [193,194].

9.2. Regression Measures

As regression models generate predictions on a continuous scale, the objective is to estimate how close the predictions were to the true values [195]. Among the studies selected, we found that regression models used Distance Metrics and Error Metrics to estimate the strength of the association or the distance between the predicted values and the true values [29,42,87,93,96,128,152]. We would like to emphasize that these metrics are used to compare the performance of the composite biomarkers rather than the performance of the individual features. The most common Distance Metrics were the correlation (also known as R) and the percentage of the variance explained (R^2). Both were used to assess the strength of the association between the predicted and true values [196]. There is no rule of thumb for interpreting the strength of R^2. While an R^2 closer to 1 can be obtained in clinical trials, a low R^2 can still be useful with respect to trends in the data. We would like to address two points of caution when using the R^2 [185,187]. First, it is not always suitable to compare R^2 across different datasets, as different clinical populations are likely to differ in their feature variance. Second, the R^2 will increase with the number of features. To compensate for this, one may use the adjusted R^2 to account for the number of features [197,198].

The Error Metrics included the Mean Absolute Error (MAE), Mean Squared Error (MSE), and Root Mean Squared Error (RMSE) [133]. The MAE measures the average absolute difference between the true and predicted values. The MAE is easy to interpret and robust to outliers. The absolute difference accounts for negative differences. The MSE squares the error instead of providing the absolute error, which gives more weight to the bigger errors. The MSE is sensitive to outliers and not easy to interpret, as the results will not have the same unit as the output. However, the RMSE provides an estimation of the error in the same units as the output while maintaining the properties of the MSE [199].

10. Model Validation

In ML, model validation refers to the process of evaluating the generalizability of a trained model on an unseen dataset. Selecting the most appropriate model validation approach depends on the size and characteristics of the datasets. Three datasets are required for model validation: the training, test, and validation datasets. In most cases, the validation dataset can be a subset of the original dataset; however, this can lead to data leakage, which could produce overly optimistic results. Another approach is to create a validation dataset from an independent (but comparable) dataset, which ensures an unbiased and independent evaluation of the ML model. However, a limitation is that the performance evaluation may reflect high variance due to the limited size of the dataset [200]. Moreover, it is crucial to highlight that a participant should only be present in a single dataset, such as the training dataset, and should not simultaneously appear in other datasets such as the testing or validation datasets. When a participant's observations are distributed across multiple datasets, data leakage can occur, compromising the accuracy estimation and its applicability to new participants [183]. As a result, cross-validation on the observation level rather than the participant level is methodologically flawed. Unfortunately, this is a common issue and needs to be accounted for in future studies [201].

Cross-validation is a popular validation method that uses resampling to train, test, and validate a model using different subsets of the data. The training dataset is used to train the ML model to learn the patterns within a dataset. The validation dataset is used to tune the hyperparameters of the model based on the performance of the ML model trained on the training dataset. The test dataset provides an unbiased estimate of the performance of the final ML model after training and validation. In the scenario when both validation and test datasets are used, the test datasets are only used to assess the model once (via hold-out validation) or multiple times (via nested cross-validation). In general, datasets need to meet two main requirements. The datasets should not have shared or overlapping observations to ensure that data leakage does not lead to bias in the estimates, and all observations must be statistically independent [202]. When applying feature engineering or feature selection with cross-validation, any transformation or selection steps should be performed within each fold of the cross-validation to prevent biasing in the training of the prediction model with information from the test dataset [203]. The overall performance of the prediction models, obtained by averaging across each iteration of the cross-validation, evaluates the effectiveness of the combined feature reduction and learning methods in estimating the label for a given dataset.

Among the selected studies, we found that the most popular cross-validation methods were k-fold cross-validation (N = 27), Leave-One-Out cross-validation (N = 16), and custom validation (N = 8). Overall, 15 studies did not report the use a validation method. K-fold cross-validation randomly splits the datasets in 'k' folds; one-fold is used for testing and the remaining folds are used for training. This step is repeated until every unique fold has been used as the test dataset, and the overall performance is based on the average of the performance of each model in each fold [204]. Leave-one-out cross-validation is a specific type of k-fold cross-validation, wherein individual observations (or participants) are the test datasets, and the remaining cases are used for training. Leave-one-out cross validation prevents data leakage across datasets, as repeated measurements of the same subjects can lead to the violation of independence assumption for ordinary cross-validation [204–206].

We would like to highlight the advantages of the nested cross-validation approach. While nested cross-validation was the least popular approach, we would argue that nested cross-validation is a more robust approach for selecting and evaluating a ML model [207]. Currently, the model section without the nested cross-validation approach uses the same data to both tune the model hyperparameters and evaluate its performance. Therefore, information is "leaked" between the training and validation of the model, which can lead to overfitting [207]. Nested cross-validation consists of an inner loop and an outer loop. The outer loop assesses the model performance, while the inner loop assesses the hyperparameter selection [207]. Each iteration of the outer loop is split into a different combination of training and test sets. The outer loop training set is used in the inner loop, which is further split into a training and validation dataset. The inner loop split is repeated over k-folds, and the best performing model across the k-folds is evaluated in the outer loop. This ensures that different data are used to optimize the models' hyperparameters and evaluate the model's performance. The final model performance represents the average and standard deviation of the model performance as selected by each of the outer loops. Without the standard deviation or confidence intervals, it is not possible to evaluate the spread or stability of the prediction error of the given models [208,209].

It is important to highlight that cross-validation is only used to approximate the generalization error of the models built and not to build the final model that will be used for making predictions [205,210]. The average prediction error across the folds gives an expected error for a single model built on the single dataset. If the variance of the prediction error is too high, then the model is considered unstable. To select a single model, it is recommended that researchers rebuild the model using the full dataset [211]. If an external validation set is available, then this validation set can be used to evaluate and compare the single prediction error to that of the cross-validation prediction error.

11. Recommendations

In this recommendation section, we address the main issues consistently identified in the selected studies and how to amend these issues for future trials (see Figure 5 for a simplified overview of these recommendations). It is important to bear in mind the regulatory implications for developing ML-derived biomarkers. Within the European Union, AI medical systems and devices are considered high risk; therefore, they are subject to stringent reviews prior to being made available on the market [212]. These review requirements emphasize the importance of achieving high levels of performance, transparency, and minimal risk in ML-derived biomarker development [213]. High performance implies that the developed ML models must be accurate, robust, and capable of reliably and consistently predicting the target outcome variable. Furthermore, transparency in ML-derived biomarker development refers to the provision of clear and adequate information to the user, including appropriate human-readable measures to minimize risks associated with the use of the system. The development of ML-derived biomarkers must also aim to minimize risks and discriminatory outcomes, which can be achieved by training the ML model on high-quality datasets that are representative of the target population and by conducting adequate risk assessment checks [214]. These considerations are critical for ensuring the safe and effective use of ML-derived biomarkers in clinical practice.

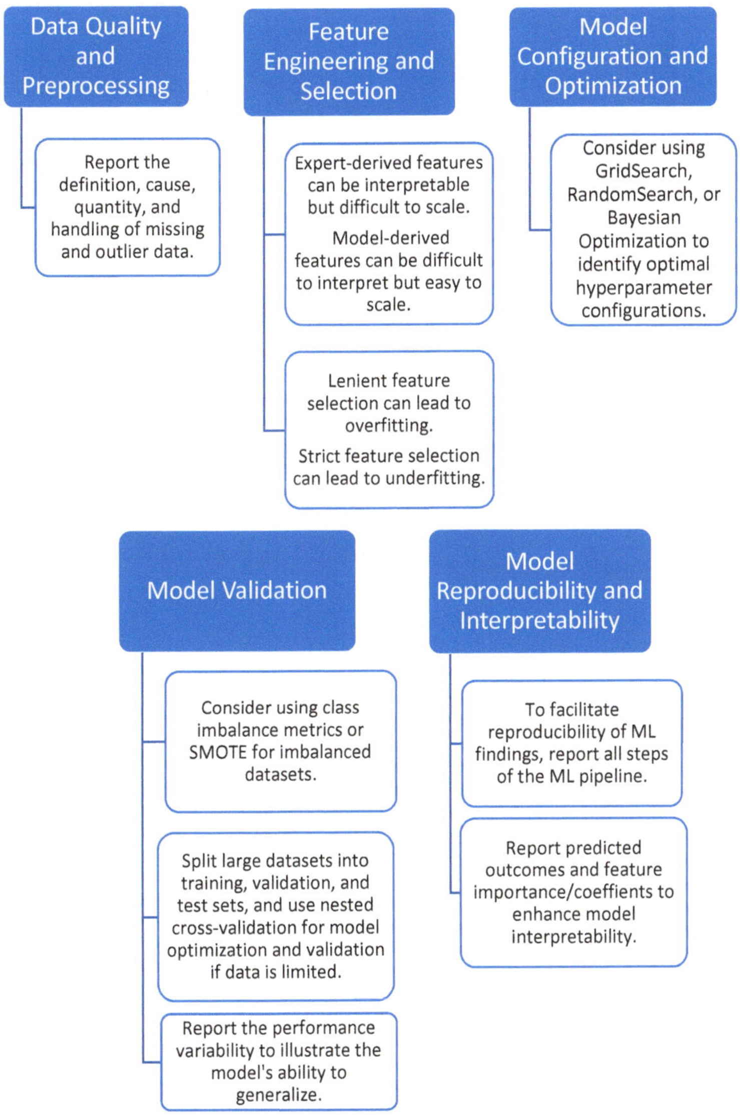

Figure 5. General recommendations for building an effective and reproducible ML pipeline.

11.1. Inclusion of Healthy Controls

When conducting a study focused on disease classification or estimation, the inclusion of control data can serve several purposes. By comparing the data from individuals with the condition of that of the healthy controls, researchers can discern whether the observed differences are specific to the condition or a result of unrelated factors. Moreover, analyzing the performance of a model on control subjects can shed light on the biomarker's effectiveness and reliability. By evaluating how well the model distinguishes between healthy controls and patients with the condition, researchers can gain a better understanding of its predictive capabilities. This evaluation can provide insights into potential false positives or false negatives that may occur when using the model in real-world settings.

It is worth noting that, when including control data, the control data should be appropriately matched with the patient population data. Having age- and gender-matched control subjects can help minimize confounding variables, improving the accuracy of the analysis. This matching process allows researchers to draw more robust conclusions about the relationship between the identified features or patterns and the disease activity while also reducing the potential impact of demographic factors on the results.

The finding that only half of the studies included healthy controls is significant as it highlights a potential gap or limitation in the existing body of research. Without the inclusion of controls, it becomes challenging to attribute identified features or patterns solely to the CNS disorder or the severity of the condition. Further, if the dataset only contains a relatively homogeneous population, it calls the reliability and predictive capabilities of the models into question. We encourage future researchers to include control subjects in their studies, as it would improve the strength of their biomarkers and the validity of their findings.

11.2. Data Quality and Preprocessing

The remote monitoring of clinical trials can generate large and complex datasets that include longitudinal data from multiple subjects and data sourced from multiple sensors, resulting in a multi-dimensional data structure. To this point, we recommend using the WHO mHealth Technical Evidence Review Groups' mHealth evidence and evidence reporting and assessment (mERA) 16-item checklist to provide transparency on which mHealth invention was used, where, and how it was implemented to support the reproducibility of the mHealth data collection [215]. To ensure the quality and reliability of the data, it is important to assess the quality of the data. This assessment includes examining the data for missing and outlier data and understanding how these factors might affect the generalizability and reproducibility of the ML model. While most studies provide detailed information on patient populations, the devices used, and the data collected, they often underreport information related to data quality and preprocessing steps. Therefore, it is important to provide sufficient details on the methods used to preprocess the data, including the quantity of missing and outlier data and the strategies employed to handle such data. This information can ensure that the data collection and preprocessing process can be reproduced, which, in turn, can enhance the credibility and generalizability of the ML model.

11.3. Feature Engineering and Selection

There is a wide variety of manual or automated techniques used for engineering and selecting features to feed a model. ML models perform best when feature engineering and selection are leveraged to formulate potentially clinically relevant features from existing data. In addition, the performance of the ML model can be optimized, and the computational time can be reduced when the redundancy across the features is reduced. While only selecting the most informative features can remove noise (therefore reducing the likelihood of overfitting), selecting too few features may reduce the strength of the (combined) signal in the dataset, making the ML model vulnerable to underfitting. Feature engineering and selection can be guided by domain expertise and/or automated statistical models, where multiple features are evaluated by their importance in predicting the outcome. While automated feature engineering techniques, such as clustering, PCA, and DL, can be used to extract a reduced set of representative features, this risks a potential decline in interpretability, which may limit its clinical application.

11.4. Model Configuration and Optimization

When selecting the ML models, there are several factors that should be considered, such as model objectives, model types, model hyperparameters, and model evaluation. Poor design choices and lenient hyperparameter tuning and validation in these steps can lead to poor model performance. We recommend that researchers carefully consider each step of building their ML pipeline by comparing multiple ML algorithms, using automated methods for assessing multiple hyperparameter configurations, and using nested cross-validation to both optimize and validate the ML models.

11.5. Model Validation

We would recommend using a minimum of three datasets to validate a ML model and train, validate, and test a dataset. At no point should the test set be used for the model configuration, which includes the data transformation, feature engineering, and selection, or the tuning of the hyperparameters. The test dataset could either be a subset of the original data (with no overlapping subjects or observations) or a separate external dataset. The use of an external dataset is ideal as this ensures that there is no influence of bias during the data collection period and that there is no data leakage between the datasets. If an external dataset is not available or if the dataset is not sufficiently large, we recommend nested cross-validation. This resampling method supports model hyperparameter tuning and performance evaluation without the risk of data leakage across the dataset.

It is crucial to report the evaluation metric results for each dataset. In the case of cross-validation reporting, we recommend that researchers report the distribution of the performance measures (e.g., the mean and standard deviation or median and 95% confidence interval) across the folds to show the average and variability of the performance of the models. As cross-validation evaluates the prediction error across multiple ML models, we would also recommend reporting the performance of the final model selected. This is achieved by re-training a ML model on the full dataset and evaluating the performance on an external dataset [207,210]. This would give insight into how well the model would perform under different circumstances. We also highly recommend using multiple evaluation metrics for assessing the model's performance. Seeing as a model might excel for one metric and fail for another, this underscores the need for comprehensive evaluation. Employing multiple metrics ensures optimal operation and reduces the likelihood of blind spots.

Once the final model has been trained, there are three approaches to choose from to apply the model to a new target dataset. The first approach is to test the model "as-is", implying that the ready-made model can be used in its original state without modifications [216]. In the second scenario, the train data and the target data may have different characteristics, which may lead to a distribution shift. The type of distribution shift between the two datasets can occur for many reasons, including different mHealth devices used for data collection, environmental noise, and sampling bias [217]. When this occurs, transfer learning can be used to finetune the ready-made model and update its weights to better suit the target dataset [216]. In the third scenario, the target dataset may have different requirements than the original training dataset [216]. As a result, the decision boundary of the classification model can be altered, such as optimizing the model for a sensitivity of 90% instead of accuracy. Whether testing the model as-is, employing transfer learning, or adjusting the decision boundary, these strategies offer flexibility in adapting the model to different settings and improving its performance for validation purposes.

11.6. Model Reproducibility and Interpretability

Equally important as the model performance are the ML models' reproducibility and interpretability. Reproducibility is a core component for ensuring that a ML model can be validated and reused by clinical researchers. Technical reproducibility involves using the same computational procedures to produce consistent model outcomes. Statistical reproducibility ensures that the model demonstrates similar statistical performance across different subsets of data. Conceptual reproducibility refers to achieving consistent results under new conditions, such as data collected from different settings [216]. Transparency regarding data quality, feature engineering and selection methods, the hyperparameters considered and selected, and the model validation protocol can help ease the ability of the scientific community to recreate the work in the published literature. Best practices for reproducibility include publishing the code on GitHub or by publishing FAIR metadata [211,218,219].

Given the potential clinical application of ML models, prior to modeling, researchers should determine the model's interpretability requirement. While ML models provide researchers with what was predicted, interpretability requires that the model can explain why it made the prediction [185]. Interpretability enables us to understand the causal relationships between the data and the ML model's predictions. There are two situations in which the interpretability of a model is required: when an inaccurate prediction can have severe or even fatal consequences for the patients (such as a misclassified diagnosis [220]) and when the interpretability can be used to identify novel relationships between clinical factors and the predicted outcome (such as factors influencing treatment outcomes [221]). There can be two situations in which interpretability is not required: situations in which incorrect predictions do not have severe consequences (such as counting the number of coughs [222]) or situations in which the ML model has been sufficiently validated in real clinical applications, even if the predictions are not perfect [223]. While black box models may offer more accurate predictions than an interpretable model, they only provide limited insight into how the predictions were made. Therefore, both interpretable and black box models have their respective merits.

There are two broad approaches towards achieving interpretability [224]. One approach is to use easy-to-interpret models, such as Linear or Logistic Regression, where the coefficients of the features can provide insight into the features' associations with the predicted outcome. The other approach is to use explanation methods for explaining complex or black box models, such as SHapley Additive exPlanations plots (SHAP), Local Interpretable Model-agnostic Explanations (LIME), or Anchors [224]. We recommend that researchers report whether their final selected model was an interpretable model or a black box [225]. If it was interpretable, we recommend discussing what interpretations can be derived from the models.

12. Conclusions

The rise and breadth of ML applications in clinical trials highlight the increasing reliance and importance of ML in the development of novel biomarkers [226]. While the advances in ML applications have demonstrated great potential for innovative biomarker development, the process of its development is not well documented, which, in turn, limits the reproducibility of these findings. This review has illustrated the steps taken to translate raw data from mHealth technologies into meaningful clinical biomarkers using ML. Given the lack of consistent reporting in the ML methods, the present review cannot provide a complete or detailed picture of the notable and generic practices. However, the authors have provided an overview of the status quo of the development and translation of ML-derived biomarkers in mHealth-focused clinical trials. The recommended checklist provided in the review could serve as a foundation for the design of future ML-derived biomarkers in conventional ML practices. By encouraging consistent and transparent reporting, researchers can accelerate the integration of novel biomarkers derived from mHealth sensors and ML pipelines into future clinical trials.

Author Contributions: Conceptualization, A.Z., R.J.D. and W.K.; methodology, A.Z.; formal analysis, A.Z.; data curation, A.Z. and A.A.d.G.; writing—original draft preparation A.Z., writing—review and editing A.Z., A.A.d.G., G.J.G., R.J.D., I.E.Y., W.K. and A.A.d.G.; visualization, A.Z.; supervision, G.J.G., R.J.D. and W.K. All authors have read and agreed to the published version of the manuscript.

Funding: This research received no external funding.

Institutional Review Board Statement: Not applicable.

Informed Consent Statement: Not applicable.

Data Availability Statement: No new data were created or analyzed in this study. Data sharing is not applicable to this article.

Conflicts of Interest: The authors declare no conflict of interest.

References

1. Au, R.; Lin, H.; Kolachalama, V.B. Tele-Trials, Remote Monitoring, and Trial Technology for Alzheimer's Disease Clinical Trials. In *Alzheimer's Disease Drug Development*; Cambridge University Press: Cambridge, UK, 2022; pp. 292–300. [CrossRef]
2. Inan, O.T.; Tenaerts, P.; Prindiville, S.A.; Reynolds, H.R.; Dizon, D.S.; Cooper-Arnold, K.; Turakhia, M.; Pletcher, M.J.; Preston, K.L.; Krumholz, H.M.; et al. Digitizing clinical trials. *NPJ Digit. Med.* **2020**, *3*, 101. [CrossRef] [PubMed]
3. Teo, J.X.; Davila, S.; Yang, C.; Hii, A.A.; Pua, C.J.; Yap, J.; Tan, S.Y.; Sahlén, A.; Chin, C.W.-L.; Teh, B.T.; et al. Digital phenotyping by consumer wearables identifies sleep-associated markers of cardiovascular disease risk and biological aging. *bioRxiv* **2019**. [CrossRef]
4. Brietzke, E.; Hawken, E.R.; Idzikowski, M.; Pong, J.; Kennedy, S.H.; Soares, C.N. Integrating digital phenotyping in clinical characterization of individuals with mood disorders. *Neurosci. Biobehav. Rev.* **2019**, *104*, 223–230. [CrossRef] [PubMed]
5. Kourtis, L.C.; Regele, O.B.; Wright, J.M.; Jones, G.B. Digital biomarkers for Alzheimer's disease: The mobile/wearable devices opportunity. *NPJ Digit. Med.* **2019**, *2*, 9. [CrossRef] [PubMed]
6. Bhidayasiri, R.; Mari, Z. Digital phenotyping in Parkinson's disease: Empowering neurologists for measurement-based care. *Park. Relat. Disord.* **2020**, *80*, 35–40. [CrossRef]
7. Prosperi, M.; Min, J.S.; Bian, J.; Modave, F. Big data hurdles in precision medicine and precision public health. *BMC Med. Inform. Decis. Mak.* **2018**, *18*, 139. [CrossRef]
8. Torres-Sospedra, J.; Ometov, A. Data from Smartphones and Wearables. *Data* **2021**, *6*, 45. [CrossRef]
9. García-Santıllán, A.; del Flóres-Serrano, S.; López-Morales, J.S.; Rios-Alvarez, L.R. Factors Associated that Explain Anxiety toward Mathematics on Undergraduate Students. (An Empirical Study in Tierra Blanca Veracruz-México). *Mediterr. J. Soc. Sci.* **2014**, *5*. [CrossRef]
10. Iniesta, R.; Stahl, D.; Mcguffin, P. Machine learning, statistical learning and the future of biological research in psychiatry. *Psychol. Med.* **2016**, *46*, 2455–2465. [CrossRef]
11. Rajula, H.S.R.; Verlato, G.; Manchia, M.; Antonucci, N.; Fanos, V. Comparison of Conventional Statistical Methods with Machine Learning in Medicine: Diagnosis, Drug Development, and Treatment. *Medicina* **2020**, *56*, 455. [CrossRef]
12. Getz, K.A.; Rafael, A.C. Trial watch: Trends in clinical trial design complexity. *Nat. Rev. Drug. Discov.* **2017**, *16*, 307. [CrossRef] [PubMed]
13. Getz, K.A.; Stergiopoulos, S.; Marlborough, M.; Whitehill, J.; Curran, M.; Kaitin, K.I. Quantifying the Magnitude and Cost of Collecting Extraneous Protocol Data. *Am. J. Ther.* **2015**, *22*, 117–124. [CrossRef] [PubMed]
14. Getz, K.A.; Wenger, J.; Campo, R.A.; Seguine, E.S.; Kaitin, K.I. Assessing the Impact of Protocol Design Changes on Clinical Trial Performance. *Am. J. Ther.* **2008**, *15*, 450–457. [CrossRef]
15. Globe Newswire. Rising Protocol Design Complexity Is Driving Rapid Growth in Clinical Trial Data Volume, According to Tufts Center for the Study of Drug Development. Available online: https://www.globenewswire.com/news-release/2021/01/12/2157143/0/en/Rising-Protocol-Design-Complexity-Is-Driving-Rapid-Growth-in-Clinical-Trial-Data-Volume-According-to-Tufts-Center-for-the-Study-of-Drug-Development.html (accessed on 12 January 2021).
16. Santos, W.M.D.; Secoli, S.R.; de Araújo Püschel, V.A. The Joanna Briggs Institute approach for systematic reviews. *Rev. Lat. Am. Enferm.* **2018**, *26*, e3074. [CrossRef]
17. Central Nervous System Diseases—MeSH—NCBI. 2023. Available online: https://www.ncbi.nlm.nih.gov/mesh?Db=mesh&Cmd=DetailsSearch&Term=%22Central+Nervous+System+Diseases%22%5BMeSH+Terms%5D (accessed on 5 January 2023).
18. Martinez, G.J.; Mattingly, S.M.; Mirjafari, S.; Nepal, S.K.; Campbell, A.T.; Dey, A.K.; Striegel, A.D. On the Quality of Real-world Wearable Data in a Longitudinal Study of Information Workers. In Proceedings of the 2020 IEEE International Conference on Pervasive Computing and Communications Workshops, PerCom Workshops 2020, Austin, TX, USA, 23–27 March 2020. [CrossRef]
19. Ruiz Blázquez, R.R.; Muñoz-Organero, M. Using Multivariate Outliers from Smartphone Sensor Data to Detect Physical Barriers While Walking in Urban Areas. *Technologies* **2020**, *8*, 58. [CrossRef]
20. Poulos, J.; Valle, R. Missing Data Imputation for Supervised Learning. *Appl. Artif. Intell.* **2018**, *32*, 186–196. [CrossRef]

21. Schafer, J.L.; Graham, J.W. Missing data: Our view of the state of the art. *Psychol. Methods* **2002**, *7*, 147–177. [CrossRef]
22. Evers, L.J.; Raykov, Y.P.; Krijthe, J.H.; de Lima, A.L.S.; Badawy, R.; Claes, K.; Heskes, T.M.; Little, M.A.; Meinders, M.J.; Bloem, B.R. Real-life gait performance as a digital biomarker for motor fluctuations: The Parkinson@Home validation study. *J. Med. Internet Res.* **2020**, *22*, e19068. [CrossRef]
23. Papadopoulos, A.; Kyritsis, K.; Klingelhoefer, L.; Bostanjopoulou, S.; Chaudhuri, K.R.; Delopoulos, A. Detecting Parkinsonian Tremor from IMU Data Collected In-The-Wild using Deep Multiple-Instance Learning. *IEEE J. Biomed. Health Inform.* **2019**, *24*, 2559–2569. [CrossRef]
24. Tougui, I.; Jilbab, A.; El Mhamdi, J. Analysis of smartphone recordings in time, frequency, and cepstral domains to classify Parkinson's disease. *Healthc. Inform. Res.* **2020**, *26*, 274–283. [CrossRef]
25. Meyerhoff, J.; Liu, T.; Kording, K.P.; Ungar, L.H.; Kaiser, S.M.; Karr, C.J.; Mohr, D.C. Evaluation of Changes in Depression, Anxiety, and Social Anxiety Using Smartphone Sensor Features: Longitudinal Cohort Study. *J. Med. Internet Res.* **2021**, *23*, e22844. [CrossRef] [PubMed]
26. Dinesh, K.; Snyder, C.W.; Xiong, M.; Tarolli, C.G.; Sharma, S.; Dorsey, E.R.; Sharma, G.; Adams, J.L. A Longitudinal Wearable Sensor Study in Huntington's Disease. *J. Huntingt. Dis.* **2020**, *9*, 69–81. [CrossRef] [PubMed]
27. Cho, C.-H.; Lee, T.; Lee, H.-J. Mood Prediction of Patients with Mood Disorders by Machine Learning Using Passive Digital Phenotypes Based on the Circadian Rhythm: Prospective Observational Cohort Study. 2019. Available online: https://www.ncbi.nlm.nih.gov/pmc/articles/PMC6492069/ (accessed on 5 January 2023).
28. Tanaka, T.; Kokubo, K.; Iwasa, K.; Sawa, K.; Yamada, N.; Komori, M. Intraday activity levels may better reflect the differences between major depressive disorder and bipolar disorder than average daily activity levels. *Front. Psychol.* **2018**, *9*, 2314. [CrossRef] [PubMed]
29. Palmius, N.; Tsanas, A.; Saunders, K.E.A.; Bilderbeck, A.C.; Geddes, J.R.; Goodwin, G.M.; De Vos, M. Detecting bipolar depression from geographic location data. *IEEE Trans. Biomed. Eng.* **2017**, *64*, 1761–1771. [CrossRef] [PubMed]
30. Abdullah, S.; Matthews, M.; Frank, E.; Doherty, G.; Gay, G.; Choudhury, T. Automatic detection of social rhythms in bipolar disorder. *J. Am. Med. Inform. Assoc.* **2016**, *23*, 538–543. [CrossRef]
31. Ramsperger, R.; Meckler, S.; Heger, T.; van Uem, J.; Hucker, S.; Braatz, U.; Graessner, H.; Berg, D.; Manoli, Y.; Serrano, J.A.; et al. Continuous leg dyskinesia assessment in Parkinson's disease -clinical validity and ecological effect. *Park. Relat. Disord.* **2016**, *26*, 41–46. [CrossRef]
32. Gossec, L.; Guyard, F.; Leroy, D.; Lafargue, T.; Seiler, M.; Jacquemin, C.; Molto, A.; Sellam, J.; Foltz, V.; Gandjbakhch, F.; et al. Detection of Flares by Decrease in Physical Activity, Collected Using Wearable Activity Trackers in Rheumatoid Arthritis or Axial Spondyloarthritis: An Application of Machine Learning Analyses in Rheumatology. *Arthritis Care Res.* **2019**, *71*, 1336–1343. [CrossRef]
33. Bai, R.; Xiao, L.; Guo, Y.; Zhu, X.; Li, N.; Wang, Y.; Chen, Q.; Feng, L.; Wang, Y.; Yu, X.; et al. Tracking and monitoring mood stability of patients with major depressive disorder by machine learning models using passive digital data: Prospective naturalistic multicenter study. *JMIR Mhealth Uhealth* **2021**, *9*, e24365. [CrossRef]
34. Schwab, P.; Karlen, W. A Deep Learning Approach to Diagnosing Multiple Sclerosis from Smartphone Data. *IEEE J. Biomed. Health Inform.* **2021**, *25*, 1284–1291. [CrossRef]
35. Aghanavesi, S. Smartphone-Based Parkinson's Disease Symptom Assessment. Licentiate Dissertation, Dalarna University, Falun, Sweden, 2017.
36. Maleki, G.; Zhuparris, A.; Koopmans, I.; Doll, R.J.; Voet, N.; Cohen, A.; van Brummelen, E.; Groeneveld, G.J.; De Maeyer, J. Objective Monitoring of Facioscapulohumeral Dystrophy During Clinical Trials Using a Smartphone App and Wearables: Observational Study. *JMIR Form. Res.* **2022**, *6*, e31775. [CrossRef]
37. Twose, J.; Licitra, G.; McConchie, H.; Lam, K.H.; Killestein, J. Early-warning signals for disease activity in patients diagnosed with multiple sclerosis based on keystroke dynamics. *Chaos* **2020**, *30*, 113133. [CrossRef] [PubMed]
38. Cho, C.H.; Lee, T.; Kim, M.G.; In, H.P.; Kim, L.; Lee, H.J. Mood prediction of patients with mood disorders by machine learning using passive digital phenotypes based on the circadian rhythm: Prospective observational cohort study. *J. Med. Internet Res.* **2019**, *21*, e11029. [CrossRef]
39. Little, R.J.A.; Rubin, D.B. *Complete-Case and Available-Case Analysis, Including Weighting Methods*; John Wiley & Sons, Ltd.: Hoboken, NJ, USA, 2014; pp. 41–58. [CrossRef]
40. Demissie, S.; LaValley, M.P.; Horton, N.J.; Glynn, R.J.; Cupples, L.A. Bias due to missing exposure data using complete-case analysis in the proportional hazards regression model. *Stat. Med.* **2003**, *22*, 545–557. [CrossRef]
41. Enders, C.K.; London, N.Y. *Applied Missing Data Analysis*; Guilford Press: New York, NY, USA, 2010.
42. Zhang, Y.; Folarin, A.A. Predicting Depressive Symptom Severity Through Individuals' Nearby Bluetooth Device Count Data Collected by Mobile Phones: Preliminary Longitudinal Study. Available online: https://www.ncbi.nlm.nih.gov/pmc/articles/PMC8367113/ (accessed on 5 January 2023).
43. Creagh, A.P.; Dondelinger, F.; Lipsmeier, F.; Lindemann, M.; De Vos, M. Longitudinal Trend Monitoring of Multiple Sclerosis Ambulation using Smartphones. *IEEE Open J. Eng. Med. Biol.* **2022**, *3*, 202–210. [CrossRef] [PubMed]
44. Wu, C.-T.; Li, G.-H.; Huang, C.-T.; Cheng, Y.-C.; Chen, C.-H.; Chien, J.-Y.; Kuo, P.-H.; Kuo, L.-C.; Lai, F. Acute exacerbation of a chronic obstructive pulmonary disease prediction system using wearable device data, machine learning, and deep learning: Development and cohort study. *JMIR Mhealth Uhealth* **2021**, *9*, e22591. [CrossRef] [PubMed]

45. Jakobsen, P.; Garcia-Ceja, E.; Riegler, M.; Stabell, L.A.; Nordgreen, T.; Torresen, J.; Fasmer, O.B.; Oedegaard, K.J. Applying machine learning in motor activity time series of depressed bipolar and unipolar patients compared to healthy controls. *PLoS ONE* **2020**, *15*, e0231995. [CrossRef]
46. Lekkas, D.; Jacobson, N.C. Using artificial intelligence and longitudinal location data to differentiate persons who develop posttraumatic stress disorder following childhood trauma. *Sci. Rep.* **2021**, *11*, 10303. [CrossRef]
47. Richman, M.B.; Trafalis, T.B.; Adrianto, I. Missing data imputation through machine learning algorithms. In *Artificial Intelligence Methods in the Environmental Sciences*; Springer: Dordrecht, The Netherlands, 2009; pp. 153–169. [CrossRef]
48. Jerez, J.M.; Molina, I.; García-Laencina, P.J.; Alba, E.; Ribelles, N.; Martín, M.; Franco, L. Missing data imputation using statistical and machine learning methods in a real breast cancer problem. *Artif. Intell. Med.* **2010**, *50*, 105–115. [CrossRef]
49. Lakshminarayan, K.; Harp, S.A.; Goldman, R.P.; Samad, T. Imputation of Missing Data Using Machine Learning Techniques. In *KDD Proceedings 1996*; AAAI Press: Palo Alto, CA, USA, 1996; Volume 96.
50. Aggarwal, C.C. *Data Mining*; Springer International Publishing: Cham, Switzerland, 2015. [CrossRef]
51. Ledolter, J.; Kardon, R.H. Does Testing More Frequently Shorten the Time to Detect Disease Progression? *Transl. Vis. Sci. Technol.* **2017**, *6*, 1. [CrossRef]
52. Bazgir, O.; Habibi, S.A.H.; Palma, L.; Pierleoni, P.; Nafees, S. A classification system for assessment and home monitoring of tremor in patients with Parkinson's disease. *J. Med. Signals Sens.* **2018**, *8*, 65–72. [CrossRef]
53. Williamson, J.R.; Telfer, B.; Mullany, R.; Friedl, K.E. Detecting Parkinson's Disease from Wrist-Worn Accelerometry in the U.K. Biobank. *Sensors* **2021**, *21*, 2047. [CrossRef] [PubMed]
54. Buda, T.S.; Khwaja, M.; Matic, A. Outliers in Smartphone Sensor Data Reveal Outliers in Daily Happiness. *Proc. ACM Interact. Mob. Wearable Ubiquitous Technol.* **2021**, *5*, 1–19. [CrossRef]
55. Buda, T.S.; Caglayan, B.; Assem, H. DeepAD: A generic framework based on deep learning for time series anomaly detection. In *Lecture Notes in Computer Science (Including Subseries Lecture Notes in Artificial Intelligence and Lecture Notes in Bioinformatics)*; Springer: Berlin/Heidelberg, Germany, 2018; pp. 577–588. [CrossRef]
56. Arora, S.; Venkataraman, V.; Zhan, A.; Donohue, S.; Biglan, K.; Dorsey, E.; Little, M. Detecting and monitoring the symptoms of Parkinson's disease using smartphones: A pilot study. *Park. Relat. Disord.* **2015**, *21*, 650–653. [CrossRef] [PubMed]
57. Guyon, I.; Elisseeff, A. An Introduction to Feature Extraction. In *Feature Extraction*; Springer: Berlin/Heidelberg, Germany, 2006; pp. 1–25. [CrossRef]
58. Raju, V.N.G.; Lakshmi, K.P.; Jain, V.M.; Kalidindi, A.; Padma, V. Study the Influence of Normalization/Transformation process on the Accuracy of Supervised Classification. In Proceedings of the 2020 Third International Conference on Smart Systems and Inventive Technology (ICSSIT), Tirunelveli, India, 20–22 August 2022; pp. 729–735. [CrossRef]
59. Dara, S.; Tumma, P. Feature Extraction by Using Deep Learning: A Survey. In Proceedings of the 2018 Second International Conference on Electronics, Communication and Aerospace Technology (ICECA), Coimbatore, India, 29–31 March 2018; pp. 1795–1801. [CrossRef]
60. Tizzano, G.R.; Spezialetti, M.; Rossi, S. A Deep Learning Approach for Mood Recognition from Wearable Data. In Proceedings of the IEEE Medical Measurements and Applications, MeMeA 2020—Conference Proceedings, Bari, Italy, 1 June–1 July 2020; Institute of Electrical and Electronics Engineers Inc.: Piscataway, NJ, USA, 2020. [CrossRef]
61. Garcia-Ceja, E.; Riegler, M.; Jakobsen, P.; Torresen, J.; Nordgreen, T.; Oedegaard, K.J.; Fasmer, O.B. Motor Activity Based Classification of Depression in Unipolar and Bipolar Patients. In Proceedings of the 2018 IEEE 31st International Symposium on Computer-Based Medical Systems (CBMS), Karlstad, Sweden, 18–21 June 2018; pp. 316–321. [CrossRef]
62. Liu, H. Feature Engineering for Machine Learning and Data Analytics. In *Feature Engineering for Machine Learning and Data Analytics*; Taylor & Francis Group: Boca Raton, FL, USA, 2018.
63. Nargesian, F.; Samulowitz, H.; Khurana, U.; Khalil, E.B.; Turaga, D. Learning feature engineering for classification. *IJCAI Int. Jt. Conf. Artif. Intell.* **2017**, 2529–2535. [CrossRef]
64. Kuhn, M.; Johnson, K. *Feature Engineering and Selection: A Practical Approach for Predictive Models*; Chapman and Hall/CRC: Boca Raton, FL, USA, 2019. [CrossRef]
65. Ronao, C.A.; Cho, S.-B. Human activity recognition with smartphone sensors using deep learning neural networks. *Expert Syst. Appl.* **2016**, *59*, 235–244. [CrossRef]
66. Nweke, H.F.; Teh, Y.W.; Al-garadi, M.A.; Alo, U.R. Deep learning algorithms for human activity recognition using mobile and wearable sensor networks: State of the art and research challenges. *Expert Syst. Appl.* **2018**, *105*, 233–261. [CrossRef]
67. Zdravevski, E.; Lameski, P.; Trajkovik, V.; Kulakov, A.; Chorbev, I.; Goleva, R.; Pombo, N.; Garcia, N. Improving Activity Recognition Accuracy in Ambient-Assisted Living Systems by Automated Feature Engineering. *IEEE Access* **2017**, *5*, 5262–5280. [CrossRef]
68. McGinnis, R.S.; Mahadevan, N.; Moon, Y.; Seagers, K.; Sheth, N.; Wright, J.A., Jr.; Dicristofaro, S.; Silva, I.; Jortberg, E.; Ceruolo, M.; et al. A machine learning approach for gait speed estimation using skin-mounted wearable sensors: From healthy controls to individuals with multiple sclerosis. *PLoS ONE* **2017**, *12*, e0178366. [CrossRef]
69. Maxhuni, A.; Muñoz-Meléndez, A.; Osmani, V.; Perez, H.; Mayora, O.; Morales, E.F. Classification of bipolar disorder episodes based on analysis of voice and motor activity of patients. *Pervasive Mob. Comput.* **2016**, *31*, 50–66. [CrossRef]

70. Yamakawa, T.; Miyajima, M.; Fujiwara, K.; Kano, M.; Suzuki, Y.; Watanabe, Y.; Watanabe, S.; Hoshida, T.; Inaji, M.; Maehara, T. Wearable epileptic seizure prediction system with machine-learning-based anomaly detection of heart rate variability. *Sensors* **2020**, *20*, 3987. [CrossRef]
71. Fuchs, C.; Nobile, M.S.; Zamora, G.; Degeneffe, A.; Kubben, P.; Kaymak, U. Tremor assessment using smartphone sensor data and fuzzy reasoning. *BMC Bioinform.* **2021**, *22*, 57. [CrossRef] [PubMed]
72. Aich, S.; Pradhan, P.M.; Park, J.; Sethi, N.; Vathsa, V.S.S.; Kim, H.C. A validation study of freezing of gait (FoG) detection and machine-learning-based FoG prediction using estimated gait characteristics with a wearable accelerometer. *Sensors* **2018**, *18*, 3287. [CrossRef] [PubMed]
73. Rodríguez-Martín, D.; Samà, A.; Pérez-López, C.; Català, A.; Arostegui, J.M.M.; Cabestany, J.; Bayés, À.; Alcaine, S.; Mestre, B.; Prats, A.; et al. Home detection of freezing of gait using Support Vector Machines through a single waist-worn triaxial accelerometer. *PLoS ONE* **2017**, *12*, e0171764. [CrossRef] [PubMed]
74. Supratak, A.; Datta, G.; Gafson, A.R.; Nicholas, R.; Guo, Y.; Matthews, P.M. Remote monitoring in the home validates clinical gait measures for multiple sclerosis. *Front. Neurol.* **2018**, *9*, 561. [CrossRef]
75. Bro, R.; Smilde, A.K. Principal component analysis. *Anal. Methods* **2014**, *6*, 2812–2831. [CrossRef]
76. Kim, J.; Lim, J. A Deep Neural Network-Based Method for Prediction of Dementia Using Big Data. *Int. J. Environ. Res. Public Health* **2021**, *18*, 5386. [CrossRef]
77. Clustering. In *Principles of Data Mining*; Springer: London, UK, 2007; pp. 221–238. [CrossRef]
78. Arabie, P.; Hubert, L.J. An Overview of Combinatorial Data Analysis. In *Clustering and Classification*; World Scientific: Singapore, 1996; pp. 5–63. [CrossRef]
79. Lu, J.; Shang, C.; Yue, C.; Morillo, R.; Ware, S.; Kamath, J.; Bamis, A.; Russell, A.; Wang, B.; Bi, J. Joint Modeling of Heterogeneous Sensing Data for Depression Assessment via Multi-task Learning. In *Proceedings of the Proceedings of the ACM on Interactive, Mobile, Wearable and Ubiquitous Technologies*; Association for Computing Machinery: New York, NY, USA, 2018; Volume 2, pp. 1–21. [CrossRef]
80. Sabatelli, M.; Osmani, V.; Mayora, O.; Gruenerbl, A.; Lukowicz, P. Correlation of significant places with self-reported state of bipolar disorder patients. In Proceedings of the 2014 4th International Conference on Wireless Mobile Communication and Healthcare-Transforming Healthcare through Innovations in Mobile and Wireless Technologies (MOBIHEALTH), Athens, Greece, 3–5 November 2014; pp. 116–119. [CrossRef]
81. Faurholt-Jepsen, M.; Busk, J.; Vinberg, M.; Christensen, E.M.; HelgaÞórarinsdóttir; Frost, M.; Bardram, J.E.; Kessing, L.V. Daily mobility patterns in patients with bipolar disorder and healthy individuals. *J. Affect. Disord.* **2021**, *278*, 413–422. [CrossRef]
82. Miotto, R.; Wang, F.; Wang, S.; Jiang, X.; Dudley, J.T. Deep learning for healthcare: Review, opportunities and challenges. *Brief. Bioinform.* **2018**, *19*, 1236–1246. [CrossRef]
83. Marx, V. The big challenges of big data. *Nature* **2013**, *498*, 255–260. [CrossRef]
84. Li, Y.; Ding, L.; Gao, X. On the decision boundary of deep neural networks. *arXiv* **2018**, arXiv:1808.05385.
85. Juen, J.; Cheng, Q.; Schatz, B. A Natural Walking Monitor for Pulmonary Patients Using Mobile Phones. *IEEE J. Biomed. Health Inform.* **2015**, *19*, 1399–1405. [CrossRef] [PubMed]
86. Cole, B.T.; Roy, S.H.; De Luca, C.J.; Nawab, S.H. Dynamical learning and tracking of tremor and dyskinesia from wearable sensors. *IEEE Trans. Neural Syst. Rehabil. Eng.* **2014**, *22*, 982–991. [CrossRef] [PubMed]
87. Peraza, L.R.; Kinnunen, K.M.; McNaney, R.; Craddock, I.J.; Whone, A.L.; Morgan, C.; Joules, R.; Wolz, R. An automatic gait analysis pipeline for wearable sensors: A pilot study in parkinson's disease. *Sensors* **2021**, *21*, 8286. [CrossRef] [PubMed]
88. Saeys, Y.; Abeel, T.; Van De Peer, Y. Robust feature selection using ensemble feature selection techniques. In *Lecture Notes in Computer Science (Including Subseries Lecture Notes in Artificial Intelligence and Lecture Notes in Bioinformatics)*; Springer: Berlin/Heidelberg, Germany, 2008; pp. 313–325. [CrossRef]
89. Cai, J.; Luo, J.; Wang, S.; Yang, S. Feature selection in machine learning: A new perspective. *Neurocomputing* **2018**, *300*, 70–79. [CrossRef]
90. Jabar, H.; Khan, R.Z. Methods to avoid over-fitting and under-fitting in supervised machine learning (comparative study). *Comput. Sci. Commun. Instrum. Devices* **2015**, *70*, 163–172.
91. Hall, M.A. Correlation-based Feature Selection for Machine Learning. Ph.D. Thesis, The University of Waikato, Hamilton, New Zealand, 1999.
92. Hall, M.A.; Smith, L.A. Feature selection for machine learning: Comparing a correlation-based filter approach to the wrapper. In Proceedings of the FLAIRS Conference 1999, Orlando, FL, USA, 1–5 May 1999; Volume 1999, pp. 235–239.
93. Galperin, I.; Hillel, I.; Del Din, S.; Bekkers, E.M.; Nieuwboer, A.; Abbruzzese, G.; Avanzino, L.; Nieuwhof, F.; Bloem, B.R.; Rochester, L.; et al. Associations between daily-living physical activity and laboratory-based assessments of motor severity in patients with falls and Parkinson's disease. *Park. Relat. Disord.* **2019**, *62*, 85–90. [CrossRef]
94. Dong, C.; Ye, T.; Long, X.; Aarts, R.M.; van Dijk, J.P.; Shang, C.; Liao, X.; Chen, W.; Lai, W.; Chen, L.; et al. A Two-Layer Ensemble Method for Detecting Epileptic Seizures Using a Self-Annotation Bracelet with Motor Sensors. *IEEE Trans. Instrum. Meas.* **2022**, *71*, 4005013. [CrossRef]
95. Creagh, A.P.; Simillion, C.; Bourke, A.K.; Scotland, A.; Lipsmeier, F.; Bernasconi, C.; van Beek, J.; Baker, M.; Gossens, C.; Lindemann, M.; et al. Smartphone-and Smartwatch-Based Remote Characterisation of Ambulation in Multiple Sclerosis during the Two-Minute Walk Test. *IEEE J. Biomed. Health Inform.* **2021**, *25*, 838–849. [CrossRef]

96. Chen, O.Y.; Lipsmeier, F.; Phan, H.; Prince, J.; Taylor, K.I.; Gossens, C.; Lindemann, M.; de Vos, M. Building a Machine-Learning Framework to Remotely Assess Parkinson's Disease Using Smartphones. *IEEE Trans. Biomed. Eng.* **2020**, *67*, 3491–3500. [CrossRef]
97. Steyerberg, E.W.; Eijkemans, M.J.C.; Habbema, J.D.F. Stepwise selection in small data sets: A simulation study of bias in logistic regression analysis. *J. Clin. Epidemiol.* **1999**, *52*, 935–942. [CrossRef]
98. Austin, P.C.; Tu, J.V. Bootstrap Methods for Developing Predictive Models. *Am. Stat.* **2004**, *58*, 131–137. [CrossRef]
99. Zimmerman, D.W. Power Functions of the Test and Mann-Whitney Test Under Violation of Parametric Assumptions. *Percept. Mot. Skills* **1985**, *61*, 467–470. [CrossRef]
100. Urbanowicz, R.J.; Meeker, M.; la Cava, W.; Olson, R.S.; Moore, J.H. Relief-based feature selection: Introduction and review. *J. Biomed. Inform.* **2018**, *85*, 189–203. [CrossRef]
101. Kira, K.; Rendell, L.A. *A Practical Approach to Feature Selection*; Elsevier: Amsterdam, The Netherlands, 1992; pp. 249–256. [CrossRef]
102. Verma, N.K.; Salour, A. Feature selection. *Stud. Syst. Decis. Control* **2020**, *256*, 175–200. [CrossRef]
103. Yaman, O.; Ertam, F.; Tuncer, T. Automated Parkinson's disease recognition based on statistical pooling method using acoustic features. *Med. Hypotheses* **2020**, *135*, 109483. [CrossRef] [PubMed]
104. Rodriguez-Molinero, A.; Samà, A.; Pérez-Martínez, D.A.; López, C.P.; Romagosa, J.; Bayes, A.; Sanz, P.; Calopa, M.; Gálvez-Barrón, C.; De Mingo, E.; et al. Validation of a portable device for mapping motor and gait disturbances in Parkinson's disease. *JMIR Mhealth Uhealth* **2015**, *3*, e9. [CrossRef]
105. Chandrashekar, G.; Sahin, F. A survey on feature selection methods. *Comput. Electr. Eng.* **2014**, *40*, 16–28. [CrossRef]
106. Goldsmith, J.; Bobb, J.; Crainiceanu, C.M.; Caffo, B.; Reich, D. Penalized functional regression. *J. Comput. Graph. Stat.* **2011**, *20*, 830–851. [CrossRef]
107. Prince, J.; Andreotti, F.; De Vos, M. Multi-Source Ensemble Learning for the Remote Prediction of Parkinson's Disease in the Presence of Source-Wise Missing Data. *IEEE Trans. Biomed. Eng.* **2019**, *66*, 1402–1411. [CrossRef]
108. Motin, M.A.; Pah, N.D.; Raghav, S.; Kumar, D.K. Parkinson's Disease Detection Using Smartphone Recorded Phonemes in Real World Conditions. *IEEE Access* **2022**, *10*, 97600–97609. [CrossRef]
109. Cakmak, A.S.; Alday, E.A.P.; Da Poian, G.; Rad, A.B.; Metzler, T.J.; Neylan, T.C.; House, S.L.; Beaudoin, F.L.; An, X.; Stevens, J.S.; et al. Classification and Prediction of Post-Trauma Outcomes Related to PTSD Using Circadian Rhythm Changes Measured via Wrist-Worn Research Watch in a Large Longitudinal Cohort. *IEEE J. Biomed. Health Inform.* **2021**, *25*, 2866–2876. [CrossRef] [PubMed]
110. Tracy, J.M.; Özkanca, Y.; Atkins, D.C.; Ghomi, R.H. Investigating voice as a biomarker: Deep phenotyping methods for early detection of Parkinson's disease. *J. Biomed. Inform.* **2020**, *104*, 103362. [CrossRef] [PubMed]
111. Abdulhafedh, A. Comparison between Common Statistical Modeling Techniques Used in Research, Including: Discriminant Analysis vs Logistic Regression, Ridge Regression vs LASSO, and Decision Tree vs Random Forest. *OAlib* **2022**, *9*, 1–19. [CrossRef]
112. Sánchez-Maroño, N.; Alonso-Betanzos, A.; Tombilla-Sanromán, M. Filter methods for feature selection—A comparative study. In *Lecture Notes in Computer Science (Including Subseries Lecture Notes in Artificial Intelligence and Lecture Notes in Bioinformatics)*; Springer: Berlin/Heidelberg, Germany, 2007; pp. 178–187. [CrossRef]
113. Porter, B.W.; Bareiss, R.; Holte, R.C. Concept learning and heuristic classification in weak-theory domains. *Artif. Intell.* **1990**, *45*, 229–263. [CrossRef]
114. Wu, C.-T.; Wang, S.-M.; Su, Y.-E.; Hsieh, T.-T.; Chen, P.-C.; Cheng, Y.-C.; Tseng, T.-W.; Chang, W.-S.; Su, C.-S.; Kuo, L.-C.; et al. A Precision Health Service for Chronic Diseases: Development and Cohort Study Using Wearable Device, Machine Learning, and Deep Learning. *IEEE J. Transl. Eng. Health Med.* **2022**, *10*, 2700414. [CrossRef] [PubMed]
115. de Lima, A.L.S.; Evers, L.J.; Hahn, T.; de Vries, N.M.; Daeschler, M.; Boroojerdi, B.; Terricabras, D.; Little, M.A.; Bloem, B.R.; Faber, M.J. Impact of motor fluctuations on real-life gait in Parkinson's patients. *Gait Posture* **2018**, *62*, 388–394. [CrossRef]
116. Pulliam, C.; Eichenseer, S.; Goetz, C.; Waln, O.; Hunter, C.; Jankovic, J.; Vaillancourt, D.; Giuffrida, J.; Heldman, D. Continuous in-home monitoring of essential tremor. *Park. Relat. Disord.* **2014**, *20*, 37–40. [CrossRef]
117. Goni, M.; Eickhoff, S.B.; Far, M.S.; Patil, K.R.; Dukart, J. Smartphone-Based Digital Biomarkers for Parkinson's Disease in a Remotely-Administered Setting. *IEEE Access* **2022**, *10*, 28361–28384. [CrossRef]
118. Livingston, E.; Cao, J.; Dimick, J.B. Tread carefully with stepwise regression. *Arch. Surg.* **2010**, *145*, 1039–1040. [CrossRef]
119. Li, F.; Yang, Y. Analysis of recursive feature elimination methods. In Proceedings of the the 28th ACM/SIGIR International Symposium on Information Retrieval 2005, Salvador, Brazil, 15–19 August 2005. [CrossRef]
120. Kuhn, M.; Johnson, K.; Kuhn, M.; Johnson, K. An Introduction to Feature Selection. In *Applied Predictive Modeling*; Springer: New York, NY, USA, 2013; pp. 487–519. [CrossRef]
121. Senturk, Z.K. Early diagnosis of Parkinson's disease using machine learning algorithms. *Med. Hypotheses* **2020**, *138*, 109603. [CrossRef]
122. Zhang, X.D. Machine Learning. In *A Matrix Algebra Approach to Artificial Intelligence*; Springer: Singapore, 2020. [CrossRef]
123. Russell, S.J.; Norvig, P. *Artificial Intelligence: A Modern Approach*, 4th ed.; Prentice Hall: Hoboken, NJ, USA, 2020.
124. Tinschert, P.; Rassouli, F.; Barata, F.; Steurer-Stey, C.; Fleisch, E.; Puhan, M.; Kowatsch, T.; Brutsche, M.H. *Smartphone-Based Cough Detection Predicts Asthma Control—Description of a Novel, Scalable Digital Biomarker*; European Respiratory Society (ERS): Lausanne, Switzerland, 2020; p. 4569. [CrossRef]

125. ZhuParris, A.; Kruizinga, M.D.; van Gent, M.; Dessing, E.; Exadaktylos, V.; Doll, R.J.; Stuurman, F.E.; Driessen, G.A.; Cohen, A.F. Development and Technical Validation of a Smartphone-Based Cry Detection Algorithm. *Front. Pediatr.* **2021**, *9*, 262. [CrossRef] [PubMed]
126. Fatima, M.; Pasha, M. Survey of Machine Learning Algorithms for Disease Diagnostic. *J. Intell. Learn. Syst. Appl.* **2017**, *9*, 1–16. [CrossRef]
127. Ensari, I.; Caceres, B.A.; Jackman, K.B.; Suero-Tejeda, N.; Shechter, A.; Odlum, M.L.; Bakken, S. Digital phenotyping of sleep patterns among heterogenous samples of Latinx adults using unsupervised learning. *Sleep. Med.* **2021**, *85*, 211–220. [CrossRef] [PubMed]
128. Ko, Y.-F.; Kuo, P.-H.; Wang, C.-F.; Chen, Y.-J.; Chuang, P.-C.; Li, S.-Z.; Chen, B.-W.; Yang, F.-C.; Lo, Y.-C.; Yang, Y.; et al. Quantification Analysis of Sleep Based on Smartwatch Sensors for Parkinson's Disease. *Biosensors* **2022**, *12*, 74. [CrossRef] [PubMed]
129. Farhan, A.A.; Yue, C.; Morillo, R.; Ware, S.; Lu, J.; Bi, J.; Kamath, J.; Russell, A.; Bamis, A.; Wang, B. Behavior vs. introspection: Refining prediction of clinical depression via smartphone sensing data. In Proceedings of the 2016 IEEE Wireless Health (WH), Bethesda, MD, USA, 25–27 October 2016. [CrossRef]
130. Derungs, A.; Schuster-Amft, C.; Amft, O. Longitudinal walking analysis in hemiparetic patients using wearable motion sensors: Is there convergence between body sides? *Front. Bioeng. Biotechnol.* **2018**, *6*, 57. [CrossRef]
131. Freedman, D.A. Statistical Models. In *Statistical Models: THeory and Practice*; Cambridge University Press: Cambridge, UK, 2009.
132. Ahmed, S.T.; Basha, S.M.; Arumugam, S.R.; Kodabagi, M.M. *Pattern Recognition: An Introduction*, 1st ed.; MileStone Research Publications: Bengaluru, India, 2021.
133. Ruppert, D. The Elements of Statistical Learning: Data Mining, Inference, and Prediction. *J. Am. Stat. Assoc.* **2004**, *99*, 567. [CrossRef]
134. Opitz, D.; Maclin, R. Popular Ensemble Methods: An Empirical Study. *J. Artif. Intell. Res.* **1999**, *11*, 169–198. [CrossRef]
135. Kosasi, S. Perancangan Prototipe Sistem Pemesanan Makanan dan Minuman Menggunakan Mobile Device. *Indones. J. Netw. Secur.* **2015**, *1*, 1–10. [CrossRef]
136. Hastie, T.; Tibshirani, R.; Friedman, J. *The Elements of Statistical Learning: Data Mining, Inference, and Prediction*; Springer Science & Business Media: New York, NY, USA, 2013.
137. San-Segundo, R.; Zhang, A.; Cebulla, A.; Panev, S.; Tabor, G.; Stebbins, K.; Massa, R.E.; Whitford, A.; de la Torre, F.; Hodgins, J. Parkinson's disease tremor detection in the wild using wearable accelerometers. *Sensors* **2020**, *20*, 5817. [CrossRef]
138. Ahmadi, M.N.; O'neil, M.E.; Baque, E.; Boyd, R.N.; Trost, S.G. Machine learning to quantify physical activity in children with cerebral palsy: Comparison of group, group-personalized, and fully-personalized activity classification models. *Sensors* **2020**, *20*, 3976. [CrossRef]
139. Faurholt-Jepsen, M.; Busk, J.; Frost, M.; Vinberg, M.; Christensen, E.M.; Winther, O.; Bardram, J.E.; Kessing, L.V. Voice analysis as an objective state marker in bipolar disorder. *Transl. Psychiatry* **2016**, *6*, e856. [CrossRef]
140. Jacobson, N.C.; Lekkas, D.; Huang, R.; Thomas, N. Deep learning paired with wearable passive sensing data predicts deterioration in anxiety disorder symptoms across 17–18 years. *J. Affect. Disord.* **2021**, *282*, 104–111. [CrossRef] [PubMed]
141. Hastie, T.; Tibshirani, R.; Friedman, J. Statistics the Elements of Statistical Learning. *Math. Intell.* **2009**, *27*, 83–85.
142. Patle, A.; Chouhan, D.S. SVM kernel functions for classification. In Proceedings of the 2013 International Conference on Advances in Technology and Engineering, ICATE 2013, Mumbai, India, 23–25 January 2013. [CrossRef]
143. Kim, H.S.; Kim, S.Y.; Kim, Y.H.; Park, K.S. A smartphone-based automatic diagnosis system for facial nerve palsy. *Sensors* **2015**, *15*, 26756–26768. [CrossRef] [PubMed]
144. Luca, S.; Karsmakers, P.; Cuppens, K.; Croonenborghs, T.; Van de Vel, A.; Ceulemans, B.; Lagae, L.; Van Huffel, S.; Vanrumste, B. Detecting rare events using extreme value statistics applied to epileptic convulsions in children. *Artif. Intell. Med.* **2014**, *60*, 89–96. [CrossRef] [PubMed]
145. Ghoraani, B.; Hssayeni, M.D.; Bruack, M.M.; Jimenez-Shahed, J. Multilevel Features for Sensor-Based Assessment of Motor Fluctuation in Parkinson's Disease Subjects. *IEEE J. Biomed. Health Inform.* **2020**, *24*, 1284–1295. [CrossRef]
146. Kramer, O. K-Nearest Neighbors. In *Dimensionality Reduction with Unsupervised Nearest Neighbors. Intelligent Systems Reference Library*; Springer: Berlin/Heidelberg, Germany, 2013; Volume 51. [CrossRef]
147. Jeon, H.; Lee, W.; Park, H.; Lee, H.J.; Kim, S.K.; Kim, H.B.; Jeon, B.; Park, K.S. Automatic classification of tremor severity in Parkinson's disease using awearable device. *Sensors* **2017**, *17*, 2067. [CrossRef]
148. Grunerbl, A.; Muaremi, A.; Osmani, V.; Bahle, G.; Ohler, S.; Troster, G.; Mayora, O.; Haring, C.; Lukowicz, P. Smartphone-based recognition of states and state changes in bipolar disorder patients. *IEEE J. Biomed. Health Inform.* **2015**, *19*, 140–148. [CrossRef]
149. Pranckevičius, T.; Marcinkevičius, V. Comparison of Naive Bayes, Random Forest, Decision Tree, Support Vector Machines, and Logistic Regression Classifiers for Text Reviews Classification. *Balt. J. Mod. Comput.* **2017**, *5*, 221–232. [CrossRef]
150. Worster, A.; Fan, J.; Ismaila, A. Understanding linear and logistic regression analyses. *Can. J. Emerg. Med.* **2007**, *9*, 111–113. [CrossRef]
151. Morrow-Howell, N. The M word: Multicollinearity in multiple regression. *Soc. Work. Res.* **1994**, *18*, 247–251. [CrossRef]
152. Schwenk, M.; Hauer, K.; Zieschang, T.; Englert, S.; Mohler, J.; Najafi, B. Sensor-derived physical activity parameters can predict future falls in people with dementia. *Gerontology* **2014**, *60*, 483–492. [CrossRef] [PubMed]
153. Lecun, Y.; Bengio, Y.; Hinton, G. Deep learning. *Nature* **2015**, *521*, 436–444. [CrossRef] [PubMed]

154. Tu, J.V. Advantages and disadvantages of using artificial neural networks versus logistic regression for predicting medical outcomes. *J. Clin. Epidemiol.* **1996**, *49*, 1225–1231. [CrossRef]
155. Mudiyanselage, T.K.B.; Xiao, X.; Zhang, Y.; Pan, Y. Deep Fuzzy Neural Networks for Biomarker Selection for Accurate Cancer Detection. *IEEE Trans. Fuzzy Syst.* **2020**, *28*, 3219–3228. [CrossRef]
156. Yagin, F.H.; Cicek, I.B.; Alkhateeb, A.; Yagin, B.; Colak, C.; Azzeh, M.; Akbulut, S. Explainable artificial intelligence model for identifying COVID-19 gene biomarkers. *Comput. Biol. Med.* **2023**, *154*, 106619. [CrossRef]
157. Wang, Y.; Lucas, M.; Furst, J.; Fawzi, A.A.; Raicu, D. Explainable Deep Learning for Biomarker Classification of OCT Images. In Proceedings of the 2020 IEEE 20th International Conference on Bioinformatics and Bioengineering (BIBE), Cincinnati, OH, USA, 26–28 October 2020; pp. 204–210. [CrossRef]
158. Fisher, J.M.; Hammerla, N.Y.; Ploetz, T.; Andras, P.; Rochester, L.; Walker, R.W. Unsupervised home monitoring of Parkinson's disease motor symptoms using body-worn accelerometers. *Park. Relat. Disord.* **2016**, *33*, 44–50. [CrossRef]
159. Frogner, J.I.; Noori, F.M.; Halvorsen, P.; Hicks, S.A.; Garcia-Ceja, E.; Torresen, J.; Riegler, M.A. One-dimensional convolutional neural networks on motor activity measurements in detection of depression. In Proceedings of the HealthMedia 2019—Proceedings of the 4th International Workshop on Multimedia for Personal Health and Health Care, Co-Located with MM 2019, Nice, France, 21–25 October 2019; pp. 9–15. [CrossRef]
160. Meisel, C.; el Atrache, R.; Jackson, M.; Schubach, S.; Ufongene, C.; Loddenkemper, T. Machine learning from wristband sensor data for wearable, noninvasive seizure forecasting. *Epilepsia* **2020**, *61*, 2653–2666. [CrossRef]
161. Matarazzo, M.; Arroyo-Gallego, T.; Montero, P.; Puertas-Martín, V.; Butterworth, I.; Mendoza, C.S.; Ledesma-Carbayo, M.J.; Catalán, M.J.; Molina, J.A.; Bermejo-Pareja, F.; et al. Remote Monitoring of Treatment Response in Parkinson's Disease: The Habit of Typing on a Computer. *Mov. Disord.* **2019**, *34*, 1488–1495. [CrossRef]
162. Weiss, K.; Khoshgoftaar, T.M.; Background, D.W. A survey of transfer learning. *J. Big Data* **2016**, *3*, 1345–1459. [CrossRef]
163. Kamishima, T.; Hamasaki, M.; Akaho, S. TrBagg: A Simple Transfer Learning Method and its Application to Personalization in Collaborative Tagging. In Proceedings of the 2009 Ninth IEEE International Conference on Data Mining, Miami, FL, USA, 6–9 December 2009; pp. 219–228. [CrossRef]
164. Fu, Z.; He, X.; Wang, E.; Huo, J.; Huang, J.; Wu, D. Personalized Human Activity Recognition Based on Integrated Wearable Sensor and Transfer Learning. *Sensors* **2021**, *21*, 885. [CrossRef] [PubMed]
165. Chen, Y.; Qin, X.; Wang, J.; Yu, C.; Gao, W. FedHealth: A Federated Transfer Learning Framework for Wearable Healthcare. *IEEE Intell. Syst.* **2020**, *35*, 83–93. [CrossRef]
166. Goschenhofer, J.; Pfister, F.M.J.; Yuksel, K.A.; Bischl, B.; Fietzek, U.; Thomas, J. Wearable-Based Parkinson's Disease Severity Monitoring Using Deep Learning. In *Lecture Notes in Computer Science (Including Subseries Lecture Notes in Artificial Intelligence and Lecture Notes in Bioinformatics)*; Springer: Berlin/Heidelberg, Germany, 2020; Volume 11908 LNAI, pp. 400–415. [CrossRef]
167. Hssayeni, M.D.; Jimenez-Shahed, J.; Burack, M.A.; Ghoraani, B. Ensemble deep model for continuous estimation of Unified Parkinson's Disease Rating Scale III. *Biomed. Eng. Online* **2021**, *20*, 1–20. [CrossRef] [PubMed]
168. Zhang, Y.; Yang, Q. Special Topic: Machine Learning An overview of multi-task learning. *Natl. Sci. Rev.* **2018**, *5*, 30–43. [CrossRef]
169. Lee, G.; Yang, E.; Hwang, S. Asymmetric multi-task learning based on task relatedness and loss. In Proceedings of the International Conference on Machine Learning 2016, New York, NY, USA, 19–24 June 2016; pp. 230–238.
170. Xin, W.; Bi, J.; Yu, S.; Sun, J.; Song, M. Multiplicative Multitask Feature Learning. *J. Mach. Learn. Res. JMLR* **2016**, *17*, 1–33.
171. Zhang, Z.; Jung, T.P.; Makeig, S.; Pi, Z.; Rao, B.D. Spatiotemporal sparse Bayesian learning with applications to compressed sensing of multichannel physiological signals. *IEEE Trans. Neural Syst. Rehabil. Eng.* **2014**, *22*, 1186–1197. [CrossRef] [PubMed]
172. Schneider, J.; Vlachos, M. Personalization of deep learning. In *Data Science–Analytics and Applications: Proceedings of the 3rd International Data Science Conference–iDSC2020*; Springer: Wiesbaden, Geramny, 2021; pp. 89–96.
173. Khademi, A.; El-Manzalawy, Y.; Buxton, O.M.; Honavar, V. Toward personalized sleep-wake prediction from actigraphy. In Proceedings of the 2018 IEEE EMBS International Conference on Biomedical and Health Informatics, BHI 2018, Vegas, NV, USA, 4–7 March 2018; Institute of Electrical and Electronics Engineers Inc.: Piscataway, NJ, USA, 2018; pp. 414–417. [CrossRef]
174. Kuhn, M.; Johnson, K. *Applied Predictive Modeling*; Springer: New York, NY, USA, 2013. [CrossRef]
175. Pal, M. Random forest classifier for remote sensing classification. *Int. J. Remote Sens.* **2005**, *26*, 217–222. [CrossRef]
176. Putin, E.; Mamoshina, P.; Aliper, A.; Korzinkin, M.; Moskalev, A.; Kolosov, A.; Ostrovskiy, A.; Cantor, C.; Vijg, J.; Zhavoronkov, A. Deep biomarkers of human aging: Application of deep neural networks to biomarker development. *Aging* **2016**, *8*, 1021–1033. [CrossRef]
177. Yang, L.; Shami, A. On hyperparameter optimization of machine learning algorithms: Theory and practice. *Neurocomputing* **2020**, *415*, 295–316. [CrossRef]
178. Waring, J.; Lindvall, C.; Umeton, R. Automated machine learning: Review of the state-of-the-art and opportunities for healthcare. *Artif. Intell. Med.* **2020**, *104*, 101822. [CrossRef]
179. Bergstra, J.; Ca, J.B.; Ca, Y.B. Random Search for Hyper-Parameter Optimization Yoshua Bengio. 2012. Available online: http://scikit-learn.sourceforge.net (accessed on 5 January 2023).
180. Beam, A.L.; Manrai, A.K.; Ghassemi, M. Challenges to the Reproducibility of Machine Learning Models in Health Care. *JAMA* **2020**, *323*, 305. [CrossRef] [PubMed]

181. Ahlrichs, C.; Samà, A.; Lawo, M.; Cabestany, J.; Rodríguez-Martín, D.; Pérez-López, C.; Sweeney, D.; Quinlan, L.R.; Laighin, G.Ò.; Counihan, T.; et al. Detecting freezing of gait with a tri-axial accelerometer in Parkinson's disease patients. *Med. Biol. Eng. Comput.* **2016**, *54*, 223–233. [CrossRef]
182. Rosenwein, T.; Dafna, E.; Tarasiuk, A.; Zigel, Y. *Detection of Breathing Sounds during Sleep Using Non-Contact Audio Recordings*; Institute of Electrical and Electronics Engineers Inc.: Piscataway, NJ, USA, 2014. [CrossRef]
183. Pérez-López, C.; Samà, A.; Rodríguez-Martín, D.; Moreno-Aróstegui, J.M.; Cabestany, J.; Bayes, A.; Mestre, B.; Alcaine, S.; Quispe, P.; Laighin, G.; et al. Dopaminergic-induced dyskinesia assessment based on a single belt-worn accelerometer. *Artif. Intell. Med.* **2016**, *67*, 47–56. [CrossRef] [PubMed]
184. Bernad-Elazari, H.; Herman, T.; Mirelman, A.; Gazit, E.; Giladi, N.; Hausdorff, J.M. Objective characterization of daily living transitions in patients with Parkinson's disease using a single body-fixed sensor. *J. Neurol.* **2016**, *263*, 1544–1551. [CrossRef] [PubMed]
185. Carvalho, D.V.; Pereira, E.M.; Cardoso, J.S. Machine Learning Interpretability: A Survey on Methods and Metrics. *Electronics* **2019**, *8*, 832. [CrossRef]
186. Zhou, J.; Gandomi, A.H.; Chen, F.; Holzinger, A. Evaluating the Quality of Machine Learning Explanations: A Survey on Methods and Metrics. *Electronics* **2021**, *10*, 593. [CrossRef]
187. Hossin, M.; Sulaiman, M.N. A Review on Evaluation Metrics for Data Classification Evaluations. *Int. J. Data Min. Knowl. Manag. Process* **2015**, *5*, 1–11. [CrossRef]
188. He, H.; Ma, Y. *Imbalanced Learning*; Wiley: Hoboken, NJ, USA, 2013. [CrossRef]
189. Wan, S.; Liang, Y.; Zhang, Y.; Guizani, M. Deep Multi-Layer perceptron classifier for behavior analysis to estimate Parkinson's disease severity using smartphones. *IEEE Access* **2018**, *6*, 36825–36833. [CrossRef]
190. Chicco, D.; Tötsch, N.; Jurman, G. The matthews correlation coefficient (Mcc) is more reliable than balanced accuracy, bookmaker informedness, and markedness in two-class confusion matrix evaluation. *BioData Min.* **2021**, *14*, 1–22. [CrossRef]
191. Jurman, G.; Riccadonna, S.; Furlanello, C. A Comparison of MCC and CEN Error Measures in Multi-Class Prediction. *PLoS ONE* **2012**, *7*, e41882. [CrossRef]
192. Faurholt-Jepsen, M.; Busk, J.; HelgaÞórarinsdóttir; Frost, M.; Bardram, J.E.; Vinberg, M.; Kessing, L.V. Objective smartphone data as a potential diagnostic marker of bipolar disorder. *Aust. N. Z. J. Psychiatry* **2019**, *53*, 119–128. [CrossRef] [PubMed]
193. Xu, Z.; Shen, D.; Nie, T.; Kou, Y. A hybrid sampling algorithm combining M-SMOTE and ENN based on Random forest for medical imbalanced data. *J. Biomed. Inform.* **2020**, *107*, 103465. [CrossRef] [PubMed]
194. Zeng, M.; Zou, B.; Wei, F.; Liu, X.; Wang, L. Effective prediction of three common diseases by combining SMOTE with Tomek links technique for imbalanced medical data. In Proceedings of the 2016 IEEE International Conference of Online Analysis and Computing Science, ICOACS 2016, Chongqing, China, 28–29 May 2016; pp. 225–228. [CrossRef]
195. Botchkarev, A. Performance Metrics (Error Measures) in Machine Learning Regression, Forecasting and Prognostics: Properties and Typology. *Interdiscip. J. Inf. Knowl. Manag.* **2018**, *14*, 45–76. [CrossRef]
196. di Bucchianico, A. Coefficient of Determination. In *Encyclopedia of Statistics in Quality and Reliability*; Wiley: Hoboken, NJ, USA, 2007. [CrossRef]
197. Piepho, H. A coefficient of determination (R^2) for generalized linear mixed models. *Biom. J.* **2019**, *61*, 860–872. [CrossRef] [PubMed]
198. Gelman, A.; Pardoe, I. Bayesian Measures of Explained Variance and Pooling in Multilevel (Hierarchical) Models. *Technometrics* **2006**, *48*, 241–251. [CrossRef]
199. Hodson, T.O. Root-mean-square error (RMSE) or mean absolute error (MAE): When to use them or not. *Geosci. Model. Dev.* **2022**, *15*, 5481–5487. [CrossRef]
200. Mezzadri, G.; Laloë, T.; Mathy, F.; Reynaud-Bouret, P. Hold-out strategy for selecting learning models: Application to categorization subjected to presentation orders. *J. Math. Psychol.* **2022**, *109*, 102691. [CrossRef]
201. Gholamiangonabadi, D.; Kiselov, N.; Grolinger, K. Deep Neural Networks for Human Activity Recognition with Wearable Sensors: Leave-One-Subject-Out Cross-Validation for Model Selection. *IEEE Access* **2020**, *8*, 133982–133994. [CrossRef]
202. Little, M.A.; Varoquaux, G.; Saeb, S.; Lonini, L.; Jayaraman, A.; Mohr, D.C.; Kording, K.P. Using and understanding cross-validation strategies. Perspectives on Saeb et al. *Gigascience* **2017**, *6*, 1–6. [CrossRef]
203. Peterson, R.A.; Cavanaugh, J.E. Ordered quantile normalization: A semiparametric transformation built for the cross-validation era. *J. Appl. Stat.* **2020**, *47*, 2312–2327. [CrossRef]
204. Zhang, Y.; Yang, Y. Cross-validation for selecting a model selection procedure. *J. Econom.* **2015**, *187*, 95–112. [CrossRef]
205. Refaeilzadeh, P.; Tang, L.; Liu, H. Cross-Validation. In *Encyclopedia of Database Systems*; Springer: Berlin/Heidelberg, Germany, 2016; pp. 1–7. [CrossRef]
206. Browne, M.W. Cross-validation methods. *J. Math. Psychol.* **2000**, *44*, 108–132. [CrossRef] [PubMed]
207. Wainer, J.; Cawley, G. Nested cross-validation when selecting classifiers is overzealous for most practical applications. *Expert Syst. Appl.* **2021**, *182*, 115222. [CrossRef]
208. Kohavi, R. A Study of Cross-Validation and Bootstrap for Accuracy Estimation and Model Selection. 1995. Available online: http://robotics.stanford.edu/~ronnyk (accessed on 5 January 2023).
209. Vanwinckelen, G.; Blockeel, H. On estimating model accuracy with repeated cross-validation. In *BeneLearn 2012: Proceedings of the 21st Belgian-Dutch Conference on Machine Learning*; Benelearn 2012 Organization Committee: Ghent, Belgium, 2012; pp. 39–44.

210. Parvandeh, S.; Yeh, H.-W.; Paulus, M.P.; McKinney, B.A. Consensus Features Nested Cross-Validation. *bioRxiv* **2020**. [CrossRef] [PubMed]
211. Goble, C.; Cohen-Boulakia, S.; Soiland-Reyes, S.; Garijo, D.; Gil, Y.; Crusoe, M.; Peters, K.; Schober, D. Fair computational workflows. *Data Intell.* **2020**, *2*, 108–121. [CrossRef]
212. Muehlematter, U.J.; Daniore, P.; Vokinger, K.N. Approval of artificial intelligence and machine learning-based medical devices in the USA and Europe (2015–20): A comparative analysis. *Lancet Digit. Health* **2021**, *3*, e195–e203. [CrossRef]
213. Beckers, R.; Kwade, Z.; Zanca, F. The EU medical device regulation: Implications for artificial intelligence-based medical device software in medical physics. *Phys. Med.* **2021**, *83*, 1–8. [CrossRef]
214. van Oirschot, J.; Ooms, G. *Interpreting the EU Artificial Intelligence Act for the Health Sector*; Health Action International: Amsterdam, The Netherlands, February 2022.
215. Agarwal, S.; LeFevre, A.; Lee, J.; L'engle, K.; Mehl, G.; Sinha, C.; Labrique, A. Guidelines for reporting of health interventions using mobile phones: Mobile health (mHealth) evidence reporting and assessment (mERA) checklist. *BMJ* **2016**, *352*, i1174. [CrossRef]
216. Yang, J.; Soltan, A.A.S.; Clifton, D.A. Machine learning generalizability across healthcare settings: Insights from multi-site COVID-19 screening. *NPJ Digit. Med.* **2022**, *5*, 69. [CrossRef]
217. Petersen, E.; Potdevin, Y.; Mohammadi, E.; Zidowitz, S.; Breyer, S.; Nowotka, D.; Henn, S.; Pechmann, L.; Leucker, M.; Rostalski, P.; et al. Responsible and Regulatory Conform Machine Learning for Medicine: A Survey of Challenges and Solutions. *IEEE Access* **2022**, *10*, 58375–58418. [CrossRef]
218. FAIR Principles—GO FAIR. Available online: https://www.go-fair.org/fair-principles/ (accessed on 16 December 2021).
219. Fletcher, R.R.; Nakeshimana, A.; Olubeko, O. Addressing Fairness, Bias, and Appropriate Use of Artificial Intelligence and Machine Learning in Global Health. *Front. Artif. Intell.* **2021**, *3*, 116. [CrossRef] [PubMed]
220. Mei, J.; Desrosiers, C.; Frasnelli, J. Machine Learning for the Diagnosis of Parkinson's Disease: A Review of Literature. *Front. Aging Neurosci.* **2021**, *13*, 633752. [CrossRef] [PubMed]
221. Chekroud, A.M.; Bondar, J.; Delgadillo, J.; Doherty, G.; Wasil, A.; Fokkema, M.; Cohen, Z.; Belgrave, D.; DeRubeis, R.; Iniesta, R.; et al. The promise of machine learning in predicting treatment outcomes in psychiatry. *World Psychiatry* **2021**, *20*, 154–170. [CrossRef] [PubMed]
222. Kruizinga, M.D.; Zhuparris, A.; Dessing, E.; Krol, F.J.; Sprij, A.J.; Doll, R.; Stuurman, F.E.; Exadaktylos, V.; Driessen, G.J.A.; Cohen, A.F. Development and technical validation of a smartphone-based pediatric cough detection algorithm. *Pediatr. Pulmonol.* **2022**, *57*, 761–767. [CrossRef] [PubMed]
223. Doshi-Velez, F.; Kim, B. Towards A Rigorous Science of Interpretable Machine Learning. *arXiv* **2017**, arXiv:1702.08608.
224. Ignatiev, A. Towards Trustable Explainable AI. 2020. Available online: https://www.kaggle.com/uciml/zoo-animal-classification (accessed on 5 January 2023).
225. Walsh, I.; Fishman, D.; Garcia-Gasulla, D.; Titma, T.; Pollastri, G.; Capriotti, E.; Casadio, R.; Capella-Gutierrez, S.; Cirillo, D.; Del Conte, A.; et al. DOME: Recommendations for supervised machine learning validation in biology. *Nat. Methods* **2021**, *18*, 1122–1127. [CrossRef]
226. Zippel, C.; Bohnet-Joschko, S. Rise of Clinical Studies in the Field of Machine Learning: A Review of Data Registered in ClinicalTrials.gov. *Int. J. Environ. Res. Public Health* **2021**, *18*, 5072. [CrossRef] [PubMed]

sensors

MDPI

Article

A Snapshot-Stacked Ensemble and Optimization Approach for Vehicle Breakdown Prediction

Reza Khoshkangini [1,2,*], Mohsen Tajgardan [3], Jens Lundström [2], Mahdi Rabbani [4] and Daniel Tegnered [5]

1 Internet of Things and People Research Center (IoTap), Department of Computer Science and Media Technology, Malmö University, 211 19 Malmö, Sweden
2 Center for Applied Intelligent Systems Research (CAISR), Halmstad University, 301 18 Halmstad, Sweden; jens.r.lundstrom@hh.se
3 Faculty of Electrical and Computer Engineering, Qom University of Technology, Qom 1519-37195, Iran; tajgardan.m@qut.ac.ir
4 Canadian Institute for Cybersecurity (CIC), University of New Brunswick (UNB), Fredericton, NB E3B 9W4, Canada; m.rabbani@unb.ca
5 Volvo Group Connected Solutions, 417 56 Gothenburg, Sweden; daniel.tegnered@volvo.com
* Correspondence: reza.khoshkangini@mau.se

Abstract: Predicting breakdowns is becoming one of the main goals for vehicle manufacturers so as to better allocate resources, and to reduce costs and safety issues. At the core of the utilization of vehicle sensors is the fact that early detection of anomalies facilitates the prediction of potential breakdown issues, which, if otherwise undetected, could lead to breakdowns and warranty claims. However, the making of such predictions is too complex a challenge to solve using simple predictive models. The strength of heuristic optimization techniques in solving np-hard problems, and the recent success of ensemble approaches to various modeling problems, motivated us to investigate a hybrid optimization- and ensemble-based approach to tackle the complex task. In this study, we propose a snapshot-stacked ensemble deep neural network (SSED) approach to predict vehicle claims (in this study, we refer to a claim as being a breakdown or a fault) by considering vehicle operational life records. The approach includes three main modules: Data pre-processing, Dimensionality Reduction, and Ensemble Learning. The first module is developed to run a set of practices to integrate various sources of data, extract hidden information and segment the data into different time windows. In the second module, the most informative measurements to represent vehicle usage are selected through an adapted heuristic optimization approach. Finally, in the last module, the ensemble machine learning approach utilizes the selected measurements to map the vehicle usage to the breakdowns for the prediction. The proposed approach integrates, and uses, the following two sources of data, collected from thousands of heavy-duty trucks: Logged Vehicle Data (LVD) and Warranty Claim Data (WCD). The experimental results confirm the proposed system's effectiveness in predicting vehicle breakdowns. By adapting the optimization and snapshot-stacked ensemble deep networks, we demonstrate how sensor data, in the form of vehicle usage history, contributes to claim predictions. The experimental evaluation of the system on other application domains also indicated the generality of the proposed approach.

Keywords: breakdown prediction; optimization; deep neural networks; ensemble learning

1. Introduction

Machine Learning-based approaches for predictive modeling have become one of the main tools in the Predictive Maintenance (PdM) strategy of the automotive industry, to improve the overall maintenance and reliability of operations, and to reduce costs and safety issues [1]. Such predictive models attempt to model and predict the breakdown of an asset (e.g., truck, bus, or car) which, if it occurred within the agreed warranty time period, could lead to a warranty claim. This set of systems is precious software technology that minimizes unexpected down-time which, in turn, improves safety and asset reliability. PdM is highly valuable for reducing costs as it means maintenance is performed only when necessary. Based on the figures reported in [2], maintenance is one of the leading vehicle operational expenses, accounting for around 59% of overall costs. Furthermore, under the PdM umbrella, claim or breakdown prediction of upcoming down-time significantly decreases the risk of human life being jeopardized in, for example, heavy-duty vehicles (in this study, the data source for the target labels is a warranty claim database and, therefore, we interchangeably use down-time, claim, breakdown and fault, such that they all point to the concept of a component defect leading to a claim in the dataset).

Heavy-duty vehicles are complex systems, with numerous possible specifications, operations and driver behaviors, wherein component down-times can originate from multiple sub-components failing for various causes. The correct prediction of breakdowns is critical and essential because failing to precisely recognize such breakdowns leads to increased maintenance costs, increased customer safety risks, decreased customer satisfaction, and lowered brand value. All of these factors add up to a significant loss for companies. Thus, correctly predicting faults that could cause customer or company loss has become an important goal in today's modern industrial strategy. Under this vision, many studies have investigated the prediction of warranty claims or component breakdowns, reliability, and degradation in the context of the vehicle industry, utilizing machine learning and statistical approaches, such as deep neural networks, recurrent neural networks, support vector machines or stochastic approaches [3–5]. Diagnostic models were developed, based on the data collected from machines (e.g, industrial machines, turbines, vehicles, trains, etc.) to estimate the performance and condition of a component. Many of the studies utilized stochastic approaches to estimate component failures [6,7]. Many studies have focused on estimating the remaining useful life (RUL) of a component, utilizing optimization or neural networks algorithms [7–10]. For example, Yang et al. [11] studied claim forecasting, where they focused on product usage rate and introduced two predictive models. In this study, the authors considered Weibull's distributed time assumption to foretell the number of breakdowns. More recently, an interesting work by Chehade et al. [12] studied the problem of data maturation in predicting warranty claims. They utilized Gaussian Mixture distribution of claim patterns from similar vehicles to forecast vehicle warranty claims as early as possible [12].

From the literature, we observed that considerable progress has been made in the context of predicting upcoming breakdowns in the automotive industry by adopting machine learning and statistical solutions. We also noticed that most of these combine and use the approaches with the age and lifetime of a particular component to estimate the component's breakdown. However, several recent studies in such areas [13,14] partially use vehicle usage to predict the need for maintenance to avoid breakdowns. Complexity, both in terms of formulating the problem and the high dimensionality of the usage data, make the task challenging, requiring further investigation.

Although the works mentioned vary widely in terms of the application context, problem to solve, and technique developed to tackle the challenge, machine learning has proved to be successful, particularly in the automotive industry [15]. However, there is still room for improvement (in terms of prediction accuracy and explainability) in existing claim prognostic systems. We believe that such forecasting operations can be improved by incorporating AI and advanced deep neural network approaches.

This motivated us to develop an ensemble solution taking advantage of hundreds of sensor signals to predict breakdowns (which often lead to warranty claims, the source of our target variable). However, the challenge of high-dimensional data needs to be dealt with. Thus, in this study, we propose an ensemble method to extract and select the best representation of the data–the most informative features–to build the predictive models for warranty claim prediction (WCP). The approach includes three modules. In the first module, that of "Data pre-processing". we developed a set of practices to integrate various sources of data, to extract relevant information and to segment the data into various time windows. In the second module, that of "Dimensionality Reduction", we designed an optimization procedure, using a Genetic Algorithm (GA) coupled with Elastic Net, to translate high-dimensional data into a representation with lower dimensionality without compromising prediction performance. GA is a method inspired by the evolutionary process and is used for solving constrained or unconstrained optimization problems [16]. The method runs several GA operators that frequently change individuals to reach the best solutions through several generations.

In the third module, that of "Snapshot-stacked Ensemble", which contains two phases, we developed an ensemble deep neural network to use the measurements (predictors, measurements, and features all refer to the same concept of sensor data) to map vehicle usage to claims. This module was developed by means of a Snapshot-stacked Ensemble method to increase the performance of the predictive model. In this fashion, several training learners are built from different perspectives to solve a complex claim prediction using vehicle usage. This is achieved by horizontally aggregating the outputs of several snapshot models, utilizing a Cyclic Cosine Annealing Schedule (CCAS) [17] concerning the principle of diversity, to improve predictive model performance. Indeed, each snapshot refers to a separate deep neural network, generated over one training process and saved, which is then used to build an ensemble model. These snapshots result in meta-learning that is capable of learning from various models (deep neural networks) to increase the accuracy of the claim prediction. In other words, we exploit the outcome of each snapshot to build an ensemble of networks for better prediction.

To the best of our knowledge, this study is the first to investigate the performance of ensemble optimization and snapshot-stacked deep neural networks to translate higher dimensional data into a lower dimension to be utilized for claim prediction. Earlier fault estimation was done through manual inspection and statistical approaches. Applying artificial intelligence techniques on sensor data automates fault detection, requiring less effort and achieving more accuracy in a timely fashion [18].

The following research questions (RQs) further elaborate the investigative objectives of our proposed approach:

- **RQ1—Measurement selection:** How could vehicle usage best be extracted for use in claim prediction, and to what extent could the extracted features contribute to the predictive model?
- **RQ2—Claim prediction, based on vehicle usage:** To what extent could ensemble deep neural networks contribute to, and improve, claim prediction in the automotive industry?

Our study aimed to address claim prediction by mapping vehicle usage to future breakdowns. First, we concentrated on selecting a subset of features that would best represent vehicle usage. Since heuristic and optimization solutions have demonstrated their capabilities to tackle np-hard problems in various domains [19,20], we hypothesized that the genetic method would be able to address this issue. Thus, to answer RQ1 we adapted the evolutionary approach to feature selection so as to select the sub-optimal features. The evaluation was performed over several generations and compared with other standard deep neural network and ensemble approaches. The contribution of the extracted features to the predictive model was evaluated by means of an A/B test, guided by the aforementioned RQ1. We report on the results of that optimization and A/B test, to show how the selected and extracted features positively affect the performance of the model. Second, we maintained that the optimization could perform better by coupling it with an ensemble learning approach. Hence, to answer RQ2, we developed a Snapshot-stacked ensemble learning system through two phases of generating multiple diverse models, which were used to build meta-learning. The meta-learning was then utilized for the final claim or breakdown prediction. The reported figures confirm that this development improves claim prediction performance compared with other ML approaches. The integration of these two methods in predictive maintenance strategy, which our research outlines, constitutes this paper's main contribution. The following are our contributions in this study:

- Whereas previous studies mostly used limited features in their prognostic systems, we used hundreds of real sensors' data (LVD) from heavy-duty trucks to model and map vehicle usage to component breakdown.
- Taking into consideration the high dimensionality of sensor data, an optimization approach was introduced to extract the best representation of the data characterizing vehicle usage in order to build an accurate predictive model for warranty claim forecasting.
- Taking into consideration the low classification accuracy of claim prediction using vehicle usage, a snapshot-stacked ensemble deep learning approach is proposed in this paper. Compared with existing algorithms, our method effectively improves classification performance.

The rest of the paper is organized as follows. Section 2 presents related studies. In Section 3, we describe the data used. In Section 4 we explain the proposed approach. Section 5 covers experimental evaluation and results, followed by a discussion and summary of the work in Section 6.

2. Related Works

Predictive maintenance techniques have previously been investigated and developed to examine and monitor the condition of tools and equipment to estimate when maintenance needs to be performed. In this regard, identifying component failures proffers opportunities for companies to take preemptive action, in the form of predictive maintenance, to minimize breakdowns and prevent more significant issues in the long term. This mitigates safety risks, lowers costs and even supports sustainable production in smart cities [21,22]. Many attempts have been made [23–25] to develop various statistical and machine learning solutions to predict component failures in a wide range of applications such as the following: manufacturing [26,27], automotive [28–31], and energy [32]. Prognostic models have been based on data collected from machines, social networks, and other sources to estimate the performance and condition of components or the number of upcoming warranty claims [33]. These models widely used statistical approaches to estimate component failures [6,7,34,35]. For example, in the early 1990s, in [34], a log-linear Poisson model was utilized to build a forecasting system based on the dates of warranty claims to predict failures. Later, Fredette et al. [5] developed a mixed Non-homogeneous Poisson Process (NHPP) to tackle a similar problem, that of component down-time prediction. An interesting remaining useful life (RUL) forecasting approach was introduced in [7], in which the Wiener process [36] with drift was utilized to model degradation in component sensors. Recently, in [33], social media data was used to improve performance in daily warranty claim prediction. To tackle the challenge of warranty data maturation, an intriguing study was developed by Chehade et al. [12]. This investigation used warranty claim data to devise a Gaussian mixture distribution of claim patterns at current and future maturation levels. They took advantage of Bayesian theories to assess the conditional posterior distribution of warranty claims at future maturation based on the available data at the current maturation level.

To the best of our knowledge, the larger part of studies focusing on forecasting problems widely use neural networks to estimate the RUL of components and the state of complex machines [37]. One example is in the context of energy production [38], wherein a self-evolving maintenance scheduler technique was proposed to detect damage in the gearbox bearings of wind turbines. The technique uses an ANN to establish a condition monitoring system (CMS) based on supervisory control and acquisition data. To tackle a similar problem, in [39] a Sparse Autoencoder-based Deep Neural Network (SAE–DNN) approach was developed to detect bearing breakdowns in multi-component systems.

A typical optimization approach was introduced in [9], using Relevance Vector Machine, with the support of ant colony optimization (ACO), to detect gearbox fault detection.

To tackle a similar problem, a neural network-based forecasting system was introduced in [40], based on a dynamic Cuckoo search optimization algorithm. Such forecasting is crucial, since it can help manufacturers lower maintenance costs, increase customer satisfaction, and safety, etc.

Lifetime distribution estimation as a practical solution has gone a long way in component fault estimation. For instance, a probabilistic model was developed by Singpurwalla et al. in [41], wherein time, and time-dependent quantity, were utilized to build a forecasting model. Later, Kaminskiy et al. [42] proposed a component failure prediction approach, based on piece-wise application of the Weibull and exponential distributions, to predict intrinsic failure period and early failure period. Similarly, in [43] two variables of age and mileage of the vehicle were considered to estimate the mean cumulative number of breakdowns.

Among the vast numbers of approaches introduced and developed for various prediction tasks in many applications, ensemble techniques have been quite successful. Ensemble machine learning approaches are techniques that integrate several base ML models to provide improved performance [44–48]. As illustrated in Figure 1, these techniques can be divided into three categories: Stacking, Bagging, and Boosting. The Stacking, or Stacked generalization, technique [49] uses and combines predictions from multiple predictive models to build a new model on the training data. Then the model is utilized for the final prediction of the test data. Bagging is a technique that seeks a mixed group of ensemble predictive models built by various parts of the training data [50]. Then, statistical approaches, such as averaging or voting, are used to combine prediction results from members. Boosting is another ensemble approach that creates a strong predictive model from several weak classifiers [51]. In this approach, the training data is used to build a predictive model. The second model then tries to correct the mistakes from the first classifier. This process continue until the maximum number of predictive models is reached or the perfect prediction acquired.

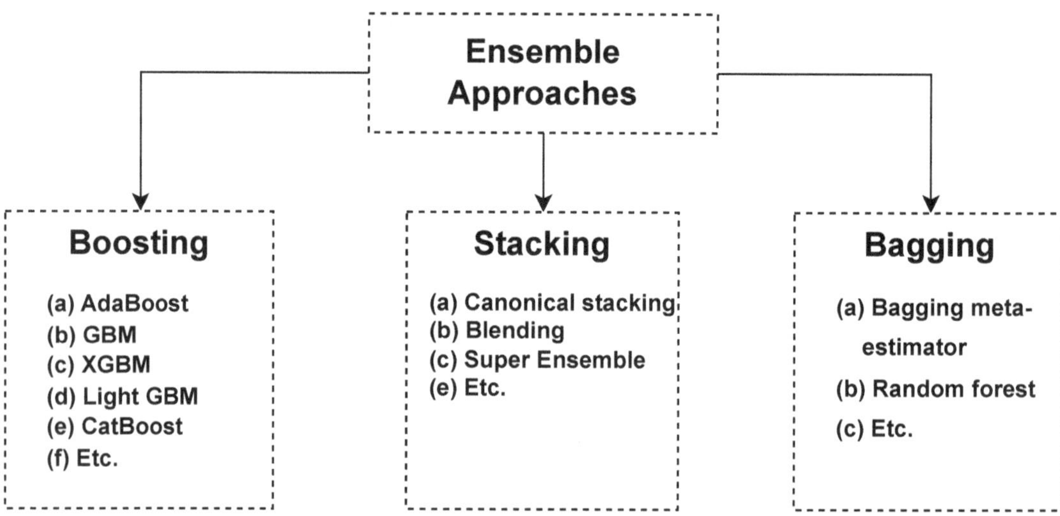

Figure 1. Taxonomy of ensemble approaches.

Researchers working on applications from different sectors also took advantage of ensemble methods [52]. In the context of claim and fault detection, we observed interesting studies in different sectors [53,54]. For example, a bagging ensemble-based approach was developed in [55] to forecast faults in a single-shaft industrial gas turbine. In this approach, the fault prognosis was based on weighted voting to improve the robustness of the fault detection technique. To tackle the same problem in a similar domain, in [56] a deep neural network ensemble approach was developed and applied to a wind turbine data series.

This work trained the network offline based on a training set of the imagination matrix that transformed the data series through segmentation technology. Then, the network was retrained by transferring learning, based on the renewed training set, to achieve enough knowledge to forecast the fault. In [57], an ensemble approach was introduced to predict the degradation of track geometry, wherein different models were developed by means of regression, deterioration, and classification. In the context of an electrical system, ref. [54] introduced an arc series ensemble fault detection approach, wherein they developed a boosting converter CPL and a buck convertor constant power load (CPL) to study different arc fault behaviors.

All in all, two main limitations can be observed in the related works presented: (1) most are limited to models based on only a few input parameters to predict warranty claims; (2) lack in use of complex real-world operational data from thousands of deployed products, which come with the challenges of noisy and missing data. Our study addresses these issues by deploying a complex ensemble-based system on high-dimensional data collected from thousands of heavy-duty trucks.

To our knowledge, the combination of evolutionary technique and ensemble deep neural network as an approach to increase warranty claim prediction accuracy, is novel. Our approach addresses the fault detection problem in the automotive industry, which is generic and applicable to a variety of fault and claim predictions in different sectors. Borrowing from the taxonomy in [56], our warranty claim prediction can be characterized as a stacked ensemble approach, since it takes into consideration the prediction of several diverse snapshot models to build meta-learning, and then the model is used for a final prediction. In the remainder of this paper, we describe the data, our proposed solution and the evaluation of its potential, both for warranty claim data and datasets from other applications.

3. Data Presentation

In this section we present the following two data sets which were used to test the proposed forecasting method: Logged Vehicle Data (LVD), including specifications, and usage, of vehicles aggregated in a cumulative fashion; Warranty Claim data (WCD), consisting of claim information reported over the vehicles' life-times during their warranty periods (24 months).

3.1. Logged Vehicle Data

The LVD used in this study were collected from commercial trucks over a four-year period, from 2016 to 2019. The LVD consisted of the aggregated usage information for a fleet of heavy-duty trucks operating in Europe. The parameters were collected using telematics, as well as through manual inspection when a vehicle visited an authorized workshop for repairs and service. In general, two types of parameters are logged in the LVD. The first type describes the configuration of the vehicles, e.g., engine type, gearbox specification, or type of pumps. The second type logs the usage of the vehicle during its operation. This data is logged in histogram and scalar formats, continuously aggregated and contains a number of different parameters, such as fuel consumption, compressor usage, gear usage, cargo load, engine ambient temperature vs. ambient pressure log, coolant temperature vs. oil temperature log, etc. An example of cumulative scalar features and its structure over one year are illustrated in Figure 2.

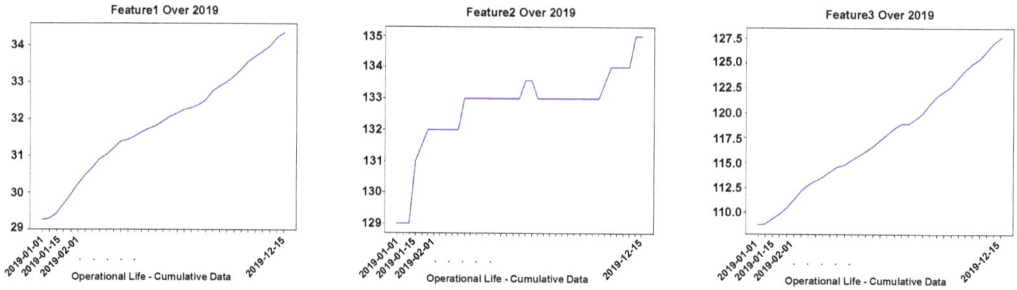

(**a**) feature1-over year 2019 (**b**) feature2-over year 2019 (**c**) feature3-over year 2019

Figure 2. Illustration of the vehicle sensor data logged as scalar features. The scalars indicate the vehicle usage over one year in 2019. As is clear, the parameter was cumulative and monotonous–in feature1 and feature3–and increased over the year.

3.2. *Warranty Claim Data*

Claim data contains information regarding a vehicle's warranty claims logged during its operation, collected by *OEM* authorized workshops in Europe. In particular, the claim database stores information about which part or component of a vehicle was repaired or changed (due to a failure) along with the date. The parts and components are identified by the normalized identification codes using four different levels of detail. This claim dataset contains various parameters, such as names of components, codes, descriptions, and dates, etc.

4. Proposed Approach

4.1. *Problem Formulation*

This section represents the formulations determined to tackle the claim forecasting problem. Our approach considers the WCD and LVD as the system's inputs and employs optimization and ensemble-based deep neural networks. This study attempted to predict, one month beforehand, defects related to the power train in heavy-duty vehicles, during their warranty period of 24 months from when they started operating in service. Thus, our intention in this claim prediction investigation could be divided into two problem formulations, as follow:

- First, we formulated the process of dimensionality reduction as an optimization task, and developed an optimization approach to extract a better set of features characterizing vehicle usage over time. The approach was carried out in a time-series fashion based on the hypothesis that vehicle behavior changes over time, so different predictors might have different impacts on claims/failures and, consequently, on performances of predictive models.
- Second, we investigated the design of an ensemble machine learning approach, formulating the task as a classification to predict the claim/defect related to a power train. Given the selected predictors from the optimization step, the ensemble approach mapped the usage to the upcoming claims one month ahead during the warranty period.

Figure 3 shows the high-level structure of the proposed claim prediction approach for the automotive industry. It contains three main modules: Data pre-processing, Feature Selection Optimization and Ensemble Modeling. The data pre-processing module consists of three processes: data integration, feature extraction, and data segmentation. The optimization module is then employed to choose the measurements in a time-series fashion. Here, the best representation of data in each segment is transferred to the ensemble modeling module. Subsequently, the ensemble method builds the predictive model over two phases to forecast warranty claims, given the vehicle usage. These modules and sub-modules are described as follow:

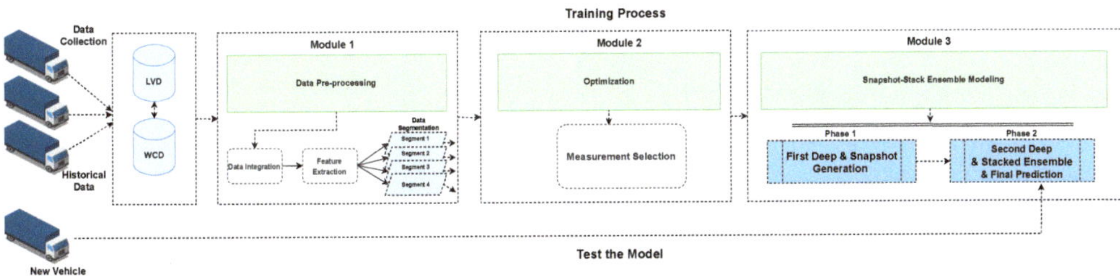

Figure 3. The conceptual view of the proposed breakdown prediction approach.

4.2. Module 1: Data Pre-Processing

Data preparation is an integral part of any machine learning problem and our fault diagnosis task is not an exception. In this section, we describe how we prepared the data to use in our optimization process.

4.2.1. Data Integration

This module aimed to merge the LVD and WCD to create an integrated dossier with the usage and failure information. We combined two data sets based on the vehicle's "Chassis id", "Date of readout, which is logged", and "Date of failure report". To merge these two datasets, we borrowed the method that we used in our previous study [18], where we took advantage of a Volvo expert's knowledge to determine a time window of one month as an interval where the symptoms of failures are most likely to be visible. Therefore, the prediction horizon was set to 30 days for the developed model. In this way, the merged data-set included a new attribute named "Target" as the target feature T. This parameter has a value of 1 for a given sample (r_i) if, and only if, failure of a specific component of interest is reported. Therefore, each vehicle may have multiple failures (for the same component) during its operational life. However, only one failure (for the same component) is considered during the desired time window. More formally, each time-window/time span is assigned a binary label according to Equation (1), where tw refers to the length of the time window that has the highest impact on failures in trucks and where τ is the current time (end of time window). Note: tw is determined based on a Volvo expert's knowledge, which differs from one type of failure to another. Thus, in this study, tw was set to one month which was considered large enough to overcome the effects of choosing a window that is too small, e.g., having a prediction horizon that is too tight or not being able to cover the whole chain of events leading to a failure. If a shorter time window would have been chosen, the target feature would have made more sense to be continuous, due to gradual transitioning into the chain of events.

$$T_{r_i} = \begin{cases} 1 & \text{if a fault happened in } [\tau - tw, \tau] \\ 0 & \text{if no fault happened in } [\tau - tw, \tau] \end{cases} \tag{1}$$

4.2.2. Feature Extraction

This sub-module computes the differences between consecutive measurements of each sensor signal accumulator and quantizes the features. Indeed, this process was used due to the accumulative nature of data. It is essential to extract the actual usage in each timestamp, so that an abnormality estimating a breakdown in the near future might be indicated.

Equation (2) shows how a new feature (from each original feature) is calculated:

$$f_{i_{new}} = f_{i_{r_{i+1}}} - f_{i_{r_i}} \tag{2}$$

$$Quarterlies = \begin{cases} Q1 & (\text{n}+1)/4 \\ Q2 & (\text{n}+1)/2 \\ Q3 & (3(\text{n}+1)/4)) \end{cases} \tag{3}$$

$$Cat = \begin{cases} low \text{ if } f_{i_{new}} < Q1 \\ medium \text{ if } \text{Q1} < f_{i_{new}} < Q3 \\ high \text{ if } f_{i_{new}} > Q3 \end{cases} \tag{4}$$

where $f_{i_{new}}$ denotes the extracted feature, and r_i and r_{i+1} represent the readouts that are collected over the time and n points to the number of samples or readouts. Basically, in the feature extraction module, we calculate the *delta* between subsequent readouts, and exploit them as new features in training the models. Indeed, these extracted features demonstrate the actual usage (and their changes) of vehicles in a bi-weekly fashion. The anomaly behavior of the vehicle hidden in the data is revealed from these changes, which might indicate potential faults in the near future. In addition, information related to each season (spring, summer, autumn and winter) is added as a new feature in the data.

Figure 4 depicts only the usage change of 10 vehicles considering 8 new features ($f_{i_{new}}$)– out of more than 600 features in total. Furthermore, we quantified the level of usage change $f_{i_{new}}$ into low, medium, and high changes. This quantization was done with the support of expert knowledge from Volvo engineers in the field. Equations (3) and (4) explain how the quartiles, and accordingly the level of changes in vehicle usage, are defined. In this way, we assume that substantial change (reduction or raise $f_{i_{new}} > Q3$) in the usage of vehicles leads to performance degradation and component failure in the future.

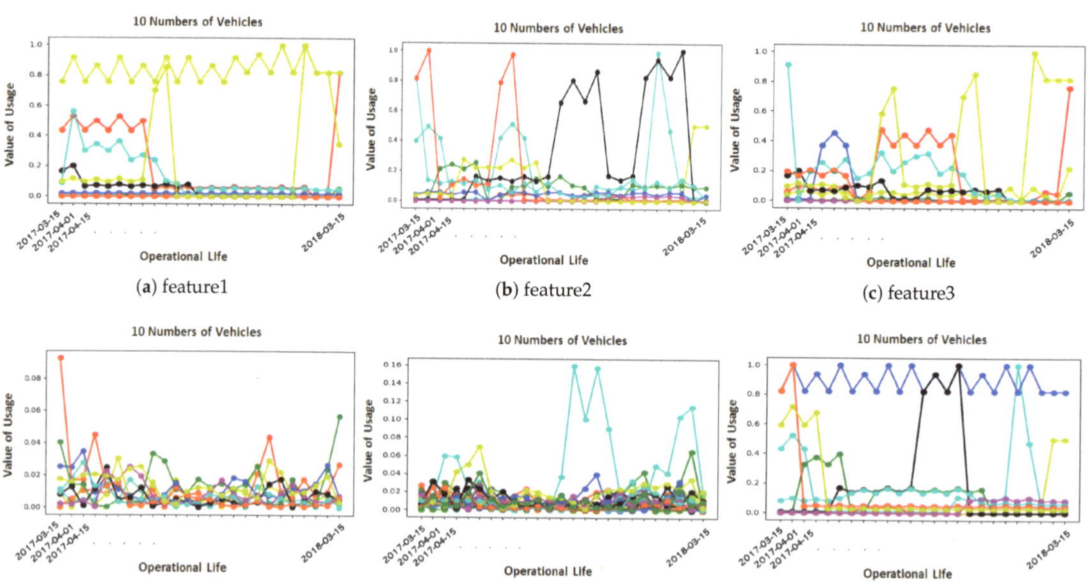

(**a**) feature1 (**b**) feature2 (**c**) feature3 (**d**) feature4 (**e**) feature5 (**f**) feature6

Figure 4. Illustration of the extracted features ($f_{i_{new}}$) from ten vehicles. The subplots (**a–f**) show significant (formulated in Equation (2)) and gradual changes bi-weekly in each parameter over one year. This means the original LVD data were logged every two weeks and collected by the authorized OEM workshops when the vehicles visited the workshops anywhere in Europe.

To show how the changes might relate to failures, we grouped the vehicles into two categories of Healthy and Unhealthy vehicles. A healthy vehicle is one that has no faults during its operational life, while an unhealthy vehicle has at least one reported claim in its history.

Figure 5 demonstrates how the changes are related to the groups of vehicles, where the y-axis explains the relative frequency of usage change in four different categories. The vehicles with high, medium and low or no numbers of significant changes are shown on the x-axis. It can be observed from the usage pattern, plotted through the sub-figures, that the proportion of significant changes in unhealthy vehicles was higher than in healthy vehicles during their life histories (except one depicted in Figure 4f). Likewise, the proportion of healthy vehicles was more than that of unhealthy vehicles once we considered no-changes to evaluate the association between them. Through this experiment, we concluded that healthy vehicles had less usage deviation than unhealthy vehicles. Thus, exploiting this information contributed to building a better predictive model to result in more accurate predictions.

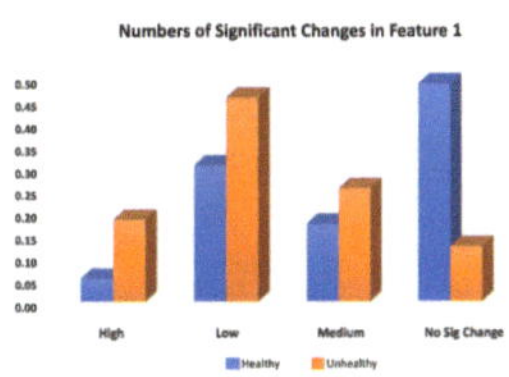

(**a**) feature1

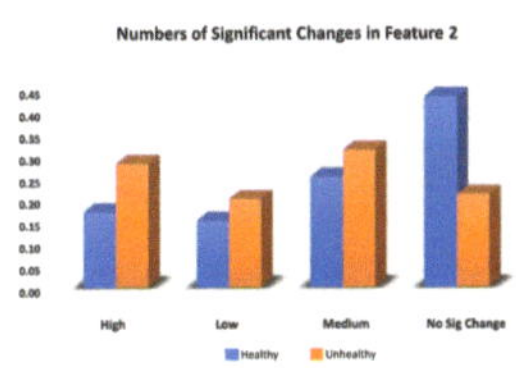

(**b**) feature2

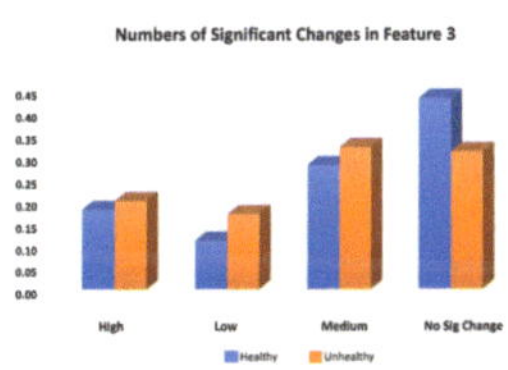

(**c**) feature3

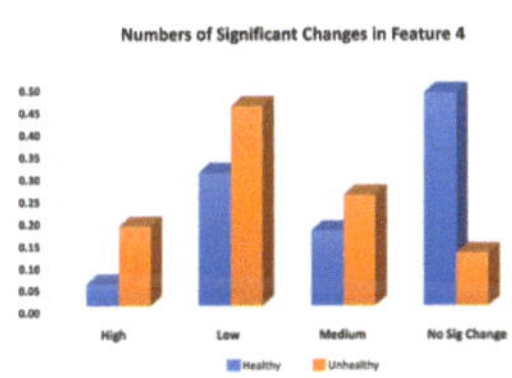

(**d**) feature4

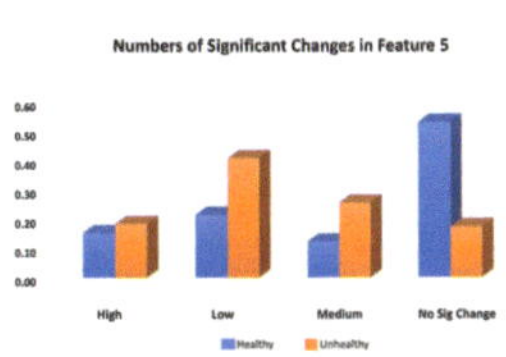

(**e**) feature5

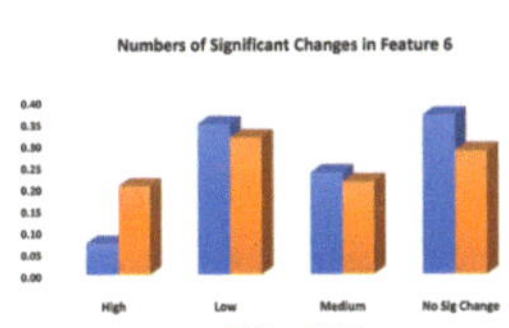

(**f**) feature6

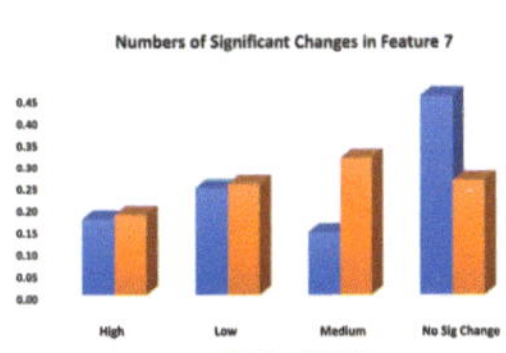

(**g**) feature7

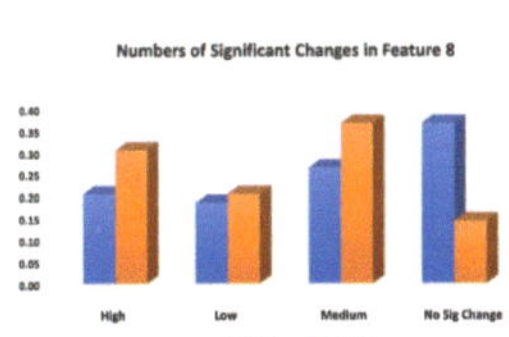

(**h**) feature8

Figure 5. Illustration of significant usage changes in healthy and unhealthy vehicles. These subplots show how the significant usage changed in eight different features (extracted features), related to the health of the vehicles. Note: we used more than 600 features in our study, and these were just eight features we randomly selected for the purpose of plotting.

4.2.3. Data Segmentation

The segmentation was conducted by having hypotheses in which vehicle demeanor differs from context to context, and the context has a certain impact on the vehicle's performance. In the domain of the automotive industry, context can be characterized as location, application domain, time, specific road conditions, etc. Furthermore, this segmentation might aid to increase the performance of the predictive model. Thus, in this study, with this low-resolution data, the operational time of the vehicles over different seasons was considered to segment the data. Basically, we partitioned the whole time-series data into four segments representing the four seasons in a year.

This time-series data set contained 337,573 samples with 375 original features (parameters/measurements), characterizing the usage style of 2412 unique vehicles over four years of operation. In addition, information related to each season was added as a new feature in the data. In this way, in each segment of time we defined three months as a season over a year, and each time segment contained the vehicles which operated in that period of time in different years (regardless of their ages or production months). Since vehicles have different ages and operational patterns over the year, each segment contained various numbers of samples/readouts as follows: Segment1 contained 86,768 readouts; Segment2 included 90,709 readouts; Segment3 held 61,364 readouts, and, finally, Segment4 contained 98,732 readouts.

Bearing in mind the assumption raised in the integration section (Section 4.2.1), to label the LVD (one-month operation may have the highest impact on component down-times), we took account a very rare case of failure on the first day of each season, so each segment also contained one-month operation from the previous season. This allowed the model to map the usage to a failure if the failure happened on the first day of the season.

4.3. *Module 2: Optimization: Measurement Selection*

This section describes the module developed to conduct the measurement selection process. We formulated this process as an optimization task, wherein we aimed to find the best representation of vehicle usage (a subset of measurements/features) to increase the predictive model's performance in forecasting component failure. In other words, our objective was to discover the optimal subset of measurements that could characterize vehicle usage and map these to failures. Thus, to this end, we constructed this module by modifying the original Genetic Optimization [58], which was successfully applied in a wide range of applications [59,60], just to cite a few. As the main contribution in this module, we utilized the Elastic Net [61] technique, by adding simple penalties (a combination of L1 and L2 regularization) into the optimization techniques (we call this ELSGA), which acted as a generation component to initiate a decent population in the first step for the optimization process. Although, data dimensionality reduction approaches, such as PCA [61,62] or LDA [63], could be implemented for this problem, in order to maintain interpretability, using such methods was avoided in this dimensionality reduction step. The optimization module itself consists of three main sub-modules: Initialization component (IC), Evaluation component (Objective function) and Criterion. These sub-modules are described as follows:

4.3.1. Initialization Component (IC)

Given the desired segment (Seg_i) from the segmentation module, IC initiates the first generation of the population (rather than randomly select the first generation of features, which is the fundamental process in the GA algorithm, we aimed to start the optimization process with better population of the measurements) using a convex integration of lasso [64] and ridge [65] methods called Elastic Net (LS) [66]. We took advantage of the LS capability, that uses L1 and L2 penalties to shrink and select features, to generate the first population of predictors. Let us suppose we have $R = (r_1, \ldots, r_n)$ readouts in the given time series Seg_i data-set with $F = (f_1, \ldots, f_m)$ features/predictors (measurements), and $T = (t_1, \ldots, t_n)$ as the target variables.The value $a_{i,j}$ refers to the value of feature f_i in readout r_i.

$$R = \begin{pmatrix} & \mathbf{f}_1 & \cdots & \mathbf{f}_m \\ \mathbf{r}_1 = & a_{1,1} & \cdots & a_{1,m} \\ \mathbf{r}_2 = & a_{2,2} & \cdots & a_{2,m} \\ \vdots & \vdots & \ddots & \vdots \\ \mathbf{r}_n = & a_{n,2} & \cdots & a_{n,m} \end{pmatrix} \times T = \begin{pmatrix} \mathbf{T} \\ t_1 \\ t_2 \\ \vdots \\ t_n \end{pmatrix}$$

Thus, R and T are our model matrix, and prediction of target variable $\hat{t}$ can be described as:

$$\hat{t} = \sum_{j=1}^{m} \beta_i f_j = \beta^T F \tag{5}$$

where the input function is the dot product of the predictors and coefficients β:

$$\sum_{j=1}^{m} = \beta_1 f_1 + \beta_2 f_{1/2} + \cdots + \beta_m f_m \tag{6}$$

Thus, in our initialization process, the task is defined as a classification problem, so β can be calculated by Equation (7):

$$Elastic(\hat{\beta}) = (tf) + \lambda\left[(1-\alpha)||\beta||^2 + \alpha||\beta||_1\right] \tag{7}$$

$$(tf) = argmin \sum_{k=1}^{n} (y_k P_k^T \beta) \tag{8}$$

$$||\beta||_1 = \sum_{j=1}^{m} |\beta_m| \tag{9}$$

$$||\beta||^2 = \sum_{j=1}^{m} \beta_m^2 \tag{10}$$

where is a margin-based loss function and $y \in \{0,1\}$. Using $||\beta||_1$ and $||\beta||^2$, which are the norm l_1 and euclidean norm, we measure the largeness of the coefficient.

λ and α are the two parameters to tune. α controls the integration of the two penalties and λ manages the amount of penalization, where α can take values between [0–1], and λ could be a positive number.

There are two alternatives for the LS: LS turns to ridge method once α gets close to 0, while LS serves as lasso when α reaches 1. Thus, to find the best α and $l1$, we tuned these parameters through walk-forward cross-validation by the learning process. This means we trained different models exploiting various values in an optimization manner to find the best setting. In this way, the algorithm overcame the drawbacks of the Lasso and Ridge functions [67] to select the best predictors. Thus, in IC we used an agent to perform Equation (11) to initialize the first population to be delivered to the Evaluation component and kept generating such decent populations to inject from the GA sector over the process.

$$IC = \begin{cases} Rand(\frac{1}{2} * LS(seg^i_{x_{j_{f_{1,\ldots,m}}}}), pop) \\ i = 1, \ldots, m \\ j = 1, \ldots, n \\ p = 1, \ldots, m \\ pop = [6, 8, 10, 12, 14] \end{cases} \tag{11}$$

where $Rand()$ is a random selection of decent initiated predictors by $LS(seg_i)$ and $Rand()$ is designed to keep the diversity within the decent selected predictors. The value pop denotes the size of the population that should be initialized at the first generation. The value of pop

was tuned with the above five numbers to find the optimal number for the population. The reasons for developing such an initialization component, rather than random generation, are as follows: (a) we hypothesized that good parents most likely generate better children than bad (or random) parents. This means randomly initializing the parent may lead to non-satisfying predictive performance in the fitness function, ignoring the cross-over step, and, accordingly, terminating the Genetic process; (b) in any optimization process, it is crucial to reach the optimal solution in the early stages (generations), particularly when the system needs to deal with a huge amount of streaming data. Thus, the random selection may lead to post-bonding to the late generations; (c) the nature of the GA approach is based on random generation and selection of a population to obtain the best solution; therefore, the approach does not concern the correlation between the features and the predictive performance. There might be highly correlated features that can contribute together to performance. Hence, the design and integration of IC support the GA operators by injecting such information to generate a better population over the optimization process. For this reason, we employed LS in this initialization component since there might be highly correlated predictors in the high dimensional data; thus, LS automatically included all the highly related predictors in the population if one of them was selected. However, it needs to be remarked that there might be cases among these highly related features wherein only one is sufficient and contributes to the predictive performance, and so GA solves such an issue.To summarize, this design with LS and GA allows the inclusion of the many correlations expected, since features coming from sensors measure correlated processes. However, the GA-enabled part of the algorithm tries to select the best subset out of such correlations and does not necessarily keep all of them. The ensemble of these two algorithms overcomes the raised concerns, particularly as the automotive sector deals with high-dimensional streaming data over time. This enables the system to decrease dimensionality and, at the same time, increase the performance of the predictive model.

4.3.2. Evaluation Component (Fitness Function)

At the core of this component, and of the GA process, a decision tree-based ensemble method [68] (XGBoost) was used as a classifier to evaluate the importance of the individuals and their impacts on the predictive model. This objective function of claim prediction $L(\theta)$, can be expressed as follows [68]:

$$L(\theta) = y_i log(logistic(y_i(\hat{y}_i^{t-1} + f_t(r_i))) + (1 - y_i) log(1 - logistic(\hat{y}_i^{t-1}) + f_t(r_i)))$$

where L is the log-likelihood function, and, thus, $logistic(y_i(\hat{y}_i^{t-1}))$ is the probability. f_t denotes the tth tree, and y^{t-1} shows the prediction of r_i sample, at $t-1$ iteration. In this experiment, we used the logistic function, as our problem was formulated as a binary classification. Indeed, the merit of each population represented as solutions are evaluated as fitness value, so the best chromosomes are ranked and selected for the next generation. We formulated our fitness function as an optimization, where individuals with higher performance are considered for the next generation. In a parameterized fashion, the four top best populations were selected, in the selection process to be given to GA operators for cross-over and to generate a new solution (offspring).

A typical crossover method, called Partially Mapped (PMX) [69], was used in this optimization problem. The main idea behind this decision was to support the GA process by having a better convergence rate and diversity and to reduce stagnation during the optimization process [70].

4.3.3. Criterion Component

Three criteria were incorporated into a composite optimization approach as follows: *(1) poor fitness value (Tr_1), (2) maximum fitness value (Tr_2),* and *(3) maximum generation (Tr_3).* The first criterion talks about the poorness of generated predictors by the IC component at the first generation. The IC is called if, and only if, the fitness value could not pass the minimum threshold.

$$Cr = \begin{cases} Tr_1 \ is \ met \ \text{ if f(x)} \geq 0.5, otherwise \ \ call \ \text{ IC} \\ Tr_2 \ is \ met \ \text{ if f(x)} \geq 0.95, then \ \ terminate \ \ GA \\ Tr_3 \ is \ met \ \text{ if itr} = \text{max gen}, then \ \ terminate \ \ GA \end{cases} \quad (12)$$

This criterion was established to ensure the first population has the potential for further generation during the optimization process. The second criterion indicates that the fitness output has been sufficiently reached for the optimization process to be terminated once it is met. As the third proxy Tr_3, we defined a maximum generation. Taking into account the first criterion Tr_1, once one of Tr_2 or Tr_3 is met, the selected operators/features are injected into the ensemble modeling (in the third module) to build a complex and general predictive model for the final claim forecasting.

4.4. Module 3: Snapshot-Stacked Ensemble Modeling (Network Construction and Training)

In this section, we describe the third module of the proposed approach that contains two main phases. In the first phase, we constructed an ANN, which was used to generate several snapshot models from a single training process. Then, we took those snapshots (models) for validation, and, in the second phase, the prediction output (class probabilities) of each snapshot model was added to the dataset for final training and prediction. Indeed, the second phase acts as a meta-learning process, where it attempts to learn from the output of the snapshots to reduce the classification errors obtained from the first network.

4.4.1. Snapshot Ensemble on Vehicle Usage

The ensemble technique and its modeling is a process in which several models are generated to forecast the outcome. This can be done by utilizing different predictive models or different training data sets. The ensemble model then combines each model's output to obtain the final prediction for the test data. The method highly supports the network to increase the predictive model's performance compared to the individual model by decreasing the generalization error. Since training multiple deep networks is computationally expensive, we utilized Snapshot Ensemble to train multiple models by a single training process [71]. We utilized this approach, which we introduced in [71], and adapted and integrated it into the optimization technique for the breakdown prediction. However, this technique has the challenge of generating similar models, resulting in similar prediction performances. This does not contribute to the predictive model at the output level unless the network generates diverse models during the training process. One way to overcome this challenge and build diverse models is to change the learning rate during the learning process. Thus, to this end, we exploited the Cyclic Cosine Annealing Schedule (CCAS) [17], that splits the training process into N cycles, each starting with a large learning rate and relatively rapidly decreasing to the minimum value before increasing quickly again. Equation (13) shows the learning rate over the iterations.

$$\alpha(t) = \frac{\alpha_0}{2}(cos(\frac{\pi \, mod(t-1, \lceil T/n \rceil)}{\lceil T/N \rceil}) + 1) \quad (13)$$

where, α denotes the learning rate at iteration t, α_0 is the initial learning rate (maximum rate), and T refers to the total number of the training iterations. Thus, the function anneals the learning rate from α_0 to $f(T/C)$ which is $\approx$0 over the course of a cycle [72]. Value N is the number of cycles in which the network is trained throughout the training process, and in which the best weights at the end of each cycle are saved as the snapshot of the model. This results in N model snapshots. In fact, in this process, the learning rate is raised periodically, reinforcing the model's convergence at the global minimum, rather than at the local minimum, over the snapshot ensemble. This leads to elimination of the need to manually find the optimal maximum learning rate. The overall snapshot process is illustrated in Figure 6, where at the end of each iteration the model with the maximum performance is saved as the snapshot. All the models are then used to predict the outcome in the first phase. The use of CCAS does not ensure diversity of the produced models, due to the stochastic nature of the algorithm. Therefore, the diversity of the snapshot models were carefully studied by computing and investigating any dissimilarities between prediction outputs. In short, in this step, we generated multiple diverse deep neural networks to build an ensemble of predictive models. This ensemble of models was then used to construct meta-learning (in the second phase) for the final breakdown prediction.

4.4.2. Training the Network

Table 1 shows the construction of the network in the first phase, wherein we trained the ANN network by incrementally creating a Sequential model. The network parameters were initialized using "He" initializer [73] (shown in Equation (14)), where G is the Gaussian probability distribution with a zero-centered and standard deviation of $\sqrt{\frac{2}{n_l}}$. In this initialization form, biases are initialized at 0 and n_l is the number of inputs to that node in layer l.

$$w_l = G(0, \sqrt{\frac{2}{n_l}}) \tag{14}$$

In order to use stochastic gradient descent to train the deep neural networks, rectified linear unit (ReLU) [74] as the–active function–was utilized to transfer the summed weighted input from the nodes into the output of the node.

$$f(x) = max(o, x) \tag{15}$$

In Equation (15), $f(x)$ is the $ReLU$ function, where it outputs 0 for $\forall\ x < 0$, while for positive inputs $x \geq 0$, it returns the same value $f(x) = x$. The Batch Normalization [75] technique is used to normalize the input of each layer to reduce the internal covariate shift problem leading to stabilizing the learning process. This is followed by a stochastic gradient descent technique called "Adam" optimization, which iteratively updates network weights during the training process [76].

We used a regularization method called "Dropout" to reduce the over-fitting issue. Basically, this led to the provision of different neural network architectures when training the networks. In the output layer, we utilized the Soft-max activation function–depicted in Equation (16)–for the probability distribution of the target variables.

$$\sigma(z_i) = \frac{e^{z_i}}{\sum_{j=1}^{K} e^{z_j}} \tag{16}$$

where $\sigma(z_i)$ are the elements of the input vector to the softmax function, K is the number of classes and $\sum_{j=1}^{K} e^{z_j}$ is the normalization term, which ensures all the output values of the function sum to 1 (each class probability is in the range (0, 1)). Finally, the output of Equation (16) is injected to the $Argmax$ function–defined in Equation (17) to predict the hard label.

$$hp_{r_i} = argmax(\sigma(z_i)) \tag{17}$$

where hp_{r_i} carries the hard label prediction of the readout r_i in the test set, expressing whether the usage corresponds to a claim/breakdown (1) or not (0).

Table 1. The architecture of the first deep networks–deep1–(left side) and the second deep network (right side). The values of the model hyper-parameters were selected experimentally in a systematic fashion.

Layers (Type)	Output Shape	# of Parameters	Layers (Type)	Output Shape	# of Parameters
Sequential()	–		Sequential()	–	
batch00 (BatchNormalization)	(None, 160)	640	batch00 (BatchNormalization)	(None, 214)	856
layer00 (Dense)	(None, 160)	25,760	layer00 (Dense)	(None, 214)	46,010
batch10 (BatchNormalization)	(None, 160)	640	batch10 (BatchNormalization)	(None, 214)	856
layer10 (Dense)	(None, 40)	6440	layer10 (Dense)	(None, 53)	11,395
batch20 (BatchNormalization)	(None, 40)	160	batch20 (BatchNormalization)	(None, 53)	212
layer20 (Dense)	(None, 13)	533	layer20 (Dense)	(None, 17)	918
batch30 (BatchNormalization)	(None, 13)	52	batch30 (BatchNormalization)	(None, 17)	68
dropout (Dropout)	(None, 13)	0	dropout (Dropout)	(None, 17)	0
layer30 (Dense)	(None, 8)	112	layer30 (Dense)	(None, 10)	180
batch40 (BatchNormalization)	(None, 8)	32	batch40 (BatchNormalization)	(None, 10)	40
layer40 (Dense)–softmax()	(None, 2)	18	layer40 (Dense)	(None, 2)	22
Total parameters:	34,387		Total parameters:	60,557	
Trainable parameters:	33,625		Trainable parameters:	59,541	
Non-trainable parameters:	762		Non-trainable parameters:	1,016	

4.4.3. Stacking and Horizontal Ensemble

Each snapshot model provides two types of outputs; Soft prediction (Equation (16)) and Hard label (Equation (17)). The former returns two values: the predicted probability of the readout belonging to a vehicle not having a claim and the predicted probability indicating the usage corresponding to a claim/fault. The second output is the hard label representing whether the component is healthy (0), given the vehicle usage, or faulty (1). Accordingly, Equation (18) shows the numbers of the new features which are generated by the Snapshots models.

$$feature_{new} = N_{Sn} \times \sum_{k=1}^{f} (C_k + 1) \quad (18)$$

where N_{S_n} refers to the snapshot numbers, and f is the number of features. C_k points to the binary class of the vehicle status (healthy–0– or unhealthy–1–) which is summed by "1" as the hard-label of that prediction. Table 2 illustrates the snapshot and stacking horizontal ensembles calculations, which are described in detail as follow:

- $r_1, \ldots, r_n$: refers to the vehicle non-stationary readouts/samples, which are collected over time.
- $f_1, \ldots, f_m$: represents the LVD parameters characterizing a vehicle's behavior in r_i.
- $P_{r_1(sp_1)} \cdots P_{r_1(sp_m)}$, indicates the output of the snapshot models–in the first phase–for the given readouts.
- p_{c_1}, p_{c_2}, represents the predicted probability of class 0 (healthy) and class 1 (unhealthy), for the given readout r_i.
- hp_{r_i} refers to the hard prediction of the given readout r_i (0 or 1).

The output of the soft predictions and the hard labels are horizontally combined with the original features (validation set, see Figure 7). This means that, by considering Equation (18), each snapshot model could add three extra parameters carrying knowledge of the previous model's performance on that specific readout. The new data–as the new training set–is then passed to the second deep net to build the meta-learning to reduce the forecasting errors obtained from the previous phase. Indeed, meta-learning is built by stacking the outputs of the snapshots, which are diverse deep neural networks and able to capture the expected complex signal relations, as compared to learning algorithms with less capacity, see Tables 5 and 6. As is described above, CCAS is utilized to construct 20 different predictive deep network models in this stacking process. To obtain the final fault prediction, we injected the test set, which was held out, into the same generated snapshots to translate the test set's structure to the new one with more dimensions, similar to the new training set (see Figure 3 left-bottom). Finally, the performance of the meta-learning is assessed by the test set. Table 1 (right side) describes the architecture of the second deep network, and Algorithm 1 shows the pseudo code of the whole process in this complex claim prediction.

Table 2. Illustration of the snapshot and stacked ensemble output.

Readouts	Original Features			Snapshot Models' Predictions		
	f_1	...	f_m	$P_{r_1(sp_1)} = (p_{c1}, p_{c2}, hp_r)$	...	$P_{r_1(sp_m)} = (p_{c1}, p_{c2}, hp_r)$
r_1	$r_{1_{f_1}}$	...	$r_{1_{f_m}}$			
r_2	.	...				.
.	.	...				.
.	.	...				.
r_n	$r_{n_{f_1}}$	...	$r_{n_{f_m}}$	$P_{r_n(sp_1)} = (p_{c1}, p_{c2}, hp_{r_n})$	...	$P_{r_n(sp_m)} = (p_{c1}, p_{c2}, hp_r)$

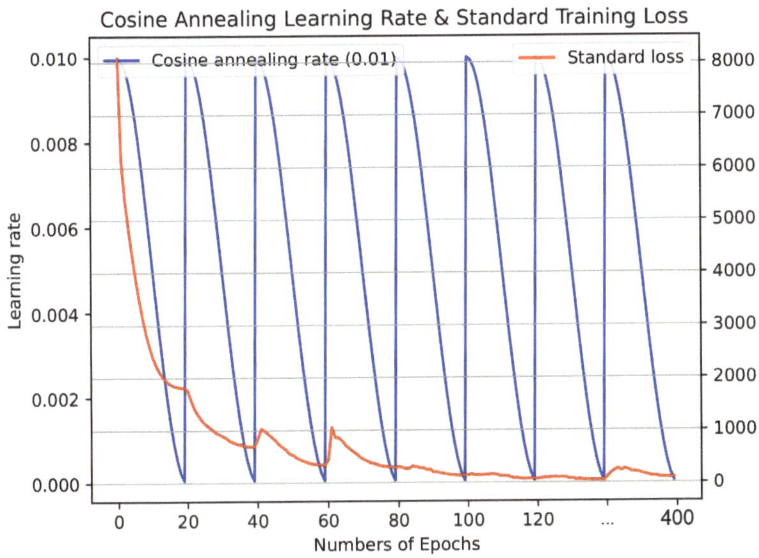

Figure 6. Visualization of the standard loss (red) and cosine annealing learning and cycles (blue). In this practice, N = 20 snapshot models were generated and saved to be used in the first and the second phases of the networks.

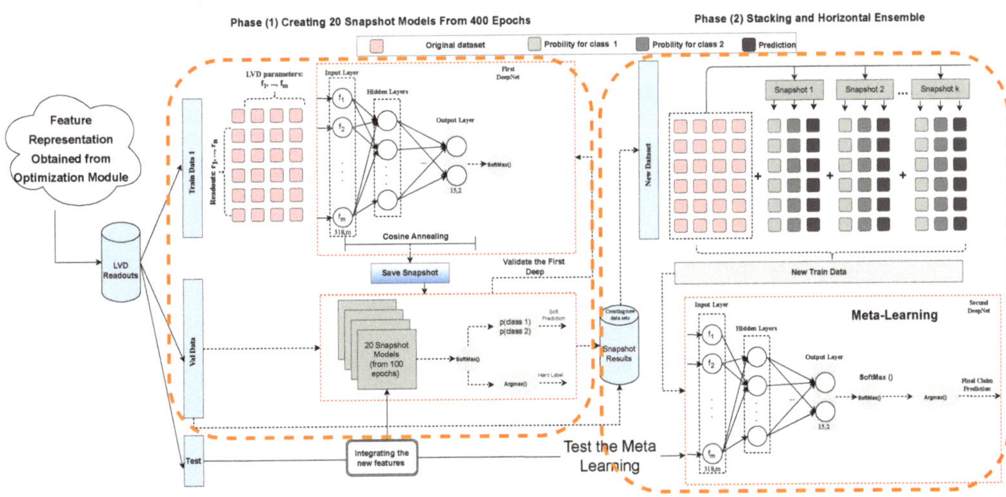

Figure 7. The schematic representation of the third module starts by partitioning the LVD data (injected from Module2) into three parts: training data, validation data, and test data. The training and validation are used in the first phase to generate the snapshot (SPs) models. In this process, CCSA is utilized to construct the snapshots where the function periodically decreases the learning rate from α_0 to ≈ 0 in a cycle; in this study, it was every 20 snapshots (see Figure 6). Then, at the end of each cycle, the model with the maximum performance is saved as the snapshot. In the second phase, the output of the SPs is horizontally added to the original set (validation data) to make a new dataset. Accordingly, the new dataset is used to build the meta-learning (with the training part) for the final prediction. Finally, the test part is utilized (with the new dimensions) to validate the meta-learning performance.

Algorithm 1 The proposed SSED approach, Module 1, 2 and 3

```
Input: R                                              ▷ readouts R={r_1,...,r_n}
Output: Fp                                            ▷ final claim prediction
Trd ∈ R                                               ▷ training data
Vld ∈ R                                               ▷ validation data
Tsd ∈ R                                               ▷ test data
S                                                     ▷ Snapshot
Ls ← {}                                               ▷ list of snapshots
Lseg ← {}                                             ▷ list of segments
Linf ← {}                                             ▷ list of informative features
Dtr ← {}                                              ▷ new dataset for training
Dts ← {}                                              ▷ new dataset for test
CCAS                                                  ▷ cosine annealing function (Equation (13))
N                                                     ▷ number of cycles, (Equation (13))
Deep1                ▷ 1st network to generate snapshots (Table 1-left side)
Deep2                ▷ 2nd network for final prediction (Table 1-right side)
Module 1 is started:
R ← label (R)                                         ▷ label R, Equation (1)
Lseg ← Segment(R)                                     ▷ partition R into four segments
Module 2 is started:
for Sg_i ∈ Lseg do
    while Cr is not met do
        Linf ← Optimization(Sg_i)                     ▷ select informative features
    end while
end for
R_new ← Union(Linf)                                   ▷ union the informative features
Module 3-phase 1 is started:
Train (Deep1 with CCAS | R_new)
while i ≤ E do
    if mod (i\N) == 0 then
        Ls ← S                                        ▷ save the snapshot in Ls
    end if
end while
Load Ls                                               ▷ load all snapshots
for S_i ∈ Ls do
    p = S_i.predict(Vld)                              ▷ validate S_i with validation set
end for
Module 3-phase 2 is started:
Creating a new data set
for S_i ∈ Ls do
    Dtr ← stack(p, Vld)                               ▷ stack the prediction to Vld
end for
for S_i ∈ Ls do
    Dts ← stack(p, Tsd)                               ▷ stack the prediction to Tsd
end for
Mm ← Train (Deep2 | Dtr)                              ▷ building meta-learning
Fp ← Mm.predict(Dts)                                  ▷ final prediction
```

5. Experimental Evaluation and Results

To facilitate the implementation of the approach, we recall the two research questions, introduced in Section 1, on which we based the evaluation of the proposed approach as follows:

5.1. RQ1—Feature Selection with Optimization Results

Bearing in mind RQ1, we constructed two evaluation tracks. Firstly, the proposed optimization approach (ELSGA) was evaluated over several generations and compared with the standard genetic algorithm (without the IC component). Secondly, the selected features (using ELSGA) were used to build models through a type of time-series k-fold cross validation called "walk-forward cross validation (WFCV)" [77] and compared with other ML, deep learning and ensemble based approaches.

The optimization process was carried out as feature selection by selecting the most informative parameters to predict a vehicle's component failures (a component related to the turbocharger). We built a forecasting system in which a supervised ML algorithm examined the contribution of the parameters to the claims. Considering results from our previous study [18], in which vehicle usage changed over time, we segmented the data into different parts. This segmentation process was done by assuming that the distribution of the vehicle usage logged in the same season over the years was relatively similar with respect to the other seasons. Thus, in each segment there might be a set of parameters that had more impact on the failures, and, accordingly, more pertinent to the prediction of failure. We designed the optimization process to be implemented in each segment in order to identify the most informative features. However, in the second tier of the evaluation, the union of the selected features (from all segments) built the predictive models, and was compared with other algorithms.

Since there is no universal configuration for such a heuristic optimization algorithm that is considered to be the best setting, GA parameters, including population size, mutation rate, number of parents inside the mating pool, and number of elements to mutate, were selected in an optimization manner. This meant the implementation code was constructed in a parameterized fashion to find the best configuration of the GA settings. Table 3 lists the parameters that our optimization system tuned to obtain the best performance.

Table 4 describes the output of the IC at initiating the first population on the whole and segmented data. The values α and $l1 - norm$ were selected through the WFCV (as the optimal values) to train the elastic model.

Table 3. The tuned parameters in the proposed optimization approach.

	Tested Parameters	Best Parameter
# of Generation	—	**100**
# of items to mutate	1–10	**3**
# of parents in mating pool	2, 4, 6, 8	**4**
Population size	6, 8, 10, 12, 14	**10**

Table 4. The Output of the Initialization Component.

	α	L1-Norm	Reduction %	# of Decent Predictors
Total Data	7.1×10^{-7}	0.050	29.79%	410
Seg1	1.1×10^{-5}	0.99	86.3%	80
Seg2	1.4×10^{-5}	0.70	82.3%	103
Seg3	$2.2 \times 10^{+5}$	0.50	77.5%	131
Seg4	3×10^{-5}	0.10	46.4%	313

The whole data-set with the lowest feature reduction ($FR = 0.29\%$) resulted in the highest number of predictors in the first population, with 410. Among the segments, 313 predictors were selected with coefficients larger than 0 in $Seg4$, which also indicated the lowest reduction proportion within the segments. Likewise, the IC initiated 80 predictors out of 577 as being decent parents for further generation in $Seg1$. The figures indicated that the IC delivered different numbers of predictors to GA operators–meaning that, in each segment/season, different parameters had different impacts on failure of further generations in each segment. It was observed the value of α in all cases was close to zero, which meant that the LS almost turned into the ridge method to initiate the first generation.

To evaluate the performance of the selected features, we formulated the objective function as a classification task, wherein the function seeks the optimal performance of the breakdowns forecasting, taking into account vehicle usage. Thus, given the vehicle usage with n readout samples and the selected predictors/features m, which were injected by GA operators, $D = (r_i, t_i)(|D| = n, r_{i_{f1...,fm}}, t_i \in (0,1))$ in each generation, the function predicted whether the usage led to a failure or not.

In this stage of the experiment, the data was partitioned into training and test data. More to the point, in each segment of the data "throughout the optimization process", we took three months of vehicle operation, with one month from the previous season, into account to train the model, and the last month was used to test the model. The statistical information of the training and test data in each segment is listed as follows:

- Segment1 (*Seg*1) contained 86,768 readouts, where 52,060 were considered for training and 34,708 for testing the model.
- Segment2 (*Seg*2) contained 90,709 readouts, where 54,421 were considered for training and 36,288 for testing the model.
- Segment3 (*Seg*3) contained 61,364 readouts, where 38,045 were considered for training and 23,319 for testing the model.
- Segment4 (*Seg*4) contained 98,732 readouts, where 75,036 were considered for training and 23,696 for testing the model.

As is mentioned in the Segmentation section, in each segment, the readout time-series samples were placed in the time order, in order to use the past data to build the model, and future data to test the model. Since these data were cumulative, and the scale of the readouts differed from year to year, data normalization–using Equation (19)–was conducted to form the data on the same scale.

$$normalize\_value = \frac{value - min}{max - min} \tag{19}$$

The three criteria, tr_1, tr_2 and tr_3, were set to 0.5, 0.95 and 100 to recall the IC or terminate the optimization process so as to provide the best predictors, respectively.

The plots in Figure 8 illustrate the superiority of the ELSGA approach in all segments compared with the GA approach without the IC component over the whole optimization process. We noticed a considerable jump in the performance provided by ELSGA that describes how well the IC adapted to the optimization process to generate decent individuals in the first population. The increase at the early stages assured that the whole optimization process could be terminated before the criterion tr_3 was met (maximum 100 generation). However, $tr_2 \geq 95\%$ was not passed due to the complexity of the problem. ELSGA performed better by reporting around 80% AUC in all segments with 3% higher than GA over the process. It was also noticeable that the AUC value obtained by ELSGA in *Segment4* was higher among the other segments, with 82% AUC vs. 79% obtained by the basic optimization. The statistical assessment of the data revealed that *Segment4* contained vehicles which had more claims compared to the vehicles located in other segments in this highly unbalanced data. This might be the reason why our model performed better, since more data and positive samples were available to build the predictive model.

The overall results suggest the segmentation formulation is compatible and positively contributes to better claim prediction performance with optimization approaches, particularly the ELSGA approach.

In the second track of the evaluation, the output of the ELSGA was taken into consideration to build the model using the XGBoost classifier [68]. In this experiment, the selected predictors, due to ELSGA, were exploited to build the predictive models and to compare with models trained with several classifiers by injecting the data. In fact, multiple classifier (including deep learning and ensemble approaches) results were selected as the baseline to compare with ELSGA. The similar WFCV method was used to build and evaluate the models with various portions of the data in a time-series fashion (this decision was made since our data was time-series data and the folds could not be randomly selected by normal k-fold cross-validation).

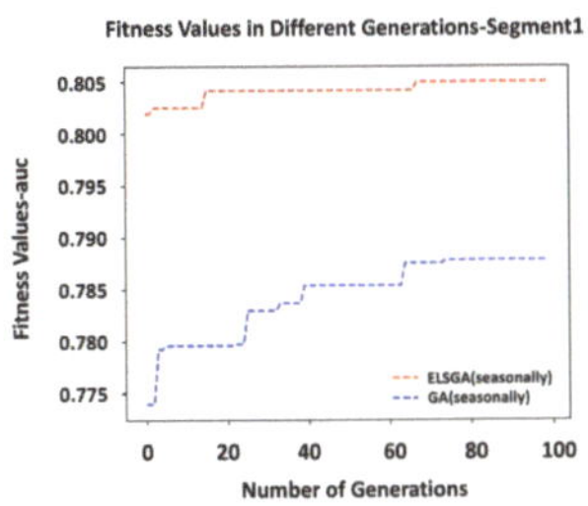

(**a**) Segment 1.

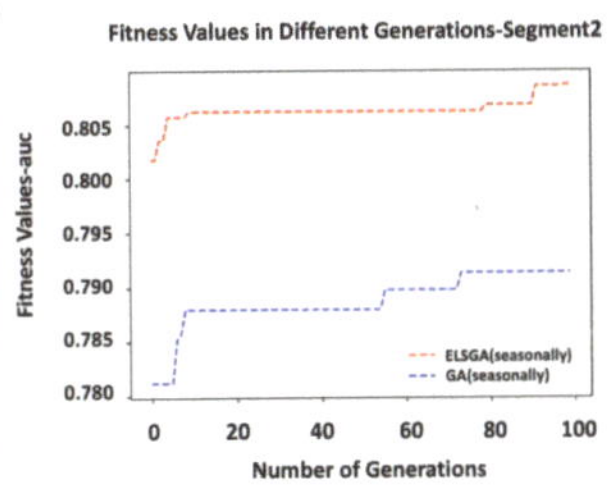

(**b**) Segment 2.

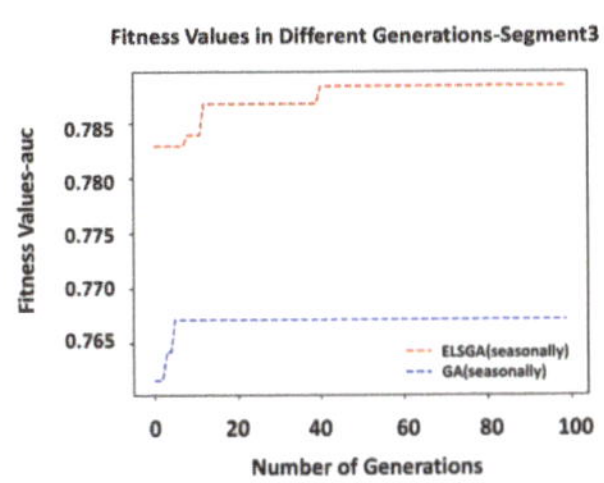

(**c**) Segment 3.

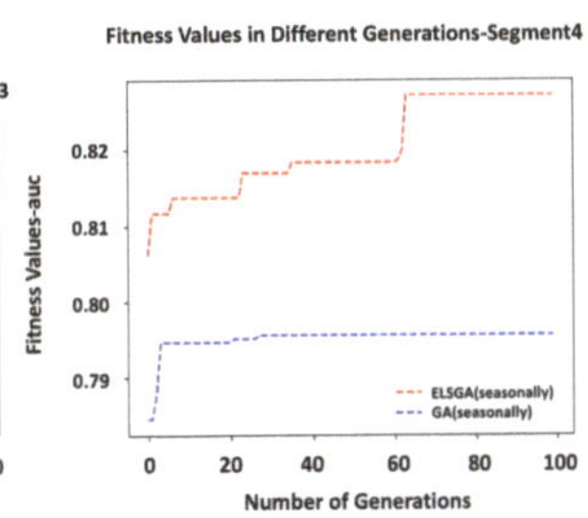

(**d**) Segment 4.

Figure 8. The performance of the proposed ELSGA and GA in each segment.

Table 5 shows the comparison of our proposed ELSGA method and multiple classifiers. The figures obtained from different predictive models demonstrate how complex the task is to predict claims by taking vehicle usage into account. As can be seen, most of the classifiers performed poorly, providing low $auc \approx 0.50$ values. Practically, these results showed that most of the linear classifiers had no discriminatory capacity in regard to this complex problem, and could not map the usage to breakdowns. In contrast, deep learning models (such as CNN, LSTM and biLSTM) showed much better performance compared to the linear classifiers, by $auc = 0.70 \pm 2$.

Among the examined predictive models, Boosting and Stacking performed close to the proposed approach by $auc = 0.75$ and 0.74 vs. $auc = 0.76$, respectively, when total data was considered. Concerning the segmentation, we can clearly observe that the proposed approach (Module2-ELSGA) significantly outperformed the other classifiers, except for Bagging. It is necessary to remark that, within deep learning approaches, CNN worked well on the last segment by providing $auc = 0.77$. To go beyond this performance assessment, we applied a $5 \times 2cv$ paired statistical t-test to evaluate how significant the results of the classifiers were. In almost all cases, the p-values were smaller than the critical value $\alpha = 0.05$, so this rejected the null hypothesis, by stating that there was a significant difference between the outcomes. The statistical tests showed that only the performance of the ensemble-based approaches were relatively close to that of the proposed approach in most cases. It is also fair to remark that the statistical test on the performance of the CNN model, in the last segment, indicated that the difference was not significant. It can be observed that in one case even Bagging performed slightly better than the proposed approach in Segment 3; however, the difference was not significant. In Segment 1, the figures indicate Stacking performed quite similarly to our proposed ELSGA approach by reporting $auc = 0.81$.

Table 5. The comparison between Module2 (ELSGA (XGB)) and different approaches (standard machine learning, Gaussian Process (GP), deep neural networks, and ensemble-based approaches, such as Bagging, Boosting and Stacking approaches) for RQ1. We used a $5 \times 2cv$ paired t-test to examine how significant the differences were between the performances of the two models. This meant that every time we compared only two models, one ELSGA vs. one classifier.

Classifiers	All Data		Segment1		Segment2		Segment3		Segment4	
	auc	p	auc	p	auc	p	auc	p	auc	p
Logreg	0.50 ± 0.00	<0.05	0.50 ± 0.00	<0.05	0.50 ± 0.00	<0.05	0.50 ± 0.00	<0.05	0.50 ± 0.00	<0.05
Kneig	0.52 ± 0.02	<0.05	0.544 ± 0.01	<0.05	0.52 ± 0.01	<0.05	0.54 ± 0.04	<0.05	0.52 ± 0.02	<0.05
SGD	0.50 ± 0.00	<0.05	0.50 ± 0.00	<0.05	0.50 ± 0.00	<0.05	0.50 ± 0.00	<0.05	0.50 ± 0.00	<0.05
QdA	0.53 ± 0.02	<0.05	0.50 ± 0.00	<0.05	0.50 ± 0.00	<0.05	0.50 ± 0.00	<0.05	0.50 ± 0.00	<0.05
SVM	0.55 ± 0.05	<0.05	0.56 ± 0.02	<0.05	0.53 ± 0.03	<0.05	0.57 ± 0.04	<0.05	0.57 ± 0.06	<0.05
DT	0.54 ± 0.01	<0.05	0.58 ± 0.01	<0.05	0.59 ± 0.02	<0.05	0.58 ± 0.02	<0.05	0.60 ± 0.03	<0.05
Ridge	0.71 ± 0.07	<0.05	0.73 ± 0.04	<0.05	0.70 ± 0.06	<0.05	0.71 ± 0.05	<0.05	0.71 ± 0.07	<0.05
SVC	0.52 ± 0.01	<0.05	0.52 ± 0.01	<0.05	0.51 ± 0.001	<0.05	0.51 ± 0.001	<0.05	0.53 ± 0.02	<0.05
GP	0.50 ± 0.00	<0.05	0.50 ± 0.00	<0.05	0.50 ± 0.00	<0.05	0.50 ± 0.00	<0.00	0.50 ± 0.00	<0.05
CNN	0.65 ± 0.08	<0.05	0.67 ± 0.07	<0.05	0.70 ± 0.06	<0.05	0.71 ± 0.041	<0.05	0.77 ± 0.11	0.14
LSTM	0.69 ± 0.05	0.2	0.74 ± 0.02	<0.05	0.71 ± 0.06	<0.05	0.71 ± 0.06	<0.05	0.72 ± 0.05	<0.05
biLSTM	0.70 ± 0.05	0.2	0.74 ± 0.03	<0.05	0.70 ± 0.07	<0.05	0.72 ± 0.05	<0.05	0.72 ± 0.10	<0.05
Blending	0.58 ± 0.01	<0.05	0.58 ± 0.004	<0.05	0.59 ± 0.03	<0.05	0.58 ± 0.01	<0.05	0.58 ± 0.01	<0.05
Bagging	0.71 ± 0.06	<0.05	0.80 ± 0.04	0.24	0.78 ± 0.06	0.30	**0.81 ± 0.06**	**0.41**	0.79 ± 0.09	0.35
Boosting	0.75 ± 0.05	0.38	0.75 ± 0.05	<0.05	0.76 ± 0.06	0.19	0.76 ± 0.04	0.21	0.75 ± 0.09	0.15
Stacking	0.74 ± 0.14	<0.05	**0.81 ± 0.01**	0.16	0.79 ± 0.06	0.6	0.78 ± 0.04	<0.05	0.81 ± 0.08	0.3
Module2 ELSGA (XGB)	**0.76 ± 0.06**		0.81 ± 0.04		**0.80 ± 0.07**		0.80 ± 0.07		**0.82 ± 0.09**	

In addition, we conducted an A/B test by comparing the performance of the models trained by the original predictors (MORGP) vs. the models built by the extracted predictors (MEXP). Both groups of predictors (original and extracted) were selected by the optimization approach. Indeed, we aimed to reveal the contribution of the extracted features and original features to predictive models.

From Figure 9, we can observe that more than 50% of the selected predictors (except in the last segment) were extracted from the extraction process, which translated to the importance of such predictors in the prediction task (Orange vs. purple bars). In fact, we considered this proportionate comparison mainly as a sanity check. However, to quantify the impact of the extracted features, several models were built through WFCV to construct and evaluate the models. The figures suggest the MEXP outperformed the MORGP by 3%, 1%, 7%, and 10% in segments one, two, three, and four, respectively (green vs. pink bars). The statistical tests also indicated how significant the differences were in segments two, three, and four by rejecting the null hypothesis (α is set to 0.05 as critical value). Although the test failed to reject the null hypothesis (by p-value = 3.32), and concluded the difference between the two models was not significant in the first segment (green and pink bars), the Extracted predictors (green) showed their valuable impacts by providing almost the same performances 0.66 with respect to the models trained by the mixed predictors (blue bar). The information extracted and used as extracted predictors in the models had greater, and more significant, impact on the predictive model's performances, compared to the original predictors.

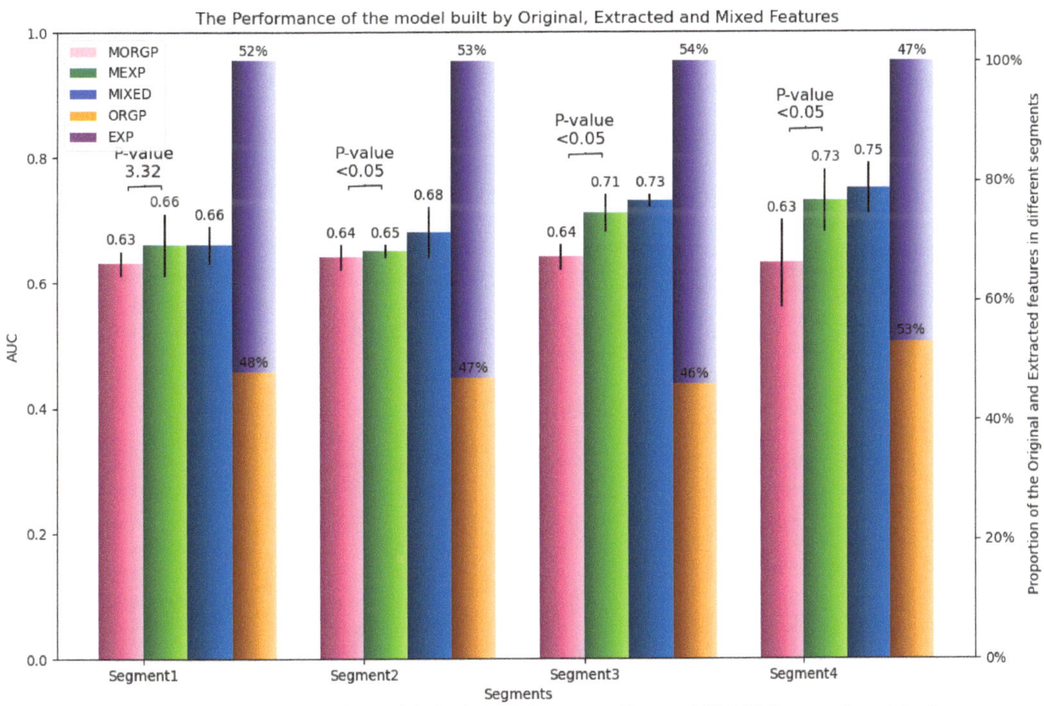

Figure 9. The outcome of the models built by different predictors. "ORGP" denotes the original predictors percentage, and "EXP" shows the proportion of the extracted predictors obtained after the optimization process.

5.2. RQ2—Snapshot-Stacked Ensemble Results

To answer RQ2, we took the output of the optimization process as the input of the ensemble module. We aimed to build a general model that could be used to predict the claims over the year, given vehicle usage. Thus, identical features from all segments were combined and injected into the first deep network to build and generate several diverse snapshot models. In addition, each season, as one additional feature, was added to the identical features to support generalization.

Given Equation (13), the number of cycles were set to $N = 20$ to generate 20 different snapshot models over the 400 epochs. Of the data, 60% were considered to train the snapshots and 30% to be the validation set to obtain the hard labels and soft predictions. Accordingly, 10% of the data was held out to test the meta-learning model at the final stage (see Figure 7 for visualization purposes). To quantify the diversity in the generated models, we used a disagreement measure, defined in Equation (20). This metric calculated the relation between the number of times the base classifiers predicted the same label, in terms of the total number of instances [78].

$$dis_{i,j} = \frac{1}{N}\sum_{k=1}^{N} fn(Sn_i(r_k) \neq Sn_j(r_k)) \tag{20}$$

where $Sn_i(r_k)$ refers to the label assigned by the snapshot (Sn_i) to readout r_k. Value $fn()$ is the function of truth predicate that counts the cases where Sn_i was correct and the Sn_j was wrong and vise-versa. Figure 10 illustrates the disagreement between the 20 snapshots. In the first phase, the hard labels predicted by the 20 snapshot models (here we took the average of the models' performances) was compared with the performance of the same algorithms used in the second module (Module2) evaluation, depicted in Table 6. The AUC values obtained over the 5-fold WFCV showed that, in almost all cases, our approach was superior in the first phase (Module3 First phase–Snapshots–, see Table 6), compared to the other approaches. This was not the case when it came to comparing the ensemble-based and Module2 performances. The classification results from all segments (0.81 vs. 0.79 in Segment 1, 0.80 vs. 0.75 in Segment 2, 0.82 vs. 0.80 in Segment 3 and 0.82 vs. 0.78 in Segment 4) and union of the features (0.76 vs. 0.75) suggested the first module performed better with respect to the first deep net (first phase). A similar observation was obtained when the first phase was compared with other ensemble approaches, such as Bagging, Boosting, and Stacking. This meant we did not see any improvement from the snapshot models in the first deep neural networks.

This motivated us to construct meta-learning with the aim to learn from the errors received in the first phase and improve the prediction in the second phase, that being the final prediction. The results of the snapshot models (20×3) were horizontally added to the data-set (the validation set in the first phase) to be trained and tested again in the second phase. This resulted in 60 extra new features in the data set. This meant the data considered as the validation set in the first deep networks (first phase/layer) became a new data set by carrying 60 extra features (in the second phase). Accordingly, the test set (10% of the whole data), which was held out, was used to assess the model. It is necessary to remark that, in each phase, we trained the deep neural network separately in one training shot. This meant that, in the first phase, we constructed 20 models in one training process using CCAS, and, in the second phase, we built a meta-learning model through one training process.

The result of the second phase is shown in Table 6 (last row), which was compared with other approaches. The figures represented in the table confirm significant improvement due to the stacking and ensemble snapshots performances generated in the first phase. We almost obtained $auc > 0.80$ in all experiments, with the exception of one in Segment 2 ($auc = 0.75$). The meta-learning built by the snapshot models could learn the errors obtained from the previous training and prediction phases. This resulted in a general predictive model that could be potentially used for claim/fault prediction in all seasons. Indeed, the diversity of the snapshots, illustrated in Figure 10, highly supported the second phase to build a generalized meta-learning model for the final prediction. Taking into account the results obtained from the above experiments, the ensemble approaches performed well, compared with other classifiers, including linear and deep networks. However, the superiority of our approach in this complex problem was very evident. This led us to examine these approaches in a different context. Thus, we conducted the proposed approach with the other three ensemble approaches on several different datasets to assess the generality and to ascertain whether the SSED performed equally to, or better than, its performance in other application domains. This is an important consideration, since it assesses the generality of the approach to deal with data from different contexts (see Table 7 for the details information of the data sets).

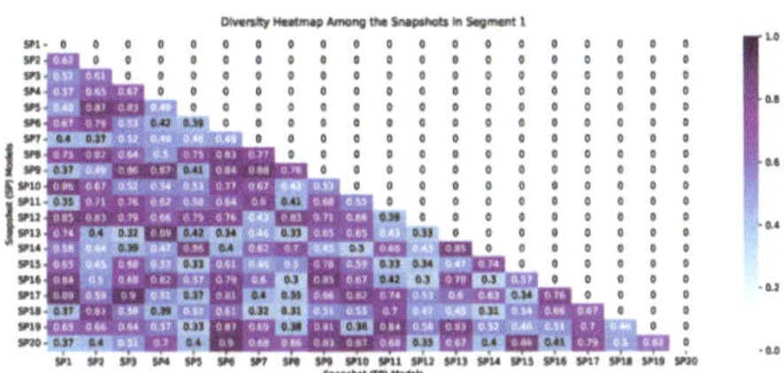

(**a**) Segment 1

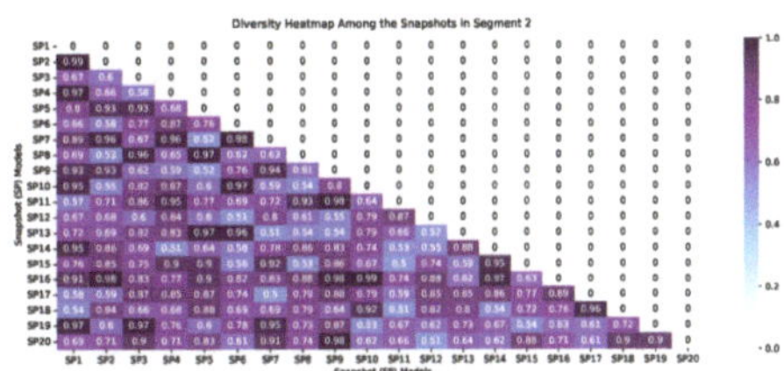

(**b**) Segment 2

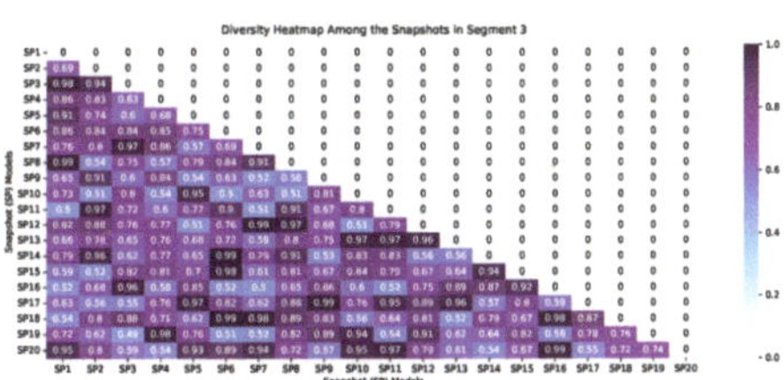

(**c**) Segment 3

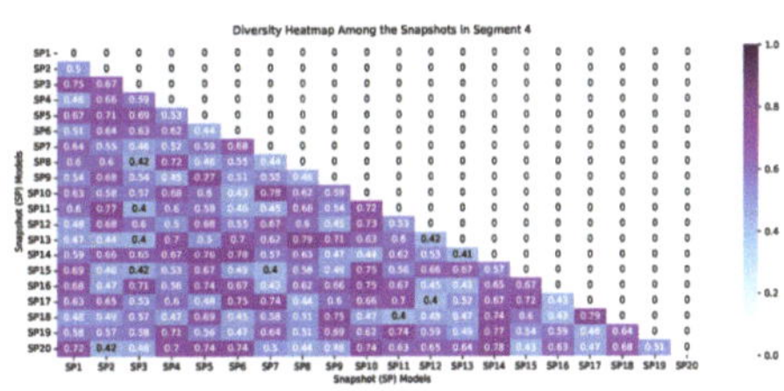

(**d**) Segment 4

Figure 10. The disagreement of 20 snapshot models generated by cosine annealing in different segments. The heat-map shows how the generated snapshots were diverse in the first phase of Module2. In these plots, each cell under the orthogonal represents the disagreement between two snapshots–e.g., SP1 vs. SP2–. The more *div* (Equation (20)) value close to 1 the more diverse the models. From the heat-maps we observed that the snapshots in Segment 2 and Segment 3 were more diverse than in Segment 1 and Segment 4.

The observations obtained from Table 8 show that SSED, in most cases, took the first rank, in terms of accuracy of performance, and outperformed other ensemble approaches. Individual comparison between SSED Vs. Bagging and Boosting indicated that the SSED almost consistently outperformed the two approaches on all datasets. However, this consistency was violated on dataset 7, where the figures show that Boosting provided slightly better results (%2). In contrast, the comparison between SSED vs. Stacking suggests that Staking had better generalization on the datasets. However, the statistical tests on the results (datasets 3, 7, 9, and 10) described the differences as not being statistically significant, when considering the critical value at $\alpha = 0.05$. Concerning the same comparison, we observed the value of the t-test on the figures obtained on datasets 1, 4, 6, 10, 11, where the SSED performed better, confirmed the differences were statistically significant, and concluded that SSED outperformed the Stacking approach.

Table 6. The comparison between Module 3 (SSED)-first phase, Module3 (SSED)-second phase and different approaches (standard machine learning, Gaussian Process (GP), deep neural networks, and ensemble based approaches, such as Bagging, Boosting and Stacking approaches) for RQ2. We used $5 \times 2cv$ paired t-test to examine how significant the differences were between the performances of the approaches. In this statistical test, only "Module3 (SSED)-second phase" was compared with other algorithm performances. Every time we compared only two models; e.g., Module3 (SSED)–second phase vs. Boosting, or Module3 (SSED)–second phase vs. Module3 (SSED)–first phase, or Module2–ELSGA (XGB).

Classifiers	Union of the Features		Segment1		Segment2		Segment3		Segment4	
	auc	p	auc	p	auc	p	auc	p	auc	p
Logreg	0.50 ± 0.00	<0.05	0.50 ± 0.00	<0.05	0.50 ± 0.00	<0.05	0.50 ± 0.00	<0.05	0.50 ± 0.00	<0.05
Kneig	0.52 ± 0.02	<0.05	0.544 ± 0.01	<0.05	0.52 ± 0.01	<0.05	0.54 ± 0.04	<0.05	0.52 ± 0.02	<0.05
SGD	0.50 ± 0.00	<0.05	0.50 ± 0.00	<0.05	0.50 ± 0.00	<0.05	0.50 ± 0.00	<0.05	0.50 ± 0.00	<0.05
QdA	0.53 ± 0.02	<0.05	0.50 ± 0.00	<0.05	0.50 ± 0.00	<0.05	0.50 ± 0.00	<0.05	0.50 ± 0.00	<0.05
SVM	0.55 ± 0.05	<0.05	0.56 ± 0.02	<0.05	0.53 ± 0.03	<0.05	0.57 ± 0.04	<0.05	0.57 ± 0.06	<0.05
DT	0.54 ± 0.01	<0.05	0.58 ± 0.01	<0.05	0.59 ± 0.02	<0.05	0.58 ± 0.02	<0.05	0.60 ± 0.03	<0.05
Ridge	0.71 ± 0.07	<0.05	0.73 ± 0.04	<0.05	0.70 ± 0.06	<0.05	0.71 ± 0.05	<0.05	0.71 ± 0.07	<0.05
SVC	0.52 ± 0.01	<0.05	0.52 ± 0.01	<0.05	0.51 ± 0.001	<0.05	0.51 ± 0.001	<0.05	0.53 ± 0.02	<0.05
GP	0.50 ± 0.00	<0.05	0.50 ± 0.00	<0.05	0.50 ± 0.00	<0.05	0.50 ± 0.00	<0.00	0.50 ± 0.00	<0.05
CNN	0.65 ± 0.08	<0.05	0.67 ± 0.07	<0.05	0.70 ± 0.06	0.12	0.71 ± 0.041	<0.05	0.77 ± 0.11	0.2
LSTM	0.69 ± 0.05	0.06	0.74 ± 0.02	<0.05	0.71 ± 0.06	0.19	0.71 ± 0.06	<0.05	0.72 ± 0.05	<0.05
biLSTM	0.70 ± 0.05	0.2	0.74 ± 0.03	<0.05	0.70 ± 0.07	<0.05	0.72 ± 0.05	<0.05	0.72 ± 0.10	<0.05
Blending	0.58 ± 0.01	<0.05	0.58 ± 0.004	<0.05	0.59 ± 0.03	<0.05	0.58 ± 0.01	<0.05	0.58 ± 0.01	<0.05
Bagging	0.71 ± 0.06	0.10	0.80 ± 0.04	0.10	**0.78 ± 0.06**	0.25	0.81 ± 0.06	0.37	0.79 ± 0.09	0.25
Boosting	0.75 ± 0.05	0.21	0.75 ± 0.05	<0.05	**0.76 ± 0.06**	0.43	0.76 ± 0.04	0.09	0.75 ± 0.09	0.09
Stacking	0.74 ± 0.14	<0.05	0.81 ± 0.01	<0.05	**0.79 ± 0.06**	0.14	0.78 ± 0.04	<0.05	0.81 ± 0.08	<0.05
Module2 ELSGA (XGB)	0.76 ± 0.06	0.28	0.81 ± 0.04	0.26	0.80 ± 0.07	0.12	0.80 ± 0.07	0.30	0.82 ± 0.09	0.38
Module3 SSED First phase (Snapshots)	0.75 ± 0.04	0.21	0.79 ± 0.03	0.05	0.75 ± 0.07	0.46	0.80 ± 0.05	0.30	0.78 ± 0.10	0.20
Module3 SSED Second phase (Stacked Ensemble)	**0.80 ± 0.11**	—	**0.83 ± 0.04**	—	0.75 ± 0.04	—	**0.82 ± 0.07**	—	**0.84 ± 0.08**	—

Table 7. The detailed information of the datasets.

#	Database	# of Samples	# of Features	# of Classes
1	VBS-fault	8712	4	2
2	Car insurance claim	10,302	27	2
3	Covtype	581,012	55	3
4	Cmiyc	27,297	22	2
5	Kdd	494,021	42	17
6	Building permit	198,900	43	8
7	Diabetes	768	9	2
8	Mnist	9271	787	10
9	Musk	6598	168	2
10	Parkinson	756	757	2
11	Spam	4400	59	2
12	Waveform	5000	41	3
13	Australia	142,193	24	2

Table 8. The comparison between SSED and three ensemble approaches on various datasets. The statistical t-test was applied to the performances of the approaches. Each time, SSED and one approach (e.g., Bagging) was compared, and the p-value reported.

#	Datasets	Accuracy						
		SSED	Bagging	p-Value	Boosting	p-Value	Stacking	p-Value
1	VBS fault	**0.95 ± 0.02**	0.93 ± 0.01	<0.05	0.75 ± 0.35	0.14	0.93 ± 0.01	<0.05
2	Car insurance claim	**0.77 ± 0.01**	0.73 ± 0.006	0.069	0.74 ± 0.004	0.27	0.73 ± 0.001	0.067
3	Covtype	0.81 ± 0.1	0.68 ± 0.009	<0.05	0.63 ± 0.002	<0.05	0.84 ± 0.005	0.32
4	Cmiyc	**0.98 ± 0.01**	0.91 ± 0.01	<0.05	0.93 ± 0.006	<0.05	0.91 ± 0.004	<0.05
5	Kdd	**0.80 ± 0.23**	0.36 ± 0.34	<0.05	0.78 ± 0.001	0.480	0.62 ± 0.43	0.288
6	Building permit	**0.93 ± 0.02**	0.89 ± 0.003	<0.05	0.94 ± 0.004	<0.05	0.91 ± 0.002	<0.05
7	Diabetes	0.73 ± 0.11	0.75 ± 0.03	0.35	0.74 ± 0.04	0.41	**0.75 ± 0.03**	0.35
8	Mnist	0.93 ± 0.04	0.85 ± 0.004	<0.05	0.86 ± 0.008	<0.05	**0.99 ± 0.004**	<0.05
9	Musk	0.97 ± 0.01	0.98 ± 0.002	0.28	0.84 ± 0.3	0.19	**0.99 ± 0.01**	0.282
10	Parkinson	**0.94 ± 0.02**	0.73 ± 0.05	<0.05	0.82 ± 0.04	<0.05	0.77 ± 0.02	<0.05
11	Spam	0.87 ± 0.15	0.82 ± 0.21	0.46	0.91 ± 0.17	0.35	**0.98 ± 0.003**	0.10
12	Waveform	**0.88 ± 0.02**	0.86 ± 0.003	0.34	0.82 ± 0.01	<0.05	0.86 ± 0.01	0.23
13	Australia	**0.99 ± 0.003**	0.81 ± 0.003	<0.05	0.99 ± 0.007	0.42	0.99 ± 0.01	0.29

6. Discussion and Conclusions and Future Work

This paper presents a two-phase snapshot-stacked ensemble approach to predict warranty claims/breakdowns in the automotive sector. The experimental evaluation, using the logged data from thousands of heavy-duty trucks, showed that the proposed approach could properly predict failures based on vehicle usage. Regarding the research questions introduced in Section 1, we evaluated the key aspects, which expressed our approach's efficiency, effectiveness, stability and generality. Considering the first objective (RQ1), figures obtained in model performance experiments showed a significant difference between the optimization and other classifiers. The results in terms of *feature contribution* were auspicious. Our A/B test experiments, illustrated in Figure 9, explain how effectively the extracted features contributed to the predictive model. Statistical tests also indicated the improvement induced by the extracted features was significantly different from the original features. It was also observed that the model constructed by the extracted features in the first segment provided similar performances compared to the model trained by both the original and extracted predictors. Therefore, we answer RQ1 by stating that: *our proposed adapted heuristic approach outperforms the other approaches to extract the best representation of vehicle usage in order to build a complex ensemble deep network for final claim prediction.*

Concerning the second objective (RQ2), the results explained how meta-learning could learn from the errors received by the snapshot models in the first phase of the ensemble approach to predict warranty claims. Indeed, the results in the second phase confirmed the stability of the meta-learning performance over the course of the year, taking into account different distributions of vehicle usage. The SSED could produce a generic predictive model that performed well on data from different seasons for claim predictions but was also promising for other less complex datasets comparable to the performance of Stacking. Therefore, we answer *RQ*2 by stating that: *the snapshot-stacked ensemble deep neural networks approach can successfully map usage to claims and outperforms the other compared algorithms. The figures obtained in the last phase also showed the potential in building a general forecasting model for this complex claim prediction problem. In addition, the results obtained on various datasets confirmed this claim.*

The findings of this work also reveal limitations of the proposed approach, which suggest new directions for future research. The first limitation pertains to the issue of modeling vehicle usage by taking into account various contexts. In this work, our concerns were time and vehicle usage to formulate the claim prediction. However, we did not include other contexts, such as location (where the vehicles were driven), which highly affect usage and predictive models. This limitation strongly motivates us to extend our approach by incorporating context knowledge (e.g., location) for a better mapping between usage and breakdowns. An interesting extension of our work could focus on vehicle profiling. We could utilize segmentation to model vehicles' behaviors in different contexts, leading to identification of good and bad driving styles as an important factor in vehicle performance. Thus, in this matter, driver behavior as an extra knowledge input could be used to characterize the style of vehicle usage in different contexts. Regarding contexts, we hypothesize that climate could impact snapshot diversity. This could also explain the dissimilarities shown in the results, which could be subject to further investigation.

Author Contributions: Conceptualization, R.K., D.T. and J.L.; Data curation, D.T.; Methodology, R.K., M.T., J.L. and M.R.; Writing—review & editing, R.K., M.T., J.L. and M.R.; Implementation, R.K. All authors have read and agreed to the published version of the manuscript.

Funding: This research was funded by Center for Applied Intelligent Systems Research (CAISR) at Halmstad University.

Institutional Review Board Statement: The authors would like to appreciate the anonymous reviewers for their advantageous comments.

Data Availability Statement: Data sharing not applicable.

Conflicts of Interest: The authors declare no conflict of interest.

References

1. Rögnvaldsson, T.; Nowaczyk, S.; Byttner, S.; Prytz, R.; Svensson, M. Self-monitoring for maintenance of vehicle fleets. *Data Min. Knowl. Discov.* **2018**, *32*, 344–384. [CrossRef]
2. Murray, D.; Glidewell, S. An Analysis of the Operational Costs of Trucking: 2019 Update. 2019. Available online: https://trid.trb.org/view/1663899 (accessed on 18 November 2019).
3. Boss, G.J.; Jones, A.R.; Lingafelt, C.S.; McConnell, K.C.; Moore, J.E. Predicting Vehicular Failures Using Autonomous Collaborative Comparisons to Detect Anomalies. U.S. Patent 10,109,120, 23 October 2018.
4. Lei, Y.; Yang, B.; Jiang, X.; Jia, F.; Li, N.; Nandi, A.K. Applications of machine learning to machine fault diagnosis: A review and roadmap. *Mech. Syst. Signal Process.* **2020**, *138*, 106587. [CrossRef]

5. Fredette, M.; Lawless, J.F. Finite-horizon prediction of recurrent events, with application to forecasts of warranty claims. *Technometrics* **2007**, *49*, 66–80. [CrossRef]
6. Ding, B.; Fang, H. Fault prediction for nonlinear stochastic system with incipient faults based on particle filter and nonlinear regression. *ISA Trans.* **2017**, *68*, 327–334. [CrossRef]
7. Man, J.; Zhou, Q. Prediction of hard failures with stochastic degradation signals using Wiener process and proportional hazards model. *Comput. Ind. Eng.* **2018**, *125*, 480–489. [CrossRef]
8. Saon, S.; Hiyama, T. Predicting remaining useful life of rotating machinery based artificial neural network. *Comput. Math. Appl.* **2010**, *60*, 1078–1087.
9. Liu, Z.; Guo, W.; Tang, Z.; Chen, Y. Multi-sensor data fusion using a relevance vector machine based on an ant colony for gearbox fault detection. *Sensors* **2015**, *15*, 21857–21875. [CrossRef]
10. Lei, Y.; Li, N.; Guo, L.; Li, N.; Yan, T.; Lin, J. Machinery health prognostics: A systematic review from data acquisition to RUL prediction. *Mech. Syst. Signal Process.* **2018**, *104*, 799–834. [CrossRef]
11. Yang, D.; He, Z.; He, S. Warranty claims forecasting based on a general imperfect repair model considering usage rate. *Reliab. Eng. Syst. Saf.* **2016**, *145*, 147–154. [CrossRef]
12. Chehade, A.; Savargaonkar, M.; Krivtsov, V. Conditional Gaussian mixture model for warranty claims forecasting. *Reliab. Eng. Syst. Saf.* **2022**, *218*, 108180. [CrossRef]
13. Nowaczyk, S.; Prytz, R.; Rögnvaldsson, T.; Byttner, S. Towards a machine learning algorithm for predicting truck compressor failures using logged vehicle data. In Proceedings of the 12th Scandinavian Conference on Artificial Intelligence, Aalborg, Denmark, 20–22 November 2013; IOS Press: Amsterdam, The Netherlands, 2013; pp. 205–214.
14. Prytz, R.; Nowaczyk, S.; Rögnvaldsson, T.; Byttner, S. Predicting the need for vehicle compressor repairs using maintenance records and logged vehicle data. *Eng. Appl. Artif. Intell.* **2015**, *41*, 139–150. [CrossRef]
15. Carvalho, T.P.; Soares, F.A.; Vita, R.; Francisco, R.d.P.; Basto, J.P.; Alcalá, S.G. A systematic literature review of machine learning methods applied to predictive maintenance. *Comput. Ind. Eng.* **2019**, *137*, 106024. [CrossRef]
16. Dennett, D.C.; Dennett, D.C. *Darwin's Dangerous Idea: Evolution and the Meanins of Life*; Number 39; Simon and Schuster: New York, NY, USA, 1996.
17. Fu, H.; Li, C.; Liu, X.; Gao, J.; Celikyilmaz, A.; Carin, L. Cyclical annealing schedule: A simple approach to mitigating kl vanishing. *arXiv* **2019**, arXiv:1903.10145.
18. Khoshkangini, R.; Mashhadi, P.S.; Berck, P.; Gholami Shahbandi, S.; Pashami, S.; Nowaczyk, S.; Niklasson, T. Early Prediction of Quality Issues in Automotive Modern Industry. *Information* **2020**, *11*, 354. [CrossRef]
19. Borboudakis, G.; Tsamardinos, I. Extending greedy feature selection algorithms to multiple solutions. *Data Min. Knowl. Discov.* **2021**, *35*, 1393–1434. [CrossRef]
20. Guidotti, R. Counterfactual explanations and how to find them: Literature review and benchmarking. *Data Min. Knowl. Discov.* **2022**, 1–55. [CrossRef]
21. Theissler, A.; Pérez-Velázquez, J.; Kettelgerdes, M.; Elger, G. Predictive maintenance enabled by machine learning: Use cases and challenges in the automotive industry. *Reliab. Eng. Syst. Saf.* **2021**, *215*, 107864. [CrossRef]
22. Vrignat, P.; Kratz, F.; Avila, M. Sustainable manufacturing, maintenance policies, prognostics and health management: A literature review. *Reliab. Eng. Syst. Saf.* **2022**, *218*, 108140. [CrossRef]
23. Xiao, Y.; Wang, H.; Zhang, L.; Xu, W. Two methods of selecting Gaussian kernel parameters for one-class SVM and their application to fault detection. *Knowl. Based Syst.* **2014**, *59*, 75–84. [CrossRef]
24. Killeen, P.; Ding, B.; Kiringa, I.; Yeap, T. IoT-based predictive maintenance for fleet management. *Procedia Comput. Sci.* **2019**, *151*, 607–613. [CrossRef]
25. Hecker, S.; Dai, D.; Van Gool, L. Failure prediction for autonomous driving. In Proceedings of the 2018 IEEE Intelligent Vehicles Symposium (IV), Changshu, China, 26–30 June 2018; pp. 1792–1799.
26. Aivaliotis, P.; Georgoulias, K.; Chryssolouris, G. The use of Digital Twin for predictive maintenance in manufacturing. *Int. J. Comput. Integr. Manuf.* **2019**, *32*, 1067–1080. [CrossRef]
27. Yu, W.; Dillon, T.; Mostafa, F.; Rahayu, W.; Liu, Y. A global manufacturing big data ecosystem for fault detection in predictive maintenance. *IEEE Trans. Ind. Inform.* **2019**, *16*, 183–192. [CrossRef]
28. Theissler, A. Detecting known and unknown faults in automotive systems using ensemble-based anomaly detection. *Knowl. Based Syst.* **2017**, *123*, 163–173. [CrossRef]
29. Wang, Q.; Bu, S.; He, Z. Achieving predictive and proactive maintenance for high-speed railway power equipment with LSTM-RNN. *IEEE Trans. Ind. Inform.* **2020**, *16*, 6509–6517. [CrossRef]
30. Liang, P.; Deng, C.; Wu, J.; Yang, Z.; Zhu, J.; Zhang, Z. Single and simultaneous fault diagnosis of gearbox via a semi-supervised and high-accuracy adversarial learning framework. *Knowl. Based Syst.* **2020**, *198*, 105895. [CrossRef]
31. Liu, J.; Xu, Y.; Wang, L. Fault information mining with causal network for railway transportation system. *Reliab. Eng. Syst. Saf.* **2022**, *220*, 108281. [CrossRef]
32. de Sousa, P.H.F.; Nascimento, N.d.M.M.e.; Almeida, J.S.; Filho, P.P.R.; Albuquerque, V.H.C.d. Intelligent Incipient Fault Detection in Wind Turbines based on Industrial IoT Environment. *J. Artif. Intell. Syst.* **2019**, *1*, 1–19. [CrossRef]
33. Shokouhyar, S.; Ahmadi, S.; Ashrafzadeh, M. Promoting a novel method for warranty claim prediction based on social network data. *Reliab. Eng. Syst. Saf.* **2021**, *216*, 108010. [CrossRef]

34. Kalbfleisch, J.; Lawless, J.; Robinson, J. Methods for the analysis and prediction of warranty claims. *Technometrics* **1991**, *33*, 273–285. [CrossRef]
35. Gao, K.; Khoshgoftaar, T.M. A comprehensive empirical study of count models for software fault prediction. *IEEE Trans. Reliab.* **2007**, *56*, 223–236. [CrossRef]
36. Ye, Z.S.; Chen, N.; Shen, Y. A new class of Wiener process models for degradation analysis. *Reliab. Eng. Syst. Saf.* **2015**, *139*, 58–67. [CrossRef]
37. Wei, J.; Dong, G.; Chen, Z. Remaining useful life prediction and state of health diagnosis for lithium-ion batteries using particle filter and support vector regression. *IEEE Trans. Ind. Electron.* **2017**, *65*, 5634–5643. [CrossRef]
38. Saufi, S.R.; Ahmad, Z.A.B.; Leong, M.S.; Lim, M.H. Challenges and opportunities of deep learning models for machinery fault detection and diagnosis: A review. *IEEE Access* **2019**, *7*, 122644–122662. [CrossRef]
39. Yang, Z.; Baraldi, P.; Zio, E. A method for fault detection in multi-component systems based on sparse autoencoder-based deep neural networks. *Reliab. Eng. Syst. Saf.* **2022**, *220*, 108278. [CrossRef]
40. Zhang, W.; Han, G.; Wang, J.; Liu, Y. A BP neural network prediction model based on dynamic cuckoo search optimization algorithm for industrial equipment fault prediction. *IEEE Access* **2019**, *7*, 11736–11746. [CrossRef]
41. Singpurwalla, N.D.; Wilson, S.P. Failure models indexed by two scales. *Adv. Appl. Probab.* **1998**, *30*, 1058–1072. [CrossRef]
42. Kaminskiy, M.P.; Krivtsov, V.V. G-renewal process as a model for statistical warranty claim prediction. In Proceedings of the Annual reliability and maintainability symposium. 2000 Proceedings. International symposium on product quality and integrity (Cat. No. 00CH37055), Los Angeles, CA, USA, 24–27 January 2000; pp. 276–280.
43. Chukova, S.; Robinson, J. Estimating mean cumulative functions from truncated automotive warranty data. *Mod. Stat. Math. Methods Reliab.* **2005**, *10*, 121.
44. Zhang, M.L.; Zhou, Z.H. Exploiting unlabeled data to enhance ensemble diversity. *Data Min. Knowl. Discov.* **2013**, *26*, 98–129. [CrossRef]
45. Li, Z.; Wu, D.; Hu, C.; Terpenny, J. An ensemble learning-based prognostic approach with degradation-dependent weights for remaining useful life prediction. *Reliab. Eng. Syst. Saf.* **2019**, *184*, 110–122. [CrossRef]
46. Large, J.; Lines, J.; Bagnall, A. A probabilistic classifier ensemble weighting scheme based on cross-validated accuracy estimates. *Data Min. Knowl. Discov.* **2019**, *33*, 1674–1709. [CrossRef] [PubMed]
47. Dong, X.; Yu, Z.; Cao, W.; Shi, Y.; Ma, Q. A survey on ensemble learning. *Front. Comput. Sci.* **2020**, *14*, 241–258. [CrossRef]
48. Fauvel, K.; Fromont, E.; Masson, V.; Faverdin, P.; Termier, A. XEM: An explainable-by-design ensemble method for multivariate time series classification. *Data Min. Knowl. Discov.* **2022**, *36*, 917–957. [CrossRef]
49. Zhou, Z.H. Ensemble learning. In *Machine Learning*; Springer: Berlin/Heidelberg, Germany, 2021; pp. 181–210.
50. González, S.; García, S.; Del Ser, J.; Rokach, L.; Herrera, F. A practical tutorial on bagging and boosting based ensembles for machine learning: Algorithms, software tools, performance study, practical perspectives and opportunities. *Inf. Fusion* **2020**, *64*, 205–237. [CrossRef]
51. Mosavi, A.; Sajedi Hosseini, F.; Choubin, B.; Goodarzi, M.; Dineva, A.A.; Rafiei Sardooi, E. Ensemble boosting and bagging based machine learning models for groundwater potential prediction. *Water Resour. Manag.* **2021**, *35*, 23–37. [CrossRef]
52. Sołtys, M.; Jaroszewicz, S.; Rzepakowski, P. Ensemble methods for uplift modeling. *Data Min. Knowl. Discov.* **2015**, *29*, 1531–1559. [CrossRef]
53. Yang, S.; Shao, Q.; Bian, C. Reliability analysis of ensemble fault tolerance for soft error mitigation against complex radiation effect. *Reliab. Eng. Syst. Saf.* **2022**, *217*, 108092. [CrossRef]
54. Le, V.; Yao, X.; Miller, C.; Tsao, B.H. Series DC arc fault detection based on ensemble machine learning. *IEEE Trans. Power Electron.* **2020**, *35*, 7826–7839. [CrossRef]
55. Mousavi, M.; Moradi, M.; Chaibakhsh, A.; Kordestani, M.; Saif, M. Ensemble-based fault detection and isolation of an industrial Gas turbine. In Proceedings of the 2020 IEEE International Conference on Systems, Man, and Cybernetics (SMC), Toronto, ON, Canada, 11–14 October 2020; pp. 2351–2358.
56. Zhang, D.; Gao, Z. An Ensemble Approach for Fault Diagnosis via Continuous Learning. In Proceedings of the 2021 IEEE 19th International Conference on Industrial Informatics (INDIN), Palma de Mallorca, Spain, 21–23 July 2021.
57. Cárdenas-Gallo, I.; Sarmiento, C.A.; Morales, G.A.; Bolivar, M.A.; Akhavan-Tabatabaei, R. An ensemble classifier to predict track geometry degradation. *Reliab. Eng. Syst. Saf.* **2017**, *161*, 53–60. [CrossRef]
58. Srinivas, M.; Patnaik, L.M. Genetic algorithms: A survey. *Computer* **1994**, *27*, 17–26. [CrossRef]
59. Whitley, D.; Sutton, A.M. Genetic algorithms-A survey of models and methods. In *Handbook of Natural Computing*; Springer: Berlin/Heidelberg, Germany, 2012; pp. 637–671.
60. Ma, C.; Hao, W.; Pan, F.; Xiang, W. Road screening and distribution route multi-objective robust optimization for hazardous materials based on neural network and genetic algorithm. *PLoS ONE* **2018**, *13*, e0198931. [CrossRef]
61. Zhou, T.; Tao, D.; Wu, X. Manifold elastic net: A unified framework for sparse dimension reduction. *Data Min. Knowl. Discov.* **2011**, *22*, 340–371. [CrossRef]
62. Abdi, H.; Williams, L.J. Principal component analysis. *Wiley Interdiscip. Rev. Comput. Stat.* **2010**, *2*, 433–459. [CrossRef]
63. Xanthopoulos, P.; Pardalos, P.M.; Trafalis, T.B. Linear discriminant analysis. In *Robust Data Mining*; Springer: Berlin/Heidelberg, Germany, 2013; pp. 27–33.
64. Fonti, V.; Belitser, E. Feature selection using lasso. *VU Amst. Res. Pap. Bus. Anal.* **2017**, *30*, 1–25.

65. Osman, H.; Ghafari, M.; Nierstrasz, O. Automatic feature selection by regularization to improve bug prediction accuracy. In Proceedings of the 2017 IEEE Workshop on Machine Learning Techniques for Software Quality Evaluation (MaLTeSQuE), Klagenfurt, Austria, 21–21 February 2017; pp. 27–32.
66. Zou, H.; Hastie, T. Regularization and variable selection via the elastic net. *J. R. Stat. Soc. Ser. B Stat. Methodol.* **2005**, *67*, 301–320. [CrossRef]
67. Tang, J.; Alelyani, S.; Liu, H. Feature selection for classification: A review. In *Data Classification: Algorithms and Applications*; CRC Press: Boca Raton, FL, USA, 2014; p. 37.
68. Chen, T.; Guestrin, C. Xgboost: A scalable tree boosting system. In Proceedings of the 22nd acm Sigkdd International Conference on Knowledge Discovery and Data Mining, San Francisco, CA, USA, 13–17 August 2016.
69. Goldberg, D.E.; Lingle, R. Alleles, loci, and the traveling salesman problem. In *Proceedings of the an International Conference on Genetic Algorithms and Their Applications*; Lawrence Erlbaum: Hillsdale, NJ, USA, 1985; Volume 154, pp. 154–159.
70. Nigam, A.; Friederich, P.; Krenn, M.; Aspuru-Guzik, A. Augmenting genetic algorithms with deep neural networks for exploring the chemical space. *arXiv* **2019**, arXiv:1909.11655.
71. Khoshkangini, R.; Mashhadi, P.; Tegnered, D.; Lundström, J.; Rögnvaldsson, T. Predicting Vehicle Behavior Using Multi-task Ensemble Learning. *Expert Syst. Appl.* **2023**, *212*, 118716. [CrossRef]
72. Huang, G.; Li, Y.; Pleiss, G.; Liu, Z.; Hopcroft, J.E.; Weinberger, K.Q. Snapshot ensembles: Train 1, get m for free. *arXiv* **2017**, arXiv:1704.00109.
73. He, K.; Zhang, X.; Ren, S.; Sun, J. Delving deep into rectifiers: Surpassing human-level performance on imagenet classification. In Proceedings of the IEEE International Conference on Computer Vision, Santiago, Chile, 11–18 December 2015; pp. 1026–1034.
74. Ramachandran, P.; Zoph, B.; Le, Q.V. Searching for activation functions. *arXiv* **2017**, arXiv:1710.05941.
75. Ioffe, S.; Szegedy, C. Batch normalization: Accelerating deep network training by reducing internal covariate shift. In Proceedings of the International Conference on Machine Learning, PMLR, Lille, France, 7–9 July 2015.
76. Kingma, D.P.; Ba, J. Adam: A method for stochastic optimization. *arXiv* **2014**, arXiv:1412.6980.
77. Falessi, D.; Narayana, L.; Thai, J.F.; Turhan, B. Preserving Order of Data When Validating Defect Prediction Models. *arXiv* **2018**, arXiv:1809.01510.
78. Kuncheva, L.I.; Whitaker, C.J. Measures of diversity in classifier ensembles and their relationship with the ensemble accuracy. *Mach. Learn.* **2003**, *51*, 181–207. [CrossRef]

cancers

Review

Recent Advances in Deep Learning and Medical Imaging for Head and Neck Cancer Treatment: MRI, CT, and PET Scans

Mathew Illimoottil [1] and Daniel Ginat [2,*]

[1] School of Medicine, University of Missouri-Kansas City, Kansas City, MO 64018, USA
[2] Department of Radiology, The University of Chicago, Chicago, IL 60637, USA
* Correspondence: dtg1@uchicago.edu

Simple Summary: Deep learning techniques have significant potential in head and neck cancer imaging, particularly in tumor detection, segmentation, and outcome prediction using magnetic resonance imaging (MRI), computed tomography (CT), and positron emission tomography (PET) scans. Advanced deep learning methods, such as convolutional autoencoders, generative adversarial networks (GANs), and transformer models, have further enhanced imaging capabilities. Comparing deep learning and traditional techniques and the advantages and limits of each reveals their complementary roles in cancer management. Integrating radiogenomics with deep learning models promises further advancements in personalized care. However, challenges such as standardization, data quality, model overfitting, and computational requirements persist. Addressing these issues, integrating multimodal and temporal information, enhancing explainability, and conducting clinical validation are crucial for implementing deep learning models in head and neck cancer diagnosis and treatment. Overcoming these obstacles will pave the way for improved patient outcomes and personalized treatment strategies in head and neck cancer management.

Abstract: Deep learning techniques have been developed for analyzing head and neck cancer imaging. This review covers deep learning applications in cancer imaging, emphasizing tumor detection, segmentation, classification, and response prediction. In particular, advanced deep learning techniques, such as convolutional autoencoders, generative adversarial networks (GANs), and transformer models, as well as the limitations of traditional imaging and the complementary roles of deep learning and traditional techniques in cancer management are discussed. Integration of radiomics, radiogenomics, and deep learning enables predictive models that aid in clinical decision-making. Challenges include standardization, algorithm interpretability, and clinical validation. Key gaps and controversies involve model generalizability across different imaging modalities and tumor types and the role of human expertise in the AI era. This review seeks to encourage advancements in deep learning applications for head and neck cancer management, ultimately enhancing patient care and outcomes.

Keywords: magnetic resonance imaging (MRI); computed tomography (CT); positron emission tomography (PET); deep learning; radiomics; radiogenomics; convolutional neural networks (CNNs); artificial intelligence (AI); generative adversarial networks (GANs); head and neck; cancer; treatment

1. Introduction

Head and neck cancers represent a heterogeneous group of malignancies that arise from various structures in the head and neck region, including the oral cavity, pharynx, larynx, nasal cavity, paranasal sinuses, and salivary glands [1]. These cancers are often associated with significant morbidity and mortality, and their management requires a multidisciplinary approach involving surgery, radiation therapy, and chemotherapy [2]. Accurate diagnosis, staging, and monitoring of head and neck cancers are crucial for determining the optimal treatment strategy, predicting patient outcomes, and assessing treatment response.

Medical imaging techniques, such as magnetic resonance imaging (MRI), computed tomography (CT), and positron emission tomography (PET) scans, have been indispensable in the assessment of head and neck cancers [3]. However, the interpretation of these images can be challenging due to the complex anatomy, diverse tumor biology, and overlapping features of malignant and benign lesions [4]. Moreover, traditional imaging techniques often rely on human expertise, which may be subjective and prone to variability.

Deep learning algorithms, particularly convolutional neural networks (CNNs), have shown potential in improving the accuracy, efficiency, and consistency of medical image analysis [5]. These advanced computational techniques can automatically learn hierarchical feature representations from raw imaging data, enabling the extraction of clinically relevant information and potentially surpassing human-level performance in certain tasks.

In this review, we delve into the limitations of traditional imaging techniques in head and neck cancer imaging and discuss advanced deep learning techniques, such as convolutional autoencoders for data compression, image denoising, and feature extraction; generative adversarial networks (GANs) for super-resolution; and transformer models such as the Vision Transformer (ViT). We also compare deep learning and traditional imaging techniques in head and neck cancer imaging, emphasizing their complementary roles and the potential benefits of integrating these methods.

In recent years, deep learning applications have expanded to include tumor detection, segmentation, classification, and response prediction, as well as the integration of multi-modal imaging data and the development of radiomics-based predictive models [6]. We explore the potential of deep learning to advance precision medicine in head and neck cancer management by enabling personalized treatment planning and outcome predictions. Additionally, we examine the emerging field of radiogenomics, which utilizes information about the relationship between imaging features and genomic information to create predictive models that can better guide medical treatment. We identify gaps in current research and propose future directions that may further enhance the impact of deep learning on the diagnosis and treatment of head and neck cancers.

This review aims to inform researchers and clinicians about the progress, potential, and obstacles in the application of deep learning techniques in head and neck cancer imaging, ultimately guiding future research toward improving patient outcomes and personalized treatment strategies. By incorporating the latest advancements and a comprehensive comparison of traditional and deep learning techniques, we provide a concise but informative understanding of the current landscape and future prospects in head and neck cancer imaging.

2. Limitations of Traditional Imaging Techniques in Head and Neck Cancer Imaging

Traditional imaging techniques, such as magnetic resonance imaging (MRI), computed tomography (CT), and positron emission tomography (PET) scans, play crucial role sin the diagnosis, staging, and monitoring of head and neck cancers [7]. However, these methods often rely on human expertise for image interpretation, which may be subjective and prone to variability [8]. In this section, we discuss the limitations of traditional imaging techniques in head and neck cancer imaging, focusing on the challenges associated with human expertise and the potential consequences for clinical decision-making.

2.1. Subjectivity and Interobserver Variability

Radiologists may assess medical images based on qualitative criteria, such as lesion shape, size, or intensity, which can lead to inconsistencies in diagnosis and staging among different observers [9]. For example, in the case of head and neck cancers, the complex anatomy and overlapping features of malignant and benign lesions can make accurate interpretation of imaging data challenging, even for experienced radiologists [10]. Additionally, manual delineation of tumor boundaries in imaging data can be time-consuming, and interobserver variability can affect the accuracy and reliability of the results [11].

2.2. Factors Affecting Human Expertise

Human interpretation of medical images may be influenced by factors such as fatigue, experience, and cognitive biases, which can contribute to discrepancies in decision-making [12]. For instance, studies have shown that radiologists' diagnostic performance can be affected by factors such as workload, time constraints, and confirmation bias, leading to potential errors in cancer detection and staging [13,14]. Furthermore, the lack of standardized guidelines and protocols for image interpretation in head and neck cancer imaging can exacerbate these issues, resulting in suboptimal clinical decision-making and patient management.

2.3. Need for Improved Imaging Techniques

Given the limitations of traditional imaging techniques and the potential impact on patient outcomes, there is a growing need for more accurate, efficient, and consistent methods for head and neck cancer imaging. Deep learning algorithms, particularly convolutional neural networks (CNNs), offer a promising solution to these challenges since they can automatically learn hierarchical feature representations from raw imaging data and potentially surpass human-level performance in certain tasks [15,16]. In the following sections, we discuss the recent advances in deep learning applications for head and neck cancer treatment, focusing on the use of MRI, CT, and PET scans.

3. Deep Learning in Medical Imaging

3.1. Deep Learning: A Brief Overview

Deep learning is a subfield of machine learning that involves the use of artificial neural networks with multiple layers, also known as deep neural networks, to automatically learn hierarchical representations of input data [15]. This advanced computational approach has shown remarkable success in various domains, including natural language processing, speech recognition, and computer vision. Convolutional neural networks (CNNs) are a type of deep learning architecture specifically designed for image analysis. They have demonstrated exceptional performance in various computer vision tasks, such as image classification, object detection, and segmentation. Figure 1 shows a lymph node metastasis segmented by a neural network algorithm. CNNs typically consist of multiple layers, including convolutional, pooling, and fully connected layers, that work together to automatically learn features and representations from raw image data, making them particularly suitable for medical image analysis, as shown in Figure 2. Some CNNs also include additional layers such as upsampling, which can be useful for image segmentation tasks.

3.2. Deep Learning in Medical Image Analysis: Performance Improvements over Traditional Methods

The application of deep learning techniques to medical image analysis has shown promising results in various domains, leading to significant advancements in the detection and diagnosis of diseases, the segmentation of anatomical structures, and the prediction of treatment outcomes [16]. Deep learning algorithms can learn complex patterns from medical images, generalizing well to new data and achieving human-level or even superior performance in many tasks [17]. This ability offers the potential to greatly improve the accuracy, efficiency, and consistency of medical image analysis, ultimately benefiting patient care and outcomes.

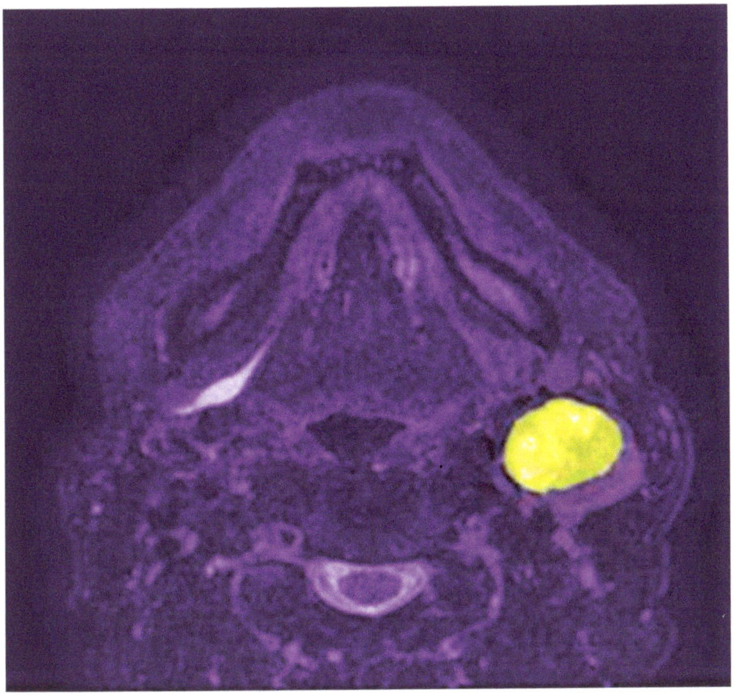

Figure 1. Axial T2-weighted MRI shows a left neck lymph node metastasis segmented by a neural network algorithm (highlighted in yellow).

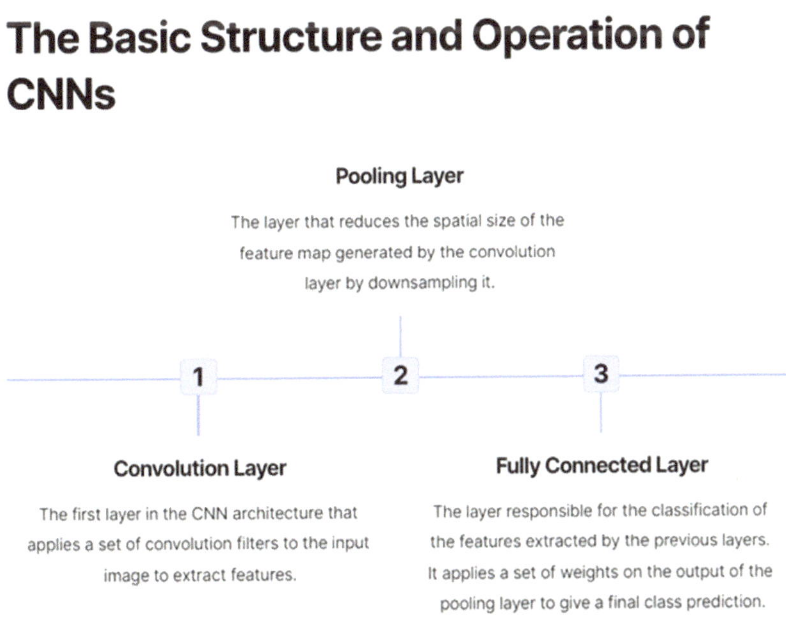

Figure 2. On the basic structure and operation of convolutional neural networks (CNNs).

Quantitative comparisons between deep learning models and traditional pre-deep learning imaging techniques have demonstrated substantial improvements in performance. For instance, in the diagnosis of diabetic retinopathy from retinal images, deep learning models have achieved sensitivity of 96.8% and specificity of 87.0%, significantly surpassing traditional methods with sensitivities of 49.3–85.5% and specificities of 71.0–93.4% [18]. Similarly, deep learning-based detection of pulmonary nodules on CT scans has exhibited higher accuracy (94.2%) than conventional computer-aided detection methods (79.1%) [19]. In the segmentation of brain tumors from MRI scans, deep learning models have achieved a Dice similarity coefficient of 0.88, outperforming traditional methods with coefficients ranging from 0.65 to 0.85 [20].

Moreover, deep learning has enabled the development of models that can integrate multi-modal and multi-scale imaging data, as well as clinical and demographic information, to generate more accurate and comprehensive predictions for patient outcomes and treatment responses [17]. This ability has contributed to the growing field of radiomics, which aims to extract and analyze high-dimensional quantitative features from medical images to build predictive models for personalized medicine [21]. As a result, deep learning is playing an increasingly important role in advancing precision medicine and improving patient care across a range of diseases and medical conditions. Figure 3 demonstrates the process by which CNNs are trained and fine-tuned.

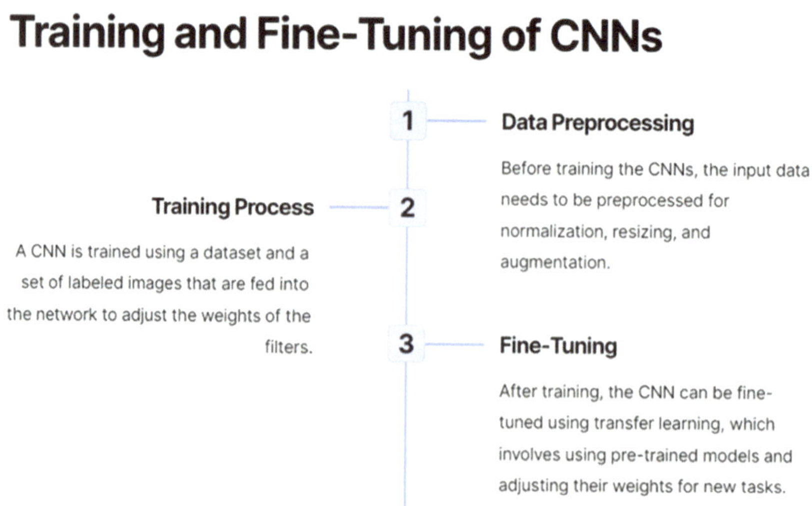

Figure 3. On the training and fine-tuning of convolutional neural networks (CNNs).

4. Deep Learning in Head and Neck Cancer Imaging

4.1. MRI

Magnetic resonance imaging (MRI) is a non-invasive imaging modality that provides excellent soft tissue contrast and high spatial resolution, making it a valuable tool for the assessment of head and neck cancers [22]. Deep learning techniques have been applied to various tasks in head and neck MRI, ranging from tumor detection and segmentation to treatment response prediction and prognosis assessment.

4.1.1. Tumor Detection and Segmentation

Convolutional neural networks (CNNs) have been employed for the automated detection and segmentation of head and neck tumors on MRI scans [23]. Recent studies have reported promising results in segmenting primary tumors and lymph nodes in patients with oropharyngeal cancer [24]. For instance, Men et al. (2017) reported a Dice similarity coefficient (DSC) of 0.85 for primary tumor segmentation using a 3D CNN, compared to a DSC of 0.74 for an atlas-based approach [25]. Advanced CNN architectures, such as U-Net and its variants, have been particularly successful in these segmentation tasks [26].

However, challenges remain in segmenting tumors with irregular shapes and heterogeneous intensity profiles, which may require the development of more sophisticated deep learning models or the incorporation of additional information, such as clinical or demographic data, to improve segmentation performance [5].

4.1.2. Treatment Response Prediction and Prognosis Assessment

Deep learning models have also shown potential in predicting treatment response in head and neck cancer patients using MRI scans. Recent studies have demonstrated the ability of CNNs to predict outcomes, such as tumor regression, local control, and overall survival, from pre- and post-treatment MRI scans for various forms of cancer [27,28]. Some research in other forms of cancer has explored the use of multi-parametric MRI, combining multiple imaging sequences to improve the accuracy of treatment response prediction [29].

Radiomics, which involves the extraction of high-dimensional features from medical images, has been combined with deep learning for improved prediction of treatment response and prognosis [6]. This approach can lead to more personalized treatment plans and better monitoring of patient outcomes. However, the clinical utility of these models remains to be established through large-scale, prospective studies.

4.2. CT

Computed tomography (CT) scans are widely used in the diagnosis and staging of head and neck cancers due to their high spatial resolution, fast acquisition times, and ability to visualize bony structures [30]. Deep learning applications in head and neck cancer CT imaging include tumor detection and segmentation, outcome prediction, and treatment planning.

4.2.1. Tumor Detection and Segmentation

Several studies have reported the use of CNNs for the detection and segmentation of head and neck tumors in CT scans [31,32]. CNNs have demonstrated high accuracy in segmenting primary tumors and lymph nodes, as well as in delineating organs at risk, such as the spinal cord and parotid glands, for radiotherapy planning [16,25,31].

Techniques such as 3D CNNs and multi-scale learning have been employed to better capture the spatial relationships between structures and to improve segmentation performance [33].

However, the performance of these models can be affected by the presence of artifacts, such as metal implants and dental fillings, which are common on head and neck CT scans [34]. Future research may explore the use of deep learning techniques for artifact reduction or may develop models that are more robust to artifacts in the image data.

4.2.2. Outcome Prediction and Treatment Planning

Deep learning models have also been developed for predicting treatment outcomes in head and neck cancer patients using CT scans. A recent study demonstrated the ability of a CNN to predict overall survival and disease-free survival from pre-treatment CT scans in patients with nasopharyngeal carcinoma [35]. Moreover, deep learning models incorporating radiomic features have shown improved prediction of treatment response and survival [33].

In addition, deep learning has been applied to treatment planning, including the optimization of radiotherapy plans, for head and neck cancer patients. These models can automatically generate dose distribution predictions and identify optimal beam configurations to minimize radiation exposure to healthy tissues while maximizing tumor coverage [36]. However, the generalizability and clinical impact of these models warrant further investigation, including multi-center studies and the evaluation of their performance in diverse patient populations.

4.3. PET

Positron emission tomography (PET) scans, which provide functional information about tumor metabolism and tissue perfusion, have gained increasing importance in the management of head and neck cancers [37]. Deep learning techniques have been applied to PET scans for tasks such as tumor detection, segmentation, and outcome prediction.

4.3.1. Tumor Detection and Segmentation

CNNs have shown promising results in the detection and segmentation of head and neck tumors on PET scans [24]. Methods such as transfer learning and the use of multi-modal data, combining PET with CT, or MRI, have been explored to improve the performance of deep learning models in this domain [38].

However, the performance of these models can be affected by the low spatial resolution and high noise levels inherent to PET imaging, as well as variations in image acquisition protocols and reconstruction algorithms [39]. Future research should focus on developing models that are more robust to these challenges and investigating strategies for harmonizing PET data from different sources.

4.3.2. Outcome Prediction and Treatment Monitoring

Deep learning models have also been explored for predicting treatment outcomes in head and neck cancer patients using PET scans. Studies have reported the potential of CNNs to predict tumor response, recurrence, and survival from pre- and post-treatment PET scans [40]. Some research has also examined the use of PET-derived radiomic features in combination with deep learning models for improved outcome prediction [41].

Furthermore, deep learning has been applied to monitor treatment response in head and neck cancer patients using PET scans. These models can potentially identify early metabolic changes that indicate treatment effectiveness, allowing for timely adjustments to treatment plans and improved patient outcomes [42]. However, further research is needed to validate these models in large, prospective cohorts and to establish their clinical utility.

5. Comparison of Deep Learning and Traditional Imaging Techniques in Head and Neck Cancer Imaging

While deep learning techniques have shown promise in various head and neck cancer imaging tasks, it is crucial to consider the specific situations in which traditional imaging techniques remain valuable and those in which deep learning tools are more appropriate.

5.1. Advantages of Traditional Imaging Techniques

1. Interpretability: Traditional imaging techniques provide interpretable results, as they are often based on well-established, handcrafted features and statistical methods [43]. This interpretability allows clinicians to better understand the rationale behind the decision-making process, which is essential for building trust in the results and ensuring appropriate clinical actions.
2. Lower computational requirements: Traditional imaging methods typically have lower computational demands compared to deep learning approaches [16], making them more accessible and easier to implement on standard workstations without the need for high-performance computing resources.
3. Robustness to variations: Traditional imaging techniques may be more robust to variations in imaging protocols and acquisition parameters [44] since they rely on well-established features that are less sensitive to changes in image quality and appearance.

5.2. Advantages of Deep Learning Techniques

1. Improved accuracy: Deep learning techniques, particularly CNNs, have demonstrated superior performance in many head and neck cancer imaging tasks, including tumor detection, segmentation, and outcome prediction, compared to traditional methods [25,27,45].
2. Automatic feature learning: Deep learning models can automatically learn relevant features from the data without the need for manual feature engineering [5], reducing the potential for human bias and enabling the discovery of novel imaging biomarkers that may not be apparent using traditional techniques.
3. Integration of multi-modal and multi-parametric data: Deep learning models can efficiently manage and integrate information from various imaging modalities (e.g., MRI, CT, and PET) and different image sequences [16], potentially providing a more comprehensive assessment of tumor characteristics and treatment response.

5.3. Guidance for Optimal Use

1. Preliminary analysis and simpler tasks: In situations in which a quick, preliminary analysis is required or for less complex tasks, traditional imaging techniques may be more suitable due to their lower computational demands and ease of implementation [16].
2. Interpretability and trust: When interpretability and trust in the decision-making process are critical, traditional methods may be more appropriate, as they provide more transparent and explainable results [46].
3. Resource-limited settings: In settings in which computational resources are limited, traditional imaging techniques may be more feasible since they generally have lower computational requirements [16].
4. Complex tasks and improved performance: For more complex tasks or when seeking improved accuracy and performance, deep learning techniques are more suitable [24,25,27,45]. This suitability includes for tasks such as tumor detection and segmentation, outcome prediction, and treatment planning in head and neck cancer imaging.

By carefully considering the specific situation and the desired outcomes, clinicians and researchers can select the most appropriate method for achieving accurate and reliable results in head and neck cancer imaging.

6. Gaps, Controversies, and Future Directions

6.1. Lack of Standardization and Benchmarking

One major challenge in the field of deep learning for head and neck cancer imaging is the lack of standardization and benchmarking. Diverse datasets, preprocessing techniques, model architectures, and training strategies have been used in different studies, making it difficult to compare the performance of various models and to assess their clinical utility [38]. Inconsistent reporting of model performance metrics, such as accuracy, sensitivity, and specificity, further complicates the comparison of results across studies. Future research should focus on establishing standardized datasets, evaluation metrics, and reporting guidelines to facilitate the benchmarking and comparison of deep learning models. In addition, promoting open science practices, such as sharing of data and code, can help to accelerate the development and validation of deep learning models for head and neck cancer imaging.

6.2. Integration of Multimodal, Temporal, and Clinical Information

Most deep learning studies in head and neck cancer imaging have focused on single-modality imaging data, such as MRI, CT, or PET scans. However, integrating multimodal information, such as combining functional and anatomical imaging data, can potentially improve the performance of deep learning models and provide a more comprehensive understanding of tumor characteristics [39]. Moreover, incorporating temporal information from longitudinal imaging data can potentially enhance the prediction of treatment response, tumor recurrence, and patient outcomes [40]. In addition to imaging data, the integration of clinical information, such as patient demographics, tumor histology, and treatment details, can further improve the performance of deep learning models in head and neck cancer management. Future research should explore the development of deep learning models that can effectively integrate multimodal, temporal, and clinical information for personalized treatment planning and prognostication.

6.3. Explainability, Interpretability, and Trustworthiness

Deep learning models, particularly CNNs, are often considered "black boxes" due to their complex architectures and the lack of transparency in the decision-making process [47]. This lack of transparency can hinder the adoption of deep learning techniques in clinical practice since clinicians may be reluctant to trust a model's predictions without understanding the underlying reasoning. Developing explainable and interpretable deep learning models is crucial to bridge this gap and promote their acceptance in the medical community [41]. Techniques such as attention mechanisms, layer-wise relevance propagation, and visualization of feature maps can help to elucidate the reasoning behind a model's predictions and build trust among clinicians. Future research should focus on incorporating explainability and interpretability into the design of deep learning models for head and neck cancer imaging, as well as on developing methods to assess the trustworthiness and robustness of these models in the face of noisy, incomplete, or adversarial data.

6.4. Clinical Implementation, Validation, and Impact Assessment

The translation of deep learning models from research to clinical practice requires rigorous validation and assessment of their impact on patient outcomes [42]. Large-scale, prospective studies would be useful to establish the performance, generalizability, and clinical utility of deep learning models for head and neck cancer imaging [25]. These studies should involve diverse patient populations and imaging data from multiple centers to ensure the robustness of the models in real-world settings. Moreover, the integration of these models into existing clinical workflows, the assessment of their cost-effectiveness, and the evaluation of their impact on patient care, such as the reduction in diagnostic errors, optimization of treatment planning, and improvement of patient outcomes, are essential steps toward their successful implementation [48]. Future research should also explore the development of user-friendly, scalable, and secure software tools and platforms that can facilitate the deployment of deep learning models in clinical settings and enable their widespread adoption in head and neck cancer management.

7. Challenges and Limitations of Deep Learning Models in Head and Neck Cancer Imaging

7.1. Data Quality and Quantity

The performance of deep learning models heavily relies on the quality and quantity of the training data [5]. Obtaining large, diverse, and well-annotated datasets for head and neck cancer imaging can be challenging due to factors such as privacy concerns, data sharing restrictions, and the time-consuming nature of manual annotation by experts [49]. Additionally, variations in imaging protocols, scanner types, and image acquisition parameters across different institutions can lead to inconsistencies in the data, affecting model performance [50]. Future research should focus on developing methods to leverage smaller or imperfect datasets, explore data augmentation techniques, and promote multi-institutional collaboration to enhance the performance of deep learning models.

7.2. Model Overfitting

Deep learning models are prone to overfitting, especially when trained on small datasets, leading to poor generalization to new data [6]. Overfitting can result in models that perform well on the training data but fail to accurately predict outcomes for new patients, limiting their clinical utility. Regularization techniques and model architectures that mitigate overfitting, such as dropout, batch normalization, and transfer learning, should be considered when developing deep learning models for head and neck cancer imaging [51]. Moreover, the use of cross-validation and external validation cohorts can help to evaluate and improve model generalizability.

7.3. Computation Requirements

The training and deployment of deep learning models often require substantial computational resources, such as graphics processing units (GPUs) and specialized hardware [5]. This need may pose a challenge for small clinical centers and researchers with limited access to high-performance computing facilities. Developing efficient model architectures, exploring strategies for model compression and acceleration, and utilizing cloud-based platforms for training and deployment can help to overcome these challenges and make deep learning models more accessible to a broader range of institutions and researchers.

7.4. Ethical Considerations

The use of deep learning models in medical imaging raises several ethical concerns, including those concerning data privacy, algorithmic fairness, and accountability [52]. Ensuring the protection of patient data is paramount, and researchers should adhere to data protection regulations and employ techniques such as data anonymization and secure data storage to safeguard patient privacy [53]. Moreover, biases in training data can lead to unfair treatment recommendations or misdiagnoses for certain patient populations, making it essential to address potential biases in the development and deployment of deep learning models. Establishing transparent and accountable practices, such as reporting model performance across diverse patient groups and involving stakeholders in the model development process can help to ensure ethical and responsible use of these technologies in head and neck cancer imaging.

8. Radiogenomics: A Promising Avenue for Deep Learning in Head and Neck Cancer Imaging

8.1. Radiogenomics: An Overview

Radiogenomics is an emerging interdisciplinary field that investigates the relationship between imaging features and genomic information, with the goal of developing predictive models that can guide personalized treatment decisions [10]. By integrating radiogenomic data with deep learning models, researchers and clinicians can gain a more comprehensive understanding of tumor biology, heterogeneity, and treatment response, ultimately leading to improved patient outcomes in head and neck cancer management.

8.2. Radiogenomic Features in Deep Learning Models

Several studies have explored the incorporation of radiogenomic features into deep learning models for head and neck cancer imaging [25,33]. Combining radiomic features extracted from MRI, CT, or PET scans with genomic data, such as gene expression profiles or mutational status, can improve the accuracy of tumor detection, segmentation, and outcome prediction [54]. Additionally, deep learning models can be designed to automatically learn radiogenomic representations from the imaging data, potentially revealing novel imaging-genomic associations that may have clinical implications [54]. This approach could lead to the identification of previously unknown biomarkers and therapeutic targets in head and neck cancers.

8.3. Challenges and Future Directions

Despite the promising potential of radiogenomics in deep learning for head and neck cancer imaging, several challenges need to be addressed. First, the acquisition of genomic data can be costly and invasive, limiting their availability in clinical settings [55]. Non-invasive methods, such as liquid biopsy, may offer a more accessible source of genomic information in the future [56]. Second, the integration of imaging and genomic data presents computational challenges since these data types have different scales and dimensions [10]. Developing novel deep learning architectures and data fusion techniques that can effectively combine and process multimodal data is a crucial area for future research. Additionally, addressing the issues of data heterogeneity, missing data, and standardization of imaging and genomic data will be essential for the successful application of radiogenomics in deep learning models [57].

8.4. Clinical Implementation and Validation

As with any novel approach in medical imaging, the translation of radiogenomic-based deep learning models to clinical practice requires rigorous validation and assessment of their impact on patient outcomes. Large-scale, multi-institutional studies would be useful to establish the performance, generalizability, and clinical utility of these models in head and neck cancer management [16]. Furthermore, the integration of radiogenomic-based models into existing clinical workflows and the assessment of their cost-effectiveness are essential steps toward their successful implementation. As radiogenomics continues to evolve, it is expected that deep learning models incorporating radiogenomic features will play an increasingly important role in guiding personalized treatment strategies for head and neck cancer patients.

9. Advanced Deep Learning Techniques in Head and Neck Cancer Imaging

9.1. Convolutional Autoencoders for Data Compression, Image Denoising, and Feature Extraction

Convolutional autoencoders (CAEs) are a type of deep learning model that can be used for unsupervised feature learning, image denoising, and data compression [58]. They consist of an encoder network, which reduces the input image to a compact representation, and a decoder network, which reconstructs the image from this representation. CAEs have been applied in various medical imaging tasks, including head and neck cancer imaging, for improving image quality and reducing noise, as well as extracting relevant features that can be used in downstream analysis [59,60].

9.2. Generative Adversarial Networks for Super-Resolution

Generative adversarial networks (GANs) are a class of deep learning models that consist of two competing networks: a generator and a discriminator [61]. The generator creates synthetic images, while the discriminator evaluates the quality of these images, aiming to distinguish between real and generated images. GANs have been used for various medical image enhancement tasks, such as super-resolution, in which the goal is to generate high-resolution images from low-resolution input data. In the context of head and neck cancer imaging, GANs can be utilized to improve image resolution, potentially leading to more accurate tumor detection and segmentation [62,63].

9.3. Transformer Models: Vision Transformer (ViT)

Transformer models, initially introduced for natural language processing tasks, have been adapted for image analysis with the development of the Vision Transformer (ViT) [64]. ViT divides an image into non-overlapping patches, linearly embeds them into a sequence of fixed-size vectors, and processes these vectors using the transformer architecture. This approach allows the model to capture long-range dependencies and spatial relationships between image regions. ViT has shown competitive performance in various medical imaging tasks, including head and neck cancer detection and segmentation, and it can be considered an alternative to traditional CNN-based methods [65,66].

10. Conclusions

Deep learning techniques have made significant strides in the field of head and neck cancer imaging, demonstrating exceptional performance in critical tasks, such as tumor detection, segmentation, and outcome prediction, using various imaging modalities, such as MRI, CT, and PET scans. Advanced deep learning techniques, including convolutional autoencoders, generative adversarial networks (GANs), and transformer models such as the Vision Transformer (ViT), have further enhanced the capabilities of these models, offering promising improvements in data compression, image denoising, super-resolution, and feature extraction. The incorporation of radiogenomics into deep learning models offers a promising avenue for further advancements, potentially enabling a more comprehensive understanding of tumor biology and heterogeneity and guiding personalized treatment strategies in head and neck cancer management.

Despite these encouraging results, a number of challenges must be addressed to facilitate the successful implementation of deep learning models in head and neck cancer diagnosis and treatment. One major obstacle is the lack of standardization in terms of datasets, preprocessing techniques, and model architectures, rendering it difficult to compare the performances of various models and assess their clinical utility. Additionally, acquiring large, diverse, and well-annotated datasets for head and neck cancer imaging poses a challenge due to privacy concerns, data sharing restrictions, and the labor-intensive nature of manual annotation by experts.

In light of the limitations of traditional imaging techniques, such as subjectivity and variability in human interpretation, the integration of deep learning and traditional imaging techniques could lead to more robust and accurate diagnosis and treatment planning. Recognizing the complementary roles of these approaches is essential for advancing head and neck cancer imaging and management.

Model overfitting, a common issue in deep learning, is another concern, particularly when models are trained on small datasets. This situation can lead to poor generalization from new data and reduced clinical utility. Moreover, the computational requirements for training and deploying deep learning models can be substantial, often necessitating specialized hardware and resources that may not be readily accessible to smaller clinical centers or researchers. Addressing these challenges, along with integrating multimodal and temporal information from different imaging modalities and timepoints, is crucial for further enhancing the performance of deep learning models. Developing explainable and interpretable models is also essential to bridge the gap between complex model architectures and clinical decision-making, fostering trust and acceptance among medical professionals.

Finally, conducting rigorous clinical validations through large-scale, prospective studies is necessary to establish the performance, generalizability, and clinical utility of deep learning models for head and neck cancer imaging. Overcoming these obstacles, while embracing the synergy between advanced deep learning techniques and traditional imaging approaches, will be instrumental in paving the way for improved patient outcomes and the realization of personalized treatment strategies in head and neck cancer management.

Author Contributions: Writing—review and editing, M.I. and D.G. All authors have read and agreed to the published version of the manuscript.

Funding: This research received no external funding.

Acknowledgments: This article was prepared in part using ChatGPT.

Conflicts of Interest: The authors declare no conflict of interest.

References

1. Chow, L.Q.M. Head and neck cancer. *N. Engl. J. Med.* **2020**, *382*, 60–72. [CrossRef] [PubMed]
2. Marur, S.; Forastiere, A.A. Head and neck squamous cell carcinoma: Update on epidemiology, diagnosis, and treatment. *Mayo Clin. Proc.* **2016**, *91*, 386–396. [CrossRef]
3. Chiesa-Estomba, C.M.; Lequerica-Fernández, P.; Fakhry, N. Imaging in head and neck cancer: United in diversity. *Cancers* **2019**, *11*, 1077.
4. Wang, J.; Wu, C.J.; Bao, M.L.; Zhang, J.; Wang, X.N.; Zhang, Y.D. Machine learning-based analysis of MR radiomics can help to improve the diagnostic performance of PI-RADS v2 in clinically relevant prostate cancer. *Eur. Radiol.* **2017**, *27*, 4082–4090. [CrossRef]

5. Litjens, G.; Kooi, T.; Bejnordi, B.E.; Setio, A.A.A.; Ciompi, F.; Ghafoorian, M.; van der Laak, J.A.W.M.; van Ginneken, B.; Sánchez, C.I. A survey on deep learning in medical image analysis. *Med. Image Anal.* **2017**, *42*, 60–88. [CrossRef] [PubMed]
6. Gillies, R.J.; Kinahan, P.E.; Hricak, H. Radiomics: Images are more than pictures, they are data. *Radiology* **2016**, *278*, 563–577. [CrossRef]
7. Pignon, J.P.; le Maître, A.; Maillard, E.; Bourhis, J. Meta-analysis of chemotherapy in head and neck cancer (MACH-NC): An update on 93 randomised trials and 17,346 patients. *Radiother. Oncol.* **2009**, *92*, 4–14. [CrossRef]
8. Alizadeh, M.; Safaei, A.; Dorri, G.; Zendehbad, S.A. Application of deep learning and machine learning methods to radiomics and texture analysis of medical images: A review. *J. Biomed. Phys. Eng.* **2018**, *8*, 405–424.
9. Petrick, N.; Sahiner, B.; Armato, I.I.I.S.G.; Bert, A.; Correale, L.; Delsanto, S.; Freedman, M.T.; Fryd, D.; Gur, D.; Hadjiiski, L.; et al. Evaluation of computer-aided detection and diagnosis systems. *Med. Phys.* **2013**, *40*, 087001. [CrossRef]
10. Lambin, P.; Rios-Velazquez, E.; Leijenaar, R.; Carvalho, S.; Van Stiphout, R.G.; Granton, P.; Zegers, C.M.; Gillies, R.; Boellard, R.; Dekker, A.; et al. Radiomics: Extracting more information from medical images using advanced feature analysis. *Eur. J. Cancer* **2012**, *48*, 441–446. [CrossRef]
11. Taha, A.A.; Hanbury, A. Metrics for evaluating 3D medical image segmentation: Analysis, selection, and tool. *BMC Med. Imaging* **2015**, *15*, 29. [CrossRef]
12. Krupinski, E.A. Current perspectives in medical image perception. *Atten. Percept. Psychophys.* **2010**, *72*, 1205–1217. [CrossRef]
13. Drew, T.; Vo ML, H.; Olwal, A.; Jacobson, F.; Seltzer, S.E.; Wolfe, J.M. Scanners and drillers: Characterizing expert visual search through volumetric images. *J. Vis.* **2013**, *13*, 3. [CrossRef] [PubMed]
14. Berbaum, K.S.; Franken, E.A.; Dorfman, D.D.; Miller, E.M.; Caldwell, R.T.; Kuehn, D.M.; Berbaum, M.L. Role of faulty deci-sion making in the satisfaction of search effect in chest radiography. *Acad. Radiol.* **2000**, *7*, 1098–1106. [CrossRef] [PubMed]
15. LeCun, Y.; Bengio, Y.; Hinton, G. Deep learning. *Nature* **2015**, *521*, 436–444. [CrossRef] [PubMed]
16. Shen, D.; Wu, G.; Suk, H.I. Deep learning in medical image analysis. *Annu. Rev. Biomed. Eng.* **2017**, *19*, 221–248. [CrossRef]
17. Esteva, A.; Kuprel, B.; Novoa, R.A.; Ko, J.; Swetter, S.M.; Blau, H.M.; Thrun, S. Dermatologist-level classification of skin cancer with deep neural networks. *Nature* **2017**, *542*, 115–118. [CrossRef]
18. Gulshan, V.; Peng, L.; Coram, M.; Stumpe, M.C.; Wu, D.; Narayanaswamy, A.; Venugopalan, S.; Widner, K.; Madams, T.; Cuadros, J.; et al. Development and validation of a deep learning algorithm for detection of diabetic retinopathy in retinal fundus photographs. *JAMA* **2016**, *316*, 2402–2410. [CrossRef]
19. Ciompi, F.; Chung, K.; Van Riel, S.J.; Setio, A.A.; Gerke, P.K.; Jacobs, C.; Scholten, E.T.; Schaefer-Prokop, C.; Wille, M.M.; Marchiano, A.; et al. Towards automatic pulmonary nodule management in lung cancer screening with deep learning. *Sci. Rep.* **2017**, *7*, 46479. [CrossRef]
20. Menze, B.H.; Jakab, A.; Bauer, S.; Kalpathy-Cramer, J.; Farahani, K.; Kirby, J.; Burren, Y.; Porz, N.; Slotboom, J.; Wiest, R.; et al. The multimodal brain tumor image segmentation benchmark (BRATS). *IEEE Trans. Med. Imaging* **2015**, *34*, 1993–2024. [CrossRef]
21. Lambin, P.; Leijenaar, R.T.H.; Deist, T.M.; Peerlings, J.; de Jong, E.E.C.; van Timmeren, J.; Sanduleanu, S.; Larue, R.T.H.M.; Even, A.J.G.; Jochems, A.; et al. Radiomics: The bridge between medical imaging and personalized medicine. *Nat. Rev. Clin. Oncol.* **2017**, *14*, 749–762. [CrossRef]
22. Czernin, J.; Allen-Auerbach, M.; Schelbert, H.R. Improvements in cancer staging with PET/CT: Literature-based evidence as of September 2006. *J. Nucl. Med.* **2007**, *48*, 78S–88S.
23. Ibragimov, B.; Xing, L. Segmentation of organs-at-risks in head and neck CT images using convolutional neural networks. *Med. Phys.* **2017**, *44*, 547–557. [CrossRef]
24. Zhong, Z.; Kim, Y.; Zhou, L.; Plichta, K. Deep-learning based, automated segmentation of head and neck cancer in PET/CT images. In Proceedings of the 2018 IEEE 15th International Symposium on Biomedical Imaging (ISBI 2018), Washington, DC, USA, 4–7 April 2018; pp. 177–181.
25. Men, K.; Dai, J.; Li, Y. Automatic segmentation of the clinical target volume and organs at risk in the planning CT for rectal cancer using deep dilated convolutional neural networks. *Med. Phys.* **2017**, *44*, 6377–6389. [CrossRef]
26. Ronneberger, O.; Fischer, P.; Brox, T. U-Net: Convolutional networks for biomedical image segmentation. In *International Conference on Medical Image Computing and Computer-Assisted Intervention*; Springer: Cham, Switzerland, 2015; pp. 234–241.
27. Lao, J.; Chen, Y.; Li, Z.C.; Li, Q.; Zhang, J. A deep learning-based radiomics model for prediction of survival in glioblastoma multiforme. *Sci. Rep.* **2017**, *7*, 10353. [CrossRef]
28. Wang, H.; Zhou, Z.; Li, Y.; Chen, Z.; Lu, P.; Wang, W.; Liu, W.; Yu, L. Comparison of machine learning methods for classify-ing mediastinal lymph node metastasis of non-small cell lung cancer from 18F-FDG PET/CT images. *EJNMMI Res.* **2019**, *9*, 15.
29. Hötker, A.M.; Tarlinton, L.; Mazaheri, Y.; Woo, K.M.; Gönen, M.; Saltz, L.B.; Goodman, K.A.; Garcia-Aguilar, J.; Gollub, M.J. Multiparametric MRI in the assessment of response of rectal cancer to neoadjuvant chemoradiotherapy: A comparison of morphological, volumetric and functional MRI parameters. *Eur. Radiol.* **2016**, *26*, 4303–4312. [CrossRef] [PubMed]
30. Som, P.M.; Curtin, H.D. *Head and Neck Imaging*, 4th ed.; Mosby: St. Louis, MO, USA, 2003.
31. Zhu, W.; Huang, Y.; Zeng, L.; Chen, X.; Liu, Y.; Qian, Z.; Du, N.; Fan, W.; Xie, X. AnatomyNet: Deep learning for fast and fully automated whole-volume segmentation of head and neck anatomy. *Med. Phys.* **2019**, *46*, 576–589. [CrossRef] [PubMed]
32. Haarburger, C.; Valentinitsch, A.; Rienmüller, A.; Kuder, T.A.; Bickelhaupt, S. Automated detection and segmentation of head and neck carcinomas using a 3D U-Net on planning CT scans. *Eur. Radiol. Exp.* **2020**, *4*, 1–10.

33. Wu, J.; Gensheimer, M.F.; Dong, X.; Rubin, D.L.; Napel, S.; Diehn, M. Robust intratumor partitioning to identify high-risk subregions in lung cancer: A pilot study. *Int. J. Radiat. Oncol. Biol. Phys.* **2017**, *98*, 647–656. [CrossRef]
34. Tajbakhsh, N.; Shin, J.Y.; Gurudu, S.R.; Hurst, R.T.; Kendall, C.B.; Gotway, M.B.; Liang, J. Convolutional neural networks for medical image analysis: Full training or fine tuning? *IEEE Trans. Med. Imaging* **2016**, *35*, 1299–1312. [CrossRef] [PubMed]
35. Lu, L.; Ehmke, R.C.; Schwartz, L.H.; Zhao, B. Assessing agreement between radiomic features computed for multiple CT imaging settings. *PloS ONE* **2015**, *10*, e0143848. [CrossRef] [PubMed]
36. Kearney, V.; Chan, J.W.; Haaf, S.; Descovich, M.; Solberg, T.D. DoseNet: A volumetric dose prediction algorithm using 3D fully convolutional networks. *Med. Phys.* **2018**, *45*, 6343–6353. [CrossRef] [PubMed]
37. Adams, M.C.; Turkington, T.G.; Wilson, J.M.; Wong, T.Z. A systematic review of the factors affecting accuracy of SUV measurements. *Am. J. Roentgenol.* **2010**, *195*, 310–320. [CrossRef] [PubMed]
38. Liu, S.; Liu, S.; Cai, W.; Che, H.; Pujol, S.; Kikinis, R.; Feng, D.; Fulham, M.J. Multimodal neuroimaging feature learning for multiclass diagnosis of Alzheimer's disease. *IEEE Trans. Biomed. Eng.* **2015**, *62*, 1132–1140. [CrossRef]
39. Eiber, M.; Weirich, G.; Holzapfel, K.; Souvatzoglou, M.; Haller, B.; Rauscher, I.; Beer, A.J.; Wester, H.J.; Gschwend, J.; Schwaiger, M.; et al. Simultaneous 68Ga-PSMA HBED-CC PET/MRI Improves the Localization of Primary Prostate Cancer. *Eur. Urol.* **2016**, *70*, 829–836. [CrossRef]
40. Zhou, M.; Scott, J.; Chaudhury, B.; Hall, L.; Goldgof, D.; Yeom, K.W.; Iv, M.; Ou, Y.; Kalpathy-Cramer, J.; Napel, S.; et al. Radiomics in Brain Tumor: Image Assessment, Quantitative Feature Descriptors, and Machine-Learning Approaches. *Am. J. Neuroradiol.* **2018**, *39*, 208–216. [CrossRef]
41. Vallières, M.; Freeman, C.R.; Skamene, S.R.; El Naqa, I. A radiomics model from joint FDG-PET and MRI texture features for the prediction of lung metastases in soft-tissue sarcomas of the extremities. *Phys. Med. Biol.* **2015**, *60*, 5471–5496. [CrossRef]
42. Vallières, M.; Kay-Rivest, E.; Perrin, L.J.; Liem, X.; Furstoss, C.; Aerts, H.J.; Khaouam, N.; Nguyen-Tan, P.F.; Wang, C.S.; Sultanem, K.; et al. Radiomics strategies for risk assessment of tumour failure in head-and-neck cancer. *Sci. Rep.* **2017**, *7*, 10117. [CrossRef]
43. Bankier, A.A.; Levine, D.; Halpern, E.F.; Kressel, H.Y. Consensus interpretation in imaging research: Is there a better way? *Radiology* **2010**, *257*, 14–17. [CrossRef]
44. Fortin, J.P.; Parker, D.; Tunç, B.; Watanabe, T.; Elliott, M.A.; Ruparel, K.; Roalf, D.R.; Satterthwaite, T.D.; Gur, R.C.; Gur, R.E.; et al. Harmonization of multi-site diffusion tensor imaging data. *Neuroimage* **2017**, *161*, 149–170. [CrossRef]
45. Men, K.; Chen, X.; Zhang, Y.; Zhang, T.; Dai, J.; Yi, J.; Li, Y. Deep deconvolutional neural network for target segmentation of nasopharyngeal cancer in planning computed tomography images. *Front. Oncol.* **2019**, *9*, 1200. [CrossRef] [PubMed]
46. Guidotti, R.; Monreale, A.; Ruggieri, S.; Turini, F.; Giannotti, F.; Pedreschi, D. A survey of methods for explaining black box models. *ACM Comput. Surv. (CSUR)* **2018**, *51*, 1–42. [CrossRef]
47. Castelvecchi, D. Can we open the black box of AI? *Nat. News* **2016**, *538*, 20–23. [CrossRef]
48. Jha, S.; Topol, E.J. Adapting to artificial intelligence: Radiologists and pathologists as information specialists. *JAMA* **2016**, *316*, 2353–2354. [CrossRef]
49. Topol, E.J. High-performance medicine: The convergence of human and artificial intelligence. *Nat. Med.* **2019**, *25*, 44–56. [CrossRef] [PubMed]
50. Zwanenburg, A.; Vallières, M.; Abdalah, M.A.; Aerts, H.J.W.L.; Andrearczyk, V.; Apte, A.; Ashrafinia, S.; Bakas, S.; Beukinga, R.J.; Boellaard, R.; et al. The image biomarker standardization initiative: Standardized quantitative radiomics for high-throughput image-based phenotyping. *Radiology* **2020**, *295*, 328–338. [CrossRef]
51. Srivastava, N.; Hinton, G.; Krizhevsky, A.; Sutskever, I.; Salakhutdinov, R. Dropout: A simple way to prevent neural networks from overfitting. *J. Mach. Learn. Res.* **2014**, *15*, 1929–1958.
52. Char, D.S.; Shah, N.H.; Magnus, D. Implementing machine learning in health care—addressing ethical challenges. *N. Engl. J. Med.* **2018**, *378*, 981–983. [CrossRef]
53. Rodrigues, J.J.; de la Torre, I.; Fernández, G.; López-Coronado, M. Analysis of the security and privacy requirements of cloud-based electronic health records systems. *J. Med. Internet Res.* **2013**, *15*, e186. [CrossRef]
54. Aerts, H.J.; Velazquez, E.R.; Leijenaar, R.T.; Parmar, C.; Grossmann, P.; Carvalho, S.; Bussink, J.; Monshouwer, R.; Haibe-Kains, B.; Rietveld, D.; et al. Decoding tumour phenotype by noninvasive imaging using a quantitative radiomics approach. *Nat. Commun.* **2014**, *5*, 4006. [CrossRef]
55. Leek, J.T.; Scharpf, R.B.; Bravo, H.C.; Simcha, D.; Langmead, B.; Johnson, W.E.; Geman, D.; Baggerly, K.; Irizarry, R.A. Tackling the widespread and critical impact of batch effects in high-throughput data. *Nat. Rev. Genet.* **2010**, *11*, 733–739. [CrossRef]
56. Siravegna, G.; Marsoni, S.; Siena, S.; Bardelli, A. Integrating liquid biopsies into the management of cancer. *Nat. Rev. Clin. Oncol.* **2017**, *14*, 531–548. [CrossRef]
57. Sollini, M.; Antunovic, L.; Chiti, A.; Kirienko, M. Towards clinical application of image mining: A systematic review on artificial intelligence and radiomics. *Eur. J. Nucl. Med. Mol. Imaging* **2018**, *45*, 2076–2089. [CrossRef]
58. Gu, J.; Wang, Z.; Kuen, J.; Ma, L.; Shahroudy, A.; Shuai, B.; Liu, T.; Wang, X.; Wang, G.; Cai, J.; et al. Recent advances in convolutional neural networks. In Proceedings of the 2015 8th International Congress on Image and Signal Processing (CISP), Shenyang, China, 14–16 October 2015; pp. 393–398.
59. Chen, H.; Zhang, Y.; Zhang, W.; Liao, P.; Li, K.; Zhou, J.; Wang, G. Low-dose CT via convolutional neural network. *Biomed. Opt. Express* **2018**, *8*, 679–694. [CrossRef] [PubMed]

60. Zhou, Z.; Siddiquee MM, R.; Tajbakhsh, N.; Liang, J. UNet++: A Nested U-Net Architecture for Medical Image Segmentation. In *Deep Learning in Medical Image Analysis and Multimodal Learning for Clinical Decision Support*; Springer: Cham, Switzerland, 2020; pp. 3–11.
61. Goodfellow, I.; Pouget-Abadie, J.; Mirza, M.; Xu, B.; Warde-Farley, D.; Ozair, S.; Courville, A.; Bengio, Y. Generative adversarial nets. In *Advances in Neural Information Processing Systems*; MIT Press: Cambridge, MA, USA, 2014; pp. 2672–2680.
62. Wolterink, J.M.; Leiner, T.; Viergever, M.A.; Isgum, I. Generative adversarial networks for noise reduction in low-dose CT. *IEEE Trans. Med. Imaging* **2017**, *36*, 2536–2545. [CrossRef] [PubMed]
63. Son, J.; Park, S.J.; Jung, K.H. Retinal vessel segmentation in fundoscopic images with generative adversarial networks. *arXiv* **2018**, arXiv:1801.00863.
64. Dosovitskiy, A.; Beyer, L.; Kolesnikov, A.; Weissenborn, D.; Zhai, X.; Unterthiner, T.; Dehghani, M.; Minderer, M.; Heigold, G.; Gelly, S.; et al. An image is worth 16x16 words: Transformers for image recognition at scale. *arXiv* **2020**, arXiv:2010.11929.
65. Chen, C.; Qin, C.; Qiu, H.; Tarroni, G.; Duan, J.; Bai, W.; Rueckert, D. Iterative residual refinement for joint optic-disc and cup segmentation with transformers. In Proceedings of the International Conference on Medical Image Computing and Computer-Assisted Intervention, Strasbourg, France, 27 September–1 October 2021; Springer: Cham, Switzerland, 2021; pp. 118–127.
66. Sharma, A.; Gupta, A.; Kumar, A.; Ray, A.; Sharma, S. Transformer-based deep learning model for multi-class segmentation of head and neck cancer. In Proceedings of the International Conference on Image Analysis and Recognition, Lecce, Italy, 23–27 May 2022; Springer: Cham, Switzerland, 2022; pp. 202–214.

sensors

MDPI

Article

Accurate Nonlinearity and Temperature Compensation Method for Piezoresistive Pressure Sensors Based on Data Generation

Mingxuan Zou [1], Ye Xu [2], Jianxiang Jin [1], Min Chu [1] and Wenjun Huang [1,*]

1 College of Control Science and Engineering, Zhejiang University, Hangzhou 310027, China
2 China Petroleum & Chemical Corporation, Beijing 100728, China
* Correspondence: wjhuang@zju.edu.cn

Abstract: Piezoresistive pressure sensors exhibit inherent nonlinearity and sensitivity to ambient temperature, requiring multidimensional compensation to achieve accurate measurements. However, recent studies on software compensation mainly focused on developing advanced and intricate algorithms while neglecting the importance of calibration data and the limitation of computing resources. This paper aims to present a novel compensation method which generates more data by learning the calibration process of pressure sensors and uses a larger dataset instead of more complex models to improve the compensation effect. This method is performed by the proposed aquila optimizer optimized mixed polynomial kernel extreme learning machine (AO-MPKELM) algorithm. We conducted a detailed calibration experiment to assess the quality of the generated data and evaluate the performance of the proposed method through ablation analysis. The results demonstrate a high level of consistency between the generated and real data, with a maximum voltage deviation of only 0.71 millivolts. When using a bilinear interpolation algorithm for compensation, extra generated data can help reduce measurement errors by 78.95%, ultimately achieving 0.03% full-scale (FS) accuracy. These findings prove the proposed method is valid for high-accuracy measurements and has superior engineering applicability.

Keywords: piezoresistive pressure sensor; multidimensional compensation; data generation; mixed polynomial kernel extreme learning machine; aquila optimization; high-accuracy measurement

1. Introduction

Piezoresistive pressure sensors based on micro-electro-mechanical systems (MEMSs) have been widely used in industrial control [1,2], health monitoring [3,4], automotive applications [5,6], and wearable equipment [7,8]. Diversified application scenarios and working conditions create lots of challenges for pressure sensors. Especially in many industrial fields such as oil refining and chemical production, pressure sensors are usually required to achieve high-accuracy (typically 0.05% FS) and high-stability measurement over a wide range of temperatures (typically from −40 to 85 degrees Celsius). Unfortunately, the inherent nonlinearity of the pressure sensor seems to be unavoidable. Material properties, structural design, manufacturing process, and process bias all contribute to this result. In addition, changes in the measurement environment (mainly temperature) exacerbate the output response error and significantly reduce the measurement accuracy. Therefore, sensor compensation considering both nonlinearity and temperature is essential to obtain high-accuracy measurement results. Previous studies have proposed many compensation methods, which generally can be divided into the hardware self-compensation approach and software algorithm compensation approach.

Hardware self-compensation approaches mitigate temperature drift and nonlinearity error primarily through built-in analog circuits [9–11], which are easy to implement and offer good real-time performance. But the built-in circuits are also susceptible to temperature changes, which limits the sensor accuracy. Particularly in some industrial applications requiring a wide temperature range, a 1% full-scale error (FSE) level will still remain after compensation [12]. In conclusion, the difference between the multiple sensor responses and the corresponding compensation hardware characteristic makes it difficult for the hardware compensation approach to attain high-accuracy measurement in batch production.

Software algorithm compensation approaches show strong robustness and demand less manufacturing technology; therefore, they have numerous applications in practical engineering. In essence, this approach is a post-compensation method. First, the pressure sensor needs to be placed in different temperature environments, applying different pressures and recording the response voltage of the sensor. These data about the input–output response relationship of the sensor are called sensor calibration data. Then, based on the calibration dataset, additional compensation algorithms are used to correct the sensor output, such as various numerical calculation methods: look-up table algorithm [13], polynomial interpolation [14], and the hybrid compensation method [15]. Over the past few decades, researchers have concentrated on developing more efficient compensation algorithms, including some machine learning techniques [16–18], such as artificial neural networks, support vector machines (SVMs), and extreme learning machines (ELMs). They also used meta-heuristic algorithms [19,20] and ensemble learning methods [21] for temperature and nonlinearity compensation.

However, existing research on software compensation has some problems, for example, ignoring the importance of calibration data and the limitations of computational resources. Calibration data are the basis for sensor compensation and increasing the amount of these data can easily improve the compensation effect. Due to the time-consuming and energy-consuming calibration process, it is not advisable to increase the amount of data by adding testing temperature points and pressure points, as it would bring more costs and reduce the production efficiency of the sensor. In the production process, one must make a trade-off between calibration costs and target accuracy, while selecting the appropriate compensation algorithm as well. Moreover, compensation algorithms are always executed by a microcontroller unit (MCU) with limited computing resources. Compared with traditional numerical fitting methods, machine learning, and intelligent optimization algorithms improve compensation accuracy and induce higher computational complexity at the same time. It forces manufacturers to endure more input and output delay times (even being unable to run) or choose more expensive microcontrollers [22,23].

In this study, we present a novel compensation method to solve the above problems. The feature of this method is that it uses the proposed aquila optimizer optimized mixed polynomial kernel extreme learning machine algorithm to learn the calibration process and generate more data. Then, it uses the bilinear interpolation algorithm for sensor compensation in the dataset containing real calibration data and extra generated data to improve the measurement accuracy. The proposed method can effectively realize high-accuracy measurements in a wide temperature range with a small amount of calibration experiment time. Notably, the training and prediction of the AO-MPKELM model described in this method are implemented by an industrial personal computer. Only the bilinear interpolation is executed by the MCU; thus, the present method has lower complexity and stronger engineering practicability than those studies that use the machine learning method as the compensation algorithm. This study is the first to propose that compensation accuracy can be improved by generating more calibration data (without increasing calibration time), which is highly innovative.

The remainder of this paper is organized as follows. In Section 2, the input–output response of the pressure sensor and calibration compensation processes are theoretically analyzed, and the specific implementation steps of the presented compensation method are described. Section 3 provides the details of the proposed calibration data generation algorithm AO-MPKELM. Section 4 analyzes and discusses the results of the calibration and compensation experiments, and the conclusions are presented in Section 5.

2. Principle and Method of Pressure Sensor Compensation

2.1. Pressure Sensor Response

The piezoresistive pressure sensor is a device that utilizes the piezoresistive effect of silicon. Figure 1 shows the measurement principle of the piezoresistive pressure sensor; it uses four silicon-doped varistors to form a Wheatstone full bridge to convert the input pressure signal to the output electrical signal. The applied pressure causes the varistor to deform, resulting in a change in resistance value and simultaneously affecting the output voltage. In addition, the ambient temperature and the usage time can also affect the sensor characteristics. Silicon is a temperature-sensitive semiconductor, which makes the pressure sensor present different characteristics at different temperatures. If an input signal p is given to a sensor s, its output voltage response u_s can be expressed as follows:

$$u_s = r_s(p, T, t) + \Delta H_s + n_s, \tag{1}$$

where $r_s(p, T, t)$ denotes the response function of the pressure sensor, p is the applied real pressure, T and t represent current ambient temperature and sensor usage time, respectively, ΔH_s is the uncertain hysteresis caused by the mechanical parts and structure materials, and n_s represents the system noise. In this study, the current ambient temperature was measured using an additional temperature sensor.

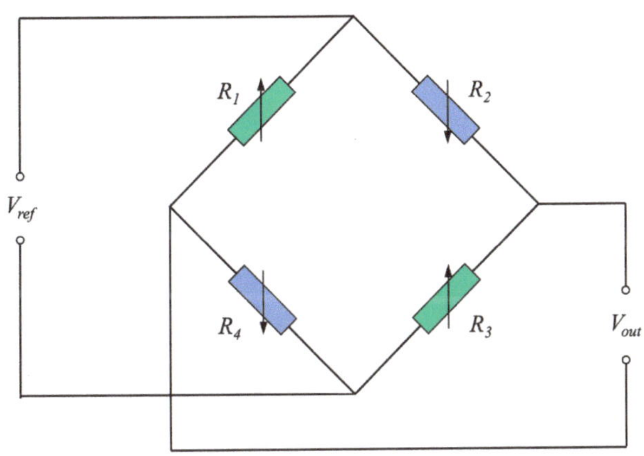

Figure 1. Measurement principle of the piezoresistive pressure sensor.

With the development of manufacturing technology, various system errors and time changes (typically in the scale of years) have less impact on sensor performance; therefore, (1) can be simplified into the following form:

$$u_s = r_s(p, T). \tag{2}$$

2.2. Calibration and Compensation

In fact, the calibration process of the sensor is the sensor response function (2). The input–output response relationship can be obtained by placing the sensor at different pressures and temperatures. In this process flow, sensor characteristics, measurement range, working temperature zone, and target accuracy should be considered simultaneously when determining the specific number and value of pressure and temperature points. Additionally, the cumulative errors of multiple processes in production will lead to differentiation between the nonlinear characteristics and the piezoresistive effect [22] of each sensor, even for sensors produced in the same batch. So, calibrating each pressure sensor itself is necessary to achieve high-accuracy measurement; that is, each sensor must use its own calibration experiment data for compensation.

Taking sensor s as an example, assuming that n_p pressure points and n_T temperature points need to be tested, the acquired calibration dataset includes the input–output response of $n_p \times n_T$ groups. This calibration experiment dataset can be mathematically expressed by

$$S_C = \{(p_i, T_j, u_{ij})\},\ i = 1, \cdots, n_p,\ j = 1, \cdots, n_T. \tag{3}$$

After obtaining the reference calibration dataset, the voltage response and the temperature value obtained by the additional temperature sensor can be used to compensate for measurement errors. This process can be expressed as follows:

$$\hat{p} = \sigma(u,\ T), \tag{4}$$

where $\hat{p}$ is the compensated pressure (i.e., the measurement result), and σ is called the compensation function [12]. As described in Section 1, many algorithms have been used to fit this function accurately. But the data are considerably more important than the learning model. Thus, more laboratory calibration data are needed to gain more accurate measurement results.

Finally, p_f represents the full-scale range of the pressure sensor. The full-scale error of the sensor measurement results can be expressed as:

$$FSE = \frac{\hat{p} - p}{p_f} \times 100\%. \tag{5}$$

2.3. Implementation Steps of the Proposed Method

Figure 2 shows the framework of the proposed compensation method based on data generation, which includes the following four steps:

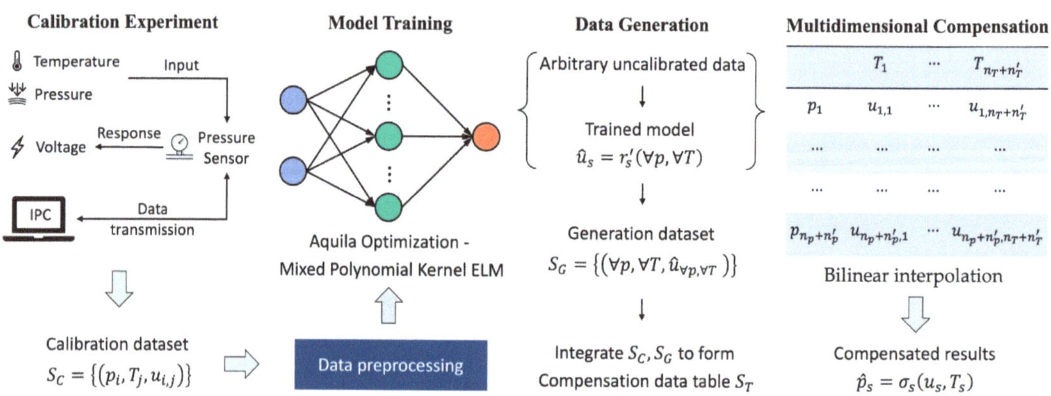

Figure 2. Framework of proposed compensation method based on data generation.

Step 1: Calibration experiment. In this experiment, we need to apply different pressures and temperatures to the sensor, obtain multiple sets of the input–output response relationships, and then transmit the real calibration dataset to an industrial personal computer for data generation.

Step 2: Model training. This step includes data preprocessing, learning the calibration experiment process using the proposed MPKELM algorithm, and the hyperparameter optimization of the learning algorithm using the AO. The detailed theoretical process will be presented in Section 3.

Step 3: Data generation. In the previous stage, we will have learned the calibration process of the sensor. The output response of pressure sensors can be predicted by inputting uncalibrated points at arbitrary temperature and pressure into the trained model. The above process is repeated to obtain the generation dataset S_G. Finally, S_C and S_G are integrated to form a compensation data table S_T, and the values are sorted from small to large. There are now $n_p + n'_p$ pressure points and $n_T + n'_T$ temperature points, where n'_p and n'_T are the number of generated pressure and temperature points, respectively.

Step 4: Multidimensional compensation. The compensation data table is sent to the MCU, and a bilinear interpolation algorithm is used to compensate for measurement errors.

3. Data Generation Algorithm

3.1. Extreme Learning Machine

The ELM [24] is a single hidden layer feedforward neural network (SLFN) learning algorithm. By randomly generating input weights and bias, the output weights of the ELM have a unique least square solution, which can be determined using the Moore–Penrose generalized inverse matrix. Consequently, the training speed of the ELM is significantly faster than traditional SLFNs with a gradient-based learning algorithm. Furthermore, it still retains the powerful nonlinear approximation ability of SLFNs [24–26] and performs well in regression and classification problems.

The structure of the ELM is the same as that of traditional SLFNs. Suppose the training set has N arbitrary samples (x_i, y_i), where $x_i = [x_{i1}, x_{i2}, \cdots, x_{in}]^{\mathrm{T}} \in R^n$ denotes the inputs, and $y_i = [y_{i1}, y_{i2}, \cdots, y_{im}]^{\mathrm{T}} \in R^m$ denotes the outputs. The weight w and bias b of each hidden layer node have been randomly generated; therefore, the ELM algorithm only needs to set the number of hidden layer nodes L to calculate the output weight β. Then, the estimated output $\hat{y}_i$ of the ELM with activation function $g(x)$ can be expressed by

$$\hat{y}_i = \sum_{j=1}^{L} \beta_j g_j(x_i) = \sum_{j=1}^{L} \beta_j g\left(w_j^{\mathrm{T}} \cdot x_i + b_j\right), i = 1, \cdots, N. \tag{6}$$

This equation can be written in matrix form as follows:

$$\hat{Y} = G\beta. \tag{7}$$

$$G = \begin{bmatrix} g(w_1^T \cdot x_1 + b_1) & \cdots & g(w_L^T \cdot x_1 + b_L) \\ \vdots & \ddots & \vdots \\ g(w_1^T \cdot x_N + b_1) & \cdots & g(w_L^T \cdot x_N + b_L) \end{bmatrix}_{N \times L}. \tag{8}$$

$$\beta = G^+ Y = \begin{bmatrix} \beta_1 \\ \vdots \\ \beta_L \end{bmatrix}_{L \times m}, \hat{Y} = \begin{bmatrix} y_1 \\ \vdots \\ y_N \end{bmatrix}_{N \times m}. \tag{9}$$

where G is the hidden layer output matrix, G^+ is the Moore–Penrose generalized inverse matrix of G, β is the unique solution of the output weight calculated by the minimum norm least square method, Y is the expected output, and $\hat{Y}$ is the estimated output.

In [27,28], the orthogonal projection method is applied to solve the Moore–Penrose generalized inverse matrix. Regularization coefficient γ and identity matrix I are introduced to enhance the stability of the ELM. Thus, the output weight β can be expressed as

$$\beta = G^+ Y = G^T \left(\gamma I + GG^T\right)^{-1} Y. \tag{10}$$

For any sample x, the output function of the ELM is

$$f_{ELM}(x) = g(x)\beta. \tag{11}$$

3.2. Mixed Polynomial Kernel ELM

The kernel ELM is an extension of the ELM [28] that further improves its generalization ability using a Mercer kernel, such as the Gaussian kernel. If we define the kernel matrix $\Omega = GG^T$, with $\Omega_{i,j} = g(x_i) \cdot g(x_j)^T = \kappa(x_i, x_j)$, then (11) can be written as follows:

$$
\begin{aligned}
f_{KELM}(x) &= g(x)\beta = g(x)G^T\left(\gamma I + GG^T\right)^{-1}Y \\
&= \begin{bmatrix} \kappa(x,\, x_1) \\ \vdots \\ \kappa(x, x_N) \end{bmatrix}^T (\gamma I + \Omega)^{-1}Y.
\end{aligned}
\tag{12}
$$

Based on the analysis of the nonlinear output response of the pressure sensor, a mixed polynomial kernel function is designed to accurately simulate the calibration process of the sensor in this study. The mixed polynomial kernel function $\kappa(x_i, x_j)$ can be defined as follows:

$$
\kappa(x_i, x_j) = \sum_{d=2}^{D} \theta_d \left(x_i^T x_j + 1\right)^d, \tag{13}
$$

where θ_d is the weight of the d_{th} degree polynomial kernel, and D is the highest order of the mixed polynomial kernel function, which needs to be set artificially according to the specific sensor type.

As mentioned above, the output nonlinearity of the sensor is not only affected by the variation of the semiconductor piezoresistive coefficient at different temperatures but also by many other factors in the manufacturing process, which may lead to different output characteristics, even in the same batch. Hence, optimizing the hyperparameters of the proposed MPKELM (including the weights θ_d of each kernel function and regularization coefficient γ) is necessary to improve the robustness of the algorithm.

3.3. *Parameter Optimization Using Aquila Optimizer*

The aquila is a large bird of prey that is active in the northern hemisphere; its agile flight and skillful methods to catch prey make it a top hunter. By observing and imitating the predation process of the aquila, Abualigah et al. [29] proposed the AO algorithm in 2021. Compared with other meta-heuristic algorithms, the AO exhibits considerable superiority in practical engineering problems [29–32]. The modeling process for the parameter optimization of MPKELM using the AO algorithm is as follows.

3.3.1. Solution Initialization

The AO is a swarm intelligence optimization algorithm. Assume a swarm with population size P, and the problem to be optimized is Q-dimensional; the possible solutions θ of the mixed polynomial kernel parameters to be optimized can be expressed as

$$
\theta = \begin{bmatrix}
\theta_{1,1} & \cdots & \theta_{1,q} & \cdots & \theta_{1,Q} \\
\vdots & \ddots & \vdots & \ddots & \vdots \\
\theta_{p,1} & \cdots & \theta_{p,q} & \cdots & \theta_{p,Q} \\
\vdots & \ddots & \vdots & \ddots & \vdots \\
\theta_{P,1} & \cdots & \theta_{P,q} & \cdots & \theta_{P,Q}
\end{bmatrix}, \tag{14}
$$

with

$$
\theta_{pq} = rand \times \left(UB_q - LB_q\right) + LB_q, p = 1, \cdots, P, q = 1, \cdots, Q, \tag{15}
$$

where $rand$ is a random number between 0 and 1, and UB_q and LB_q indicate the q_{th} upper and lower bounds of the parameters to be optimized, respectively.

3.3.2. Global Exploration (θ_1)

When $\tau \leq \frac{2}{3}T$, the AO algorithm starts the exploration phase, where τ and T denote the current and maximum iteration numbers, respectively. In the first strategy, global exploration, the aquila bird seeks the best prey areas at high altitudes by taking off vertically. The mathematical model of this behavior can be expressed as follows:

$$
\theta_1(\tau+1) = \theta_b(\tau) \times \left(1 - \frac{\tau}{\mathrm{T}}\right) + (\theta_M(\tau) - \theta_b(\tau) * rand), \tag{16}
$$

with

$$\theta_M(\tau) = \frac{1}{P}\sum_{p=1}^{P}\theta_p(\tau),\ \forall q = 1,\cdots,Q, \tag{17}$$

where $\theta_1(\tau+1)$ is the solution for the next iteration of τ produced by θ_1, $\theta_b(\tau)$ denotes the best solution until the τth iteration, and $\theta_M(\tau)$ represents the average solution of P populations at the τth iteration.

3.3.3. Local Exploration (θ_2)

In the second strategy, local exploration, when the aquila bird has identified a prey area, it circles the target prey, looking for the best time to land and prepare to attack. This behavior is mathematically expressed as follows:

$$\theta_2(\tau+1) = \theta_b(\tau) \times Levy(D) + \theta_R(\tau) + (n-m) * rand, \tag{18}$$

where $\theta_2(\tau+1)$ is the solution for the next iteration of τ produced by θ_2, $Levy(D)$ is the levy flight distribution function, which is expressed by (19), and $\theta_R(\tau)$ represents a solution that is randomly selected from P populations at the τth iteration.

$$Levy(D) = \zeta \times \frac{u \times \varepsilon}{|v|^{\frac{1}{\lambda}}}, \tag{19}$$

where ζ is a fixed constant 0.01, u and v are random numbers between 0 and 1, and ε is expressed as follows:

$$\varepsilon = \left(\frac{\Gamma(1+\lambda) \times \sin\left(\frac{\pi\lambda}{2}\right)}{\Gamma\left(\frac{1+\lambda}{2}\right) \times \lambda \times 2^{\left(\frac{\lambda-1}{2}\right)}}\right), \tag{20}$$

where λ is a fixed constant 1.5. m and n in (18) can be calculated using the following formulas:

$$m = r \times \sin(\varphi) \tag{21}$$

$$n = r \times \cos(\varphi), \tag{22}$$

with

$$r = r_1 + U \times Q_1 \tag{23}$$

$$\varphi = -\omega \times Q_1 + \frac{3 \times \pi}{2}, \tag{24}$$

where r_1 is a random number between 1 and 20, U is a fixed constant 0.00565, Q_1 represents an integer number between 1 and the optimization problem dimension Q, and ω is a constant fixed at 0.005.

3.3.4. Global Exploitation (θ_3)

When $\tau > \frac{2}{3}T$, the AO algorithm starts the exploitation phase. In the third strategy, global exploitation, when the aquila bird has identified a prey area and is ready to land, it descends vertically and launches a tentative attack to see how the prey responds. This behavior can be mathematically expressed as follows:

$$\theta_3(\tau+1) = (\theta_b(\tau) - \theta_M(\tau)) \times \alpha - rand + ((UB - LB) \times rand + LB) \times \delta, \tag{25}$$

where $\theta_3(\tau+1)$ is the solution for the next iteration of τ produced by θ_3, and α and δ are both adjustment parameters that are equal to 0.1.

3.3.5. Local Exploitation (θ_4)

In the fourth strategy, when the aquila bird approaches its prey, it will attack based on the stochastic movement of the prey. The mathematical model of this behavior can be expressed as follows:

$$\theta_4(\tau+1) = QF \times \theta_b(\tau) - F_1 \times \theta(t) \times rand - F_2 \times Levy(D) + rand \times F_1, \quad (26)$$

with

$$QF(t) = \tau^{\frac{2\times rand-1}{(1-T)^2}} \quad (27)$$

$$F_1 = 2 \times rand - 1 \quad (28)$$

$$F_2 = 2 \times \left(1 - \frac{\tau}{T}\right), \quad (29)$$

where $\theta_4(\tau+1)$ is the solution for the next iteration of τ produced by θ_4, QF is a tuning function used to balance different iterative search strategies, which is calculated by (27), F_1 represents the attack trajectory of the aquila based on the stochastic movement of the prey, and F_2 denotes the flight slope during the AO algorithm, which decreases from 2 to 0 as the number of iterations increases.

3.3.6. Performance Measurement

In the parameter optimization process of MPKELM, we use the root mean square error (RMSE) for performance measurement. The fitness function can be expressed as

$$E = \sqrt{\frac{1}{m}\sum_{i=1}^{m}(\hat{y}_i - y_i)^2}, \quad (30)$$

where $\hat{y}_i$ is the predicted value of the MPKELM, y_i is the true value, and m represents the number of samples contained in the training set.

3.3.7. Parameter Optimization Flow

Figure 3 shows the optimization flow of the MPKELM using the AO algorithm, and the specific process can be summarized in the following four steps:

Step 1: (15) is used to initialize the candidate solutions θ of the mixed polynomial kernel function to start the improvement procedure, and the relevant parameters are set in the AO algorithm, such as the maximum number of iterations T and the adjustment parameters α and δ.

Step 2: The current population fitness value is calculated using fitness function (30). Then, the best solution and average solution are selected based on the calculation results.

Step 3: At this step, the iteration begins. When $\tau \leq \frac{2}{3}T$, the AO algorithm enters the exploration phase and randomly selects between global exploration and local exploration to update the best solution. When $\tau > \frac{2}{3}T$, the AO algorithm enters the exploitation phase and randomly selects between global exploitation and local exploitation to update the best solution.

Step 4: When the current number of iterations τ meets the maximum number of iterations T, the optimization process ends, and the optimal parameter solution and fitness value of the MPKELM are output; otherwise, Step 3 is repeated.

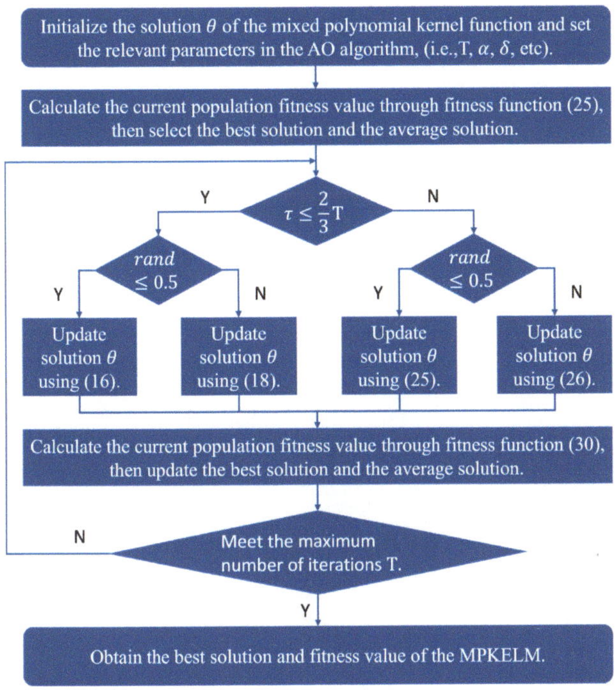

Figure 3. Parameter optimization flow of MPKELM using the AO algorithm.

4. Analysis of Experiments and Results

The silicon piezoresistive differential pressure sensor selected in the experiment has a measurement range of −250 to +250 kPa and an operating temperature of −40 to +85 °C. We use a modular pressure controller (GE Druck PACE5000, Baker Hughes Company, Houston, TX, USA) to control the pressure precisely. The accuracy of this device can reach 0.001% FS (including nonlinearity, hysteresis, repeatability, and temperature effects over the calibrated temperature range) within the required working conditions. The experimental temperature environment is determined by a high–low temperature test chamber (ET0440, Guangzhou-GWS Environmental Equipment Company, Guangzhou, China). The set temperature of this equipment ranges from −40 to +150 °C, and the ambient temperature fluctuation is less than 0.5 °C. Figure 4 shows the actual calibration experiment system, where the pressure transducer is the device packaging the pressure sensor and the additional temperature sensor described in Section 2.

First, we carry out a detailed calibration of the sensor equipment to analyze the nonlinearity and temperature drift characteristics without compensation. The pressure calibration points are selected in an arithmetic sequence with a range of −250 to 250 kPa, tested every 31.25 kPa for a total of 17 pressure points. Temperature calibration points are also selected in an arithmetic sequence with a range of −40 to 85 °C, tested every 5 °C for a total of 26 temperature points, and 442 groups of data are finally recorded. A low-noise 24-bit analog-to-digital converter (AD7799) collects the output voltage and transfers the data to the industrial personal computer. Each voltage data point is collected and averaged several times after stability to reduce the measurement uncertainty (the RMS noise of this device is approximately 27 nV).

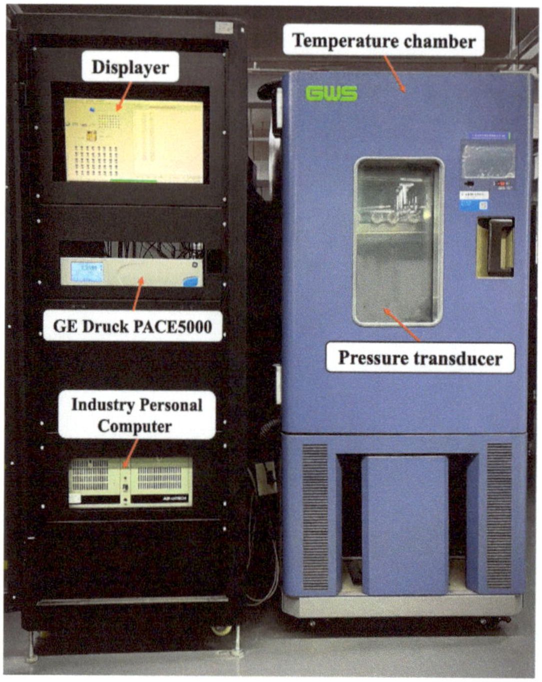

Figure 4. Actual calibration experiment system.

After collecting the real calibration data, the input–output relationship of the sensor is considered as a simple first-order linear function to obtain the response deviation without compensation. Taking two points, 250 kPa and −250 kPa at room temperature (25 °C), as the benchmark, the uncompensated measurement error can be obtained through calculation.

Figure 5 is the pressure bias surface before compensation, from which we can find that the measurement error of the silicon piezoresistive pressure sensor increases with the applied pressure increase. This is due to the inherent nonlinear effect between the change in resistance of the silicon-doped resistor and the applied pressure, which becomes more obvious when the input signal varies widely. Meanwhile, as the deviation between ambient temperature and the room temperature (25 °C) increases, the measurement error increases, which is related to the changes in temperature-induced sensitivity. When the sensor is subjected to positive pressure, the measured pressure rises at high temperature and falls at low temperature, and the opposite is true when subjected to negative pressure. The rising temperature decreases semiconductor silicon resistivity, thus reducing sensor sensitivity. Without compensation, the maximum error of this pressure sensor occurs at (−250 kPa, −40 °C) and (250 kPa, −40 °C), which is 46.92 kPa and −42.07 kPa, respectively. We can use (5) to calculate the FSE of these two points as 9.38% and −8.41%, respectively. The results of the experiments demonstrate that the output of the silicon piezoresistive pressure sensor exhibits some nonlinearity and is affected by temperature drift significantly. Consequently, the measurement error of this sensor is considerable without compensation and cannot meet the demands of engineering and scientific applications.

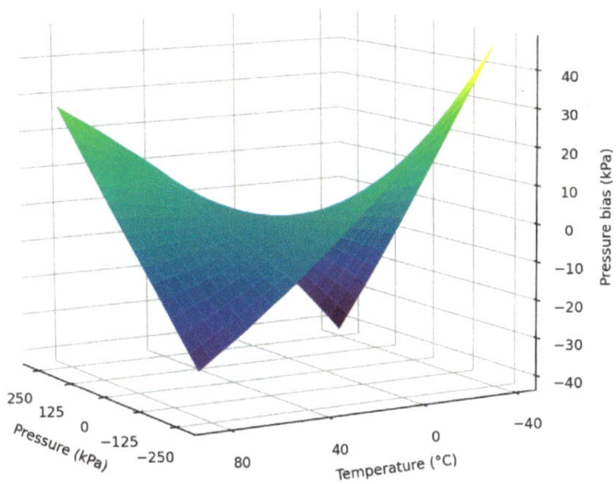

Figure 5. Pressure bias surface before compensation.

4.1. Calibration Experiment

Next, we compare the real calibration data and the generated data to verify the effectiveness of the proposed method. In this experiment, we select five temperature points, −40, −10, 25, 50, and 85 °C, and nine pressure points, −250, −187.5, −125, −62.5, 0, 62.5, 125, 187.5, and 250 kPa, as the calibration dataset for algorithm training, with a total 45 groups of input–output relationships. The remaining 397 groups of experimental data will serve as the test set and will be compared with the generated data to measure the performance of the proposed calibration generation method. Actually, in the batch application of this method, the quality of the generated data does not need to be verified; only the correlation points of the training dataset must be tested. The distribution of the training data is relatively uniform and centralized; therefore, we use (26), the min–max normalization, to normalize different input features x_i to the range of [0, 1] to eliminate the influence of data units and improve the convergence speed and accuracy of the model:

$$x_i = \frac{x_i - \min(x_i)}{\max(x_i) - \min(x_i)}. \tag{31}$$

In the proposed algorithm, the highest order D of the mixed polynomial kernel function is a parameter that needs to be artificially set to better simulate the calibration process of the sensor.

In Table 1, we compare the performance of the AO-MPKELM models under different highest mixing orders, including the RMSE and mean absolute error (MAE) of the training set and test set, as well as the time required by the algorithm. When $D = 4$, the model exhibits the best performance in the test set, and its RMSE and MAE are 1.66×10^{-4} and 1.23×10^{-4}, respectively. When $D = 5$, the model exhibits the highest accuracy on the training set; however, its performance on the test set is inferior to that of the fourth-order AO-MPKELM, which indicates that the introduction of the fifth-order polynomial kernel function has caused certain overfitting problems in the model.

Table 1. Performance of the AO-MPKELM models under different highest mixing orders.

D	RMSE		MAE		Time (s)
	Training	Testing	Training	Testing	
2	4.78×10^{-3}	4.08×10^{-3}	3.62×10^{-3}	3.16×10^{-3}	0.68
3	4.87×10^{-4}	4.63×10^{-4}	3.98×10^{-4}	3.48×10^{-4}	0.78
4	8.76×10^{-5}	1.66×10^{-4}	6.26×10^{-5}	1.23×10^{-4}	0.97
5	6.54×10^{-5}	1.80×10^{-4}	4.81×10^{-5}	1.34×10^{-4}	1.16

Owing to limited data, the optimization process of the AO is conducted on the training set. The larger the value of D, the more likely overfitting will occur; thus, selecting the correct parameter D is important in the proposed method. Both theoretical and experimental results show that when the model becomes more complex (D increases), the running time of the algorithm will also increase. However, compared with a calibration time of tens of hours, a difference in calculation time of a few seconds between different D levels can be ignored. All the above experiments were conducted using the Python3.7 environment, and the test device used a 10th-generation Intel Core i5 processor and 16 GB memory. There is no doubt that the commonly used industrial computer can meet the above calculation load.

Table 2 shows the set hyperparameter range and optimization results of the AO under different highest mixing orders. In this study, the population number and iteration times of the AO are set at 15 and 20, respectively.

Table 2. The set hyperparameter range and optimization results of the AO under different highest mixing orders.

Parameter	Range	Result			
		D = 2	D = 3	D = 4	D = 5
γ	$[1 \times 10^3, 1 \times 10^7]$	7.52×10^5	2.90×10^5	1.91×10^4	3.80×10^3
θ_2	$[1, 1 \times 10^2]$	92.31	6.88	21.86	57.16
θ_3	$[1, 1 \times 10^2]$	0	70.01	13.14	27.49
θ_4	$[1, 1 \times 10^2]$	0	0	27.52	64.79
θ_5	$[1, 1 \times 10^2]$	0	0	0	37.57

In practical application, we only tested the data of the training set and did not know the real voltage value of the test set; thus, we could not select the hyperparameters according to the results of the test set. Therefore, in the subsequent analysis and experiment, we adopted the AO-MPKELM model ($D = 5$), which performed best on the training set, to express the application effect in the actual project objectively.

Figure 6 shows the voltage deviation between generated and real values at each temperature and pressure point. In most cases, the voltage deviation is less than 0.5 mV, and its maximum deviation is 0.71 mV under all working conditions. Through several calibration experiments, we observed millivolt voltage deviations between calibration data in the same working conditions under the influence of superimposed errors by the GE Druck PACE5000, temperature chamber, and pressure sensor. It proves the authenticity and consistency of the generated data and indicates that we can correctly simulate the calibration process of the pressure sensor by modeling its input–output response relationship to generate more calibration data.

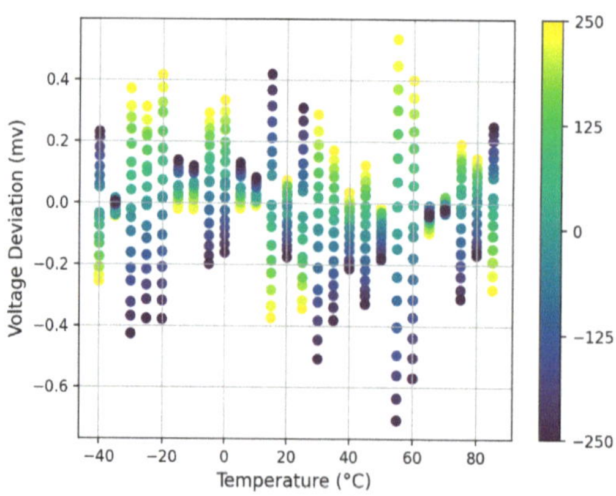

Figure 6. Voltage deviation between generated and real values at each temperature and pressure point.

4.2. Compensation Experiment

After verifying the quality of the generated data, we conducted a series of ablation and comparison experiments to demonstrate the superiority of the proposed method. A variety of compensation algorithms were selected for comparison in the real calibration dataset, including bilinear interpolation (BI), polynomial regression (PR), back-propagation neural networks (BPNNs), support vector machines (SVMs), and kernel extreme learning machine (KELM), all of which have better compensation effects in this field and are currently more suitable for implementation in microprocessors. Although the above algorithms can achieve online compensation, the training process of machine learning still consumes lots of resources, and the BI is undoubtedly the most convenient and fastest algorithm for practical applications. Therefore, in the proposed compensation method based on data generation, BI is adopted as the multidimensional compensation algorithm, and the compensation dataset contains both generated data by AO-MPKELM and real calibration data. We use DG-BI as an abbreviation for this method; the DG represents the data generation process using AO-MPKELM, and BI represents the final compensation algorithms. Some setting parameters of these algorithms are shown in Table 3.

Table 3. The setting parameters of different compensation algorithms.

Algorithm	Parameter	Value
PR	Highest order	5
BP	Hidden layer number	4
	Optimizer	Adam
	Learning rate	1×10^{-3}
SVM	Penalty parameter C	1×10^{6}
	Tolerance	1×10^{-3}
	Epsilon	1×10^{-3}
KELM	Penalty parameter C	1×10^{6}
	Variance σ of gaussian kernel	1

Table 4 shows the compensation results of different compensation methods, including the maximum positive full-scale error, maximum negative full-scale error, average full-scale error (absolute value), and standard deviation. Detailed compensation results of different algorithms under different temperatures and pressures are presented in Figure 7. The full-scale error surface of bilinear interpolation is smoother than that of the polynomial regression algorithm, with a smaller average FSE and standard deviation, and the maximum error occurs at a room temperature of 25 °C. The polynomial regression performs poorly at 60 to 80 °C, and the excessively high-dimensional nature caused overfitting in the high-temperature region, so its maximum full-scale error reached 0.123%. As the most popular regression model in recent years, BP has been proven to have a strong generalization ability in various tasks. But in this experiment, the performance is not satisfactory. Insufficient training data hindered us in obtaining optimal weights and bias through gradient descent. The two machine learning methods suitable for small data (SVM and KELM) are both based on the kernel function theory and thus perform similarly, outperforming the models mentioned above. However, there is still nearly 0.06% FSE under extreme working conditions. The proposed DG-BI exhibits the most robust performance and provides the most stable compensation effect over the full-scale and temperature region. Compared to the control group for the ablation experiment that used the BI algorithm only in the real calibration dataset, its measurement error is reduced by 78.95%, which proves the correctness and superiority of the presented theory in this paper.

Table 4. Compensation results of different methods.

Method	Max Positive FSE	Max Negative FSE	Average	Std
BI	1.33×10^{-3}	-1.15×10^{-3}	3.36×10^{-4}	4.45×10^{-4}
PR	1.23×10^{-3}	-9.32×10^{-4}	4.39×10^{-4}	5.61×10^{-4}
BPNN	2.82×10^{-3}	-4.93×10^{-3}	8.72×10^{-4}	1.17×10^{-3}
SVM	5.48×10^{-4}	-7.82×10^{-4}	1.77×10^{-4}	2.05×10^{-4}
KELM	5.79×10^{-4}	-6.11×10^{-4}	1.83×10^{-4}	2.25×10^{-4}
DG-BI	$\boldsymbol{2.36 \times 10^{-4}}$	$\boldsymbol{-2.80 \times 10^{-4}}$	$\boldsymbol{5.90 \times 10^{-5}}$	$\boldsymbol{7.81 \times 10^{-5}}$

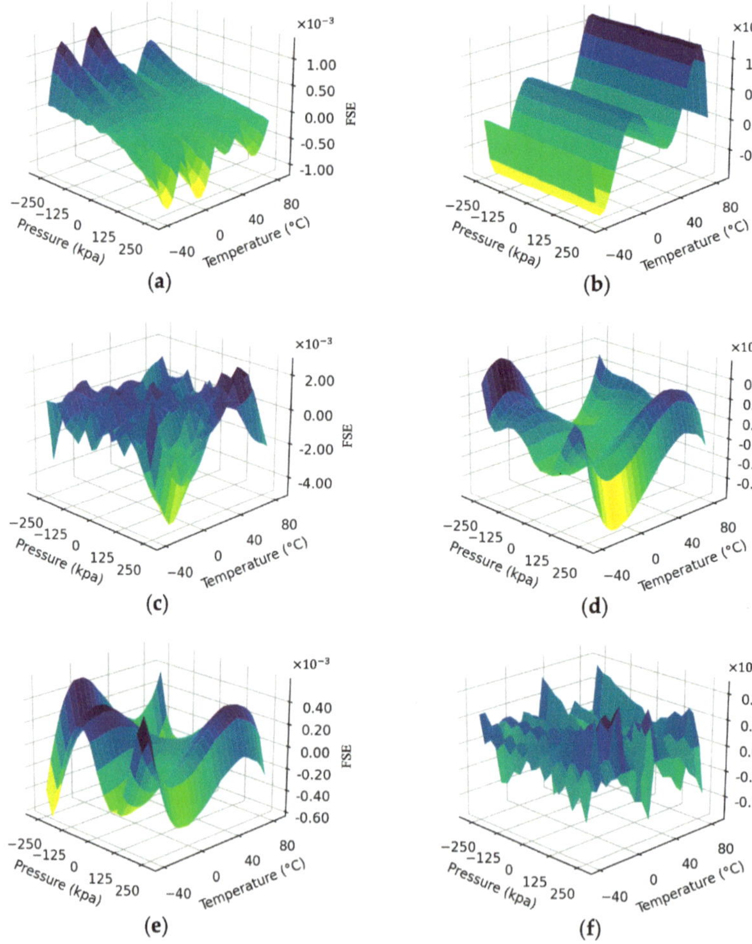

Figure 7. Full-scale error surfaces after multidimensional compensation using different methods: (**a**) bilinear interpolation, (**b**) polynomial regression, (**c**) back-propagation neural network, (**d**) support vector machines, (**e**) kernel extreme learning machine, (**f**) proposed DG-BI method.

In this measurement system, the achievable accuracy after compensation is about 0.03% FS, which is better than other methods. The maximum positive full-range error and maximum negative full-range error of the sensor are 0.024% and −0.028%, respectively. Contrasted with the corresponding values before compensation, the errors are reduced to 0.25% and 0.33%. These findings verify that the compensation method in this study can significantly decrease the measurement bias. Moreover, the complexity of this method is lower than that of BP and other learning algorithms; therefore, this method is conducive to achieving high-accuracy compensation in practical engineering applications.

5. Conclusions

In this study, we presented a novel multidimensional compensation method, DG-BI, for piezoresistive pressure sensors, which combined the AO and proposed MPKELM to generate more calibration data and used BI for compensation to improve measurement accuracy. Experimental results show that the AO-MPEKLM accurately simulated the calibration process of the pressure sensor and can further improve the compensation effect by using additionally generating high-quality data. The maximum voltage deviation between generated data and actual calibration data is 0.71 mV, and the full-scale error after compensation is 0.028%, only 0.33% of that before compensation. Through the ablation experiment, we found that the measurement accuracy was improved by nearly five times compared to that using only the real calibration data and BI algorithm. This compensation method has a guiding effect on existing compensation research. The high-accuracy measurement can be achieved by using more generated data rather than complex compensation algorithms, which makes software compensation more suitable for engineering applications.

Author Contributions: Conceptualization, M.Z. and W.H.; methodology, M.Z. and J.J.; funding acquisition, W.H. and J.J.; resources, W.H. and M.C.; project administration, J.J. and Y.X.; investigation, Y.X. and M.Z.; data curation, M.Z. and M.C.; visualization, M.C. and Y.X.; formal analysis, M.Z. and M.C.; software, M.Z. and J.J.; validation, Y.X. and M.Z.; supervision W.H. and Y.X.; Writing—original draft, M.Z. and M.C.; Writing—review and editing M.Z. and J.J. All authors have read and agreed to the published version of the manuscript.

Funding: This research was funded by the National Key Research and Development Program of China (2022YFB4201603).

Institutional Review Board Statement: Not applicable.

Informed Consent Statement: Not applicable.

Data Availability Statement: Not applicable.

Acknowledgments: Technical support and experimental material for this study were supported by Zhejiang Supcon Sensing Technology Co., Ltd.

Conflicts of Interest: The authors declare no conflict of interest.

References

1. Song, P.; Ma, Z.; Ma, J.; Yang, L.; Wei, J.; Zhao, Y.; Zhang, M.; Yang, F.; Wang, X. Recent Progress of Miniature MEMS Pressure Sensors. *Micromachines* **2020**, *11*, 56. [CrossRef] [PubMed]
2. Jena, S.; Gupta, A. Review on Pressure Sensors: A Perspective from Mechanical to Micro-Electro-Mechanical Systems. *Sens. Rev.* **2021**, *41*, 320–329. [CrossRef]
3. Xu, T.; Wang, H.; Xia, Y.; Zhao, Z.; Huang, M.; Wang, J.; Zhao, L.; Zhao, Y.; Jiang, Z. Piezoresistive Pressure Sensor with High Sensitivity for Medical Application Using Peninsula-Island Structure. *Front. Mech. Eng.* **2017**, *12*, 546–553. [CrossRef]
4. Kang, K.; Park, J.; Kim, K.; Yu, K.J. Recent Developments of Emerging Inorganic, Metal and Carbon-Based Nanomaterials for Pressure Sensors and Their Healthcare Monitoring Applications. *Nano Res.* **2021**, *14*, 3096–3111. [CrossRef]
5. Qian, J.; Kim, D.-S.; Lee, D.-W. On-Vehicle Triboelectric Nanogenerator Enabled Self-Powered Sensor for Tire Pressure Monitoring. *Nano Energy* **2018**, *49*, 126–136. [CrossRef]
6. Soy, H.; Toy, İ. Design and Implementation of Smart Pressure Sensor for Automotive Applications. *Measurement* **2021**, *176*, 109184. [CrossRef]
7. Li, W.; Lu, W.; Sha, X.; Xing, H.; Lou, J.; Sun, H.; Zhao, Y. Wearable Gait Recognition Systems Based on MEMS Pressure and Inertial Sensors: A Review. *IEEE Sens. J.* **2022**, *22*, 1092–1104. [CrossRef]
8. Cui, X.; Huang, F.; Zhang, X.; Song, P.; Zheng, H.; Chevali, V.; Wang, H.; Xu, Z. Flexible Pressure Sensors via Engineering Microstructures for Wearable Human-Machine Interaction and Health Monitoring Applications. *iScience* **2022**, *25*, 104148. [CrossRef]
9. Wang, L.; Zhu, R.; Li, G. Temperature and strain compensation for flexible sensors based on thermosensation. *ACS Appl. Mater. Interfaces* **2019**, *12*, 1953–1961. [CrossRef]

10. Aryafar, M.; Hamedi, M.; Ganjeh, M.M. A Novel Temperature Compensated Piezoresistive Pressure Sensor. *Measurement* **2015**, *63*, 25–29. [CrossRef]
11. Devi, R.; Gill, S.S. Performance Investigation of Carbon Nanotube Based Temperature Compensated Piezoresistive Pressure Sensor. *Silicon* **2022**, *14*, 3931–3938. [CrossRef]
12. Pieniazek, J.; Ciecinski, P. Temperature and Nonlinearity Compensation of Pressure Sensor with Common Sensors Response. *IEEE Trans. Instrum. Meas.* **2020**, *69*, 1284–1293. [CrossRef]
13. Altinoz, B.; Unsal, D. Look up Table Implementation for IMU Error Compensation Algorithm. In Proceedings of the 2014 IEEE/ION Position, Location and Navigation Symposium (PLANS 2014), Monterey, CA, USA, 5–8 May 2014; pp. 259–261.
14. Ali, I.; Asif, M.; Shehzad, K.; Rehman, M.R.U.; Kim, D.G.; Rikan, B.S.; Pu, Y.; Yoo, S.S.; Lee, K.-Y. A Highly Accurate, Polynomial-Based Digital Temperature Compensation for Piezoresistive Pressure Sensor in 180 nm CMOS Technology. *Sensors* **2020**, *20*, 5256. [CrossRef]
15. Guo, Z.; Lu, C.; Wang, Y.; Liu, D.; Huang, M.; Li, X. Design and Experimental Research of a Temperature Compensation System for Silicon-on-Sapphire Pressure Sensors. *IEEE Sens. J.* **2017**, *17*, 709–715. [CrossRef]
16. Kayed, M.O.; Balbola, A.A.; Lou, E.; Moussa, W.A. Hybrid Smart Temperature Compensation System for Piezoresistive 3D Stress Sensors. *IEEE Sens. J.* **2020**, *20*, 13310–13317. [CrossRef]
17. Ma, Z.; Wang, G.; Rui, X.; Yang, F.; Wang, Y. Temperature Compensation of a PVDF Stress Sensor and Its Application in the Test of Gun Propellant Charge Compression Stress. *Smart Mater. Struct.* **2019**, *28*, 025018. [CrossRef]
18. Li, J.; Hu, G.; Zhou, Y.; Zou, C.; Peng, W.; Alam, J. SM Study on temperature and synthetic compensation of piezo-resistive differential pressure sensors by coupled simulated annealing and simplex optimized kernel extreme learning machine. *Sensors* **2017**, *17*, 894. [CrossRef] [PubMed]
19. Ruan, Y.; Yuan, L.; Yuan, W.; He, Y.; Lu, L. Temperature Compensation and Pressure Bias Estimation for Piezoresistive Pressure Sensor Based on Machine Learning Approach. *IEEE Trans. Instrum. Meas.* **2021**, *70*, 1008610. [CrossRef]
20. Liang, H.; Chen, H.; Lu, Y. Research on Sensor Error Compensation of Comprehensive Logging Unit Based on Machine Learning. *J. Intell. Fuzzy Syst.* **2019**, *37*, 3113–3123. [CrossRef]
21. Li, J.; Zhang, C.; Zhang, X.; He, H.; Liu, W.; Chen, C. Temperature compensation of piezo-resistive pressure sensor utilizing ensemble AMPSO-SVR based on improved AdaBoost. RT. *IEEE Access* **2020**, *8*, 12413–12425. [CrossRef]
22. Barlian, A.A.; Park, W.-T.; Mallon, J.R.; Rastegar, A.J.; Pruitt, B.L. Review: Semiconductor Piezoresistance for Microsystems. *Proc. IEEE* **2009**, *97*, 513–552. [CrossRef] [PubMed]
23. dos Santos Pereira, R.; Cima, C.A. Thermal Compensation Method for Piezoresistive Pressure Transducer. *IEEE Trans. Instrum. Meas.* **2021**, *70*, 9510807.
24. Huang, G.-B.; Zhu, Q.-Y.; Siew, C.-K. Extreme Learning Machine: Theory and Applications. *Neurocomputing* **2006**, *70*, 489–501. [CrossRef]
25. Zheng, Y.; Chen, B.; Wang, S.; Wang, W.; Qin, W. Mixture Correntropy-Based Kernel Extreme Learning Machines. *IEEE Trans. Neural Netw. Learn. Syst.* **2022**, *33*, 811–825. [CrossRef] [PubMed]
26. Shi, X.; Kang, Q.; An, J.; Zhou, M. Novel L1 Regularized Extreme Learning Machine for Soft-Sensing of an Industrial Process. *IEEE Trans. Ind. Inf.* **2022**, *18*, 1009–1017. [CrossRef]
27. Deng, W.; Zheng, Q.; Chen, L. Regularized Extreme Learning Machine. In Proceedings of the 2009 IEEE Symposium on Computational Intelligence and Data Mining, Nashville, TN, USA, 30 March–2 April 2009; pp. 389–395.
28. Huang, G.B.; Zhou, H.; Ding, X.; Zhang, R. Extreme Learning Machine for Regression and Multiclass Classification. *IEEE Trans. Syst. Man Cybern. B* **2012**, *42*, 513–529. [CrossRef]
29. Abualigah, L.; Yousri, D.; Abd Elaziz, M.; Ewees, A.A.; Al-qaness, M.A.A.; Gandomi, A.H. Aquila Optimizer: A Novel Meta-Heuristic Optimization Algorithm. *Comput. Ind. Eng.* **2021**, *157*, 107250. [CrossRef]
30. Abd Elaziz, M.; Dahou, A.; Alsaleh, N.A.; Elsheikh, A.H.; Saba, A.I.; Ahmadein, M. Boosting COVID-19 Image Classification Using MobileNetV3 and Aquila Optimizer Algorithm. *Entropy* **2021**, *23*, 1383. [CrossRef]
31. Guo, Z.; Yang, B.; Han, Y.; He, T.; He, P.; Meng, X.; He, X. Optimal PID Tuning of PLL for PV Inverter Based on Aquila Optimizer. *Front. Energy Res.* **2022**, *9*, 812467. [CrossRef]
32. Wang, S.; Ma, J.; Li, W.; Khayatnezhad, M.; Rouyendegh, B.D. An Optimal Configuration for Hybrid SOFC, Gas Turbine, and Proton Exchange Membrane Electrolyzer Using a Developed Aquila Optimizer. *Int. J. Hydrog. Energy* **2022**, *47*, 8943–8955. [CrossRef]

electronics

Article

YOLO-Xray: A Bubble Defect Detection Algorithm for Chip X-ray Images Based on Improved YOLOv5

Jie Wang [1,†], Bin Lin [2,†], Gaomin Li [1], Yuezheng Zhou [1], Lijun Zhong [1], Xuan Li [3] and Xiaohu Zhang [1,*]

1 School of Aeronautics and Astronautics, Sun Yat-sen Unviersity, Guangzhou 510725, China; wangj688@mail2.sysu.edu.cn (J.W.); zhonglj9@mail.sysu.edu.cn (L.Z.)
2 College of Photonic and Electronic Engineering, Fujian Normal University, Fuzhou 350117, China; linb@fjnu.edu.cn
3 China Aerospace Components Engineering Center, Beijing 100094, China
* Correspondence: zhangxiaohu@mail.sysu.edu.cn
† These authors contributed equally to this work.

Abstract: In the manufacturing of chips, the accurate and effective detection of internal bubble defects of chips is essential to maintain product reliability. In general, the inspection is performed manually by viewing X-ray images, which is time-consuming and less reliable. To solve the above problems, an improved bubble defect detection model YOLO-Xray based on the YOLOv5 algorithm for chip X-ray images is proposed. First, the chip X-ray images are preprocessed by image segmentation to construct the chip X-ray defect dataset, namely, CXray. Then, in the input stage, the K-means++ algorithm is used to re-cluster the CXray dataset to generate the anchors suitable for our dataset. In the backbone network, a micro-scale detection head is added to improve the capabilities for small defect detection. In the neck network, the bi-direction feature fusion idea of BiFPN is used to construct a new feature fusion network based on the improved backbone to fuse the semantic features of different layers. In addition, the Quality Focal Loss function is used to replace the cross-entropy loss function to solve the imbalance of positive and negative samples. The experimental results show that the mean average precision (mAP) of the YOLO-Xray algorithm on the CXray dataset reaches 93.5%, which is 5.1% higher than the original YOLOv5. Meanwhile, the YOLO-Xray algorithm achieves state-of-the-art detection accuracy and speed compared with other mainstream object detection models. This shows the proposed YOLO-Xray algorithm can provide technical support for bubble defect detection in chip X-ray images. The CXray dataset is also open and available at CXray.

Keywords: bi-direction feature fusion; bubble defect detection; deep learning; X-ray images; YOLOv5

1. Introduction

With the widespread use of semiconductor chips across a wide field of applications, the quality of chips is increasingly demanding. During the packaging process of chips, bubble defects frequently occur. Excessive bubbles might compromise the airtightness of the circuit and the stability of the cover, hence decreasing the reliability of the products. Therefore, internal bubble inspection of chips is a crucial component of the manufacturing process, playing a crucial role in regulating chip quality. At present, the detection of bubble defects in chip X-ray images mainly relies on manual operation, which has low accuracy, low efficiency, and expensive labor costs. Therefore, accurately and effectively detecting bubble defects in chip X-ray images is crucial.

Earlier works on chip defect detection have focused on traditional image processing methods. They use hand-crafted features, such as geometric features [1], color features [2], texture features [3], scale-invariant characteristic transform (SIFT) [4], local binary patterns (LBPs) [5], and other features to detect defects. For example, Chen et al. [6] proposed an inspection system for integrated circuit (IC) molding surfaces based on surface texture features, including normalization, shrinking, segmenting, and Fourier-based image restoration and defect identification, achieving a high accuracy rate of 94.2%. Lin et al. [7] proposed a region segmentation search-based defect detection method for complex structure chips, which incorporates the canny operator, contour interpolation, and shape interference elimination. Zhong et al. [8] proposed a three-stage technique for defect detection in flexible integrated circuit package substrates (FICSs), including image contrast enhancement, standard template acquisition, and featuring probability calculation, which outperforms existing methods in the time–accuracy tradeoff. However, on the one hand, hand-crafted features are selected heavily relying on rich expert experience and require a large number of parameters to be adjusted. On the other hand, hand-crafted features are sensitive to complex backgrounds, noise, and uneven illumination, resulting in poor robustness. In addition, X-ray chip images have some challenges in bubble defect detection: (1) The overall gray value of the X-ray chip image is low, and the contrast between defects and background is low. (2) There is considerable interference from noise and complex background structures. (3) The scale of defects varies greatly, with irregular shapes and more defects of a small dimension. Therefore, it is difficult for traditional methods to accurately detect defects of different scales and shapes.

In recent years, with the development of computer vision technology, object detection networks based on deep learning have been successfully used in the field of defect detection, including fabric [9], metallic [10,11], railway [12], optical components [13], and so on. For instance, He et al. [11] proposed an end-to-end defect detection network for steel plate defects by combining multi-level features, achieving promising results with an accuracy of 99.67% for defect classification and 82.3% mAP for defect detection, and also contributed a defect detection dataset NEU-DET for fine-tuning. Lin et al. [14] proposed a convolutional neural network combined with class activation mapping for detecting light-emitting diode (LED) chip defects, reaching a classification error of 5.04% while also contributing an LCD chip dataset. Luo et al. [15] proposed a decoupled two-stage object detection model for FPCB surface defect detection, creating two distinct features for the localization and classification tasks, reaching an mAP of 94.15%, and built an FPBC surface defect detection dataset called FPCB-DET. Li et al. [16] proposed WearNet, a lightweight CNN, for surface scratch detection in contact sliding, achieving an impressive classification accuracy of 94.16% and successfully deploying the model on an embedded system, as well as gathering a large-scale surface scratch dataset. Object detection networks are mainly classified into two categories: one-stage and two-stage. Two-stage detection algorithms represented by R-CNN [17], Faster R-CNN [18], Cascade R-CNN [19], and Mask R-CNN [20] are divided into two steps. First, region proposals are created based on feature extraction, and then, these proposals are classified and regressed. One-stage detection algorithms represented by You Only Look Once (YOLO) series [21–24], Single-Shot Detector (SSD) [25], and RetinaNet [26] directly regress the classification and localization of targets based on feature extraction, which makes them meet real-time object detection tasks, but their accuracy is inferior to that of two-stage detection algorithms.

Currently, the YOLOv5 algorithm is frequently used for industrial defect detection because it has a considerable advantage in terms of detection accuracy and detection efficiency, resulting in major advancements in certain industrial circumstances. Zhang et al. [27] developed SOD-YOLO based on YOLOv5 for wind turbine blade (WTB) surface flaws, which improved the mAP by 7.82% when compared to YOLOv5. Du et al. [28] presented BV-YOLOv5S, a lightweight model designed for the detection of pavement defects. Compared to deep network learning models such as YOLOv3-Tiny, YOLOv5S, and B-YOLOv5S, BV-YOLOv5S demonstrated significant improvements of 4.1%, 3%, and 0.9%, respectively. Zhang et al. [29] proposed an enhanced YOLOv5-based solar cell defect detection method by incorporating deformable convolution and an ECA-Net attention mechanism, thereby increasing the mAP by 7.85% in comparison to YOLOv5. Shi et al. [30] proposed an improved YOLOv5 for the detection of steel surface defects by incorporating attention mechanisms and re-clustering anchor boxes, demonstrating a significant improvement in average precision, especially for small targets and targets with extreme aspect ratios. Wang et al. [31] suggested an improved MS-YOLOv5 model based on the YOLOv5 algorithm for detecting surface defects in aluminum profiles by using PE-Neck and multi-stream network components, reaching an AP of over 80% for each defect. The YOLOv7 algorithm [32], a YOLOv5-based enhanced network model proposed in 2022, surpasses existing target detectors on the MS COCO dataset in terms of speed and accuracy. In our practical implementations of chip bubble defect detection, YOLOv5 outperforms the state-of-the-art YOLOv7 model. Therefore, the YOLOv5 model is selected as the baseline. Nevertheless, there are several obstacles to overcome when applying YOLOv5 directly to the issue of bubble defect identification, including tiny target detection, imbalanced samples, and limited defective samples. Considering the above challenges, we develop a high-precision model, YOLO-Xray, based on YOLOv5 for chip X-ray image analysis. The main contributions of the paper are as follows:

1. We construct a dataset (CXray) as the research basis for bubble defects detection by foreground segmentation and homography transformation.
2. Various optimizations, which include the K-means++ clustering algorithm, the incorporation of micro-scale detection heads, the integration of the idea of BiFPN, and QFocal Loss function are introduced to improve the performance of the original YOLOv5 model.
3. Numerous experiments on the CXray dataset have verified the effectiveness of our proposed YOLO-Xray model.

The remainder of the article is organized as follows: Section 2 describes the acquisition and preprocessing of our CXray dataset. Section 3 presents detailed information on the improved YOLOv5. Section 4 elaborates on the experimental results and discussions. Section 5 draws conclusions and discusses future works.

2. Dataset

2.1. Data Acquisition

Nowadays, X-ray technology has been commonly used in nondestructive testing applications for chip package detection. Because X-rays can penetrate the chips to visualize the inner defects that are invisible to the naked eye without damaging them. X-ray images of chips used in this study were generated by X-ray inspection equipment from a chip testing factory, as shown in Figure 1. In this study, a typical type of chip was used as the research object. Additionally, the defect that this study focuses on the most is bubbles, which are typically seen in X-ray pictures of chips.

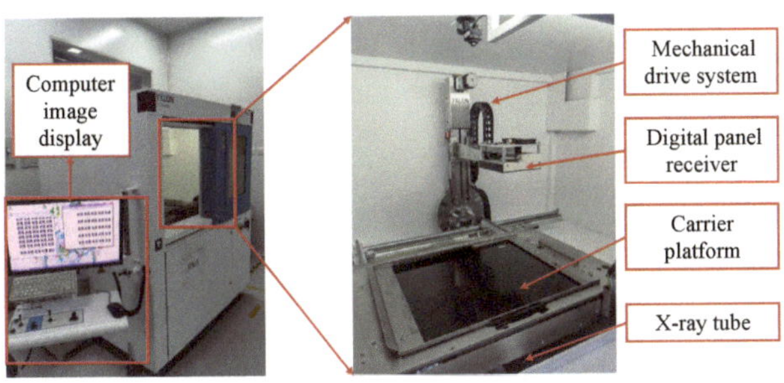

Figure 1. X-ray image inspection equipment.

2.2. Data Preprocessing and Annotation

As shown in Figure 2a, the original resolution of chip X-ray images collected by X-ray inspection equipment is 1004 × 620 pixels. However, the original image contains multiple chips, and only the chip in the image's central region must be identified, as illustrated in the red-boxed part of Figure 2a. Since the chips may be placed obliquely, they are tilted in the image. Therefore, we develop a combination of foreground segmentation and homography transformation to extract and normalize the central chip.

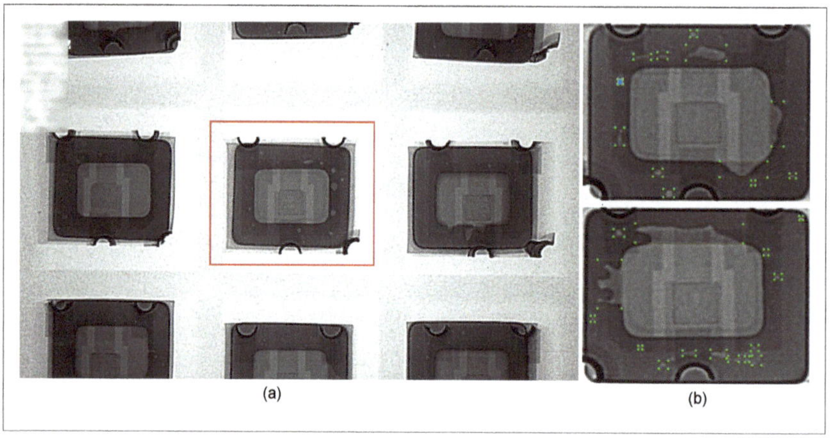

Figure 2. (**a**) Raw images captured by X-ray inspection equipment. (**b**) Samples of X-ray images with annotations of bubble defects.

The steps of the algorithm are described as follows:

1. The original images are center-cropped to 512×512 pixels.
2. Gaussian filtering of a kernel size of 5×5 is employed to remove Gaussian noise in the X-ray image, such as particle noise on the fluorescent screen.
3. The OTSU algorithm [33] is adopted to determine the threshold value for binarization segmentation.
4. Morphological opening and closing operations are applied to the image after segmentation for noise points filtering and hole filling.
5. Two-pass algorithms are used to label the contours of all connected regions and filter out the central chip based on the prior geometrical information.
6. The Homography transformation matrix, H, is applied to map an image from different scales to a uniform scale. Suppose that (x, y) and (x', y') are the pixel coordinates of the original image and transformed image, respectively. The form of transformation is shown in the following equation:

$$\begin{bmatrix} x' \\ y' \\ 1 \end{bmatrix} = \begin{bmatrix} h_1 & h_2 & h_3 \\ h_4 & h_5 & h_6 \\ h_7 & h_8 & 1 \end{bmatrix} \begin{bmatrix} x \\ y \\ 1 \end{bmatrix} \quad (1)$$

where $h_1, \ldots, h_8$ are transformation parameters. The H matrix can be solved through four feature points at least. We first solve the minimum area bounding rectangle (MABR) of the target contour based on an approximation algorithm. Then, we select four corner points of MABR as reference points(u, v), and (u', v') is the corresponding coordinate of four static boundary points of the ideal image.

After image preprocessing, we obtained an image containing only a single chip. We randomly selected 1000 images from real industrial environments with varying noise situations and lighting attributes. All images were annotated carefully into the VOC dataset [34] format by professionals using the labeling software called LabelImg. At the same time, the class names and coordinates of the bounding box of each bubble defect are stored in corresponding XML files. The CXray dataset holds 1000 images with around 10,000 bubble defects. The dataset is randomly split into the training set, the validation set, and the test set, according to a ratio of 7:2:1. Some samples of the CXray dataset are shown in Figure 2b.

3. Method

3.1. YOLO-Xray Algorithm

In the previous section, we constructed a dataset of chip bubble defect X-ray images called CXray. The following are some of the characteristics of the CXray dataset: (1) The defects may appear at any location in the image. (2) There may be a large number of tiny defects exhibiting modest and dense distribution features. (3) There are large-scale long strips and irregular bubble defects of varying sizes, as shown in Figure 2b.

As one of the most advanced single-stage object detectors, the YOLOv5 algorithm [35] possesses advantageous characteristics such as rapid detection speed, high accuracy, and adaptability. However, when bubble defect identification is explicitly applied, there are still several obstacles to overcome. It has difficulty detecting small and irregular defects, resulting in a greater rate of missed detection.

To address the above problems, we propose an improved YOLOv5 model YOLO-Xray for detecting small defects and irregular defects in the CXray dataset. A schematic diagram of the structure of the YOLO-Xray model is shown in Figure 3, and the composition of each component is shown in Figure 4. The model is mainly composed of four parts: the input module, backbone network, feature extraction network, and detection network. In the input module, the K-means++ clustering algorithm was used to regenerate the optimal anchors suitable for our dataset. In the detection network, based on the original three detection heads, a micro-scale detection head was designed to improve the detection performance of small-sized targets. In the feature extraction network, the idea of bi-directional feature fusion was introduced to fuse feature information from multiple scales for better learning of irregularly shaped defect features. In the classification loss and confidence loss, the Quality Focal Loss function was used to replace the original cross-entropy loss function to solve the imbalance of positive and negative samples.

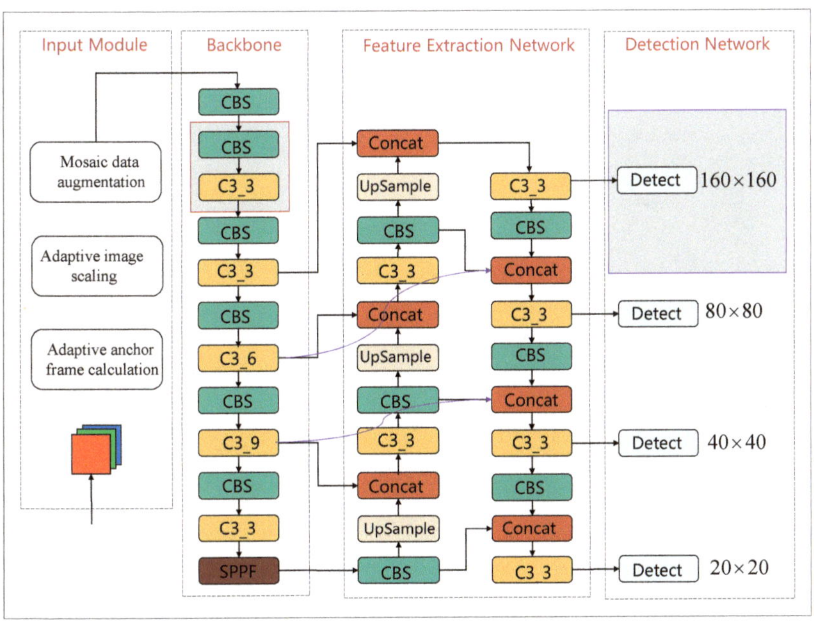

Figure 3. Schematic diagram of the structure of the YOLO-Xray model.

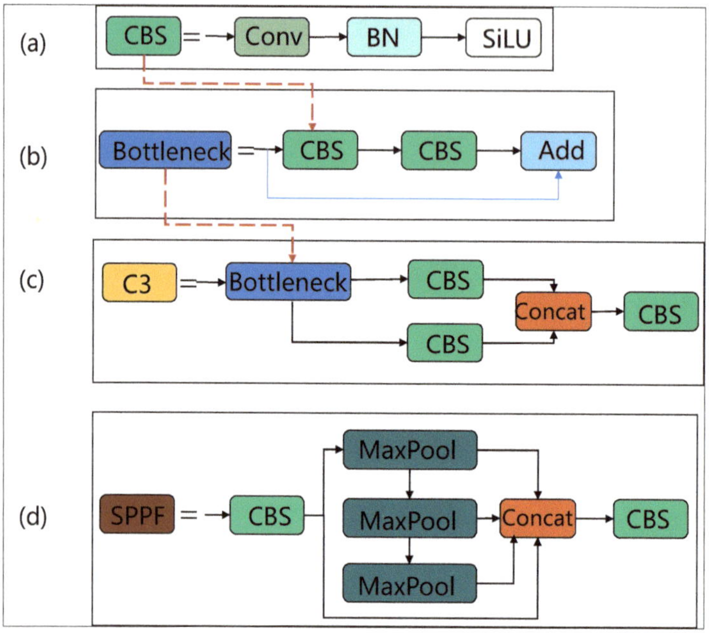

Figure 4. Schematic diagram of each functional module: (**a**) CBS; (**b**) bottleneck; (**c**) C3; (**d**) SPPF.

3.2. Anchor Design

The YOLOv5 model for object detection relies on the generation of a set of anchor boxes on the input image, with the selection of anchor boxes having a direct effect on the model's accuracy. In YOLOv5, when detecting a single category, predefined anchor boxes are utilized. However, the parameters of the predefined anchor boxes are derived from the COCO dataset [36], which has a large gap with the CXray dataset. Figure 5 depicts the size distribution of bubble flaws in the CXray dataset, revealing that our dataset contains a higher proportion of smaller-sized objects and larger aspect ratios. Considering that the CXray dataset focuses solely on detecting bubbles as defects, it becomes necessary to re-cluster the actual annotated bounding boxes.

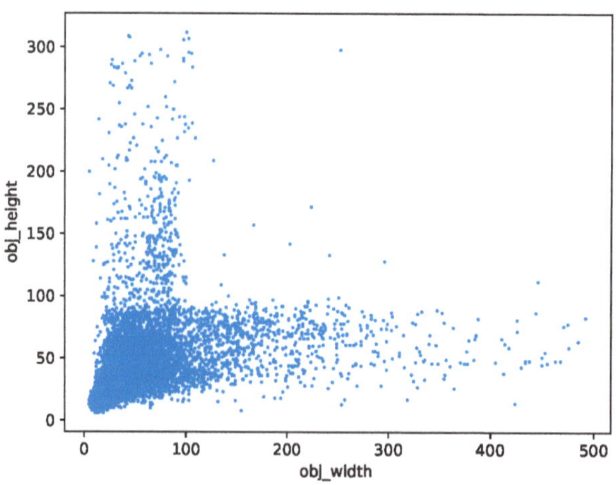

Figure 5. Size scatter distribution plots of bubble defects.

Therefore, the K-means++ algorithm [37] is used to re-cluster the prior anchor boxes that are suitable for our datasets. Table 1 displays the predefined anchor boxes and clustered anchor boxes.

The following is a description of how the K-means++ algorithm works in terms of its flow:

Table 1. Comparison of Anchor Boxes Before and After Improvement.

Feature Map Scale	Before	After
Large-scale	[116,90, 156,198, 373,326]	[38,34, 61,51, 142,57]
Medium-scale	[30,61, 62,45, 59,119]	[16,15, 21,19, 27,24]
Small-scale	[10,13, 16,30, 33,23]	[10,11, 14,12, 12,15]

Input: Read the CXray dataset and retrieve a set B of the widths and heights of all ground truth target boxes. Set K to the number of clustering centers.

Output: K anchor boxes.

1. Randomly select a value from the set, B, as the initial clustering center, C.
2. Calculate the minimum IoU distance, $d(x)$, between each sample, x, and the existing clustering centers, C. Determine the probability, denoted by $p(x)$, that each sample will serve as the subsequent clustering center. Select the next clustering center using the roulette wheel method, where samples with greater probabilities have a greater chance of being chosen.
3. Repeat step 2 until there are K clustering centers.
4. Calculate the IOU distance to the K clustering centers for each sample in the dataset and assign it to the cluster corresponding to the clustering center with the shortest distance.
5. Calculate the average width and height of each cluster and use those values as the new K clustering centers.
6. Repeat steps 4 and 5 until there is no longer any change in the positions of the clustering centers, resulting in the desired anchor boxes.

$d(x)$ and $p(x)$are defined as follows:

$$d(x) = 1 - IoU(x, C) \tag{2}$$

$$p(x) = \frac{d(x)^2}{\sum_{x \in B} d(x)^2} \tag{3}$$

where $x = (w1, h1)$, $C = (w2, h2)$, and IoU represents the intersection-over-union ratio between a sample point, x, and a clustering center, C, ranging from 0 to 1. If they are more similar, the IoU will be larger. IoU is defined as follows:

$$IoU = \frac{x \cap C}{x \cup C} \tag{4}$$

3.3. Micro-Scale Detection Head

The original YOLOv5 algorithm detects targets of different sizes using $8\times$, $16\times$, and $32\times$ downsampled feature maps as feature layers. Assuming 640×640 image data as input, the three output detection scales of size 20×20, 40×40, and 80×80 were utilized to identify targets of large, medium, and small sizes, respectively. However, the CXRay dataset contains a large number of defects that represent only a small portion of the whole image. The model is required to have a strong ability for detecting the small object. While the original YOLOv5 primarily employs the feature layer after $8\times$ downsampling to detect the small target, it neglected to utilize the target location information in the lower feature layers, resulting in the loss of the small target location. To solve the problem, we added a micro-scale small object detector compared with YOLOv5. The improved backbone network and detection network structure is shown in Figure 6.

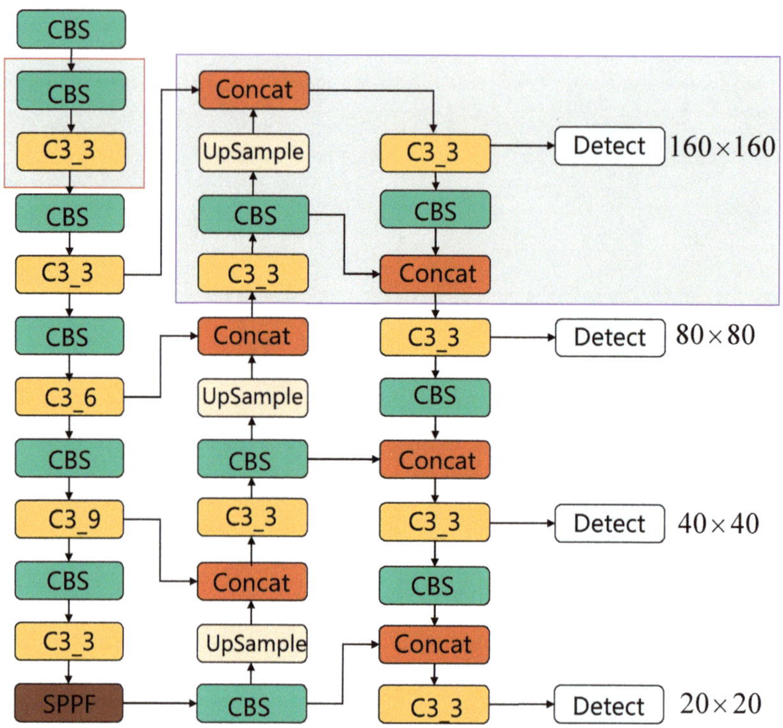

Figure 6. The structure of improved backbone and detection network.

Instead of using a 4× downsampling feature layer for feature fusion based on the original backbone network, we inserted a module (as depicted by the red box in Figure 6) between the 2× and 4× downsampling feature layers to improve the feature extraction capability and make better use of the micro-scale feature layer. The fusion module continues to employ the PANet [38] and FPN [39] networks for the feature fusion of different layers. The final output size for the small target detection layer is 160 × 160 (as shown in the purple box in Figure 6).

3.4. Bi-Directional Feature Fusion Module

The shallow features focus more on details and location information, which is good for localization, while deep features have stronger semantic information, which is good for classification. In the neck module, the YOLOv5 adopts the FPN structure and PAN structure to combine features of different layers. The FPN structure adopts a top-down structure, which can maintain the deeper semantic information but does not conceal the usage of shallow location and detail information, as shown in Figure 7a. The PAN structure adds a bottom–top downsampling path to the FPN structure, which can maintain the deeper semantic information and shallower location information, as shown in Figure 7b. However, in the CXRay dataset, there are long strip and irregular bubble defects with a limited sample size, and the majority of bubble defects only account for a small percentage of the overall image. To address the above-mentioned characteristics of defects and improve the defect detection accuracy, we updated the four scaled feature layers of the backbone network and referred to the concept of bi-directional feature fusion to build a new YOLOv5 feature fusion network.

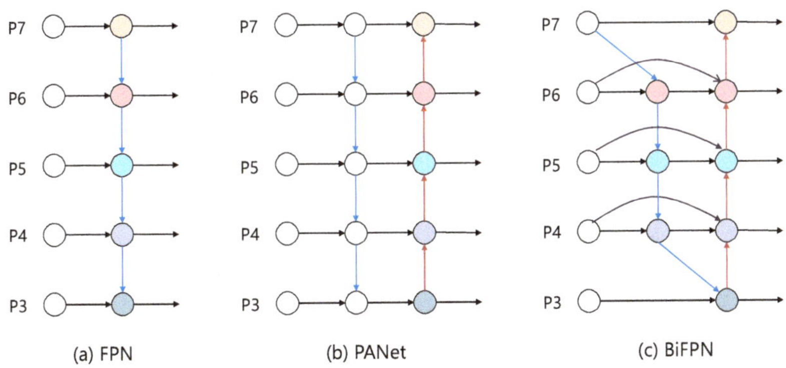

Figure 7. Structure diagrams of FPN, PANet, and BiFPN.

The BiFPN structure is a new feature fusion module suggested by Google for the EfficientDet network [40] that is based on PANet's architecture, as shown in Figure 7c. There are some significant design differences between the BiFPN and the PAN structure: (1) For the nodes with single input and output edges, because they are not involved in feature fusion, they contribute less to the network, so we removed the intermediate nodes of P3 and P7 in the PANet structure in Figure 7b. (2) Adding jump connections between input and output nodes at the same scale, as they are at the same layer, enables the fusion of more features without incurring an excessive computational expense.

Figure 8 illustrates the topology of the enhanced feature fusion module, in which the feature fusion technique is still merged by the number of channels. Feature layers acquired from the customized backbone network at 4×, 8×, 16×, and 32× downsampling are shown as $P_{2,1} - P_{5,1}$ in Figure 8. We eliminated the feature fusion layers $P_{2,2}$ and $P_{5,2}$ in layers 2 and 5, respectively, and introduced two jump connection lines that went from $P_{3,1}$ to $P_{3,3}$ and $P_{4,1}$ to $P_{4,3}$, respectively.

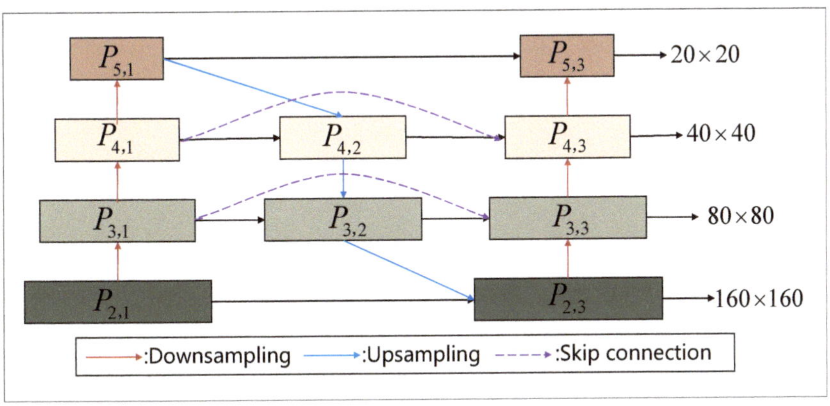

Figure 8. Improved feature extraction network.

3.5. Loss Function

The loss function of YOLOv5 consists of three parts: target confidence loss, localization loss, and category loss. The category loss and confidence loss are both binary cross entropy (BCE) losses, while the localization loss is the CIoU loss.

The BCE loss function is as follows:

$$L_{BCE}(y,\hat{p}) = -ylog(\hat{p}) - (1-y)log(1-\hat{p}) \tag{5}$$

where $\hat{p}$ is the probability of the predicted sample, and y is the true label.

Nevertheless, when detecting the CXray dataset, the model generates a large number of candidate frames, but these candidate frames contain fewer regions of the target (positive samples), and the majority of them are backgrounds that do not contain the target (negative samples), which can lead to the problem of positive and negative sample imbalance. The negative samples cannot be used to train a network, and a significant quantity of simple negative examples will degrade the model's performance during training, preventing the network from acquiring relevant information to effectively identify the target. In addition to positive and negative samples, there are also difficult and simple samples to consider. To solve the above problems, He et al. [26] suggested the focal loss for object detection networks, and this loss function may provide weight to each sample according to its predicted probability. The expression is as follows:

$$L_{focal} = \begin{cases} -\alpha(1-\hat{p})^{\gamma}log(\hat{p}), & \text{if } y=1; \\ -(1-\alpha)\hat{p}^{\gamma}log(\hat{p}), & \text{if } y=0. \end{cases} \tag{6}$$

where $\hat{p}$ is the probability of the predicted sample, y is the true label, α is the weight of a balanced positive and negative sample, and $(1-\hat{p})^{\gamma}$ and $\hat{p}^{\gamma}$ are the modulation coefficients.

However, focal loss only allows discrete category labels like 0 and 1, and confidence scores and classification scores are produced individually during the training process but combined during the inference step. Due to increased crossover and overlap between candidate frames when the targets are crowded, certain candidates with exact locations but low confidence are suppressed, which, in turn, affects the final detection results. To solve the above problems, Quality Focal Loss (QFocal Loss) [41] was used to replace the BCE Loss as the confidence loss and category loss. QFocal Loss is an extended form of focal loss on continuous label values, which can effectively balance both positive and negative as well as difficult and easy samples, and can also be adapted to the IOU-based supervision of continuous probability distributions. The expression is as follows:

$$L_{Qfocal} = -\alpha_t * |y-p|^{\beta}[(1-y)log(1-p) + ylog(p)] \tag{7}$$

where $\hat{p}$ is the probability of the predicted sample, y is the 0~1 label after smoothing, $\alpha_t = y*\alpha + (1-y)*(1-\alpha)$ is used to balance positive and negative samples, and $|y-p|^{\beta}$ is used to balance difficult and easy samples.

In this paper, we keep the original CIoU loss adopted by the YOLOv5 as the bounding box loss function. The formula of the CIoU loss function is shown below:

$$L_{CIoU} = 1 - IoU + \frac{\rho^2(b,b^{gt})}{C^2} + \alpha v \tag{8}$$

$$\alpha = \frac{v}{(1-IoU)+v} \tag{9}$$

$$v = \frac{4}{pi}(artan\frac{w^{gt}}{h^{gt}} - artan\frac{w}{h})^2 \tag{10}$$

where ρ^2 represents the Euclidean distance between the prediction box's center point and the ground truth box's center point. c is the diagonal length of the minimal enclosing box that encompasses both the prediction box and the ground truth box. w^{gt} and h^{gt} represent the width and height of the ground truth box, respectively, and w and h represent the width and height of the prediction box, respectively.

4. Experiments and Discussion

In this part, we first introduce the implementation detail and evaluation criterion. Then, we design two sets of experiments on the CXray dataset, the first of which conducts the ablation experiments, and the other compares the performance of the improved YOLO-Xray with the mainstream one-stage and two-stage object detection models. Finally, we visualize the qualitative results of chip determination.

4.1. Implementation

We conducted bubble defect detection experiments using the CXray dataset. The configuration of the experimental environment is described as follows: Ubuntu18.04, 128 G RAM, Intel Core i9-10920X, and NVIDIA RTX 3090Ti with 24 G memory. The model training and testing are conducted under PyTorch1.8.0 and CUDA11.1. The training hyperparameters are as follows: the input size was 640 × 640, the training epoch was set to 300, the batch size was set to 16, and the learning rate was set to 0.001.

4.2. Evaluation Criterion

Precision (P), recall (R), and $F1$-score are commonly used as evaluation metrics to assess the performance of the model. Precision is the model's capacity to precisely identify the detected objects. Recall denotes the model's capacity to identify and capture all instances of the target objects. $F1$-score takes into consideration both precision and recall and can be viewed as their harmonic mean. They are defined as follows:

$$P = \frac{TP}{TP + FP} \tag{11}$$

$$R = \frac{TP}{TP + FN} \tag{12}$$

$$F1 = 2 \times \frac{P \times R}{P + R} \tag{13}$$

where TP is the number of bubble defects that are correctly detected; FN is the number of bubble defects that are not detected; and FP is the number of incorrectly bubble defects. The mAP is the mean value of the AP of all classes.

The precision–recall (P–R) curve depicts the relationship between precision and recall graphically. Precision is represented on the vertical axis (y-axis), and recall is plotted on the horizontal axis (x-axis) in the P–R curve. The average precision (AP) is defined as the area under the P–R curve. The mean average precision (mAP) represents the average accuracy across all classes. These are defined as follows:

$$AP = \int_0^1 P(R)dR \tag{14}$$

$$mAP = \frac{1}{n}\sum_{i=1}^{n} AP_i \tag{15}$$

where n is the number of classes.

In engineering practical applications, the speed of detection and the size of the model are crucial factors. The frames per second (FPS) metric is used to evaluate the speed of the model during the testing stage. The model size is the quantity of storage space needed to store the model.

Therefore, the mean average precision (mAP), the frames per second (FPS), and the model size are used as metrics to assess the detection performance of the experiments.

4.3. Ablation Experiments

To verify the effectiveness of the proposed different improvements on the performance of the previous YOLOv5 algorithm, the five ablation studies were designed to evaluate the impact of K-means++, micro-scale detection layer, BiFPN, and QFocal Loss. A detailed comparison of different configurations is listed in Table 2, where ✓and ✕ represent the corresponding improvement strategies used and not used in the experiments, respectively.

Table 2. Experimental Results of Different Improvements.

Method	K-Means++	Micro-Scale Head	BiFPN	QFocal Loss	mAP0.5	Detection Speed (FPS)	Model Size (Mb)
A (the original YOLOv5)	✕	✕	✕	✕	0.884	**156.25**	**6.70**
B (A+ K-means++)	✓	✕	✕	✕	0.905	156.25	6.70
C (B+ micro-scale head)	✓	✓	✕	✕	0.912	138.89	8.34
D (C+ BiFPN)	✓	✓	✓	✕	0.919	119.05	9.56
E (YOLO-Xray)	✓	✓	✓	✓	**0.935**	119.05	9.56

The values in bold show the best results.

From Table 2, it can be observed that the baseline (original YOLOv5) achieved an mAP0.5 of 88.4% and the detection speed was 156.25 FPS.

A → B: Method B was to use the K-means++ technique to establish the priori boxes for the original three detection heads before the training step, with the same amount of prior frames as in the baseline model, resulting in no increase in model size or inference time. Because the updated priori boxes were better suited to the goal size of the dataset, method B outperformed the baseline model by 2.1% while having no influence on model size or detection speed, demonstrating the efficiency of the K-means++ clustering algorithm.

B → C: Method C was used to better extract the feature of tiny targets, and a micro-scale target detection head was introduced to the backbone of YOLOv5. The feature fusion still uses a combination of FPN and PANet to output a new detection head for tiny target prediction. As a consequence, the upgraded model was able to detect more tiny targets that were previously difficult to detect, boosting the mAP by 0.7%. However, because the new detection head increased the number of parameters, the detection speed was slower than previously.

C → D: Method D introduced the bi-directional feature fusion concept of BiFPN, which provides efficient multi-scale feature fusion via cross-level connections and sibling jump connections. Method D enhanced the detection of irregularly shaped and dense objects, the mAP was increased by 0.7%, and the detection speed was 119.05 FPS.

D → E: Method E used QFocal Loss as the loss function to fix the problem of an uneven number of positive and negative samples, as well as hard and easy ones, and the mAP increased by 1.6%. The loss function was optimized during the training phase, without changing the model size or detection speed. Finally, the mAP of the improved YOLO-Xray model reached 93.6%, which is 5.1% higher than the baseline model, proving the effectiveness of the improvement. When compared to the original YOLO5, the detection speed of YOLO-Xray is 37.20 FPS slower, yet it can still detect bubble defects in real time.

The results demonstrate the effectiveness of our improvements in the following four stages: input module, backbone network, neck module, and loss function. The improved YOLO-Xray can accurately detect bubble defects, including tiny targets and incomplete targets.

4.4. Performance Comparison of Different Object Detection Algorithms

To further evaluate the detection performance of the improved YOLO-Xray proposed in this study, the algorithm was compared with mainstream object detection models, which include two-stage networks (Faster R-CNN [18]) and one-stage networks (EfficientDet [40],

YOLOv4 [24], YOLOv5 [35], and the latest YOLOv7 [32]) under the same environment configuration. All the models were trained and tested on the CXray dataset. The mAP and FPS were used to evaluate and compare each detection algorithm, and a comparison of the experimental results are shown in Table 3.

Table 3. Performance Comparison between Mainstream Detection Models.

Model	mAP0.5	Detection Speed (FPS)
Faster R-CNN	0.907	30
EfficientDet	0.845	73
YOLOv4	0.865	57
YOLOv5	0.884	156
YOLOv7	0.867	122
YOLO-XRay	**0.935**	**119**

The values in bold show the best results.

According to Table 3, it can be seen that among the detection results, our YOLO-Xray model is the best in terms of overall performance. Although YOLOv7 outperformed many object detectors on the MS COCO dataset, with an map50 of 69.7%, the detection performance on the CXray dataset was inferior to that of the original YOLOv5. Compared to YOLOv7, our YOLO-Xray has higher detection accuracy, and the detection speed is close to that of YOLOv7. Faster R-CNN has a better detection accuracy of over 90% mAP, but the worst detection speed at only 30 FPS. The YOLO-Xray achieved 93.5% mAP, 2.8% higher than Faster R-CNN, 9% higher than EfficientDet, 7% higher than YOLOV4, 5.1% higher than YOLOV5, and 6.8% higher than YOLOv7. At the same time, the efficiency of YOLO-Xray is much better than other models, reaching 119 FPS, which is 4, 1.6, and 2.1 times faster than Faster R-CNN, EfficientDet, and YOLOV4, respectively. Based on the extensive analysis presented above, our suggested YOLO-Xray model has the greatest performance in terms of detection accuracy and speed for bubble defect detection, demonstrating the efficacy of our model.

4.5. Visualization of Defect Detection Results

Figure 9 illustrates some detection results of the original YOLOv5 and YOLO-Xray for the CXray dataset under various noise and contrast circumstances.

By comparing Figure 9, it can be found that in the first and second lines, there are some irregular bubble defects connected to the contour of the inner frame. The original YOLOv5 suffers from missed detection. While the improved YOLO-Xray accurately detects these bubble defects with high confidence. In the third and fourth lines of the detection results, there are several small and inconspicuous bubble defects. The original network is not ideal for detecting such bubble defects, and the leakage problem is serious. The improved model enhances the capability of detecting the types of defects and improves the detection of tiny bubbles. The results demonstrate the strong robustness and effectiveness of our YOLO-Xray model in complex environments.

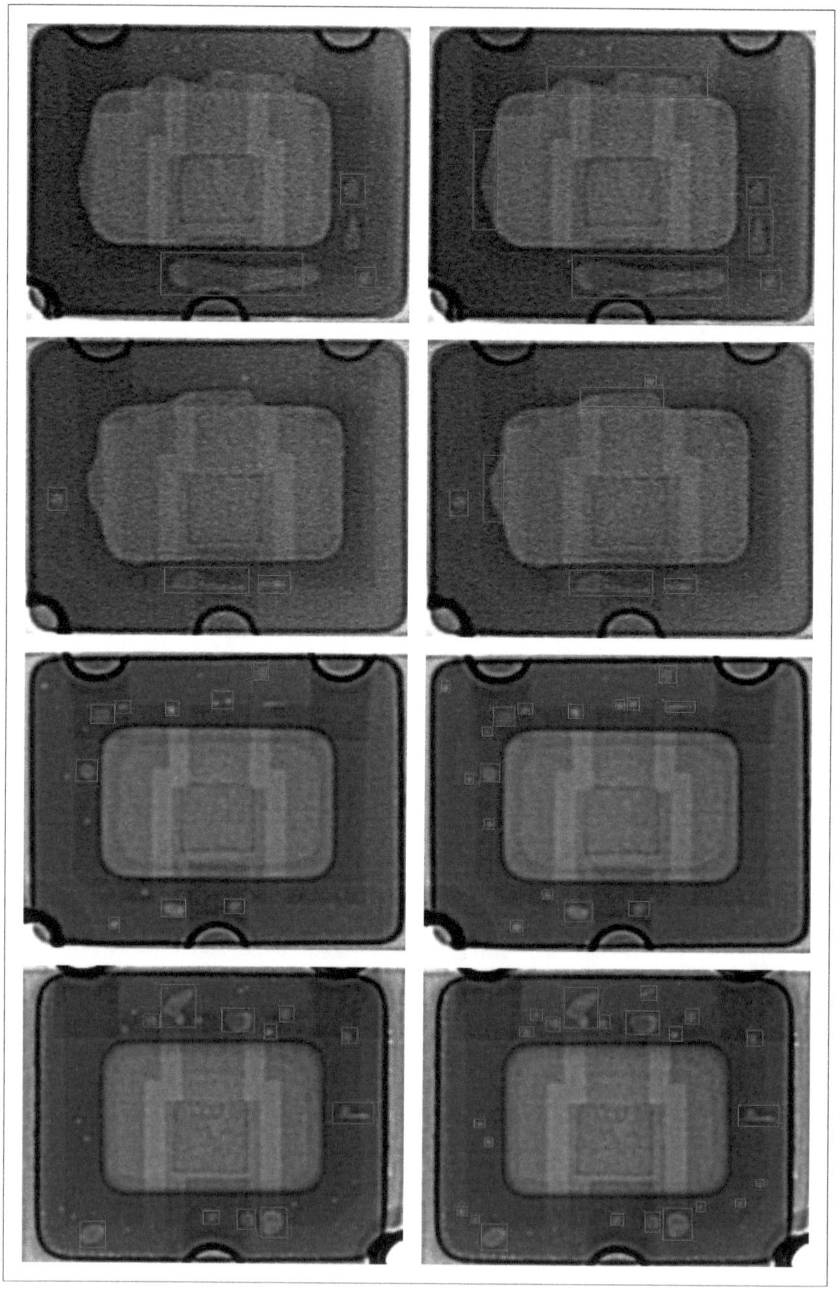

(a) (b)

Figure 9. Comparison of detection results between the original YOLOv5 and the improved YOLO-Xray. (**a**) YOLOv5, (**b**) YOLO-Xray.

5. Conclusions

In this paper, we proposed a bubble defect detection model based on YOLOv5 to address the problem of poor accuracy and low efficiency of traditional methods. Firstly, we constructed a dataset called CXray as the research basis for bubble defect detection, which is freely downloadable for research purposes. The dataset contains 1000 images with around 10,000 annotations. Various optimizations, which include the K-means++ clustering algorithm, the incorporation of micro-scale detection heads, BiFPN, and QFocal loss, were introduced to improve the performance of the original YOLOv5 model, which can effectively detect small and irregular bubble defects. The experimental results show that the improved YOLO-Xray increased the mAP from 88.4% to 93.5% compared to the original YOLOv5 and outperformed mainstream object detection models in terms of detection accuracy and speed on the CXray dataset. For chip manufacturers, our proposed YOLO-Xray model offers substantial benefits. It can considerably enhance the precision of defect detection, resulting in higher product quality and lower failure rates. In turn, this serves to reduce the costs associated with manufacturing defects. In addition, YOLO-Xray's high frames per second (FPS) performance reduces inspection time, resulting in increased efficiency and meeting the industry's detection requirements.

However, there are some future research areas for improvements:

1. Training the YOLO-Xray model requires a large quantity of annotated datasets, which can be both costly and time-consuming to acquire. Future research should investigate data augmentation techniques and the application of generative adversarial networks to generate synthetic defect samples. These methods can reduce the need for manually annotated data and accelerate the training process.
2. The YOLO-Xray model is based on enhancements made to the original network. In light of the need for lightweight models, additional research should investigate strategies for lightweight network architectures that reduce the parameter size of the model. This would allow its deployment on embedded devices, thereby expanding its applicability.

Lastly, we will continue to optimize our YOLO-Xray model and explore its application to different product types, aiming to expand its scope and versatility.

Author Contributions: Conceptualization, X.Z. and J.W.; methodology, J.W.; software, J.W.; validation, J.W., B.L., G.L., Y.Z. and L.Z.; formal analysis, J.W.; investigation, J.W.; resources, X.L.; data curation, J.W.; writing—original draft preparation, J.W.; writing—review and editing, J.W.; visualization, J.W.; supervision, X.Z.; project administration, X.Z.; funding acquisition, X.Z. All authors have read and agreed to the published version of the manuscript.

Funding: This research received no external funding.

Institutional Review Board Statement: Not applicable.

Informed Consent Statement: Not applicable.

Data Availability Statement: The CXray dataset is open and available at https://github.com/EudicL/CXray (accessed on 12 July 2023).

Acknowledgments: The authors wish to acknowledge China Aerospace Components Engineering for providing the X-ray images of chips used in this study.

Conflicts of Interest: The authors declare no conflict of interest.

References

1. Choi, K.J.; Lee, Y.H.; Moon, J.W.; Park, C.K.; Harashima, F. Development of an Automatic Stencil Inspection System Using Modified Hough Transform and Fuzzy Logic. *IEEE Trans. Ind. Electron.* **2007**, *54*, 604–611. [CrossRef]
2. Peng, X.; Chen, Y.; Yu, W.; Zhou, Z.; Sun, G. An online defects inspection method for float glass fabrication based on machine vision. *Int. J. Adv. Manuf. Technol.* **2008**, *39*, 1180–1189. [CrossRef]
3. Chetverikov, D.; Hanbury, A. Finding defects in texture using regularity and local orientation. *Pattern Recognit.* **2002**, *35*, 2165–2180. [CrossRef]
4. Lowe, D. Object recognition from local scale-invariant features. In Proceedings of the Seventh IEEE International Conference on Computer Vision, Corfu, Greece, 20–27 September 1999; Volume 2, pp. 1150–1157. [CrossRef]
5. Ojala, T.; Pietikainen, M.; Maenpaa, T. Multiresolution gray-scale and rotation invariant texture classification with local binary patterns. *IEEE Trans. Pattern Anal. Mach. Intell.* **2002**, *24*, 971–987. [CrossRef]

6. Chen, S.H.; Perng, D.B. Automatic optical inspection system for IC molding surface. *J. Intell. Manuf.* **2016**, *27*, 915–926. [CrossRef]
7. Lin, B.; Wang, J.; Yang, X.; Tang, Z.; Li, X.; Duan, C.; Zhang, X. Defect Contour Detection of Complex Structural Chips. *Math. Probl. Eng.* **2021**, *2021*, 5518675. [CrossRef]
8. Zhong, Z.; Ma, Z. A Novel Defect Detection Algorithm for Flexible Integrated Circuit Package Substrates. *IEEE Trans. Ind. Electron.* **2022**, *69*, 2117–2126. [CrossRef]
9. Kumar, A. Computer-Vision-Based Fabric Defect Detection: A Survey. *IEEE Trans. Ind. Electron.* **2008**, *55*, 348–363. [CrossRef]
10. Ferguson, M.; Ak, R.; Lee, Y.T.T.; Law, K.H. Automatic localization of casting defects with convolutional neural networks. In Proceedings of the 2017 IEEE International Conference on Big Data (Big Data), Boston, MA, USA, 11–14 December 2017; pp. 1726–1735. [CrossRef]
11. He, Y.; Song, K.; Meng, Q.; Yan, Y. An End-to-End Steel Surface Defect Detection Approach via Fusing Multiple Hierarchical Features. *IEEE Trans. Instrum. Meas.* **2020**, *69*, 1493–1504. [CrossRef]
12. Chen, J.; Liu, Z.; Wang, H.; Núñez, A.; Han, Z. Automatic Defect Detection of Fasteners on the Catenary Support Device Using Deep Convolutional Neural Network. *IEEE Trans. Instrum. Meas.* **2018**, *67*, 257–269. [CrossRef]
13. Tang, S.; He, F.; Huang, X.; Yang, J. Online PCB Defect Detector on a New PCB Defect Dataset. *arXiv* **2019**, arXiv:1902.06197.
14. Lin, H.; Li, B.; Wang, X.; Shu, Y.; Niu, S. Automated defect inspection of LED chip using deep convolutional neural network. *J. Intell. Manuf.* **2019**, *30*, 2525–2534. [CrossRef]
15. Luo, J.; Yang, Z.; Li, S.; Wu, Y. FPCB Surface Defect Detection: A Decoupled Two-Stage Object Detection Framework. *IEEE Trans. Instrum. Meas.* **2021**, *70*, 1–11. [CrossRef]
16. Li, W.; Zhang, L.; Wu, C.; Cui, Z.; Niu, C. A new lightweight deep neural network for surface scratch detection. *Int. J. Adv. Manuf. Technol.* **2022**, *123*, 1999–2015. [CrossRef]
17. Girshick, R.; Donahue, J.; Darrell, T.; Malik, J. Rich Feature Hierarchies for Accurate Object Detection and Semantic Segmentation. In Proceedings of the 2014 IEEE Conference on Computer Vision and Pattern Recognition, Columbus, OH, USA, 23–28 June 2014; pp. 580–587. [CrossRef]
18. Ren, S.; He, K.; Girshick, R.; Sun, J. Faster R-CNN: Towards Real-Time Object Detection with Region Proposal Networks. *Adv. Neural Inf. Process. Syst.* **2015**, *28*. [CrossRef]
19. Cai, Z.; Vasconcelos, N. Cascade R-CNN: Delving Into High Quality Object Detection. In Proceedings of the 2018 IEEE/CVF Conference on Computer Vision and Pattern Recognition, Salt Lake City, UT, USA, 18–22 June 2018; pp. 6154–6162. [CrossRef]
20. He, K.; Gkioxari, G.; Dollár, P.; Girshick, R.B. Mask R-CNN. In Proceedings of the IEEE International Conference on Computer Vision, Venice, Italy, 22–29 October 2017.
21. Redmon, J.; Divvala, S.; Girshick, R.; Farhadi, A. You Only Look Once: Unified, Real-Time Object Detection. In Proceedings of the 2016 IEEE Conference on Computer Vision and Pattern Recognition (CVPR), Las Vegas, NV, USA, 27–30 June 2016; pp. 779–788. [CrossRef]
22. Redmon, J.; Farhadi, A. YOLO9000: Better, Faster, Stronger. In Proceedings of the 2017 IEEE Conference on Computer Vision and Pattern Recognition (CVPR), Honolulu, HI, USA, 21–26 July 2017; pp. 6517–6525. [CrossRef]
23. Redmon, J.; Farhadi, A. YOLOv3: An Incremental Improvement. *arXiv* **2018**, arXiv:1804.02767.
24. Bochkovskiy, A.; Wang, C.; Liao, H.M. YOLOv4: Optimal Speed and Accuracy of Object Detection. *arXiv* **2020**, arXiv:2004.10934.
25. Liu, W.; Anguelov, D.; Erhan, D.; Szegedy, C.; Reed, S.E.; Fu, C.; Berg, A.C. SSD: Single Shot MultiBox Detector. *arXiv* **2015**, arXiv:1512.02325.
26. Lin, T.Y.; Goyal, P.; Girshick, R.; He, K.; Dollár, P. Focal Loss for Dense Object Detection. In Proceedings of the 2017 IEEE International Conference on Computer Vision (ICCV), Venice, Italy, 22–29 October 2017; pp. 2999–3007. [CrossRef]
27. Zhang, R.; Wen, C. SOD-YOLO: A Small Target Defect Detection Algorithm for Wind Turbine Blades Based on Improved YOLOv5. *Adv. Theory Simul.* **2022**, *5*, 2100631. [CrossRef]
28. Du, F.J.; Jiao, S.J. Improvement of lightweight convolutional neural network model based on YOLO algorithm and its research in pavement defect detection. *Sensors* **2022**, *22*, 3537. [CrossRef]
29. Zhang, M.; Yin, L. Solar cell surface defect detection based on improved YOLO v5. *IEEE Access* **2022**, *10*, 80804–80815. [CrossRef]
30. Shi, J.; Yang, J.; Zhang, Y. Research on steel surface defect detection based on YOLOv5 with attention mechanism. *Electronics* **2022**, *11*, 3735. [CrossRef]
31. Wang, T.; Su, J.; Xu, C.; Zhang, Y. An intelligent method for detecting surface defects in aluminium profiles based on the improved YOLOv5 algorithm. *Electronics* **2022**, *11*, 2304. [CrossRef]
32. Wang, C.Y.; Bochkovskiy, A.; Liao, H.Y.M. YOLOv7: Trainable bag-of-freebies sets new state-of-the-art for real-time object detectors. *arXiv* **2022**, arXiv:2207.02696.
33. Otsu, N. A Threshold Selection Method from Gray-Level Histograms. *IEEE Trans. Syst. Man Cybern.* **1979**, *9*, 62–66. [CrossRef]
34. Everingham, M.; Gool, L.V.; Williams, C.K.I.; Winn, J.M.; Zisserman, A. The Pascal Visual Object Classes (VOC) Challenge. *Int. J. Comput. Vis.* **2010**, *88*, 303–338. [CrossRef]
35. Ultralytics. YOLOv5. 2022. Available online: https://github.com/ultralytics/yolov5 (accessed on 20 June 2023).
36. Lin, T.Y.; Maire, M.; Belongie, S.; Hays, J.; Perona, P.; Ramanan, D.; Dollar, P.; Zitnick, C.L. Microsoft COCO: Common Objects in Context. In Proceedings of the 13th European Conference on Computer Vision (ECCV), COMPUTER VISION–ECCV 2014, PT V, Zurich, Switzerland, 6–12 September 2014; Fleet, D., Pajdla, T., Schiele, B., Tuytelaars, T., Eds.; Lecture Notes in Computer Science; Volume 8693, pp. 740–755. [CrossRef]

37. Vincent, O.; Makinde, A.; Salako, O.; Oluwafemi, O. A self-adaptive k-means classifier for business incentive in a fashion design environment. *Appl. Comput. Inform.* **2018**, *14*, 88–97. [CrossRef]
38. Liu, S.; Qi, L.; Qin, H.; Shi, J.; Jia, J. Path Aggregation Network for Instance Segmentation. *arXiv* **2018**, arXiv:1803.01534. [CrossRef]
39. Lin, T.Y.; Dollár, P.; Girshick, R.; He, K.; Hariharan, B.; Belongie, S. Feature Pyramid Networks for Object Detection. In Proceedings of the 2017 IEEE Conference on Computer Vision and Pattern Recognition (CVPR), Honolulu, HI, USA, 21–26 July 2017; pp. 936–944. [CrossRef]
40. Tan, M.; Pang, R.; Le, Q.V. EfficientDet: Scalable and Efficient Object Detection. *arXiv* **2019**, arXiv:1911.09070. [CrossRef]
41. Li, X.; Wang, W.; Wu, L.; Chen, S.; Hu, X.; Li, J.; Tang, J.; Yang, J. Generalized Focal Loss: Learning Qualified and Distributed Bounding Boxes for Dense Object Detection. *arXiv* **2020**, arXiv:2006.04388.

sensors

MDPI

Article

LPO-YOLOv5s: A Lightweight Pouring Robot Object Detection Algorithm

Kanghui Zhao [1], Biaoxiong Xie [1], Xingang Miao [1,2,*] and Jianqiang Xia [2]

[1] Beijing Engineering Research Center of Monitoring for Construction Safety, Beijing University of Civil Engineering and Architecture, Beijing 100044, China; 2108550022112@stu.bucea.edu.cn (K.Z.)
[2] Anhui Hengli Additive Manufacturing Technology Co., Ltd., Wuhu 241200, China
* Correspondence: miaoxingang@bucea.edu.cn

Abstract: The casting process involves pouring molten metal into a mold cavity. Currently, traditional object detection algorithms exhibit a low accuracy and are rarely used. An object detection model based on deep learning requires a large amount of memory and poses challenges in the deployment and resource allocation for resource limited pouring robots. To address the accurate identification and localization of pouring holes with limited resources, this paper designs a lightweight pouring robot hole detection algorithm named LPO-YOLOv5s, based on YOLOv5s. First, the MobileNetv3 network is introduced as a feature extraction network, to reduce model complexity and the number of parameters. Second, a depthwise separable information fusion module (DSIFM) is designed, and a lightweight operator called CARAFE is employed for feature upsampling, to enhance the feature extraction capability of the network. Finally, a dynamic head (DyHead) is adopted during the network prediction stage, to improve the detection performance. Extensive experiments were conducted on a pouring hole dataset, to evaluate the proposed method. Compared to YOLOv5s, our LPO-YOLOv5s algorithm reduces the parameter size by 45% and decreases computational costs by 55%, while sacrificing only 0.1% of mean average precision (mAP). The model size is only 7.74 MB, fulfilling the deployment requirements for pouring robots.

Keywords: casting process; object detection; YOLOv5s; DSIFM

1. Introduction

Casting is the basic process of the equipment manufacturing industry, and pouring is a crucial step in the casting process.Detection of pouring hole positions belongs to the object detection task in computer vision, which finds application in various industrial fields. Traditional object detection algorithms typically rely on manually designed feature extractors, such as SIFT [1], HOG [2], and others. However, these handcrafted features may not be suitable for different tasks and datasets, necessitating frequent adjustments and modifications. This increases the complexity and labor costs of the algorithms. Furthermore, traditional object detection algorithms are relatively slow and fail to meet real-time requirements. They also exhibit insensitivity to complex scenes and target occlusions, often resulting in inaccurate detection of partial targets or difficulties in distinguishing between targets. In contrast, deep-learning-based object detection algorithms have gained wide adoption in various intelligent robotics applications, due to their higher accuracy and faster speed. In deep learning, object detection can be categorized into two types: two-stage detection algorithms, and one-stage detection algorithms [3]. Two-stage detection algorithms generally achieve a higher localization and target recognition accuracy, but they are slower in terms of detection speed. Common two-stage algorithms include R-CNN [4], Fast R-CNN [5], Faster R-CNN [6], and Mask R-CNN [7]. In the field of unmanned aerial vehicles (UAVs), Yun et al. [8] proposed an improved algorithm for landslide detection in UAV imagery based on Mask R-CNN. By introducing attention mechanisms and modifying the feature pyramid network's structure, they were able to enhance the accuracy of landslide information extraction. On the other hand, one-stage algorithms offer a faster detection speed but may have a lower detection accuracy, such as SSD [9] and YOLO [10–12]. In the transportation domain, Lai et al. [13] developed a small traffic sign detection network (STC-YOLO) based on YOLOv5, for complex environments. By introducing a multi-head attention module and improving the loss function, they achieved a 5.9% improvement in mean average precision (mAP), yielding promising results. For the detection of small objects, Xiao et al. [14] proposed a context-enhanced and feature-refined network specifically designed for detecting small objects. Additionally, in many practical studies, the large model size of deep learning poses efficiency challenges when running on resource-limited embedded devices. Consequently, researchers have focused on lightweight model designs. Prominent lightweight models include SqueezeNet [15], MobileNet [16–18], ShuffleNet [19,20], and GhostNet [21], which have a reduced number of parameters and computational complexity to some extent without significant performance degradation. Jia et al. [22] proposed a lightweight and high-performance rice pest identification model, by replacing the backbone network of YOLOv7 with MobileNetv3 and combining this with an attention mechanism. It is worth mentioning that compared to lightweight networks such as ShuffleNetv2, MobileNetv3 can better balance efficiency and accuracy when completing real-world image classification tasks on mobile terminals and has therefore been widely adopted [23].

Sampling on feature maps is a very important step in object detection. The CARAFE [24] operator has a large receptive field and sufficient semantic information, without adding too many parameters, and its performance is better than transpose convolution [25] and other upsampling operations. Zhang et al. proposed a novel quad feature pyramid network (Quad-FPN) combined with a deformable convolutional FPN (DE CO FPN), content aware feature reassembly FPN (CA FR FPN), a path aggregation space attention FPN (PA SA FPN), and a balance scale global attention FPN (BS GA FPN), to achieve real-time monitoring of ship images [26].

At present, there are two main types of head modules for object detection: coupling heads, and decoupling heads. Among them, the detection effect of the decoupling head is higher than that of the coupling head in most cases. Muralidhara et al. [27] increased the upper limit of current video object detection methods by introducing DyHead detection heads into faster R-CNN, combining scale, space, and task aware attention, and achieving good results.

In recent years, with the rapid development of deep learning, more and more researchers are committed to research in the field of object detection (i.e., backbone, neck, and head). Li et al. proposed omnidimensional dynamic convolution (ODConv) [28], ODConv is a more generalized, yet elegant, dynamic convolution design. It utilizes novel multidimensional attention mechanisms and parallel strategies, to learn the complementary attention of convolutional kernels in all dimensions of the kernel space. Yu et al. proposed PoolFormer [29] by replacing the attention module in transformers with a spatial pooling operation, which achieved a good detection performance. Yang et al. proposed focal modulation networks (FocalNet) [30]; this network can modulate automatically and gradually converge to the object area that induces the recognition of the category. Yu et al. proposed a new network called InceptionNeXt [31] by combining inception with ConvNeXt, which not only had a high throughput but also maintained competitive performance. Generalized FPN [32] refines and integrates high-level semantic features and low-level spatial features. Based on GFPN, Xu et al. [33] proposed a novel Efficient-RepGFPN to achieve the design of real-time object detection. Li et al. designed a new neck network, named slim neck [34], to achieve a higher computational cost-effectiveness of detectors. Zhuang et al. proposed a novel task-specific context decoupling (TSCODE) [35] head, which further disentangles the feature encoding for classification and positioning tasks. Li et al. [36] adopted a hybrid-channel strategy, to build a more efficient decoupled head, which accelerated the inference speed of the network. However, a key issue that almost all of the above modules face is the significant increase in the number of network parameters and computational complexity, making them unsuitable for lightweight research on networks.

Regrettably, there has been no research on the visual recognition of pouring robots conducted by researchers. Therefore, the purpose of this study was to design a lightweight object detection algorithm suitable for pouring robots, which could be used to solve the problem of large-scale neural network models deployed in devices with limited resources and seizing other logical and peripheral resources. Inspired by the aforementioned works, this paper introduces the LPO-YOLOv5s algorithm, which stands for lightweight pouring hole detection algorithm based on YOLOv5s. To achieve this, we propose utilizing the MobileNetv3 network as the backbone network. By doing so, we can significantly reduce the model's parameters and computational complexity. However, reducing parameters often leads to a certain loss in performance. In order to minimize this loss, we designed a depthwise separable information fusion module (DSIFM) and incorporated it between the backbone layer and the neck layer of the algorithm. This addition does not result in a significant increase in parameters. In the neck network, we also considered using CARAFE, due to its lightweight features. Additionally, DyHead was employed in the network prediction stage to fine-tune the model's performance. The lightweight model proposed in this paper achieved excellent detection results, with fewer parameters and a smaller overall size.

The following are the main contributions of this article:

1. A lightweight object detection algorithm for pouring tasks is proposed, with MobileNetv3 as the backbone network, greatly reducing the parameter number and computational complexity;
2. A DSIFM module is designed, to enable the network to obtain different levels of receptive field, and enrich and fuse the contextual feature information, so as to carry out effective feature extraction;
3. During the upsampling process, a CARAFE operator is used to better focus on the relevant point information in the local area and obtain stronger semantic information;
4. In the model detection phase, a plug and play DyHead detection head is used to improve the detection performance of the algorithm;
5. A dataset of pouring hole locations has been established, to provide assistance and a reference for future research.

2. Methods

2.1. Basic Module

YOLOv5 is a one-stage object detection algorithm that utilizes multiple layers of convolutional neural networks to extract the position and category information of objects. The network structure of YOLOv5 consists of four parts: input, backbone, neck, and prediction. The input part preprocesses the dataset and includes operations such as mosaic data augmentation, adaptive anchor box calculation, and adaptive image scaling. The backbone is comprised of Conv modules and C3 modules. The Conv module includes Conv2D, Batch Normalization, and SiLU activation function. The C3 module consists of three standard convolutions and multiple bottleneck modules, which are interconnected through residual connections to extract features from the feature map. The SPPF module, located at the end of the backbone, is based on the SPP [37] (spatial pyramid pooling) module and aims to extract high-level semantic features by performing multiple max pooling operations. The neck part adopts the structures of a feature pyramid network (FPN) [38] and pyramid attention network (PAN) [39]. The FPN propagates high-level semantic features in a top-down manner, through upsampling and fusion, while the PAN conveys strong localization features of shallow-level feature maps in a bottom-up manner. By combining FPN and PAN, the network obtains both localization and semantic information, enhancing the feature fusion capability of the network. In the prediction phase, all generated prediction boxes are filtered using the complete intersection over union (CIoU-loss) [40] and non-maximum suppression (NMS) [41] algorithms. The prediction boxes with the highest confidence scores are outputted, completing the object detection process. The specific flowchart is shown in Figure 1.

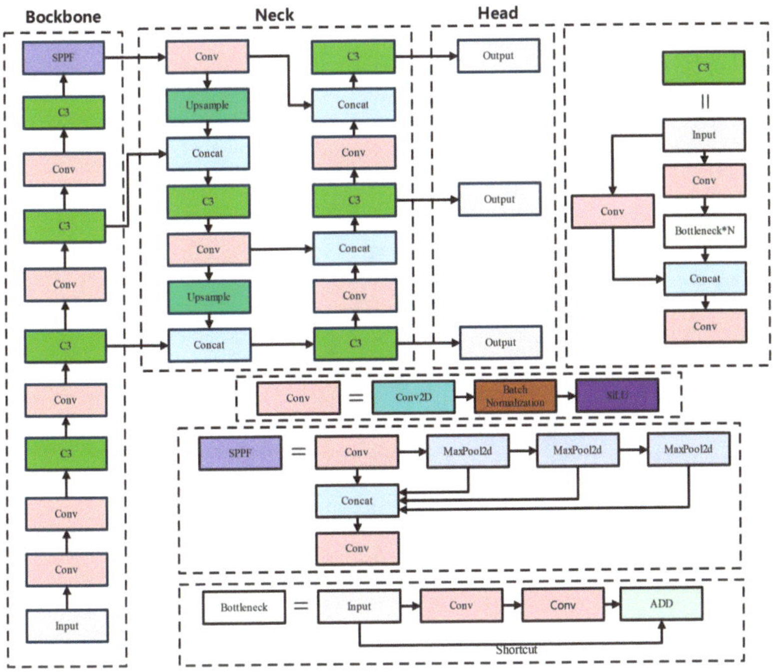

Figure 1. Structures of YOLOv5s network.

2.2. *LPO-YOLOv5s*

The LPO-YOLOv5 algorithm represents an improvement over YOLOv5. The overall structure of the network is similar, comprising an input, backbone network, neck network, and head network, as depicted in Figure 2. To achieve a lightweight model and reduce the parameter quantity and network complexity, we employ a lightweight MobileNetv3 network in the backbone network for feature extraction. Addressing the issue of imbalanced information distribution between high-level and low-level feature maps, we introduce DSIFM to replace the SPPF module. In the neck network, we utilize the CARAFE module to effectively aggregate contextual information. Furthermore, the DyHead detection head in the head network serves as a plug-and-play module for detecting different perspectives, thereby improving the detection performance of pouring hole positions.

2.2.1. Lightweight Backbone: MobileNetv3

The MobileNetv3 network retains the effective depthwise separable convolution of MobileNetv1 and the linear bottleneck inverted residual structure of MobileNetv2. Additionally, it introduces a lightweight attention module called an SE structure. Moreover, the network incorporates a novel non-linear activation function called h-swish, which combines with the ReLU activation function to reduce memory access, improve network accuracy, and minimize latency costs.

In Figure 3, the module begins by expanding the input features through a 1×1 convolution, to enhance the information content of the image. Next, the expanded features undergo a 3×3 convolution for depthwise separable convolution, which reduces the parameter computation. The resulting features from the depthwise separable convolution are then passed through the SE attention module. Within this module, the features are globally average pooled and processed by two fully connected layers, automatically assigning weights to the feature maps. This weight allocation amplifies valuable feature channels, while suppressing irrelevant ones. It is worth noting that the activation function for the first fully connected layer is ReLU, while the second fully connected layer employs h-swish. Finally, the reweighted features are dimensionally reduced through a 1×1 convolution and added back to the original input feature map, resulting in a new feature map.

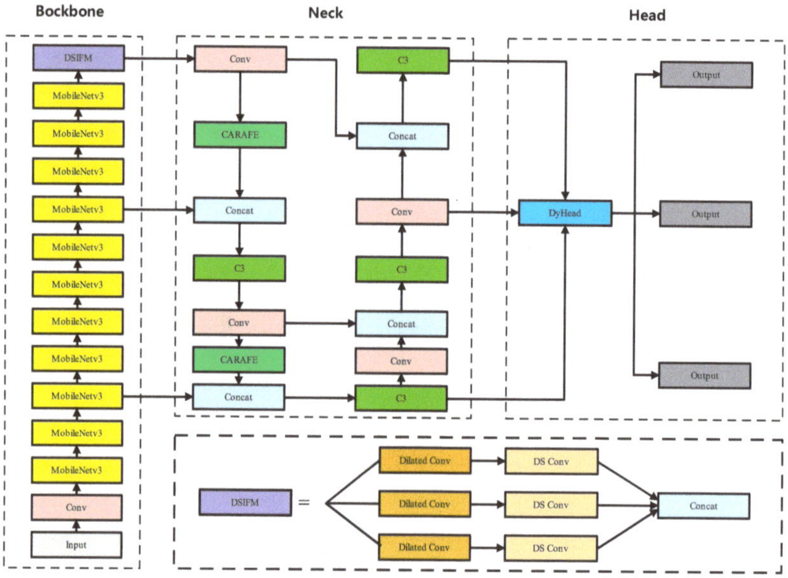

Figure 2. Structure of the LPO-YOLOv5s network.

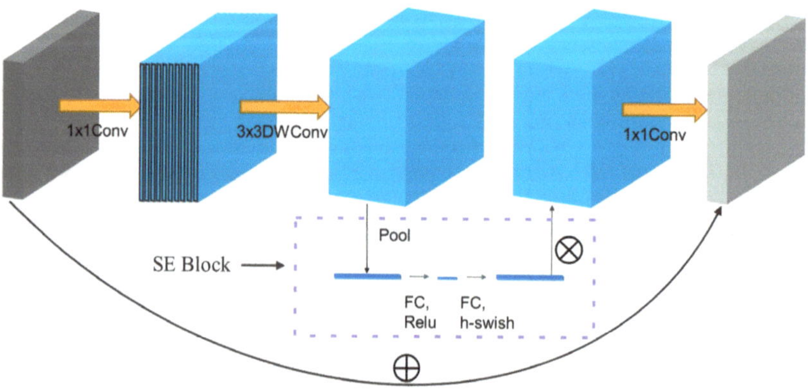

Figure 3. Structure of the lightweight backbone: MobileNetv3.

2.2.2. CARAFE: Lightweight Upsampling Operator

Feature upsampling plays a critical role in the feature pyramid, particularly in object detection tasks. YOLOv5 incorporates the nearest-neighbor interpolation method [42] as its feature upsampling operator. This method employs grayscale values of the nearest pixels to guide the upsampling process, utilizing the same upsampling kernel for every position in the feature map. While this algorithm is computationally efficient and fast, it neglects the influence of neighboring pixels, resulting in noticeable discontinuities in the resampled grayscale values and a significant loss of image features. To address these limitations, CARAFE introduces a universal lightweight upsampling operator that performs upsampling based on the input content, incorporating a larger receptive field, to efficiently aggregate contextual information. CARAFE is lightweight, requires no additional parameters or computational cost, and provides increased processing speeds.

Figure 4 illustrates the module structure of CARAFE. Assuming the size of the input feature map is C × H × W, the upsampling factor is σ, and the size of the upsampling kernel is k_{up}. First, the given feature map is compressed along the channel dimension using a 1 × 1 convolution, resulting in a compressed feature map of size H × W × C_m. Further channel compression reduces the computational burden of the network and accelerates the processing speed. Next, a convolutional layer with a size of $k_{encoder} \times k_{encoder}$ is used to encode and predict the upsampling kernel based on the compressed feature map. The input channel is C_m, and the output channel is $\sigma^2 k_{up}^2$. The predicted upsampling kernel is then unfolded in the spatial dimension, resulting in an upsampling convolutional kernel of shape σH × σW × k_{up}^2. Finally, the Softmax function is applied to normalize the upsampling kernel, ensuring that the sum of its weights is 1.

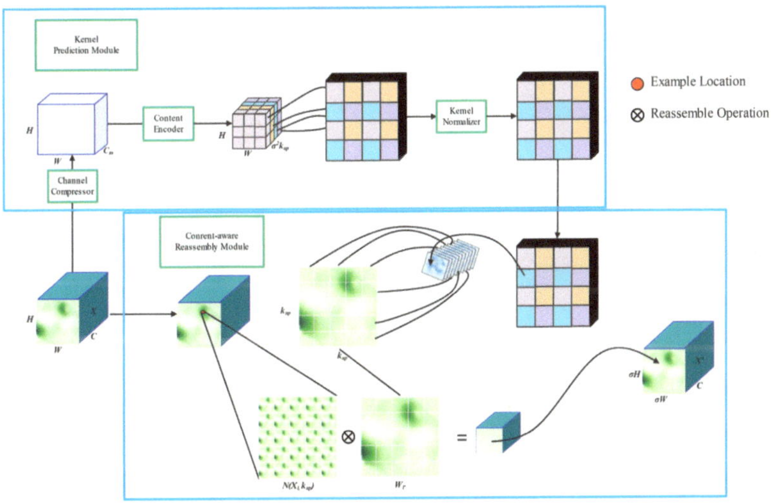

Figure 4. Lightweight upsampling operator: CARAFE. Note: H, height of the pouring hole feature map; W, width of the pouring hole feature map; C_m, length of the compressed pouring hole feature map; C, length of the original pouring hole feature map; X, original pouring hole feature map; σ, up-sampling multiplier; k_{up}, reorganization kernel size; X', reorganization pouring hole feature map; $W_{l'}$, feature reorganization kernel; N, square region; X_l, feature map at l.

Through the action of each recombined convolutional kernel $W_{l'}$, the content-aware reassembly module recombines features within local regions using a weighted summation operator. The recombination process for the target position l' and the corresponding square region N($X_{l'}$,k_{up}) centered at $l = (i,j)$ is shown in Equation (1), where $r = \lfloor k_{up}/2 \rfloor$.

$$X'_{l'} = \sum_{n=-r}^{r} \sum_{m=-r}^{r} W_{l'(n,m)} \cdot X_{(i+n,j+m)} \tag{1}$$

The feature map after the CARAFE operator can better focus on relevant point information within local regions compared to the nearest-neighbor interpolation sampling method, enabling the capture of stronger semantic information.

2.2.3. Depthwise Separable Information Fusion Module: DSIFM

Background contextual information plays a crucial role in object detection tasks and should also be considered in the pouring task. In the benchmark model YOLOv5, the SPPF module connects the backbone feature extraction network and the neck feature fusion network, partially addressing the challenge of the network's difficulty in extracting high-level semantic features. However, the max pooling layer in the module leads to the loss of detailed information in the feature map. This lost information cannot be recovered through subsequent upsampling operations, making it challenging for the network to acquire sufficient contextual information. As a result, the algorithm's performance in complex tasks is suboptimal. To overcome this limitation, we propose a deep separable information fusion module (DSIFM) to replace the SPPF module. Our DSIFM aims to obtain contextual information with different receptive fields, and a model structure diagram is shown in Figure 5.

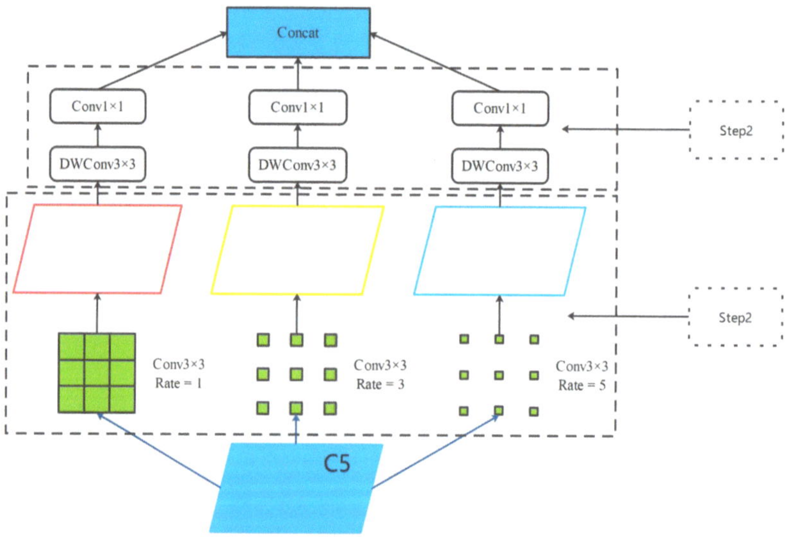

Figure 5. The overall framework of the DSIFM. This module mainly consists of two steps: a multi-scale dilated convolution module, and depth separable convolution module.

To enrich the contextual feature information and facilitate effective feature extraction, we applied dilated convolution operations to the feature maps that underwent five down-sampling operations. We used dilation rates of 1, 3, and 5, as features with both large and small receptive fields, to complement each other. By incorporating different dilation rates, the object detection network can acquire multiple levels of receptive fields. This approach enhances contextual feature information. Moreover, dilated convolution introduces zeros between convolutional kernels, resulting in a relatively smaller number of additional parameters. This aligns with our objective of designing a lightweight network. Then, the feature maps outputted from the dilated convolution module were each subjected to a depthwise separable convolution and finally combined for enhanced network robustness.

2.2.4. Dynamic Head: Unifying with Attentions

Due to the variations in pouring tasks, the size of pouring holes can differ significantly. Consequently, the detection head of the network must possess the capability to detect pouring holes of varying scales. Furthermore, considering the robot's positional movement during operation, the detection head needs to incorporate spatial perception. To address these challenges, we introduced DyHead as the detection head algorithm during the network's prediction stage. DyHead is a multi-attention unified detection head method proposed by Dai et al. [43]. It treats the input of the backbone network as distinct attention mechanisms deployed at the scale level, spatial level, and channel level. This approach achieves a unified attention mechanism for the detection head. Whether used in a one-stage or two-stage object detection algorithm, DyHead, as a plug-and-play lightweight module, can be embedded to enhance the detection performance of the algorithm.

Before applying the DyHead detection head, it is necessary to scale the feature pyramid outputted by the backbone network to a consistent scale. The feature maps of the scaled pyramid can be represented as a four-dimensional tensor $\boldsymbol{F} \in R^{L\times H\times W\times C}$, where L denotes the number of levels in the feature pyramid, and H and W represent the height and width of the feature map, respectively. The channel dimension is denoted by C, indicating the number of channels in the feature map. As H and W represent spatial features, they are replaced with S, defined as S = H × W. This abstraction allows us to view the feature map as a three-dimensional tensor, with dimensions L, S, and C focusing on scale perception, spatial perception, and channel perception, respectively.

Figure 6 illustrates the principle of the DyHead. The feature maps undergo mapping and sequentially pass through the scale perception module (π_L), spatial perception module (π_S), and task perception module (π_C), to achieve a unified attention mechanism for the detection head. In the scale perception module, the L dimension of the feature map undergoes global average pooling, followed by feature extraction using a 1 × 1 convolutional layer and ReLU activation function. The result is then passed through a hard sigmoid activation function and element-wise multiplied with the input feature map, concluding the scale perception module. For the spatial perception module, the input tensor is processed using a 3 × 3 convolutional layer, to obtain the offset values of the feature map, as well as the weight terms of the feature map offsets. Spatial perception is achieved through the utilization of deformable convolutional [44]. In the task perception module, the tensor undergoes global average pooling on the L × S dimension to reduce the dimensionality and parameter count. It then proceeds through two fully connected layers and a normalization layer. Dynamic ReLU [45] is applied, which outputs different channel values based on different tasks, completing the task perception of the feature map.

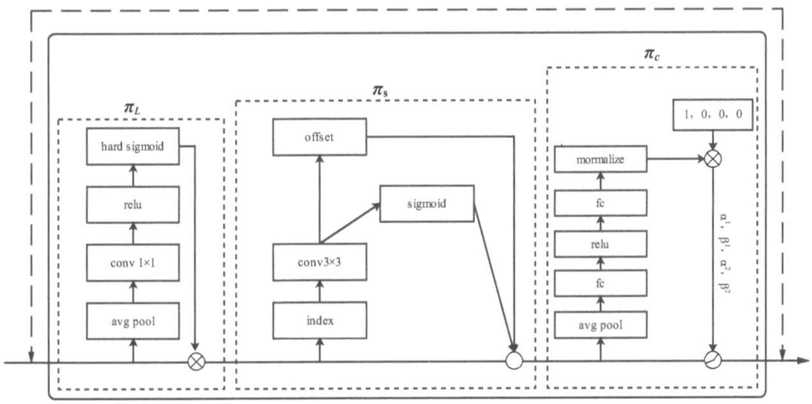

Figure 6. A specific introduction to DyHead.

3. Expression Results

3.1. Experimental Setting

This study utilized Python version 3.8, PyTorch version 1.11.0, CUDA version 11.2, and a GeForce RTX 3060 VENTUS 2X 12G OC graphics card. The chosen hyperparameters included hyp.scratch-low, a batch size of 32, and 300 rounds of training. In order to ensure the rigor and smooth progress of the experiment, the pre-training weights and other parameters all adopted the default values of YOLOv5s.

3.2. Experimental Dataset

3.2.1. Data Acquisition

Due to the limited number of researchers studying pour hole detection, a publicly available complete dataset is currently unavailable. Therefore, all the data in this study were captured in the natural working environment of the pouring workshop at Anhui Hengli Additive Manufacturing Technology Co., Ltd., in Wuhu City, Anhui Province, China, using a smartphone. To ensure the quality and effectiveness of the dataset, we aimed to cover different shooting scenarios as comprehensively as possible. During the data collection process, we observed the impact of lighting conditions on image quality. Hence, we made efforts to include various lighting conditions in the collection, encompassing both natural and artificial light. Additionally, we considered the diversity of shooting angles and distances, to ensure sufficient coverage in the dataset. As pouring tasks in the factory may involve unrelated personnel obstructing the pour holes, we also incorporated data with partially obscured pour holes, to enhance the dataset's robustness. Through a series of data augmentation operations, such as translation and flipping, the final dataset consisted of 2790 images. These images were divided into training, testing, and validation sets in an 8:1:1 ratio. The resolution of the data ranged from 1080 pixels × 1920 pixels to 2080 pixels × 4608 pixels. Sample images from the dataset are shown in Figure 7.

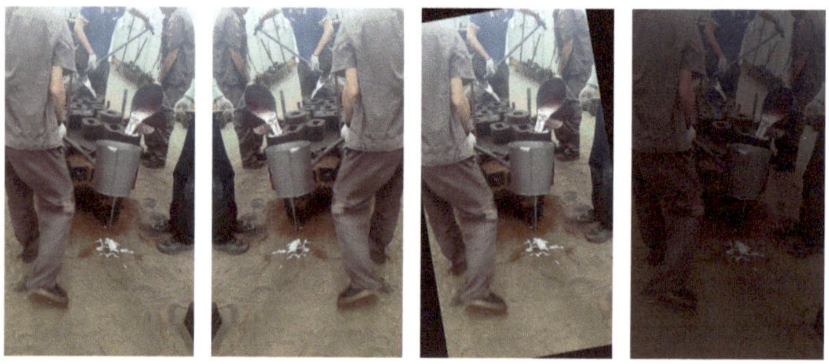

(**a**) original (**b**) flip (**c**) rotate (**d**) dark treatment

Figure 7. Data set of hole location detection in the pouring environment.

3.2.2. Data Annotation

In addition to dataset acquisition, the annotation task is also crucial. We employed the Labelimg tool for the manual annotation of the dataset. Labelimg can generate annotation data in two formats: VOC format and YOLO format. For the convenience of experimentation, all image data annotations were conducted with the YOLO format. Through manual annotation, we ensured that each photo accurately marked the position and size of the pouring hole. Considering and addressing all these details contributed to the quality and reliability of the dataset, enhancing the accuracy and generalization capability of the model, and providing effective data support for the pouring hole detection task.

3.3. Evaluation Metrics

This paper evaluated the LPO-YOLOv5s model using widely recognized evaluation metrics in the field of object detection. The evaluation metrics employed included the precision rate (P), recall rate (R), mAP, F1-score, and weight size. Since this article presents a lightweight object detection algorithm, the parameter count was the most crucial metric. The definitions of the various evaluation metrics are as follows:

$$P = \frac{TP}{TP + FP} \tag{2}$$

$$R = \frac{TP}{TP + FN} \tag{3}$$

In Equations (2) and (3), TP represents the number of correctly detected hole positions, where the detection results and the labels in the images matched. Similarly, FP indicates the number of false detections, where the detection results identified hole positions, but the ground truth labels were backgrounds. FN represents the number of missed hole positions, where the model's detection results identified backgrounds, but the model labeled them as hole positions, indicating the rate of missed hole positions.

$$F1\text{-}score = 2 \times \frac{P \times R}{P + R} \tag{4}$$

In Equation (4), the *F1-score* is used to balance the weights of P and R, representing the value when both reach their highest points simultaneously.

$$AP = \int_0^1 P(R)dR \tag{5}$$

$$mAP = \frac{1}{N}\sum_{i=1}^{N} AP_i \tag{6}$$

Here, AP represents the area under the P-R curve, N represents the total number of classes, and mAP is the average value of AP across all classes. In this paper, mAP is reported as mAP@0.5, indicating the average precision when the IoU threshold was set to 0.5, In the experiments, a higher mAP value signified a better model performance, indicating the algorithm's effectiveness.

3.4. Results

This paper conducted numerous experiments using the pouring hole detection dataset, and the metrics obtained during the model training process are shown in Figure 8.

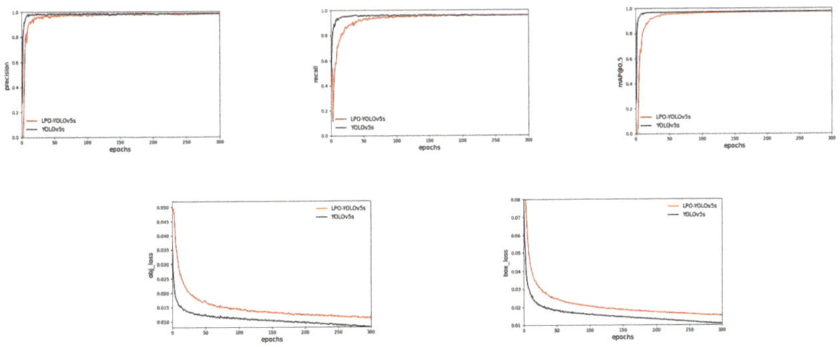

Figure 8. The model training results for LPO-YOLOv5s and YOLOv5s.

From Figure 8, it is evident that, as the number of model training iterations was increased, both the accuracy, recall rate, and mAP@0.5 value of the original YOLOv5s algorithm and the LPO-YOLOv5s algorithm gradually improved. Although replacing the backbone network resulted in significant fluctuations and a slower convergence speed of the LPO-YOLOv5s algorithm in the first 50 epochs, when the number of training iterations reached 100 epochs, the metrics of both algorithms gradually stabilized and almost converged to the same value. We also found that, compared to the yolov5s algorithm, our method had a higher median value in the model loss curve, but ultimately converged to a relatively small value, which was in line with the experimental expectations.

Additionally, according to Table 1, we can see that our algorithm, compared to YOLOv5s, experienced only a slight decrease of 0.1% in mAP@0.5, while reducing the parameter count by 45% and lowering the computational complexity by 55%. This makes it easier to deploy in hardware and meets the requirements for pouring robots. The detection results of the two algorithms are shown in the Figure 9.

Table 1. The results of the LPO-YOLOv5s and YOLOv5s models in the validation.

Model	P	R	F1	mAP@0.5	GFLOPs	Parameters	FPS	Weight Size
YOLOv5s	97.8%	96.4%	0.97	97.3%	15.8	7,012,822	91	13.6 (MB)
Ours	98.2%	96.0%	0.97	97.2%	7.0	3,874,924	80	7.74 (MB)

3.4.1. Comparative Experiment

To further demonstrate the performance advantages of the LPO-YOLOv5s algorithm, this paper conducted experimental comparisons on the pouring hole dataset using several mainstream object detection algorithms. Since the algorithm proposed in this paper is an improvement on one-stage algorithms, only faster R-CNN, a two-stage algorithm, was selected for comparison, while the rest were single-stage detection models such as SSD, YOLOX [46], YOLOv7 [47], YOLOv8, and so on. The experimental results are shown in Table 2.

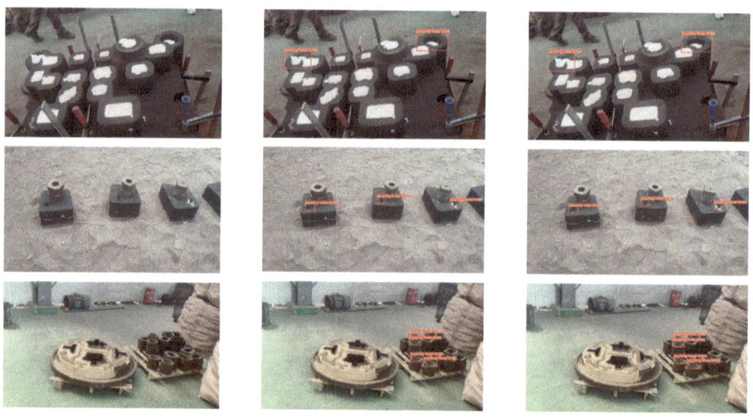

(**a**) Original data (**b**) YOLOv5s (**c**) LPO-YOLOv5s

Figure 9. The detection performance of the two algorithms before and after improvement.

Table 2. Comparative experimental results of the various models on the hole detection dataset.

Model	P	R	F1	mAP@0.5	FPS	Weight Size
SSD	96.8%	97.7%	0.87	95.7%	62	90.6 (MB)
Faster-RCNN	88.1%	98.3%	0.88	98.1%	11	108 (MB)
YOLOv3	96.9%	92.4%	0.95	97.3%	63	235 (MB)
YOLOX-S	96.8%	95.9%	0.96	98.7%	82	34.3 (MB)
YOLOv7-tiny	97.8%	95.3%	0.96	96.3%	85	11.6 (MB)
YOLOv8s	97.4%	97.0%	0.97	97.2%	131	21.4 (MB)
Ours	98.2%	96.0%	0.97	97.2%	80	7.74 (MB)

According to Table 2, it can be observed that the proposed algorithm achieved the highest precision among the algorithms in the table, with a value of 98.2%. This precision value was 10.1% higher than that of the Faster R-CNN algorithm and 0.4% higher than that of the YOLOv5s algorithm. Moreover, the proposed algorithm also performed well in terms of the F1 score, reaching 0.97, which was the highest value among all the algorithms. The faster R-CNN algorithm had the highest recall rate among all the algorithms. However, due to its large parameter size and lower values in other metrics, it did not meet the requirements for pouring robot operation. It is worth noting that the YOLOX-s algorithm had the highest mAP@0.5 value. However, its model size was relatively large, making it unsuitable for deployment in resource-limited embedded devices. Similarly, YOLOv8 had the fastest inference speed, reaching 131 FPS, but its model size was about three times that of the algorithm in this paper. Additionally, during the experiments, we found that the recall rate of the proposed method was relatively low compared to the other state-of-the-art algorithms, and there was no significant improvement compared to the original YOLOv5s. However, in the context of pouring tasks, precision is of greater importance. Due to the nature of pouring tasks involving human operators, a slight decrease in recall rate would not significantly affect the progress of pouring tasks. On the contrary, a higher precision value indicates a more accurate detection and better performance in completing pouring tasks.

3.4.2. Ablation Experiment

To verify the effectiveness of each module of the proposed LPO-YOLOv5s algorithm in pouring hole detection, this study conducted ablation experiments under the same experimental conditions, using the precision and mAP@0.5 as evaluation metrics. From Table 3, it can be observed that the MobileNetv3 module was primarily introduced to lighten the model, which indeed resulted in some performance loss. However, with the integration of other modules, the performance of the improved model gradually improved. Among them, the DSIFM module improved the accuracy of the model by 0.4%, while the CARAFE module primarily affected the recall rate of the model. Finally, it was found that the addition of the DyHead module on top of these improvements made the model even more prominent, with significant improvements in the precision and mAP@0.5 values. The experimental results fully demonstrate the effectiveness of the designed LPO-YOLOv5s algorithm presented in this paper.

Table 3. The results of the ablation experiment.

Model	MobileNetv3	DSIFM	CARAFE	DyHead	Precision	mAP@0.5
YOLOv5s	×	×	×	×	0.978	0.973
A	√	×	×	×	0.974	0.965
B	√	√	×	×	0.979	0.966
C	√	√	√	×	0.979	0.968
D	√	√	√	√	0.982	0.972

4. Conclusions

The total number of foundries in China has exceeded 20,000, with an average annual production of 1777 tons of castings per factory, which is significantly lower than the top-ranked country Germany, producing 8818 tons. China ranks last among the top-ten countries in global casting production. This is primarily due to the low level of automation in the foundry industry. To tackle this challenge, as well as enhance the intelligence and meet the demands of digitized factories, this paper proposed a lightweight pouring hole detection algorithm for pouring robots. In terms of reducing parameters, we introduced the MobileNetV3 network to replace the backbone of YOLOv5s. However, through experiments, we found that the performance of the algorithm declined to a certain extent. Therefore, between the backbone network and the neck network of YOLOv5s, we designed a DSIFM module, to strengthen the information fusion between the network layers. Experiments proved that the DSIFM module improved the detection accuracy, and the CARAFE upsampling operator improved the algorithm's recall rate. At the end of the algorithm, a DyHead attention detection head was used, and the detection advantages of the algorithm in this paper were shown through testing with a pouring dataset. While minimizing performance loss, the algorithm achieved significant reductions in parameters and computational complexity, with the model size being only 7.74 MB, a 43% reduction compared to the unimproved scale. Through experimental comparisons with other algorithms, the proposed algorithm achieved a highest accuracy of 98.2%, meeting the requirements for pouring robots, in terms of both detection accuracy and parameter scale. In the future, we will further optimize the pouring robot to accurately identify and locate targets. At the same time, we will also consider improving the robot's autonomous navigation ability for independently planning paths, to accelerate the automation process of the casting industry.

Author Contributions: Conceptualization, X.M. and K.Z.; methodology, K.Z.; software, K.Z.; validation, K.Z.; formal analysis, K.Z.; investigation, K.Z., B.X. and X.M.; resources, J.X.; data curation, K.Z.; writing—original draft preparation, K.Z.; writing—review and editing, K.Z. and X.M.; visualization, K.Z.; supervision, X.M.; project administration, X.M. and J.X.; funding acquisition, X.M. All authors have read and agreed to the published version of the manuscript.

Funding: Funded by the Open Research Fund Program of the Beijing Engineering Research Center of Monitoring for Construction Safety (grant no. BJC2020K009).

Data Availability Statement: Not applicable.

Conflicts of Interest: The authors declare no conflict of interest.

References

1. Lowe, D.G. Distinctive image features from scale-invariant keypoints. *Int. J. Comput. Vis.* **2004**, *60*, 91–110. [CrossRef]
2. Dalal, N.; Triggs, B. Histograms of oriented gradients for human detection. In Proceedings of the 2005 IEEE Computer Society Conference on Computer Vision and Pattern Recognition (CVPR'05), San Diego, CA, USA, 20–26 June 2005; IEEE: Piscataway Township, NJ, USA, 2005; Volume 1, pp. 886–893.
3. Zou, Z.; Chen, K.; Shi, Z.; Guo, Y.; Ye, J. Object detection in 20 years: A survey. *Proc. IEEE* **2023**, *111*, 257–276. [CrossRef]
4. Girshick, R.; Donahue, J.; Darrell, T.; Malik, J. Rich feature hierarchies for accurate object detection and semantic segmentation. In Proceedings of the IEEE Conference on Computer Vision and Pattern Recognition, San Francisco, CA, USA, 18–20 June 1996; pp. 580–587.
5. Girshick, R. Fast r-cnn. In Proceedings of the IEEE International Conference on Computer Vision, Cambridge, MA, USA, 20–23 June 1995; pp. 1440–1448.
6. Ren, S.; He, K.; Girshick, R.; Sun, J. Faster r-cnn: Towards real-time object detection with region proposal networks. *Adv. Neural Inf. Process. Syst.* **2015**, *28*, 91–99. [CrossRef]
7. He, K.; Gkioxari, G.; Dollár, P.; Girshick, R. Mask r-cnn. In Proceedings of the IEEE International Conference on Computer Vision, Cambridge, MA, USA, 20–23 June 1995; pp. 2961–2969.
8. Yun, L.; Zhang, X.; Zheng, Y.; Wang, D.; Hua, L. Enhance the accuracy of landslide detection in UAV images using an improved Mask R-CNN Model: A case study of Sanming, China. *Sensors* **2023**, *23*, 4287. [CrossRef]
9. Liu, W.; Anguelov, D.; Erhan, D.; Szegedy, C.; Reed, S.; Fu, C.Y.; Berg, A.C. Ssd: Single shot multibox detector. In Proceedings of the Computer Vision—ECCV 2016: 14th European Conference, Amsterdam, The Netherlands, 11–14 October 2016; Proceedings, Part I 14; Springer: Berlin/Heidelberg, Germany, 2016; pp. 21–37.
10. Redmon, J.; Divvala, S.; Girshick, R.; Farhadi, A. You only look once: Unified, real-time object detection. In Proceedings of the IEEE Conference on Computer Vision and Pattern Recognition, San Francisco, CA, USA, 18–20 June 1996; pp. 779–788.
11. Redmon, J.; Farhadi, A. YOLO9000: Better, faster, stronger. In Proceedings of the IEEE Conference on Computer Vision and Pattern Recognition, San Francisco, CA, USA, 18–20 June 1996; pp. 7263–7271.
12. Redmon, J.; Farhadi, A. Yolov3: An incremental improvement. *arXiv* **2018**, arXiv:1804.02767.
13. Lai, H.; Chen, L.; Liu, W.; Yan, Z.; Ye, S. STC-YOLO: Small Object Detection Network for Traffic Signs in Complex Environments. *Sensors* **2023**, *23*, 5307. [CrossRef]
14. Xiao, J.; Zhao, T.; Yao, Y.; Yu, Q.; Chen, Y. Context augmentation and feature refinement network for tiny object detection. In Proceedings of the ICLR 2022 Conference, Virtual, 25–29 April 2022.
15. Iandola, F.N.; Han, S.; Moskewicz, M.W.; Ashraf, K.; Dally, W.J.; Keutzer, K. SqueezeNet: AlexNet-level accuracy with 50x fewer parameters and <0.5 MB model size. *arXiv* **2016**, arXiv:1602.07360.
16. Howard, A.G.; Zhu, M.; Chen, B.; Kalenichenko, D.; Wang, W.; Weyand, T.; Andreetto, M.; Adam, H. Mobilenets: Efficient convolutional neural networks for mobile vision applications. *arXiv* **2017**, arXiv:1704.04861.

17. Sandler, M.; Howard, A.; Zhu, M.; Zhmoginov, A.; Chen, L.C. Mobilenetv2: Inverted residuals and linear bottlenecks. In Proceedings of the IEEE Conference on Computer Vision and Pattern Recognition, San Francisco, CA, USA, 18–20 June 1996; pp. 4510–4520.
18. Howard, A.; Sandler, M.; Chu, G.; Chen, L.C.; Chen, B.; Tan, M.; Wang, W.; Zhu, Y.; Pang, R.; Vasudevan, V. Searching for mobilenetv3. In Proceedings of the IEEE/CVF International Conference on Computer Vision, Seoul, Republic of Korea, 27 October–2 November 2019; pp. 1314–1324.
19. Zhang, X.; Zhou, X.; Lin, M.; Sun, J. Shufflenet: An extremely efficient convolutional neural network for mobile devices. In Proceedings of the IEEE Conference on Computer Vision and Pattern Recognition, San Francisco, CA, USA, 18–20 June 1996; pp. 6848–6856.
20. Ma, N.; Zhang, X.; Zheng, H.T.; Sun, J. Shufflenet v2: Practical guidelines for efficient cnn architecture design. In Proceedings of the European Conference on Computer Vision (ECCV), Tel Aviv, Israel, 23–27 October 2022; pp. 116–131.
21. Han, K.; Wang, Y.; Tian, Q.; Guo, J.; Xu, C.; Xu, C. Ghostnet: More features from cheap operations. In Proceedings of the IEEE/CVF Conference on Computer Vision and Pattern Recognition, Seattle, WA, USA, 14–19 June 2020; pp. 1580–1589.
22. Jia, L.; Wang, T.; Chen, Y.; Zang, Y.; Li, X.; Shi, H.; Gao, L. MobileNet-CA-YOLO: An Improved YOLOv7 Based on the MobileNetV3 and Attention Mechanism for Rice Pests and Diseases Detection. *Agriculture* **2023**, *13*, 1285. [CrossRef]
23. Qian, S.; Ning, C.; Hu, Y. MobileNetV3 for image classification. In Proceedings of the 2021 IEEE 2nd International Conference on Big Data, Artificial Intelligence and Internet of Things Engineering (ICBAIE), Nanchang, China, 26–28 March 2021; IEEE: Piscataway Township, NJ, USA, 2021; pp. 490–497.
24. Wang, J.; Chen, K.; Xu, R.; Liu, Z.; Loy, C.C.; Lin, D. Carafe: Content-aware reassembly of features. In Proceedings of the IEEE/CVF International Conference on Computer Vision, Seoul, Republic of Korea, 27 October–2 November 2019; pp. 3007–3016.
25. Dumoulin, V.; Visin, F. A guide to convolution arithmetic for deep learning. *arXiv* **2016**, arXiv:1603.07285.
26. Zhang, T.; Zhang, X.; Ke, X. Quad-FPN: A novel quad feature pyramid network for SAR ship detection. *Remote Sens.* **2021**, *13*, 2771. [CrossRef]
27. Muralidhara, S.; Hashmi, K.A.; Pagani, A.; Liwicki, M.; Stricker, D.; Afzal, M.Z. Attention-guided disentangled feature aggregation for video object detection. *Sensors* **2022**, *22*, 8583. [CrossRef]
28. Li, C.; Zhou, A.; Yao, A. Omni-dimensional dynamic convolution. *arXiv* **2022**, arXiv:2209.07947.
29. Yu, W.; Luo, M.; Zhou, P.; Si, C.; Zhou, Y.; Wang, X.; Feng, J.; Yan, S. Metaformer is actually what you need for vision. In Proceedings of the IEEE/CVF Conference on Computer Vision and Pattern Recognition, New Orleans, LA, USA, 18–26 June 2022; pp. 10819–10829.
30. Yang, J.; Li, C.; Dai, X.; Gao, J. Focal modulation networks. *Adv. Neural Inf. Process. Syst.* **2022**, *35*, 4203–4217.
31. Yu, W.; Zhou, P.; Yan, S.; Wang, X. Inceptionnext: When inception meets convnext. *arXiv* **2023**, arXiv:2303.16900.
32. Jiang, Y.; Tan, Z.; Wang, J.; Sun, X.; Lin, M.; Li, H. GiraffeDet: A heavy-neck paradigm for object detection. *arXiv* **2022**, arXiv:2202.04256.
33. Xu, X.; Jiang, Y.; Chen, W.; Huang, Y.; Zhang, Y.; Sun, X. Damo-yolo: A report on real-time object detection design. *arXiv* **2022**, arXiv:2211.15444.
34. Li, H.; Li, J.; Wei, H.; Liu, Z.; Zhan, Z.; Ren, Q. Slim-neck by GSConv: A better design paradigm of detector architectures for autonomous vehicles. *arXiv* **2022**, arXiv:2206.02424.
35. Zhuang, J.; Qin, Z.; Yu, H.; Chen, X. Task-Specific Context Decoupling for Object Detection. *arXiv* **2023**, arXiv:2303.01047.
36. Li, C.; Li, L.; Jiang, H.; Weng, K.; Geng, Y.; Li, L.; Ke, Z.; Li, Q.; Cheng, M.; Nie, W.; et al. YOLOv6: A single-stage object detection framework for industrial applications. *arXiv* **2022**, arXiv:2209.02976.
37. He, K.; Zhang, X.; Ren, S.; Sun, J. Spatial pyramid pooling in deep convolutional networks for visual recognition. *IEEE Trans. Pattern Anal. Mach. Intell.* **2015**, *37*, 1904–1916. [CrossRef]
38. Lin, T.Y.; Dollár, P.; Girshick, R.; He, K.; Hariharan, B.; Belongie, S. Feature pyramid networks for object detection. In Proceedings of the IEEE Conference on Computer Vision and Pattern Recognition, San Francisco, CA, USA, 18–20 June 1996; pp. 2117–2125.
39. Li, H.; Xiong, P.; An, J.; Wang, L. Pyramid attention network for semantic segmentation. *arXiv* **2018**, arXiv:1805.10180.
40. Zheng, Z.; Wang, P.; Liu, W.; Li, J.; Ye, R.; Ren, D. Distance-IoU loss: Faster and better learning for bounding box regression. In Proceedings of the AAAI Conference on Artificial Intelligence, New York, USA, 7–12 February 2020; Volume 34, pp. 12993–13000.
41. Neubeck, A.; Van Gool, L. Efficient non-maximum suppression. In Proceedings of the 18th International Conference on Pattern Recognition (ICPR'06), Hong Kong, China, 20–24 August 2006; IEEE: Piscataway Township, NJ, USA, 2006; Volume 3, pp. 850–855.
42. Rukundo, O.; Cao, H. Nearest neighbor value interpolation. *arXiv* **2012**, arXiv:1211.1768.
43. Dai, X.; Chen, Y.; Xiao, B.; Chen, D.; Liu, M.; Yuan, L.; Zhang, L. Dynamic head: Unifying object detection heads with attentions. In Proceedings of the IEEE/CVF Conference on Computer Vision and Pattern Recognition, Online, 19–25 June 2021; pp. 7373–7382.
44. Dai, J.; Qi, H.; Xiong, Y.; Li, Y.; Zhang, G.; Hu, H.; Wei, Y. Deformable convolutional networks. In Proceedings of the IEEE International Conference on Computer Vision, Cambridge, MA, USA, 20–23 June 1995; pp. 764–773.
45. Chen, Y.; Dai, X.; Liu, M.; Chen, D.; Yuan, L.; Liu, Z. Dynamic relu. In Proceedings of the Computer Vision—ECCV 2020: 16th European Conference, Glasgow, UK, 23–28 August 2020; Proceedings, Part XIX 16; Springer: Berlin/Heidelberg, Germany, 2020; pp. 351–367.

46. Ge, Z.; Liu, S.; Wang, F.; Li, Z.; Sun, J. Yolox: Exceeding yolo series in 2021. *arXiv* **2021**, arXiv:2107.08430.
47. Wang, C.Y.; Bochkovskiy, A.; Liao, H.Y.M. YOLOv7: Trainable bag-of-freebies sets new state-of-the-art for real-time object detectors. In Proceedings of the IEEE/CVF Conference on Computer Vision and Pattern Recognition, Vancouver, BC, Canada, 18–22 June 2023; pp. 7464–7475.

electronics

Article

Skeleton-Based Human Action Recognition Based on Single Path One-Shot Neural Architecture Search

Yujian Jiang [1,2,3,4,*], Saisai Yu [1,2,3,4], Tianhao Wang [1,2,3,4], Zhaoneng Sun [1,2,3,4] and Shuang Wang [1,2,3,4]

1 State Key Laboratory of Media Convergence and Communication, Communication University of China, Beijing 100024, China; sai_yss@cuc.edu.cn (S.Y.); wangtianhao@cuc.edu.cn (T.W.); sunzhaoneng@cuc.edu.cn (Z.S.); wangshuang@cuc.edu.cn (S.W.)
2 Key Laboratory of Acoustic Visual Technology and Intelligent Control System, Ministry of Culture and Tourism, Communication University of China, Beijing 100024, China
3 Beijing Key Laboratory of Modern Entertainment Technology, Communication University of China, Beijing 100024, China
4 School of Information and Communication Engineering, Communication University of China, Beijing 100024, China
* Correspondence: yjjiang@cuc.edu.cn

Abstract: Skeleton-based human action recognition based on Neural Architecture Search (NAS.) adopts a one-shot NAS strategy. It improves the speed of evaluating candidate models in the search space through weight sharing, which has attracted significant attention. However, directly applying the one-shot NAS method for skeleton recognition requires training a super-net with a large search space that traverses various combinations of model parameters, which often leads to overly large network models and high computational costs. In addition, when training this super-net, the one-shot NAS needs to traverse the entire search space of the complete skeleton recognition task. Furthermore, the traditional method does not consider the optimization of the search strategy. As a result, a significant amount of search time is required to obtain a better skeleton recognition network model. A more efficient weighting model, a NAS skeleton recognition model based on the Single Path One-shot (SNAS-GCN) strategy, is proposed to address the above challenges. First, to reduce the model search space, a simplified four-category search space is introduced to replace the mainstream multi-category search space. Second, to improve the model search efficiency, a single-path one-shot approach is introduced, through which the model randomly samples one architecture at each step of the search training optimization. Finally, an adaptive Covariance Matrix Adaptation Evolution Strategy (CMA-ES) is proposed to obtain a candidate structure of the perfect model automatically. With these three steps, the entire network architecture of the recognition model (and its weights) is fully and equally trained significantly. The search and training costs will be greatly reduced. The search-out model is trained by the NTU-RGB + D and Kinetics datasets to evaluate the performance of the proposed model's search strategy. The experimental results show that the search time of the proposed method in this paper is 0.3 times longer than that of the state-of-the-art method. Meanwhile, the recognition accuracy is roughly comparable compared to that of the SOTA NAS-GCN method.

Keywords: neural architecture search; graph convolution neural network; skeleton action recognition; search space

1. Introduction

Skeleton-based human action recognition has grown in popularity as a research topic in computer vision in recent years. It has been extensively employed in various domains, including human–computer interaction and stage performance arts. A Graph Convolutional Network (GCN) [1,2] is the mainstream method for skeleton recognition, which excels in handling non-Euclidean data and has produced outstanding outcomes in human skeleton recognition. However, when creating a 10-layer network that alternates between a spatial GCN and temporal GCN, it is typically necessary to use a GCN for the skeletal recognition directly. To find the perfect network layout and training settings, and to achieve an adequate recognition accuracy, a significant amount of manual training and intricate ablation experiments are frequently needed. Some researchers have suggested using the GCN and the NAS (Neural Architecture Search) to automatically determine the best network layout and parameters to increase the model's recognition effectiveness. For instance, the literature (Pérez-Rúa, Juan-Manuel et al.) [3] proposed a multi-modal skeleton recognition model based on neural architecture search, which achieved a high recognition accuracy in the skeleton recognition task. However, to introduce NAS into the skeleton recognition task, factors such as the number of layers of the skeleton recognition model, the selection of the optimization modules in the model, and the categories and weights of the dataset need to be considered. As a result, NAS should initially offer a vast search space for the skeleton identification operation. The optimal model parameters, however, take a while to converge because there are too many candidate operations in the search space.

Training thousands of models is difficult or impossible for a traditional machine learning task. To solve the problem of architecture search for models, the researchers propose the idea of sharing weights between models: instead of the traditional method of training thousands of individual models from scratch, a super network is proposed, which can be trained to simulate any architecture in the search space. To make the search more flexible, instead of deciding whether a particular layer is convolutional, average pooling, or maximum pooling, the method changes the search space to mix all of the above choices in one search process. The search time is reduced by assigning weights to each component of the search space when training the skeleton model (Peng, Wei et al., 2020) [4]. A simple example of a search space is shown in Figure 1, where one can apply a 3×3 convolution, 5×5 convolution, or max-pooling layer at specific locations in the network. The search space contains three different operations; the one-shot model adds their outputs together. The implementation idea is to treat all operations, such as convolution, average pooling, and maximum pooling, as channels and allow the controller to select a mask over these channels. It is possible to train a single model containing all three operations, rather than training three separate models. By allowing the parameters to be shared between all the architectures in the search space, thus avoiding the need to train each architecture from scratch, it is more computationally efficient than a standard NAS.

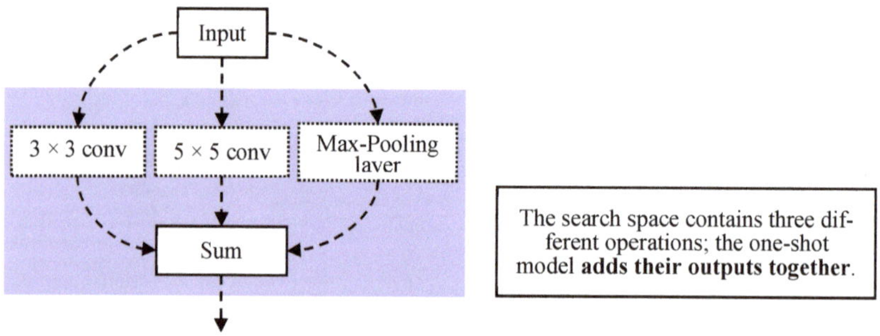

Figure 1. One-shot NAS method.

However, the weight-sharing method based on NAS must traverse the whole network's search space, which causes the search process to be overtrained in the NAS search area, and results in the absence of optimization and an automated optimization strategy. Therefore, the NAS has to thoroughly search the ideal network topology and training parameters for the skeleton identification challenge. This thesis considers the field of a GCN based on a Single Path One-Shot Neural Architecture Search (SNAS-GCN) for human skeleton recognition as the main subject of its study. In this paper, a simplified, four-category NAS search space for skeleton recognition tasks is proposed. Subsequently, to reduce the search training time of the super-net and obtain an excellent skeleton recognition model in a relatively short period, the weight-sharing search is replaced by a single path random search, which is a single path one-shot search strategy. Applying SNAS-GCN to the skeletal identification job is not simple. Skeleton recognition jobs have complicated and bloated search spaces due to the different types of models (RNN, CNN, and GCN) and data inputs (skeleton data and RGB data). A challenging problem in this domain is optimizing the search process in the NAS space. In addition, many combinations of network layers and parameters are applied to the skeleton recognition task. Obtaining the optimal combination and optimal network structure using an optimization-seeking algorithm is another challenge. To quickly identify the ideal neural network in the search space, the Covariance Matrix Adaptation Evolution Strategy (CMA-ES) is presented. With the proposed SNAS-GCN strategy, a graph convolutional network for skeleton recognition is constructed rapidly. In evaluating the performance of the search strategy proposed in this paper, the model approach took 18 h less than the NAS-GCN model to train a skeleton recognition network with a roughly comparable accuracy on the NTU-RGB + D [5] and Kinetics [6] datasets.

The contributions of this paper include the following two points:

- To improve the search efficiency and reduce the NAS search time, the optimization operation of the NAS search space is simplified. Secondly, a single-path one-shot weight-sharing model is proposed to replace the original weight-sharing strategy.
- To automatically sample candidate networks from the super-net, a new covariance matrix adaptation evolution strategy is employed.

2. Related Work

2.1. Skeleton Recognition Based on GCN

Human action recognition based on skeleton data has attracted more attention due to its robustness to changes in human scale, viewpoint, and background. With the development of deep learning, traditional methods use recurrent neural networks (RNN) [5,7,8] and convolutional neural networks (CNN) [9–12] to compute and analyze skeleton data. However, the method based on RNNs and CNNs has a high complexity, and the model's ability to deal with the skeleton structure needs to be improved [13]. A graph convolution network (GCN) can better capture the space–time relationship between bodies and extract the advanced features of the human skeleton. Therefore, in recent years, researchers have proposed a skeleton recognition model based on a graph convolution network (GCN). (Yan, Sijie, 2018) [1] proposed the spatiotemporal graph convolutional network (ST-GCN) model, which extracts human body feature information by introducing the Spatio-temporal map convolution network and solving the problem of the previous model only being able to deal with temporal features, but not extract spatial features. (Shi et al., 2019) proposed a two-stream adaptive graph convolutional network (2s-AGCN) [2] model based on the ST-GCN model, which improves performance by collecting the dual information of joints and bones. (Liu, Ziyu et al., 2020) [14] proposed an MS-G3D model based on multi-scale expansion. By eliminating the dependence between distances, it can directly model cross Spatio-temporal joints, which solves the problem of the previous methods not being able to capture complex Spatio-temporal relationships.

The above methods all adopt the skeleton recognition model based on a GCN. Although the GCN model has achieved good advantages in processing non-Euclidean data, implementing an efficient neural network usually requires the manual setting of the network parameters and a long training time. (Peng, Wei et al., 2020) [4] proposed the GCN-NAS model based on neural architecture search. The shortcoming of the previous GCN model only setting parameters manually is solved by introducing a neural architecture search of a dynamic graph. The limitation of the previously fixed graph is broken and the accuracy of the skeleton recognition is improved. By building an automatic search strategy for neural architectures, this paper hopes to solve the problem of the amount of time one needs to spend designing models for skeleton recognition networks.

2.2. Neural Architecture Search

Because neural architecture search can automatically find the optimal network model, this method provides a new possibility to avoid manually designing the network structure in the field of image vision [15–19]. For example, NAS algorithms based on reinforcement learning (RL) (Zoph, Barret, 2016) [19] can replace manually designed networks. However, such methods need to consume hundreds of GPUs for their search training. To reduce the search training time of traditional NAS methods, researchers have proposed a one-shot neural architecture search strategy [20]. For example, Brock et al. (2017) [15] proposed a SMASH model based on one-shot neural architecture search, which generates suboptimal weights by introducing an auxiliary network. The overall search speed is improved because the generated suboptimal weights are related to the weights of normal training in accuracy. In addition, (Pham et al., 2018) [18] proposed a neural structure search (ENAS) method based on sub-model weight sharing, which finds the neural network structure by introducing a controller to search the optimal sub-network in an ample search space. (Luo et al., 2018) [21] proposed an automatic neural network design method based on continuous optimization based on ENAS. This method maps the neural architecture to a constant vector space and solves the problems that could not be optimized in the continuous space in the past. (Zhu, Hui et al., 2019) [22] also proposed the EENAS method based on ENAS, which accelerates the search process by introducing a pre-learning strategy, thereby reducing the amount of computation. (Chu, Xiangxiang. 2021) [23] presented a uniformly sampled FairNAS technique, whose sampling and training procedures may completely use the search space's potential and deliver the best results among similar one-off models. (Liang, Tingting et al., 2021) [24] proposed a new search space in which each candidate is closely connected by a directed acyclic graph. Therefore, the effective method has been excellent and the method's mobility has been dramatically improved compared to the previous methods, and advanced results can be obtained on multiple datasets. In addition, this method proposes an efficient one-shot search algorithm to find the optimal path structure.

The above methods are all one-shot neural architecture search methods based on super-nets. This class of methods must first train a super-net and then use evolutionary algorithms to find the optimal candidate paths. However, the super-net needs to design a complex search space, which leads to a too-large network model and a long search training time. To quickly find the optimal neural network under the framework of one-shot neural architecture search, Guo [25] proposed a single-path sampling strategy to train the super network, which solves the problem of the search space of the previous model being too complex, and further reduces the search training time. (Bender et al., 2018) [26] introduced the dropout strategy, which solves the problem of the previous super network being too complicated by randomly deleting operations with low weights in the training process.

2.3. Graph Convolutional Networks Based on Neural Architecture Search

With the rapid development of neural architecture search, the time required for the automatic design of a graph neural network is significantly reduced. Recently, (Gao, Yang et al., 2020) [27] proposed a GraphNAS model based on neural architecture search and designed the best graph neural architecture by introducing reinforcement learning strategies. Furthermore, by introducing a novel parameter-sharing strategy in (Zhou, Kaixiong et al.) [28], an automatic graph neural network (AGNN) framework is proposed. The above methods show the possibility of applying neural architecture search to a graph neural network. However, due to the overly complex design of the search space, the search efficiency of traditional graph neural networks needs to be further optimized. To further improve the search efficiency of neural architecture search in the field of GCNs, (Ding, Yuhui et al., 2021) [29] proposed a differentiable search DiffMG model, which solves this problem by introducing a novel and efficient search algorithm. (CAI, Shaofei et al., 2021) [30] proposed a graph neural structure (GNAS) based on a gradient search strategy to obtain a higher performance. (Li, Guohao et al.) [31] proposed a greedy algorithm SGAS based on neural structure search, which can find the best architecture and reduce the search cost.

Inspired by the above research work, to improve the efficiency of neural architecture search and better apply it to the automatic search of a skeleton recognition model, this work designs a simplified search space and proposes a covariance adaptive improvement strategy based on an evolutionary algorithm to find the best architecture.

3. Methods

In this section, the single-path one-shot NAS-strategy-based methods for optimizing the GCN structure for skeleton recognition are proposed. First, the GCN-based skeleton recognition task is described in Section 3.1. Then, the one-shot single-path NAS strategy is described in Section 3.2. Next, the search space for the neural architecture search is defined in Section 3.3. Finally, Section 3.4 introduces how to automatically optimize the best GCN network from all the candidate architectures using the covariance matrix adaptive evolutionary algorithm (CMA-ES).

The framework of the SNAS-GCN is shown in Figure 2. Three functional modules are abstracted based on the graph neural architecture search process. The search space module contains a predefined search space for the GCN architecture, which can be customized by the user for a specific task. The GCN module is then used in the search space to implement the GCN model building and training for different downstream graph tasks on known graph data and configuration parameters. The search module implements the search function and search management for the GCN architecture, using a search algorithm to sample the GCN architecture for the best structure.

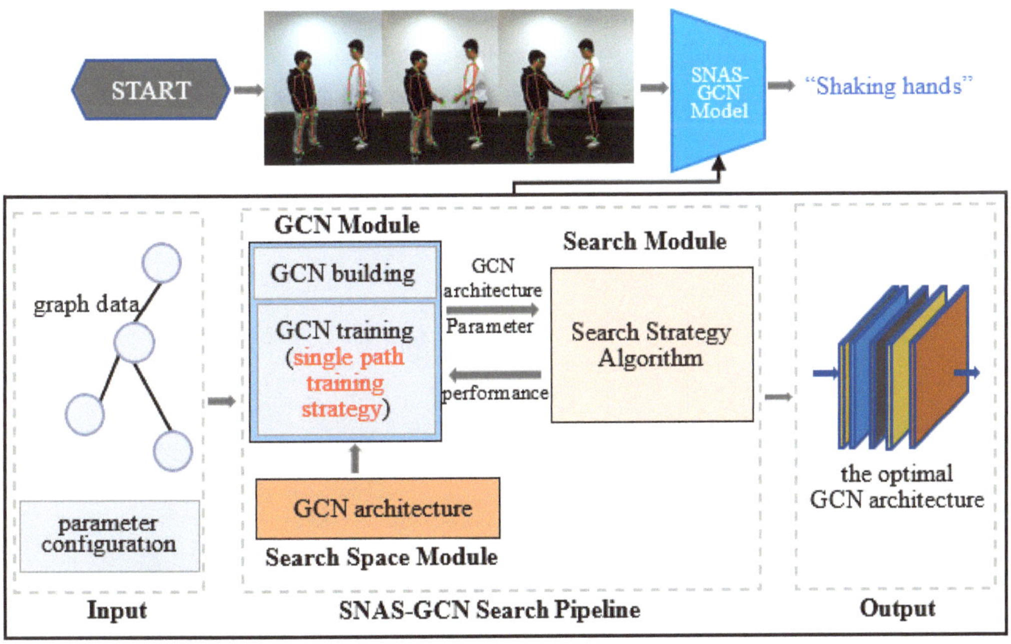

Figure 2. Graphical model architecture.

3.1. GCN-Based Skeleton Recognition Network

Motion capture devices or posture estimation techniques in a video can be used to gather skeleton data. Typically, these data consist of many frames. The joint coordinates that make up each frame of the data are included. As a result, the two-dimensional or three-dimensional coordinates of the human joints in each frame serve as typical representations of the skeleton sequence. An undirected spatiotemporal graph $\mathcal{G} = \{\mathcal{V}, \mathcal{E}, \mathcal{A}\}$ is built using a skeleton sequence in this study. The nodes and edges of this spatiotemporal graph represent the skeleton's joints and bones, respectively, where $\mathcal{V}$ represents all the joints in the skeleton sequence and $|\mathcal{E}|$ represents the edge connection. An adjacency matrix $\mathcal{A} \in \mathcal{R}^{n \times n}$ represents the connection of the joints.

Spatial graph convolution and temporal graph convolution are the two main components of the spatio-temporal graph based on the skeleton recognition model. An example of a constructed spatio-temporal skeleton diagram, including spatial and temporal dimensions, is shown in Figure 3, where the joints are represented as vertices and their natural connections in the body are represented as spatial edges. For the temporal dimension, the corresponding joints between two adjacent frames are connected to temporal edges. Among these, one-dimensional convolution modeling is used in the temporal domain, while graph convolution modeling is used in the spatial domain. The model constructs a spatial–temporal map of the skeletal sequence in two steps. First, the joints within a frame are connected to the edges according to the connectivity of the human body structure. Second, in each frame, each joint will be connected to the same joint. The connections in this setting are naturally defined and do not need to be manually assigned. A supervised learning problem using graph data may be used to frame the skeletal recognition issue. The robust representation of $\mathbb{G}$ will be learned using a GCN to improve the action class prediction. The GCN model is built by using neural structure search, which automatically improves the skeletal recognition model.

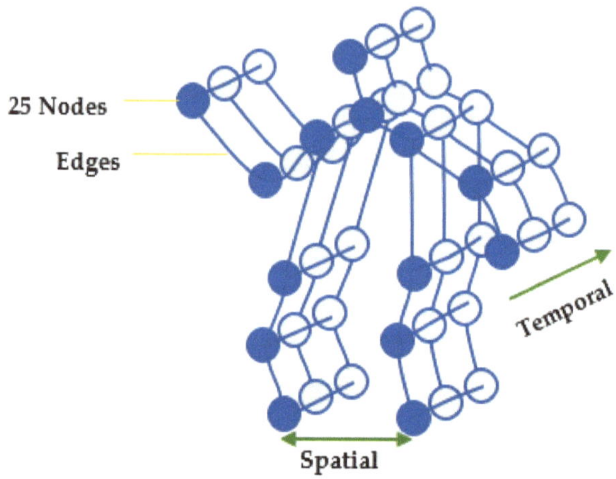

Figure 3. Spatio-temporal graph of the skeleton sequence.

The research inserts 10 GCN blocks into the network for the search and training to be compatible with existing state-of-the-art GCN approaches, as illustrated in Figure 4. The spatial module for each GCN block comprises channel convolution filters, which are convolution filters with a kernel size of 9×1 that are applied along the temporal axis to record temporal data. The graph is projected onto a feature space with channel 64 by the network's initial GCN block. The outputs of three GCN layers with 64-dimensional channels follow. The output channels of the three layers are then multiplied by two to obtain a total of 128 dimensions. The finished three-layer network includes 256 output channels for different dimensions. Each GCN block is subjected to the ResNet technique, similar to (Yan, and Xong, 2018) [1]. The collected characteristics are then used to make a final prediction in a fully linked layer.

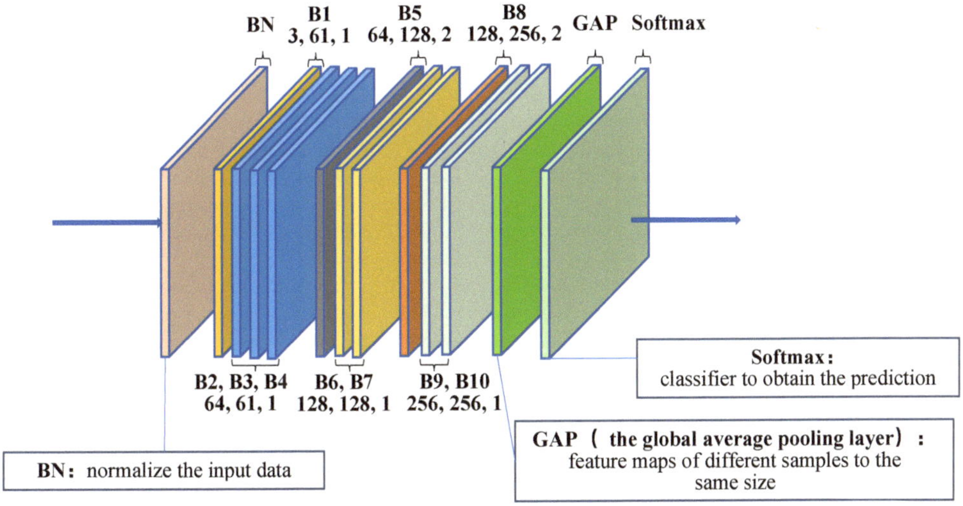

Figure 4. Neural network structure.

3.2. A Single Path One-Shot NAS Method

3.2.1. One-Shot Weight Sharing Method

Without a loss of generality, the architecture search space $\mathcal{A}$ is represented by a directed acyclic graph (D.A.G.). The network architecture is $a \in \mathcal{A}$, denoted as $\mathcal{N}(a, w)$, with a weight w.

The neural architecture search aims to solve two related problems. The first one is weight optimization, as shown in Equation (1):

$$w_a = \underset{w}{\operatorname{argmin}} \mathcal{L}_{\text{train}}(\mathcal{N}(a, w)) \tag{1}$$

where $\mathcal{L}_{\text{train}}(\cdot)$ is the loss function on the training set.

The second is architecture optimization. It finds architectures trained on the training set and has the best accuracy on the validation set, as shown in Equation (2):

$$a* = \underset{a \in \mathcal{A}}{\operatorname{argmax}} \text{A.C.C.}_{\text{val}}(\mathcal{N}(a, w_a)) \tag{2}$$

where $ACC_{\text{val}}(\cdot)$ is the accuracy of the validation set.

Traditional NAS methods perform these two optimization problems in a nested fashion. Many architectures are sampled from the $\mathcal{A}$ system and trained from scratch, as shown in Equation (1). The cost of each training is high, and only a small dataset and small search space (such as a single block) can complete this training in a short time.

To alleviate the above problems, the NAS method adopts a weight-sharing strategy. The architectural search space $\mathcal{A}$ is encoded in the super-net, denoted as $\mathcal{N}(\mathcal{A}, \mathcal{W})$, where $\mathcal{W}$ is the weight in the super-net. The super-net is trained only once. The weights are directly inherited by all the architectures from $\mathcal{W}$. Therefore, they share weights in common graph nodes. The architecture can be fine-tuned as needed, but it does not need to be trained from scratch. This is achieved by dividing the training and architecture search of the super network into two consecutive steps. Therefore, the architecture search speed is improved.

In general, the formula for these two consecutive steps of the super-net is as follows: First, the weights of the super-net are optimized, as shown in Equation (3):

$$W_{\mathcal{A}} = \underset{W}{\operatorname{argmin}} \mathcal{L}_{\text{train}}(\mathcal{N}(\mathcal{A}, W)) \tag{3}$$

Second, the search architecture of the super-net is optimized, as shown in Equation (4):

$$a* = \underset{a \in \mathcal{A}}{\text{argmax}} \text{A.C.C.}_{\text{val}}(\mathcal{N}(a, W_{\mathcal{A}}(a))) \tag{4}$$

During the search process, each sampled architecture a inherits its weights from $W_{\mathcal{A}}$ to $W_{\mathcal{A}}(a)$. The main differences between Equations (1), (2) and (4) are that the architecture weights have been trained in advance, and the evaluation of $ACC_{\text{val}}(\cdot)$ only requires inference and does not need to retrain new Weights. Generally, the one-shot weight-sharing method is essentially one-shot training and multiple inferences. Therefore, the search efficiency is improved.

However, compared to traditional methods, the super network reduces the cost of the architecture search by an order of magnitude, but still needs to train a large enough super network. The super network should contain enough search space, resulting in a too-large network model, which increases the calculation cost, and the search efficiency still needs to be improved.

Recent one-shot approaches have attempted to use a "path dropout" strategy to address the problem of oversized super-network models [26]. Each edge in the super-net graph is randomly eliminated in Equation (3), and the dropout rate parameter controls the randomness. In the above way, the joint adaptation of the node weights is reduced during the training, thereby reducing the search time [32]. The dropout rate parameter, however, significantly impacts this strategy's training methodology. Because of this, the issue of super-net training is challenging and still not fully resolved.

3.2.2. Single Path One-Shot Algorithm

The single path one-shot technique is introduced by revisiting the basic idea behind the concept of weight sharing. The effectiveness of the architectural search in Equation (4) critically depends on the fact that the inherited weights $W_{\mathcal{A}}(a)$ do not require fine-tuning. Second, $W_{\mathcal{A}}(a)$ can correctly forecast the architecture on the validation set. The weights $W_{\mathcal{A}}(a)$ should, ideally, be close to the ideal weights w_a in (1). The value of the approximation depends on how much the training loss $\mathcal{L}_{train}(\mathcal{N}(a, W_{\mathcal{A}}(a)))$ is minimized, which demands that the weights $W_{\mathcal{A}}$ of the super-net be improved in a way that simultaneously improves all of the designs in the search space, as demonstrated in Equation (5):

$$W_{\mathcal{A}} = \underset{W}{\text{argmin}} \mathbb{E}_{a \sim \Gamma(\mathcal{A})}[\mathcal{L}_{\text{train}}(\mathcal{N}(a, W(a)))] \tag{5}$$

where $\Gamma(\mathcal{A})$ is the prior distribution of $a \in \mathcal{A}$ during the training process. (Guo, Zichao et al.) [25] found that a uniformly constrained sampling method can better extract the ideal architectures from the search space. Equation (3) is realized explicitly in Equation (5). Thus, just one weight $W(a)$ is enabled and updated at a time throughout each optimization phase while an architecture a is randomly selected. Memory use is effective, and being embodied as a random super-net, the super-net is no longer effective.

A single path super-net structure is proposed to reduce the cooperative adaptation between the node weights and achieve fast search results. Each architecture is a path, as shown in Figure 5. As the choice block does not recognize branches, a random path must be kept in this case. Therefore, by randomly selecting a sub-network, its validation accuracy is evaluated during the training phase.

As shown in Figure 5, the choice block is made up of several candidate structure alternatives. In Section 3.3, the choice block in the search space is described in detail as consisting of a Chebyshev choice block and feature structure choice block. In a single-path super-net, each choice block is executed one option at a time. By sampling every option block, a single path may be found.

The simplicity of this method is used to find different architectural elements by defining several choice blocks. To facilitate complex search spaces, two additional choice blocks are particularly recommended.

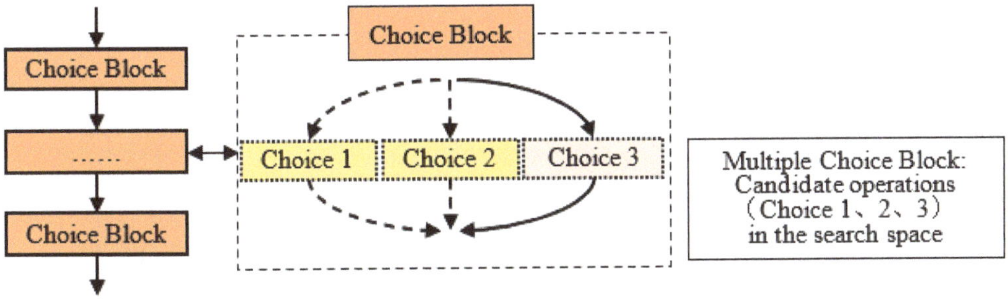

Figure 5. Single path neural architecture search structure diagram.

3.3. GCN Search Space Design

Designing a search space for single-path one-shot architecture search is a challenging problem, because the following competing requirements must be balanced. First, the search space should be reasonably designed and expressive enough to capture a variety of helpful candidate architectures. Secondly, the accuracy of the validation set generated by the one-shot model must be able to predict the accuracy generated by the independent model training. Finally, the one-shot model must be small enough to use limited computing resources (i.e., memory and time) for the search training [33].

To enrich the search space, this paper designs two choice blocks: the Chebyshev choice block and the feature structure choice block, as shown in Figure 6.

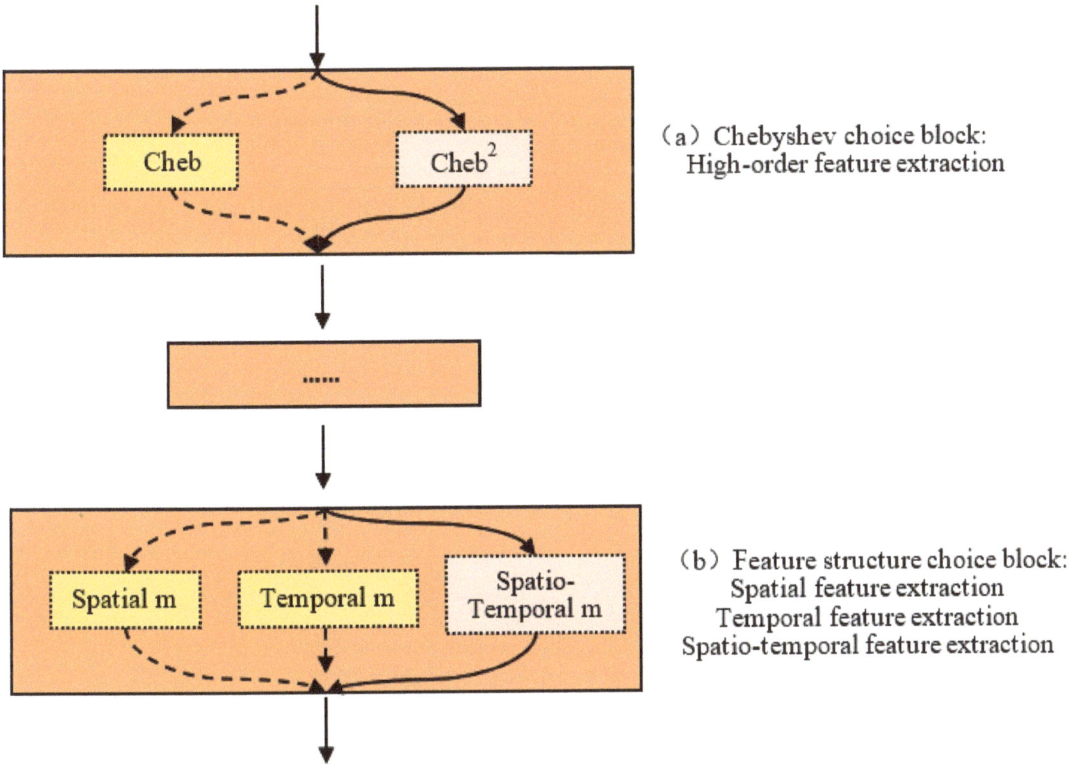

Figure 6. Design of choice blocks in the GCN search space.

Figure 6a shows the Chebyshev choice block module, composed of a first- and second-order Chebyshev polynomial function. Figure 6b shows the characteristic structural choice block module. It consists of a spatial feature neural operation (spatial m), a temporal feature neural operation (temporal m), and a spatio temporal feature neural operation (spatio temporal m). The entire search space consists of a number of the two aforementioned choice blocks, which are sampled by a single path to search for a better skeleton recognition network.

3.3.1. Feature Structure Choice Blocks

As shown in Figure 3, human actions can be recognized and resolved through the temporal and spatial sequences of human joint positions. By forming a high-level representation of the skeletal sequence through the spatio-temporal map, the recognition efficiency and accuracy can be further improved. Thus, the feature structure choice block includes spatial feature neural operations, temporal feature neural operations, and spatio-temporal feature neural operations.

The spatial features are extracted based on the structural correlation of the spatial node connections. To determine the connection strength between two nodes, (Shi. 2019) [2] applied the normalized Gaussian function on the graph nodes and calculated the similarity score as the correlation of the nodes, as shown in Equation (6):

$$\forall i,j \in \mathcal{V}, A_D(i,j) = \frac{e^{\phi(h(x_i))\otimes\psi(h(x_j))}}{\sum_{j=1}^{n} e^{\phi(h(x_i))\otimes\psi(h(x_j))}} \tag{6}$$

According to the $h(x_i)$ and $h(x_j)$ of the nodes and their corresponding representations, the correlation score $A_D(i,j)$ between them is calculated. $\phi(\cdot)$ and $\psi(\cdot)$ are two projection functions, called *conv_s* in Figure 7, which can be implemented by channel convolution filters. This way, the correlation between the nodes can be captured, which is the spatial feature "Spatial m" in Figure 7.

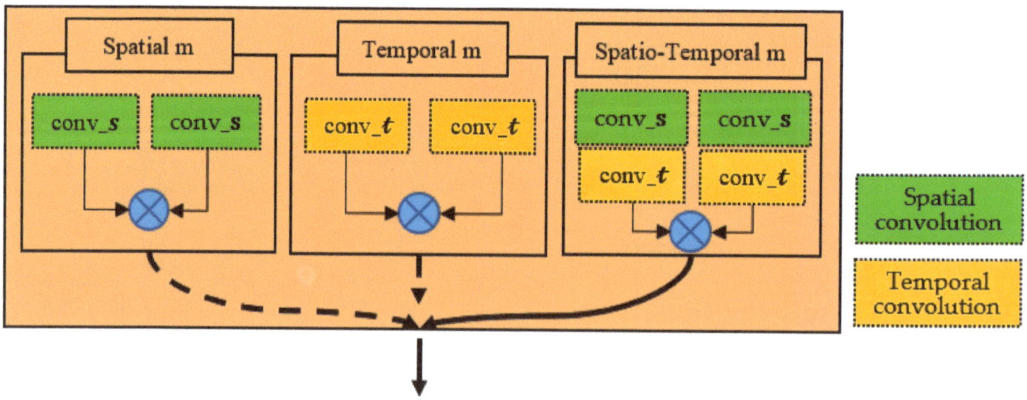

Figure 7. Choice block of feature structure.

The topology of the spatial feature map is the most intuitive. However, when temporal correlations are ignored, hidden joint connections are lost. For example, if there is no time information in the NTU-RGB + D dataset, it is difficult to tell whether a person is touching his head or waving his hand. As a result, including temporal information in action recognition models improves their accuracy. First, using Equation (6), the temporal feature introduces a Gaussian function, which calculates the node correlation. Second, to extract information from the temporal of each node, the functions $\phi(\cdot)$ and $\psi(\cdot)$ are implemented by two temporal convolutions *conv_t*, as shown in Figure 7. "Temporal m" is the name of this module.

Previous GCN approaches have been based on predefined graph structures constrained by temporal and spatial structures, while lacking a discussion of spatio-temporal correlations, thus ignoring the implied joint associations. However, different layers contain different semantic information, and therefore layer-specific mechanisms should be designed to construct a spatio-temporal graph.

The Spatio-temporal module, which is denoted as "Spatio-Temporal m" in Figure 7, can be directly constructed using "Spatial m" and "Temporal m." After the spatial feature neural operations and temporal feature neural operations have been formulated, the Spatio-temporal feature neural operations within the skeleton sequence must be modelled. Through constructing the graph, the temporal dimension of the graph is built by connecting the identical joints, and the graph is constructed by connecting the similar joints in consecutive frames, which allows us to define a very simple strategy for extending the spatial graph into the spatio-temporal domain. The same Gaussian function sampling is required to complete the convolution operation on the spatio-temporal graph, as in the spatial-only or time-only cases. In this way, a good convolution operation is performed on the constructed spatio-temporal graph.

3.3.2. Chebyshev Choice Block

Chebyshev polynomials provide high-order connections to GCN networks and can obtain high-level graph features. Therefore, (Deferrard et al.) [34] introduced a new spectral domain graph convolution network, which accomplishes quick localization and a low complexity, to overcome the shortcomings of the early spectral domain graph convolution network. The convolution kernel $\}_\theta$ in a spectral domain graph can be approximated by Chebyshev polynomials of order R, as shown in Equation (7):

$$Y = \sum_{r=0}^{R} \theta'_r T_r\left(\hat{L}\right) X \tag{7}$$

where $R = 1, \theta'_r$ denotes the Chebyshev coefficient. $X \in \mathcal{R}^n$ is the input representation of $\mathcal{G}$ and its n elements. The Chebyshev polynomial $T_r\left(\hat{L}\right)$ is defined recursively as Equation (8):

$$T_r\left(\hat{L}\right) = 2\hat{L}T_{r-1}\left(\hat{L}\right) - T_{r-2}\left(\hat{L}\right) \tag{8}$$

where, $T_0 = 1$ and $T_1 = \hat{L}$. Here, $\hat{L} = 2L/\lambda_{max} - I_n$ is normalized to $[-1, 1]$, $\lambda_{max} = 2$. The graph Laplacian L, of which the normalized definition is $L = I_n - D^{-1/2}AD^{-1/2}$ and $D_{ii} = \sum_j A_{ij}$, is used for Fourier transform. Chebyshev polynomial functions of the first or second order are constructed on different network layers in the search space, as shown in Figure 6. With a maximum order of 2, the function module can be built from Equation (8).

3.4. *Search Strategy Algorithm*

Random search for the architecture search in Equation (4) is adopted in the traditional search strategy algorithm. However, this has a limited effect on tight search spaces. This paper uses an evolutionary algorithm. The Covariance Matrix Adaptive Evolutionary Strategies (CMA-ES) is one of the most powerful evolutionary algorithms in the field of real-valued optimization, with many successful applications [35–37]. The invariance of CMA-ES, which is attained by carefully thought-out mutation and selection operators, and its successful adaptability to the mutation distribution, are its key benefits. The architectural parameter ⊣ is described by a Gaussian distribution in the CMA-ES method. Then, to update the distribution of architectures, the CMA-ES algorithm examines a collection of designs and chooses significant samples based on their performance. From the architectural dispersion, the optimal architecture can eventually be found.

The CMA-ES algorithm is divided into three parts, as shown in Figure 8.

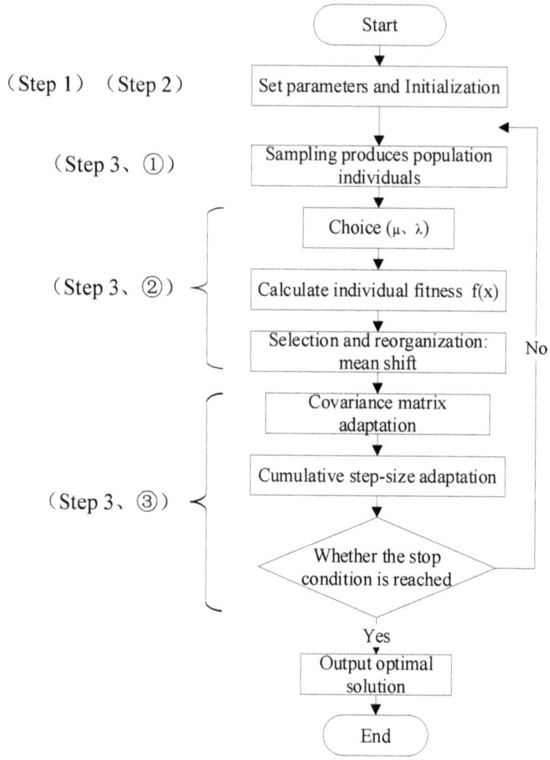

Figure 8. CMA-ES algorithm flow.

Step 1: Parameter setting

This includes the number of children λ, the number of parents μ, the recombination weight $w_i = 1 \ldots$ μ, the cumulative learning rate c_σ controlled by the step size, the decay parameter d_σ of the step size update, the cumulative learning rate c_c of the rank-one update of the covariance matrix, the covariance matrix, the learning rate c_1 of the variance matrix rank-one update, and the learning rate c_u of the covariance matrix rank μ update.

Step 2: Initialization

This includes choosing the distribution mean and step size. The evolutionary path is set as: $p_\sigma = 0, p_c = 0$. The covariance matrix is set as $C = I$ and the number of current iterations is $\} = 0$.

Step 3: Loop until the termination condition is reached

This includes ① sampling from the population, ② a reselection and recombination of the samples based on their fitness, and ③ updating the internal state variables based on the reordered samples.

This section describes, in detail, steps②and③of step 3 above. Then, in the step of selecting reorganization, the new mean value of the search distribution is selected from the sample μ, the weighted average value of the points, through which the recombination of the best offspring is achieved to calculate the new parental status, as shown in Equations (9) and (10):

$$\langle \boldsymbol{y} \rangle_\mathrm{w} = \sum_{i=1}^{\mu} w_i \boldsymbol{y}_{i:\lambda} \ \ where \ \sum_{i=1}^{\mu} w_i = 1, w_i > 0 \tag{9}$$

$$\boldsymbol{m} \leftarrow \boldsymbol{m} + \sigma \langle \boldsymbol{y} \rangle_\mathrm{w} = \sum_{i=1}^{\mu} w_i \boldsymbol{x}_{i:\lambda} \tag{10}$$

where m is the mean and $\mu \leq \lambda$ is the parent population size, i.e., the number of selected samples. $w_{i=1 \ldots \mu} \in \mathcal{R}+$: the positive weight coefficient for recombination, e.g., $w_{i=1 \ldots \mu} = 1/\mu$. Equation (10) is equivalent to calculating the mean of μ selection points.

The fitness $f(x)$ is calculated for each new individual, as shown in Equation (11). The reselection and reorganization are performed according to the fitness.

$$f(x_{1:\lambda}) \leq f(x_{2:\lambda}) \leq \cdots \leq f(x_{i:\lambda}) \tag{11}$$

where $x_{i:\lambda}$: is the i-th optimal individual. According to the fitness ranking, the top $\mu < \lambda$ individuals are intercepted for parameter updating.

An evolutionary path is the sequence of consecutive steps taken over many generations for the update process in step 3, ③. A series of consecutive sums of steps can be used to represent an evolutionary path, and these sums are referred to as cumulative sums. Exponential smoothing is used to create the evolutionary path p_c, and $p_c^{(0)} = 0$, as seen in Equation (12):

$$p_c \leftarrow (1 - c_c)p_c + h_\sigma \sqrt{c_c(2 - c_c)\mu_{\text{eff}}} \langle y \rangle_w \tag{12}$$

For each step of selection, the covariance matrix adaptation increases the scale in only one direction. The same technique as that in Equation (12) is adopted to construct the evolutionary path p_σ, as shown in Equation (13). Unlike Equation (12), however, Equation (13) constructs conjugate evolutionary paths, because the expected length of evolutionary path p_c in Equation (12) is determined by the path's direction. Initializing $p_\sigma^{(0)} = 0$, the conjugate evolutionary path is shown in Equation (13):

$$p_\sigma \leftarrow (1 - c_\sigma)p_\sigma + \sqrt{c_\sigma(2 - c_\sigma)\mu_{\text{eff}}} C^{-\frac{1}{2}} \langle y \rangle_w \tag{13}$$

The step control is updated, as shown in Equation (14):

$$\sigma \leftarrow \sigma \times \exp\left(\frac{c_\sigma}{d_\sigma}\left(\frac{\|p_\sigma\|}{E\|\mathcal{N}(0,I)\|} - 1\right)\right). \tag{14}$$

The covariance matrix is updated, as shown in Equation (15):

$$C \leftarrow (1 - c_1 - c_\mu)C + c_1(p_c p_c^T + \delta(h_\sigma)C) + c_\mu \sum_{i=1}^{\mu} w_i y_{i:\lambda} y_{i:\lambda}^T \tag{15}$$

The above parameters are continuously updated until the optimal network structure is obtained. More details of the CMA algorithm can be found in Algorithm 1 below.

Algorithm 1: CMA—ES algorithm

set

{\displaystyle\lambda}

$\lambda, \mu, w_i = 1 \ldots \mu, c_\sigma, d_\sigma, c_c, c_1$ *and* c_u // number of samples per iteration, at least two, generally >4

initialize $m, \sigma, C = I, p_\sigma = 0, p_c = 0$

{\displaystyle m}

// initialize state variables

while not terminate **do** // iterate

for i

Algorithm 1: *Cont.*

{\displaystyle i}

in $\{1 \ldots \lambda\}$

{\displaystyle \{1\ldots \lambda \}}

do // sample

{\displaystyle \lambda }

λ new solutions and evaluate them
$x_i =$

{\displaystyle x_{i}={}}

sample_multivariate_normal(mean = m

{\displaystyle {}=m}

, covariance_matrix = $\sigma^2 C$

{\displaystyle {}=\sigma ^{2}C}

)

{\displaystyle f_{i}=\operatorname {fitness} (x_{i})}

$f_i = \text{fitness}(x_i)$

{\displaystyle x_{1\laots \lambda }}

$x_{1\ldots\lambda} \leftarrow x_{s(1)\ldots s(\lambda)}$

{\displaystyle x_{s(1)\ldots s(\lambda)}}

with

{\displaystyle s(i)=\operatorname {argsort} (f_{1\ldots \lambda },i)}

$s(i) = \text{argsort}(f_{1\ldots\lambda}, i)$// sort solutions

{\displaystyle m'=m}

$m' = m$ // need later

Algorithm 1: *Cont.*

{\displaystyle m-m′}

$m - m'$ and $x_i - m'$

{\displaystyle x_{i}-m′}

{\displaystyle m}

$m \leftarrow$ update_m$(x_1, \ldots, x_\lambda)$

{\displaystyle (x_{1},\ldots,x_{\lambda })}

// move means to better solutions

{\displaystyle p_{\sigma }}

$p_\sigma \leftarrow$ update_ps$\left(p_\sigma, \sigma^{-1} C^{-\frac{1}{2}} (m - m')\right)$

{\displaystyle (p_{\sigma },\sigma ^{-1}C^{-1/2}(m-m′))}

// update isotropic evolution path

{\displaystyle p_{c}}

$p_c \leftarrow$ update_pc$(p_c, \sigma^{-1}(m - m'), \| p_\sigma \|)$

{\displaystyle (p_{c},\sigma ^{-1}(m-m′),\|p_{\sigma }\|)}

// update anisotropic evolution path

{\displaystyle C}

$C \leftarrow$ update_C$(C, p_c, (x_1 - m')/\sigma, \ldots, (x_\lambda - m')/\sigma)$

{\displaystyle (C,p_{c},(x_{1}-m′)/\sigma,\ldots,(x_{\lambda }-m′)/\sigma)}

// update covariance matrix

{\displaystyle \sigma }

$\sigma \leftarrow$ update_sigma

Algorithm 1: *Cont.*

{\displaystyle (\sigma,\ | p_{\sigma }\ |)}

$\sigma, \| p_\sigma \|)$ // update step-size using isotropic path length

return m

{\displaystyle m}

or x_1

{\displaystyle x_{1}}

3.5. Summary

The combination of an efficient search space design, a single-path one-shot super-net strategy, and an evolutionary structural search algorithm enables the efficient and flexible search of the skeleton recognition network. Therefore, this model is easy to train for search, occupies little memory, and is highly competitive on large datasets. To validate the model's efficiency, the approach proposed by the model is evaluated in Section 4.

4. Experiments

In this section, the effectiveness and superiority of the single-path search strategy are evaluated using experiments. NTU-RGB + D and dynamics are used in the experiment.

4.1. Dataset and Evaluation Metrics

The NTU-RGB+D dataset is a significant public human motion recognition dataset captured by the second-generation Kinect. The dataset collects all the data patterns that the Kinect camera can provide, including depth maps, 3D joint information, RGB frames, and R.IR sequences. The data consist of 56,000 action sequences and 4 million frames in 60 categories of actions, 60 of which were demonstrated by 40 volunteers aged between 10 and 35. The 60 types of movements are divided into three main categories: 40 daily activities, 9 health-related activities, and 11 pairwise interactive actions.

Kinetics is a video-based action recognition dataset that only provides raw video clips without skeletal data. To flatten the joint positions in the dataset, all the videos were first resized to a resolution of 340 × 256, and then the frame rate was converted into 30 fps. Secondly, each frame's bones in the video were extracted through Openpose to generate the Kinetics skeleton data (7.5 GB). The Kinetics dataset includes 400 human action categories, each of which has at least 400 video clips taken from different Youtube videos, with a duration of about ten seconds. The categories of the dataset are mainly divided into three categories: human–object interaction, such as playing a musical instrument; and human–human interaction, such as shaking hands, hugging, and sports, etc. They are named person, person–person, and person–object.

4.2. Experimental Details

The experiments in this section were implemented on PyTorch (Paszke et al., 2017) [38] with an NVIDIA RTX 3090 (with 24 G RAM) GPU. The experimental environment is consistent with the current state-of-the-art GCN methods (Yan, Xiong, and Lin 2018; Shi et al., 2019) [1,2].

The joint NTU-RGB + D data were searched experimentally during the search to find the best structure. During the training process, the stochastic gradient descent (S.G.D.) with a momentum of 0.9 was used as the optimization algorithm for the network. Cross entropy loss was selected as the loss function of the recognition task. During the search and training, the weight decay was set as 0.0001 or 0.0006, respectively. For the NTU-RGB + D dataset, there could be up to two individuals in each dataset sample. If the number of entities in the sample was less than 2, the second entity was filled by 0. The maximum number of frames per sample was 300, and for samples with less than 300 frames, these were repeated until 300 frames were reached. The learning rate was set to 0.1 and divided by 10 at periods 30, 45, and 60. The training process ended at the 61st calendar time.

4.2.1. Ablation Study

To test the efficacy of the single-path neural structure search and confirm that the model could search for the best GCN network, the following experience was conducted on NTU-RGB + D using the cross-view dataset as a baseline. Each experiment showed the time (mins) results for searching one epoch of search time. A series of baselines were established for the experiments as a point of comparison: (1) There was a fixed choice of one or more candidate options. Only Temporal m (T), Spatio-Temporal m (ST), and Spatial m + Temporal m + Spatio-Temporal m (S + T + ST) were in the search space; (2) there was the random selection of an option from the search space. An option could be chosen and combined with cheb^2. The single path search lowered the search cost as much as was practicable while still being able to guarantee accuracy when combined with the experimental findings, which are shown in Table 1. Second, the usefulness of including second-order Chebyshev polynomials was demonstrated by the fact that adding cheb^2 to the search space increased the accuracy compared to the conventional approach, while maintaining a relatively constant search time. The overall experimental findings support the method's efficacy and efficiency (Single Path One-Shot, SNAS-GCN).

Table 1. Performance comparison of NTU-RGB + D CV evaluation.

Methods	Joint (%)	Search Time per Epoch (mins)
Ours (T)	93.8	20.46
Ours (ST)	93.8	16.26
Ours (S + T + ST)	93.7	24.19
Ours (T + Cheb^2)	94.2	20.53
Ours (ST + Cheb^2)	94.2	16.28
Ours (S + T + ST + Cheb^2)	94.1	24.22
Ours(SNAS-GCN)	94.3	17.43

The results also show that the temporal feature neural operation (T) took longer to search than the Spatio-temporal feature neural operation (ST). First, the temporal module involved interactions outside the same frame. In contrast, the spatial feature interactions were limited to features of the same dimension at the same time step, resulting in a long (T) operation search time. Second, the search time of the (S + T + ST + Cheb^2) experimental module demonstrated that, while the super-net ensured a better accuracy by simultaneously searching all the modules, it also took longer.

4.2.2. Search Cost Analysis

The memory cost and total time cost of training a super network in search space were adopted to measure the model's performance. All the super networks underwent 61 iterations of training, with a batch size of 16, and were trained using an NVIDIA RTX 3090 24 G GPU. Table 2 shows the search cost of the search space; and Table 3 shows that the modeling approach in this paper clearly used less search time than the baseline model.

1. Cost analysis of search space

The experimental results in Table 2 show that searching based on a single-path architecture is efficient compared to the search cost of the traditional method (GCN-NAS) [4] (experiments conducted in the same experimental environment). This is because searching within a single path with only a few layers is far more efficient than searching the entire architecture with many layers, and then defining the overall architecture by stacking units.

Table 2. Cost analysis of search space.

Methods	Joint (%)	Search Time per Epoch (mins)
Ours(S + T + ST + Cheb^2)	94.1	24.22
GCN-NAS(S + T + ST + Cheb^2)	94.3	25.1
Ours(7 Categories)	94.4	25.2
GCN-NAS(7 Categories)	94.6	26.1
Ours (single path NAS)	94.3	17.43

Table 3. Search costs for each baseline model.

Method	Training Time	Search Time	Total Time	CV (Joint) (%)
MFAS [3]	--	150.91 h	--	93.46
GCN-NAS [4]	24 h	46.2 h	70.2 h	94.6
SAR-NAS [39]	--	29 h	--	94.3
Ours (Joint)	17.7 h	34.4 h	52.1	94.3

The experimental results in Table 2 show that, compared to the search costs of traditional methods (GCN-NAS) [4] (experiments conducted in the same experimental environment), a search based on a single-path architecture is effective. This is because searching in a single path with only a few layers is much more efficient than searching in an entire architecture with many layers, and then defining the whole architecture through stacked units.

This study demonstrates that a simplified search space using single-path search can produce good search results in a GCN architecture search. The model in this paper was based on single-path grid search, but with simplified operations and fast optimization seeking.

The above comparison indicates that using a simplified search space with a single-path search can produce good search results in GCN architecture searches. It can be seen that the model proposed in this article has a simple operation and fast optimization speed.

2. Comparison of search costs with the baseline model

The GCN-NAS model was compared with the model proposed in this paper in the same empirical setting with the same experimental parameters, such as the number of epochs and batch sizes. The experimental results showed that the model approach of the article was less time-consuming and had approximately the same accuracy.

Due to the different search spaces and data types, as well as the fact that the source code of the papers is not publicly available, the MFAS and SAR-NAS methods could not be reproduced in this experiment. Various papers have found that the MFAS method takes 150.9 h to search on four NVIDIA Tesla P100 16 GB GPUs, while the SAR-NAS method takes 29 h to search on one NVIDIA TitanXP 12 G GPU. By analogy, the search times required for these two experiments are also too long.

4.2.3. Comparison with State-of-the-Art (SOTA)

This section first searches the network on the NTU-RGB + D dataset to obtain a better performance. The initial learning rate was 0.1 and the learning rate was updated at 30, 45, and 60 epochs, respectively. An NVIDIA RTX 3090 24 G GPU was used for the training. A data expansion was performed using clipping and rotation angles to reduce the network overfitting. The searched models were compared with seven SOTA skeleton-based action recognition methods, including a CNN-based method, a GCN-based method, and a GCN- and NAS-based method. Table 4 shows the performance results of the method in this paper for NTU-RGB + D.

Table 4. Comparison of SNAS-GCN with other methods on NTU-RGB + D60 in terms of accuracy, search time consumed.

Architecture	CS (%)	CV (%)	Time-Consuming Search
HCN [40]	86.5	91.1	--
ST-GCN [1]	81.5	88.3	--
AS-GCN [41]	86.8	94.2	--
DARTS [42]	83.9	92.0	--
SAR-NAS [39]	86.4	94.3	29 h+
MFAS [3]	--	93.46	150.91 h+
2S-AGCN [2]	88.5	95.1	--
NAS-GCN [4]	89.4	95.7	70.2 h
2s HA-GCN [43]	91.5	96.6	--
Ours (Joint)	87.1	94.3	52.1 h
Ours (Bone)	86.0	94.0	--
Ours(Joint + Bone)	89.0	95.0	--

Regarding the NTU-RGB + D and Kinetics skeleton datasets, Tables 4 and 5 show the results for these two datasets, respectively. As can be seen from Tables 4 and 5, the search model achieved the best results on both datasets for all the evaluation metrics, which demonstrates the effectiveness of the approach proposed in the article.

Table 5. Performance of SNAS-GCN versus other methods on the Kinetics dataset.

Model	Top-1 (%)	Top-5 (%)
ST-GCN [1]	30.7	52.8
AS-GCN [41]	34.8	56.5
DARTS [42]	32.1	54.0
2S-AGCN [2]	36.1	58.7
SAR-NAS [39]	33.6	56.3
NAS-GCN (Joint) [4]	35.5	57.9
NAS-GCN (Bone) [4]	34.9	57.1
NAS-GCN [4]	37.1	60.1
2s HA-GCN [43]	37.4	60.5
Ours (Joint)	35.6	57.9
Ours (Bone)	34.8	57.0
Ours (Joint + Bone)	37.0	60.0

Table 4 shows that the MFAS modelling approach used more than 150 hours for searching, but the recognition accuracy still has a disadvantage compared to the model in this paper. Compared to the SAR-NAS method, the model in this paper shows a substantial improvement in recognition accuracy while maintaining search efficiency. Table 5 shows that the models trained by a single-path neural architecture search are well ahead of traditional methods and slightly below the latest techniques in terms of their recognition accuracy.

5. Conclusions

This work studied the skeleton action recognition task based on neural architecture search. It explored finding the best model quickly and automatically through neural architecture search. The main contributions of this paper include: firstly, a simplified four-category search space was constructed instead of a traditional eight-category search space. The simplified search space reduced the complexity of the super network. Secondly, a single-path one-shot weight-sharing strategy was proposed to reduce the search time of neural architecture search in the super network, thus reducing the computational cost. Finally, an evolutionary strategy algorithm was proposed, which could automatically select the best architecture from all the training architectures. The NTU-RGB + D and Kinetics datasets' experimental results verified the proposed method's effectiveness. Compared to the latest NAS-GCN method, the search time of the proposed method in this paper was 0.3 times longer than that of the SOTA method, while the recognition accuracy was approximately the same.

It is hoped to deploy this model in embedded systems and apply it to various application scenarios, such as stage performances and human–computer interaction. In addition, it is hoped to continue simplifying the evolutionary algorithm and improving the search speed of the NAS.

Author Contributions: Conceptualization, Y.J., S.Y. and Z.S.; methodology, Y.J. and Z.S.; software, Z.S. and T.W.; validation, Z.S., S.W. and S.Y.; formal analysis, Y.J. and Z.S.; investigation, Y.J. and S.W.; resources, Y.J. and Z.S.; data curation, Z.S. and S.Y.; writing—original draft preparation, Y.J. and S.Y.; writing—review and editing, Y.J., S.Y. and T.W.; visualization, Y.J. and S.Y.; supervision, Y.J.; project administration, Y.J.; funding acquisition, Y.J. All authors have read and agreed to the published version of the manuscript.

Funding: This work was supported by Funds for the National Key R&D Program of China, Ministry of science and technology of China (2021YFF0307603), the National cultural and tourism science and technology innovation project of the Ministry of culture and Tourism (GJWLKJCXGC-095) and the Fundamental Research Funds for the Central Universities: CUC22WH001.

Data Availability Statement: The data sets used in this paper are public, free, and available at NTU RGB + D 60: https://rose1.ntu.edu.sg/dataset/actionRecognition/ (accessed on 21 July 2021). Kinetics: https://www.deepmind.com/open-source/kinetics (accessed on 1 November 2021).

Conflicts of Interest: The authors declare that they have no conflict of interest regarding the publication of this paper.

References

1. Sijie, Y.; Xiong, Y.; Lin, D. Spatial-temporal graph convolutional networks for skeleton-based action recognition. In Proceedings of the Thirty-Second AAAI Conference on Artificial Intelligence, New Orleans, LA, USA, 2–7 February 2018.
2. Shi, L.; Zhang, Y.; Cheng, J.; Lu, H. Two-stream adaptive graph convolutional networks for skeleton-based action recognition. In Proceedings of the IEEE/CVF Conference on Computer Vision and Pattern Recognition, Long Beach, CA, USA, 15–20 June 2019.
3. Perez-Rua, J.-M.; Vielzeuf, V.; Pateux, S.; Baccouche, M.; Jurie, F. Mfas: Multimodal fusion architecture search. In Proceedings of the IEEE/CVF Conference on Computer Vision and Pattern Recognition, Long Beach, CA, USA, 15–20 June 2019.
4. Peng, W.; Hong, X.; Chen, H.; Zhao, G. Learning graph convolutional network for skeleton-based human action recognition by neural searching. *Proc. AAAI Conf. Artif. Intell.* **2020**, *34*, 2669–2676. [CrossRef]
5. Shahroudy, A.; Liu, J.; Ng, T.-T.; Wang, G. Ntu rgb+ d: A large scale dataset for 3d human activity analysis. In Proceedings of the IEEE Conference on Computer Vision and Pattern Recognition, Las Vegas, NV, USA, 26 June–1 July 2016.
6. Kay, W.; Carreira, J.; Simonyan, K.; Zhang, B.; Hillier, C.; Vijayanarasimhan, S.; Viola, F.; Green, T.; Back, T.; Natsev, P.; et al. The kinetics human action video dataset. *arXiv* **2017**, arXiv:1705.06950.
7. Song, S.; Lan, C.; Xing, J.; Zeng, W.; Liu, J. An end-to-end spatio-temporal attention model for human action recognition from skeleton data. In Proceedings of the AAAI Conference on Artificial Intelligence, San Francisco, CA, USA, 4–9 February 2017; Volume 31.
8. Zhang, P.; Lan, C.; Xing, J.; Zeng, W.; Xue, J.; Zheng, N. View adaptive recurrent neural networks for high performance human action recognition from skeleton data. *Proc. IEEE Int. Conf. Comput. Vis.* **2017**, *41*, 1963–1978.
9. Soo, K.T.; Reiter, A. InterpreTable 3d human action analysis with temporal convolutional networks. In Proceedings of the 2017 IEEE Conference on Computer Vision and Pattern Recognition Workshops (CVPRW), Honolulu, HI, USA, 21–26 July 2017.
10. Liu, M.; Liu, H.; Chen, C. Enhanced skeleton visualization for view invariant human action recognition. *Pattern Recognit.* **2017**, *68*, 346–362. [CrossRef]
11. Ullah, K.I.; Afzal, S.; Lee, J.W. Human activity recognition via hybrid deep learning based model. *Sensors* **2022**, *22*, 323. [CrossRef] [PubMed]
12. Kang, J.; Shin, J.; Shin, J.; Lee, D.; Choi, A. Robust human activity recognition by integrating image and accelerometer sensor data using deep fusion network. *Sensors* **2021**, *22*, 174. [CrossRef] [PubMed]
13. Monti, F.; Boscaini, D.; Masci, J.; Rodolà, E.; Svoboda, J.; Bronstein, M.M. Geometric deep learning on graphs and manifolds using mixture model cnns. In Proceedings of the IEEE Conference on Computer Vision and Pattern Recognition, Honolulu, HI, USA, 21–26 July 2017.
14. Liu, Z.; Zhang, H.; Chen, Z.; Wang, Z.; Ouyang, W. Disentangling and unifying graph convolutions for skeleton-based action recognition. In Proceedings of the IEEE/CVF Conference on Computer Vision and Pattern Recognition, Seattle, WA, USA, 13–19 June 2020.

15. Brock, A.; Lim, T.; Ritchie, J.M.; Weston, N. Smash: One-shot model architecture search through hypernetworks. *arXiv* **2017**, arXiv:1708.05344.
16. Liu, C.; Zoph, B.; Neumann, M.; Shlens, J.; Hua, W.; Li, L.-J.; Fei-Fei, L.; Yuille, A.; Huang, J.; Murphy, K. Progressive neural architecture search. In Proceedings of the European Conference on Computer Vision (ECCV), Munich, Germany, 8–14 September 2018.
17. Perez-Rua, J.-M.; Baccouche, M.; Pateux, S. Efficient progressive neural architecture search. *arXiv* **2018**, arXiv:1808.00391.
18. Pham, H.; Guan, M.Y.; Zoph, B.; Le, Q.V.; Dean, J. Efficient neural architecture search via parameters sharing. In Proceedings of the International Conference on Machine Learning, PMLR, Stockholm, Sweden, 10–15 July 2018.
19. Barret, Z.; Le, Q.V. Neural architecture search with reinforcement learning. *arXiv* **2016**, arXiv:1611.01578.
20. Yao, Y.; Liu, R.; Zhang, J.; Zhong, W.; Fan, X.; Luo, Z. Hardware-Aware Low-Light Image Enhancement via One-Shot Neural Architecture Search with Shrinkage Sampling. In Proceedings of the 2021 IEEE International Conference on Multimedia and Expo (ICME), Shenzhen, China, 5–9 July 2021.
21. Luo, R.; Tian, F.; Qin, T.; Chen, E.; Liu, T.Y. Neural architecture optimization. *Adv. Neural Inf. Process. Syst.* **2018**, *31*, 1–12.
22. Zhu, H.; An, Z.; Yang, C.; Xu, K.; Zhao, E.; Xu, Y. EENA: Efficient evolution of neural architecture. In Proceedings of the IEEE/CVF International Conference on Computer Vision Workshops, Seoul, Republic of Korea, 27–28 October 2019.
23. Chu, X.; Zhang, B.; Xu, R. Fairnas: Rethinking evaluation fairness of weight sharing neural architecture search. In Proceedings of the IEEE/CVF International Conference on Computer Vision, Montreal, QC, Canada, 10–17 October 2021.
24. Liang, T.; Wang, Y.; Tang, Z.; Ling, H. Opanas: One-shot path aggregation network architecture search for object detection. In Proceedings of the IEEE/CVF Conference on Computer Vision and Pattern Recognition, Nashville, TN, USA, 20–25 June 2021.
25. Guo, Z.; Zhang, X.; Mu, H.; Heng, W.; Liu, Z.; Wei, Y.; Sun, J. Single path one-shot neural architecture search with uniform sampling. In *European Conference on Computer Vision*; Springer: Cham, Switzerland, 2020.
26. Bender, G.; Kindermans, P.J.; Zoph, B.; Vasudevan, V.; Le, Q. Understanding and simplifying one-shot architecture search. In Proceedings of the International Conference on Machine Learning, PMLR, Stockholm, Sweden, 10–15 July 2018.
27. Gao, Y.; Yang, H.; Zhang, P.; Zhou, C.; Hu, Y. Graphnas: Graph neural architecture search with reinforcement learning. *arXiv* **2019**, arXiv:1904.09981.
28. Zhou, K.; Song, Q.; Huang, X.; Hu, X. Auto-gnn: Neural architecture search of graph neural networks. *arXiv* **2019**, arXiv:1909.03184. [CrossRef] [PubMed]
29. Ding, Y.; Yao, Q.; Zhao, H.; Zhang, T. Diffmg: Differentiable meta graph search for heterogeneous graph neural networks. In Proceedings of the 27th ACM SIGKDD Conference on Knowledge Discovery & Data Mining, Singapore, 14–18 August 2021.
30. Cai, S.; Li, L.; Deng, J.; Zhang, B.; Zha, Z.-J.; Su, L.; Huang, Q. Rethinking graph neural architecture search from message-passing. In Proceedings of the IEEE/CVF Conference on Computer Vision and Pattern Recognition, Nashville, TN, USA, 20–25 June 2021.
31. Li, G.; Qian, G.; Delgadillo, I.C.; Müller, M.; Thabet, A.; Ghanem, B. Sgas: Sequential greedy architecture search. In Proceedings of the IEEE/CVF Conference on Computer Vision and Pattern Recognition, Seattle, WA, USA, 13–19 June 2020.
32. Zhang, H.; Jin, Y.; Hao, K. Evolutionary search for complete neural network architectures with partial weight sharing. *IEEE Trans. Evol. Comput.* **2022**, *26*, 1072–1086. [CrossRef]
33. Xia, X.; Xiao, X.; Wang, X.; Zheng, M. Progressive Automatic Design of Search Space for One-Shot Neural Architecture Search. In Proceedings of the IEEE/CVF Winter Conference on Applications of Computer Vision, Waikoloa, HI, USA, 3–8 January 2022.
34. Michaël, D.; Bresson, X.; Vandergheynst, P. Convolutional neural networks on graphs with fast localized spectral filtering. *Adv. Neural Inf. Process. Syst.* **2016**, *29*, 1–9.
35. Zhang, C.; Xiao, C.; Guo, X. Covariance Matrix Evolutionary Preference-based Policy Search for Robot Confrontation. In Proceedings of the 2022 7th Asia-Pacific Conference on Intelligent Robot Systems (ACIRS), Waikoloa, HI, USA, 3–8 January 2022.
36. Nilotpal, S.; Chen, K.-W. Neural Architecture Search using Covariance Matrix Adaptation Evolution Strategy. *arXiv* **2021**, arXiv:2107.07266.
37. Zhang, J.; Qi, H.; Ji, Y.; Ren, Y.; He, M.; Su, M.; Cai, X. Nonlinear acoustic tomography for measuring the temperature and velocity fields by using the covariance matrix adaptation evolution strategy algorithm. *IEEE Trans. Instrum. Meas.* **2021**, *71*, 4500214. [CrossRef]
38. Paszke, A.; Gross, S.; Chintala, S.; Chanan, G.; Yang, E.; DeVito, Z.; Lerer, A. Automatic differentiation in pytorch. In Proceedings of the 31st Conference on Neural Information Processing Systems (NIPS 2017), Long Beach, CA, USA, 4 December 2017.
39. Zhang, H.; Hou, Y.; Wang, P.; Guo, Z.; Li, W. SAR-NAS: Skeleton-based action recognition via neural architecture searching. *J. Vis. Commun. Image Represent.* **2020**, *73*, 102942. [CrossRef]
40. Li, C.; Zhong, Q.; Xie, D.; Pu, S. Co-occurrence feature learning from skeleton data for action recognition and detection with hierarchical aggregation. *arXiv* **2018**, arXiv:1804.06055.
41. Li, M.; Chen, S.; Chen, X.; Zhang, Y.; Wang, Y.; Tian, Q. Actional-structural graph convolutional networks for skeleton-based action recognition. In Proceedings of the IEEE/CVF Conference on Computer Vision and Pattern Recognition, Long Beach, CA, USA, 15–20 June 2019.

42. Liu, H.; Simonyan, K.; Yang, Y. Darts: Differentiable architecture search. *arXiv* **2018**, arXiv:1806.09055.
43. Xing, H.; Burschka, D. Skeletal human action recognition using hybrid attention based graph convolutional network. In Proceedings of the 2022 26th International Conference on Pattern Recognition (ICPR), Montreal, QC, Canada, 21–25 August 2022.

sensors

MDPI

Article

Smart Contract Vulnerability Detection Based on Deep Learning and Multimodal Decision Fusion

Weichu Deng [1], Huanchun Wei [2,*], Teng Huang [1], Cong Cao [1], Yun Peng [1] and Xuan Hu [3,4]

1 Institute of Artificial Intelligence and Blockchain, Guangzhou University, Guangzhou 510006, China; 2112106114@e.gzhu.edu.cn (W.D.); huangteng1220@gzhu.edu.cn (T.H.); 2112206107@e.gzhu.edu.cn (C.C.)
2 School of Beidou, Guangxi University of Information Engineering, Nanning 530299, China
3 Information Security Research Center, CEPREI Laboratory, Guangzhou 510610, China
4 Key Laboratory of Ministry of Industry and Information Technology, Guangzhou 510610, China
* Correspondence: whc191021@163.com

Abstract: With the rapid development and widespread application of blockchain technology in recent years, smart contracts running on blockchains often face security vulnerability problems, resulting in significant economic losses. Unlike traditional programs, smart contracts cannot be modified once deployed, and vulnerabilities cannot be remedied. Therefore, the vulnerability detection of smart contracts has become a research focus. Most existing vulnerability detection methods are based on rules defined by experts, which are inefficient and have poor scalability. Although there have been studies using machine learning methods to extract contract features for vulnerability detection, the features considered are singular, and it is impossible to fully utilize smart contract information. In order to overcome the limitations of existing methods, this paper proposes a smart contract vulnerability detection method based on deep learning and multimodal decision fusion. This method also considers the code semantics and control structure information of smart contracts. It integrates the source code, operation code, and control-flow modes through the multimodal decision fusion method. The deep learning method extracts five features used to represent contracts and achieves high accuracy and recall rates. The experimental results show that the detection accuracy of our method for arithmetic vulnerability, re-entrant vulnerability, transaction order dependence, and Ethernet locking vulnerability can reach 91.6%, 90.9%, 94.8%, and 89.5%, respectively, and the detected AUC values can reach 0.834, 0.852, 0.886, and 0.825, respectively. This shows that our method has a good vulnerability detection effect. Furthermore, ablation experiments show that the multimodal decision fusion method contributes significantly to the fusion of different modalities.

Keywords: multimodal fusion; smart contract; vulnerability detection; deep learning

1. Introduction

A smart contract is the technical core of blockchain 2.0 [1], first proposed by computer scientist Nick Szabo in 1995. It is a contract clause established in digital form, with self-verification and automatic execution characteristics. Unlike traditional contracts, smart contracts have high efficiency, low costs, and high security, and no "denial" occurs. Smart contracts are designed to perform contracts securely and efficiently without a trusted third party. Previously limited by technological developments, smart contracts have been widely used with the rapid development of blockchains in recent years. Ethereum [2], as the first blockchain platform to introduce smart contracts, supports the Turing-complete programming language Solidity [3].

In essence, the smart contract is a piece of code that continuously runs on the blockchain [4], which cannot be modified once deployed and is automatically executed when a predetermined condition is triggered. Smart contracts running on the blockchain can achieve reliable information exchange, value transfer, and asset management, so smart contracts are usually bound to assets. Once security problems occur, they can cause substantial economic losses [5]. Smart contract security problems have followed the increase in smart contracts. According to statistics, 89% of smart contracts have security vulnerabilities.

The security of smart contracts [6] has also attracted the attention of hackers, who have been stealing illegal assets by exploiting smart contract vulnerabilities. In 2016, attackers used a re-entrant vulnerability in TheDAO's crowdfunding contract to attack it, resulting in about USD 60 million in losses [7]. In 2017, a vulnerability in the Parity wallet [8] contract led to USD 31 million in user losses. In 2018, hackers took advantage of the integer overflow vulnerability in the Ethereum ERC-20 smart contract to transfer a large number of BEC tokens issued by the United States Chain company out of thin air, resulting in its market value dropping to almost zero [9]. According to statistics from SlowMist's website [10], attacks against Ethereum smart contracts have caused cumulative losses of more than USD 3.1 billion as of 2023.

With the frequent occurrence of blockchain security incidents and the increasing number of assets involved in smart contracts, the vulnerability problem of smart contracts has also attracted the attention of more scholars [11]. Due to the characteristics of smart contracts being tamper-evident and automatically executed, stolen assets are difficult to recover once security problems occur in smart contracts. Therefore, there is an urgent need to detect potential security vulnerabilities before deployment to ensure the security of smart contracts and the safety of the asset interests of each participant.

Current smart contract vulnerability detection methods [12] are mainly based on traditional vulnerability detection methods of conventional programs (such as C++ and Java), usually using static analysis [13] and dynamic execution techniques [14] to detect vulnerabilities. This method often relies on expert experience and has problems such as low automation, low efficiency, and insufficient flexibility. In recent years, with the rapid development of deep learning, the use of deep learning technology for vulnerability detection of smart contracts has become a research hotspot [15]. Detection technologies based on deep learning can make up for the problems of low automation, low efficiency, and dependence on expert knowledge in traditional detection methods. However, most of the current methods based on deep learning [16] consider a single data feature and are unable to fully mine the vulnerability information of smart contracts. Additionally, there is poor scalability and a lack of consideration for multimodal characteristics.

1.1. Goals and Contributions

In order to address the above challenges, this paper proposes a smart contract vulnerability detection method based on deep learning and multimodal decision fusion, which does not rely on expert knowledge and supports a wide range of vulnerability types. Firstly, multiple features that can be used for vulnerability detection are extracted from smart contracts. In this paper, the source code (SC) [17], operation code (OP) [18], and control-flow graph (CFG) [19] of smart contracts are extracted. Secondly, the extracted features are used to train the neural network classifier, and the decision of each modal data is obtained. Finally, the decisions of these classifiers are integrated with the stacking-decision fusion method as the final decision. This paper experimentally evaluates the proposed approach on a publicly available dataset for four common vulnerability types. The experimental results show that, compared with existing methods, the proposed method can achieve higher accuracy and a higher AUC value, and better affects the vulnerability detection of smart contracts. The approach presented in this paper covers a more comprehensive type of smart contract vulnerabilities. The main contributions of this paper include:

- The proposal of a multimodal decision fusion vulnerability detection method for smart contract vulnerability detection. Experiments show that the proposed method is superior to other methods in detecting smart contract vulnerabilities.
- A comparison of the effectiveness of three modalities—source code, operation code, and control-flow graph—on the vulnerability detection task, revealing the effectiveness of the different modalities.

1.2. Layout

This paper is organized as follows. Section 2 reviews the existing research, including traditional vulnerability detection techniques for smart contracts, deep learning-based vulnerability detection techniques, and multimodal fusion techniques. Section 3 describes the flow of the proposed detection method for multimodal decision fusion vulnerability. Section 4 presents the experiments and conclusions. Section 5 discusses the experimental results and the advantages and disadvantages of the proposed method. Section 6 summarizes the work in this paper and presents the conclusion.

2. Review

This section compares and analyzes existing smart contract vulnerability detection and multimodal fusion techniques, identifies the advances and shortcomings of the current research, and further elaborates on the advantages of our approach.

2.1. Vulnerability Types

Smart contracts, as a piece of program code, are inescapably vulnerable, and these vulnerabilities continue to increase. The SWC vulnerability repository [20] alone has flagged as many as 136 types of vulnerabilities. No single tool can cover all the known and unknown vulnerabilities that may appear. Therefore, to verify the effectiveness of our proposed approach, we experimentally validate it against several common and damaging vulnerabilities. These include arithmetic [21], re-entrant [22], transaction-ordering dependence [23], and locked-ether [24] vulnerabilities. In this paper, these vulnerabilities are, respectively, abbreviated as follows for ease of representation: ARTHM, RENT, TimeO, and LE.

- ARTHM: Smart contracts are usually executed by virtual machines (such as EVMs), where integers have a range of maximum and minimum values that they can represent in the virtual machine. When an integer exceeds this range, an overflow occurs. For example, if a number is stored as a uint8, it means that the number is stored as an 8-bit unsigned number with values ranging from 0 to $2^8 - 1$. Hackers exploit the integer overflow vulnerability and make it easy for ordinary developers to make mistakes when writing smart contracts, resulting in integer overflow in smart contracts.
- RENT: In Ethereum, since smart contracts can call external contracts or send ether, these operations require the contract to submit external calls. Attackers can use these calls outside the contract to cause attack hijacking, allowing the attacked contract can be re-executed at any location. This bypasses the restrictions in the original code, resulting in re-entrant attacks. A re-entrant vulnerability triggered the DAO attack [25]. Ethereum was forced to implement a hard fork to recover the damage, and due to the anonymity of the Ethereum network, the attackers are still at large.
- TimeO: A race condition vulnerability occurs when the outcome of a contract's operation depends on the order of transactions submitted to it. In Ethereum, transactions are grouped into blocks, and the block confirmation time is approximately 17 s. After receiving the transaction request, the miner selects the transaction to be included in the block based on who has paid a high enough gas price. So, miners can anticipate which transactions will occur before they are finalized. For example, in a smart contract of the information reward type, the contract will reward the person who submits the answer first. In this case, a malicious node running an Ethereum client intercepts the reward by submitting a request at a higher price after receiving a transaction request from another node.
- LE: A locked-ether vulnerability means that ether is at risk of being frozen. The reason for ether locking is that smart contract developers do not consider the transfer function of ether when developing smart contracts. They only consider the receive function of ether. This will result in the smart contract account being able only to receive ether and not being able to transfer it out, resulting in the received ether becoming frozen.

2.2. Smart Contract Vulnerability Detection

From the perspective of technology-driven development, smart contract vulnerability detection can be divided into traditional methods and deep learning methods, where traditional methods are driven by expert experience in development. In contrast, deep learning methods are driven by data.

2.2.1. Traditional Methods

Traditional methods for detecting vulnerabilities in smart contracts draw heavily from conventional methods for detecting vulnerabilities in programming languages such as C\C++ and Java, which typically use static analysis and dynamic execution techniques to detect vulnerabilities. For example, Maian [26] uses a symbolic execution approach, which detects vulnerabilities by exploring the invocation paths of the contract; however, the number of paths to be explored grows exponentially with the branching state. Securify [11] and Zeus [27] propose a formal verification approach to vulnerability detection, achieving a low false positive rate of detection; however, these methods rely on expert auditing. ContractFuzzer [28] uses fuzzy testing for vulnerability detection and can successfully detect the RENT vulnerability and Parity wallet vulnerability that led to the DAO contract event.

Although traditional vulnerability detection methods are still widely popular today, they generally suffer from low automation, low efficiency, and long detection times due to their reliance on expert experience, and detecting smart contract vulnerabilities is still time-consuming and labor-intensive.

2.2.2. Deep Learning Methods

In recent years, with the rapid development of deep learning, vulnerability detection of smart contracts using deep learning techniques has become a hot research topic. Based on the data-driven deep learning approach, the key to the effectiveness of vulnerability detection for smart contracts lies in the extraction of features and the amount of data used for training. For example, Gao et al. [29] proposed an automated method based on word embedding to learn the features of smart contracts in Solidity. Zhang et al. [30] proposed a vulnerability detection method that combines information graphs and integrated learning using information graph embedding to extract features from smart contracts. Sendner et al. [31] proposed the first migration learning-based vulnerability detection method for smart contracts, which uses a generic feature extractor to learn the bytecode semantics of smart contracts and uses independent branches to learn the features of each vulnerability type. Zhuang et al. [32] first proposed a contract graph to represent the syntactic and semantic structures of smart contracts and then used graph convolutional neural networks to analyze vulnerabilities built on top of the contract graph.

Vulnerability detection technologies based on deep learning can address the problems of low automation, low efficiency, and dependence on expert knowledge of traditional detection methods. However, most of the current methods based on deep learning consider limited data characteristics, are unable to fully mine the vulnerability information of smart contracts, have poor scalability, and lack consideration for multimodal characteristics.

2.3. Multimodal Fusion

A modality is a description of something from a particular perspective. Multimodality usually contains two or more modalities and refers to a description of things from multiple perspectives. Usually, to utilize data from multiple modalities, it is necessary to fuse these modal data. According to the different fusion stages, multimodal fusion techniques can be divided into feature-based, decision-based, and hybrid fusion approaches. Currently, multimodal data fusion is one of the most effective ways to efficiently utilize large amounts of multisource data.

2.3.1. Feature-Based Fusion

The process of feature-based fusion is shown in Figure 1a. Feature-based fusion usually involves extracting features from different modalities and then fusing them directly, which is suitable for cases where there is a high correlation between modalities. Yang et al. [33] used feature-level fusion for audio and video features for speech recognition, but it suffers from the deficiency of challenging feature extraction. Besides deep learning-based feature extraction, the fuzzy inference method and generative rule method are standard methods for feature-level fusion. The generated high-dimensional feature vector limits feature-based fusion, and the method cannot model complex relationships because of the difficulty in obtaining the cross-correlation between different features by directly fusing multiple modal features.

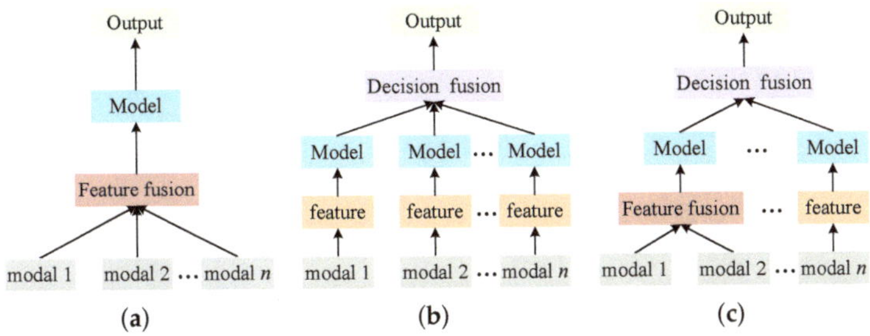

Figure 1. Comparison of the three fusion methods. (**a**) Feature-based fusion. (**b**) Decision-based fusion. (**c**) Hybrid Fusion.

2.3.2. Decision-Based Fusion

The process of decision-based fusion is shown in Figure 1b. Decision-based fusion [34] considers the variability of different modalities by using the features extracted from different modalities to train sub-models individually After obtaining decisions based on each modality, a combination of these decisions (e.g., vote, sum, average, etc.) is used as the final result. Decision-based fusion is a way to fuse the local decision information obtained from the data after further evaluation or inference [35], and each fusion modality is assumed to be independent of the others. Decision-based fusion is commonly used in human–computer interaction scenarios [36] such as emotion recognition [37]. Makiuchi et al. [38] proposed a new cross-representational speech model for emotion recognition. The model is based on a Transformer model to extract features and uses a CNN model to make decisions. Finally, based on the results from speech and text, the two modalities are fused using a fractional fusion approach, resulting in good outcomes. Xie et al. [39] fully exploited the similarity information between graphical texts and used a multilayer semantic decision-level fusion model to classify the sentiment of graphical texts in social media. Decision-based fusion can fuse heterogeneous data but suffers from the drawback of easy information loss.

2.3.3. Hybrid Fusion

The process of hybrid fusion is shown in Figure 1c. The hybrid fusion approach combines feature- and decision-based fusion. As a compromise between the two fusion methods, the limitations of feature- and decision-level fusion are improved by considering the modality's characteristics and the learning task's impact. Lin et al. [40] used feature- and decision-level fusion for audio and video, respectively, applied to a sentiment analysis study, and demonstrated that multimodal fusion outperformed unimodal fusion. Additionally, the decision-level fusion results were slightly better compared to the feature-level fusion results. Although hybrid fusion combines the advantages of feature- and decision-level fusion, it also makes the model more complex and makes learning more difficult.

2.3.4. Discussion

Figure 1 shows the flows of the three fusion methods. Each of the three fusion methods has advantages and disadvantages. Feature-based fusion captures the relationship between features better but is prone to overfitting. Decision-based fusion is less prone to overfitting, and different modalities have less influence on each other, but it does not allow the classifier to train all the data simultaneously. The hybrid fusion method is flexible to use, and while it synthesizes the advantages of both, it also increases the model's structural complexity and training difficulty.

In real-world smart contract vulnerability detection, the data distribution of vulnerability samples is often unbalanced [40]. On unbalanced datasets, the method based on feature fusion is more likely to overfit. In addition, when it is unclear whether a modality is suitable for smart contract vulnerability detection, the feature-based fusion method needs to consider the impact that adding features of a modality may have on the overall detection performance. In contrast, the decision-based fusion method can use all possible experiment modalities. The correlation between the three modalities of smart contracts used in this paper is small, and all modalities can be independently applied to the vulnerability detection of smart contracts. Therefore, the feature-based fusion method is not considered to avoid interference between different modalities. Last but not least, decision-based fusion methods are more interpretable.

We compared the scope and examples of applications of these three modal fusion approaches. The results of this comparison are shown in Table 1.

Table 1. Comparison of the three fusion methods.

Method	Stages	Scope of Applicability	Examples
Feature-based	Early	Emotion recognition	[41–43]
Decision-based	Terminal	Emotion recognition	[44,45]
Hybrid Fusion	Early and Terminal	Event detection	[46–48]

3. Our Method

This section describes the proposed approach, including the general structure and specific process.

3.1. Overview

3.1.1. Structure

The overall structure of the proposed approach is shown in Figure 2, and the whole process is divided into subclassifier learning and multimodal decision fusion. The subclassifiers make decisions based on the characteristics of the different modalities of the smart contract and determine whether there is a vulnerability. On the other hand, multimodal decision fusion takes each subclassifier's decision output as input and learns decision fusion strategies based on the decision output and the true labels.

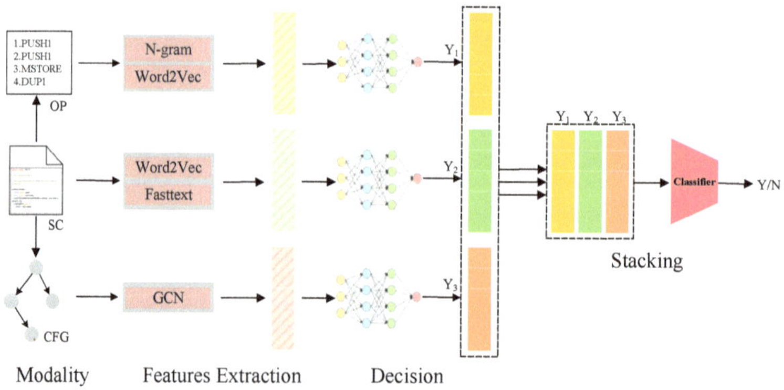

Figure 2. Multimodal fusion vulnerability detection scheme graph. Source code is the initial form in which a smart contract is written out and stored as a sequence of text. The operation code is the product of the compiler's compilation and records each EVM operation step when executing the contract. The control-flow graph shows how each code block is invoked when the contract is running.

3.1.2. Procedure

As shown in Figure 3, the procedure of our method is divided into four stages according to the order of execution.

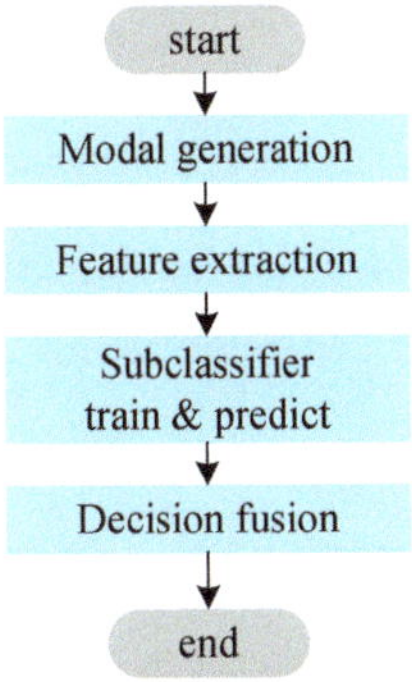

Figure 3. The procedure of our method. Our method is divided into four main stages: modal generation, feature extraction, subclassifier training and prediction, and decision fusion.

1. Modal generation. This stage generates different modal representations of the smart contract data based on the source code. This stage inputs the source code and outputs both the OP and CFG modalities.
2. Feature extraction. This stage extracts different features from the modalities for deep learning model training. This stage inputs the modal data and outputs multiple features.
3. Subclassifier training and prediction. The features extracted from each modality are used to train a deep neural network classifier to make a decision for a particular modality. This stage inputs the features of the modality and outputs the predicted values.
4. Decision fusion. This stage uses the stacking method to fuse the outputs of the individual subclassifiers. A final classifier is trained using the outputs of the subclassifiers as features. This classifier outputs the final prediction.

3.1.3. Discussion

This subsection discusses the feasibility of our proposed methodology. The source code contains the most primitive information about the smart contract [49] and is the primary basis for identifying vulnerabilities [50]. Compared with the source code, the operation code records more detailed operations of the EVM [51] when executing smart contracts, and contract violations can be detected through the operation code. The operation code records semantic information about the contract, whereas the control-flow graph records structural information about the contract. Some vulnerabilities tend to have anomalous structures. Table 2 shows the application of the three modalities for smart contract vulnerability detection.

Table 2. Examples of the three modalities used for vulnerability detection.

Modality	**Author**	**Year**	**Outcome**
Source code	Jiang et al. [52]	2022	High detection performance
	Gao et al. [29]	2020	Semantic clone detection
	Jeon et al. [53]	2021	High detection performance
Operation code	Albert et al. [54]	2022	Smart contract optimization
	Zhang et al. [55]	2023	Short detection time
	Ashizawa et al. [56]	2021	Detects both code clones and vulnerabilities
Control-flow graph	Liu et al. [57]	2023	High detection performance
	Agarwal et al. [58]	2022	High detection accuracy rates
	Liu et al. [59]	2023	Enables fine-grained vulnerability detection

In summary, the three modalities are feasible for smart contract vulnerability detection.

3.2. Modality Generation

The complete life cycle of a smart contract, from writing to execution, can be divided into three layers: the source code layer, build layer, and execution layer. This paper corresponds to these three layers using three modalities for vulnerability detection: source code (SC), operation code (OP), and control-flow graph (CFG). The features of the different modalities of smart contracts are first extracted. Then, different classifiers are trained using these features to make good use of the attributes of a single modality for vulnerability detection. Since the modal data of the smart contract's operation code and control-flow graph are usually not directly available, the operation code and control-flow graph need to be generated based on the source code.

3.2.1. Operation Code Generation

The operation code records the specific operation of each step of the virtual machine when the smart contract is executed, which is equivalent to a system instruction. Common operation codes and their corresponding meanings are shown in Table 3.

The solc compiler usually generates the operation code for smart contracts. A fragment of operation code generated using the example contract *0x0ce6d5a093d4166237c7a9ff8e0553b0293214a1.sol* is shown in Figure 4.

```
PUSH1 0x80 PUSH1 0x40 MSTORE PUSH8
0xDE0B6B3A7640000 PUSH1 0x0 SSTORE PUSH10
0x3F870857A3E0E3800000 PUSH1 0x1 SSTORE PUSH1
0x0 PUSH1 0x2 SSTORE CALLVALUE DUP1 ISZERO
```

Figure 4. An operation code fragment.

Table 3. Operation codes and their meanings.

Type	Name	Meaning
Stack operations	PUSHx POPx SWAPx	Push x-byte values on the stack Pop x-byte values on the stack Swap the top of the stack with the value of the xth position
Logic commands	AND OR NOT	Logical operations: AND Logical operations: OR Logical operations: NOT
Arithmetic commands	ADD MUL SUB	Arithmetic operations: ADD Arithmetic operations: MUL Arithmetic operations: SUB
Address and wallet commands	ADDRESS BALANCE ORIGIN	Obtains the address Obtains the balance Obtains the execution start address of the current contract
Predictable variable commands	TIMESTAMP NUMBER DIFFICULTY	Obtains the timestamp Obtains the block number Obtains the mining difficulty

Given that the meaning of the operation code is independent of the address, the operation code needs to be normalized to prevent the extracted features from having too large a dimension. The processing method used in this paper is as follows:

- The instructions PUSH1-PUSHx, POP1-POPx, etc., are unified as PUSH and POP;
- Remove the operands 0x80, 0x40, 0x0, etc.;

- Remove the operation addresses, such as 0xDE0B6B3A7640000.

The processed operation codes are shown in Figure 5, and the normalized operation codes proceed to the next stage of feature extraction.

```
PUSH PUSH MSTORE PUSH PUSH SSTORE PUSH
PUSH SSTORE PUSH PUSH SSTORE CALLVALUE
DUP ISZERO
```

Figure 5. An operation code fragment specification.

3.2.2. Control-Flow Graph Generation

The control-flow graphs record the variables and functions that each part of the system calls when the EVM executes a smart contract. This paper uses the Slither [60] tool to generate the control-flow graphs. Control-flow graphs generated using the sample contract *0x00000000000da14c27c155bb7c1ac9bd7519eb3b.sol* are shown in Figure 6.

As can be seen in Figure 6, the control-flow graph of a smart contract is a kind of directed acyclic graph, where the nodes of the graph are the various functions/variables in the contract, and the edges of the graph represent the invocation relationships between the functions or variables. The control-flow graphs show the structural information of the smart contract.

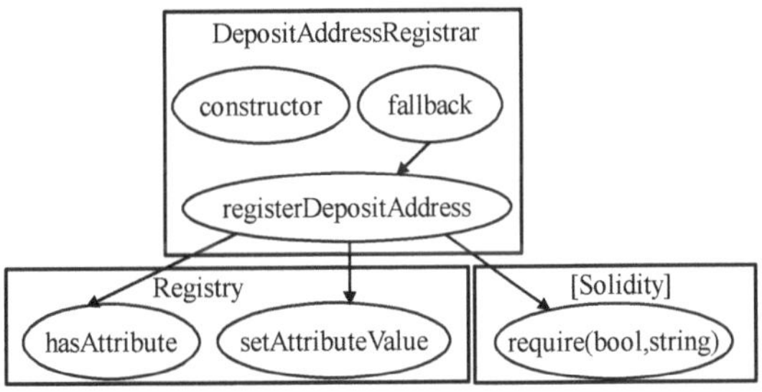

Figure 6. Example of a control-flow graph. The source code for this contract can be found in Appendix A. Combining the control-flow graph with the source code clearly shows the relationship between the code calls.

3.3. *Modal Features Extraction*

In Section 3.2, we obtained three different modal representations of the same smart contract. In this subsection, we extract features from these modalities to obtain a more computationally friendly form and to remove redundant information.

3.3.1. Source Code Embedding

The source code of a smart contract consists of lines of characters that cannot be used directly in a machine learning model and must be transformed into a numerical vector, i.e., text embedding. Keywords and variables with linguistic information from the smart contract in the source code are embedded into the high-dimensional space by performing text embedding on the source code. This paper embeds the source code using the Word2Vec model [61] and the Fasttext model [62].

Word2Vec is the process of mapping a sparse word vector of one-hot form into an n-dimensional dense vector using a single implicit layer, where words with similar meanings are mapped to similar positions in the vector space. Using neural networks allows flexible modeling of context, and it is one of the most popular approaches to text embedding. Word2Vec is based on the idea that the context of a word determines the semantics of that word. According to the semantic decision, Word2Vec is divided into the CBOW and Skip-gram models. Among them, the CBOW model predicts the central word based on the words around it, whereas the Skip-gram model predicts the surrounding words based on the central word.

Fasttext is a Facebook open source word vector and text classification tool that provides a simple and efficient method for learning text classification and representation with high performance and speed. The Fasttext model is a word vector-based text classification model. It uses character-level n-gram-based features to represent words in the text, thus avoiding the traditional bag-of-words model, which must consider all possible word sequences. In Fasttext's training process, each word is represented as a fixed-length vector. These vectors are combined into a vector representation of the text, and finally, a softmax activation function is used for classification.

The process of embedding source code using the Word2Vec and Fasttext models is as follows. The first step is to clean the text. It may contain various characters unrelated to the code's semantics (e.g., comments), so the source code text needs to be cleaned before training with the model, including removing deactivated words, comments, and useless punctuation. The next step is model training. After obtaining the text of the source code after text cleaning, the text is processed into a suitable form for the input of the Word2Vec and Fasttext models, and then the model is trained using the text. The final step is text-vector generation. The trained model performs text embedding for all contracts to obtain text vectors containing semantic information and supporting mathematical operations.

3.3.2. Operation Code Vectorization

Before the smart contract runs, the source code must be compiled into the form of operation code, which contains the basic operations of the program's execution (such as pop stack, push stack, take operand, etc.). Compared to the source code, the operation information contained in the operation code is more detailed and accurate. Similar to the source code, operation code sequences cannot be directly applied to machine learning models. This paper uses simultaneous N-gram and Word2Vec models to vectorize the operation code.

N-gram is an algorithm based on a statistical language model [63]. The basic idea is to process the content within the text using bytes within a sliding window of size N, forming a sequence of byte fragments of length N. Each byte fragment is called a gram, and the frequency of all grams is counted and filtered according to a pre-set threshold to form a list of crucial grams. This list constitutes the vector feature space of this text, with each kind of gram in this list representing a feature vector dimension. In order to prevent the generated vector dimensions from becoming too large, this paper uses a sliding window of N = 2 to extract features from the normalized operand sequence, ultimately generating 5183-dimensional features through experimentation.

The process of vectorizing operand sequences using the Word2Vec model is similar to the embedding process of the source code. First, the text is pre-processed to obtain the training corpus for the Word2Vec model. Then, the Word2Vec model is trained using the corpus, and the text vectors of individual characters are computed sequentially after the training is completed. Finally, the text vectors in all operands are summed to obtain the operand vectors of the whole contract.

3.3.3. Graph Convolution of Control-Flow Graphs

The graph neural network, a deep learning-based graph data processing method with good performance and interpretation, is widely used in various fields. The graph convolution network (GCN) [64] is a feature extractor similar to convolutional neural networks, except that its object is graph data. The graph convolution generally consists of three processes:

1. Send: Each node sends its feature information to its neighbor nodes after transformation. This step involves extracting and transforming the feature information of the nodes.
2. Receive: Each node aggregates the feature information of neighboring nodes. This step involves fusing the local structure information of the nodes.
3. Transform: Perform nonlinear transformation after aggregating the previous information to increase the expressiveness of the model.

In this paper, the features of the control-flow graph are extracted using a graph convolutional neural network, and the final experiment generates 64-dimensional features that are used in later classifier decisions.

3.4. Multimodal Decision Fusion

After obtaining the features from the different modalities, a neural network classifier is trained for each modality, and this classifier is used to apply the features of the different modalities. After obtaining the decisions from the neural network for each modality, *stacking* is applied to fuse the decisions and obtain an optimal decision that considers all the modalities.

The *stacking* decision fusion uses the output of each subclassifier as input. Suppose the classifiers obtained from each modality training are $\{L_1(x), L_2(x), \ldots, L_n(x)\}$, and the corresponding decision outputs are $\{y_1, y_2, \ldots, y_n\}$. Then, the single input of the fusion model is $\{y_1, y_2, \ldots, y_n, Y\}$, where Y is the true label. Various machine learning models can be used as the decision-fusion model, but this paper uses logistic regression as the fusion model. The reasons are as follows:

- There are not many kinds of modalities, and the dimensionality of the input of the fusion model is not large, so it is not easy to overfit by using simple models.
- The logistic regression model is highly interpretable, and the importance of each modality on the final decision can be obtained to judge the effectiveness of each modality on contract vulnerability detection.

In this paper, we use the binomial logistic regression model. As a classification model, the binomial logistic regression model is represented by a conditional probability distribution $P(Y = 1 \mid x)$. The random variable X takes the value of a real number, and the random variable Y takes the value of $\{0, 1\}$. The logistic regression model is defined in Equation (1). In the logistic regression model, the log odds of the output $Y = 1$ form a linear function of the input x. In model learning, for a given dataset T, the parameters of the model can be estimated using the maximum likelihood method.

$$\log \frac{P(Y = 1 \mid x)}{1 - P(Y = 1 \mid x)} = w \cdot x \tag{1}$$

In Equation (1), $x \in R^{n+1}$ is the input, $Y \in \{0, 1\}$ is the output, and $w \in R^{n+1}$ is the parameter known as the weight.

The basic idea of using *stacking* decision fusion in this thesis is to combine the decisions from each modality to form the optimal decision. Assuming that the effect of a particular modality is insignificant and the decision holds no value, it is also possible to set voting weights, etc., so that the final decision of that modality has the lowest impact on the final decision. The worst effect of decision-level fusion will not be lower than the effect of the best single modality.

4. Experiments

In this section, we empirically evaluate our proposed approach using a publicly available dataset. In order to evaluate the performance of the proposed method, the following research questions were designed:

- RQ1: Can the proposed method effectively detect common vulnerabilities in smart contracts, and is its vulnerability detection better than that of existing methods?
- RQ2: Does adding new features within the same modality help improve model performance?
- RQ3: Do different strategies of cross-modal fusion affect model performance?

The following experiments address each of these research questions.

4.1. Experiment Settings

4.1.1. Dataset

In this section, we conduct tests on a publicly available dataset and analyze the effectiveness of the proposed approach from multiple perspectives. We used the ScrawlD dataset [65], a real Ethereum smart contract dataset with vulnerability tags, to test against four common contract vulnerability types: ARTHM, TimeO, LE, and RENT. We employed a crawler to obtain the source code of the corresponding contracts from Ethscan. Considering the unbalanced nature of the data, we resampled a few classes of categories using the ADASYN algorithm [66]. A total of 9252 smart contract data were used in the experiments.

4.1.2. Experiment Environment

The vulnerability detection model proposed in this paper mainly consists of four parts: modal generation, which compiles source code into an operation code and generates a control-flow graph; feature extraction, which is used to extract the features of each mode; a neural network classifier, which is used to classify samples according to each modal feature; and decision fusion, which is used to fuse the output of each modal classifier. The modal generation tools were mainly implemented using Python and SIF tools [67]. Feature extraction was implemented through text embedding and graph convolution models. The neural network classifiers and decision fusion were implemented using Pytorch and SkLearn. The system environment versions included Ubuntu 18.04, Python 3.10, Scikit-learn 1.2.2, and Pytorch 1.13.1.

All experiments were conducted on a computer equipped with an Intel Xeon Gold 6240R CPU @ 2.6 GHz, GPU Tesla V100S-32 GB, and 64 GB RAM.

4.1.3. Parameter Setting

The input type of the graph convolutional neural network contained the following three parts: the dimensionality of each node in the graph (uniformly coded as 128 dimensions in this paper), the representation of the adjacency matrix corresponding to the nodes in the graph, and the label corresponding to the graph. The output was the number of graph categories. The hidden layer was $128 \times 128 \times 128 \times 128 \times 128 \times 64$, and the *batch* size was set to 256. The neural network classifier was a fully connected feedforward network with $200 \times 100 \times 100 \times 50$ hidden layers. There is no straightforward method for determining the number of layers of the neural network, and we determined a better layer configuration based on experience and experiments.

The activation function of both networks was *Relu*, a binary cross-entropy loss function. The *Relu* activation function is nonlinear and less prone to gradient vanishing. The network parameters were adjusted using the *Adam* optimizer, which is commonly used in deep learning. The learning rate for the graph convolutional neural network was set to 0.005, and the learning rate of the fully connected feedforward network was set to 0.001. A lower learning rate prevents neural networks from converging to local optimal points.

Decision fusion was implemented using a logistic regression model, and to prevent overfitting, L_2 regularization was applied to constrain the model parameters. A grid search method was applied to explore the optimal regularization parameters within the range of $[0.001, 20]$, with 500 equidistant points. L_2 regularization was chosen because there were fewer features, and L_2 regularization prevents overfitting.

We randomly selected 70% of the dataset as the training set and the remaining 30% as the test set. and report the average results from multiple training runs.

4.1.4. Evaluation Metrics

We applied the following four widely used evaluation metrics to measure the effectiveness of our method against other methods. Accuracy (ACC) is the percentage of all samples correctly detected. Precision (P) is the ratio of correctly detected vulnerable samples to all detected vulnerable samples. Recall (R) is the ratio of correctly detected vulnerable samples to all vulnerable samples. The F1 score (F1) is the summed average of precision and recall, used as a score to measure the overall effectiveness. The AUC value (i.e., the area under the ROC curve) can be used to compare the performance of different classifiers. The exact formulas for the evaluation metrics are as follows:

$$\mathrm{ACC} = \frac{\mathrm{TP} + \mathrm{TN}}{\mathrm{TP} + \mathrm{FP} + \mathrm{TN} + \mathrm{FN}} \tag{2}$$

$$\mathrm{R} = \frac{\mathrm{TP}}{\mathrm{TP} + \mathrm{FN}} \tag{3}$$

$$\mathrm{P} = \frac{\mathrm{TP}}{\mathrm{TP} + \mathrm{FP}} \tag{4}$$

$$\mathrm{F1} = 2 \times \frac{\mathrm{P} \times \mathrm{R}}{\mathrm{P} + \mathrm{R}} \tag{5}$$

4.2. Comparison Experiment (Addressing RQ1)

In order to verify the effectiveness of the proposed method, a comparative test was conducted using existing smart contract vulnerability detection tools on the same dataset. These tools included Mythril [68], Smartcheck [69], and Slither [60]. In addition, we selected TMP [32] as a similar method for comparison. Table 4 shows the comparison of these methods and our proposed method, these methods are described as follows:

- Mythril: A free security analysis tool provided by the Ethereum open source community. Mythril, one of the most well-known smart contract security tools, can detect security vulnerabilities in Solidity smart contracts and perform in-depth analysis. Mythril is a static vulnerability analysis tool that relies simply on symbolic execution [70] and the SMT method to perform vulnerability detection. It currently supports the detection of vulnerabilities including integer-overflow, timestamp-dependency, and re-entry attacks.
- Smartcheck: A scalable static analysis tool for detecting vulnerabilities or code issues in Solidity smart contracts. Smartcheck converts smart contracts into a structured XML form and then identifies smart contract vulnerabilities by matching Xpath patterns. Although Smartcheck is a simple and efficient tool for detecting smart contract vulnerabilities, the types of vulnerabilities supported for detection are limited.
- Slither: The first open source static analysis framework for the Solidity language. Slither can find vulnerabilities within seconds as a tool dedicated to the security analysis of Solidity smart contracts. Slither is user-friendly and provides numerous APIs for developers to use. However, the tool suffers from underreporting of severe issues, such as not being very sensitive to detecting integer-overflow vulnerabilities and requiring manual code auditing.
- TMP: A novel temporal information propagation network that learns vulnerability features in a normalized contract graph using graph convolution. A contract graph is a graph that represents data and control dependencies between program statements.

The comparison experiments were performed on the same dataset using tools including Mythril, Smartcheck, Slither, TMP, and the method proposed in this paper. The experimental results are shown in Table 5. As can be seen from the experimental results, the method proposed in this paper can support more types of vulnerabilities because it does not rely on expert knowledge. Compared to existing mature, smart contract vulnerability detection tools, the method proposed in this paper achieved higher detection accuracy and AUC values on various types of vulnerabilities, as well as better overall detection performance.

Table 4. Comparison of existing methods and our proposed method.

Method	Year	Technical Category	Vulnerability Types
Mythril	2017	symbolic execution	medium
Smartcheck	2018	static analysis	few
Slither	2019	static analysis	few
TMP	2020	deep learning	many
Our method	-	deep learning	many

The traditional approach relies on expert experience and requires constant updating of the characteristics of smart contract vulnerabilities. Often, even experienced experts make mistakes and omit some vital information that can be used for vulnerability detection, which is the root cause of the poor effectiveness of this type of smart contract vulnerability detection method. TMP is a vulnerability detection method based on the concept of contract graphs that was proposed in [32]. The graph-based detection method is often suitable for security vulnerabilities caused by contract inter-call. In Table 5, it can be seen that TMP achieved the best detection accuracy for the RENT vulnerability. At the same time, vulnerabilities such as ARTHM may appear in a line of code and cannot be represented graphically. Traditional and deep learning-based approaches use limited information about the characteristics of contracts.

Our method uses as much contractual characterization information as possible and combines this information organically. Therefore, our method's vulnerability detection is better than existing methods.

Table 5. Experimental comparison with other methods.

Method	ARTHM		RENT		LE		TimeO	
	ACC	AUC	ACC	AUC	ACC	AUC	ACC	AUC
Mythril	0.681	0.689	0.716	0.843	-	-	0.879	0.5
Smartcheck	0.564	0.5	-	-	0.925	0.798	-	-
Slither	-	-	0.729	0.851	0.927	0.804	-	-
TMP	0.628	0.561	0.745	0.669	0.841	0.507	0.772	0.684
Our method	0.916	0.834	0.909	0.852	0.895	0.825	0.948	0.886

The "-" symbols in the table mean that the vulnerability detection tool does not support this type of contract.

4.3. Ablation Experiments

In this subsection, we verify the effectiveness of each part of the proposed method. Firstly, an experimental comparison between features within a single modality and feature fusion is performed. Secondly, the effectiveness of cross-modal fusion is verified.

4.3.1. Comparison between Features within a Single Modality and Feature Fusion (Addressing RQ2)

After extracting the features from the modalities using the feature extractor, the neural network classifier was trained directly, and the effects were compared. The ARTHM vulnerability was used as an example to compare the effectiveness of different types of features within the different modalities for vulnerability detection, and the experimental results are shown in Table 6.

In Table 6, it can be seen that the addition of new features of a modality can effectively improve the performance of the training model in cases where the detection performance of the original single intra-modal feature is poor.

Table 6. Intra-modal compar. Using ARTHM as an example.

Modal	Feature	ACC	AUC	R	P	F1
SC	FastText	0.741	0.739	0.724	0.674	0.699
	Word2Vec	0.839	0.824	0.728	0.871	0.793
	Merge	0.851	0.844	0.787	0.861	0.823
OP	N-gram	0.858	0.846	0.758	0.898	0.822
	Word2Vec	0.847	0.835	0.741	0.887	0.807
	Merge	0.856	0.843	0.74	0.914	0.818
CFG	GCN	0.745	0.733	0.648	0.730	0.687

4.3.2. Comparison of Inter-Modal Features and Multimodal Decision Fusion (Addressing RQ3)

Based on the information in Section 4.3.1, the neural network classifier was trained using the features obtained through intra-modal feature fusion, and the experimental results of the different modalities were compared. Table 7 shows the experimental results of the different modalities for four vulnerability types, where Concat indicates that the features of these modalities were stitched together to form a new feature and used as the input of the neural network classifier. In Table 7, it can be seen that the features of the OP outperformed those of the CFG and SC on the vulnerability detection task. The direct fusion of cross-modal features resulted in a slight performance improvement compared to the optimal modality.

Table 7. Inter-modal comparison.

Type	Use_Data	AUC	ACC	P	R	F1
ARTHM	CFG	0.727	0.735	0.704	0.667	0.685
	OP	0.837	0.840	0.815	0.814	0.815
	SC	0.781	0.790	0.785	0.710	0.745
	Concat	0.829	0.829	0.785	0.832	0.808
LE	CFG	0.703	0.738	0.371	0.647	0.472
	OP	0.817	0.846	0.553	0.772	0.644
	SC	0.644	0.770	0.383	0.447	0.413
	Concat	0.823	0.839	0.535	0.798	0.641
RENT	CFG	0.703	0.601	0.207	0.836	0.332
	OP	0.885	0.867	0.469	0.909	0.619
	SC	0.775	0.722	0.278	0.845	0.418
	Concat	0.876	0.871	0.476	0.881	0.618
TimeO	CFG	0.703	0.601	0.207	0.836	0.332
	OP	0.885	0.867	0.469	0.909	0.619
	SC	0.775	0.722	0.278	0.845	0.418
	Concat	0.876	0.871	0.476	0.881	0.618

We obtained the decision outputs from each modality, and the results of decision fusion using logistic regression are shown in Table 8. The modalities corresponding to the weights are OP, SC, CFG, and Concat. The magnitude of the weights of logistic regression can be regarded as the importance of the influence of the different modalities on the results, and the positive and negative signs can be interpreted as the decision propensities of the modalities.

As shown in Table 8, the performance improvement of sampling for unbalanced samples was significant, especially for the ACC and AUC values. However, the improvement in the accuracy rate was not significant and can make the model more inclined to classify the samples into a few categories. Figure 7 further visualizes the results shown in Tables 7 and 8.

Table 8. The multimodal decision fusion results.

Type	Sampling	AUC	ACC	P	R	F1	Weights
ARTHM	0	0.843	0.904	0.834	0.808	0.82	7.663, −2.486, 1.420, 4.740
	1	0.834	0.916	0.779	0.853	0.814	3.622, −1.596, 0.018, 4.876
LE	0	0.816	0.908	0.729	0.689	0.708	8.361, 0.390, 1.643, 2.933
	1	0.825	0.895	0.556	0.788	0.652	4.483 , −1.898, 2.285, 3.356
RENT	0	0.636	0.884	0.560	0.290	0.382	5.451, −1.134, 2.180, 2.675
	1	0.852	0.909	0.280	0.890	0.426	2.016, 1.390, 1.050, 5.184
TimeO	0	0.844	0.948	0.64	0.745	0.688	13.704, 3.089, 3.664, 1.858
	1	0.886	0.948	0.507	0.888	0.645	7.540, 0.767, 1.707, 1.727

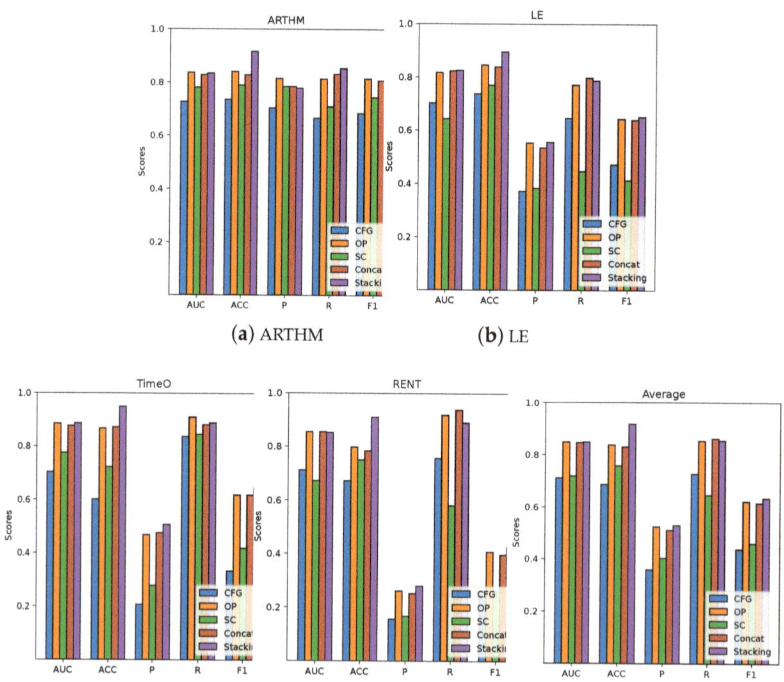

(**a**) ARTHM (**b**) LE

(**c**) TimeO (**d**) RENT (**e**) Averaged

Figure 7. Performance evaluation of different modalities for different types of vulnerability detection. In the figure, the OX-axis represents the different evaluation indicators; the OY-axis represents the score of a particular evaluation indicator, which takes a range of 0∼1. In addition to the AUC, the other indicators can also be interpreted as percentages. (**a**–**d**) Experimental results for different vulnerabilities. (**e**) Averaged results.

The detection method using logistic regression models for multimodal models performed better for the detection task of four common smart contract vulnerabilities. The ACC and AUC values of the decision fusion model improved significantly compared to the method that directly fused the features of each modality.

5. Results

The experimental results show that the smart contract vulnerability detection method proposed in this paper outperforms existing methods, achieving high accuracy and a high recall of vulnerability detection. Ablation experiments show that adding new features within the same modality can improve the performance of the vulnerability detection model. In the smart contract vulnerability detection task, the decision fusion of multiple modalities contributes more to improving detection accuracy. In addition, using the resampling method to address the dataset's unbalanced vulnerability types can somewhat mitigate the imbalanced category problem. However, the detection accuracy rate cannot be improved. In the vulnerability detection task, losses from missed detections are often more significant than those from false positives, so achieving a high recall rate is more important than a high accuracy rate.

6. Conclusions

In this paper, from a multimodal perspective, three modalities—source code, operation code, and control-flow graph—with five features were extracted for vulnerability detection from the combined smart contract life cycle. The multimodal decision fusion approach can fully utilize the semantic and structural features of smart contracts. The results of the comparison experiments also verify the proposed approach's effectiveness and superiority.

Compared with existing smart contract vulnerability detection tools, the proposed approach has many advantages. Firstly, the method does not rely on expert experience and is completely data-driven. Secondly, the method is highly flexible. Finally, the method covers more types of vulnerabilities and achieves a higher inspection accuracy rate. The method proposed in this paper can be used for the initial screening of smart contract vulnerabilities to increase the efficiency of smart contract developers and auditors in identifying smart contract vulnerabilities. Based on this paper, adding new modalities can further improve the performance of vulnerability detection.

Some data imbalances in the experiment led to low detection accuracies. Future work can consider using small samples and unsupervised learning to address this problem. Furthermore, the lack of a standard unified smart contract vulnerability dataset remains an urgent problem that needs to be addressed.

Author Contributions: Conceptualization, W.D. and T.H.; investigation, W.D. and C.C.; supervision, T.H., Y.P. and H.W.; writing—original draft, W.D.; writing—review and editing, W.D., C.C., X.H. and T.H. All authors have read and agreed to the published version of the manuscript.

Funding: This work was supported by the National Natural Science Foundation of China (62002074), NSF of Guangdong Province (2023A1515030273), and the Key Laboratory, Ministry of Industry and Information Technology, China.

Institutional Review Board Statement: Not applicable.

Informed Consent Statement: Not applicable.

Data Availability Statement: Not applicable.

Conflicts of Interest: The authors declare no conflict of interest.

Appendix A. An Example of Smart Contract Source Code

```
pragma solidity ^0.4.23;

interface Registry {
function setAttributeValue
(address who, bytes32 what, uint val) external;
function hasAttribute(address _who, bytes32 _attribute)
external view returns(bool);}

contract DepositAddressRegistrar {
Registry public registry;
bytes32 public constant IS_DEPOSIT_ADDRESS
= "isDepositAddress";
event DepositAddressRegistered
(address registeredAddress);
constructor(address _registry) public {
registry = Registry(_registry);}

function registerDepositAddress() public {
address shiftedAddress = address(uint(msg.sender) >> 20);
require(!registry.hasAttribute
(shiftedAddress, IS_DEPOSIT_ADDRESS),
"deposit address already registered");
registry.setAttributeValue
(shiftedAddress, IS_DEPOSIT_ADDRESS, uint(msg.sender));
emit DepositAddressRegistered(msg.sender);}

function() external payable {
registerDepositAddress();
msg.sender.transfer(msg.value);}}
```

References

1. Deng, W.; Huang, T.; Wang, H. A Review of the Key Technology in a Blockchain Building Decentralized Trust Platform. *Mathematics* **2023**, *11*, 101. [CrossRef]
2. Buterin, V. A Next-Generation Smart Contract and Decentralized Application Platform. *White Paper* **2014**, *3*, 1–36. Available online: https://blockchainlab.com/pdf/Ethereum_white_paper-a_next_generation_smart_contract_and_decentralized_application_platform-vitalik-buterin.pdf (accessed on 1 July 2023).
3. Dannen, C. *Introducing Ethereum and Solidity*; Springer: Berlin/Heidelberg, Germany, 2017; Volume 1.
4. Zheng, Z.; Xie, S.; Dai, H.N.; Chen, W.; Chen, X.; Weng, J.; Imran, M. An overview on smart contracts: Challenges, advances and platforms. *Future Gener. Comput. Syst.* **2020**, *105*, 75–491. [CrossRef]
5. Zhao, P.; Li, C.; Fu, Y.; Hui, Y.; Zhang, Y.; Cheng, N. Blockchain-Enabled Conditional Decentralized Vehicular Crowdsensing System. *IEEE Trans. Intell. Transp. Syst.* **2022**, *23*, 18937–18950. [CrossRef]
6. Lin, S.; Zhang, L.; Li, J.; Ji, L.; Sun, Y. A survey of application research based on blockchain smart contract. *Wirel. Netw.* **2022**, *28*, 635–690. [CrossRef]
7. Atzei, N.; Bartoletti, M.; Cimoli, T. A survey of attacks on Ethereum Smart Contracts. Cryptology ePrint Archive, Paper 2016/1007. 2016. Available online: https://eprint.iacr.org/2016/1007 (accessed on 1 August 2023).
8. The Parity Wallet Hack Explained. 2023. Available online: https://blog.openzeppelin.com/on-the-parity-wallet-multisig-hack-405a8c12e8f7 (accessed on 1 August 2023).
9. Peng, K.; Li, M.; Huang, H.; Wang, C.; Wan, S.; Choo, K.K.R. Security Challenges and Opportunities for Smart Contracts in Internet of Things: A Survey. *IEEE Internet Things J.* **2021**, *8*, 12004–12020. [CrossRef]
10. Slowmist. 2023. Available online: https://hacked.slowmist.io/ (accessed on 14 May 2023).
11. Tsankov, P.; Dan, A.; Drachsler-Cohen, D.; Gervais, A.; Buenzli, F.; Vechev, M. Securify: Practical security analysis of smart contracts. In Proceedings of the 2018 ACM SIGSAC Conference on Computer and Communications Security, Toronto, ON, Canada, 15–19 October 2018; pp. 67–82.
12. Kushwaha, S.S.; Joshi, S.; Singh, D.; Kaur, M.; Lee, H.N. Systematic review of security vulnerabilities in ethereum blockchain smart contract. *IEEE Access* **2022**, *10*, 6605–6621. [CrossRef]

13. Ghaleb, A.; Pattabiraman, K. How effective are smart contract analysis tools? evaluating smart contract static analysis tools using bug injection. In Proceedings of the 29th ACM SIGSOFT International Symposium on Software Testing and Analysis, Los Angeles, CA, USA, 18–22 July 2020; pp. 415–427.
14. Wang, H.; Liu, Y.; Li, Y.; Lin, S.W.; Artho, C.; Ma, L.; Liu, Y. Oracle-supported dynamic exploit generation for smart contracts. *IEEE Trans. Dependable Secur. Comput.* **2020**, *19*, 1795–1809. [CrossRef]
15. Qian, P.; Liu, Z.; He, Q.; Zimmermann, R.; Wang, X. Towards Automated Reentrancy Detection for Smart Contracts Based on Sequential Models. *IEEE Access* **2020**, *8*, 19685–19695. [CrossRef]
16. He, D.; Wu, R.; Li, X.; Chan, S.; Guizani, M. Detection of Vulnerabilities of Blockchain Smart Contracts. *IEEE Internet Things J.* **2023**, *10*, 12178–12185. [CrossRef]
17. Hwang, S.J.; Choi, S.H.; Shin, J.; Choi, Y.H. CodeNet: Code-Targeted Convolutional Neural Network Architecture for Smart Contract Vulnerability Detection. *IEEE Access* **2022**, *10*, 32595–32607. [CrossRef]
18. Huang, J.; Zhou, K.; Xiong, A.; Li, D. Smart Contract Vulnerability Detection Model Based on Multi-Task Learning. *Sensors* **2022**, *22*, 1829. [CrossRef] [PubMed]
19. Yadav, K.; Naval, S. CFG Analysis for Detecting Vulnerabilities in Smart Contracts. In Proceedings of the International Conference on Smart Computing and Communication, Jaipur, India, 24–25 January 2023; Springer: Berlin/Heidelberg, Germany, 2023; pp. 753–763.
20. Smart Contract Weakness Classification Registry. 2023. Available online: https://swcregistry.io/ (accessed on 30 May 2023).
21. CVE. 2023. Available online: https://nvd.nist.gov/vuln/detail/CVE-2018-10769 (accessed on 31 May 2023).
22. Ethereum Smart Contract Best Practices. 2023. Available online: https://consensys.github.io/smart-contract-best-practices/attacks/reentrancy/ (accessed on 31 May 2023).
23. Medium. 2023. Available online: https://medium.com/coinmonks/solidity-transaction-ordering-attacks-1193a014884e (accessed on 31 May 2023).
24. Perez, D.; Livshits, B. Smart Contract Vulnerabilities: Vulnerable Does Not Imply Exploited. In Proceedings of the 30th USENIX Security Symposium, USENIX Security 2021, Virtual, 11–13 August 2021; Bailey, M., Greenstadt, R., Eds.; USENIX Association: Berkeley, CA, USA, 2021; pp. 1325–1341.
25. N'Da, A.A.K.; Matalonga, S.; Dahal, K. Characterizing the Cost of Introducing Secure Programming Patterns and Practices in Ethereum. In Proceedings of the Trends and Innovations in Information Systems and Technologies, Budva, Montenegro, 7–10 April 2020; Rocha, Á., Adeli, H., Reis, L.P., Costanzo, S., Orovic, I., Moreira, F., Eds.; Springer: Cham, Switzerland, 2020; pp. 25–34.
26. Nikolic, I.; Kolluri, A.; Sergey, I.; Saxena, P.; Hobor, A. Finding The Greedy, Prodigal, and Suicidal Contracts at Scale. In Proceedings of the 34th Annual Computer Security Applications Conference, ACSAC 2018, San Juan, PR, USA, 3–7 December 2018; ACM: New York, NY, USA, 2018; pp. 653–663. [CrossRef]
27. Kalra, S.; Goel, S.; Dhawan, M.; Sharma, S. ZEUS: Analyzing Safety of Smart Contracts. In Proceedings of the 25th Annual Network and Distributed System Security Symposium, NDSS 2018, San Diego, CA, USA, 18–21 February 2018; The Internet Society: Reston, VA, USA, 2018.
28. Jiang, B.; Liu, Y.; Chan, W.K. ContractFuzzer: Fuzzing smart contracts for vulnerability detection. In Proceedings of the 33rd ACM/IEEE International Conference on Automated Software Engineering, ASE 2018, Montpellier, France, 3–7 September 2018; Huchard, M., Kästner, C., Fraser, G., Eds.; ACM: New York, NY, USA, 2018; pp. 259–269. [CrossRef]
29. Gao, Z.; Jiang, L.; Xia, X.; Lo, D.; Grundy, J. Checking Smart Contracts With Structural Code Embedding. *IEEE Trans. Softw. Eng.* **2021**, *47*, 2874–2891. [CrossRef]
30. Zhang, L.; Wang, J.; Wang, W.; Jin, Z.; Zhao, C.; Cai, Z.; Chen, H. A Novel Smart Contract Vulnerability Detection Method Based on Information Graph and Ensemble Learning. *Sensors* **2022**, *22*, 3581. [CrossRef] [PubMed]
31. Sendner, C.; Chen, H.; Fereidooni, H.; Petzi, L.; König, J.; Stang, J.; Dmitrienko, A.; Sadeghi, A.; Koushanfar, F. Smarter Contracts: Detecting Vulnerabilities in Smart Contracts with Deep Transfer Learning. In Proceedings of the 30th Annual Network and Distributed System Security Symposium, NDSS 2023, San Diego, CA, USA, 27 February–3 March 2023; The Internet Society: Reston, VA, USA, 2023.
32. Zhuang, Y.; Liu, Z.; Qian, P.; Liu, Q.; Wang, X.; He, Q. Smart Contract Vulnerability Detection using Graph Neural Network. In Proceedings of the Twenty-Ninth International Joint Conference on Artificial Intelligence—IJCAI 2020, Yokohama, Japan, 11–17 July 2020; pp. 3283–3290. [CrossRef]
33. Choi, W.Y.; Song, K.Y.; Lee, C.W. Convolutional Attention Networks for Multimodal Emotion Recognition from Speech and Text Data. In Proceedings of the Grand Challenge and Workshop on Human Multimodal Language (Challenge-HML), Melbourne, Australia, 20 July 2018; pp. 28–34.
34. Tang, K.; Ma, Y.; Miao, D.; Song, P.; Gu, Z.; Tian, Z.; Wang, W. Decision Fusion Networks for Image Classification. *IEEE Trans. Neural Netw. Learn. Syst.* **2022**, 1–14. [CrossRef] [PubMed]
35. Gandhi, A.; Adhvaryu, K.; Poria, S.; Cambria, E.; Hussain, A. Multimodal sentiment analysis: A systematic review of history, datasets, multimodal fusion methods, applications, challenges and future directions. *Inf. Fusion* **2022**, 91, 424–444. [CrossRef]
36. Li, M. An analysis of multimodal emotional braincomputer interface. In Proceedings of the 2021 International Conference on Intelligent Computing, Automation and Systems (ICICAS), Chongqing, China, 29–31 December 2021; IEEE: Piscataway, NJ, USA, 2021; pp. 407–411.

37. Hu, J.; Liu, Y.; Zhao, J.; Jin, Q. MMGCN: Multimodal Fusion via Deep Graph Convolution Network for Emotion Recognition in Conversation. In Proceedings of the 59th Annual Meeting of the Association for Computational Linguistics and the 11th International Joint Conference on Natural Language Processing (Volume 1: Long Papers), Virtual Event, 1–6 August 2021; pp. 5666–5675.
38. Makiuchi, M.R.; Uto, K.; Shinoda, K. Multimodal Emotion Recognition with High-Level Speech and Text Features. In Proceedings of the 2021 IEEE Automatic Speech Recognition and Understanding Workshop (ASRU), Cartagena, Colombia, 13–17 December 2021; pp. 350–357.
39. Xie, H.; Mao, J.; Li, G. Sentiment classification of image-text information with multi-layer semantic fusion. *Data Anal. Knowl. Discov.* **2021**, *5*, 103–114. [CrossRef]
40. Ling, S.; Zhang, X.; Guo, M.; Zhang, W. Method Research on Multimodal Emotion Recognition Based on Audio and Video. *J. Signal Process.* **2021**, *37*, 1889–1898. [CrossRef]
41. Fan, W.; He, Z.; Xing, X.; Cai, B.; Lu, W. Multi-modality Depression Detection via Multi-scale Temporal Dilated CNNs. In Proceedings of the 9th International on Audio/Visual Emotion Challenge and Workshop, AVEC@MM 2019, Nice, France, 21–25 October 2019; Ringeval, F., Schuller, B.W., Valstar, M.F., Cummins, N., Cowie, R., Pantic, M., Eds.; ACM: New York, NY, USA, 2019; pp. 73–80. [CrossRef]
42. Yin, S.; Liang, C.; Ding, H.; Wang, S. A Multi-Modal Hierarchical Recurrent Neural Network for Depression Detection. In Proceedings of the 9th International on Audio/Visual Emotion Challenge and Workshop, AVEC@MM 2019, Nice, France, 21–25 October 2019; Ringeval, F., Schuller, B.W., Valstar, M.F., Cummins, N., Cowie, R., Pantic, M., Eds.; ACM:New York, NY, USA, 2019; pp. 65–71. [CrossRef]
43. Ray, A.; Kumar, S.; Reddy, R.; Mukherjee, P.; Garg, R. Multi-level Attention Network using Text, Audio and Video for Depression Prediction. In Proceedings of the 9th International on Audio/Visual Emotion Challenge and Workshop, AVEC@MM 2019, Nice, France, 21–25 October 2019; Ringeval, F., Schuller, B.W., Valstar, M.F., Cummins, N., Cowie, R., Pantic, M., Eds.; ACM: New York, NY, USA, 2019; pp. 81–88. [CrossRef]
44. Meng, H.; Huang, D.; Wang, H.; Yang, H.; Al-Shuraifi, M.; Wang, Y. Depression recognition based on dynamic facial and vocal expression features using partial least square regression. In Proceedings of the 3rd ACM international workshop on Audio/Visual Emotion Challenge, AVEC@ACM Multimedia 2013, Barcelona, Spain, 21 October 2013; Schuller, B.W., Valstar, M.F., Cowie, R., Krajewski, J., Pantic, M., Eds.; ACM: New York, NY, USA, 2013; pp. 21–30. [CrossRef]
45. Gaw, N.; Yousefi, S.; Gahrooei, M.R. Multimodal data fusion for systems improvement: A review. *IISE Trans.* **2022**, *54*, 1098–1116. [CrossRef]
46. Shalu, H.; P, H.; CN, H.S.; Das, A.; Majumder, S.; Datar, A.; MS, S.M.; Das, A.; Kadiwala, J. Depression Status Estimation by Deep Learning based Hybrid Multi-Modal Fusion Model. *arXiv* **2020**, arXiv:2011.14966.
47. Alghowinem, S.; Goecke, R.; Cohn, J.F.; Wagner, M.; Parker, G.; Breakspear, M. Cross-cultural detection of depression from nonverbal behaviour. In Proceedings of the 11th IEEE International Conference and Workshops on Automatic Face and Gesture Recognition, FG 2015, Ljubljana, Slovenia, 4–8 May 2015; IEEE Computer Society: Washington, DC, USA, 2015; pp. 1–8. [CrossRef]
48. Morales, M.R.; Scherer, S.; Levitan, R. A Linguistically-Informed Fusion Approach for Multimodal Depression Detection. In Proceedings of the Fifth Workshop on Computational Linguistics and Clinical Psychology: From Keyboard to Clinic, CLPsych@NAACL-HTL, New Orleans, LA, USA, 5 June 2018; Loveys, K., Niederhoffer, K., Prud'hommeaux, E., Resnik, R., Resnik, P., Eds.; Association for Computational Linguistics: Stroudsburg, PA, USA, 2018; pp. 13–24. [CrossRef]
49. Shakya, S.; Mukherjee, A.; Halder, R.; Maiti, A.; Chaturvedi, A. Smartmixmodel: Machine learning-based vulnerability detection of solidity smart contracts. In Proceedings of the 2022 IEEE International Conference on Blockchain (Blockchain), Espoo, Finland, 22–25 August 2022; IEEE: Piscataway, NJ, USA, 2022; pp. 37–44.
50. Akter, M.S.; Shahriar, H.; Bhuiya, Z.A. Automated vulnerability detection in source code using quantum natural language processing. In Proceedings of the Inernational Conference on Ubiquitous Security, Zhangjiajie, China, 28–31 December 2022; Springer: Berlin/Heidelberg, Germany, 2022; pp. 83–102.
51. Wu, C.; Xiong, J.; Xiong, H.; Zhao, Y.; Yi, W. A review on recent progress of smart contract in blockchain. *IEEE Access* **2022**, *10*, 50839–50863. [CrossRef]
52. Jiang, F.; Cao, Y.; Xiao, J.; Yi, H.; Lei, G.; Liu, M.; Deng, S.; Wang, H. VDDL: A deep learning-based vulnerability detection model for smart contracts. In Proceedings of the International Conference on Machine Learning for Cyber Security, Guangzhou, China, 2–4 December 2022; Springer: Berlin/Heidelberg, Germany, 2022; pp. 72–86.
53. Jeon, S.; Lee, G.; Kim, H.; Woo, S.S. Smartcondetect: Highly accurate smart contract code vulnerability detection mechanism using bert. In Proceedings of the KDD Workshop on Programming Language Processing, Virtual, 14–18 August 2021.
54. Albert, E.; Gordillo, P.; Hernández-Cerezo, A.; Rubio, A.; Schett, M.A. Super-optimization of smart contracts. *ACM Trans. Softw. Eng. Methodol. (TOSEM)* **2022**, *31*, 1–29. [CrossRef]
55. Zhang, H.; Zhang, W.; Feng, Y.; Liu, Y. SVScanner: Detecting smart contract vulnerabilities via deep semantic extraction. *J. Inf. Secur. Appl.* **2023**, *75*, 103484. [CrossRef]
56. Ashizawa, N.; Yanai, N.; Cruz, J.P.; Okamura, S. Eth2Vec: Learning contract-wide code representations for vulnerability detection on ethereum smart contracts. In Proceedings of the 3rd ACM International Symposium on Blockchain and Secure Critical Infrastructure, Hong Kong, China, 7–11 June 2021; pp. 47–59.

57. Liu, Z.; Jiang, M.; Zhang, S.; Zhang, J.; Liu, Y. A Smart Contract Vulnerability Detection Mechanism Based on Deep Learning and Expert Rules. *IEEE Access* **2023**, *11*, 77990–77999. [CrossRef]
58. Agarwal, S.; Godboley, S.; Krishna, P.R. Cyclomatic Complexity Analysis for Smart Contract Using Control Flow Graph. In Proceedings of the International Conference on Computing, Communication and Learning, Warangal, India, 27–29 October 2022; Springer: Berlin/Heidelberg, Germany, 2022; pp. 65–78.
59. Liu, H.; Fan, Y.; Feng, L.; Wei, Z. Vulnerable Smart Contract Function Locating Based on Multi-Relational Nested Graph Convolutional Network. *J. Syst. Softw.* **2023**, *204*, 111775. [CrossRef]
60. Feist, J.; Grieco, G.; Groce, A. Slither: A static analysis framework for smart contracts. In Proceedings of the 2019 IEEE/ACM 2nd International Workshop on Emerging Trends in Software Engineering for Blockchain (WETSEB), Montreal, QC, Canada, 27 May 2019; IEEE: Piscataway, NJ, USA, 2019; pp. 8–15.
61. Mikolov, T.; Chen, K.; Corrado, G.; Dean, J. Efficient Estimation of Word Representations in Vector Space. In Proceedings of the 1st International Conference on Learning Representations, ICLR 2013, Scottsdale, Arizona, USA, 2–4 May 2013.
62. Kuyumcu, B.; Aksakalli, C.; Delil, S. An automated new approach in fast text classification (fastText) A case study for Turkish text classification without pre-processing. In Proceedings of the 2019 3rd International Conference on Natural Language Processing and Information Retrieval, Tokushima, Japan, 28–30 June 2019; pp. 1–4.
63. Wang, H.; He, J.; Zhang, X.; Liu, S. A short text classification method based on N-gram and CNN. *Chin. J. Electron.* **2020**, *29*, 248–254. [CrossRef]
64. Kipf, T.N.; Welling, M. Semi-Supervised Classification with Graph Convolutional Networks. In Proceedings of the 5th International Conference on Learning Representations, ICLR 2017, Toulon, France, 24–26 April 2017.
65. Sujeet Yashavant, C.; Kumar, S.; Karkare, A. ScrawlD: A Dataset of Real World Ethereum Smart Contracts Labelled with Vulnerabilities. *arXiv* **2022**, arXiv:2202.11409.
66. He, H.; Bai, Y.; Garcia, E.A.; Li, S. ADASYN: Adaptive synthetic sampling approach for imbalanced learning. In Proceedings of the 2008 IEEE International Joint Conference on Neural Networks (IEEE World Congress on Computational Intelligence), Hong Kong, China, 1–6 June 2008; IEEE: Piscataway, NJ, USA, 2008; pp. 1322–1328.
67. Peng, C.; Akca, S.; Rajan, A. SIF: A Framework for Solidity Contract Instrumentation and Analysis. In Proceedings of the 2019 26th Asia-Pacific Software Engineering Conference (APSEC), Putrajaya, Malaysia, 2–5 December 2019; IEEE: Piscataway, NJ, USA, 2019; pp. 466–473.
68. Mythril. 2023. Available online: https://github.com/ConsenSys/mythril (accessed on 31 May 2023).
69. Tikhomirov, S.; Voskresenskaya, E.; Ivanitskiy, I.; Takhaviev, R.; Marchenko, E.; Alexandrov, Y. SmartCheck: Static Analysis of Ethereum Smart Contracts. In Proceedings of the 1st IEEE/ACM International Workshop on Emerging Trends in Software Engineering for Blockchain, WETSEB@ICSE 2018, Gothenburg, Sweden, 27 May–3 June 2018; ACM: New York, NY, USA, 2018; pp. 9–16. [CrossRef]
70. Baldoni, R.; Coppa, E.; D'Elia, D.C.; Demetrescu, C.; Finocchi, I. A Survey of Symbolic Execution Techniques. *ACM Comput. Surv.* **2018**, *51*, 50:1–50:39. [CrossRef]

mathematics

MDPI

Article

AI-Enabled Condition Monitoring Framework for Outdoor Mobile Robots Using 3D LiDAR Sensor

Sathian Pookkuttath *, Povendhan Arthanaripalayam Palanisamy and Mohan Rajesh Elara

Engineering Product Development Pillar, Singapore University of Technology and Design (SUTD), Singapore 487372, Singapore; povendhan_palanisamy@sutd.edu.sg (P.A.P.); rajeshelara@sutd.edu.sg (M.R.E.)
* Correspondence: sathian_pookkuttath@sutd.edu.sg

Abstract: An automated condition monitoring (CM) framework is essential for outdoor mobile robots to trigger prompt maintenance and corrective actions based on the level of system deterioration and outdoor uneven terrain feature states. Vibration indicates system failures and terrain abnormalities in mobile robots; hence, five vibration threshold classes for CM in outdoor mobile robots were identified, considering both vibration source system deterioration and uneven terrain. This study proposes a novel CM approach for outdoor mobile robots using a 3D LiDAR, employed here instead of its usual use as a navigation sensor, by developing an algorithm to extract the vibration-indicated data based on the point cloud, assuring low computational costs without losing vibration characteristics. The algorithm computes cuboids for two prominent clusters in every point cloud frame and sets motion points at the corners and centroid of the cuboid. The three-dimensional vector displacement of these points over consecutive point cloud frames, which corresponds to the vibration-affected clusters, are compiled as vibration indication data for each threshold class. A simply structured 1D Convolutional Neural Network (1D CNN)-based vibration threshold prediction model is proposed for fast, accurate, and real-time application. Finally, a threshold class mapping framework is developed which fuses the predicted threshold classes on the 3D occupancy map of the workspace, generating a 3D CbM map in real time, fostering a Condition-based Maintenance (CbM) strategy. The offline evaluation test results show an average accuracy of vibration threshold classes of 89.6% and consistent accuracy during real-time field case studies of 89%. The test outcomes validate that the proposed 3D-LiDAR-based CM framework is suitable for outdoor mobile robots, assuring the robot's health and operational safety.

Keywords: AI; 3D LiDAR; point cloud; vibration; condition monitoring; 1D CNN; outdoor mobile robot; condition-based maintenance; operational safety

MSC: 68T40; 68T01; 68T07; 93C85

1. Introduction

The rapid advancements in sensor technologies and the advent of Deep Learning and Computer Vision techniques have enhanced the development of outdoor mobile robots and anticipated high market need for various applications, such as pavement sweeping, landscaping, pest control, inspection, logistics, and security. As per ABI research, the expected shipment of outdoor mobile robots by 2030 is 350,000 units, compared with 40,000 units in 2021, a compound annual growth rate (CAGR) of 27% [1]. The state-of-the-art technology for outdoor mobile robots is more focused on operational performance, for instance, understanding the outdoor scene and environmental perception [2], pose estimation and localization [3], visual navigation [4], motion control design [5], path planning [6], and pedestrian safety [7]. Since outdoor mobile robots are exposed to uneven terrain and weather extremities, the system deterioration rate in outdoor mobile robots is higher compared with those for indoor applications. Terrainability in outdoor mobile robots, the ability to negotiate various terrain irregularities, has been studied mainly to classify typical terrain types, such as tiles, stone, gravel, sand, asphalt, and grass. It is generally conducted by two approaches [8]: a wheel–terrain interactive vibration proprioceptive traversability analysis (retrospective techniques) and geometry or appearance-based exteroceptive analysis (prospective methods) using onboard sensors and adopting suitable Artificial Intelligence (AI) techniques for accurate classification. Some examples include terrain classification studies using an Inertial Measurement Unit (IMU) in [9,10], camera-based classification in [11], and a combination of camera and IMU-based classification in [12]. Similarly, a traversability analysis using 3D LiDAR is conducted in [13], and a comparative study showed better terrain classification under different lighting conditions using a 3D LiDAR sensor than a camera in [14]. These approaches will help to classify undesired terrain features; however, they are not an ideal solution for monitoring system deterioration caused by long-term usage or to alert if exposed to any unstructured terrain which causes an accelerated system failure or potential hazards, allowing to opt for a suitable maintenance strategy. Hence, an automated condition monitoring (CM) method is vital for outdoor mobile robots to ensure the robot's health and operational safety. CM is an enabler to predict potential failure so that maintenance or corrective actions can be promptly triggered [15,16]. This can foster a Condition-based Maintenance (CbM) strategy for mobile robots' deployments, i.e., trigger a proper action based on the health condition of the robot or workspace terrain features only when needed, assuring productivity and safety [17].

The CM frameworks for mobile robots exploring the onboard typical proprioceptive and exteroceptive sensors, considering both the robot's internal and external conditions, are an unexplored research area. The existing research works are focused mainly on autonomous and functional enhancements. This is crucial because of the vast demand for mobile robots, optimized maintenance resources, and assured operational safety. Our research focuses on utilizing the typical onboard sensors for CM applications, other than their usual uses for perception, navigation, obstacle avoidance, path planning, etc. Towards this research objective, we have been developing suitable CM frameworks for indoor mobile robots and presented the works based on an IMU sensor [18] and camera [19]. Developing an appropriate CM framework for an outdoor mobile robot is more challenging considering the hostile, dynamic, unpredictable, and highly uneven terrain and climate changes in the outdoor workspace environment, especially for fumigation or fogging, lawn mowing, and pavement sweeping applications. Hence, in the proposed work, we explore a 3D LiDAR sensor, an inevitable sensor for mapping and navigation in outdoor autonomous robots, for the CM application of outdoor mobile robots addressing various challenges.

1.1. Problem Statement

It is crucial to monitor and ensure the health conditions of an outdoor mobile robot and that its workspace terrain features are safe or at acceptable levels during its entire course of operation, as here the robot is exposed to undetected and uneven terrain features and adverse weather conditions, which causes an accelerated mode of system degradation resulting unexpected downtime, catastrophic failure, high maintenance cost, operational hazards, and customer dissatisfaction. The terrainability studies mentioned in [9–14] will help to plan paths, especially for last mail delivery or security robots, where the end destination or point-to-point travel can be executed with a suitable path planner, minimizing the traverse through unstructured terrain. However, such plans may need to be revised often as the terrain features may keep changing in an outdoor environment. Besides, this approach is not an ideal solution for the robots deployed for outdoor cleaning and maintenance applications, such as pavement sweeping, fogging or fumigation, and lawn mowing or pruning branches because, here, the robot has to cover a predefined area, completely following a traverse pattern which may include uneven terrain such as curbs, cliffs, exposed tree-roots, broken or missing paved stones, pits, gutters, raised utility hole lids, drainage system, etc. Such uneven terrain features, which cause adverse effects on the robot's health, will not be captured as obstacles in the environment mapping due to their small size. This scenario is the worst in lawn mowing and fumigation or fogging robots, as grass fields usually cover such undesired terrain imperfections. Hence, the passable or impassable exteroceptive terrain classification method is also impractical. Here, due to continuous operation of such sweeping, fumigation, and lawn mowing tasks on uneven areas, the robot degrades at a faster rate, which eventually causes damage to wheels and their alignment, structural deformation and instability, quick wear out of sweeping brushes, looseness of perception sensors assembly, etc., affecting functional performance, safety, and maintenance cost. The system will also degrade naturally and fail due to long-term continuous deployment, even if the terrain is structured and well-retained. Hence, depending on the robot's design capabilities, applications, terrain types, and changes in climate conditions, the rate of system degradation and potential hazards varies, and a periodic maintenance strategy may not be an ideal solution. As per our knowledge, there is no automated real-time monitoring system to evaluate how good or bad the health condition of the robot and the given workspace for deployment, especially for outdoor environment maintenance robots, other than the human visual inspection method currently followed. The aforementioned facts reveal that an automated real-time CM system is crucial for outdoor mobile robots considering system degradation, terrain flaws, and safety, which supports a CbM strategy to optimize the maintenance resources.

The key challenges involved in the proposed 3D-LiDAR-based CM framework include integrating a 3D Lidar as a reliable vibration indicator sensor posing technical difficulties, as its primary function is for mapping and navigation. Developing an algorithm extracting the vibration-indicated data from the numerous point cloud data for a fast and low computational cost for real-time application is a complex task. Identifying different internal and external factors to assess the states of the robot and extreme outdoor terrain features including severity levels for classification to fit outdoor CM applications is crucial and needs a representative ruggedized outdoor robot for real-world data acquisition, and much trial and error is involved. Adapting a suitable AI model to detect the unique features of vibration-indicated data from dynamically changing outdoor environments for each class with high accuracy is challenging. As the proposed study is based on 3D LiDAR point clouds, the dynamically changing outdoor features and range of 3D LiDAR are tough for performance accuracy, unless a robust and sophisticated data processing algorithm is developed.

1.2. Contributions

- This research proposed an automated CM framework for outdoor mobile robots to alert the system deterioration and terrain-related flaws of the workspace causing accelerated degradation or threat to decide whether the robot can easily traverse, traverse with caution, or avoid continuous operation.
- Vibration are used as a key indicator of system degradation or uneven terrain; the classified vibration threshold levels in an outdoor robot are safe, moderately safe, and unsafe. Further, the causes of abnormal vibrations are classified as system-generated or terrain-induced to assist proper maintenance or corrective actions.
- We explored a novel approach of using a 3D LiDAR sensor for condition monitoring applications in outdoor mobile robots based on point cloud data instead of its usual application for navigation.
- The 3D LiDAR point cloud data affected by changes in the linear and angular motion of the robot due to vibration threshold classes are used as vibration-indicated data.
- A data extracting algorithm was developed, assuring low computational cost without losing vibration characteristics, by forming cuboids for two prominent clusters in each point cloud frame with computed motion points at its corners and centroid. The vector displacements of these points on three axes, based on the vibration-affected clusters, are compiled from consecutive frames and used as vibration indication data for training each class.

- A simply structured 1D-CNN-based vibration threshold class prediction model was developed for fast, accurate, and real-time execution.
- A 3D CbM map was generated, developing a framework to map the predicted classes on the 3D occupancy map of the workspace in real time, enabling the maintenance team to trigger prompt actions based on the robot's condition or workspace, thereby fostering a CbM strategy.

The remainder of this paper is structured as follows: Section 2 briefs the related published works. Section 3 overviews the proposed framework, including vibration-indicated data modeling. Section 4 elaborates on data acquisition, 1D CNN model training, results, and discussion. Real-time field case studies and observations are presented in Section 5. Finally, the work is summarized in Section 6.

2. Related Works

The vibration data for fault detection and prognosis of equipment is typically measured using micro-electro-mechanical systems (MEMS) or piezoelectric accelerometers [20–23], mainly for bearings, motors, and machinery. Accelerometers are also used for terrain classification in outdoor mobile robot applications, for instance, classifying indoor floor, paving, asphalt, gravel, boule court, and grass in [24], as well as in [25] for a four-wheeled planetary exploration rover to classify sand, gravel, and clay. Though CNN-based Deep Learning models are generally used for image pixel-based data, there are many studies where a 1D CNN is used for feature learning on vibration-signal-based raw sensor data for multinomial classification. For instance, a human activity recognition and classification task was conducted using accelerometer and gyroscope sensor raw data applying a 1D CNN model in [26]. One-dimensional CNN models are widely used to to extract vibration-affected features effectively to classify and predict various defects or severity conditions accurately in real-time, mainly due to their simple structure, low computational complexity, and easy deployability [27]. For instance, vibration and 1D-CNN-based CM or fault diagnosis and prognostics studies are conducted for machine components [20,28], structural systems [29], and industrial robots [30]. However, such vibration-based CM research works are not much explored for assuring the health and safety of autonomous indoor and outdoor mobile robots even though there is great research scope by utilizing typical onboard sensors, as well as considering their vast market demand for various applications. To fill such research gaps, two vibration and 1D-CNN-based CM frameworks were proposed previously for indoor mobile cleaning robots, mainly to predict the internal and external sources of abnormal vibrations which cause accelerated system deterioration or threats, thereby enabling a CbM strategy. Firstly, there was an IMU sensor-based study [18] where the vibration-affected change in linear acceleration, angular velocity, and angular acceleration of the robot was modeled as vibration data. In the second work, a monocular camera sensor [19] was used, and the vibration was modeled as sparse optical flow. Here, the change in optical flow vector displacement over consecutive frames due to the robot's vibrations are used as vibration indication data for training different abnormal vibration source classes.

An outdoor mobile robot's operational conditions and system degradation differ from indoor robots due to its extreme workspace nature. The current studies mainly focus on terrain classification using various sensors to classify terrain types using suitable AI models to conclude passable or impassable terrain. For example, in the study in [9], terrain classification of mainly asphalts, mud, gravel, and grass, based on magnitude frequency response on X and Y axis rotation and Z axis acceleration rate, is captured through an IMU sensor in an autonomous ground vehicle, using a Probabilistic Neural Network (PNN). A similar study using IMU sensor data on a wheeled mobile robot was conducted to classify typical outdoor terrain types adapting MultiLayer Perceptron (MLP) neural networks in [10]. Here, the authors also indicated a lower prediction accuracy of classes when using the accelerometer and gyroscope separately. Camera-based visual terrain classification for such terrain features was conducted in [11] using various texture-based descriptors and applying Random Forest (RF) as the classifier. An IMU-and-camera-based sensor fusion method was applied in the study in [12] to classify 14 types of terrain classes adapting Support Vector Machine (SVM), which also mentioned a faster classifier is needed for real-time prediction. Semisupervised learning with a positive Naive Bayes classifier was conducted in [13] using 3D LiDAR data for traversability analysis in outdoor mobile robots by defining the feature vectors with different dimensions. The study in [14] shows that 3D-LiDAR-based outdoor terrain classification is better under different lighting conditions due to its illumination-invariant property compared with the camera study, and both used RF as the classifier. These studies are limited to the terrainability or traversability of the robot without considering the internal system degradation level or assessing the safe threshold levels to continue the operation in a given workspace, which includes uneven and unseen terrain features. Also, the existing works do not consider ground-level negative obstacles, such as minor pits and gutters, which affect traversability and the robot's health, which is crucial for CM, too. Besides, to assess the performance of the terrainability/traversability for extreme, uneven outdoor terrains, a ruggedized platform representative of real-world outdoor applications is necessary; however, it mostly uses a small, lightweight simple prototype with the required sensors mounted.

Three-dimensional LiDAR is an inevitable sensor in an outdoor autonomous robot for navigation based on its point cloud data. Recently, many algorithms have been developed using 3D LiDAR point cloud data for fast and accurate prediction of the 3D motion of points in dynamic environments, mainly scene flow estimation, enabling various real-time applications for outdoor robots and autonomous vehicles. For example, FlowNet3D, an architecture explained in [31], specifies the motion between two frames estimating translation vector flow for each point in the frame, assisting 3D scan registration and motion segmentation applications, enabling better point feature learning and predicting the flow of original points. Any Motion Detector (AMD-Net), a class agnostic scene dynamics estimation architecture proposed in [32], outputs a static and dynamic segmentation of each cell and its velocity. Instead of handling object detection and scene flow estimation separately, the PillaFlowNet network [33] takes both simultaneously, providing fast and accurate results. A Fast Hierarchical Network (FH-Net) method is proposed in [34], minimizing the cost and latency issues of typical scene flow methods and directly computing the key points flow from the source point to its target points. These studies show that the recent point-cloud-based works are more focused on dynamic object detection, motion segmentation, and tracking to assist autonomous outdoor robots or vehicles. Hence, there is a research scope to explore for CM applications analyzing vibration-affected point cloud data to assure outdoor mobile robots' health and operational safety.

3. Overview of the Proposed Framework

The overview of the proposed 3D LiDAR point-cloud-based CM framework for outdoor mobile robots, predicting the vibration threshold classes in real-time to trigger proper maintenance or corrective action promptly, is illustrated in Figure 1 and elaborated in the following sections.

3.1. Outdoor Mobile Robot for Data Acquisition and Trials

An in-house-designed autonomous wheeled mobile robot is used in this work for vibration-affected point cloud data acquisition for different threshold classes for 1D CNN training and real-time field case studies. The robot base platform is developed in a way to easily mount various modular payloads for diverse outdoor applications, such as pavement sweeping, fumigation or fogging (pest control), and landscaping applications, as shown in Figure 2, i.e., a vacuum motor–brush–bin assembly for pavement sweeping (Figure 2(2.1), a sprayer with tank assembly for fumigation (Figure 2(2.2)), and a grass cutting wire (lawn mowing), and a robotic biarm assembly (pruning branches) for landscaping applications (Figure 2(2.3)) are integrated for specific outdoor applications. However, for the extensive data acquisition and field trials of this study, we used the fumigation robot (Figure 2(2.2)), as this robot has to traverse both pavements and grass fields to serve its purpose, ensuring all the potential extreme outdoor terrain features are covered and studied.

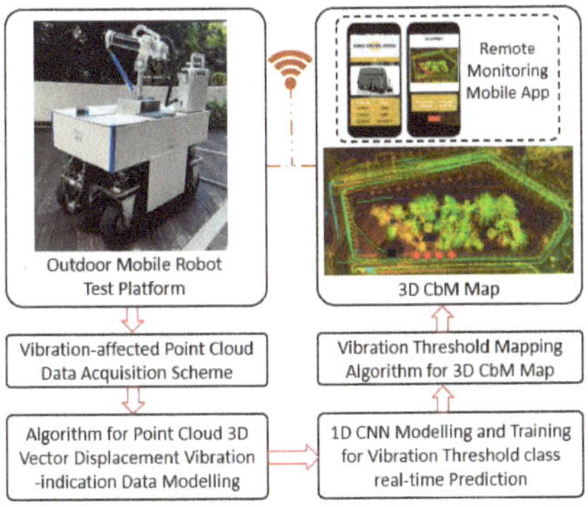

Figure 1. Overview of the proposed CM framework for outdoor mobile robots.

Figure 2. Test platform: outdoor mobile robot for various applications.

The overall size and weight of the robot without payload are 1.6 (L) × 0.85 (W) × 1 (H) m and 300 Kg, respectively, and designed for an additional payload of up to 200 Kg. A ruggedized chassis is built with the weldment of stainless-steel hollow sections and plates, machined stainless steel shafts, aluminum brackets for mounting all critical parts, and a double-wishbone suspension system to fit outdoor driving environment applications. Four hub wheel motors with 250 W of power and a wheel size of 38 (outer diameter) × 15 (width) cm were used for transmission, with 100 W steering motors designed as independent steering drive wheels for omnidirectional mobility. A 3D LiDAR Velodyne VLP-16 is used for perception and localization, and subsequently in autonomous navigation, and a Vectornav VN-100 IMU sensor and wheel encoders are included for dead reckoning to estimate the robot's position. A 300 Ah 48 V DC lithium iron phosphate (LFP) battery is used to power the system, including payloads. An Industrial PC Anewtech IPC970-10WF5E1A is integrated for steering the entire operation and data storage, and the Robot Operating System (ROS) middleware in the Ubuntu Noetic operating system is used to execute all communications. A TP-Link LTE 4G Router is added for Internet and networking, including for the robot's remote health monitoring.

3.2. Vibration Threshold Classes for Condition Monitoring

An outdoor wheeled mobile robot is generally designed in a ruggedized manner, anticipating uneven terrain, compared with an indoor mobile robot. However, the chances of system degradation or risk of hazardous operations are still high because of the unexpected and uncontrolled changes in terrain features, as explained in the problem statement section, especially while deploying the robot in autonomous mode. A wheeled mobile robot produces aberrant vibration as an early failure symptom, and its amplitude level depends on the system deterioration severity or the extremity of the terrain defects or undetected features. Hence, a vibration-level-based assessment or threshold level classification is a prerequisite to executing an automated CM system. Based on the robot design and application use cases furnished, the vibration threshold levels are broadly classified as safe, moderately safe, and unsafe. Further, the moderately safe and unsafe classes are classified based on the causes of vibration, either system-degradation-generated or terrain-induced. Hence, a total of five vibration-threshold-level classes are identified in this study for outdoor mobile robots, as illustrated in Figure 3. Here, the safe class is the acceptable vibration level, considering the robot is healthy and deployed at a smooth/plain terrain workspace. Moderately safe (M) is an acceptable class, but it is monitored as the condition may worsen faster. The unsafe (U) class advises not to continue the operations long term due to high chances of total system failure and operational hazards. The moderately safe vibration generated by the system (M-S) represents system deterioration, such as wheels wobbling, worn out or punctured wheels, etc., including unbalanced loading, and here, the exposed terrain is considered even and well-maintained. The moderately safe vibration due to poor terrain (M-T) class is due to uneven and undetected terrain features, which cause accelerated system degradation. Similarly, the unsafe system-generated (U-S) and unsafe terrain-induced (U-T) classes comparatively represent severe vibrations due to these distinguished causes.

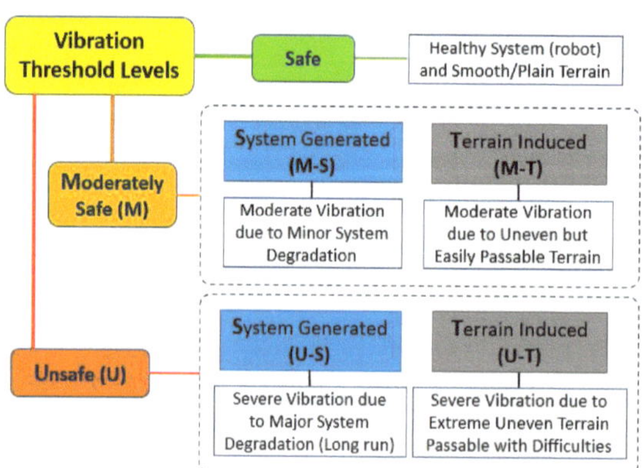

Figure 3. Vibration threshold classification for condition monitoring.

3.3. Point-Cloud-Based Vibration-Indicated Data Modeling

A mobile robot vibrates or jerks moderately or severely in angular and linear ways, caused by system deterioration or undesired terrain features. As 3D LiDAR is a prevalent sensor for outdoor autonomous mobile robots, we exploited the behavior of 3D LiDAR point cloud data across the five vibration threshold classes to develop a methodology to model the vibration-affected point cloud data as vibration indication data for CM applications in outdoor mobile robots. The onboard 3D LiDAR Velodyne VLP-16 sensor is set at a subscription rate of 10 Hz, and the change in the three-dimensional motion field of points in a point cloud due to vibration classes is analysed and tracked through consecutive frames, similar to scene flow techniques. The 3D LiDAR sensor set-up and the three-dimensional vector displacement of the motion point in point cloud space between two consecutive frames are illustrated in Figure 4, including visualization of a point cloud frame captured through ROS graphical interface RVIZ. The 3D LiDAR sensor Velodyne VLP-16 was firmly mounted directly to the metal chassis at the front side of the robot on the top–middle area, and other payload mounting configurations were considered as well.

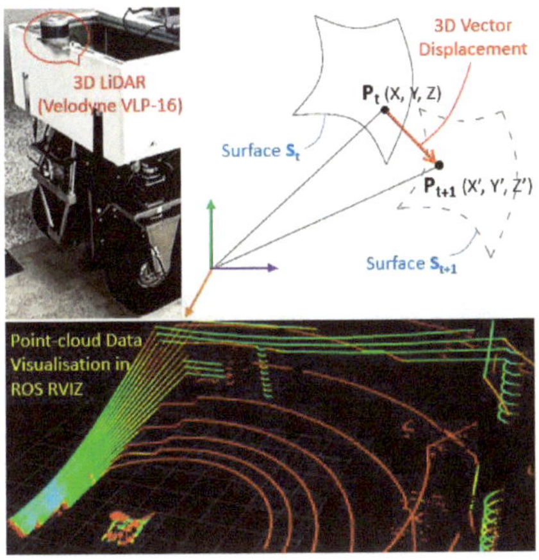

Figure 4. Three-dimensional LiDAR set up and point cloud 3D vector displacement.

The change in motion point displacement on three axes due to various vibration threshold levels of the robot is modeled as 3D vector displacement data, as represented in Equation (1), which is used as the vibration-indicated data (Vib_{data}). As illustrated in Figure 4, the motion point P is represented as (X, Y, Z) in 3D space at time t, and (X', Y, Z') is the displaced point in the consecutive frame at time $(t+1)$.

$$Vib_{data} = [(X' - X), (Y' - Y), (Z' - Z)] = [\delta X, \delta Y, \delta Z] \quad (1)$$

The Velodyne VLP-16 generates 3D images of the environment in 360° through 16 channels with a vertical field of view +15° to −15° by acquiring around 300,000 points per second. It is not computationally recommended to consider every point for modeling as vibration-indicated data for real-time deployment. The point cloud for any prominent objects in the environment will generally be denser than its background or small objects. Hence, we developed a fast and computationally low-cost method by creating a vibration-indicated data modeling algorithm without losing the vibration characteristics. Firstly, the algorithm set a boundary around the 3D LiDAR position, starting from 1.5 m and limited to 10 m for every scene and detected all object clusters within this range/boundary. Then, the two most prominent clusters were selected, and all other points adding a filter layer were removed. A bounding box (cuboid) is set for each cluster based on size. The bounding box motion corresponds to the cluster motion, which is directly related to the linear and angular motion of the robot due to changes in vibration threshold levels. Next, a point is computed at each corner and the centroid of the cuboid as motion points and the three-dimensional vector displacement of these points is tracked through every consecutive frame. As the motion of these points in 3D space corresponds to the vibration-affected robot's motion, finally, the vector displacement elements of these nine points on the three axes from each cluster are extracted and modeled as vibration-indication data, as represented in Equation (1). The vibration-indicated data are extracted as $[n \times 3]$, where n is the number of motion points, and 3 is for the vector displacement data on three axes. Figure 5 depicts the steps to extract the vibration-indicated data from the point cloud, assuring low computation cost and representing vibration threshold classes.

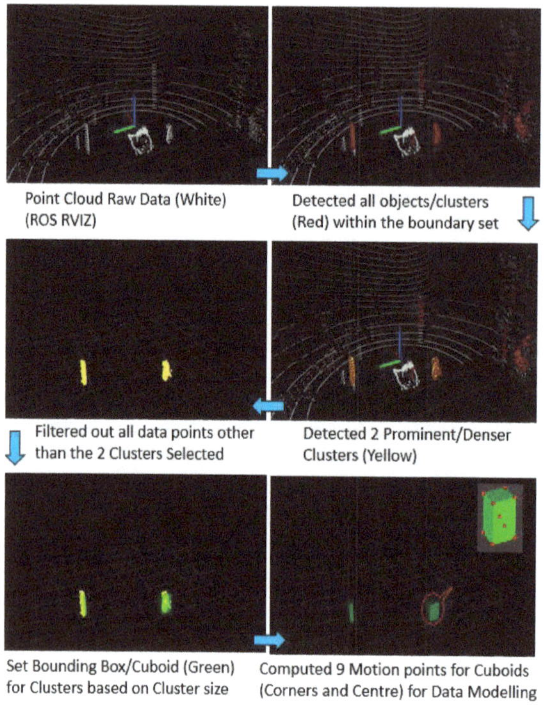

Figure 5. Point cloud data filtering and vibration-indicated data modeling.

Further, the algorithm compiles the vibration-indicated data as follows. The 3D LiDAR data are subscribed at 10 Hz, i.e., every 0.1 s a new frame is generated. A total of 18 displacement points (9 points $\times$ 2 clusters) are formed from every two consecutive frames, each representing three axes elements. The sequential data collected from 30 frames, i.e., over 3 s, are treated as one sample of data for training and during real-time prediction, which is [540 $\times$ 3] and saved as a CSV file, as illustrated in Figure 6. The CSV files for each vibration threshold class are labeled for preparing the training dataset.

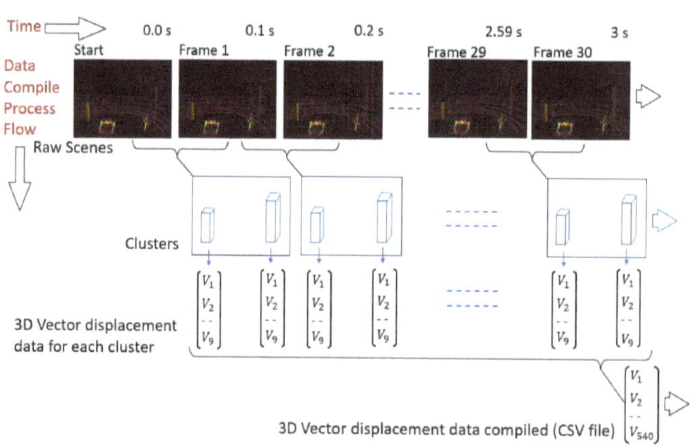

Figure 6. Vibration-indicated data compiling for training and prediction.

3.4. One-Dimensional CNN Modeling for Threshold Level Classification

A good understanding of Deep Learning and convolutional operation is essential to adopt a suitable classification model based on the input data. It is also essential to develop the structure, mainly the convolutional layers, for fast, higher-accuracy, and real-time application. Convolutional Neural Networks can be used based on process data such as 1D signal, 2D image, and 3D video [35–39]. In the proposed study, a 1D Convolutional Neural Network (1D CNN) model is adopted to classify the vibration threshold levels and for real-time prediction due to its simple structure and low computation cost, as well as from considering previous CM work [18,19] results compared with other AI models. A four-layer 1D CNN model is structured, and it works based on the convolution operations on data vectors, as mentioned in Equations (2) and (3) [40]. Here, an input vector to output layer **c** of length $(N - L + 1)$ is convolved with a filter vector by applying a nonlinear activation function, where, x = input data vector of length N, ω = filter vector of length L, and b = bias to best fit for given data. After the convolution layers, the max pooling output vector is used to distinguish the key characteristics and decrease the number of parameters. Here, $m \times 1$ = kernel size, u = window function, and s = filter moving stride over the input vector **c**.

$$\text{Output layer } \mathbf{c}(j) = f\left(\sum_{i=0}^{L-1} \omega(i)x(j-i) + b\right), \quad j = 0, 1, \ldots, N-1 \tag{2}$$

$$\text{Max Pooling output vector } \mathbf{d} = max(u(m \times 1, s)\mathbf{c}) \tag{3}$$

As explained in the data modeling section, the temporal-vibration-affected compiled data for every 30 frames form an array of $[N \times 540 \times 3]$, where N = the total number of samples. This data structure is flattened into $[1 \times 1620]$ for each sample and fed into the 1D convolution layer as input data. A suitable standardization, filter size, activation function, and other measures are applied for fast and accurate classification. A batch normalization layer is used for the first two convolution layers for independent and higher learning rates, tested with different filter configurations and finalized as 64 for the first two layers and 32 for the rest for effective feature detection. A kernel (convolution window) size of 3 is opted for, moving in one direction over the input data, reducing the complexity of the model. We noticed a Rectified Linear Activation Unit (ReLU) helps to learn the nonlinear pattern of these input data; hence, we applied it for each convolutional layer. A dropout layer at a rate of 0.2 is added to prevent any overfitting of the training data. From the second convolutional layer onward, a max pooling layer of stride size two is applied, which helps lower computation time. Next, in the output layer, a fully connected layer is generated where the pooled feature is flattened, creating a 1D array. Finally, a Softmax function is applied as the final activation function in the output layer to predict the probability of vibration threshold classes.

For better prediction accuracy, the training is conducted on two levels. Level I classifies the three higher classes—safe, moderately safe (M), and unsafe (U), saved as Model-A. Level II training is to find the cause of the undesired vibration, i.e., whether it is system-generated (S) or terrain-induced (T). Hence, for level II training, two models are developed; Model-B is trained to find the causes of the moderately safe class vibration, i.e., moderately safe due to system-generated/deteriorated vibration (M-S) or moderately safe due to terrain-induced vibration (M-T), and Model-C is for finding the causes of the unsafe class vibration, predicting U-S or U-T. Hence, based on the Model-A-predicted class (other than for the safe class), the algorithm will call either Model-B or C. More details of this two-level training approach are furnished in Section 4.2. The proposed 1D-CNN-based model structure to fit with the given point-cloud-based vector displacement data is illustrated in Figure 7.

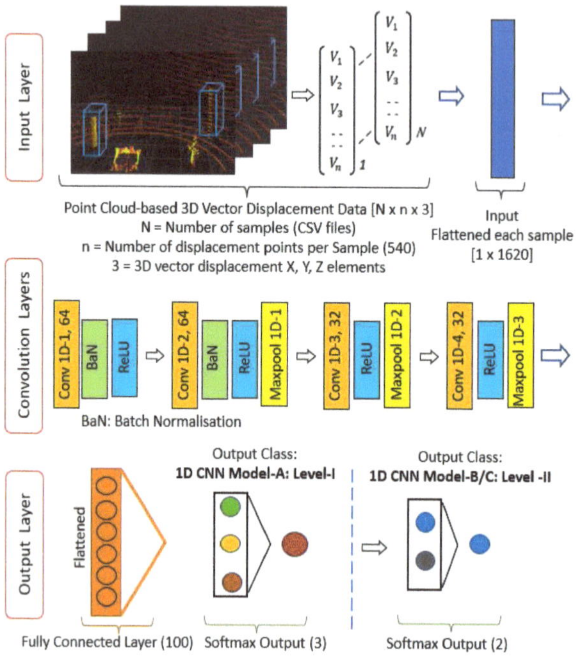

Figure 7. One-dimensional CNN structure for vibration threshold class training.

3.5. Three-Dimensional CbM Map for Condition Monitoring

A vibration threshold mapping framework is developed to generate a 3D CbM map to visually track the robot's health states and terrain-related workspace issues in real time. It is carried out in two stages; firstly, a 3D occupancy grid map is generated for the robot's workspace using a 3D LiDAR sensor and a Simultaneous Localization and Mapping (SLAM) system [41] based on the HDL-graph-SLAM algorithm [42,43]. Then, a vibration threshold class mapping algorithm is developed which fuses the threshold classes that need either attention or immediate actions, i.e., classes other than safe, on the 3D occupancy grid map. The mapping algorithm tracks the robot's position based on the robot's localization data, receives the predicted class, and fuses instantly at the geometrical coordinates of the robot's footprint, with a square dot of unique color for each class, including a timestamp. The 3D CbM map maintains a top-down view, and the undesired classes are color-coded: orange, green, blue, and red for M-S, M-T, U-S, and U-T, respectively. Figure 8 illustrates this framework generating a 3D CbM map, including the photograph of the robot and workspace. This is pavement at the SUTD campus, where we introduced some aluminum profiles around 4–6 cm in height to induce an unsafe-terrain-related class (U-T). Accordingly, the framework predicted the U-T class as the robot ran through this simulated uneven terrain and marked the class on the 3D occupancy grid map, matching the spot where the profiles are inserted. The maintenance team can monitor the system-generated health conditions of the robot or terrain-induced workspace features from the 3D CbM map in real time as required, enabling maintenance or corrective actions promptly based on the conditions of the robot or workspace, i.e., fostering a CbM strategy for productivity and safety.

Figure 8. Three-dimensional CbM map by fusing threshold class on 3D occupancy map.

4. Experiments and Results

This section elaborates on the set-up for training data acquisition of each threshold level class, including modification of the robot and different outdoor workspaces selected; the methodology followed to train and evaluate the 1D CNN model and the results and discussions are furnished.

4.1. Dataset Preparation for Training and Evaluation

The training dataset preparation and compiling, especially for the abnormal threshold classes representing real-world scenarios of outdoor applications, is crucial in this study. The outdoor fumigation robot is selected for all data acquisition tests, as it has to traverse both pavements and grass fields, set the linear speed of 0.3 m/s and operate in a straight and zig-zag pattern, as per the application intended. We assured the Velodyne VLP-16 3D LiDAR sensor was mounted firmly with the metal chassis and set a point cloud data subscription to 10 Hz for recording the vibration-indicated data from the point cloud by following the data modeling methodology explained in Section 3.3.

For the safe class data acquisition, firstly, the robot's health is assured to be good and deployed at different well-maintained plain areas, such as uniformly paved tiled ground, smooth concrete pavements, and plain grass fields without any hidden pits, stones, or roots. We monitored the robot's actions closely, ensured there were no noticeable vibrations, and that all data captured were labeled as safe class. Next, for the system-generated vibration classes, the robot was modified by loosening the wheel couplings and mounting brackets, puncturing one or more wheels alternatively, asymmetrically loading payloads and other heavy items such as the battery, and lowering the performance of the suspension system. Initially, we set all these modifications at a minor level and drove on the same plain terrain area tested for the safe class. An obvious vibration is observed compared withthe safe class trials; however, the robot can traverse easily, and the vibration-affected data were labeled as moderately safe in the system-generated (M-S) class. Next, the above modifications worsened for the unsafe class, causing a higher level of vibration than the M-S class, and it was labeled as unsafe in the system-generated (U-S) class. The U-S class vibration is threatening and has chances of total failure, hence the intermittent stopping and readjusting before continuing for training data collection purposes with all safety precautions, including wireless handheld emergency stop and close monitoring.

For the terrain-induced moderately safe class (M-T), the robot is exposed to different-sized obstacles of 2–4 cm in height, representing stones and roots, as well as pockets of around the same size to capture the vibration levels at small pits and gutters. Such terrain features are introduced to one or more wheels individually and simultaneously and then data labeled as M-T class for training and evaluation were acquired. Finally, data were captured for the terrain-induced unsafe (U-T) class by exposing the positive and negative obstacles of 4–7 cm in height, which shows hard-hitting vibration and chances for falling, tipping hazards while traversing through such ground-level obstacles, or the robot becoming stuck and continuing to jerk to overcome the obstacles. The same precautions explained for the U-S class were followed to avoid total failure or hazards while collecting all of the training data. These robot health states and terrain features used for the data acquisition phase are shown in Figures 9 and 10, respectively. As explained in the data modeling Section 3.3, a total of 1200 samples (each sample is a CSV file of [540 × 3]) are compiled for each class for the 1D CNN model training and validation. Additionally, 500 dataset samples were collected for each class to evaluate the trained model's accuracy.

Figure 9. Robot modification for system-based training data acquisition.

Figure 10. Robot exposure for terrain-based training data acquisition.

4.2. One-Dimensional CNN Model Training and Evaluation

A supervised learning strategy was applied for training the proposed 1D CNN model using the unique point-cloud-based dataset acquired through our in-house-developed outdoor mobile robot. The dataset, with a total of 1200 samples for each class, was split into 80% for training and 20% for model validation. K-fold cross-validation was used to assure dataset quality, avoid overlearning, and improve generalization, and the applied K value was 5. The training framework was developed using Nvidia GeForce GTX 1080 Ti-powered workstation and Tensorflow DL-library [44]. This training uses a momentum with gradient descent optimizing strategy for faster learning and to avoid getting stuck with local minima. An adaptive moment optimization (Adam) optimizer [45] with three parameters, a learning rate of 0.001, an exponential decay rate for the first moment of 0.9, and a second of 0.999, is applied. A categorical cross-entropy loss function is used to assure minimum loss while compiling for better classification accuracy. The other hyperparameters finalized during the training were a batch size of 32 and an epoch of 100 after testing with different values.

We conducted the training in two approaches with the same dataset to decide on a better approach concerning the prediction accuracy of the five vibration threshold classes, following the same 1D CNN structure and parameters explained earlier. In the first approach (Approach-1), the model is trained using all five classes together in a single step, as usual practice. The second approach (Approach-2) is carried out by developing three models (Models A, B, and C), and the way the training dataset is complied for each model in the respective classes is explained as follows. (1) Model-A: All datasets are compiled for the three higher-level classes—safe, moderately safe (M), and unsafe (U). Here, all moderately safe threshold class datasets due to system or terrain are compiled together and labeled under M-class. Similarly, for U-class, all the unsafe threshold class datasets due to system or terrain are compiled together and labeled as U-class. (2) Model-B: Model-B only deals with the moderately safe class (M) training. The moderately safe threshold class datasets acquired from system or terrain causes are compiled separately and labeled as moderately safe–system-generated (M-S) and moderately safe–terrain-induced (M-T), respectively, for binary classification. (3) Model-C: Similarly, Model-C deals with training the unsafe class (U) to predict whether the unsafe class is due to the system (U-S) or terrain (U-T) by compiling and labeling only these particular datasets. For Approach-2, as per the prediction result of Model-A, the algorithm will call either Model-B or Model-C for subsequent prediction for the specific causes of vibration. Here, the final effective prediction accuracy is calculated based on the prediction probability of both Model A and Model B or C. Also, the total number of dataset samples used for each class for all the models during the respective training ensured the same thing. The training approaches, models, and configurations explored for selecting a better prediction approach are illustrated in Figure 11. Also, all the hyperparameters (as explained earlier) used during training for both approaches were the same, and the loss and accuracy plots for the selected approach (Approach-2 as detailed below) are shown in Figure 12.

The prediction accuracy of this proposed 1D CNN model is evaluated with 500 samples which were not used in any of the configurations during the training phase. The statistical metrics—Accuracy, Precision, Recall, and F1 Score—and Equations (4)–(7) [46] are used to assess the approaches and models. Here, TP, TN, FP, and FN are True Positive, True Negative, False Positive, and False Negative. The prediction accuracy of both approaches (Approach-1 and 2) based on the evaluation samples are presented in Table 1. The final effective accuracy in Approach-2 for the M-S, M-T, U-S, and M-T is listed, accounting for the probability predicted for each model involved in a particular class. Still, Approach-2 has a significantly better average accuracy of 89.6% than the single-model Approach-1 (80.8%). The detailed statistical metrics results are listed for the selected Approach-2 in Table 2. The inference time for Approach-1 (8.3 ms) is faster than the multimodel Approach-2 (12.6 ms). However, the higher inference time for Approach-2 does not affect the real-time applications, as the model needs to wait 3 s to compile a new sample dataset for each prediction due to data subscription and the number of frames for each sample set. The

Approach-2-trained model can also be used to predict the three higher-level classes only—safe, moderately safe, and unsafe—without calling for Models B and C. This is designed in case the maintenance team wants to take action to stop or proceed to ensure operational safety. We also found a higher accuracies of 94%, 95%, and 97% for the safe, moderately safe, and unsafe classes, respectively, and an inference time of 5.2 ms.

$$Accuracy = \frac{TP + TN}{TP + FP + TN + FN} \tag{4}$$

$$Precision = \frac{TP}{TP + FP} \tag{5}$$

$$Recall = \frac{TP}{TP + FN} \tag{6}$$

$$F1\ Score = \frac{2 \times Precision \times Recall}{Precision + Recall} \tag{7}$$

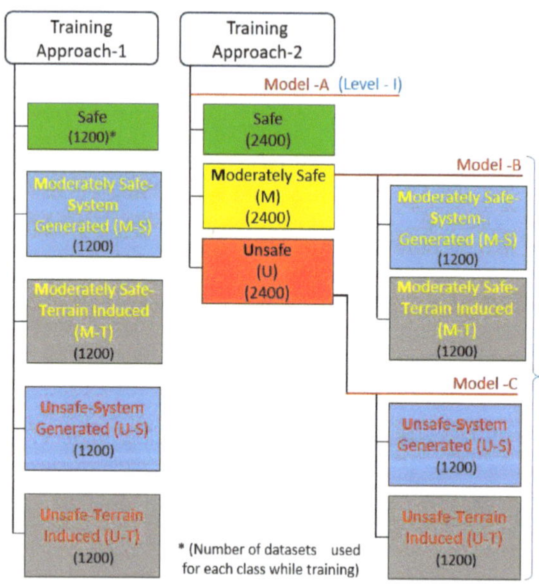

Figure 11. Two training approaches and datasets configurations.

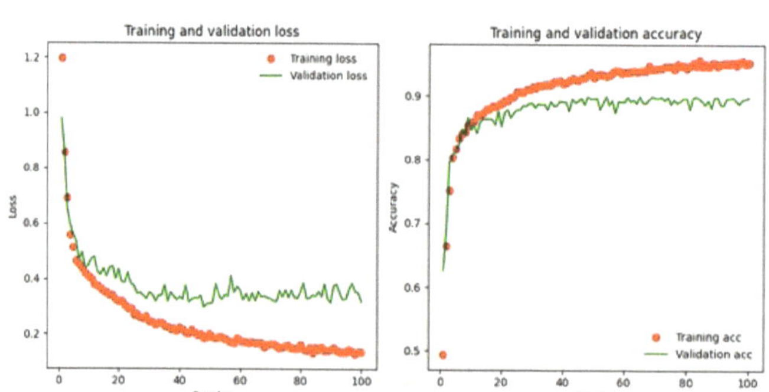

Figure 12. Loss and accuracy plots for training and validation.

Table 1. Assessment results of training Approaches 1 and 2.

Vibration Threshold Class	Safe	M-S	M-T	U-S	U-T
Approach-1 Accuracy (%)	82	77	81	79	85
Approach-2 Accuracy (%)	94	85	89	88	92

Table 2. Evaluation results for the threshold classes: Approach-2.

Vib Threshold Class	Precision	Recall	F1 Score	Accuracy
Safe	0.95	0.94	0.93	0.94
M-S	0.83	0.85	0.86	0.85
M-T	0.88	0.90	0.90	0.89
U-S	0.86	0.89	0.89	0.88
U-T	0.92	0.92	0.91	0.92

5. Real-Time Field Case Studies

The proposed 3D LiDAR point-cloud-based CM framework is validated by conducting three real-time field case studies using the outdoor fumigation robot in different outdoor areas at the SUTD campus, including a grass field, representing real outdoor robot's workspaces with varying terrain features. The areas selected and the ground-level obstacles or features introduced for these trials differed from those used during the training data acquisition phase. The onboard 3D LiDAR Velodyne VLP-16 sensor, its mounting position, and the 10 Hz subscription rate were kept the same as in the training phase. The healthy state of the robot was assured before starting each field trial. An Inference Engine (Algorithm 1) was developed based on the training Approach-2, covering levels I and II, and the models run on the TensorFlow platform to apply their knowledge for inferring a new data sample received in real time. The engine appends point cloud frames for a batch size of thirty and extracts the vibration-indicated 3D vector displacement data from the consecutive frames. Here, the temporal buffer (*TempBuffer*) is a 2D array of a [540, 3] placeholder for a single dataset, which is used as input to the 1D CNN model that returns the threshold class. The inference buffer(*InfBuffer*) is a placeholder for the inferred class (*InfClass*) for each class predicted, which will be sent to the mapping algorithm.

A 3D-occupancy grid map (*map*) was generated for each test area, selected as a pre-work for field trials. The vibration threshold class mapping algorithm, as illustrated in Algorithm 2, is used to fuse the abnormal threshold classes predicted (*PC*) instantly on the 3D occupancy map. The algorithm tracks the robot's position on the map and receives the predicted class from Algorithm 1 in real time. Each class received will fuse/mark at the center point of the robot's footprint (*RFP*) on the map concurrently with the timestamp and following the unique color for each class, thus generating a 3D CbM map.

The first case study was conducted in an open-ground area with uniformly paved tiles, which include some pits under maintenance and drainage manholes with raised covers. For testing, we selected an area of around 12×22 m^2 and conducted a complete area coverage in a zig-zag travel pattern. The threshold class prediction and the 3D CbM map generation follow Algorithms 1 and 2, respectively. The terrain-induced moderately safe and unsafe classes were predicted when the robot overcomes the uneven spots mentioned. The tested area and the 3D CbM map fusing the moderately safe terrain-induced class (M-T) in green color and unsafe terrain-induced class (U-T) in red are shown in Figure 13.

Algorithm 1 Inference Engine

```
while Frame is not empty do
    TempBuffer = call VectorDisp(Frame)
    Format TempBuffer to have shape (n, 30)
    Append TempBuffer to InfBuffer
    counterFrame = counterFrame + 1
    if counterFrame == 30 then
        InfClass = Call modelA
        if InfClass == M then
            InfClass = call modelB
            counterFrame = 0
            Clean InfBuffer
        else if InfClass == U then
            InfClass = call modelC
            counterFrame = 0
            Clean InfBuffer
        else if  then
            counterFrame = 0
            Clean InfBuffer
        end if
    end if
end while
```

Algorithm 2 Vibration Threshold Levels Mapping

```
PC ← [M-S, M-T, U-S, U-T]
map ← 3Dgridmap
RFP ← [x, y]
DT ← [Date, Time]
if PC is Safe then Donothing
else if PC is Abnormal then
    if PC is M-S then
        Color ← Orange
    else if PC is M-T then
        Color ← Green
    else if PC is U-S then
        Color ← Blue
    else if PC is U-T then
        Color ← Red
        map[RFP.x][RFP.y] ← color
    end if
end if
saveMap ← map, DT
```

An open grass field at the campus was selected for the second case study, as well as an area of around 20×35 m^2 for trial runs. The ground is almost fully covered by grass with regions of slops, small hidden pits, many ground-level stones of different sizes and shapes, and long, short, small, and big tree root sizes are spread out on the ground and mostly covered by grass. The robot was tested with a zig-zag pattern and we observed the robot's behavior closely. The 3D CbM map initially showed an intermittent M-T and U-T class while traversing, especially on pits and roots. Whenever the robot hit bigger roots, jerks were generated and it tried to deviate from its planned path; simultaneously, the robot keept predicting the U-T classes. The power tripped once due to power surges from the motors while overcoming roots. As more unsafe classes due to terrain were predicted, we stopped the robot after one hour to avoid any permanent damage to the robot or hazardous events. The threshold classes fused based on some last sample data selected are shown in the 3D CbM map together with the test environment photo in Figure 14.

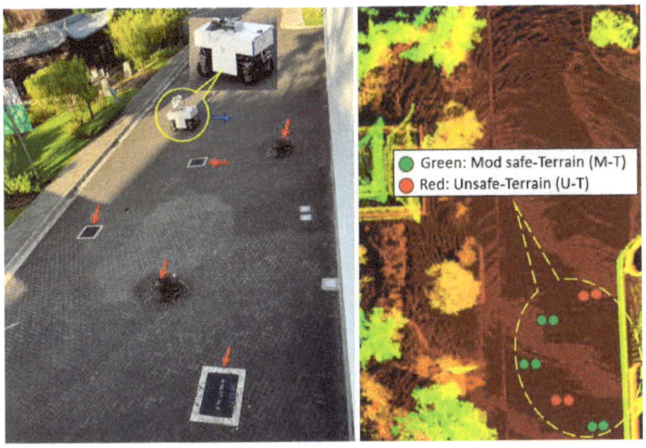

Figure 13. Case study-1: Terrain features inducing abnormal classes.

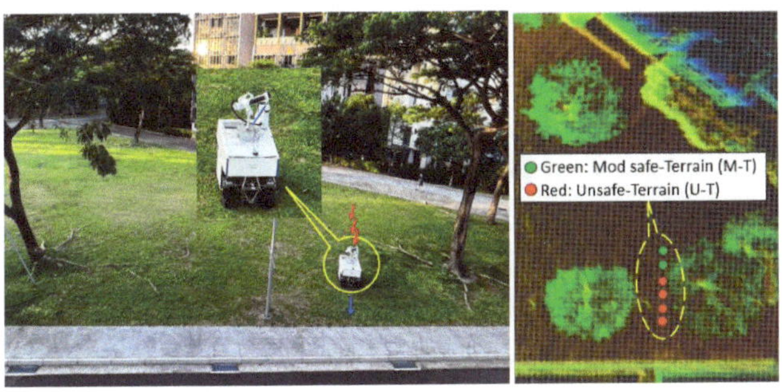

Figure 14. Case study-2: Grass field with roots and hidden terrain features.

The final case study was focused mainly on representing the system-generated threshold classes. Hence, we selected minimally used pavement with uniformly paved tiles and an area of 10×25 m^2 for deployment. This area also includes uneven and unstructured spots such as missing or broken tiles, drainage cover at different heights, curbs, small pits, cliffs, gutters, etc. Additionally, some common hard trashes on the road were introduced, such as small stones, rods, bottles, cans, etc., to simulate the actual pavements. The robot was driven in the predefined hard-coded zig-zag path pattern for area coverage in a closed loop. The 3D LiDAR map generated during the prework of the experiment was used along with the hdl-localization algorithm for localization. An intermittent prediction of M-T and U-T classes was observed whenever the robot was exposed to medium and severe terrain features and fused on the 3D-occupancy grid map in real time. This test was planned on average four hours per day, and we continued testing for ten days with no corrective actions on terrain and noticed M-S classes fused occasionally and observed the robot's vibrations when it traversed through the plain and even terrain. The robot was allowed to continue its operation without any maintenance actions, exposing it to the same unstructured terrain features. Further, the algorithm predicted system-generated unsafe classes (U-S) on the 15th day, and we noticed an obvious vibration in the robot. For the safety of the robot and environment, we stopped the experiment and discovered the hub wheel couplings assembly was getting looser than the original settings, and one of the wheels was punctured. The 3D CbM map with random samples from the last few fused vibration threshold classes is shown in Figure 15, including the test environment photo.

Figure 15. Case study-3: Long-term field trial—terrain and system issues.

We randomly collected 500 samples of each class from the above three case studies, observing the ground truth during the robot's operation to assess the real-time prediction accuracy. Accordingly, the average accuracy found was 89%, and the individual classes' accuracies are listed in Table 3, which are close to the offline evaluation results; hence, the feasibility and consistent accuracy of the proposed CM framework for outdoor mobile robots is validated for real-time applications.

Table 3. Real-time prediction accuracy of threshold classes.

Threshold Vibration Classes	Safe	M-S	M-T	U-S	U-T
Prediction Accuracy (%)	93	84	89	87	92

Other than helping maintenance engineers to trigger prompt maintenance or corrective actions based on the health condition of the robot or uneven terrain conditions through real-time monitoring, the proposed CM framework also helps robot deployment contract companies assess how good a given area is for deployment, assessing the terrain-induced classes and their moderate and unsafe class populations using the proposed 3D CbM map. This facilitates the revision of their robot's maintenance and deployment rental strategies. As per our knowledge, this assessment is currently practiced manually, which is not representative, especially for outdoor fumigation or lawn mowing robots, as most of the ground-level terrain features are covered by grass. Other impacts of this work/3D CbM map are for robot manufacturers to improve the design and assembly processes based on the system generated and the effects of terrain classes predicted, mainly considering faster loosening of critical assembly, deterioration, wear and tear rate of components, unbalanced structure, etc. Compared with a camera sensor, a 3D LiDAR sensor is not affected by outdoor light conditions; also, a long range of up to 100 m is feasible with the 3D LiDAR used in this work, enabling it for the outdoor environment. However, the limitations and improvements needed for the proposed work include enhanced real-time prediction accuracy to avoid any False Positive predictions, performance degradation issues of 3D LiDAR sensor in bad weather conditions, especially heavy rain, fog, and snow, limitations to detect proper clusters if no surrounding objects, and 3D LiDAR is expensive.

The robustness of the proposed system, mainly the capability of the system to maintain performance, accuracy, and functionality under varying conditions and uncertainties, is important, i.e., good performance even in the worst internal or external scenarios [47,48]. Here, the 3D-LiDAR-based CM approach's robustness refers to its capability to capture the vibration indicators and detect the system deterioration and the terrain conditions affecting the robot's health across varying outdoor environments, which is validated through the real-time field case studies explained. For future works, the system's resilience can be improved, i.e., reducing the risks involved due to inevitable disruptions, with the intrinsic capability of the system to recover and continue its functional performance [49,50]. The resilience of our proposed CM system can be improved using the onboard IMU sensor or integrating a camera sensor as a redundant CM sensor by modeling and training the relevant vibration-indicated data. We will explore such heterogeneous data from these additional independent sensors to improve the resilience of this 3D-LiDAR-based CM approach for future work.

6. Conclusions

An automated condition monitoring (CM) strategy for outdoor mobile robots is essential to assure productivity and safety, considering extreme outdoor terrain features and faster system deterioration. Vibration is a key indicator of system failure and terrain abnormalities, especially in mobile robots. Hence, in this work, we proposed and validated a novel CM framework using a 3D LiDAR sensor, applying it differently than its typical use for navigation, by developing an algorithm to extract the vibration-indicated data from the point cloud, assuring low computation cost without losing vibration characteristics. We identified five vibration threshold classes for CM in outdoor robots considering both system deterioration and terrain conditions, namely safe, moderately safe due to system (M-S) and terrain (M-T), and unsafe due to system (U-S) and terrain (U-T). The algorithm computed cuboids for two prominent clusters in every point cloud frame, and the 3D vector displacement of the points set at the corners and centroid of the cuboid, which correspond to the vibration-affected clusters, are compiled as vibration indication data. A simply structured 1D-CNN-based vibration threshold prediction model is developed for fast, accurate, and real-time execution. Further, a 3D CbM map is generated by developing a framework to map the predicted anomalous threshold classes on the 3D occupancy map of the workspace in real time, which allows monitoring of the robot's health conditions and uneven terrain features.

The proposed CM framework for an outdoor mobile robot is unique in terms of the vibration classes that consider both the robot's internal health state and external terrain features affected, including the threshold levels, 3D LiDAR sensor used for vibration-based CM application, the methodology developed for extracting vibration-indicated data, and the data acquisition and real-world scenario validation using the most representative outdoor robot; it achieved an average real-time prediction accuracy of 89%. The accurate and consistent prediction results of real-time field trials using the in-house-developed outdoor robot show that the proposed 3D-LiDAR-based CM framework is robust, which facilitates the maintenance team in triggering prompt maintenance or corrective actions based on the robot's condition or its workspace, fostering a CbM strategy. This work also helps revise rental policies to fit intended workspaces and improves robot design and assembly processes. Further, we plan to explore more heterogeneous sensors to complement each other and sensor fusion approaches to improve the prediction accuracy of the vibration threshold classes, as well as enabling system resilience in case one sensor fails, for condition monitoring in outdoor mobile robots.

Author Contributions: Conceptualization, S.P. and M.R.E.; methodology, S.P. and P.A.P.; software, P.A.P. and S.P.; validation, S.P.; formal analysis, S.P., M.R.E. and P.A.P.; investigation, S.P. and M.R.E.; resources, M.R.E.; data, S.P. and P.A.P.; writing—original draft preparation, S.P. and P.A.P.; supervision, M.R.E.; project administration, M.R.E.; funding acquisition, M.R.E. All authors have read and agreed to the published version of the manuscript.

Funding: This research is supported by the National Robotics Programme under its National Robotics Programme (NRP) BAU, Ermine III: Deployable Reconfigurable Robots, Award No. M22NBK0054 and also supported by A*STAR under its "RIE2025 IAF-PP Advanced ROS2-native Platform Technologies for Cross-sectorial Robotics Adoption (M21K1a0104)" programme.

Data Availability Statement: Data will be made available on request.

Acknowledgments: The authors would like to thank, the National Robotics Programme, the Agency for Science, Technology and Research and SUTD for their support. Thanks to Abdullah Aamir Hayat for his inputs.

Conflicts of Interest: The authors declare no conflict of interest.

References

1. ABIResearch. Shipments of Outdoor Mobile Robots to 350,000 by 2030. Available online: https://www.abiresearch.com/press/labor-shortages-and-workplace-safety-concerns-propel-shipments-of-outdoor-mobile-robots-to-350000-by-2030/ (accessed on 10 January 2023).
2. Zhang, F.S.; Ge, D.Y.; Song, J.; Xiang, W.J. Outdoor scene understanding of mobile robot via multi-sensor information fusion. *J. Ind. Inf. Integr.* **2022**, *30*, 100392. [CrossRef]
3. Yang, L.; Wang, L. A semantic SLAM-based dense mapping approach for large-scale dynamic outdoor environment. *Measurement* **2022**, *204*, 112001. [CrossRef]
4. Liang, Z.; Fang, T.; Dong, Z.; Li, J. An Accurate Visual Navigation Method for Wheeled Robot in Unstructured Outdoor Environment Based on Virtual Navigation Line. In Proceedings of the The International Conference on Image, Vision and Intelligent Systems (ICIVIS 2021), Changsha, China, 15–17 June 2021; Springer: Singapore, 2022; pp. 635–656.
5. Liu, F.; Li, X.; Yuan, S.; Lan, W. Slip-aware motion estimation for off-road mobile robots via multi-innovation unscented Kalman filter. *IEEE Access* **2020**, *8*, 43482–43496. [CrossRef]
6. Dong, Y.; Liu, S.; Zhang, C.; Zhou, Q. Path Planning Research for Outdoor Mobile Robot. In Proceedings of the 12th International Conference on CYBER Technology in Automation, Control, and Intelligent Systems (CYBER), Baishan, China, 27–31 July 2022; IEEE: New York, NY, USA, 2022; pp. 543–547.
7. Manikandan, N.; Kaliyaperumal, G. Collision avoidance approaches for autonomous mobile robots to tackle the problem of pedestrians roaming on campus road. *Pattern Recognit. Lett.* **2022**, *160*, 112–121. [CrossRef]
8. Papadakis, P. Terrain traversability analysis methods for unmanned ground vehicles: A survey. *Eng. Appl. Artif. Intell.* **2013**, *26*, 1373–1385. [CrossRef]
9. Dupont, E.M.; Moore, C.A.; Collins, E.G.; Coyle, E. Frequency response method for terrain classification in autonomous ground vehicles. *Auton. Robot.* **2008**, *24*, 337–347. [CrossRef]
10. Csík, D.; Odry, Á.; Sárosi, J.; Sarcevic, P. Inertial sensor-based outdoor terrain classification for wheeled mobile robots. In Proceedings of the IEEE 19th International Symposium on Intelligent Systems and Informatics (SISY), Subotica, Serbia, 16–18 September 2021; IEEE: New York, NY, USA, 2021; pp. 159–164.
11. Khan, Y.N.; Komma, P.; Bohlmann, K.; Zell, A. Grid-based visual terrain classification for outdoor robots using local features. In Proceedings of the IEEE Symposium on Computational Intelligence in Vehicles and Transportation Systems (CIVTS), Paris, France, 11–15 April 2011; IEEE: New York, NY, USA, 2011; pp. 16–22.
12. Weiss, C.; Tamimi, H.; Zell, A. A combination of vision-and vibration-based terrain classification. In Proceedings of the IEEE/RSJ International Conference on Intelligent Robots and Systems, Nice, France, 22–26 September 2008; IEEE: New York, NY, USA, 2008; pp. 2204–2209.
13. Suger, B.; Steder, B.; Burgard, W. Traversability analysis for mobile robots in outdoor environments: A semi-supervised learning approach based on 3D-lidar data. In Proceedings of the IEEE International Conference on Robotics and Automation (ICRA), Seattle, WA, USA, 26–30 May 2015; IEEE: New York, NY, USA, 2015; pp. 3941–3946.
14. Laible, S.; Khan, Y.N.; Bohlmann, K.; Zell, A. 3d lidar-and camera-based terrain classification under different lighting conditions. In Proceedings of the Autonomous Mobile Systems 2012, Stuttgart, Germany, 26–28 September 2012; Springer: Berlin/Heidelberg, Germany, 2012; pp. 21–29.
15. Ahmed, N.; Day, A.; Victory, J.; Zeall, L.; Young, B. Condition monitoring in the management of maintenance in a large scale precision CNC machining manufacturing facility. In Proceedings of the IEEE International Conference on Condition Monitoring and Diagnosis, Bali, Indonesia, 23–27 September 2012; IEEE: New York, NY, USA, 2012; pp. 842–845.
16. Davies, A. *Handbook of Condition Monitoring: Techniques and Methodology*; Springer Science & Business Media: Dordrecht, The Netherlands, 2012.
17. Prajapati, A.; Bechtel, J.; Ganesan, S. Condition based maintenance: A survey. *J. Qual. Maint. Eng.* **2012**, *18*, 384–400. [CrossRef]
18. Pookkuttath, S.; Rajesh Elara, M.; Sivanantham, V.; Ramalingam, B. AI-enabled predictive maintenance framework for autonomous mobile cleaning robots. *Sensors* **2021**, *22*, 13. [CrossRef]
19. Pookkuttath, S.; Gomez, B.F.; Elara, M.R.; Thejus, P. An optical flow-based method for condition-based maintenance and operational safety in autonomous cleaning robots. *Expert Syst. Appl.* **2023**, *222*, 119802. [CrossRef]
20. Eren, L.; Ince, T.; Kiranyaz, S. A generic intelligent bearing fault diagnosis system using compact adaptive 1D CNN classifier. *J. Signal Process. Syst.* **2019**, *91*, 179–189. [CrossRef]
21. Janssens, O.; Slavkovikj, V.; Vervisch, B.; Stockman, K.; Loccufier, M.; Verstockt, S.; Van de Walle, R.; Van Hoecke, S. Convolutional neural network based fault detection for rotating machinery. *J. Sound Vib.* **2016**, *377*, 331–345. [CrossRef]
22. Kumar, P.; Shankar Hati, A. Convolutional neural network with batch normalisation for fault detection in squirrel cage induction motor. *IET Electr. Power Appl.* **2021**, *15*, 39–50. [CrossRef]
23. Abdeljaber, O.; Sassi, S.; Avci, O.; Kiranyaz, S.; Ibrahim, A.A.; Gabbouj, M. Fault detection and severity identification of ball bearings by online condition monitoring. *IEEE Trans. Ind. Electron.* **2018**, *66*, 8136–8147. [CrossRef]
24. Weiss, C.; Frohlich, H.; Zell, A. Vibration-based terrain classification using support vector machines. In Proceedings of the IEEE/RSJ international conference on intelligent robots and systems, Beijing, China, 9–13 October 2006; IEEE: New York, NY, USA, 2006; pp. 4429–4434.

25. Brooks, C.A.; Iagnemma, K. Vibration-based terrain classification for planetary exploration rovers. *IEEE Trans. Robot.* **2005**, *21*, 1185–1191. [CrossRef]
26. Brownlee, J. 1D convolutional neural network models for human activity recognition. *Mach. Learn. Mastery* **2018**, *26*, 2021.
27. Kiranyaz, S.; Avci, O.; Abdeljaber, O.; Ince, T.; Gabbouj, M.; Inman, D.J. 1D convolutional neural networks and applications: A survey. *Mech. Syst. Signal Process.* **2021**, *151*, 107398. [CrossRef]
28. Ince, T.; Kiranyaz, S.; Eren, L.; Askar, M.; Gabbouj, M. Real-time motor fault detection by 1-D convolutional neural networks. *IEEE Trans. Ind. Electron.* **2016**, *63*, 7067–7075. [CrossRef]
29. Abdeljaber, O.; Avci, O.; Kiranyaz, S.; Gabbouj, M.; Inman, D.J. Real-time vibration-based structural damage detection using one-dimensional convolutional neural networks. *J. Sound Vib.* **2017**, *388*, 154–170. [CrossRef]
30. Wang, J.; Wang, D.; Wang, X. Fault diagnosis of industrial robots based on multi-sensor information fusion and 1D convolutional neural network. In Proceedings of the 39th Chinese Control Conference (CCC), Shenyang, China, 27–29 July 2020; IEEE: New York, NY, USA, 2020; pp. 3087–3091.
31. Liu, X.; Qi, C.R.; Guibas, L.J. Flownet3d: Learning scene flow in 3d point clouds. In Proceedings of the IEEE/CVF Conference on Computer Vision and Pattern Recognition, Long Beach, CA, USA, 15–20 June 2019; pp. 529–537.
32. Filatov, A.; Rykov, A.; Murashkin, V. Any motion detector: Learning class-agnostic scene dynamics from a sequence of lidar point clouds. In Proceedings of the IEEE International Conference on Robotics and Automation (ICRA), Paris, France, 31 May–31 August 2020; IEEE: New York, NY, USA, 2020; pp. 9498–9504.
33. Duffhauss, F.; Baur, S.A. PillarFlowNet: A real-time deep multitask network for LiDAR-based 3D object detection and scene flow estimation. In Proceedings of the IEEE/RSJ International Conference on Intelligent Robots and Systems (IROS), Las Vegas, NV, USA, 24 October–24 January 2020; IEEE: New York, NY, USA, 2020; pp. 10734–10741.
34. Ding, L.; Dong, S.; Xu, T.; Xu, X.; Wang, J.; Li, J. Fh-net: A fast hierarchical network for scene flow estimation on real-world point clouds. In Proceedings of the European Conference on Computer Vision, Tel Aviv, Israel, 23–27 October 2022; Springer: Cham, Switzerland, 2022; pp. 213–229.
35. Zhang, W.; Yang, G.; Lin, Y.; Ji, C.; Gupta, M.M. On Definition of Deep Learning. In Proceedings of the World Automation Congress (WAC), Stevenson, WA, USA, 3–6 June 2018; pp. 1–5. [CrossRef]
36. LeCun, Y.; Bengio, Y.; Hinton, G. Deep learning. *Nature* **2015**, *521*, 436–444. [CrossRef]
37. Albawi, S.; Mohammed, T.A.; Al-Zawi, S. Understanding of a convolutional neural network. In Proceedings of the International Conference on Engineering and Technology (ICET), Antalya, Turkey, 21–23 August 2017; IEEE: New York, NY, USA, 2017; pp. 1–6.
38. Kiranyaz, S.; Ince, T.; Abdeljaber, O.; Avci, O.; Gabbouj, M. 1-D convolutional neural networks for signal processing applications. In Proceedings of the ICASSP 2019—IEEE International Conference on Acoustics, Speech and Signal Processing (ICASSP), Brighton, UK, 12–147 May 2019; IEEE: New York, NY, USA, 2019; pp. 8360–8364.
39. Chen, S.; Yu, J.; Wang, S. One-dimensional convolutional auto-encoder-based feature learning for fault diagnosis of multivariate processes. *J. Process. Control.* **2020**, *87*, 54–67. [CrossRef]
40. Mitiche, I.; Nesbitt, A.; Conner, S.; Boreham, P.; Morison, G. 1D-CNN based real-time fault detection system for power asset diagnostics. *IET Gener. Transm. Distrib.* **2020**, *14*, 5766–5773. [CrossRef]
41. Hess, W.; Kohler, D.; Rapp, H.; Andor, D. Real-time loop closure in 2D LIDAR SLAM. In Proceedings of the IEEE International Conference on Robotics and Automation (ICRA), Stockholm, Sweden, 16–21 May 2016; IEEE: New York, NY, USA, 2016; pp. 1271–1278.
42. Koide, K.; Miura, J.; Menegatti, E. A portable three-dimensional LIDAR-based system for long-term and wide-area people behavior measurement. *Int. J. Adv. Robot. Syst.* **2019**, *16*, 1729881419841532. [CrossRef]
43. Koide, K. HDL-Graph-SLAM Algorithm. Available online: https://github.com/koide3/hdl_graph_slam (accessed on 10 August 2022).
44. Abadi, M.; Barham, P.; Chen, J.; Chen, Z.; Davis, A.; Dean, J.; Devin, M.; Ghemawat, S.; Irving, G.; Isard, M.; et al. TensorFlow: A system for Large-Scale machine learning. In Proceedings of the 12th USENIX Symposium on Operating Systems Design and Implementation (OSDI 16), Savannah, GA, USA, 2–4 November 2016; pp. 265–283.
45. Kingma, D.P.; Ba, J. Adam: A method for stochastic optimization. *arXiv* **2014**, arXiv:1412.6980.
46. Grandini, M.; Bagli, E.; Visani, G. Metrics for multi-class classification: An overview. *arXiv* **2020**, arXiv:2008.05756.
47. Baker, J.W.; Schubert, M.; Faber, M.H. On the assessment of robustness. *Struct. Saf.* **2008**, *30*, 253–267. [CrossRef]
48. Kouvelis, P.; Yu, G. *Robust Discrete Optimization and Its Applications*; Springer Science & Business Media: New York, NY, USA, 2013; Volume 14.
49. Hosseini, S.; Barker, K.; Ramirez-Marquez, J.E. A review of definitions and measures of system resilience. *Reliab. Eng. Syst. Saf.* **2016**, *145*, 47–61. [CrossRef]
50. Hollnagel, E.; Woods, D.D.; Leveson, N. *Resilience Engineering: Concepts and Precepts*; Ashgate Publishing, Ltd.: Farnham, UK, 2006.

mathematics

MDPI

Article

AI-Enabled Condition Monitoring Framework for Indoor Mobile Cleaning Robots

Sathian Pookkuttath *, Prabakaran Veerajagadheswar and Mohan Rajesh Elara

Engineering Product Development Pillar, Singapore University of Technology and Design (SUTD), Singapore 487372, Singapore; prabakaran@sutd.edu.sg (P.V.); rajeshelara@sutd.edu.sg (M.R.E.)
* Correspondence: sathian_pookkuttath@sutd.edu.sg

Abstract: Autonomous mobile cleaning robots are ubiquitous today and have a vast market need. Current studies are mainly focused on autonomous cleaning performances, and there exists a research gap on monitoring the robot's health and safety. Vibration is a key indicator of system deterioration or external factors causing accelerated degradation or threats. Hence, this work proposes an artificial intelligence (AI)-enabled automated condition monitoring (CM) framework using two heterogeneous sensor datasets to predict the sources of anomalous vibration in mobile robots with high accuracy. This allows triggering proper maintenance or corrective actions based on the condition of the robot's health or workspace, easing condition-based maintenance (CbM). Anomalous vibration sources are classified as induced by uneven Terrain, Collision with obstacles, loose Assembly, and unbalanced Structure, which causes accelerated system deterioration or potential hazards. Here, an unexplored heterogeneous sensor dataset using inertial measurement unit (IMU) and current sensors is proposed for effective recognition across different vibration classes, resulting in higher-accuracy prediction. A simple-structured 1D convolutional neural network (1D CNN) is developed for training and real-time prediction. A 2D CbM map is generated by fusing the predicted classes in real time on an occupancy grid map of the workspace to monitor the conditions of the robot and workspace remotely. The evaluation test results of the proposed method show that the usage of heterogeneous sensors performs significantly more accurately (98.4%) than previous studies, which used IMU (92.2%) and camera (93.8%) sensors individually. Also, this model is comparatively fast, fit for the environment, and ideal for real-time applications in mobile robots based on field trial validations, enhancing mobile robots' productivity and operational safety.

Keywords: AI; vibration; condition monitoring; IMU; current sensor; 1D CNN; condition-based maintenance; operational safety; mobile cleaning robot

MSC: 68T40; 68T01; 68T07; 93C85

1. Introduction

The demand for autonomous professional indoor cleaning robots keeps increasing due to frequent cleaning and disinfection of large workplaces, especially after the COVID-19 pandemic, and due to labor issues such as availability, cost, efficiency, and infection risks. In [1], it is predicted that 48.6 million cleaning robot units will be sold in 2023. Also, [2] shows a growth projection of the cleaning robot market to USD 25.9 billion by 2027 from USD 9.8 billion in 2022. As a support to this industry, many technological enhancements works have been conducted recently to improve autonomous and functional efficiency, primarily perception [3,4], path planning [5,6], area coverage [7,8], energy saving [9,10], and cleaning performance assessment [11,12]. However, a significant research gap observed for mobile cleaning robots is that little work has been conducted on developing an automated condition monitoring (CM) framework considering both internal and external factors that affect the robot's health and operational safety. Such a CM framework can enable maintenance strategies such as condition-based maintenance (CbM), i.e., maintenance only when needed, based on the robot's condition or workspace. This CM framework in addition to CbM is crucial to avoid catastrophic failure, malfunction, environmental hazards, downtime, under-utilization of components, higher maintenance cost, and customer dissatisfaction. The execution of such an automated CM framework is also paramount by considering the enormous requirements of cleaning robots and opting for a suitable rental strategy for manufacturers or cleaning contract companies, followed by a fixed rental policy [13] for deployment irrespective of workspace conditions. Inertial measurement unit (IMU) sensors are versatile, especially for vibration-based applications, not only for robots but for human motion studies as well [14,15]. However, heterogeneous data are preferred for effective CM application in mobile robots with high-accuracy prediction, which complement each other during different scenarios affecting the robot's health and operational safety. Hence, this research proposes an unexplored heterogeneous sensor dataset-based automated CM framework using IMU and current sensors for indoor mobile robots facilitating real-time CbM and applying deep learning (DL) techniques.

1.1. Problem Statement

All mobile robots deteriorate naturally after long-term continuous usage, mainly due to wear and tear, loosening subassemblies such as wheel couplings, sensors, etc., and fail in their operational and functional capabilities, even if deployed in the intended environment. However, other reasons that accelerate the system degradation rate include unbalanced structural loading or deployment on uneven terrains, such as tactile pavers, pebble pathways, and damaged tiles. Collision with obstacles due to sensor limitations, faults, and inaccuracies also causes accelerated system failure or becomes hazardous, for example, LiDAR sensors' limit to detecting glass doors, as mentioned in [16,17], and also sometimes miss detecting thin metal legs of tables or unseen low-height table bases. Here, vibration is the common decisive element for system degradation and threats, and such anomalous vibration sources are mainly terrain, collision, assembly, and structure induced. Also, one vibration source may lead to another unless prompt maintenance action is taken. For instance, continuous drive on an uneven terrain causes the loose assembly of system components. Early identification of such abnormal vibration sources will help to assess and fix system level deterioration promptly, avoiding catastrophic failure, isolate or redirect the robot from the hazardous workspaces, plan for a robot-friendly workspace, improve design and assembly, etc. Hence, detecting the characteristics of vibration types using onboard sensors and predicting the sources of such anomalous vibrations using an easily executable DL technique is critical in mobile robots. Towards this, we introduced a vibration-based automated CM framework enabling CbM in previous works using an IMU sensor [18] and monocular camera sensor [19] individually. The camera-based study was better than the IMU-based study in real-time prediction time and accuracy. However, camera-based work is not recommended for dark/shadow environments; also, the camera lens's glass cover cleanliness is to be assured in a dusty cleaning environment. In both studies, the accuracy of vibration source prediction needs to improve significantly to ensure the correct maintenance action is placed; otherwise, it raises additional downtime and maintenance costs. We observed the robot's behavior during the IMU and camera-based studies, and it was found that that, sometimes, there was no apparent vibration in the Collision and Structure classes, being wrongly predicted as the Normal class. At the same time, the robot kept trying to overcome the affected/abnormal load or obstacles. Therefore, IMU and camera sensors are often limited to accurately predicting these classes. Hence, more studies are essential for a higher-accuracy CM framework to execute in indoor mobile robots, assuring the robot's health and operational safety by exploring heterogeneous sensors with unique data features.

1.2. Related Works

In the literature, vibration-based health monitoring studies by analyzing the characteristics of vibration data for fault detection, classification, and prediction have been mainly conducted for machine components or systems such as bearings [20], motors [21], machines [22,23], and structural systems [24]. Condition monitoring and fault diagnosis studies were also conducted for industrial robots, for example, to reduce downtime in a wafer transfer robot [25], predicting failure in a packaging robot [26], safe stop prediction in a collaborative robot [27], etc. The advent of artificial intelligence (AI) has enhanced fault diagnosis and prediction, and out of various techniques, the 1D convolutional neural network (1D CNN) was found prominent due to its simple structure, fast inferencing, computationally lower cost, and better prediction accuracy, making it suitable for real-time applications [28]. The advantage of the 1D CNN model has also been verified in previous works [18,19], showing a better accuracy and inference time compared to other common approaches such as long short-term memory (LSTM), CNN-LSTM, multilayer perceptron (MLP), and support vector machine (SVM). The type of sensor generally used for collecting vibration data in various systems are micro-electromechanical systems (MEMS) or piezoelectric accelerometers [20,29–31]. The studies described above show that vibration-based CM research has been scarcely conducted for wheeled mobile autonomous robots, especially for cleaning applications. Also, the different onboard typical sensors and their configurations are to be explored more for high-accuracy real-time condition monitoring applications in mobile robots.

Competitive (independent measurement of the same property), complementary (different measures complement each other, resolving the shortcomings of each sensor), and cooperative (originate data from two sensors) sensor fusion techniques are used in mobile robots for state estimation, localization, and navigation [32]. For instance, a camera visual system and robot odometry fusion method are proposed in [33] using the Kalman filter (KF), resulting in better localization accuracy in mobile robots compared to tests conducted with each sensor data used separately. In [34], the authors simulated fusing odometry and sonar sensor data for mobile robot navigation to improve position estimation and correct systematic error using the extended Kalman filter (EKF) and the adaptive fuzzy logic system (AFLS). A sensor fusion method to reduce position errors and better target tracking for mobile robots is proposed in [35] using an encoder, an inertial sensor, and active beacon sensors and applying the unscented Kalman filter (UKF). However, such sensor fusion techniques mostly needed extensive system modeling and calibration processes of statistical filtering techniques. To avoid such expensive approaches, recently, techniques have been applied, such as the long short-term memory (LSTM) model adapted in [36] for learning fused camera pose and IMU data. Another DL method, VINet architecture, is proposed by fusing camera and IMU data in [37] for visual–inertial odometry motion estimation. Similarly, a hybrid framework, recurrent convolutional neural network (RCNN), is developed in [38] for higher accuracy and robustness in the localization of mobile robots, fusing a 2D laser scanner and IMU sensor data.

In the literature, existing terrain classification works for mobile robots have been been conducted to detect the terrain type for traversability analysis, using onboard proprioceptive and exteroceptive sensors, including wheel–terrain interactive and noninteractive works. These are primarily conducted using IMU sensor-based vibration data and suitable classifiers. For instance, adaptive terrain classification using an IMU sensor to classify grass, small and large gravel, and hard grounds at different operating speeds with the random forest classifier in [39] observed an average accuracy of 93%. Five types of typical indoor terrains, including different tiles and carpets, are classified based on IMU acceleration and angular velocity data in [40] using a linear Bayes normal classifier, and a prediction accuracy of around 89% was found. Similarly, in [41], the authors used the magnitude frequency response on X- and Y-axis rotation and Z-axis acceleration data captured through an IMU sensor to classify dirt, gravel/grass, asphalts, and mud using the classifier probabilistic neural network (PNN) and found an accuracy around 84% considering different driving speeds as well. Similarly, different terrain types such as concrete, pebble, grass, paving stone, sand, and synthetic running track are classified using IMU gyroscope and accelerometer data and applying the multilayer perceptron (MLP) neural network in [42], resulted in a prediction accuracy of 97.18 to 99.7 with different processing window sizes. In [43], the authors used a combination of IMU vibration and camera-image-based data for the classification of fourteen types of outdoor terrains using an SVM classifier and found an average accuracy of 87.33%. These terrain classification studies may help CM applications for the external terrain-induced vibration source class. However, other factors/classes we proposed for CM, such as collision, loose assembly, and unbalanced structure affecting the robot's failure or potential hazards, are still unsolved.

Current sensors are the least exploited sensors for mobile robot applications other than energy consumption or energy-efficient strategy studies for navigation in different trajectories, as addressed in [44,45]. Motor current signature analysis (MCSA) is one of the earliest fault diagnosis approaches in motor-driven equipment by monitoring the changes in current readings [46]. However, as per our knowledge, integrating a current sensor either individually or with other onboard sensors for condition monitoring applications in mobile robots has not been explored.

1.3. Contributions

The key contributions of this study are listed below:

1. An unexplored and compelling heterogeneous sensor dataset-based CM framework is developed for mobile robots using IMU and current sensors, resolving the shortcomings of individual sensors in detecting anomalous vibration class features.
2. Vibration-affected data are modeled by formulating an algorithm to ensure the temporal data from IMU and current sensors are subscribed and fused concurrently.
3. Vibration data are trained by developing a simple, fast, and low-computation-cost 1D CNN model for real-time prediction of the internal and external anomalous vibration sources that cause system degradation and operational hazards.
4. A vibration source mapping framework is developed, which fuses the predicted vibration classes on the 2D occupancy grid of the environment in real time, generating a 2D CbM map that enables prompt CbM or corrective actions for the robot's workspace.
5. Based on the case study results, the proposed heterogeneous sensor-based CM framework is ideal for mobile robots for real-time CbM applications, with a significant improvement in prediction accuracy compared to previous works where IMU and camera sensors were used individually.

The remainder of this paper is organized as follows. Section 2 shows an overview of the proposed work. Section 3 elaborates on the experiments and results, including a detailed comparison with previous works. Section 4 explains the field case studies for real-time application, and a conclusion is presented in Section 5.

2. Overview of the Proposed Framework

The overview of the proposed heterogeneous sensor dataset-based anomalous vibration source prediction framework for the CM of autonomous mobile indoor cleaning robots enabling CbM is illustrated in Figure 1, and the details of each module are explained as follows.

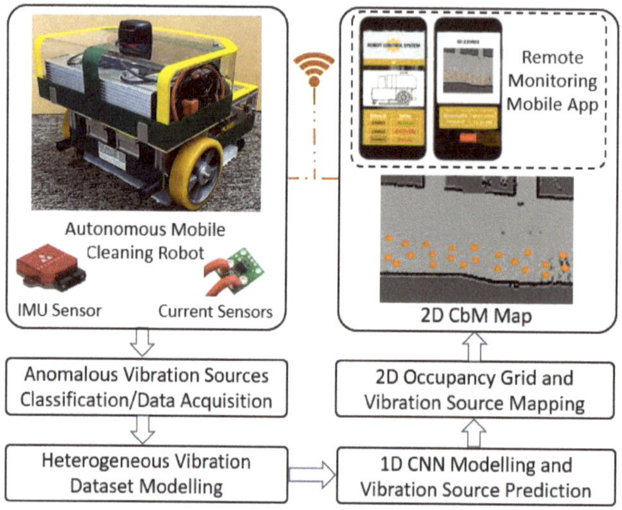

Figure 1. Overview of the heterogeneous sensor dataset-based CM framework for mobile robots.

2.1. Mobile Robot Test Platform and Vibration Source Classes

An in-house-developed autonomous steam mopping and disinfection robot suitable for large indoor workplaces is used in this study for data acquisition and real-time case studies. The photos of this wheeled mobile robot with critical components are shown in Figure 2. The maximum size of the robot is 45(L) $\times$ 40(W) $\times$ 38(H) cm and around 20 kg weight, including the cleaning payload. The robot chassis is built using an aluminum plate and extruded channels for a robust structure, and all the electromechanical and sensor assemblies are fastened to the chassis for smooth and stable locomotion across all the typical indoor floors. A differential wheel drive mechanism is used with two caster wheel supports. Two Oriental motors of 24 V and 1.9 N-m torque brushless DC geared motors (BLHM015K-50) are used for each drive wheel with a closed-loop control function. An RPLiDAR A2 scanner is mounted at the top of the robot with 360° coverage for localization and mapping. A Vectornav VN-100 IMU sensor is firmly mounted on the aluminum base plate at the center portion of the two-wheel axis, enabling navigation. A 24 V 40 Ah Li-ion battery is used to power the whole system, and an NVIDIA Jetson AGX Xavier computer is used to steer the entire operation and data storage. The robot operating system (ROS) melodic framework is used and acts as the middleware to establish all communications. More details of this robot are published in the system architecture paper [47]. A D-Link 4G/LTE mobile router is used for monitoring the robot's health remotely. A temporary, clear acrylic cover is used to observe the system deterioration during testing.

A wheeled mobile robot generates aberrant vibration as an early symptom of failure. This could be due to various internal factors such as wear and tear, loose assembly, parts displacement or misalignment, unbalanced weight, etc., raised from the robot itself by continuous use or poor design and assembly. External factors are mainly unintended or uneven and damaged terrain, collision with undetected or unseen objects, or forced external interaction. Hence, by monitoring such abnormal vibration sources, as classified in Figure 3, appropriate maintenance actions can be triggered promptly, avoiding total system failure or incidents. Here, the Normal class is represented as the robot in healthy condition and deployed at its intended smooth operational environments, and any minor natural vibration generated is accepted.

Figure 2. Test platform. Autonomous steam mopping robot.

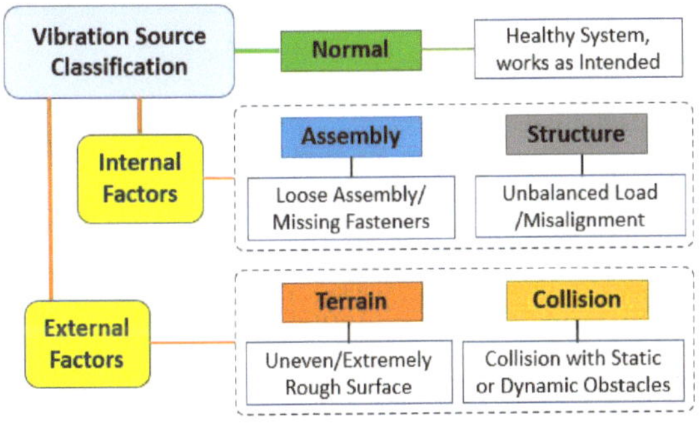

Figure 3. Vibration source classes in a wheeled mobile robot.

2.2. *Vibration-Affected Heterogeneous Sensors Dataset Modeling*

A wheeled mobile robot vibrates at different rates linearly or in angular motion due to various vibration sources explained in the previous section. Similarly, a change in power required is observed when the robot is exposed to such undesired vibration sources, resulting in a higher current consumption compared to Normal class operation. As the vibration-affected linear or angular-motion-related data from the IMU sensor and the current consumption data from current sensors are heterogeneous, the final dataset helps to fill the gaps of individual sensors in extracting the relevant vibration-affected characteristics during various health states and exposures of the robot, which is vibration source classes. The onboard Vectornav VN-100 IMU sensor is used to extract the rate of linear and angular motion of the robot, which provides linear acceleration and angular velocity from its built-in accelerometer and gyroscope. We also added angular acceleration, calculated from angular velocity, to further strengthen the dataset features for training. A low-cost ACHS-7124 Hall-effect-based isolated linear current sensor is used with left and right drive motors and connected with an Arduino Mega 2560 microcontroller to measure the current rate consumed. These proprioceptive sensors do not depend on each other, and they are unaffected by the environment/light conditions, enabling higher accuracy in vibration class prediction. The dataset acquisition scheme, including the circuit diagram for the proposed approach, is depicted in Figure 4. (Only one current sensor motor assembly is shown for illustration.)

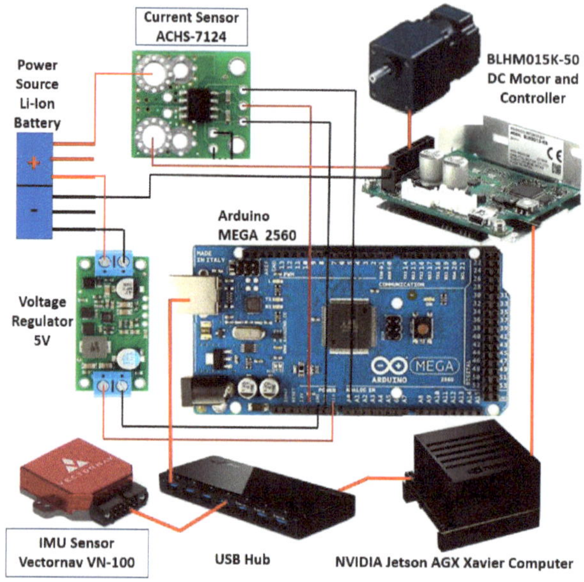

Figure 4. Heterogeneous sensor dataset acquisition scheme.

Here, the vibration-indicated data, mainly linear acceleration, angular velocity, and angular acceleration based on the IMU readings and the current consumed data measured from the current sensor connected with left and right drive wheel motors, all affected by the various vibration sources, are expressed in (1). As two types of sensors with different behavior are used in this study, the sequential data subscription rate is critical, and we set 80 Hz for IMU and set the same rate for the current sensor data by lowering the delay in the Arduino program and confirmed the total number of samples data collected at any particular time is same. This higher-frequency IMU subscription rate helps to reduce the vibration class prediction time to half compared to the previous individual IMU-based study without compromising the accuracy due to the complementary current sensor. The linear acceleration is measured in the X-, Y-, and Z-axes of the robot and the corresponding angular velocity and acceleration in roll (X), pitch (Y), and yaw (Z). Including the left and right motor current sensor data, a total of 11 data features are considered, and each feature is compiled with 128 temporal data (processing window size) over 1.6 s. Hence, one sample data for the 1D CNN training and prediction is set as an array of [128 × 11], as given in Expression (2).

$$Vib_{data} = [A_{lin}, V_{ang}, A_{ang}, I_L, I_R] \tag{1}$$

$$\begin{bmatrix} Acc_{Lin}(X)_{t1}, Acc_{Lin}(X)_{t2}, \ldots Acc_{Lin}(X)_{t128} \\ Acc_{Lin}(Y)_{t1}, Acc_{Lin}(Y)_{t2}, \ldots Acc_{Lin}(Y)_{t128} \\ Acc_{Lin}(Z)_{t1}, Acc_{Lin}(Z)_{t2}, \ldots Acc_{Lin}(Z)_{t128} \\ Vel_{Ang}(X)_{t1}, Vel_{Ang}(X)_{t2}, \ldots Vel_{Ang}(X)_{t128} \\ Vel_{Ang}(Y)_{t1}, Vel_{Ang}(Y)_{t2}, \ldots Vel_{Ang}(Y)_{t128} \\ Vel_{Ang}(Z)_{t1}, Vel_{Ang}(Z)_{t2}, \ldots Vel_{Ang}(Z)_{t128} \\ Acc_{Ang}(X)_{t1}, Acc_{Ang}(X)_{t2}, \ldots Acc_{Ang}(X)_{t128} \\ Acc_{Ang}(Y)_{t1}, Acc_{Ang}(Y)_{t2}, \ldots Acc_{Ang}(Y)_{t128} \\ Acc_{Ang}(Z)_{t1}, Acc_{Ang}(Z)_{t2}, \ldots Acc_{Ang}(Z)_{t128} \\ I(L)_{t1}, I(L)_{t2}, \ldots I(L)_{t128} \\ I(R)_{t1}, I(R)_{t2}, \ldots I(R)_{t128} \end{bmatrix} \tag{2}$$

2.3. 1D Convolutional Neural Network Modeling

A simple, effective, fast, and computationally low-cost neural network is crucial for real-time condition monitoring and predicting the vibration source classes. Hence, we adapted a 1D convolutional neural network model following the convolution operations on data vectors as expressed in Equations (3) and (4) [48], simplified the model structure with a minimum number of convolution layers, and opted for suitable hyperparameters to fit with our fused dataset.

$$\text{Output layer } \mathbf{c}(j) = f\left(\sum_{i=0}^{L-1} \omega(i)x(j-i) + b\right), \quad j = 0, 1, \ldots, N-1 \tag{3}$$

This nonlinear activation function convolves the input vector to the output layer using a filter vector. Here, x (of length N) = input data vector, ω (of length L) = filter vector, term b = bias to best fit for given data, and $\mathbf{c}$ = output layer (of length $N - L + 1$).

$$\text{Max-pooling output vector } \mathbf{d} = max(u(m \times 1, s)\mathbf{c}) \tag{4}$$

The max-pooling output vector **d** is applied after convolution layers to detect the key characteristics and reduce the number of parameters. Here, $m \times 1$ is kernel size, u is a window function, and s is a filter moving stride over input vector **c**.

The input vibration data for 1D CNN training is compiled into an array of [n $\times$ 128 $\times$ 11], where n is the total number of samples. Hence, the compiled/fused dataset is flattened into a 1D array [1 $\times$ 1408] and fed into 1D convolution. The convolution structure is made simple with only four layers and uses 64 filters for the first two layers and 32 for the rest. All layers used a convolution window (kernel) size of 3. The rectified linear activation unit (ReLU) is applied to each convolutional layer to learn the complex nonlinear pattern of vibration-affected heterogeneous sensor data. After each convolution layer, a max-pooling layer with a stride size of 2 is added to reduce the computation time. Further, a dropout layer is added at a rate of 0.2 for each layer to prevent over-fitting while training. Finally, the pooled feature maps are converted to a 1D array using a flattening function in the output layer to predict the multinomial probability of vibration source classes. The structure of this 1D CNN-based model proposed is illustrated in Figure 5.

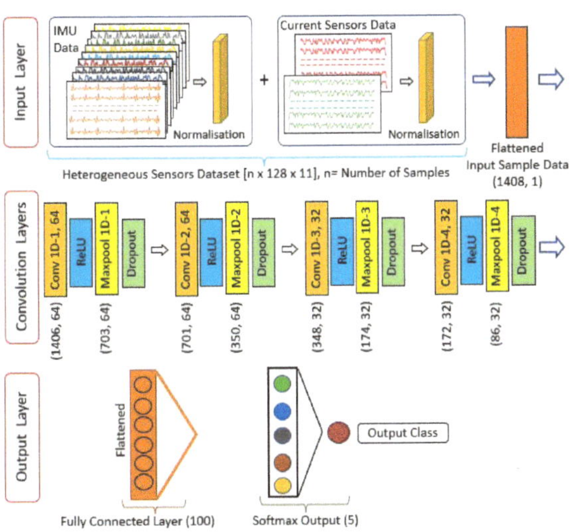

Figure 5. One-dimensional (1D) CNN structure and data shape modeled for training.

2.4. 2D CbM Map for Real-Time Condition Monitoring

A 2D CbM map is generated by developing a vibration source class mapping framework to visually track the internal- and external-based anomalies in real time. This is mainly achieved by preparing a 2D environment occupancy grid map for the robot' deployed area using the onboard RPLidar scanner and applying a grid-based mapping algorithm cartographer simultaneous localization and mapping (SLAM) with the Ceres-based scan matcher [49]. The vibration source mapping algorithm extracts each anomalous vibration class predicted and fused on the 2D occupancy grid map, as illustrated in Figure 6. Here, the robot moves from a concrete floor to a small-pebble-paved area in a corridor, and the system predicts the Terrain class continuously, which is fused on the 2D grid map instantly at the exact location where it is detected, i.e., at the geometrical coordinates of robot's footprint, generating a 2D CbM map. The abnormal predicted classes are marked on the grid map with a unique color for each class: blue, green, orange and red for the Assembly, Structure, Terrain and Collision classes, respectively. The real-time position of the robot's footprint is obtained from the localization data. The maintenance team can monitor remotely and trigger prompt maintenance or corrective actions based on the robot's or workspace's condition using a mobile app connected through the message queuing telemetry transport (MQTT) protocol.

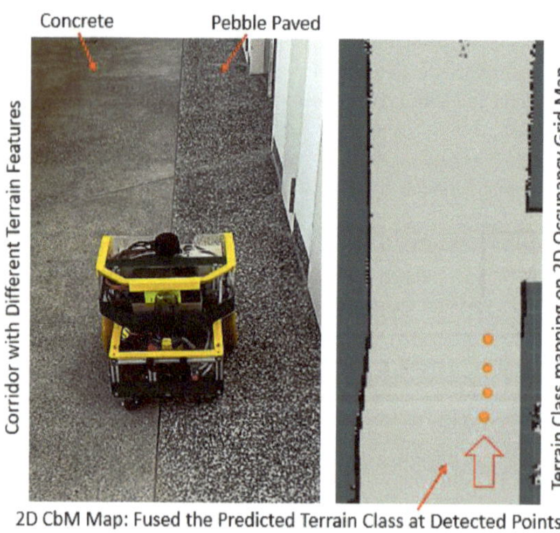

Figure 6. Anomalous class prediction and 2D CbM map generation.

3. Experiments, Results, and Discussion

This section elaborates on the data acquisition and visualization, 1D CNN training, evaluation and results, and comparative study with previous works and discussions.

3.1. Training Dataset Preparation and Visualization

Preparing a qualitative and quantitative training dataset representing each vibration source class for 1D CNN training and evaluation is critical in this study. Hence, we modified and exposed the robot inducing the four abnormal vibration classes. For Terrain class data, we drove the robot through typical but undesired terrain features for an indoor robot, such as small-pebble-paved and uneven/unstructured/damaged indoor floors. Similarly, exposed colliding with potentially unseen/undetected common static obstacles such as table bases, thin legs of tables, and glass doors, also externally applied forces such as collision with dynamic objects, to collect the collision-induced data. Further, various critical components were loosened for the Assembly class, such as wheel couplings, mounting brackets, and mopping heads. Finally, the robot was modified by shifting the heavy components, such as the battery, boiler assembly, etc., from their designed location, also causing an unbalanced system with a more worn-out wheel for collecting unstable Structure class vibration-affected data. The Normal class data were collected, deploying the robot in good condition in terms of the robot's health and deployed workspace. Some of the modified and exposed cases for anomalous classes are illustrated in Figure 7. As the robot is used for steam mopping and disinfection, all the tests were performed by driving at the intended linear speed of 0.05 m/s and angular speed of 0.4 rad/s with typical cleaning patterns, mainly straight and zig-zag.

Figure 7. Robot set-up for vibration-affected training data acquisition for abnormal classes.

The raw data collected from both IMU and current sensors are visually represented, plotting time–amplitude and time–current graphs for each vibration source class, as shown in Figures 8 and 9. This depicts the change in linear and angular motion and current consumed across the different classes. For IMU data features, the three axes' data are plotted in the same graph, and here, the linear acceleration in the Z-axis is brought to the same scale for plotting and visualization. Similarly, current data readings from the left and right motor are plotted in the same graph. In both plots, randomly selected sample data, i.e., 128 temporal data acquired over 1.6 s, are used.

Data normalization preprocessing was performed to bring the data x into a standard scale without losing information for better convergence during 1D CNN training. Based on the data characteristics the IMU data were normalized into -1 to $+1$ and for the current sensor 0 to 1 using Equations (5) and (6), respectively. A total of 2500 samples [$2500 \times 128 \times 11$] for each vibration source class were collected and split into 80% for training and 20% for validation of the 1D CNN model. Additionally, 500 data samples were recorded for each class to evaluate the trained model.

$$\text{IMU Dataset}: x_{Normalised} = 2\frac{x - min(x)}{max(x) - min(x)} - 1 \tag{5}$$

$$\text{Current Sensor Dataset}: x_{Normalised} = \frac{x - min(x)}{max(x) - min(x)} \tag{6}$$

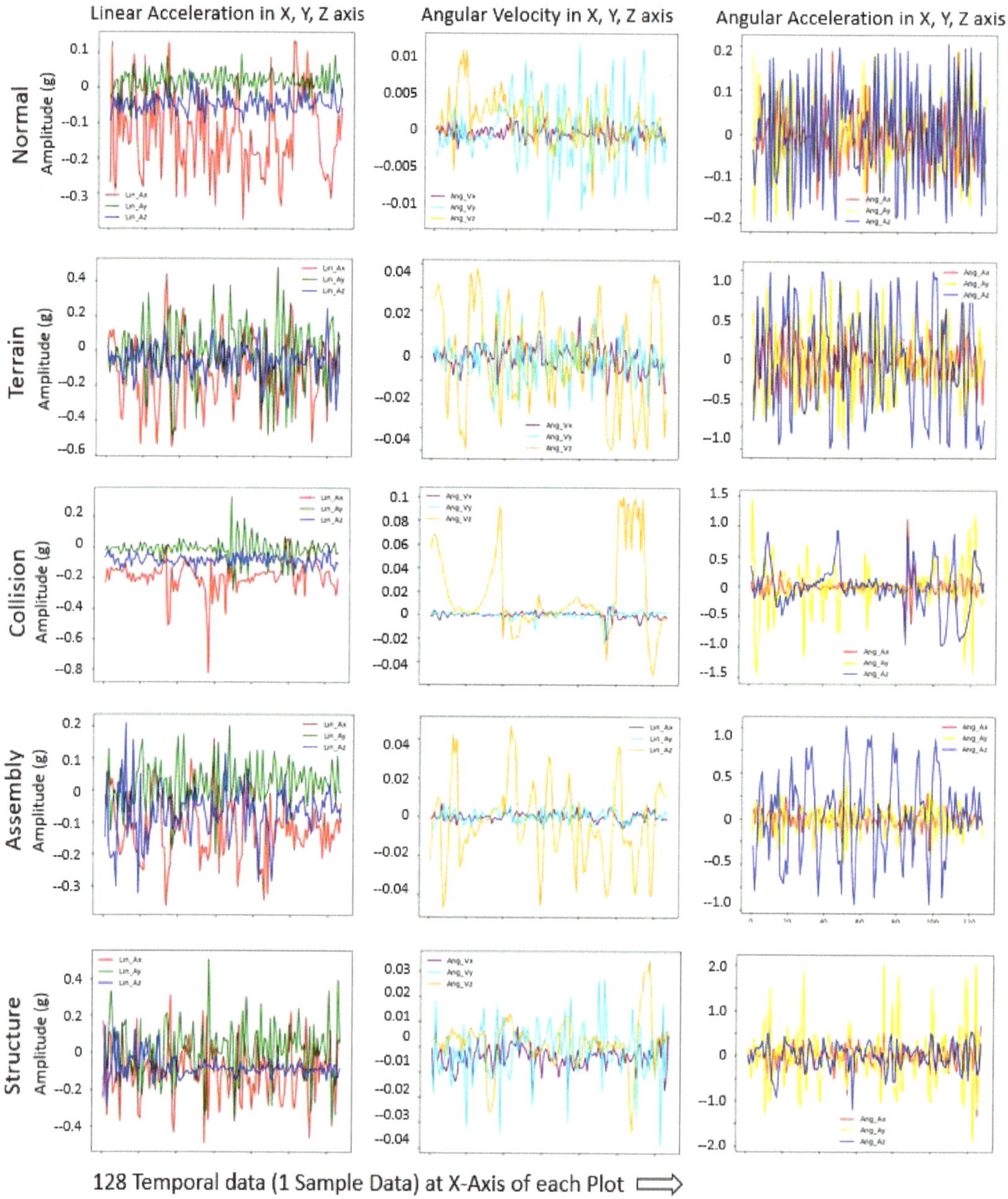

Figure 8. Vibration-affected data from IMU sensor for each class.

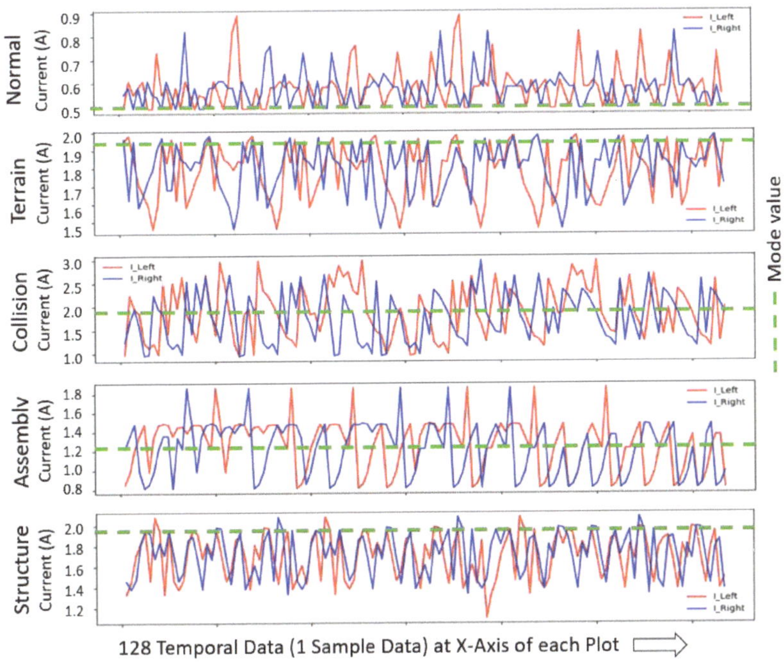

Figure 9. Vibration-affected data from current sensor for each class.

3.2. 1D CNN Training and Evaluation

A supervised learning strategy was applied to train the unique dataset, compiled as explained in the previous sections, using the Tensorflow DL-library [50] and the Nvidia GeForce GTX 1080 Ti-powered workstation. A K-fold (K = 5) cross-validation technique was applied for generalization and to avoid over-learning to ensure the dataset quality. Here, the datasets were split into k subsets, where k − 1 subsets are to train the model and the remaining to evaluate the model's performance. Momentum with gradient descent optimization strategy was applied to speed up learning and help not to get stuck with local minima. We tested with different hyperparameters for training for optimum results and opted as follows. An adaptive moment optimization (Adam) optimizer [51] with a learning rate of 0.001 and an exponential decay rate of 0.9 for the first and 0.999 for the second moment was used. A categorical cross-entropy loss function was applied to assure minimum loss while compiling the model, which fits better for this multinomial classification and probability prediction. The model with a batch size of 32 and an epoch size of 80 showed better performance, as plotted in Figure 10, showing the loss and accuracy curve graphs for training and validation.

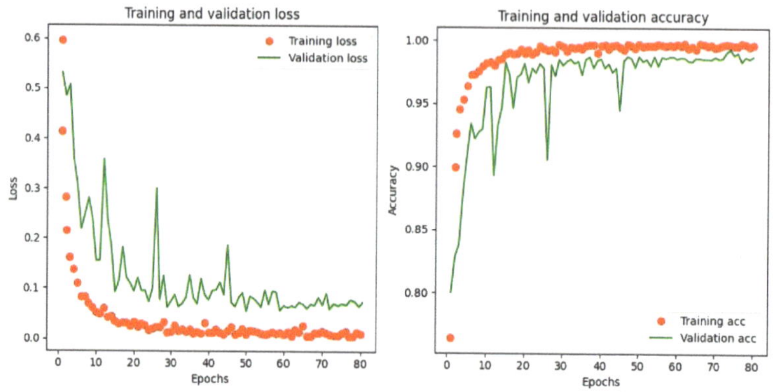

Figure 10. Loss and Accuracy curves for Training and Validation.

Further, the prediction accuracy of the 1D CNN model was evaluated with 500 samples that were not used during training. The assessment was performed based on the statistical metrics Accuracy, Precision, Recall, and F1 Score, as per Equations (7)–(10) [52], following the confusion matrix. Here, TP = true positive, TN = true negative, FP = false positive, and FN = false negative. This offline test statistical measure results are listed in Table 1, with an average accuracy measured as 98.4%. The approximate inference time to predict a given sample's class measured is 0.196 milliseconds. We also tested training only with the current sensor data to understand the performance of the current sensors in predicting the vibration classes instead of the heterogeneous sensor dataset. However, the model did not converge effectively, ending with inferior accuracy of an average of 79.6% only. The offline test result (Accuracy) of each class using the current-sensor-based training is listed in Table 2.

$$Accuracy = \frac{TP + TN}{TP + FP + TN + FN} \tag{7}$$

$$Precision = \frac{TP}{TP + FP} \tag{8}$$

$$Recall = \frac{TP}{TP + FN} \tag{9}$$

$$F1Score = \frac{2 \times Precision \times Recall}{Precision + Recall} \tag{10}$$

Table 1. One-dimensional (1D) CNN model evaluation result: IMU and current sensor heterogeneous dataset based.

Vibration Class	Precision	Recall	F1 Score	Accuracy
Normal	0.99	0.98	0.99	0.99
Terrain	0.98	0.96	0.98	0.98
Collision	0.99	0.99	0.98	0.99
Assembly	0.99	0.96	0.97	0.98
Structure	0.98	0.97	0.99	0.98

Table 2. One-dimensional (1D) CNN model evaluation result: only current sensor based.

Vibration Class	Normal	Terrain	Collision	Assembly	Structure
Accuracy Prediction (%)	78	85	81	74	80

As mentioned in the related works, the studies other than our previous IMU and camera-based CM studies were mainly for terrain classification, i.e., not considering the collision-related abnormality and internal health degradation, such as loose assembly and unbalanced structure. Hence, a direct comparison with such IMU-based existing terrainability studies is not appropriate. However, if only the Terrain class results of the proposed research are chosen, the prediction accuracy (98%) is much better than the works mentioned in existing terrain classification studies [39–43].

Beyond, we conducted a comparison study to assess the performance of the proposed 1D CNN model with other AI models, mainly SVM, MLP, LSTM, and CNN-LSTM, in terms of prediction accuracy and inference time. The same training and evaluation dataset and processing resources applied for the 1D CNN model are used here for a fair comparison. The TensorFlow library is used for training MLP, LSTM, and CNN-LSTM, whereas the Scikitlearn package is used for the SVM model. Other than for SVM, the key hyperparameter functions/values such as the Adam optimizer, a learning rate of 0.001, and categorical cross-entropy loss function are used the same as the 1D CNN model. The key parameters for the SVM model are set as C = 100 and gamma = 0.01 and applied radial basis function (RBF) kernel. The test results, prediction accuracy (average of five classes) in %, and inference time to process one sample data in milliseconds for each model, including the proposed 1D CNN model, are listed in Table 3. The comparison studies show that the selection of the optimum model depends on the application, characteristics of the data for each class identified, how the data are compiled for training, the necessity of real-time prediction, etc. Hence, a 1D CNN model is ideal for this proposed IMU–current sensor heterogeneous data-based CM framework.

Table 3. Comparison with other AI models.

Model	Accuracy (%)	Inference Time (ms)
1D CNN	98.4	0.196
CNN-LSTM	94.2	0.368
LSTM	90.6	0.314
MLP	84.3	0.205
SVM	82.9	1.985

3.3. Comparative Study with Our Previous CM Studies and Discussion

This section provides a performance comparison of the proposed IMU–current sensor heterogeneous dataset-based CM framework with our previous two similar but individual sensor-based CM works developed for mobile indoor robots, i.e., using a proprioceptive sensor IMU [18] and an exteroceptive sensor monocular camera [19]. The comparison study was performed based on three factors: the average prediction accuracy of the five vibration source classes, the total time from collecting a new sample data until predicting the class, and the suitability to fit with the deployed environmental conditions, mainly light and dust.

Regarding the prediction accuracy, the IMU sensor-based work found the lowest accuracy of 92.2% due to poor prediction of the Collision and Structure classes. This will raise a minor probability of triggering a wrong maintenance action, adversely affecting cost and downtime. This approach was found computationally less complex with the fastest inference time (0.162 ms) due to fewer data features and processes involved. However, the IMU subscription rate was finally set to 40 Hz to improve prediction accuracy because a higher subscription rate further reduced the accuracy. Hence, it is required to wait for a more extended period of 3.2 s to compile one data set for real-time prediction. The performance of this approach is independent of environmental conditions.

The monocular camera sensor-based study was a novel approach in terms of optical flow application for condition monitoring in mobile robots. Here, the vibration source prediction accuracy was slightly better (93.8%) than that of the IMU. However, there are still risks involved in real-time deployment, considering the few chances of triggering improper maintenance action and the consequences. This approach is computationally more complex in deriving the displacement vector data from each image frame and thereby a higher inference time from the given data sample (7.9 ms). However, the total time taken for real-time data recording and predicting a class, with a recording rate of 30 frames per second (fps), was observed the fastest around 0.2 s only. Another factor noted was the accuracy of the camera approach is affected by poor light and a dusty environment. The system fails to collect accurate vector data representing the particular class at an illuminance of less than 50 lux. Hence, the camera-based approach is not advised for places with dark shades and poorly illuminated rooms/confined areas. Also, the lens glass cover cleanliness should be assured while the robot operates in a dusty environment.

In the proposed IMU–current sensor heterogeneous dataset approach, the current sensors complement the IMU sensor's shortcomings in accuracy and total prediction time. The distinctive higher current consumption while the robot was trying to overcome uneven terrain, obstacles, and unbalancing factors helped to extract the vibration-affected features of classes more effectively and improved the prediction accuracy of all the vibration classes. This heterogeneous dataset approach also helped to increase the IMU subscription rate twice (80 Hz) without compromising the accuracy; hence, the total time to predict a class from recording a data sample is reduced to half (1.6 s) compared to the individual IMU sensor model. Though the number of features added with two current values, the inference time (0.196 ms) is still close to the previous IMU work. Also, as both IMU and current sensors are proprioceptive, environmental changes such as light and dust are unaffected. Hence, the proposed IMU–current sensor heterogeneous sensor dataset-based CM framework is more promising for real-time deployment, especially regarding accuracy and environmental conditions. These observations and discussions are summarized in Table 4. Here, Sensor type is the sensor used indicating a single sensor or multiple sources; Accuracy is the average accuracy of five classes from model evaluation; Time taken is the approximate time consumed to predict a class, including data recording time for one sample; and Environment fit is the suitability of the approach irrespective of light and dust.

Table 4. Performance comparison with previous works proposed.

Sensor Type	Accuracy	Time Taken	Environment Fit
IMU + current sensor	98.4%	1.6 s	Excellent
Monocular camera [19]	93.8%	0.2 s	Fair
IMU [18]	92.2%	3.2 s	Excellent

The proposed combination of IMU and current-sensor-based study shows a significant improvement in prediction accuracy of the vibration source classes compared to our previous individual IMU and camera-based studies. The higher prediction accuracy is vital for the CM applications to avoid false positive predictions and trigger improper maintenance or corrective actions. Though the camera-based work can make eight times more predictions in 1.6 s, due to the different data subscriptions and methodology developed, the close ground truth comparison study shows that such predictions are the same as the single prediction of the proposed study. Also, compared to the robot's operating speed of 0.05 m/s, the 1.6 s for a single prediction had less impact and was acceptable from the real-time case studies conducted. Though cost-wise, the IMU sensor is expensive. It is an inevitable sensor for autonomous robots; hence, this additional application is proposed as a CM sensor together with current sensors.

4. Real-Time Field Case Studies

The proposed IMU–current-sensor-based CM framework is validated by conducting real-time case studies at four different types of indoor environments at the SUTD campus, which are the canteen and pantry area, glass-walled lobby hallway, pebble-paved indoor corridor, and our robotics lab area. It is also assured that none of these areas was used during training data sample collection.

There are two pre-works conducted for these field trials. The first is developing an Inference Engine to apply the 1D CNN model's knowledge for inferring every new data sample in real time. Inference Engine Algorithm 1 shows the way 128 temporal data (*Data*) and a total of 11 sensor features (*Feature*) are acquired, forming a new dataset [128 × 11] for prediction. Here, a placeholder *TemporalBuffer* is used while collecting the dataset, and *InferenceBuffer* holds one complete dataset as 1D CNN input and returns the predicted vibration source class.

Algorithm 1 Inference Engine

```
while Feature, Data are not empty do
    TemporalBuffer = call Sensor(Feature, Data)
    Format TemporalBuffer to have shape (11,128)
    Append TemporalBuffer to InferenceBuffer
    counterFeature = counterFeature + 1
    counterData = counterData + 1
    if counterFeature == 11, counterData == 128 then
        InferredClass = Call 1DCNN(InferenceBuffer)
        counterFeature = 0
        counterData = 0
        Clean InferenceBuffer
    end if
end while
```

Secondly, a cartographer SLAM-based 2D occupancy grid map was created for each test area. Then, a vibration source mapping algorithm (Algorithm 2) was developed, which maps the predicted abnormal vibration classes (*PC*) with the unique color at the center of the robot's footprint (*RFP*) on the 2D grid map (*map*) where the class was predicted, and saved the real-time 2D CbM map with time stamp for each class fused. All the field case studies were conducted in autonomous mode at the intended speed of 0.05 m/s in a zig-zag travel pattern, ensuring a sufficiently charged battery; also, each test started verifying the robot was healthy.

Algorithm 2 Vibration Source Mapping

```
PC ← [Terrain, Collision, Assembly, Structure]
map ← 2Dgridmap
RFP ← [x, y]
DT ← [Date, Time]
if PC is Normal then Donothing
else if PC is Abnormal then
    if PC is Terrain then
        Color ← Orange
    else if PC is Collision then
        Color ← Red
    else if PC is Assembly then
        Color ← Blue
    else if PC is Structure then
        Color ← Green
        map[RFP.x][RFP.y] ← color
    end if
end if
saveMap ← map, DT
```

The first case study was conducted considering a moderately crowded place with different furniture and floor types and selected SUTD canteen and an office pantry room. The canteen floor is built with various flooring materials and features such as smooth concrete, tactile pavements, uneven wooden floors, and tables with flat bases. The canteen area was messed up with temporary cable routing, misplaced furniture, etc. Two classes of abnormal vibrations were noticed after driving through the canteen space: terrain and collision induced. The Terrain class was mainly due to the cable routing, tactile pavement, and uneven wooden floor, while the unseen table base caused the Collision class. Next, it was tested in the office pantry area with a smooth tiled floor and tall and thin metal-legged chairs. During this testing, collision-induced vibrations were predicted when the robot collided with a chair as the RPLiDAR could not detect the thin long metal leg of the chair. The robot tried to push the chair further but was blocked by a pillar with a higher current consumption, matching the training characteristics of the Collision class, though no obvious angular motion vibrations were noticed. Some of these classes fused on the SLAM map generating a 2D CbM map with the relevant trial run photos are shown in Figure 11.

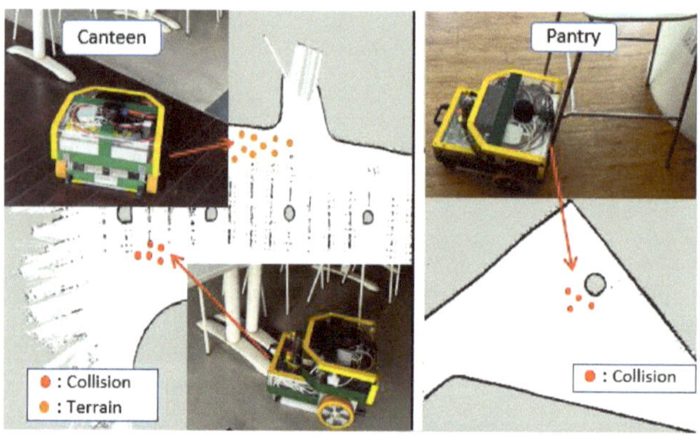

Figure 11. Two-dimensional (2D) CbM map. Canteen and Pantry area: Case study 1.

The second case study was tested in a carpeted lobby hallway where one wall was built using transparent rugged glass. During the robot's drive near the glass wall, the RPLiDAR did not precisely detect the transparent glass wall, resulting in occasionally hitting the wall. We stopped the robot to avoid damage to the robot, glass wall, and other potential hazards. Next, the robot was driven in a mixed concrete and small-pebble-paved corridor for the third field trial. During the trial runs, the Terrain-induced vibration class is predicted and fused on the map whenever the robot moves to the pebble-paved floors. As it was an isolated area with no workspace hazards, we continued the testing on the same pebble-paved floor without any maintenance action. However, after six days, intermittent Assembly-induced vibration classes were observed along with the Terrain class, mainly due to the loose assembly of wheel couplings with motors. Figure 12 shows the 2D CbM maps for case studies at the lobby and corridor, indicating some of the randomly selected fused vibration classes in real-time towards the end of the test, also the workspace photos.

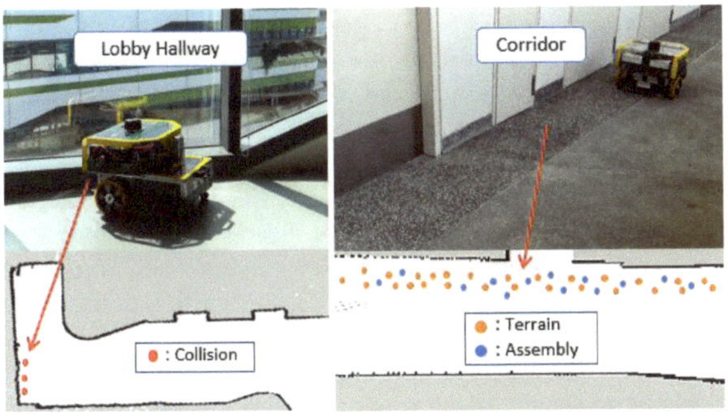

Figure 12. Two-dimensional (2D) CbM maps. Lobby and Corridor area: Case Studies 2 and 3.

Finally, we conducted a long-term test drive by isolating a small area (around $9 \times 6\ \text{m}^2$) of the robotics lab with a vinyl floor and removing all obstacles, i.e., tested with no potential external factors. The robot ran continuously with timely recharging for 21 days without any abnormal classes predicted. However, apparent wheel wobbling was later observed, and Assembly-induced vibrations were fused on the SLAM map concurrently. We continued the test without any maintenance; hence, the condition worsened. Detaching the battery bracket and mopping pad assembly resulted in parts shifting and Structure-induced class prediction. These intermittent Assembly and Structure classes predicted/fused for the last few samples before stopping the robot are shown in the 2D CbM map Figure 13 with the lab test area photo.

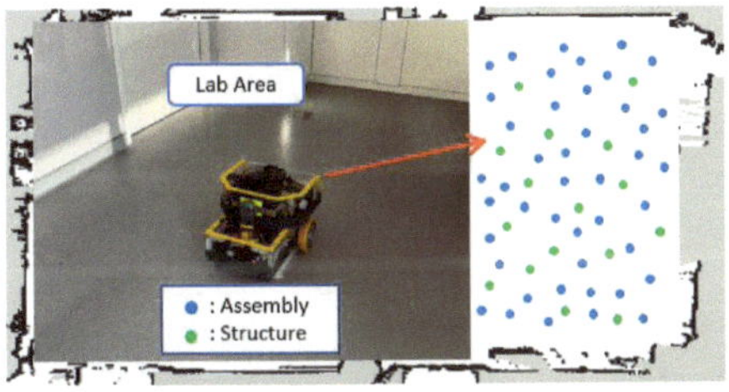

Figure 13. Two-dimensional (2D) CbM map. Robotics Lab area: Case Study 4.

We closely observed the robot's behavior during these field trials, especially when it started predicting abnormal classes, and randomly collected 400 samples for each class, i.e., during Normal and all Abnormal classes, for assessing real-time prediction accuracy. The results show an average accuracy of 98.0 %. The individual class accuracy is listed in Table 5. This indicates very close prediction accuracy as the offline evaluation test results. The consistent and high prediction accuracy during the field trials shows the fitness of this proposed method for real-time deployment in professional cleaning robots for health monitoring, promptly triggering the right maintenance action and ensuring operational safety.

Table 5. Real-time prediction accuracy of vibration sources.

Vibration Class	Normal	Terrain	Collision	Assembly	Structure
Accuracy Prediction (%)	99	98	99	98	96

5. Conclusions

An automated maintenance strategy for mobile cleaning robots is essential to assure productivity and safety. The proposed heterogeneous sensor dataset-based condition monitoring framework using IMU and current sensors fills such research gaps by predicting the anomalous internal and external vibration sources, which cause system degradation or potential hazards, with high accuracy for real-time deployment. The distinctive data features of these two proprioceptive sensors were modeled as vibration data, which complement each other to sort out the flaws of individual use, resulting in a higher accuracy prediction. Also, the proposed simple, fast and low computation cost four-layer 1D CNN model was validated through four field case studies as an ideal neural network model for real-time condition monitoring applications in mobile robots by accurately predicting the abnormal vibration sources, using the proposed temporal fused dataset. Compared to similar previous works developed, the proposed work excels in accuracy, time, and fit for the environment. The IMU–current sensor heterogeneous dataset model results in 6.2% higher accuracy, and the prediction time is reduced to half compared to individual IMU sensor work. Similarly, it is 4.6% more accurate and does not affect any environment light and dust compared to the camera sensor-based study. The 2D CbM map fusing the predicted abnormal vibration source class on a SLAM-generated environment map helps the maintenance team track the system degradation of the robot, identify any operational hazards in advance, and trigger prompt maintenance or corrective actions, enabling a real-time CbM strategy. The proposed framework will also help robot manufacturers and cleaning contract companies review their maintenance and rental strategies, design and assembly processes, and plan for robot-friendly workspaces. We plan automated CM frameworks for outdoor mobile robots to extend this research.

Author Contributions: Conceptualization, S.P. and M.R.E.; methodology, S.P. and P.V.; software, P.V. and S.P.; validation, S.P.; formal analysis, S.P., M.R.E. and P.V.; investigation, S.P. and M.R.E.; resources, M.R.E.; data, S.P. and P.V.; writing–original draft preparation, S.P. and P.V.; supervision, M.R.E.; project administration, M.R.E.; funding acquisition, M.R.E. All authors have read and agreed to the published version of the manuscript.

Funding: This research is supported by the National Robotics Programme under its National Robotics Programme (NRP) BAU, Ermine III: Deployable Reconfigurable Robots, Award No. M22NBK0054 and also supported by A*STAR under its "RIE2025 IAF-PP Advanced ROS2-native Platform Technologies for Cross-sectorial Robotics Adoption (M21K1a0104)" programme.

Data Availability Statement: Data will be made available on request.

Acknowledgments: The authors would like to thank, the National Robotics Programme, the Agency for Science, Technology and Research and SUTD for their support. Thanks to Abdullah Aamir Hayat for his inputs.

Conflicts of Interest: The authors declare no conflict of interest.

References

1. Robotics, W. World Robotics 2020 Report. Available online: http://reparti.free.fr/robotics2000.pdf (accessed on 10 August 2021).
2. MarketsandMarkets. Cleaning Robot Market. Available online: https://marketsandmarkets.com/Market-Reports/cleaning-robot-market-22726569.html (accessed on 10 March 2023).
3. Le, A.V.; Ramalingam, B.; Gomez, B.F.; Mohan, R.E.; Minh, T.H.Q.; Sivanantham, V. Social density monitoring toward selective cleaning by human support robot with 3D based perception system. *IEEE Access* **2021**, *9*, 41407–41416.
4. Wang, Z.; Liao, H.; Jia, Z.; Wu, J. Semantic Mapping Based on Visual SLAM with Object Model Replacement Visualization for Cleaning Robot. In Proceedings of the 2022 IEEE International Conference on Robotics and Biomimetics (ROBIO), Jinghong, China, 5–9 December 2022; IEEE: Piscataway, NJ, USA, 2022; pp. 569–575.
5. Ravankar, A.; Ravankar, A.A.; Watanabe, M.; Hoshino, Y.; Rawankar, A. Multi-robot path planning for smart access of distributed charging points in map. *Artif. Life Robot.* **2021**, *26*, 52–60. [CrossRef]
6. Moon, W.; Park, B.; Nengroo, S.H.; Kim, T.; Har, D. Path planning of cleaning robot with reinforcement learning. In Proceedings of the 2022 IEEE International Symposium on Robotic and Sensors Environments (ROSE), Abu Dhabi, United Arab Emirates, 14–15 November 2022; IEEE: Piscataway, NJ, USA, 2022; pp. 1–7.
7. Prabakaran, V.; Elara, M.R.; Pathmakumar, T.; Nansai, S. Floor cleaning robot with reconfigurable mechanism. *Autom. Constr.* **2018**, *91*, 155–165. [CrossRef]
8. Muthugala, M.V.J.; Samarakoon, S.B.P.; Elara, M.R. Design by robot: A human-robot collaborative framework for improving productivity of a floor cleaning robot. In Proceedings of the 2022 International Conference on Robotics and Automation (ICRA), Philadelphia, PA, USA, 23–27 May 2022; IEEE: Piscataway, NJ, USA, 2022; pp. 7444–7450.

9. Khan, A.; Noreen, I.; Habib, Z. An energy efficient coverage path planning approach for mobile robots. In *Intelligent Computing, Proceedings of the 2018 Computing Conference, London, UK, 10–12 July 2018*; Springer: Berlin/Heidelberg, Germany, 2019; Volume 2, pp. 387–397.
10. Manimuthu, A.; Le, A.V.; Mohan, R.E.; Veerajagadeshwar, P.; Huu Khanh Nhan, N.; Ping Cheng, K. Energy consumption estimation model for complete coverage of a tetromino inspired reconfigurable surface tiling robot. *Energies* **2019**, *12*, 2257. [CrossRef]
11. Zheng, K.; Chen, G.; Cui, G.; Chen, Y.; Wu, F.; Chen, X. Performance metrics for coverage of cleaning robots with mocap system. In *Intelligent Robotics and Applications, Proceedings of the 10th International Conference, ICIRA 2017, Wuhan, China, 16–18 August 2017*; Proceedings, Part III 10; Springer: Berlin/Heidelberg, Germany, 2017; pp. 267–274.
12. Pathmakumar, T.; Elara, M.R.; Gómez, B.F.; Ramalingam, B. A reinforcement learning based dirt-exploration for cleaning-auditing robot. *Sensors* **2021**, *21*, 8331. [CrossRef]
13. Chng, G. Softbank Robotics Launches First Rent-a-Robot Offering for Cleaning Services in Singapore. Available online: https://www.techgoondu.com/2019/09/25/softbank-robotics-launches-first-rent-a-robot-offering-for-cleaning-services-in-singapore/ (accessed on 28 November 2021).
14. Cardarelli, S.; di Florio, P.; Mengarelli, A.; Tigrini, A.; Fioretti, S.; Verdini, F. Magnetometer-free sensor fusion applied to pedestrian tracking: A feasibility study. In Proceedings of the 2019 IEEE 23rd International Symposium on Consumer Technologies (ISCT), Ancona, Italy, 19–21 June 2019; IEEE: Piscataway, NJ, USA, 2019; pp. 238–242.
15. Yang, S.; Garg, N.P.; Gao, R.; Yuan, M.; Noronha, B.; Ang, W.T.; Accoto, D. Learning-Based Motion-Intention Prediction for End-Point Control of Upper-Limb-Assistive Robots. *Sensors* **2023**, *23*, 2998. [CrossRef]
16. Tibebu, H.; Roche, J.; De Silva, V.; Kondoz, A. Lidar-based glass detection for improved occupancy grid mapping. *Sensors* **2021**, *21*, 2263. [CrossRef]
17. Foster, P.; Sun, Z.; Park, J.J.; Kuipers, B. Visagge: Visible angle grid for glass environments. In Proceedings of the 2013 IEEE International Conference on Robotics and Automation, Karlsruhe, Germany, 6–10 May 2013; IEEE: Piscataway, NJ, USA, 2013; pp. 2213–2220.
18. Pookkuttath, S.; Rajesh Elara, M.; Sivanantham, V.; Ramalingam, B. AI-Enabled Predictive Maintenance Framework for Autonomous Mobile Cleaning Robots. *Sensors* **2022**, *22*, 13. [CrossRef]
19. Pookkuttath, S.; Gomez, B.F.; Elara, M.R.; Thejus, P. An optical flow-based method for condition-based maintenance and operational safety in autonomous cleaning robots. *Expert Syst. Appl.* **2023**, *222*, 119802. [CrossRef]
20. Eren, L.; Ince, T.; Kiranyaz, S. A generic intelligent bearing fault diagnosis system using compact adaptive 1D CNN classifier. *J. Signal Process. Syst.* **2019**, *91*, 179–189. [CrossRef]
21. Ince, T.; Kiranyaz, S.; Eren, L.; Askar, M.; Gabbouj, M. Real-time motor fault detection by 1-D convolutional neural networks. *IEEE Trans. Ind. Electron.* **2016**, *63*, 7067–7075. [CrossRef]
22. Tiboni, M.; Remino, C.; Bussola, R.; Amici, C. A review on vibration-based condition monitoring of rotating machinery. *Appl. Sci.* **2022**, *12*, 972. [CrossRef]
23. Luo, B.; Wang, H.; Liu, H.; Li, B.; Peng, F. Early fault detection of machine tools based on deep learning and dynamic identification. *IEEE Trans. Ind. Electron.* **2018**, *66*, 509–518. [CrossRef]
24. Abdeljaber, O.; Avci, O.; Kiranyaz, S.; Gabbouj, M.; Inman, D.J. Real-time vibration-based structural damage detection using one-dimensional convolutional neural networks. *J. Sound Vib.* **2017**, *388*, 154–170. [CrossRef]
25. Kim, H.G.; Yoon, H.S.; Yoo, J.H.; Yoon, H.I.; Han, S.S. Development of predictive maintenance technology for wafer transfer robot using clustering algorithm. In Proceedings of the 2019 International Conference on Electronics, Information, and Communication (ICEIC), Auckland, New Zealand, 22–25 January 2019; IEEE: Piscataway, NJ, USA, 2019; pp. 1–4.
26. Onur, K.; Kaymakci, O.T.; Mercimek, M. Advanced Predictive Maintenance with Machine Learning Failure Estimation in Industrial Packaging Robots. In Proceedings of the 2020 International Conference on Development and Application Systems (DAS), Suceava, Romania, 21–23 May 2020; IEEE: Piscataway, NJ, USA, 2020; pp. 1–6.
27. Aliev, K.; Antonelli, D. Proposal of a monitoring system for collaborative robots to predict outages and to assess reliability factors exploiting machine learning. *Appl. Sci.* **2021**, *11*, 1621. [CrossRef]
28. Kiranyaz, S.; Avci, O.; Abdeljaber, O.; Ince, T.; Gabbouj, M.; Inman, D.J. 1D convolutional neural networks and applications: A survey. *Mech. Syst. Signal Process.* **2021**, *151*, 107398. [CrossRef]
29. Janssens, O.; Slavkovikj, V.; Vervisch, B.; Stockman, K.; Loccufier, M.; Verstockt, S.; Van de Walle, R.; Van Hoecke, S. Convolutional neural network based fault detection for rotating machinery. *J. Sound Vib.* **2016**, *377*, 331–345. [CrossRef]
30. Kumar, P.; Shankar Hati, A. Convolutional neural network with batch normalisation for fault detection in squirrel cage induction motor. *IET Electr. Power Appl.* **2021**, *15*, 39–50. [CrossRef]
31. Abdeljaber, O.; Sassi, S.; Avci, O.; Kiranyaz, S.; Ibrahim, A.A.; Gabbouj, M. Fault detection and severity identification of ball bearings by online condition monitoring. *IEEE Trans. Ind. Electron.* **2018**, *66*, 8136–8147. [CrossRef]
32. Alatise, M.B.; Hancke, G.P. A review on challenges of autonomous mobile robot and sensor fusion methods. *IEEE Access* **2020**, *8*, 39830–39846. [CrossRef]
33. Mikulová, Z.; Duchoň, F.; Dekan, M.; Babinec, A. Localization of mobile robot using visual system. *Int. J. Adv. Robot. Syst.* **2017**, *14*, 1729881417736085. [CrossRef]

34. Sasiadek, J.; Hartana, P. Sensor data fusion using Kalman filter. In Proceedings of the Third International Conference on Information Fusion, Sun City, South Africa, 1–4 November 2020; IEEE: Piscataway, NJ, USA, 2000; Volume 2, pp. WED5–WED19.
35. Shaikh, M.M.; Bahn, W.; Lee, C.; Kim, T.i.; Lee, T.J.; Kim, K.S.; Cho, D. Mobile robot vision tracking system using unscented Kalman filter. In Proceedings of the 2011 IEEE/SICE International Symposium on System Integration (SII), Kyoto, Japan, 20–22 December 2011; IEEE: Piscataway, NJ, USA, 2011; pp. 1214–1219.
36. Rambach, J.R.; Tewari, A.; Pagani, A.; Stricker, D. Learning to fuse: A deep learning approach to visual-inertial camera pose estimation. In Proceedings of the 2016 IEEE International Symposium on Mixed and Augmented Reality (ISMAR), Merida, Yucatan, Mexico, 19–23 September 2016; IEEE: Piscataway, NJ, USA, 2016; pp. 71–76.
37. Clark, R.; Wang, S.; Wen, H.; Markham, A.; Trigoni, N. Vinet: Visual-inertial odometry as a sequence-to-sequence learning problem. In Proceedings of the AAAI Conference on Artificial Intelligence, San Francisco, CA, USA, 4–9 February 2017; Volume 31.
38. Li, C.; Wang, S.; Zhuang, Y.; Yan, F. Deep sensor fusion between 2D laser scanner and IMU for mobile robot localization. *IEEE Sens. J.* **2019**, *21*, 8501–8509. [CrossRef]
39. Wang, M.; Ye, L.; Sun, X. Adaptive online terrain classification method for mobile robot based on vibration signals. *Int. J. Adv. Robot. Syst.* **2021**, *18*, 17298814211062035. [CrossRef]
40. Tick, D.; Rahman, T.; Busso, C.; Gans, N. Indoor robotic terrain classification via angular velocity based hierarchical classifier selection. In Proceedings of the 2012 IEEE International Conference on Robotics and Automation, St Paul, MN, USA, 14–18 May 2012; IEEE: Piscataway, NJ, USA, 2012; pp. 3594–3600.
41. Dupont, E.M.; Moore, C.A.; Collins, E.G.; Coyle, E. Frequency response method for terrain classification in autonomous ground vehicles. *Auton. Robot.* **2008**, *24*, 337–347. [CrossRef]
42. Csík, D.; Odry, Á.; Sárosi, J.; Sarcevic, P. Inertial sensor-based outdoor terrain classification for wheeled mobile robots. In Proceedings of the 2021 IEEE 19th International Symposium on Intelligent Systems and Informatics (SISY), Subotica, Serbia, 16–18 September 2021; IEEE: Piscataway, NJ, USA, 2021; pp. 159–164.
43. Weiss, C.; Tamimi, H.; Zell, A. A combination of vision-and vibration-based terrain classification. In Proceedings of the 2008 IEEE/RSJ International Conference on Intelligent Robots and Systems, Nice, France, 22–26 September 2008; IEEE: Piscataway, NJ, USA, 2008; pp. 2204–2209.
44. Constantin, G.; Maroșan, I.A.; Crenganiș, M.; Botez, C.; Gîrjob, C.E.; Biriș, C.M.; Chicea, A.L.; Bârsan, A. Monitoring the Current Provided by a Hall Sensor Integrated in a Drive Wheel Module of a Mobile Robot. *Machines* **2023**, *11*, 385. [CrossRef]
45. Rapalski, A.; Dudzik, S. Energy Consumption Analysis of the Selected Navigation Algorithms for Wheeled Mobile Robots. *Energies* **2023**, *16*, 1532. [CrossRef]
46. Kryter, R.; Haynes, H. *Condition Monitoring of Machinery Using Motor Current Signature Analysis*; Technical Report; Oak Ridge National Lab.: Oak Ridge, TN, USA, 1989.
47. Pookkuttath, S.; Elara, M.R.; Mohan Rayguru, M.; Saldi, Z.S.; Sivanantham, V.; Ramalingam, B. Snail: An Eco-Friendly Autonomous Steam Mopping Robot for Cleaning and Disinfection of Floors. *Mathematics* **2023**, *11*, 1086. [CrossRef]
48. Mitiche, I.; Nesbitt, A.; Conner, S.; Boreham, P.; Morison, G. 1D-CNN based real-time fault detection system for power asset diagnostics. *IET Gener. Transm. Distrib.* **2020**, *14*, 5766–5773. [CrossRef]
49. Hess, W.; Kohler, D.; Rapp, H.; Andor, D. Real-time loop closure in 2D LIDAR SLAM. In Proceedings of the 2016 IEEE International Conference on Robotics and Automation (ICRA), Stockholm, Sweden, 16–21 May 2016; IEEE: Piscataway, NJ, USA, 2016; pp. 1271–1278.
50. Abadi, M.; Barham, P.; Chen, J.; Chen, Z.; Davis, A.; Dean, J.; Devin, M.; Ghemawat, S.; Irving, G.; Isard, M.; et al. Tensorflow: A system for large-scale machine learning. In Proceedings of the Osdi, Savannah, GA, USA, 2–4 November 2016; Volume 16, pp. 265–283.
51. Kingma, D.P.; Ba, J. Adam: A method for stochastic optimization. *arXiv* **2014**, arXiv:1412.6980.
52. Grandini, M.; Bagli, E.; Visani, G. Metrics for multi-class classification: An overview. *arXiv* **2020**, arXiv:2008.05756.

remote sensing

MDPI

Article

Single Object Tracking in Satellite Videos Based on Feature Enhancement and Multi-Level Matching Strategy

Jianwei Yang [1,2,3], Zongxu Pan [1,2,3,*], Yuhan Liu [1,2], Ben Niu [1,2,3] and Bin Lei [1,2,3]

1 Aerospace Information Research Institute, Chinese Academy of Sciences, Beijing 100094, China; yangjianwei20@mails.ucas.ac.cn (J.Y.)
2 Key Laboratory of Technology in Geo-Spatial Information Processing and Application System, Chinese Academy of Sciences, Beijing 100190, China
3 School of Electronic, Electrical and Communication Engineering, University of Chinese Academy of Sciences, Beijing 100049, China
* Correspondence: zxpan@mail.ie.ac.cn

Abstract: Despite significant advancements in remote sensing object tracking (RSOT) in recent years, achieving accurate and continuous tracking of tiny-sized targets remains a challenging task due to similar object interference and other related issues. In this paper, from the perspective of feature enhancement and a better feature matching strategy, we present a tracker SiamTM specifically designed for RSOT, which is mainly based on a new target information enhancement (TIE) module and a multi-level matching strategy. First, we propose a TIE module to address the challenge of tiny object sizes in satellite videos. The proposed TIE module goes along two spatial directions to capture orientation and position-aware information, respectively, while capturing inter-channel information at the global 2D image level. The TIE module enables the network to extract discriminative features of the targets more effectively from satellite images. Furthermore, we introduce a multi-level matching (MM) module that is better suited for satellite video targets. The MM module firstly embeds the target feature map after ROI Align into each position of the search region feature map to obtain a preliminary response map. Subsequently, the preliminary response map and the template region feature map are subjected to the Depth-wise Cross Correlation operation to get a more refined response map. Through this coarse-to-fine approach, the tracker obtains a response map with a more accurate position, which lays a good foundation for the prediction operation of the subsequent sub-networks. We conducted extensive experiments on two large satellite video single-object tracking datasets: SatSOT and SV248S. Without bells and whistles, the proposed tracker SiamTM achieved competitive results on both datasets while running at real-time speed.

Keywords: satellite video; object tracking; siamese network; feature enhancement; matching strategy

1. Introduction

Single object tracking (SOT) refers to the task of predicting the position and size of a given target in the subsequent frames of a video, given its initial information in the first frame [1]. It is an important research area in computer vision and has been utilized in various applications, such as motion object analysis [2], automatic driving [3], and human–computer interactions [4]. As deep learning demonstrates outstanding performance in various visual fields, more and more researchers have introduced deep learning into the field of single object tracking, achieving many astonishing results. Tao et al. [5] proposed the SINT network, the first to apply the Siamese network to single-object tracking fields. SINT learns a matching function through a Siamese network using the first frame of target information as a template, and calculates the matching score between the template and all subsequent frames sampled. The highest score represents the position of the target in the current frame. SiamFC [6] abstracts the tracking process into a similarity learning problem. By learning a function $f(Z, X)$ to compare the similarity between the template image Z and the search image X, the target's location can be predicted based on the position with the highest score on the response map. Building upon SiamFC, SiamRPN [7] introduces the RPN module from Faster R-CNN [8], which eliminates the time-consuming multi-scale testing process, further improving performance and speeding up the system. Li et al. [9] proposed the SiamRPN++ network, which introduces the ResNet-50 [10] network to improve feature extraction capability. Additionally, they proposed Depth-wise Cross Correlation to replace the previous Up-Channel Cross Correlation, drastically reducing the number of parameters and enhancing the overall stability of the training. Xu et al. [11] proposed an anchor-free tracking algorithm called SiamFC++. The algorithm enhances the original SiamFC tracker with added position regression, quality score, and multiple joint training losses, resulting in a significantly improved tracking performance. SiamCAR [12] is also a novel anchor-free fully convolutional Siamese tracking network, which decomposes visual tracking tasks into two sub-problems: pixel-wise classification and target bounding box regression, and solves end-to-end visual tracking problems in a pixel-wise manner. Guo et al. [13] proposed a simple and perceptible Siamese graph network, SiamGAT, for generic object tracking. The tracker establishes part-to-part correspondences between the target and search regions using a full bipartite graph and applies a graph attention mechanism to propagate target information from template features to search features. Yang et al. [14] proposed a dedicated Siamese network, SiamMDM, designed for single object tracking in satellite videos. This network addresses the challenge of weak features exhibited by typical targets in satellite videos by incorporating feature map fusion and introducing a dynamic template branch. Additionally, the network suggests an adaptive fusion of both motion model predictions and Siamese network predictions to alleviate issues commonly encountered in satellite videos, such as partial or full occlusions.

As Transformer [15] has shown great potential in the field of object detection, more and more researchers are trying to introduce Transformer into the field of object tracking. Due to the utilization of only spatial features in Siamese algorithms, they may not be particularly suitable for scenarios involving target disappearance or significant object variations. To address this limitation, Yan et al. [16] proposed the incorporation of a transformer architecture, which combines spatial and temporal characteristics, effectively resolving the issue of long-range interactions in sequence modeling. Chen et al. [17] pointed out that correlation operation is a simple way of fusion, and they proposed a Transformer tracking method based on the attention fusion mechanism called TransT. SparseTT [18] has designed a sparse attention mechanism that allows the network to focus on target information in the search area, and proposed a Double-Head approach to improve classification and regression accuracy. The ToMP [19] tracker also replaces traditional optimization-based model predictors with transformers. This tracker incorporates two novel encoding methods that include both target position information and range information. In addition, a parallel two-stage tracking method is proposed to decouple target localization and bounding box regression, achieving a balance between accuracy and efficiency. In order to fully leverage the capabilities of self-attention, Gui et al. [20] introduced a novel tracking framework known as MixFormer, which deviates from traditional tracking paradigms. Additionally, they proposed the MAM module, which employs attention mechanisms to perform feature extractions and feature interactions simultaneously. This renders MixFormer remarkably concise without the need for additional fusion modules. Ye et al. [21] introduced OSTrack, a concise and efficient one-stream one-stage tracking framework. This tracker leverages the prior knowledge of similarity scores obtained in the early stages and proposes an in-network early candidate elimination module, thereby reducing inference time. To address the issue of inhibited performance improvements due to independent correlation calculations in attention mechanisms, AiATrack [22] introduces an Attention in Attention (AiA) module. This module enhances appropriate correlations and suppresses erroneous correlations by seeking consensus among all relevant vectors. SwinTrack [23] employs Transformer for feature extraction and fusion, enabling full interaction between the template region and the search region for tracking. Moreover, SwinTrack extensively investigates various strategies for feature fusion, position encoding, and training loss to enhance performance.

In recent years, with the advancement of remote sensing and image processing technology [24–26], the resolution of satellite video has been continuously improving, enabling the tracking of trains, ships, airplanes, and even ordinary cars from a remote sensing perspective. Compared with general optical object tracking, remote sensing single object tracking (RSOT) [27–30] faces several challenges that lead to significant performance degradation when directly applying deep learning-based single-object tracking methods to RSOT. Taking the common vehicle target in the tracking dataset as an example, we conducted a comparative analysis involving vehicle targets captured in natural image scenes, aerial views from unmanned aerial vehicles (UAV), and satellite perspectives. The results are depicted in Figure 1. The vehicle depicted in Figure 1a is selected from the Car24 video sequence in the OTB [31] dataset. In the displayed image frame, the vehicle target occupies 1848 pixels, accounting for 2.41% of the entire image. Despite the limited resolution of this image, the rich feature information at the rear of the vehicle provides sufficient discriminative cues for the tracker's decision-making process. The car depicted in Figure 1b is extracted from the car4 video sequence in the UAV123 [32] dataset. The car target occupies a total of 1560 pixels, which accounts for approximately 0.17% of the entire image. Despite the diminished proportion of the vehicle target within the entire image when viewed from the aerial perspective of a UAV, the contour of the vehicle target remains distinctly discernible. The car in Figure 1c is selected from the car_04 video sequence in the SatSOT [33] dataset. In complete contrast to the previous two images, the car target in this image occupies only 90 pixels, taking up only 0.01% of the whole image.

(a) Car in OTB100 dataset (b) Car in UAV123 dataset (c) Car in SatSOT dataset

Figure 1. Vehicle targets from different shooting angles. From left to right: car in natural image scene; car in drone aerial perspective; car in satellite capture perspective. Compared to the previous two, the proportion of the cars in the satellite images is very low, and the cars themselves are tiny in size, lacking distinct and identifiable features.

The comparison of the three images in Figure 1 reveals that the minuscule size of the vehicle targets in satellite videos results in a reduced amount of feature information. Therefore, it is essential to employ corresponding feature enhancement techniques to bolster the features of small objects in satellite videos. Additionally, it is unreasonable to rely solely on the Depth-wise Cross Correlation matching approach from the general object tracking domain, given the aforementioned scarcity of features in small objects. In summary, applying methods from the natural tracking field to the RSOT field primarily confronts two critical issues, as follows.

(1) **Weak target feature:** Due to the altitude of the satellite and the spatial resolution of satellite videos, the target in the satellite video usually occupies only a tiny percentage of the entire image [34]. Compared to generic optical targets, targets to be tracked in satellite video are generally too small in size, resulting in insufficient feature information for trackers to exploit [35].
(2) **Inappropriate matching strategy:** The common Depth-wise Cross Correlation finds the best match within the search map based on the target appearance and texture information provided by the template map. However, due to the tiny size of objects in satellite videos, which usually exist in clusters or lines in images and lack noticeable contour features, the Depth-wise Cross Correlation matching strategy is not fully applicable to RSOT.

In summary, how to effectively suppress the background information around the target and enhance the intensity of the target's own features has become an unavoidable problem in the study of target tracking in the RSOT field. At the same time, it is essential to design a matching method suitable for target tracking in satellite videos. Based on the above thinking, we have designed a tracker SiamTM specifically for satellite video target tracking based on feature enhancement and coarse-to-fine matching strategies.

The main contributions of this paper can be summarized as follows.

- Firstly, we propose a novel target information enhancement module that can capture the direction and position-aware information from both the horizontal and vertical dimensions, as well as inherent channel information from a global perspective of the image. The target information enhancement module embeds position information into channel attention to enhance the feature expression of our proposed SiamTM algorithm for small targets in satellite videos.
- Secondly, we have designed a multi-level matching module that is better suited for satellite video targets' characteristics. It combines coarse-grained semantic abstraction information with fine-grained location detail information, effectively utilizing template information to accurately locate the target in the search area, thereby improving the network's continuous tracking performance of the target in various complex scenarios.
- Finally, extensive experiments have been conducted on two large-scale satellite video single-object tracking datasets, SatSOT and SV248S. The experimental results show that the proposed SiamTM algorithm achieved state-of-the-art performance in both success and precision metrics, while having a tracking speed of 89.76 FPS, exceeding the standard of real-time tracking.

2. Related Work

In this chapter, we will introduce relevant works from two aspects. Firstly, we will introduce some typical feature enhancement methods in single-object tracking. Secondly, we will discuss the contributions of previous research in changing the matching methods between the template feature map and the search feature map.

2.1. Feature Enhancement Methods in Single Object Tracking

As one of the most crucial steps in a Siamese-based tracker, the quality of the feature maps extracted from the backbone directly influences the tracking performance and robustness. Therefore, numerous researchers have been exploring methods to obtain feature maps with richer feature information and emphasize focus on key regions. In early Siamese-based trackers (e.g., SINT [5] and SiamFC [6]), they employed a modified version of AlexNet [36] as the feature extraction network. Due to the shallow architecture of AlexNet, it lacked strong feature representation capabilities, leading to limited performance in early trackers. To further enhance network performance, subsequent works such as SiamRPN++ [9] and SiamDW [37] introduced a deeper backbone such as ResNet [10] to replace the original shallow AlexNet network. As a result of ResNet's simplicity and powerful feature extraction capabilities, it has become the default backbone extraction network in Siamese-based trackers [38]. Given a fixed backbone, exploring methods to further enhance the network's feature representation capabilities has naturally become a topic of great interest among researchers.

Fan et al. [39] proposed a multi-stage tracking framework called Siamese Cascaded Region Proposal Network (C-RPN), which consists of a series of cascaded RPNs in the Siamese network from deep high-level to shallow low-level layers. By introducing a novel Feature Transfer Block (FTB), C-RPN effectively utilizes multi-level features for each RPN, thereby enhancing its ability to leverage both high-level semantic information and low-level spatial information. Yu et al. [40] introduced a novel Siamese attention mechanism that computes deformable self-attention features and cross-attention features. The deformable self-attention features capture rich contextual information in the spatial domain and selectively enhance the interdependencies between channel features. On the other hand, the cross-attention features aggregate and communicate abundant information between the template and search regions, thereby improving the discriminative power of the features. Cao et al. [41] proposed an Attentional Aggregation Network (AAN) that leverages attention mechanisms to enhance the expressive power of features. The AAN utilizes both a Self-Attention Aggregation Network (Self-AAN) and a Cross-Attention Aggregation Network (Cross-AAN) to aggregate attention. By incorporating self-attention and cross-attention mechanisms, the AAN effectively captures the dependencies and relationships between different regions of the input, leading to improved feature representation. Similarly, Xie et al. [42] introduced a target-dependent feature network. By incorporating deep cross-image feature correlations into multiple layers of the feature network, this novel approach effectively suppresses non-target features and possesses the ability to extract instance-variant features. Chan et al. [1] first proposed a fine feature aggregation module to integrate low-level and high-level features for a more robust feature representation. They then utilized a Compound Attention Module to independently encode the local key information of template features and the global contextual information of search features.

In addition to the aforementioned methods for enhancing the feature extraction capabilities of trackers in the general object tracking domain, researchers have also proposed targeted approaches to address the inherent challenges posed by weak target features, susceptibility to background interference, and occlusion in satellite videos.

Cao et al. [43] proposed the utilization of a three-branch Siamese network structure. In addition to using the first frame as the template branch, they also incorporated a template branch based on the previous frame. Furthermore, in order to fully leverage deep and shallow features, multiple attention mechanisms were introduced after various stages of the backbone network to achieve significant representation of object features. Song et al. [29] proposed an attention-based tracker using a Siamese network architecture. By jointly optimizing multiple attention modules, they achieved information filtering and focused on key regions, thereby enhancing robustness to weak features and background noise. Zhang et al. [44] proposed an architectural framework designed specifically for satellite video object tracking, known as ThickSiam. ThickSiam replaces the original residual modules in the backbone network with Thickened Residual Blocks, aiming to extract robust semantic features. Nie et al. [45] proposed the information compensation module called Dim-Aware to enhance the representation of object features. This module utilizes high-frequency and crucial information to enhance the localization of small objects.

2.2. Matching Methods between Template Region and Search Region

Since the target tracking model structure based on the Siamese network was determined as the basic framework, the matching method between the target region and search region has undergone multiple iterations and improvements. In the early tracker SiamFC, the Siamese network used the naive correlation matching method. SiamFC considered the feature map corresponding to the search region as a convolutional kernel and performed a correlation operation on the feature map corresponding to the template region, obtaining a single-channel response map. The target position was determined based on the location of the maximum value in the response map. Subsequently, in the SiamRPN network, the authors proposed an Up-Channel Cross Correlation matching method to facilitate the subsequent operation of the classification and regression headers. However, Up-Channel Cross Correlation can lead to excessive parameterization. Therefore, in the SiamPRN++ network, the authors proposed the Depth-wise Cross Correlation matching method to address the abovementioned issue. It dramatically simplifies parameterization, balances the training of both branches, and stabilizes the training process, resulting in better network convergence. Alpha-Refine [46] believes that the key to improving the fine-tuning performance is to extract and maintain detailed spatial information as much as possible. Therefore, the network proposed uses a pixel-wise correlation instead of traditional correlation operations and deep correlation for high-quality feature representation, ensuring that each correlation map encodes local information of the target and avoids an extremely large correlation window. PG-Net [47] proposed that traditional similarity measurement methods introduce a lot of background noise during the tracking prediction process. Therefore, a pixel-to-global matching method is proposed to reduce the impact of noise. Zhou et al. [48] proposed a fine-grained saliency mining module to capture local saliency and a saliency-association modeling module to associate the captured salient regions and learn the global correlation between the target template and search image for state estimation. Zhang et al. [49] believe that a single matching operator is difficult to ensure stable tracking in challenging environments. Therefore, the authors proposed six different matching operators to replace the traditional cross-correlation operation. These operators are combined to explore complementary features, and a structural search method is used to select the most suitable combination of operators.

3. Proposed Approach

In this section, we introduce the proposed tracker SiamTM network in detail. Firstly, in Section 3.1, we expound the proposed SiamTM single-object tracking network from a holistic perspective. In order to extract more valuable and discriminative features from the template feature map and the search feature map, we creatively introduce a target information enhancement (TIE) module, and a detailed explanation of this module is presented in Section 3.2. Furthermore, we propose a multi-level matching (MM) module that integrates target information into the search feature map to improve tracking performance in Section 3.3.

3.1. Overall Architecture

The overall framework of the proposed SiamTM network is shown in Figure 2. The SiamTM network consists of three subnetworks: the feature extraction subnetwork, the feature enhancement and matching subnetwork, and the target prediction subnetwork. Similar to SiamCAR [12], the feature extraction network of SiamTM comprises two parts, the template branch and the search branch. The size of the template image is $3 \times 127 \times 127$, while that of the search image is $3 \times 255 \times 255$. Among them, the first dimension represents the number of channels in the image, and the last two dimensions represent the height and width of the image. The template region image and the search region image are fed into the modified ResNet50 [9] network with weight parameter sharing at the same time. After feature extraction, we obtain a $256 \times 15 \times 15$ template feature map and a $256 \times 31 \times 31$ search feature map. To enhance localization and discern foreground from background more effectively, we leverage the features extracted from the final three residual blocks of the backbone in reference to SiamCAR. Extensive literature [9,12,50] has substantiated that the joint utilization of low-level and high-level feature maps significantly contributes to improved accuracy in tracking. This is because low-level features encompass a multitude of informative cues facilitating precise positioning, such as the target's edge details and color attributes. On the other hand, high-level features encompass a greater wealth of semantic information that aids in effectively differentiating between foreground and background.

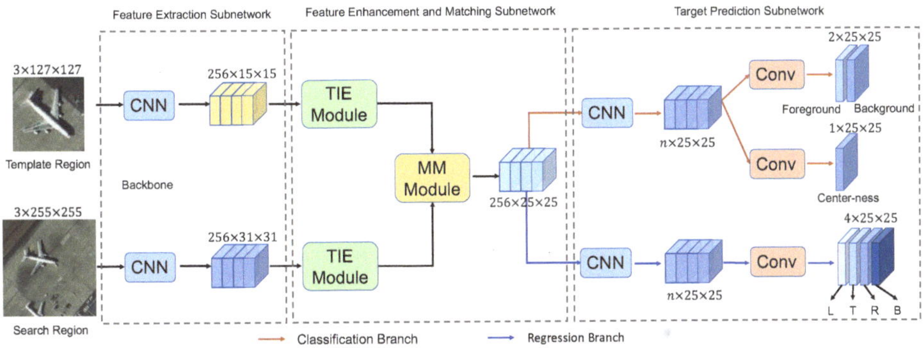

Figure 2. Architecture of the proposed SiamTM network. The SiamTM network consists of three subnetworks: the feature extraction subnetwork, the feature enhancement and matching subnetwork, and the target prediction subnetwork.

In order to extract more beneficial and distinguishable features from the original feature maps, and thereby facilitate the subsequent processing of the network, the template region feature map and the search region feature map are separately fed into the TIE module. The above operations outputs new feature maps that focus more on the information of the target itself, while maintaining the same size as the original feature maps. Compared to targets in natural images, targets in satellite videos are smaller in size and are more easily interfered by other objects in the background. In response to these characteristics of satellite video targets, a multi-level matching (MM) module was designed. The new template feature map and the search feature map are fed into the MM matching module. First, the template feature is embedded into each position of the search feature map to obtain a more accurate center position of the target, resulting in a preliminary rough response map. Then, through coarse-grained semantic abstraction information matching, the target outline is determined, resulting in a more refined final response map after matching.

Finally, the response map is fed into the target prediction subnetwork. In this subnetwork, the classification branch is used to distinguish the foreground and background in the current frame and to perform center-ness calculation, while the regression branch is used to determine the predicted bounding box. For the classification branch, it first performs a feature transformation on the response map $F_{response}$ output from the Feature Enhancement and Matching Subnetwork through a four-layer CNN structure to get the feature map F_{cls}. Each layer in the CNN structure consists of a Convolution layer, a GroupNorm layer, and a ReLU layer successively. The feature map F_{cls} is subsequently channel transformed by two independent single-layer Convolution layers to get the output feature maps $R_{cls} \in \mathbb{R}^{2\times25\times25}$ and $R_{cen} \in \mathbb{R}^{1\times25\times25}$. R_{cls} is used to differentiate the foreground from the background of the input image, whereas R_{cen} denotes the center-ness score of each position. Similar to the classification branch, the regression branch first performs a feature transformation on $F_{response}$ through an identical but independent CNN structure to obtain the feature map $F_{reg} \in \mathbb{R}^{n\times25\times25}$. Subsequently, a channel transformation is performed on F_{reg} through a Convolution layer to obtain the output of the regression branch $R_{reg} \in \mathbb{R}^{4\times25\times25}$. Each point in the feature map R_{reg} respectively represents the distance from the corresponding position to the four sides of the bounding box in the search region.

3.2. Target Information Enhancement Module

Compared to objects in natural scenarios, objects in remote sensing images are smaller and contain less feature information [51,52]. Therefore, how to effectively extract distinctive features from objects in remote sensing images has become one of the critical issues affecting subsequent tracking performance. Studies [53] on lightweight networks have shown that channel attention can significantly improve model performance. However, channel attention often ignores the crucial positional information in visual tasks that capture target structures [54]. Therefore, it is necessary to consider how to embed positional information into channel attention. Based on the above problems and corresponding considerations, we designed a feature enhancement module to capture inter-channel information at the global 2D image level, while also capturing direction and position-aware information along the two spatial directions. The TIE module is shown in Figure 3.

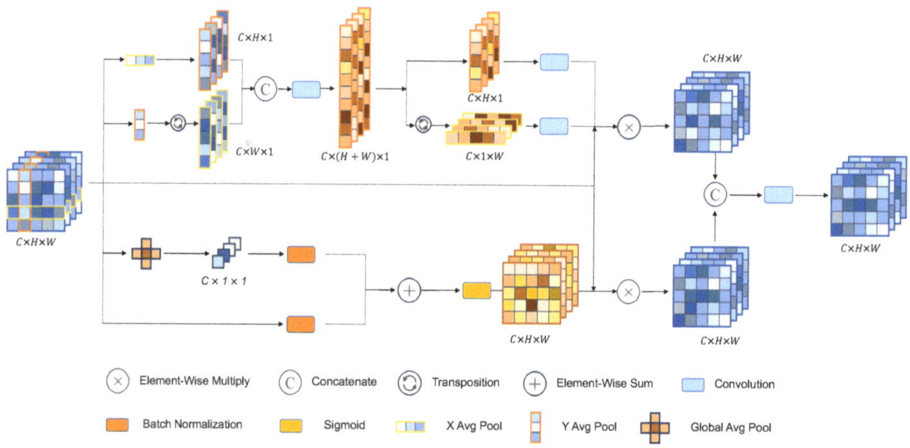

Figure 3. Architecture of the Target Information Enhancement Module. The meanings of each symbol are shown below the image.

Given a feature map $X \in \mathbb{R}^{C\times H\times W}$, first, a one-dimensional feature-encoding operation is performed on each channel using pooling kernels of size $(H, 1)$ and $(1, W)$ along the horizontal and vertical directions, respectively. The output expression of the c-th channel with a height of h is shown below:

$$e_c^h = \frac{1}{W}\sum_{0\leq i\leq W} x_c(h, i), \tag{1}$$

where c and h represent the channel and height of the current operation respectively, while W represents the width of the image.

Similarly, the output expression of the c-th channel with a width of w is shown as follows:

$$e_c^w = \frac{1}{H}\sum_{0\leq j\leq H} x_c(j, w), \tag{2}$$

where H represents the height of the image.

Then, concatenate the feature map $X_{horizontal} \in \mathbb{R}^{C\times H\times 1}$ obtained through vertical average pooling with the feature map $X_{vertical} \in \mathbb{R}^{C\times W\times 1}$ obtained through horizontal average pooling and transpose and transform it into an intermediate feature map $X_{inter} \in \mathbb{R}^{C\times (H+W)\times 1}$ through a 1×1 convolution kernel. The intermediate feature map retains spatial information from the original feature map in both vertical and horizontal directions and also captures relationships between channels.

After that, the intermediate feature map X_{inter} is divided into two separate tensors, $X'_{horizontal} \in \mathbb{R}^{C\times H\times 1}$ and $X'_{vertical} \in \mathbb{R}^{C\times 1\times W}$, along the spatial dimension. After undergoing a 1×1 convolution operation, attention weights W_h and W_w are obtained. The final output expression for the 1D part is

$$Y_1 = X \times C_1(X'_{horizontal}) \times C_1(X'_{vertical}), \tag{3}$$

where $C_1(\cdot)$ denotes the 1×1 convolution layer, $C_1(X'_{horizontal})$ and $C_1(X'_{vertical})$ denote the attention weights along the horizontal direction and the attention weights along the vertical direction, respectively.

In addition, in order to capture more distinctive target information from a global perspective and highlight more effective target features, we started from the 2D level of the image and performed global average pooling on the original feature map to obtain the feature map $X_{global} \in \mathbb{R}^{C\times 1\times 1}$. We also generated another intermediate feature map, whose output expression is:

$$Y_2 = \sigma(X_{global} \oplus X) \otimes X, \tag{4}$$

where $\oplus$ means the broadcasting addition, $\otimes$ means the element-wise multiplication, and σ denotes the Sigmoid operation.

Finally, we concatenate and reduce the one-dimensional feature map Y_1 and the two-dimensional feature map Y_2 obtained from feature encoding operations, and obtain the final output feature map.

3.3. Multi-Level Matching Module

Feature matching in Siamese networks refers to integrating feature maps obtained from the template branch and the search branch, calculating the similarity between each region of the template feature map and the search feature map, and finally, outputting a response map. The output response map is then sent to subsequent target prediction subnetworks for classification, regression, and other operations. As one of the most essential steps in the single-object tracking network, the quality of the feature matching directly determines the tracking performance. In existing works, since SiamRPN++ introduced Depth-wise Cross Correlation to the Siamese tracking network, most tracking networks have used this correlation operation as their feature matching subnetwork. Few networks modify the feature matching module for the characteristics of remote sensing targets. However, the Depth-wise Cross Correlation uses the target template as a spatial filter to convolve over the search area, emphasizing coarse-grained semantic abstractions such as target contours and appearance, while ignoring position information. Therefore, we propose a multi-level matching (MM) module, and the architecture of the MM module is shown in Figure 4.

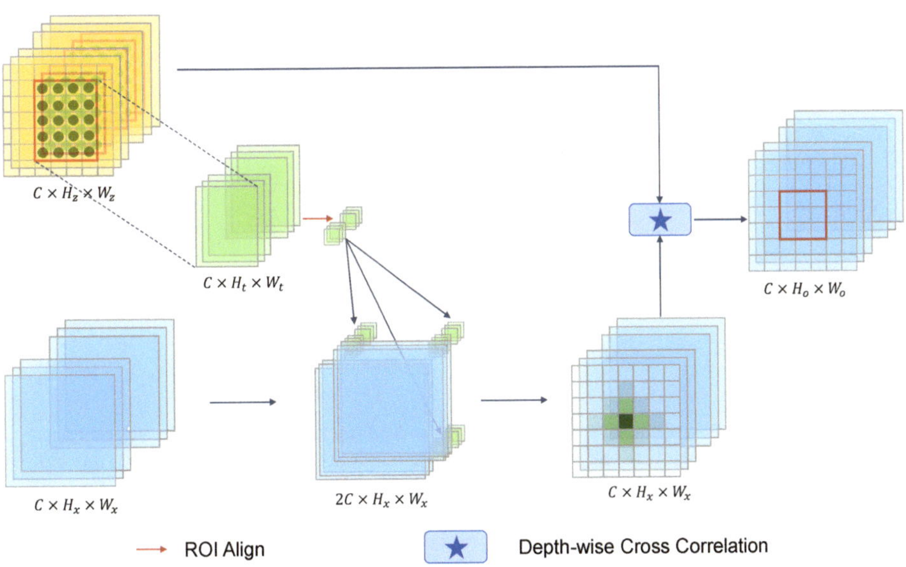

Figure 4. Architecture of the Multi-level Matching Module.

In order to achieve a more precise localization and tracking of tiny-sized objects in remote sensing videos, the proposed MM Module employs a coarse-to-fine feature matching fashion. Initially, the template feature map, denoted as X_z, is utilized to extract the segment containing solely the features of the target, followed by the application of the ROI Align operation. The resultant feature map X_t^p is then embedded into the feature map of the search region, denoted as X_s, yielding an initial response map. Subsequently, a Depth-wise Cross Correlation operation is conducted between the template feature map X_z and the preliminary response map X_s^*, yielding the ultimate response map. Within this multi-level matching mechanism, each position within the preliminary response map X_s^* assimilates information from the template features, thus rendering higher response values for pixels corresponding to the object's location within the response map. Moreover, the Depth-wise Cross Correlation operation imparts further constraints upon the object's contours. The MM Module aims to enhance the performance of the tracker by employing a two-step approach: initially coarsely localizing the target's position and subsequently refining the determination of the target's bounding box. This methodology leads to the acquisition of a more refined response map, consequently improving the tracking performance. The specific steps of the MM Module are delineated as follows.

Firstly, the feature map of the template region and the feature map of the search region, both of which have undergone feature extraction and enhancement, are, respectively, represented as $X_z \in \mathbb{R}^{C \times H_z \times W_z}$ and $X_s \in \mathbb{R}^{C \times H_x \times W_x}$. Then, according to the label information, extract the feature map that only contains the information of the target itself from the template area feature map, represented by $X_t \in \mathbb{R}^{C \times H_t \times W_t}$, ignoring the interfering background information in the template area feature map. Next, we use ROI Align to transform feature map $X_t \in \mathbb{R}^{C \times H_t \times W_t}$ containing only target information into feature map $X_t^p \in \mathbb{R}^{C \times 1 \times 1}$. Feature map X_t^p will be embedded into the corresponding position of each pixel in the search area feature map, resulting in a 2C feature map $X_s^* \in \mathbb{R}^{2C \times H_x \times W_x}$. This ensures that every position in the feature map contains target features for later processing. To efficiently control the channel dimension of the feature map and avoid complex matrix operations, we use a 1×1 convolution to reduce the dimensionality of the feature map and obtain a response map with only C channels. After these operations, the network obtains a preliminary matching result, which is more focused on the localization of the target center rather than confirming the outline of the target rectangle.

Subsequently, a Depth-wise Cross Correlation operation is performed between the template area feature map and the preliminarily matched response map, which emphasizes more on locating the target bounding box, and the final response map obtained is fed into the target prediction subnetwork. Through this multi-level matching method, more precise positioning of the center point location and target rectangle box prediction can be achieved. The entire formula expression for the MM module is shown as follows.

$$F_{response} = X_z \star Concat(ROIAlign_1(X_t), X_s), \tag{5}$$

where $F_{response}$ denotes the final output response map, $\star$ denotes the Depth-wise Cross Correlation, $Concat$ denotes the Concatenation operation described above to embed X_t^p into X_s, and $ROIAlign_1$ denotes the process of converting the feature map X_t to the feature map X_t^p of size 1×1.

4. Evaluation

To test the performance of the proposed SiamTM tracker, a series of comparative and ablation experiments were carried out. The following section consists of three parts around the experimental content. Section 4.1 introduces the details of the experimental setup, including the dataset used in the experiment and evaluation methods. Section 4.2 analyzes in detail the role of each module in the proposed method through ablation experiments. In addition, in Section 4.3, we compare the proposed method with 12 other state-of-the-art object tracking algorithms to demonstrate the superiority of our algorithm.

4.1. Experimental Setup

4.1.1. Introduction to Experimental Datasets

In this paper, we conducted experiments using two large-scale satellite video single-object tracking datasets named SatSOT [33] and SV248S [55]. We performed an ablation study on the SatSOT dataset to analyze the effectiveness of each module. Furthermore, we compared the proposed SiamTM tracker with 12 other single-object tracking methods on both SatSOT and SV248S tracking datasets to validate the effectiveness and superiority of the proposed method.

SatSOT is a dedicated dataset focused on satellite video single-object tracking. The dataset consists of 105 satellite video sequences comprising a total of 27,664 frames and covering four typical moving targets in satellite videos, namely vehicles, trains, airplanes, and ships. The vehicle targets appear in 65 video sequences, train targets appear in 26 video sequences, while airplane and ship targets appear in 9 and 5 video sequences, respectively. The average length of the videos in the SatSOT dataset is 263 frames, with over 70% of the bounding box sizes in the sequence being less than 1000 pixels. The average number of pixels occupied by cars is the least, at only 112.4 pixels, while trains occupy the most, a whopping 39,566.3 pixels. To indicate the characteristics and challenges faced by each satellite video sequence and help better analyze the strengths and weaknesses of trackers, the dataset lists 11 challenge attributes in the satellite videos, and the corresponding abbreviations and definitions for each attribute are shown in Table 1.

Table 1. The 11 challenge attributes in the SatSOT dataset with their abbreviations and definitions.

Attribute	Abbreviation	Definition
Background Clutter	BC	The background has similar appearance as the target.
Illumination Variation	IV	The illumination of the target region changes significantly.
Low Quality	LQ	The image is low quality and the target is difficult to be distinguished.
Rotation	ROT	The target rotates in the video.
Partial Occlusion	POC	The target is partially occluded in the video.
Full Occlusion	FOC	The Target is fully occluded in the video.
Tiny Object	TO	At least one ground truth bounding box has less than 25 pixels.
Similar Object	SOB	There are objects of similar shape or same type around the target.
Background Jitter	BJT	Background jitter brings by the shaking of satellite camera.
Aspect Ratio Change	ARC	The ratio of the bounding-box aspect ratio of the first and the current frame is outside the range [0.5, 2].
Deformation	DEF	Non-rigid object deformation.

SV248S is a satellite video single-object tracking dataset composed of 248 video sequences. The dataset focuses on tiny objects in satellite videos. It includes four categories of objects: ships, vehicles, large vehicles, and airplanes, but no trains, which is slightly different from the SatSOT dataset. In SV248S, 202 video sequences track vehicles, 37 video sequences track large vehicles, 6 video sequences track airplanes, and the remaining 3 video sequences track ships. Compared to the SatSOT dataset, the video sequences in SV248S have a longer average length, with all sequences ranging from 500 frames to 753 frames. Additionally, the targets in the SV248S dataset are smaller, with an average of less than 65.6 pixels occupying the frames in over 81.45% of all video sequences. The largest target type airplane only occupies an average of approximately 2284.8 pixels. With sufficient video sequence numbers and a focus on tiny targets in satellite videos, the SV248S dataset is suitable for the comprehensive evaluation of a tracker's performance on satellite video targets, especially tiny and weak targets.

4.1.2. Evaluation Criteria

To quantitatively analyze the results of satellite video single-object tracking, we have adopted two standard evaluation methods: success score and precision score. In addition, we have introduced the FPS indicator to evaluate the inference speed of each tracker.

The success score represents the Intersection over Union (IoU) between the predicted bounding boxes and the ground truth bounding boxes, and is represented by the following formula.

$$S = \frac{|B_p \cap B_{gt}|}{|B_p \cup B_{gt}|}, \tag{6}$$

where B_p and B_{gt} represent the predicted bounding box and the ground truth bounding box, respectively, while $\cap$ and $\cup$ represent intersection and union. $|\cdot|$ represents the total number of pixels in the corresponding area. For a specific frame in the tracking task, if the IoU between the predicted box and the actual bounding box in the current frame exceeds a certain threshold $t_s \in (0, 1]$, it is considered that the target is successfully tracked in the current frame. Otherwise, the tracker is considered to have failed in tracking the target in the current frame.

The precision score represents the center location error (CLE) between the center point of the predicted bounding box and the center point of the ground truth bounding box, expressed as follows:

$$CLE = \sqrt{(x_1 - x_2)^2 + (y_1 - y_2)^2}, \tag{7}$$

where (x_1, y_1) represents the center coordinate of the predicted bounding box, and (x_2, y_2) represents the center coordinate of the ground truth bounding box. For a specific frame, if the CLE between the predicted bounding box and the ground truth bounding box is less than a certain threshold $t_{CLE} \in [0, +\infty)$, it is considered that the tracker has successfully tracked the object in the current frame. Otherwise, it is considered that the tracker has failed to follow the target in the current frame accurately. In natural image single-object tracking datasets such as OTB [31] and GOT10k [56], the threshold is usually set to $t_{CLE} = 20$ pixels. However, typical targets in remote sensing images, such as vehicles, ships, and airplanes, are smaller in size. In order to accurately measure the true tracking performance of trackers, we set the threshold to $t_{CLE} = 5$ pixels in this paper.

Frames per second (FPS) measures how many frames the tracker can complete its tracking task per second. The higher the number, the more capable the tracker is of processing tasks per second, and the better the real-time performance of the tracker. We use this metric to reflect and compare the inference speeds of different trackers.

4.1.3. Implementation Details

We implemented the proposed tracking network SiamTM on the open-source deep learning library PyTorch 2.0.0, and implemented it on a 64-bit Ubuntu 20.04 workstation with 24 GB memory GeForce RTX4090 GPU. We used a modified ResNet-50 as the same in [9] as the backbone for the Siamese network feature extraction subnetwork, which was pre-trained on the ImageNet [57] dataset. The size of the initial frame template was set to 127 × 127 pixels, and the search region was set to 255 × 255 pixels. During the training phase, the batch size was set to 48, and the network was trained end-to-end for 20 epochs using the SGD method. The starting learning rate, momentum, and weight decay for SGD were set to 0.005, 0.9, and 0.0001, respectively. Consistent with SiamCAR [12], during the first 10 epochs, the parameters of the feature extraction subnetwork of the Siamese network were completely frozen, and only the following part was trained. In the subsequent 10 epochs, the last three blocks of the ResNet-50 were unfrozen for joint training. We trained our SiamTM network using data from VISO [58].

4.2. Ablation Study

The proposed SiamTM method consists of two novel components: (1) the TIE module, and (2) the MM module. To validate the effectiveness of each module in the SiamTM method and clarify the roles each module plays in improving tracking performance, we conducted a series of ablation experiments on the SatSOT dataset.

The performance of four trackers with different combinations of proposed modules is presented in Table 2. In this context, "T" abbreviates the TIE module, while "M" stands for the MM module.

Table 2. Ablation experiments of the proposed tracker on the SatSOT dataset.

Trackers	TIE Module	MM Module	Prec. (%)	Succ. (%)	Speed (FPS)
Baseline	-	-	56.4	44.6	158.41
SiamCAR+T	✓	-	60.2	46.6	113.65
SiamCAR+M	-	✓	59.5	46.7	109.17
SiamCAR+T+M	✓	✓	60.8	47.5	89.76

The performance comparison of SiamCAR, SiamCAR+T, SiamCAR+M, and the proposed SiamTM tracker can be readily observed in Table 2. Notably, the Precision and Success scores of the SiamTM tracker surpass those of SiamCAR, SiamCAR+T, and SiamCAR+M in a pronounced manner. In the case of the SiamCAR+T tracker, the utilization of the TIE module enables the tracker to effectively filter out pertinent feature information on top of the original feature extraction. Consequently, compared to the Baseline tracking network, there has been an improvement of 3.8 percentage points in precision score and 2.0 percentage points in success score. Furthermore, for the SiamCAR+M tracker, a more suitable MM module specifically designed for satellite video target tracking has been employed. The MM module allows for improved matching between the template region feature map and the search region feature map, resulting in a more accurate response map. In terms of the obtained results, compared to the baseline SiamCAR tracking network, the tracker SiamCAR+M exhibits an increase of 3.1 percentage points in precision score and 2.1 percentage points in success score. The SiamTM tracker proposed in this study introduces a TIE module to enhance the features of both the template region and the search region from both a 1D directional aspect and a 2D global aspect. Additionally, a MM module specifically tailored for satellite video targets is utilized. Effectively combining the proposed two modules can enhance the tracker's tracking performance in terms of feature extraction and feature matching dimensions. Ultimately, this approach achieves competitive results on the SatSOT dataset, with a precision score of 60.8 and a success score of 47.5. However, there is no such thing as a free lunch, and as a consequence, the inference speed of the tracker decreased by 68.64 FPS compared to the baseline, reaching 89.76 FPS. Considering that this speed is still significantly higher than the standard threshold of real-time tracking at 24 FPS, it is acceptable to trade off some inference speed in exchange for improved tracking performance.

In order to qualitatively analyze the results before and after the addition of the modules, the classification (Cls) maps of the baseline SiamCAR tracker and the proposed SiamTM tracker are visualized, and the visualization results are shown in Figure 5.

Comparing the last two columns of Figure 5, it is easy to see that the proposed SiamTM tracker is more capable of suppressing similar distractors in the search area map when distinguishing between foreground and background than the baseline SiamCAR tracker, while SiamTM is more focusing on the information of the object itself. In summary, thanks to the TIE module and MM module, the SiamTM tracker shows excellent robustness and tracking performance in the face of Tiny Object and Similar Object challenge attributes.

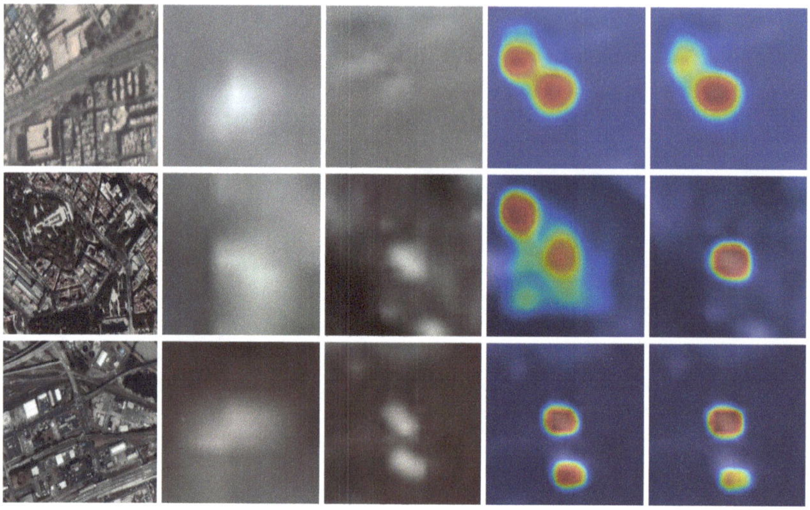

Figure 5. Comparison of classification (Cls) map between baseline and our proposed method. The first column is the original image, the second and third columns are the template region image and the search region image, respectively. The fourth and fifth columns are the Cls maps corresponding to the baseline method and the proposed method, respectively.

4.2.1. Target Information Enhancement Module

The TIE module integrates 1D information along the spatial direction dimension as well as 2D information across the image global dimension. To validate the effectiveness of fusing the 1D and 2D information, we conducted an ablation study on SatSOT dataset. We compared four different scenarios: solely employing the baseline model SiamCAR, employing solely 1D information on top of the baseline, employing solely 2D information on top of the baseline, and employing the proposed TIE module that integrates 1D and 2D information on top of the baseline. We evaluated the precision score and success score, and the experimental results are presented in Table 3.

Table 3. Performance comparison of TIE module with different structures.

Baseline	1D	1D†	2D	Prec. (%)	Succ. (%)	Speed (FPS)
✓	-	-	-	56.4	44.6	158.41
✓	✓	-	-	58.1	45.5	131.48
✓	-	✓	-	57.7	45.0	142.94
✓	-	-	✓	57.4	44.4	151.93
✓	✓	-	✓	60.2	46.6	113.65

In Table 3, "✓" indicates that the module is used in the tracker, while "-" indicates that the module is not used. From Table 3, it can be observed that the baseline SiamCAR exhibits the highest inference speed, reaching 158.41 FPS. However, it achieves the lowest levels of precision score (56.4%) and success rate score (44.6%) among the four comparisons. After incorporating only 1D or 2D information on top of the baseline, the precision score of the tracking network has shown improvements compared to the baseline. Specifically, the inclusion of 1D information improves the tracker's precision score by 1.7 percentage points compared to the baseline, while the inclusion of 2D information improves the tracker's precision score by 1.0 percentage points compared to the baseline. As for the Success metric, after adding 1D information, the tracker improves by 0.9 percentage points compared to the baseline, while the tracker with 2D information declines by 0.2 percentage points compared to the baseline, remaining almost equal to the baseline. When only adding

2D information, the network emphasizes discriminative features and the center of the target, so the tracker rises in the Precision metric. However, the Success metric places emphasis on the degree of overlap between the predicted bounding box and ground truth bounding box, which is more dependent on the shape and size of the box, and the 2D information contributes less to this. Hence, the tracker remains essentially unchanged in the Success metrics. To provide a more comprehensive demonstration of the effectiveness of our proposed module in exploiting 1D information, we conducted a comparison between two methodologies: one involves encoding spatial information along the horizontal and vertical directions using an intermediate feature map X_{inter}, followed by attention map generation (represented as "1D" in Table 3), while the other directly perform two groups of 1×1 convolution operations on $X_{horizonal}$ and $X_{vertical}$ to generate the attention weight (represented as "$1D^{\dagger}$" in Table 3). As can be seen from Table 3, compared to the baseline, directly performing two groups of 1×1 convolution operations on $X_{horizonal}$ and $X_{vertical}$ to generate the attention weight is still effective in improving the overall tracking performance of the tracker. The use of the "$1D^{\dagger}$" attention generation method resulted in a 1.3 percentage point improvement in precision score and a 0.4 percentage point improvement in success score. However, better results can be achieved by first encoding the spatial information along both the horizontal and vertical directions through an intermediate feature map X_{inter}. The use of "1D" attention generation method resulted in a 1.7 percentage point improvement in precision score and a 0.9 percentage point improvement in success score. This demonstrates that the approach of first encoding spatial information together and then generating attention maps is superior in terms of effectiveness to the approach of directly generating attention maps. However, this enhancement comes at the expense of varying degrees of reduction in inference speed. When applying the TIE module on the baseline, this module concatenates and reduces the obtained 1D and 2D information along the channel dimension, resulting in a feature map enriched with enhanced features. With the integration of the target information enhancement module, the tracking network exhibits a significant improvement of 3.8 percentage points in precision metric and 2.0 percentage points in success metric, demonstrating the effectiveness of the proposed target information enhancement module.

4.2.2. Multi-Level Matching Module

As previously mentioned, the MM Module is composed of a cascaded Concatenation matching operation and a Depth-wise Cross Correlation operation. To compare the performance difference between using the MM Module and a single matching operation, we conducted ablation experiments on the matching operation component of the network. The experimental results are presented in Table 4.

Table 4. Performance comparison of different matching methods.

Concatenation	Depth-Wise Cross Correlation	Multi-Level Matching Module	Prec. (%)	Succ. (%)	Speed (FPS)
✓	-	-	57.9	42.0	111.13
-	✓	-	56.4	44.6	158.41
-	-	✓	59.5	46.7	109.17

In Table 4, "✓" indicates the matching method used in the tracker. As indicated in Table 4, employing a MM module results in a precision score of 59.5% and success score of 46.7% for the tracking network, which stand as the highest values among the compared methods. Both the Concatenation matching operation and the Depth-wise Cross Correlation matching operation possess distinct advantages. The accuracy score attained by the Concatenation matching operation amounts to 57.9%, demonstrating a superiority of 1.5% over the Depth-wise Cross Correlation approach in this evaluation metric. This advantage primarily arises from the fact that the Concatenation matching operation embeds the information solely contained within the target itself into each position of the search feature map. As a result, the tracking network exhibits an enhanced ability to determine the target's location. From another perspective, Depth-wise Cross Correlation employs the template feature map as a convolution kernel, conducting correlation operation on the search feature map. This process effectively constrains the target range by utilizing information such as the target's contour and texture, enabling the tracking network to predict the target's bounding box with higher success, which is not achievable by concatenation matching operation. As a result, the Depth-wise Cross Correlation operation outperforms the Concatenation matching operation by a margin of 2.6 percentage points in terms of success score. The MM module combines the Concatenation matching operation with the Depth-wise Cross Correlation operation. Firstly, it embeds the target's own information into each position of the search feature map using concatenation, resulting in an intermediate response map. Subsequently, the Depth-wise Cross Correlation operation is performed on the template feature map and the search feature map to better delineate the target's range, yielding the final feature map. The results of ablation experiments validate the effectiveness of the proposed MM Module.

4.3. Comparison with State-of-the-Art

4.3.1. Evaluated Trackers

To validate the competitiveness of our algorithm SiamTM, we compared the proposed SiamTM algorithm with 12 state-of-the-art target tracking algorithms, namely SiamFC [6], SiamRPN [7], ATOM [59], SiamRPN++ [9], SiamRPN++_lt [9], SiamMask [60], DiMP18 [61], DiMP50 [61], SiamCAR [12], PrDiMP18 [62], ToMP50 [19], and ToMP101 [19].

4.3.2. Overall Performance on SatSOT and SV248S

Tables 5 and 6 present the quantitative results of precision score, success score, and speed for the proposed SiamTM tracker and the selected trackers on the SatSOT and SV248S datasets, respectively. The precision score and success score of the SiamTM tracker outperform those of the selected comparative trackers on both two satellite video single-object tracking datasets. These findings demonstrate the effectiveness of the proposed SiamTM tracker in satellite video object tracking. In terms of speed, the top three performers on both datasets are SiamRPN, SiamFC, and ATOM. Due to the presence of the TIE module and the MM module, the speed of SiamTM on SatSOT and SV248S is 89.76 FPS and 72.50 FPS, respectively, which shows a slight decrease compared to the Baseline SiamCAR. However, it still surpasses the real-time tracking benchmark of 24 FPS. This trade-off, sacrificing some speed in exchange for higher tracking performance, is considered acceptable.

Taking the SV248S dataset as an example, we examine the performance of various trackers on this dataset. In terms of precision metric, the SiamTM tracker achieves the highest score of 75.3%, surpassing the second-ranked SiamCAR by an improvement of 5.2 percentage points and the third-ranked SiamRPN++ by an improvement of 9.7 percentage points. This remarkable performance is primarily attributed to the superior matching paradigm of the MM module. After enhancing the extracted feature maps with the TIE module, the MM module first obtains a rough intermediate response map through Concatenation. This response map emphasizes the proximity between the centers of objects rather than contour matching. Subsequently, the final response map is obtained based on Depth-wise Cross Correlation on this intermediate response map. By adopting this approach of initial center-point coarse matching followed by contour fine matching, SiamTM achieves reduced centroid error between the predicted target bounding boxes and the ground truth, as well as a higher overlap rate between contours. Furthermore, it is noteworthy that the ToMP series, which is based on Transformer for tracking, has achieved state-of-the-art performance in natural scene tracking. However, its results in satellite video object tracking are rather mediocre. Unlike objects in natural scenes, targets in satellite videos exhibit lower inter-class separability [63]. Therefore, although ToMP has been trained extensively on natural scene images, directly applying it to RSOT is inappropriate, as experiment results have also confirmed.

Table 5. Overall performance on SatSOT. The best three performances are respectively highlighted with red, green, and blue colors. (For features, ConvFeat/CF:Convolutional Feature, TF:Transformer).

Trackers	Features	Backbone	Prec. (%)	Succ. (%)	Speed (FPS)
SiamFC	ConvFeat	AlexNet	49.8	41.3	298.3
SiamRPN	ConvFeat	AlexNet	49.4	38.5	436.6
ATOM	ConvFeat	ResNet-18	52.9	42.4	249.2
SiamRPN++	ConvFeat	ResNet-50	55.4	41.5	172.0
SiamRPN++_lt	ConvFeat	ResNet-50	52.5	38.4	148.2
SiamMask	ConvFeat	ResNet-50	55.6	40.2	173.5
DiMP18	ConvFeat	ResNet-18	52.8	42.6	115.9
DiMP50	ConvFeat	ResNet-50	51.3	42.0	92.7
SiamCAR	ConvFeat	ResNet-50	56.4	44.6	158.41
PrDiMP18	ConvFeat	ResNet-18	46.0	39.7	69.5
ToMP50	CF+TF	ResNet-50	49.2	38.8	58.1
ToMP101	CF+TF	ResNet-101	46.7	36.9	52.4
SiamTM(Ours)	ConvFeat	ResNet-50	60.8	47.5	89.76

Table 6. Overall performance on SV248S. The best three performances are respectively highlighted with red, green, and blue colors. (For features, ConvFeat/CF:Convolutional Feature, TF:Transformer).

Trackers	Features	Backbone	Prec. (%)	Succ. (%)	Speed (FPS)
SiamFC	ConvFeat	AlexNet	63.4	39.4	402.0
SiamRPN	ConvFeat	AlexNet	34.4	14.7	625.6
ATOM	ConvFeat	ResNet-18	62.8	36.4	262.0
SiamRPN++	ConvFeat	ResNet-50	65.6	40.5	180.7
SiamRPN++_lt	ConvFeat	ResNet-50	56.8	21.5	167.6
SiamMask	ConvFeat	ResNet-50	55.9	21.9	179.6
DiMP18	ConvFeat	ResNet-18	58.3	35.3	111.5
DiMP50	ConvFeat	ResNet-50	61.4	36.7	82.6
SiamCAR	ConvFeat	ResNet-50	70.1	44.8	176.16
PrDiMP18	ConvFeat	ResNet-18	57.6	36.4	68.2
ToMP50	CF+TF	ResNet-50	38.7	16.5	65.4
ToMP101	CF+TF	ResNet-101	37.1	16.0	56.0
SiamTM(Ours)	ConvFeat	ResNet-50	75.3	48.7	72.50

4.3.3. Attribute-Based Evaluation

To analyze the tracking performance of the trackers on different challenge attributes, we plotted the precision score plots and success score plots of the trackers on each challenge attribute of the SatSOT dataset, respectively. The plotting results are shown in Figures 6 and 7.

In terms of precision score, the proposed SiamTM tracker outperforms the other 12 tracking algorithms in eight challenge attributes, including the overall attribute. Specifically, SiamTM achieves the highest performance in the overall attribute, Background Clutter, Low Quality, Rotation, Partial Occlusion, Tiny Object, Similar Object, and Deformation. Notably, SiamTM delivers exceptional performance in the Tiny Object challenge attribute. It achieves a precision score of 71.6%, which is 11.9 percentage points higher than the second-ranked SiamCAR tracker in this attribute. Figure 6 clearly demonstrates that, compared to other trackers, the SiamTM tracker excels in exploring and enhancing feature information within small targets, resulting in superior tracking of such objects. This improvement is particularly significant for satellite video tracking datasets predominantly composed of small-sized objects.

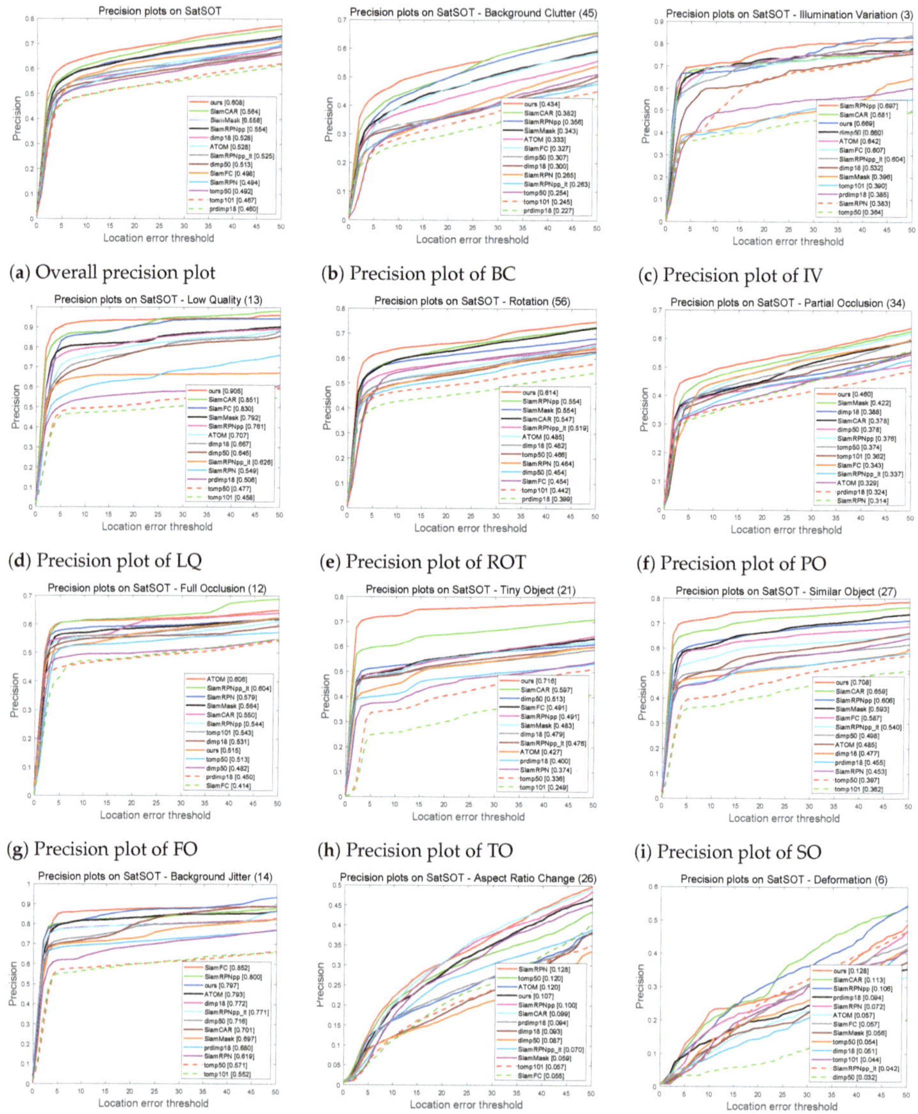

(**a**) Overall precision plot (**b**) Precision plot of BC (**c**) Precision plot of IV

(**d**) Precision plot of LQ (**e**) Precision plot of ROT (**f**) Precision plot of PO

(**g**) Precision plot of FO (**h**) Precision plot of TO (**i**) Precision plot of SO

(**j**) Precision plot of BJ (**k**) Precision plot of ARC (**l**) Precision plot of DEF

Figure 6. Precision plots of 13 trackers under overall attributes and different challenge attributes.

Regarding success score, compared to the other 12 tracking algorithms, the proposed SiamTM tracker achieves the highest success scores in 9 out of the 12 challenge attributes. For the remaining three challenge attributes, namely Illumination Variation, Full Occlusion, and Scale Variation, the trackers with the highest scores are SiamFC, ATOM, and ToMP50, respectively. Similar to the precision score metric, SiamTM outperforms the second-ranked tracker by 5.6 percentage points in the Tiny Object (TO) attribute. Hence, the success score plots across various challenge attributes demonstrate that the SiamTM tracker can deliver a robust tracking performance in scenes with different challenges.

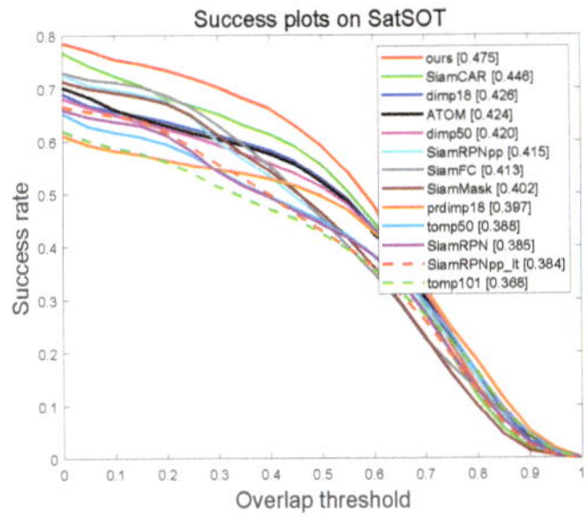

(**a**) Overall success plot

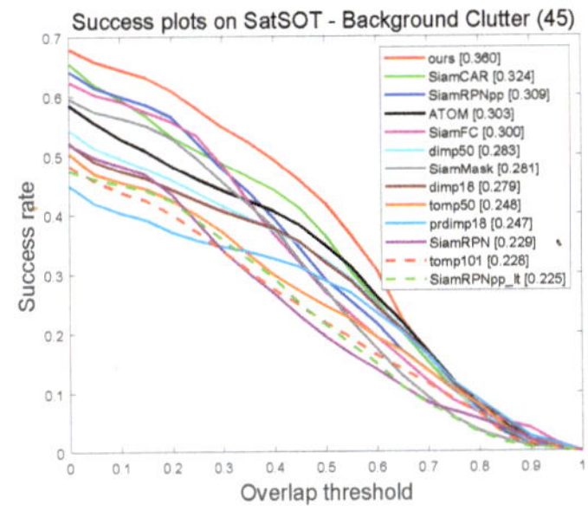

(**b**) Success plot of BC

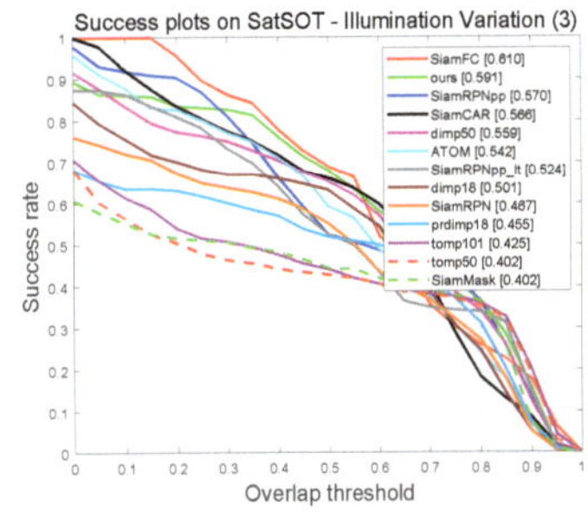

(**c**) Success plot of IV

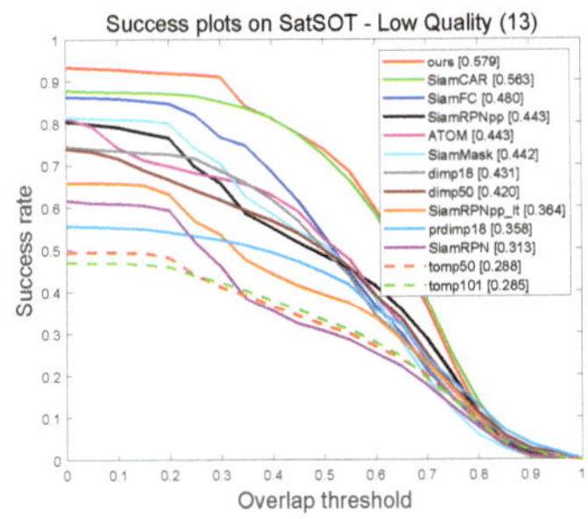

(**d**) Success plot of LQ

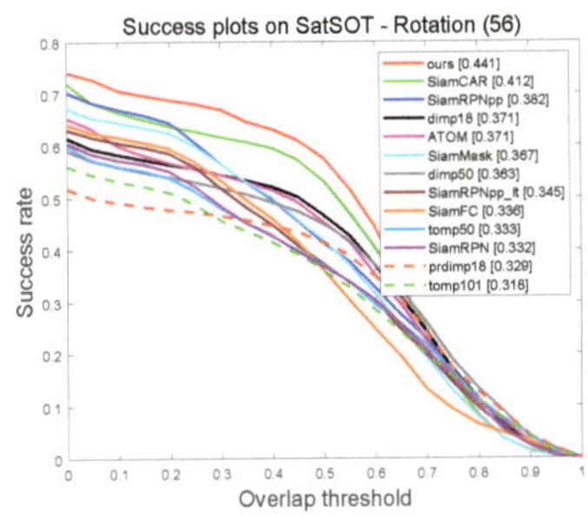

(**e**) Success plot of ROT

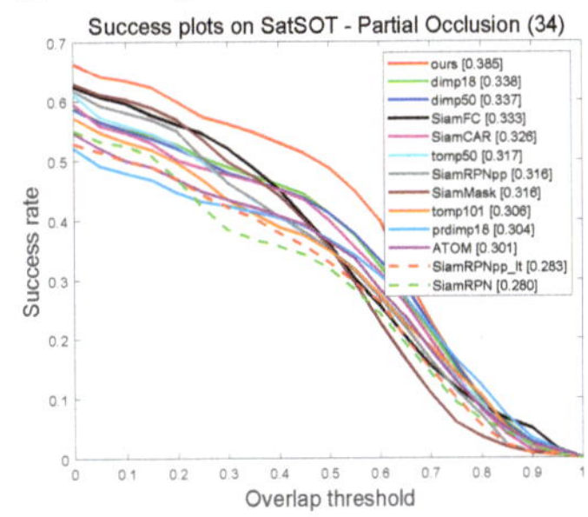

(**f**) Success plot of PO

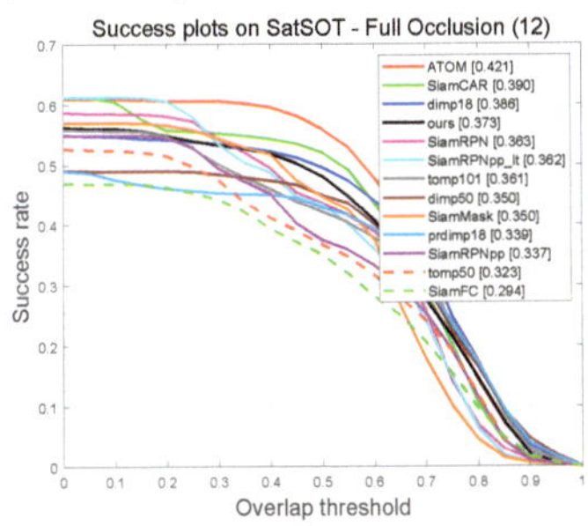

(**g**) Success plot of FO

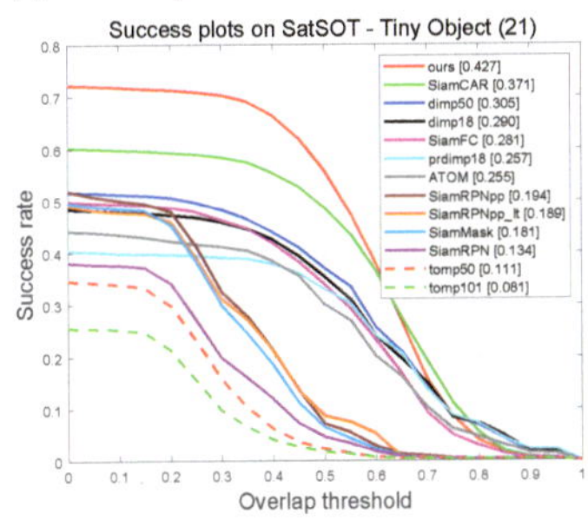

(**h**) Success plot of TO

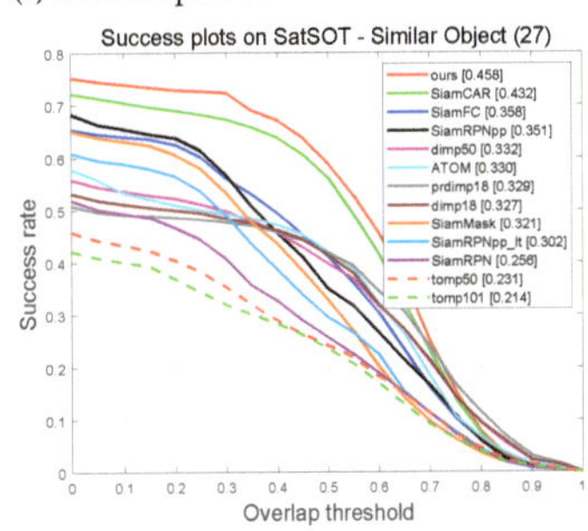

(**i**) Success plot of SO

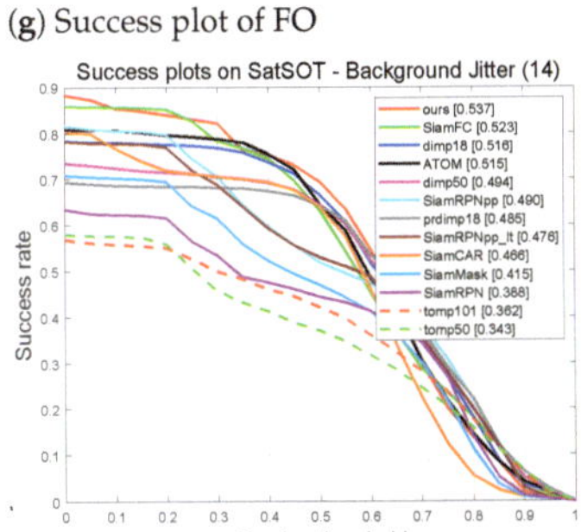

(**j**) Success plot of BJ

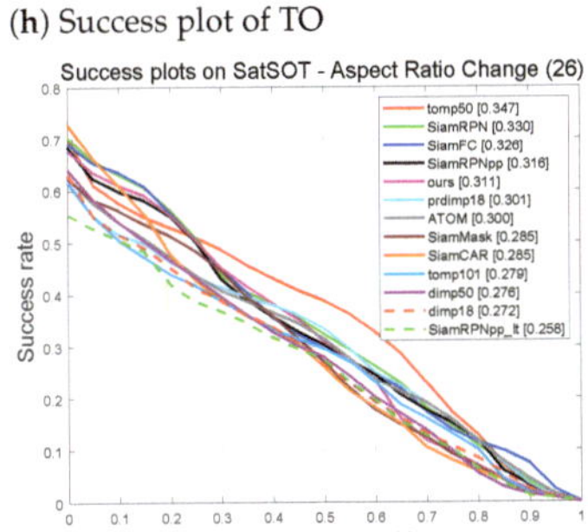

(**k**) Success plot of ARC

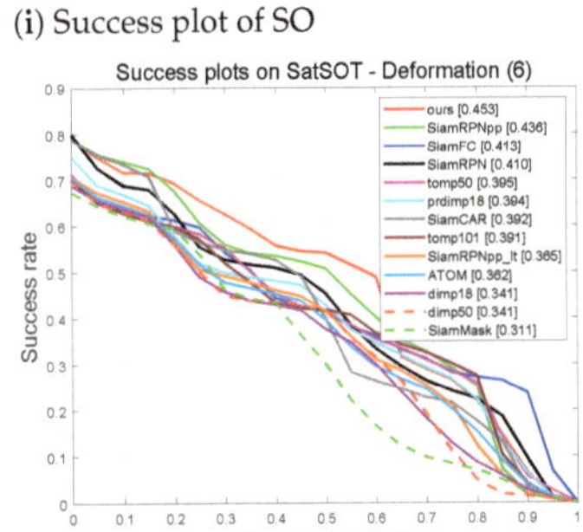

(**l**) Success plot of DEF

Figure 7. Success plots of 13 trackers under overall attributes and different challenge attributes.

4.3.4. Visual Analysis

To qualitatively analyze the tracking performance of the proposed SiamTM tracker, we visualized its tracking results alongside the tracking results of four other trackers that achieved the highest accuracy scores on the SatSOT dataset, as well as the ground truth. The visualization results are shown in Figure 8. In order to comprehensively analyze the tracking performance of the trackers under different challenging attributes, we selected four video sequences from SatSOT, each with distinct challenging attributes. From top to bottom in Figure 8, the sequences are Car_06, Car_07, Car_27, and Car_45.

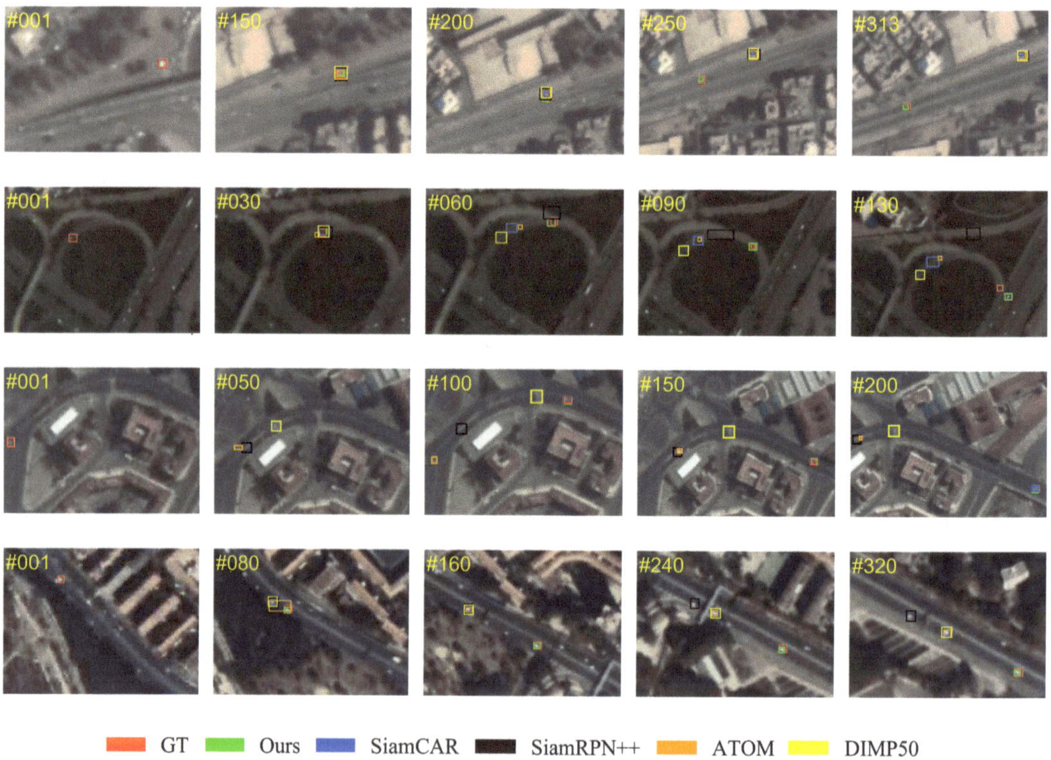

Figure 8. Qualitative comparison of five trackers and groundtruth in different videos with different challenging attributes.

The Car_06 sequence exhibits the attributes of low quality (LQ) and tiny object (TO), allowing us to visually compare the tracking performance of different trackers in terms of tracking low-quality and small objects. In the early frames of the Car_06 sequence, the target itself has relatively prominent features, enabling clear distinction between the target and its surrounding background. As a result, all selected trackers demonstrate good tracking performance in frames 1, 150, and 200, with SiamTM yielding tighter predicted bounding boxes. In the later frames of the Car_06 sequence, the discriminative features between the target and the background gradually diminish, limiting the information provided by the target itself. From frames 250 and 313, it can be observed that while other trackers experience tracking drift, the proposed SiamTM tracker maintains correct and consistent tracking. These results indicate that the SiamTM tracker exhibits robust tracking capability, allowing it to track small targets in low-quality satellite images.

The video sequences named Car_07 and Car_27 face similar types of challenges. Both sequences primarily encounter challenges such as background clutter (BC), rotation (ROT), target occlusion (TO), and low quality (LQ). Throughout the Car_07 video image sequence, the tracked target consistently resides in a complex environment with low lighting conditions and undergoes rotation, which significantly amplifies the difficulty of tracking small targets in such environments. When observing small targets in this scenario, it was noticed at the 30th frame that the ATOM tracker exhibited tracking drift during the early stages of the video image sequence. At the 60th and 90th frames of the video, all other trackers, except for the proposed SiamTM tracker, encountered tracking drift or failures. Similarly, when examining frames 50, 100, 150, and 200 of the Car_27 image sequence, it becomes evident that SiamRPN++, ATOM, and DIMP experienced tracking failures at various stages, while only SiamCAR and the proposed SiamTM achieved consistent and accurate target tracking. These findings indicate that the SiamTM tracker, relative to other trackers, better handles the challenges of tracking small targets amidst complex backgrounds and rotation.

In the video sequence Car_45, in addition to the common challenge attributes found in the previous three video sequences (TO, BC, ROT, LQ), it also exhibits unique attributes such as partial occlusion (POC), similar object (SOB), and background jitter (BJT). Upon observing frames 80 and 160 of Car_45, it can be observed that due to the presence of similar objects in proximity, both SiamCAR and SiamRPN experienced tracking drift, while DIMP50 also encountered difficulties. However, SiamTM maintained robustness and achieved accurate tracking of the vehicle target. Even in the presence of short-term occlusion (when the vehicle passes under a bridge), SiamTM is able to predict the target's position accurately once the occlusion ends. This demonstrates that the SiamTM tracker exhibits excellent tracking performance in complex scenarios involving multiple challenges, such as partial occlusion, similar object, and background jitter.

The above visualization results demonstrate that the proposed SiamTM tracker is capable of effectively handling various challenge attributes in satellite video tracking scenarios, particularly excelling in challenges such as tiny object (TO) and low quality (LQ). This provides strong evidence of the effectiveness of SiamTM in satellite video object tracking.

4.3.5. Failure Case Analysis

While our approach has yielded highly competitive results in single-object tracking on satellite videos, inevitably, our proposed SiamFM method has encountered tracking drift or failure in some video sequences. Taking the second video sequence in Figure 8 as an example, during the initial stages of tracking, our tracker successfully distinguishes the target of interest from the surrounding background interference, thanks to the assistance provided by the TIE module and MM module. This enables the tracker to maintain continuous tracking of the target. However, at frame 130 of the video sequence, the tracker experienced tracking drift, where the predicted results deviated towards a similar distractor located very close to the target's position. The reason for the tracking drift is as follows: due to various factors such as the satellite video's resolution and background jitter, the target's distinguishable features become scarce at frame 130 of the video sequence and are almost overwhelmed by the background information. Conversely, within the same search area, a distractor appears that is very similar to the target. This distractor possesses highly similar features to the target's template in the first frame. For the proposed tracker SiamFM, which relies on the appearance features of the target for foreground-background discrimination, it is unable to discern the distinction between the target and the interfering object in such scenarios. Consequently, it erroneously assigns a higher classification confidence to the interfering object. This, in turn, leads to the tracking drifting from the correct target to the surrounding similar distractor. In the future, integrating feature enhancement with solutions such as super-resolution or motion-aware approaches holds promise for addressing tracking failure cases of this nature.

5. Conclusions

In this paper we present SiamTM, a single-object tracking algorithm based on the target information enhancement (TIE) module and the multi-level matching (MM) module, especially for typical moving targets in satellite video. First, to address the problem of indistinct features caused by the tiny size of objects in satellite video, we propose a TIE module to extract more effective and discriminative features from the original feature maps corresponding to the images, laying a good foundation for subsequent matching and target prediction tasks. Furthermore, In order to solve the problem that the matching method of the common Siamese network algorithm cannot fully release the tracker performance and better match the template feature map and the search feature map, we propose a MM module that is more suitable for satellite video objects. We conduct comprehensive experiments on two dedicated satellite video single-object tracking datasets, namely SatSOT and SV248S. The results of ablation experiments show that the proposed two modules proposed in this article effectively improve the accuracy of single-object tracking in satellite videos. When compared with other 12 competitive methods, our proposed method SiamTM achieved state-of-the-art tracking results. The propoed SiamTM method selects and enhances discriminative features by means of attention, while acquiring response maps by moving from coarse to fine, which inevitably increases the computational complexity of the tracker. In the future, we will further improve the proposed method by reducing the computational complexity and maintaining the accuracy and robustness of the method for satellite video object tracking.

Author Contributions: Conceptualization, J.Y. and Z.P.; methodology, J.Y., Z.P and Y.L.; resources, Z.P. and Y.L.; writing—original draft preparation, J.Y. and B.N.; and supervision, Z.P. and B.L. All authors have read and agreed to the published version of the manuscript.

Funding: This work was supported by the Youth Innovation Promotion Association, CAS under number 2022119.

Data Availability Statement: The SatSOT dataset is available at http://www.csu.cas.cn/gb/jggk/kybm/sjlyzx/gcxx_sjj/sjj_wxxl/ (accessed on 17 July 2023). The SV248S dataset is available at https://github.com/xdai-dlgvv/SV248S (accessed on 17 July 2023).

Conflicts of Interest: The authors declare no conflicts of interest.

References

1. Chan, S.; Tao, J.; Zhou, X.; Bai, C.; Zhang, X. Siamese implicit region proposal network with compound attention for visual tracking. *IEEE Trans. Image Process.* **2022**, *31*, 1882–1894. [CrossRef] [PubMed]
2. Shao, J.; Du, B.; Wu, C.; Zhang, L. Tracking objects from satellite videos: A velocity feature based correlation filter. *IEEE Trans. Geosci. Remote Sens.* **2019**, *57*, 7860–7871. [CrossRef]
3. Lee, K.H.; Hwang, J.N. On-road pedestrian tracking across multiple driving recorders. *IEEE Trans. Multimed.* **2015**, *17*, 1429–1438. [CrossRef]
4. Dong, X.; Shen, J.; Wu, D.; Guo, K.; Jin, X.; Porikli, F. Quadruplet network with one-shot learning for fast visual object tracking. *IEEE Trans. Image Process.* **2019**, *28*, 3516–3527. [CrossRef] [PubMed]
5. Tao, R.; Gavves, E.; Smeulders, A.W. Siamese instance search for tracking. In Proceedings of the IEEE Conference on Computer Vision and Pattern Recognition, Las Vegas, NV, USA, 27–30 June 2016; pp. 1420–1429. [CrossRef]
6. Bertinetto, L.; Valmadre, J.; Henriques, J.F.; Vedaldi, A.; Torr, P.H. Fully-convolutional siamese networks for object tracking. In Proceedings of the Computer Vision–ECCV 2016 Workshops, Amsterdam, The Netherlands, 8–10 and 15–16 October 2016; Proceedings, Part II 14; Springer: Berlin/Heidelberg, Germany, 2016; pp. 850–865. [CrossRef]
7. Li, B.; Yan, J.; Wu, W.; Zhu, Z.; Hu, X. High performance visual tracking with siamese region proposal network. In Proceedings of the IEEE Conference On Computer Vision and Pattern Recognition, Salt Lake City, UT, USA, 18–23 June 2018; pp. 8971–8980. [CrossRef]
8. Ren, S.; He, K.; Girshick, R.; Sun, J. Faster R-CNN: Towards Real-Time Object Detection with Region Proposal Networks. *IEEE Trans. Pattern Anal. Mach. Intell.* **2017**, *39*, 1137–1149. [CrossRef]
9. Li, B.; Wu, W.; Wang, Q.; Zhang, F.; Xing, J.; Yan, J. Siamrpn++: Evolution of siamese visual tracking with very deep networks. In Proceedings of the IEEE/CVF Conference On Computer Vision and Pattern Recognition, Long Beach, CA, USA, 16–20 June 2019; pp. 4282–4291. [CrossRef]
10. He, K.; Zhang, X.; Ren, S.; Sun, J. Deep residual learning for image recognition. In Proceedings of the IEEE Conference On Computer Vision and Pattern Recognition, Las Vegas, NV, USA, 26 June–1 July 2016; pp. 770–778. [CrossRef]

11. Xu, Y.; Wang, Z.; Li, Z.; Yuan, Y.; Yu, G. Siamfc++: Towards robust and accurate visual tracking with target estimation guidelines. In Proceedings of the AAAI Conference On Artificial Intelligence, Online, 7–12 February 2020; Volume 34, pp. 12549–12556. [CrossRef]
12. Guo, D.; Wang, J.; Cui, Y.; Wang, Z.; Chen, S. SiamCAR: Siamese fully convolutional classification and regression for visual tracking. In Proceedings of the IEEE/CVF Conference On Computer Vision and Pattern Recognition, Seattle, WA, USA, 14–19 June 2020; pp. 6269–6277. [CrossRef]
13. Guo, D.; Shao, Y.; Cui, Y.; Wang, Z.; Zhang, L.; Shen, C. Graph attention tracking. In Proceedings of the IEEE/CVF Conference On Computer Vision and Pattern Recognition, Nashville, TN, USA, 20–25 June 2021; pp. 9543–9552. [CrossRef]
14. Yang, J.; Pan, Z.; Wang, Z.; Lei, B.; Hu, Y. SiamMDM: An Adaptive Fusion Network with Dynamic Template for Real-time Satellite Video Single Object Tracking. *IEEE Trans. Geosci. Remote Sens.* **2023**, *61*, 1–19. [CrossRef]
15. Vaswani, A.; Shazeer, N.; Parmar, N.; Uszkoreit, J.; Jones, L.; Gomez, A.N.; Kaiser, Ł.; Polosukhin, I. Attention is all you need. *Adv. Neural Inf. Process. Syst.* **2017**, *30*, 5998–6008. [CrossRef]
16. Yan, B.; Peng, H.; Fu, J.; Wang, D.; Lu, H. Learning spatio-temporal transformer for visual tracking. In Proceedings of the IEEE/CVF International Conference on Computer Vision, Nashville, TN, USA, 19–25 June 2021; pp. 10448–10457. [CrossRef]
17. Chen, X.; Yan, B.; Zhu, J.; Wang, D.; Yang, X.; Lu, H. Transformer tracking. In Proceedings of the IEEE/CVF Conference on Computer Vision and Pattern Recognition, Nashville, TN, USA, 19–25 June 2021; pp. 8126–8135. [CrossRef]
18. Fu, Z.; Fu, Z.; Liu, Q.; Cai, W.; Wang, Y. SparseTT: Visual Tracking with Sparse Transformers. In Proceedings of the Thirty-First International Joint Conference on Artificial Intelligence—IJCAI-22, Vienna, Austria, 23–29 July 2022; pp. 905–912. [CrossRef]
19. Mayer, C.; Danelljan, M.; Bhat, G.; Paul, M.; Paudel, D.P.; Yu, F.; Van Gool, L. Transforming model prediction for tracking. In Proceedings of the IEEE/CVF Conference on Computer Vision and Pattern Recognition, New Orleans, LA, USA, 18–24 June 2022; pp. 8731–8740. [CrossRef]
20. Cui, Y.; Jiang, C.; Wang, L.; Wu, G. Mixformer: End-to-end tracking with iterative mixed attention. In Proceedings of the IEEE/CVF Conference on Computer Vision and Pattern Recognition, New Orleans, LA, USA, 18–24 June 2022; pp. 13608–13618. [CrossRef]
21. Ye, B.; Chang, H.; Ma, B.; Shan, S.; Chen, X. Joint feature learning and relation modeling for tracking: A one-stream framework. In *European Conference on Computer Vision*; Springer Nature: Cham, Switzerland, 2022; pp. 341–357. [CrossRef]
22. Gao, S.; Zhou, C.; Ma, C.; Wang, X.; Yuan, J. Aiatrack: Attention in attention for transformer visual tracking. In *European Conference on Computer Vision*; Springer Nature: Cham, Switzerland, 2022; pp. 146–164. [CrossRef]
23. Lin, L.; Fan, H.; Zhang, Z.; Xu, Y.; Ling, H. Swintrack: A simple and strong baseline for transformer tracking. *Adv. Neural Inf. Process. Syst.* **2022**, *35*, 16743–16754. [CrossRef]
24. Wang, J.; Shao, Z.; Huang, X.; Lu, T.; Zhang, R.; Li, Y. From artifact removal to super-resolution. *IEEE Trans. Geosci. Remote Sens.* **2022**, *60*, 1–15. [CrossRef]
25. Wang, J.; Shao, Z.; Huang, X.; Lu, T.; Zhang, R.; Cheng, G. Pan-sharpening via deep locally linear embedding residual network. *IEEE Trans. Geosci. Remote Sens.* **2022**, *60*, 1–13. [CrossRef]
26. Wang, J.; Shao, Z.; Huang, X.; Lu, T.; Zhang, R. A Dual-Path Fusion Network for Pan-Sharpening. *IEEE Trans. Geosci. Remote Sens.* **2022**, *60*, 1–14. [CrossRef]
27. Shao, J.; Du, B.; Wu, C.; Zhang, L. Can we track targets from space? A hybrid kernel correlation filter tracker for satellite video. *IEEE Trans. Geosci. Remote Sens.* **2019**, *57*, 8719–8731. [CrossRef]
28. Feng, J.; Zeng, D.; Jia, X.; Zhang, X.; Li, J.; Liang, Y.; Jiao, L. Cross-frame keypoint-based and spatial motion information-guided networks for moving vehicle detection and tracking in satellite videos. *ISPRS J. Photogramm. Remote Sens.* **2021**, *177*, 116–130. [CrossRef]
29. Song, W.; Jiao, L.; Liu, F.; Liu, X.; Li, L.; Yang, S.; Hou, B.; Zhang, W. A joint siamese attention-aware network for vehicle object tracking in satellite videos. *IEEE Trans. Geosci. Remote Sens.* **2022**, *60*, 1–17. [CrossRef]
30. Lin, B.; Bai, Y.; Bai, B.; Li, Y. Robust Correlation Tracking for UAV with Feature Integration and Response Map Enhancement. *Remote Sens.* **2022**, *14*, 4073. [CrossRef]
31. Wu, Y.; Lim, J.; Yang, M.H. Online object tracking: A benchmark. In Proceedings of the IEEE Conference on Computer Vision and Pattern Recognition, Portland, OR, USA, 23–28 June 2013; pp. 2411–2418. [CrossRef]
32. Mueller, M.; Smith, N.; Ghanem, B. A benchmark and simulator for uav tracking. In Proceedings of the Computer Vision–ECCV 2016: 14th European Conference, Amsterdam, The Netherlands, 11–14 October 2016; Proceedings, Part I 14; Springer: Berlin/Heidelberg, Germany, 2016; pp. 445–461. [CrossRef]
33. Zhao, M.; Li, S.; Xuan, S.; Kou, L.; Gong, S.; Zhou, Z. SatSOT: A benchmark dataset for satellite video single object tracking. *IEEE Trans. Geosci. Remote Sens.* **2022**, *60*, 1–11. [CrossRef]
34. Wu, D.; Song, H.; Fan, C. Object Tracking in Satellite Videos Based on Improved Kernel Correlation Filter Assisted by Road Information. *Remote Sens.* **2022**, *14*, 4215. [CrossRef]
35. Wu, J.; Pan, Z.; Lei, B.; Hu, Y. FSANet: Feature-and-spatial-aligned network for tiny object detection in remote sensing images. *IEEE Trans. Geosci. Remote Sens.* **2022**, *60*, 1–17. [CrossRef]
36. Krizhevsky, A.; Sutskever, I.; Hinton, G.E. Imagenet classification with deep convolutional neural networks. *Adv. Neural Inf. Process. Syst.* **2012**, *25*, 1097–1105. [CrossRef]

37. Zhang, Z.; Peng, H. Deeper and wider siamese networks for real-time visual tracking. In Proceedings of the IEEE/CVF Conference On Computer Vision and Pattern Recognition, Long Beach, CA, USA, 15–20 June 2019; pp. 4591–4600.
38. Javed, S.; Danelljan, M.; Khan, F.S.; Khan, M.H.; Felsberg, M.; Matas, J. Visual object tracking with discriminative filters and siamese networks: A survey and outlook. *IEEE Trans. Pattern Anal. Mach. Intell.* **2022**, *45*, 6552–6574. [CrossRef]
39. Fan, H.; Ling, H. Siamese cascaded region proposal networks for real-time visual tracking. In Proceedings of the IEEE/CVF Conference On Computer Vision and Pattern Recognition, Long Beach, CA, USA, 15–20 June 2019; pp. 7952–7961.
40. Yu, Y.; Xiong, Y.; Huang, W.; Scott, M.R. Deformable siamese attention networks for visual object tracking. In Proceedings of the IEEE/CVF Conference On Computer Vision and Pattern Recognition, Seattle, WA, USA, 14–19 June 2020; pp. 6728–6737.
41. Cao, Z.; Fu, C.; Ye, J.; Li, B.; Li, Y. SiamAPN++: Siamese attentional aggregation network for real-time UAV tracking. In Proceedings of the 2021 IEEE/RSJ International Conference on Intelligent Robots and Systems (IROS), IEEE, Prague, Czech Republic, 27 October–1 October 2021; pp. 3086–3092.
42. Xie, F.; Wang, C.; Wang, G.; Cao, Y.; Yang, W.; Zeng, W. Correlation-aware deep tracking. In Proceedings of the IEEE/CVF Conference on Computer Vision and Pattern Recognition, New Orleans, LA, USA, 18–24 June 2022; pp. 8751–8760.
43. Cao, J.; Song, C.; Song, S.; Xiao, F.; Zhang, X.; Liu, Z.; Ang, M.H., Jr. Robust object tracking algorithm for autonomous vehicles in complex scenes. *Remote Sens.* **2021**, *13*, 3234. [CrossRef]
44. Zhang, X.; Zhu, K.; Chen, G.; Liao, P.; Tan, X.; Wang, T.; Li, X. High-resolution satellite video single object tracking based on thicksiam framework. *GISci. Remote Sens.* **2023**, *60*, 2163063. [CrossRef]
45. Nie, Y.; Bian, C.; Li, L. Object tracking in satellite videos based on Siamese network with multidimensional information-aware and temporal motion compensation. *IEEE Geosci. Remote Sens. Lett.* **2022**, *19*, 1–5. [CrossRef]
46. Yan, B.; Zhang, X.; Wang, D.; Lu, H.; Yang, X. Alpha-refine: Boosting tracking performance by precise bounding box estimation. In Proceedings of the IEEE/CVF Conference on Computer Vision and Pattern Recognition, Nashville, TN, USA, 20–25 June 2021; pp. 5289–5298. [CrossRef]
47. Liao, B.; Wang, C.; Wang, Y.; Wang, Y.; Yin, J. Pg-net: Pixel to global matching network for visual tracking. In Proceedings of the Computer Vision–ECCV 2020: 16th European Conference, Glasgow, UK, 23–28 August 2020; Proceedings, Part XXII 16; Springer: Berlin/Heidelberg, Germany, 2020; pp. 429–444. [CrossRef]
48. Zhou, Z.; Pei, W.; Li, X.; Wang, H.; Zheng, F.; He, Z. Saliency-associated object tracking. In Proceedings of the IEEE/CVF International Conference on Computer Vision, Montreal, BC, Canada, 11–17 October 2021; pp. 9866–9875. [CrossRef]
49. Zhang, Z.; Liu, Y.; Wang, X.; Li, B.; Hu, W. Learn to match: Automatic matching network design for visual tracking. In Proceedings of the IEEE/CVF International Conference on Computer Vision, Montreal, BC, Canada, 11–17 October 2021; pp. 13339–13348. [CrossRef]
50. Chen, Z.; Zhong, B.; Li, G.; Zhang, S.; Ji, R. Siamese box adaptive network for visual tracking. In Proceedings of the IEEE/CVF Conference On Computer Vision and Pattern Recognition, Seattle, WA, USA, 13–19 June 2020; pp. 6668–6677. [CrossRef]
51. Chen, S.; Wang, T.; Wang, H.; Wang, Y.; Hong, J.; Dong, T.; Li, Z. Vehicle tracking on satellite video based on historical model. *IEEE J. Sel. Top. Appl. Earth Obs. Remote Sens.* **2022**, *15*, 7784–7796. [CrossRef]
52. Wang, Z.; Yang, J.; Pan, Z.; Liu, Y.; Lei, B.; Hu, Y. APAFNet: Single-Frame Infrared Small Target Detection by Asymmetric Patch Attention Fusion. *IEEE Geosci. Remote Sens. Lett.* **2022**, *20*, 1–5. [CrossRef]
53. Hu, J.; Shen, L.; Sun, G. Squeeze-and-excitation networks. In Proceedings of the IEEE Conference on Computer Vision and Pattern Recognition, Salt Lake City, UT, USA, 18–22 June 2018; pp. 7132–7141. [CrossRef]
54. Hou, Q.; Zhou, D.; Feng, J. Coordinate attention for efficient mobile network design. In Proceedings of the IEEE/CVF Conference On Computer Vision and Pattern Recognition, Nashville, TN, USA, 20–25 June 2021; pp. 13713–13722. [CrossRef]
55. Li, Y.; Licheng, J.; Huang, Z.; Zhang, X.; Zhang, R.; Song, X.; Tian, C.; Zhang, Z.; Liu, F.; Shuyuan, Y.; et al. Deep learning-based object tracking in satellite videos: A comprehensive survey with a new dataset. *IEEE Geosci. Remote Sens. Mag.* **2022**. [CrossRef]
56. Huang, L.; Zhao, X.; Huang, K. Got-10k: A large high-diversity benchmark for generic object tracking in the wild. *IEEE Trans. Pattern Anal. Mach. Intell.* **2019**, *43*, 1562–1577. [CrossRef]
57. Russakovsky, O.; Deng, J.; Su, H.; Krause, J.; Satheesh, S.; Ma, S.; Huang, Z.; Karpathy, A.; Khosla, A.; Bernstein, M.; et al. Imagenet large scale visual recognition challenge. *Int. J. Comput. Vis.* **2015**, *115*, 211–252. [CrossRef]
58. Yin, Q.; Hu, Q.; Liu, H.; Zhang, F.; Wang, Y.; Lin, Z.; An, W.; Guo, Y. Detecting and tracking small and dense moving objects in satellite videos: A benchmark. *IEEE Trans. Geosci. Remote Sens.* **2021**, *60*, 1–18. [CrossRef]
59. Danelljan, M.; Bhat, G.; Khan, F.S.; Felsberg, M. Atom: Accurate tracking by overlap maximization. In Proceedings of the IEEE/CVF Conference on Computer Vision and Pattern Recognition, Long Beach, CA, USA, 15–20 June 2019; pp. 4660–4669. [CrossRef]
60. Wang, Q.; Zhang, L.; Bertinetto, L.; Hu, W.; Torr, P.H. Fast online object tracking and segmentation: A unifying approach. In Proceedings of the IEEE/CVF Conference on Computer Vision and Pattern Recognition, Long Beach, CA, USA, 15–20 June 2019; pp. 1328–1338. [CrossRef]
61. Bhat, G.; Danelljan, M.; Gool, L.V.; Timofte, R. Learning discriminative model prediction for tracking. In Proceedings of the IEEE/CVF International Conference on Computer Vision, Seoul, Republic of Korea, 27 October–2 November 2019; pp. 6182–6191. [CrossRef]

62. Danelljan, M.; Gool, L.V.; Timofte, R. Probabilistic regression for visual tracking. In Proceedings of the IEEE/CVF Conference On Computer Vision and Pattern Recognition, Virtual, 14–19 June 2020; pp. 7183–7192. [CrossRef]
63. Zhang, T.; Zhang, X.; Zhu, P.; Jia, X.; Tang, X.; Jiao, L. Generalized few-shot object detection in remote sensing images. *ISPRS J. Photogramm. Remote Sens.* **2023**, *195*, 353–364. [CrossRef]

mathematics

MDPI

Article

AI-Enabled Vibrotactile Feedback-Based Condition Monitoring Framework for Outdoor Mobile Robots

Sathian Pookkuttath *, Raihan Enjikalayil Abdulkade, Mohan Rajesh Elara and Prabakaran Veerajagadheswar

Engineering Product Development Pillar, Singapore University of Technology and Design (SUTD), Singapore 487372, Singapore; raihan_abdulkader@sutd.edu.sg (R.E.A.); rajeshelara@sutd.edu.sg (M.R.E.); prabakaran@sutd.edu.sg (P.V.)

* Correspondence: sathian_pookkuttath@sutd.edu.sg

Abstract: An automated Condition Monitoring (CM) and real-time controlling framework is essential for outdoor mobile robots to ensure the robot's health and operational safety. This work presents a novel Artificial Intelligence (AI)-enabled CM and vibrotactile haptic-feedback-based real-time control framework suitable for deploying mobile robots in dynamic outdoor environments. It encompasses two sections: developing a 1D Convolutional Neural Network (1D CNN) model for predicting system degradation and terrain flaws threshold classes and a vibrotactile haptic feedback system design enabling a remote operator to control the robot as per predicted class feedback in real-time. As vibration is an indicator of failure, we identified and separated system- and terrain-induced vibration threshold levels suitable for CM of outdoor robots into nine classes, namely Safe, moderately safe system-generated, and moderately safe terrain-induced affected by left, right, and both wheels, as well as severe classes such as unsafe system-generated and unsafe terrain-induced affected by left, right, and both wheels. The vibration-indicated data for each class are modelled based on two sensor data: an Inertial Measurement Unit (IMU) sensor for the change in linear and angular motion and a current sensor for the change in current consumption at each wheel motor. A wearable novel vibrotactile haptic feedback device architecture is presented with left and right vibration modules configured with unique haptic feedback patterns corresponding to each abnormal vibration threshold class. The proposed haptic-feedback-based CM framework and real-time remote controlling are validated with three field case studies using an in-house-developed outdoor robot, resulting in a threshold class prediction accuracy of 91.1% and an effectiveness that, by minimising the traversal through undesired terrain features, is four times better than the usual practice.

Keywords: AI; condition monitoring; vibration; vibrotactile haptic feedback; wearable device; IMU; current sensor; 1D CNN; outdoor mobile robot; remote control; operational safety

MSC: 68T40; 68T01; 68T07; 93C85

1. Introduction

The rapid developments of robotics fundamental techniques and Artificial Intelligence (AI) models have enabled versatile outdoor robots to reduce human efforts in performing dull, dirty, and repetitive tasks, thereby resolving labour shortage issues and workplace-related safety infection risks. There is a vast market demand for diverse outdoor applications such as pavement sweeping, fumigation and fogging, landscaping, inspection, logistics, and security. Based on ABI Research, an expected shipment volume of outdoor mobile robots is 350,000 units by 2030, compared to 40,000 units in 2021 [1]. Much research has been published on autonomous performance improvement to fit the outdoor environment; for example, environmental perception [2], visual navigation [3], path planning [4], localisation [5], motion control [6], and pedestrian safety [7]. Though the system degradation in an outdoor robot is faster due to its exposure to extreme terrain and weather changes, focus on Condition Monitoring (CM) studies are still not commonly used to assess health states and terrain-related flaws, causing accelerated deterioration and potential hazards. The above mentioned research works are mainly focused on performing operations autonomously. However, there is still a long way to go to make these robots operate without human guidance by considering the open, hostile, and unsafe outdoor environment. Also, local regulations and certifications are to be checked and complied with for deploying robots autonomously on public pavements [8].

In this context, shared autonomy, where the robot's autonomous system and operator input work collaboratively to make decisions and achieve common goals [9,10], is a safe and feasible option for outdoor robot deployment. Also, studies on haptic-feedback-based motion control of robots [11–14] have enabled effective remote operation. An environment-independent sensor selection for data acquisition and a fast and accurate classifier model is crucial for real-time CM frameworks for the outdoor environment. There are numerous AI-enabled and vibration-based terrainability studies for outdoor robots using different sensors to classify tile, stone, sand, asphalt, grass, and gravel [15–19], which will help to decide the traversability of the robot, but there is not a proper CM approach for assessing the health condition of the robot and level of terrain-induced concerns that accelerate system degradation. Hence, an automated CM and feasible remote controlling framework involving a shared autonomy and haptic feedback system is imperative for outdoor robots assisting in either stopping or minimising the exposure to extreme terrain conditions based on the internal health and external terrain states of the robot. Towards this effort, this paper proposes and discusses vibration-based anomalies prediction using Inertial Measurement Unit (IMU) and current sensors data and a 1D Convolutional Neural Network (1D CNN) as the classifier, along with real-time robot control, by developing a wearable vibrotactile haptic feedback system, assuring productivity and operational safety in outdoor mobile robots.

1.1. Problem Statement

The deployment of mobile robots in a dynamically changing and unsafe outdoor environment is still a challenging problem, and condition monitoring studies performed to assess the health conditions of the robot and the uneven terrain features affecting accelerated deterioration is an unexplored research area, resulting in catastrophic failure, a high maintenance cost, unexpected downtime, operational hazards, and customer dissatisfaction. The aforesaid terrainability studies may help the last mail delivery or outdoor security robots in path planning, avoiding terrain-related flaws. However, in fumigation, lawn mowing, and pavement sweeping applications, the robot is supposed to cover a given area for the intended function. In particular, traversing through grass fields is more challenging where exteroceptive techniques are impractical as most ground-level imperfections are grass-covered. The system also degrades naturally due to continuous usage, even when deployed in a plain area, or there may be accelerated deterioration due to the terrain flaws depending on the workspace deployed. Hence, periodic maintenance is not advised for outdoor robots. The manual procedure is generally practised to assess the robot's condition and how good the given area is for deployment, which is labour-oriented and results in high maintenance costs. Hence, a study on automated CM strategy for outdoor mobile robots is vital for addressing this problem.

Selecting proper sensors to extract the characteristics of both system degradation and terrain-induced concerns is critical, and it should be environment-independent for outdoor robots. Vibration is widely accepted for fault detection, and most of the studies used an IMU sensor for vibration-indicated data modelling based on the change in linear and angular motions. However, sometimes robots are stopped by ground-level obstacles where no abnormal vibrations are noticed, so more suitable sensors must be integrated and explored. A fast and computationally low-cost neural network model with higher prediction accuracy is mandated for real-time applications to avoid false positive actions.

The shared autonomy and haptic-feedback-based approach is a feasible approach for CM in outdoor robots; however, the current studies are mainly focused on controlling robot arms, motion control, and safe navigation, avoiding obstacles in mobile robots and drones in their heading direction, and do not explore CM applications in outdoor mobile robots. Since CM covers internal and external factors and different threshold levels are involved, a multi-level haptic feedback patterns system is essential for effectively monitoring and controlling the robot. The proposed study analyses the abovementioned problems for a feasible solution for CM applications in outdoor mobile robots.

1.2. Related Works

Abnormal vibration is an indicator of system failure; hence, many fault detection and prognosis studies are conducted based on vibration-indicated data collected through suitable sensors, mainly accelerometers [20–22]. The advent of deep learning models, especially 1D CNN, is found to be accurate in extracting the unique features in order to classify and assess the severity of anomalies, and they are used for real-time applications due to their simple structure, low computational complexity, and easy deployability [23]. The vibration and 1D-CNN-based works are common for the CM and fault assessment of machine components [24,25], structural systems [26], and industrial robots [27]. Other CM and fault diagnosis works for industrial robots include minimising downtime in a wafer transfer robot [28], failure prediction in a packaging robot [29], and a safe stop for a collaborative robot [30]. However, such CM works are not extended for outdoor mobile robots, though a great research scope using the typical onboard sensors and a necessity because of their vast market demand and safety. Towards this CM effort, we introduced two vibration-based approaches in previous works for indoor mobile robots, considering both internal and external sources of vibration. In an IMU sensor-based work [31], the change in angular velocity and linear acceleration data due to vibration sources is mainly modelled as vibration data and, in the second work, a monocular camera sensor [32] is used and the change in optical flow 2D vector displacement data is modelled as vibration-indicated data for training. In both works, a 1D CNN was adapted as a classifier and compared with other models, such as a Support Vector Machine (SVM), Multilayer Perceptron (MLP), Long Short-Term Memory (LSTM), and CNN-LSTM, showing better accuracy and inference time.

The CM studies for outdoor robots are more challenging and unexplored, and the existing studies are limited to terrain classification using different sensors and AI models to assess traversability only; for instance, a terrain type study using IMU and a Probabilistic Neural Network (PNN) for autonomous ground vehicles in [15], classifying asphalts, gravel, mud, and grass. A terrain feature study based on texture-based descriptors is conducted in [17] using a camera and Random Forest (RF) classifier. A sensor fusion approach using IMU and camera is used in [18] for terrain classification, adapting an SVM as the classifier and mentioning that a faster model is needed for real-time application. A 3D LiDAR sensor is used for traversability analysis in [19] with the positive naive Bayes classifier by defining the feature vectors with different dimensions. Regarding the wheel–terrain tough interactions due to ground-level imperfections and obstacles, a higher torque is applied to overcome the flaws and, naturally, a higher current consumption occurs. However, the current sensors are more used for energy efficient strategy studies in mobile robots [33,34] rather than CM applications, except an earlier study in [35] for fault diagnosis in motor-driven equipment by monitoring current readings.

The vibrotactile haptic feedback approach is used in many systems, conveying information to the user or operator through vibration patterns, mainly for human health monitoring or assistance, by developing wearable devices, such as the health monitoring of an offshore operator in [36] and wheelchair navigation for disability people in [37]. Recently, these techniques have been applied to remotely controlled industrial and mobile robots. For instance, haptic-feedback-based teleoperation using a Phantom Omni haptic feedback master device and a 6-DOF robot arm was used with a force sensor as a slave, enabling the operator to adjust the position and direction of the arm safely from a distance based on the distance from the haptic feedback to the manipulator [38], and a similar study was conducted in [39] to stop the manipulator avoiding any collision. An inexpensive electro-tactile-based haptic feedback method is presented in [40] for obstacle avoidance in a cluttered workspace of a teleoperated robot arm. A wearable hand gesture recognition and vibrotactile system is developed in [41] to control drone navigation, where the obstacle information in the heading direction is fed back to the operator. A haptic feedback and shared autonomy-based teleoperation for the safe navigation of mobile robots is explained in [11], where obstacle range information is fed back to the operator as a force through a haptic probe. Similar force feedback studies for mobile robots were conducted, including [12–14]. These haptic-feedback-based works are mainly undertaken for indoor robots for smooth navigation and avoiding obstacles in the heading direction. However, to our knowledge, there are no such works for CM applications in outdoor robots, feeding back the system deterioration stage and terrain-related features to ensure the robot's health and operational safety.

1.3. Contributions

1. An unexplored vibration-based Condition Monitoring (CM) framework is introduced for outdoor mobile robots considering both the system deterioration state and uneven terrain features.
2. Eight abnormal states are identified for an outdoor wheeled robot set at different threshold levels for CM in view of real-world cases.
3. The vibration-indicated data are modelled based on IMU and current sensors data, which complement each other for system-generated/degraded and wheel–terrain interaction-related robot behaviour.
4. A simple 1D CNN-based model is structured for the fast, accurate, and computationally low cost enabling of the real-time prediction of vibration threshold classes.
5. A novel wearable vibrotactile haptic feedback device architecture is presented for CM applications in outdoor robots and configured with unique tactile feedback patterns for each threshold class.
6. The real-time field case studies conducted using an in-house-developed outdoor fumigation robot show the impacts of the proposed research framework, enhancing the robot's health and operational safety.
7. The proposed AI-enabled vibrotactile feedback-based CM framework for real-time remote controlling is the first of its kind for outdoor mobile robots.

The remainder of this paper is structured as follows. Section 2 presents an overview and methodology of the proposed work. Section 3 elaborates on the framework evaluation and results. Section 4 explains field case studies conducted to validate the proposed method for real-time application, and Section 5 concludes with a summary of the work.

2. Overview of the Vibrotactile Feedback-Based Condition Monitoring Framework

An overview of the proposed vibrotactile feedback-based condition monitoring and real-time remote controlling framework for health monitoring and avoiding uneven terrain features in outdoor mobile robots is depicted in Figure 1. It mainly consists of two systems: a health status and uneven terrain-induced threshold classes publishing system at the robot side and a haptic-feedback-based real-time remote controlling system using a vibrotactile haptic feedback device at the operator side. The data transfer between the robot (publisher) and the vibrotactile device (subscriber) is accomplished by setting a topic through a Message Queuing Telemetry Transport (MQTT) broker protocol. The design and methodology followed to develop and test these two systems for the proposed framework are elaborated as follows.

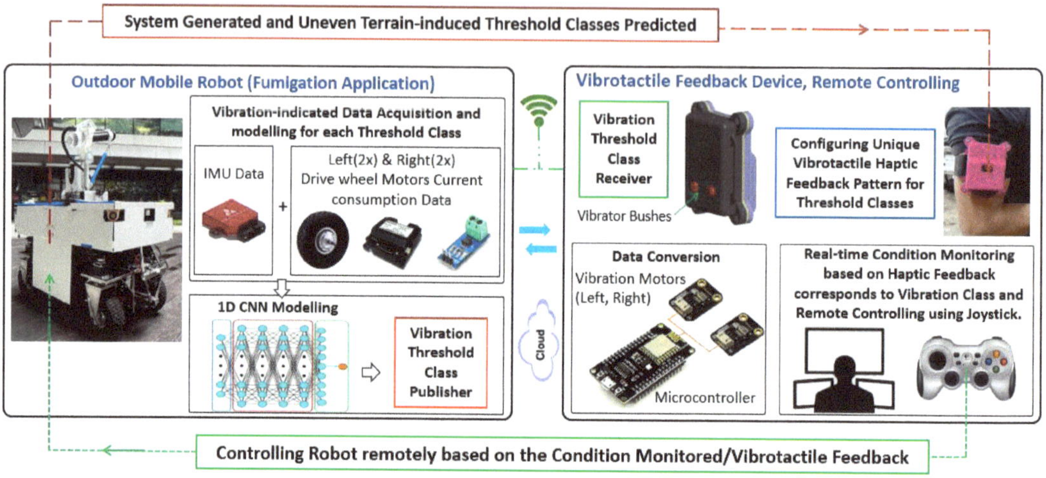

Figure 1. Overview of the vibrotactile feedback-based condition monitoring and remote controlling.

2.1. Predicting the Anomalous Cases for Condition Monitoring—Robot Side

This section explains the methodology followed for identifying and predicting the anomalous cases at the robot side, covering the details of the outdoor robot platform used for the data acquisition and field trials, the system-generated/deteriorated and uneven terrain-induced vibration threshold classes, vibration-indicated data modelling, and the 1D CNN model structured for classification and prediction.

2.1.1. Outdoor Mobile Robot Test Platform

An in-house-developed wheeled outdoor mobile robot is used in this study for vibration-indicated data acquisition and real-time field case studies. The base platform of this robot is designed with interchangeable payloads for outdoor applications such as fumigation (pest control), pavement sweeping, and lawn mowing by mounting a sprayer with a tank assembly, sweeping brushes, and grass-cutting wire module, respectively. Here, we used the fumigation robot as it is intended to travel on both pavements and grass fields, exposing maximum undesired terrain features, and is hence an ideal test platform for this study. The size of the robot is 1.6 (L) × 0.85 (W) × 1 (H) m and it has an approximate weight of 300 Kg. The robot's locomotion unit consists of four hub wheel motors of 250 W and steering motors of 100 W power, and the wheel size is 38 (outer diameter) × 15 (width) cm. An omnidirectional mobility is designed with independent steering drive wheels enabling better manoeuvrability. The robot is built with a robust chassis using the welding of stainless-steel sections, plates, shafts, and brackets for mounting all critical sub-assemblies and sensors, including a double wishbone suspension system to fit driving for outdoor applications. A 300 Ah 48V DC Lithium iron phosphate (LFP) battery is used to power the system and payloads. An Industrial PC Anewtech IPC970-10WF5E1A is integrated for entire operation and data storage. The Robot Operating System (ROS) middleware in the Ubuntu Noetic operating system is used to execute all communications. A TP-Link LTE 4G Router is added for remote monitoring and control, establishing Internet and wireless networking.

A shared autonomy framework is executed; hence, the robot's autonomous behaviour is combined with human input as needed to monitor and control the robot, assuring the robot's health and operational safety. The onboard computer processes a low-level autonomy stack using the perception sensors integrated to avoid large obstacles. The sensors include four ultrasonic sensors (UCC4000) mounted at four corners of the robot and two 2D LiDARs (SICK TIM581) mounted at the front and rear sides. A stereo camera (ZED 2) is mounted at the front towards the heading direction. For live streaming, four cameras are mounted: one main camera facing towards the heading direction, one at the rear side for any reverse motion, and the other two cameras at the left and right sides. This shared autonomy framework integrated into this fumigation robot was based on the previous works from our research lab [10], and a briefing of this shared autonomy framework is as follows. The four ultrasonic sensors detect obstacles, including transparent glass walls (when the robot traverses nearby buildings) and moving objects at all sides. The two 2D LiDAR sensors enable the robot to mainly detect any prominent cliffs, kerbs, or steps. The stereo camera detects the presence of pedestrians, bicycles, animals, and vehicles at the front, and measures pedestrian density as well. The four cameras help with live streaming. Additionally, a decision-making process based on multi-layer fuzzy logic is integrated into the robot as additional safety layers, considering the operator's attention level and switching the operation modes, accordingly avoiding hazardous events. However, this is applicable only for the large obstacles in the heading direction of the robot, and any unseen and undetected ground-level obstacles are still harmful to the robot's health and operational safety. The fumigation outdoor robot, as explained, used as a test platform in this study, is shown in Figure 2, including the perception sensors integrated for shared autonomy. Depending on the remote location, the human–robot interface/shared autonomy layer is connected through a network protocol, either using a local area network or cloud computing.

Figure 2. Shared autonomy integrated outdoor fumigation robot test platform.

2.1.2. Vibration Threshold Classes for Condition Monitoring

A mobile robot produces unusual vibrations as early symptoms of failure and potential hazards. The sources of such vibrations could be due to internal system issues like component deterioration, wear and tear, loose assembly, and an unbalanced structure, or external factors like uneven terrain and undetected ground-level small obstacles/features. Here, the severity of vibration depends on the various levels of such internal and external factors. Hence, a vibration threshold level-based classification used to distinguish the vibration sources and their acceptance assessment is essential to trigger prompt and proper corrective or maintenance action. For example, if the robot's vibration is moderate, close attention is needed to proceed or change the heading direction. Suppose that the vibrations are severe due to internal or external factors: the operator should stop and fix the issues as soon as possible or change the heading direction to avoid ground-level obstacles and hazards. Hence, this study proposes nine vibration threshold classes for condition monitoring suitable for a four-wheeled outdoor mobile robot by considering both system-generated vibrations due to internal concerns and terrain-flaws-related external circumstances, namely Safe (no action required), Moderately safe due to System-generated (MS), Moderately safe due to Terrain-induced Left wheels affected (MTL), Moderately safe due to Terrain-induced

Right wheels affected (MTR), Moderately safe due to Terrain-induced where both Left and Right wheels are affected (MT), Unsafe due to System-generated (US), Unsafe due to Terrain-induced Left wheels affected (UTL), Unsafe due to Terrain-induced Right wheels affected (UTR), and Unsafe due to Terrain-induced where both Left and Right wheels are affected (UT). Here, the MS class need to be kept in view to ensure that the condition is not worsening. The MTL, MTR, and MT classes need attention while proceeding or the heading direction may change. For the US class, the robot should be stopped and the issues should be fixed with proper maintenance actions. For the UTL, UTR, and UT classes, it is better to stop and change the heading angle and proceed only if it is safe. These nine vibration threshold classes tested and validated for condition monitoring in outdoor robots are depicted in Figure 3.

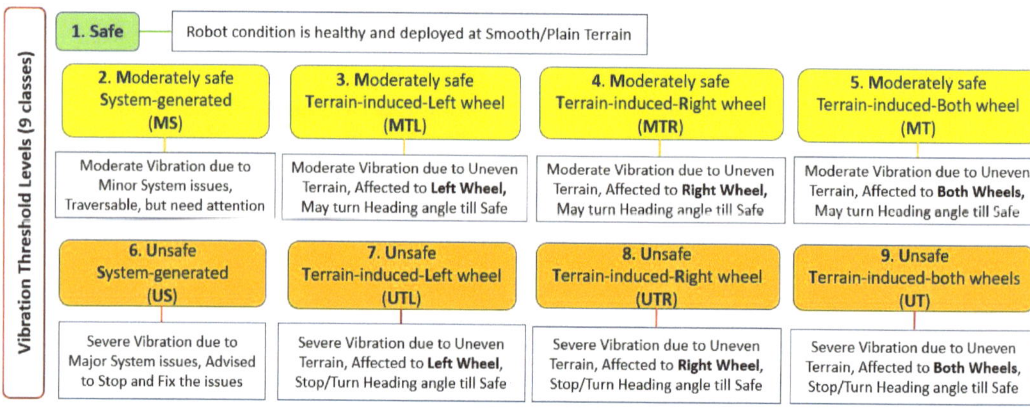

Figure 3. Vibration threshold classification for condition monitoring.

2.1.3. Vibration-Indicated Dataset Modelling

A wheeled outdoor mobile robot vibrates at different rates linearly or in angular motion (roll, pitch, yaw) due to being system-generated or terrain-induced, as explained in the previous section. Such abnormal movements of the robot could be assessed through IMU sensor data. Sometimes, the robot may not show an apparent vibration/angular motion when stuck with ground-level obstacles; instead, it keeps trying to overcome the obstacle, resulting in a higher torque required and higher current consumption. Hence, adding the current consumed data from each wheel motor will enhance the terrain-related threshold classes assessment specific to the left or right side affected for corrective action. Thus, these heterogeneous IMU and current sensors data support filling the gaps of individual sensors in extracting the relevant vibration-affected features. In this study, we used the onboard Vectornav VN-100 IMU sensor to extract the rate of the linear and angular motion of the robot at different vibration threshold classes, i.e., linear acceleration and angular velocity, from the accelerometer and gyroscope integrated with IMU. An ACS712 current sensor module is integrated with each wheel drive motor to tap the current consumption using an Arduino Mega 2560 microcontroller, enabling better prediction accuracy for terrain-induced classes specific to left and right. The calculated angular acceleration from the angular velocity data is added for further dataset strengthening. These proprioceptive sensors do not depend on each other and are unaffected by the environment or light conditions, facilitating higher-accuracy prediction. As the robot locomotion is controlled by four independent steering wheels (two drive wheels at both the front and rear) with holonomic mobility, the average current value of the front-left wheel motor and rear-left wheel motor is recorded as the left current value, and similarly for the Right. The IMU current sensors data acquisition scheme is depicted in Figure 4.

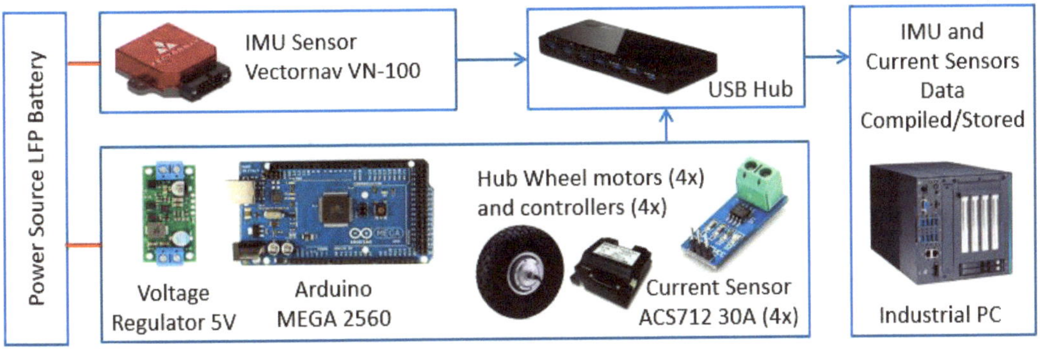

Figure 4. Dataset acquisition scheme using IMU and current sensors.

In this work, the vibration-indicated data (Vib_{data}) are the changes in the linear acceleration, angular velocity, angular acceleration measured through the IMU sensor, and data of the average current consumed by left-side pair motors and right-side pair motors measured using the current sensors, due to the various vibration threshold classes, which is modelled and represented as in (1). Here, the two types of sensors used have different behaviour; hence, the sequential data subscription rate is critical to represent all data recorded for the particular class in the given period. This is assured by setting the IMU subscription rate at 80 Hz and lowering the delay in the Arduino program to the same rate as the current sensor data subscription. It also confirmed that the total number of samples data collected at any particular time is the same. The data features for linear acceleration, angular velocity, and angular acceleration are recorded in three axes—a total of nine data from IMU and two data features from the robot's left and right-side motors pair—hence, a total of 11 data features are used. Further, each data feature is compiled with 128 temporal data, forming one sample datum for training acquired in 1.6 s. Hence, a sample datum is an array of [128 × 11] as represented in (2). The IMU and current sensors data are normalised into standard scale without losing information for better convergence during training, and then compiled into an array of [n × 128 × 11] as training and evaluation datasets for each class for the 1D CNN training, where n is the total number of samples.

$$Vib_{data} = [A_{lin}, V_{ang}, A_{ang}, I(L)_{\frac{(Left\_Front)+(Left\_Rear)}{2}}, I(R)_{\frac{(Right\_Front)+(Right\_Rear)}{2}}] \quad (1)$$

$$\begin{bmatrix} Acc_{Lin}(X)_{t1}, Acc_{Lin}(X)_{t2}, \ldots Acc_{Lin}(X)_{t128} \\ Acc_{Lin}(Y)_{t1}, Acc_{Lin}(Y)_{t2}, \ldots Acc_{Lin}(Y)_{t128} \\ Acc_{Lin}(Z)_{t1}, Acc_{Lin}(Z)_{t2}, \ldots Acc_{Lin}(Z)_{t128} \\ Vel_{Ang}(X)_{t1}, Vel_{Ang}(X)_{t2}, \ldots Vel_{Ang}(X)_{t128} \\ Vel_{Ang}(Y)_{t1}, Vel_{Ang}(Y)_{t2}, \ldots Vel_{Ang}(Y)_{t128} \\ Vel_{Ang}(Z)_{t1}, Vel_{Ang}(Z)_{t2}, \ldots Vel_{Ang}(Z)_{t128} \\ Acc_{Ang}(X)_{t1}, Acc_{Ang}(X)_{t2}, \ldots Acc_{Ang}(X)_{t128} \\ Acc_{Ang}(Y)_{t1}, Acc_{Ang}(Y)_{t2}, \ldots Acc_{Ang}(Y)_{t128} \\ Acc_{Ang}(Z)_{t1}, Acc_{Ang}(Z)_{t2}, \ldots Acc_{Ang}(Z)_{t128} \\ I(L)_{t1}, I(L)_{t2}, \ldots I(L)_{t128} \\ I(R)_{t1}, I(R)_{t2}, \ldots I(R)_{t128} \end{bmatrix} \quad (2)$$

2.1.4. 1D CNN Modelling for Training and Prediction of Threshold Classes

A simple structured 1D-CNN-based model is proposed to train threshold classes for fast, accurate prediction and real-time CM applications. The model follows the convolution operations on data vectors using a nonlinear activation function that convolves the input vector to the output layer using a filter vector, as in Equations (3) and (4) [42]. Here, x (of length N) is the input data vector, ω (of length L) is the filter vector, the term b is the bias to the best fit for given data, and $\mathbf{c}$ is the output layer (of length $N - L + 1$). For detecting the key characteristics and reducing the number of parameters, a max pooling output vector $\mathbf{d}$ is added after convolution layers, where $m \times 1$ is the kernel size, u is a window function, and s is the filter moving stride over input vector $\mathbf{c}$.

$$\text{Output layer } \mathbf{c}(j) = f\left(\sum_{i=0}^{L-1} \omega(i)x(j-i) + b\right), \text{where } j = 0, 1, \ldots, N-1 \tag{3}$$

$$\text{Max pooling output vector } \mathbf{d} = max(u(m \times 1, s)\mathbf{c}) \tag{4}$$

The dataset [128 × 11] prepared for each class is flattened into a 1D array [1 × 1408] to be fed into convolution layers. The CNN structure for this study is made simple with four layers, and uses sixty-four filters for the first two layers, and then thirty-two for the remaining. A convolution window (kernel) size of 3 is used for each layer. For effective learning of the non-linear vibration-indicated heterogeneous sensors data pattern, a Rectified Linear activation Unit (ReLU) is applied to each convolutional layer, followed by a max pooling layer with a stride size of 2, enabling a reduction in the computation time. To avoid over-fitting while training, a dropout layer is added at a rate of 0.2 for each layer. Finally, with a flattening function, the pooled feature maps are converted into a 1D array in the output layer, which predicts the multinomial probability of threshold classes. The proposed 1D-CNN-based model structure is illustrated in Figure 5.

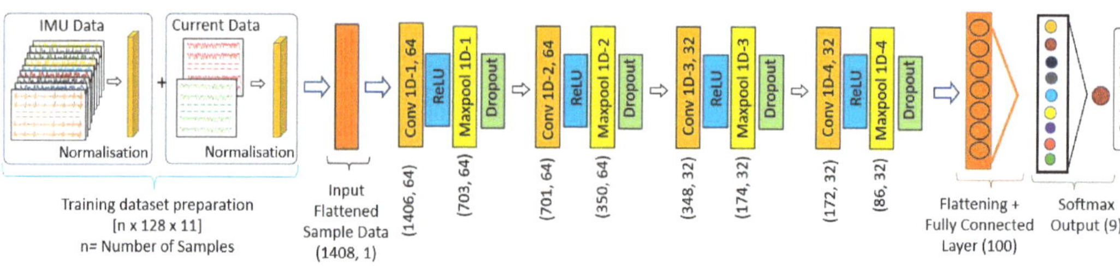

Figure 5. One-dimensional CNN model structure and data shape for training.

2.2. *Vibrotactile Feedback System for Condition Monitoring and Remote Controlling—Operator Side*

This section explains the vibrotactile haptic feedback device developed and the methodology followed to receive and convert the predicted abnormal classes to unique haptic feedback patterns corresponding to the abnormal threshold classes, enabling prompt actions by the remote operator.

2.2.1. Vibrotactile Haptic Feedback Device Architecture

A vibrotactile haptic feedback device is developed to understand the anomalous cases related to the robot's system/deterioration level and uneven terrain features causing accelerated degradation or hazards, thereby switching the robot's control to operator mode and taking necessary actions in real-time. Here, the anomalous cases are the eight abnormal threshold classes (MS, MTL, MTR, MT, US, UTL, UTR, and UT) identified and predicted by the robot (publisher). The haptic device comprises an ESP8266 ESP-12E microcontroller, two vibration modules/motors (DFR0440), a 9 V alkaline battery, a 5 V voltage regulator, and a power switch. A 3D-printed middle bracket for holding/guiding the components and top and bottom protective enclosures are used and fastened with screws. The DFR0440 vibration module is built with an Eccentric Rotating Mass (ERM) vibration motor; hence, the module vibrates when the motor shaft with offset mass rotates. The vibration modules are vertically guided independently and floated with low-tension compression springs. A hemispherical plastic bush is attached to each vibration module, acting as an actuator for tactile stimulation. The two vibration modules and the vibrator bushes named left and right also represent the robot's left and right wheels in the cases of left- and right-wheel-induced vibration classes. The vibration modules stimulate based on the vibration pattern configured for each threshold class and effectively deliver tactile feedback to the operator through the left and right vibrator bushes. As it is a wearable device, various ergonomics and safety factors have been considered while designing the device; for instance, a compact and minimum overall size while at the same time keeping the left and right bushes as far as possible to distinct the two bushes' feedback, a lower weight, being easy to wear with a single hand, being adjustable to fit with the operator's upper arm size, and having no sharp corners, especially regarding the vibrator bushes, which are in contact with the operator, for mechanical safety. In addition, the electrical circuits and battery connection are carefully wired and isolated for electrical safety. The device is compact, with a size of $80 \times 68 \times 38$ mm, and weighs 100 gm. A single armband with a velcro fastener is used to firmly hold the device with the remote operator's upper arm, and the spring-loaded left and right vibrator bushes ensure contact with the operator. The design and assembly details of the haptic feedback device proposed for the CM application of an outdoor mobile robot are depicted in Figure 6.

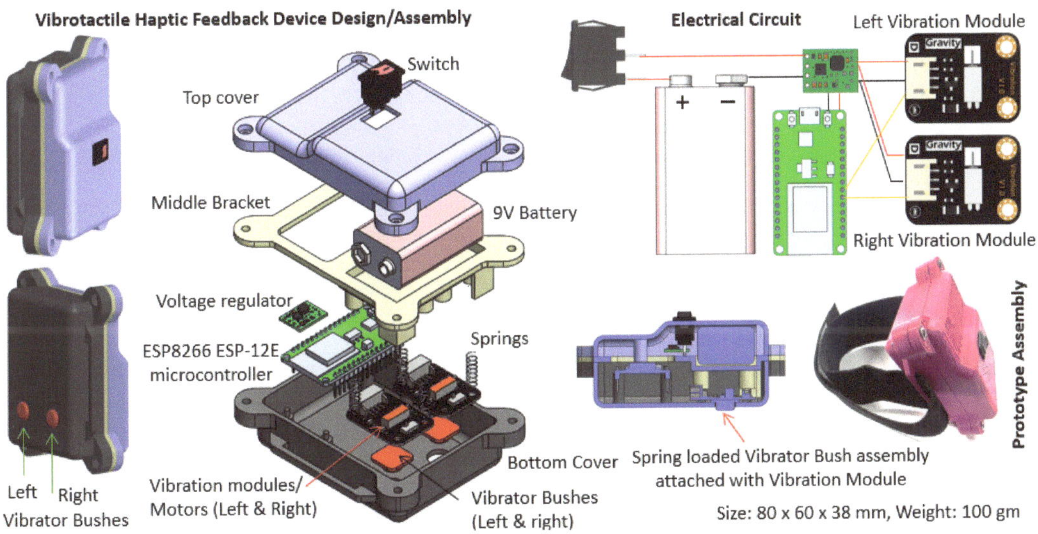

Figure 6. Vibrotactile haptic feedback device architecture.

2.2.2. Vibrotactile Haptic Feedback Patterns for Threshold Classes and Encoding

The vibration threshold classes predicted by the 1D CNN model at the robot side are published to a topic via MQTT Broker, which is subscribed by the microcontroller at the haptic feedback device. The vibrotactile feedback pattern for each threshold class should be easy to perceive and intuitive to avoid additional cognitive attempts from the operator. Hence, distinctive vibration feedback patterns are planned that correspond to each threshold class by changing the vibration intensity of the module (ERM speed), setting to ON/OFF and continuous ON modes, and utilising the left and right vibration modules/bushes. The vibration intensities of the module are set by changing the duty cycle of the Pulse Width Modulation (PWM) signal sent to the vibration module. The left, right, or both vibration modules will be stimulated for the terrain-induced classes specific to the affected left-, right-, or both-side wheels. For all Moderately safe classes (MS, MTL, MTR, and MT), the PWM duty cycle is set as 50% and, for all Unsafe classes (US, UTL, UTR, and UT), it is set to 100%, resulting in moderate and the highest vibration intensity, respectively. Further, to distinguish between system-generated and terrain-induced, the vibration pattern is set as ON and OFF mode for System-generated classes (MS and US) with an off time of 300 ms. And, for Terrain-induced classes (MTL, MTR, MT, UTL, UTR, and UT), the vibration mode is set as a continuous pattern. The predicted vibration threshold classes' conversion to specific vibrotactile feedback patterns is illustrated in Figure 7. Here, 1.6 s is used for the vibration pattern, considering each prediction based on every new dataset.

As there are a total of eight abnormal classes (excluding the safe class, which needs no action) in our proposed study, designing a unique pattern for each class was one of the challenges in the work, which we realised by the combination of the mechanical design of the haptic device, hardware, and software architecture. The mechanical design and hardware setup were made simple with only two actuators (left and right vibrator tactile bush/vibrator module) as more actuators/vibrator modules raise complexity in ergonomics and assessment. Similarly, the programming simplified the design by limiting the Pulse Width Modulation (PWM) duty cycles to only two settings (50% and 100%). Based on these minimum number of features/ parameters, the unique haptic feedback pattern for eight classes is made simple and intuitive.

In summary, for all Moderate classes (MS, MTL, MTR, and MT), the haptic feedback intensity (vibration intensity of the vibrator module/motor) will be moderate, and for the Unsafe classes (US, UTL, UTR, and UT) the haptic feedback intensity is set at maximum (not harmful) to distinguish between these two threshold levels. Further, the haptic feedback is set ON and OFF for the System-generated vibration MS and US classes (internal system degradation), i.e., the device will be OFF for 300 ms after a 500 ms feedback period. These vibration patterns designed are intuitive or need less learning effort; hence, the operator can easily make decisions based on this vibration intensity difference and ON/OFF vibration pattern configurations for the specific class.

As per the haptic feedback received through the vibrator bushes, the operator can trigger the corrective actions by changing the robot's heading angle, moving left or right, thereby minimising the traversal through uneven terrain or planning to stop the robot if the robot persistently predicts unsafe classes either due to system-generated or terrain-induced flaws. The distinctive haptic feedback is evaluated, simulating the eight classes, and is tested by different users. No haptic feedback is programmed for safe class prediction.

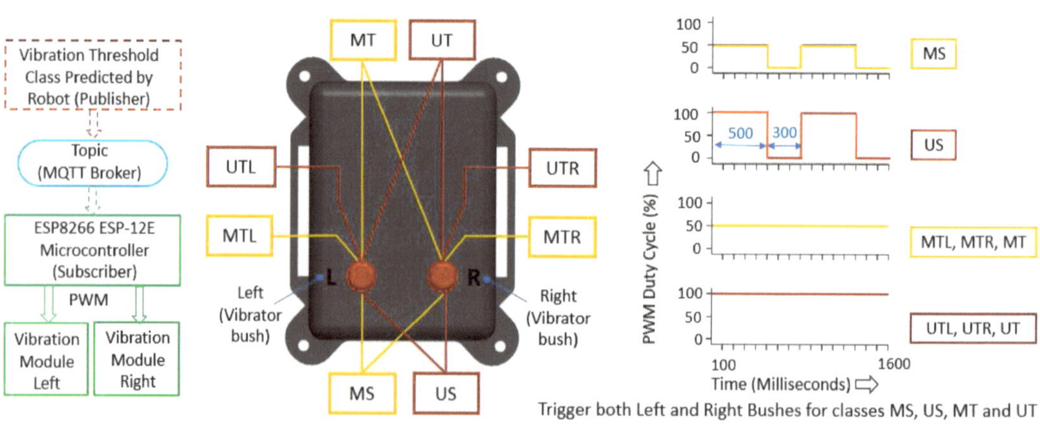

Figure 7. Vibrotactile feedback patterns for abnormal threshold classes.

3. Training and Evaluation of 1D-CNN-Based Threshold Class Prediction Model

This section composes data acquisition methods representing the nine classes based on the two heterogeneous sensors and their visual representation, followed by the 1D CNN model training for optimum performance and evaluation results.

3.1. Training Dataset Preparation for the Vibration Threshold Classes

The data acquisition representing the nine classes for training the 1D CNN model, using the two heterogeneous sensors integrated with the fumigation robot, is critical. Hence, it is firstly confirmed that the Vectonav VN-100 IMU sensor and the four ACS712 30 A current sensors are mounted firmly with the chassis and wheel motors, and the data subscription rate (80 Hz) and dataset collection algorithm work properly, compiling the data in an array of [n × 128 × 11] as explained in Section 2.1.3. The robot's linear speed is set at 0.3 m/s, which is the same as the operating speed, and the robot is driven in a zig-zag pattern while collecting the training data. For the safe class dataset preparation, the robot is tested, assuring good health, and is operated on plain and well-maintained pavements and grass fields, where no apparent vibrations are observed. The dataset collected during this trial is saved and labelled as the safe class. The data collection for the abnormal threshold classes involves robot modification representing system-generated classes and running through uneven and simulated ground-level obstacle environments, including positive and negative obstacles/features and unstructured grass fields. Some of these setups used to capture the internal system-generated classes (health issues) and external uneven/undetected ground-level obstacles/features for terrain-induce classes data collection are shown in Figure 8.

Figure 8. Robot modification and terrain exposure for threshold classes training data acquisition.

The robot was modified on a minor level for the Moderately safe System-generated class (MS). This includes loosening the components assembly mounting brackets, setting the air pressure of one or more wheels as low, reducing the suspension system performance, and placing offset the heavier items such as the battery. Then, the robot was driven on the same plain terrains as the safe class and a minor level of vibration was noticed compared to the safe class, which is purely due to the degraded system causes. The data were collected and labelled as the MS class. Further, for the Unsafe System-generated class (US), the modifications above intensified more with close observations and safety measures, such as an emergency stop, resulting in a higher vibration of the robot compared to the MS class. The data collected during these trials were labelled as the US class.

The left and right wheels of the robot are kept 0.7 m apart; hence the spot ground-level obstacles are exposed to the particular wheel only, and the deterioration rate is affected by the specific wheel and related assembly only. In this view, if the robot is exposed to the spot obstacles at left-side wheels (front and rear), it will be labelled as MTL or UTL depending on the size of the ground-level obstacles, and, similarly for right-side wheels, MTR or UTR. The advantage of this approach is that the operator can easily avoid the terrain-induced vibration, detecting left or right spot obstacles by moving left and right, utilising the holonomic mobility of the robot as well. Alternatively, if the obstacle is spread like a long root or a large uneven pebbled area, both left/right wheels undergo the same experience; hence, such data are labelled as MT or UT.

Next, for collecting moderately safe terrain-induced classes data, the robot is exposed to typical pavements and grass fields with uneven/unstructured surfaces, damaged/missing tiles pavement, stones, roots, pits, utility hole raised lids, and gutters, including simulated obstacles. Here such selected imperfections had a moderate size of around 2–4 cm, both positive and negative. As explained above, these Moderate Terrain-induced data are recorded as MTL, MTR, and MT. The same exercise and approach are repeated for Unsafe Terrain-induced classes (UTL, UTR, and UT), driving through the undesired terrain features of sizes of around 4-6 cm height, and the data were labelled accordingly.

For all Unsafe classes (US, UTL, UTR, and UT), as the robot vibrates more severely, the deterioration rate become faster, and chances of total failure and hazards are high. Hence, during such trials, the robot stopped intermittently and adjusted to continue trials as an unsafe class. Also, all safety measures like close observation and an emergency stop button were taken if needed. For each threshold class, a total of 2500 samples, forming an array of [$2500 \times 128 \times 11$], were saved and split into 80% and 20% for training and validation, respectively, for training the 1D CNN model. Additionally, a total of 500 dataset samples were recorded for each class to evaluate the accuracy of the trained model. The vibration-indicated IMU and current sensors data across the nine classes found different values/patterns, enabling a faster convergence in extracting the unique characteristics of each class and a higher prediction accuracy. For a visual representation of this unique dataset for each class, we randomly collected a single dataset of 128 elements and used only five features here for plotting, i.e., features from IMU (angular velocity in X, Y, and Z axes) and the current sensor (left and right wheel current consumption) as shown in Figures 9 and 10.

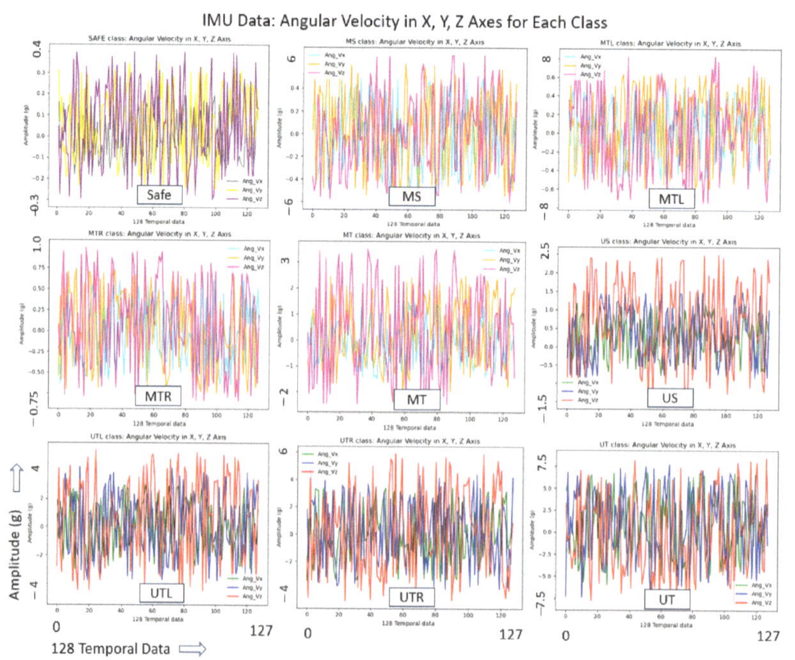

Figure 9. Vibration-indicated data from IMU sensor for each class.

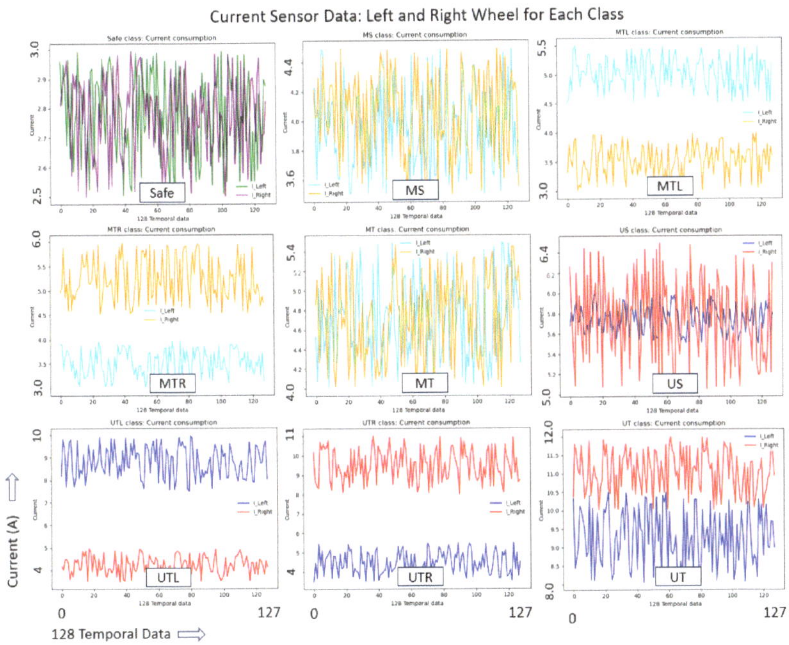

Figure 10. Vibration-indicated data from current sensors for each class.

As the data features are in different scales, especially the current sensors values, normalisation pre-processing is conducted for the standard scale, supporting faster convergence during the 1D CNN training. Here, IMU data were normalised into −1 to +1 and 0 to 1 for current consumption data following Equations (5) and (6), respectively.

$$\text{Dataset\_IMU Sensor} : x_{Normalised} = 2\frac{x - min(x)}{max(x) - min(x)} - 1 \quad (5)$$

$$\text{Dataset\_Current sensor} : x_{Normalised} = \frac{x - min(x)}{max(x) - min(x)} \quad (6)$$

3.2. 1D CNN Model Training and Evaluation Results

This section provides the 1D CNN model training methodology and the results. The heterogeneous sensors-based unique dataset compiled is trained following a supervised learning strategy using an Nvidia GeForce GTX 1080 Ti-powered workstation and Tensorflow DL-library [43]. To avoid over-learning and for generalisation, a K-fold cross-validation technique is applied, where k = 5, assuring the dataset quality. A momentum with gradient descent optimising strategy is adapted for faster learning and in order not to be stuck with local minima. After tuning with different functions and values, the following hyperparameters are used for the optimum performance of the model. An adaptive learning rate approach optimisation algorithm—an adaptive moment optimisation (Adam) optimiser [44] is applied with a learning rate of 0.001. Here, the exponential decay rates of 0.9 and 0.999 are used for the first and second moment, respectively. Considering this model's multinomial classification and probability prediction, a categorical cross-entropy loss function is used, ensuring minimum loss while compiling the model. The model performed better when set with a batch size of 32 and an epoch size of 100, as plotted in Figure 11, depicting the loss and accuracy curve graphs during training and validation.

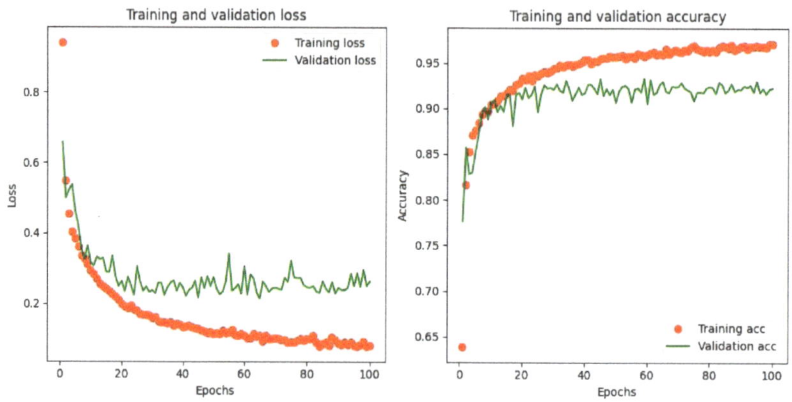

Figure 11. Loss and accuracy curves during training and validation.

The prediction accuracy of the trained model is assessed based on the statistical metrics precision, recall, F1-score, and accuracy following Equations (7)–(10) [45], where TP, TN, FP, and FN are True Positive, True Negative, False Positive, and False Negative, respectively. The additional 500 samples recorded for each class during data acquisition, which were not part of training, were used for the 1D CNN model evaluation, resulting in an average prediction accuracy of 92.6% with an inference time of 0.238 ms. Table 1 lists the detailed results of this offline evaluation test for each class. Hence, this proposed model is suitable for accurately predicting the vibration threshold classes and condition monitoring applications in outdoor mobile robots

$$Precision = \frac{TP}{TP + FP} \tag{7}$$

$$Recall = \frac{TP}{TP + FN} \tag{8}$$

$$F1Score = \frac{2 \times Precision \times Recall}{Precision + Recall} \tag{9}$$

$$Accuracy = \frac{TP + TN}{TP + FP + TN + FN} \tag{10}$$

Table 1. One-dimensional CNN model evaluation result.

Vibration Threshold Classes	Precision	Recall	F1 Score	Accuracy
Safe	0.94	0.92	0.93	0.93
Mod safe System-generated (MS)	0.87	0.89	0.88	0.88
Mod safe Terrain-induced—Left (MTL)	0.90	0.89	0.92	0.91
Mod safe Terrain-induced—Right (MTR)	0.91	0.94	0.92	0.92
Mod safe-Terrain induced-Left & Right (MT)	0.95	0.96	0.93	0.95
Unsafe System-generated (US)	0.90	0.93	0.89	0.90
Unsafe Terrain-induced—Left (UTL)	0.93	0.94	0.91	0.93
Unsafe Terrain-induced—Right (UTR)	0.94	0.94	0.92	0.94
Unsafe Terrain-induced—Left & Right (UT)	0.97	0.98	0.95	0.97

A summary of the various challenges and fixes during the 1D-CNN-based classifier training and an evaluation are as follows. The training data acquisition and compilation representing all nine classes were challenging; the data subscription rate for both types of sensors selected (IMU and current sensors) should be the same, and a higher data subscription rate is preferred for a maximum number of predictions per unit of time, where each data feature (a total of 11 features from two types of sensors) with a process window size (128 temporal data) is to be confirmed, forming a single dataset for training. This was realised by developing a data subscription and compiling algorithm, lowering the delay in the Arduino program during the current consumption data subscription for the maximum data subscription rate, and ensuring that the dataset was prepared for all the features with a uniform processing window size. As the proposed IMU and current heterogeneous sensor data fall on different scales, there are difficulties in converging speed during training; hence, the data are brought to a common scale by normalisation. A fast and low computation cost and higher prediction accuracy are crucial for the CM in a mobile robot; hence, the model is structured with a minimum number (four) of convolution layers and tested with different filter sizes and convolution windows in order to choose ideal parameters to fit the given dataset. We opted for a suitable activation function (ReLU) for this classification model, tested with different configurations of pooling and dropout layers to reduce the computation time and overfitting, and finalised the structure of the 1D CNN model for a fast and accurate prediction model. Similarly, the challenges faced during the training for fast convergence and better accuracy were fixed by exploring the various hyperparameters of the model. We used the accuracy and loss curve tool to monitor how well the proposed 1D CNN model fits training data (training loss) and new data (validation loss) and updated the hyperparameters accordingly to finalise a fast and accurate model.

4. Real-Time Experiments and Framework Validation

The proposed haptic-feedback-based real-time condition monitoring and remote controlling framework to assess system-generated issues mainly deterioration and minimise traverse through terrain-related flaws reducing the chances of accelerated deterioration and hazards, is validated by conducting three field case studies. Prior to the case studies, a threshold class inference and haptic feedback trigger engine is developed as illustrated in the flowchart, Figure 12. Here, the engine applies the 1D CNN model's knowledge for every new data sample for inferencing in real-time. A total of 128 temporal data (*Data*) and 11 features (*Feature*) form a new dataset [128 $\times$ 11] for prediction. A placeholder *TemporalBuffer* is used during dataset collection, and *InferenceBuffer* holds one complete dataset as input to the 1D CNN model, runs on the TensorFlow platform, and returns the predicted threshold class. The algorithm also switches the predicted class (*PC*) to the corresponding vibration pattern, as illustrated. Next, the robot's shared autonomy and control operations over cloud computing, including the class inference and haptic feedback engine performance, are verified by running the robot on a plain pavement and intruding rods of different sizes, testing safe, MTL, MTR, MT, UTL, UTR, and UT classes as shown in Figure 13.

Field case studies are planned to assess all the threshold classes' real-time prediction accuracy and how the vibrotactile haptic feedback system enables operators to stop, avoid/minimise ground-level obstacles, and ensure a safe class. Two engineers are trained to learn vibration patterns for each threshold class and are assigned these field trials to monitor and switch over to manual control and operate based on haptic feedback threshold class patterns. The robot is driven in a zig-zag pattern to cover the given test area at a linear speed of 0.3 m/s. The IMU and current sensors data subscription ensure the same at 80 Hz. Before starting each trial, it is also ensured that the robot's health conditions are good.

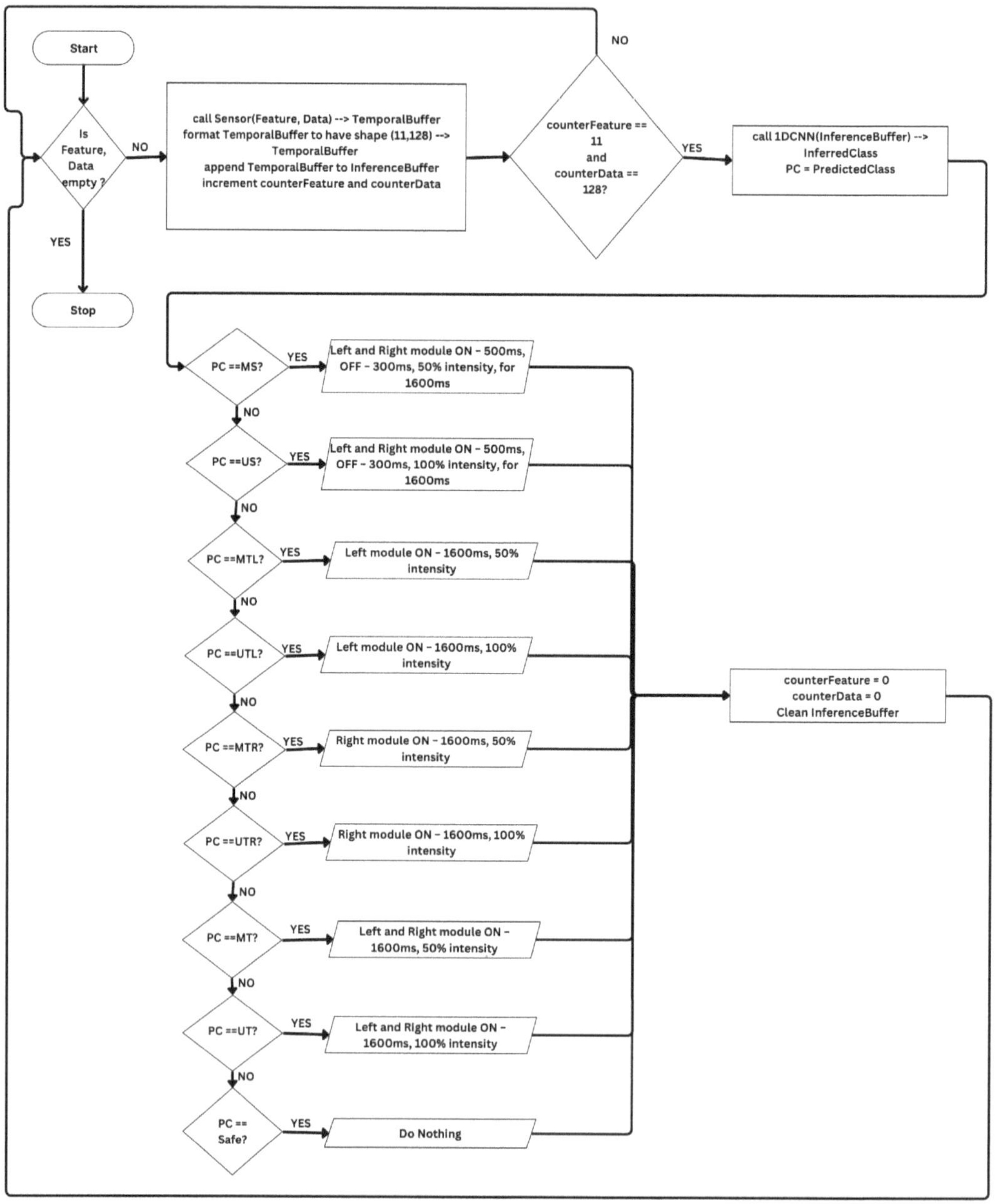

Figure 12. Flowchart: class inference and haptic feedback engine.

Figure 13. Pre-trial set up and framework testing.

A plain ground with uniformly paved tiles is identified for the first case study and a total area of 10×12 m^2 is selected for testing. There are few drainage utility holes with raised covers as ground-level obstacles and also some aluminium profiles of different sizes are kept, randomly inducing terrain threshold classes. Secondly, an uneven grass field of a test area of 15×20 m^2 is selected, where many natural surface level flaws of different sizes are observed, such as small pits, stones, and tree roots spread over the ground, that are either passable or blocking. Most of these imperfections are covered by grass so that they can be detected and controlled only through this proposed retrospective technique. Finally, a pavement area is identified with ground-level obstacles, such as missing or broken tiles, small pits, cliffs, kerbs, gutters, and drainage cover at varied heights, and more obstacles are added, which is the same as the first case study, simulating a poorly maintained pavement. Here, a total area of 8×16 m^2 was selected and isolated from other users, considering further testing as a long-term trial for system-generated classes. The test areas for these three field case studies are shown in Figure 14.

Figure 14. Three case studies conducted for haptic-feedback-based CM framework validation.

We planned three operation modes for each case study to validate the effectiveness of this proposed haptic-feedback-based condition monitoring and robot controlling framework. The first mode is without executing the proposed haptic feedback system and observing the total number of abnormal threshold classes predicted by the system. This means that the robot is exposed to each test area with uneven terrain, which will cause an accelerated system deterioration or become hazardous if driven continuously. The second and third modes run with the proposed haptic feedback system, and the robot is controlled remotely by operators 1 and 2, respectively. Here, we mainly tested how many ground-level obstacles are avoided based on the haptic feedback when the robot hit the terrain obstacles and control, avoiding the severe features or being limited to the moderate class, or helping not to repeat this for the rear wheels by adequately changing the heading direction. So, here, we mainly checked the total number of abnormal threshold classes reduced compared to without the haptic feedback and controlling system. Also, with two operators, the repetitiveness of the system was tested and validated. The robot drove on the three field areas selected for each mode and covered the whole area once in a zig-zag pattern for each test. The test results summary of these three case studies (CS 1-3) for the three modes of operation is plotted in Figure 15.

Here, for instance, in case study 1, without integrating the haptic feedback system for real-time control, which is mode 1, the robot encountered 62 abnormal terrain features. At the same time, the operator could control the robot for the same test environment, avoiding such terrain anomalies based on the proposed haptic feedback, and encountered only 17 terrain features (as shown in mode 2), reducing the chances of accelerated system degradation and operational hazards.

Accordingly, the total number of abnormal classes exposed without haptic feedback and hence the chances of system deterioration and hazards is four times more than with the proposed system. Also, there is not much difference between modes 2 and 3, i.e., the ease of operation in detecting the feedback pattern and controlling is validated. Additionally, for mode 1 without haptic-feedback-based operations, a safety operator on the spot was involved whenever the robot got stuck, especially when encountering MT, UTL, UTR, and UT classes. The total number of instances of manual involvement needed to resume the operations as per mode 1 for case studies 1–3 was 4, 13, and 5, respectively. This also shows additional time and effort for a robot without integrating such a proposed CM framework, other than the system degradation and chances of hazardous events.

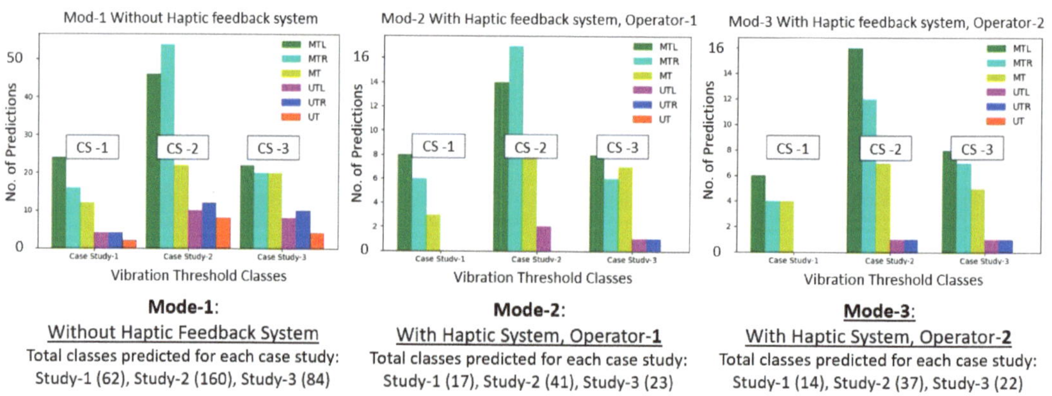

Figure 15. Threshold class predicted/haptic feedback summary: 3 case studies, 3 operation modes.

Here, no system-deteriorated classes are observed as the robot was tested in healthy condition and the test was completed in the short term to cover the given area once. So, for testing system-generated classes, the third case study was planned to continue for the long term (average 4 h a day) by removing the comparatively bigger ground-level obstacles, allowing the robot to pass over the obstacles limited to moderated safe terrain-induced classes. The robot was driven in a closed loop and zig-zag pattern in the predefined area. The prediction classes and haptic feedback were tested daily. On the 20th day, an intermittent prediction of the MS class and corresponding haptic feedback pattern was observed. We continued this state for a few hours, closely monitoring the ground truth with safety precautions such as e-stop, and, further, one of the wheels became punctured and generated wobbling. The system occasionally predicted the US class as haptic feedback whenever the wobbling effect was severe. To avoid total failure or hazards, we stopped the operations and triggered maintenance actions to fix the loose assembly and replace the wheel.

We randomly collected 300 samples of each class from the above case studies and long-term tests, closely observing the ground truth and actual haptic feedback to assess the real-time prediction accuracy. Accordingly, the average accuracy was 91.1%, as listed in Table 2 for each class, close to the offline evaluation results. This shows the feasibility and repetitive accuracy of the proposed vibrotactile haptic-feedback-based condition monitoring framework for outdoor mobile robots for real-time applications.

Table 2. Real-time prediction accuracy of threshold classes.

Haptic Feedback Classes	Safe	MS	MTL	MTR	MT	US	UTL	UTR	UT
Prediction Accuracy (%)	92	86	88	89	90	92	94	93	96

The proposed study is distinct from the existing techniques, such as shared autonomy, haptic feedback, and terrain classification. The usual shared autonomy works mainly to switch to operator control due to the limitation of safe navigation, i.e., mainly avoiding obstacles in its heading direction. The haptic feedback devices are generally used for human health monitoring and to collect information about the obstacles in mobile robots. The terrain classification intends to detect the type of terrain in order to assess the traversability. However, our proposed work is for CM applications in outdoor mobile robots detecting both the robot's degradation level and terrain extremities, causing accelerated deterioration and operational hazards. Hence, this is an additional protection other than shared autonomy capabilities.

5. Conclusions

The deployment of a mobile robot in a public outdoor workspace exposed to extreme terrain features is still challenging due to the lack of a proper robot health monitoring system, real-time control that avoids ground-level obstacles, limitations of technology readiness, and other local regulations. This paper discussed and proposed a feasible and novel approach for the real-time condition monitoring and remote control of an outdoor robot by developing an AI-enabled haptic-feedback-based condition monitoring framework that assesses the robot's health state and enables remote operators to minimise traversing through undesired terrain flaws. We proposed eight abnormal vibration threshold levels for a typical four-wheeled outdoor robot for condition monitoring, including the system health/deterioration status and the wheel–terrain-related undesired effects due to extreme terrain features. The vibration-indicated data for the threshold classes were modelled based on the onboard IMU and current sensors data, which are independent of environment conditions and complement each other, representing the robot's behaviour for each class. A simple structured 1D CNN model was developed and trained on this heterogeneous sensor data for fast, accurate, and real-time execution. A wearable vibrotactile haptic feedback device architecture suitable for the condition monitoring of outdoor robots and configured multi-level vibrotactile patterns unique to each threshold class is presented, enabling the real-time controlling of the robot. A shared autonomy integrated in-house-developed outdoor fumigation robot was used as a test platform in this work to validate the proposed framework by conducting three field case studies, including a long-term test. Accordingly, the threshold class prediction accuracy was found to be 91.1%, and the effectiveness of the real-time haptic device operation was evaluated to be four times better than the usual practice by assisting either in stopping or minimising the exposure to extreme terrain features. Hence, the proposed AI-enabled vibrotactile feedback-based condition monitoring and remote control significantly impact the robot's health and operational safety. We will further improve the prediction accuracy by exploring additional sensors as future work suitable for outdoor mobile robots.

Author Contributions: Conceptualisation, S.P., P.V. and M.R.E.; methodology, S.P., R.E.A. and P.V.; software, R.E.A. and S.P.; validation, S.P. and R.E.A.; formal analysis, S.P., M.R.E. and R.E.A.; investigation, S.P. and R.E.A.; resources, M.R.E.; data, S.P. and R.E.A.; writing—original draft preparation, S.P., R.E.A. and P.V.; supervision, M.R.E.; project administration, M.R.E.; funding acquisition, M.R.E. All authors have read and agreed to the published version of the manuscript.

Funding: This research is supported by the National Robotics Programme under its National Robotics Programme (NRP) BAU, Ermine III: Deployable Reconfigurable Robots, Award No. M22NBK0054 and also supported by A*STAR under its "RIE2025 IAF-PP Advanced ROS2-native Platform Technologies for Cross-sectorial Robotics Adoption (M21K1a0104)" programme.

Data Availability Statement: Data will be made available on request.

Acknowledgments: The authors would like to thank the National Robotics Programme, the Agency for Science, Technology and Research, and SUTD for their support.

Conflicts of Interest: The authors declare no conflict of interest.

References

1. ABIResearch. Labor Shortages and Workplace Safety Concerns Propel Shipments of Outdoor Mobile Robots to 350,000 by 2030. Available online: https://www.abiresearch.com/press/labor-shortages-and-workplace-safety-concerns-propel-shipments-of-outdoor-mobile-robots-to-350000-by-2030/ (accessed on 10 January 2023)
2. Zhang, F.S.; Ge, D.Y.; Song, J.; Xiang, W.J. Outdoor scene understanding of mobile robot via multi-sensor information fusion. *J. Ind. Inf. Integr.* **2022**, *30*, 100392. [CrossRef]
3. Liang, Z.; Fang, T.; Dong, Z.; Li, J. An Accurate Visual Navigation Method for Wheeled Robot in Unstructured Outdoor Environment Based on Virtual Navigation Line. In Proceedings of the International Conference on Image, Vision and Intelligent Systems (ICIVIS 2021), Changsha, China, 15–17 June 2021; pp. 635–656.
4. Dong, Y.; Liu, S.; Zhang, C.; Zhou, Q. Path Planning Research for Outdoor Mobile Robot. In Proceedings of the 2022 12th International Conference on CYBER Technology in Automation, Control, and Intelligent Systems (CYBER), Jilin City, China, 27–31 July 2022; pp. 543–547.
5. Yang, L.; Wang, L. A semantic SLAM-based dense mapping approach for large-scale dynamic outdoor environment. *Measurement* **2022**, *204*, 112001. [CrossRef]
6. Liu, F.; Li, X.; Yuan, S.; Lan, W. Slip-aware motion estimation for off-road mobile robots via multi-innovation unscented Kalman filter. *IEEE Access* **2020**, *8*, 43482–43496. [CrossRef]
7. Manikandan, N.; Kaliyaperumal, G. Collision avoidance approaches for autonomous mobile robots to tackle the problem of pedestrians roaming on campus road. *Pattern Recognit. Lett.* **2022**, *160*, 112–121. [CrossRef]
8. CETRAN. Centre of Excellence for Testing & Research of Autonomous Vehicles—NTU. Available online: https://cetran.sg/ (accessed on 10 November 2022)
9. Reddy, S.; Dragan, A.D.; Levine, S. Shared autonomy via deep reinforcement learning. *arXiv* **2018**, arXiv:1802.01744.
10. Azcarate, R.F.; Daniela, S.; Hayat, A.; Yi, L.; Muthugala, M.V.J.; Tang, Q.; Povendhan, A.; Leong, K.; Elara, M. Shared Autonomy for Safety Between a Self-reconfigurable Robot and a Teleoperator Using Multi-layer Fuzzy Logic. In Proceedings of the 2022 IEEE/RSJ International Conference on Intelligent Robots and Systems (IROS), Kyoto, Japan, 23–27 October 2022; pp. 141–148.
11. Lee, S.; Sukhatme, G.S.; Kim, G.J.; Park, C.M. Haptic control of a mobile robot: A user study. In Proceedings of the IEEE/RSJ International Conference on Intelligent Robots and Systems, Lausanne, Switzerland, 30 September–4 October 2002; Volume 3, pp. 2867–2874.
12. Farkhatdinov, I.; Ryu, J.H.; An, J. A preliminary experimental study on haptic teleoperation of mobile robot with variable force feedback gain. In Proceedings of the 2010 IEEE Haptics Symposium, Waltham, MA, USA, 25–26 March 2010; pp. 251–256.
13. Rösch, O.J.; Schilling, K.; Roth, H. Haptic interfaces for the remote control of mobile robots. *Control. Eng. Pract.* **2002**, *10*, 1309–1313. [CrossRef]
14. Diolaiti, N.; Melchiorri, C. Obstacle avoidance for teleoperated mobile robots by means of haptic feedback. In Proceedings of the IEEE 1st International Workshop on Advances in Service Robotics, Online, 31 March 2003.
15. Dupont, E.M.; Moore, C.A.; Collins, E.G.; Coyle, E. Frequency response method for terrain classification in autonomous ground vehicles. *Auton. Robot.* **2008**, *24*, 337–347. [CrossRef]
16. Csík, D.; Odry, Á.; Sárosi, J.; Sarcevic, P. Inertial sensor-based outdoor terrain classification for wheeled mobile robots. In Proceedings of the 2021 IEEE 19th International Symposium on Intelligent Systems and Informatics (SISY), Subotica, Serbia, 16–18 September 2021; pp. 159–164.
17. Khan, Y.N.; Komma, P.; Bohlmann, K.; Zell, A. Grid-based visual terrain classification for outdoor robots using local features. In Proceedings of the 2011 IEEE Symposium on Computational Intelligence in Vehicles and Transportation Systems (CIVTS) Proceedings, Paris, France, 11–15 April 2011; pp. 16–22.

18. Weiss, C.; Tamimi, H.; Zell, A. A combination of vision-and vibration-based terrain classification. In Proceedings of the 2008 IEEE/RSJ International Conference on Intelligent Robots and Systems, Nice, France, 22–26 September 2008; pp. 2204–2209.
19. Suger, B.; Steder, B.; Burgard, W. Traversability analysis for mobile robots in outdoor environments: A semi-supervised learning approach based on 3D-lidar data. In Proceedings of the 2015 IEEE International Conference on Robotics and Automation (ICRA), Seattle, WA, USA, 26–30 May 2015; pp. 3941–3946.
20. Janssens, O.; Slavkovikj, V.; Vervisch, B.; Stockman, K.; Loccufier, M.; Verstockt, S.; Van de Walle, R.; Van Hoecke, S. Convolutional neural network based fault detection for rotating machinery. *J. Sound Vib.* **2016**, *377*, 331–345. [CrossRef]
21. Kumar, P.; Shankar Hati, A. Convolutional neural network with batch normalisation for fault detection in squirrel cage induction motor. *IET Electr. Power Appl.* **2021**, *15*, 39–50. [CrossRef]
22. Abdeljaber, O.; Sassi, S.; Avci, O.; Kiranyaz, S.; Ibrahim, A.A.; Gabbouj, M. Fault detection and severity identification of ball bearings by online condition monitoring. *IEEE Trans. Ind. Electron.* **2018**, *66*, 8136–8147. [CrossRef]
23. Kiranyaz, S.; Avci, O.; Abdeljaber, O.; Ince, T.; Gabbouj, M.; Inman, D.J. 1D convolutional neural networks and applications: A survey. *Mech. Syst. Signal Process.* **2021**, *151*, 107398. [CrossRef]
24. Eren, L.; Ince, T.; Kiranyaz, S. A generic intelligent bearing fault diagnosis system using compact adaptive 1D CNN classifier. *J. Signal Process. Syst.* **2019**, *91*, 179–189. [CrossRef]
25. Ince, T.; Kiranyaz, S.; Eren, L.; Askar, M.; Gabbouj, M. Real-time motor fault detection by 1-D convolutional neural networks. *IEEE Trans. Ind. Electron.* **2016**, *63*, 7067–7075. [CrossRef]
26. Abdeljaber, O.; Avci, O.; Kiranyaz, S.; Gabbouj, M.; Inman, D.J. Real-time vibration-based structural damage detection using one-dimensional convolutional neural networks. *J. Sound Vib.* **2017**, *388*, 154–170. [CrossRef]
27. Wang, J.; Wang, D.; Wang, X. Fault diagnosis of industrial robots based on multi-sensor information fusion and 1D convolutional neural network. In Proceedings of the 2020 39th Chinese Control Conference (CCC), Shenyang, China, 27–29 July 2020; pp. 3087–3091.
28. Kim, H.G.; Yoon, H.S.; Yoo, J.H.; Yoon, H.I.; Han, S.S. Development of predictive maintenance technology for wafer transfer robot using clustering algorithm. In Proceedings of the 2019 International Conference on Electronics, Information, and Communication (ICEIC), Auckland, New Zealand, 22–25 January 2019; pp. 1–4.
29. Onur, K.; Kaymakci, O.T.; Mercimek, M. Advanced Predictive Maintenance with Machine Learning Failure Estimation in Industrial Packaging Robots. In Proceedings of the 2020 International Conference on Development and Application Systems (DAS), Suceava, Romania, 21–23 May 2020; pp. 1–6.
30. Aliev, K.; Antonelli, D. Proposal of a monitoring system for collaborative robots to predict outages and to assess reliability factors exploiting machine learning. *Appl. Sci.* **2021**, *11*, 1621. [CrossRef]
31. Pookkuttath, S.; Rajesh Elara, M.; Sivanantham, V.; Ramalingam, B. AI-Enabled Predictive Maintenance Framework for Autonomous Mobile Cleaning Robots. *Sensors* **2022**, *22*, 13. [CrossRef]
32. Pookkuttath, S.; Gomez, B.F.; Elara, M.R.; Thejus, P. An optical flow-based method for condition-based maintenance and operational safety in autonomous cleaning robots. *Expert Syst. Appl.* **2023**, *222*, 119802. [CrossRef]
33. Constantin, G.; Maroșan, I.A.; Crenganiș, M.; Botez, C.; Gîrjob, C.E.; Biriș, C.M.; Chicea, A.L.; Bârsan, A. Monitoring the Current Provided by a Hall Sensor Integrated in a Drive Wheel Module of a Mobile Robot. *Machines* **2023**, *11*, 385. [CrossRef]
34. Rapalski, A.; Dudzik, S. Energy Consumption Analysis of the Selected Navigation Algorithms for Wheeled Mobile Robots. *Energies* **2023**, *16*, 1532. [CrossRef]
35. Kryter, R.; Haynes, H. *Condition Monitoring of Machinery Using Motor Current Signature Analysis*; Technical Report; Oak Ridge National Lab.: Oak Ridge, TN, USA, 1989.
36. Sanfilippo, F.; Pettersen, K.Y. A sensor fusion wearable health-monitoring system with haptic feedback. In Proceedings of the 2015 11th International Conference on Innovations in Information Technology (IIT), Dubai, United Arab Emirates, 1–3 November 2015; pp. 262–266.
37. Devigne, L.; Aggravi, M.; Bivaud, M.; Balix, N.; Teodorescu, C.S.; Carlson, T.; Spreters, T.; Pacchierotti, C.; Babel, M. Power wheelchair navigation assistance using wearable vibrotactile haptics. *IEEE Trans. Haptics* **2020**, *13*, 52–58. [CrossRef]
38. Song, G.; Guo, S.; Wang, Q. A Tele-operation system based on haptic feedback. In Proceedings of the 2006 IEEE International Conference on Information Acquisition, Weihai, China, 20–23 August 2006; pp. 1127–1131.
39. Dekker, I.; Kellens, K.; Demeester, E. Design and Evaluation of an Intuitive Haptic Teleoperation Control System for 6-DoF Industrial Manipulators. *Robotics* **2023**, *12*, 54. [CrossRef]
40. Pamungkas, D.S.; Ward, K. Tele-operation of a robot arm with electro tactile feedback. In Proceedings of the 2013 IEEE/ASME International Conference on Advanced Intelligent Mechatronics, Wollongong, Australia, 9–12 July 2013; pp. 704–709.
41. Lee, J.W.; Yu, K.H. Wearable Drone Controller: Machine Learning-Based Hand Gesture Recognition and Vibrotactile Feedback. *Sensors* **2023**, *23*, 2666. [CrossRef] [PubMed]
42. Mitiche, I.; Nesbitt, A.; Conner, S.; Boreham, P.; Morison, G. 1D-CNN based real-time fault detection system for power asset diagnostics. *IET Gener. Transm. Distrib.* **2020**, *14*, 5766–5773. [CrossRef]
43. Abadi, M.; Barham, P.; Chen, J.; Chen, Z.; Davis, A.; Dean, J.; Devin, M.; Ghemawat, S.; Irving, G.; Isard, M.; et al. Tensorflow: A system for large-scale machine learning. In Proceedings of the Osdi, Savannah, GA, USA, 2–4 November 2016; Volume 16, pp. 265–283.

44. Kingma, D.P.; Ba, J. Adam: A method for stochastic optimization. *arXiv* **2014**, arXiv:1412.6980.
45. Grandini, M.; Bagli, E.; Visani, G. Metrics for multi-class classification: An overview. *arXiv* **2020**, arXiv:2008.05756.

mathematics

MDPI

Article

Machine Learning-Based Model Predictive Control of Two-Time-Scale Systems

Aisha Alnajdi [1], Fahim Abdullah [2], Atharva Suryavanshi [2] and Panagiotis D. Christofides [1,2,*]

1 Department of Electrical and Computer Engineering, University of California, Los Angeles, CA 90095, USA; aishaaln2@gmail.com
2 Department of Chemical and Biomolecular Engineering, University of California, Los Angeles, CA 90095, USA; fa2@g.ucla.edu (F.A.); atharvasurya99@g.ucla.edu (A.S.)
* Correspondence: pdc@seas.ucla.edu; Tel.: +1-(310)-794-1015

Abstract: In this study, we present a general form of nonlinear two-time-scale systems, where singular perturbation analysis is used to separate the dynamics of the slow and fast subsystems. Machine learning techniques are utilized to approximate the dynamics of both subsystems. Specifically, a recurrent neural network (RNN) and a feedforward neural network (FNN) are used to predict the slow and fast state vectors, respectively. Moreover, we investigate the generalization error bounds for these machine learning models approximating the dynamics of two-time-scale systems. Next, under the assumption that the fast states are asymptotically stable, our focus shifts toward designing a Lyapunov-based model predictive control (LMPC) scheme that exclusively employs the RNN to predict the dynamics of the slow states. Additionally, we derive sufficient conditions to guarantee the closed-loop stability of the system under the sample-and-hold implementation of the controller. A nonlinear chemical process example is used to demonstrate the theory. In particular, two RNN models are constructed: one to model the full two-time-scale system and the other to predict solely the slow state vector. Both models are integrated within the LMPC scheme, and we compare their closed-loop performance while assessing the computational time required to execute the LMPC optimization problem.

Keywords: two-time-scale systems; machine learning; recurrent neural networks; long short-term memory recurrent neural networks; feedforward neural network; process control; model predictive control; nonlinear systems; singular perturbations

MSC: 93B45; 93C10; 93C70

1. Introduction

Various applications in the field of chemical engineering involve systems that exhibit different time scales. Instances of such systems include biochemical processes, catalytic reactors, and distillation columns. Furthermore, other scientific sectors, such as power electronics, communication networks, and biological systems, also feature systems with dynamics evolving in disparate time scales. Specifically, in the context of chemical engineering, researchers have utilized the method of singular perturbation to decouple two-time-scale systems into reduced-order subsystems, each associated with a distinct time scale (e.g., [1]). This approach simplifies the analysis of the ordinary differential equations governing the system, allowing for the design of a suitable well-conditioned control law that stabilizes the system. In the context of control system design, model predictive control (MPC) is widely recognized as one of the leading and most convenient approaches employed in stabilizing different types of nonlinear systems. MPC is essentially an optimization problem formulated with a well-defined objective function aimed at enhancing the performance of the system while meeting constraints associated with the system's physical structure and closed-loop stability. The fact that MPC accounts for multiple inputs, outputs, and constraints within its framework is what makes it a suitable choice for designing control systems to stabilize chemical processes. However, in singularly perturbed systems, the presence of distinct time scales can lead to challenges when employing MPC without accounting for the state evolution in the different time scales. Neglecting this aspect can result in degradation of the closed-loop performance, long time consumption, or, in more critical situations, instability and stiffness in the control system due to issues with the controller's effectiveness and the presence of ill-conditioning [2].

However, if the time-scale multiplicity is accounted for, MPC has proven to be an efficient control strategy when it comes to situations involving time-scale separation. For instance, in [3], the design of a composite control system was presented, which included two MPC designs: one for the slow subsystem and one for the fast subsystem. Using stability analysis and the theory of singular perturbations, the authors investigated and guaranteed the closed-loop stability for the full, nonlinear two-time-scale system under Lyapunov-based tracking MPC. To additionally consider process economics, a similar strategy was proposed in [4], which focused on the design of a composite controller involving two MPC problems. Specifically, for the stability of the fast subsystem (and full two-time-scale system as a result), a Lyapunov-based tracking MPC was used for the fast states based on the error between the fast states and the slow manifold using a quadratic cost function, while a Lyapunov-based economic MPC was used for the slow subsystem to optimize the economic considerations and achieve any other desirable closed-loop stability properties using a non-quadratic, economic cost function. The authors of [3,4] focused on first-principles-based controllers, and in [5], a two-time-scale system was separated into slow and fast subsystems using a singular perturbation strategy and then modeled using only data from the system via the method of sparse identification for nonlinear dynamics. Subsequently, an MPC was designed based on the reduced-order slow subsystem modeled using sparse identification, and closed-loop stability guarantees were derived. Furthermore, numerical simulations were used to demonstrate the reduced computational cost of the reduced-order sparse-identified model.

Due to the vast advancements in the chemical industry and the multitude of complex reactions occurring in real-life chemical process applications, engineers encounter challenges when constructing first-principles models for these chemical processes. As a result, in recent years, machine learning (ML) models have been utilized to approximate the dynamics of chemical processes. These models are highly beneficial when designed carefully, as they offer a reliable and efficient alternative to classical first-principles models and can then be incorporated into MPC frameworks. Many works have been conducted in this field. For instance, the authors of [6] offered essential insights and a fundamental theory regarding the integration of machine learning techniques into MPC schemes. Their work focused on designing a recurrent neural network (RNN) model to approximate the nominal nonlinear system and integrated the model within an MPC framework to stabilize the system. In [7], polynomial nonlinear autoregressive with exogenous inputs (NARX) models were used to build nonlinear MPC, with the relevant polynomial terms to retain selected via sparse regression. The proposed MPC was applied to a multi-input multi-output chemical reactor, and an algebraic modeling language was used to reduce computational time.

Another research direction termed "approximate MPC" involves the replacement of the MPC optimization problem with a closed-form expression that approximates the MPC control actions without having to solve an optimization problem. This is meant to address the computational challenges inherent in classical MPC at the expense of lower accuracy and generalizability. In [8], an approximate MPC was designed for heating, ventilation, and air-conditioning (HVAC) systems due to the stringent computational resources available for building control systems. Specifically, the MPC optimization problem was imitated using a recurrent neural network with a structure of a nonlinear autoregressive network with exogenous inputs trained with data from 30 days of operation in a closed loop under MPC.

While both MPC and approximate MPC greatly reduced the cooling consumption of the HVAC system when tested, approximate MPC solved for the control actions over 100 times faster throughout the testing period, with only a 12% degradation in performance compared to MPC. An alternate research direction to reduce the computational burden of MPC is the design of "explicit MPC", in which the optimal control actions are computed offline using multiparametric programming methods, allowing the online component of MPC to only have to search a table of linear gains and compute a single function evaluation. A summary of studies investigating explicit MPC is provided in [9]. Machine learning has also been investigated in the context of explicit MPC. In [10], for example, an explicit control law was learned for discrete-time linear time-invariant systems using machine learning. Specifically, the key challenge addressed was that of high dimensionality, which has been attempted to be solved in the existing literature by identifying suboptimal polytope partitions of the state space and designing control laws based on such regions. Ref. [10] focused on tackling high dimensionality in a similar manner but by using reinforcement learning and additionally investigated constraint satisfaction and feasibility guarantees for explicit MPC. Specifically, by using deep neural networks with rectified linear units as the activation function and a modified policy gradient algorithm based on prior knowledge, the objectives were met and demonstrated via three numerical examples of varying levels of dimensionality and complexity. A potential limitation of the approximate and explicit MPC approaches is that they are practically applicable to only smaller-scale systems, even though they are applicable to systems with very fast sampling times [9]. However, for general nonlinear systems and nonlinear MPC, due to the highly nonconvex optimization problem, methods such as the use of suboptimal polytopes [10] may not be readily extended to the nonlinear case.

While ML techniques have shown great success when incorporating them into MPC schemes, the practical application of machine learning-based MPC schemes to real-life processes poses significant challenges. The fact that the ML model is developed using a finite number of data samples makes it challenging to assess the accuracy of the model when considering generalized scenarios. Therefore, researchers have developed methods to quantify the error of the process model used in MPC and study its effects on closed-loop performance and stability. For example, ref. [11] proposed the use of Bayesian neural networks (BNN) to quantify the plant-model mismatch and subsequently create adaptive scenarios in real time for scenario-based MPC using the BNN. Based on the simulation results from a cold atmospheric plasma system, the proposed approach outperformed scenario-based MPC without adaptive scenarios and adaptive scenario-based MPC with Gaussian process regression. On more theoretical fronts, researchers have also studied the concept of generalization error bounds, which are crucial to study to ensure the construction of reliable models that are capable of performing accurate predictions on unseen or new data. Specifically, a low generalization error bound guarantees the effectiveness and high performance of the ML model and, consequently, the MPC scheme utilizing the designed ML model. To address this, ref. [12] derived a generalization error bound for a fully connected RNN (FCRNN) model, discussed the main factors that affect the bound, and integrated the FCRNN model into an MPC scheme to stabilize a nonlinear process. Moreover, the authors of [13] focused on deriving a generalization error bound for a feedforward neural network (FNN) that was used to construct a control Lyapunov-barrier function. The authors of [14] studied the generalization error bound for both a partially connected RNN (PCRNN) and an LSTM-RNN and subsequently carried out a comparison study between the generalization performance of an FCRNN and a PCRNN. Additionally, all the aforementioned works incorporated the designed machine learning models into Lyapunov-based MPC schemes to demonstrate the ability of the LMPCs to stabilize a chemical process. To the best of our knowledge, the application of generalization error bounds to neural networks modeling two-time-scale systems has not yet been investigated.

In light of the above considerations, in this article, we use the theory of statistical machine learning to construct the generalization error bounds for neural networks modeling two-time-scale systems. Moreover, the computational time for a neural network-based MPC is known to be an important practical consideration. However, the computational time for an RNN-based MPC with and without decomposition into lower-order subsystems has not yet been studied. Therefore, we introduce two LMPC frameworks, one designed based on an RNN model that predicts the slow state vector and the other designed based on an RNN model that models the full two-time-scale system, and compare the two LMPC schemes in terms of the closed-loop stability and computational time. The rest of this article is structured as follows. Section 2 introduces the essential preliminaries, as well as the general class of two-time-scale systems considered in this work. Section 3 discusses the generalization error bounds of neural networks modeling two-time-scale systems. In Section 4, we present a machine learning-based Lyapunov-based MPC using an RNN that predicts the slow state vector, followed by a closed-loop stability analysis and the corresponding results. In Section 5, a chemical process example is used to demonstrate the effectiveness of the designed controller. Section 6 summarizes the key findings obtained in this study.

2. Preliminaries

2.1. Notations

The Euclidean norm of a vector is denoted by $|\cdot|$. The transpose of the vector x is given by x^T. Given a matrix B, its Frobenius norm is denoted as $||B||_F$. Moreover, the infinity norm of the 1-norms of the columns of B is denoted as $||B||_{1,\infty} = \max_j \sum_i |B_{i,j}|$. The expression $L_F V(x)$ denotes the standard Lie derivative $L_F V(x) := \frac{\partial V(x)}{x} f(x)$. Set subtraction is denoted by '\', i.e., $A \backslash B := \{x \in \mathbb{R}^n | x \in A, x \notin B\}$. A function $f(x)$ is of class $\mathcal{C}^1$ if it is continuously differentiable in its domain. A continuous function $\alpha : [0, a) \to [0, \infty)$ belongs to class $\mathcal{K}$ if it is strictly increasing and is zero only when evaluated at zero. A function $f : \mathbb{R}^n \to \mathbb{R}^m$ is said to be L-Lipschitz continuous if there exists $L \geq 0$, such that $|f(a) - f(b)| \leq L|a - b|$ for all $a, b \in \mathbb{R}^n$. A continuous function $\beta : [0, a) \times [0, \infty) \to [0, \infty)$ belongs to class $\mathcal{KL}$ if, for each constant value of t, the function $\beta(\cdot, t)$ is of class $\mathcal{K}$, and for each constant value of s, the function $\beta(s, \cdot)$ is decreasing and approaches zero as $s \to \infty$. The probability that the event A will occur is denoted as $\mathbb{P}(A)$. Additionally, the expected value of a random variable X is denoted as $\mathbb{E}[X]$.

2.2. Class of Systems

The following family of ordinary differential equations describes the general class of two-time-scale continuous-time nonlinear systems with k states considered in this work:

$$\dot{x} = f_1(x, z, u, \epsilon) \tag{1a}$$
$$\epsilon \dot{z} = f_2(x, z, \epsilon) \tag{1b}$$

where $x \in \mathbb{R}^n$ is the slow state vector, $z \in \mathbb{R}^p$ is the fast state vector, and $n + p = k$. $u \in \mathbb{R}^{q_1}$ is the bounded manipulated input vector. The input vector u is constrained by $u \in U := \{|u_i| \leq u_{1_{\max,i}}, i = 1, ..., q_1\}$. We assume the vector functions $f_1(x, z, u, \epsilon)$ and $f_2(x, z, \epsilon)$ to be sufficiently smooth vector functions in $\mathbb{R}^n$ and $\mathbb{R}^p$, respectively. Moreover, we assume that the closed-loop stability region of the nonlinear system in Equation (1) is defined by the region Ω_{ρ_F}, where $\Omega_{\rho_F} := \{x \in D | V(x, z) \leq \rho_F\}$, where $\rho_F > 0$ and D is an open neighborhood around the origin. The speed ratio of the slow to the fast dynamics of the system is represented by the small positive parameter ϵ. In Equation (1b), we observe that the speed ratio ϵ pre-multiplies the derivative of the fast state vector z, which enables us to utilize a well-known strategy called "reduced-order modeling" via singular perturbations. In this approach, we will be able to decompose the system in Equation (1) into two different reduced-order subsystems. We follow the strategy thoroughly illustrated in [2].

The following equations are obtained by setting $\epsilon = 0$ in Equation (1):

$$\dot{\bar{x}} = f_1(\bar{x}, \bar{z}, \bar{u}, 0) \tag{2a}$$
$$0 = f_2(\bar{x}, \bar{z}, 0) \tag{2b}$$

where $\bar{x}$ and $\bar{z}$ are the slow state vector and the fast state vector associated with the system in Equation (1) under the case where $\epsilon = 0$, respectively. Additionally, the following assumption is considered to be fundamental in the theory of singular perturbations.

Assumption 1. *Equation (2b) has an isolated solution, which is determined by*

$$\bar{z} = \bar{f}_2(\bar{x}) \tag{3}$$

where $\bar{z}$ is a quasi-steady state for the fast state z, and the function $\bar{f}_2 : \mathbb{R}^n \rightarrow \mathbb{R}^p$ is continuously differentiable.

We substitute Equation (3) into Equation (2a) to obtain the reduced-order slow subsystem,

$$\dot{\bar{x}} = f_1(\bar{x}, \bar{f}_2(\bar{x}), \bar{u}, 0) \tag{4}$$

Furthermore, to obtain the dynamics of the reduced-order fast subsystem, we introduce a fast timescale $\tau = t/\epsilon$ and a new coordinate $\bar{\bar{\bar{z}}} = z - \bar{z}$ to re-write Equation (1b) with respect to the newly introduced variables $\bar{\bar{\bar{z}}}$ and τ. Then, the fast subsystem can be expressed as the derivative of $\bar{\bar{\bar{z}}}$ with respect to τ while setting $\epsilon = 0$,

$$\frac{\mathrm{d}\bar{\bar{\bar{z}}}}{\mathrm{d}\tau} = f_2(\bar{x}, \bar{\bar{\bar{z}}} + \bar{f}_2(\bar{x}), 0) \tag{5}$$

Starting at t_0, the initial conditions of the state vectors x and z are given by the vectors x_0 and z_0, respectively. Based on Equations (1)–(5), and by utilizing the basic theoretical concepts of two-time-scale systems in [2], the following equation will hold for all $t \in [t_p, T]$, where $t_p > t_0$

$$z(t) = \bar{z}(t) + \mathcal{O}(\epsilon). \tag{6}$$

where $\bar{z}(t)$ reflects the slow transient of z, and $\mathcal{O}(\epsilon)$ is an error of order epsilon. A variable, $z(t)$, is $\mathcal{O}(\epsilon)$, where ϵ is a positive constant if there exists a positive constant $\bar{k}$ (independent of ϵ), such that $|z(t)| \leq \bar{k}\epsilon$. Furthermore, if we were to consider the time interval where $t \in [t_0, T]$, Equation (6) would be slightly modified, such that the approximation of the state z is given by the following:

$$z(t) = \bar{z}(t) + \bar{\bar{\bar{z}}}(t) + \mathcal{O}(\epsilon), \tag{7}$$

where $\bar{z}(t)$ and $\bar{\bar{\bar{z}}}(t)$ are the slow and fast transients of z, respectively. Additionally, the approximation of state x by $\bar{x}$ is given by the following equation for all $t \in [t_0, T]$

$$x(t) = \bar{x}(t) + \mathcal{O}(\epsilon). \tag{8}$$

The fast subsystem in Equation (5) needs to satisfy certain stability properties for the above closeness of the solution estimates to hold true, which are described by the following assumption.

Assumption 2. *The equilibrium $\bar{\bar{\bar{z}}}(\tau) = 0$ in Equation (5) demonstrates uniform asymptotic stability in x_0 and t_0. In addition, $z_0 - \bar{z}(t_0)$ resides within its domain of attraction. As a result, for all $\tau \geq 0$, $\bar{\bar{\bar{z}}}(\tau)$ exists.*

If Assumption 2 holds true, then

$$\lim_{\tau \to \infty} \bar{\bar{z}}(\tau) = 0, \tag{9}$$

This implies that, at some time $t_p > t_0$, z will approach close to its quasi-steady state $\bar{z}$. The local stability of the equilibrium of the fast subsystem holds if the following verifiable condition holds.

Assumption 3. *All the eigenvalues of* $\frac{\partial f_2}{\partial z}$*, computed for* $\epsilon = 0$*, along* $\bar{x}(t), \bar{z}(t)$*, exhibit real parts less than a constant negative value, i.e.,*

$$\Re \, \lambda \left\{ \frac{\partial f_2}{\partial z} \right\} \leq -c < 0. \tag{10}$$

Utilizing the aforementioned assumptions, we establish the well-known "Tikhonov's theorem" in the following theorem.

Theorem 1 (c.f. Theorem 3.1 in [2])**.** *Consider that Assumptions 2 and 3 hold true. Then, for all* $t \in [t_0, T]$*, Equations (7) and (8) are valid. In addition, there exists a specific time instant* $t_p \geq t_0$*, such that Equation (6) is valid for all* $t \in [t_p, T]$*.*

Several sources such as [15] are useful for analyzing slightly varied formulations of Theorem 1 and reviewing its proof. At this point, we have investigated several essential concepts of two-time-scale systems. We will assume that the error between x and $\bar{x}$ is not greater than $\mathcal{O}(\epsilon)$. This assumption allows us to simplify the analysis by approximating $\bar{x}$ as x. Similarly, the same principle applies to z and $\bar{z}$, with our focus on capturing the slow transient of the fast state dynamics z, that is, $\bar{z}$. Therefore, based on this assumption, we express the reduced-order slow subsystem in Equation (4) as the following throughout this manuscript,

$$\dot{x} = F(x, u) := f(x) + g(x)u, \qquad x(t_0) = x_0, \tag{11}$$

where $f(\cdot)$ and $g(\cdot)$ are sufficiently smooth vector functions of dimensions $n \times 1$ and $n \times q_1$, respectively. Without loss of generality, we assume that the initial time $t_0 = 0$ and that $f(0) = 0$; hence, the steady state of the nonlinear system in Equation (11) is the origin. Finally, we will assume that the fast subsystem is globally asymptotically stable, which is necessary for establishing the stability of the closed-loop system under model predictive control using machine learning models in Section 4.

Assumption 4. *The origin of the closed-loop fast subsystem described by Equation (5) exhibits global asymptotic stability uniformly in* x*. This implies that there exists a function* $\beta_{\bar{\bar{z}}}$ *of class* $\mathcal{KL}$*, such that for any* $\bar{\bar{z}}(0) \in \mathbb{R}^p$*,*

$$|\bar{\bar{z}}(t)| \leq \beta_{\bar{\bar{z}}}\left(|\bar{\bar{z}}(0)|, \frac{t}{\epsilon} \right) \qquad \forall t \geq 0 \tag{12}$$

2.3. Stabilizability Assumption via Control Lyapunov Function

Regarding the dynamics of the slow subsystem described in Equation (11), we assume that there exists a locally Lipschitz feedback controller, $\Phi(x) \in U$, which renders the origin of the slow subsystem in Equation (11) asymptotically stable, i.e., there exists a $\mathcal{C}^1$ Lyapunov function $V(x)$ that is continuously differentiable and meets the following set of inequalities:

$$a_1(|x|) \leq V(x) \leq a_2(|x|), \tag{13a}$$

$$\frac{\partial V(x)}{\partial x} F(x, \Phi(x)) \leq -a_3(|x|), \tag{13b}$$

$$\left|\frac{\partial V(x)}{\partial x}\right| \leq a_4(|x|) \tag{13c}$$

where a_1, a_2, a_3, and a_4 are class $\mathcal{K}$ functions for all $x \in \mathbb{R}^n \subset D$. Additionally, given that the system $F(x,u)$ has a Lipschitz property and the input u is constrained by the set U, then there exist positive constants M, L_x, and $L_x^{'}$, such that the subsequent inequalities hold for all $x, x' \in D$, and $u \in U$:

$$|F(x,u)| \leq M \tag{14a}$$

$$|F(x,u) - F(x',u)| \leq L_x|x - x'| \tag{14b}$$

$$\left|\frac{\partial V(x)}{\partial x}F(x,u) - \frac{\partial V(x')}{\partial x}F(x',u)\right| \leq L_x^{'}|x - x'| \tag{14c}$$

Moreover, the region $\Omega_\rho := \{x \in D | V(x) \leq \rho\}, \rho > 0$, is defined as the stability region for the slow subsystem in Equation (11).

Now we discuss the incorporation of machine learning models, particularly neural networks, in process systems engineering. The first step in developing any type of neural network model is generating a comprehensive data set that effectively captures the input–output relation. This data set is used to train the network and enables it to learn patterns present in the data and make accurate predictions. To generate a data set that captures the dynamics of the nonlinear system in Equation (1) within the region Ω_ρ, we follow the data generation technique described in [6]. The first step is to carry out various open-loop simulations of the nonlinear system in Equation (1), covering a wide range of initial conditions (i.e., $x_0 \in \Omega_\rho$ and $z_0 \in \Omega_\rho$) and valid input signals under sample-and-hold implementation (i.e., the input is applied to the system as piecewise constant functions, $u = u(t_k) \in U, \forall t \in [t_k, t_{k+1})$, where $t_{k+1} := t_k + \Delta$, Δ is the sampling period). To integrate the nonlinear system in Equation (1), the well-established forward Euler approach is used with a sufficiently small integration time step $h_c \ll \Delta$. As a result, an extensive data set containing the dynamics of the system is generated, which is then utilized to train a neural network employing the Keras library.

2.4. Recurrent Neural Networks

Consider designing a recurrent neural network (RNN) model that predicts only the slow state vector x in Equation (1), or in other words, to approximate the dynamics of the slow subsystem in Equation (11). The network is trained using m data points $(\mathbf{x}_{i,t}, \mathbf{y}_{i,t})$ of T-time length, where $i = 1, ..., m$ and $t = 1, ..., T$. $\mathbf{x}_{i,t} \in \mathbb{R}^{d_x}$ and $\mathbf{y}_{i,t} \in \mathbb{R}^{d_y}$ are the RNN input and output, respectively. Additionally, d_x and d_y are the dimensions of the RNN input and output, respectively. It is worth mentioning that the notations used for the RNN inputs and outputs are represented in bold to establish a clear distinction between the RNN notations and the notations used to describe the nonlinear two-time-scale system in Equation (1). The RNN input $\mathbf{x}_{i,t}$ comprises the current slow state measurement, along with the manipulated inputs applied through the time steps $t = 1, ..., T$, where the manipulated inputs are generated randomly within the set U. The RNN output $\mathbf{y}_{i,t}$ comprises the slow state predicted through the time steps $t = 1, ..., T$. As a result, the RNN model is designed to make predictions of the slow states over one sampling period with $T = \frac{\Delta}{h_c}$ time steps. The data set utilized to develop the RNN model is constructed by generating m independent data sequences obtained from an underlying distribution over $\mathbb{R}^{d_x \times T} \times \mathbb{R}^{d_y \times T}$.

With the aim of simplification, we introduce a single-hidden-layer RNN model, given that $\mathbf{h}_i \in \mathbb{R}^{d_h}$ are the hidden states and can be evaluated as follows:

$$\mathbf{h}_{i,t} = \sigma_h(U\mathbf{h}_{i,t-1} + W\mathbf{x}_{i,t}) \tag{15}$$

where the element-wise nonlinear activation function is denoted by σ_h. $U \in \mathbb{R}^{d_h \times d_h}$ is the weight matrix associated with the hidden states, whereas $W \in \mathbb{R}^{d_h \times d_x}$ is the weight matrix associated with the input vector. Furthermore, the following equation evaluates the output layer $y_{i,t}$ of the RNN model:

$$\mathbf{y}_{i,t} = \sigma_y(\mathcal{V}\mathbf{h}_{i,t}) \tag{16}$$

where the element-wise activation function of the output layer is denoted by σ_y, and $\mathcal{V} \in \mathbb{R}^{d_y \times d_h}$ is the weight matrix associated with the output layer. In this particular application, the input of the RNN model will be the current slow state measurement, $x \in \mathbb{R}^n$, as well as the manipulated input $u \in \mathbb{R}^{q_1}$ for the next sampling period. The output of the RNN model will be the predicted slow states x for at least one sampling period ahead. The basic structure of the RNN is represented in Figure 1.

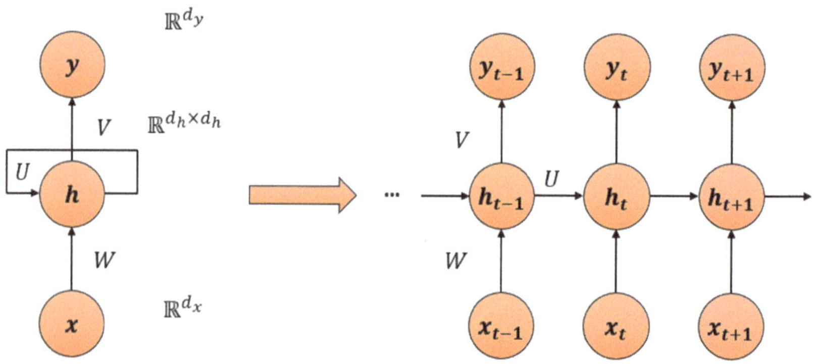

Figure 1. Recurrent neural network structure.

Let $\check{\mathbf{y}}$ be the predicted value and $\mathbf{y}$ be the actual value. We can then define the loss function $L(\mathbf{y}, \check{\mathbf{y}})$ that computes the squared difference between the actual and predicted values. More precisely, we will consider the L_2 (i.e., mean squared error) loss function. Without loss of generality, we establish the following standard assumptions that are required for the study of the generalization error bounds and are reviewed for completeness:

Assumption 5. *The inputs of the RNN are bounded (i.e., $\|\mathbf{x}_{i,t}\| \leq B_X$ for all $i = 1, ..., m$ and $t = 1, ..., T$).*

Assumption 6. *The Frobenius norms of the weight matrices are bounded (i.e., $\|\mathcal{V}\|_F \leq B_{\mathcal{V},F}$, $\|W\|_F \leq B_{W,F}$, $\|U\|_F \leq B_{U,F}$).*

Assumption 7. *σ_h is a positive homogeneous and 1-Lipschitz continuous nonlinear activation function (i.e., $\sigma_h(\alpha z) = \alpha\sigma_h(z)$ holds $\forall\alpha \geq 0$ and $z \in \mathbb{R}$).*

Assumption 8. *The data sets used for training, testing, and validation are drawn from the same distribution.*

Remark 1. *Although the RNN model is developed to approximate the dynamics of the slow subsystem described by Equation (11), we emphasize the fact that the data used to train this RNN model are constructed using the nonlinear two-time-scale system in Equation (1). The reason is that in most practical scenarios, engineers are limited to working with data obtained from the original nonlinear two-time-scale system in Equation (1). Additionally, it is possible to design an RNN model that predicts the full two-time-scale dynamics of Equation (1) using full-state measurements.*

Remark 2. *Assumption 5 takes into account the bounded nature of the RNN inputs. This aligns with the observation that the states x and the inputs u are restricted within certain limits, where $x \in \Omega_\rho$ and $u \in U$. Assumption 6 requires that the weight matrices of the RNN are bounded. This requirement can be met while training the RNN since the search for the optimal RNN parameters* *is limited to a finite class of neural network hypotheses. Assumption 8 indicates that the RNN model constructed using data derived from industrial operations will be utilized in the exact same process under the condition that the data distribution remains unchanged. It is important to point out that in the context of assessing the generalization performance for machine learning models, Assumption 8 is regarded as essential. Several well-known activation functions can be used to construct the RNN model that satisfy Assumption 7; the rectified linear unit (ReLu) activation function is one such example.*

2.5. Feedforward Neural Networks

Using the RNN model described in Section 2.4, we are able to predict the slow state vector x. However, it is important to devise a methodology to predict the values of the fast states. In line with the framework proposed in [5], we consider the design of a feedforward neural network (FNN) that predicts the fast states using the previously predicted slow states from the RNN model. More precisely, the FNN input will be the slow states x, and its output will be the fast states z. The FNN model is trained using m independent data points from an underlying distribution over $\mathbb{R}^{d_x^F} \times \mathbb{R}^{d_y^F}$, where d_x^F and d_y^F are the dimensions of the FNN input and output, respectively. In this particular application, given that the FNN input is the slow state, $x \in \mathbb{R}^n$, and its output is the fast state $z \in \mathbb{R}^p$, we have $d_x^F = n$ and $d_y^F = p$. The general form of an FNN model can be formulated as follows:

$$\mathbf{y}^F = \sigma_d(Q_d\sigma_{d-1}(Q_{d-1}\sigma_{d-2}(...\sigma_1(Q_1\mathbf{x}^F)))) \tag{17}$$

where d is the total number of FNN layers, $\mathbf{y}^F \in \mathbb{R}^{d_y^F}$ is the predicted output of the FNN, and $\mathbf{x}^F \in \mathbb{R}^{d_x^F}$ is the FNN's input. For each FNN layer l, where $l = 1, \ldots, d$, the weight parameter matrix is denoted as Q_l, and the activation function is represented as σ_l. The depth of the network is represented by the number of layers d. However, the width of the network is the maximal number of neurons in a hidden layer and is represented by $h_{\max}$. Given that h_l denotes the number of neurons in the lth layer, the width of the network can be expressed as $h_{\max} = \max_{l=1,\ldots,d}\{h_l\}$. The general FNN structure is shown in Figure 2.

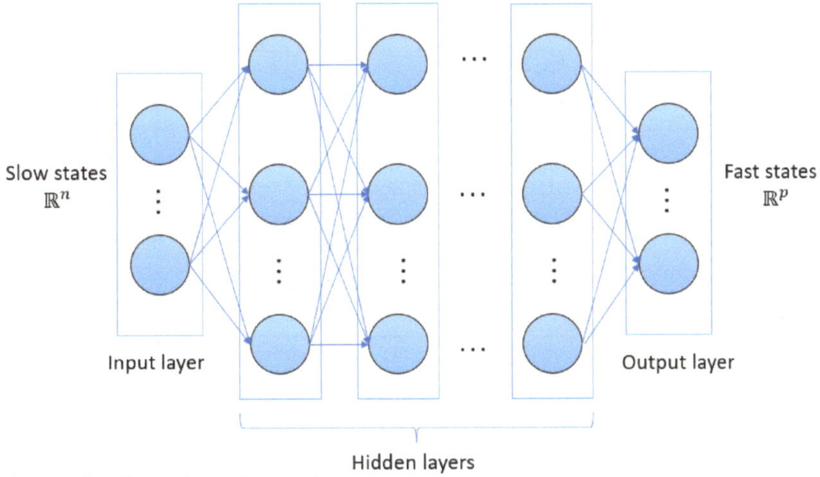

Figure 2. Feedforward neural network structure.

If $\check{\mathbf{y}}^F$ is the predicted value and $\mathbf{y}^F$ is the actual value, we can define the loss function $L(\mathbf{y}^F, \check{\mathbf{y}}^F)$ that computes the squared difference between the actual and the predicted values. More precisely, we will consider the L_2 (i.e., mean squared error) loss function. Similar to the discussion on RNNs, we establish the following assumptions for the developed FNN model:

Assumption 9. *The inputs of the FNN are bounded (i.e., $||\mathbf{x}_i^F|| \leq B_X$ for all $i = 1, ..., m$).*

Assumption 10. *The norms of the weight matrices are bounded in the sense that the maximal 1-norm of the rows of the weight matrices in the hidden layers and output is bounded (i.e., $||Q||_{1,\infty} \leq B_Q$).*

Assumption 11. *The activation function σ_l is 1-Lipschitz continuous and satisfies $\sigma_l(0) = 0$, where $l = 1, ..., d$ (e.g., $\tanh(\cdot)$).*

Assumption 9 is an implication of the assumption that the RNN output (slow states) is bounded. Assumptions 9–11 follow the same reasoning as Assumptions 5–7, as explained in Remark 2. Building upon the previous assumptions, it should be noted that Assumption 8 remains applicable to the FNN model.

3. Generalization Error Bounds of Neural Networks Modeling Two-Time-Scale Systems

The primary goal of constructing any neural network model is to make accurate predictions. Therefore, analyzing the generalization error bounds for various types of neural networks modeling different types of nonlinear systems enables us to improve the construction and design of these network models. Typically, the assessment of neural networks is conducted using a finite set of training samples. Hence, it is essential to investigate the generalization error bound for the neural network model, which allows us to evaluate the performance of the model in accurately predicting outcomes for new, unseen data. In other words, considering new data from the same distribution, the generalization error bound quantifies the ability of a neural network model to make accurate predictions for unseen data that the neural network has not encountered before.

In this work, we consider an RNN that predicts the dynamics of the slow states. Then, an FNN is utilized to predict the fast states using the predicted slow states as the input for the FNN. The generalization error bounds of the neural networks modeling two-time-scale systems are investigated. Particularly, in this section, we discuss the generalization error bounds of both the RNN (which predicts the slow states' dynamics) and the FNN (which predicts the fast states). The generalization error bound is derived by utilizing the basic concepts of statistical machine learning theory. Hence, in the upcoming subsection, we outline the fundamental concepts and definitions used to derive the generalization error bounds. Additionally, it is worth mentioning that in the following subsection, the same principles that apply to $\mathbf{x}$,$\mathbf{y}$, and $\check{\mathbf{y}}$ also extend to $\mathbf{x}^F$,$\mathbf{y}^F$, and $\check{\mathbf{y}}^F$, respectively. In the same vein, the principles that apply to d_x and d_y are also true for d_x^F and d_y^F, respectively. Let $\mathcal{H}$ be the hypothesis class of RNN functions $h(\cdot)$ that map the d_x-dimensional input $\mathbf{x} \in \mathbb{R}^{d_x}$ to the d_y-dimensional output $\check{\mathbf{y}} \in \mathbb{R}^{d_y}$. Similarly, let $\mathcal{H}^F$ be the hypothesis class of FNN functions $h^F(\cdot)$ that map the d_x-dimensional input $\mathbf{x}^F \in \mathbb{R}^{d_x}$ to the d_y-dimensional output $\check{\mathbf{y}}^F \in \mathbb{R}^{d_y}$. We also note that in the following subsection, the principles that apply to $\mathcal{H}$ and $h(\cdot)$ are also valid for $\mathcal{H}^F$ and $h^F(\cdot)$, respectively. It should be noted that the dimensions d_x and d_y differ based on whether the network is an RNN or an FNN.

3.1. Generalization Error Bound Preliminaries

Definition 1. *Consider a function, h, which makes predictions of the output $\mathbf{y}$ corresponding to the input $\mathbf{x}$, along with an underlying distribution D. The generalization error or the expected loss is formulated as follows:*

$$\mathbb{E}[L(h(\mathbf{x}), \mathbf{y})] = \int_{X \times Y} L(h(\mathbf{x}), \mathbf{y})\, \rho(\mathbf{x}, \mathbf{y})\, \mathrm{d}\mathbf{x}\, \mathrm{d}\mathbf{y}. \tag{18}$$

where $\rho(\mathbf{x}, \mathbf{y})$ represents the joint probability distribution for $\mathbf{x}$ and $\mathbf{y}$, whereas X and Y represent the vector space for all possible inputs and outputs, respectively, and L denotes the loss function.

We introduce the following definition of the empirical error as an approximation of the expected loss due to the fact that the joint probability distribution ρ is unknown in many scenarios.

Definition 2. *Considering a data set consisting of m data samples $S = (s_1, ..., s_m)$, with each $s_i = (\mathbf{x}_i, \mathbf{y}_i)$, the empirical error or risk can be expressed as*

$$\hat{\mathbb{E}}_S[L(h(\mathbf{x}), \mathbf{y})] = \frac{1}{m} \sum_{i=1}^{m} L(h(\mathbf{x}_i), \mathbf{y}_i) \tag{19}$$

We note that the m data samples are gathered from the same data distribution. In this work, the loss function $L(\mathbf{y}, \check{\mathbf{y}})$ is chosen as the mean squared error (i.e., L_2 loss function). Furthermore, the generalization error bounds are derived using "Rademacher complexity", a widely recognized concept in the field of machine learning theory that is used to determine the complexity and richness of a class of functions. The Rademacher complexity is defined as follows:

Definition 3. *Given a data set $S = \{s_1, ..., s_m\}$ with m samples, and a hypothesis class $\mathcal{F}$ of real-valued functions, the empirical Rademacher complexity of $\mathcal{F}$ can be defined as follows:*

$$\mathcal{R}_S(\mathcal{F}) = \mathbb{E}_\epsilon \left[\sup_{f \in \mathcal{F}} \frac{1}{m} \sum_{i=1}^{m} \epsilon_i f(s_i) \right] \tag{20}$$

where $\boldsymbol{\epsilon} = (\epsilon_1, ..., \epsilon_m)^T$ consists of the Rademacher random variables ϵ_i. These variables are independent and identically distributed (i.i.d.), satisfying the condition that $\mathbb{P}(\epsilon_i = -1) = \mathbb{P}(\epsilon_i = 1) = 0.5$.

Given that $\mathcal{G}_t$ is the class of loss functions associated with the function class $\mathcal{H}$ and is defined as follows

$$\mathcal{G}_t = \{g_t : (\mathbf{x}, \mathbf{y}) \to L(h(\mathbf{x}), \mathbf{y}), h \in \mathcal{H}\}, \tag{21}$$

where $\mathbf{x}$ and $h(\mathbf{x})$ are the model's input and output, respectively, whereas $\mathbf{y}$ denotes the actual value of the output, the upper bound of the generalization error can be computed using the Rademacher complexity $\mathcal{R}_S(\mathcal{G}_t)$, as shown in the following Lemma.

Lemma 1 (c.f. Theorem 3.3 in [16]). *Given a data set $S = (s_i)$, $i = 1, ..., m$, that consists of m i.i.d. data samples, where $s_i = (\mathbf{x}_{i,t}, \mathbf{y}_{i,t})_{t=1}^T$ for an RNN and $s_i = (\mathbf{x}_i, \mathbf{y}_i)$ for an FNN, with probability $1 - \delta$, the following inequality holds for all $g_t \in \mathcal{G}_t$ over the data samples S:*

$$\mathbb{E}[g_t(\mathbf{x}, \mathbf{y})] \leq \frac{1}{m} \sum_{i=1}^{m} g_t(\mathbf{x}_i, \mathbf{y}_i) + 2\mathcal{R}_S(\mathcal{G}_t) + 3\sqrt{\frac{\log(\frac{2}{\delta})}{2m}} \tag{22}$$

For a comprehensive proof of Lemma 1, interested readers can refer to [12,16]. By observing Equation (22), clearly, the upper bound of the generalization error relies on three terms. The first term is the empirical loss ($\frac{1}{m}\sum_{i=1}^{m} g_t(\mathbf{x}_i, \mathbf{y}_i)$). The second term is the Rademacher complexity ($2\mathcal{R}_S(\mathcal{G}_t)$), whereas the third term depends on the data set size m, along with the confidence level δ. At this stage, the goal is to be able to effectively quantify the generalization error bound using predefined and measurable values. This can be achieved by imposing a further upper bound on the Rademacher complexity term. Having established the background and basic concepts of the generalization error bound, our next discussion focuses on examining the generalization error bounds for neural networks modeling two-time-scale systems.

3.2. RNN Generalization Error Bound

In this subsection, we investigate the generalization error bound for the RNN model that predicts the slow dynamics of the two-time-scale system defined in Equation (1). We start by introducing the following Lemma, which upper-bounds the Rademacher complexity in terms of the RNN parameters.

Lemma 2. *Let $\mathcal{H}_{k,t}$, $k = 1, ..., d_y$ represent the class of real-valued functions associated with the kth component of the RNN output at the tth time step, with the activation functions and weight matrices satisfying Assumptions 5–8. Given a data set $S = (\mathbf{x}_{i,t}, \mathbf{y}_{i,t})_{t=1}^{T}, i = 1, ..., m$, consisting of m i.i.d. data samples, the following inequality holds for the Rademacher complexity:*

$$\mathcal{R}_S(\mathcal{H}_{k,t}) \leq \frac{M(\sqrt{2\log(2)t}+1)B_X}{\sqrt{m}} \tag{23}$$

where $M = B_{V,F}B_{W,F}\frac{B_{U,F}^{t}-1}{B_{U,F}^{-1}}$, and B_X is the upper bound for the RNN inputs.

For a comprehensive and detailed proof of Lemma 2, interested readers may refer to [12]. By applying the results derived and proven in [12], the following Lemma specifically addresses the generalization error bound for the RNN model that predicts the slow dynamics of the two-time-scale system defined in Equation (1).

Lemma 3. *Consider the general class of the two-time-scale continuous-time nonlinear systems described in Equation (1) under the assumption that the slow and fast states in Equation (1) are stable and the perturbation parameter ϵ is sufficiently small. An RNN satisfying Assumptions 5–8 is constructed to predict the slow states of the system at the tth time step. Given the L_r–Lipschitz loss function that belongs to the family of loss functions $\mathcal{G}_t$ associated with the RNN function of class $\mathcal{H}_t$ and a data set $S = (\mathbf{x}_{i,t}, \mathbf{y}_{i,t})_{t=1}^{T}$, $i = 1, ..., m$, with m i.i.d. data samples, the following inequality holds true with a probability of at least $1-\delta$ over S for all $t \in [t_0, T]$:*

$$\mathbb{E}[g_t^{RNN}(\check{x}, x)] \leq \frac{1}{m}\sum_{i=1}^{m} g_t(\check{x}_i, x_i) + 3\sqrt{\frac{\log(\frac{2}{\delta})}{2m}} + \mathcal{O}\left(L_r d_y \frac{MB_X(1+\sqrt{2\log(2)t})}{\sqrt{m}}\right) \tag{24}$$

where $M = B_{V,F}B_{W,F}\frac{B_{U,F}^{t}-1}{B_{U,F}^{-1}}$, and B_X is the upper bound for the RNN inputs.

Equation (24) indicates that the generalization error bound of the RNN that predicts the slow state vector x depends on a number of factors, such as the number of training samples m, the time length t of the RNN's inputs, the upper bound on the input vector B_X, and the complexity hypothesis class in terms of the weight matrices M, as well as the empirical loss ($\frac{1}{m}\sum_{i=1}^{m} g_t(\check{x}_i, x_i)$). We emphasize that the actual slow state is denoted as x, and the slow state predicted by the RNN model is $\check{x}$. Additionally, based on Equation (8), the generalization error bound in Equation (24) holds for all $t \in [t_0, T]$ under the assumption that the slow and fast states are stable, and the perturbation parameter ϵ is sufficiently small.

Now we discuss the implementation of the RNN's generalization error bound for a specific loss function. A locally Lipschitz continuous loss function is used to optimize the RNN's weights and biases. To be more precise, we employ the MSE loss function of the form,

$$L = \frac{1}{m}\sum_{i=1}^{m}(\check{\mathbf{y}}_i, \mathbf{y}_i)^2, \tag{25}$$

where $\check{\mathbf{y}}$ and $\mathbf{y}$ are the predicted and actual values, respectively. Since the loss function L is locally Lipschitz continuous, it satisfies the following inequality,

$$|L(\mathbf{y}, \check{\mathbf{y}}_2) - L(\mathbf{y}, \check{\mathbf{y}}_1)| \leq L_r|\check{\mathbf{y}}_2 - \check{\mathbf{y}}_1| \tag{26}$$

where L_r is the local Lipschitz constant for the loss function L. Since the RNN model predicts the slow states of Equation (1), the loss function is computed over the actual and predicted slow states, x and $\check{x}$, respectively. Moreover, the expected loss of L is upper-bounded by the following inequality, with a probability of at least $1-\delta$:

$$\mathbb{E}[L(\check{x},x)] \leq \frac{1}{m}\sum_{i=1}^{m} L(\check{x}_i,x_i) + 3\sqrt{\frac{\log(\frac{2}{\delta})}{2m}} + \mathcal{O}\left(L_r d_y \frac{MB_X(1+\sqrt{2\log(2)t})}{\sqrt{m}}\right) \tag{27}$$

Given that the MSE loss function L computes the error between the RNN's output x and the RNN's predicted output $\check{x}$, the upper bound on $|\check{x}-x|$ can be written as follows:

$$|\check{x}-x| \leq \sqrt{\frac{1}{m}\sum_{i=1}^{m} L(\check{x}_i,x_i) + 3\sqrt{\frac{\log(\frac{2}{\delta})}{2m}} + \mathcal{O}\left(L_r d_y \frac{MB_X(1+\sqrt{2\log(2)t})}{\sqrt{m}}\right)} \tag{28}$$

3.3. FNN Generalization Error Bound

In this subsection, we investigate the generalization error bound for the FNN model that predicts the fast states of the two-time-scale system defined in Equation (1). We start by introducing the following Lemma, which upper-bounds the Rademacher complexity in terms of the FNN parameters.

Lemma 4 (c.f. Theorem 2 in [17]). *Given a class of scalar-valued functions $\mathcal{H}_k^F$ and a neural network with depth d, where $\|Q_1\|_{1,\infty} \leq B_Q$ for all $l = 1,\ldots,d$ and Assumptions 8–11 are satisfied, the following inequality holds for the Rademacher complexity:*

$$\mathcal{R}_S(\mathcal{H}_k^F) \leq \frac{2B_X(B_Q)^d\sqrt{d+1+\log(d_x)}}{\sqrt{m}} \tag{29}$$

For a comprehensive and detailed proof of Lemma 4, interested readers may refer to [17]. By applying the results derived and proven in [13], the following Lemma specifically addresses the generalization error bound for the FNN model that predicts the fast states using the slow states of the two-time-scale system defined in Equation (1).

Lemma 5. *Consider the general class of two-time-scale continuous-time nonlinear systems described in Equation (1) under the assumption that the slow and fast states are stable, and the perturbation parameter ϵ is sufficiently small. An FNN satisfying Assumptions 8–11 is constructed to predict the fast states using the slow states. Given the L_r–Lipschitz loss functions associated with the vector-valued FNN hypothesis class $\mathcal{H}^F$ and a data set S consisting of m i.i.d. data samples, the following inequality holds true, with a probability of at least $1-\delta$ over S for all $t \in [t_p, T]$, where t_p is defined in Theorem 1:*

$$\mathbb{E}[g_t^{FNN}(\check{z},z)] \leq \frac{1}{m}\sum_{i=1}^{m} g_t(\check{z}_i,z_i) + 3\sqrt{\frac{\log(\frac{2}{\delta})}{2m}} + \mathcal{O}\left(L_r d_y \frac{B_X(B_Q)^d\sqrt{d+1+\log(d_x)}}{\sqrt{m}}\right) \tag{30}$$

Equation (30) indicates that the generalization error bound of the FNN modeling the fast subsystem depends on several factors, such as the number of training samples m, the upper bound on the input vector B_X, the complexity hypothesis class in terms of the upper bound of the weight matrices B_Q, and the number of layers d, as well as the empirical loss ($\frac{1}{m}\sum_{i=1}^{m} L(\check{z}_i,z_i)$). We emphasize that the actual fast state is denoted as z, and the fast state predicted by the FNN model is $\check{z}$. Therefore, the generalization error bound of Equation (30) holds for all $t \in [t_p, T]$ under the assumption that the slow and fast states are stable, and the perturbation parameter ϵ is sufficiently small.

Now we discuss the implementation of the FNN's generalization error bound specifically on the MSE loss function. A locally Lipschitz continuous loss function such as the

MSE is used to optimize the FNN's weights and biases. The MSE loss function satisfies the inequality of Equation (26). Since the FNN model predicts the fast states of Equation (1), the loss function is computed over the actual and predicted fast states, z and $\check{z}$, respectively. Moreover, the expected loss of L is upper-bounded by the following inequality, with a probability of at least $1-\delta$:

$$\mathbb{E}[L(\check{z},z)] \leq \frac{1}{m}\sum_{i=1}^{m} L(\check{z}_i,z_i) + 3\sqrt{\frac{\log(\frac{2}{\delta})}{2m}} + \mathcal{O}\left(L_r d_y \frac{B_X (B_Q)^d \sqrt{d+1+\log(d_x)}}{\sqrt{m}}\right) \tag{31}$$

Given that the MSE loss function L computes the error between the RNN's output z and the RNN's predicted output $\check{z}$, the upper bound on $|\check{z}-z|$ can be written as follows:

$$|\check{z}-z| \leq \sqrt{\frac{1}{m}\sum_{i=1}^{m} L(\check{z}_i,z_i) + 3\sqrt{\frac{\log(\frac{2}{\delta})}{2m}} + \mathcal{O}\left(L_r d_y \frac{B_X (B_Q)^d \sqrt{d+1+\log(d_x)}}{\sqrt{m}}\right)} \tag{32}$$

4. Machine Learning-Based LMPC Using an RNN That Approximates the Slow Subsystem

Having investigated the generalization error bounds for neural networks modeling two-time-scale systems, the rest of this manuscript focuses on the design and performance of LMPC using RNN models with and without time-scale decomposition. Specifically, the objective is to design an RNN to predict only the slow state vector x, following the approach outlined in Section 2.4, and then incorporate the designed RNN model within an LMPC scheme to show that it is sufficient to stabilize the full two-time-scale system in Equation (1). In this section, we present a stability analysis for the proposed LMPC framework. For simplicity, in this section, we present the RNN model that predicts the slow state vector x as a continuous-time nonlinear system of the following form:

$$\dot{\hat{x}} = F_{nn}(\hat{x},u) := A\hat{x} + \Theta^T \hat{z} \tag{33}$$

where $\hat{x} \in \mathbb{R}^n$ is the RNN's state vector and $u \in \mathbb{R}^{q_1}$ is the manipulated input vector. The weight matrices are denoted as A and Θ. The diagonal coefficient matrix A has negative diagonal entries. Θ is defined as $\Theta = [\theta_1,...,\theta_n] \in \mathbb{R}^{(q_1+n)\times n}$, with entries $\theta_i = b_i[w_{i1},...,w_{i(q_1+n)}], i = 1,...,n$, given that w_{ij} represents the weight associated with the connection between the ith neuron and the jth output, with i ranging from 1 to n and j ranging from 1 to (q_1+n). The vector $\hat{z} = [\hat{z}_1,...,\hat{z}_n,\hat{z}_{n+1},...,\hat{z}_{n+q_1}] = [\sigma(\hat{x}_1)...\sigma(\hat{x}_n),u...u_{q_1}] \in \mathbb{R}^{n+q_1}$ consists of both the network states $\hat{x}$ and inputs u. Additionally, a nonlinear activation function $\sigma(\cdot)$ is applied to the network states $\hat{x}$ when constructing the vector $\hat{z}$. It is worth mentioning that the weight matrices and activation functions satisfy Assumptions 5–8. For simplicity, we assume a single-hidden-layer RNN model represented by Equation (33), which does not explicitly include bias terms. Nevertheless, it should be emphasized that the results obtained in this section are not limited to single-hidden-layer RNN models, but can be generalized to include deep RNN models that consist of multiple hidden layers.

4.1. Lyapunov-Based Control Using an RNN Model

Taking into account the RNN model designed to predict the values of the slow states, we assume that there exists a stabilizing control law $\Phi_{nn}(x) \in U$ (e.g., a P-controller, a Lyapunov-based controller) that can render the origin of the RNN model in Equation (33) asymptotically stable in an open neighborhood around the origin denoted as $\hat{D}$. Hence, there exists a continuously differentiable Lyapunov function $\hat{V} : \mathbb{R}^n \to \mathbb{R}_{\geq 0}$ that satisfies the following inequalities:

$$\hat{a}_1(|x|) \leq \hat{V}(x) \leq \hat{a}_2(|x|) \tag{34a}$$

$$\frac{\partial \hat{V}(x)}{\partial x} F_{nn}(x,\Phi_{nn}(x)) \leq -\hat{a}_3(|x|) \tag{34b}$$

$$\left|\frac{\partial \hat{V}(x)}{\partial x}\right| \leq \hat{\alpha}_4(|x|) \tag{34c}$$

where $\hat{\alpha}_i$ are class $\mathcal{K}$ functions for all $x \in \mathbb{R}^n \subset \hat{D}$. The Lyapunov function $\hat{V}$ can be designed using a quadratic formulation, which satisfies the required criteria, or, alternately, using learning-based approaches, as conducted using a linear parameter-varying (LPV) framework in [18]. In the simulation example presented in our work below, a quadratic Lyapunov function is used, and extensive open-loop and closed-loop simulations are conducted to verify that the asymptotic stability assumption is met. The Lyapunov level set that ensures the stability of the RNN model is defined as $\Omega_{\hat{\rho}} := \{x \in \hat{D} \mid \hat{V}(x) \leq \hat{\rho}\}$, where $\hat{\rho} > 0$. This implies that the closed-loop stability region of the RNN model in Equation (33) is denoted as $\Omega_{\hat{\rho}}$. Moreover, there exist positive constants M_{nn} and L_{nn}, such that the following inequalities are satisfied for all $x, x' \in \Omega_{\hat{\rho}}$, and $u \in U$:

$$|F_{nn}(x,u)| \leq M_{nn} \tag{35a}$$

$$\left|\frac{\partial \hat{V}(x)}{\partial x}F_{nn}(x,u) - \frac{\partial \hat{V}(x')}{\partial x}F_{nn}(x',u)\right| \leq L_{nn}|x - x'| \tag{35b}$$

The feedback controller $u = \Phi_{nn}(x) \in U$ can stabilize the dynamics of the slow subsystem in Equation (11) under a sufficiently small modeling error between the nominal slow subsystem and the RNN model. This result is shown in the following proposition.

Proposition 1. *Consider the RNN model in Equation (33) that satisfies the stabilizability conditions of Equation (34) and is rendered asymptotically stable around the origin under the control law $u = \Phi_{nn}(x) \in U$ for all $x \in \Omega_{\hat{\rho}}$. Then, the origin of the slow subsystem in Equation (11) is asymptotically stable if there exists a positive real number ν_m, where $\nu_m < \hat{\alpha}_3(|x|)/\hat{\alpha}_4(|x|)$, such that the modeling error between the slow subsystem in Equation (11) and the RNN model in Equation (33) (i.e., $\nu = |F(x,u) - F_{nn}(x,u)|$) is upper-bounded by ν_m for all $x \in \Omega_{\hat{\rho}}$.*

Proof. We follow the strategy outlined in [5] to prove Proposition 1. The main objective is to show that, under the control law $u = \Phi_{nn}(x) \in U$, the slow subsystem in Equation (11) is asymptotically stable around the origin for all $x \in \Omega_{\hat{\rho}}$ if $\Phi_{nn}(x)$ renders the RNN designed to approximate the slow subsystem asymptotically stable with a bounded modeling error between the nominal slow subsystem and the RNN model. In other words, we show that $\dot{\hat{V}}(x) \leq 0$ for the slow subsystem in Equation (11) under the controller $u = \Phi_{nn}(x) \in U$ based on the RNN model. We utilize the inequalities in Equations (34b) and (34c) and compute $\dot{\hat{V}}(x)$ as follows:

$$\begin{aligned}
\dot{\hat{V}} &= \frac{\partial \hat{V}(x)}{\partial x}F(x,\Phi_{nn}(x)) \\
&= \frac{\partial \hat{V}}{\partial x}(F_{nn}(x,\Phi_{nn}(x)) + F(x,\Phi_{nn}(x)) - F_{nn}(x,\Phi_{nn}(x)) \\
&\leq -\hat{\alpha}_3(|x|) + \hat{\alpha}_4(|x|)(F(x,\Phi_{nn}(x)) - F_{nn}(x,\Phi_{nn}(x))) \\
&\leq -\hat{\alpha}_3(|x|) + \nu_m\hat{\alpha}_4(|x|)
\end{aligned} \tag{36}$$

By assigning $\nu_m < \frac{\hat{\alpha}_3(|x|)}{\hat{\alpha}_4(|x|)}$, we obtain $\dot{\hat{V}}(x) \leq -\tilde{\alpha}_3(|x|) \leq 0$, where $\tilde{\alpha}_3(|x|) = -\hat{\alpha}_3(|x|) + \nu_m\hat{\alpha}_4(|x|) > 0$ since $\hat{\alpha}_3$ and $\hat{\alpha}_4$ are known functions chosen in such a way as to ensure that the above result holds. To show that if the general class $\mathcal{K}$ functions are chosen as $\hat{\alpha}_3 = a_3|x|$ and $\hat{\alpha}_4 = a_4|x|$, where a_3 and a_4 are constants, then $\nu_m = \frac{a_3}{a_4}$. Hence, we guarantee the closed-loop stability of the slow subsystem in Equation (11) around the origin under the control law $\Phi_{nn}(x) \in U$ for all $x \in \Omega_{\hat{\rho}}$. □

The RNN model designed in Equation (33) is incorporated into an LMPC scheme, where the control actions computed by the LMPC are implemented in a sample-and-hold fashion. Moreover, this sample-and-hold implementation of the LMPC control law $u = \Phi_{nn}(x) \in U$ establishes certain properties. These properties are studied in the following two propositions. Specifically, the following proposition shows that the error between the state of the slow subsystem in Equation (11) and the state predicted by the RNN model in Equation (33) is bounded.

Proposition 2. *Consider the RNN model in Equation (33) and the slow subsystem in Equation (11), starting with the same initial condition $x_0 = \hat{x}_0 \in \Omega_{\hat{\rho}}$. There exists a class $\mathcal{K}$ function $f_w(\cdot)$ and a positive constant κ, such that the following inequalities are satisfied for all $x, \hat{x} \in \Omega_{\hat{\rho}}$:*

$$|x(t) - \hat{x}(t)| \leq f_w(t) := \frac{\nu_m}{L_x}\left(e^{L_x t} - 1\right) \tag{37a}$$

$$\hat{V}(x) \leq \hat{V}(\hat{x}) + \hat{a}_4(\hat{a}_1^{-1}(\hat{\rho}))|x - \hat{x}| + \kappa|x - \hat{x}|^2 \tag{37b}$$

Proof. Considering that $e(t) = x(t) - \hat{x}(t)$ represents the error vector between the state of the slow subsystem in Equation (11) and the state of the RNN model in Equation (33), the bound for the time derivative of $e(t)$ is given as follows:

$$\begin{aligned} |\dot{e}(t)| &= |F(x, u) - F_{nn}(\hat{x}, u)| \\ &\leq |F(x, u) - F(\hat{x}, u)| + |F(\hat{x}, u) - F_{nn}(\hat{x}, u)| \end{aligned} \tag{38}$$

Using Equation (14b), for all $x, \hat{x} \in \Omega_{\hat{\rho}}$, the first term in Equation (38) can be bounded as follows:

$$\begin{aligned} |F(x, u) - F(\hat{x}, u)| &\leq L_x|x(t) - \hat{x}(t)| \\ &\leq L_x|x(t) - \hat{x}(t)| \end{aligned} \tag{39}$$

The term $|F(\hat{x}, u) - F_{nn}(\hat{x}, u)|$ in Equation (39) denotes the modeling error, and it is upper-bounded by ν_m for all $\hat{x} \in \Omega_{\hat{\rho}}$. Hence, the term $\dot{e}(t)$ can be further bounded by utilizing the bound of the modeling error and the bound of Equation (39), as follows:

$$\begin{aligned} |\dot{e}(t)| &\leq L_x|x(t) - \hat{x}(t)| + \nu_m \\ &\leq L_x|e(t)| + \nu_m \end{aligned} \tag{40}$$

Taking into account the zero initial condition (i.e., $e(0) = 0$), we integrate the inequality in Equation (40) and obtain the following upper bound for the error vector for all $x, \hat{x} \in \Omega_{\hat{\rho}}$:

$$|e(t)| = |x(t) - \hat{x}(t)| \leq \frac{\nu_m}{L_x}\left(e^{L_x t} - 1\right) \tag{41}$$

Equation (37b) can be derived using the Taylor series expansion of $\hat{V}(x)$ around $\hat{x}$ for all $x, \hat{x} \in \Omega_{\hat{\rho}}$ as follows:

$$\hat{V}(x) \leq \hat{V}(\hat{x}) + \frac{\partial \hat{V}(\hat{x})}{\partial x}|x - \hat{x}| + \kappa|x - \hat{x}|^2 \tag{42}$$

where κ is a positive real number. Furthermore, Equations (34a) and (34b) are used to further upper-bound $\hat{V}(x)$ as follows:

$$\hat{V}(x) \leq \hat{V}(\hat{x}) + \hat{a}_4(\hat{a}_1^{-1}(\hat{\rho}))|x - \hat{x}| + \kappa|x - \hat{x}|^2 \tag{43}$$

□

The following proposition demonstrates that the closed-loop state of the slow subsystem described in Equation (11) is maintained within the stability region $\Omega_{\hat{\rho}}$ at all times.

Additionally, utilizing the Lyapunov-based controller $u = \Phi_{nn}(x) \in U$ through sample-and-hold implementation ensures that the closed-loop state of the slow subsystem can be ultimately bounded within a small region around the origin $\Omega_{\rho_{\min}}$.

Proposition 3 (c.f Proposition 3 in [5]). *Consider the slow subsystem in Equation (11) under the controller $\Phi_{nn}(\hat{x}) \in U$ that satisfies the conditions in Equation (34) and is implemented in a sample-and-hold fashion (i.e., $\Phi_{nn}(\hat{x}(t_k)), \forall t \in [t_k, t_{k+1})$, where $t_{k+1} := t_k + \Delta$) to stabilize the RNN model in Equation (33). Then, there exist $\epsilon_w > 0$, $\Delta > 0$ and $\hat{\rho} > \rho_{\min} > \rho_{nn} > \rho_s$ that satisfy*

$$-\hat{a}_3(\hat{a}_2^{-1}(\rho_s)) + L_{nn} M_{nn} \Delta \leq -\epsilon_s \tag{44a}$$

$$-\tilde{a}_3(\hat{a}_2^{-1}(\rho_s)) + L_x' M_F \Delta \leq -\epsilon_w \tag{44b}$$

and

$$\rho_{nn} := \max\{\hat{V}(\hat{x}(t+\Delta)) \mid \hat{x}(t) \in \Omega_{\rho_s}, u \in U\} \tag{45a}$$

$$\rho_{\min} \geq \rho_{nn} + -\hat{a}_4(\hat{a}_1^{-1}(\hat{\rho})) f_w(\Delta) + \kappa(f_w(\Delta))^2 \tag{45b}$$

such that, for any $x(t_k) \in \Omega_{\hat{\rho}} \backslash \Omega_{\rho_s}$, there exists a class $\mathcal{KL}$ function β_x and a class $\mathcal{K}$ function $\bar{\gamma}$, such that the following inequality is satisfied:

$$|x(t)| \leq \beta_x(|x(0)|, t) + \bar{\gamma}(\rho_{\min}) \tag{46}$$

and the state $x(t)$ of the nominal slow subsystem in Equation (11) is bounded in $\Omega_{\hat{\rho}}$ for all times and ultimately bounded in $\Omega_{\rho_{\min}}$.

Proof. Assuming that $x(t_k) = \hat{x}(t_k)$, the first part of establishing this proof involves showing that $\hat{V}(\hat{x})$ is decreasing under the control law $u(t) = \Phi_{nn}(x(t_k)) \in U$ for $t \in [t_k, t_{k+1})$, where $x(t_k)$ and $\hat{x}(t_k)$ are the state of the slow subsystem in Equation (11) and the state of the RNN model in Equation (33), respectively. The time derivative of $\hat{V}(\hat{x})$ for all $t \in [t_k, t_{k+1})$ is calculated as follows:

$$\begin{aligned} \dot{\hat{V}}(\hat{x}(t)) &= \frac{\partial \hat{V}(\hat{x}(t))}{\partial \hat{x}} F_{nn}(\hat{x}(t), \Phi_{nn}(\hat{x}(t_k))) \\ &= \frac{\partial \hat{V}(\hat{x}(t_k))}{\partial \hat{x}} F_{nn}(\hat{x}(t_k), \Phi_{nn}(\hat{x}(t_k))) \\ &\quad + \frac{\partial \hat{V}(\hat{x}(t))}{\partial \hat{x}} F_{nn}(\hat{x}(t), \Phi_{nn}(\hat{x}(t_k))) \\ &\quad - \frac{\partial \hat{V}(\hat{x}(t_k))}{\partial \hat{x}} F_{nn}(\hat{x}(t_k), \Phi_{nn}(\hat{x}(t_k))) \end{aligned} \tag{47}$$

By utilizing the inequalities in Equations (34a) and (34b), we further bound $\dot{\hat{V}}(\hat{x}(t))$ as follows:

$$\begin{aligned} \dot{\hat{V}}(\hat{x}(t)) \leq & -\hat{a}_3(\hat{a}_2^{-1}(\rho_s)) + \frac{\partial \hat{V}(\hat{x}(t))}{\partial \hat{x}} F_{nn}(\hat{x}(t), \Phi_{nn}(\hat{x}(t_k))) \\ & - \frac{\partial \hat{V}(\hat{x}(t_k))}{\partial \hat{x}} F_{nn}(\hat{x}(t_k), \Phi_{nn}(\hat{x}(t_k))) \end{aligned} \tag{48}$$

By applying the Lipschitz inequalities in Equation (35), we proceed with bounding $\dot{\hat{V}}(\hat{x}(t))$ as follows:

$$\begin{aligned} \dot{\hat{V}}(\hat{x}(t)) &\leq -\hat{a}_3(\hat{a}_2^{-1}(\rho_s)) + L_{nn}|\hat{x}(t) - \hat{x}(t_k)| \\ &\leq -\hat{a}_3(\hat{a}_2^{-1}(\rho_s)) + L_{nn} M_{nn} \Delta \end{aligned} \tag{49}$$

Therefore, if Equation (44a) is satisfied, the following inequality holds for all $\hat{x}(t_k) \in \Omega_{\hat{\rho}} \backslash \Omega_{\rho_s}$ and $t \in [t_k, t_{k+1})$:

$$\dot{\hat{V}}(x(t)) \leq -\epsilon_s \tag{50}$$

Upon integrating the aforementioned differential equation over the time interval $t \in [t_k, t_{k+1})$, it is derived that $\hat{V}(\hat{x}(t_{k+1})) \leq \hat{V}(\hat{x}(t_k)) - \epsilon_s \Delta$. Therefore, when Equation (44a) is satisfied, this leads to the fact that $\dot{\hat{V}}(x(t))$ is negative for any $\hat{x}(t_k) \in \Omega_{\hat{\rho}} \backslash \Omega_{\rho_s}$. As a result, utilizing the control law $u = \Phi_{nn}(\hat{x})$ in a sample-and-hold fashion ensures that the closed-loop state of the RNN model in Equation (33) is bounded within the region $\Omega_{\hat{\rho}}$ and moves toward the origin. Nevertheless, it should be observed that Equation (50) may not hold true in cases where $x(t_k) = \hat{x}(t_k) \in \Omega_{\rho_s}$, which indicates that the state $\hat{x}(t_k)$ may leave the region Ω_{ρ_s} within one sampling period. Hence, we introduce the region $\Omega_{\rho_{nn}}$ in Equation (45a) to guarantee that the closed-loop state $\hat{x}(t_k)$ of the RNN model will be bounded in the region $\Omega_{\rho_{nn}}$ within one sampling period for all $t \in [t_k, t_{k+1})$, $u \in U$ and $\hat{x}(t_k) \in \Omega_{\rho_s}$. Consider the case where $\hat{x}(t_{k+1})$ leaves the region Ω_{ρ_s}. Then, the controller $u = \Phi_{nn}(x(t_{k+1}))$ reactivates to derive the states into the region Ω_{ρ_s}, and Equation (50) will be satisfied again at $t = t_{k+1}$. Up to this point, it can be concluded that the state of the RNN system in Equation (33) is ultimately bounded in $\Omega_{\rho_{nn}}$ for all $x_0 \in \Omega_{\hat{\rho}}$.

The second part of this proof is to demonstrate that the controller $u = \Phi_{nn}(x) \in U$, implemented in a sample-and-hold fashion, effectively bounds the states of the slow subsystem in Equation (11) in $\Omega_{\hat{\rho}}$ and can ultimately derive the states to a small region around the origin. This requires showing that $\hat{V}(x)$ for the slow subsystem in Equation (11) is decreasing under the control law $u(t) = \Phi_{nn}(x(t_k))$ for $t \in [t_k, t_{k+1})$ and $x(t_k) = \hat{x}(t_k) \in \Omega_{\hat{\rho}} \backslash \Omega_{\rho_s}$. The derivative of $\hat{V}(x(t))$ with respect to time is computed as follows:

$$\begin{aligned}
\dot{\hat{V}}(x(t)) &= \frac{\partial \hat{V}(x(t))}{\partial x} F(x(t), \Phi_{nn}(x(t_k))) \\
&= \frac{\partial \hat{V}(x(t_k))}{\partial x} F(x(t_k), \Phi_{nn}(x(t_k))) \\
&\quad + \frac{\partial \hat{V}(x(t))}{\partial x} F(x(t), \Phi_{nn}(x(t_k))) \\
&\quad - \frac{\partial \hat{V}(x(t_k))}{\partial x} F(x(t_k), \Phi_{nn}(x(t_k)))
\end{aligned} \tag{51}$$

Using the inequality in Equation (36), which holds for all $x \in \Omega_{\hat{\rho}} \backslash \Omega_{\rho_s}$, the first term in Equation (51) can be further bounded as follows:

$$\begin{aligned}
\dot{\hat{V}}(x(t)) \leq &- \tilde{a}_3(\hat{a}_2^{-1}(\rho_s)) + \frac{\partial \hat{V}(x(t))}{\partial x} F(x(t), \Phi_{nn}(x(t_k)), \xi) \\
&- \frac{\partial \hat{V}(x(t_k))}{\partial x} F(x(t_k), \Phi_{nn}(x(t_k)), 0)
\end{aligned} \tag{52}$$

where $\tilde{a}_3(\cdot)$ was previously defined in the proof of Proposition 1. By using the Lipschitz condition stated in Equation (14), we derive the following upper bound for $\dot{\hat{V}}(x(t))$:

$$\begin{aligned}
\dot{\hat{V}}(x(t)) &\leq \tilde{a}_3(\hat{a}_2^{-1}(\rho_s)) + L_x^{'} |x(t) - x(t_k)| \\
&\leq \tilde{a}_3(\hat{a}_2^{-1}(\rho_s)) + L_x^{'} M_F \Delta
\end{aligned} \tag{53}$$

Therefore, if Equation (44b) holds true, the following inequality is satisfied for all $x(t_k) \in \Omega_{\hat{\rho}} \backslash \Omega_{\rho_s}$ and for all $t \in [t_k, t_{k+1})$:

$$\dot{\hat{V}}(x(t)) \leq -\epsilon_w \tag{54}$$

By integrating the above differential equation over the time interval $t \in [t_k, t_{k+1})$ between any two arbitrary points within the previous time interval, it can be shown that for all $x(t_k) \in \Omega_{\hat{\rho}} \backslash \Omega_{\rho_S}$, the following results are obtained:

$$\hat{V}(x(t_{k+1})) \leq V(x(t_k)) - \epsilon_w \Delta \tag{55}$$

$$\hat{V}(x(t)) \leq \hat{V}(x(t_k)), \forall t \in [t_k, t_{k+1}) \tag{56}$$

Based on Equation (54), it can be observed that $\dot{\hat{V}}(x(t))$ is negative for all $x(t_k) \in \Omega_{\hat{\rho}} \backslash \Omega_{\rho_S}$. As a result, the state of the slow subsystem in Equation (11) remains continuously bounded within the region $\Omega_{\hat{\rho}}$ and can be ultimately driven toward the origin in each sampling period by implementing the control law $u = \Phi_{nn}(x)$. In addition, if $x(t_k) \in \Omega_{\rho_s}$, the state of the RNN model in Equation (33) is bounded within the region $\Omega_{\rho_{nn}}$ for one sampling period. This outcome has been demonstrated in the first part of the proof. Taking onto account the bounded modeling error between the state of the RNN, as described in Equation (33), and the state of the slow subsystem in Equation (11), a compact set $\Omega_{\rho_{\min}} \supset \Omega_{\rho_{nn}}$ exists and satisfies Equation (45b). The set $\Omega_{\rho_{\min}}$ is introduced to ensure that the state of the slow subsystem in Equation (11) is bounded within the region $\Omega_{\rho_{\min}}$ during one sampling period, provided that the state of the RNN represented by Equation (33) is bounded within the region $\Omega_{\rho_{nn}}$. In the case where the state $x(t)$ enters the set $\Omega_{\rho_{\min}} \backslash \Omega_{\rho_S}$, it has been shown that Equation (56) is satisfied. Therefore, under the control law $u = \Phi_{nn}(x)$, the state $x(t)$ will be driven toward the origin in the next sampling period, ultimately bounding the state of the slow subsystem within a small region around the origin $\Omega_{\rho_{\min}}$. Hence, considering the continuity property of the Lyapunov function $\hat{V}$, it follows that there exists a class $\mathcal{KL}$ function β_x and a class $\mathcal{K}$ function $\bar{\gamma}$, such that if $x_0 \in \Omega_{\hat{\rho}}$, then $x(t) \in \Omega_{\hat{\rho}}$ for all $t \leq t_0$:

$$|x(t)| \leq \beta_x(|x(0)|, t) + \bar{\gamma}(\rho_{\min}) \tag{57}$$

□

4.2. Machine Learning-Based LMPC Formulation

In this subsection, we introduce the machine learning-based LMPC formulation that utilizes the RNN model designed in Equation (33) to predict future states over a certain future horizon. Furthermore, the LMPC is essentially an optimization problem consisting of an objective function and a number of constraints. This optimization problem is solved repeatedly to compute the optimal input trajectory. The following shows the formulation of the machine learning-based LMPC optimization problem:

$$\mathcal{J} = \min_{u \in S(\Delta)} \int_{t_k}^{t_{k+N}} L_{MPC}(\tilde{x}(t), u(t))dt \tag{58a}$$

$$\text{s.t.} \quad \dot{\tilde{x}}(t) = F_{nn}(\tilde{x}(t), u(t)) \tag{58b}$$

$$u(t) \in U, \ \forall\, t \in [t_k, t_{k+N}) \tag{58c}$$

$$\tilde{x}(t_k) = x(t_k) \tag{58d}$$

$$\dot{\hat{V}}(x(t_k), u) \leq \dot{\hat{V}}(x(t_k), \Phi_{nn}(x(t_k))), \text{ if } x(t_k) \in \Omega_{\hat{\rho}} \backslash \Omega_{\rho_{nn}} \tag{58e}$$

$$\hat{V}(\tilde{x}(t)) \leq \rho_{nn}, \ \forall\, t \in [t_k, t_{k+N}), \text{ if } x(t_k) \in \Omega_{\rho_{nn}} \tag{58f}$$

where the predicted slow state trajectory is denoted as $\tilde{x}$. The set of piecewise constant functions with period Δ is denoted as $S(\Delta)$, and the number of sampling periods in the prediction horizon is denoted by N. The optimal control action $u^*(t)$ is calculated by repeatedly solving the machine learning-based LMPC optimization problem over the entire prediction horizon $t \in [t_k, t_{k+N})$. Then, the optimal control action $u^*(t_k)$ computed at the first sampling period is transmitted by the controller to be applied to the process. Subsequently, the resulting real-time state, $x(t_k)$, is fed back to the machine learning-based

LMPC optimization problem to compute the optimal input trajectory for the next sampling time. The optimization problem aims to minimize the time integral of $L_{MPC}(\tilde{x}(t), u(t))$ over the prediction horizon, which is represented in the cost function in Equation (58a). As for the constraints of the optimization problem, the first constraint demonstrated in Equation (58b) is essentially the RNN model utilized to approximate the states of the slow subsystem's dynamics. Equation (58c) represents the input constraints that may be applied to the process over the entire prediction horizon. The initial condition needed to solve Equation (58b) is essentially the slow state measurement at $t = t_k$, which is specified in the constraint in Equation (58d). In the case where $x(t_k) \in \Omega_{\hat{\rho}} \backslash \Omega_{\rho_{nn}}$, the constraint Equation (58e) ensures that the closed-loop state converges toward the origin. Furthermore, if the state $x(t_k)$ enters the region $\Omega_{\rho_{nn}}$, it is necessary to guarantee that the states predicted by the RNN model remain bounded within the region $\Omega_{\rho_{nn}}$ throughout the entire prediction horizon. This is accomplished by employing Equation (58f).

4.3. Closed-Loop Stability

Assuming that certain conditions are met, the following theorem establishes the closed-loop stability of a singularly perturbed system described by Equation (1) when the machine learning-based LMPC of Equation (58) is implemented.

Theorem 2. *Consider the singularly perturbed system in Equation (1) in a closed loop, with the optimal control action u^* calculated using the machine learning-based LMPC in Equation (58) based on the controller $\Phi_{nn}(x)$ that meets the conditions of Equation (34). Additionally, assuming that Propositions 1–3 and Assumptions 1 and 4 hold true, then there exist class $\mathcal{KL}$ functions β_x and β_z, a pair of positive real numbers (δ, d), and $\epsilon^* > 0$, such that if* $\max\{|x(0), |z(0)|\} \leq \delta$ *and $\epsilon \in (0, \epsilon^*]$, then, for all $t \geq 0$,*

$$|x(t)| \leq \beta_x(|x(0)|, t) + \bar{\gamma}(\rho_{\min}) + d \tag{59}$$

$$|z(t)| \leq \beta_z\left(|z(0)|, \frac{t}{\epsilon}\right) + d \tag{60}$$

Proof. We follow the proof analysis in [5]. Moreover, we consider that by substituting the optimal control action u^* into Equation (1), the closed-loop system can be represented as follows:

$$\dot{x} = f_1(x, z, u^*, \epsilon) \tag{61a}$$
$$\epsilon\dot{z} = f_2(x, z, \epsilon) \tag{61b}$$

We set $\epsilon = 0$ and obtain the following,

$$\dot{x} = f_1(x, z, u^*, 0) \tag{62a}$$
$$0 = f_2(x, z, 0) \tag{62b}$$

where the assumption that the error between z and $\bar{z}$ is not greater than $\mathcal{O}(\epsilon)$ enables us to simplify the analysis by approximating $\bar{z}$ as z. By solving Equation (62b) for z, we obtain a unique, isolated root $z = \bar{f}(x)$. We substitute the root into Equation (62a) and obtain the following for the reduced-order slow subsystem,

$$\dot{x} = f_1(x, \bar{f}_2(x), u^*, 0) \tag{63}$$

By observing the LMPC equation in Equation (58), in the case where $x(t_k) \in \Omega_{\hat{\rho}} \backslash \Omega_{\rho_{nn}}$, Equation (58e) is activated, such that the Lyapunov function $\hat{V}$ under the calculated control law u is decreased. According to the finding in Proposition 3 and considering a sufficiently small modeling error, the state of the slow subsystem in Equation (61a) will gradually converge to the origin and enter the region $\Omega_{\rho_{nn}}$ within a finite number of sampling time steps. Equation (58f) is activated once the state $x(t_k)$ enters the region $\Omega_{\rho_{nn}}$, where it ensures that the predicted state is bounded within $\Omega_{\rho_{nn}}$ throughout the prediction horizon. Additionally, by ensuring that the modeling error and the sampling time are sufficiently small, Proposition 3 establishes that the actual state in Equation (61a) can be maintained within a slightly expanded set $\Omega_{\rho_{min}}$, encompassing $\Omega_{\rho_{nn}}$. Hence, LMPC guarantees that for any initial state $x_0 \in \Omega_{\hat{\rho}}$, the state $x(t)$ of the closed-loop slow subsystem described by Equation (61a) is maintained bounded within the region $\Omega_{\hat{\rho}}$ at all times and fulfills the bound of Equation (57) that was established in Proposition 3.

Additionally, by introducing a fast time scale $\tau = t/\epsilon$, a new coordinate $\bar{\bar{z}} = z - \bar{f}_2(x)$, and considering $\epsilon = 0$, the closed-loop fast subsystem is given by:

$$\frac{d\bar{\bar{z}}}{d\tau} = f_2(x, \bar{\bar{z}} + \bar{f}_2(x), 0) \tag{64}$$

According to Assumption 4, global asymptotic stability is achieved for the origin of the closed-loop fast subsystem in Equation (64), satisfying Equation (12) for any $\bar{\bar{z}}(0) \in \mathbb{R}^p$. As a result, the closed-loop system in Equation (61) meets all the assumptions required for Theorem 1 in [19] to hold true. Hence, there exist class $\mathcal{KL}$ functions β_x and β_z, a pair of positive real numbers (δ, d), and $\epsilon^* > 0$, such that if $\max\{|x(0), |z(0)|\} \leq \delta$ and $\epsilon \in (0, \epsilon^*]$, the closed-loop states of the slow and fast systems are bounded by Equations (59) and (60). □

Remark 3. *It is possible to define the general class of two-time-scale systems with an additional input associated with the fast subsystem as follows:*

$$\dot{x} = f_1(x, z, u, \epsilon) \tag{65a}$$

$$\epsilon\dot{z} = f_2(x, z, u_2, \epsilon) \tag{65b}$$

where $u_2 \in \mathbb{R}^{q_2}$ is the bounded manipulated input vector associated with the fast subsystem. The input vector u_2 is constrained by $u_2 \in U_2 := \{|u_i| \leq u_{2_{\max,i}}, i = 1, ..., q_2\}$. However, if there exists a stabilizing control law $u_2 = h_{stable}(x, z)$ that stabilizes the dynamics of the fast subsystem, then the general class of two-time-scale systems defined by Equation (65) can be reduced to the class of systems presented in Equation (1), which is the primary focus of this work. Furthermore, there are relevant works that have addressed the stability analysis of classes of systems defined in Equation (65), which share similar concepts but incorporate certain modifications. Interested readers may refer to [3] to further explore this subject.

5. Example of Application to a Chemical Process

In this section, we apply the proposed reduced-order machine learning strategy to a chemical process example. Specifically, we build two ML models—a reduced-order ML model to approximate the slow subsystem, and another ML model to capture the dynamics of the full two-time-scale system. Subsequently, each ML model is incorporated into an LMPC scheme, and the performance of the controllers is compared in terms of the closed-loop properties and computational efficiencies.

We consider a perfectly mixed, non-isothermal CSTR, in which an irreversible and endothermic reaction that transforms chemical A to product B ($A \rightarrow B$) occurs. Figure 3 depicts the CSTR, where C_A represents the concentration of chemical A, and T_r represents the temperature of the reactor's contents. C_{A0} denotes the inlet molar concentration of chemical A, which is fed into the reactor at flow rate F and temperature T_{A0}. Assuming that the vessel maintains a constant holdup, V_r denotes the volume of liquid present in the reactor. Due to the occurrence of an endothermic reaction within the reactor, a heating jacket with volume V_j is used to provide the energy as required. T_{j0} is the inlet temperature of the heat-transfer fluid provided to the jacket with a flow rate F_j. The reactor's contents and the heat-transfer fluid maintain constant densities, denoted as ρ_m and ρ_j, respectively. In addition, they both have constant heat capacities, denoted as $c_{p,m}$ and $c_{p,j}$, respectively. The enthalpy of the reaction is ΔH_r, the heat-transfer coefficient is U, and the heat-transfer contact area between the reactor and the jacket is A_r. Given that k_0 is the pre-exponential constant, R is the ideal gas constant, and E is the activation energy of the reaction, the time-varying rate constant of the reaction is denoted by k and is given by the following equation:

$$k = k_0 \exp\left(\frac{-E}{RT_r}\right) \tag{66}$$

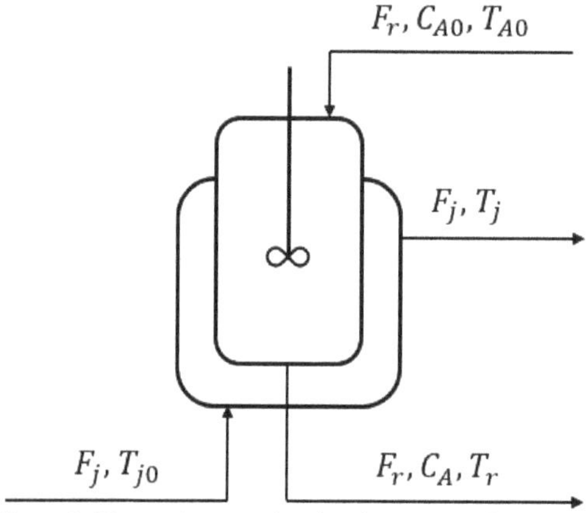

Figure 3. The continuous-stirred tank reactor with jacket.

The first-principles equations of the CSTR are described by the following dynamic material and energy balance equations:

$$V_r \frac{\mathrm{d}C_A}{\mathrm{d}t} = F_r(C_{A0} - C_A) - k_0 \exp\left(\frac{-E}{RT_r}\right) C_A V_r \tag{67a}$$

$$V_r \frac{\mathrm{d}T_r}{\mathrm{d}t} = F_r(T_{A0} - T_r) + \frac{-\Delta H_r}{\rho_m C_{p,m}} k_0 \exp\left(\frac{-E}{RT_r}\right) C_A V_r + \frac{UA_r}{\rho_m C_{p,m}}(T_j - T_r) \tag{67b}$$

$$V_j \frac{\mathrm{d}T_j}{\mathrm{d}t} = F_j(T_{j0} - T_j) - \frac{UA_r}{\rho_j C_{p,j}}(T_j - T_r) \tag{67c}$$

where the values of the process parameters are listed in Table 1. Additionally, we consider the constant $\epsilon = \frac{V_j}{V_r}$ to be the singular perturbation parameter. Hence, the system of Equation (67) can be written in the standard singularly perturbed form of Equation (1), which implies that the concentration of chemical A (C_A) and the temperature of the reactor (T_r) are the slow states, whereas the temperature of the heating jacket (T_j) can be considered as the fast state. The input of the system in Equation (67) is the feed concentration of chemical A (C_{A0}). The control objective is to drive the system to its stable steady state $(C_{As}, T_{rs}, T_{js}) = (2.54 \text{ kmol m}^{-3}, 274.4 \text{ K}, 303.3 \text{ K})$, corresponding to the steady-state input value of $C_{A0s} = 3.75 \text{ kmol m}^{-3}$, while the input is permitted to vary within the range $\Delta C_{A0} = [-3.5, 3.5] \text{ kmol m}^{-3}$. To shift the steady state of the system in Equation (67) to the origin, we express the states and the input in terms of deviation variables and rewrite the system in Equation (67). Specifically, the deviation variables are defined as follows:

$$\Delta C_A = C_A - C_{As} \tag{68a}$$

$$\Delta T_r = T_r - T_{rs} \tag{68b}$$

$$\Delta T_j = T_j - T_{js} \tag{68c}$$

$$\Delta C_{A0} = C_{A0} - C_{A0s} \tag{68d}$$

Therefore, to write the CSTR system of Equation (67) in the standard form of Equation (1), we define $x^T = [\Delta C_A \quad \Delta T_r]$, $z = \Delta T_j$, and the input $u = \Delta C_{A0}$. The explicit Euler method is utilized to simulate the CSTR system, i.e., Equation (67) is numerically integrated with a sufficiently small time step $h_c = 10^{-4}$ h. The open-source software package IPOPT [20] is employed to solve the ML-based LMPC optimization problem in Equation (58), with a sampling time of $\Delta = 0.03$ h.

Table 1. Notations and parameter values of the CSTR.

$V_r = 1.0\ \text{m}^3$	$E = 8.0 \times 10^3\ \text{kcal kg}^{-1}$
$V_j = 0.08\ \text{m}^3$	$U = 1000.0\ \text{kcal h}^{-1}\text{m}^{-2}\text{K}^{-1}$
$A_r = 6.0\ \text{m}^3$	$k_0 = 3.36 \times 10^6\ \text{h}^{-1}$
$C_{A0s} = 3.75\ \text{kmol m}^{-3}$	$\rho_m = 900.0\ \text{kg m}^{-3}$
$C_{As} = 2.54\ \text{kmol m}^{-3}$	$\rho_j = 800.0\ \text{kg m}^{-3}$
$R = 1.987\ \text{kcal kmol}^{-1}\text{K}^{-1}$	$c_{p,m} = 0.231\ \text{kcal kg}^{-1}\text{K}^{-1}$
$F_r = 3.0\ \text{m}^3\text{h}^{-1}$	$c_{p,j} = 0.200\ \text{kcal kg}^{-1}\text{K}^{-1}$
$F_j = 20.0\ \text{m}^3\text{h}^{-1}$	$\Delta H_r = 5.4 \times 10^4\ \text{kcal mol}^{-1}$
$T_{rs} = 274.4\ \text{K}$	$T_{j0} = 357.5\ \text{K}$
$T_{js} = 303.3\ \text{K}$	$T_{A0} = 310.0\ \text{K}$

5.1. Data Generation and Development of RNN Models

The first step in building the RNN models was to generate a comprehensive data set that captured the dynamics of the system in Equation (67). To implement the explicit Euler algorithm, we first needed to initialize the process model using a set of initial conditions $(C_A(0), T_r(0), T_j(0))$, where $C_A(0) \in [0, 9]$ mol/m^3, $T_r(0) \in [280, 370]$ K and $T_j(0) \in [300, 390]$ K. We conducted open-loop simulations by integrating the CSTR system defined by Equation (67) using a sufficiently small time step h_c for 10^6 random initial conditions within the specified ranges. We applied random input signals $C_{A0} \in [0.5, 7.5]$ mol/m^3 over a period of one sampling period Δ, which yielded a data set of 10^6 trajectories, each with a length equal to Δ. The data set was then divided into three sets—training, validation, and testing. The training of the models was performed using the open-source library Keras [21]. The first RNN model was designed to predict the dynamics of the full two-time-scale system in Equation (67) (i.e, it predicts the dynamics of the slow and fast states). In other words, it utilizes the current state measurements $x(t_k)$ and $z(t_k)$, and the manipulated input for the subsequent sampling period $u(t) \in [t_k, t_{k+1})$ to predict the slow and fast state measurements of $x(t)$ and $z(t)$ $\forall t \in [t_k, t_{k+1}]$. For simplicity, we denote the model that predicts the dynamics of the full two-time-scale system in Equation (67) as RNN_F, where the subscript "F" denotes "full" and not "fast". This model was designed using a single layer of 20 long short-term memory recurrent neural network (LSTM-RNN) units, and the validation and testing errors achieved were 1.434×10^{-6} and 1.432×10^{-6}, respectively. In addition, the total number of learning parameters for training this model was 2063. We note that the network was designed in a step-wise manner. Initially, we started with a simple structure, consisting of a single-layer network with one LSTM-RNN unit but the validation error was high. Gradually, we increased the network's complexity by adding more LSTM-RNN units. This progression was continued until we achieved a satisfactory level of validation loss. In this particular case, the optimal outcome was achieved using a single-layer network with 20 LSTM-RNN units. Further increasing the number of LSTM-RNN units beyond 20 resulted in an increase in the validation loss. Therefore, RNN_F was constructed using 20 LSTM-RNN units. On the other hand, the second RNN, denoted as RNN_S, was designed to predict the dynamics of only the slow states in Equation (67). Specifically, this RNN utilizes the current slow state measurement $x(t_k)$ and the manipulated input for the subsequent sampling period $u(t) \in [t_k, t_{k+1})$ to predict the slow state measurements of $x(t)$ $\forall t \in [t_k, t_{k+1}]$. This model was designed using a single layer of five LSTM-RNN units corresponding to a total of 192 learning parameters following the same step-wise manner strategy used to design RNN_F, with a validation error of 2.67×10^{-4} and a testing error of 2.68×10^{-4}. The results of both models are summarized in Table 2. The low testing errors for both models indicate that the modeling error, as defined in Proposition 1, is sufficiently small over a wide range of open-loop trajectories. Due to having fewer learning parameters, the RNN_S model is significantly less complex compared to the RNN_F model, possibly leading to a lower computational cost and making it a more practical choice for control applications. Therefore, our aim is to investigate whether the LMPC scheme utilizing RNN_S can sufficiently meet the desired control objective more efficiently compared to the LMPC scheme using RNN_F, which is demonstrated via closed-loop simulations of the CSTR in Equation (67) under both LMPC designs. However, before proceeding, it is essential to define the Lyapunov and objective functions used in both LMPC schemes. For the RNN_F-based LMPC, since the controller design was based on the dynamics of the full two-time-scale system in Equation (67), the Lyapunov function is defined as:

$$V_F(x,z) = \begin{bmatrix} x - x_S \\ z - z_S \end{bmatrix}^T P_F \begin{bmatrix} x - x_S \\ z - z_S \end{bmatrix} \tag{69}$$

where the matrix P_F is given by

$$P_F = \begin{bmatrix} 91.6 & 2.9 & 4.3 \\ 2.9 & 0.8 & 0.2 \\ 4.3 & 0.2 & 0.7 \end{bmatrix}$$

The objective function for the RNN_F-based LMPC is given by $L_F(x,z,u) = |x|^2_{Q_{1F}} + |z|^2_{Q_{2F}} + |u|^2_{Q_{3F}}$, where $Q_{1F} = \begin{bmatrix} 20 & 0 \\ 0 & 0.05 \end{bmatrix}$, $Q_{2F} = 6$, and $Q_{3F} = 0.1$. On the other hand, the RNN_S-based LMPC was designed based on the dynamics of the slow subsystem of the CSTR, specifically Equations (67a) and (67b) expressed in the standard form. Therefore, the Lyapunov function of the slow subsystem used in the RNN_S-based LMPC is given by:

$$V(x) = (x - x_s)^T P (x - x_s) \tag{70}$$

where the matrix P is defined as

$$P = \begin{bmatrix} 84.9 & 1.2 \\ 1.2 & 0.4 \end{bmatrix}$$

The objective function is given by $L(x,u) = |x|^2_{Q_1} + |u|^2_{Q_2}$, where $Q_1 = \begin{bmatrix} 20 & 0 \\ 0 & 0.05 \end{bmatrix}$ and $Q_2 = 0.05$.

Remark 4. *In this particular application, we used long short-term memory recurrent neural network (LSTM-RNN) units to construct the models. LSTM-RNNs are a special type of RNN known for their ability to overcome the common problem of vanishing/exploding gradients in RNNs, resulting in generally better performance. For more information about LSTM-RNNs, interested readers may refer to [22]. However, we can always utilize classical RNNs that are well-suited for this particular application by choosing appropriate structures and carefully tuning the hyperparameters to obtain the desired objective, as well as an acceptable level of performance.*

Table 2. Specifications of the constructed recurrent neural network models.

RNN Model	RNN_F	RNN_S
Number of units used	20	5
Testing error	1.432×10^{-6}	2.68×10^{-4}
Validation error	1.434×10^{-6}	2.67×10^{-4}
Number of learning parameters	2063	192

5.2. Simulation Results

In this subsection, we conduct closed-loop simulations of the CSTR system described by Equation (67) under the two LMPC schemes, one with each RNN model. We first assess the ability of both LMPCs to achieve the control objective adequately compared to LMPCs that utilize the respective first-principles systems as their process models. Subsequently, we compare the RNN-based LMPCs to each other in terms of closed-loop performance and the computational time required for the simulations.

To establish the satisfactory performance of both RNN-based LMPCs, we compared their closed-loop performance to first-principles-based LMPCs, with the latter serving as the baseline for the best possible performance achievable under LMPC. Specifically, the LMPC in Equation (58) was considered in two scenarios, with the process model of Equation (58e) being different between each scenario:

- Scenario 1: The LMPC utilizing RNN_F as the process model was compared to an LMPC employing the first-principles model in Equation (67), denoted as FP_F, as its process model.
- Scenario 2: The LMPC utilizing RNN_S as the process model was compared to an LMPC employing the first-principles slow subsystem in Equations (67a) and (67b), denoted as FP_S, as its process model. In this case, for the first-principles-based LMPC, we note that the full CSTR system in Equation (67) was integrated, but only the slow states C_A and T_r from Equations (67a) and (67b) were used in calculating the LMPC cost function of Equation (58a) and the Lyapunov function V.

For both scenarios, we started the simulations from the initial condition $IC_{\text{main}} = (\Delta C_A(0), \Delta T_r(0), \Delta T_j(0)) = (1, 30, 40)$. The LMPC prediction horizon was set to $N = 3$, and the remaining controller parameters were described in Section 5.1. The state and input trajectories under the LMPCs of scenarios 1 and 2 are shown in Figures 4 and 5, respectively. To quantitatively compare the performance of the different LMPCs, the time integral of the cost function of the LMPC, $\int_{t=0}^{t_f} L(x(\tau), u(\tau))\, d\tau$, was calculated over the entire simulation duration, $t_f = 3$ h. For scenario 1, the cost function values were 3458 and 3485 for the FP_F-based LMPC and the RNN_F-based LMPC, respectively, whereas for scenario 2, the values of the cost function were 176 and 179 for the FP_S-based LMPC and the RNN_S-based LMPC, respectively. For both scenarios, we noticed that the value of the cost function when utilizing the corresponding RNN model closely aligned with that achieved when employing the respective first-principles model, demonstrating the reliable predictive performance of the designed RNN models and their ability to drive the system to the steady state when incorporated into an LMPC scheme.

Next, we demonstrated the computational efficiency of the RNN_S-based LMPC due to its lower complexity while still achieving the desired controller performance. For IC_{main}, the computational time for the RNN_F-based LMPC shown in Figure 4 was 5578 s , whereas, for the RNN_S-based LMPC shown in Figure 5, it was significantly lower at 2170 s, which was 61% lower. This substantial difference in computational time can make RNN_S a more practical choice for real-time application of MPC, where computations must be completed within a sampling period to be sent to the actuator. Hence, to rigorously investigate the impact of integrating RNN_F and RNN_S into an LMPC framework in terms of stability and computational demand, we conducted closed-loop simulations from several different initial conditions in addition to IC_{main}.

The initial conditions chosen for further investigation are outlined in Table 3. Ten different initial conditions (IC_i where $i = 1, ..., 10$) within the operating region were selected, along with IC_{main}. For each initial condition, we simulated the CSTR system described in Equation (67) under the RNN_F-based LMPC, as well as the RNN_S-based LMPC. The state and input profiles under each LMPC for four representative initial conditions are depicted in Figures 6–9, which indicate that both LMPC schemes successfully drove the process in Equation (67) to its steady state. The closed-loop behavior was similar for the remaining six initial conditions in terms of convergence to the steady state. Additionally, Table 3 provides the computational times required in terms of CPU time for the entire simulation duration of $t_f = 3$ h when applying either LMPC scheme for the 11 different initial conditions. For all the tested initial conditions, the computational time required to execute the RNN_S-based LMPC was less than that required for the RNN_F-based LMPC. Among all the initial conditions in Table 3, the largest relative difference in computational time was observed for IC_3, with a percentage difference of 92% between the two LMPC schemes. Hence, the incorporation of RNN_S in the LMPC framework has been demonstrated to be more practical and computationally efficient compared to employing RNN_F, without compromising stability and closed-loop performance.

The computational time of an MPC optimization problem can be influenced by several factors. As mentioned in [23], MPC computational times are generally affected by the horizon length N, the number of constraints in the MPC optimization problem, and the number of states and control actions of the system. The computational time is also directly related to the complexity of the process model. Hence, given the lower computational times across all 11 initial conditions tested, incorporating RNN_S into an LMPC framework is a more practical approach compared to the use of RNN_F. In the study of [24], it was observed that the CPU time of the IPOPT optimization problem increased as the complexity of the problem increased. Additionally, several works have discussed computational complexity evaluations of neural networks. For instance, the authors of [25] offered a comprehensive and systematic analysis for quantifying and comparing the computational complexity of neural network layers. The authors proposed and computed three computational metrics per layer for different types of neural networks and presented the exact mathematical expressions and derivations, providing a deep understanding of various networks' complexities. Upon analyzing the results derived in [25], it is evident that a neural network's complexity is proportional to its hyperparameters. For example, factors like the number of neurons, the number of hidden units, the input time sequence size, the number of output neurons, and other parameters associated with the structure, size, and design of the network, all of which affect the complexity and performance of a network, as well as the time required to solve the ML-based MPC optimization problem. As RNN_S consisted of only 5 LSTM-RNN units while RNN_F consisted of 20 LSTM-RNN units, the complexity of RNN_F was much higher than that of RNN_S, leading to a correspondingly complex MPC optimization problem. Hence, over the entire simulation duration of $t_f = 3$ h, the total CPU time required to solve all 100 MPC optimization problems over the 100 sampling periods was significantly less for the RNN_S-based LMPC than the RNN_F-based LMPC.

Remark 5. *Although the state and input trajectories in Figure 6 have not completely settled, the process is stabilized under both the RNN_F-based LMPC and the RNN_S-based LMPC. However, the process requires simulation for a longer duration, exceeding 3 h, for the states and input to ultimately converge to their steady-state values and remain close to the origin. The reason for limiting the simulation time is to ensure fair and clear comparisons between the computational times for both controllers from various initial conditions, as listed in Table 3.*

Remark 6. *In this particular application, RNN_S was designed to utilize the current slow state measurement and the manipulated input for the next sampling period. As a result, the output of RNN_S will be the predicted slow states x for only one sampling period ahead (1Δ). Consequently, in this application, since the prediction horizon was chosen to be an integer multiple of the sampling* *time Δ (i.e., $N = 3$ implies that $t_{k+N} = t_k + 3\Delta$ in Equation (58a)), based on the design of RNN_S, RNN_S is required to be invoked at least three times to be able to make predictions for the entire prediction horizon. Additionally, an alternative method involves designing an RNN model, denoted as RNN_{new}, with the same goal of predicting the slow states. However, RNN_{new} utilizes not only the current slow state measurements and the manipulated input for the next sampling period but also the current fast state measurements to predict the slow states one sampling period Δ ahead, i.e., $x(t_{k+1})$. To obtain predictions for the entire prediction horizon ($N = 3$), an FNN takes as its input the output of RNN_{new} and predicts the fast state z at the next sampling period, i.e., the FNN predicts $z(t_{k+1})$ from $x(t_{k+1})$, following the design described in Section 2.5. Subsequently, for the next sampling period, the outputs of both RNN_{new} and the FNN are given as inputs to RNN_{new}, enabling it to predict the slow states $x(t_{k+2})$ for the next sampling period in the prediction horizon.*

Table 3. Computational times for the RNN_F-based LMPC and the RNN_S-based LMPC over the simulation duration of $t_f = 3$ h starting from various initial conditions within the operating region.

Index	Initial Condition	Computational Time (s)	
	$(\Delta C_A(0), \Delta T_r(0), \Delta T_j(0))$	RNN_F-Based LMPC	RNN_S-Based LMPC
IC_{main}	(1, 30, 40)	5578	2170
IC_1	(−1, 50, 40)	16,059	1807
IC_2	(−1, −10, −3)	4801	2667
IC_3	(−3, 30, 5)	31,884	2417
IC_4	(−2, −10, 100)	17,896	1921
IC_5	(1, 90, 10)	6078	1990
IC_6	(1, 20, 60)	5795	2404
IC_7	(3, −6, 20)	21,231	2411
IC_8	(1, 50, 50)	5161	2225
IC_9	(−2, 30, 60)	10,275	2194
IC_{10}	(−1, 10, 80)	18,146	2106

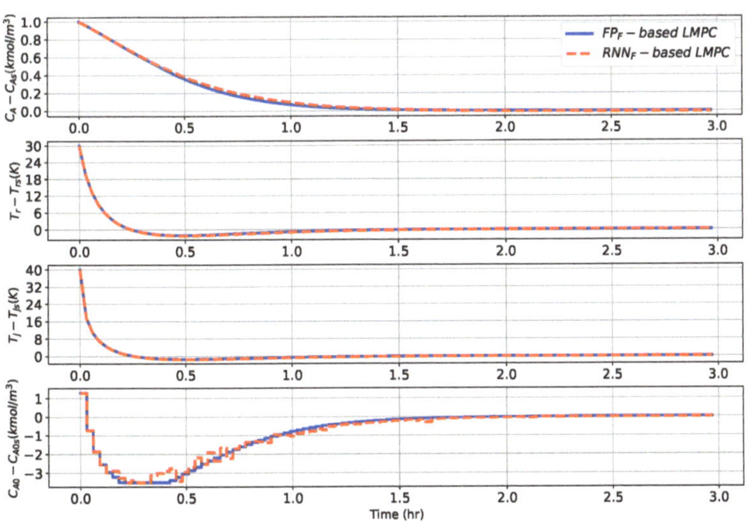

Figure 4. States and input trajectories of the CSTR under the Lyapunov-based MPC scheme using the first-principles model of the full process (FP_F-based LMPC, blue line) and RNN_F (RNN_F-based LMPC, red dashed line).

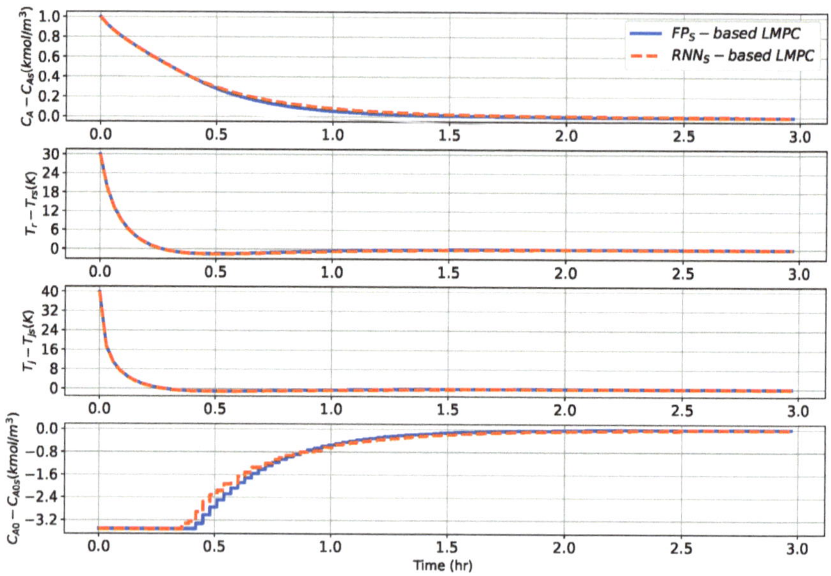

Figure 5. States and input trajectories of the CSTR under the Lyapunov-based MPC scheme using the first-principles model of the slow-subsystem (FP_S-based LMPC, blue line) and RNN_S (RNN_S-based LMPC, red dashed line).

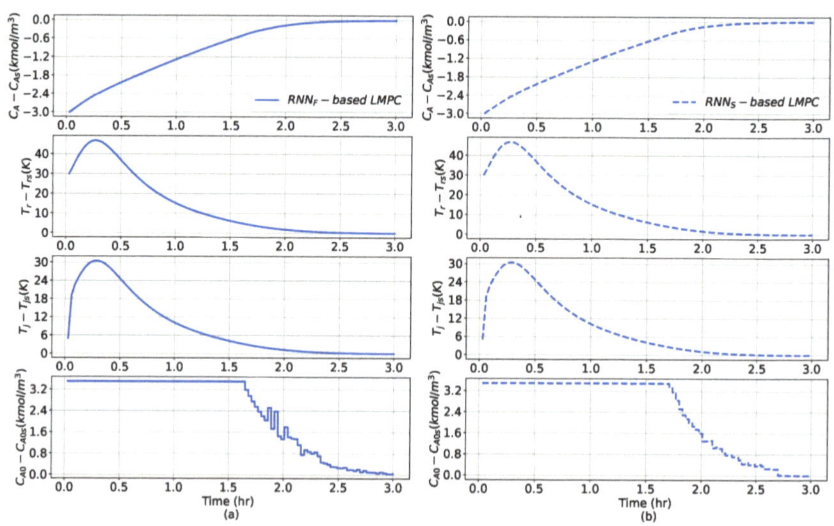

Figure 6. Considering the initial condition $IC_3 = (-3, 30, 5)$, (**a**) illustrates the time-varying profiles of the states and the input under the RNN_F-based LMPC (solid line), whereas (**b**) shows the time-varying profiles of the states and the input under the RNN_S-based LMPC (dashed line).

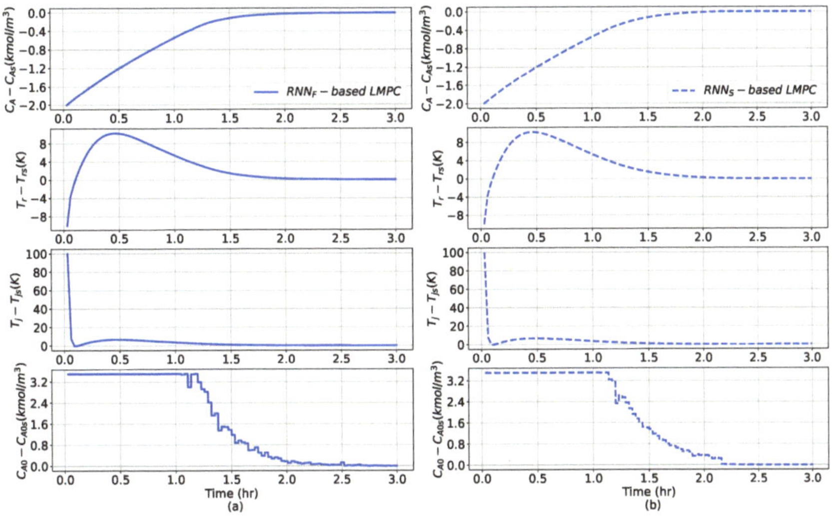

Figure 7. Considering the initial condition $IC_4 = (-2, -10, 100)$, (**a**) illustrates the time-varying profiles of the states and the input under the RNN_F-based LMPC (solid line), whereas (**b**) shows the time-varying profiles of the states and the input under the RNN_S-based LMPC (dashed line).

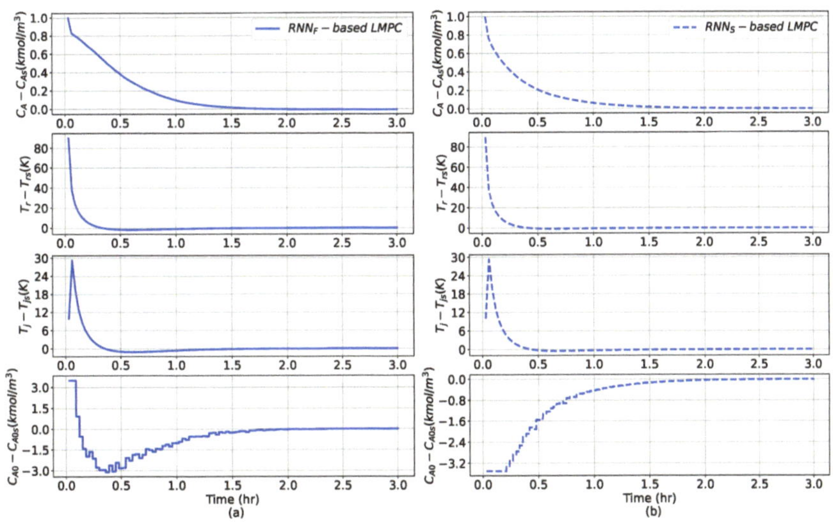

Figure 8. Considering the initial condition $IC_5 = (1, 90, 10)$, (**a**) illustrates the time-varying profiles of the states and the input under the RNN_F-based LMPC (solid line), whereas (**b**) shows the time-varying profiles of the states and the input under the RNN_S-based LMPC (dashed line).

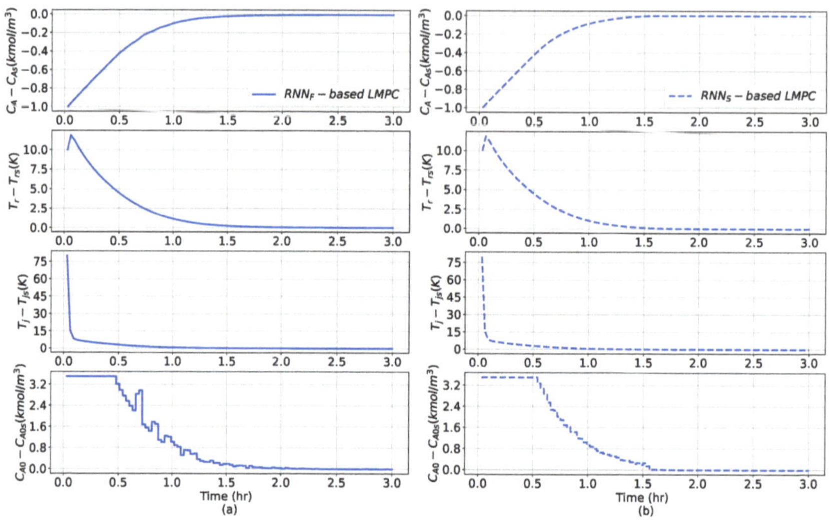

Figure 9. Considering the initial condition $IC_{10} = (-1, 10, 80)$, (**a**) illustrates the time-varying profiles of the states and the input under the RNN_F-based LMPC (solid line), whereas (**b**) shows the time-varying profiles of the states and the input under the RNN_S-based LMPC (dashed line).

6. Conclusions

In this work, we introduced a general class of nonlinear two-time-scale systems, where the two distinct time scales can be decoupled into slow subsystem and fast subsystem dynamics using singular perturbation analysis. Machine learning was employed to approximate the dynamics of both subsystems. In particular, an RNN was used to predict the slow state vector or, in other words, to approximate the dynamics of the slow subsystem, whereas an FNN was used to approximate the dynamics of the fast subsystem. To draw inferences about the performance of the neural network models on unseen data, the generalization error bounds for both neural networks when modeling two-time-scale systems were derived. Subsequently, the RNN modeling of the slow states (RNN_S) was incorporated into an LMPC framework and sufficient conditions to guarantee closed-loop stability were derived under the sample-and-hold implementation of the LMPC. Finally, a chemical process example was used to demonstrate the efficiency of the LMPC scheme using an RNN to predict the slow states compared to first-principles-based LMPC schemes and an LMPC scheme using an RNN to model the full two-time-scale system. Through closed-loop simulations, while both RNN-based LMPCs were able to achieve the desired control objective of driving the system to the steady state, due to the significantly lower complexity of the RNN approximating only the slow subsystem, the LMPC based on RNN_S required significantly less computational time in all simulations compared to the LMPC using the RNN modeling the full singularly perturbed system, which supported the use of RNN_S in real-time MPC applications.

Author Contributions: Conceptualization, A.A., F.A., A.S. and P.D.C.; methodology, A.A., F.A., A.S. and P.D.C.; software, A.A., F.A. and A.S.; validation, A.A., F.A. and A.S.; formal analysis, A.A., F.A. and A.S.; investigation, A.A., F.A. and A.S.; resources, P.D.C.; data curation, A.A., F.A. and A.S.; writing—original draft preparation, A.A., F.A. and A.S.; writing—review and editing, A.A., F.A., A.S. and P.D.C.; supervision, P.D.C.; project administration, P.D.C.; funding acquisition, P.D.C. All authors have read and agreed to the published version of the manuscript.

Funding: Financial support from the National Science Foundation, CBET-CBET-2140506, and the Department of Energy, through the Office of Energy Efficiency and Renewable Energy (EERE) under the Advanced Manufacturing Office Award Number DEEE0007613, is gratefully acknowledged. The first author acknowledges the support of Kuwait University via the KU scholarship program.

Data Availability Statement: Data is available upon request to the corresponding author.

Conflicts of Interest: The authors declare that they have no conflicts of interest regarding the publication of this research article.

References

1. Christofides, P.D.; Daoutidis, P. Feedback control of two-time-scale nonlinear systems. *Int. J. Control* **1996**, *63*, 965–994.
2. Kokotović, P.; Khalil, H.K.; O'reilly, J. *Singular Perturbation Methods in Control: Analysis and Design*; SIAM: Philadelphia, PA, USA, 1999.
3. Chen, X.; Heidarinejad, M.; Liu, J.; Christofides, P.D. Composite fast-slow MPC design for nonlinear singularly perturbed systems. *AIChE J.* **2012**, *58*, 1802–1811. [CrossRef]
4. Ellis, M.; Heidarinejad, M.; Christofides, P.D. Economic model predictive control of nonlinear singularly perturbed systems. *J. Process Control* **2013**, *23*, 743–754. [CrossRef]
5. Abdullah, F.; Wu, Z.; Christofides, P.D. Sparse-identification-based model predictive control of nonlinear two-time-scale processes. *Comput. Chem. Eng.* **2021**, *153*, 107411. [CrossRef]
6. Wu, Z.; Tran, A.; Rincon, D.; Christofides, P.D. Machine Learning-Based Predictive Control of Nonlinear Processes. Part II: Computational Implementation. *AIChE J.* **2019**, *65*, e16734. [CrossRef]
7. Nikolakopoulou, A.; Braatz, R.D. Polynomial NARX-based nonlinear model predictive control of modular chemical systems. *Comput. Chem. Eng.* **2023**, *177*, 108272. [CrossRef]
8. Yang, S.; Wan, M.P.; Chen, W.; Ng, B.F.; Dubey, S. Experiment study of machine-learning-based approximate model predictive control for energy-efficient building control. *Appl. Energy* **2021**, *288*, 116648. [CrossRef]
9. Alessio, A.; Bemporad, A. A survey on explicit model predictive control. In *Nonlinear Model Predictive Control: Towards New Challenging Applications*; Springer: Berlin/Heidelberg, Germany, 2009; pp. 345–369.
10. Chen, S.; Saulnier, K.; Atanasov, N.; Lee, D.D.; Kumar, V.; Pappas, G.J.; Morari, M. Approximating explicit model predictive control using constrained neural networks. In Proceedings of the 2018 Annual American Control Conference (ACC), Milwaukee, WI, USA, 27–29 June 2018; pp. 1520–1527.
11. Bao, Y.; Chan, K.J.; Mesbah, A.; Velni, J.M. Learning-based adaptive-scenario-tree model predictive control with improved probabilistic safety using robust Bayesian neural networks. *Int. J. Robust Nonlinear Control* **2023**, *33*, 3312–3333. [CrossRef]
12. Wu, Z.; Rincon, D.; Gu, Q.; Christofides, P.D. Statistical machine learning in model predictive control of nonlinear processes. *Mathematics* **2021**, *9*, 1912. [CrossRef]
13. Chen, S.; Wu, Z.; Christofides, P.D. Statistical Machine-Learning-based Predictive Control Using Barrier Functions for Process Operational Safety. *Comput. Chem. Eng.* **2022**, *163*, 107860. [CrossRef]
14. Alhajeri, M.S.; Alnajdi, A.; Abdullah, F.; Christofides, P.D. On generalization error of neural network models and its application to predictive control of nonlinear processes. *Chem. Eng. Res. Des.* **2023**, *189*, 664–679. [CrossRef]
15. Hoppensteadt, F. Properties of solutions of ordinary differential equations with small parameters. *Commun. Pure Appl. Math.* **1971**, *24*, 807–840. [CrossRef]
16. Mohri, M.; Rostamizadeh, A.; Talwalkar, A. *Foundations of Machine Learning*; MIT Press: Cambridge, MA, USA, 2018.
17. Golowich, N.; Rakhlin, A.; Shamir, O. Size-independent sample complexity of neural networks. In Proceedings of the Conference on Learning Theory, Stockholm, Sweden, 5–9 July 2018; pp. 297–299.
18. Bao, Y.; Abbas, H.S.; Velni, J.M. A learning- and scenario-based MPC design for nonlinear systems in LPV framework with safety and stability guarantees. *Int. J. Control* 2023, *in press*. [CrossRef]
19. Christofides, P.D.; Teel, A.R. Singular perturbations and input-to-state stability. *IEEE Trans. Autom. Control* **1996**, *41*, 1645–1650. [CrossRef]
20. Wächter, A.; Biegler, L.T. On the implementation of an interior-point filter line-search algorithm for large-scale nonlinear programming. *Math. Program.* **2006**, *106*, 25–57. [CrossRef]
21. Chollet, F. Keras. 2015. Available online: https://keras.io (accessed on 11 August 2023).
22. Hochreiter, S.; Schmidhuber, J. Long short-term memory. *Neural Comput.* **1997**, *9*, 1735–1780. [CrossRef] [PubMed]
23. Rawlings, J.B.; Mayne, D.Q.; Diehl, M. *Model Predictive Control: Theory, Computation, and Design*; Nob Hill Publishing: Madison, WI, USA, 2017; Volume 2.
24. Pöri, L. Comparison of Two Interior Point Solvers in Model Predictive Control Optimization. Master's Thesis, Aalto University, Espoo, Finland, 2016.
25. Freire, P.J.; Srivallapanondh, S.; Napoli, A.; Prilepsky, J.E.; Turitsyn, S.K. Computational complexity evaluation of neural network applications in signal processing. *arXiv* **2022**, arXiv:2206.12191.

sensors

MDPI

Article

PSBF: *p-adic* Integer Scalable Bloom Filter

Wenlong Yi, Chuang Wang, Qiliang Xie, Yingding Zhao and Jing Jia *

School of Software, Jiangxi Agricultural University, Nanchang 330045, China; yiwenlong@jxau.edu.cn (W.Y.); wangchuang@stu.jxau.edu.cn (C.W.); xieqiliangg@outlook.com (Q.X.); zhaoyingding@jxau.edu.cn (Y.Z.)
* Correspondence: kevinjia0120@jxau.edu.cn

Abstract: Given the challenges associated with the dynamic expansion of the conventional bloom filter's capacity, the prevalence of false positives, and the subpar access performance, this study employs the algebraic and topological characteristics of *p-adic* integers to introduce an innovative approach for dynamically expanding the *p-adic* Integer Scalable Bloom Filter (PSBF). The proposed method involves converting the target element into an integer using a string hash function, followed by the conversion of said integer into a *p-adic* integer through algebraic properties. This process automatically establishes the topological tree access structure of the PSBF. The experiment involved a comparison of access performance among the standard bloom filter, dynamic bloom filter, and scalable bloom filter. The findings indicate that the PSBF offers advantages such as avoidance of a linear storage structure, enhanced efficiency in element insertion and query, improved storage space utilization, and reduced likelihood of false positives. Consequently, the PSBF presents a novel approach to the dynamic extensibility of bloom filters.

Keywords: *p-adic*; number theory; access optimization; bloom filter; scalable

1. Introduction

The advent of contemporary information technologies, including big data, the Internet of Things, and cloud computing, has resulted in a substantial increase in data volume, posing significant challenges to computer systems in terms of data processing, storage, and retrieval. The bloom filter, a well-established probabilistic data structure, serves the purpose of determining the presence of a data element within a collection. This data structure offers notable advantages such as efficient space utilization and expedited query speed, rendering it extensively employed in various domains such as big data analysis [1], optimization of data storage in the Internet of Things [2], and safeguarding data security in cloud computing [3]. However, the conventional bloom filter exhibits a false positive probability during member queries, wherein an element that does not actually belong to the set is erroneously identified as a member of the set. As depicted in Figure 1, the bloom filter essentially comprises a binary vector V of a specific size m and multiple hash mapping functions H. The size of the bloom filter needs to be predetermined based on the number of data elements E to be stored and the desired probability of false positives, which poses challenges in accurately estimating the size of the stored elements in a system. The increase in false positives is observed in a bloom filter upon surpassing a specific threshold of stored elements. Fixed-size bloom filters are prone to significant storage space wastage or elevated false positives. Consequently, the expandability of bloom filters holds substantial research importance.

To address the aforementioned issues, previous studies have put forth over 60 enhanced bloom filters with the aim of enhancing their adaptability and performance [4]. These enhanced bloom filters exhibit enhanced flexibility and performance benefits in various application scenarios, including content caching [5–10], data storage [11–15], and other domains. By reducing false positives, optimizing implementation, augmenting collection diversity, and enriching functionalities, these enhanced bloom filters offer improved adaptability and efficacy, thereby rendering bloom filters a more potent tool for handling extensive datasets.

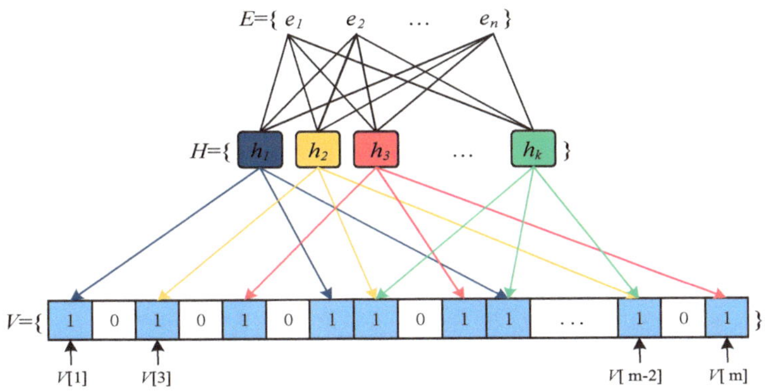

Figure 1. Data structure of a standard bloom filter.

Despite the enhanced effectiveness of bloom filters in mitigating false positives, their ability to address computational overhead and additional space consumption remains limited. To tackle this issue, dynamic capacity expansion serves as a mechanism that facilitates automatic expansion of the bloom filter's capacity in line with the growth of data volume. This approach not only significantly enhances the scalability of processing extensive datasets but also diminishes the rate of misjudgment. A robust dynamic expansion is crucial for enhancing the applicability of bloom filters and is a pivotal consideration in the development of extensive data processing systems.

The *p-adic* number serves as a valuable parameterization tool for computations within hyper-dimensional spaces [16]. Not only does it possess commendable algebraic properties, but it also forms a finite commutative ring with addition and multiplication operations while exhibiting favorable topological characteristics. The authors utilize the algebraic and topological characteristics of *p-adic* numbers to present a pioneering and scalable approach to bloom filters. The primary contributions are as follows:

1. The proposal of a bloom filter automatic expansion technique founded on *p-adic* integers. This technique enables the bloom filter to dynamically adapt its size in accordance with the actual number of stored elements, thereby resolving the problems associated with inefficient storage utilization and elevated false positive rates encountered in bloom filters employing fixed-size binary vectors;
2. Algorithms for the insertion and querying of a bloom filter, utilizing *p-adic* integers, have been developed, and their effectiveness in terms of storage and query efficiency has been validated through experimental analysis;
3. The algorithms address the limitation of conventional bloom filters in terms of expandability when the number of stored elements is uncertain, thereby enhancing the practicality and adaptability of bloom filters.

This paper is composed of five main sections. Section 1, Introduction, provides an in-depth exploration of the research purpose, existing issues, and the authors' contributions in the field of bloom filters. Section 2, Related Works, critically analyzes the current research status and proposes research directions aimed at optimizing the performance of bloom filtering. Section 3, Methods, elaborates on the design ideas and algorithms employed in the *p*-adic Integer Scalable Bloom Filter (PSBF) in a comprehensive manner. Section 4, Results, presents the execution of the selection experiment conducted to determine the suitable hash function and the optimal prime p for the PSBF. Section 5, Discussion, presents a comparative and analytical examination of the performance tests conducted on PSBF and three other bloom filters. Additionally, Section 6, Conclusion, provides a summary of the advantages exhibited by PSBF.

2. Related Works

In contrast to conventional bloom filters, scalable bloom filters possess the ability to adjust their size dynamically in order to accommodate expanding data scales. Almeida et al. [17] addressed the issue of capacity limitations by proposing the utilization of multiple bloom filters to construct an expandable bloom filter. Nonetheless, this approach presents potential limitations, primarily including the division of data into multiple partitions and the requirement for query operations to be conducted across multiple partitions, thereby augmenting the query overhead. Additionally, the maintenance of multiple bloom filters necessitates additional storage overhead.

The dynamic bloom filter, as proposed by Guo et al. [18], is a resizable bloom filter that adjusts its size based on the cardinality of the designated dataset, resulting in a reduced rate of misjudgment. However, the dynamic resizing process of the bloom filter may incur additional time overhead due to memory allocation and data rearrangement. Furthermore, the inclusion of counters or bitmaps for maintaining the bloom filter introduces storage overhead. In response to these challenges, Patgir et al. [19] introduced rFilter, a member filter that offers excellent expandability and efficient utilization of space. In contrast to the conventional bloom filter, rFilter diminishes the likelihood of false positives; however, it is constrained in its capacity to retain bit-sharing information to only twice, thereby limiting its ability to share such information.

Kleyko et al. [20] conducted an analysis on the boundaries of false positives and true positives in the counting bloom filter and introduced a technique known as the Autoscaling Bloom Filter (ABF). The capacity of ABF can be adjusted based on the probability boundary of false positives and true positives, thereby enhancing the extensibility of conventional bloom filters by relaxing the need for a perfect true positive rate. ABF achieves this by sacrificing a non-zero true positivity rate in exchange for a reduced false positivity rate, which restricts the potential application scenarios where a non-zero true positivity rate is acceptable.

Kim et al. [21] employed the cyclic displacement method to enhance the operational speed of the standard bloom filter for both "add" and "query" operations while maintaining the same false positive rate. However, it should be noted that this method solely focuses on improving the computational efficiency of the standard bloom filter and does not address the persistently high false positive rate observed in the fixed-length bloom filter when processing extensive data. In contrast, Rottenstreich et al. [22] utilized Orthogonal Latin Square codes, and polynomials defined modulo primes to represent data elements, thereby leveraging the linear memory dependence of the dataset to accommodate a larger number of data elements. Despite its ability to prevent false positives in cases where the number of collection elements is below the set threshold, this method fails to account for the dynamic nature of the dataset. In response to this limitation, Nayak et al. [23] introduced a bloom filter known as RobustBF, which enhances the selection of hash functions and incorporates them into the 2D bloom filter [24] to minimize false positives and optimize memory usage. However, as RobustBF is currently implemented as a single-threaded system, its performance may require further enhancements to effectively handle high-throughput or low-latency data streams.

Dayan et al. [25] successfully incorporated InfiniFilter into the existing quotient filter [26]. This implementation utilizes a hash table to store the fingerprint of each entry, and its adaptable hash slot format ensures consistent functionality for "insert", "query", and "delete" operations. However, as the data volume increases, InfiniFilter's performance in the "insert" operation falls short compared to the original quotient filter. Cohen et al. [27] introduced the concept of Spectral Bloom Filters (SBF), wherein the bit vector of the conventional bloom filter is extended to incorporate a counter vector. The functionality of querying and deleting SBF elements is facilitated by this counter operation. It is worth noting that both SBF and standard bloom filters are constrained by the dimensions of the bit vector. Kiss et al. [28] introduced the EGH filter, a data structure that enables bloom filter operations while ensuring error-free operations within a constrained set and a limited number of stored elements. However, the design limits the capability to handle dynamic collections.

These observations highlight the presence of challenging obstacles in the dynamic expansion mechanism of the bloom filter. The urgent need for comprehensive research lies in determining the optimal balance between the computing and memory overhead resulting from expansion, as well as devising strategies to minimize the occurrence of false positives in the filter.

3. Methods

This study proposes a solution to the issues associated with fixed-size bloom filters in previous works, namely the excessive waste of storage space and high false positive rates. Drawing inspiration from the design of the quotient filter [26], the authors introduce the PSBF. By increasing the prime number p and depth d, the PSBF dynamically expands to accommodate the storage elements, thereby optimizing the utilization of storage space.

As depicted in Figure 2, the data structure of the PSBF primarily consists of a tree logical structure comprising a linked list and three key operations: "initialize"; "insert"; and "query". Initially, a Hash Function is employed to map any data element e to an integer e^*. Subsequently, this integer is transformed into a *p-adic* integer using a prime number p. Furthermore, the depth d of the bloom filter is determined based on the number of bits present in the converted *p-adic* integer. Consequently, the storage structure of the PSBF adopts a tree structure, thereby circumventing the linear storage structure employed in previous methodologies. This novel approach facilitates more efficient element management.

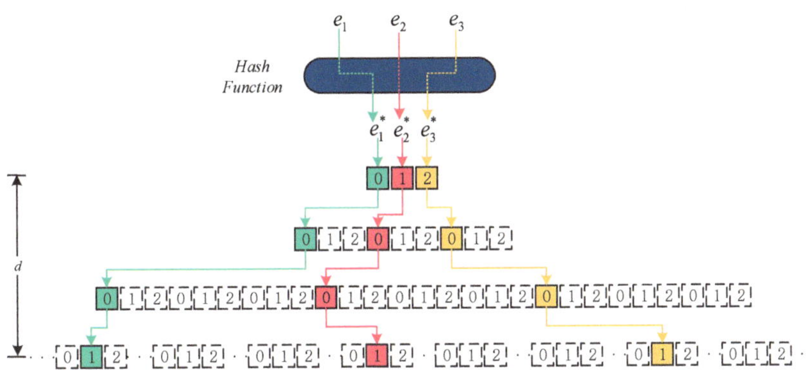

Figure 2. Data structure of *3-adic* Integer Scalable Bloom Filter.

1. The *p-adic* integer algebraic representation is defined as follows: Let α be an element in the number domain Q_p. It can be expressed as Equation (1).

$$\alpha = \sum_{i=n}^{\infty} a_i p^i \tag{1}$$

where p is a prime number, n is an integer, and $a_i \in \mathbf{Z}$. If $i = 0$ and $a_i \in \{0,1,\ldots,p-1\}$, then α is an integer, and there exists a unique *p-adic* decomposition for α in the case of modulo p. The *p-adic* expanded power polynomial is depicted in Equation (2).

$$\alpha = a_n p^n + a_{n+1} p^{n+1} + a_{n+2} p^{n+2} + \ldots \tag{2}$$

It is utilized to represent α. Specifically, the *p-adic* integer is chosen in the PSBF, with n set to 0. Consequently, each coefficient is proposed to form a representation of its corresponding coefficient, as illustrated in Equation (3).

$$\alpha = a_n a_{n+1} \ldots a_{n-2} a_{n-1} . a_0 a_1 a_2 \ldots \tag{3}$$

where the point between the coefficients a_{n-1} and a_0 is referred to as the *p-adic* point, which exclusively represents the symbol for the n value, while a_i is denoted as the *p-adic* number.

2. Coefficient extraction operation. Taking the example of *3-adic*, the *p-adic* integer representation is obtained. Let an integer $e = 11$ be given, and the *3-adic* expanded form of the number e is calculated using Equation (2), where $n = 0$. The process involves three steps:

step 1 $\alpha \equiv a_0 (\text{mod } 3^0) \rightarrow a_0 = 2$;
step 2 $\alpha \equiv a_1 (\text{mod } 3^1) \rightarrow a_1 = 0$;
and step 3 $\alpha \equiv a_2 (\text{mod } 3^2) \rightarrow a_2 = 1$.
After completing these steps, the result is $11 = 2 \times 3^0 + 0 \times 3^1 + 1 \times 3^2$, yielding the *3-adic* number α as 0.201.

3. Calculation of the probability of false positives. The PSBF involves determining the depth d and prime number p of the *p-adic* expansion. The number of slices from layer 1 to layer d is represented by $p, p^2, p^3, \ldots, p^d$. The probability of a false positive for a single number P_f is calculated using Equation (4).

$$P_f = \prod_{i=1}^{d} \frac{1}{p^i} \tag{4}$$

while the probability that the number does not have a false positive P_t is denoted by Equation (5).

$$P_t = 1 - \prod_{i=1}^{d} \frac{1}{p^i} \tag{5}$$

Furthermore, the probability that n numbers do not have false positives P_t^n is denoted by Equation (6).

$$P_t^n = \left(1 - \prod_{i=1}^{d} \frac{1}{p^i}\right)^n \tag{6}$$

The calculation process for determining the probability P of false positives for n numbers is illustrated in Equation (7).

$$P = 1 - \left(1 - \prod_{i=1}^{d} \frac{1}{p^i}\right)^n \tag{7}$$

Hence, the likelihood of encountering false positives in a bloom filter is inversely related to the magnitude of the expanded prime number p and depth d while being directly proportional to the number n of stored elements.

4. PSBF operation. During the initialization stage, the algorithm "InitBloom" is executed to generate a *bt* bloom filter. As depicted in Algorithm 1, a *"tree"* array is initialized. Each data element comprises two attributes, namely *"Id"* and *"Index"*, where *"Id"* represents the index and *"Index"* denotes the array of its own structure.

Algorithm 1 InitBloom

Input: none
Output: *bt*
1.**function** InitBloom
2. *bt*⇐create an empty array *tree;*
3. **return** *bt*
4.**end function**

In the stage of inserting data elements, an element must first undergo mapping using a string hash function. The hash function employed in this study is BKDRHash [29], which is capable of transforming any string into an integer. The algorithm used for this conversion is Algorithm 2.

Algorithm 2 BKDRHash

```
Input: str
Output: hashCode
function BKDRHash (str)
    seed⇐131;
    hash⇐0;
    for i⇐0 to len(str) - 1 do
        hash⇐(hash×seed) + unit64(str[i]);
    end for
    return hashCode&0x7FFFFFFF
end function
```

Algorithm 3 delineates the sequential procedures entailed in the conversion of integers to *p-adic* integers. The process commences with the establishment of an empty string "*new_num_str*", designated for the storage of the converted outcome. Furthermore, the variables "reminder" and "*reminder_string*" are introduced to accommodate the retention of the residual value of the present bit and its corresponding string representation, correspondingly. Subsequently, the loop should be executed until the input integer "*num*" reaches 0, signifying the termination of the loop. Within each iteration, the remainder of the expanded prime number p divided by "*num*" should be obtained to determine the current bit's remainder. It is imperative to ascertain whether the remainder falls within the range of 10 to 75. If it does, the corresponding character representation in the predefined mapping table "*tenToAny*" should be identified. Conversely, if the remainder does not fall within this range, it should be converted into a string representation. The resulting current bit string, "*remainder_string*", should then be appended to the end of "*new_num_str*". The prime number n is used to divide "*num*", resulting in an updated value of "*num*". Upon completion of the loop, the converted string representation of a *p-adic* integer is stored in "*new_num_str*" and returned.

Algorithm 3 DecimalToAny

```
Input: num, p
Output: new_num_str
function DecimalToAny(num,p)
    initialize variables new_num_str, remainder, remainder_string;
    while num≠0 do
        remainder⇐ num mod p;
        if remainder > 9 and remainder < 76 then
            remainder_string⇐tenToAny[remainder];
        else
            remainder_string⇐str(remainder);
        end if
        new_num_str⇐ tetany + remainder_string;
        num⇐num/p;
    end while
    return new_num_str
end function
```

PSBF element inserted. Let elements $e_1 = 27$, $e_2 = 28$, $e_3 = 29$, and $e_4 = 30$, then Algorithm 3 is executed to obtain their respective *3-adic* integers: 0.0001; 0.1001; 0.2001; and 0.0101. The specific steps for inserting these elements into the bloom filter are illustrated in Figure 3.

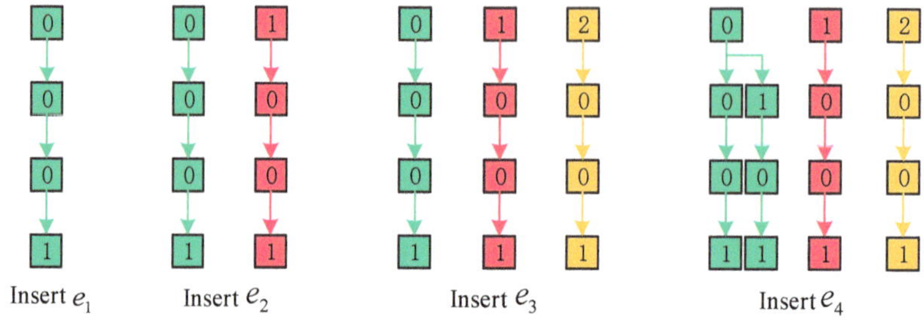

Figure 3. Storage status of the PSBF.

The initial digit of the *p-adic* integer of the element e_1 is determined to be 0. In the event that no slice with an index of 0 is present in the first layer, a new slice will be appended and assigned an index of 0. Similarly, the second digit is also determined to be 0. If no slice with an index of 0 is found in the second layer, a new slice will be added, and the corresponding index will be assigned as 0. The third digit is likewise determined to be 0. In the event that no slice with an index of 0 is found in the third layer, a new slice will be appended and assigned an index of 0. Finally, the fourth digit is determined to be 1. If a slice with an index of 1 is not located in the fourth layer, a new slice will be appended, and the corresponding index will be assigned a value of 1. The verification process confirms the input of all bits, thereby completing the storage of the element. Similarly, the storage of elements e_2 and e_3 can be accomplished. For the *p-adic* integer of the elemente_4, the initial position is 0. In the first layer, a slice with an index of 0 is identified, prompting a transition to the subsequent layer of the slice. In the second position, where the index is 1, no slice is present in the second layer of the slice. Consequently, a new slice will be appended, and the corresponding index will be assigned as 1. Similarly, in the third position, where the index is 0, no slice is detected in the third layer. As a result, a new slice will be appended, and the corresponding index will be assigned as 0. In the fourth position, where the index is 1, no slice is found in the fourth layer. Therefore, a new slice will be appended, and the corresponding index will be assigned as 1. Verify and ensure that all bits have been inputted, and finalize the *p-adic* integer storage of the elemente_4. The precise methodology is illustrated in Algorithm 4.

Algorithm 4 InsertElement

```
Input: e, bt
Output: newBt
function InsterElement(e, bt)
    initialize variables path, flag, num, newBt;
    for i⇐0 to len(e)—1 do
        for h⇐0 to len(bt))—1 do
            if string(e[i]) == bt[h].Id then
                flag⇐true;
                if i =- 0 then
                    path⇐strconv.Itoa(h) + ".Index"
                else
                    path⇐path + "."+ strconv.Itoa(h) + ".Index";
                end if
                bt⇐bt[h].Index;
                break
        else
            flag⇐false;
        end if
```

Algorithm 4 *Cont.*

```
        end for
        if !flag then
            if i == 0 then
                create a new node a of the tree, add a to the end of newBt, record path, and bt⇐newBt [len(newBt)-1].Index;
            else
                create a new node a of the tree, add a to the end of bt, update newBt, and bt⇐bt [len(bt)-1].Index;
                    if num == 0 then
                        path⇐path + "."+ strconv.Itoa(v) + ".Index";
                        num++;
                    else
                        path⇐ path + ".0.Index";
                    end if
                end if
            end if
    end for
    return s
end function
```

The "SearchElement" algorithm, which is a component of the element query algorithm, bears a resemblance to the "InsertElement" algorithm. It employs the same hash function utilized in the stored procedure to convert the queried element into an integer. The selection of the expanded prime number *p* remains consistent with the one used during the element storage process, and the integer resulting from the element mapping is expanded to derive the *p-adic* integer of the element. The detailed implementation of this process is presented in Algorithm 5.

Algorithm 5 SearchElement

```
Input: e, bt
Output: true or false
function SearchElement(e, bt)
    for i ⇐0 to len(e)—1 do
        if len(bt) < 1 then
            return false
        end if
        found⇐ false;
        for h⇐0 to len(bt)—1 do
            if string(e[i]) = bt[h].Id then
                found⇐true;
                bt⇐ bt[h].Index;
                break
            end if
        end for
        if !found then
            return false
        end if
    end for
    return true
end function
```

4. Results

All experiments conducted in this study were performed within a consistent environment, utilizing the following main hardware configuration: Intel(R) Core(TM) i5-8250 U CPU @ 1.60 GHz, 1.80 GHz, and 12 GB RAM. The operating system employed wazs Windows 11 64-bit, and the programming language utilized was Golang 1.18.4. In order to achieve optimal performance of the PSBF, various prime numbers p and hash functions were chosen for experimentation in this study.

4.1. Selection of the Optimal Prime Number p

For this study, the hash function's output result length was set at 32, thus limiting the prime number p within the range of 2 to 31. The experiment employed a comparative analysis of insertion and query operations to determine the minimum prime number p that yields the optimal performance in a PSBF. Figure 4a,b demonstrates the impact of selecting prime number p on the insertion and query capabilities of the bloom filter when inserting 1000 to 20,000 data elements. As the number of elements increases, both insertion and query times gradually rise, albeit with slight variations in the performance observed across different prime numbers p. In the context of insertion and query operations, the prime number p with a value of 31 demonstrated optimal performance and exhibited a brief execution time. Conversely, prime numbers such as 2, 3, 5, 7, etc., tended to exhibit a prolonged execution time, particularly when the number of elements was substantial, thereby resulting in a more pronounced degradation in performance. Consequently, this study advocates for the utilization of the expanded prime number p = 31 within the bloom filter as it yields superior performance and efficiency.

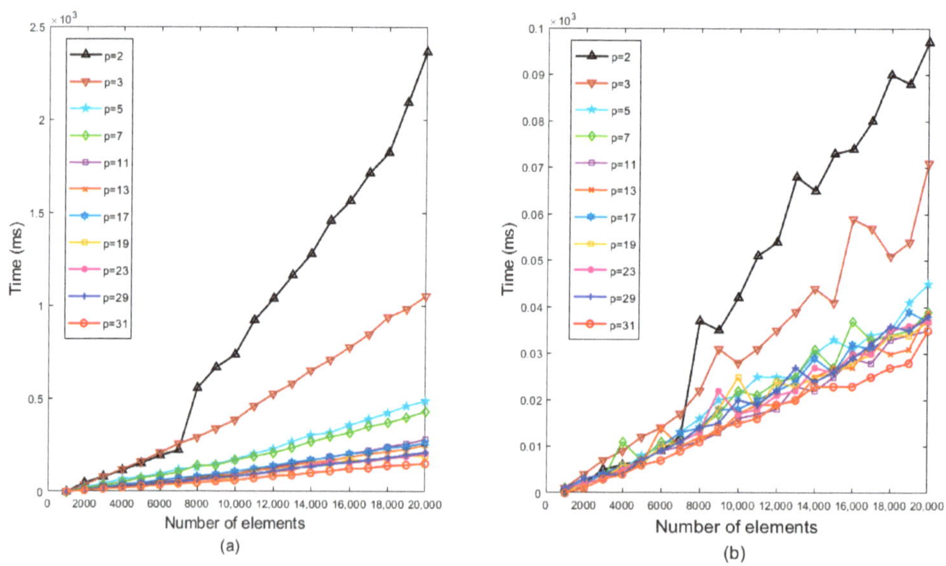

Figure 4. Time comparison of selecting different expansion prime numbers p from 2 to 31 intervals. (**a**) Insertion operation; (**b**) query operation.

4.2. Selection of a Suitable Hash Function

The determination of the most suitable number of hash functions for conventional bloom filters is contingent upon the quantity of inserted elements and the dimensions of the bit array employed by said filters [30]. In contrast, the hash function employed by the PSBF diverges from that of traditional bloom filters. Specifically, the hash function in the PSBF transforms the elements to be manipulated into integers of fixed length. As a result of the incorporation of *p-adic* numbers in the PSBF, it becomes solely necessary to decompose the fixed-length integers into *p-adic* numbers based on the optimal prime number p. The PSBF does not internally utilize a hash function. In essence, the utilization of the PSBF does not necessitate the consideration of the number of elements or the dimensions of the bit array within the bloom filter. The selection of a suitable hash function for the PSBF pertains to the efficiency of operations performed on elements that are mapped to fixed-length sequences of integers.

The experiment involved the insertion and query of 1000–20,000 identical data elements using eight string hash functions, namely BKDRHash [29], APHash [31], DJBHash [32], JSHash [33], RSHash [34], SDBMHash [35], PJWHash [36], and ELFHash [37]. The results of these experiments are illustrated in Figure 5a and 5b, respectively. BKDRHash, APHash, JSHash, SDBMHash, and PJWHash exhibited favorable performance in terms of insertion and query time, displaying relatively consistent performance. However, DJBHash was characterized by a prolonged insertion and query time, rendering it unsuitable for large-scale data insertion and query operations in the PSBF. RSHash demonstrated an even higher query time, with a significant decline in performance when dealing with substantial data volumes. Although ELFHash exhibited relatively high insertion and query times, its growth rate was comparatively gradual.

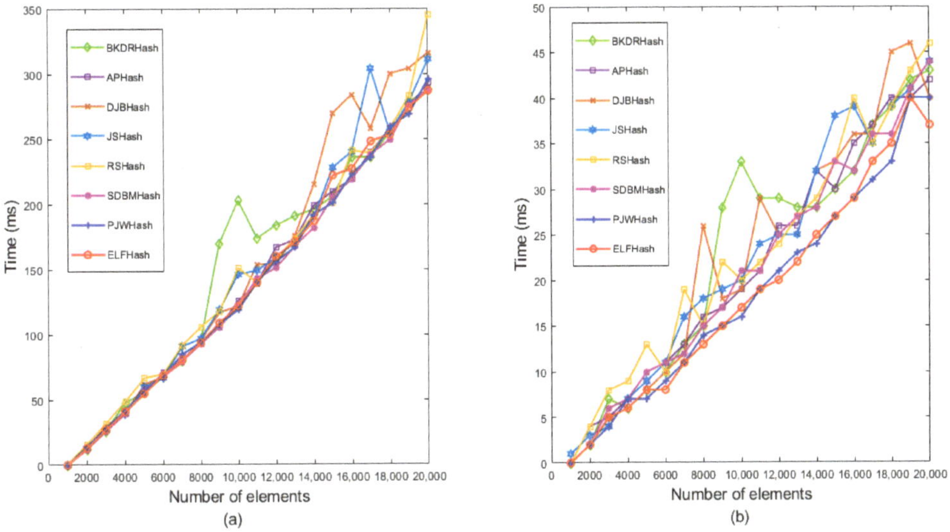

Figure 5. Time comparison of eight various hash functions. (**a**) Insertion operation; (**b**) query operation.

5. Discussion

It is worth noting that the conventional bloom filter exhibits a false positive probability during member queries, wherein an element that is not present in the set is erroneously identified as being part of the set. Furthermore, the size of the bloom filter necessitates predefinition based on the desired probability of false positives and the number of elements to be stored. Estimating the accurate size of the stored elements in a system poses a challenge. When the number of elements stored in a bloom filter exceeds a specific threshold, the occurrence of false positives will escalate. The utilization of fixed-size bloom filters will inevitably result in significant storage wastage or a high rate of false positives. To ascertain the superiority of the PSBF, this study undertakes a comparative analysis of insertion time, query time, space utilization rate, and misjudgment count in relation to standard bloom filters, scalable bloom filters [17], and dynamic bloom filters [18].

5.1. Analysis of Insertion and Query Performance

Regarding the insertion operation, as depicted in Figure 6a, the insertion time of the four distinct filters exhibited a gradual increase as the number of elements grew. Nonetheless, this increase remained relatively low, with the time being consistently below 0.14 ms. As the data elements increased, both the PSBF and expandable bloom filters demonstrated more consistent performance, while standard bloom filters and dynamic bloom filters exhibited comparatively higher insertion times. Furthermore, the insertion performance of the system exhibited significant fluctuations with an increase in the number of inserted elements. Regarding query operations, as depicted in Figure 6b, the query time for the four bloom filters gradually increased as the number of elements grew. Dynamic bloom filters demonstrated relatively high query times, while the PSBF and scalable bloom filters exhibited relatively stable query performance. However, the PSBF outperformed the other three bloom filters in terms of both query performance and time.

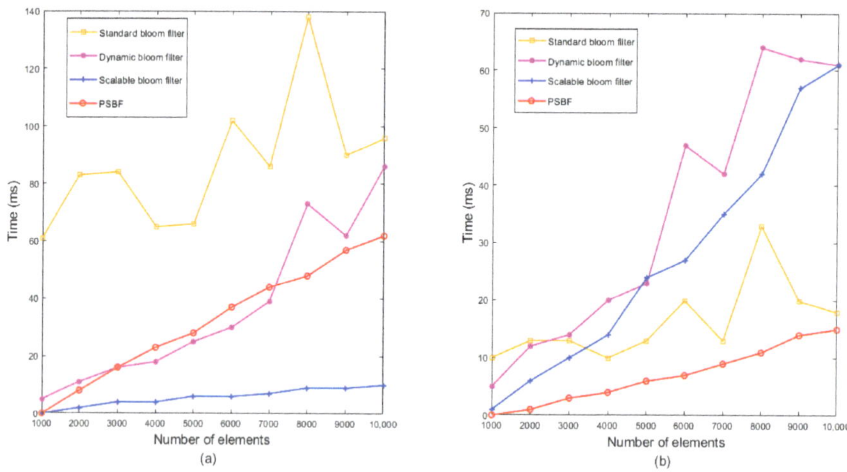

Figure 6. Time comparison of the different bloom filters. (**a**) Insertion operation; (**b**) query operation.

5.2. Analysis of Space Utilization and Misjudgment Number

In relation to space utilization, as depicted in Figure 7, the space utilization of both the standard bloom filter and the dynamic bloom filter progressively rose with the augmentation of inserted elements. Nevertheless, upon reaching the predetermined value of inserted elements, the dynamic bloom filter, in order to maintain a false positive probability lower than the anticipated threshold, generated a bloom filter with greater space and allocated a substantial amount of free space, consequently leading to a sudden decrease in space utilization. The space utilization rate of the expandable bloom filter will exhibit a gradual decline as the number of inserted elements increases. Once the space utilization rate surpasses a specific threshold, a larger bloom filter will be generated and stored at the forefront of the bloom filter set, serving as an active bloom filter [4]. Consequently, this leads to increased available space for expandable bloom filters and a reduction in space utilization.

This study employed the PSBF method to initially convert the target element into an integer, followed by the selection of an expanded prime number p to convert the integer into a p-adic integer further. The depth d of the resulting bloom filter was determined based on the number of bits in the converted *p-adic* integer. The storage structure of the PSBF was transformed into a tree, deviating from the linear storage structure employed in current technologies. This modification enhances the efficiency of element insertion and query operations. Furthermore, the PSBF employs a selection process to determine the expanded prime number p based on the given integer, which is then utilized to convert the integer into a *p-adic* integer. Subsequently, the depth d of the bloom filter is determined based on the number of bits in the converted *p-adic* integer. This approach allows the PSBF to circumvent the need to consider the size of the bloom filter and the stored data, enabling automatic expansion of capacity and storage of any number of elements based on the storage elements. Consequently, the PSBF effectively utilizes the available storage space.

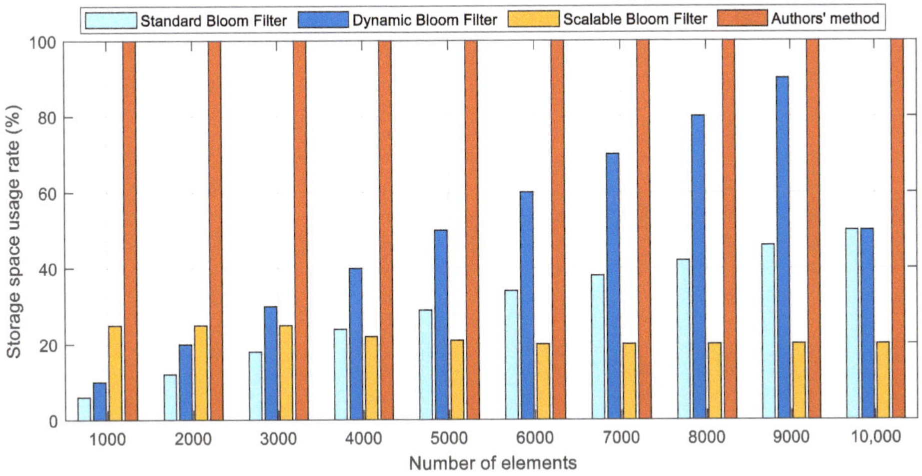

Figure 7. Storage space usage rate of four various bloom filters.

Based on the findings presented in Figure 8, it is evident that the standard bloom filter lacks the capability of automatic expansion. Consequently, when the space utilization rate is high, the likelihood of false positives in the bloom filter increases significantly. Therefore, in order to maintain a lower-than-anticipated probability of false positives, the standard bloom filter must compromise its space utilization. The dynamic bloom filter consistently recorded zero misjudgments, whereas the expandable bloom filter exhibited a gradual linear increase in misjudgments. The PSBF algorithm selected an appropriate expanded prime number p by considering the integer obtained from the hash function, thereby minimizing false positives. When the *p-adic* integer depth of all data conversions was equal, it became possible to attain a probability of zero false positives.

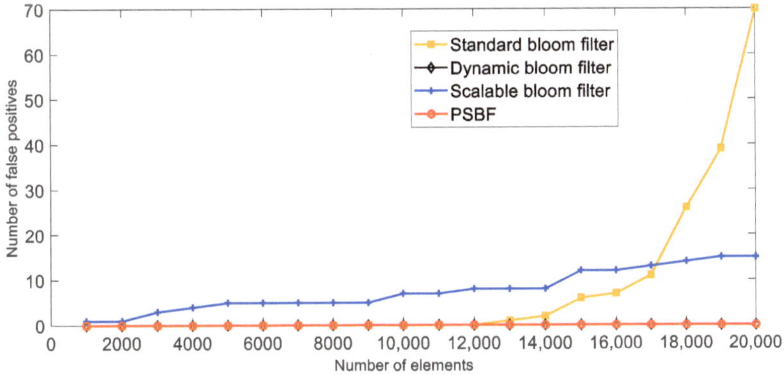

Figure 8. Number of misjudgments of four different bloom filters.

As previously stated, the PSBF demonstrates clear advantages in terms of query time, space utilization, and number of misjudgments. It consistently exhibited shorter query times, better space utilization, and fewer misjudgments when compared to the scalable bloom filter, regardless of the size of the element set or the number of elements. Although the insertion operation time of the PSBF was slightly higher than that of the scalable bloom filter, in practical application scenarios, the insertion operation time of the bloom filter was typically significantly smaller than the query operation time. Consequently, the overall performance of the PSBF surpasses that of the scalable bloom filter.

6. Conclusions

To address the limitations of capacity in the existing bloom filter, it becomes unfeasible to reprocess extensive data for filtration. Moreover, when the data to be filtered approaches the upper limit of capacity during the reprocessing process, the false positive rate significantly rises, leading to a substantial decline in the efficacy of data filtration. Consequently, this gives rise to technical challenges that diminish the accuracy of the filtering process. This study introduced a bloom filter with an automatic expansion tree structure, utilizing the algebraic and topological properties of *p-adic* integers. The proposed approach offers improved efficiency in terms of element insertion and query operations. Moreover, it eliminates the need for parameter setting to accommodate capacity expansion, thereby mitigating space wastage and reducing the likelihood of false positives.

Author Contributions: Conceptualization, W.Y.; methodology, W.Y. and J.J.; software, C.W., Q.X. and J.J.; validation, Y.Z.; formal analysis, W.Y. and J.J.; investigation, W.Y.; resources, W.Y.; data curation, C.W.; writing—original draft preparation, W.Y. and C.W.; writing—review and editing, W.Y.; visualization, Y.Z. and J.J.; supervision, W.Y.; project administration, W.Y.; funding acquisition, W.Y. All authors have read and agreed to the published version of the manuscript.

Funding: The study was financially supported by the National Key Research and Development Program of China (Grant No. 2022YFD1600601); the Natural Science Foundation of Jiangxi Province (Grant No. 20212BAB202015); Jiangxi Provincial Special Program 03 and 5G Projects (Grant No. 20232ABC03A18); and the 2023 Jiangxi Provincial Special Fund Projects for Graduate Student Innovation (Grant No. YC2023-S413).

Institutional Review Board Statement: Not applicable.

Informed Consent Statement: Not applicable.

Data Availability Statement: The data from this study can be obtained upon request from the first or corresponding author.

Conflicts of Interest: The authors declare no conflict of interest.

References

1. Singh, A.; Garg, S.; Kaur, K.; Batra, S.; Kumar, N.; Choo, K.-K.R. Fuzzy-folded bloom filter-as-a-service for big data storage in the cloud. *IEEE Trans. Ind. Inform.* **2018**, *15*, 2338–2348. [CrossRef]
2. Singh, A.; Garg, S.; Batra, S.; Kumar, N.; Rodrigues, J.J. Bloom filter based optimization scheme for massive data handling in IoT environment. *Future Gener. Comput. Syst.* **2018**, *82*, 440–449. [CrossRef]
3. Yang, C.; Tao, X.; Zhao, F.; Wang, Y. Secure data transfer and deletion from counting bloom filter in cloud computing. *Chin. J. Electron.* **2020**, *29*, 273–280. [CrossRef]
4. Luo, L.; Guo, D.; Ma, R.T.; Rottenstreich, O.; Luo, X. Optimizing bloom filter: Challenges, solutions, and comparisons. *IEEE Commun. Surv. Tutor.* **2018**, *21*, 1912–1949. [CrossRef]
5. Hou, R.; Zhang, L.; Wu, T.; Mao, T.; Luo, J. Bloom-filter-based request node collaboration caching for named data networking. *Clust. Comput.* **2019**, *22*, 6681–6692. [CrossRef]
6. Ni, J.; Zhang, K.; Vasilakos, A.V. Security and privacy for mobile edge caching: Challenges and solutions. *IEEE Wirel. Commun.* **2020**, *28*, 77–83. [CrossRef]
7. He, P.; Xue, K.; Yang, J.; Xia, Q.; Liu, J.; Wei, D.S. FASE: Fine-grained accountable and space-efficient access control for multimedia content with in-network caching. *IEEE Trans. Netw. Serv. Manag.* **2021**, *18*, 4462–4475. [CrossRef]
8. Cohen, I.; Einziger, G.; Scalosub, G. False negative awareness in indicator-based caching systems. *IEEE/ACM Trans. Netw.* **2022**, *30*, 2674–2687. [CrossRef]
9. Zhang, L.; Bai, Z.; Cui, B.; Wu, Z. Search mechanism for data contents based on bloom filter and tree hybrid structure in system wide information management. *IET Commun.* **2023**, *17*, 1262–1273. [CrossRef]
10. Zeng, Z.; Xiao, R.; Lin, X.; Luo, T.; Lin, J. Double locality sensitive hashing Bloom filter for high-dimensional streaming anomaly detection. *Inf. Process. Manag.* **2023**, *60*, 103306. [CrossRef]
11. Liu, Y.; Zhang, L.Y.; Li, J. Fast detection of maximal exact matches via fixed sampling of query K-mers and Bloom filtering of index K-mers. *Bioinformatics* **2019**, *35*, 4560–4567. [CrossRef]
12. Patgiri, R.; Nayak, S.; Borgohain, S.K. Passdb: A password database with strict privacy protocol using 3d bloom filter. *Inf. Sci.* **2020**, *539*, 157–176. [CrossRef]
13. Byun, H.; Lim, H. Learned FBF: Learning-Based Functional Bloom Filter for Key–Value Storage. *IEEE Trans. Comput.* **2021**, *71*, 1928–1938. [CrossRef]

14. Lemane, T.; Medvedev, P.; Chikhi, R.; Peterlongo, P. Kmtricks: Efficient and flexible construction of bloom filters for large sequencing data collections. *Bioinform. Adv.* **2022**, *2*, vbac029. [CrossRef]
15. Liang, Y.; Ma, J.; Miao, Y.; Kuang, D.; Meng, X.; Deng, R.H. Privacy-Preserving Bloom Filter-Based Keyword Search over Large Encrypted Cloud Data. *IEEE Trans. Comput.* **2023**. [CrossRef]
16. Yi, W.; Gerasimov, I.; Kuzmin, S.; He, H. A stream processing approach to distance measurement of integers in p-adic metric space. In Proceedings of the 2017 XX IEEE International Conference on Soft Computing and Measurements (SCM), Saint Petersburg, Russia, 24–26 May 2017; pp. 617–620.
17. Almeida, P.S.; Baquero, C.; Preguiça, N.; Hutchison, D. Scalable bloom filters. *Inf. Process. Lett.* **2007**, *101*, 255–261. [CrossRef]
18. Guo, D.; Wu, J.; Chen, H.; Yuan, Y.; Luo, X. The dynamic bloom filters. *IEEE Trans. Knowl. Data Eng.* **2009**, *22*, 120–133. [CrossRef]
19. Patgiri, R.; Borgohain, S.K.; Bhattacharjee, A. rFilter: A scalable and space-efficient membership filter. In Proceedings of the 2018 5th International Conference on Signal Processing and Integrated Networks (SPIN), Noida, India, 22–23 February 2018; pp. 478–485.
20. Kleyko, D.; Rahimi, A.; Gayler, R.W.; Osipov, E. Autoscaling bloom filter: Controlling trade-off between true and false positives. *Neural Comput. Appl.* **2020**, *32*, 3675–3684. [CrossRef]
21. Kim, M.-K.; Kim, S.-R. Modifications using Circular Shift for a Better Bloom Filter. In Proceedings of the International Conference on Research in Adaptive and Convergent Systems, Gwangju, Korea, 13–16 October 2020; pp. 149–154.
22. Rottenstreich, O.; Reviriego, P.; Porat, E.; Muthukrishnan, S. Constructions and applications for accurate counting of the bloom filter false positive free zone. In Proceedings of the Symposium on SDN Research, San Jose, CA, USA, 3 March 2020; pp. 135–145.
23. Nayak, S.; Patgiri, R. robustBF: A high accuracy and memory efficient 2d bloom filter. *arXiv* **2021**, arXiv:2106.04365.
24. Patgiri, R.; Nayak, S.; Borgohain, S.K. rDBF: A r-dimensional bloom filter for massive scale membership query. *J. Netw. Comput. Appl.* **2019**, *136*, 100–113. [CrossRef]
25. Dayan, N.; Bercea, I.; Reviriego, P.; Pagh, R. InfiniFilter: Expanding Filters to Infinity and Beyond. *Proc. ACM Manag. Data* **2023**, *1*, 1–27. [CrossRef]
26. Bender, M.A.; Farach-Colton, M.; Johnson, R.; Kuszmaul, B.C.; Medjedovic, D.; Montes, P.; Shetty, P.; Spillane, R.P.; Zadok, E. Don't thrash: How to cache your hash on flash. In Proceedings of the 3rd Workshop on Hot Topics in Storage and File Systems (HotStorage 11), Portland, OR, USA, 14 June 2011.
27. Cohen, S.; Matias, Y. Spectral Bloom Filters. In Proceedings of the 2003 ACM SIGMOD International Conference on Management of Data, San Diego, CA, USA, 10–12 June 2003; pp. 241–252.
28. Kiss, S.Z.; Hosszu, É.; Tapolcai, J.; Rónyai, L.; Rottenstreich, O. Bloom filter with a false positive free zone. *IEEE Trans. Netw. Serv. Manag.* **2021**, *2*, 2334–2349. [CrossRef]
29. Zhu, X.; Zhang, Q.; Cheng, T.; Liu, L.; Zhou, W.; He, J. DLB: Deep Learning Based Load Balancing. In Proceedings of the 2021 IEEE 14th International Conference on Cloud Computing (CLOUD), Chicago, IL, USA, 5–10 September 2021; pp. 648–653.
30. Kim, K.; Jeong, Y.; Lee, Y.; Lee, S. Analysis of counting bloom filters used for count thresholding. *Electronics* **2019**, *8*, 779. [CrossRef]
31. Saxena, M.; Saurabh, P.; Verma, B. A new hashing scheme to overcome the problem of overloading of articles in Usenet. In Proceedings of the Advances in Computer Science, Engineering & Applications: Proceedings of the Second International Conference on Computer Science, Engineering and Applications (ICCSEA 2012), New Delhi, India, 25–27 May 2012; Volume 1, pp. 967–975.
32. Bernstein, D.J. DJB Hash. Available online: http://www.partow.net/programming/hashfunctions/#DJBHashFunction (accessed on 7 August 2023).
33. Wu, W.-Q.; Xue, M.-T.; Xing, Q.-J.; Yu, F. High-Parallelism Hash-Merge Architecture for Accelerating Join Operation on FPGA. *IEEE Trans. Circuits Syst. II: Express Briefs* **2021**, *68*, 2650–2654. [CrossRef]
34. Shi, Y.; Huang, S.; Lou, J. A characteristic standardization method for circuit input vectors based on Hash algorithm. *J. Ambient. Intell. Humaniz. Comput.* **2022**, *13*, 1505–1513. [CrossRef]
35. Jain, S.; Pandey, M. Hash table based word searching algorithm. *Int. J. Comput. Sci. Inf. Technol.* **2012**, *3*, 4385–4388.
36. Lin, Y.; Huang, Z.; Li, Y. Learning hash index based on a shallow autoencoder. *Appl. Intell.* **2023**, *53*, 14999–15010. [CrossRef]
37. Trimoska, M.; Ionica, S.; Dequen, G. Time-memory analysis of parallel collision search algorithms. *IACR Trans. Cryptogr. Hardw. Embed. Syst.* **2021**, 254–274. [CrossRef]

remote sensing

Article

Long-Time Coherent Integration for Marine Targets Based on Segmented Compensation

Zhenfang Zhao [1], Yisong Zhang [1,2], Wenguang Wang [1,*], Ben Liu [3] and Wei Wu [4]

1 School of Electronic and Information Engineering, Beihang University, Beijing 100191, China; zhaozhenfang_0509@buaa.edu.cn (Z.Z.); gsjrzys_hq@bank-of-china.com (Y.Z.)
2 Bank of China, Beijing 100032, China
3 Institute of Deep-Sea Science and Engineering, Chinese Academy of Sciences, Sanya 572000, China; liub@idsse.ac.cn
4 College of Weaponry Engineering, Naval University of Engineering, Wuhan 430030, China; 1712021043@nue.edu.cn
* Correspondence: wwenguang@buaa.edu.cn; Tel.: +86-10-8231-7240

Abstract: Long-time coherent integration is an effective method for dim target detection from heavy sea clutter. To detect dim targets, a novel long-time coherent integration method based on segmented compensation is proposed in this paper. The method models the complex motion of a marine target as the combination of multi-stage uniformly accelerated motions. According to the difference of energy distribution in Doppler frequency domain, this method can suppress sea clutter and detect the regions of interest (ROIs). Using time–frequency domain energy analysis, the potential target can be extracted. After estimating the parameters and segmentation, for the potential targets, the phase compensation factor can be used to eliminate the Doppler frequency modulation caused by the complex motion. Finally, for the compensated signal, long-time coherent integration is performed to realize the target detection and discrimination under low signal-to-clutter ratio. To verify the effectiveness of the proposed method, we apply simulation data and measured CSIR data in the experiments. The results show that the proposed method can integrate the energy of target more effectively than MTD and RFrFT, and the novel method has better detection performance for complex moving targets under low signal-to-clutter ratio situation.

Keywords: parameter estimation; motion segmentation; phase compensation; coherent integration

1. Introduction

Target detection in the background of sea clutter is always an important problem in radar applications. Sea clutter, affected by many factors such as grazing angle, wind, polarization, and so on, has a very large dynamic range. Both the low signal-to-clutter ratio (SCR) caused by the heavy sea clutter and the complex motion of the marine targets are the factors reducing the detection performance. In addition to being affected by their own power, the motion of small marine targets is also affected by other factors such as sea waves and wind blowing, resulting in strong time variability. The stable detection of maneuvering marine targets has always been a challenging problem in water traffic surveillance, anti-smuggling, maritime search and rescue, and so on.

Long-time coherent integration is an important method to effectively increase the SCR of dim targets to improve the detection probability [1]. Long-time integration is mainly divided into two types according to the coherence of the signal, namely incoherent integration and coherent integration. Typical incoherent integration includes Hough transform [2,3], Radon transform [4], tracking before detection [5,6], and so on. Incoherent integration is just the superposition of echo in amplitude without considering the phase of signal, so the integration performance is not so good at low SCR. Compared with incoherent integration, coherent integration takes into account both the amplitude and phase, and can obtain higher SCR than that of incoherent integration. During the long observation, the envelope of the target echo will migrate across different range bins. Moreover, the complex motion of the target will cause the Doppler frequency spectrum expansion. So, the correction of the range migration (RM) and the compensation of the Doppler frequency are the main problems in the long-time coherent integration. Keystone transform (KT) is a range migration correction method widely used in synthetic aperture radar imaging. The core of KT is to scale the time axis of the echo in the range–frequency domain [7] so as to correct the linear range migration caused by the target motion [8,9]. But KT is not effective for complex movement because it cannot correct the range curving caused by the high-order movement of the target, nor can it eliminate the Doppler ambiguity [10]. Ref. [11] proposed an algorithm which applied azimuth dechirping and the second-order keystone transform to eliminate the Doppler ambiguity to correct the range curvature. RFT [12] is another method widely used for target range migration correction. It can perform coherent integration along a model-based track on the range–time plane [13], which can correct the range curvature caused by the higher-order motion of the target. However, it depends on the search of the motion parameters of different models, and the calculation complexity is relatively high [14]. To compensate the Doppler frequency modulation (DFM) caused by the high-order motion of the target, many methods have been proposed successively to improve the integration effect of the target echo, such as the time reversal transform [15,16], fractional Fourier transform (FrFT) [17,18], Lv's distribution (LVD) [19], etc. Among them, fractional Fourier transform eliminates Doppler frequency modulation through the rotation of the time–frequency plane, which needs to search the acceleration of the moving target [20]. LVD gives the natural representation of LFM signal in the center frequency-chirp rate (CFCR) domain [21], which transforms the signal from the time–frequency domain to the CFCR domain by scaling the parametric symmetric instantaneous autocorrelation function (PSIAF). Ref. [22] proposed a fast non-searching method based on the adjacent cross-correlation function (ACCF) and LVD, which can remove the range migration and reduce the order of Doppler frequency modulation to realize the coherent integration, target detection and parameter estimation for maneuvering target. The method in Ref. [23] is based on frequency spectrum segment processing (FSSP) and the segmental Lv's distribution (SLVD), where the FSSP divides the received signal into several subband signals and expands the range resolution of the subband signals to eliminate the RM and SLVD is applied to achieve the coherent integration of the subband signals. By combining Radon transform with FrFT, RFrFT can compensate Doppler frequency modulation while correcting the range migration [24]. To achieve long-time coherent integration of uniformly accelerated targets, the method in Ref. [25] extracts signals along the track on the range–time plane through a multidimensional parameter search, and converts the correction of range migration and the phase compensation into the matching of motion parameters. Based on RFrFT, W. Wang proposed a two-stage RFrFT long-time coherent integration for marine targets [26] to address the problem of excessive computation caused by multidimensional parameters search. The method includes two steps: the first step detects regions of interest (ROIs) based on the spectral similarity between adjacent range bins, and then performs out RFrFT integration and target discrimination within ROI regions. It reduces the calculation cost of RFrFT and has good detection performance at the same time. However, due to the diversity and complexity of the influencing factors, the marine target always has a complex motion, which is difficult to accurately describe during the long-time integration by using a single uniformly accelerated model. As a result, it will be difficult to completely eliminate the problem of Doppler frequency modulation caused by the target motion. Aiming at the complex motions of marine targets, this paper proposes a novel long-time coherent integration method based on segmented phase compensation. On the basis of extracting potential targets with low threshold, the motion of the targets is decomposed into a combination of multi-stage uniformly accelerated motions. After estimating the motion parameters of potential targets using the symmetric instantaneous autocorrelation function (SIAF), the phase compensation is performed to eliminate Doppler frequency modulation for each segment. Finally, the complex moving targets can be detected using long-time coherent integration.

The contributions of this paper are as follows:

1. Aiming at the problem of mismatch between the complex motion and the single motion model, this paper presents a new modeling method that decomposes the complex motion of the target into the combination of multiple uniformly accelerated motions to achieve a simplified description.
2. For each segment, the parameters under low SCR are estimated under the model constraints, and then the compensation factor is constructed according to the parameter estimation to compensate the secondary order phase to eliminate the Doppler frequency modulation caused by the complex motion.
3. To eliminate the false alarms that may exist in the detection results, a target discrimination method based on the 3 dB spectrum width of the symmetric instantaneous autocorrelation function is proposed, which can effectively distinguish the false alarm caused by the sea clutter.

The structure of this paper is as follows: In Section 2, the echo model of moving target and some classic coherent integration methods are introduced. In Section 3, a long-time coherent integration method based on segment compensation is presented, including detection of ROIs, parameter estimation and motion segmentation of potential targets, long-time coherent integration and target discrimination. In Section 4, the effectiveness of the proposed method is verified based on simulation data and measured data. Finally, the work of this paper is summarized and some useful conclusions are given.

2. Signal Processing Models

2.1. Marine Target Echo Model

In order to obtain higher range resolution, linear frequency modulation (LFM) is widely used as radar transmitting signal; its form can be expressed as Equation (1).

$$s(t) = rect(\frac{t}{T_p})\exp[j2\pi(f_c t + \frac{kt^2}{2})] \tag{1}$$

where $rect(u) = \begin{cases} 1, |u| \leq \frac{1}{2} \\ 0, |u| > \frac{1}{2} \end{cases}$, f_c is the center frequency, $k = B/T_p$ is the chirp rate of the LFM signal whose bandwidth is B and pulse width is T_p.

The motion of the marine target will cause the radial distance between the target and the radar to change at any time. If the distance between the target and the radar at the time t_m is $R(t_m)$, the echo delay will be $\tau(t_m) = 2R(t_m)/C$, where C denotes the speed of light and the received echo of the radar can be expressed as Equation (2).

$$s_r(t, t_m) = \sigma_r rect(\frac{t - \tau(t_m)}{T_p})\exp[-j2\pi f_c \tau(t_m)]\exp\{j\pi k[t - \tau(t_m)]^2\} \tag{2}$$

where σ_r is the complex scattering coefficient of the target. After pulse compression, the form of the signal is as Equation (3).

$$s_{pc}(t, t_m) = A_0 \sin c\{B[t - \tau(t_m)\}\exp[-j2\pi f_c \tau(t_m)] \tag{3}$$

In Equation (3), A_0 is related to σ_r, which can be approximated as a constant. If the changing of the range in long-time integration exceeds the range resolution of the radar, the motion of the target will cause migration of the echo envelope, and the integration based on a single range bin will not be very effective.

The motion of the marine target is complex and time-varying, but the acceleration of the target is relatively stable in a short time, which means the target motion can still be regarded as the uniform acceleration movement in a short time. For the marine target, if the initial radial velocity is v_0, the acceleration is a, then the instantaneous velocity $v(t) = v_0 + at$, and the instantaneous Doppler frequency $f_d(t) = 2(v_0 + at)/\lambda$, so the second-order phase caused by the acceleration can be expressed as Equation (4).

$$\phi(t) = 2\pi \int_0^t f_d dt = \frac{4\pi}{\lambda}(v_0 t + \frac{1}{2}at^2) \tag{4}$$

To simplify the expression, the motion of marine targets can be modeled as a combination of multi-stage uniformly accelerated motions. The radial distance corresponding to the ith segment can be expressed as Equation (5).

$$R_i(t_m) = R_{i-1} + v_{i-1}(t_m - \sum_{j=0}^{i-1} t_j) + \frac{1}{2}a_i(t_m - \sum_{j=0}^{i-1} t_j)^2 \tag{5}$$

In Equation (5), v_{i-1} is the radial initial velocity of the ith segment, a_i is the acceleration, and the duration of the jth segment is t_j and $t_0 = 0$, then $\sum_{j=0}^{i-1} t_j$ is the total duration before the ith segment.

Due to the non-uniform motion of the target, its Doppler frequency changes with time, which will cause the problem of Doppler frequency modulation. As a result, the Doppler spectrum will disperse in a wide range of Doppler domains. The Doppler frequency of the echo generated by the movement of the ith segment is shown in Equation (6), and the Doppler spectrum width is proportional to a_i.

$$f_{di}(t_m) = \frac{2v_{i-1} + 2a_i t_m}{\lambda} \tag{6}$$

The echo of the ith segment can be compressed as Equation (7).

$$s_{pci}(t, t_m) = A_0 \sin c\{B[t - \frac{2R_i(t_m)}{C}]\} \exp[-j\frac{4\pi}{\lambda}R_i(t_m)] \tag{7}$$

If the complex motion of the target can be segmented into a combination of N segments of uniformly accelerated motion, then the compressed echo of the marine target can be expressed as Equation (8).

$$s_{pc}(t, t_m) = \sum_{i=1}^{N} s_{pci}(t, t_m) \tag{8}$$

2.2. Coherent Integration

MTD is a widely used target detection method; the core of MTD is to improve the SCR of moving targets by using coherent integration, so it is also regarded as a set of Doppler filter banks [27], which can be defined as Equation (9).

$$X_{MTD} = \int s(t, r)e^{-j2\pi ft}dt \tag{9}$$

To perform MTD, the key processing is the Fourier transform (FT) to the echo in the same range bin [28]. Owing to the RM and DFM, the integration result is influenced by the motion of the target. In addition, the MTD is just a type of short-time integration procedure.

To solve the coupling between the range walking and slow time, RFT is proposed to realize long-time coherent integration. Different from the existing Radon transformation or FT, the RFT may be quite helpful for the detection of moving targets due to the correction of range migration [29]. The standard RFT combining the Fourier transform and the Radon transform is defined as Equation (10).

$$X_{RFT} = \int s(t, r_0 - v_0 t)e^{-j2\pi ft}dt \tag{10}$$

where $s(t, r_0 - v_0 t)$ is a two-dimension complex function defined in (t, r) plane, and $r = r_0 - v_0 t$. Therefore, Equation (9) is a special case of Equation (10) when the target is very slow and there is no range migration. When the target moves uniformly and crosses the range gates, neither MTD nor RFT can integrate the target energy effectively. As a generalization of the FT, FrFT [30] is proposed for signal processing, which is defined as Equation (11).

$$X_\alpha(u) = \int x(t) K_\alpha(t, u) dt \tag{11}$$

where α is the transformation angle; the value range of α is 0 to π, which can be calculated by $\alpha = p\pi/2$. p is the transformation order, which can be determined by searching the acceleration of the moving target after dimensional normalization. The relationship between the acceleration and transformation order is shown in Equation (12).

$$p = -\frac{2\,\mathrm{arccot}(2a_i S^2 \lambda)}{\pi} + 2 \tag{12}$$

In (12), S is the scale factor, depending on the sampling frequency f_S and integral time T_n, $S = \sqrt{T_n / f_S}$. The transform kernel K_α is defined as follows:

$$K_\alpha(t, u) = \begin{cases} A e^{(j\frac{u^2+t^2}{2}\cot\alpha - jut\csc\alpha)}, \alpha \neq n\pi \\ \delta(t-u), \alpha = 2n\pi \\ \delta(t+u), \alpha = (2n \pm 1)\pi \end{cases} \tag{13}$$

where $A = \sqrt{(1 - j\cot\alpha)/2\pi}$, FrFT can be seen as a rotation of the time–frequency plane at an angle of α. When $\alpha = \pi/2$, FrFT is the Fourier transform. Compared with Fourier transform, FrFT can compensate Doppler frequency modulation caused by the uniformly accelerated motion effectively, and is widely used for the long-time coherent integration of dim marine targets.

By combining FrFT and RFT, RFrFT [31] is proposed to achieve long-time coherent integration of maneuvering targets. RFrFT is defined as Equation (14).

$$X_{RFrFT} = \int s(t, r_0 - v_0 t - \frac{1}{2}at^2) K_\alpha(t, u) dt \tag{14}$$

where the range $r = r_0 - v_0 t - \frac{1}{2}at^2$ represents the accelerated or high-order motion of the target, which is used to search the echo of moving target in the plane (t, r).

RFrFT can extract the echoes of the target on the range–time plane according to the motion model. In order to detect dim targets from noise and heavy sea clutter, RFrFT must search for different parameter combinations of $[r_0, v_0, a]$; this results in high computation. RFrFT can be regarded as a kind of extension of the MTD, FrFT and standard RFT, which can be used effectively in target detection from very low SCNR and even from the situation with range migration and Doppler frequency modulation.

3. Long-Time Coherent Integration Based on Segmented Compensation

Based on the models in Section 2, to the complex moving target, such as the variable acceleration moving, a long-time coherent integration method based on segmented compensation is proposed in this section. The proposed long-time coherent integration includes the processing of ROI detection, motion estimation and segmentation, phase compensation and coherent integration, and target discrimination. Below is the framework of the proposed method, which is shown in Figure 1. To obtain the potential targets, firstly, the ROIs are detected in the range–Doppler domain. Then, the velocity and acceleration of each ROI is estimated to realize the motion segmentation. Based on the parameters estimation and motion segmentation, the phase factor is constructed to eliminate Doppler frequency modulation caused by the movement of the target. For the compensated signal, long-time coherent integration is performed and target detection is realized. After target discrimination, the false alarms caused by sea clutter can be eliminated.

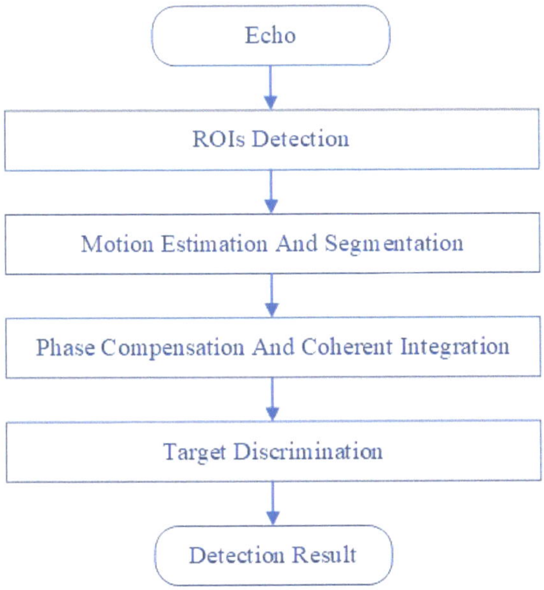

Figure 1. Framework of long-time coherent integration based on segmented compensation.

3.1. ROI Detection

This section includes three modules: wavelet packet decomposition, ROI detection and potential target signal extraction.

In radar surveillance of the sea, sea clutter is the main component of the echo; small marine targets are often submerged in them. In radar echoes, the energy of sea clutter is distributed in all range bins, while the target only appears in specific range bins. Because the motion of the target is subject to specific constraints, its Doppler spectrum characteristics are different from those of sea clutter. The difference in energy distribution between the target and the sea clutter can be used to distinguish the target from the sea clutter.

Wavelet packet decomposition (WPD) can decompose the signal into the components with different frequencies, which can characterize the energy distribution. WPD can decompose the signal into two parts using a set of orthogonal wavelet bases: high frequency and low frequency, and then each part is used as input, and will be decomposed again. In this way, the corresponding components $c_j(j = 1, 2, \ldots, 2^i)$ with different frequency bands and satisfied $\sum_{j=1}^{2^i} c_j(n) = s(n)$ can be obtained at the ith layer.

To each range bin, the WPD provides the energy distribution with frequency. And the energy in the jth frequency band of the mth range bin can be expressed as Equation (15).

$$E_{m,j} = \sum_{n} [c_{m,j}(n)]^2 \tag{15}$$

In (15), $m = 1, 2, \ldots, M$ is the index number of the range bin. After extracting the energy distribution, the echo is decomposed into $M \cdot 2^i$ frequency units $P_{m,j}$ in the frequency domain, and the energy of each frequency unit is $E_{m,j}$.

The high-energy units can be detected by comparison with the threshold T_m, defined as Equation (16).

$$T_m = mean(E_{m,j}) + \eta_m \cdot std(E_{m,j}) \tag{16}$$

where η_m is the threshold coefficient, then a frequency unit can be judged whether it is a high-energy unit.

After the high-energy unit detection, the higher unit numbers in all range bins will be used to analyze whether a frequency unit belongs to sea clutter or a potential target. If most range bins are judged to be high-energy units, a frequency unit will be considered to belong to sea clutter, because it is impossible for a marine target to pass through a lot of range bins and show high energy in many range bins in a short time. The sea clutter can be partly suppressed by using the high-energy unit numbers to weight each frequency unit, such as Equation (17).

$$E_{m,j}{}' = (1 - \frac{M'}{M})E_{m,j} \tag{17}$$

where M' is the unit numbers that are judged as high-energy in the M range bins. To the sea clutter, its M' is close to M, but to the target, M' is much smaller than M. This difference makes the potential targets in the scene can be extracted. The region where the potential target is located is called the region of interest (ROI).

Let the extracted Kth ROI be $ROI_k(k = 1, 2, \ldots, K)$ with the range R_k, and the Doppler frequency D_k. If the R_k includes multiple adjacent range bins, we should select the the right bin to provide echo. To solve this problem, the signals in R_k are divided into L segments in time domain. To each segment, the signal is transformed into the Doppler frequency domain, then the bin with the highest energy of each segment in D_k will be extracted. The extracted segments are connected successively to complete the signal extraction of potential targets.

3.2. Motion Estimation and Segmentation

In motion estimation and segmentation, the initial velocity of the signal in each short period is estimated, and the acceleration between segments is estimated according to the initial velocity between adjacent segments. Finally, the motion is segmented according to the estimated parameters.

The motion of the marine target is not only controlled by its own power, but also affected by environmental factors such as sea waves and wind. In a long-time observation, the variation of the influencing factors leads to the complex motion of the marine target, which is difficult to describe using a certain model. To model the complex movements, the motion can be decomposed into multiple fragments. In each short period, the target acceleration can be considered as relatively stable and can be approximated to a constant value, so the complex motion of the marine target can be decomposed into several short-time uniformly accelerated motions. The parameters of the uniformly accelerated motion of each stage are estimated respectively, and the phase compensation factor is constructed to eliminate the Doppler frequency modulation caused by the uniformly accelerating motion of the target.

In the extracted ROIs, the movement of the ith segment can be simplified as Equation (18) in the slow time dimension.

$$s_i(t_m) = A_0 \exp[j2\pi f_{di}(t_m - \sum_{j=0}^{i-1} t_j) + j\pi k_i{}'(t_m - \sum_{j=0}^{i-1} t_j)^2] \tag{18}$$

where $f_{di} = \frac{2v_{i-1}}{\lambda}$ is the Doppler center frequency of the signal in the ith segment, which can be calculated from the initial velocity in the ith segment; $k_i{}' = \frac{2a_i}{\lambda}$ is the Doppler modulation rate of the signal in this segment; and a_i is the acceleration in the segment.

Then, the symmetric instantaneous autocorrelation function (SIAF) [32] of $s_i(t_m)$ at time t_{0i} can be obtained as Equation (19).

$$f_i(t_{0i}, \tau) = s_i(t_{0i} + \frac{\tau}{2})s_i{}^*(t_{0i} - \frac{\tau}{2}) = A_0{}^2 \exp[j2\pi f_{di}\tau + j2\pi k_i{}'(t_{0i} - \sum_{j=0}^{i-1} t_j)\tau] \tag{19}$$

where $(\cdot)^*$ denotes the complex conjugation, for a given t_0; $f_i(t_0, \tau)$ is a one-dimensional function $f_i(\tau)$, its Fourier transform has the form as Equation (20).

$$F_i(f_\tau) = A_0{}^2 \sin c\{f_\tau - [f_{di} + k_i'(t_{0i} - \sum_{j=0}^{i-1} t_j)]\} \approx A_0{}^2 \sin c(f_\tau - f_{di}) \tag{20}$$

Due to the short duration of each motion segment and the limited small acceleration of the marine target, the item $k_i'(t_{0i} - \sum_{j=0}^{i-1} t_j)$ can be negligible to the target initial velocity estimation. The Doppler frequency $f_{\max i} = \arg\max_{f_\tau} [F_i(f_\tau)]$ corresponds to the peak of the transformed spectrum. Then, the estimated Doppler frequency $\hat{f}_{di}$ for the motion in the ith segment can be calculated as Equation (21).

$$\hat{f}_{di} = \frac{\sum_{f_j = f_{\max i} - \Delta f}^{f_{\max i} + \Delta f} f_j \cdot F_i(f_j)}{\sum_{f_j = f_{\max i} - \Delta f}^{f_{\max i} + \Delta f} F_i(f_j)} \tag{21}$$

where Δf is the Doppler window width, then the initial velocity of the ith segment can be estimated by $\hat{v}_{i-1} = \frac{\hat{f}_{di}\lambda}{2}$. The acceleration estimation, as shown in Equation (22), can be obtained from the velocity of two adjacent segments.

$$\hat{a}_i = \frac{\hat{v}_i - \hat{v}_{i-1}}{t_{0i} - t_{0(i-1)}} \tag{22}$$

3.3. Phase Compensation and Long-Time Coherent Integration

According to the estimated motion parameters, the phase compensation factor can be constructed. The core of phase compensation is to eliminate the quadratic phase term caused by the uniform acceleration motion of the target. For the ith segment, the phase compensation factor is ϕ_i, which has the form of Equation (23). After the phase compensation, the coherent integration is performed to improve the SCR of the echo.

$$\phi_i(t_m) = \frac{1}{\lambda}[2(\hat{v}_{i-1} - \hat{v}_0)(t_m - \sum_{j=0}^{i-1} t_j) + \hat{a}_i(t_m - \sum_{j=0}^{i-1} t_j)^2] \tag{23}$$

After phase compensation, the phase of the signal in ith segment will not contain t^2 term anymore, in the form of Equation (24).

$$s_i'(t_m) = s_i(t_m) \cdot \exp[-j2\pi\phi_i(t_m)] = A_0 \exp[j\frac{4\pi}{\lambda} v_0 (t_m - \sum_{j=0}^{i-1} t_j)] \tag{24}$$

In theory, to the compensated signal $s'(t) = \{s_1'(t), s_2'(t), \ldots, s_N'(t)\}$, the Doppler frequency modulation caused by the quadratic phase term will be eliminated; that means the long-time coherent integration can be realized to obtain higher SCR.

3.4. Target Discrimination

The extracted ROIs contain both moving targets and false alarms caused by sea clutter, so the target discrimination is needed to eliminate false alarms. For a target with a certain motion constraints, its spectrum $F(f_\tau)$ of the SIAF of a certain segment will have the form of $\sin c$ function that the peak is located at the center of Doppler frequency as shown in Equation (20), while that of sea clutter has stronger randomness, and the energy will be distributed in a wider range on the frequency domain. Therefore, the spectrum width of the SIAF between the target and the sea clutter is obviously different, so the spectral width can be used for target discrimination. For the spectrum of the Kth ROI, the 3 dB spectrum width can be calculated, denoted as W_k. For a ROI, if its W_k satisfies inequality (25), it can be determined as the target.

$$W_k \leq \eta \cdot W_{3dB} \tag{25}$$

where η is the threshold factor; W_{3dB} is the 3 dB spectrum width of the $\sin c$ function; $W_{3dB} = 0.89 \cdot W_{Rayleigh}$ [33], where $W_{Rayleigh}$ is the Rayleigh bandwidth of the $\sin c$ function, which is related to the length of the time series. For a signal with a duration of τ s, its Rayleigh bandwidth is $W_{Rayleigh} = \frac{1}{\tau}$ Hz [34].

4. Experimental Verification

In order to verify the performance of the proposed long-time coherent integration based on segmented compensation, in this section, we organize the experiments based on simulated data and measured data, respectively. The simulation data are obtained by adding simulated low-SCR targets into the measured CSIR sea clutter. The measured data are collected by CSIR on the southwest coast of South Africa, which includes a moving boat.

4.1. Dim Targets Detection

Two simulated marine dim targets with very low SCR are added to the measured sea clutter to obtain the simulation data. The measured sea clutter is collected by CSIR on the southwest coast of South Africa in 2006, which is numbered as CFC17-001. The parameters of the radar observation are shown in Table 1.

Table 1. Parameters of radar observation.

Parameters	Value
Frequency (GHz)	9
PRF (KHz)	5
Initial distance (m)	3000.63
Range resolution (m)	15
Grazing angle (deg)	0.853–1.27
Wind speed (m/s)	7.97

For the simulated targets, the echo model of them uses the model which is introduced in Section 2.1. The motion of target 1 consists of two uniformly accelerated motions; the initial speed is 6 m/s; the accelerations of the two motions are -2 m/s^2 and 1 m/s^2, respectively. It crosses the 21st and the 22nd range bins during the observation. The motion of target 2 includes three stages of uniformly accelerated motions with the initial speed of 12 m/s; the accelerations of the three motions are 4 m/s^2, -3 m/s^2 and 2 m/s^2, respectively. It crosses the 51st and the 52nd range bins during the observation. The detailed motion parameters of the simulated targets are shown in Table 2.

Table 2. Motion parameters of simulated marine targets.

Parameters	Target 1	Target 2
Initial distance (m)	310	755
Initial velocity (m/s)	6	12
Accelerations (m/s^2)	−2, 1	4, −3, 2
Duration (s)	0.5, 0.5	0.3, 0.3, 0.4
SCR (dB)	−15	−17

The range–time image of simulated data is shown in Figure 2. The different colors represent different intensities of echoes. The simulated data consist of 96 range bins, and we select the 5000 pulses in 0–1 s for the experiment. Due to the low SCR, the targets are submerged in the sea clutter.

For the simulated data, after the five-layer wavelet packet decomposition is performed to each range bin, the frequency unit with high energy can be detected. The detection results are shown in Figure 3 where yellow represents the extracted high-energy units, while blue represents the lower-energy parts. It can be seen that the high-energy units are mainly concentrated in the low-frequency region.

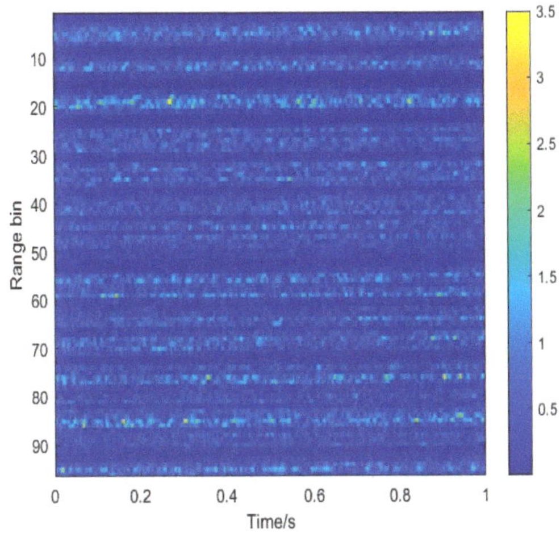

Figure 2. Range–time image of the simulation data.

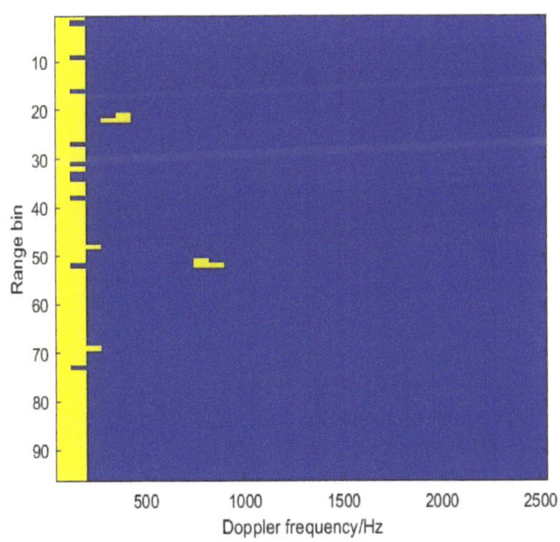

Figure 3. Detection result of the high–energy units.

After clutter suppression using Equation (18), the extracted ROIs are shown in Figure 4, which are labeled as 1, 2, 3, 4 according to the radial distance from radar, respectively. The range bins and Doppler frequency of the four ROIs are listed in Table 3. Among them, ROI 2 and ROI 4 are in the same Doppler frequency range; ROI 1 and ROI 3 cross different range bins, and ROI 2 and ROI 4 locate in the 48th range bin and the 66th range bin, respectively during the observation. For ROIs across different range bins, the time–frequency domain energy analysis described in Section 3.1 is used to select and extract the echo of the potential target.

Table 3. The location of the ROIs.

ROI	Range Bin	Doppler Frequency/Hz
ROI 1	21, 22	312.5~390.6
ROI 2	48	156.3~234.4
ROI 3	51, 52	781.3~859.4
ROI 4	69	156.3~234.4

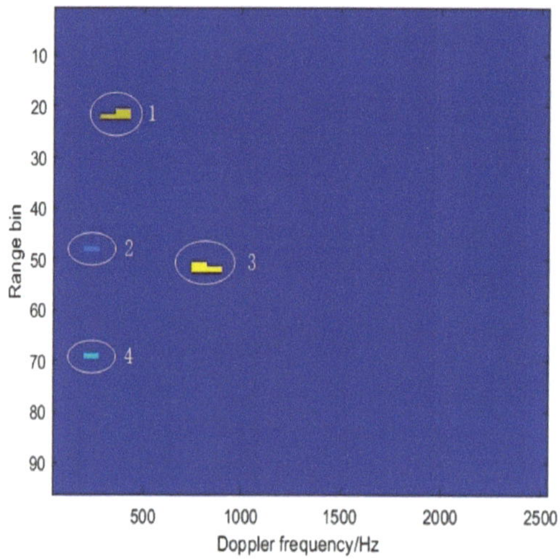

Figure 4. The extracted ROIs after clutter suppression.

Using the parameter estimation and motion segmentation described in Section 3.2, the parameters of the initial velocity and acceleration of each ROI are obtained. According to the parameters estimation, the motion of ROI 1 can be regarded as a combination of two stages of uniformly accelerated motion, from 0 to 0.5 s; the estimated initial velocity and acceleration are 5.92 m/s and -1.84 m/s^2, respectively; and in the range 0.5 to 1 s, the estimated initial velocity and acceleration are 5 m/s and 1.16 m/s^2, respectively. For ROI 2, its motion can be divided into three stages of uniformly accelerated motion, that is, 0 to 0.3 s, 0.3 to 0.4 s and 0.4 to 1 s; the estimated initial velocities are 2.92 m/s, 2.58 m/s and -0.17 m/s, respectively, the estimated acceleration are -1.13 m/s^2, -27.5 m/s^2, 4.87 m/s^2, respectively. The motion of ROI 3 can be segmented into three uniformly accelerated motions, that is, 0 to 0.3 s, 0.3 to 0.6 s and 0.6 to 1 s; the initial speed of the three motion segments is 12.17 m/s, 13.08 m/s, 12.17 m/s; and the acceleration is 3.03 m/s^2, -3.03 m/s^2, 2.9 m/s^2, respectively. The motion of ROI 4 can be divided into three stages of uniformly accelerated motion, from 0 to 0.6 s; the estimated initial velocity and acceleration are 2.75 m/s and -3.05 m/s^2, 0.6 to 0.8 s; its estimated initial velocity and acceleration are 0.92 m/s and 10.4 m/s^2; and from 0.8 to 1 s, its estimated initial velocity and acceleration are 3 m/s and -5.85 m/s^2, respectively.

Noise, data segmentation, clutter interference, etc., can affect the parameter estimation. The mean square error (MSE) is used to measure the accuracy of parameter estimation of the targets, which is calculated as $MSE = \frac{1}{n}\sum_{i=1}^{n}(Y_i - \hat{Y}_i)^2$, where $\hat{Y}_i$ is the estimation of Y_i. The MSE of velocity estimation and acceleration estimation of target 1 are 6.2×10^{-3} m/s and 2.56×10^{-2} m/s^2, respectively; The MSE of velocity estimation and acceleration estimation of target 2 are 4.9×10^{-2} m/s and 0.58 m/s^2, respectively. It can be seen that the estimation result is relatively accurate. It should be noted that the excessive parameter estimation errors can lead to significant phase mismatch, which will cause

Doppler spectrum expansion cannot be effectively eliminated. Because the Doppler frequency of sea clutter is affected by many environmental factors, it has strong temporal variability, which is difficult to describe its mechanism by specific model. Therefore, according to the parameter estimation and segmentation results of ROI 2 and ROI 4 shown in Figure 5b,d, the changes of velocity and acceleration have strong randomness and are significantly different from those of marine targets with specific motion models.

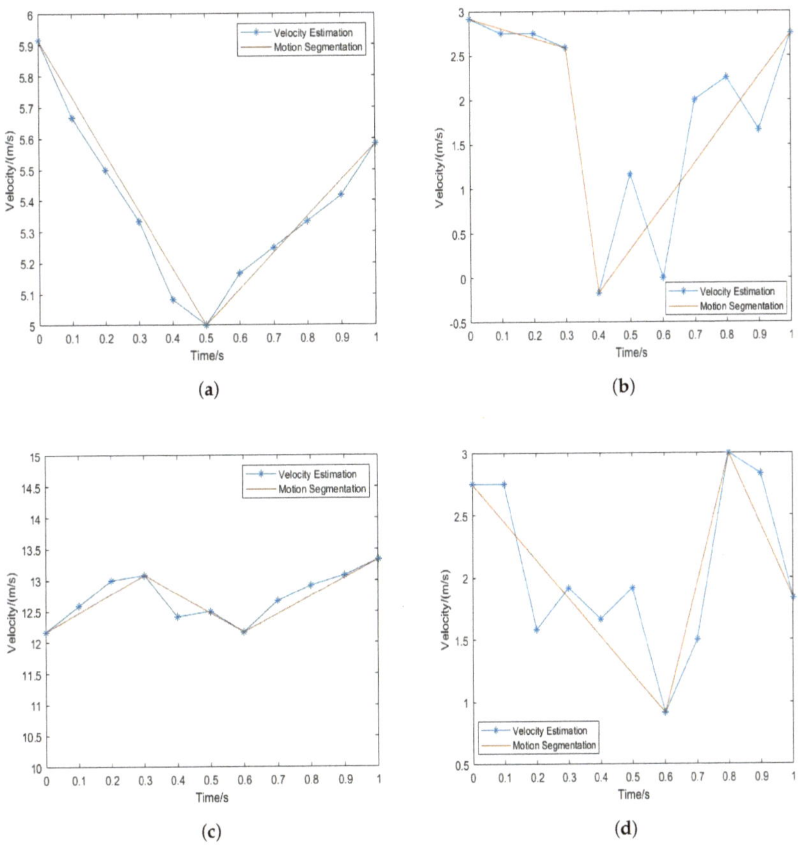

Figure 5. Parameter estimation and motion segmentation results of 4 ROIs of simulation data, where (**a**–**d**) correspond to ROI 1-ROI 4, respectively.

According to the estimation of motion parameters and the segmentation, the extracted echo of four ROIs can be compensated in each segment, respectively. Then the long-time coherent integration can be performed to obtain high SCR. Taking ROI 1 as an example, the segmented compensation integration results are shown in Figure 6a. MTD is a typical target detection method, which is based on coherent integration; as a comparison, the integration effect of MTD of ROI 1 is also given in Figure 6b. As can be seen from Figure 6, the proposed segmented coherent integration can well compensate the Doppler frequency modulation caused by the variable acceleration movement of the marine target, so that the target can be accumulated in a small range of frequency domain. To finely describe the integration results, the SCR is calculated as the ratio between the maximum power of the target and the average power of sea clutter. The SCR after segmented compensation integration is 19.7 dB, which is significantly higher than that of the coherent integration of MTD, which is 14.7 dB. In Figure 6, both the sea clutter and the target correspond to peaks, to distinguish them clear, the target is circled in red.

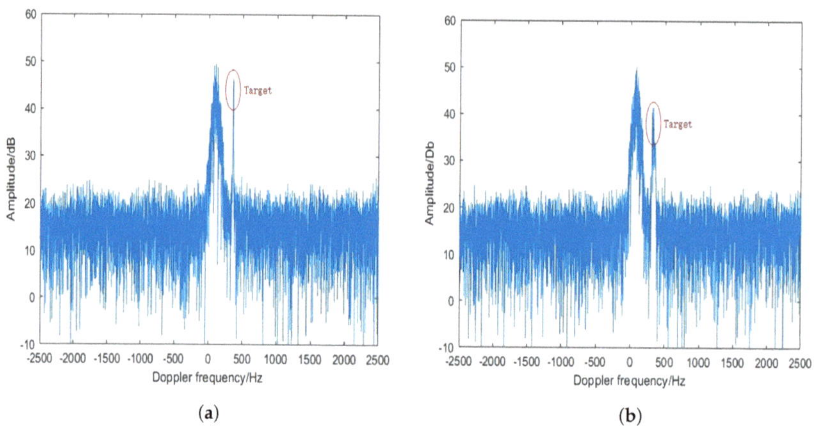

Figure 6. Comparison of MTD and proposed method in ROI 1. (**a**) Integration result of proposed method. (**b**) Integration result of MTD.

The detection of ROIs is performed under low SCR, and there will be false alarms caused by sea clutter, so it is necessary to discriminate the extracted ROIs. For each ROI, the 3 dB spectral width of SIAF is calculated. Based on the motion segmentation of four ROIs, the segment with the longest duration of each ROI is selected for target discrimination. The selected segments of ROI 1-ROI 4 are 0 s to 0.5 s, 0.4 s to 1 s, 0.6 s to 1 s and 0 s to 0.6 s, respectively. For the selected segments, the standard 3 dB spectral width of SIAF are about 1.78 Hz, 1.48 Hz, 2.225 Hz and 1.48 Hz, respectively, taking the coefficient $\eta = 1.5$, the spectral width thresholds are calculated. The target discrimination results are shown in Table 4, where ROI 1 and 3 are the areas where the targets exist, which is consistent with the setting of the simulation. By using long-time coherent integration, the very low SCR targets are detected successfully.

Figure 7 shows the spectrum of the SIAF of the four ROIs, where the red line denotes the 3 dB attenuation of the maximum value. As can be seen from Figure 7, due to the irregular motion of the sea clutter, the 3 dB spectral width of the SIAF is wider than that of the marine targets with clear motion constraints.

Table 4. Target discrimination of ROIs.

ROI	3 dB Band Width/Hz	Threshold/Hz	Discrimination
ROI 1	2	2.67	T
ROI 2	70.8	2.23	F
ROI 3	2.5	3.34	T
ROI 4	64.3	2.23	F

In this experiment, the proposed long-time coherent integration based on segmented compensation achieves the energy integration and target detection of two targets with complex motion under very low SCR, and effectively removes the false alarms, indicating that the proposed method has the ability of marine dim target detection with complex motion.

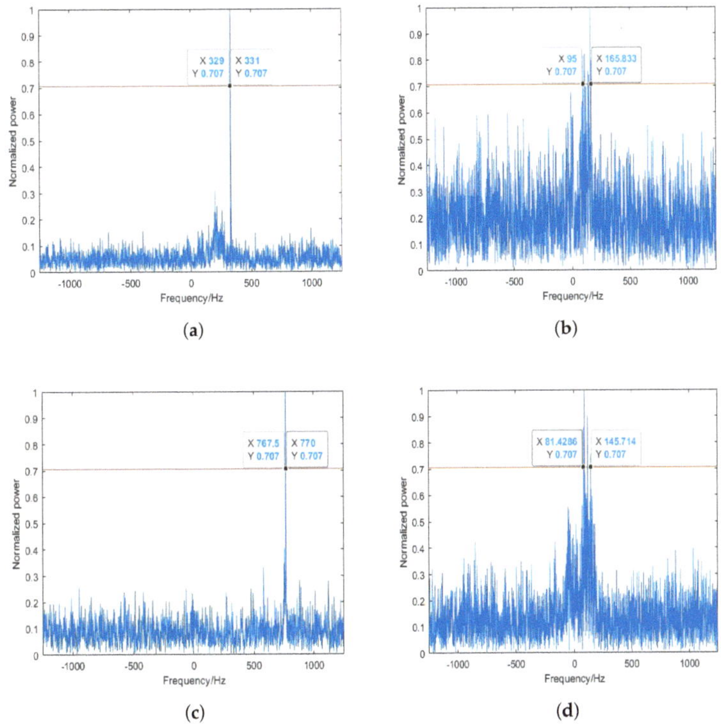

Figure 7. The spectrum of symmetric instantaneous autocorrelation function of the 4 ROIs in simulated data, where (**a**–**d**) correspond to the spectrum of symmetric instantaneous autocorrelation function of ROI 1-ROI 4, respectively.

4.2. Detection Performance Simulation

Target detection from sea clutter is affected by many factors in order to verify the performance of the proposed method. Monte Carlo simulation under different SCR is used. The sea clutter can be simulated using K-distribution model, in which a variable acceleration target is randomly inserted. The initial velocity of the target takes a random value in the range of $0 \sim 10$ m/s, and the acceleration takes a random value in the range of $-5 \sim 5$ m/s^2. Each target includes two stages of uniformly accelerated motions.

In the experiments, to compare the detection performance, four methods of coherent integration: segmented compensation integration, RFrFT, RFT and MTD, are used to realize the coherent integration; CFAR detection was performed to detect the targets. The false alarm rate of CFAR detection was set to 10^{-4}, and the times of the Monte Carlo experiments were set to 10^4. The detection probability under each SCR was calculated, and the ROC curve is shown in Figure 8. According to the simulation results, under the same SCR, the detection probability of the proposed method is higher than those of RFrFT, RFT and MTD. To obtain the same detection probability, the required SCR of the proposed method is about 4.5 dB less than that of the RFrFT, and about 6 dB less than that of the RFT. The simulation results show that the proposed method can effectively improve the dim target detection performance under heavy sea clutter, and the detection probability is higher than that of RFrFT, RFT and MTD under the same conditions, especially to lower SCR.

If there are multiple targets in the actual scene and relatively close, then the threshold of the CA-CFAR detector will be affected by the neighboring targets, and the detection performance will decrease.

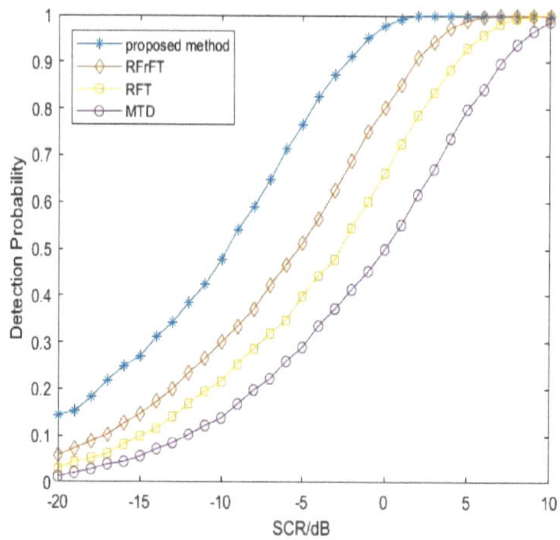

Figure 8. Simulation results of Pd-SCR.

4.3. Target Detection Based on Measured Data

In order to verify the detection effect of the proposed method on the actual scene, the measured CSIR data TFC17-006, which were also collected from the southwest coastline of South Africa in 2006, were used for the experiment. The radar was operating in X-band, with a range resolution of 15m and a pulse repetition frequency of 5000 Hz. During the radar observation, the wind speed was 11.96 m/s and the grazing angle was from 0.501 to 0.560 degree. Like Ref. [26], we also selected 5000 pulses between 80.5 to 81.5 s for the experiment. A boat, as the target, was located in the 23rd and 24th range bins in the data. The range–time image of the measured data is shown in Figure 9.

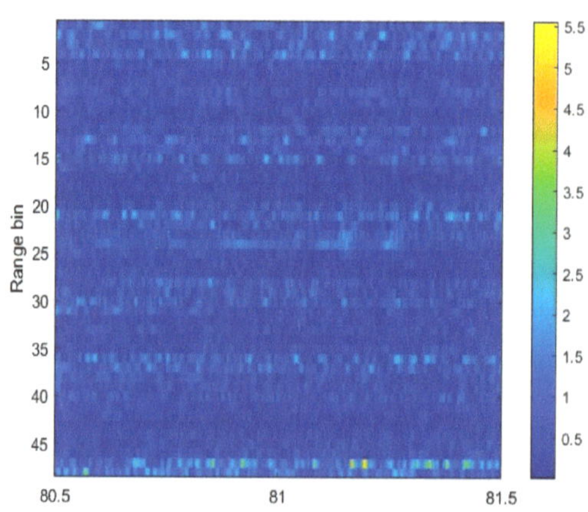

Figure 9. Range–time image of the measured CSIR data.

Same as the Ref. [26], four ROIs can be extracted from this experimental scenario, as shown in Figure 10. The four ROIs are labeled as 1, 2, 3 and 4, respectively, according to the radial distance from radar. Among them, ROI 1 is located in the 17th range bin; ROI 2 is distributed in the 23rd and 24th range bins; ROI 3 and ROI 4 are located in the 28th range bin and 44th range bin, respectively. The Doppler frequency of the four ROIs range from 312.5 Hz to 468.75 Hz. For the extracted potential targets, the method described in Section 3.2 is used for motion parameter estimation and segmentation.

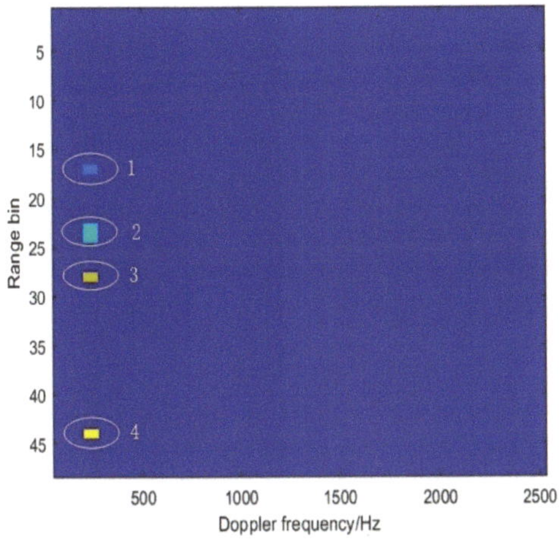

Figure 10. The extracted ROIs from the measured data.

To the extracted ROIs, the velocity estimation and motion segmentation results are shown in Figure 11. Among them, ROI 2 is caused by a real boat. Just like Figure 11b, the movement of the boat is divided into 3 segments. ROI 1, ROI 3 and ROI 4 are false alarms caused by sea clutter, which shows the irregular movement. The parameters of velocity and acceleration in each segment can be estimated to construct the phase compensation factor.

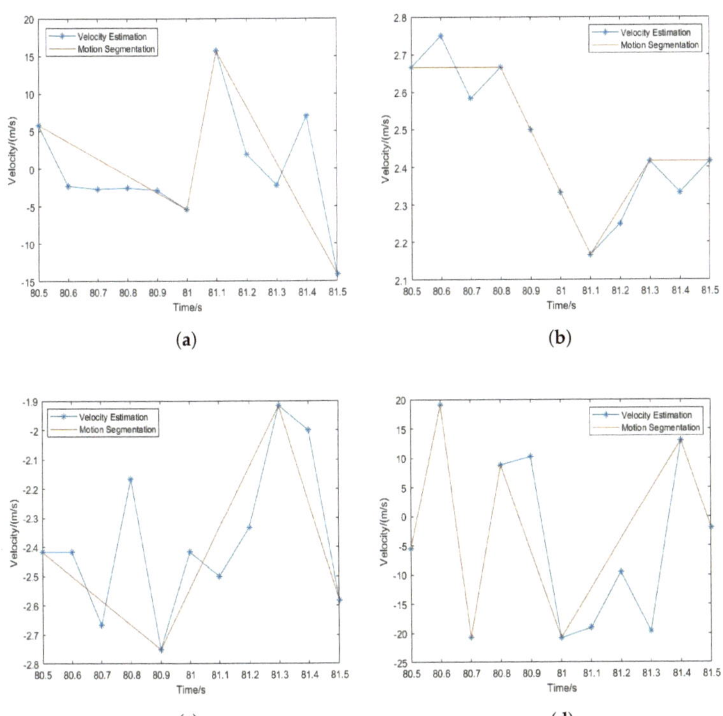

Figure 11. Parameter estimation and motion segmentation results of 4 ROIs of measured data, where (**a**–**d**) correspond to the results of ROI 1-ROI 4, respectively.

Figure 12 shows the spectrum of the SIAF of the 4 ROIs; the red line in the Figure 12 shows the 3 dB attenuation of the maximum value. The selected motion segments of ROI 1-ROI 4 are 80.5 s to 81 s, 80.8 s to 81.1 s, 80.9 s to 81.3 s and 81 s to 81.4 s, respectively. The target discrimination of the four ROIs is performed based on the 3 dB spectral width of the SIAF of the selected segments. The 3 dB spectrum width of the four ROIs are 22 Hz, 3.33 Hz, 68.75 Hz and 50 Hz, respectively. Taking the coefficient $\eta = 1.5$, the spectral width thresholds are 2.67 Hz, 4.45 Hz, 3.34 Hz, 3.34 Hz, respectively. For ROI 2, its 3 dB spectral width of the SIAF is less than the spectral width threshold, which is discriminated as the target. ROI 1, ROI 3 and ROI 4 are identified as false alarms caused by sea clutter. The result of target discrimination is consistent with the ground truth.

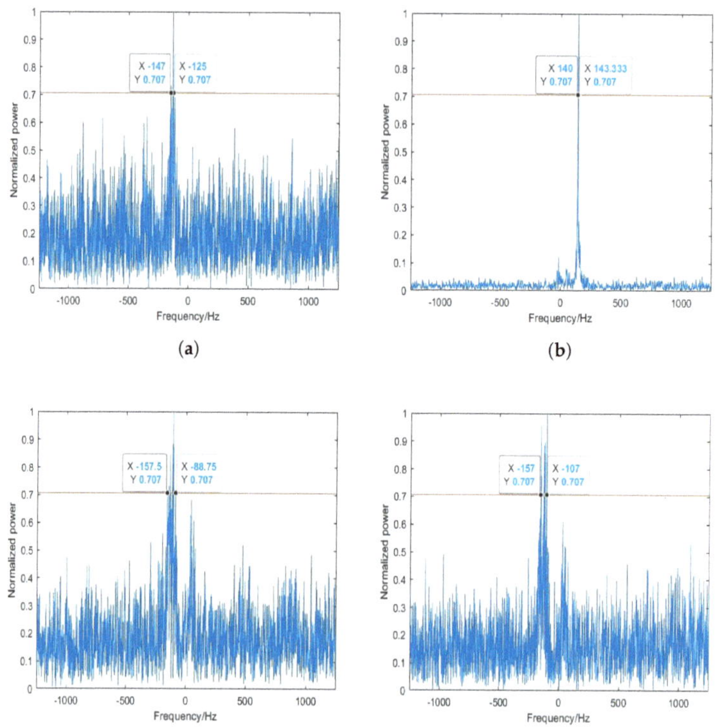

(**a**) (**b**) (**c**) (**d**)

Figure 12. The spectrum of symmetric instantaneous autocorrelation function of the 4 ROIs in measured data, where (**a**–**d**) correspond to the spectrum of symmetric instantaneous autocorrelation function of ROI 1-ROI 4, respectively.

After segmented compensation, the integration results are shown in Figure 13. In Figure 13, the integrated energy of target is higher than the energy of sea clutter. Experiment results indicate that the long-time coherent integration based on segmented compensation can effectively compensate the Doppler frequency modulation caused by the complex motion of the marine target, and can realize the efficient integration of target energy. As a comparison, the integration effect of MTD in the 24th range bin and that of RFrFT in ROI 2 are also given in Figure 14a,b. As can be seen from Figures 13 and 14, the proposed method can significantly improve the SCR after coherent integration.

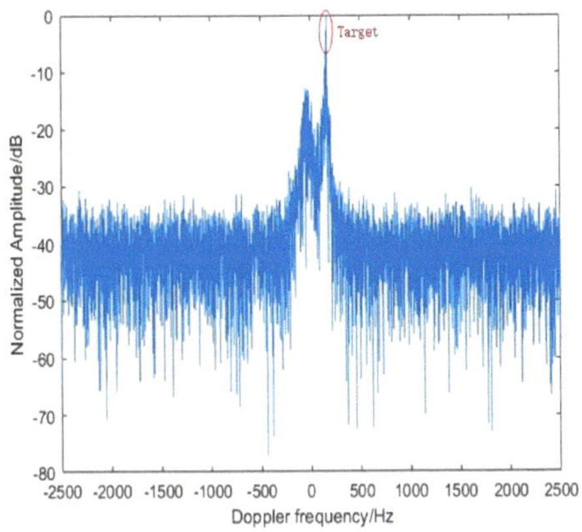

Figure 13. Integration result of proposed method.

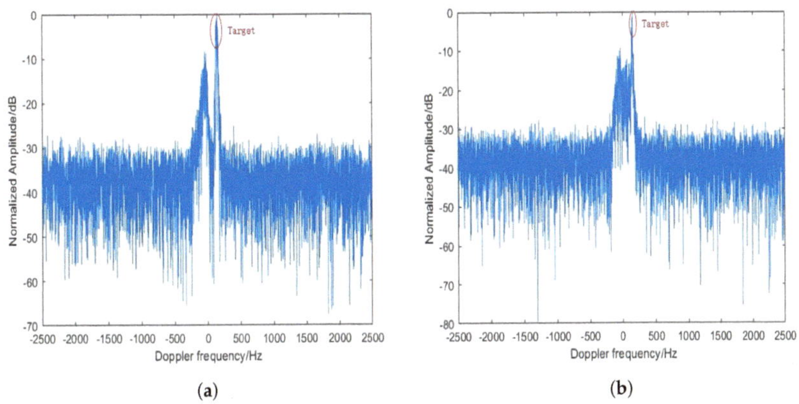

Figure 14. Integration results of MTD and RFrFT. (**a**) Integration result of MTD in 24th range bin. (**b**) Integration result of RFrFT in ROI 2.

In order to show the integration effect of the proposed method on the boat in measured data more clearly, Table 5 lists the integration effect of MTD, RFrFT and the proposed method. It can be seen that the proposed method can obtain the highest SCR, which is about 4.5 dB higher than that of the representative RFrFT method. The experimental results of the measured data verify that the proposed method can effectively improve the SCR of complex moving target, and is an effective dim target detection method.

Table 5. The SCR of different integration methods.

Methods	SCR/dB
MTD of 23rd range bin	25.9
MTD of 24th range bin	26.3
RFrFT	28.5
Proposed method	33.0

5. Conclusions

In this paper, a long-time coherent integration method based on segmented compensation is proposed for the maneuvering target detection from sea clutter. The proposed method models the complex motion of the maritime target as a multi-stage uniformly accelerated motion with different parameters. By estimating the moving parameters of each stage, the Doppler frequency modulation caused by the acceleration will be eliminated. And finally, the long-time coherent integration is realized, which will increase the detection performance of dim targets with complex movement.

Based on multi-stage motion modeling, the quadratic phase of the complex moving target can be compensated through parameter estimation in each stage. The processing of segmented compensation provides a new method to realize long-time coherent integration to a complex moving target.

Maritime target detection is affected by various factors, especially at low SCR. To eliminate false alarms, a novel threshold based on the 3 dB bandwidth of the spectrum of symmetric instantaneous autocorrelation function is constructed for target discrimination. The threshold is effective for determining whether an ROI is a target or sea clutter.

The target detection experiment based on simulated data show that the proposed method can extract ROIs, estimate moving parameters and detect targets successfully from sea clutter under very low SCR of −15 dB and −17 dB. At the same time, it can obtain the best detection performance than the representative detection methods, such as MTD, RFT and RFrFT. Under the SCR of −5 dB, the detection probability of the proposed method is about 0.3 higher than that of RFrFT, 0.4 higher than that of RFT and 0.5 higher than that of MTD. The result indicates that the proposed method has good detection performance for weak and small targets with very low SCR. The experiment based on the measured CSIR data show that the proposed method can effectively integrate target energy to obtain higher SCR than the RFrFT of 4.5 dB and detect the real boat from sea clutter. The experimental results show that the proposed long-time coherent integration method based on segmented compensation can effectively integrate complex moving targets' energy to detect them from sea clutter.

Author Contributions: Conceptualization, Z.Z. and W.W. (Wenguang Wang); methodology, Z.Z. and Y.Z.; software, Z.Z. and Y.Z.; validation, W.W. (Wenguang Wang), B.L. and W.W. (Wei Wu); formal analysis, Z.Z. and Y.Z.; investigation, Z.Z.; resources, W.W. (Wenguang Wang); data curation, Z.Z. and Y.Z.; writing—original draft preparation, Z.Z.; writing—review and editing, W.W. (Wenguang Wang) and B.L.; visualization, Z.Z.; supervision, W.W. (Wenguang Wang); project administration, W.W. (Wenguang Wang); funding acquisition, W.W. (Wei Wu). All authors have read and agreed to the published version of the manuscript.

Funding: This research was funded by the National Natural Science Foundation of China, Grant No. 62073334 and Grant No. 61771028.

Data Availability Statement: Not applicable.

Conflicts of Interest: The authors declare no conflict of interest.

References

1. Richards, M. Coherent integration loss due to white Gaussian phase noise. *IEEE Signal Process. Lett.* **2003**, *10*, 208–210. [CrossRef]
2. Carlson, B.; Evans, E.; Wilson, S. Search radar detection and track with the Hough transform. I. system concept. *IEEE Trans. Aerosp. Electron. Syst.* **1994**, *30*, 102–108. [CrossRef]
3. Huang, P.; Liao, G.; Yang, Z.; Xia, X.G.; Ma, J.; Zheng, J. Ground Maneuvering Target Imaging and High-Order Motion Parameter Estimation Based on Second-Order Keystone and Generalized Hough-HAF Transform. *IEEE Trans. Geosci. Remote Sens.* **2017**, *55*, 320–335. [CrossRef]
4. Kong, Y.K.; Cho, B.L.; Kim, Y.S. Ambiguity-free Doppler centroid estimation technique for airborne SAR using the Radon transform. *IEEE Trans. Geosci. Remote Sens.* **2005**, *43*, 715–721. [CrossRef]
5. Ebenezer, S.P.; Papandreou-Suppappola, A. Generalized Recursive Track-Before-Detect with Proposal Partitioning for Tracking Varying Number of Multiple Targets in Low SNR. *IEEE Trans. Signal Process.* **2016**, *64*, 2819–2834. [CrossRef]
6. Kim, D.Y.; Ristic, B.; Guan, R.; Rosenberg, L. A Bernoulli Track-Before-Detect Filter for Interacting Targets in Maritime Radar. *IEEE Trans. Aerosp. Electron. Syst.* **2021**, *57*, 1981–1991. [CrossRef]
7. Wang, M.; Li, X.; Zhang, Z.; Cui, G.; Yeo, T.S. Coherent Integration and Parameter Estimation for High-Speed Target Detection with Bistatic MIMO Radar. *IEEE Trans. Geosci. Remote Sens.* **2023**, *61*, 5107915. [CrossRef]
8. Perry, R.P.; DiPietro, R.C.; Fante, R.L. Coherent Integration with Range Migration Using Keystone Formatting. In Proceedings of the 2007 IEEE Radar Conference, Waltham, MA, USA, 17–20 April 2007; pp. 863–868. [CrossRef]

9. Zhu, D.; Li, Y.; Zhu, Z. A Keystone Transform without Interpolation for SAR Ground Moving-Target Imaging. *IEEE Geosci. Remote Sens. Lett.* **2007**, *4*, 18–22. [CrossRef]
10. Huang, P.; Liao, G.; Yang, Z.; Xia, X.G.; Ma, J.T.; Ma, J. Long-Time Coherent Integration for Weak Maneuvering Target Detection and High-Order Motion Parameter Estimation Based on Keystone Transform. *IEEE Trans. Signal Process.* **2016**, *64*, 4013–4026. [CrossRef]
11. Yang, J.; Zhang, Y. An Airborne SAR Moving Target Imaging and Motion Parameters Estimation Algorithm with Azimuth-Dechirping and the Second-Order Keystone Transform Applied. *IEEE J. Sel. Top. Appl. Earth Obs. Remote Sens.* **2015**, *8*, 3967–3976. [CrossRef]
12. Xu, J.; Yu, J.; Peng, Y.N.; Xia, X.G. Radon-Fourier Transform for Radar Target Detection, I: Generalized Doppler Filter Bank. *IEEE Trans. Aerosp. Electron. Syst.* **2011**, *47*, 1186–1202. [CrossRef]
13. Ding, Z.; Zhang, T.; Li, Y.; Li, G.; Dong, X.; Zeng, T.; Ke, M. A Ship ISAR Imaging Algorithm Based on Generalized Radon-Fourier Transform with Low SNR. *IEEE Trans. Geosci. Remote Sens.* **2019**, *57*, 6385–6396. [CrossRef]
14. Lang, P.; Fu, X.; Dong, J.; Yang, J. An Efficient Radon Fourier Transform-Based Coherent Integration Method for Target Detection. *IEEE Geosci. Remote Sens. Lett.* **2023**, *20*, 3501905. [CrossRef]
15. Li, X.; Huang, L.; Zhang, S.; Wang, W.Q. Phase Compensation and Time-Reversal Transform for High-Order Maneuvering Target Detection. *IEEE Geosci. Remote Sens. Lett.* **2022**, *19*, 4025905. [CrossRef]
16. Lin, L.; Sun, G.; Cheng, Z.; He, Z. Long-Time Coherent Integration for Maneuvering Target Detection Based on ITRT-MRFT. *IEEE Sens. J.* **2020**, *20*, 3718–3731. [CrossRef]
17. Yang, J.; Liu, X.; Yang, B.; Lu, J.; Liao, G. Detection and Speed Estimation of Moving Target Based on Phase Compensation and Coherent Accumulation Using Fractional Fourier Transform. *Sensors* **2020**, *20*, 1410. [CrossRef] [PubMed]
18. Chen, X.; Guan, J.; He, Y.; Zhang, J. Detection of low observable moving target in sea clutter via fractal characteristics in fractional Fourier transform domain. *IET Radar, Sonar Navig.* **2013**, *7*, 635–651. [CrossRef]
19. Lv, X.; Bi, G.; Wan, C.; Xing, M. Lv's Distribution: Principle, Implementation, Properties, and Performance. *IEEE Trans. Signal Process.* **2011**, *59*, 3576–3591. [CrossRef]
20. Yu, X.; Chen, X.; Huang, Y.; Guan, J. Fast Detection Method for Low-Observable Maneuvering Target via Robust Sparse Fractional Fourier Transform. *IEEE Geosci. Remote Sens. Lett.* **2020**, *17*, 978–982. [CrossRef]
21. Luo, S.; Bi, G.; Lv, X.; Hu, F. Performance analysis on Lv distribution and its applications. *Digit. Signal Process.* **2013**, *23*, 797–807. [CrossRef]
22. Li, X.; Cui, G.; Kong, L.; Yi, W. Fast Non-Searching Method for Maneuvering Target Detection and Motion Parameters Estimation. *IEEE Trans. Signal Process.* **2016**, *64*, 2232–2244. [CrossRef]
23. Xiang, J.; Cui, W.; Tian, J. Coherent Integration Algorithm for a Maneuvering Target Based on Frequency Spectrum Segment Processing. *IEEE Access* **2020**, *8*, 115646–115654. [CrossRef]
24. Gao, C.; Tao, R.; Kang, X. Weak Target Detection in the Presence of Sea Clutter Using Radon-Fractional Fourier Transform Canceller. *IEEE J. Sel. Top. Appl. Earth Obs. Remote Sens.* **2021**, *14*, 5818–5830. [CrossRef]
25. Chen, X.; Guan, J.; Liu, N.; He, Y. Maneuvering Target Detection via Radon-Fractional Fourier Transform-Based Long-Time Coherent Integration. *IEEE Trans. Signal Process.* **2014**, *62*, 939–953. [CrossRef]
26. Wang, W.; Chen, C.; Qu, G.; Zuo, Y.; Zeng, G. Low-observable target detection using two-stage RFRFT. *IET Radar Sonar Navig.* **2019**, *13*, 653–663. [CrossRef]
27. Barton, D.K.; Leonov, S.A. *Radar Technology Encyclopedia*; Artech House Inc.: Norwood, MA, USA, 1997
28. Levanon, N.; Mozeson, E.; Signals, R.; Levanon, C. Radar Signals. *IEEE Aerosp. Electron. Syst. Mag.* **2013**, *20*, 32–33. [CrossRef]
29. Hussain, M.; Ahmed, R.; Cheema, H.M. Segmented Radon Fourier Transform for Long-Time Coherent Radars. *IEEE Sens. J.* **2023**, *23*, 9582–9594. [CrossRef]
30. Sun, H.B.; Liu, G.S.; Gu, H.; Su, W.M. Application of the fractional Fourier transform to moving target detection in airborne SAR. *IEEE Trans. Aerosp. Electron. Syst.* **2002**, *38*, 1416–1424. [CrossRef]
31. Pei, J.; Huang, Y.; Guan, J.; Dong, Y.; Chen, X. Long-time coherent integration method combining pulse compression and Radon fractional Fourier transform. In Proceedings of the 2021 CIE International Conference on Radar (Radar), Haikou, China, 15–19 December 2021; pp. 1266–1269. [CrossRef]
32. Claasen, T.A.C.M.; Mecklenbrauker, W.F.G. The Wigner Distribution—A Tool for Time-Frequency Signal Analysis Part Ill: Relations with Other Time-Frequency Signal Transformations. *Philips J. Res.* **1980**, *35*, 372–389.
33. Miller, R. Fundamentals of Radar Signal Processing (Richards, M.A.; 2005) [Book review]. *IEEE Signal Process. Mag.* **2009**, *26*, 100–101. [CrossRef]
34. Cumming, I.G.; Wong, F.H. *Digital Processing of Synthetic Aperture Radar Data: Algorithms and Implementation*; Artech House: Boston, MA, USA, 2005.

cancers

MDPI

Article

A Study on Survival Analysis Methods Using Neural Network to Prevent Cancers

Chul-Young Bae [1], Bo-Seon Kim [1], Sun-Ha Jee [2], Jong-Hoon Lee [3] and Ngoc-Dung Nguyen [3,*]

[1] Mediage Research Center, Seongnam-si 13449, Republic of Korea
[2] Department of Epidemiology and Health Promotion, Institute for Health Promotion, Graduate School of Public Health, Yonsei University, Seoul 03722, Republic of Korea
[3] Moadata AI Labs, Seongnam-si 13449, Republic of Korea
* Correspondence: nguyen@moadata.ai; Tel.: +82-70-4099-5131

Simple Summary: According to cancer statistics published in 2020, there were 19.3 million new cancer cases and almost 10.0 million cancer deaths worldwide. This suggests that cancer is one of the main health threats worldwide. Survival analysis, combining the exponential distribution with regression analysis, can predict the time when a specific event will occur. This nationwide follow-up study aims to present survival deep learning models to assist in the early identification of cancer incidence. Our models consistently achieved high performance in 10 types of cancer. By applying the techniques outlined in this paper, clinical biomarkers, demographic, and anthropometric data can be utilized to predict the risk of cancer occurrence.

Abstract: Background: Cancer is one of the main global health threats. Early personalized prediction of cancer incidence is crucial for the population at risk. This study introduces a novel cancer prediction model based on modern recurrent survival deep learning algorithms. **Methods:** The study includes 160,407 participants from the blood-based cohort of the Korea Cancer Prevention Research-II Biobank, which has been ongoing since 2004. Data linkages were designed to ensure anonymity, and data collection was carried out through nationwide medical examinations. Predictive performance on ten cancer sites, evaluated using the concordance index (c-index), was compared among nDeep and its multitask variation, Cox proportional hazard (PH) regression, DeepSurv, and DeepHit. **Results:** Our models consistently achieved a c-index of over 0.8 for all ten cancers, with a peak of 0.8922 for lung cancer. They outperformed Cox PH regression and other survival deep neural networks. **Conclusion:** This study presents a survival deep learning model that demonstrates the highest predictive performance on censored health dataset, to the best of our knowledge. In the future, we plan to investigate the causal relationship between explanatory variables and cancer to reduce cancer incidence and mortality.

Keywords: survival analysis; recurrent neural network; LSTM; Cox PH; biomarkers; cancer; cohort; follow-up

1. Introduction

Cancer stands as a leading cause of premature death in upper-middle- and high-income countries [1]. Specifically, female breast, lung, colorectal, prostate, stomach, and liver cancers account for the highest incidence and mortality rates [1,2]. Early personalized prediction of cancer incidence holds paramount importance for the population at risk. Substantial evidence regarding the risk factors of noncommunicable diseases, including cancer, has primarily been drawn from large-scale follow-up studies, with many of them employing survival analysis.

Survival analysis is centered on determining the probability of surpassing a certain period without encountering the event, as well as relevant outcomes. It has demonstrated its robustness in comparing risks among subgroups within the targeted population and has found applications in diverse fields, spanning from economics to healthcare [3]. Consequently, survival analysis has been utilized in the quest to identify factors associated with cancer and predict cancer survival [4].

The Cox proportional hazard (Cox PH) regression model [5] has emerged as a popular tool within the realm of survival analysis due to its interpretability and predictive ability. However, it rests upon strong assumptions regarding covariates. To mitigate these limitations, numerous models have been developed based on the Cox PH model, incorporating artificial intelligence techniques [4]. One such machine learning approach, the random survival forest, has gained prominence. Certain studies [6–8] were among the first to introduce a single hidden layer network for survival analysis. Building upon similar concepts, Katzman [9] further developed a deep survival network for personalized treatment recommendations. The predictive efficacy of survival networks was markedly enhanced by the application of multitask learning and residual connections, as proposed in the model known as DeepHit [10].

Recent research has suggested that survival neural networks outperform their machine learning and traditional statistical counterparts [10,11]. Leveraging data from a large-scale blood-based cohort, we conducted a comparative assessment of the performance of various survival deep networks, using Cox PH as the benchmark, to predict ten cancers that rank among the most common globally [2,12]. Furthermore, we introduced nDeep—a model that, to the best of our knowledge, achieves the highest possible performance on real censored health datasets. In an attempt to interpret the effects of those biomarkers in each model, feature importance was also calculated.

2. Materials and Methods

2.1. Subjects

A total of 160,407 observations (95,229 men and 62,169 women) were collected from a blood-based cohort within The Korean Cancer Prevention Study-II Biobank, an ongoing study initiated in 2004 through nationwide medical examinations [13]. Participants underwent annual follow-ups utilizing personal identification numbers, enabling linkage with the National Cancer Center registry, hospital admission records, and death registers. The integrity of all linkages was guaranteed with anonymity safeguards. During routine health check-ups, individuals diagnosed with cancer or other severe conditions (such as organ failure, diabetes, malignant hypertension), as well as those who died, were subsequently excluded from the study.

2.2. Characteristics, Biomarkers, and Events

Anthropometry (height and weight) was recorded by InBody (Biospace, Cheonan-si, Korea) when participants wore light clothes. For waist circumference, the thinnest area between the iliac crest and the inferior part of the lowest rib was measured in an upright position. Forced vital capacity and forced expiratory volume in 1 second were recorded by an electronic spirometer in standing position, while the better results out of two records were taken.

Seated blood pressure was then taken by a registered nurse or technician using a sphygmomanometer after resting 5 min. Fasting blood glucose, total cholesterol, and other biomarkers were measured in the laboratory of the hospital by a COBAS INTEGRA 800 (Roche, Berlin, Germany) and a 7600 Analyzer (Hitachi, Tokyo, Japan). More details can be found at [13]. For our final analysis, these features were included as covariates:

- Demographic: AGE—age, years; SEX—sex, 1 as male and 2 as female.
- Metabolic characteristics: WT—weight, kilograms; HT—height, centimeters; and BMI—body mass index = weight/height^2 (kg/m^2); WC—waist circumference, centimeters; SBP—systolic blood pressure, mmHg; DBP—diastolic blood pressure, mmHg; CHO total cholesterol, mg/dL; HDL—high-density cholesterol, mg/dL; TG—triglycerides,

mg/dL; LDL—low-density cholesterol, mg/dL; GLU—glucose, mg/dL; PP—pulse pressure = systolic blood pressure − diastolic blood pressure, mmHg.

- Liver function: ALB—albumin, g/dL; GLOBULIN—globulin, g/dL; AGR—albumin globulin rate; BIL—total bilirubin, mg/dL; DBIL—direct bilirubin, mg/dL; ALP—alkaline phosphatase, units/L; AST—aspartate aminotransferase, units/L; ALT—alanine aminotransferase, units/L; GGTP—gamma-glutamyl transpeptidase, units/L.
- Kidney function: CREAT—creatinine, mg/dL; BUN—blood urea nitrogen, mg/dL; SG—specific gravity; PH—urine pH; CCR—creatinine clearance rate = (140 − Age) × Weight/(Creatinine × 72) for Men OR ((140 − Age) × Weight/(Creatinine × 72)) × 0.85 for Women.
- Pancreas function: AMYLASE—amylase, units/L.
- Pulmonary function: FVC—forced lung capacity, measured in liters; FEV1—forced expiratory volume, measured in liters.

The study sample is representational regarding age. The number of men accounts for 60% of the total sample (Table 1).

Table 1. Basic information on demography, anthropometry, and clinical characteristics of study population.

	Mean ± SD	Min	Max	Male Mean ± SD	Female Mean ± SD
AGE	41.62 ± 10.53	9.00	88.00	42.04 ± 9.95	40.98 ± 11.31
HT	166.80 ± 8.40	130.00	198.00	171.75 ± 5.95	159.31 ± 5.52
WT	65.79 ± 12.00	32.00	137.00	72.12 ± 9.93	56.23 ± 7.78
WC	80.58 ± 9.45	41.00	127.00	84.80 ± 7.65	74.21 ± 8.26
BMI	23.53 ± 3.17	12.91	42.36	24.42 ± 2.88	22.18 ± 3.10
SBP	117.65 ± 14.11	61.00	199.00	121.23 ± 12.87	112.25 ± 14.19
DBP	74.12 ± 10.04	30.00	134.00	76.55 ± 9.56	70.45 ± 9.63
CHO	188.83 ± 33.02	3.68	382.00	192.04 ± 32.86	183.99 ± 32.68
HDL	52.27 ± 10.70	2.00	118.00	48.80 ± 9.01	57.50 ± 10.92
TG	132.82 ± 83.35	30.00	690.00	155.29 ± 90.44	98.96 ± 56.51
LDL	112.13 ± 31.03	1.00	299.60	115.23 ± 31.25	107.47 ± 30.11
FVC	26.93 ± 30.82	0.04	158.00	25.33 ± 30.39	29.16 ± 31.33
FEV1	28.64 ± 33.96	0.42	178.00	26.58 ± 33.25	31.56 ± 34.81
ALB	4.53 ± 0.25	2.20	5.90	4.58 ± 0.24	4.45 ± 0.24
GLOBULIN	2.76 ± 0.17	1.80	3.80	2.74 ± 0.17	2.78 ± 0.16
AGR	1.64 ± 0.13	0.85	2.39	1.67 ± 0.13	1.59 ± 0.11
BIL	0.87 ± 0.33	0.10	2.93	0.94 ± 0.34	0.75 ± 0.28
DBIL	0.33 ± 0.12	0.04	1.07	0.36 ± 0.12	0.28 ± 0.10
ALP	133.52 ± 50.56	10.00	445.00	142.70 ± 50.77	119.69 ± 46.93
AST	22.69 ± 8.90	1.00	135.00	24.65 ± 9.55	19.75 ± 6.85
ALT	24.43 ± 17.23	1.00	180.00	29.46 ± 18.95	16.87 ± 10.68
GGTP	34.23 ± 32.45	1.90	390.00	44.58 ± 36.75	18.67 ± 14.83
GLU	90.56 ± 15.19	10.00	208.00	92.37 ± 16.18	87.83 ± 13.07
AMYLASE	70.98 ± 19.00	3.00	195.00	69.63 ± 18.65	73.04 ± 19.35
CREAT	0.98 ± 0.18	0.38	2.40	1.09 ± 0.14	0.83 ± 0.12
BUN	13.74 ± 3.25	0.45	59.00	14.50 ± 3.13	12.57 ± 3.07
SG	1.02 ± 0.01	1.00	1.05	1.02 ± 0.01	1.02 ± 0.01
PH	5.68 ± 0.76	0.50	9.00	5.66 ± 0.75	5.72 ± 0.77
PP	43.54 ± 9.64	0.00	101.00	44.69 ± 9.50	41.80 ± 9.59
CCR	92.92 ± 20.02	20.14	213.37	91.66 ± 19.86	94.85 ± 20.12

Events of interest are ten cancer sites that are among the cancers with top incidence and top mortality rates: thyroid, gastric, breast, colorectal, lung, prostate, liver, kidney, uterine-cervical, and lymphoma cancers.

2.3. Project Pipeline

For the raw dataset, observations that are left-censored or contain values that are 6SD away from the mean were excluded (3.3% of the sample). Features with a missing rate of less than 25% were retained and imputed. The time variable was defined by subtracting the year of study enrolment from the year of getting diagnosed with any of the above cancers (event case) or finishing/quitting the study before any event (censored case). The processed data were then fed into each of the five models so that cancer risk and feature importance could be calculated (Figure 1). Our analysis process are released at https://github.com/ngocdung03/nDeep.

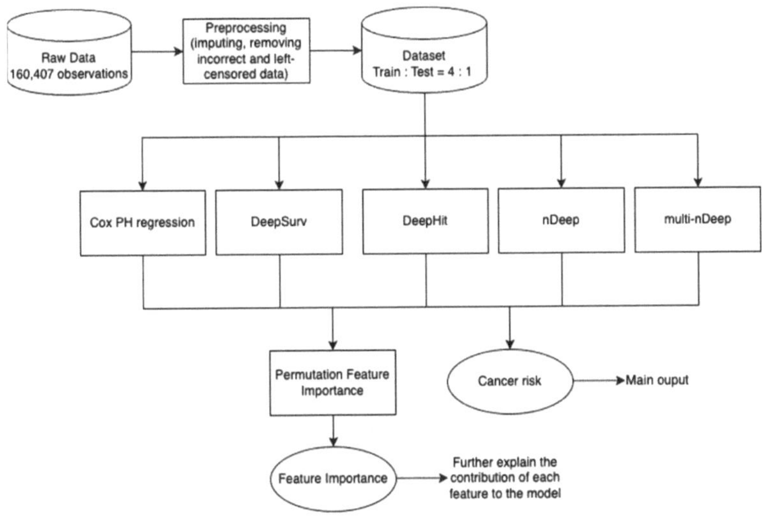

Figure 1. Project pipeline.

2.4. Model Description

Four neural networks and one Cox PH model were included for predictive performance comparison.

2.4.1. Cox PH Regression

Developed in 1972, Cox PH is a semi-parametric regression model for determining the association between predictors (covariates) and the rate of an event happening at a specific time point (hazard rate).

$$h(t) = h_0(t)exp(\beta \times x_i)$$

where,

- t is the survival time
- $h(t)$ is the hazard function
- x_i is the vector of covariates
- $h_0(t)$ is the baseline hazard when all the xi are equal to zero.
- β is the vector of coefficients of x_i

By maximizing this partial likelihood, the vector of β is estimated:

$$L_c(\beta) = \prod_{i:E_i=1} \frac{exp(\beta \times x_i)}{\sum_{j \in R(T_i)} exp(\beta \times x_j)}$$

where,

- E_i, T_i, and x_i are event indicator, survival time, and baseline covariates of the ith observation.
- The likelihood is defined in non-censored observations (E_i = 1).
- $R(t)$ = {i: T_i >= t} is the set of participants at risk of event at time t.

2.4.2. DeepSurv

DeepSurv is a deep neural network (DNN) that outputs the estimate of hazard function $\hat{h_\theta}(x)$ with θ as the weights of the network [9]. Unlike the model of [8], it applied modern techniques for training DNN, including weight decay regularization, batch normalization, Rectified Linear Units (ReLU), dropout, and gradient descent optimization.

2.4.3. DeepHit

DeepHit is a multitask neural network that is formed by shared layer(s) and K task-specific sub-networks (K is the number of tasks). There is a residual connection from input to each task-specific sub-network. Deep residual learning [14] is a technique for overcoming the accuracy degradation when increasing the depth of the network. In addition, with multitask learning, DeepHit can alleviate the competing risk issue, which is common in survival analysis and studies of cancer.

In this study, we performed two multitask learning DeepHit models on these groups of tasks: Group 1—thyroid, gastric, breast, colorectal, and lung cancers; and Group 2—prostate, liver, kidney, uterus-cervical, and lymphoma cancers.

2.4.4. nDeep

This model contains three layers of long short-term memory (LSTM) and one fully connected layer that outputs predicted hazard. LSTM is a type of recurrent network (RNN) that has feedback connections that can address vanishing or exploding gradient issues of RNN [15]. Its architecture transforms survival data into sequences of n time steps, which can then be processed and analyzed. A basic LSTM unit includes a cell, an input gate, an output gate, and a forget gate (Figure 2). The equations for a forward pass of an LSTM cell are as follows:

$$f_t = \sigma_g\left(W_f x_t + U_f h_{t-1} + b_f\right)$$
$$i_t = \sigma_g(W_i x_t + U_i h_{t-1} + b_i)$$
$$o_t = \sigma_g(W_o x_t + U_o h_{t-1} + b_o)$$
$$\tilde{c}_t = \sigma_c(W_c x_t + U_c h_{t-1} + b_c)$$
$$c_t = f_t \odot c_{t-1} + i_t \odot \tilde{c}_t$$
$$h_t = o_t \odot \sigma_h(c_t)$$

where,

- The initial values $c_0 = 0$ and $h_0 = 0$
- $x_t \in R^d$: vector of input
- $f_t \in (0,1)^h$: vector of forget gate's activation
- $i_t \in (0,1)^h$: vector of input/update gate's activation
- $o_t \in (0,1)^h$: vector of output gate's activation
- $h_t \in (-1,1)^h$: vector of hidden state, also output
- $\tilde{c}_t \in (-1,1)^h$: vector of cell input activation
- $c_t \in R^h$: vector of cell state
- $W \in R^{hxd}$, $U \in R^{hxd}$ and $b \in R^h$: vectors of weight matrices and bias
- $\odot$: element-wise product (Hadamard product)

2.4.5. Multi-nDeep

Being inspired by the combination of multitask learning and residual connection, multi-nDeep has a shared block of fully connected layers and K task-specific blocks of nDeep network (Figure 3). The residual is connected from the input to each task-specific block. Models of multitask-nDeep were trained on these groups of tasks: Group 1—thyroid and gastric; Group 2—breast, colorectal, and lung; Group 3—prostate and liver; and Group 4—kidney, uterus-cervical, and lymphoma cancers.

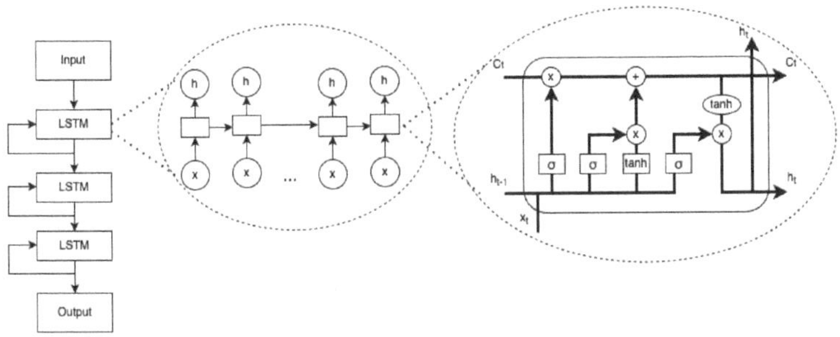

Figure 2. Architecture of nDeep.

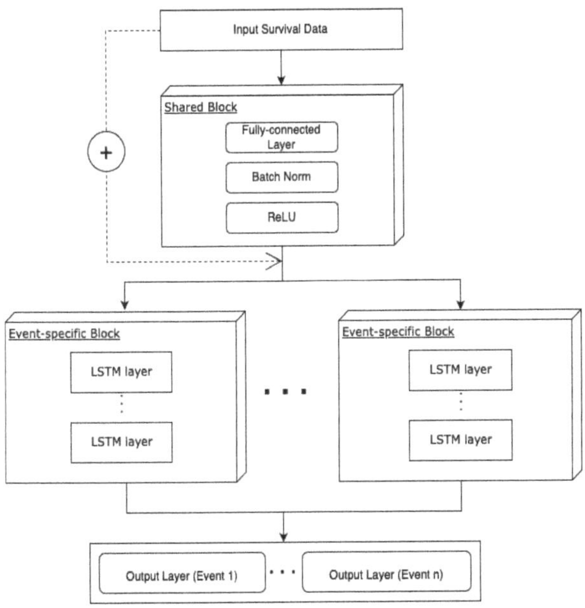

Figure 3. Architecture of mtl-nDeep.

2.4.6. Hyperparameters of DNN Models

Hyperparameters of each DNN model are shown in Supplementary Table S1. Their optimization progress is depicted in Supplementary Figure S1.

2.5. Model Learning

Four networks were involved in two training loops. The outer loop started with the dataset being randomly divided into a train set and a test set (ratio 4:1). In the inner loop, the valid set was further sampled on the train set with the probability of 0.2 so that hyperparameters were optimized by an automatic framework called Optuna with 60 iterations. This framework employs a Tree-structured Parzen Estimator (TPE), a state-of-the-art algorithm based on Bayesian optimization, which is faster and more efficient than grid search (Supplementary Figure S1). The tuned hyperparameters were used for training parameters of each neural network. The inner loop returned a learned model, which was run on the test set 30 times to output predicted cancer hazard risks and averaged c-index (Figure 4).

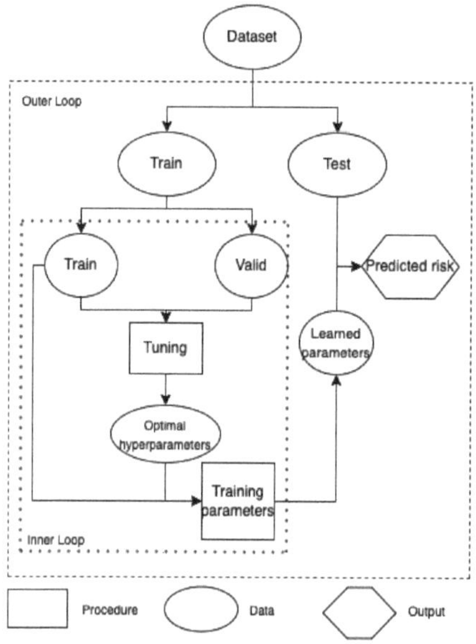

Figure 4. The workflow for optimizing hyperparameters of the models. The inner loop is a process to find the optimal hyperparameters during a training process. Furthermore, the outer loop is a process to evaluate the optimal hyperparameters that were obtained by training with test sets once again.

The loss function used for each DNN is calculated by the sum of L1 + L2 where: L1 is the negative log partial likelihood:

$$L1 = -\sum_{i=1}^{N} I_i \left[\hat{h_\theta}(x_i) - log \sum_{j \in R(t_i)} e^{\hat{h_\theta}(x_j)} \right]$$

where,

- N: number of subjects
- I: indicator function (1: death, 0: otherwise)
- θ: weights of the network
- $\hat{h_\theta}$: estimated hazard function
- x_i, x_j: vectors of covariates
- $R(t_i)$: set of subjects at risk at time t_i

L2 is the ranking loss function:

$$L2 = \sum_{i \neq j} I_{i,j} \cdot \eta\left(\hat{F}\left(s^{(i)} \middle| x^{(i)}\right), \hat{F}\left(s^{(j)} \middle| x^{(j)}\right)\right)$$

where,

- $I_{i,j}$: indicator function of acceptable pairs (i,j) [16]
- $\eta(x,y) = exp\left(\frac{-(x-y)}{\sigma}\right)$: convex loss function
- $\hat{F}$: estimated cumulative incidence function
- s: time
- x: covariate

2.6. Performance Metric

Model accuracy was evaluated by Harrell's concordance index (c-index) [16]. A c-index of 1 represents perfect accuracy. Contrary, a c-index which is 0.5 or less suggests the model performs no better than a naïve model. Five-fold cross-validation was performed 30 times for each task, with the test set being sampled with replacement. The final c-index was achieved by averaging the c-indices of those 30 trials.

2.7. Feature Importance

For each model, permutation feature importance (PFI) was calculated by the module permutation importance of *sklearn* version 1.0.2—a Python library. PFI scores were evaluated based on the c-index. The values of PFI are presented in figures and grouped by a separation line for interpretation. The groups of features represent demographic, metabolic, liver function, pulmonary function, kidney function, and pancreas function.

2.8. Ethical Considerations

Informed consent of participants was received under the support of the Metabolic Syndrome Research Program of the Seoul Metropolitan Government in December 2005. This study was approved by the Yonsei University Institutional Review Board (IRB Approval No. 4-2011-0277).

The current study is approved by the Korea Institute of Bioethics Policy Electromagnetic Concern Committee to be exempt from examination. All methods were used based on ethical regulations and guidelines.

3. Results

Table 2 indicates the number of cases on each cancer site and the corresponding rate over the total number of samples. Cases/total sample rates were small, suggesting an imbalance in data.

Table 2. Number and rate of cancer cases on each site.

Cancer Site	Number of Cases	Cases/Total Cancer Cases	Cases/Total Sample
Thyroid	3385	0.33120	0.02189
Gastric	1499	0.15109	0.00969
Breast	1169	0.11783	0.00756
Colorectum	1031	0.10392	0.00667
Lung	757	0.07630	0.00489
Prostate	697	0.07026	0.00451
Liver	439	0.04425	0.00284
Kidney	352	0.03548	0.00228
Uterus-cervical	306	0.03084	0.00198
Lymphoma	286	0.02883	0.00185

3.1. Comparing c-Index of Each Model

Figure 5 shows the c-index of five models on different tasks of predicting cancer. DeepHit and Cox PH regression models attained the equivalent pattern of c-index. Although c-indices of DeepHit on many cancers are slightly lower as compared to Cox PH, DeepHit, a multitasking model, reached c-index as high as 0.8962 and 0.8822 for prostate and liver cancers. DeepSurv did not indicate any predictive power with the c-index of about 0.5 for every model. On the other hand, nDeep and multi-nDeep achieved the c-indices, which are consistently more than 0.8. Although there was a slight improvement in multi-nDeep as compared to nDeep, the effects were not remarkable. In Supplementary Table S2, the best c-index of each cancer is presented in bold. For reference purpose, c-indices of variants of Cox PH were included in Table S3.

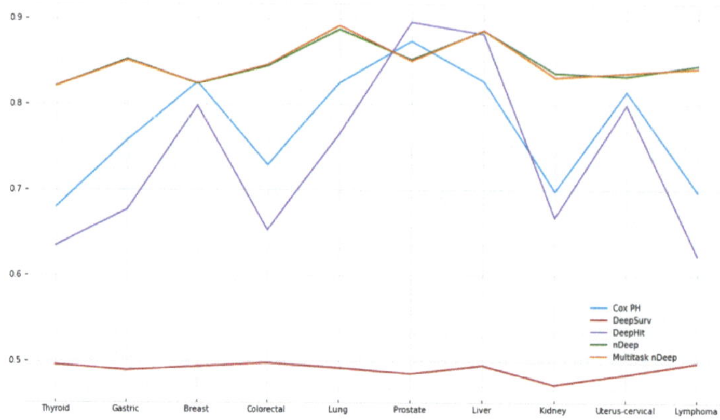

Figure 5. Best c-index of each model. The y-axis represents c-index values.

3.2. Feature Importance

For Cox PH, the most important features are demographic ones (Age and Sex), followed by liver function (Figure 6A). Age is the most decisive factor in the majority of Cox PH models, while sex applies to thyroid cancer, breast cancer, colorectal, and uterus-cervical cancers (particularly, women have a higher risk of these cancers as compared with men in this study). In addition, direct bilirubin, a liver function metric, was also a crucial determinant of thyroid cancer. Metabolic characteristics even showed higher importance than demographic factors for kidney and lymphoma cancers, where waist circumference and weight were the main predictors. Kidney function characteristics showed moderate importance for some cancers (creatinine clearance rate for gastric, prostate, and liver cancers, and blood urea nitrogen for kidney cancer). Amylase, a pancreas function index, and creatinine clearance rate were moderately important for liver cancer.

Similar patterns were observed for DeepHit on gastric, lung, prostate, liver, kidney, and lymphoma cancers, although there were more contributions from factors across categories other than the dominant factor (Figure 6C). For kidney and lymphoma cancer, metabolic characteristics were still the most important but attributed to other metrics (weight and low-density lipoprotein, and triglyceride, respectively). Regarding colorectal cancer, age is the main determinant rather than sex. PFIs, according to thyroid and breast cancers, are relatively equitably distributed across categories. Uterus-cervical cancer is mostly attributed to height.

The figure of feature importance for DeepSurv (Figure 6B) was almost uninformative, with rather random distributions around zero. In Figure 6D,E, with PFI score <0.001, the effect of each feature on nDeep models was minor. Generally, the patterns of multitask-nDeep were similar to nDeep.

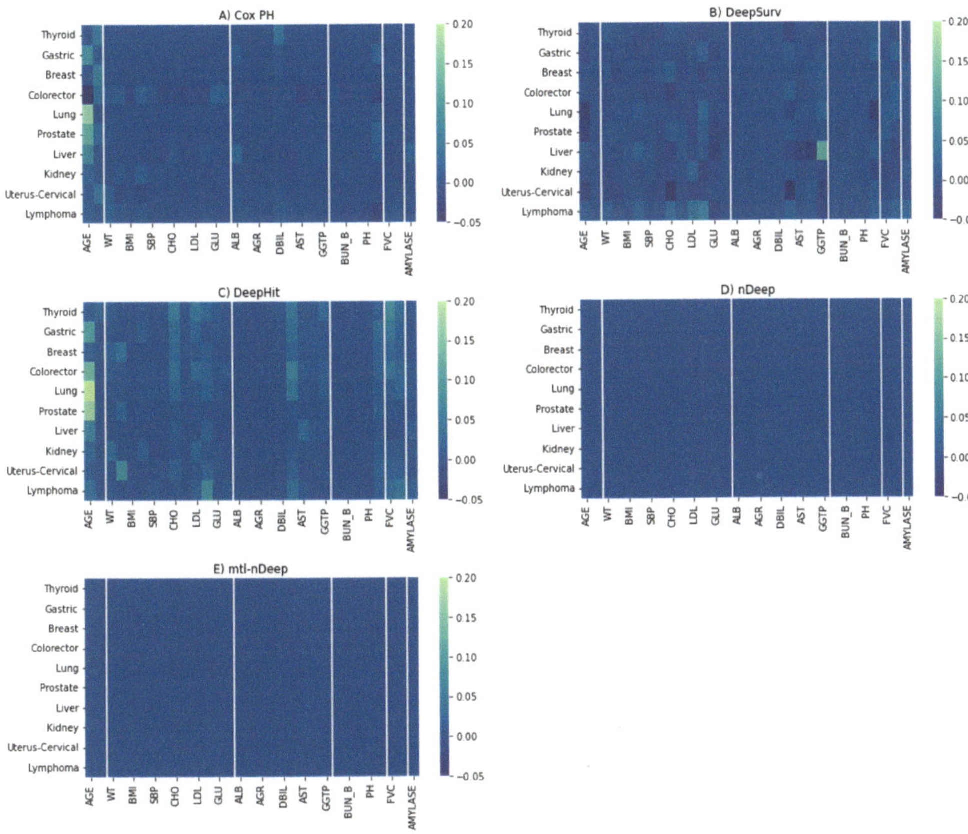

Figure 6. The heatmaps of PFI for five models. Differences in the c-index after removing a feature were represented in a manner where the brighter the color, the more important the feature is. (**A**) Cox PH. (**B**) DeepSurv. (**C**) DeepHit. (**D**) nDeep. (**E**) mtl-nDeep.

4. Discussion

Only two models, nDeep and multitask nDeep, outperformed the traditional Cox PH regression model. On the other hand, DeepSurv exhibited no improvement over a naïve model. This is inconsistent with the results presented in reference [9], where the c-index of DeepSurv can be more than 0.7. Similarly, DeepHit showed comparable c-indices with Cox PH in our study, while its performance was significantly better in generated data [10]. That signifies the disparities in nature between the simulated and real censored data. One of the persistent challenges in follow-up health data is the low rate of uncensored cases, especially when the event is a rare disease such as cancer, which poses significant hurdles for the accuracy of survival analysis [17]. Furthermore, in our study, DeepSurv is the only neural network without residual connections, which could explain its underperformance. In contrast, other than the well-known Cox PH, nDeep emerges as the best model in terms of both accuracy and cost-effectiveness in the current study. It seems that the absence of a particular feature did not hinder the performance of nDeep and its variants.

There are two main factors that affect nDeep performance. Firstly, drawing inspiration from natural language processing, we treated the observed data for a participant and a visit as a single sequence with n time steps and one subsequence with m time steps (where $m < n$), respectively. Secondly, our model architectures incorporated the benefits of residual connections and multiple gates in an LSTM memory cell to effectively capture the dynamic nature of covariates. Moreover, finding the optimal models was eased by applying techniques including TPE algorithm.

The connection between age and cancer is well-established [18]. In fact, our analysis based on Cox PH regression models indicates that age has an impact on nearly all types of cancers. Aging decreases the effectiveness of autophagy, a process responsible for damaged organelles removal, leading to uncontrollable cellular homeostasis and cancer genesis [19]. Some cancers, such as breast and uterus-cervical cancers, are strongly affected by sex, which is expected as these cancers predominantly or exclusively affect females [2,12]. Additionally, our study reveals a higher prevalence of colorectal cancer in women compared to men. This is possibly due to both genetic and environmental factors, including dietary habits [20]. Similarly, the female predominance in thyroid cancer cases can be attributed to factors such as sex hormones, genetics, and the immune system [21–24].

Neural network like DeepHit helps learn the non-linear effects of factors beyond age and sex in cancer risk assessment. It was observed that metabolic metrics, liver function, pulmonary function, and kidney function indices serve as predictors for certain cancer types, including thyroid, gastric, breast, colorectal, kidney, and lymphoma cancers.

Obesity-related metabolic disorders have long been recognized as risk factors for thyroid cancer [25]. Specifically, WC is a strong indicator of the association between overweight/obesity and thyroid cancer [26]. Obesity also highly promotes the progression of various other cancers through mechanisms involving hormones, insulin and insulin-like growth factors, sex steroids, and obese inflammation [27,28]. Increased inflammation in adipose tissue, a consequence of obesity, results in changes in adipokine secretion patterns and the release of pro-inflammatory cytokines, leading to insulin resistance rises in tissues with active metabolism [29]. Overexpression of insulin-like growth factors may have neoplastic effects by promoting cell cycle progression and inhibiting apoptosis, either directly or indirectly, through interaction with established oncogenic systems such as steroid hormones and integrins [30]. Moreover, excess peripheral adipose tissue contributes to steroid aromatization, leading to elevated estrogen levels. Androgens and androgenic antecedents are changed over to estradiol by the chemical aromatase. Within the setting of weight and abundance of fat tissue, aromatase expanded action leads to higher transformation rates coming about in higher levels of estrogens [31]. Estrogen advances tumorigenesis in endometrial tissue by incitement of cell multiplication and hindrance of apoptosis [32].

A recent study demonstrated a link between metabolic dysfunction-associated fatty liver and thyroid cancer, even after adjusting for WC [33]. The relationship between thyroid cancer, a hormone-sensitive cancer, and metabolic dysfunction was found to be modified by menopausal status in females. Higher levels of liver function markers were pointed out to be associated with a reduced risk of colorectal cancer [34]. In the ileum, bilirubin undergoes deconjugation through bacterial and mucosal processes, potentially possessing antioxidant and anti-inflammatory properties [35]. Considering that inflammation and oxidative stress are pivotal contributors to cancer development, this association is noteworthy [36]. The link between pulmonary function and cancer is primarily limited to lung cancer [37]. Likewise, kidney function has been found to be associated with kidney cancer mainly.

This study represents the first attempt to compare different neural network models and traditional models in survival analysis using clinical biomarkers on a large-scale cohort. Our findings align with prior evidence indicating that deep neural networks outperform other methods in healthcare survival analysis. Furthermore, we have clarified the role of different factors in predicting cancer incidence. However, according to the PFI heatmaps, while Cox PH regression heavily relies on information regarding sex and age, cancers appear to be more evenly affected by all the features in our models. Therefore, in the future, we should investigate advanced models capable of distinguishing key factors clearly and incorporate additional biomarkers for more accurate cancer prediction. Additionally, exploring the causal relationship between key factors and specific cancers is crucial, as it can help identify causal factors and facilitate cancer prevention efforts.

5. Conclusions

We have introduced nDeep and its multitasking version for predicting cancer risks based on clinical biomarkers. These algorithms transform health follow-up data into sequences that can be fed into recurrent neural networks, enabling high predictive performance. Our model can assist health care providers in determining people at risk of ten different cancers.

Supplementary Materials: The following supporting information can be downloaded at: https://www.mdpi.com/article/10.3390/cancers15194757/s1, Figure S1: Hyperparameters optimization of 4 models bases on TPE algorithms; Table S1: Hyperparameters of DNN models; Table S2: C-index of each model and imputation method; Table S3: C-index of variants of Cox PH models.

Author Contributions: C.-Y.B. conceptualized the study, designed the research. B.-S.K. performed the literature review and critically revised the article for important intellectual content. S.-H.J. supervised the study. J.-H.L. conducted the data collection and analysis. N.-D.N. interpreted the results and drafted the manuscript. All authors have read and agreed to the published version of the manuscript.

Funding: This research received no external funding.

Institutional Review Board Statement: The current study is approved by the Korea Institute of Bioethics Policy Electro-magnetic Concern Committee to be exempt from examination. All methods were conducted based on ethical regulations and guidelines.

Informed Consent Statement: Informed consents of participants were received under the support of the Metabolic Syndrome Research Program of the Seoul Metropolitan Government in December 2005. This study was approved by the Yonsei University Institutional Review Board (IRB Approval No. 4-2011-0277).

Data Availability Statement: The raw data supporting the conclusions of this article will be made available by the authors, without undue reservation.

Conflicts of Interest: The authors declare no conflict of interest.

References

1. World Health Organization—The Top 10 Causes of Death. Available online: https://www.who.int/news-room/fact-sheets/detail/the-top-10-causes-of-death (accessed on 2 December 2022).
2. Sung, H.; Ferlay, J.; Siegel, R.L.; Laversanne, M.; Soerjomataram, I.; Jemal, A.; Bray, F. Global cancer statistics 2020: {GLOBOCAN} estimates of incidence and mortality worldwide for 36 cancers in 185 countries. *CA A Cancer J. Clin.* **2021**, *71*, 209–249. [CrossRef]
3. Emmert-Streib, F.; Dehmer, M. Introduction to survival analysis in practice. *Mach. Learn. Knowl. Extr.* **2019**, *1*, 1013–1038. [CrossRef]
4. Deepa, P.; Gunavathi, C. A systematic review on machine learning and deep learning techniques in cancer survival prediction. *Biophys. Mol. Biol.* **2022**, *174*, 62–71.
5. Cox, D. Regression models and life-tables. *J. R. Stat. Soc. Ser. B* **1972**, *34*, 187–202. [CrossRef]
6. Ishwaran, H.; Kogalur, U.B. Random survival forests for R. *R News* **2007**, *7*, 25–31.
7. Ishwaran, H.; Kogalur, U.B.; Blackstone, E.H.; Lauer, M.S. Random survival forests. *Ann. Appl. Stat.* **2008**, *2*, 841–860. [CrossRef]
8. Faraggi, D.; Simon, R. A neural network model for survival data. *Stat. Med.* **1995**, *14*, 73–82. [CrossRef] [PubMed]
9. Katzman, J.L.; Shaham, U.; Cloninger, A.; Bates, J.; Jiang, T.; Kluger, Y. DeepSurv: Personalized treatment recommender system using a Cox proportional hazards deep neural network. *BMC Med. Res. Methodol.* **2018**, *18*, 1–12. [CrossRef] [PubMed]
10. Lee, C.; Zame, W.; Yoon, J.; Van Der Schaar, M. Deephit: A deep learning approach to survival analysis with competing risks. In Proceedings of the AAAI Conference on Artificial Intelligence, New Orleans, LA, USA, 2–7 February 2018; Volume 32.
11. Kim, S.; Song, H.; Kim, S.; Kim, B.; Lee, J.-G. Revisit Prediction by Deep Survival Analysis. In Proceedings of the Pacific-Asia Conference on Knowledge Discovery and Data Mining, Chengdu, China, 16–19 May 2022; pp. 514–526.
12. Wild, C.; Weiderpass, E.; Stewart, B. *World Cancer Report: Cancer Research for Cancer Prevention*; International Agency for Research on Cancer: Lyon, France, 2020.
13. Jee, Y.H.; Emberson, J.; Jung, K.J.; Lee, S.J.; Lee, S.; Back, J.H.; Hong, S.; Kimm, H.; Sherliker, P.; Jee, S.H.; et al. Cohort profile: The korean cancer prevention study-II (KCPS-II) Biobank. *Int. J. Epidemiol.* **2018**, *47*, 385–386f. [CrossRef]
14. He, K.; Zhang, X.; Ren, S.; Sun, J. Deep residual learning for image recognition. In Proceedings of the IEEE Conference on Computer Vision and Pattern Recognition, Las Vegas, NV, USA, 27–30 June 2016; pp. 770–778.
15. Hochreiter, S.; Schmidhuber, J. Long short-term memory. *Neural Comput.* **1997**, *9*, 1735–1780. [CrossRef]
16. Harrell Jr, F.E.; Lee, K.L.; Califf, R.M.; Pryor, D.B.; Rosati, R.A. Regression modelling strategies for improved prognostic prediction. *Stat. Med.* **1984**, *3*, 143–152. [CrossRef] [PubMed]

17. Maalouf, M.; Trafalis, T.B. Rare events and imbalanced datasets: An overview. *Int. J. Data Min. Model. Manag.* **2011**, *3*, 375–388. [CrossRef]
18. SEER*Explorer SEER Incidence Rates by Age at Diagnosis, 2015–2019. Available online: https://seer.cancer.gov/statistics-network/explorer/application.html?site=1&data_type=1&graph_type=3&compareBy=sex&chk_sex_3=3&chk_sex_2=2&rate_type=2&race=1&advopt_precision=1&advopt_show_ci=on&hdn_view=0#graphArea (accessed on 2 December 2022).
19. Tabibzadeh, S. Role of autophagy in aging: The good, the bad, and the ugly. *Aging Cell* **2022**, e13753. [CrossRef] [PubMed]
20. Kim, S.-E.; Paik, H.Y.; Yoon, H.; Lee, J.E.; Kim, N.; Sung, M.-K. Sex-and gender-specific disparities in colorectal cancer risk. *World J. Gastroenterol. WJG* **2015**, *21*, 5167. [CrossRef]
21. Suteau, V.; Munier, M.; Briet, C.; Rodien, P. Sex Bias in Differentiated Thyroid Cancer. *Int. J. Mol. Sci.* **2021**, *22*, 12992. [CrossRef]
22. Shobab, L.; Burman, K.D.; Wartofsky, L. Sex Differences in Differentiated Thyroid Cancer. *Thyroid* **2022**, *32*, 224–235. [CrossRef]
23. Rahbari, R.; Zhang, L.; Kebebew, E. Thyroid cancer gender disparity. *Future Oncol.* **2010**, *6*, 1771–1779. [CrossRef]
24. Morganti, S.; Ceda, G.; Saccani, M.; Milli, B.; Ugolotti, D.; Prampolini, R.; Maggio, M.; Valenti, G.; Ceresini, G. Thyroid disease in the elderly: Sex-related differences in clinical expression. *J. Endocrinol. Investig.* **2005**, *28*, 101–104.
25. Yin, D.-T.; He, H.; Yu, K.; Xie, J.; Lei, M.; Ma, R.; Li, H.; Wang, Y.; Liu, Z. The association between thyroid cancer and insulin resistance, metabolic syndrome and its components: A systematic review and meta-analysis. *Int. J. Surg.* **2018**, *57*, 66–75. [CrossRef]
26. Nguyen, D.N.; Kim, J.H.; Kim, M.K. Association of Metabolic Health and Central Obesity with the Risk of Thyroid Cancer: Data from the Korean Genome and Epidemiology Study. *Cancer Epidemiol. Biomark. Prev.* **2022**, *31*, 543–553. [CrossRef]
27. Osorio-Costa, F.; Rocha, G.Z.; Dias, M.M.; Carvalheira, J.B. Epidemiological and molecular mechanisms aspects linking obesity and cancer. *Arq. Bras. Endocrinol. Metabol.* **2009**, *53*, 213–226. [PubMed]
28. Avgerinos, K.I.; Spyrou, N.; Mantzoros, C.S.; Dalamaga, M. Obesity and cancer risk: Emerging biological mechanisms and perspectives. *Metabolism* **2019**, *92*, 121–135. [PubMed]
29. Gallagher, E.J.; LeRoith, D. Obesity and diabetes: The increased risk of cancer and cancer-related mortality. *Physiol. Rev.* **2015**, *95*, 727–748. [PubMed]
30. Moschos, S.J.; Mantzoros, C.S. The role of the IGF system in cancer: From basic to clinical studies and clinical applications. *Oncology* **2002**, *63*, 317–332. [CrossRef]
31. Crosbie, E.J.; Zwahlen, M.; Kitchener, H.C.; Egger, M.; Renehan, A.G. Body Mass Index, Hormone Replacement Therapy, and Endometrial Cancer Risk: A Meta-Analysis. *Cancer Epidemiol. Biomark. Prev.* **2010**, *19*, 3119–3130.
32. Shaw, E.; Farris, M.; McNeil, J.; Friedenreich, C. Obesity and endometrial cancer. *Obes. Cancer* **2016**, 107–136.
33. Liu, Z.; Lin, C.; Suo, C.; Zhao, R.; Jin, L.; Zhang, T.; Chen, X. Metabolic dysfunction-associated fatty liver disease and the risk of 24 specific cancers. *Metabolism* **2022**, *127*, 154955. [CrossRef]
34. He, M.-M.; Fang, Z.; Hang, D.; Wang, F.; Polychronidis, G.; Wang, L.; Lo, C.-H.; Wang, K.; Zhong, R.; Knudsen, M.D.; et al. Circulating liver function markers and colorectal cancer risk: A prospective cohort study in the UK Biobank. *Int. J. Cancer* **2021**, *148*, 1867–1878. [CrossRef]
35. Stocker, R.; Yamamoto, Y.; McDonagh, A.F.; Glazer, A.N.; Ames, B.N. Bilirubin is an antioxidant of possible physiological importance. *Science* **1987**, *235*, 1043–1046. [CrossRef]
36. Horsfall, L.J.; Rait, G.; Walters, K.; Swallow, D.M.; Pereira, S.P.; Nazareth, I.; Petersen, I. Serum bilirubin and risk of respiratory disease and death. *JAMA* **2011**, *305*, 691–697. [CrossRef]
37. Sarna, L.; Evangelista, L.; Tashkin, D.; Padilla, G.; Holmes, C.; Brecht, M.L.; Grannis, F. Impact of respiratory symptoms and pulmonary function on quality of life of long-term survivors of non-small cell lung cancer. *Chest* **2004**, *125*, 439–445. [CrossRef] [PubMed]

remote sensing

MDPI

Article

HFCC-Net: A Dual-Branch Hybrid Framework of CNN and CapsNet for Land-Use Scene Classification

Ningbo Guo [1], Mingyong Jiang [1,*], Lijing Gao [2], Kaitao Li [1], Fengjie Zheng [1], Xiangning Chen [1] and Mingdong Wang [1]

[1] Space Information Academic, Space Engineering University, Beijing 101407, China; sxguonb@163.com (N.G.)
[2] State Key Laboratory of Remote Sensing Science, Aerospace Information Research Institute, Chinese Academy of Sciences, Beijing 100101, China; gaolj200869@aircas.ac.cn
* Correspondence: jiangmingyong2010@163.com; Tel.: +86-176-8325-3692

Abstract: Land-use scene classification (LUSC) is a key technique in the field of remote sensing imagery (RSI) interpretation. A convolutional neural network (CNN) is widely used for its ability to autonomously and efficiently extract deep semantic feature maps (DSFMs) from large-scale RSI data. However, CNNs cannot accurately extract the rich spatial structure information of RSI, and the key information of RSI is easily lost due to many pooling layers, so it is difficult to ensure the information integrity of the spatial structure feature maps (SSFMs) and DSFMs of RSI with CNNs only for LUSC, which can easily affect the classification performance. To fully utilize the SSFMs and make up for the insufficiency of CNN in capturing the relationship information between the land-use objects of RSI, while reducing the loss of important information, we propose an effective dual-branch hybrid framework, HFCC-Net, for the LUSC task. The CNN in the upper branch extracts multi-scale DSFMs of the same scene using transfer learning techniques; the graph routing-based CapsNet in the lower branch is used to obtain SSFMs from DSFMs in different scales, and element-by-element summation achieves enhanced representations of SSFMs; a newly designed function is used to fuse the top-level DSFMs with SSFMs to generate discriminant feature maps (DFMs); and, finally, the DFMs are fed into classifier. We conducted sufficient experiments using HFCC-Net on four public datasets. The results show that our method has better classification performance compared to some existing CNN-based state-of-the-art methods.

Keywords: remote sensing; scene classification; image interpretation

1. Introduction

As remote sensing technology continues to develop, LUSC plays an increasingly important role in the field of remote sensing [1]. For instance, the support of LUSC is needed in the fields of ecological environment construction, urban construction planning, disaster analysis, and disaster relief resource dispatching [2,3]. Most importantly, with the continuous improvement of the decision-making system and management level in various fields of the country, people's needs and requirements for surface target information are becoming higher and higher, and the technology of LUSC has inevitably become a key technical subject of wide concern in the field of remote sensing [4,5].

In land-use scene images (LUSIs), there is a large difference between images of the same category and a high degree of similarity between images of different categories. As shown in Figure 1, the semantic expressions of images in the categories of "Resident" and "Parking" in the RSSCN7 dataset have a high degree of similarity, and those of "Forest" and "Mountain" in the UCM dataset do; moreover, the semantic expressions of images in the categories of "Medium Residential" in the OPTIMAL dataset and "Park" in the SIRI dataset have a high degree of difference. In addition, affected by the diversity, multi-resolution, and complex spatial content distribution of LUSIs, the LUSC task of automation and high accuracy is, subsequently, full of uncertainties. Therefore, it has become a challenging task to adequately express the detailed features of different land-use category images and achieve high accuracy classification results.

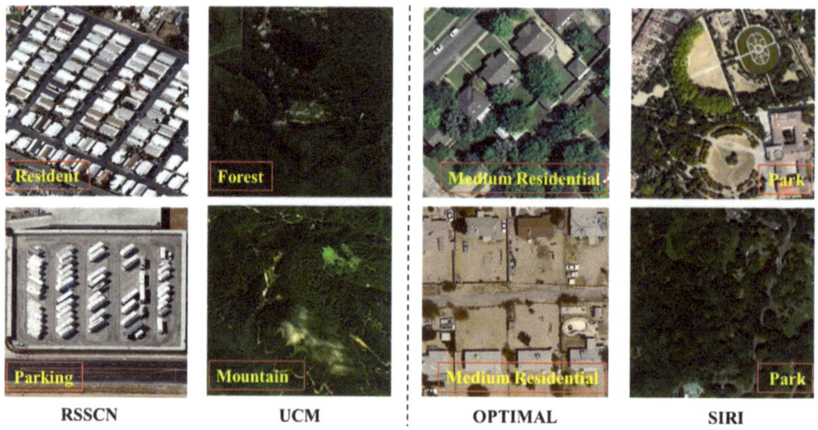

Figure 1. Schematic of land-use scene confusion.

The traditional technique is to use manual design of features [6]. One class is pixel-based classification: each pixel is considered as the smallest classification unit, and the spectral, morphological, textural, and spatial information of the pixel is first extracted, from which the features that can represent the different classes are then selected and classified using different classifiers. This type of approach is susceptible to image local heterogeneity and noise in high-resolution images [7]. The other category is object-based classification: segmentation based on a specific target in the image and its spatial structural relationship, combining multiple pixel points into an object with similar features, so as to achieve the purpose of feature extraction, analysis, and classification. In such methods, the classification results are limited by the goodness of the segmentation results [8]. Moreover, traditional techniques need to rely on experts' empirical analyses and a lot of manually designed features, which are not only time-consuming and labor-intensive, but also easy to fall into local extremes [9]. These characteristics lead to the fact that when using traditional methods to deal with the task of classifying a large number of LUSIs based on RSI data, not only is the efficiency of the classification is not high, but also the accuracy of the classification cannot be stable all the time.

The latest techniques for LUSC are based on deep learning methods, which automatically learn image features with multiple characteristics through the network, and use a large number of labeled samples to quickly fit an image classification model [10]. Owing to the ability to automatically learn image features and the high classification accuracy, CNN-based methods have become the most popular methods for LUSC. First, a convolutional layer is used to filter the input image with multiple channels to capture information such as different textures, shapes, and colors in the image; second, the pooling layer is used to reduce the size and computation of the feature map and improve the computational efficiency; and last, the fully connected layer is used to map the features output from the pooling layer to the corresponding category labels to achieve the goal of LUSC. To prevent overfitting and improve the generalization ability of the model, the method often employs some regularization techniques such as dropout and batch normalization, which can help reduce the number of parameters in the network, improve the robustness of the model, and make the network easier to train and optimize. However, during the process of pooling RSIs, not only is a lot of effective information lost, but also the spatial resolution of the image is reduced; moreover, the noise on the image and the effect of image rotation and scaling due to the change of viewpoints have a relatively large impact on the accuracy of target category interpretation. All these are also major issues that affect the further development of CNN-based techniques in LUSC [11].

CapsNet was proposed by Hinton et al. in 2017 [12], and was mainly proposed to address the limitations of traditional CNNs in tasks such as pose estimation, and its core idea is to replace neurons in traditional neural networks with capsules, and to capture spatial relationships and hierarchical structures between objects by introducing dynamic routing algorithms, so as to improve the robustness and generalization ability of the model. In recent years, it has been successfully applied to tasks such as brain tumor image classification [13], gait recognition [14], iris recognition [15], and LUSC [16], with good results. Compared with CNN-based classification methods, CapsNet-based methods are more expressive in feature extraction and representation of images, which is manifested in three aspects: 1. Pose invariance: it can better learn and express the pose information in the input data, which enables it to have a significant advantage in recognizing objects with different poses; 2. Hierarchical representation: features are represented as vectors in the capsule through a dynamic routing algorithm, and this hierarchical representation can better capture the spatial relationship and contextual information between features; 3. Robustness: with a certain degree of robustness, it can better cope with the problems of noise, deformation, and interference in the input data [17]. Although CapsNets have shown potential advantages in a number of tasks, some shortcomings still exist at present. For example, the introduction of a large number of dynamic routing operations has led to an increase in the complexity of the network and slower training and inference, and the current architecture is still relatively simple, which needs to be further optimized and improved to accommodate larger image data [18].

For the purpose of further improving the classification accuracy of LUSIs, especially for RSIs with relatively high image resolution, large heterogeneity of image content, and certain noise, features extraction and interpretation ability of commonly used network models on LUSIs need to be further improved, which requires the use of the spatial structure information of RSIs in the meantime, and, thus, reduces the loss of important feature maps in the process of DSFM extraction and the issue of category misclassification during scene classification. Therefore, we propose to design a novel scene classification model, called HFCC-Net, which combines the strengths of image object location and pose perception based on the CapsNet approach together with the advantages of DSFM extraction capability of the CNN-based method, to provide richer feature representations and more accurate scene category determination results for the LUSC task. The main contributions are as follows:

1. We designed a novel LUSC method named HFCC-Net with hybrid CapsNet and CNN, which could fully utilize the global semantic information and local spatial structure information of RSIs to effectively represent global and local feature maps, and obtained competitive level of classification compared with the advanced ones;
2. We propose an algorithm for generating discriminative feature maps, which could fuse the global DSFMs obtained through CNN and the local SSFMs obtained by the graph routing-based CapsNet, and not only achieves the maximization of the information content for feature representation, but also improves the classification accuracy;
3. We conducted full experiments on four typical LUSI datasets, and not only successfully verified the advancement of HFCC-Net, but also analyzed in detail the main factors affecting the classification accuracy.

The rest of the paper is organized as follows: in Section 2, a description of the relevant studies is given; Section 3 presents the data and evaluation indicators; Section 4 discusses the proposed methodology; Section 5 conducts experiments and discusses the factors influencing; and finally, conclusions are drawn in Section 6.

2. Related Work

2.1. CapsNet-Based for Classification

In recent years, researchers have conducted extensive research and in-depth exploration of CapsNet-based image classification techniques, and the related results are mainly reflected in three aspects. First, the routing algorithm is improved. Sabour et al. [12] proposed an adaptive routing algorithm, which regards the number of routing times as the hyper-parameters of the model and dynamically adjusts the number of routing times according to the performance in the training process, so that the model can better learn the distributional characteristics of the data and improve the classification accuracy. Hinton et al. [19] use a 4×4 matrix to express the pose parameters of the object and switch to an expectation–maximization routing algorithm, which reduces the transformation matrix and reduces the training parameters and computational effort. Li et al. [20] treat the capsules in each layer as nodes of a graph and use bidirectional graph routing to learn the internal relationships between capsules in the same layer. Second, the architectural design of capsule network is improved. Tao et al. [21] propose using an adaptive capsule layer instead of the main capsule layer of CapsNet, thus, effectively utilizing the potential spatial relationships between capsule vectors and, thus, improving the classification accuracy. Phaye et al. [22] designed a dense CapsNet using dense layers instead of convolutional layers, and a diverse CapsNet using a hierarchical architecture to learn capsules, both with improved model performance. Xiong et al. [23] added a convolutional capsule layer and a capsule pooling layer to the original network, which reduced the number of model parameters and improved the experimental efficiency. Jia et al. [24] constructed multi-scale master capsules using residual convolutional layers and positional dot products, and used a sigmoid function to determine the weight coefficients between capsules, achieving better recognition performance. Zhou et al. [25] utilize twin capsule networks to solve the problem of information loss in the processing of remote sensing images by convolutional networks. Third, capsule attention mechanism is introduced. Hoogi et al. [26] added a capsule attention mechanism, which can adaptively focus on the important features in the image, enhance the important capsule at the same time, can inhibit the influence of irrelevant capsule, and improve the accuracy of the model. Gu et al. [27] replace the original dynamic routing method with an image pooling approach using multiple heads of attention, which not only makes the model computationally smaller, but also has better classification performance and adversarial robustness. Yu et al. [28] add an attention module for channel and spatial feature calibration, which results in higher quality target features learned by the network, sharper semantic information, and further improvement in the final classification accuracy. Although the above methods are able to achieve an improvement in image classification efficiency and accuracy, these methods are basically applied to image datasets with small samples, and have not been attempted in LUSI datasets. Furthermore, the architecture of CapsNet is still relatively simple, which makes it difficult to achieve better results in the classification of higher resolution and larger size RSIs for the time being.

2.2. CNN-Based for Classification

Among the latest scene classification approaches, CNN has become a classical neural network for processing image structured data. According to the means of classification feature map extraction, CNN-based classification methods can be classified into two categories: pure convolutional classification methods and classification methods in which convolution is fused with other different networks. For example, Xia et al. [29] processed RSI data using VGG network, GoogLeNet network, and AlexNet network, and achieved a high classification accuracy with the participation of pre-trained models. Sun et al. [30] fuse the DSFMs extracted from different convolutional layers of VGG-16 and select the best combination, which realizes the full utilization of complementary information between multi-layer DSFMs. Yu et al. [31] utilize two CNN networks to extract the original image and significantly enhanced image, respectively, and finally the two feature maps are fused for classification. Zhang et al. [32] enhance the model classification rate by adding channels and spatial attention to MobileNet V2. Yang et al. [33] propose the use of twinned CNNs for scene classification. Anwer et al. [34] propose using the ResNet network to extract the RGB features and texture encoding features of the model, and realize feature fusion by constructing a two-branch feature extraction architecture, and then achieve the effect of classification. Gao et al. [35] augmented the DSFMs extracted by CNN using the attention mechanism in the channel and spatial branches, and used the augmented fused features for scene classification. Liu et al. [36] improve classification performance by fusing different layers of feature maps from a single CNN. Wu et al. [37] build on the research of CNNs and propose stacking multiple columns of encoders for classification. Although methods such as combining different DSFMs or increasing the attention mechanism can reach a high scene classification accuracy, they do not solve the loss of information problem in the process of RSI feature map learning. In addition, researchers have proposed fusing CNN with other different networks in the hope of achieving complementary use of the strengths of different networks. For instance, Wang et al. [38] combine CNN and LSTM and add the attention mechanism, which speeds up the convergence and improves the accuracy. Zhang et al. [16] utilize VGG-16 and Inception-v3 in tandem with CapsNet to form a new network for classification, and achieve some results, but the convolutional network at the front of the whole network is still lossy, resulting in insufficient classification accuracy. Peng et al. [39] combine GNN and CNN to construct the spatial and topological relations of RSIs, and the classification accuracy reaches an advanced level. Obviously, this type of method can achieve higher classification accuracy, but it also increases the computational amount of feature learning, and in addition to the complexity of the combined model being higher, more importantly, it does not solve the loss messages of the feature maps and classification errors existing in the CNN itself.

2.3. Fusions for Classification

Along with the continuous in-depth study of the LUSC task, researchers began to try to use CapsNet to make up for the shortcomings of CNN-based techniques in the image classification task, and gradually achieved better classification results. Divided according to the manner in which these two network frameworks are combined, this fusion scene classification method can be divided into two types: one is fusion in the form of a sequence. The CNN first performs feature extraction on the input image and then feeds the extracted features into CapsNet for further processing. This approach allows the model to capture more complex patterns with CapsNet on top of the CNN. For example, Xu et al. [14] implement a network framework for gait image recognition by connecting convolutional and capsule modules in series. Phaye et al. [40] combine DenseNet and CapsNets to formulate better master capsules, creating an efficient recognition framework. Xiang et al. [41] use CFMs of different scales as inputs to the CapsNet to learn features and obtain a rich representation to achieve high recognition accuracy. Jampour et al. [42] use the deep features extracted from the second residual block of ResNet as the input data to CapsNet to obtain recognition results. Wang et al. [43] use ResNet to extract LiDAR data features and then use CapsNet for recognition, which solves the problem of information loss in ResNet network to some extent. Yousra et al. [44] extract features from the VGG19 model trained on the ImageNet dataset and input them into the newly designed CapsNet to obtain recognition results. In contrast to the above approach, Zhang et al. [45] input the features acquired through the CapsNet into MobileNetV2, which achieves the lightweight and accurate recognition requirements. This method is still based on CapsNet in essence, the difference is that the original input image data are replaced by the DSFM after the convolution operation; although it achieves certain image classification requirements, in the process of convolution of the image, part of the information is also lost. Another is fusion in a parallel form. The CNN and CapsNet are connected in parallel and each processes the input image and then the respective outputs are merged and classified. This approach can simultaneously utilize the excellent feature extraction capabilities of CNN and the powerful structure-aware capabilities of CapsNet. For example, Wang et al. [46] design a dual-channel network framework that fuses CNN and CapsNet, extracts convolutional and capsule features simultaneously and separately for the input data, then fuses the two features to form new classification features with more discriminative information, and finally inputs the fused features into a classifier to achieve classification. This type of method utilizes only single-scale DSFMs of the input image for the feature learning process in the CapsNet branch, and the two-branch feature fusion approach is simple and prone to produce redundant features and, furthermore, there is no improvement to CapsNet and CNN.

3. Materials

3.1. LUSI Datasets for LUSC

We used four typical LUSI datasets to conduct experiments and validate our proposed algorithm about scene classification. The specific parameters of the datasets are shown in Table 1, and Figure 2 illustrates the samples for each land-use category.

Table 1. Relevant parameters of the LUSI datasets.

Land Use Datasets	RSSCN	SIRI	UCM	OPTIMAL
Category number	7	12	21	31
Number per category	400	200	100	60
Total number	2800	2400	2100	1860
Image size	256 × 256	200 × 200	256 × 256	256 × 256
Source	Google Earth	GF-1 et al.	USGS	Google Earth

(1) RSSCN [47]: The dataset contains images from seven different scenes with high spatial resolution, and the images of each scene come from different viewpoints, lighting conditions, and seasonal variations, which not only provide more detailed information about the features, but also simulate the real RSI scenes;

(2) SIRI [48]: The dataset is organized from data acquired by satellite sensors such as GF-1 and GF-2, as well as a number of global and domestic open datasets. These images cover images from different geographic locations, light conditions and seasons, including farmland, forests, cities, etc., which can provide diversity and richness of scene types and help to carry out refined RSI processing and analysis;

(3) UCM [49]: The dataset provides images with high spatial resolution, which can provide more detailed and clear RSI information, which is very important for conducting refined RSI processing and analysis. Moreover, both the categories of geographic scenes and the size of the dataset satisfy the need to utilize experiments to validate the algorithms;

(4) OPTIMAL [38]: The dataset is a multimodal RSI dataset that contains many types of remote sensing data, such as optical, radar and infrared images. This enables the dataset to reflect the diversity and richness of RSI data more comprehensively. In addition, the images in it not only cover rich geographic scenes, but also have a high data scale.

3.2. Evaluation Metrics for LUSC

(1) Scene classification accuracy (SCA): It is an intuitive evaluation metric, which is a benchmark for comparing the performance of different classification algorithms or different datasets, and can directly tell us how well the algorithms classify the whole dataset. Assuming that land-use category of $scene_i$ in the categorized dataset is $prediciton_i$, land-use category kind is $(0, 1, 2, \cdots, j)$, and the function to predict the category of LUSI is f, the SCA is calculated as follows:

$$SCA = \frac{\sum_{i=0}^{j+1} (f(scene_i) = prediction_i)}{j+1} \tag{1}$$

Figure 2. *Cont.*

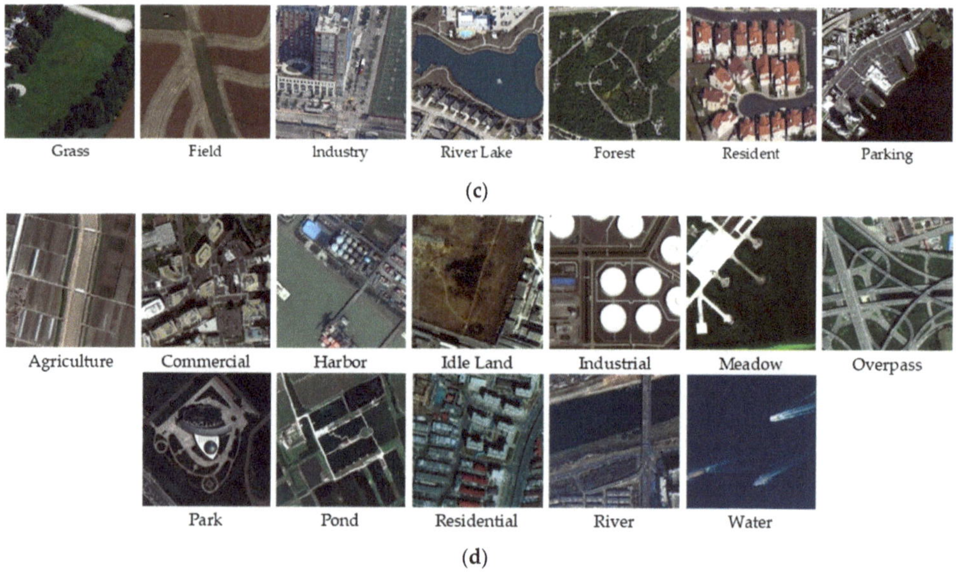

Figure 2. Instances of the scenes within LUSI datasets. (**a**) Sample examples of the UCM; (**b**) sample examples of the OPTIMAL; (**c**) sample examples of the RSSCN; (**d**) sample examples of the SIRI.

(2) Confusion matrix (CM): It can reflect the mutual misclassification probability between images of various categories. Formally, it is a two-dimensional matrix, with rows and columns representing the true and predicted labels of the images, respectively, and the values of the elements on the diagonal lines indicate the probability that the samples of the corresponding categories are correctly classified, while the values of the elements on the off-diagonal lines indicate the probability that the samples with the true labels are misclassified.

Shown in Table 2, for a scene classification task containing j categories, the CM is a $j \times j$ matrix, where the i-th row and j-th column denote the probability that a target of category i is classified into category j. We show the CM obtained by the best classifier, and, as can be seen, the probability values on the diagonal line are the largest.

Table 2. Schematic representation of the CM.

Probability Value		**True Category Labels**				
		$scene_0$	$scene_1$	...	$scene_j$	$scene_{j+1}$
Predicted Category Labels	$scene_0$	1	0	0	0	0
	$scene_1$	0	1	0	0	0
	...	0	0	1	0	0
	$scene_j$	0	0	0	1	0
	$scene_{j+1}$	0	0	0	0	1

4. Methodology

4.1. Overall Framework of HFCC-Net

HFCC-Net is a novel scene classifier with a two-branch network architecture (see Figure 3), which can classify the deep semantic and spatial structure features extracted by CNN and CapsNet at the same time after fusion, and has an outstanding performance in dealing with LUSC with complex background contents. The overall structure of HFCC-Net consists of three parts: the DSFM extraction branch built by the transfer-learning-based CNN, the SSFM extraction branch built by the graph routing-based CapsNet, and the DFM and prediction module generated by fusing the two branches.

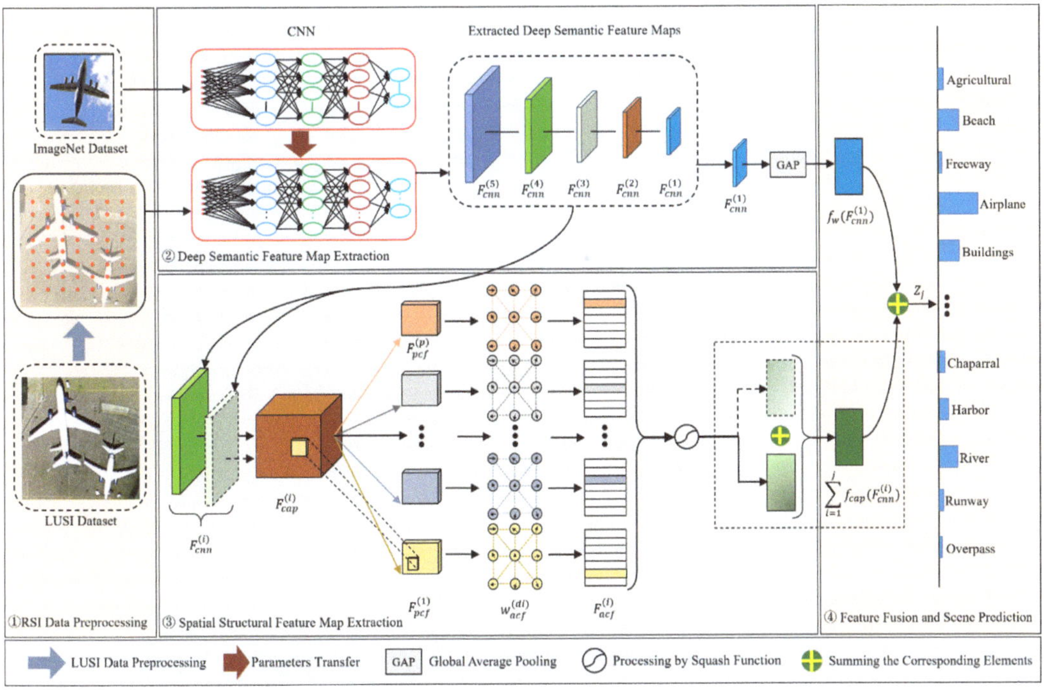

Figure 3. The structure of HFCC-Net.

Before inputting the RSI data into the network, the data are first processed to be classified by the model. That is, it mainly consists of three steps: the first step is to crop the image to match the input size of the model (we set the size of the image as 224 × 224); the second step is to decode the image data into tensor data; and the third step is to convert the pixel values into floating point data type and perform normalization.

In HFCC-Net, considering that the training cost of utilizing CapsNet is higher than utilizing CNN, for example, utilizing CapsNet requires more computational resources and training time when dealing with a large number of high-resolution RSIs, in order to achieve the task of model training faster and more accurately, instead of choosing to utilize the original image data directly for the computation, we chose to use the DSFMs extracted by CNNs in the upper branch as the input data to the lower branch of the CapsNet. Inspired by the literature [27], we integrated the graph structure into the CapsNet in the lower branch, and utilized the multi-head attention graph-based pooling operation to replace the routing operation of the traditional method, which further reduced the training volume of HFCC-Net. Meanwhile, we utilized a transfer learning technique in computing DSFM, using pre-trained models from the ImageNet dataset.

In addition, we designed a new feature fusion algorithm to generate the DFM for improving the utilization efficiency of DSFM and SSFM. Finally, the DFM was fed into the classifier consisting of the SoftMax function to obtain the predicted probabilities.

4.2. Deep Semantic Feature Map Extraction

The deep semantic feature extraction module of the upper branch is based on CNN. To extract the rich semantic information of the LUSI data, a deep CNN, which is Xception [50], was used. As is known, CNN is used to extract semantic information by performing convolutional operations on the data obtained from the input layer. The traditional convolution is that the convolution kernel traverses the input data according to the step size, and each traversal multiplies the input value with the corresponding position of the convolution kernel and then adds the operation, and the feature matrix is obtained after traversal in turn. However, Xception is different. As shown in Figure 4, it designs the traditional convolution as a deeply separable convolution, i.e., point-by-point convolution and deep convolution. First, a convolution kernel of size 1 × 1 is used for point-by-point convolution to reduce the computational complexity; then, a 3 × 3 deep convolution is applied to disassemble and reorganize the feature map; simultaneously, ReLUs are not added to ensure that data are not corrupted. The whole network structure is stacked by multiple deep separable convolutions.

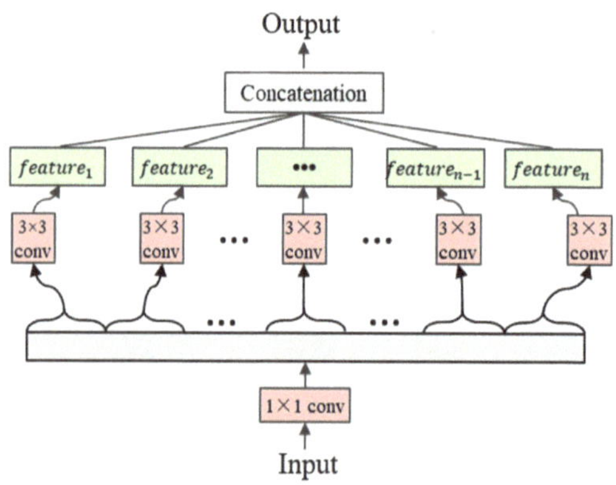

Figure 4. Examples of deeply separable convolution.

Compared to the traditional structure, Xception has four advantages: first, it uses cross-layer connections in the deep network to avoid the problem of degradation of the deep network, to make the network easier to train, and to improve the accuracy of the network; second, it uses the maximum pooling layer, which reduces the loss of information and improves the accuracy of the network; third, it uses the depth-separable convolutional layer, which reduces the number of parameters and improves the computational efficiency; and fourth, different sized images are used for training during the training process, which improves the robustness of the network.

Figure 5 shows a schematic of the flow of Xception to extract deep semantic features from the input data. The size of each DSFM is set to $h \times w \times d$, where h denotes the height, w donates the width of the DSFM, and d denotes the channel dimension of the DSFM. Furthermore, we modified the last two layers of the Xception to use the new global average pooling (GAP) and fully connected layer (FCL).

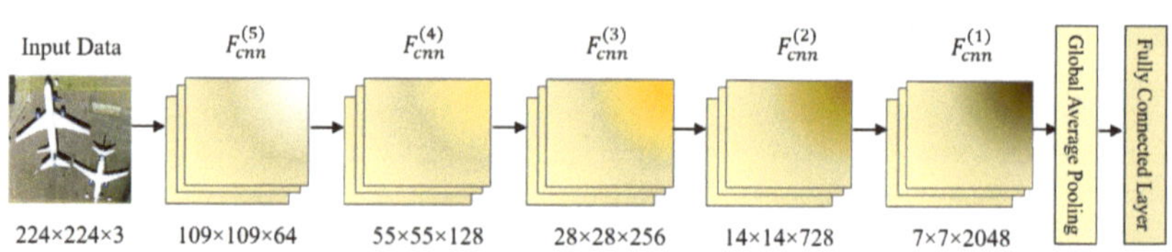

Figure 5. DSFMs, modified from Xception.

In the structure of Xception, the convolutional layer and the pooling layer generally appear at the same time. There are two main approaches to pooling: maximum pooling and average pooling. Maximum pooling retains the maximum value of each region as the result of the calculation, while average pooling is calculated by calculating the average value of each region as the result of the calculation. As shown in Figure 6, the 4 × 4 convolutional feature traversal is computed using the 2 × 2 filter according to the size of the step size of 2. Retaining the maximum value gives the maximum pooled feature map, and retaining the average value gives the average pooled feature map. Xception usually chooses a filter of size 3 × 3, traversed in 2 steps, to achieve maximum pooling.

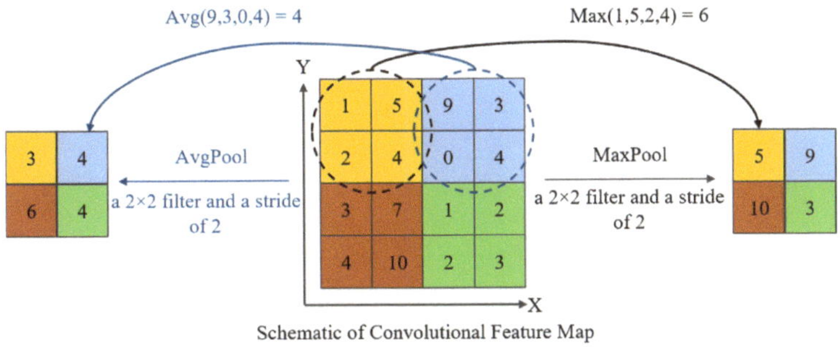

Figure 6. Schematic of the pooling process.

The last DSFM processed by GAP is converted into an n-dimensional vector for easy classification by FCL. Its calculation is:

$$y_{out}(x_{in}) = f(w^T x_{in} + b) \tag{2}$$

where y_{out} is the output vector, f is the activation function, x_{in} is the input data with flattening the pooled DSFM to one-dimensional form, w is the weight vector, and b is the offset vector.

4.3. Spatial Structural Feature Map Extraction

As is known, CapsNet consists of an encoder and a decoder. The encoder consists of three layers, which serves to perform feature extraction on the input data; in addition, a routing process for computation between capsule layers and an activation function for capsule feature classification are included. The decoder is mainly composed of three FCLs and is used to ensure that the encoded information is reduced to the original input features. For the LUSC task, we mainly utilized the encoder part with adaptations.

For the rapid extraction of the rich structural information of the LUSI data while reducing the network parameters, inspired by the literature [27], the SSFM extraction module of the lower branch was built by using the graph routing-based CapsNet.

Specifically, it includes three important parts. The first part is to modify the original input layer to process the DSFMs of the upper branch. As shown in Figure 7, the shallow semantic feature map $F_{cnn}^{(i)}$ extracted by CNN are fine-tuned in size with convolutional operations to fit the input requirements of the CapsNet. The specific method is to use 256 convolution kernels of size 3 × 3 to perform convolution operations on the shallow semantic feature map in a traversal manner by step size 1. The ReLU function is also utilized to retain the values of the elements in the output features that are greater than 0, and the values of the other elements are set to 0. Finally, the 256 feature maps are stitched together by the cat function to form the input feature map $F_{cap}^{(i)}$ of the next layer.

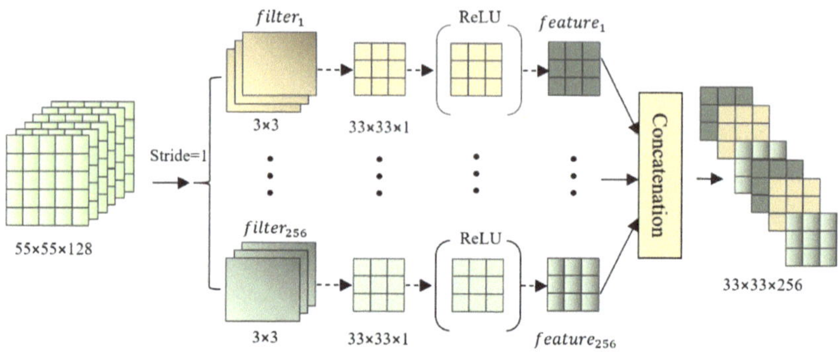

Figure 7. Convolutional operation process for image data.

The second part is to compute the primary capsule feature $F_{pcf}^{(i)}$ using the output feature map $F_{cap}^{(i)}$ of the previous layer. First, N_{filter} convolution kernels of size 3×3 are utilized to carry out convolution operations in a traversal manner with a step size of 2 to obtain feature maps $F_{cap}^{(i_1)}$, where $N_{filter} = 512/d_{incap}$; then, the output feature maps $F_{cap}^{(i_1)}$ are converted into a feature matrix $F_{cap}^{(i_2)}$ of the shape of (b, p, d_{incap}), where d_{incap} denotes the number of primary capsules, the value of p is the product of the square of the size of the feature after the convolution and the value of N_{filter}. Finally, the squash function is utilized to perform normalization on the new form of feature map $F_{cap}^{(i_2)}$ to obtain the main capsule feature map $F_{pcf}^{(i)}$, which is calculated as follows:

$$F_{pcf}^{(i)} = \frac{\left\|F_{cap}^{(i_2)}\right\|^2}{1 + \left\|F_{cap}^{(i_2)}\right\|^2} \frac{F_{cap}^{(i_2)}}{\left\|F_{cap}^{(i_2)}\right\|} \tag{3}$$

When the modulus length of $F_{cap}^{(i_2)}$ tends to positive infinity, $F_{pcf}^{(i)}$ tends to 1; when the modulus length of $F_{cap}^{(i_2)}$ tends to 0, $F_{pcf}^{(i)}$ tends to 0. $F_{cap}^{(i_2)} / \left\|F_{cap}^{(i_2)}\right\|$ denotes the unit vector, which compresses the range of values of the modulus length of $F_{pcf}^{(i)}$ between [0, 1], keeping the feature direction unchanged, i.e., keeping the attributes of the image features represented by the feature unchanged.

The third part is the calculation of advanced capsule feature maps $F_{acf}^{(i)}$ using primary capsule feature maps $F_{pcf}^{(i)}$. After adding a dimension to the third dimension of feature map $F_{pcf}^{(i)}$, we obtain $F_{pcf}^{(i_1)} = (b, p, 1, d_{incap})$, then we perform matrix multiplication operation with randomly initialized weight parameters $w_{pcf}^{(i_1)} = (b, p, d_{outcap})$ to get $F_{pcf}^{(i_2)} = (b, p, 1, d_{outcap})$, where $d_{outcap} = 2 \times d_{incap}$, and then, we convert the feature map $F_{pcf}^{(i_2)}$ to $F_{pcf}^{(d_i)} = (b, p, d_{outcap})$, and finally input $F_{pcf}^{(d_i)}$ into the multiple graphs pooling module to obtain $F_{acf}^{(i)}$.

The multiple graphs pooling module serves to transform the primary capsule feature map $F_{pcf}^{(i)}$ into advanced capsule feature map $F_{acf}^{(i)}$, and the key technique is to calculate the weight coefficients $w_{acf}^{(i)}$ of each primary capsule feature map. Figure 8 shows a simple relationship between a two-dimensional array and an image.

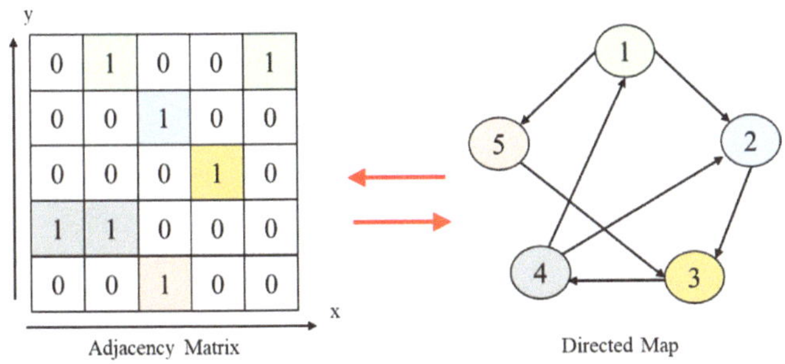

Figure 8. Schematic diagram of directed graph with adjacency matrix.

The first step is to compute the nodes and edges of feature map $F_{pcf}^{(d_i)}$. Consider feature map $F_{pcf}^{(d_i)} = (b, p, d_{outcap})$ as d_{outcap} graphs, where each single graph consists of p nodes, and the vector p can be seen as the feature map of the corresponding node.

The second step is to calculate the attention coefficients of the d_{outcap} unigraph. Matrix multiplication of the adjacency matrix of the $d_{outcap}^{(i)}$ unigram is performed with the node features of the $d_{outcap}^{(i)}$ unigram, and then multiplied with the random initialization weight parameter w, and then the calculation result is inputted into the softmax function to obtain the attention coefficient $w_{acf}^{(d_i)}$ of the feature $F_{pcf}^{(d_i)}$.

$$w_{acf}^{(d_i)} = softmax(e^{-\frac{\left\|F_{pcf}^{(d_{ix})}\right\| - \left\|F_{pcf}^{(d_{iy})}\right\|}{2\delta^2}} \cdot F_{pcf}^{(d_i)} \cdot w) \tag{4}$$

where $F_{pcf}^{(d_i)}$ denotes the feature map of the $d_{outcap}^{(i)}$ dimension, and d_{ix} and d_{iy} denote the index of the feature map $F_{pcf}^{(d_i)}$ node, respectively.

The third step is to use multiply the attention coefficient with the node feature to obtain the attention feature, then the squash function is used to perform the normalization disposal to obtain the advanced capsule feature map $F_{acf}^{(i)}$.

$$F_{acf}^{(i)} = squash\left(\frac{1}{p}\sum\left(\left(w_{acf}^{(d_i)}\right)^T \cdot F_{pcf}^{(d_i)}\right)\right) \tag{5}$$

4.4. Feature Map Fusion and Scene Prediction

To obtain both DSFMs and SSFMs of the LUSIs, we designed a simple two-branch fusion module with different characteristics, where the feature maps extracted from the upper-branch CNN and the lower-branch CapsNet are integrated and inputted to the recognizer. The computational formulae used in this fusion module are as follows:

$$Z_j = f_w(F_{cnn}^{(1)}) \oplus \left(\lambda \cdot \left(\sum_{i=1}^{j} f_{cap}(F_{cnn}^{(i)})\right)\right) \tag{6}$$

where Z_j denotes the fused feature, called DFM, $F_{cnn}^{(1)}$ denotes the last layer of semantic feature map extracted by the CNN, f_w denotes the mean GAP and FCL, $\oplus$ denotes the use of summing according to the values of the corresponding elements one by one, λ is the control coefficient, which denotes the key parameter controlling the deep fusion of the two-branch feature maps, and $\sum_{i=1}^{j} f_{cap}(F_{cnn}^{(i)})$ denotes the new spatial structure feature map formed by the CNN extracted feature maps input to the CapsNet after summing the corresponding elements.

After obtaining the DFM Z_j, it is first fed into the softmax function to calculate the probability of the image belonging to each class, and then the obtained probability values are used to calculate the predicted loss together with the probability values of the data labels after the image has been encoded by one-hot coding. While training the model using the HFCC-Net method, we used the cross-entropy loss function to obtain the minimum loss in both the training and validation datasets, achieving the task of optimizing the weight parameters of the model. Equation (7) lists the loss values obtained for a batch size of land-use data after softmax function and cross-entropy loss calculation.

$$loss(b,c) = -\frac{1}{b}\sum_{i=1}^{b}\sum_{j=1}^{c} y_{ij}\log\left(e^{z_i} \Big/ \sum_{j=1}^{c} e^{z_{ij}}\right) \quad (7)$$

where b denotes the number of the LUSI, c denotes the number of categories of land-use scenes, j denotes the true probability value of the i-th scene data label (if the true category of the i-th scene data label is equal to j then, y_{ij} is equal to 1, otherwise 0), and y_{ij} denotes the predicted probability that the i-th scene image data belongs to category j-th.

5. Results

5.1. Experimental Conditions

5.1.1. Platform Settings

To verify the outstanding performance of our proposed method in the LUSC task, we used Python3.7 language to build the HFCC-Net network framework under Pytorch framework, and fully trained on a Windows 10 operating system using each of the four classification datasets talked about in Section 3 to verify the advancement of HFCC-Net. Specifically, we use NVIDIA GPU acceleration during the training process to increase the speed of model training and inference. The specific environment configuration is shown in Table 3 below:

Table 3. Parameters for the platforms.

Platforms		Parameters
Hardware	Processor	Intel(R) Xeon(R) W-2245 CPU
	RAM	128 GB
	Graphics card	NVIDIA RTX A6000
	Memory	48 GB
Software	Operating system	Windows 10
	Operational environment	Python 3.7
	Deep learning framework	Pytorch

5.1.2. Training Details

To facilitate the comparison and analysis of the classification performance of HFCC-Net, we randomly selected 20%, 50%, and 80% of the four datasets as the training data, and the corresponding 80%, 50%, and 20% of the data as the test data, respectively. When inputting the data into the HFCC-Net model, we set the size of the images to 224 × 224, and the number of images inputted in each batch was set to 64, and each dataset was trained five times, with 200 rounds each time, and the mean and standard deviation of SCA were counted.

In the training process, we choose the cross-entropy loss function to describe the difference between the image category results predicted by HFCC-Net and the real image categories; in order to effectively deal with the gradient noise and non-smoothness in the learning process, we used Adam's algorithm to update the parameters of the HFCC-Net network. In addition, for speeding up the convergence speed while avoiding overfitting, the learning rate takes the value of 0.0001 and the weight decay coefficient takes the value of 0.0005.

5.2. Experimental Results

5.2.1. Results on the RSSCN

Table 4 shows the SCA obtained by 10 different algorithms trained on the RSSCN dataset. In order to achieve the best classification results, we set the structural parameters of HFCC-Net, i.e., j takes the value of 2, λ takes the value of 1, d_{incaps} takes the value of 32, and $d_{outcaps}$ takes the value of 64. The experimental data obtained show that when 50% of the training data are used for learning, HFCC-Net can achieve 95.30% of the average accuracy for 50% of the test data; when 20% of the training data are used for learning, HFCC-Net can achieve 93.61% of the average accuracy for 80% of the test data. Comparing the SCA values under similar experimental conditions, it can be seen that the SCAs obtained by HFCC-Net with different training data are all significantly competitive. Specifically, the average accuracy using HFCC-Net is improved by 0.66 compared with the method [51] that only fuses different DSFMs, and the SCA of HFCC-Net is relatively high when LUSI is used for training with relatively less data compared to the method that enhances DSFMs with attentional mechanisms [52].

Table 4. SCA on the RSSCN.

Practical Method	Classification Results (SCA, %)	
	50% for Training	20% for Training
GoogleNet [29]	85.84 ± 0.92	82.55 ± 1.11
Two-stage fusion [36]	92.37 ± 0.72	-
VGG-VD-16 [29]	87.18 ± 0.94	83.98 ± 0.87
TEX-Net-LF [34]	94.00 ± 0.55	92.45 ± 0.45
Dual-attention [35]	93.25 ± 0.28	91.07 ± 0.65
CaffeNet [29]	88.25 ± 0.62	85.57 ± 0.95
Deep filter banks [37]	90.40 ± 0.60	-
LCNN-BFF [51]	94.64 ± 0.21	-
EfficientNetB3-Attn-2 [52]	96.17 ± 0.23	93.30 ± 0.19
EfficientNetB3-Basic [52]	94.39 ± 0.10	92.06 ± 0.39
HFCC-Net	95.30 ± 0.24	93.61 ± 0.47

To further analyze the ability of HFCC-Net to classify images of each category, we utilize the best model obtained from training data of the RSSCN to classify the test data and view the misclassification between images of each category.

Figure 9a shows the image category misclassification obtained by classifying the test data with a 50% share of the data using the best model obtained after training in the data with a 50% share of the data. Among the seven image categories of the RSSCN, the most accurately classified category is 'Forest', which has a classification probability of 0.96; 'Industry' and 'parking' because of their similar structure. 'Industry' is classified with a probability of 0.88 because of its similar structure; however, the average probability of the overall classification accuracy of the images in the other categories is greater than 0.90; Figure 9b illustrates the best model obtained by using the model trained on 20% of the test data, and the best model obtained by using the model trained on 80% of the test data. The most accurately classified category is 'Forest', with a probability of accuracy of 1. 'Grass' and 'Field' have similar structure, which causes the model to classify 'Grass' as grass, however, the average probability of other categories of images being classified accurately is greater than 0.95, which also proves the sophistication of our method.

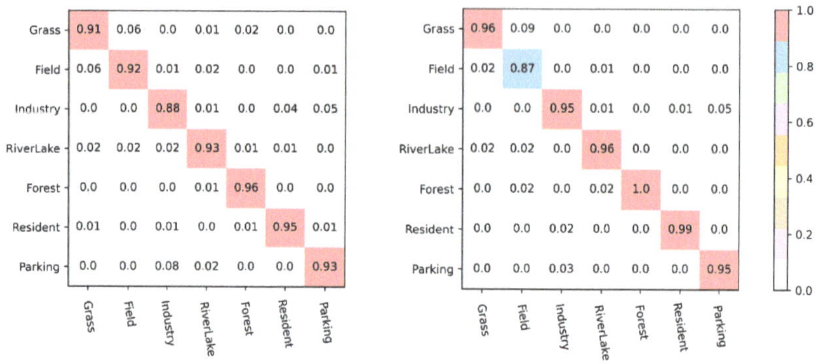

Figure 9. CMs on the RSSCN. (**a**) 50% for training. (**b**) 20% for training.

5.2.2. Results on the SIRI

Table 5 shows the SCA obtained by 11 different algorithms trained on the SIRI dataset. In order to achieve the best classification results, we set the structural parameters of HFCC-Net, i.e., j takes the value of 2, λ takes the value of 1, d_{incaps} takes the value of 32, and $d_{outcaps}$ takes the value of 64. While using 50% of the training data for learning, HFCC-Net can achieve an average accuracy of 96.72% with 50% of the testing data; comparing the values of SCA, it can be seen that our method has higher accuracy under similar experimental conditions. e.g., 0.97 improvement over the average accuracy obtained by using the Siamese ResNet-50 [33], 2.35 improvement over the average accuracy obtained with the Siamese CapsNet [25], etc. When learning with 80% of the training data, HFCC-Net can achieve an average accuracy of 97.78% with 20% of the test data. Similarly, our method achieves better classification results, e.g., 0.28 improvement in average accuracy over the Siamese ResNet-50 [33], 0.79 improvement in average accuracy over the Siamese CapsNet [25], and better SCA for HFCC-Net compared to the state-of-the-art methods [53,54].

Table 5. SCA on the SIRI.

Practical Method	Classification Results (SCA, %)	
	50% for Training	80% for Training
ResNet-50 [25]	94.67	95.63
AlexNet [25]	82.50	88.33
VGG-16 [25]	94.42	96.25
Fine-tuning MobileNetV2 [32]	95.77 ± 0.16	96.21 ± 0.31
Siamese ResNet-50 [33]	95.75	97.50
Siamese AlexNet [33]	83.25	88.96
Siamese VGG-16 [33]	94.50	97.30
Siamese CapsNet [25]	94.37	96.99
DenseNet + DenseNet [54]	-	96.37
DenseNet + VGG-16 [54]	-	94.16
MSAA-Net [53]	-	95.21 ± 0.65
HFCC-Net	96.72 ± 0.43	97.78 ± 0.41

Figure 10a shows the misclassification of image categories obtained by classifying the test data with a 50% share of the data using the best model obtained after training in the data with a 50% share of the data. Among the 12 image categories of the SIRI, the most accurately classified are "Overpass" and "residential", whose classification probability is 1; the average probability of overall classification accuracy for all categories of images is not less than 0.92; Figure 9b shows the average probability of overall classification accuracy for all categories, 0.92; Figure 10b shows the image category misclassification obtained by classifying 20% of the test data using the best model obtained after training on 80% of the data. The most accurately classified categories are "idle_land", "industrial", "overpass", "park", "pond", "residential" and "water", the probability of accuracy is 1; the average probability of an image being accurately classified for all categories is not less than 0.93.

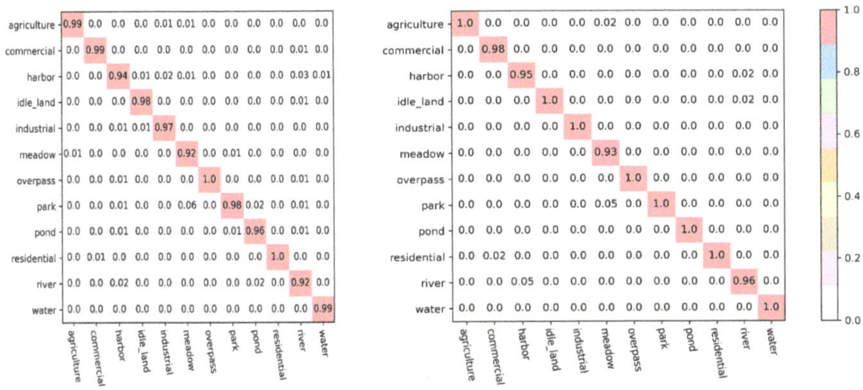

(a) (b)

Figure 10. CMs on the SIRI. (**a**) 50% for training. (**b**) 80% for training.

5.2.3. Results on the UCM

Table 6 shows the SCA obtained by 12 different algorithms trained on the UCM dataset. To achieve the best classification results, we set the structural parameters of HFCC-Net, i.e., j takes the value of 2, λ takes the value of 1, d_{incaps} takes the value of 32, and $d_{outcaps}$ takes the value of 64. The experimental data obtained show that when using 50% of the training data for learning, HFCC-Net in 50% of the test data can achieve an average accuracy of 98.38%; a comparison of the values of SCA shows that under similar experimental conditions, our method has a higher SCA, e.g., an increase of 0.79 over the average accuracy obtained using Inception-v3-CapsNet [16] and the average accuracy improvement over the average accuracy obtained using VGG-16-CapsNet [16] is 3.05. When learning with 80% of the training data, HFCC-Net achieves 99.20% with 20% of the test data; similarly, our method improves the average accuracy over Inception-v3-CapsNet [16] by 0.15, and over VGG-16-CapsNet [16] by average accuracy by 0.39. In addition, HFCC-Net is more competitive compared to state-of-the-art methods [51,52] in classification.

Table 6. SCA on the UCM.

Practical Method	Classification Results (SCA, %)	
	50% for Training	80% for Training
VGG-VD-16 [29]	94.14 ± 0.69	95.21 ± 1.20
GoogLeNet [29]	92.70 ± 0.60	94.31 ± 0.89
GBNET [30]	95.71 ± 0.19	96.90 ± 0.23
GBNET+Global feature [30]	97.05 ± 0.19	98.57 ± 0.48
CaffeNet [29]	93.98 ± 0.67	95.02 ± 0.81
Two-stream fusion [31]	96.97 ± 0.75	98.02 ± 1.03
ARCNET-VGG16 [38]	96.81 ± 0.14	99.12 ± 0.40
Inception-v3-CapsNet [16]	97.59 ± 0.16	99.05 ± 0.24
LCNN-BFF [51]	-	99.29 ± 0.24
EfficientNetB3-Attn-2 [52]	97.90 ± 0.36	99.21 ± 0.22
EfficientNetB3-Basic [52]	97.63 ± 0.06	98.73 ± 0.20
VGG-16-CapsNet [16]	95.33 ± 0.18	98.81 ± 0.22
HFCC-Net	98.38 ± 0.18	99.20 ± 0.46

The best model obtained using the training data of the UCM was utilized to classify the test data and to see the misclassification between each category of images. Figure 11a shows the misclassification of image categories obtained by classifying the test data with a 50% share of the data using the best model obtained after training in the data with a 50% share of the data. Among the 21 image categories of the UCM, "baseball diamond", "buildings", "dense residential", "golf course", "intersection", "medium residential", "sparse residential", and "storage tanks" are accurately classified with a probability close to 1, while the other categories are classified with a probability of 1. Figure 11b shows the best model obtained using the data trained on 80% of the data. Figure 11b shows the image category misclassification obtained by classifying the test data with 20% of the data using the best model obtained after training on 80% of the data. The probability of accurately classifying "buildings" and "intersection" is close to 1, while the probability of classifying the other categories is 1.

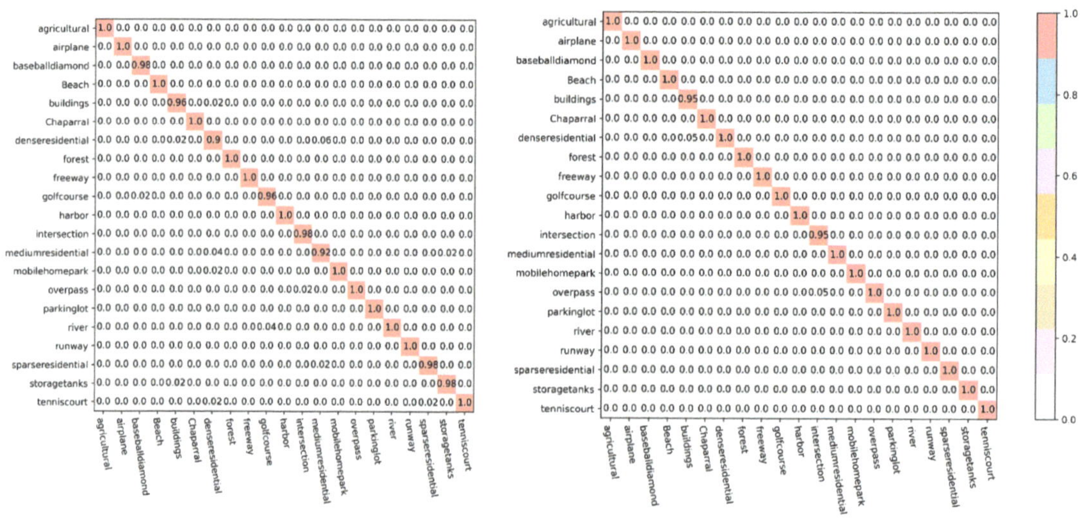

Figure 11. CMs on the UCM. (**a**) 50% for training. (**b**) 80% for training.

5.2.4. Results on the OPTIMAL

Table 7 shows the SCA obtained by 13 different algorithms trained on the OPTIMAL dataset. To achieve the best classification results, we set the structural parameters of HFCC-Net, i.e., j takes the value of 2, λ takes the value of 0.2, d_{incaps} takes the value of 32, and $d_{outcaps}$ takes the value of 64. The obtained experimental data show that HFCC-Net can achieve an average accuracy of 94.80% with 20% of the test data when learning with 80% of the training data. Comparing the values of SCA, it can be seen that under similar experimental conditions, our method has a higher SCA, e.g., HFCC-Net improves the average accuracy by 1.43 over the average accuracy obtained by SopNet-GCN-ResNet50 [39], and 1.25 over the average accuracy obtained by using SopNet-GCN-ResNet50 [39], and even reaches the level of a CNN-based state-of-the-art method [52].

Using the best model obtained from the training data of the OPTIMAL, we classify the test data and see the misclassification between each category of images. Figure 12 shows the misclassification of image categories obtained by using the best model obtained after training in the data with a share of 80% to classify the test data with a share of 20%. Among the 31 image categories of the OPTIMAL, "intersection" and "commercial area" have a similar structure, resulting in some "intersection" being classified as "commercial area"; "mountain" and "desert" have similar structures, resulting in some "mountain" being classified as "desert"; "railway" and "runway" have similar structures, resulting in some "railway" being classified as "runway". Surprisingly enough, the probability of classification for the other 28 categories is close to 1.

Table 7. SCA on the OPTIMAL.

Practical Method	Classification Results (SCA, %)
	80% for Training
Fine-tuning VGGNet16 [29]	87.45 ± 0.45
Fine-tuning GoogLeNet [29]	82.57 ± 0.12
GBNET [30]	91.40 ± 0.27
GBNET+Global feature [30]	93.28 ± 0.27
Fine-tuning AlexNet [38]	81.22 ± 0.19
VGG-VD-16 [29]	87.45 ± 0.45
ARCNET-ResNet34 [38]	91.28 ± 0.45
ARCNET-VGGNet16 [38]	92.70 ± 0.35
ARCNET-AlEXNET [38]	85.75 ± 0.35
SopNet-GCN-ResNet50 [39]	93.37 ± 0.68
SopNet-GAT-ResNet50 [39]	93.55 ± 0.74
EfficientNetB3-Attn-2 [52]	95.86 ± 0.22
EfficientNetB3-Basic [52]	94.76 ± 0.26
HFCC-Net	94.80 ± 0.89

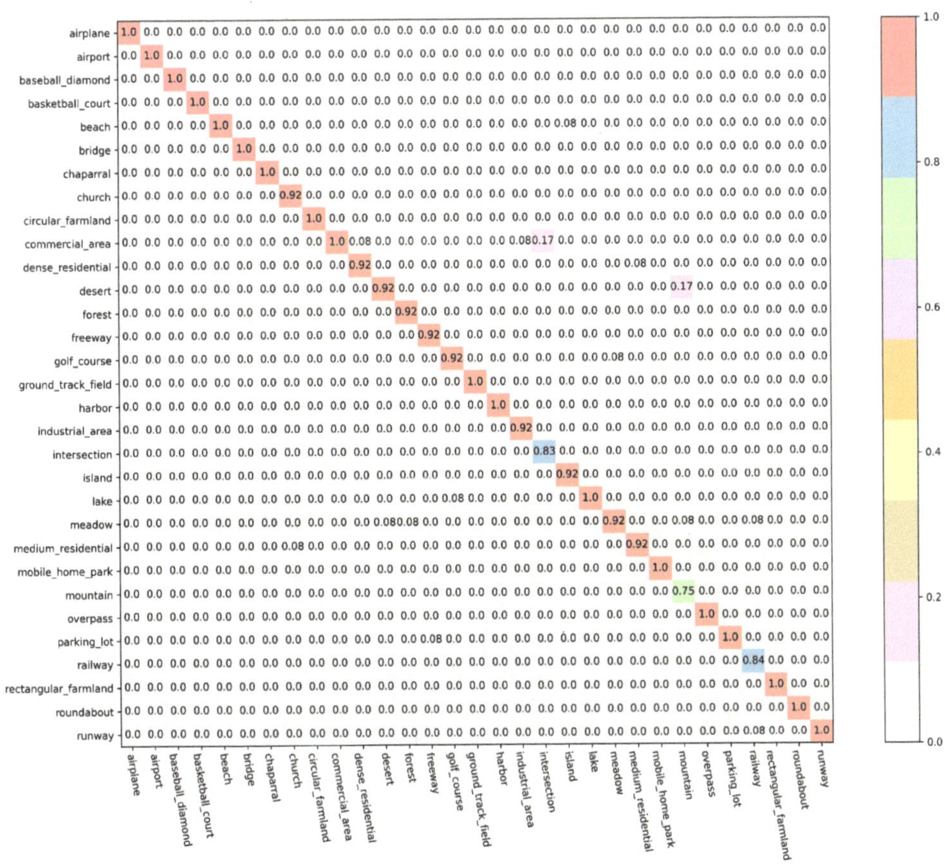

Figure 12. CM on the OPTIMAL, 80% for training.

5.3. Discussion

5.3.1. Combinatorial Patterns of DSFMs

As can be seen from Figure 2, the SSFMs of the lower branch of HFCC-Net are obtained by inputting the deep semantic features of the upper branch into the graph routing-based CapsNet. In order to further verify the outstanding contribution of our selected DSFMs to the classification performance of HFCC-Net, we conducted an ablation study on the inputs of the lower-branch CapsNet. The specific idea is to select one, two, three, and four feature maps as the input feature maps of the lower branch. Considering that the increasing number of convolutional layers tends to cause layer-by-layer loss of image information and that the current CapsNet is more suitable for small-size image features, we focus on the middle- and high-level features of the upper branch when selecting semantic feature maps. The specific input feature maps are shown in Table 8.

Table 8. Input feature maps for the lower branch.

Total	DSFMs	Discriminative Feature Maps
1	$F_{cnn}^{(4)}$	$Z_1 = f_w(F_{cnn}^{(1)}) \oplus (\lambda \cdot f_{cap}(F_{cnn}^{(4)}))$
2	$F_{cnn}^{(4)}, F_{cnn}^{(3)}$	$Z_2 = f_w(F_{cnn}^{(1)}) \oplus (\lambda \cdot f_{cap}(F_{cnn}^{(4)}, F_{cnn}^{(3)}))$
3	$F_{cnn}^{(4)}, F_{cnn}^{(3)}, F_{cnn}^{(2)}$	$Z_3 = f_w(F_{cnn}^{(1)}) \oplus (\lambda \cdot f_{cap}(F_{cnn}^{(4)}, F_{cnn}^{(3)}, F_{cnn}^{(2)}))$
4	$F_{cnn}^{(4)}, F_{cnn}^{(3)}, F_{cnn}^{(2)}, F_{cnn}^{(1)}$	$Z_4 = f_w(F_{cnn}^{(1)}) \oplus (\lambda \cdot f_{cap}(F_{cnn}^{(4)}, F_{cnn}^{(3)}, F_{cnn}^{(2)}, F_{cnn}^{(1)}))$

Under the premise that other experimental conditions remain unchanged, we utilized four kinds of the LUSI to start training separately. Based on the experience, the value of λ is tentatively set to 1. By selecting the best training model under the four inputs, we performed classification prediction on the test data, and obtained the accuracy of HFCC-Net for the four datasets under four different inputs of DSFMs. As shown in Figure 13, when the input method of $j = 2$, better classification results can be obtained.

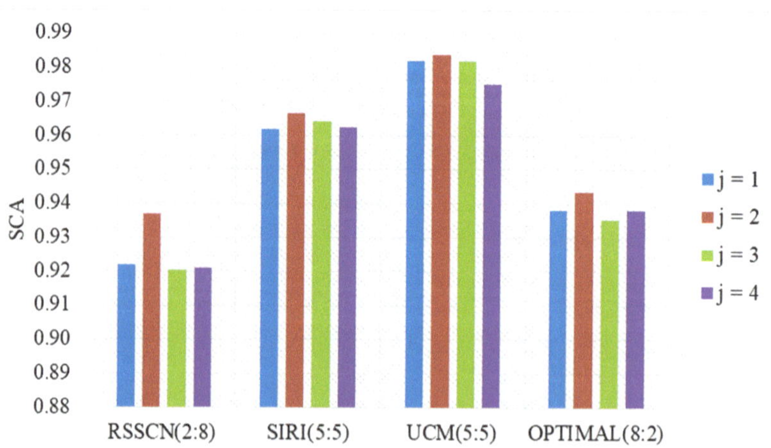

Figure 13. Results of SCA for CapsNet with various inputs.

5.3.2. Control Coefficient for Dual-Branch Feature Fusion

There are three main fusion methods for DSFMs: one is to splice different convolutional feature maps in channel dimension; one is to add different convolutional feature maps element-by-element; and one is to multiply two convolutional feature maps element-by-element. After extensive experiments using HFCC-Net in different fusion ways, we find that the element-by-element summation has an important contribution to the classification of the model. In addition, in order to speed up the network convergence and maximize the advantages of the DSFM and SSFM, we discuss the performance of the HFCC-Net model with different fusion control coefficients.

When performing the fusion of the corresponding elements of discriminative feature maps, considering that the SSFMs are special and the values of the elements are large relative to the DSFMs, we discuss how the different control coefficients λ of all the elements of the SSFMs affect the fusion effect. Specifically, the discussion process is divided into two steps: first the order of magnitude of λ is determined, in accordance with the multiples of 10 to deflate, and we found that when the control coefficient takes 1, a higher classification accuracy can be obtained; and then we determined more specific control coefficients, and following with the law of the equivariant series, we chose to carry out experiments on the three special values of 0.8, 0.5, and 0.2.

The optimal fusion control coefficients for HFCC-Net differ across the four datasets because of the variability in the image content of the land-use scenes and because of the differences in the amount of data in the categories of the scene images. Under the premise that other experimental conditions remain unchanged, the SCA we obtained on the predicted datasets of the four datasets are shown in Figure 14, which shows that when λ takes the value of 1, the effect of feature map fusion can promote HFCC-Net to achieve the better classification performance on the RSSCN, SIRI, and UCM datasets; when the value is taken as 0.2, the effect of feature map fusion is more suitable for the OPTIMAL dataset.

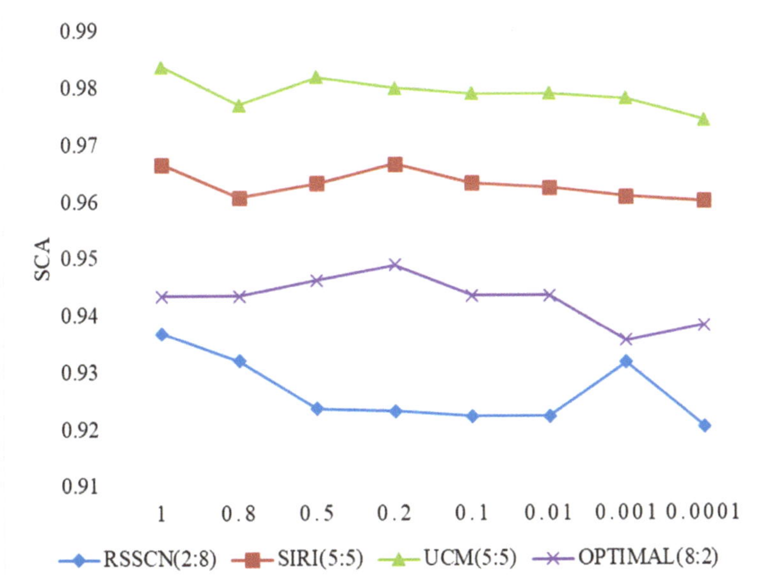

Figure 14. Results of SCA with different control coefficients.

5.3.3. Input and Output Dimensions of Capsules

A capsule is the basic unit of CapsNet, similar to a neuron, which is able to extract richer feature representation. Specifically, each capsule represents a feature map, and increasing the number of capsules increases the expressive power of the network, allowing it to capture more details and complex feature maps. However, the increase also increases the computational and storage overhead of the network. Too many capsules may cause the network to overfit the training data, reducing its ability to generalize to new data. In addition, if the number of capsules is not set properly, it may lead to training difficulties such as gradient vanishing or gradient explosion.

We have designed four different combinations of capsule sizes in HFCC-Net based on the characteristics of the down-branch network and the input data. Specifically, the dimensions of the input capsules d_{incaps} are sequentially designed as 8, 16, 32, and 64, and the dimensions of the output capsules $d_{outcaps}$ are correspondingly designed as 16, 32, 64, and 128. Other conditions are kept unchanged, and the SCA of the HFCC-Net model is obtained after training on four LUSI datasets as shown in Figure 15. It can be seen that HFCC-Net can achieve a better classification performance when the dimension of the input capsules d_{incaps} is set to 32 and the dimension of the output capsules $d_{outcaps}$ is set to 64.

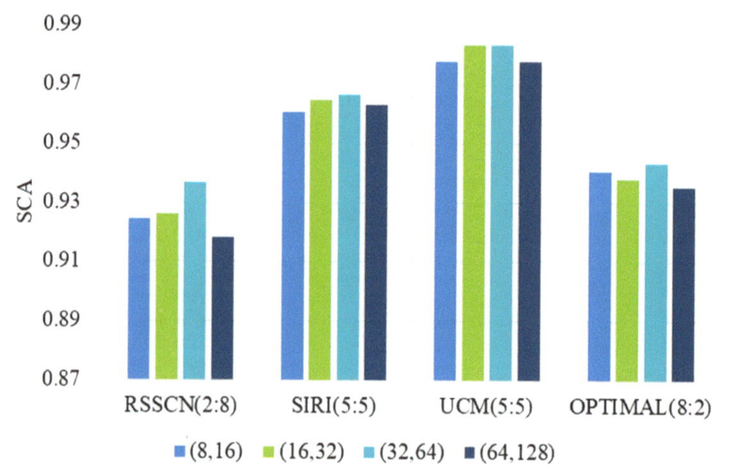

Figure 15. Results of SCA with various dimensions.

5.3.4. Different Backbone Network Architecture

The backbone network can capture local information such as spatial structure, texture, and edges of the input data, as well as higher-level semantic and contextual information, and the goodness of the backbone network is directly related to the performance and performance of the whole deep learning system.

To analyze the goodness of the HFCC-Net backbone network, we conducted a comparative experiment. Specifically, we first selected two powerful feature extraction CNNs as the upper-branch backbone networks. The first time, we extracted the semantic features of the conv2_x and conv3_x layers using ResNet-50 [55], and used the two features as the input information of the HFCC-Net under-branch network; the second time, we extracted the semantic features of the conv3_256 and conv4_512 layers using VGG-16 [56], and used them as the HFCC-Net under-branch network's inputs; other experimental conditions are kept constant, and our obtained SCAs on the four datasets are shown in Figure 16. It can be seen that compared with the two typical backbones, the upper branching network of HFCC-Net has a more obvious promotion effect on SCA. Similarly, we selected the most popular CapsNet as the lower-branch backbone network to compare and verify the feature extraction ability of the lower branch of HFCC-Net. Different from the method adopted by HFCC-Net, CapsNet [12] uses a dynamic routing mechanism as the information transfer method. We take $F_{cnn}^{(3)}$ and F_{cnn}^{4} of the Xception network as the input of CapsNet, and obtain the spatial structure features of the image through the dynamic routing mechanism. After the experiments, it can be seen that compared with CapsNet, the lower branch network of HFCC-Net promotes SCA more obviously in the LUSI datasets.

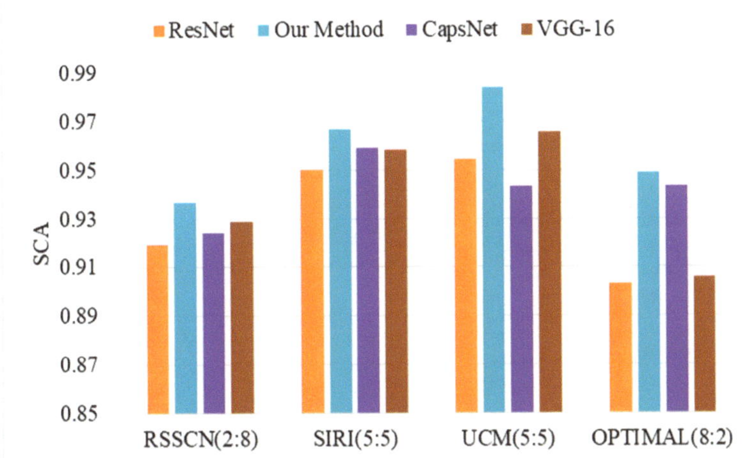

Figure 16. Results of SCA with different backbones.

6. Conclusions

CNN-based techniques for the LUSC are techniques that have emerged in recent years, which could autonomously learn information-rich features from the original pixels of RSI, and automatically complete the classification with high accuracy. However, the pooling layer of CNN leads to more data loss the more convolutional layers it has. Moreover, CNN does not utilize the spatial structure features between individual objects in the image content. Therefore, it is extremely challenging to improve the classification accuracy of CNN for complex LUSI. CapsNet is a novel neural network that better captures the structural relationships between the contents in an image through capsule units. To improve the SCA of the LUSC and make up for the shortcomings of CNN methods, we propose a dual-branch hybrid framework, HFCC-Net, which makes more complete use of the LUSI's global semantic and local structural information. DSFM of RSI is extracted by using the transfer learning technique, SSFM is obtained by using the shallow DSFM as the input of CapsNet, and the DFMs of the two branches are fused in turn by using the newly designed fusion function and finally the DFMs are input into the classifier to obtain the predictive probability of the LUSI. The experimental results on four public datasets prove the effectiveness of HFCC-Net, and in future work, we plan to improve the CNN, such as adding the attention mechanism, to further improve the classification accuracy of the model.

Author Contributions: N.G. and M.J.: methodology, software, writing—original draft; L.G., K.L., F.Z. and X.C.: supervision; M.W.: validation, investigation. All authors have read and agreed to the published version of the manuscript.

Funding: This research was funded by Internal Parenting Program (grant number: 145AXL250004000X).

Institutional Review Board Statement: Not applicable.

Informed Consent Statement: Not applicable.

Data Availability Statement: RSSCN7 dataset: https://aistudio.baidu.com/datasetdetail/52117 (accessed on 15 August 2023); SIRI-WHU dataset: https://figshare.com/articles/dataset/SIRI_WHU_Dataset/8796980 (accessed on 15 August 2023); UC-Merced dataset: http://weegee.vision.ucmerced.edu/datasets/landuse.html (accessed on 15 August 2023); OPTIMAL-31 dataset: https://aistudio.baidu.com/datasetdetail/51798 (accessed on 15 August 2023).

Conflicts of Interest: The authors declare no conflict of interest.

Abbreviations

The following abbreviations are used in this manuscript:

LUSC	Land-use scene classification
RSI	Remote sensing image
LUSI	Land-use scene image
CNN	Convolutional neural network
CapsNet	Capsule network
DSFM	Deep semantic feature map
SSFM	Spatial structure feature map
DFM	Discriminative feature map
GAP	Global average pooling
FCL	Fully connected layer
SCA	Sence classification accuracy
CM	Confusion matrix

References

1. Dutta, S.; Das, M. Remote sensing scene classification under scarcity of labelled samples—A survey of the state-of-the-arts. *Comput. Geosci.* **2023**, *171*, 105295. [CrossRef]
2. Cheng, G.; Xie, X.; Han, J.; Guo, L.; Xia, G.-S. Remote sensing image scene classification meets deep learning: Challenges, methods, benchmarks, and opportunities. *IEEE J. Sel. Top. Appl. Earth Obs. Remote Sens.* **2020**, *13*, 3735–3756. [CrossRef]
3. Wang, J.; Li, W.; Zhang, M.; Tao, R.; Chanussot, J. Remote Sensing Scene Classification via Multi-Stage Self-Guided Separation Network. *IEEE Trans. Geosci. Remote Sens.* **2023**, *61*, 5615312.
4. Huang, X.; Liu, F.; Cui, Y.; Chen, P.; Li, L.; Li, P. Faster and Better: A Lightweight Transformer Network for Remote Sensing Scene Classification. *Remote Sens.* **2023**, *15*, 3645. [CrossRef]
5. Zhang, J.; Zhao, H.; Li, J. TRS: Transformers for remote sensing scene classification. *Remote Sens.* **2021**, *13*, 4143. [CrossRef]
6. Thapa, A.; Horanont, T.; Neupane, B.; Aryal, J. Deep Learning for Remote Sensing Image Scene Classification: A Review and Meta-Analysis. *Remote Sens.* **2023**, *15*, 4804. [CrossRef]
7. Khatami, R.; Mountrakis, G.; Stehman, S.V. A meta-analysis of remote sensing research on supervised pixel-based land-cover image classification processes: General guidelines for practitioners and future research. *Remote Sens. Environ.* **2016**, *177*, 89–100. [CrossRef]
8. Kavzoglu, T.; Tonbul, H. An experimental comparison of multi-resolution segmentation, SLIC and K-means clustering for object-based classification of VHR imagery. *Int. J. Remote Sens.* **2018**, *39*, 6020–6036. [CrossRef]
9. Maurya, K.; Mahajan, S.; Chaube, N. Remote sensing techniques: Mapping and monitoring of mangrove ecosystem—A review. *Complex Intell. Syst.* **2021**, *7*, 2797–2818. [CrossRef]
10. Ma, L.; Liu, Y.; Zhang, X.; Ye, Y.; Yin, G.; Johnson, B.A. Deep learning in remote sensing applications: A meta-analysis and review. *ISPRS J. Photogramm. Remote Sens.* **2019**, *152*, 166–177. [CrossRef]
11. Song, J.; Gao, S.; Zhu, Y.; Ma, C. A survey of remote sensing image classification based on CNNs. *Big Earth Data* **2019**, *3*, 232–254. [CrossRef]
12. Sabour, S.; Frosst, N.; Hinton, G.E. Dynamic routing between capsules. In Proceedings of the Advances in Neural Information Processing Systems 30 (NIPS 2017), Long Beach, CA, USA, 4–9 December 2017; Volume 30.
13. Afshar, P.; Mohammadi, A.; Plataniotis, K.N. Brain tumor type classification via capsule networks. In Proceedings of the 2018 25th IEEE International Conference on Image Processing (ICIP), Athens, Greece, 7–10 October 2018; pp. 3129–3133.
14. Xu, Z.; Lu, W.; Zhang, Q.; Yeung, Y.; Chen, X. Gait recognition based on capsule network. *J. Vis. Commun. Image Represent.* **2019**, *59*, 159–167. [CrossRef]
15. Zhao, T.; Liu, Y.; Huo, G.; Zhu, X. A deep learning iris recognition method based on capsule network architecture. *IEEE Access* **2019**, *7*, 49691–49701. [CrossRef]
16. Zhang, W.; Tang, P.; Zhao, L. Remote sensing image scene classification using CNN-CapsNet. *Remote Sens.* **2019**, *11*, 494. [CrossRef]
17. Patrick, M.K.; Adekoya, A.F.; Mighty, A.A.; Edward, B.Y. Capsule networks–a survey. *J. King Saud Univ.-Comput. Inf. Sci.* **2022**, *34*, 1295–1310.
18. Goceri, E. Analysis of capsule networks for image classification. In Proceedings of the International Conference on Computer Graphics Visualization, Computer Vision and Image Processing, Online, 21–23 July 2021.
19. Hinton, G.E.; Sabour, S.; Frosst, N. Matrix capsules with EM routing. In Proceedings of the International Conference on Learning Representations, Vancouver, BC, Canada, 30 April–3 May 2018.
20. Li, Y.; Zhao, W.; Cambria, E.; Wang, S.; Eger, S. Graph routing between capsules. *Neural Netw.* **2021**, *143*, 345–354. [CrossRef]
21. Tao, J.; Zhang, X.; Luo, X.; Wang, Y.; Song, C.; Sun, Y. Adaptive capsule network. *Comput. Vis. Image Underst.* **2022**, *218*, 103405. [CrossRef]

22. Phaye, S.S.R.; Sikka, A.; Dhall, A.; Bathula, D. Dense and diverse capsule networks: Making the capsules learn better. *arXiv* **2018**, arXiv:1805.04001.
23. Xiong, Y.; Su, G.; Ye, S.; Sun, Y.; Sun, Y. Deeper capsule network for complex data. In Proceedings of the 2019 International Joint Conference on Neural Networks (IJCNN), Budapest, Hungary, 14–19 July 2019; pp. 1–8.
24. Jia, B.; Huang, Q. DE-CapsNet: A diverse enhanced capsule network with disperse dynamic routing. *Appl. Sci.* **2020**, *10*, 884. [CrossRef]
25. Zhou, S.; Zhou, Y.; Liu, B. Using Siamese capsule networks for remote sensing scene classification. *Remote Sens. Lett.* **2020**, *11*, 757–766. [CrossRef]
26. Hoogi, A.; Wilcox, B.; Gupta, Y.; Rubin, D.L. Self-attention capsule networks for object classification. *arXiv* **2019**, arXiv:1904.12483.
27. Gu, J. Interpretable graph capsule networks for object recognition. In Proceedings of the AAAI Conference on Artificial Intelligence, Online, 2–9 February 2021; pp. 1469–1477.
28. Yu, Y.; Liu, C.; Guan, H.; Wang, L.; Gao, S.; Zhang, H.; Zhang, Y.; Li, J. Land cover classification of multispectral lidar data with an efficient self-attention capsule network. *IEEE Geosci. Remote Sens. Lett.* **2021**, *19*, 1–5. [CrossRef]
29. Xia, G.-S.; Hu, J.; Hu, F.; Shi, B.; Bai, X.; Zhong, Y.; Zhang, L.; Lu, X. AID: A benchmark data set for performance evaluation of aerial scene classification. *IEEE Trans. Geosci. Remote Sens.* **2017**, *55*, 3965–3981. [CrossRef]
30. Sun, H.; Li, S.; Zheng, X.; Lu, X. Remote sensing scene classification by gated bidirectional network. *IEEE Trans. Geosci. Remote Sens.* **2019**, *58*, 82–96. [CrossRef]
31. Yu, Y.; Liu, F. A two-stream deep fusion framework for high-resolution aerial scene classification. *Comput. Intell. Neurosci.* **2018**, *2018*, 8639367. [CrossRef]
32. Zhang, B.; Zhang, Y.; Wang, S. A lightweight and discriminative model for remote sensing scene classification with multidilation pooling module. *IEEE J. Sel. Top. Appl. Earth Obs. Remote Sens.* **2019**, *12*, 2636–2653. [CrossRef]
33. Yang, L.; Zhang, R.-Y.; Li, L.; Xie, X. Simam: A simple, parameter-free attention module for convolutional neural networks. In Proceedings of the International Conference on Machine Learning, Online, 18–24 July 2021; pp. 11863–11874.
34. Anwer, R.M.; Khan, F.S.; Van De Weijer, J.; Molinier, M.; Laaksonen, J. Binary patterns encoded convolutional neural networks for texture recognition and remote sensing scene classification. *ISPRS J. Photogramm. Remote Sens.* **2018**, *138*, 74–85. [CrossRef]
35. Gao, Y.; Shi, J.; Li, J.; Wang, R. Remote sensing scene classification with dual attention-aware network. In Proceedings of the 2020 IEEE 5th International Conference on Image, Vision and Computing (ICIVC), Beijing, China, 10–12 July 2020; pp. 171–175.
36. Liu, Y.; Liu, Y.; Ding, L. Scene classification based on two-stage deep feature fusion. *IEEE Geosci. Remote Sens. Lett.* **2017**, *15*, 183–186. [CrossRef]
37. Wu, H.; Liu, B.; Su, W.; Zhang, W.; Sun, J. Deep filter banks for land-use scene classification. *IEEE Geosci. Remote Sens. Lett.* **2016**, *13*, 1895–1899. [CrossRef]
38. Wang, Q.; Liu, S.; Chanussot, J.; Li, X. Scene classification with recurrent attention of VHR remote sensing images. *IEEE Trans. Geosci. Remote Sens.* **2018**, *57*, 1155–1167. [CrossRef]
39. Peng, F.; Lu, W.; Tan, W.; Qi, K.; Zhang, X.; Zhu, Q. Multi-output network combining GNN and CNN for remote sensing scene classification. *Remote Sens.* **2022**, *14*, 1478. [CrossRef]
40. Phaye, S.S.R.; Sikka, A.; Dhall, A.; Bathula, D.R. Multi-level dense capsule networks. In Proceedings of the Computer Vision–ACCV 2018: 14th Asian Conference on Computer Vision, Perth, Australia, 2–6 December 2018; Revised Selected Papers, Part V 14. pp. 577–592.
41. Xiang, C.; Zhang, L.; Tang, Y.; Zou, W.; Xu, C. MS-CapsNet: A novel multi-scale capsule network. *IEEE Signal Process. Lett.* **2018**, *25*, 1850–1854. [CrossRef]
42. Jampour, M.; Abbaasi, S.; Javidi, M. CapsNet regularization and its conjugation with ResNet for signature identification. *Pattern Recognit.* **2021**, *120*, 107851. [CrossRef]
43. Wang, A.; Wang, M.; Wu, H.; Jiang, K.; Iwahori, Y. A novel LiDAR data classification algorithm combined capsnet with resnet. *Sensors* **2020**, *20*, 1151. [CrossRef]
44. Yousra, D.; Abdelhakim, A.B.; Mohamed, B.A. A novel model for detection and classification coronavirus (COVID-19) based on Chest X-Ray images using CNN-CapsNet. In Proceedings of the Sustainable Smart Cities and Territories, Doha, Qatar, 27–29 April 2021; pp. 187–199.
45. Zhang, J.; Yu, X.; Lei, X.; Wu, C. A novel CapsNet neural network based on MobileNetV2 structure for robot image classification. *Front. Neurorobotics* **2022**, *16*, 1007939. [CrossRef]
46. Wang, P.; Wang, J.; Li, Y.; Li, P.; Li, L.; Jiang, M. Automatic classification of breast cancer histopathological images based on deep feature fusion and enhanced routing. *Biomed. Signal Process. Control* **2021**, *65*, 102341. [CrossRef]
47. Zou, Q.; Ni, L.; Zhang, T.; Wang, Q. Deep learning based feature selection for remote sensing scene classification. *IEEE Geosci. Remote Sens. Lett.* **2015**, *12*, 2321–2325. [CrossRef]
48. Zhao, B.; Zhong, Y.; Xia, G.-S.; Zhang, L. Dirichlet-derived multiple topic scene classification model for high spatial resolution remote sensing imagery. *IEEE Trans. Geosci. Remote Sens.* **2015**, *54*, 2108–2123. [CrossRef]
49. Yang, Y.; Newsam, S. Bag-of-visual-words and spatial extensions for land-use classification. In Proceedings of the 18th SIGSPATIAL International Conference on Advances in Geographic Information Systems, San Jose, CA, USA, 2–5 November 2010; pp. 270–279.

50. Chollet, F. Xception: Deep learning with depthwise separable convolutions. In Proceedings of the IEEE Conference on Computer Vision and Pattern Recognition, Honolulu, HI, USA, 21–26 July 2017; pp. 1251–1258.
51. Shi, C.; Wang, T.; Wang, L. Branch feature fusion convolution network for remote sensing scene classification. *IEEE J. Sel. Top. Appl. Earth Obs. Remote Sens.* **2020**, *13*, 5194–5210. [CrossRef]
52. Alhichri, H.; Alswayed, A.S.; Bazi, Y.; Ammour, N.; Alajlan, N.A. Classification of remote sensing images using EfficientNet-B3 CNN model with attention. *IEEE Access* **2021**, *9*, 14078–14094. [CrossRef]
53. Li, L.; Liang, P.; Ma, J.; Jiao, L.; Guo, X.; Liu, F.; Sun, C. A multiscale self-adaptive attention network for remote sensing scene classification. *Remote Sens.* **2020**, *12*, 2209. [CrossRef]
54. Khan, S.D.; Basalamah, S. Multi-Branch Deep Learning Framework for Land Scene Classification in Satellite Imagery. *Remote Sens.* **2023**, *15*, 3408. [CrossRef]
55. He, K.; Zhang, X.; Ren, S.; Sun, J. Deep residual learning for image recognition. In Proceedings of the IEEE Conference on Computer Vision and pattern Recognition, Las Vegas, NV, USA, 26 June–1 July 2016; pp. 770–778.
56. Tammina, S. Transfer learning using vgg-16 with deep convolutional neural network for classifying images. *Int. J. Sci. Res. Publ. IJSRP* **2019**, *9*, 143–150. [CrossRef]

remote sensing

MDPI

Article

Local Differential Privacy Based Membership-Privacy-Preserving Federated Learning for Deep-Learning-Driven Remote Sensing

Zheng Zhang *, Xindi Ma and Jianfeng Ma

The School of Cyber Engineering, Xidian University, Xi'an 710071, China; xdma@xidian.edu.cn (X.M.); jfma@mail.xidian.edu.cn (J.M.)
* Correspondence: zhengzhang1994@stu.xidian.edu.cn

Abstract: With the development of deep learning, image recognition based on deep learning is now widely used in remote sensing. As we know, the effectiveness of deep learning models significantly benefits from the size and quality of the dataset. However, remote sensing data are often distributed in different parts. They cannot be shared directly for privacy and security reasons, and this has motivated some scholars to apply federated learning (FL) to remote sensing. However, research has found that federated learning is usually vulnerable to white-box membership inference attacks (MIAs), which aim to infer whether a piece of data was participating in model training. In remote sensing, the MIA can lead to the disclosure of sensitive information about the model trainers, such as their location and type, as well as time information about the remote sensing equipment. To solve this issue, we consider embedding local differential privacy (LDP) into FL and propose LDP-Fed. LDP-Fed performs local differential privacy perturbation after properly pruning the uploaded parameters, preventing the central server from obtaining the original local models from the participants. To achieve a trade-off between privacy and model performance, LDP-Fed adds different noise levels to the parameters for various layers of the local models. This paper conducted comprehensive experiments to evaluate the framework's effectiveness on two remote sensing image datasets and two machine learning benchmark datasets. The results demonstrate that remote sensing image classification models are susceptible to MIAs, and our framework can successfully defend against white-box MIA while achieving an excellent global model.

Keywords: remote sensing image classification; local differential privacy; deep learning; federated learning; membership inference attack

1. Introduction

Remote sensing is a technology that uses sensors (e.g., satellites, airplanes) to detect physical features in a non-contact, long-distance manner [1,2]. In recent years, with the increasing deployment of remote sensing satellites, the magnitude of remote sensing data has also grown dramatically. To analyze and utilize these remote sensing image data, more scholars apply state-of-the-art deep learning techniques for remote sensing image data analysis, such as scene classification [3,4] and object detection [5]. As we know, deep learning models significantly benefit from the size and quality of the training dataset. However, in practice, on the one hand, remote sensing data need to be collected from different companies [6]. On the other hand, remote sensing data cannot directly be shared because they are usually sensitive and contain trade secrets, which may leak information about sensitive infrastructure or military facilities [7]. This is one of the major problems affecting the practical application of deep learning algorithms in remote sensing. To solve this problem, federated learning has been researched and applied to remote sensing [8–11].

Federated learning allows participants to collaboratively train an efficient global model by sharing the parameters of a local model instead of sensitive data. However, in recent years, scholars have found that federated learning is vulnerable to various inference attacks launched against model parameters, such as adversarial attacks [12,13], backdoor attacks [14], and membership inference attacks [15]. In this paper, we focus on the membership inference attack, the goal of which is to infer whether a specific data point is participating in the model's training process. Suppose an attacker successfully launches a membership inference attack against a participant, the attacker can infer the location, time, and equipment type of the company's remote sensing equipment from these membership data. This will be a serious secret disclosure for a remote sensing company.

Recently, scholars have proposed some approaches to defend against MIAs, which can be summarized as confidence score masking [16–18] (hiding the target model's true confidence scores), regularization [16,18–21] (reducing the overfitting of the target model), knowledge distillation [22,23] (transferring the private large model's knowledge to the small one through an unlabeled public dataset), and differential privacy [15,24,25] (adding noise to parameters during the model training). For defending against white-box MIAs in FL, researchers have mainly focused on regularization and differential privacy [15,26–29]. However, the majority of them struggle to balance the privacy preservation and model utility for the deep neural network with high dimensional parameters. Some scholars attempt to encrypt the model parameters with homomorphic encryption [11], which can guarantee the model's effectiveness while taking into account the privacy and security of the model. However, for complex remote sensing images and classification models, it leads to huge computational and communication costs.

To solve the abovementioned problems, we propose local differential privacy federated learning (LDP-Fed) in this paper. It is a private FL system that allows participants to train DNN models jointly under the protection of LDP. The main inspiration for our LDP-Fed is that adversaries launch membership inference attacks by collecting a set of local or global model parameters in FL. Based on this point, it is different from the central differential privacy, which adds noise during local model training [24]. We perform a piecewise mechanism (PM) on the local model parameters before they are delivered to the central server so that participants satisfy local privacy guarantees. As far as we know, there have been some works [28,29] that applied LDP mechanisms to FL to protect local model parameters. They added the same level of LDP noise to the parameters of each dimension separately. Hence, to improve the model performance, they must either increase the overall privacy budget or the number of federal learning participants, which may reduce the framework usability or raise the risk of privacy disclosure. To show the effectiveness of LDP-Fed, we evaluated LDP-Fed on various datasets. The experimental results demonstrate that LDP-Fed could defend against white-box MIAs while having a minor impact on the utility of complex DNN models. The contributions are summarized as follows:

- We propose and implement LDP-Fed to defend against white-box MIAs in the federated learning system on remote sensing, especially for global attacks launched from the central server. It allows participants to collaboratively train DNN models with formal LDP guarantees.
- To achieve optimal privacy-utility trade-offs, we optimize the noise addition method according to the characteristics of white-box MIAs and apply the piecewise mechanism (PM) that is more suitable for the parameters of the DNN models in LDP-Fed. It gives our framework more utility than others while resisting white-box membership inference attacks.
- We extensively evaluate LDP-Fed on various datasets to show the advanced trade-offs of the local participants' privacy and model utility. For the remote sensing image dataset NWPU-RESISC45 with VGG, it reduced the adversary advantage Adv of global attacks from 46.2% to 10.6% while decreasing the model accuracy by 5.0%.

The rest of our paper is organized as follows: Section 2 presents some related works and background techniques related to the subject of this paper, then an overview of the LDP-Fed framework and details of the LDP-Fed framework are presented in Section 3. Sections 4 and 5 present the experimental setup and experimental results with analysis and comparison. Section 6 concludes this paper.

2. Related Works and Background Techniques

2.1. Related Works

The development of deep learning algorithms, especially image recognition algorithms, has led to their application in remote sensing, such as scene recognition and object detection. With the increase in remote sensing equipment, the distribution of remote sensing data has become decentralized, with multiple entities (remote sensing companies) holding a small volume of data. To enhance the model's effectiveness while guaranteeing the security of remote sensing data, scholars have widely used federated learning in remote sensing [30–32]. However, it still suffers from some security issues, such as adversarial attacks, backdoor attacks, and especially membership inference attacks. These attacks target the model's parameters and pose serious security problems for federated learning. There are some typical works on traditional FL applications for these attacks but less research on remote sensing. Therefore, we propose the LDP-Fed framework for MIA in FL on remote sensing. We will describe related work in three aspects: the application of machine learning in remote sensing, membership inference attacks, and defense against membership inference attacks.

2.1.1. Machine Learning in Remote Sensing

The research on remote sensing image data classification using state-of-the-art machine learning models has been studied by scholars for many years [33,34]. Geiß et al. [35] followed the idea of learning invariant decisions function to address the problem of the efficient classification of remote sensing images under sparse data. They proposed a virtual support vector machine based on self-learning (VSVM-SL). In the same period, Wang et al. [36] used the efficient random forest (RF) algorithm to classify remote sensing image data, and the results showed that RF can achieve a superior and more stable classification performance than SVM on the land-cover dataset. With the development of deep learning algorithms, convolutional neural networks (CNN) have emerged as the most promising technique for remote sensing image classification. Zhang et al. [37] proposed the model architecture named CNN-CapsNet to improve the performance of the scene classification model for remote sensing images. Their model utilizes the convolutional layer of a pre-trained convolutional model for feature extraction from the image and then uses CapsNet to classify the intermediate features. In contrast, Li et al. [38] used a convolutional neural network for remote sensing image multi-target scene classification tasks. They proposed MLRSSC-CNN-GNN. The main idea is to generate high-level representation features with the CNN model and combine them with the graph attention network model to fully exploit the scene graph's spatial-topological relationship. Tang et al. [39] proposed a new CNN model attentional consistent network (ACNet), which can improve the model classification performance by emphasizing the local features of an image. Chen et al. [40] proposed CNSPN, a method that better solves the problem of few-shot remote sensing image classification by combining the semantic information of image class names. Along with the successful applications of deep neural networks in remote sensing image recognition, some scholars have focused on the security concerns posed by the models, such as backdoor attacks [11,14] and adversarial attacks [12,13].

2.1.2. Membership Inference Attack

We categorize membership inference attacks into white-box and black-box MIAs according to the attacker's information obtained from the target model.

Black-box membership inference attack. Shokri et al. [16] first presented the study of MIAs on the classification model. They trained an inference attack model based on the difference in model prediction between trained and untrained data. With the black box setting, they need to train a series of shadow models to simulate the target model, significantly impacting the attack model. Additionally, it is shown in [16] that overfitting was highly correlated with MIAs, and the techniques that can mitigate over-fittings, such as L2 regularization and dropout, could weaken the effect of an MIA. Yeom et al. [41]

proposed another inference attack model based on quantitatively analyzing the difference in the model's loss between the training and the testing dataset to simplify the MIA. Salem et al. [21] attempted to decrease the shadow model's number in [16] and proposed a lighter-weight MIA strategy. These schemes rely on black-box features of the target model, such as model prediction, to launch the attack.

White-box membership inference attack. Unlike the previous works, Nasr et al. [27] proposed a comprehensive white-box MIA framework. They assumed that attackers held some knowledge about the training dataset. To obtain the training dataset for the attack model, they use forward and backward propagation to obtain all the model outputs at the data points (e.g., gradients of the data, neuron outputs, and model losses). A one-hot value is added, representing whether the data point participates in the target model training. Their attacks achieve a better performance than the black-box setting. Nasr et al. [27] pointed out that the MIA effect is improved in the FL scenario due to the frequent communication enabling the attacker to acquire more helpful knowledge.

2.1.3. Defense Mechanism against Membership Inference Attack

As MIAs have demonstrated efficiency on various DNN models, researchers have proposed many protection solutions to defend against MIAs. Here, we summarize the existing schemes into four categories, i.e., confidence score masking, regularization, knowledge distillation, and differential privacy. We introduce them separately. In Table 1, we compare various aspects of the existing schemes with those of LDP-Fed.

Table 1. Comparison summary.

Function/Method	[16]	[20]	[22]	[24]	**LDP-Fed**
Data Type	Arbitrary	Image	Arbitrary	Arbitrary	Arbitrary
Federated Learning	×	✓	✓	✓	✓
Defend against white-box MIA	×	×	✓	✓	✓
Defend against black-box MIA	✓	✓	✓	✓	✓
Defense Mechanism	Confidence Masking	Mixup Regularization	Knowledge Distillation	DP-SGD	LDP
Model Performance	High	High	High	Low	High
Efficiency of Training	High	Low	Low	High	High

Confidence score masking. The motivation for confidence score masking is that defenders could hide some prediction information from the attacked model to mitigate black-box MIAs. Shokri et al. [16] proposed sharing the top-k confidence scores with the attacker instead of providing the complete prediction vector. Jia et al. [17] proposed a defense method called MemGuard using the adversarial machine learning technique. However, follow-up works [23,42] proved that neither could completely defend against membership inference attacks.

Regularization. The overfitting problem of machine learning is an important cause of membership inference attacks [16,43,44], and regularization can effectively mitigate overfitting. Therefore, many scholars have analyzed the effectiveness of various regularization methods to defend against MIA (e.g., L2-norm regularization, dropout, early stopping, and label smoothing [45]). Apart from that, regularization methods are specifically used to defend against MIAs. Nasr et al. [19] proposed adversarial regularization. Li et al. [20] proposed Mixup+MMD (Maximum Mean Discrepancy) to mitigate MIAs. Kaya et al. [45] applied data augmentation to machine learning and compared it with classical regularization methods to demonstrate that data augmentation can effectively defend against MIA experimentally.

Knowledge distillation. Knowledge distillation is often applied as an efficient model compression and acceleration technique in machine learning model deployment. It trains a small "student model" that is transferable by extracting the knowledge from an extensive deep neural network "teacher mode" with high generalization ability since the output vector of the teacher model may contain much potential information about itself [46]. Based on knowledge distillation, Shejwalkar et al. [22] proposed the DMP (Distillation for Membership Privacy) method to defend against MIAs. Unlike the conventional model training process, DMP first trains an unprotected teacher model on their private dataset and then uses the teacher model to classify an unlabeled public dataset as their soft labels. Finally, training the student model with a portion of the soft-labeled dataset. Because DMP requires additional public datasets, Zheng et al. [47] proposed complementary knowledge distillation (CKD) and pseudo complementary knowledge distillation (PCKD) that allows the process of knowledge distillation without the need for additional public datasets.

Differential privacy. Differential privacy [48] is an excellent privacy protection framework that strictly quantifies privacy by the privacy budget ϵ and provides different mechanisms to suit all data types. Since Abadi et al. [24] proposed DP-SGD (differential privacy stochastic gradient descent) for DL, DP has been widely employed in DL. The main idea of DP-SGD is to clip the gradients to obtain a bounded local sensitivity and add Gaussian noise to it. Abadi et al. [24] calculated the privacy cost by applying "moments accountant" to reduce the maximum privacy upper bound. DNN models trained with DP-SGD can effectively defend against MIAs but suffer a significant loss of model utility.

2.2. *Background Techniques*

2.2.1. Federated Learning

FL was proposed in the form of collaborative learning by Google in 2017 [49]. The importance of FL is that it enables participants to train high-quality machine learning models without sharing the original dataset directly, which may pose privacy concerns. First, N participants with the same private dataset structure agree on training a typical machine learning model with the same architecture. At each iteration, local participants receive the weights from the central server, perform local training with their private dataset, and finally submit their gradients or local model weights to the server. After that, the central server gathers and averages the gradients or model weights from clients. Afterward, the global model's parameters are distributed to the participants again, and the next iteration starts. This process will iterate until the global model converges or completes a predetermined number of iterations.

2.2.2. Local Differential Privacy

As a variant of DP, local differential privacy [50] aims to enhance the privacy protection of local participants. It allows each participant to perturb sensitive data before being uploaded to the central server. Consequently, the aggregator has no access to the original dataset from the participants. It provides better privacy protection for the participants. The definition of ϵ-LDP is given below:

Definition 1 (ϵ-Local Differential Privacy). *A randomized algorithm $\mathcal{M}$ satisfies ϵ-LDP, if and only if for any two input values v and v' in the domain of $\mathcal{M}$, and any possible output Y of $\mathcal{M}$, we have:*

$$Pr[\mathcal{M}(v) \in Y] \leq e^{\epsilon} \cdot Pr[\mathcal{M}(v') \in Y], \tag{1}$$

where ϵ is the privacy budget, which controls the privacy guarantee of mechanism $\mathcal{M}$, a smaller ϵ means stronger privacy protection, and $Pr[\cdot]$ denotes probability.

In addition, similar to DP, LDP also holds two widely-used properties [51]: post-processing and sequential composition. The former property states that it is always privacy-guaranteed to perform arbitrary computations on the output of a differentially private mechanism. The latter property offers the ability to bound the total privacy cost of releasing multiple results of differentially private mechanisms on the same input data.

Theorem 1 (Posting-Processing). *Let $\mathcal{M}(x)$ be a ϵ-LDP mechanism, where x is the input of $\mathcal{M}$, then for any (deterministic or randomized) function G, $G(\mathcal{M}(x))$ satisfies ϵ-LDP.*

Theorem 2 (Sequential Composition). *Given n mechanisms $\{\mathcal{M}_1(x), \mathcal{M}_2(x), ..., \mathcal{M}_n(x)\}$ satisfy ϵ_i-LDP, respectively, where x is the same input to all of the mechanisms, then for a new mechanism $\mathcal{A}(x) = (\mathcal{M}_1(x), \mathcal{M}_2(x), ..., \mathcal{M}_n(x))$ satisfies $\sum_{i=1}^{n} \epsilon_i$-LDP.*

2.2.3. Piecewise Mechanism

Piecewise mechanism (PM) [50] is a random algorithm that privately analyzes user data's mean value and frequency estimate. Meanwhile, it also ensures that the user side satisfies the ϵ-LDP. An aggregator will calculate the mean value over all n users in this setting. However, instead of raw data t, users send the data t^* perturbed by PM, and PM guarantees that t is an unbiased estimate of t^*.

Theorem 3. *Let $S = \frac{1}{n}\sum_{i=1}^{n} t_i^*$ and $\bar{S} = \frac{1}{n}\sum_{i=1}^{n} t_i$, with at least $1 - \beta$ probability,*

$$|S - \bar{S}| = O(\frac{\sqrt{log(1/\beta)}}{\epsilon\sqrt{n}}). \quad (2)$$

Different from the Laplace mechanism [52] and Duchi et al.'s method [53] for one-dimensional data, PM will adjust the probability destiny function according to the input value, which leads to a lower worst-case variance $\frac{4e^{\epsilon/2}}{3(e^{\epsilon/2}-1)^2}$ compared to Duchi et al.'s $\frac{(e^{\epsilon}+1)^2}{(e^{\epsilon}-1)^2}$ and Laplace mechanism's $\frac{8}{\epsilon^2}$. It is easy to prove that the PM has a lower variance than Duchi et al.'s method when $\epsilon > 1.29$. Additionally, the worst-case variance of the Laplace mechanism and PM will decrease dramatically as ϵ increases. However, PM's worst-case variance is still lower than that of the Laplace mechanism, as illustrated in Figure 1.

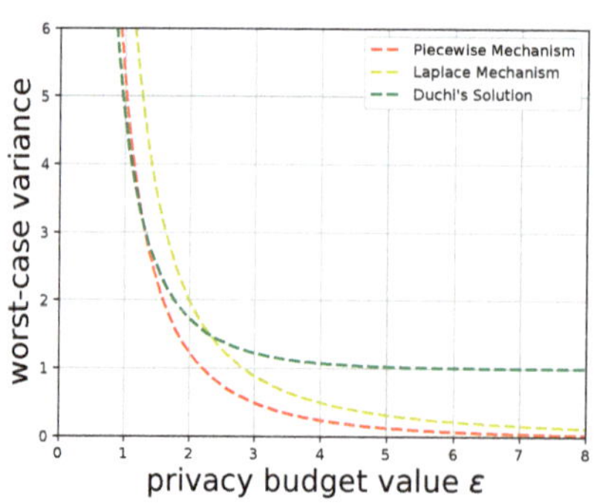

Figure 1. Different LDP mechanisms' worst-case noise variances for one-dimensional numeric data versus privacy budget ϵ.

2.2.4. Membership Inference Attack in Federated Learning

As for the MIA, the primary goal of the attack is to detect whether a data point (x, y) belongs to the training dataset of the target model $f(x, \omega)$. We separate this type of attack into white-box and black-box MIAs depending on the attacker's information about the attacked model [27]. For the white-box setting, an attacker can obtain all information (i.e., model parameters, model architecture, loss function) about the target model and use these features to train its attack model. However, an attack can only access the model through queries for the black-box attack. This paper focuses on the white-box MIA due to the FL setting.

Unfortunately, federated learning is easily attacked by MIAs. As shown in Figure 2, the adversary may be one of the participants or the CS. In this paper, we set a curious parameter server that can receive the gradients or parameters from each participant and perform a passive white-box MIA on all the local participants, respectively. The local side attacker can only observe a fraction of updated global models, so it achieves a lower attack accuracy than the global attacker. Meanwhile, MIA accuracy declines as the number of participants decreases [27].

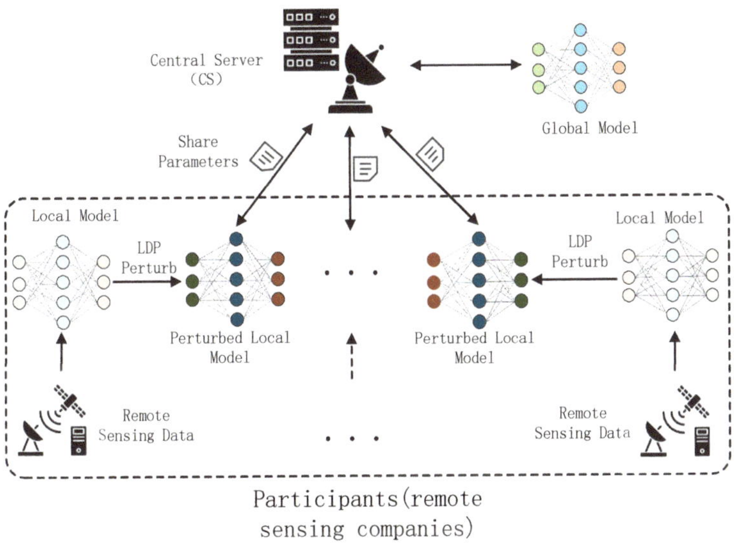

Figure 2. Overview of the LDP-Fed .

3. Methodology

3.1. Overview of LDP-Fed

The rapid development of machine learning techniques has motivated the broad application of deep learning-based image recognition in remote sensing. To improve the model's effectiveness while ensuring the security of remote sensing data, federated learning has also been widely used in remote sensing. However, frequent transmission of local model parameters in federated learning can lead to membership inference attacks on participants' remote sensing data, which are launched against the parameters of a model. Suppose an attacker launches membership attacks on a remote sensing company participating in federated learning. In that case, it can infer the location, type, and time information about the company's satellites through the membership information, resulting in a severe security risk to their satellites. Therefore, this paper proposes LDP-Fed, which aims to train an excellent global model while guaranteeing the security of participants' sensitive data.

3.1.1. Main Motivation of LDP-Fed

The target model's generalization gap (i.e., loss gap) between the model on the training dataset and the test dataset is the fundamental reason for the membership inference attack [22,54,55], which is due to the continuous fitting of the training dataset by the deep learning model during the training process. However, in the FL scenario, attackers can obtain more information about the model, enhancing the MIA's effectiveness. Hence, the methods generally used in the black-box MIA scenario for a defender cannot apply to

FL. Local differential privacy is the best option to solve this problem. Participants use local differential privacy algorithms to perturb the parameters of the local model during interaction to reduce the privacy risk. However, the direct use of LDP brings a huge model utility loss, so we designed LDP-Fed for MIA attacks. We chose more appropriate LDP algorithms and allocated a reasonable privacy budget to achieve a better trade-off between member privacy and model utility, and we improved the practicality of the LDP-Fed framework.

3.1.2. Framework of LDP-Fed

As shown in Figure 2, our LDP-Fed mainly contains two parts: (1) Central Server (CS) and (2) Participants. We apply the local differential privacy to FL to protect the parameters of local models and prevent attackers from obtaining information about the local models' training dataset through parameters.

(1) CS is responsible for communicating with local participants to collect and aggregate local models' parameters. After aggregating the local models into a global model, CS releases the model information to the chosen participants to perform local training and start the next iteration.

(2) Participants are remote sensing companies involved in federated learning. These companies collect remote sensing image data via private satellites and intend to train a more efficient global model collaboratively. At the beginning of FL, participants initialize their local model based on the parameters distributed by the CS (i.e., global model parameters). Then, the participants start to perform local training using their private remote sensing dataset. Then, participants perturb their local model's parameters through the LDP algorithm, upload the perturbation parameters to CS, and wait for the next iteration.

Different from the traditional FL framework, which transfers local models' parameters directly, we perturb the parameters by PM that make the FL process satisfy local differential privacy and can defend against white-box MIAs effectively.

3.1.3. Adversary Model

This paper assumes that the CS and participants are honest but curious. Even though they strictly perform local model training, they remain interested in obtaining the private training dataset of other remote sensing companies to infer the private information of their satellites. Under this assumption, we present the passive adversary $\mathcal{A}$, which has auxiliary data points; a part of that is sampled from the training dataset of the victims. The remaining parts have the same distribution as the training dataset but have not been involved in the model training and launch a white-box MIA to infer whether a data point appears in the training dataset of participants. $\mathcal{A}$ has the following capabilities:

(1) $\mathcal{A}$ can compromise CS to spy on the participants' local training process, collect all the local models from any participants, and launch the passive white-box MIA.

(2) $\mathcal{A}$ can compromise one remote sensing company, different from the CS situation. It can only observe a series of global models and launch a white-box MIA.

3.1.4. Privacy Requirement

In LDP-Fed, we attempt to protect the private data in each participant's private remote sensing data from disclosure. In this paper, we mainly focus on the MIA, which is a highly prevalent and efficient attack in federated learning. Therefore, we require our LDP-Fed framework to be defensible against passive white-box MIA launched from the CS and the participant sides during the process of FL training. It means that the adversary fails to distinguish whether a specific data point belongs to the training dataset of a local participant.

3.2. Details of LDP-Fed

Here, we first describe the steps of LDP-Fed illustrated in Algorithm 1, and we presente key details of LDP-Fed as shown in Figure 3.

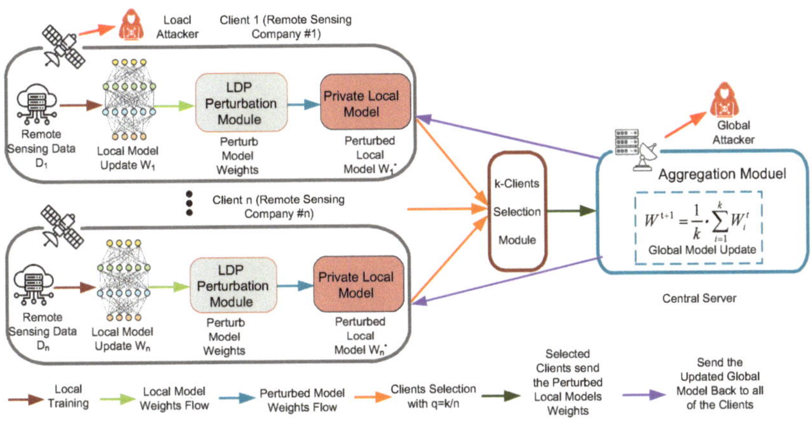

Figure 3. Private federated learning with LDP-Fed.

3.2.1. Federated Learning with LDP

In federated learning, we consider K participants with the same remote sensing data structure who want to train a remote sensing scene classification model collaboratively. During the training process of FL, each participant performs the local model training by its private remote sensing image dataset and shares the local model's weights with the CS. However, to defend against membership inference attacks, we apply PM to FL and propose LDP-Fed. As shown in Algorithm 1, we present our LDP-Fed training process from the perspective of the central server and the local participants separately.

Cloud update. The central server side is similar to the traditional FL framework. First, it generates the target model with the initial parameters θ_0 and sends it to each client with a privacy budget ϵ. Then, the server waits for responses from the clients selected by the k-clients selection module. At every iteration in FL, the k-clients selection module will select $k(k \leq n)$ clients with probability $q = k/n$, where n is the total number of the local clients. Upon getting responses, the central server aggregates and averages all the updated parameters from the local clients and sends the results back to the clients. After that, the central server moves on to the next iteration.

Clients Update. For the clients, they are participants in federated learning, and each client is a remote sensing company with a private image dataset. At each communication round r, the selected clients will update their local models using weights θ_r sent from the central server. Next, they will use their private remote sensing dataset to perform the local training by the SGD (stochastic gradients descent) [56] in parallel. Then, the clients will send the updated local models' weights θ_r' back to the central server and wait for the next iteration. However, they integrate the PM into this process for client-level security concerns. Before the clients send the updated models' weights back to the central server, each parameter will be clipped and then perturbed θ_r' by PM with the privacy budget ϵ as shown in Algorithm 2. They allocate privacy budgets based on the model's different layers, unlike the previous methods [28,29] of allocating privacy budgets equally. The privacy budget allocation is described in Section 3.2.4.

Algorithm 1: LDP-Fed.

Input: K is the total client number. T is the communication rounds number. B is the mini-batch size of the local model. E is the epoch size of the local model. η is the learning rate of the client's training process. ϵ is the privacy budget of LDP.

Sever executes:

Initialize: ω_0 and send ϵ to clients. ; // initialize model weights and send the privacy budget ϵ to clients

for *each communication round t = 1, 2...* **do**

  $k_t \leftarrow$ random set of k_t clients ; // cloud side randomly select k_t clients from K clients with uniformly $q = k_t/K$

  for *each client* $k \in k_t$ *in parallel* **do**

    $\omega_{t+1}^k \leftarrow ClientUpdate(k, \omega_t)$;

  end

  $W \leftarrow \{\omega_{t+1}^k\}_{k \in k_t}$; // Gather k_t clients' local updated model's weights

  for *each* $id \in W$ **do**

    $\omega_{t+1}[id] = \frac{1}{k_t}\sum_{i=1}^{k_t} W_{id}^i$; // average the weights of local models and update the central model

  end

end

SendWeightsToClients(ω_{t+1}) ; // send updated weights back and update all of the local models

Client executes:

function ClientUpdate(k,ω_t):

$\omega_k \leftarrow \omega_t$;

for *each local Epoch i = 1, 2, 3...E* **do**

  for *batch* $b \in B$ **do**

    $\omega_k = \omega_k - \eta \bigtriangledown L(\omega_k, b)$; // mini batch gradient descent and clients local model update

  end

end

for *each* ω *in* ω_k **do**

  $\omega = \omega / max(1, |\omega|)$;

end

ω_k^* = PM_Perturbation(ω_k, ϵ);

return ω_k^*;

Algorithm 2: PM_Perturbation.

Input: model weights $\omega \in [-1, 1]$, privacy budget ϵ

Output: perturbed model weights $\omega_k^* \in [-C, C]$

Sample x uniformly at random from $[0, 1]$;

if $x < \frac{e^{\epsilon/2}}{e^{\epsilon/2}+1}$ **then**

  Sample ω^* uniformly at random from $[l(\omega), r(\omega)]$;

else

  Sample ω^* uniformly at random from $[-C, l(\omega)) \cup (r(\omega), C]$;

end

This process between the central server and local clients will continue until the target model converges or reaches maximum communication rounds (iterations).

3.2.2. Parameter Norm Clipping

Considering the LDP-Fed system's overall privacy budget and communication costs, we transfer the local model's weights, which the local private dataset has trained for three epochs. Therefore, the input value of Algorithm 2 is likely not in the range of $[-1,1]$ compared with the FL system, which transfers the model gradients. To satisfy the local differential privacy of Algorithm 1, we have to bound the perturbed local model weights ω_r into $[-1,1]$. Thus, in Algorithm 1, we have to clip each parameter θ in ω_r before they are perturbed by Algorithm 2, i.e., for a parameter θ replaced by $\theta/max(1,|\theta|)$. The clipping ensures that if $|\theta| \leq 1$, then the parameter θ is preserved, whereas if $|\theta| > 1$ gets scaled down to be 1 or -1.

3.2.3. Parameter Perturbation

After norm clipping, we subject the parameters to perturbation based on the PM. Specifically, Given a weight ω_i of local model $\omega_i \in [-1,1]$ from local client n_i, PM outputs a perturbed value $\omega_i^* \in [-C,C]$, where

$$C = \frac{e^{\epsilon/2}+1}{e^{\epsilon/2}-1}$$

and the PM's probability density function (pdf) is a piecewise constant function as follows:

$$pdf(\omega_i^* = x|\omega_i)\begin{cases} p, & x \in [l(\omega_i), r(\omega_i)], \\ \frac{p}{e^{\epsilon}}, & x \in [-C, l(\omega_i)) \cup (r(\omega_i), C]. \end{cases} \quad (3)$$

where

$$p = \frac{e^{\epsilon} - e^{\epsilon/2}}{2e^{\epsilon/2}+2},$$

$$l(\omega_i) = \frac{C+1}{2} * \omega_i - \frac{C-1}{2},$$

$$r(\omega_i) = l(\omega_i) + C - 1.$$

In Algorithm 2, we show the pseudo-code of PM. We apply PM in our LDP-Fed because, compared with previous mechanisms (the Laplace mechanism and Duchi et al.'s method), PM performs better when dealing with neural network parameters, which have high precision (5–10 digits after decimal points). Given numeric data t to the LDP mechanism, the algorithm returns perturbed data t^*, and t^* is the unbiased estimator of the original data t. However, different LDP mechanisms have variances for t^*. The variances of PM, Duchi's solution, and the Laplace mechanism are $\frac{t^2}{e^{\epsilon/2}-1} + \frac{e^{\epsilon/2}+3}{3(e^{\epsilon/2}-1)^2}$, $(\frac{e^{\epsilon}+1}{e^{\epsilon}-1})^2 - t^2$, $\frac{8}{\epsilon^2}$, respectively. PM has a smaller variance when dealing with smaller values, while Duchi et al. 's has a larger variance.

3.2.4. Privacy Budget Allocation of LDP-Fed

The privacy budget allocation is vital for applying PM to the FL system. PM is a design for the single numerical attribute collection, so we have to use PM for each parameter of the target model. As shown in Algorithm 1, we assume the FL system has a total of E iterations, n clients, and an overall privacy budget ϵ. At each iteration, randomly select k clients for target model updating. Based on the composition property of LDP [57], to guarantee ϵ-LDP, Truex et al. [28] and Sun et al. [29] split the privacy budget ϵ into E small pieces and hold $\epsilon = \sum_{i=0}^{E-1}\epsilon_i$. At each iteration $i < E$, the small piece of the privacy budget ϵ_i will be allocated equally to every dimension of the model parameters, just as $\epsilon_p = \frac{\epsilon_i}{|\theta|}$, where the $|\theta|$ represents the total numbers of the model parameter uploaded to the central server. However, this type of privacy budget allocation assumes that the parameters are identically sensitive to LDP noise. This may lead to a less accurate DNN model or be ineffective in defending against white-box MIAs in FL. Our experiments found that the DNN model's former layers are more noise-sensitive. As the layers increase, the robustness of the parameters increases. This is due to the former layers extracting features from the training dataset and generalizing better than the later layers. Thus, a slight change could be reflected in the DNN model's accuracy. Another important point is that, in [27], their experiments showed that among the layers of the target model, the latter leak more information about the training dataset than the former layers, especially for the last layer. As shown in Theorem 3, the asymptotically optimal error bound of PM is $O(\frac{\sqrt{log(1/\beta)}}{\epsilon\sqrt{n}})$. To increase the target model utility, it has to increase the privacy budget ϵ for each parameter when we fix clients' number k, which may result in privacy leakage from the later layers' parameters when k is small.

Combining the twofold, we suggest that the defender allocates more privacy budget to the former layers of their model when using LDP to defend against white-box MIAs in FL. Specifically, given a total privacy budget ϵ in our LDP-Fed, we allocate it equally to each iteration. In each iteration, we allocate the privacy budget for all parameters in the last layer as ϵ_L. Moreover, for the parameters of layer l, we allocate the privacy budget $\epsilon_L + (L - l) * s$. In our experiments, we set $s = 1$ uniformly.

4. Experimental Setup

In this section, we briefly introduce the experimental setup. It contains the dataset information, the target model's architecture and hyperparameters in FL, the white-box inference attack model setup, and the metrics for model performance.

4.1. Datasets

To evaluate our experiments, we used four datasets: two famous remote sensing image datasets, EuroSAT and NWPU-RESISC45, and two standard image recognition benchmark datasets, Fashion-MNIST and CIFAR10.

EuroSAT was proposed by Helber et al. [58], which is a satellite image dataset based on Sentinel-2. EuroSAT provides 27,000 non-overlapping 64 × 64 color pixels images and is classified into ten categories, as shown in Figure 4. Researchers often use it as a benchmark dataset for land use and land cover issues in remote sensing image recognition.

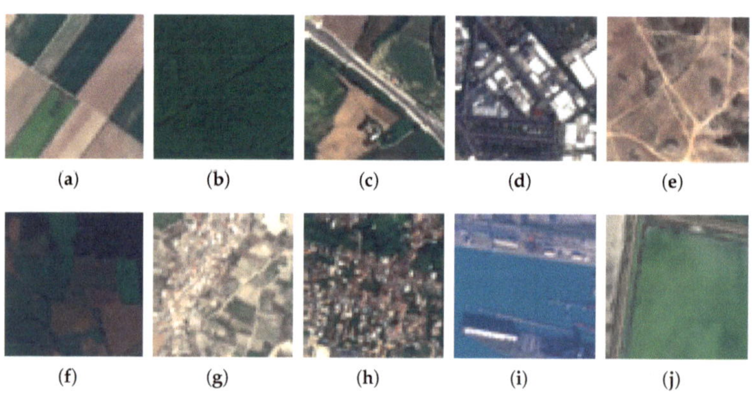

Figure 4. EuroSAT dataset sample image presentations in ten categories. (**a**) AnnualCrop. (**b**) Forest. (**c**) Highway. (**d**) Industrial. (**e**) HerbaceousVegetation. (**f**) Pasture. (**g**) PermanentCrop. (**h**) Residential. (**i**) River. (**j**) SeaLake.

NWPU-RESISC45 [59] was created by Northwestern Polytechnical University (NWPU) as a benchmark dataset for remote sensing image scene classification. It consists of 31,500 remote sensing images that cover 45 scene categories, with 700 images in each category. Due to its large-scale scene categories and data quantity, it is widely used in remote sensing image classification tasks. In Figure 5, we present part of the scene images of NWPU-RESISC45.

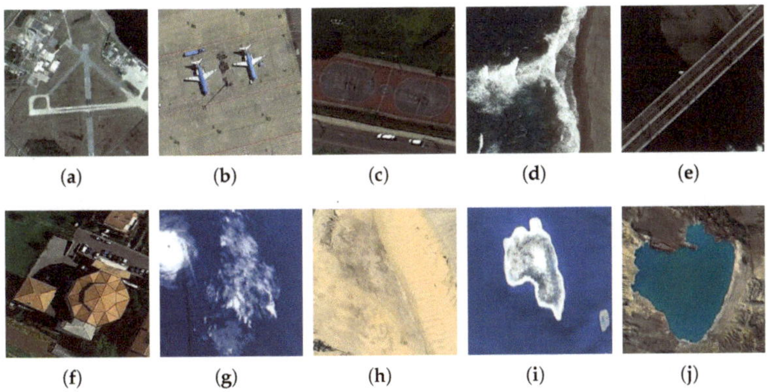

Figure 5. Presentation of a portion of the NWPU-RESISC45 sample, it contains 45 categories of remote sensing scene images. (a) Airport. (b) Airplane. (c) Basketball. (d) Beach. (e) Bridge. (f) Church. (g) Cloud. (h) Desert. (i) Island. (j) Lake.

Fashion-MNIST [60] contains 60,000 training images and 10,000 testing images containing 28 × 28 pixels gray-level (0 to 255). A classification task is trained to recognize the ten clothing labels of the input images (e.g., t-shirt, trousers, and dress).

CIFAR10 [61] is a popular benchmark dataset used to evaluate image recognition algorithms. The dataset consists of 32 × 32 color pixels and contains 50,000 training and 10,000 validation images drawn from 10 classes.

4.2. Target Model Setting

We investigated LDP-Fed on the datasets mentioned previously. For EuroSAT, we used a three convolutional layers and a two fully connected layers convolutional neural network (CNN). As for Fashion-MNIST, we used a six-layer (1024,512,256,128) fully connected neural network (FCN), which had been used by Nasr et al. [27] for a target model to be attacked. The features and label size of the dataset determine the size of the input and output layers. For the CIFAR10 and NWPU-RESISC45, we used the Alexnet [62] and VGG [63] models, respectively. We trained our model with the Adam [64] optimizer with a 0.001 learning rate.

We used two federated learning settings to verify the robustness of the LDP-Fed and the effectiveness against the white-box MIA. Specifically, we chose the parameters averaging the aggregation method in FL. At every iteration in FL, each selected participant sends the updated local model's parameter to the central server after training three epochs. We uniformly used the same dataset size for all the participants. For the former, we used four datasets, and the sizes are shown in Table 2. We averaged all the data points for all of the participants, which do not overlap between various participants. We continued the experiment's setup in [27] for the latter. We set four participants in FL, and the sizes of the datasets are shown in Table 3.

Table 2. The size of the dataset used to train and test the LDP-Fed classification models and the architecture of the models.

Dataset	Architecture	Training Size	Testing Size
EuroSAT	Convolutional Neural Network	15,000	12,000
NWPU-RESISC45	Convolutional Neural Network(VGG)	20,000	10,000
Fashion-MNIST	Fully Connected Network	60,000	10,000
CIFAR10	Convolutional Neural Network(Alexnet)	50,000	10,000

Table 3. Datasets sizes of the target and attack models in FL experiments.

Datesets	Client's Datasets Size		Attack Model Datasets Size			
	Training	Testing	Training Member	Training Non-Member	Testing Member	Testing Non-Member
EuroSAT	5000	5000	2500	2500	2500	2500
NWPU-RESISC45	10,000	10,000	5000	5000	5000	5000
Fashion-MNIST	5000	10,000	2500	2500	2500	2500
CIFAR10	15,000	10,000	10,000	5000	5000	5000

4.3. Threat Model Setting

We implemented the MIA with TensorFlow and Keras. The architecture of our threat model is the same as Nasr et al.'s [27] supervised learning part. When a global attacker launches a white-box MIA against a participant, the attacker first actively collects the local models uploaded by the participant. Then, it obtains all the helpful information, just like the model's output, hidden layers' output, and the gradients of the loss to the parameters by forward and backward passes as the input features of the attack model. Finally, a binary classification attack model is trained by supervised learning. However, the target model is changed to the global model when the attacker is a local participant. An attacker can launch an MIA at any epoch during the FL training process. In this paper, we uniformly set it to 120, which may lead to a better attack performance [27].

4.4. Metrics

To quality the model's utility in FL, we used its top-1 accuracy on the testing dataset, $\boldsymbol{Acc}$. We tested our attack model's accuracy on a dataset consisting of half members and half non-members. However, on the dataset, a random guessing strategy has an accuracy of 50%. Thus, to quantify the performance of the attack model, we followed [20,41,45] to use the adversary advantage metric $\boldsymbol{Adv}$, defined as its accuracy over a balanced 1/2 as a percentage, i.e., $Adv = (Acc - 50\%) * 2$.

5. Experimental Results and Discussion

Here, we present our experimental results for the LDP-Fed defense effectiveness to the white-box MIA in remote sensing, followed by results for analyzing the effect of parameters in LDP-Fed on its performance.

5.1. Defende against White-Box MIA

Unconstrained attack. We presented the performance of the membership inference attack without deployment defense in Table 4. All four datasets suffered from white-box MIAs to varying degrees, especially for the more complex remote sensing image datasets EuroSAT and NWPU-RESISC45. It indicates that remote sensing image classification models are more vulnerable to membership inference attacks. Further, we can conclude that a global attacker is more effective than a local attacker (e.g., 46.4% vs. 25.8% attack Adv on EuroSAT and 44.6% vs. 26.3% on CIFAR10). This is because the model updated by the local clients contains more information about the clients' training data than the averaged model in an iteration of the FL.

LDP against white-box MIAs in FL. We first have to provide good utility for the FL model when $n = 4$. Thus, we chose different ϵ values for the four datasets based on the results of the previous experiment. As shown in Table 4, we set $\epsilon = 3.0$ for EuroSAT, $\epsilon = 2.0$ for CIFAR10, $\epsilon = 5.0$ for NWPU-RESISC45, and $\epsilon = 3.0$ for Fashion-MNIST. They achieved $Acc = 82.5\%$, $Acc = 59.6\%$, $Acc = 77.9\%$, and $Acc = 83.6\%$, respectively. Our LDP-Fed can significantly reduce the white-box MIA's effectiveness while maintaining a high testing accuracy on global models. Under this LDP setting, a passive global attacker Adv is decreased dramatically for all four datasets. For EuroSAT, CIFAR10, and NWPU-RESISC45, LDP reduced attack Adv by more than 70% over the baseline attack Adv, and for Fashion-

MNIST, by almost 82%. In Table 4, we also presented the results of the LDP defending against a local passive attacker. We observed that LDP reduces the risk slightly compared to a global attacker: on EuroSAT, CIFAR10, Fashion-MNIST, and NWPU-RESISC45, it reduced the attack *Adv* by nearly 30%. Fortunately, the efficiency of local attackers decreases quickly and naturally as the number of participants increases, because averaging in the FL scenario may decrease the effects of each party [27].

Table 4. Performance of applying LDP in FL against white-box MIA on EuroSAT, CIFAR10, NWPU-RESISC45, and Fashion-MNIST.

Defense	Dateset	Privacy Budget ϵ	Acc.	Global Att.*Adv*	Local Att.*Adv*
No Defense	EuroSAT	-	89.2%	46.4%	25.8%
	CIFAR10	-	62.1%	44.6%	26.3%
	NWPU-RESISC45	-	83.5%	46.2%	28.4%
	Fashion-MNIST	-	86.5%	24.5%	12.5%
Defense with LDP	EuroSAT	3.0	82.5%	13.8%	17.6%
	CIFAR10	2.0	59.6%	12.2%	18.4%
	NWPU-RESISC45	5.0	77.9%	10.6%	18.6%
	Fashion-MNIST	2.0	83.6%	4.6%	8.2%

Why does LDP-Fed work? In Figure 6, we recorded the performance of the local and global models on four datasets during LDP-Fed's training process. Specifically, we recorded the updated local models' training and testing accuracy, the testing accuracy of the local models after LDP perturbation, and the testing accuracy of the aggregated global models at 30, 60, 90, 120, and 150 epochs, respectively. The experimental results showed that the local and global models converge after 30 epochs, and the perturbation effect of LDP on the local models is evident. It seriously reduced the testing accuracy of the local models by around 40–50 percentage points for all datasets, while the aggregated global model still performs well after aggregation. LDP perturbation caused dramatic changes in the local models' parameters, leading to significant inaccuracies in a global attacker's training dataset. That is why the LDP-Fed is so effective in defending against attacks launched from the global side. The global model received by the client still maintained a good performance, resulting in LDP-Fed being less effective than the global attacker in defending against the local attackers.

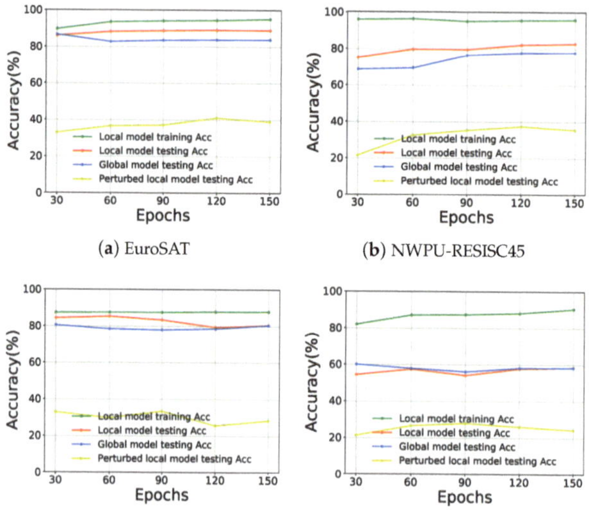

(**a**) EuroSAT (**b**) NWPU-RESISC45

(**c**) Fashion-MNIST (**d**) CIFAR10

Figure 6. Compare the effect of LDP to the local and global models' accuracy on four different datasets.

5.2. Hyperarameter of LDP-Fed Analysis

As shown in Figure 7, to evaluate the relationship between the client's number n and the global model testing accuracy in FL, we fixed the privacy budget ϵ. We changed the client's number n in four different datasets. We observed that the LDP-Fed models perform better with the increase of n, even as close as the undefended models. However, after $n = 10$, they almost hold the same performance distance between LDP-Fed and undefended models no matter the change of n. This is because, as shown in Theorem 3, the error between the mean of the original data and the perturbed data by PM is constrained by the privacy budget ϵ and the number of participants n. The error decreases rapidly as the number of participants increases. In practice, the number of clients n still should be ten or more to get an acceptable model performance in our LDP-Fed.

Another important hyperparameter to the LDP-Fed is the privacy budget ϵ. As shown in Figure 8, we analyzed the effect of c on model performance. Specifically, we set n to 20 in FL and presented the accuracy results with ϵ from 1 to 7 for all four datasets. It showed that the accuracy would increase as ϵ increases. We observed four different ϵ as an inflection point on EuroSAT ($\epsilon = 3.0$), Fashion-MNIST ($\epsilon = 2.0$), CIFAR10 ($\epsilon = 3.0$), and NWPU-RESISC45 ($\epsilon = 5.0$). The model's accuracy dropped rapidly when ϵ was less than the point. On the contrary, the accuracy increased slowly and was close to the undefended model's performance. Furthermore, more complex datasets or DNN models require more privacy costs. This is because of the more sophisticated neural network and more model weights.

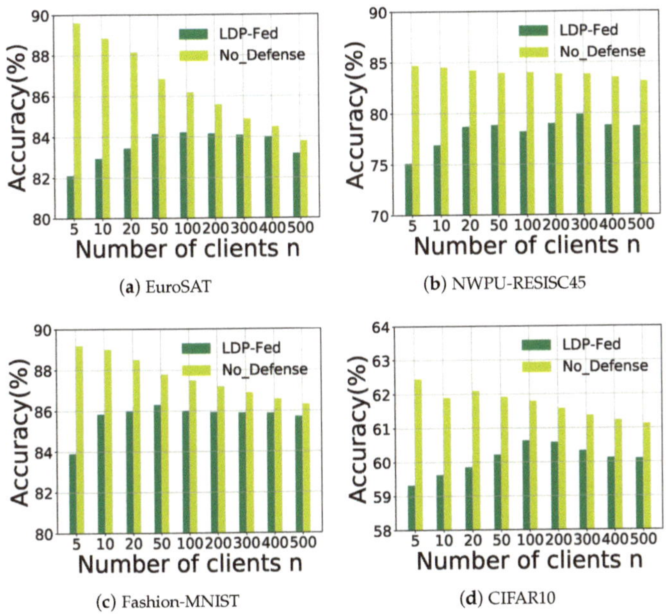

(a) EuroSAT (b) NWPU-RESISC45 (c) Fashion-MNIST (d) CIFAR10

Figure 7. Impact of client number n on the training accuracy in four different datasets.

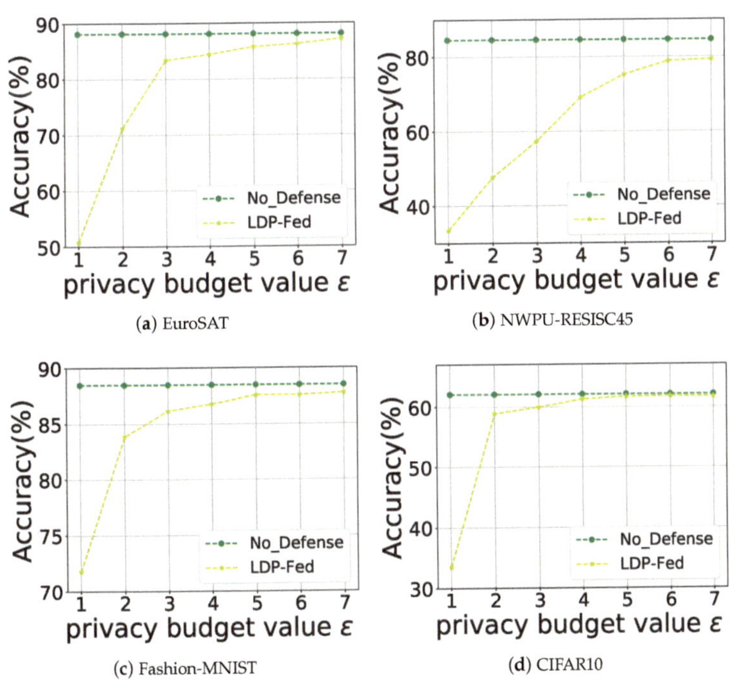

(a) EuroSAT (b) NWPU-RESISC45 (c) Fashion-MNIST (d) CIFAR10

Figure 8. Impact of privacy budget ϵ on training accuracy in four different datasets.

In addition, we evaluated the model convergence rate, which determines the communication costs between clients and the central server in LDP-Fed. As shown in Figure 9, for Fashion-MNIST, the models converged within ten communication rounds. For the rest of the dataset, limited by model complexity and sensitivity, they need more than 15 communication rounds to achieve a better performance. Our LDP-Fed is convenient for training an excellent model within an acceptable communication cost.

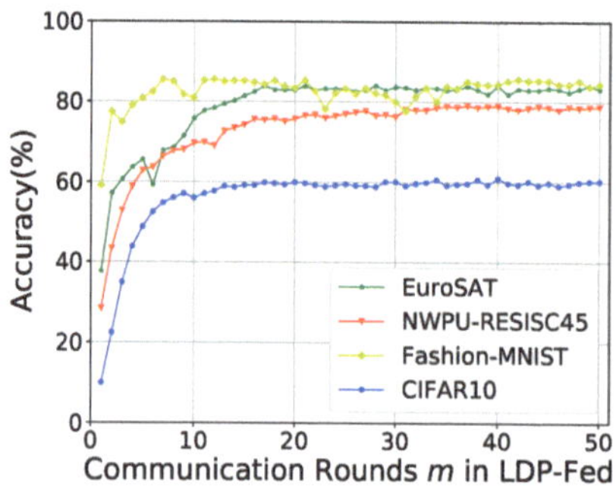

Figure 9. Analyze the relationship between the model accuracy and communication rounds.

5.3. Performance Comparison

Comparison with DP-SGD. As a gold standard, DP provides theoretical privacy guarantees. Abadi et al. [24] proposed DP-SGD to protect against privacy leakage in machine learning. Essentially, DP-SGD clipped and added Gaussian noise to the gradients computed on the private dataset during model training to limit the influence of a single sample on the target model. In this paper, we followed the method of Naseri et al. [15], which applied the DP-SGD to the local models training process of the FL. We compared LDP-Fed and DP-SGD using the empirically observed trade-offs between membership privacy and global model accuracy.

We used CIFAR10 with Alexnet and chose two different privacy budgets for the DP-SGD, $\epsilon = 2.0$ (same with LDP-Fed privacy budget setting), $\epsilon = 8.6$ (suggested by [15]), and $\delta = 10^{-5}$ in both cases. We presented the experimental results in Table 5. As shown in Table 5, our LDP-Fed and DP-SGD did mitigate the white-box MIA for a global attacker. However, LDP-Fed could achieve a better defensive performance while obtaining a target model with higher testing accuracy. When we set the privacy budget of DP-SGD to 2.0, both global and local attackers were almost ineffective. However, this accompanied a loss of more than 31% of model accuracy. As we anticipated, DP-SGD performed better against the local attackers. Unlike DP-SGD, which adds noise during local model training, LDP-Fed-aggregated perturbed parameters were unbiased estimates of the original parameters, meaning the global LDP-Fed models have much less noise compared with DP-SGD.

Table 5. Compare the performance of different defense mechanisms against a white-box MIA in FL using Alexnet on CIFAR10.

Defense Method	Privacy Budget ϵ	Model Acc.	Global Att.Adv	Local Att.Adv
No Defense	-	62.1%	44.6%	26.3%
LDP-Fed	2.0	59.6%	12.2%	18.4%
DP-SGD	2.0	42.3%	5.8%	4.2%
	8.6	54.6%	13.6%	12.8%

Comparison with other LDP-FL. In Figure 8, we set $n = 20$ for all the datasets. Our LDP-Fed achieved 73.4% accuracy with $\epsilon = 3.0$, 77.9% accuracy with $\epsilon = 5.0$, 83.8% accuracy with $\epsilon = 2.0$, and 59.3% accuracy with $\epsilon = 2.0$ on EuroSAT, NWPU-RESISC45,

Fashion-MNIST, and CIFAR10, respectively. To our knowledge, the results are very competitive with previous works. Geyer et al. [65] first applied DP under Gaussian Mechanism to federated learning to preserve clients' level privacy. While they only achieved 78%,92%, and 96% accuracy with $(\epsilon, m, n) = (8, 11, 100), (8, 54, 1000), (8, 12, 10,000)$ on MNIST with differential privacy, where (ϵ, m, n) represented the privacy budget, communication rounds, and clients number, respectively. LDP was first exploited for federated learning by Bhowmick et al. [66]. However, due to the high variance of their LDP algorithm, to improve the convolutional neural network model's accuracy to 10 percent of the original value, they required a relatively large privacy budget and communication rounds, i.e., CIFAR10 $(\epsilon = 1000, CR = 200)$. Truex et al. [28] utilized condensed local differential privacy (CLDP) to FL and up to $Acc = 86.93\%$ for Fashion-MNIST. Unlike the DP privacy budget ϵ, α-CLDP's privacy budget is controlled by the parameter α. However, α-CLDP achieved the accuracy by requiring a high privacy budget $\epsilon = \alpha \cdot 2c \cdot p$ (e.g., $\alpha = 1, c = 1, p = 10$), which may cause a weak LDP guarantee. Meanwhile, they did not apply α-CLDP to a more complex DNN model or dataset. Sun et al. [29] proposed an advanced LDP-Fed recently, which obtained high accuracy while protecting the privacy of the model, i.e., Fashion-MNIST $(\epsilon = 4, Acc = 86.26\%, n = 200)$, CIFAR10 $(\epsilon = 10, Acc = 61.46\%, n = 500)$. However, they need more than 100 clients in FL because they added the same noise to all parameters. Thus, the number of clients must be constantly increased to reduce the noise's impact on model performance. In our LDP-Fed, we designed a special noise addition method against MIAs. We can efficiently train excellent complex DNN models with fewer clients than in previous works.

6. Conclusions

As the most promising solution for remote sensing institutions to maximize the potential of data-driven models, FL has been widely used in remote sensing, such as in FL-based image scene classification. However, FL is vulnerable to white-box MIAs, which may lead to the serious disclosure of the participants' secret satellite sensitive information when we apply FL in remote sensing. To solve the problem, we presented LDP-Fed, a federated learning framework with LDP to defend against white-box MIAs in FL. Unlike conventional federated learning, our system does not require a trusted aggregation server due to local participants perturbing their model parameters with LDP to provide privacy protection. To efficiently train a complex DNN model with LDP guaranteed, we have allocated a reasonable privacy budget according to the network layers. We conducted comprehensive experiments on four datasets, including two benchmark datasets for remote sensing scene classification. Our experimental results demonstrate that remote sensing image classification models are susceptible to MIAs. Furthermore, our framework can defend against membership inference attacks while guaranteeing the utility of the models.

Author Contributions: Conceptualization, Z.Z. and X.M.; methodology, Z.Z.; software, Z.Z.; validation, Z.Z.; formal analysis, Z.Z.; investigation, Z.Z.; writing—original draft preparation, Z.Z.; writing—review and editing, X.M.; supervision, J.M.; All authors have read and agreed to the published version of the manuscript.

Funding: This work was supported in part by Key R&D Program of Shandong Province, China (Grant No. 2023CXPT05); Shenzhen Science and Technology Program (CJGJZD20220517142005013); the National Key Research and Development Program of China under Grant 2021YFB3101100; the National Natural Science Foundation of China under Grant U21A20464, 62220106004, 62072352, and 62261160651; the Key Research and Development Program of Shaanxi under Grant 2023-ZDLGY-52, and Grant 2022GY-029; the Fundamental Research Funds for the Central Universities under Grant ZYTS23167.

Data Availability Statement: Not applicable.

Conflicts of Interest: The authors declare no conflict of interest.

References

1. Yuan, Q.; Shen, H.; Li, T.; Li, Z.; Li, S.; Jiang, Y.; Xu, H.; Tan, W.; Yang, Q.; Wang, J.; et al. Deep learning in environmental remote sensing: Achievements and challenges. *Remote Sens. Environ.* **2020**, *241*, 111716. [CrossRef]
2. Thapa, A.; Horanont, T.; Neupane, B.; Aryal, J. Deep Learning for Remote Sensing Image Scene Classification: A Review and Meta-Analysis. *Remote. Sens.* **2023**, *15*, 4804. [CrossRef]
3. Gadamsetty, S.; Ch, R.; Ch, A.; Iwendi, C.; Gadekallu, T.R. Hash-based deep learning approach for remote sensing satellite imagery detection. *Water* **2022**, *14*, 707. [CrossRef]
4. Ma, D.; Wu, R.; Xiao, D.; Sui, B. Cloud Removal from Satellite Images Using a Deep Learning Model with the Cloud-Matting Method. *Remote Sens.* **2023**, *15*, 904. [CrossRef]
5. Devi, N.B.; Kavida, A.C.; Murugan, R. Feature extraction and object detection using fast-convolutional neural network for remote sensing satellite image. *J. Indian Soc. Remote Sens.* **2022**, *50*, 961–973. [CrossRef]
6. Tam, P.; Math, S.; Nam, C.; Kim, S. Adaptive resource optimized edge federated learning in real-time image sensing classifications. *IEEE J. Sel. Top. Appl. Earth Obs. Remote Sens.* **2021**, *14*, 10929–10940. [CrossRef]
7. Li, K.; Wan, G.; Cheng, G.; Meng, L.; Han, J. Object detection in optical remote sensing images: A survey and a new benchmark. *ISPRS J. Photogramm. Remote Sens.* **2020**, *159*, 296–307. [CrossRef]
8. Ruiz-de Azua, J.A.; Garzaniti, N.; Golkar, A.; Calveras, A.; Camps, A. Towards federated satellite systems and internet of satellites: The federation deployment control protocol. *Remote Sens.* **2021**, *13*, 982. [CrossRef]
9. Büyüktaş, B.; Sumbul, G.; Demir, B. Learning Across Decentralized Multi-Modal Remote Sensing Archives with Federated Learning. *arXiv* **2023**, arXiv:2306.00792.
10. Jia, Z.; Zheng, H.; Wang, R.; Zhou, W. FedDAD: Solving the Islanding Problem of SAR Image Aircraft Detection Data. *Remote Sens.* **2023**, *15*, 3620. [CrossRef]
11. Zhu, J.; Wu, J.; Bashir, A.K.; Pan, Q.; Wu, Y. Privacy-Preserving Federated Learning of Remote Sensing Image Classification with Dishonest-Majority. *IEEE J. Sel. Top. Appl. Earth Obs. Remote Sens.* **2023**,*16*, 4685–4698. [CrossRef]
12. Xu, Y.; Du, B.; Zhang, L. Assessing the threat of adversarial examples on deep neural networks for remote sensing scene classification: Attacks and defenses. *IEEE Trans. Geosci. Remote Sens.* **2020**, *59*, 1604–1617. [CrossRef]
13. Bai, T.; Wang, H.; Wen, B. Targeted universal adversarial examples for remote sensing. *Remote Sens.* **2022**, *14*, 5833. [CrossRef]
14. Brewer, E.; Lin, J.; Runfola, D. Susceptibility & defense of satellite image-trained convolutional networks to backdoor attacks. *Inf. Sci.* **2022**, *603*, 244–261.
15. Naseri, M.; Hayes, J.; De Cristofaro, E. Local and central differential privacy for robustness and privacy in federated learning. *arXiv* **2020**, arXiv:2009.03561.
16. Shokri, R.; Stronati, M.; Song, C.; Shmatikov, V. Membership inference attacks against machine learning models. In Proceedings of the 2017 IEEE symposium on security and privacy (SP), IEEE, San Jose, CA, USA, 22–24 May 2017; pp. 3–18.
17. Jia, J.; Salem, A.; Backes, M.; Zhang, Y.; Zhenqiang Gong, N. MemGuard: Defending against Black-Box Membership Inference Attacks via Adversarial Examples. *arXiv* **2019**, arXiv:1909.10594.
18. Choquette-Choo, C.A.; Tramer, F.; Carlini, N.; Papernot, N. Label-only membership inference attacks. In Proceedings of the International Conference on Machine Learning, PMLR, Virtual, 18–24 July 2021; pp. 1964–1974.
19. Nasr, M.; Shokri, R.; Houmansadr, A. Machine learning with membership privacy using adversarial regularization. In Proceedings of the the 2018 ACM SIGSAC Conference on Computer and Communications Security, Toronto, ON, Canada, 15–19 October 2018, pp. 634–646.
20. Li, J.; Li, N.; Ribeiro, B. Membership inference attacks and defenses in classification models. In Proceedings of the the Eleventh ACM Conference on Data and Application Security and Privacy, Virtual, 22 March 2021; pp. 5–16.
21. Salem, A.; Zhang, Y.; Humbert, M.; Berrang, P.; Fritz, M.; Backes, M. ML-Leaks: Model and Data Independent Membership Inference Attacks and Defenses on Machine Learning Models. In Proceedings of the 26th Annual Network and Distributed System Security Symposium (NDSS), San Diego, CA, USA, 24–27 February 2019.
22. Shejwalkar, V.; Houmansadr, A. Membership privacy for machine learning models through knowledge transfer. In Proceedings of the the AAAI Conference on Artificial Intelligence, Virtual, 2–9 February 2021; Volume 35, pp. 9549–9557.
23. Tang, X.; Mahloujifar, S.; Song, L.; Shejwalkar, V.; Nasr, M.; Houmansadr, A.; Mittal, P. Mitigating membership inference attacks by self-distillation through a novel ensemble architecture. *arXiv* **2021**, arXiv:2110.08324.
24. Abadi, M.; Chu, A.; Goodfellow, I.; McMahan, H.B.; Mironov, I.; Talwar, K.; Zhang, L. Deep learning with differential privacy. In Proceedings of the the 2016 ACM SIGSAC Conference on Computer and Communications Security, Vienna, Austria, 24–28 October 2016; pp. 308–318.
25. Jayaraman, B.; Evans, D. Evaluating differentially private machine learning in practice. In Proceedings of the 28th USENIX Security Symposium (USENIX Security 19), Santa Clara, CA, USA, 14–16 August 2019; pp. 1895–1912.
26. Xie, Y.; Chen, B.; Zhang, J.; Wu, D. Defending against Membership Inference Attacks in Federated learning via Adversarial Example. In Proceedings of the 2021 17th International Conference on Mobility, Sensing and Networking (MSN) IEEE, Exeter, UK 13–15 December 2021; pp. 153–160.
27. Nasr, M.; Shokri, R.; Houmansadr, A. Comprehensive privacy analysis of deep learning: Passive and active white-box inference attacks against centralized and federated learning. In Proceedings of the 2019 IEEE Symposium on Security and Privacy (SP), IEEE, Santa Clara, CA, USA, 20–22 May 2019; pp. 739–753.

28. Truex, S.; Liu, L.; Chow, K.H.; Gursoy, M.E.; Wei, W. LDP-Fed: Federated learning with local differential privacy. In Proceedings of the the Third ACM International Workshop on Edge Systems, Analytics and Networking, Heraklion, Greece, 27 April 2020; pp. 61–66.
29. Sun, L.; Qian, J.; Chen, X. Ldp-fl: Practical private aggregation in federated learning with local differential privacy. *arXiv* **2020**, arXiv:2007.15789.
30. Fadlullah, Z.M.; Kato, N. On smart IoT remote sensing over integrated terrestrial-aerial-space networks: An asynchronous federated learning approach. *IEEE Netw.* **2021**, *35*, 129–135. [CrossRef]
31. Chhikara, P.; Tekchandani, R.; Kumar, N.; Tanwar, S. Federated learning-based aerial image segmentation for collision-free movement and landing. In Proceedings of the the 4th ACM MobiCom Workshop on Drone Assisted Wireless Communications for 5G and Beyond, Virtual, 29 October 2021; pp. 13–18.
32. Lee, W. Federated reinforcement learning-based UAV swarm system for aerial remote sensing. *Wirel. Commun. Mob. Comput.* **2022**, *2022*, 4327380. [CrossRef]
33. Cheng, G.; Xie, X.; Han, J.; Guo, L.; Xia, G.S. Remote sensing image scene classification meets deep learning: Challenges, methods, benchmarks, and opportunities. *IEEE J. Sel. Top. Appl. Earth Obs. Remote Sens.* **2020**, *13*, 3735–3756. [CrossRef]
34. Zhang, L.; Zhang, L. Artificial intelligence for remote sensing data analysis: A review of challenges and opportunities. *IEEE Geosci. Remote Sens. Mag.* **2022**, *10*, 270–294. [CrossRef]
35. Geiß, C.; Pelizari, P.A.; Blickensdörfer, L.; Taubenböck, H. Virtual support vector machines with self-learning strategy for classification of multispectral remote sensing imagery. *ISPRS J. Photogramm. Remote Sens.* **2019**, *151*, 42–58. [CrossRef]
36. Wang, X.; Gao, X.; Zhang, Y.; Fei, X.; Chen, Z.; Wang, J.; Zhang, Y.; Lu, X.; Zhao, H. Land-cover classification of coastal wetlands using the RF algorithm for Worldview-2 and Landsat 8 images. *Remote Sens.* **2019**, *11*, 1927. [CrossRef]
37. Zhang, W.; Tang, P.; Zhao, L. Remote sensing image scene classification using CNN-CapsNet. *Remote Sens.* **2019**, *11*, 494. [CrossRef]
38. Li, Y.; Chen, R.; Zhang, Y.; Zhang, M.; Chen, L. Multi-label remote sensing image scene classification by combining a convolutional neural network and a graph neural network. *Remote Sens.* **2020**, *12*, 4003. [CrossRef]
39. Tang, X.; Ma, Q.; Zhang, X.; Liu, F.; Ma, J.; Jiao, L. Attention consistent network for remote sensing scene classification. *IEEE J. Sel. Top. Appl. Earth Obs. Remote Sens.* **2021**, *14*, 2030–2045. [CrossRef]
40. Chen, J.; Guo, Y.; Zhu, J.; Sun, G.; Qin, D.; Deng, M.; Liu, H. Improving Few-Shot Remote Sensing Scene Classification with Class Name Semantics. *IEEE Trans. Geosci. Remote Sens.* **2022**, *60*, 1–12. [CrossRef]
41. Yeom, S.; Giacomelli, I.; Fredrikson, M.; Jha, S. Privacy Risk in Machine Learning: Analyzing the Connection to Overfitting. In Proceedings of the 2018 IEEE 31st Computer Security Foundations Symposium (CSF), Oxford, UK, 9–12 July 2018; pp. 268–282.
42. Song, L.; Mittal, P. Systematic evaluation of privacy risks of machine learning models. In Proceedings of the 30th USENIX Security Symposium (USENIX Security 21), Virtual, 11–13 August 2021; pp. 2615–2632.
43. Carlini, N.; Chien, S.; Nasr, M.; Song, S.; Terzis, A.; Tramer, F. Membership inference attacks from first principles. In Proceedings of the 2022 IEEE Symposium on Security and Privacy (SP), IEEE, Santa Clara, CA, USA, 23–25 May 2022; pp. 1897–1914.
44. Liu, P.; Xu, X.; Wang, W. Threats, attacks and defenses to federated learning: issues, taxonomy and perspectives. *Cybersecurity* **2022**, *5*, 4.
45. Kaya, Y.; Dumitras, T. When does data augmentation help with membership inference attacks? In Proceedings of the International Conference on Machine Learning, PMLR, Virtual, 18–24 July 2021; pp. 5345–5355.
46. Hinton, G.; Vinyals, O.; Dean, J. Distilling the knowledge in a neural network. *arXiv* **2015**, arXiv:1503.02531.
47. Zheng, J.; Cao, Y.; Wang, H. Resisting membership inference attacks through knowledge distillation. *Neurocomputing* **2021**, *452*, 114–126. [CrossRef]
48. Cynthia, D. Differential privacy. In *Automata, Languages and Programming*; Springer: Berlin/Heidelberg, Germany, 2006; pp. 1–12.
49. McMahan, B.; Moore, E.; Ramage, D.; Hampson, S.; Aguera y Arcas, B. Communication-efficient learning of deep networks from decentralized data. In Proceedings of the Artificial Intelligence and Statistics, PMLR, Fort Lauderdale, FL, USA, 20–22 April 2017; pp. 1273–1282.
50. Wang, N.; Xiao, X.; Yang, Y.; Zhao, J.; Hui, S.C.; Shin, H.; Shin, J.; Yu, G. Collecting and Analyzing Multidimensional Data with Local Differential Privacy. *arXiv* **2019**, arXiv:1907.00782.
51. Dwork, C.; Roth, A. The algorithmic foundations of differential privacy. *Found. Trends Theor. Comput. Sci.* **2014**, *9*, 211–407. [CrossRef]
52. Dwork, C.; McSherry, F.; Nissim, K.; Smith, A. Calibrating noise to sensitivity in private data analysis. In Proceedings of the Theory of Cryptography Conference, New York, NY, USA, 4–7 March 2006; Springer: Berlin/Heidelberg, Germany, 2006; pp. 265–284.
53. Duchi, J.C.; Jordan, M.I.; Wainwright, M.J. Minimax optimal procedures for locally private estimation. *J. Am. Stat. Assoc.* **2018**, *113*, 182–201. [CrossRef]
54. Sablayrolles, A.; Douze, M.; Schmid, C.; Ollivier, Y.; Jégou, H. White-box vs black-box: Bayes optimal strategies for membership inference. In Proceedings of the International Conference on Machine Learning, PMLR, Long Beach, CA, USA, 9–15 June 2019, pp. 5558–5567.
55. Chen, D.; Yu, N.; Fritz, M. Relaxloss: Defending membership inference attacks without losing utility. *arXiv* **2022**, arXiv:2207.05801.
56. Robbins, H.; Monro, S. A stochastic approximation method. *Ann. Math. Stat.* **1951**, *22*, 400–407. [CrossRef]

57. McSherry, F.; Talwar, K. Mechanism design via differential privacy. In Proceedings of the 48th Annual IEEE Symposium on Foundations of Computer Science (FOCS'07), IEEE, Washington, DC, USA, 21–23 October 2007; pp. 94–103.
58. Helber, P.; Bischke, B.; Dengel, A.; Borth, D. Eurosat: A novel dataset and deep learning benchmark for land use and land cover classification. *IEEE J. Sel. Top. Appl. Earth Obs. Remote Sens.* **2019**, *12*, 2217–2226. [CrossRef]
59. Cheng, G.; Han, J.; Lu, X. Remote sensing image scene classification: Benchmark and state of the art. *Proc. IEEE* **2017**, *105*, 1865–1883. [CrossRef]
60. Xiao, H.; Rasul, K.; Vollgraf, R. Fashion-mnist: a novel image dataset for benchmarking machine learning algorithms. *arXiv* **2017**, arXiv:1708.07747.
61. Krizhevsky, A. Learning Multiple Layers of Features from Tiny Images. Master's Thesis, University of Tront: Toronto, ON, Canada, 2009.
62. Krizhevsky, A.; Sutskever, I.; Hinton, G.E. Imagenet classification with deep convolutional neural networks. *Adv. Neural Inf. Process. Syst.* **2012**, *25*. [CrossRef]
63. Simonyan, K.; Zisserman, A. Very deep convolutional networks for large-scale image recognition. *arXiv* **2014**, arXiv:1409.1556.
64. Kingma, D.P.; Ba, J. Adam: A method for stochastic optimization. *arXiv* **2014**, arXiv:1412.6980.
65. Geyer, R.C.; Klein, T.; Nabi, M. Differentially private federated learning: A client level perspective. *arXiv* **2017**, arXiv:1712.07557.
66. Bhowmick, A.; Duchi, J.; Freudiger, J.; Kapoor, G.; Rogers, R. Protection against reconstruction and its applications in private federated learning. *arXiv* **2018**, arXiv:1812.00984.

cancers

MDPI

Article

Prediction of Radiation Treatment Response for Locally Advanced Rectal Cancer via a Longitudinal Trend Analysis Framework on Cone-Beam CT

Zirong Li [1], Ann C. Raldow [2], Joanne B. Weidhaas [2], Qichao Zhou [1,*] and X. Sharon Qi [2,*]

[1] Manteia Medical Technologies Co., Milwaukee, WI 53226, USA; lizirong@manteiatech.com
[2] Department of Radiation Oncology, University of California Los Angeles, Los Angeles, CA 90095, USA; araldow@mednet.ucla.edu (A.C.R.); jweidhaas@mednet.ucla.edu (J.B.W.)
* Correspondence: zhouqc@manteiatech.com (Q.Z.); xqi@mednet.ucla.edu (X.S.Q.)

Simple Summary: Locally advanced rectal cancer (LARC) is commonly treated with neoadjuvant chemoradiation (CRT) followed by total mesorectal excision (TME). Patients respond to the CRT differently due to inter- and intra- patient variability in radiosensitivity. 15–27% of patients completely respond to the CRT and achieve pathologic complete response (pCR), while a large proportion of patients have a partial or poor response. TME is a highly invasive procedure, which can cause overtreatment, leading to morbidity and even mortality for the LARC patients who had a pCR. This work introduces an integrated framework to inform clinical decision making through systematical analysis of longitudinal Cone Beam Computed Tomography (CBCT) during a treatment course. The effectiveness and robustness of the response prediction model, based on quantitative imaging features derived from targeted volume from pre- and during-treatment images, were validated on a retrospective cohort of LARC patients towards personalized treatment.

Abstract: Locally advanced rectal cancer (LARC) presents a significant challenge in terms of treatment management, particularly with regards to identifying patients who are likely to respond to radiation therapy (RT) at an individualized level. Patients respond to the same radiation treatment course differently due to inter- and intra-patient variability in radiosensitivity. In-room volumetric cone-beam computed tomography (CBCT) is widely used to ensure proper alignment, but also allows us to assess tumor response during the treatment course. In this work, we proposed a longitudinal radiomic trend (LRT) framework for accurate and robust treatment response assessment using daily CBCT scans for early detection of patient response. The LRT framework consists of four modules: (1) Automated registration and evaluation of CBCT scans to planning CT; (2) Feature extraction and normalization; (3) Longitudinal trending analyses; and (4) Feature reduction and model creation. The effectiveness of the framework was validated via leave-one-out cross-validation (LOOCV), using a total of 840 CBCT scans for a retrospective cohort of LARC patients. The trending model demonstrates significant differences between the responder vs. non-responder groups with an Area Under the Curve (AUC) of 0.98, which allows for systematic monitoring and early prediction of patient response during the RT treatment course for potential adaptive management.

Keywords: locally advanced rectal cancer; cone-beam computed tomography; longitudinal trending analysis; radiomics

1. Introduction

Colorectal cancer is the fourth most common cancer diagnosed in the United States and has the second-highest mortality rate among all cancer types [1]. Locally advanced rectal cancer (LARC) is commonly treated with neoadjuvant chemoradiation (CRT) followed by total mesorectal excision (TME) and adjuvant chemotherapy [2,3]. While all patients with the locally advanced disease receive neoadjuvant radiation, not all patients benefit equally.

Approximately 15–27% [4,5] of patients completely respond to the neoadjuvant CRT and achieve a pathologic complete response (pCR), while a large proportion of patients have a partial or poor response. Given that TME is a highly invasive procedure, the current one-size-fits-all approach can cause overtreatment [6–9], leading to morbidity and even mortality for those LARC patients who had a pCR. For non-responders, neoadjuvant CRT may not provide enough benefit, and alternative or additional strategies may be indicated.

A major challenge for LARC management is distinguishing patients who may respond to RT from those who may not on an individual level. In recent years, advanced cross-sectional and functional imaging enables non-invasive characterization of the tumor, making it possible to establish pre-treatment biomarkers for the accurate prediction or early assessment of post-RT response to improve prognostication and patient stratification towards an optimal outcome. Indeed, comprehensive cancer characterization based on deep learning (such as tumor heterogeneity) derived from medical imaging could improve patient risk classification and optimize management [10–12]. Quantitative imaging features have been widely utilized to extract patient-specific information from medical images to predict the effectiveness of radiotherapy [13,14]. More and more research has shown the potential predictive power of CT-based radiomic signatures in identifying patients with postoperative pathological good responses [15–18].

In-room volumetric cone-beam computed tomography (CBCT) is widely used to ensure proper alignment but also allows us to assess tumor response during the entire treatment course. The CBCT-based radiomic provides a large number of potential features, as well as feature changes in those features, thus offering additional predictive values of tumor response [19–22]. Many researchers have begun to pay attention to daily images during treatment, and many publications have proved the values of delta-radiomic research during treatment [23–27]. Plautz et al. demonstrated that changes in features extracted from daily images could be caused by treatment and proved their potential in the early assessment of treatment response [28]. Nasief et al. proposed a prognosis analysis based on delta-radiomic features for a cohort of pancreas patients (AUC = 0.94) [29]. Boldrini et al. demonstrated that delta-radiomic and imaging features variation during magnetic resonance-guided radiotherapy treatment may represent a cCR prediction tool in locally advanced rectal cancer [30].

Traverso et al. reviewed 41 literatures and evaluated the methodological details in each study, they concluded that the repeatability and reproducibility of radiomic features are sensitive to image quality and software used to extract radiomic features. Only a few features are reproducible, some features seem impossible to be reproducible, which makes it difficult to replicate others' conclusions [31]. Delta-radiomic features, involving feature variations at different acquisition time points, offer additional potentials and benefits for experimental reproducibility. Nardone et al. demonstrated that delta-radiomic features are more robust than traditional radiomic features [32]. The feasibility of CBCT radiomic features largely depends on the quality of the radiomic features. Delgadillo et al. conducted research on the repeatability and reproducibility of CBCT, quantifying repeatability by correlating the radiomic features obtained from 20 repeat CBCT scans within 15 min. However, these radiomic features were reproducible in more than the 9.8% of reconstruction and preprocessing methods [33]. Despite of great potentials, the utilization of a longitudinal CBCT-based radiomic analysis for early response prediction has not been widely adopted at clinic possibly due to (1) tedious steps requiring additional resources and manpower, and (2) the lack of an integrated framework that truly benefits clinical decision-making.

In this work, we proposed a longitudinal radiomic trend (LRT) framework to monitor and forecast potential treatment response based on series of image features derived from daily CBCT during a treatment course. To this end, we developed an automated platform to facilitate clinical implementation. The LRT framework consists of four modules: (1) Automated registration of CBCT scans to planning CT and evaluation; (2) Feature extraction and normalization; (3) Longitudinal trending analyses; (4) Feature reduction and model creation. The proposed framework allows for more robust treatment response assessment through daily CBCT analyses for personalized medical advice.

2. Materials and Methods

2.1. Overview of Longitudinal Radiomic Trend (LRT) Framework

We developed a framework to facilitate clinical decision-making that involves daily CBCT images during the entire treatment course, alongside the localization CT image and initial plan contour during treatment. Figure 1 shows the schematic process for the proposed framework. The process and all the details are explained further below.

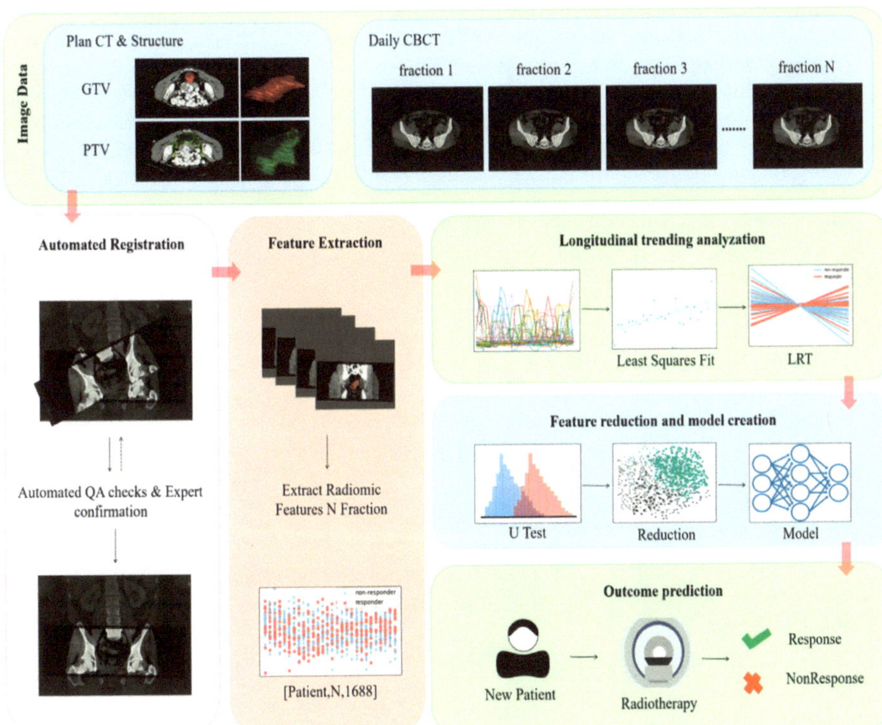

Figure 1. The schematic workflow of longitudinal radiomic trend framework (LRT).

2.1.1. Automatic Registration and Evaluation

Daily pre-treatment CBCT images are subjected to rigid registration with the planning CT images. In our framework, we used rigid transformation based on mutual information and automatically evaluated the consistency of the displacement matrix of each daily CBCT registration for each patient. Unexpected registration results were re-checked and registered manually. The automatic consistency assessment is based on manually checking the accuracy of the registration of the first day's images and then comparing the subsequent registration deformation matrix with the first day's deformation matrix. If the difference is higher than 5%, a warning is issued requiring further manual verification. Finally, all registration results were reviewed to guarantee the reliability of the registration results.

2.1.2. Feature Extraction and Normalization

The gross tumor volume (GTV) and planning tumor volume (PTV) were delineated on planning CT, with guidance of MRI and/or PET for treatment planning purposes. The original GTV and PTV contours were then transferred to the fused CBCT images, from which radiomic features were extracted separately for the GTV and PTV. While shape features were extracted from the original image without transformation, resulting in a total of 14 features. The images underwent 7 transformations for other feature extractions. These transformations include Laplacian of Gaussian, Wavelet, Gradient, SquareRoot, Logarithm, Exponential, and Square. Among them, Laplacian of Gaussian was performed 4 times with sigma values of 0.6, 1.0, 3.0, and 5.0. Wavelet was combined with high-pass and low-pass filters 8 times, resulting in 18 types of images including the original image. Each transformed image contained 93 features, including 18 first-order features, 24 GLCM features, 16 GLRLM features, 16 GLSZM features, 14 GLDM features, and 5 NGTDM features. For patients who received N fractional treatments, N groups of feature sets were calculated based on the GTV area and PTV from the daily CBCT, resulting in a total of 1688 imaging features in each group. Feature extraction was performed using PyRadiomics (version 3.0.1) [34]. For each patient, we investigated the trend of each feature's changes, rather than focusing on the magnitude of the values. To this end, we performed Z-score tests using each patient's daily feature mean and variance to standardize, and the absolute value of 3 was imposed as the outlier boundary [35].

2.1.3. Longitudinal Radiomic Feature Trending

Quantitative image features, captured by daily CBCT scans, can be useful to assess the change of radiomic features in a longitudinal pattern, which, in turn, provides early prediction to identify, quantify, and potentially predict therapy-induced changes over the course of treatment. For each fraction of j (j = 1, 2, . . ., N) of each patient, the feature (i = 1, 2, . . ., 1688) extracted from each CBCT scan, let X_j denote the number of delivered fractions.

The least-squares method [36,37] was employed to fit the features over the course of treatment, focusing on the trending of feature changes to avoid atypical features and noise in the images. For each patient, assuming a linear relationship. Equation (1) is used to capture the correlation between each radiomic feature RF_{ij} and the number of fractions X_j.

$$RF_{ij} = LRT_i \cdot X_j + \beta_i \tag{1}$$

where the slope LRT_i can be calculated in Equation (2), and the intercept of the fitted feature line β_i can be calculated by Equation (3). Calculate the fitting error for each feature by Equation (4).

$$LRT_i = \frac{\sum_j^n X_j \cdot RF_{ij} - nX \cdot RF_i}{\sum_j^n X_j^2 - nX^2} \tag{2}$$

$$\beta_i = RF_i - LRT_i \cdot X \tag{3}$$

$$ERR_i = \frac{\sum_j^n \left| RF_{ij} - \left(LRT_i \cdot X_j + \beta_i\right)\right|}{n} \tag{4}$$

X is the mean of the number of treatments. RF_{ij} denotes the features extracted from the j-th daily CBCT scan, RF_i is the mean of the N daily CBCT feature. Each feature of each patient can be characterized with a feature-specific LRT_i and β_i.

Then, substitute the specific LRT_i and β_i into Equation (5) as the LRT feature (LRTF) of each feature for patients. Subsequent feature selection and modeling will be based on LRTF.

$$LRTF_i = LRT_i + \beta_i \tag{5}$$

2.1.4. Feature Reduction and Modeling

Feature selection was performed by univariate Mann–Whitney U-test between the responder and non-responder groups, with the retained features resulting in significant LRTFs with p-values less than 0.05 in the U-test. The U-test ensures that there is a difference in the features between the patients who responded to RT. For the LRTFs retained in the

U-test, univariate logistic regression models were utilized for dimensionality reduction, keeping LRTFs that were correctly classified in the validation set. This step makes the retained features tend to have high metrics in the validation set. Final retained LRTFs are tested and modeled using a random forest (RF) algorithm to create a model that monitors and predicts early patient response during the radiation treatment delivery course. We used the predicted probability of responding patients as the RF score and the predicted probability comes from the results of the RF model; the RF model was performed using scikit-learn (version 0.24.2) [38].

We build a statistical weight (SW) model to interpret the RF results for clinical reference. The SW model was carried out by directly weighting the LRT and β obtained after fitting all features. The results of the univariate Mann–Whitney U-test reserved features to calculate the mean of the fitting features LRT and β; each LRT and β are weighted by univariate U-test p-values, weight formula (6).

$$w = \frac{0.05 - p}{0.05} \tag{6}$$

After obtaining the mean value of LRT and β and substituting them into Formula (5), we also obtain the LRTF. We calculated the accuracy for the two categories of LRTFs using a threshold of 0; a score greater than 0 suggests a potential good responder, while less than 0 suggests a poor responder patient.

For each patient, the LRT framework intervenes and start to work from the second fraction. For a treatment course of 28 fractions, 27 pairs of independent experiments, involving image features of the $1st$ fraction and kth fraction, were conducted to assess the treatment response of the patient. Leave-one-out cross-validation (LOOCV) for each experiment was performed to confirm the stability and potential of early treatment prediction of a therapeutic effect on the patient. Such experiments were performed on the image features based on the GTV region and the PTV region derived from CBCT scans, respectively.

The LRT framework was compared to radiomic models. Three radiomic models were created to predict post-RT treatment response: (1) radiomic model based on pre-treatment radiomic features; (2) during-treatment radiomic model based on 28 daily CBCT scans; (3) combined model using pre- and during-treatment radiomic features. In the radiomic models, the same imaging pre-processing, extraction, and reduction were used for training and prediction on the same patient cohort.

To evaluate model performance, Area Under Curve (AUC) and accuracy were utilized. We introduced a consistency index (CI) to evaluate the robustness of the LRT model. The CI is defined as whether it was correctly classified in the LOOCV experiments, regardless of whether the patient is included in the training data. The experiment with correct classification is recorded as 1, otherwise it is recorded as 0, and the mean of all experiments is calculated as the consistency of 6 patients.

2.2. Validation

We evaluated the proposed framework using a cohort of 30 patients (21 male and 9 female patients) diagnosed with LARC at stage II–IV. The average age (and standard deviation) of the patient cohort is 55.9 $\pm$ 11 (yr). The patient demographic information is summarized in Table 1. All patients underwent conventional neo-adjuvant fractionated radiation treatment with a prescription dose of 50.4 Gy (28 fractions) on a C-arm accelerator. The gross tumor volume (GTV) was delineated on the planning CT with the guidance of MRI or PET images. The clinical tumor volume (CTV) is the rectum and mesentery, including the perirectal, pre-sacral, and internal iliac nodes and the posterior portion of the obturator nodes. The CTV was expanded 1 cm into the posterior bladder (for bladder variation above vagina/prostate). The planning target volume (PTV) was generated using a 0.5–0.7 cm margin on the CTV. Prior to each treatment delivery, a daily kilovoltage CBCT image was acquired to ensure proper patient alignment, resulting in a total of 840 CBCT scans for the patient cohort. Four to eight weeks after the completion of CRT, all patients underwent total mesorectal excision surgery. Patients were separated into good-responder (GR) and non-GR groups according to the postoperative pathology report, MRI, or colonoscopy. The GR group (14 patients) consisted of patients with either complete clinical response (cCR) or partial response (PR), and the non-GR group consisted of patients with stable disease (SD) and progressive disease (PD).

Table 1. Patient demographic information.

Clinical Characteristics	Good Responder (GR)	Poor Responder (PR)
# of cases	14	16
Gender, n (%)		
Male	9 (64.3%)	12 (75%)
Female	5 (35.7%)	4 (25%)
Average age (standard deviation) (yr)	52.3 ± 11.7	59.2 ± 9.3
Stage, n (%)		
II	3 (21.4%)	4 (25%)
III	8 (57.2%)	9 (56.3%)
IV	3 (21.4%)	1 (6.3%)
Unknown		2 (12.4%)
Response scores, n (%)		
0	7 (50%)	
1	7 (50%)	
2		11 (68.8%)
3		5 (31.2%)

3. Results

Figure 2 shows the selected 42 features for a sample patient. Each subplot shows the fitting of each selected imaging feature when all N fractions of the CBCT scans, i.e., N = 5, 10, 15, 20, 28, were considered, respectively. Most of the features in the figure show an increasing or decreasing relationship in the overall trend of value transformation.

Random forest model yielded AUCs of 0.98 (95% CI [0.97, 0.99]) and 0.9867 (95% CI [0.969, 1.0]), respectively, for GTV and PTV regions through LOOCV in 27 experiments. Figure 3a,b are the box plots of RF scores in 27 experiments for the GTV and PTV regions, respectively. The red box are responders, and the blue box are non-responders. Significant differences were observed in RF scores among responders and non-responders. Figure 3c visualizes the number of features retained after the U-test. Figure 3d illustrates the trend of the accuracy rate change with the increase in the number of treatments with every five treatments used as a group statistically. Before the 15th fraction, the accuracy increases with increasing fractions, after the 15th fraction, the predictive accuracies becomes more stable. Although the GTV regions preserve more features than the PTV regions, the overall accuracy of the GTV area is lower than that of the PTV area, as it can be seen from Figure 3d.

Figure 4 shows the trending patterns of the LRT derived from PTV features using the SW model for all 30 patients, and the model results for N fractions are shown separately, where N = 2–28. The red lines denote trending plots for the responders, the blue lines are for non-responder patients, and the classification accuracy of the SW model uses 0 as the threshold. The classification model is directly weighted and summed based on the U-test feature screening results. The average accuracy of the GTV and PTV regions in 13 experiments before the first 15 treatments (N: 2–14) is 0.53 and 0.62, respectively. The average accuracy of six experiments with features from 15 to 20 treatments (N: 15–20) is 0.662 and 0.738, respectively, while the average accuracy of features from more than 20 treatments in eight experiments (N: 21–28) is 0.671 and 0.795, respectively.

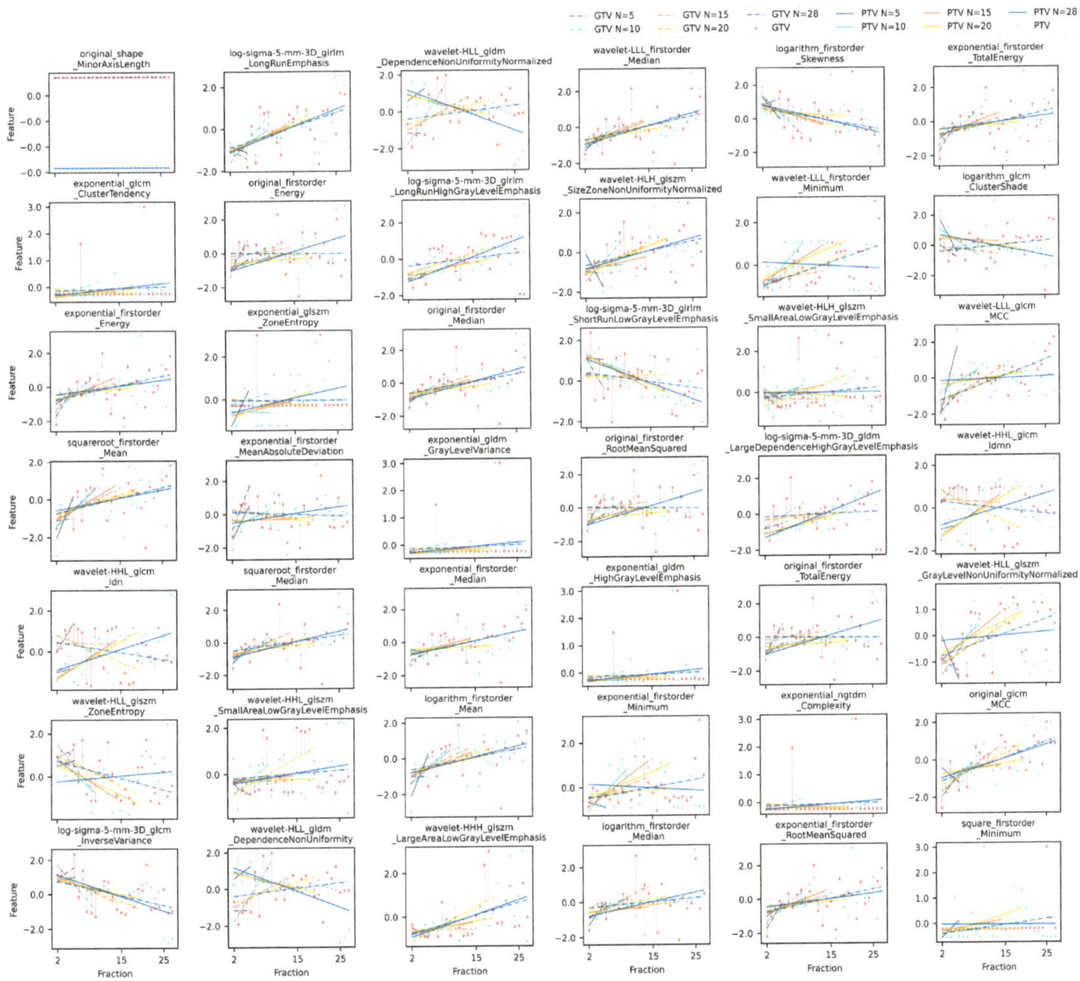

Figure 2. The least-squares fitting results for 42 preserved imaging features as a function of # of fractions for a representative patient; x-axis is # of fractions, and y-axis is the selected feature values derived from the GTV and PTV. The blue triangle points in the figure represent the feature values in the PTV of CBCT images during 28 treatments, the red circular points represent feature values of the GTV, the dotted line represents the straight line fitted by the features of the GTV area, and the solid line represents the straight line fitted by the features of the PTV. The distance between the point and the line represents the fitting error. The lines of five colors in the figure, DarkGray, MediumTurquoise, Orange, Gold, and Medium Blue, respectively, represent five groups of experiments with # of fractions of 5, 10, 15, 20, and 28.

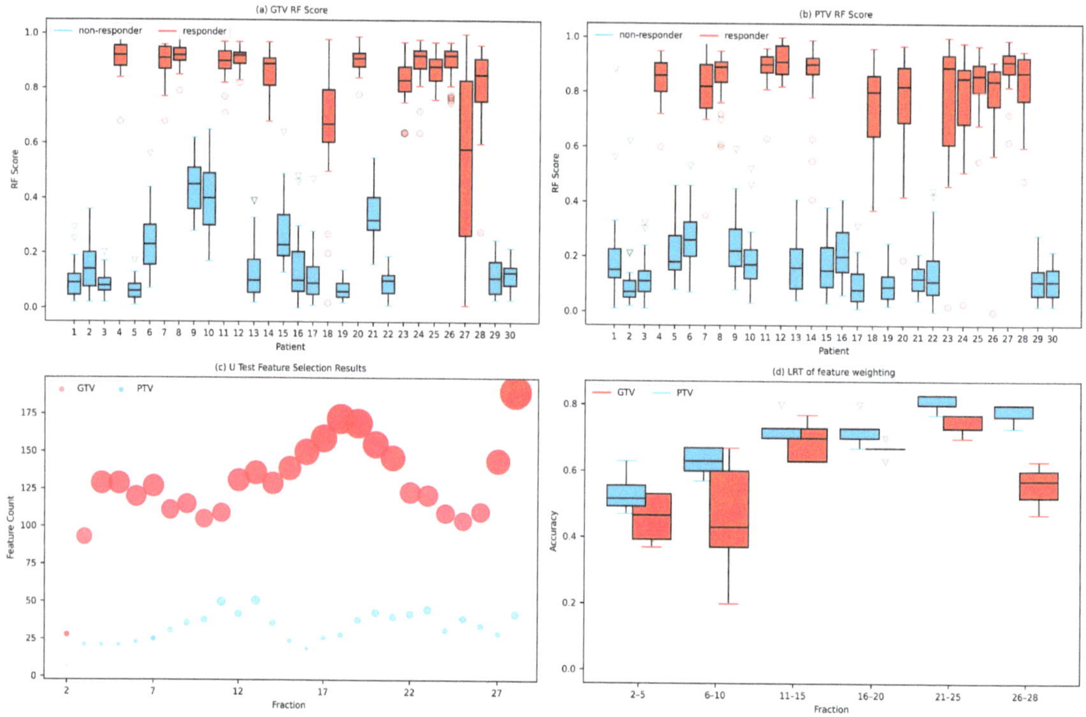

Figure 3. Box plots of patient RF scores using features derived from GTV (**a**) and PTV regions (**b**) for the cohort of LARC patients. The red boxes are responders, and the blue boxes are non-responder patients. (**c**) Number of features retained after the U-test. The red circles are the feature counts from the GTV, and the blue circles are the features derived from PTV. Larger size of circles means more features are retained. (**d**) Boxplot of model predictive accuracy using CBCT scans of previous n fractions. The red boxes are the GTV, and the blue boxes are the PTV.

Figure 5 shows the consistency of the LOOCV experiments and the least-squares fit error mean of all features at Nth fractions, where N = 2–28, for each patient. To better interpret the results, the patients were divided into two groups based on the consistency mean after 15 fractions; the consistency means greater than 0.8 are defined as the high-consistency group, and the consistency means smaller than 0.8 are defined as the low-consistency group. For patients in the low-consistency group, a certain degree of overfitting was likely caused by the small sample size, especially for those patients with response scores of 1 and 2. Consistency analysis allows us to be more cautious about high indicators that appear in machine learning models. The group of patients with high consistency increased significantly after 15 factions, the mean consistency value increased from 0.71 to 0.92, and the least-squares fit error declined from 2.69 to 0.715, suggesting an improved classification ability when 15 or more factions are considered.

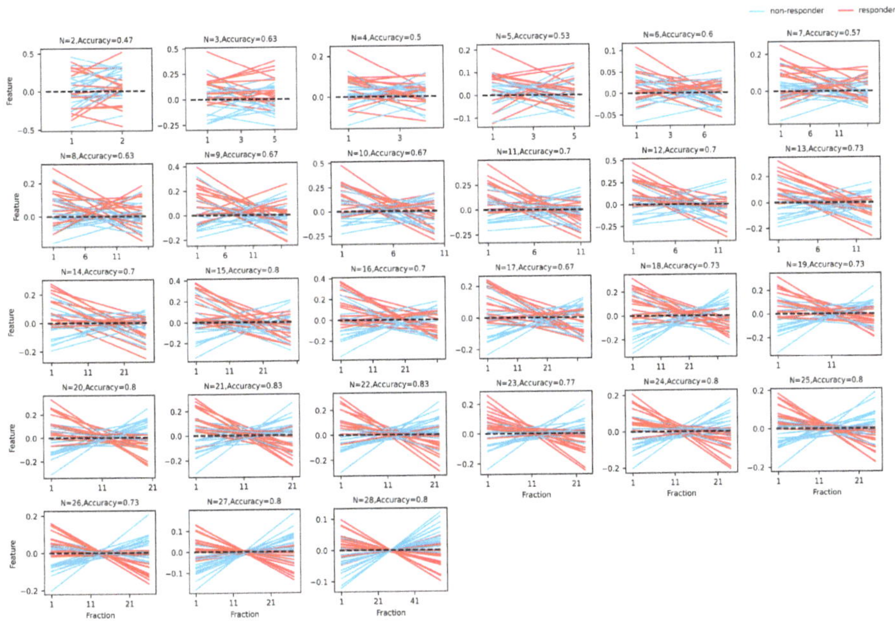

Figure 4. Trending patterns of the LRT derived from PTV features using SW model for all 30 patients. The red lines are the responders, the blue lines are the non-responder patients, with 0 as the threshold for the classification accuracy of the SW model.

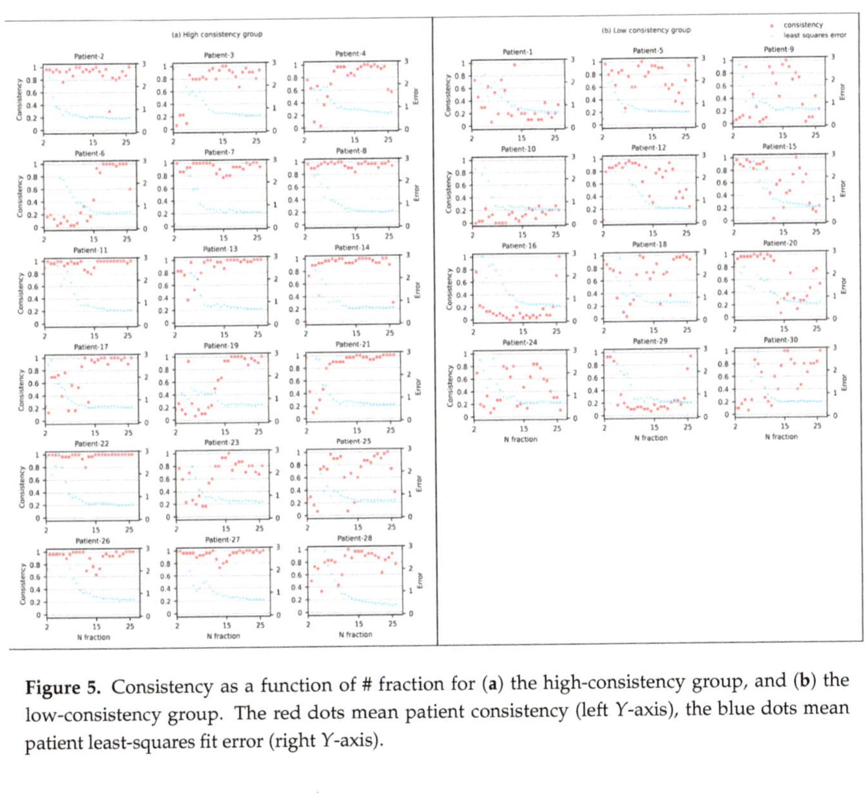

Figure 5. Consistency as a function of # fraction for (**a**) the high-consistency group, and (**b**) the low-consistency group. The red dots mean patient consistency (left Y-axis), the blue dots mean patient least-squares fit error (right Y-axis).

4. Discussion

Accurate and early prediction of patient response is critical for subsequent treatment management and adaptive planning. A longitudinal radiomic trend framework was proposed to systematically assess treatment response using CBCT in the work. This platform automates the process for the early assessment of patient response for routine clinical practice.

In the current trending analysis, we assumed that the image feature changes during the treatment course of LARC patients could be mapped to the final treatment effect through a linear relationship. Given the patients' heterogeneity, the simple linear relationship may or may not be true for all patients. Despite being a simplified linear model, our model achieved great predictive accuracy and consistency for the cohort of LARC patients in the study. Specifically, accurate predictions could be achieved as early as the 15th fraction for the patient cohort (Figure 3d). We expect that the current platform can be utilized to access non-linear relationships between longitudinal imaging series and clinical endpoints for other disease sites. Artificial intelligence (AI)/deep learning (DL) is expected to be a good alternative for predicting non-linear feature transformation trends [39,40]. However, DL approaches generally require large data sets to train and evaluate the predictive models. Given the limited data size, overfitting and other issues are likely to occur. The proposed framework is designed to be flexible to incorporate both linear and non-linear trending models for outcome prediction for different cancer types.

In the comparative experiment of imaging analyses from the GTV area versus those from the PTV area, the predictive models created based on the GTV area were less predictive than the models created from the PTV area for trending analyses and radiomic experiments using either daily fractional CT solely or a combination of daily CBCT and pre-treatment CT. This can be understood in that the predictive models, when nodal area was considered, will better capture comprehensive underlying characteristics and therefore offer a better performance.

We evaluated robustness of the LRT by analyzing the consistency of 27 sets of LOOCV experiments for each patient. A higher consistency index demonstrates that the patient's characteristics display more stability; caution should be exercised in evaluating the high accuracy achieved in machine learning models with patients who exhibit lower consistency. Compared to the machine learning models, we achieved the accuracy of 0.8 via the proposed SW models, which suggests that the LRT can separate the responders from the non-responders with decent performance. Analysis based on clinical data shows that the staging system and clinical features cannot differentiate between responders and non-responders.

Advantages of the proposed trending methods are more predictive than conventional ML-based delta-radiomic methods given the limited amount of data. In pre-treatment CT-based radiomic prediction models, the AUCs of the GTV and PTV radiomic feature sets were 0.43 and 0.62, respectively. In the CBCT-based radiomic prediction model, the AUCs ranged from 0.15 to 0.58 using the series of image features derived from the GTV region, and the AUCs ranged from 0.42 to 0.62 using the features derived from the PTV region. In the combined model, the AUCs ranged from 0.07 to 0.72 using the Image features from the GTV region, and the AUCs ranged from 0.27 to 0.65 using the features from the PTV region. The ML-based radiomic prediction models suggested serious overfitting due to the relatively small number of the data sample. The LRT may be considered similarly to project high-dimensional spatial features to low-dimensional space, which makes complex feature variables simple and useful. The proposed LRT framework, focusing on trending analysis throughout the entire course, yields more effective and robust results and maintains a good generalization ability even for relatively small data sets.

The proposed automated framework LRT platform aims to streamline the clinical utilization of longitudinal quantitative images for treatment response prediction. The platform achieved good performance for the cohort of LARC patients. The target region is currently obtained through registration from the planning GTV/PTV to the CBCT scans during the course of radiation treatment, which may or may not capture actual target volume changes. Yamashita et al. reported a tendency of rectal volume reduction during weekly CBCT through neoadjuvant concurrent chemoradiation therapy for rectal cancer [41]. Zumre et al. demonstrated that the mesorectum moved 20 mm laterally and in the anterior/posterior direction and up to 10 mm in the inferior/superior direction [42]. Although certain target volume changes are eliminated through image registration, accurate segmentation is important for accurate predictions. Furthermore, it is known that CBCT images are subject to poor quality and artifacts, which may be a big challenge for stable radiomic features. Our future work will focus on improvement of CBCT image quality to achieve high performance of automated segmentation for the targets. Lastly, the proposed LRT can alleviate the instability brought by image quality to some extent, which mayfurther reduce fitting error of the LRT and increase its confidence level.

5. Conclusions

We developed a framework that integrates image registration, quantitative feature extraction, and longitudinal imaging trending analysis to predict post-treatment clinical outcomes. The effectiveness of the framework was validated for a retrospective cohort of LARC patients with superior performance, allowing for systematic monitoring and the early assessment of patient response as early as the 15th fraction during the RT treatment course for potential adaptive management.

Author Contributions: Conceptualization, Z.L. and X.S.Q.; methodology, Z.L., X.S.Q. and Q.Z.; software, Z.L.; validation, Z.L.; formal analysis, Z.L.; investigation, Z.L.; resources, A.C.R., J.B.W., X.S.Q. and Q.Z.; data curation, X.S.Q., A.C.R. and J.B.W.; writing—original draft preparation, Z.L.; writing—review and editing, X.S.Q., A.C.R., J.B.W. and Q.Z.; visualization, Z.L.; supervision, X.S.Q. and Q.Z.; project administration, X.S.Q. and Q.Z.; funding acquisition, none. All authors have read and agreed to the published version of the manuscript.

Funding: J.B.W. reported a grant from the National Cancer Institute of the National Institutes of Health under Award R01 CA 238998. X.S.Q. reported a grant from Varian Medical Systems.

Institutional Review Board Statement: UCLA IRB#23-001021 waived.

Informed Consent Statement: Patient consent was waived under the UCLA IRB.

Data Availability Statement: The data presented in this study are available upon request from the corresponding author.

Conflicts of Interest: J.B.W. has created IP that was licensed to a company she co-founded, Miradx, by Yale university. Z.L., A.C.R., Q.Z., X.S.Q. declare no conflict of interest.

References

1. Siegel, R.L.; Miller, K.D.; Wagle, N.S.; Jemal, A. Cancer statistics, 2023. *CA A Cancer J. Clin.* **2023**, *73*, 17–48. [CrossRef] [PubMed]
2. Breugom, A.J.; Swets, M.; Bosset, J.-F.; Collette, L.; Sainato, A.; Cionini, L.; Glynne-Jones, R.; Counsell, N.; Bastiaannet, E.; Broek, C.B.M.v.D.; et al. Adjuvant chemotherapy after preoperative (chemo) radiotherapy and surgery for patients with rectal cancer: A systematic review and meta-analysis of individual patient data. *Lancet Oncol.* **2015**, *16*, 200–207. [CrossRef]
3. Ferrari, L.; Fichera, A. Neoadjuvant chemoradiation therapy and pathological complete response in rectal cancer. *Gastroenterol. Rep.* **2015**, *3*, 277–288. [CrossRef] [PubMed]
4. Janjan, N.A.; Khoo, V.S.; Abbruzzese, J.; Pazdur, R.; Dubrow, R.; Cleary, K.R.; Allen, P.K.; Lynch, P.M.; Glober, G.; Wolff, R.; et al. Tumor downstaging and sphincter preservation with preoperative chemoradiation in locally advanced rectal cancer: The MD anderson cancer center experience. *Int. J. Radiat. Oncol. Biol. Phys.* **1999**, *44*, 1027–1038. [CrossRef] [PubMed]
5. Maas, M.; Nelemans, P.J.; Valentini, V.; Das, P.; Rödel, C.; Kuo, L.-J.; Calvo, F.; García-Aguilar, J.; Glynne-Jones, R.; Haustermans, K.; et al. Long-term outcome in patients with a pathological complete response after chemoradiation for rectal cancer: A pooled analysis of individual patient data. *Lancet Oncol.* **2010**, *11*, 835–844. [CrossRef]
6. Rombouts, A.J.M.; Al-Najami, I.; Abbott, N.L.; Appelt, A.; Baatrup, G.; Bach, S.; Bhangu, A.; Garm Spindler, K.L.; Gray, R.; Handley, K.; et al. Can we save the rectum by watchful waiting or TransAnal microsurgery following (chemo) radiotherapy versus total mesorectal excision for early REctal cancer (STAR-TREC study)?: Protocol for a multicentre, randomised feasibility study. *BMJ Open* **2017**, *7*, e019474. [CrossRef]

7. Park, S.H.; Cho, S.H.; Choi, S.H.; Jang, J.K.; Kim, M.J.; Kim, S.H.; Lim, J.S.; Moon, S.K.; Park, J.H.; Seo, N.; et al. MRI assessment of complete response to preoperative chemoradiation therapy for rectal cancer: 2020 guide for practice from the korean society of abdominal radiology. *Korean J. Radiol.* **2020**, *21*, 812. [CrossRef]
8. Gambacorta, M.A.; Masciocchi, C.; Chiloiro, G.; Meldolesi, E.; Macchia, G.; van Soest, J.; Peters, F.; Collette, L.; Gérard, J.P.; Ngan, S.; et al. Timing to achieve the highest rate of pCR after preoperative radiochemotherapy in rectal cancer: A pooled analysis of 3085 patients from 7 randomized trials. *Radiother. Oncol.* **2021**, *154*, 154–160. [CrossRef]
9. López-Campos, F.; Martín-Martín, M.; Fornell-Pérez, R.; García-Pérez, J.C.; Die-Trill, J.; Fuentes-Mateos, R.; López-Durán, S.; Domínguez-Rullán, J.; Ferreiro, R.; Riquelme-Oliveira, A.; et al. Watch and wait approach in rectal cancer: Current controversies and future directions. *World J. Gastroenterol.* **2020**, *26*, 4218. [CrossRef]
10. Lin, Z.; Cai, W.; Hou, W.; Chen, Y.; Gao, B.; Mao, R.; Wang, L.; Li, Z. CT-guided survival prediction of esophageal cancer. *IEEE J. Biomed. Health Inform.* **2021**, *26*, 2660–2669. [CrossRef]
11. Cui, S.; Ten Haken, R.K.; El Naqa, I. Integrating multiomics information in deep learning architectures for joint actuarial outcome prediction in non-small cell lung cancer patients after radiation therapy. *Int. J. Radiat. Oncol. Biol. Phys.* **2021**, *110*, 893–904. [CrossRef] [PubMed]
12. Lou, B.; Doken, S.; Zhuang, T.; Wingerter, D.; Gidwani, M.; Mistry, N.; Ladic, L.; Kamen, A.; Abazeed, M.E. An image-based deep learning framework for individualising radiotherapy dose: A retrospective analysis of outcome prediction. *Lancet Digit. Health* **2019**, *1*, e136–e147. [CrossRef] [PubMed]
13. Lambin, P.; Rios-Velazquez, E.; Leijenaar, R.; Carvalho, S.; van Stiphout, R.G.P.M.; Granton, P.; Zegers, C.M.L.; Gillies, R.; Boellard, R.; Dekker, A.; et al. Radiomics: Extracting more information from medical images using advanced feature analysis. *Eur. J. Cancer* **2012**, *48*, 441–446. [CrossRef] [PubMed]
14. Gillies, R.J.; Kinahan, P.E.; Hricak, H. Radiomics: Images are more than pictures, they are data. *Radiology* **2016**, *278*, 563–577. [CrossRef]
15. Meng, Y.; Zhang, Y.; Dong, D.; Li, C.; Liang, X.; Zhang, C.; Wan, L.; Zhao, X.; Xu, K.; Zhou, C.; et al. Novel radiomic signature as a prognostic biomarker for locally advanced rectal cancer. *J. Magn. Reson. Imaging* **2018**, *48*, 605–614. [CrossRef] [PubMed]
16. Cusumano, D.; Dinapoli, N.; Boldrini, L.; Chiloiro, G.; Gatta, R.; Masciocchi, C.; Lenkowicz, J.; Casà, C.; Damiani, A.; Azario, L.; et al. Fractal-based radiomic approach to predict complete pathological response after chemo-radiotherapy in rectal cancer. *La Radiol. Med.* **2018**, *123*, 286–295. [CrossRef]
17. Horvat, N.; Veeraraghavan, H.; Khan, M.; Blazic, I.; Zheng, J.; Capanu, M.; Sala, E.; Garcia-Aguilar, J.; Gollub, M.J.; Petkovska, I. MR imaging of rectal cancer: Radiomics analysis to assess treatment response after neoadjuvant therapy. *Radiology* **2018**, *287*, 833–843. [CrossRef]
18. Bonomo, P.; Socarras Fernandez, J.; Thorwarth, D.; Casati, M.; Livi, L.; Zips, D.; Gani, C. Simulation CT-based radiomics for prediction of response after neoadjuvant chemo-radiotherapy in patients with locally advanced rectal cancer. *Radiat. Oncol.* **2022**, *17*, 84. [CrossRef]
19. Rosen, B.S.; Hawkins, P.G.; Polan, D.F.; Balter, J.M.; Brock, K.K.; Kamp, J.D.; Lockhart, C.M.; Eisbruch, A.; Mierzwa, M.L.; Haken, R.K.T.; et al. Early changes in serial CBCT-measured parotid gland biomarkers predict chronic xerostomia after head and neck radiation therapy. *Int. J. Radiat. Oncol. Biol. Phys.* **2018**, *102*, 1319–1329. [CrossRef]
20. Du, F.; Tang, N.; Cui, Y.; Wang, W.; Zhang, Y.; Li, Z.; Li, J. A novel nomogram model based on cone-beam CT radiomics analysis technology for predicting radiation pneumonitis in esophageal cancer patients undergoing radiotherapy. *Front. Oncol.* **2020**, *10*, 596013. [CrossRef]
21. Posiewnik, M.; Piotrowski, T. A review of cone-beam CT applications for adaptive radiotherapy of prostate cancer. *Phys. Med.* **2019**, *59*, 13–21. [CrossRef] [PubMed]
22. Le, W.T.; Bang, C.; Cordelle, P.; Markel, D.; Nguyen-Tan, P.F.; Bahig, H.; Kadoury, S. Comparing 3D deformations between longitudinal daily CBCT acquisitions using CNN for head and neck radiotherapy toxicity prediction. *arXiv* **2023**, arXiv:230303965.
23. Nardone, V.; Reginelli, A.; Grassi, R.; Boldrini, L.; Vacca, G.; D'ippolito, E.; Annunziata, S.; Farchione, A.; Belfiore, M.P.; Desideri, I.; et al. Delta radiomics: A systematic review. *La Radiol. Med.* **2021**, *126*, 1571–1583. [CrossRef] [PubMed]
24. Lin, P.; Yang, P.-F.; Chen, S.; Shao, Y.-Y.; Xu, L.-M.; Wu, Y.; Teng, W.; Zhou, X.-Z.; Li, B.-H.; Luo, C.; et al. A delta-radiomics model for preoperative evaluation of neoadjuvant chemotherapy response in high-grade osteosarcoma. *Cancer Imaging* **2020**, *20*, 7. [CrossRef]
25. Crombé, A.; Périer, C.; Kind, M.; De Senneville, B.D.; Le Loarer, F.; Italiano, A.; Buy, X.; Saut, O. T2-based MRI delta-radiomics improve response prediction in soft-tissue sarcomas treated by neoadjuvant chemotherapy. *J. Magn. Reson. Imaging* **2019**, *50*, 497–510. [CrossRef]
26. Shayesteh, S.; Nazari, M.; Salahshour, A.; Sandoughdaran, S.; Hajianfar, G.; Khateri, M.; Yaghobi Joybari, A.; Jozian, F.; Fatehi Feyzabad, S.H.; Arabi, H.; et al. Treatment response prediction using MRI-based pre-, post-, and delta-radiomic features and machine learning algorithms in colorectal cancer. *Med. Phys.* **2021**, *48*, 3691–3701. [CrossRef]
27. Alahmari, S.S.; Cherezov, D.; Goldgof, D.B.; Hall, L.O.; Gillies, R.J.; Schabath, M.B. Delta radiomics improves pulmonary nodule malignancy prediction in lung cancer screening. *IEEE Access* **2018**, *6*, 77796–77806. [CrossRef]
28. Plautz, T.E.; Zheng, C.; Noid, G.; Li, X.A. Time stability of delta-radiomics features and the impact on patient analysis in longitudinal CT images. *Med. Phys.* **2019**, *46*, 1663–1676. [CrossRef]
29. Nasief, H.; Zheng, C.; Schott, D.; Hall, W.; Tsai, S.; Erickson, B.; Li, X.A. A machine learning based delta-radiomics process for early prediction of treatment response of pancreatic cancer. *NPJ Precis. Oncol.* **2019**, *3*, 25. [CrossRef]

30. Boldrini, L.; Cusumano, D.; Chiloiro, G.; Casà, C.; Masciocchi, C.; Lenkowicz, J.; Cellini, F.; Dinapoli, N.; Azario, L.; Teodoli, S.; et al. Delta radiomics for rectal cancer response prediction with hybrid 0.35 t magnetic resonance-guided radiotherapy (MRgRT): A hypothesis-generating study for an innovative personalized medicine approach. *La Radiol. Med.* **2019**, *124*, 145–153. [CrossRef]
31. Traverso, A.; Wee, L.; Dekker, A.; Gillies, R. Repeatability and reproducibility of radiomic features: A systematic review. *Int. J. Radiat. Oncol. Biol. Phys.* **2018**, *102*, 1143–1158. [CrossRef] [PubMed]
32. Nardone, V.; Reginelli, A.; Guida, C.; Belfiore, M.P.; Biondi, M.; Mormile, M.; Buonamici, F.B.; Di Giorgio, E.; Spadafora, M.; Tini, P.; et al. Delta-radiomics increases multicentre reproducibility: A phantom study. *Med. Oncol.* **2020**, *37*, 38. [CrossRef] [PubMed]
33. Delgadillo, R.; Spieler, B.O.; Ford, J.C.; Kwon, D.; Yang, F.; Studenski, M.; Padgett, K.R.; Abramowitz, M.C.; Pra, A.D.; Stoyanova, R.; et al. Repeatability of CBCT radiomic features and their correlation with CT radiomic features for prostate cancer. *Med. Phys.* **2021**, *48*, 2386–2399. [CrossRef] [PubMed]
34. Van Griethuysen, J.J.M.; Fedorov, A.; Parmar, C.; Hosny, A.; Aucoin, N.; Narayan, V.; Beets-Tan, R.G.H.; Fillion-Robin, J.C.; Pieper, S.; Aerts, H.J.W.L. Computational radiomics system to decode the radiographic phenotype. *Cancer Res.* **2017**, *77*, e104–e107. [CrossRef] [PubMed]
35. Kannan, K.S.; Manoj, K.; Arumugam, S. Labeling methods for identifying outliers. *Int. J. Stat. Syst.* **2015**, *10*, 231–238.
36. York, D. Least-squares fitting of a straight line. *Can. J. Phys.* **1966**, *44*, 1079–1086. [CrossRef]
37. Johnson, M.L. Why, when, and how biochemists should use least squares. *Anal. Biochem.* **1992**, *206*, 215–225. [CrossRef]
38. Pedregosa, F.; Varoquaux, G.; Gramfort, A.; Michel, V.; Thirion, B.; Grisel, O.; Blondel, M.; Prettenhofer, P.; Weiss, R.; Dubourg, V.; et al. Scikit-learn: Machine Learning in Python. *J. Mach. Learn. Res.* **2011**, *12*, 2825–2830.
39. Pan, X.; Feng, T.; Liu, C.; Savjani, R.R.; Chin, R.K.; Sharon Qi, X. A survival prediction model via interpretable machine learning for patients with oropharyngeal cancer following radiotherapy. *J. Cancer Res. Clin. Oncol.* **2023**, *149*, 6813–6825. [CrossRef]
40. Fu, J.; Zhong, X.; Li, N.; Van Dams, R.; Lewis, J.; Sung, K.; Raldow, A.C.; Jin, J.; Qi, X.S. Deep Learning-based Radiomic Features for Improving Neoadjuvant Chemoradiation Response Prediction in Locally Advanced Rectal Cancer. *Phys. Med. Biol.* **2020**, *65*, 075001. [CrossRef]
41. Yamashita, H.; Takenaka, R.; Sakumi, A.; Haga, A.; Otomo, K.; Nakagawa, K. Analysis of motion of the rectum during preoperative intensity modulated radiation therapy for rectal cancer using cone-beam computed tomography. *Radiat. Oncol.* **2015**, *10*, 2. [CrossRef] [PubMed]
42. Alickikus, Z.A.; Kuru, A.; Aydin, B.; Akcay, D.; Gorken, I.B. The importance of mesorectum motion in determining PTV margins in rectal cancer patients treated with neoadjuvant radiotherapy. *J. Radiat. Res.* **2020**, *61*, 335–342. [CrossRef] [PubMed]

sensors

MDPI

Article

Deep Network-Assisted Quality Inspection of Laser Welding on Power Battery

Dong Wang, Yongjia Zheng, Wei Dai, Ding Tang * and Yinghong Peng

State Key Laboratory of Mechanical Systems and Vibration, Shanghai Jiao Tong University, Shanghai 200240, China; wdshop@sjtu.edu.cn (D.W.)
* Correspondence: tangding@sjtu.edu.cn

Abstract: Reliable quality control of laser welding on power batteries is an important issue due to random interference in the production process. In this paper, a quality inspection framework based on a two-branch network and conventional image processing is proposed to predict welding quality while outputting corresponding parameter information. The two-branch network consists of a segmentation network and a classification network, which alleviates the problem of large training sample size requirements for deep learning by sharing feature representations among two related tasks. Moreover, coordinate attention is introduced into feature learning modules of the network to effectively capture the subtle features of defective welds. Finally, a post-processing method based on the Hough transform is used to extract the information of the segmented weld region. Extensive experiments demonstrate that the proposed model can achieve a significant classification performance on the dataset collected on an actual production line. This study provides a valuable reference for an intelligent quality inspection system in the power battery manufacturing industry.

Keywords: power battery; laser welding; two-branch network; coordinate attention; Hough transform; quality inspection

1. Introduction

A power battery is one of the key components of new energy vehicles, and its quality determines the reliability and safety of the vehicle to a large extent. Laser welding is widely used in power battery manufacturing due to its advantages of high energy density, high precision, and precise control over the heat input [1,2]. However, on large-scale automatic production lines, on-site uncertainties such as material inhomogeneity, residual impurities, and parameter fluctuations increase the welding instability and easily lead to welding defects, which will seriously affect the quality and performance of power batteries [3,4]. At present, most of the post-welding quality evaluation of power batteries is mainly carried out by manual visual inspection, which is bound to cause low detection efficiency and high labor costs, making it difficult to meet the requirements of modern welding production for high efficiency and high quality. Therefore, an effective method is urgently needed to achieve automatic inspection of laser welding quality on power batteries.

In the past decades, traditional image processing methods have been broadly used in welding quality inspection [5–7]. Traditional visual inspection tends to manually extract geometric and textural features and build a learning-based classifier to predict the welding quality according to these features. For example, contour-based and OTSU threshold segmentation methods were used to extract keyhole features and weld width, and a back propagation neural network (BPNN) was trained to evaluate welding defects [8]. Cai et al. [9] extracted seven morphology features of keyholes and molten pools and compared the accuracy of several machine learning models (e.g., BPNN, radial basis function neural network, support vector regression) on welding quality prediction. However, such methods usually require strong expert knowledge and complex threshold settings, which are susceptible to environmental interference and have poor generalization performance and robustness.

Due to the powerful representation learning capability of convolutional neural networks (CNNs), deep learning-based visual inspection has shown excellent performance advantages in both accuracy and speed [10–12]. Visual quality inspection is generally divided into three stages: defect classification, defect localization, and defect segmentation. Defect classification models integrate feature learning and discriminative classifiers in one network to directly predict the welding quality category of the input image. Most of these models are established based on the improvement of existing classical networks, such as VGG [13], ResNet [14], MobileNet [15], etc. For instance, the pre-trained optimized VGG-16 model was established to accurately predict the welding defects on the safety vent of power batteries [16]. Furthermore, a lightweight model, SqueezeNet, was proposed to realize efficient and accurate detection [17], which is more suitable for scenarios with limited computational power. Transfer learning-based ResNet model was used to predict six spot welding defects in the automobile production site [18]. Defect localization models can obtain both accurate location and category information of welding defects, which can generally be divided into two-stage models represented by Faster R-CNN [19] and one-stage models represented by YOLO [20,21]. One-stage models pursue speed, while two-stage models emphasize accuracy. A Faster R-CNN model combining feature pyramid network was proposed, which can effectively detect small welding defects with weak contrast under complex backgrounds [22]. A bidirectional cross-scale fusion pyramid structure was introduced in YOLOv3 to facilitate the flow of information and enhance the fusion of multi-scale features for location and quality detection of small spot welding [23]. Compared to classification and localization, defect segmentation models enable a fine pixel-level localization of the defect region to obtain a more detailed description of the defect. Zhu et al. [24] presented a lightweight semantic segmentation method that can accurately segment laser welding defects of different shapes in real-time. A compressed U-shape network [25] was used for the quality inspection of laser welding in the battery production process, which has significant advantages in both speed and detection accuracy. By learning and fusing multi-scale semantic features of defects, the FPN-ResNet-34 network was proposed to predict the complete information, including defect category, boundary, and location, to achieve weld quality evaluation [26].

The success of deep learning stems from the construction of large annotated datasets. However, the amount of training data is difficult to meet for power battery production lines with strict qualification rate control. Therefore, to alleviate the problem of poor generalization performance due to insufficient training data, a two-branch network architecture is proposed for quality inspection. By learning two related tasks, weld segmentation and classification, the model can make full use of the useful supervisory information in images, thus reducing the data requirements for training the model. In addition, many defects and normal welds have high similarity in appearance, and it is difficult to find discriminative features in the images by using ordinary CNNs directly. The feature learning modules in the network should pay more attention to spatially distinguishable subtle features. Inspired by the visual attention mechanism, channel attention module [27], convolutional block attention module (CBAM) [28], and coordinate attention module (CAM) [29] have been proposed successively, which can learn critical information by assigning different weights to different regions in the images. In this paper, CAMs are introduced into the two-branch network architecture, which enables the model to effectively focus on the subtle features in the defective images that can represent its category. The accuracy and effectiveness of the proposed model are verified by the image dataset collected from the actual power battery production lines. Finally, edge detection and circular Hough transform are used to post-process the segmented weld image to obtain the weld information. In summary, the main contributions of the paper are outlined as follows:

(1) A two-branch network architecture is proposed, and weld region segmentation as a secondary task can effectively improve the classification accuracy of the model and reduce the demand for training data;

(2) The coordinate attention mechanism is introduced in the network to learn the more discriminative features between normal and defect;

(3) A framework combining deep network and conventional image processing is established to achieve efficient and accurate detection of laser welding quality.

The rest of the paper is organized as follows: The welding images dataset collected on the power battery production line is presented in Section 2. The proposed quality inspection framework is described elaborately in Section 3. In Section 4, the experimental results and discussions of the proposed model are comprehensively validated. Finally, the conclusion will be given in Section 5.

2. Overview of Welding Image Dataset

Power batteries are generally connected in series or parallel to form packs to obtain high capacity. When batteries are packaged, the connection between the battery poles and the adapter block needs to be implemented by laser welding. However, due to the high reflectivity of aluminum to the laser and the high tendency of parts to retain impurities such as stamping oil and cleaning agents, it is easy to lead to defects during the laser welding process. To carry out the model evaluation experiments, raw images of laser welding on battery packs are collected from an actual production line. To obtain high-quality images, an optical inspection system is embedded in the laser welder on the production line, consisting of an industrial camera and an LED-stabilized light source. Batteries are clamped on the assembly line by a bracket, and the light source is placed vertically above the assembly line. As the battery is welded and comes to rest in the camera's field of view, its image is captured.

To acquire enough defective data for training, the expensive time cost is spent to construct the dataset. The dataset contains a total of 1268 images, of which 1040 are qualified welds, and 208 are defective welds. The resolution of raw images is 823×1129. To match the downsampling requirements of the network, zero padding is used for the border of images, and the image size is adjusted to 832×1152. In addition, the image resolution should be moderate; too large will affect the model inference speed, while too small will easily lead to reduced or even lost weld defect information. Finally, the image resolution is resized to 416×576 to obtain a trade-off between accuracy and speed. The proposed model follows a supervised paradigm that requires labeled data for training. In this paper, the open-source tool LabelMe is used to provide pixel-level annotations of the weld region and the presence or absence of weld defects for each image. Some examples of welding images with annotation masks are illustrated in Figure 1. It can be noticed that the contrast between welds and the background is not obvious, and the defects and qualified welds are relatively similar in appearance, which poses some challenges for quality prediction.

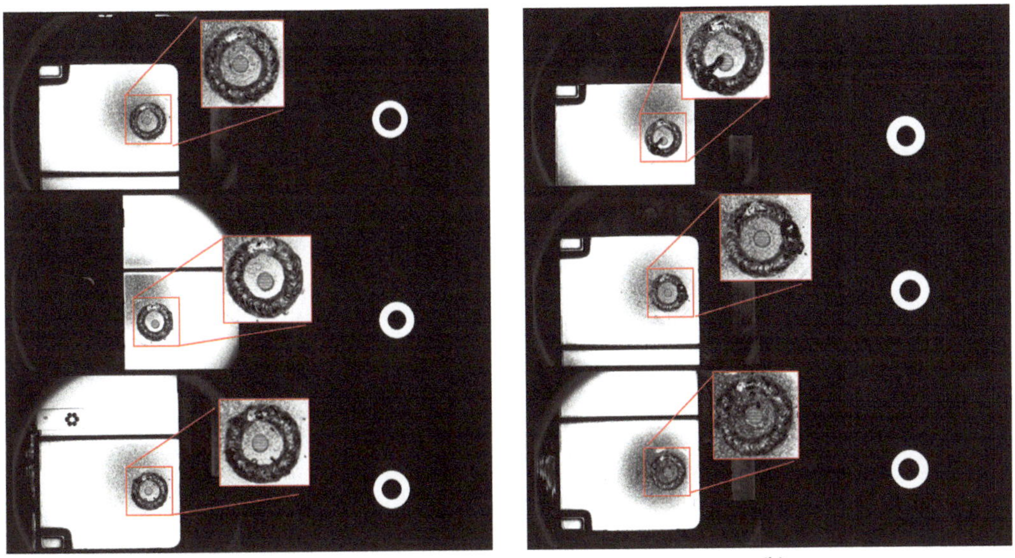

(a) (b)

Figure 1. Some images with annotation masks in the dataset. (**a**) Qualified welds; (**b**)defective welds.

3. Proposed Framework

In actual production, besides the weld appearance quality, the weld width is also an important indicator. To this end, a quality inspection framework combining deep learning and conventional image processing is developed to identify weld defects while outputting weld information such as inner radius, outer radius, and width, as shown in Figure 2. In the framework, an end-to-end two-branch network is designed to alleviate the problem of large training sample size requirements for deep learning. The two-branch network is a multi-task learning method [30–32], which enhances model generalization performance by sharing feature expressions among related tasks. The proposed network includes a segmentation network for pixel-level localization of welded regions and a classification network for quality prediction. The segmentation network contains an encoder and a decoder, where the encoder extracts multi-scale features from the original image, and the decoder aggregates the feature maps and predicts the weld segmentation results. The classification network uses the feature maps obtained by the weld region-aware encoder to achieve accurate quality prediction. In this way, the two-branch network can make full use of the supervised information in each sample, increasing the effective number of training samples and reducing the risk of overfitting. Moreover, each network branch uses CAMs for feature learning to extract more discriminative features for subtle differences in weld defects. Finally, the segmented weld region is post-processed to extract the weld information by using edge detection and circular Hough transform [33,34].

3.1. Coordinate Attention Module

The proposed network uses CAMs instead of convolution modules in traditional CNNs, which allows better learning of weld detailed features and thus improves the detection accuracy of the model in welding defects. Unlike channel attention [27] and CBAM [28], CAM uses two average pooling operations to aggregate the input feature maps into two direction-aware features along the horizontal and vertical directions (Figure 3), thus alleviating the loss of location information caused by the direct use of two-dimensional global pooling. In this way, CAM can capture long-term dependencies along one spatial direction while maintaining accurate location information along the other direction. The two attention maps are then multiplied with the input feature map to enhance the representation of the object of interest; this allows for more accurate localization of the weld defect, which helps the model to better identify quality categories. The diagram of the CAM is illustrated in Figure 3.

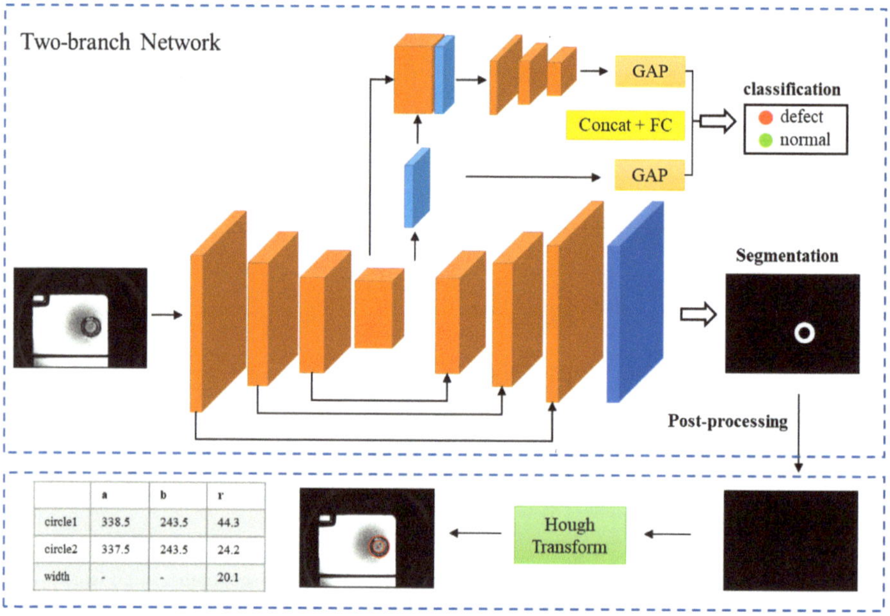

Figure 2. The proposed framework combining the two-branch network and conventional image processing. Note: (a,b) denote circle center coordinates and r denotes circle radius.

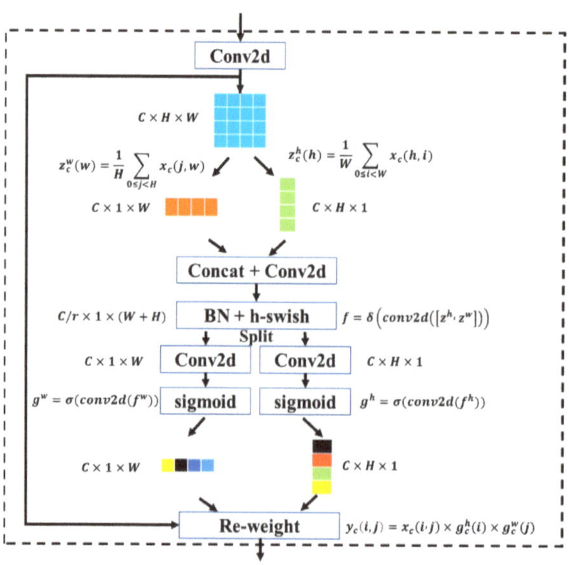

Figure 3. Diagram of the coordinate attention module (CAM). Note: the different colors in the bar below reflect the different weight values.

Specifically, given the feature maps after two-dimensional convolution, represented as $X = [x_1, x_2, \ldots, x_c]$. The average pooling of kernel size $(H,\ 1)$ and $(1,\ W)$ are then used to encode each channel along the horizontal and vertical directions, respectively. Thus, for the cth channel, the output at height h and width w can be expressed as:

$$z_c^h(h) = \frac{1}{W} \sum_{0 \le i < W} x_c(h, i) \tag{1}$$

$$z_c^w(w) = \frac{1}{H} \sum_{0 \le j < H} x_c(j, w) \tag{2}$$

The generated feature maps z_c^h and z_c^w are concatenated, combining the encoded information from the horizontal and vertical directions. Then, the concatenated feature maps are fed into a 1×1 convolution transform function, which applies a set of filters to extract relevant information from the input. Additionally, the incorporation of a Batch Normalization (BN) [29] layer following the convolution operation can effectively stabilize the input distribution of each layer within the network. This results in an improvement in training speed and a reduction in the overfitting issue.

$$f = \delta\left(Conv2d\left(\left[z^h, z^w\right]\right)\right) \tag{3}$$

where δ is the nonlinear activation function, and the h-swish function is chosen here. $f \in \mathbb{R}^{C/r \times (H+W)}$ is the intermediate feature map, and r is the reduction ratio. Then, f is split into two independent tensors along the spatial dimension, $f^h \in \mathbb{R}^{C/r \times H}$ and $f^w \in \mathbb{R}^{C/r \times W}$. By using two 1×1 two-dimensional convolution transformations, f^h and f^w are converted into tensors with the same channel of the input X, respectively. The activation function σ used here is the sigmoid function.

$$g^h = \sigma\left(Conv2d\left(f^h\right)\right) \tag{4}$$

$$g^w = \sigma(Conv2d(f^w)) \tag{5}$$

The outputs g^h and g^w are extended and used as attention weights, respectively. Finally, the output of the CAM is denoted as $Y = [y_1, y_2, \ldots, y_c]$, where each element $y_c(i,j)$ is reweighted by multiplying the original input feature $x_c(i,j)$ with the corresponding attention weights, resulting in the following:

$$y_c(i,j) = x_c(i,j) \times g_c^h(i) \times g_c^w(j) \tag{6}$$

3.2. Two-Branch Network Architecture

The detailed architecture of the two-branch network is presented in Figure 4, which consists of two parallel branch networks, a segmentation network, and a classification network. The segmentation network adopts a U-Net [35] similar structure, which consists of an encoder and a decoder, to effectively extract the multi-scale features and restore the image resolution. This branch segments the input image into weld and background regions and outputs the probability that each pixel belongs to the weld region. Specifically, the encoder contains convolutional layers, CAMs, and max-pooling layers. Each convolutional layer is followed by a batch normalization (BN) [36] layer and ReLU nonlinear function, where the combination of BN and ReLU enables more robust learning. After each 2×2 maxpooling operation, the image resolution becomes half of the original one. More importantly, several CAMs are used after the convolutional layer to enhance the model's ability to perceive weld details. The main parameters in CAMs are the number of convolutional kernels and the reduction ratio r. Detailed parameters and the number of modules are shown in Figure 4. The encoder extracts feature maps with different resolutions and information levels from the image, while the decoder continuously increases the resolution of the feature maps by upsampling. The long connection from the encoder to the decoder allows the multi-scale feature maps in the encoder to be merged into the decoder, resulting in the more accurate performance of weld region segmentation. Finally, the original resolution and channel are restored by 1×1 convolution, where the sigmoid activation determines the probability that each pixel belongs to the weld region.

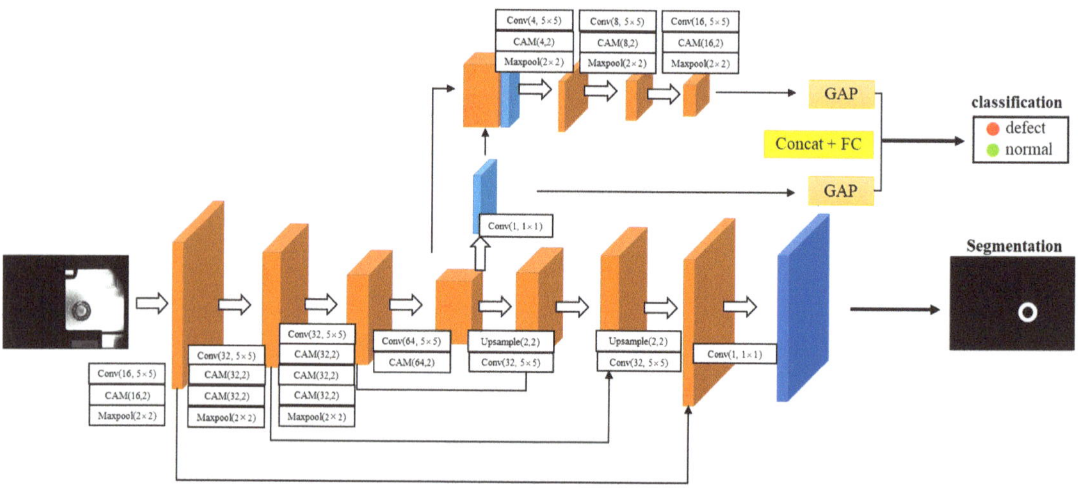

Figure 4. Detailed architecture of the two-branch network.

The classification branch and segmentation branch share the weight parameters of the encoder. Feature maps obtained from the encoder contain abundant weld region information, which effectively avoids the subsequent quality prediction being influenced by background. The weld quality features are further enriched by concatenating the output of the encoder with the single-channel mask feature maps. These feature maps are then fed into a series of combinatorial modules, each of which contains a convolutional layer with a 5×5 kernel size, a CAM, and a maximum pooling layer. The convolutional layers are followed by BN and ReLU, similarly, and the number of the convolutional kernel in each combinatorial module is 4, 8, and 16 in turn. As can be seen from Figure 4, the image resolution is further reduced, and the image is abstracted into a higher-level feature representation in the layer-by-layer operation of the classification network. The global average pooling (GAP) [37] layer reduces the redundancy and dimensionality of the features by calculating the average of all pixels within each channel of feature maps, thus reducing the overfitting possibility of the deep neural network. Finally, the GAP acts on both the high-level feature maps and single-channel mask, and their outputs are concatenated to 17 neurons, which are combined with a fully connected (FC) layer to obtain the final classification results.

The loss function of the two-branch network is a pivotal aspect of training the model effectively. In the proposed approach, the loss function consists of two distinct components, each serving a specific purpose, as presented in Equation (7). The first component employs the mean squared error (MSE) loss, which is utilized to measure the discrepancy between the predicted output of the weld region segmentation and the ground truth segmentation, as presented in Equation (8). The MSE loss ensures that the network accurately delineates the boundaries of the weld regions within the input data, minimizing the difference between the predicted and actual segmentations. The second component of the loss function incorporates the cross-entropy (CE) loss, as shown in Equation (9). Unlike the MSE loss, which assesses the similarity between continuous values, the CE loss is specifically designed for classification tasks. In our case, it addresses the quality classification aspect of the network's output. By employing the CE loss, the network is trained to accurately classify the quality of the weld regions, assigning them to different classes based on predefined criteria (e.g., good quality, bad quality). The CE loss optimizes the network's ability to correctly classify the quality of weld regions by penalizing incorrect predictions and encouraging accurate classification across different quality levels. By combining the MSE and CE losses within the overall loss function, the network's performance can be simultaneously optimized in both weld region segmentation and quality classification. This multi-component loss function helps the network learn and adapt to the complex and interrelated tasks of accurately segmenting weld regions and classifying their quality, ultimately leading to improved performance and overall system effectiveness.

$$L_{total} = L_{mse} + \lambda L_{ce} \tag{7}$$

$$L_{mse} = \frac{1}{N}\sum_{i=1}^{N}(y_i - x_i)^2 \tag{8}$$

$$L_{ce} = -\frac{1}{N}\sum_{i=1}^{N}(y_i log x_i + (1 - y_i)\log(1 - x_i)) \tag{9}$$

where x_i and y_i are the ground truth and prediction results, respectively. λ is the trade-off between two losses.

3.3. Post-Processing

The post-processing method utilizes edge detection and Hough transforms for circle detection in the input-segmented images. Edge detection is applied to extract the weld region contour using techniques like Canny [38] and Sobel [39]. Hough transform [33,34] maps the contour pixels from the image space to the circle parameter space, employing vote accumulation to identify the peak response in the parameter space; this allows for the determination of the circle parameters, which can be further employed to calculate the weld width and complete the evaluation of weld quality.

The flowchart of the Hough transform for circle detection is summarized in the algorithm (Algorithm 1). The algorithm initializes an accumulator array, $A[a, b, r]$, with zeros to store the center points (a, b) and radius (r) of circles. The Canny operator is then used to detect the edge points in the segmented image, which indicates regions with significant changes in pixel intensity. Next, a nested loop is executed: the outer loop iterates from $r = 0$ to the diagonal length of the image in order to detect circles with different radii. For each edge pixel (x, y) in the image, the algorithm enters an inner loop that iterates over angles θ from 0 to 360 degrees, covering all possible angles around the edge pixel. Inside the inner loop, the algorithm calculates the potential coordinates for the circle center, a and b, using the current edge pixel (x, y) and radius r. These calculations involve subtracting the product of the radius and the sine or cosine of the angle θ from the edge pixel coordinates. Subsequently, the algorithm increments the accumulator array $A[a, b, r]$ by 1 for each potential circle center and radius combination. This accumulation process counts the number of times a specific circle parameter combination passes through edge pixels in the image. After completing all iterations, the algorithm identifies the largest and second-largest values in the accumulator array, $A[a, b, r]$. These values correspond to the inner and outer circles of the weld. The parameters associated with these largest values represent the center coordinates (a, b) and the radius (r) of the detected circles.

Algorithm 1: Circle detection using Hough transform

Input: the segmented images, image coordinates (x, y).
Output: circle center coordinates (a, b), circle radius (r).
Initialize: $A[a, b, r] = 0$.
Detect all edge points in the segmented image using the Canny operator.
for $r = 0$ to the diagonal length of the image
    for each edge pixel (x, y) in the image
        for $\theta = 0$ to 360
            $a = x - r \times cos\theta$
            $b = y - r \times sin\theta$
            $A[a, b, r] = A[a, b, r] + 1$
Find the largest and second largest values of $A[a, b, r]$, whose parameters correspond to the inner and outer circles of the weld.

4. Experiments and Discussion

This section provides a thorough exploration of various study aspects, including implementation details, evaluation metrics, experiment results analysis, and post-processing results. The section begins by discussing implementation specifics, offering readers a comprehensive insight into employed methodologies and techniques. It then focuses on evaluation metrics utilized to measure model performance. Subsequently, analysis of experimental results, including ablation experiments and sensitivity analysis of training samples, will be presented to demonstrate the performance advantages of the model in detecting weld quality. Lastly, the section concludes by analyzing post-processing results. This section presents a concise analysis and discussion, offering valuable insights into research methodology and outcomes.

4.1. Implementation Details

Experiments are conducted on a computer with i7-9700k CPU and NVIDIA GeForce GTX 1070ti GPU, and models are implemented based on the open-source Tensorflow deep learning framework. The input images and annotated masks are normalized before training, and the pre-processed dataset is divided into training and testing sets according to 4:1, where the category ratio is kept consistent. Moreover, to improve the generalization performance of the model, data augmentation is performed on the training samples using simple geometric transformations, including random rotation and flip. There is no scientific conclusion on the hyperparameter setting of the deep network. In this paper, the best hyperparameters are selected by using five-fold cross-validation, in which the training set is equally divided into five subsets, and each subset is used as the validation set in turn, and the remaining samples are used for training. The model performance is the average of the results of five training sessions and is used as the basis for hyperparameter selection. In this paper, the hyperparameter selection includes the number of CAM, parameters of the convolution layer, and so on. The weight λ of the two components in the loss function is set to 2 because the correct identification of defects is more important compared to weld segmentation in industrial scenarios. Adam [40] is used as the optimizer for model training with an initial learning rate of 0.001 and β_1 and β_2 of 0.9. A cosine scheduler with a warm-up is adopted to adjust the learning rate [41]. In the warm-up phase, the learning rate is gradually increased to a preset initial value to avoid the instability of the model parameters due to the large initial learning rate, after which the learning rate is decayed by the cosine annealing function, and the final learning rate is stabilized at 1×10^{-5}. The training epoch and batch size are 200 and 8, respectively. The model parameters with the best performance on the validation set during the training are saved.

4.2. Evaluation Metrics

The classification branch mainly classifies images into two categories: defective and non-defective. Based on the confusion matrix in Table 1, metrics that are of concern to the production line are selected to quantify the classification performance, such as accuracy (Acc), miss alarm rate (MAR), and false alarm rate (FAR). Acc measures the overall performance of the model for defect and normal weld identification. MAR refers to the percentage of true defective welds that are missed by the model, which tends to lead to the outflow of defective welds. FAR indicates the percentage of welds predicted as defective that are misjudged by the model, which often results in wasted labor costs. The ideal situation is that the established quality evaluation method has high Acc, low MAR, and low FAR.

$$Acc = \frac{TP + TN}{TP + FP + FN + TN} \tag{10}$$

$$MAR = \frac{FN}{FN + TP} \tag{11}$$

$$FAR = \frac{FP}{FP + TP} \tag{12}$$

Table 1. Confusion Matrix.

Confusion Matrix		Prediction	
		Normal	**Defect**
True	Normal	True Negative (*TN*)	False Positive (*FP*)
	defect	False Negative (*FN*)	True Positive (*TP*)

For the segmentation branch, mean intersection over union (*mIoU*) is used as an evaluation metric, which quantifies the percentage of overlapping regions between predicted and actual annotation masks.

$$mIoU = \frac{|A \cap B|}{|A| + |B| - |A \cap B|} \tag{13}$$

where $|A|$ and $|B|$ denote the number of pixels in predicted and ground truth images while $|A \cap B|$ denotes the number of common pixels in both images.

4.3. Experiment Results

The experimental results analysis consisted of an ablation experiment and sensitivity analysis of training samples. Ablation experiments are conducted, systematically removing study components to observe their impact on overall results. Additionally, the section investigates the model's sensitivity to varying training sample sizes and examines how such variations affect performance.

4.3.1. Ablation Experiment

In this section, ablation experiments are conducted for each component of the proposed network, and the effectiveness of each component is analyzed by comparing its performance on the test set. The detailed results are presented in Table 2. The advantages of the two-branch structure are first investigated. From the comparison results in Table 2, it can be found that the performance indicators of the two-branch structure are significantly improved compared to the single-branch. For example, compared to the single-branch model with CAMs, the proposed model improves by 1.3%, 5.2%, and 2.9% in *Acc*, *MAR*, and *FAR*, respectively; this indicates that the model can learn a more accurate feature representation of weld through the joint training of two related tasks. In addition, the effect of coordinate attention on classification performance is also studied. After introducing CAMs into the feature learning module of the proposed network, the Acc increased by 0.4% compared to the two-branch CNN, while *MAR* and *FAR* decreased by 1.6% and 1.0%, respectively. Similarly, compared with the original CNN, the model with CAMs has improved in all metrics. These results indicate that CAMs have a stronger ability to capture subtle features of defects compared to ordinary CNN.

Table 2. The comparison results from different combinations of CAM and two-branch (TB).

CAM	TB	*Acc*	*MAR*	*FAR*	*mIoU*
—	—	0.964	0.069	0.154	—
√	—	0.966	0.063	0.144	—
—	√	0.975	0.027	0.125	0.926
√	√	0.979	0.011	0.115	0.957

Note: "—" means that the component does not exist in the network, while "√" means that the component is introduced.

More importantly, in terms of the magnitude of the performance improvement, the two-branch structure is better than coordinated attention. For example, the performance improvement is 0.2%, 0.6%, and 1.0% after introducing CAMs into CNN, while the improvement can reach 1.1%, 4.2%, and 2.9% when applying the two-branch structure to CNN. By establishing two related tasks, segmentation and classification, the two-branch network can highlight the weld semantic information in the training process under the full use of supervised information in the images. This structure provides better results than the direct use of coordinated attention, and the combination of both enhances the classification performance of the weld quality.

4.3.2. Sensitivity to the Number of Training Samples

As mentioned above, it is very expensive to collect a large amount of defect data on an actual production line. Therefore, the impact of a smaller training sample size on model performance is also investigated. The training set is randomly sampled for the experimental study by 95%, 90%, 80%, 70%, 60%, and 50%, respectively. The same category ratio is maintained as the original dataset in the sampling process. The rest of the settings follow the same training and testing procedure as the previous experiments. The detailed results are presented in Figure 5. It can be found that compared with the single-branch network, the two-branch structure model is not sensitive to the number of training samples. The performance of the proposed network model still remains 95.76% *Acc*, 5.20% *MAR*, and 21.15% *FAR* when the number of training samples is 70% of the original. When using a smaller dataset (50%), the Acc decreases by 6.55% compared to the original one, while *MAR* and *FAR* increase by 22.02% and 19.75% in the test set. On the contrary, the impact of decreasing the training sample size on the single-branch model is significant, with a rapid decrease in all performance metrics. The performance is poor when the number of training samples is small. For example, *Acc*, *MAR*, and *FAR* of CNN are 87.35%, 35.63%, and 46.16%, respectively. Compared to the original model, its *Acc* decreased by 9.05%, while *MAR* and *FAR* increased by 28.73% and 30.76%, respectively. Such a comparison demonstrates that the two-branch network has a significant effect on alleviating the training sample demand. Overall, the experimental results show that the method can still maintain a good, stable performance with a small number of training samples.

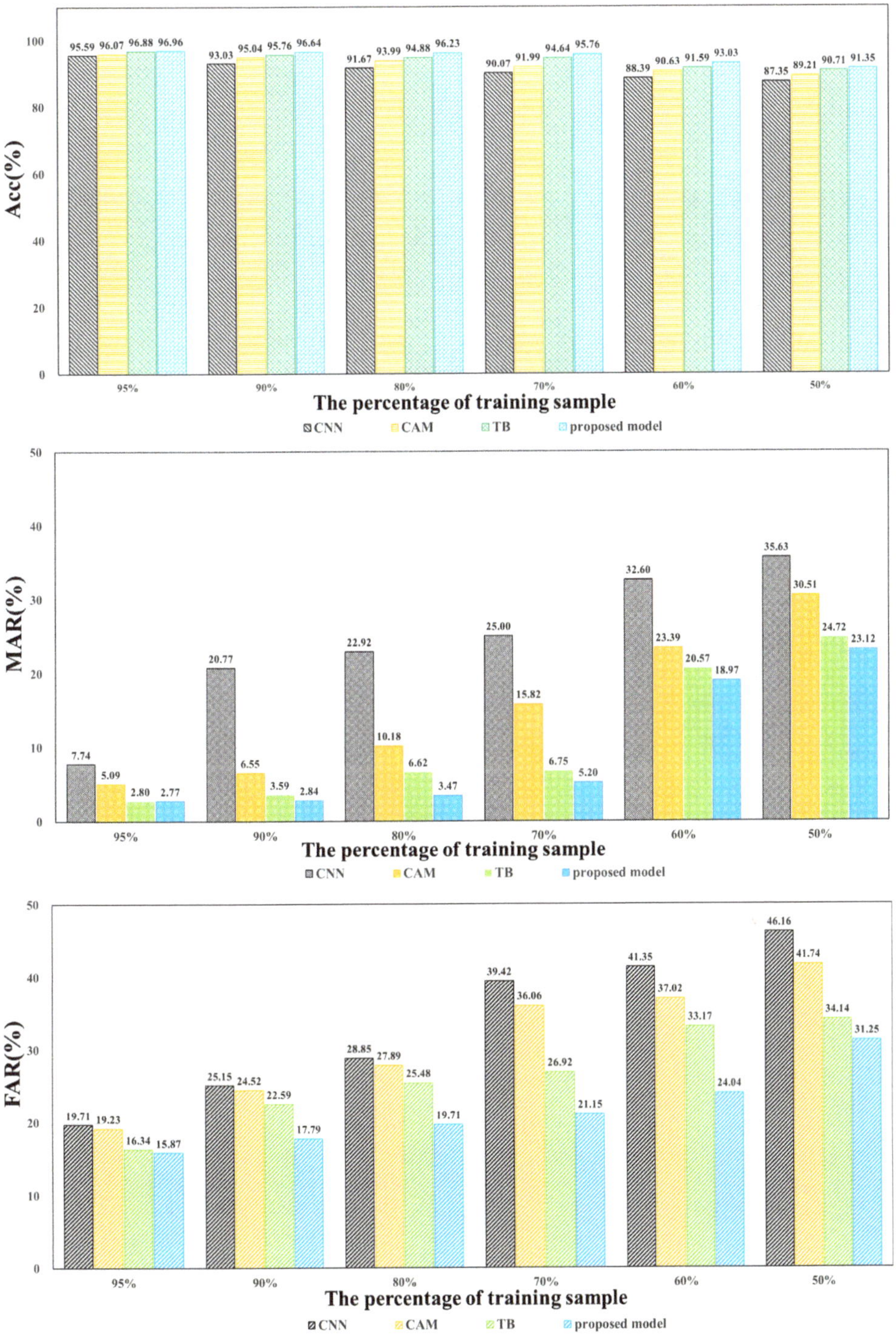

Figure 5. Performance metrics at a varying number of training samples.

4.4. Post-Processing Results

Figure 6 shows the weld segmentation results obtained by the proposed two-branch network and the corresponding annotation masks. It can be intuitively illustrated that the model can achieve the pixel-level localization of the weld relatively well. The *mIoU* results in Table 2 also quantitatively show that the model can segment the weld region accurately from the raw welding images.

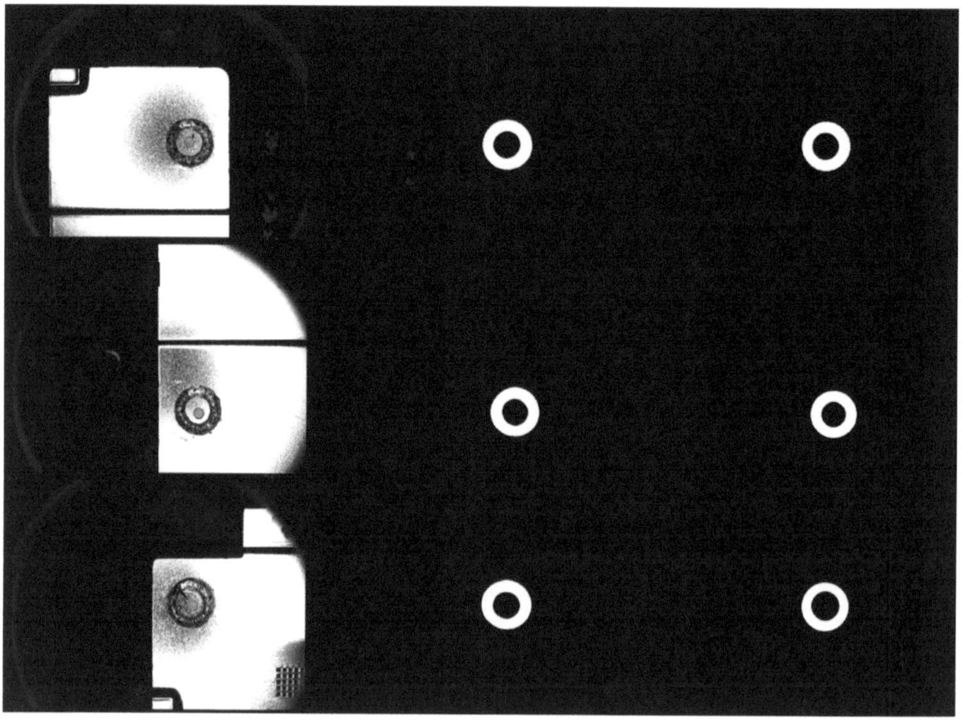

Figure 6. Some examples of segmentation results of the proposed model. From left to right are the raw welding images, the segmentation results of the proposed network, and the ground truth.

After obtaining the binary segmentation images, the post-processing method combines the Canny and circular Hough transform to achieve the extraction of the internal and external circular parameters. The weld width information can be easily calculated from the inner and outer circle parameters, which are measured in pixel lengths. Some results of post-processing methods are shown in Figure 7. It can be seen that the framework proposed in this paper can effectively extract the weld region parameters from the welding images on power batteries. In addition, the accuracy of the welding parameter extraction relies heavily on the results of the segmentation model in the previous section. When the segmentation model does not perform well, it will seriously deteriorate the effect of the post-processing algorithm.

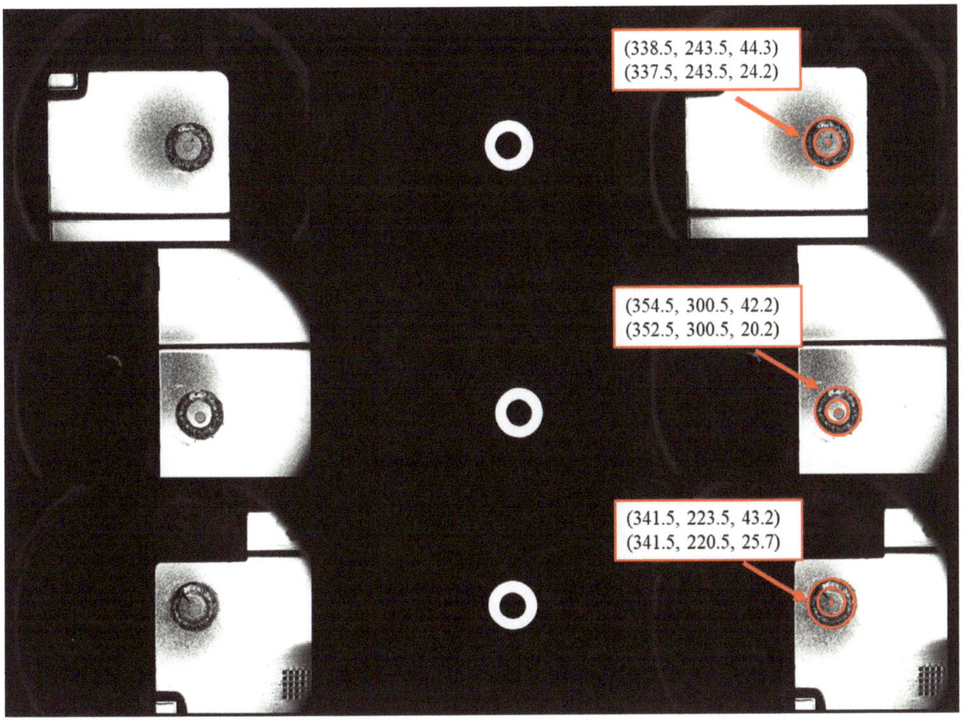

Figure 7. Some examples of the post-processing method. From left to right are the raw images, segmented images, and extracted parameters (a, b, r).

5. Conclusions

The automatic detection of laser welding quality in power batteries is crucial for ensuring the safety performance of new energy vehicles. This paper proposes a framework that combines deep network and conventional image processing techniques to achieve efficient and accurate detection of laser welding quality.

(1) The framework consists of a two-branch network, comprising a segmentation network trained with pixel-level labels and a decision network built on top of it to predict the presence of weld defects. The joint learning of these two tasks effectively enhances the accuracy of the features and the generalization ability of the model;

(2) To effectively learn the distinguishing features between normal welds and defective welds, the detection network incorporates a coordinate attention module. Additionally, a post-processing method based on the Hough transform is employed to extract geometric information about the segmented weld region, including the position of the weld's center and its width;

(3) An extensive evaluation of the model is conducted on a dataset collected from a real production line. The evaluation encompasses ablation experiments and investigates the impact of the training sample size. The experimental results demonstrate that both the coordinate attention module and the two-branch structure significantly improve the model's performance for weld detection, as indicated by metrics such as Acc, FDR, and MAR.

This study contributes a new approach to quality prediction with promising implications for battery pack welding. Although the present work is focused on a specific quality inspection application, it can be believed that the underlying principles and methodologies can be adapted and applied to similar scenarios. For instance, in body spot weld quality inspection, where the defect rate is low on the production line, the accuracy of the model in recognizing defective welds can be improved by utilizing the defective area of spot welds as an auxiliary task. In future work, the feasibility and effectiveness of extending the proposed framework will be explored for other manufacturing quality inspection tasks that have similar dataset constraints.

Author Contributions: D.W.: data collection, investigation, methodology, and writing. W.D.: methodology and writing. Y.Z.: data collection and investigation. D.T.: methodology, review, and supervision. Y.P.: review and supervision. All authors have read and agreed to the published version of the manuscript.

Funding: This work was supported by the National Natural Science Foundation of China (No. 52075325), Shanghai Jiao Tong University Medical Engineering Intersection Project (No. 23X010300732) and Shanghai Aerospace Science Joint Fund (No. USCAST2022-31).

Institutional Review Board Statement: Not applicable.

Informed Consent Statement: Not applicable.

Data Availability Statement: Not applicable.

Conflicts of Interest: The authors declare no conflict of interest.

References

1. Tsoukantas, G.; Salonitis, K.; Stournaras, A.; Stavropoulos, P.; Chryssolouris, G. On optical design limitations of generalized two-mirror remote beam delivery laser systems: The case of remote welding. *Int. J. Adv. Manuf. Technol.* **2007**, *32*, 932–941. [CrossRef]
2. Sadeghian, A.; Iqbal, N. A review on dissimilar laser welding of steel-copper, steel-aluminum, aluminum-copper, and steel-nickel for electric vehicle battery manufacturing. *Opt. Laser Technol.* **2021**, *146*, 107595. [CrossRef]
3. Cao, X.; Wallace, W.; Poon, C.; Immarigeon, J.-P. Research and progress in laser welding of wrought aluminum alloys. i. laser welding processes. *Mater. Manuf. Process.* **2003**, *18*, 1–22. [CrossRef]
4. Stavridis, J.; Papacharalampopoulos, A.; Stavropoulos, P. Quality assessment in laser welding: A critical review. *Int. J. Adv. Manuf. Technol.* **2017**, *94*, 1825–1847. [CrossRef]
5. Xu, F.; Xu, Y.; Zhang, H.; Chen, S. Application of sensing technology in intelligent robotic arc welding: A review. *J. Manuf. Process.* **2022**, *79*, 854–880.
6. Cai, W.; Wang, J.; Jiang, P.; Cao, L.; Mi, G.; Zhou, Q. Application of sensing techniques and artificial intelligence-based methods to laser welding real-time monitoring: A critical review of recent literature. *J. Manuf. Syst.* **2020**, *57*, 1–18.
7. Fan, X.; Gao, X.; Liu, G.; Ma, N.; Zhang, Y. Research and prospect of welding monitoring technology based on machine vision. *Int. J. Adv. Manuf. Technol.* **2021**, *115*, 3365–3391.
8. Gao, Y.; Zhong, P.; Tang, X.; Hu, H.; Xu, P. Feature Extraction of Laser Welding Pool Image and Application in Welding Quality Identification. *IEEE Access* **2021**, *9*, 120193–120202. [CrossRef]
9. Cai, W.; Wang, J.; Cao, L.; Mi, G.; Shu, L.; Zhou, Q.; Jiang, P. Predicting the weld width from high-speed successive images of the weld zone using different machine learning algorithms during laser welding. *Math. Biosci. Eng.* **2019**, *16*, 5595–5612. [CrossRef]
10. Zhang, Z.; Wen, G.; Chen, S. Weld image deep learning-based on-line defects detection using convolutional neural networks for Al alloy in robotic arc welding. *J. Manuf. Process.* **2019**, *45*, 208–216. [CrossRef]
11. Xia, C.; Pan, Z.; Fei, Z.; Zhang, S.; Li, H. Vision based defects detection for Keyhole TIG welding using deep learning with visual explanation. *J. Manuf. Process.* **2020**, *56*, 845–855. [CrossRef]
12. Wang, Q.; Jiao, W.; Wang, P.; Zhang, Y. A tutorial on deep learning-based data analytics in manufacturing through a welding case study. *J. Manuf. Process.* **2020**, *63*, 2–13.
13. Simonyan, K.; Zisserman, A. Very deep convolutional networks for large-scale image recognition. *arXiv* **2014**, arXiv:1409.1556.
14. He, K.; Zhang, X.; Ren, S.; Sun, J. Deep residual learning for image recognition. In Proceedings of the IEEE Computer Society Conference on Computer Vision and Pattern Recognition (CVPR), Las Vegas, NV, USA, 27–30 June 2016; pp. 770–778.
15. Howard, A.; Sandler, M.; Chen, B.; Wang, W.; Chen, L.; Tan, M.; Chu, G.; Vasudevan, V.; Zhu, Y.; Pang, R.; et al. Searching for MobileNetV3. In Proceedings of the 2019 IEEE/CVF International Conference on Computer Vision (ICCV), Seoul, Republic of Korea, 27 October–2 November 2019; pp. 1314–1324.
16. Yang, Y.; Pan, L.; Ma, J.; Yang, R.; Zhu, Y.; Yang, Y.; Zhang, L. A High-Performance Deep Learning Algorithm for the Automated Optical Inspection of Laser Welding. *Appl. Sci.* **2020**, *10*, 933. [CrossRef]
17. Yang, Y.; Yang, R.; Pan, L.; Ma, J.; Zhu, Y.; Diao, T.; Zhang, L. A lightweight deep learning algorithm for inspection of laser welding defects on safety vent of power battery. *Comput. Ind.* **2020**, *123*, 103306. [CrossRef]
18. Dai, W.; Li, D.; Tang, D.; Wang, H.; Peng, Y. Deep learning approach for defective spot welds classification using small and class-imbalanced datasets. *Neurocomputing* **2022**, *477*, 46–60. [CrossRef]

19. Ren, S.; He, K.; Girshick, R.; Sun, J. Faster R-CNN: Towards real-time object detection with region proposal networks. *IEEE Trans. Pattern Anal. Mach. Intell.* **2017**, *39*, 1137–1149. [CrossRef]
20. Redmon, J.; Divvala, S.; Girshick, R.; Farhadi, A. You only look once: Unified, real-time object detection. In Proceedings of the 2016 IEEE Conference on Computer Vision and Pattern Recognition (CVPR), Las Vegas, NV, USA, 27–30 June 2016; pp. 779–788.
21. Redmon, J.; Farhadi, A. Yolov3: An incremental improvement. *arXiv* **2018**, arXiv:1804.02767.
22. Ji, C.; Wang, H.; Li, H. Defects detection in weld joints based on visual attention and deep learning. *NDT E Int.* **2023**, *133*, 102764.
23. Dai, W.; Li, D.; Tang, D.; Jiang, Q.; Wang, D.; Wang, H.; Peng, Y. Deep learning assisted vision inspection of resistance spot welds. *J. Manuf. Process.* **2020**, *62*, 262–274.
24. Zhu, Y.; Yang, R.; He, Y.; Ma, J.; Guo, H.; Yang, Y.; Zhang, L. A Lightweight Multiscale Attention Semantic Segmentation Algorithm for Detecting Laser Welding Defects on Safety Vent of Power Battery. *IEEE Access* **2021**, *9*, 39245–39254. [CrossRef]
25. Zhang, H.; Di, X.; Zhang, Y. Real-Time CU-Net-Based Welding Quality Inspection Algorithm in Battery Production. *IEEE Trans. Ind. Electron.* **2020**, *67*, 10942–10950. [CrossRef]
26. Xu, H.; Yan, Z.H.; Ji, B.W.; Huang, P.F.; Cheng, J.P.; Wu, X.D. Defect detection in welding radiographic images based on semantic segmentation methods. *Measurement* **2021**, *188*, 110569. [CrossRef]
27. Hu, J.; Shen, L.; Albanie, S.; Sun, G.; Wu, E. Squeeze-and-excitation networks. *IEEE Trans. Pattern Anal. Mach. Intell.* **2017**, *42*, 2011–2023. [CrossRef]
28. Woo, S.; Park, J.; Lee, J.; Kweon, I.S. CBAM: Convolutional Block Attention Module. In Proceedings of the European Conference on Computer Vision (ECCV), Munich, Germany, 8–14 September 2018; pp. 3–19.
29. Hou, Q.; Zhou, D.; Feng, J. Coordinate Attention for Efficient Mobile Network Design. In Proceedings of the IEEE/CVF Conference on Computer Vision and Pattern Recognition, Nashville, TN, USA, 20–25 June 2021; pp. 13708–13717.
30. Zhang, Y.; Yang, Q. A Survey on Multi-Task Learning. *IEEE Trans. Knowl. Data Eng.* **2022**, *34*, 5586–5609.
31. Vafaeikia, P.; Namdar, K.; Khalvati, F. A Brief Review of Deep Multi-task Learning and Auxiliary Task Learning. *arXiv* **2020**, arXiv:2007.01126.
32. Xie, Z.; Chen, J.; Feng, Y.; Zhang, K.; Zhou, Z. End to end multi-task learning with attention for multi-objective fault diagnosis under small sample. *J. Manuf. Syst.* **2021**, *62*, 301–316. [CrossRef]
33. Duda, R.O.; Hart, P.E. Use of the hough transformation to detect lines and curves in pictures. *Commun. ACM* **1972**, *15*, 11–15. [CrossRef]
34. Ballard, D. Generalizing the hough transform to detect arbitrary shapes. *Pattern Recognit.* **1981**, *13*, 111–122. [CrossRef]
35. Ronneberger, O.; Fischer, P.; Brox, T. U-Net: Convolutional networks for biomedical image segmentation. In *Medical Image Computing and Computer-Assisted Intervention–MICCAI 2015, Proceedings of the 18th International Conference, Munich, Germany, 5–9 October 2015*; Springer International Publishing: Berlin/Heidelberg, Germany, 2015; pp. 234–241.
36. Ioffe, S.; Szegedy, C. Batch normalization: Accelerating deep network training by reducing internal covariate shift. *arXiv* **2015**, arXiv:1502.03167.
37. Lin, M.; Chen, Q.; Yan, S. Network in network, ICLR 2014. *arXiv* **2013**, arXiv:1312.4400.
38. Canny, J. A computational approach to edge detection. *IEEE Trans. Pattern Anal. Mach. Intell.* **1986**, *8*, 679–698. [CrossRef] [PubMed]
39. Sobel, I.; Feldman, G. A 3 × 3 isotropic gradient operator for image processing. Presented at a talk at the Stanford Artificial Project. In *Pattern Classification and Scene Analysis*; Duda, R., Hart, P., Eds.; John Wiley & Sons: Hoboken, NJ, USA, 1968; pp. 271–272.
40. Kingma, D.P.; Ba, J. Adam: A method for stochastic optimization. *arXiv* **2014**, arXiv:1412.6980.
41. Loshchilov, I.; Hutter, F. Sgdr: Stochastic gradient descent with warm restarts. *arXiv* **2016**, arXiv:1608.03983.

cancers

MDPI

Article

Identifying Lymph Nodes and Their Statuses from Pretreatment Computer Tomography Images of Patients with Head and Neck Cancer Using a Clinical-Data-Driven Deep Learning Algorithm

Sheng-Yao Huang [1,2], Wen-Lin Hsu [2,3,4], Dai-Wei Liu [1,2,3,4], Edzer L. Wu [5], Yu-Shao Peng [5], Zhe-Ting Liao [5] and Ren-Jun Hsu [1,3,4,*]

[1] Institute of Medical Science, Tzu Chi University, Hualien 970374, Taiwan; martin12345m@gmail.com (S.-Y.H.); dwliu5177@yahoo.com.tw (D.-W.L.)
[2] Department of Radiation Oncology, Hualien Tzu Chi General Hospital, Buddhist Tzu Chi Medical Foundation, Hualien 970473, Taiwan; wlhsu@tzuchi.com.tw
[3] Cancer Center, Hualien Tzu Chi Hospital, Buddhist Tzu Chi Medical Foundation, Hualien 970473, Taiwan
[4] School of Medicine, Tzu Chi University, Hualien 970374, Taiwan
[5] DeepQ Technology Corp, New Taipei City 242062, Taiwan; edzer_wu@deepq.com (E.L.W.); ys_peng@htc.com (Y.-S.P.); zt_liao@htc.com (Z.-T.L.)
* Correspondence: hsurnai@gmail.com; Tel./Fax: +886-3-8561825

Simple Summary: We proposed a deep learning algorithm to detect lymph nodes and classify them in the head and neck region on computed tomography. We further analyzed the inference result from the model and found that the size of the lymph nodes may be a characteristic for the model to classify them. This finding is consistent with current clinical aspects. We will deploy the model in clinical practice and hope to assist clinicians in finding out the lesions more correctly and efficiently.

Abstract: Background: Head and neck cancer is highly prevalent in Taiwan. Its treatment mainly relies on clinical staging, usually diagnosed from images. A major part of the diagnosis is whether lymph nodes are involved in the tumor. We present an algorithm for analyzing clinical images that integrates a deep learning model with image processing and attempt to analyze the features it uses to classify lymph nodes. Methods: We retrospectively collected pretreatment computed tomography images and surgery pathological reports for 271 patients diagnosed with, and subsequently treated for, naïve oral cavity, oropharynx, hypopharynx, and larynx cancer between 2008 and 2018. We chose a 3D UNet model trained for semantic segmentation, which was evaluated for inference in a test dataset of 29 patients. Results: We annotated 2527 lymph nodes. The detection rate of all lymph nodes was 80%, and Dice score was 0.71. The model has a better detection rate at larger lymph nodes. For those identified lymph nodes, we found a trend where the shorter the short axis, the more negative the lymph nodes. This is consistent with clinical observations. Conclusions: The model showed a convincible lymph node detection on clinical images. We will evaluate and further improve the model in collaboration with clinical physicians.

Keywords: head and neck cancer; computed tomography; deep learning; semantic segmentation; image processing

1. Introduction

Head and neck cancers have remained among the ten leading causes of cancer-related death in Taiwan for a long time [1]. They include oral cavity, oropharynx, hypopharynx, larynx, and nasopharynx cancers. Most head and neck cancers are associated with life habits, such as smoking, drinking alcohol, and chewing betel nuts.

The head and neck region has abundant lymphatic drainage [2]. The status of lymph node metastasis is critical to prognosis [3], including the location and number of cancer-involved lymph nodes and presentation of extranodal extension (ENE) [4].

Most head and neck cancers are treated surgically if eradicable, even at an advanced stage [5]. Tumors with involved lymph nodes that are completely removed have better prognoses than others [6]. Typically, surgeons conduct a clinical workup before operating to gather additional information for selecting appropriate surgical techniques, which may involve dissecting the cervical lymph nodes. The workup also affects the treatment choice in neoadjuvant, adjuvant, or even definitive treatment settings.

Medical imaging plays a crucial role in the clinical workup, with various tools available, including computer tomography (CT), magnetic resonance imaging (MRI), and positron emission tomography (PET) [5,7]. Each imaging tool offers unique advantages. For example, CT effectively finds bony invasion, while MRI excels at delineating soft tissue involvement. In contrast, PET can provide a comprehensive diagnosis of locoregional and distant metastasis by measuring cell activity using F-18 [8]. The efficacy of using medical images as clinical diagnostic tools has been evaluated [4,9]. Some features of lymph nodes related to morphology or enhancement on images may indicate tumor involvement: a shorter axis of a lymph node of >1 cm, heterogeneous enhancement, or rough border of a lymph node, which might be a sign of ENE [10]. However, even when interpreted by well-experienced clinical physicians or radiologists, the sensitivity and specificity of CT images were 72% and 83%, respectively, while the area under the receiver operating characteristic curve (AUC) was 0.65–0.69 [11,12]. In contrast, the sensitivity and specificity with MRI were 0.7–0.8 and 0.5–0.7, respectively [13,14].

Efficient and correct identification and delineation of lymph nodes is crucial for clinical diagnosis, surgical techniques, and other treatments. Traditionally, radiologists or clinical physicians such as otolaryngologists would have to view CT images to obtain information on clinical diagnosis and make treatment decisions. In Taiwan, it is common for an experienced otolaryngologist to have more than a hundred patients in one outpatient clinic, and serve more than twenty in inpatient at the same time. Even with the resident's assistance, this is still overwhelming. Although not all patients are diagnosed with head and neck diseases, it creates time pressure for physicians to read images, make decisions, and discuss with patients. In Hualien Tzu Chi Hospital, head and neck CT examinations are generally carried out within a week; most of them should be reported and submitted within 2 weeks, which is also exhausting for the staff. For junior residents, it would take more time and effort to complete image reading. An automated assistant for clinical diagnosis might relieve the loading.

There has been a recent trend toward using digital systems to assist clinical diagnosis. These systems analyze data from laboratory tests, medical records, and images to generate results for clinical needs, such as establishing clinical impressions, alerting for emergencies, or risk stratification. Among these digital systems, deep learning-based computer vision techniques have made significant progress in analyzing medical images [15].

Convolutional neural networks (CNNs) have been widely used in deep learning for computer vision tasks, including classification, object detection, and semantic segmentation [16]. Models derived from ResNet [17] or VGG [18] were used for classifying regions of interest by human experts. Fully convolutional networks like Unet [19], on the other hand, can segment the targets from medical images. Such models have been used to study lymph node status, or segmentation tasks at head and neck region. Using a CNN model, Kann et al. classified lymph nodes segmented by experts as normal or tumor-involved in CT images, achieving an impressive AUC of 0.91 [20]. The model was composed of a 3D model and a size-invariant model and was able to extract features while preventing itself from overfitting [20]. Another study examined segmentation for head and neck lymphatic drainage areas [21], which can be applied to contouring in radiotherapy. In this study, a fully convolutional neural network was proposed to deal with segmentation for head and neck lymphatic drainage area.

However, lymph nodes' inconsistent morphology and size make determining their status and delineation challenging. Lymph nodes can range in diameter from being almost invisible in medical images to >10 cm. Moreover, their 2D projections can appear with similar textures to other structures in image slices, such as vessels, muscles, or salivary glands. Another challenge in this task is annotation, a time-consuming process for clinical physicians to segment and label the lymph nodes for classification.

The most challenging issue is deploying a model in the clinical field. While having sufficient data can increase the likelihood of constructing a well-performing model, additional factors must also be considered for successful deployment. A performance gap between training and real-world data has been reported [22], and factors such as the examination settings, presentation of inference results, and the specific needs of clinicians and other healthcare professionals can all affect the model's effectiveness.

To address these challenges, we present a novel approach combining deep learning models, image processing algorithms, and domain knowledge segment and classify lymph nodes. The proposed method is further evaluated based on clinical knowledge to assess the reliability of the inference results.

2. Materials and Methods

2.1. Study Cohort

We retrospectively enrolled patients diagnosed with oral cavity, oropharynx, hypopharynx, or larynx cancers at Hualien Tzu Chi General Hospital between 1 January 2008, and 31 December 2018. These patients had confirmed diagnoses from biopsies carried out at our hospital and should receive surgery as definitive treatment. We collected presurgery contrast-enhanced head and neck CT images and surgical pathology reports. We collected pretreatment CT images if a patient received definitive concurrent chemoradiotherapy without surgery. Patients diagnosed or treated due to other cancers before would be excluded.

Supplement Table S1 shows the patients enrolled in this study. After excluding CT images with low resolutions or poorly identified targets, the final dataset included 271 patients with 274 CT image series (Figure 1). These patients were randomly divided into training ($n = 213$), validation ($n = 30$), and testing ($n = 28$) sets. The numbers of patients and lymph nodes are reported in Supplement Table S1.

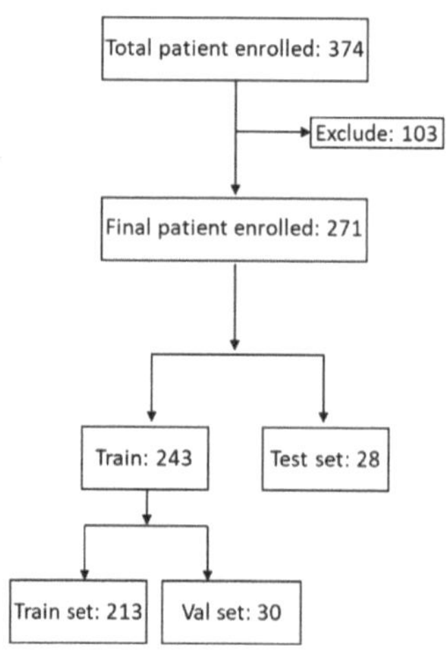

Figure 1. Flowchart of study enrollment. We initially enrolled 374 patients. One hundred and three cases were excluded due to poor resolution of images. Finally, images from 271 patients were included, with 243 in the train set and 28 in the test set.

2.2. Image Prepare and Annotations

Two clinical physicians and a radiologist reviewed the CT images, after which the clinical physicians segmented and classified the lymph nodes on the CT images. The radiologist provided advice to the physicians in case of any uncertainty. Two pathologists reviewed the pathology reports, and annotations were created to classify lymph node status based on the pathology report, which will serve as the ground truth. We used the DeepQ AI platform (https://www.deepq.ai/?lang=en, accessed on 20 September 2023) (from DeepQ, New Taipei City, Taiwan) to annotate images, which were deidentified before upload.

2.3. Model and Training Methodology

2.3.1. Model

The nn-UNet model integrates most state-of-the-art semantic medical image segmentation techniques [23]. It extracts features from images in different spacing in two stages to prevent the model from losing complete picture information when training in image patches. Computing resources can be preserved by training in patches. The model is trained by self-adjusting hyperparameters based on data features (i.e., sample size, image size, spacing, and modalities). The framework automatically defines the batch size, number of epochs, model architecture, and learning rate. However, if necessary, the user can manually modify them based on past experiences or competition on different open datasets. We constructed a model based on nnUNet to fit our situations.

First, we preprocessed images. We set pixel spacing as 0.89×0.48 mm according to the value obtained from the dicom file. Windowing and intensity were normalized by window level and width.

We chose a 3D network from clinical aspects. The morphology of lymph nodes may be confused with other structures around them in 2D projections, such as vessels, glands, muscles, or other soft tissues, that become distinct in 3D projections. We expected that the result would be better in a 3D network. The nn-UNet network will automatically adjust its architecture to handle spacing anisotropy between axes. Specifically, the network applies convolution and pooling operations to high-resolution axes until the resolution factor between axes becomes <2. This approach ensures that the model extract feature is unaffected by the varying resolution between axes. We used a patch size in the network of 160×192 in the first stage and 192×224 in the second stage. We inherited the loss function in nnUNet, which combines Dice loss and cross-entropy. Finally, post-processing was carried out to filter prediction masks with pixel numbers below the threshold. The details for model settings were summarized in Supplementary Figure S1.

We evaluated the model's inference results using two metrics: Dice score and detection rate. The Dice score (s) was calculated as follows:

$$s = \frac{2|P + GT|}{|P| + |GT|} \tag{1}$$

where P represents prediction, and GT represents ground truth (label). The detection rate (d) was calculated as follows:

$$d = \frac{TP}{TP + FP} \tag{2}$$

where TP represents true positive, and FP represents false positive. Both evaluation metrics were calculated by each image slice and averaged to represent each study.

2.3.2. Training Method

We trained the model for 300 epochs with a mini-batch size of 2. We used stochastic gradient descent as optimization with an initial learning rate of 0.01 and gradually decaying the learning rate as training progressed. An oversample technique was used to address class imbalances. Specifically, we ensured that >33.3% of the patches contained at least one positive mask. In addition to using the default data augmentation provided by nnU-Net [23], we used random translation to ensure the mask was distributed uniformly within the patches, which can improve model robustness and generalization. All experiments were trained on a single GeForce GTX 1080 Ti graphics processing unit. The training, validation, and inference were performed with Pytorch (version 1.11.0) in Python 3.9.

3. Results

3.1. Basic Image Features

This study included 2527 lymph nodes annotated from 271 patients. Supplement Table S1 shows that most lymph nodes were annotated as negative.

Distribution of Lymph Node Size and Intensity

We analyzed the lymph nodes according to their short axis size (Table 1a), finding a trend where the shorter the short axis, the more negative the lymph node. The pixel intensity in lymph node regions showed greater diversity in negative than in other lymph nodes.

Table 1. (a)The size (represented by a short axis) of lymph nodes by classes on the dataset. (b) The size of the lymph node by class from the model's inference result.

(a)									
	Train (%)			Valid (%)			Test (%)		
	<1 cm	>1 cm	Total	<1 cm	>1 cm	Total	<1 cm	>1 cm	Total
N	1470 (96)	61 (4)	1531	270 (96)	11 (4)	281	221 (96)	10 (4)	231
P	125 (45)	153 (55)	278	14 (45)	16 (55)	30	18 (60)	12 (40)	30
ENE	19 (13)	126 (87)	145	2 (13)	14 (87)	16	0	10	10

(b)			
Num (%)	<1 cm	>1 cm	Total
N	167 (96)	9 (4)	176
P	17 (60)	11 (40)	28
ENE	0	9	9

3.2. D Model Performance

3.2.1. Performance Evaluation

First, we examined the relationship between the threshold of the portion overlapped and the detection rate (Supplementary Figure S3). We found that the detection rate reached >60% at a threshold of 50%. We then evaluated the model using two settings: >0% and >50%.

We trained models with different manipulations: combine all three annotation classes, combine only the positive and ENE classes, and separate classes. First, we evaluated the ability to detect lymph nodes in three models (Supplement Table S2). The three models showed consistent Dice scores and detection rates. Then, we evaluated the inference of the model trained on separate classes (Supplement Table S3). The detection rates for negative, positive, and positive with ENE lymph nodes were 76%, 73%, and 90%, respectively. The average lymph node detection rate was 80%, while the Dice score was 0.71.

3.2.2. Inference Analysis

The Model Can Size Classify Lymph Nodes

We analyzed lymph nodes detected by the model in the test set to determine the clinical characteristics it may capture. Table 1b shows that, of the 176 identified negative lymph nodes, 167 had a short axis <1 cm. In contrast, all identified positive lymph nodes with ENE had a short axis >1 cm. Figure 2 shows some of the model's accurate predictions.

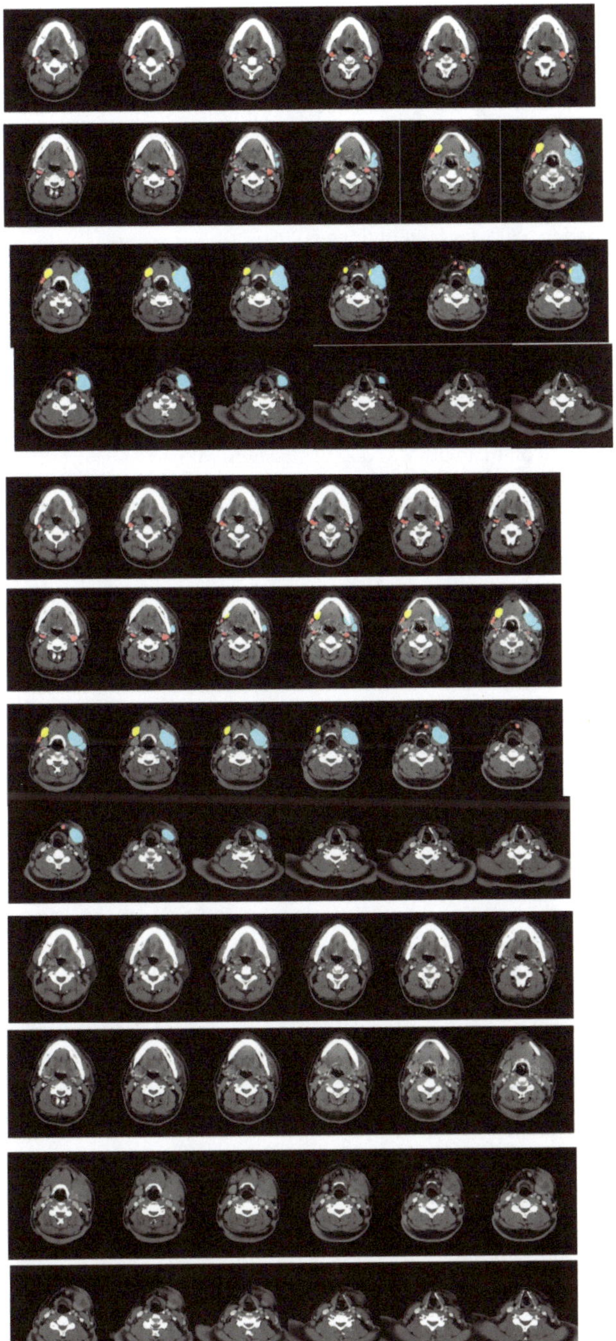

Figure 2. Examples of inference results. (Upper: inference result; middle: ground truth; lower: original image) The inference results compared with ground truth annotation. Red: negative; yellow: positive; light blue: extranodal extension.

False Negative/Positive Inferences

We found that Dice scores decreased when inferring separated classes, especially for the positive and ENE classes. We analyzed accuracy at the pixel level to determine possible reasons for this. Supplement Table S4 shows the confusion matrix. Most of the misclassified pixels were recognized as background (Supplement Table S4b). The model also predicted some background pixels as lymph nodes (Supplement Table S4c).

We also found a correlation between the detection rate and lymph node size (Supplementary Figure S4). The detection rate was 45% for lymph nodes with a short axis <5 mm. The detection rate increased to >80% for larger lymph nodes.

Misclassification of Positive and ENE Lymph Nodes

Supplement Table S4c shows that the model classified 11% of pixels labeled as positive lymph nodes as ENE and 11% labeled as ENE as positive lymph nodes. We evaluated the performance of models on the test set with combined P and ENE classes to determine how misclassification affected Dice scores (Table 2). The models trained on all separate classes or combined P and ENE classes showed better Dice scores and detection rates on the test set with combined classes than on that with P and ENE classes separated.

Table 2. Performance of model on test dataset with P and ENE classes combined. (a) Model trained in separate classes. (b) Model trained in P and ENE classes combined.

3D Metric(%)	(a)		(b)	
	ENE+P	N	ENE+P	N
Detection rate (>0)	87.50	76.58	92.50	75.68
Detection rate (>50)	80.00	59.46	87.50	61.71
FP/image	1.29	4.64	1.11	5.46

4. Discussion

4.1. *Applying a Deep Learning Model in Classifying Lymph Mode Metastasis in Head and Neck Cancer*

There is limited research on applying machine learning or deep learning algorithms to lymph nodes in patients with head and neck cancers. One reason for this is data collection. There are open datasets for medical imaging, including contouring and classification of organs and lesions [16], but almost none for head and neck cancers. A dataset is the basis for training and evaluating a model, and establishing such a novel dataset would be time- and labor-intensive, especially for clinical practitioners. Our study enrolled more patients than previous studies [20,21], and the annotated lymph nodes are also comparable. To our knowledge, this is the first study to identify lymph nodes and statuses using semantic segmentation. We hope that a model trained using such a volume of data could be helpful clinically.

4.2. *Model Inference*

4.2.1. Detection Rate and Dice Score

This study aimed to assist clinicians in detecting lymph nodes from medical images to make diagnoses and decide on treatments, but not to replace clinicians or screen through serial images. Clinical physicians should still review the images to check the model's suggestions and the primary tumor's extension. It marks potentially involved lymph nodes and reminds physicians while reviewing image slices. Therefore, we are more concerned with classification accuracy than detailed object contouring. However, since classifying an entire image based on the presence of a tumor-involved lymph node is meaningless in clinical practice, object detection remains necessary. We assigned the task to semantic segmentation due to its visual result presentation. A questionnaire survey in our hospital showed that 82% of physicians preferred lymph nodes to be presented as segmentation masks rather than bounding boxes, mainly due to visualization. When clustered or serial lymph nodes are present, segmentation masks could be easier to read than stacked bounding boxes.

Evaluating lymph node segmentation is challenging. While tumor-involved lymph nodes can be >5–6 cm, most objects are <1 cm (Table 1a). In addition, the anatomy and structure are complicated in head and neck lymphatic drainage regions. They contain many vessels and glands whose size, texture, and even intensity are similar to lymph nodes, which can interfere with the model recognizing lymph nodes.

Our model had higher detection rates than Dice scores. These metrics are quite different: the Dice score considers false negatives and positives, while the detection rate considers whether the model "captures" the object, meaning its marking of pixels labeled as ground truth. The detection rate calculation may underestimate the false positive effect. We evaluated the false positive rate (Supplement Table S3), finding 1–4 false positive components per case, depending on the class. This false positive rate should be tolerable for clinical practice, although further evaluation after deployment is necessary.

Some studies have evaluated model performance in identifying lymph nodes from medical images by detection rate [24,25], reporting a detection rate of 0.7–0.8 and a false positive rate of 10 per volume. Our study showed a better overall detection rate of 0.8 and a false positive rate of 2.36 per volume. Among classes, the best detection rate was for ENE (0.9). Among sizes, there was a better detection rate for lymph nodes with a short axis >5 mm (>0.7; Supplementary Figure S4), which was also better than in a previous study (0.62). Our improved results may be due to nnUNet's comprehensive feature extraction, especially for spatial information.

We examined the relationship between the threshold and detection rate, which is the fraction of the ground truth mask overlapped by the inference mask (i.e., true positive; Supplementary Figure S3). We obtained a detection rate of >0.6 even at a threshold of >0.5. The model could determine the location and size of those detected lymph nodes. We believe our model will be sufficiently robust as an alarm system in clinical practice.

4.2.2. Effects of Clinical Features on Inference

The model might classify lymph nodes according to their size. Table 1b shows that smaller lymph nodes tended to be classified as negative, and those with a short axis >1 cm were more likely to be tumor-involved. The short axis of all identified lymph nodes with ENE was >1 cm, consistent with current clinical experience that one feature indicating malignant lymph node changes is size, usually defined as a short axis >1 cm. It can be referred to as an interpretability of the model from a clinical point of view, and convinces the clinical physicians when the model alarms at a specific lymph node during practice.

4.2.3. The Potential of a Model Trained on Images Generated Using Different Protocols at Different Timepoints

We retrospectively collected images over 10 years. During these years, computed tomography machines, settings for examination, and protocols have changed several times. Even with image preprocesses such as intensity normalization and clipping, intensity enhancement, and contrast remained confusing (Supplementary Figure S2). The model could obtain convincing inferences on test datasets. The images might be heterogeneous in real-world data from different examination machines and protocols. It is common to see gaps in model performance between model training datasets and deployment [22,26]. One reason for this gap is the variance between training and real-world data. Several studies have examined approaches to address this problem, such as transfer learning [27] or domain adaptations [28]. However, labeled data are still necessary, and the effect is not always promising. The model trained on a heterogenous dataset may have better adaptability after deployment but not reach the perfect performance shown in their original studies.

4.2.4. P and ENE Class Misclassification

We found the model confused pixels labeled positive and positive with ENE lymph nodes. Since both classes are tumor-involved, they may share some common features from a clinical perspective, such as larger size and central necrosis. The status of these two classes makes a difference in staging and prognosis but not treatment choice. Since they are clinically suspected of malignancy, dissection during surgery or dose escalation during radiotherapy will be preferred. Therefore, it can still be valuable to classify lymph nodes as tumor-involved or not as an alarm system for clinical practice.

4.3. Limitations

4.3.1. No Consistent Image Examination Protocol

A consistent image examination protocol is still necessary to improve accuracy. Those protocols are established in clinical practice according to modalities, examination aims, targets, and clinical needs. The aim and target were specific in our case, but image quality varied over time. Further evaluation and model adjustment showed that brightness and contrast should be clipped in a range to maintain consistent intensity for corresponding structures in the images, which may lead to stable inference results.

4.3.2. Improved Classification Ability for P and ENE Classes

Future work will aim to reinforce the model's ability to classify P and ENE lymph nodes, potentially by increasing the number of ENE annotations since they were much fewer than for the other two classes (Supplement Table S1).

5. Conclusions

We present a model trained with semantic segmentation to identify lymph nodes and their tumor-involved statuses. Our model had a satisfactory detection rate, but its Dice score could be improved. After deployment, we will evaluate and further improve our model in collaboration with clinical physicians, including model adjustments and image examination protocols. Figure 3 shows our expectations for our model's clinical contribution after deployment.

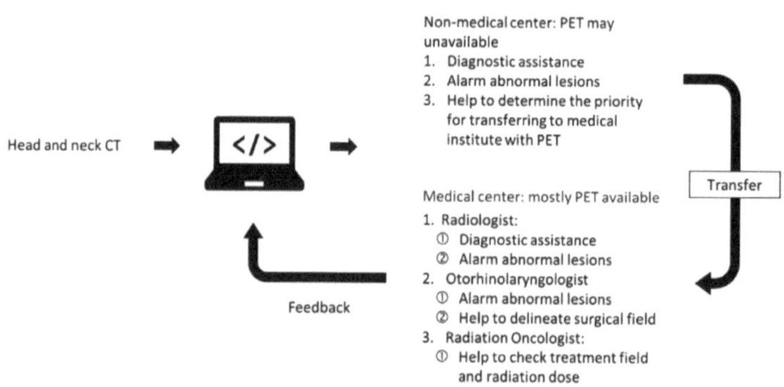

Figure 3. The anticipation of the clinical deployment. We hope the algorithm to fill the gap due to uneven medical examination resources between medical institutes.

In the future, we would like to explore more about the effect of data heterogenicity on model performance. The size of the lymph nodes will be recorded to confirm if the trend that smaller lymph nodes tend to be classified as negative is consistent. We will establish a protocol for CT examinations to obtain stable images, and analysis about intensity or other radiomics features could be applied. We hope that the result of further research could make the model more useful in clinical practice, and most importantly convincible.

Supplementary Materials: The following are available online at https://www.mdpi.com/article/10.3390/cancers15245890/s1, Figure S1: Details for training settings, Figure S2: The intensity of lymph nodes by class, Figure S3: The relationship between threshold (i.e., the portion of ground truth overlapped by prediction mask) and detection rate, Figure S4: The relationship between detection rate and size of lymph node, Table S1: patient and lymph node characteristics, Table S2: Inference result of models on test dataset with all annotation classes combined. The models were trained in different settings: (a) train dataset with three separated classes, (b) train dataset with positive and extranodal extension classes combined, (c) all three classes combined, Table S3: Inference result of model trained in separated classes. Table S4: Pixel-level confusion matrix. (a) Count by pixel. (b) Calculated by longitudinal axis. Most of the misclassified pixels for ground truth were classified as background. (c) Calculated by vertical axis. Most of the misclassified pixels for label were classified as background, but P and ENE have more cross-mistake.

Author Contributions: D.-W.L., S.-Y.H. and R.-J.H. thought of the concept and designed the architecture of the article. W.-L.H. reviewed and provided comments on material articles. S.-Y.H., E.L.W., Z.-T.L. and Y.-S.P. designed the image processing, training process, and data analysis. S.-Y.H. enrolled patients and annotated images. All authors collected and assembled material articles, wrote, and made final approval of the manuscript. All authors have read and agreed to the published version of the manuscript.

Funding: This research was funded by Hualien Tzu Chi Hospital, Buddhist Tzu Chi Medical Foundation, grant numbers TCRD-110-15, IMAR-110-01-08, TCRD112-032, TCRD112-047, and Buddhist Tzu Chi Medical Foundation, grant number TCMF-IMC 112-02.

Institutional Review Board Statement: The study was conducted according to the guidelines of the Declaration of Helsinki, and approved by the Institutional Review Board of Hualien Tzu Chi General Hospital, number IRB111-070-B.

Informed Consent Statement: Informed consent was obtained from all subjects involved in the study.

Data Availability Statement: The data were collected from Hualien Tzu Chi General Hospital under the supervision of the Institutional Review Board, and were not allowed to be available to the public.

Acknowledgments: Special thanks to Yun, Liu for assisting annotations and checking patient lists. Thanks to Yu-Tang, Chen for offering viewpoints on statistics.

Conflicts of Interest: Edzer L. Wu, Yu-Shao Peng and Zhe-Ting Liao are employed by company DeepQ Technology Corp. The remaining authors declare that the research was conducted in the absence of any commercial or financial relationships that could be construed as a potential conflict of interest.

References

1. Health Promotion Administration, Ministry of Health and Welfare, and Taiwan. *Cancer Registry Annul Report 2020 Taiwan*; Taiwan Cancer Registry: Taiwan, 2022.
2. Grégoire, V.; Levendag, P.; Ang, K.K.; Bernier, J.; Braaksma, M.; Budach, V.; Chao, C.; Coche, E.; Cooper, J.S.; Cosnard, G.; et al. CT-based delineation of lymph node levels and related CTVs in the node-negative neck: DAHANCA, EORTC, GORTEC, NCIC,RTOG consensus guidelines. *Radiother. Oncol.* **2003**, *69*, 227–236. [CrossRef] [PubMed]
3. Pisani, P.; Airoldi, M.; Allais, A.; Valletti, P.A.; Battista, M.; Benazzo, M.; Briatore, R.; Cacciola, S.; Cocuzza, S.; Colombo, A.; et al. 107th Congress of the Italian Society of Otorhinolaryngology Head and Neck Surgery Official report. *Acta Otorhinolaryngol Ital.* **2020**, *40* (Supp. S1), S1–S2. [CrossRef] [PubMed]
4. Khan, R. Lymph Node Disease and Advanced Head and Neck Imaging: A Review of the 2013 Literature. In *Current Radiology Reports*; Springer New York LLC: Berlin/Heidelberg, Germany, 2014. [CrossRef]
5. Cognetti, D.M.; Weber, R.S.; Lai, S.Y. Head and Neck Cancer an Evolving Treatment Paradigm. *Cancer* **2008**, *113*, 1911–1932. [CrossRef] [PubMed]
6. Bernier, J.; Cooper, J.S.; Pajak, T.F.; Van Glabbeke, M.; Bourhis, J.; Forastiere, A.; Ozsahin, E.M.; Jacobs, J.R.; Jassem, J.; Ang, K.-K.; et al. Defining Risk Levels in Locally Advanced Head and Neck Cancers: A Comparative Analysis of Concurrent Postoperative Radiation plus Chemotherapy Trials of the EORTC (#22931) and RTOG (# 9501). *Head Neck* **2005**, *27*, 843–850. [CrossRef]
7. National Comprehensive Cancer Network. *NCCN Guidelines Version 1.2024*; National Comprehensive Cancer Network: Fort Washington, PA, USA, 2023.

8. Cerfolio, R.J.; Ojha, B.; Bryant, A.S.; Raghuveer, V.; Mountz, J.M.; Bartolucci, A.A. The Accuracy of Integrated Pet-CT Compared with Dedicated Pet Alone for the Staging of Patients with Nonsmall Cell Lung Cancer. *Ann. Thorac. Surg.* **2004**, *78*, 1017–1023. [CrossRef] [PubMed]
9. Sun, J.; Li, B.; Li, C.J.; Li, Y.; Su, F.; Gao, Q.H.; Wu, F.L.; Yu, T.; Wu, L.; Li, L.J. Computed Tomography versus Magnetic Resonance Imaging for Diagnosing Cervical Lymph Node Metastasis of Head and Neck Cancer: A Systematic Review and Meta-Analysis. In *OncoTargets and Therapy*; Dove Medical Press Ltd.: Princeton, NJ, USA, 2015. [CrossRef]
10. Hoang, J.K.; Vanka, J.; Ludwig, B.J.; Glastonbury, C.M. Evaluation of Cervical Lymph Nodes in Head and Neck Cancer with CT and MRI: Tips, Traps, and a Systematic Approach. *Am. J. Roentgenol.* **2013**, *200*, W17–W25. [CrossRef] [PubMed]
11. Merritt, R.M.; Williams, M.F.; James, T.H.; Porubsky, E.S. Detection of Cervical Metastasis: A Meta-Analysis Comparing Computed Tomography with Physical Examination. *JAMA Otolaryngol. Neck Surg.* **1997**, *123*, 149–152. [CrossRef]
12. Schwartz, D.L.; Ford, E.; Rajendran, J.; Yueh, B.; Coltrera, M.D.; Virgin, J.; Anzai, Y.; Haynor, D.; Lewellyn, B.; Mattes, D.; et al. FDG-PET/CT Imaging for Preradiotherapy Staging of Head-and-Neck Squamous Cell Carcinoma. *Int. J. Radiat. Oncol.* **2005**, *61*, 129–136. [CrossRef]
13. de Bondt, R.; Nelemans, P.; Hofman, P.; Casselman, J.; Kremer, B.; van Engelshoven, J.; Beets-Tan, R. Detection of Lymph Node Metastases in Head and Neck Cancer: A Meta-Analysis Comparing US, USgFNAC, CT and MR Imaging. *Eur. J. Radiol.* **2007**, *64*, 266–272. [CrossRef]
14. Van den Brekel, M.W.M.; Castelijns, J.A.; Stel, H.V.; Golding, R.P.; Meyer, C.J.L.; Snow, G.B. Originals Oto-Rhino-Laryngology Modern Imaging Techniques and Ultrasound-Guided Aspiration Cytology for the Assessment of Neck Node Metastases: A Prospective Comparative Study. *Eur. Arch. Otorhinolaryngol.* **1993**, *250*, 11–17. [CrossRef]
15. Esteva, A.; Chou, K.; Yeung, S.; Naik, N.; Madani, A.; Mottaghi, A.; Liu, Y.; Topol, E.; Dean, J.; Socher, R. Deep Learning-Enabled Medical Computer Vision. *Npj Digit. Med.* **2021**, *4*, 5. [CrossRef] [PubMed]
16. Huang, S.-Y.; Hsu, W.-L.; Hsu, R.-J.; Liu, D.-W. Fully Convolutional Network for the Semantic Segmentation of Medical Images: A Survey. *Diagnostics* **2022**, *12*, 2765. [CrossRef] [PubMed]
17. He, K.; Zhang, X.; Ren, S.; Sun, J. Identity mappings in deep residual networks. *Lect. Notes Comput. Sci. Incl. Subser. Lect. Notes Artif. Intell. Lect. Notes Bioinform.* **2016**, *9908 LNCS*, 630–645.
18. Simonyan, K.; Zisserman, A. A Very Deep Convolutional Networks for Large-Scale Image Recognition. *arXiv* **2014**, arXiv:1409.1556.
19. Ronneberger, O.; Fischer, P.; Brox, T. U-net: Convolutional networks for biomedical image segmentation. *Lect. Notes Comput. Sci. (Incl. Subser. Lect. Notes Artif. Intell. Lect. Notes Bioinform.* **2015**, *9351*, 234–241. [CrossRef]
20. Kann, B.H.; Aneja, S.; Loganadane, G.V.; Kelly, J.R.; Smith, S.M.; Decker, R.H.; Yu, J.B.; Park, H.S.; Yarbrough, W.G.; Malhotra, A.; et al. Pretreatment Identification of Head and Neck Cancer Nodal Metastasis and Extranodal Extension Using Deep Learning Neural Networks. *Sci. Rep.* **2018**, *8*, 14306. [CrossRef] [PubMed]
21. Men, K.; Chen, X.; Zhang, Y.; Zhang, T.; Dai, J.; Yi, J.; Li, Y. Deep Deconvolutional Neural Network for Target Segmentation of Nasopharyngeal Cancer in Planning Computed Tomography Images. *Front. Oncol.* **2017**, *7*, 315. [CrossRef]
22. Beede, E.; Baylor, E.; Hersch, F.; Iurchenko, A.; Wilcox, L.; Ruamviboonsuk, P.; Vardoulakis, L.M. A Human-Centered Evaluation of a Deep Learning System Deployed in Clinics for the Detection of Diabetic Retinopathy. In *Conference on Human Factors in Computing Systems—Proceedings*; Association for Computing Machinery: Melbourne, Australia, 2020. [CrossRef]
23. Isensee, F.; Jaeger, P.F.; Kohl, S.A.A.; Petersen, J.; Maier-Hein, K.H. NnU-Net: A Self-Configuring Method for Deep Learning-Based Biomedical Image Segmentation. *Nat. Methods* **2021**, *18*, 203–211. [CrossRef]
24. Iuga, A.-I.; Carolus, H.; Höink, A.J.; Brosch, T.; Klinder, T.; Maintz, D.; Persigehl, T.; Baeßler, B.; Püsken, M. Automated Detection and Segmentation of Thoracic Lymph Nodes from CT Using 3D Foveal Fully Convolutional Neural Networks. *BMC Med Imaging* **2021**, *21*, 1–12. [CrossRef]
25. Oda, H.; Bhatia, K.K.; Roth, H.R.; Oda, M.; Kitasaka, T.; Iwano, S.; Homma, H.; Takabatake, H.; Mori, M.; Natori, H.; et al. Dense Volumetric Detection and Segmentation of Mediastinal Lymph Nodes in Chest CT Images. In *Medical Imaging 2018: Computer-Aided Diagnosis*; Mori, K., Petrick, N., Eds.; SPIE: Bellingham, WA, USA, 2018. [CrossRef]
26. Gulshan, V.; Peng, L.; Coram, M.; Stumpe, M.C.; Wu, D.; Narayanaswamy, A.; Venugopalan, S.; Widner, K.; Madams, T.; Cuadros, J.; et al. Development and Validation of a Deep Learning Algorithm for Detection of Diabetic Retinopathy in Retinal Fundus Photographs. *JAMA J. Am. Med. Assoc.* **2016**, *316*, 2402–2410. [CrossRef]
27. Zhuang, F.; Qi, Z.; Duan, K.; Xi, D.; Zhu, Y.; Zhu, H.; Xiong, H.; He, Q. A Comprehensive Survey on Transfer Learning. *Proc. IEEE* **2021**, *109*, 43–76. [CrossRef]
28. Gabriela Csurka. Domain Adaptation for Visual Applications: A Comprehensive Survey. 2017. Available online: http://arxiv.org/abs/1702.05374 (accessed on 20 September 2023).

sensors

Article

A Binocular Vision-Based Crack Detection and Measurement Method Incorporating Semantic Segmentation

Zhicheng Zhang [1], Zhijing Shen [1], Jintong Liu [1], Jiangpeng Shu [1] and He Zhang [1,2,*]

[1] College of Civil Engineering and Architecture, Zhejiang University, Hangzhou 310058, China; jszzc@zju.edu.cn (Z.Z.); zjshen@zju.edu.cn (Z.S.); lfyer1213@outlook.com (J.L.); jpeshu@zju.edu.cn (J.S.)
[2] Center for Balance Architecture, Zhejiang University, Hangzhou 310058, China
* Correspondence: zjuzhanghe@zju.edu.cn

Abstract: The morphological characteristics of a crack serve as crucial indicators for rating the condition of the concrete bridge components. Previous studies have predominantly employed deep learning techniques for pixel-level crack detection, while occasionally incorporating monocular devices to quantify the crack dimensions. However, the practical implementation of such methods with the assistance of robots or unmanned aerial vehicles (UAVs) is severely hindered due to their restrictions in frontal image acquisition at known distances. To explore a non-contact inspection approach with enhanced flexibility, efficiency and accuracy, a binocular stereo vision-based method incorporating full convolutional network (FCN) is proposed for detecting and measuring cracks. Firstly, our FCN leverages the benefits of the encoder–decoder architecture to enable precise crack segmentation while simultaneously emphasizing edge details at a rate of approximately four pictures per second in a database that is dominated by complex background cracks. The training results demonstrate a precision of 83.85%, a recall of 85.74% and an F1 score of 84.14%. Secondly, the utilization of binocular stereo vision improves the shooting flexibility and streamlines the image acquisition process. Furthermore, the introduction of a central projection scheme achieves reliable three-dimensional (3D) reconstruction of the crack morphology, effectively avoiding mismatches between the two views and providing more comprehensive dimensional depiction for cracks. An experimental test is also conducted on cracked concrete specimens, where the relative measurement error in crack width ranges from −3.9% to 36.0%, indicating the practical feasibility of our proposed method.

Keywords: non-contact measurement; crack width; deep learning; image processing; binocular vision

1. Introduction

Visible cracks in concrete facilitate the unimpeded infiltration of environmental chemicals, such as water, carbon dioxide and chloride ions, thereby promoting corrosion and carbonation [1,2]. When coupled with external loads [3], these durability considerations may exacerbate the occurrence of cracking and result in material discontinuities as well as a localized reduction in structural stiffness [4–7]. To prevent the functional deterioration of the bridge structure and to mitigate potential safety hazards, periodic crack inspections are essential in assessing the condition of each component and developing appropriate maintenance strategies.

Conventional inspection methods typically involve the use of handheld tools, such as a crack gauge, to detect cracks through direct contact. However, once the inspecting area becomes inaccessible (e.g., the bottom of a beam), heavy machinery like a bridge inspection vehicle is required to provide an operational platform. This entire process is characterized by a high demand for labor, extensive time consumption and substantial costs, while the detected results are susceptible to the inspector's subjectivity [8–10]. To improve this circumstance, several studies have implemented non-destructive testing

(NDT) techniques to assist manual inspection. Huston et al. [11], for instance, were able to successfully detect concrete cracks with a width as narrow as 1 mm using a ground penetrating radar (GPR) equipped with a good impedance matching antenna (GIMA). Chen et al. [12] deployed a three-dimensional laser radar, also referred to as 3D LiDAR, to quantify the length of cracking on bridge components, while Valenca et al. [13] incorporated terrestrial laser scanning (TLS) to characterize large-scale structural cracks. In recent years, there has been a growing interest in the utilization of advanced nanomaterials to achieve the self-monitoring of concrete cracks [14,15]. Roopa et al. [16] conducted a study where they incorporated carbon fiber (CF) and multiwalled carbon nanotubes (MWCNT) as nanofillers in the cementitious matrix, aiming to develop self-sensing sensors. These sensors exhibit piezoelectric properties that correspond to the structural response, enabling them to autonomously detect damage. At the microscale, the nanocomposite sensors demonstrate exceptional sensitivity to small cracks, thereby facilitating real-time monitoring of crack formation and propagation. However, it is important to note that this method is relatively susceptible to environmental factors such as temperature and humidity, which can impact its performance. Additionally, while the self-monitoring methods based on nanomaterials can provide estimates of crack width and location, it cannot provide precise information on crack morphology. In general, the exorbitant cost and limited applicability of these abovementioned methods impede their promotion, rendering it arduous to satisfy the demand for crack detection in huge-volume concrete bridges.

Over the past two decades, non-contact, high-precision and low-cost machine vision-based NDT methods have emerged as the potentially viable alternative to manual visual inspection. In this context, camera-mounted unmanned aerial vehicles (UAVs) or robots can function as image sensing-based inspection platforms [17–20]. The automatic crack detection in large volumes of acquired image data thus poses a significant challenge. Previously, researchers have utilized traditional image processing techniques (IPTs) for crack extraction, proposing hybrid approaches that integrate thresholding, morphological operators or filter concepts [21–27], as well as approaches based on mathematical transformations [28–32]. A considerable proportion of crack measurements in these studies were conducted on binary images, which can be broadly categorized into three distinct groups. The first group adopts pixel count as a quantitative metric for representing cracks. Payab et al. [33] expressed the crack area and length values in pixel numbers of crack region and skeleton, respectively, and took the ratio of the two as the average crack width. The second type entails a scale factor to convert the output of the first group into actual physical dimensions. After detecting thermal cracks on fire-affected concrete via wavelet transform, Andrushia et al. [34] adopted the unit pixel size, i.e., pixel resolution, to convert the morphological characteristics from pixel units to physical units. The final category achieves measurement by means of crack reconstruction. Liu et al. [35] employed the structure from motion (SFM) algorithm to conduct 3D reconstruction, enabling not only the acquisition of crack width but also the integration of cracks from multiple perspectives into a unified 3D scene.

The attainment of anticipated outcomes through IPT-based methods suitable for simple cracks (i.e., high contrast and good continuity) is a challenging task due to the presence of diverse noises in actual inspection data, necessitating further enhancement in their robustness [36]. Therefore, modified solutions in combination with machine learning (ML) have been proposed. Specifically, the image features extracted by IPTs pass through the supervised learning-based classifier to determine whether they are indicative of a crack. The study conducted by Prasanna et al. [37] focused on the detection of noise-robust line segment features that accurately fit cracks. They employed support vector machines, Adaboost and random forests as classifiers, utilizing spatially tuned multi-feature appearance vectors. The performance of various feature combinations was evaluated, demonstrating that integrating multiple design features into a single appearance vector yields superior classification results. Peng et al. [38] developed a cascade classifier for determining the positivity and negativity of crack detection windows by extending diverse Haar-like features and employed a monocular vision technique, which belongs to the second category of measurement methods, to calculate the actual crack width. While the incorporation of ML into such methodologies strengthens their adaptability to real-world scenarios, it is inevitable that the results will still be influenced by IPTs.

Deep learning (DL) is an emerging and powerful alternative to the above methods, with the advantage of not depending on expert-dominated heuristic thresholds or hand-designed feature descriptors, thereby greatly enhancing the accuracy and robustness of feature extraction [39]. During recent years, a multitude of researchers have extensively investigated the potential of DL-based models, particularly convolutional neural networks (CNNs), for concrete crack detection. The aforementioned studies demonstrated successful applications of CNNs in image classification [40] and object identification tasks, specifically pertaining to crack detection at both the image level/patch level [41–44] and object level [45–47]. However, neither the grid-like detected results nor the bounding boxes with class labels provide a precise description of the crack topology. In contrast, semantic segmentation categorizes each pixel into a possible class (e.g., crack or background), offering the highest level of detail in features. To detect cracks at the pixel level, Li et al. [48] trained a CNN-based local pattern predictor for coarse analysis on crack pixels. Kim et al. [49] adopted Mask R-CNN for instance segmentation of concrete cracks but not complete semantic segmentation, hence having limited precision. Zhang et al. [50] developed CrackNet-R, an effective semantic segmentation network for detecting cracks in asphalt pavement but also prone to technical isolation in practice.

With the widespread adoption of the encoder–decoder architecture in semantic segmentation, various CNNs have been proposed for pixel-level crack detection based on different variations of this structure, including fully convolutional network (FCN) [51,52], U-Net [53–56], SegNet [57–59], DeepLab series [60,61] and ResNets [62,63]. These architectures consist of two components, namely the encoder module responsible for extracting multi-scale features and the decoder module dedicated to restoring the feature information. On the one hand, the decoders upscale the final output of the encoder network to match the original input size, thereby facilitating the orientation of crack pixels. On the other hand, the encoders supply the local information during the decoding process to minimize loss of details from the input. Although the mentioned classical neural networks demonstrate proficiency in executing fundamental segmentation operations, they remain confronted with difficulties in achieving precise object edge segmentation and addressing class imbalance. Consequently, researchers have started integrating various cutting-edge methods to optimize the performance of segmentation models. In light of the requirement for both semantic understanding and fine-grained detail in segmentation tasks, a suite of attention-based methodologies [64,65] have been developed. These methods are designed to assimilate multi-scale and global contextual information, thereby enhancing the accuracy of defect identification. Chen et al. [66] have demonstrated impressive recognition accuracy in identifying different types of cracks by incorporating the Convolutional Block Attention Module (CBAM) into MobileNetV3 as the backbone network. Du et al. [67] have proposed an Attention Feature Pyramid Network that enhances the precise segmentation of road cracks within the YOLOv4 model. Similarly, Yang et al. [68] introduced a multi-scale, tri-attention network, termed MST-NET. Other advanced computational modules, such as separable convolution [69] and deformable convolution [70], have been introduced to further enhance model performance. Recognizing that the training of semantic segmentation models heavily relies on accurately annotated data, numerous researchers have also begun exploring approaches to enhance the generalization and adaptability of segmentation methods from the perspective of dataset optimization and learning strategies. For instance, Que et al. [71] have proposed a crack dataset expansion method based on generative adversarial networks (GANs), resulting in higher recall rates and F1 scores for the same model. Nguyen et al. [72] have introduced the Focal Tversky loss function to tackle class imbalance issues in crack segmentation, shedding light on the role of loss functions during model training. Furthermore, Weng et al. [73] have devised an unsupervised adaptive framework for crack detection, effectively mitigating domain shift problems among various civil infrastructure crack images.

On this basis, the first category of crack measurements was completed by Yang et al. [51], Ji et al. [60] and Kang et al. [74]. Regrettably, these results are inadequately cited for crack evaluation purposes. To make sense of the measure values, Li et al. [36] and Chen et al. [65] employed a monocular vision technique to accurately quantify the crack indicators such as area, max width and length. However, these methods rely on calibrated pixel resolution and the similar triangle relationship for unit conversion, which necessitates frontal photography of the target crack at known distances with a monocular device. As a result, restricted shooting postures increase the difficulty of remotely manipulating inspection platforms, leading to complications in image acquisition and unstable measurements.

The third category of binocular stereo vision-based measurement emerges as a promising solution to tackle the aforementioned challenges. In contrast to monocular vision, which calculates physical dimensions mapped on pixels, binocular stereo vision reconstructs the 3D coordinates of a crack in a datum coordinate system based on internal imaging geometries and the external relative posture of two cameras, as well as matching relations between two captured images. This enables a more comprehensive and reliable quantification of morphological characteristics. Furthermore, binocular vision is not constrained by a fixed photogrammetric geometry and offers greater flexibility in capturing cracks within its depth of field. Previously, Guan et al. [56] designed a vehicle-mounted binocular photography system to generate 3D pavement models and precisely estimated the volume of pavement potholes by integrating pixel-level predictions of a U-Net but failed to further quantify the segmented cracks. Yuan et al. [75] and Kim et al. [76] upgraded the automation of non-contact inspection through a robot and a UAV equipped with binocular devices, respectively, despite their crack predictions not being derived from semantic segmentation networks. Recently, Chen et al. [77] optimized DeeplabV3+ to deliver a detailed crack morphology for measurement based on binocular stereo vision, resulting in satisfactory outcomes.

In this paper, a novel non-contact crack detection and measurement method in combination with an encoder–decoder FCN and binocular stereo vision is proposed for efficient and accurate evaluation of concrete cracks in bridge structures. The proposed method not only enhances the flexibility of crack data acquisition but also enables rapid and precise extraction of crack morphology, which facilitates 3D reconstruction in the form of spatial discrete points, thereby obtaining a more comprehensive set of dimensional information regarding cracks. The limitations on shooting attitude imposed by the monocular measurement method are thus effectively addressed, along with the issues related to accuracy and robustness in traditional crack detection methods. Moreover, in contrast to conventional binocular vision-based 3D reconstruction methods that rely heavily on feature matching prior to point cloud computation, the proposed method employs projective reconstruction, which significantly alleviates computational expenses and eliminates potential mismatches between the two views.

2. Methodology

2.1. Overview

The proposed method consists of three parts, as depicted in Figure 1, which illustrates the overall workflow schematically. (I) Crack data acquisition: a tailored binocular system is constructed for capturing visible cracks from multiple angles at flexible distances, rendering it ideal for UAV-aided crack inspection. The captured image pairs subsequently serve as primary data to detect cracks. (II) Crack pixel-level detection: to achieve precise segmentation of cracks in the main images from primary data, a semantic segmentation network (i.e., the encoder–decoder FCN) is constructed with a VGG19-based encoder network and a decoder network featuring the deconvolution layer as its core. The resulting binary image is further exploited to extract pixels that characterize the morphology of the crack. (III) Crack quantitative assessment: at this stage, a binocular vision-based projection reconstruction model is employed for spatial localization of the cracked concrete surface and subsequent 3D crack reconstruction by projecting pixels extracted in the previous stage onto it. Finally, the morphological characteristics of cracks are quantitatively calculated based on the discrete reconstructed points. A detailed description of each part is presented below.

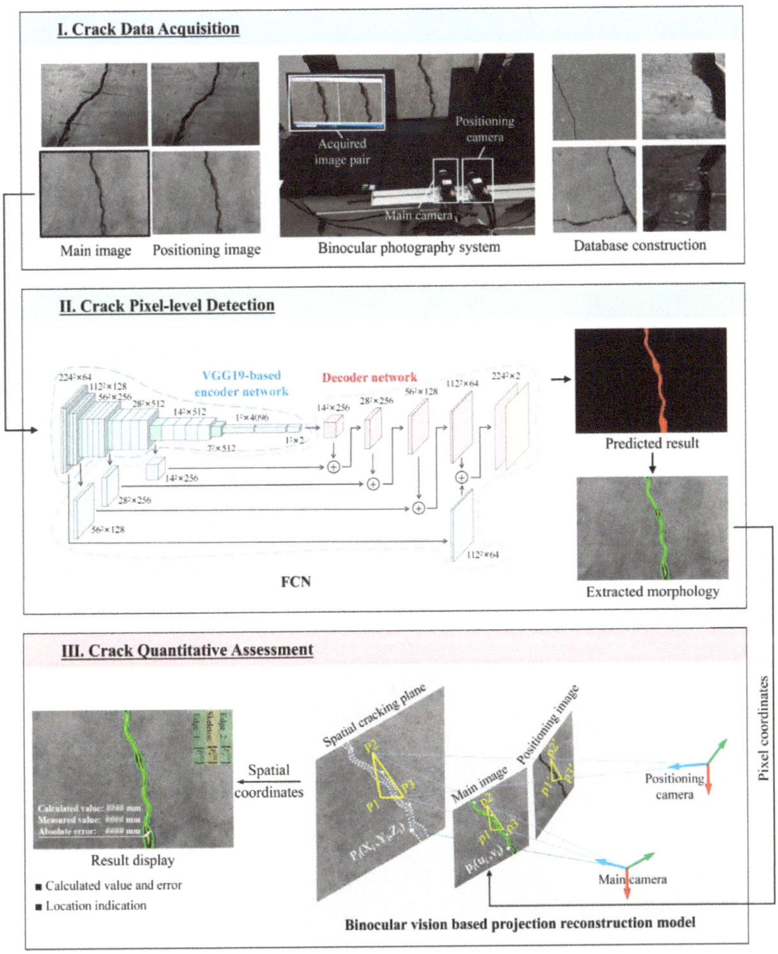

Figure 1. The overall workflow of the method. (The # represents the specific numerical results for different cracks.).

2.2. Crack Data Acquisition

To facilitate the UAV assistance, a pair of identical industrial charge-coupled device (CCD) cameras from Microvision, a supplier specialized in visual products, are rigidly assembled for a lightweight and compact binocular photography system. The specifications for each component are comprehensively presented in Table 1, where the outgoing focal length f is 16 mm, with a pixel size $\Delta u \cdot \Delta v$ of 3.75×3.75 μm^2. According to the pinhole model depicted in Figure 2a, the resolution of a single camera at an operating distance D of 200 ± 50 mm is approximately 0.047 ± 0.012 mm/pixel, which is adequate for capturing crack details. Moreover, to take into account the public field of view (Figure 2b), the relative pose of two cameras is adjusted with a narrow baseline (denoted as B and set to 5 cm) and the intersecting optical axes (realized by a left deviation of the right camera at angle θ of roughly 20°), as shown in Figure 2c. For the subsequent description, the left camera is designated as the main camera along the shooting direction, while the right camera is designated as the positioning camera. These two cameras capture images of target cracks synchronously to form stereo image pairs, which are then transmitted in real time to the inspector's laptop.

Table 1. Detailed specifications of the binocular system.

Component	Model	Specification
CCD grayscale camera@2	MV-EM120M	Sensor resolution: 1280 × 960 pixels Pixel size: 3.75 × 3.75 (μm) Size: 29 × 35 × 48.9 (mm) Weight: 50 g
Industrial fixed-focus lens@2	BT-118C1620MP5	Focal length: 16 mm Size: φ27.2 × 26.4 (mm) Weight: 75 g

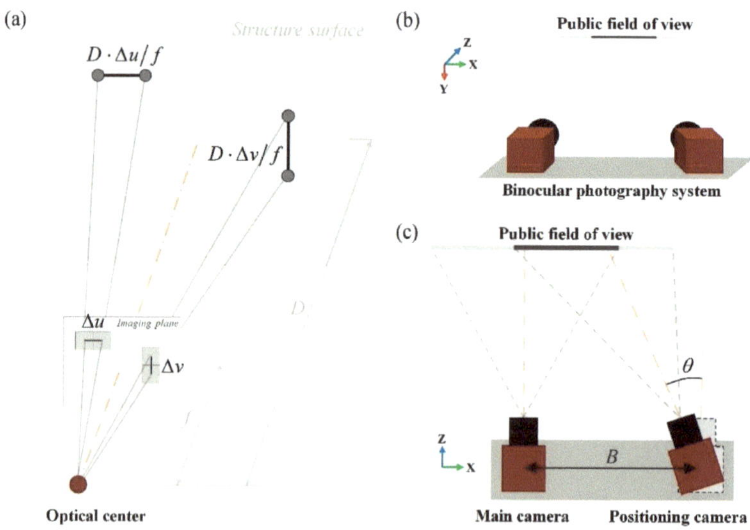

Figure 2. Considerations of the binocular system: (**a**) a pinhole model for resolution and distance trade-off; (**b**) public field of view of two specifically mounted cameras; and (**c**) overhead perspective of (**b**).

2.3. Crack Pixel-Level Detection

The accurate and efficient characterization of crack morphology is a prerequisite for real-time image measurement of concrete cracks. To accomplish this, a specialized encoder–decoder FCN is developed for detecting cracks at the pixel level. Subsequently, an integrated computer vision (CV) program is written to enable rapid extraction of the edges and skeletons that characterize the crack morphology from the FCN predictions.

2.3.1. FCN for Crack Segmentation

The state-of-the-art CNNs, such as VGG16 [52], ResNet [62] and DenseNet [36], which serve as the encoder of FCNs for robust feature extraction in crack images, directly inspire the construction of the FCN framework in this study. Among these classical CNN models, the VGG series, including VGG16 and depth-increased VGG19, are extensively applied for large-scale image detection tasks due to their good transferability. Considering that employing transfer learning [78,79] based on pre-trained parameters of VGG can not only significantly reduce the overall training time of the FCN model but also effectively enhance its performance in scenarios with limited training data, the VGG19-based encoder network is adopted to extract essential features for semantic segmentation. As shown in Figure 3a, the encoder network is topologically identical to the first 16 layers of VGG19, consisting of five convolutional blocks (also referred to as encoders in this paper) that include all convolutional layers, nonlinear activation layers utilizing the ReLU function and pooling layers. Since the encoder module does not involve neuron classification, the final softmax layer of VGG19 is excluded, while the fully connected layers are replaced by convolutional layers with two dropout layers added in between to prevent overfitting.

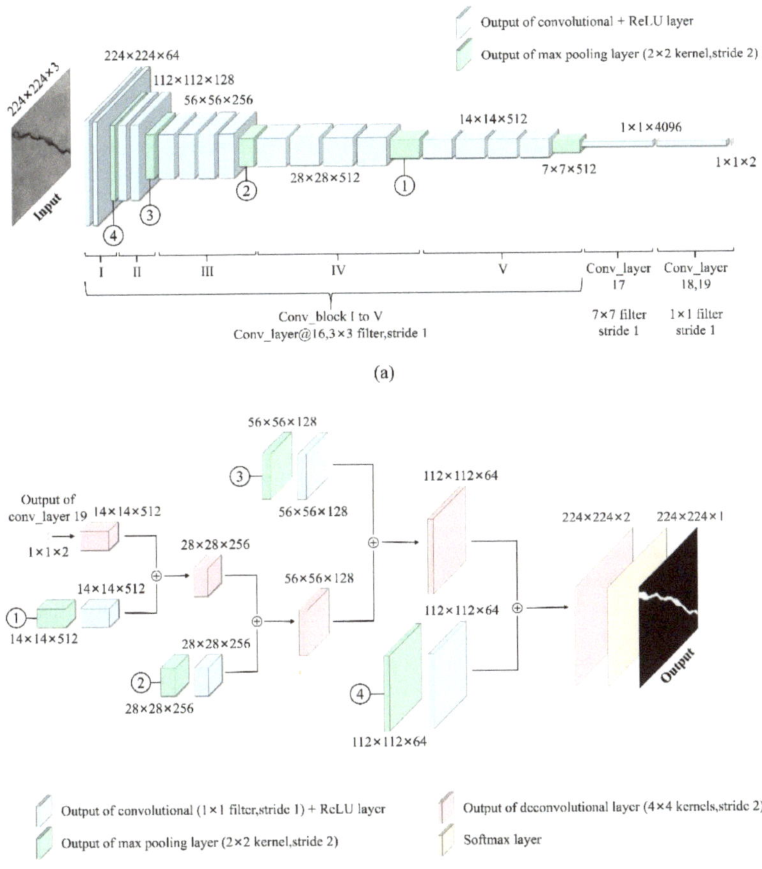

Figure 3. (**a**) Encoder network and (**b**) decoder network of FCN.

Inheriting the strengths of VGG19, each encoder conducts convolution operations through the stacking of 3×3 filters (i.e., convolution kernels) with a fixed stride length of 1 pixel, which ensures the equivalent receptive field as larger-size filters, while extracting higher-level features with fewer parameters of the convolution kernel. Moreover, ReLU activation is applied following each convolution to introduce nonlinearity, thereby enhancing the nonlinear fitting capability of the encoder network. To eliminate redundant information and to accelerate computational speed, the max pooling operation is subsequently performed on a 2×2 pixel window with a stride of 2, which results in downsampling of the output by a factor of 2. It is noteworthy that the outputs of the first four max pooling layers, numbered ④, ③, ② and ①, will also be recycled by the decoder network. Due to the three newly substituted convolution layers, namely Conv_layer 17,18 and 19, the final output is transformed from the initial class probabilities into a low-resolution feature map that characterizes the crack, which is subsequently fed into the decoder module.

The decoder network employs deconvolutional upsampling to generate a dense output and rescales the data to the original input size. To minimize the loss of details during the decoding process, the skip connection structure proposed by Bang et al. [62] is adopted to facilitate the flow of feature maps from the upstream encoders to their corresponding downstream counterparts, which enables effective integration of multi-scale and multi-level local information. Specifically, each decoder selectively fuses the local feature map with the upstream feature map at the expense of increased memory consumption.

Referring to the decoder network depicted in Figure 3b, the max pooling outputs labeled as ①, ②, ③ and ④ are initially individually convolved with a 1 × 1 kernel for densification purposes. The subsequent outputs are considered to hold local information originating from the upstream network (i.e., the encoder network) and are then arithmetically added (represented by "⊕" in Figure 3b) to the upsampling results of identical resolution obtained through deconvolution with a 4 × 4 kernel with a two-pixel stride. The entire decoder network integrates the outputs from the final layer and the first four max pooling layers of the encoder network, wherein each fused feature map undergoes a doubling in resolution through upsampling with a stride of 2. After five upsamplings, the output of conv_layer 19 is expanded to match the dimensions of the original input and then proceeds through the softmax layer, where the softmax function value determines the probability of a single pixel belonging to either the "crack" or "background" categories. Ultimately, a binary image is exported as the final prediction, with "crack" pixels assigned a value of 1, while the "background" pixels assigned a value of 0.

2.3.2. Extraction of Crack Edges and Skeletons

The CV procedure for crack edge and skeleton extraction consists of three stages: region pre-processing, edge extraction and skeleton optimization (Figure 4a). Firstly, the FCN prediction shown in Figure 4b is subjected to global segmentation using a fixed threshold of 180 as an empirical value. This procedure successfully eliminates isolated data points outside the cracks. In addition, a morphological optimization technique is employed, which entails the sequential application of dilation and erosion. After this step, marginal burrs and internal holes caused by misjudgment of the proposed FCN can be effectively eliminated. Figure 4c presents the optimized crack region. Secondly, the contour extraction technique in OpenCV is subsequently applied to acquire the single-pixel-wide crack edges. Given that the image boundary truncates the crack and forms a closed contour along with its edges, it becomes imperative to exclude the boundary pixels within this contour. The specific solution is to identify the difference set between the crack region and the pixel border of the image. Next, the connected component is calculated, and the remaining regional contours are divided into the two crack edges (Figure 4d).

Finally, the skeleton of the crack region is extracted and optimized using the fast parallel thinning algorithm proposed by Zhang et al. [80]. During this process, the superfluous branches of the original crack skeleton are pruned through deburring treatment. This involves identifying branch nodes and calculating the number of path pixels, which removes branches that fall below a preset threshold and thus retains only the longest path, i.e., the backbone portion of the skeleton. To further mitigate the issue of tail ends of the crack skeleton converging towards the cusp in the crack region, resulting in incongruity with the actual crack topology, as indicated by the red end in Figure 4e, an end trimming treatment is implemented, in which any skeleton part that falls within 20 pixels (based on experience) from the image boundary will be cropped. The final outputs, as presented in Figure 4f, are stored as pixel coordinates.

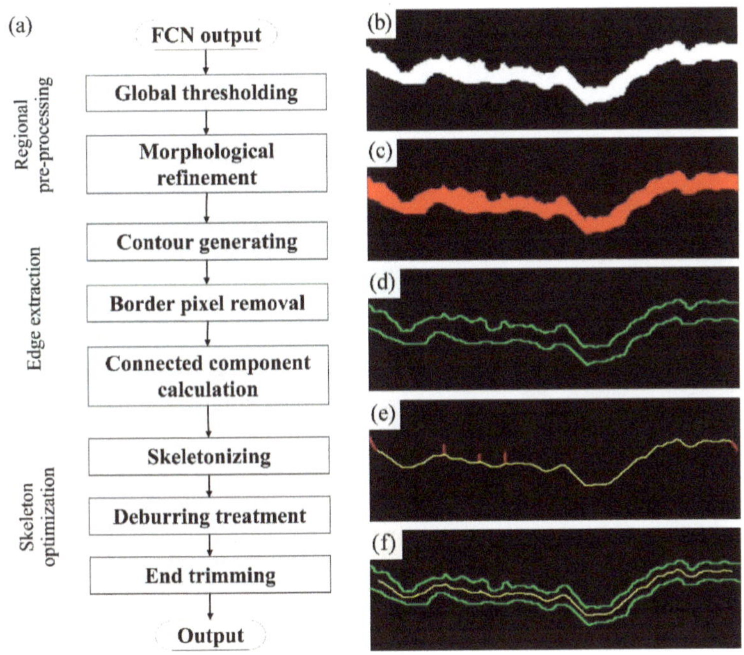

Figure 4. Procedures for crack edge and skeleton extraction: (**a**) flow chart; (**b**) FCN prediction; (**c**) refined crack region; (**d**) crack edges; (**e**) original crack skeleton (The red lines represent the pruned excess crack branches and the yellow lines represent the crack skeletons.); and (**f**) outputs of crack morphology.

2.4. Crack Quantitative Assessment

The proposed projection reconstruction model consists of a binocular vision model for locating the spatial crack plane and a central projection model for reconstructing the crack morphology. Based on the discrete reconstruction points, the dimensions of concrete cracking in bridge structures can be quantitatively assessed.

2.4.1. Binocular Vision for Crack Location

Our crack location approach is illustrated in Figure 5. First, the points of interest in a stereo image pair (Figure 5a) are extracted and matched using the correspondence search techniques, as indicated by the red dots in Figure 5b. Each pair of matching points is considered the projection of a specific point on the cracked structure onto both imaging planes, which is connected by a green line in Figure 5c. The next step involves randomly selecting three non-colinear feature points (p_1, p_2 and p_3) in one image, along with their corresponding matching points (p_1', p_2' and p_3', respectively) in another image, to form a three-point pair for the purpose of planar location. Herein, to avoid the selected points falling into the crack region, the contour is dilated by five pixels as the boundary for pre-filtering the internal feature points. Consequently, only feature points located on the background of the image remain. Finally, the binocular vision model depicted in Figure 5d is utilized to calculate the non-collinear spatial location points (P_1, P_2 and P_3) corresponding to the aforementioned three-point pair for achieving the precise localization of the flat concrete surface.

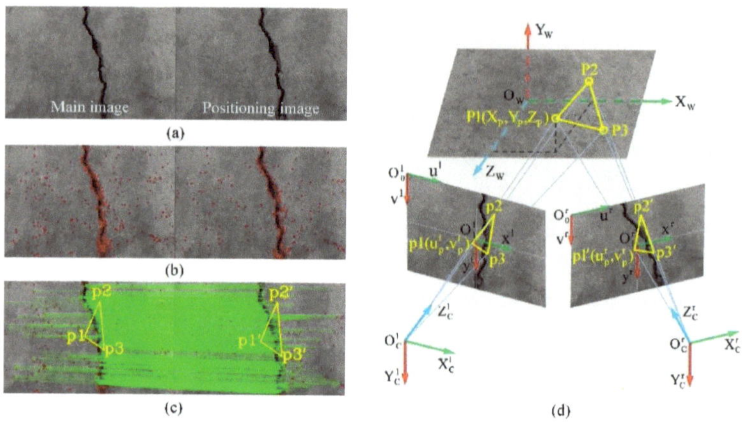

Figure 5. Crack plane location: (**a**) stereo image pair; (**b**) feature point extraction; (**c**) feature point matching with randomly selected three-point pair; and (**d**) binocular vision model to calculate the spatial location points.

Previously, the scale-invariant feature transform (SIFT) algorithm proposed by Lowe [81] was successfully applied to extract features from crack images [56,82], showcasing its robustness to rotation and translation, as well as its capability to handle variations in lighting conditions and viewpoints. Our approach employs the SIFT algorithm for scale space filtering of stereo image pairs, facilitating the detection of feature points across multiple scales. For the kth stereo image pair $\boldsymbol{I}^{(k)} = \left\{\boldsymbol{I}_1^{(k)}, \boldsymbol{I}_2^{(k)}\right\}$, with $\boldsymbol{I}_1^{(k)}$ and $\boldsymbol{I}_2^{(k)}$ representing the k main image and the positioning image, respectively, the extracted feature point sets are denoted as $\boldsymbol{F}_1^{(k)} = \left\{(p_{1,i}^{(k)}, f_{1,i}^{(k)})|i = 1 \ldots P\right\}$ and $\boldsymbol{F}_2^{(k)} = \left\{(p_{2,j}^{(k)}, f_{2,j}^{(k)})|j = 1 \ldots Q\right\}$, where $f_{1,i}^{(k)}$ and $f_{2,j}^{(k)}$ are the local feature descriptors corresponding to feature point positions $\mathrm{p}_{1,i}^{(k)}$ and $\mathrm{p}_{2,j}^{(k)}$, respectively. On this basis, the first two nearest neighbors of $(\mathrm{p}_1^{(k)}, f_1^{(k)}) \in \boldsymbol{F}_1^{(k)}$ are searched with Euclidean distance in the query set $\boldsymbol{F}_2^{(k)}$ by applying the nearest neighbor algorithm. The optimal matches are then obtained through a threshold of 0.5 to the ratio between the Euclidean distances of the nearest and second-nearest neighbors. The matching result is a set of feature point pairs, i.e., $\left\{(p_1^{(k)}, p_2^{(k)})\middle|p_1^{(k)} \in \boldsymbol{I}_1^{(k)}, p_2^{(k)} \in \boldsymbol{I}_2^{(k)}\right\}$, from which three pairs of location points are randomly selected.

The binocular photography system is simplified into a binocular vision model, as illustrated in Figure 5d. Here, $O_C^l - X_C^l Y_C^l Z_C^l$ represents the main camera coordinate system (m-CCS), $O_1^l - x^l y^l$ and $O_0^l - u^l v^l$ denote the physical and pixel coordinate systems on the main image, respectively; the positioning camera coordinate system (p-CCS), i.e., $O_C^r - X_C^r Y_C^r Z_C^r$, is situated on the right side with the two corresponding image coordinate systems $O_1^l - x^l y^l$ and $O_0^l - u^l v^l$; and $p1(u_p^l, v_p^l)$ and $p1'(u_p^r, v_p^r)$ represent the projected pixels of a specific point $P1(X_P, Y_P, Z_P)$ on the crack plane in world coordinate system $O_W - X_W Y_W Z_W$ (WCS), as captured by the two imaging planes, respectively.

Taking point $P1$ as an example for calculation, assuming WCS coincides with m-CCS, the projection relationship between $P1(X_P, Y_P, Z_P)$ and $p1(u_p^l, v_p^l)$ is given by the following:

$$Z_P \begin{pmatrix} u_p^l \\ v_p^l \\ 1 \end{pmatrix} = \boldsymbol{A}_1 \begin{bmatrix} \boldsymbol{I}_3 & \boldsymbol{O}_{3\times1} \end{bmatrix} \begin{pmatrix} X_P \\ Y_P \\ Z_P \\ 1 \end{pmatrix} = \begin{pmatrix} f^l/k^l & \gamma_1 & u_0^l \\ 0 & f^l/l^l & v_0^l \\ 0 & 0 & 1 \end{pmatrix} \begin{pmatrix} X_P \\ Y_P \\ Z_P \end{pmatrix} \tag{1}$$

where $\boldsymbol{A}_1$ is the intrinsic matrix of the main camera, with f^l the focal length, (u_0^l, v_0^l) the pixel coordinates of the principal point O_1^l, as well as k^l and l^l the physical length of the pixel unit along the u^l-axis and v^l-axis directions, respectively; γ_1 is the parameter characterizing the skew of the two image axes, which is typically zero; $\boldsymbol{I}_3$ denotes the 3×3 unit matrix, while $\boldsymbol{O}_{3\times1}$ represents the 3×1 zero vector.

The projection formula from $P1(X_P, Y_P, Z_P)$ to $p1'(u_p^r, v_p^r)$ is simultaneously established by utilizing the relative pose of the two cameras, as demonstrated below:

$$Z_{P1}\begin{pmatrix} u_{P1}^r \\ v_{P1}^r \\ 1 \end{pmatrix} = \boldsymbol{A}_2\begin{bmatrix} \boldsymbol{R} & \boldsymbol{t} \end{bmatrix}\begin{pmatrix} X_{P1} \\ Y_{P1} \\ Z_{P1} \\ 1 \end{pmatrix} = \bar{\boldsymbol{A}}_2\begin{pmatrix} f^r R_{11} & f^r R_{12} & f^r R_{13} & f^r t_x \\ f^r R_{21} & f^r R_{22} & f^r R_{23} & f^r t_y \\ R_{31} & R_{32} & R_{33} & t_z \end{pmatrix}\begin{pmatrix} X_{P1} \\ Y_{P1} \\ Z_{P1} \\ 1 \end{pmatrix} \tag{2}$$

where $\boldsymbol{A}_2$ represents the positioning camera intrinsic matrix, which is structurally and parametrically equivalent to $\boldsymbol{A}_1$; $\bar{\boldsymbol{A}}_2 = \boldsymbol{A}_2 \times \boldsymbol{diag}(1/f^r, 1/f^r, 1)$, with $\boldsymbol{diag}$ symbolizing the diagonal matrix; and $\boldsymbol{R} = [R_{ij}]_{3\times3}$ and $\boldsymbol{t} = [t_x, t_y, t_z]^T$ are the rotation matrix and translation vector, respectively, of the main camera relative to the positioning camera in the binocular system, serving as its external parameters.

From Equations (1) and (2), the spatial coordinates of the point $P1$ can be obtained:

$$X_P = Z_P\frac{x_p^l}{f^l} \tag{3}$$

$$Y_P = Z_P\frac{y_p^l}{f^l} \tag{4}$$

$$\begin{aligned} Z_P &= \frac{f^l(f^r t_x - x_p^r t_z)}{x_p^r(x_p^l R_{31} + y_p^l R_{32} + f^l R_{33}) - f^r(x_p^l R_{11} + y_p^l R_{12} + f^l R_{13})} \\ &= \frac{f^l(f^r t_y - y_p^r t_z)}{y_p^r(x_p^l R_{31} + y_p^l R_{32} + f^l R_{33}) - f^r(x_p^l R_{21} + y_p^l R_{22} + f^l R_{23})} \end{aligned} \tag{5}$$

where (x_p^l, y_p^l) and (x_p^r, y_p^r) are the physical coordinates of the projected pixels $p1(u_p^l, v_p^l)$ and $p1'(u_p^r, v_p^r)$, respectively, which can be expressed as follows:

$$\begin{bmatrix} x_p^l \\ y_p^l \\ x_p^r \\ y_p^r \\ 1 \end{bmatrix} = \begin{bmatrix} k^l & 0 & 0 & 0 & -k^l u_0^l \\ & l^l & 0 & 0 & -l^l v_0^l \\ & & k^r & 0 & -k^r u_0^r \\ & & & l^r & -l^r v_0^r \\ & & & & 1 \end{bmatrix}\begin{bmatrix} u_p^l \\ v_p^l \\ u_p^r \\ v_p^r \\ 1 \end{bmatrix} \tag{6}$$

According to Equations (5) and (6), the mapping relationship between a pair of homologous pixels to its spatial source point is established. With the internal and external parameters obtained from calibration, the location of the cracking plane can be determined in m-CCS.

2.4.2. Central Projection for Crack Reconstruction

The binocular vision model enables spatial point reconstruction, contingent upon feature matching to establish the correspondence between the two views. To alleviate computational expenses and reconstruction errors resulting from mismatches, a projection reconstruction scheme is proposed.

The central projection model is constructed by taking the origin of the main camera model, namely the optical center O_C^l, as the projection center; the determined spatial cracking plane as the easel plane; and the pixels of crack edges and skeleton extracted from the main image as the points to be projected, as shown in Figure 6a. The model achieves 3D reconstruction by mapping pixels from the main imaging plane onto the cracked concrete surface. Prior to this, the reference systems, or the main camera coordinates of target pixels need to be standardized. According to the properties of pinhole camera model, the location of the main imaging plane depicted in Figure 6b under the main camera coordinate system is as follows:

$$Z_C^l = f^l(-\frac{W}{2} - \Delta u \leq X_C^l \leq \frac{W}{2} - \Delta u, -\frac{H}{2} - \Delta v \leq Y_C^l \leq \frac{H}{2} - \Delta v) \tag{7}$$

where W and H represent the width and height of the main image, respectively, and $(\triangle u, \triangle v)$ denotes the deviation of the calibrated principal point $O_1^l(u_0, v_0)$ from the image center. Therefore, the Z_C^l-coordinates of all pixels to be projected are numerically equal to the focal length f^l. Since $O_1^l - x^l y^l$ can be regarded as the projection of the X_C^l- and Y_C^l-axes on the main imaging plane, the corresponding camera coordinates of $p_i(u_i, v_i)$ also represent the physical coordinates of (x_i, y_i), which can be interconverted by the scale factors k^l and l^l in the directions of the u^l- and v^l-axes, respectively, as well as the origin $O_1^l(u_0, v_0)$, as indicated by Equation (6). The transformation of the target pixel onto the main camera coordinate system is thus given by the following:

$$f : (u_i, v_i) \to (x_i, y_i, z_i) = ((u_i - u_0)k^l, (v_i - v_0)l^l, f^l) \tag{8}$$

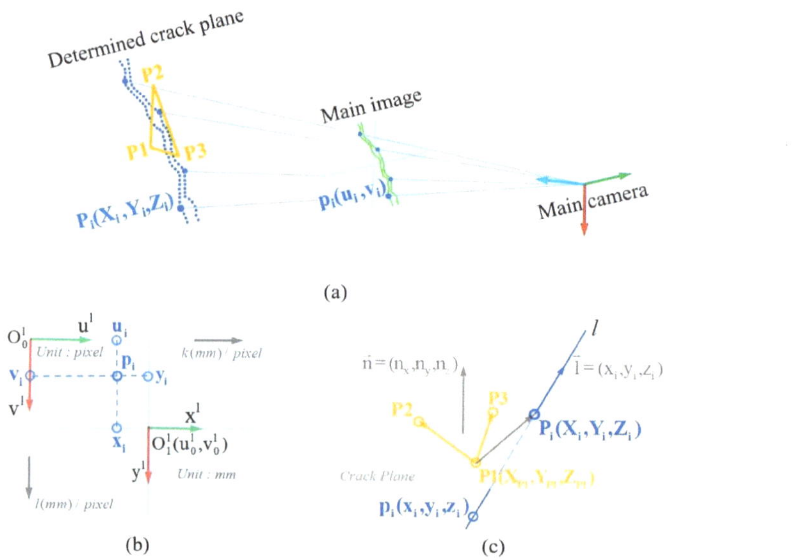

Figure 6. Central projection for crack reconstruction: (**a**) central projection model; (**b**) coordinate transformation on the main image; and (**c**) projection point calculation.

After establishing a unified reference system with Equation (8), the projection points on the easel plane are calculated. As shown in Figure 6c, $\vec{n} = (n_x, n_y, n_z)$ is the normal vector of the spatial cracking plane, determined by vectors $\overrightarrow{P1, P2}$ and $\overrightarrow{P1, P3}$; the crack pixel $p_i(x_i, y_i, z_i)$ serves as a particular point on the projection line l_i, while $\vec{l}_i = (x_i, y_i, z_i)$ is the direction vector of l_i, pointing from the projection center O_C^l to p_i; and $P_i(X_i, Y_i, Z_i)$ is the desired projection point. The equation for the intersection point is as follows:

$$\begin{cases} \overrightarrow{P1, P_i} \perp \vec{n} \\ \overrightarrow{p_i, P_i} \,//\, \vec{l} \end{cases} \Rightarrow \begin{cases} (X_i - X_{P1}, Y_i - Y_{P1}, Z_i - Z_{P1}) \cdot (n_x, n_y, n_z) = 0 \\ \dfrac{X_i - x_i}{x_i} = \dfrac{Y_i - y_i}{y_i} = \dfrac{Z_i - z_i}{z_i} = \lambda \end{cases} \tag{9}$$

where λ is the scale factor. Let $F = x_i n_x + y_i n_y + z_i n_z$, $F \neq 0$; the coordinates of the projection points obtained from the above equation are as follows:

$$X_i = (x_i(X_{P1}n_x + n_y(Y_{P1} - y_i) + n_z(Z_{P1} - z_i)) + x_i(y_i n_y + z_i n_z))/F \quad (10)$$

$$Y_i = (y_i(n_x(X_{P1} - x_i) + Y_{P1}n_y + n_z(Z_{P1} - z_i)) + y_i(x_i n_x + z_i n_z))/F \quad (11)$$

$$Z_i = (z_i(n_x(X_{P1} - x_i) + n_y(Y_{P1} - y_i) + Z_{P1}n_z) + z_i(x_i n_x + y_i n_y))/F \quad (12)$$

The 3D reconstruction of crack edges and skeletons is accomplished through the utilization of Equations (10)–(12). The morphological length of the crack is determined by calculating the cumulative Euclidean distance between adjacent skeleton points, while the width at each skeleton point is obtained by computing the Euclidean distance between the pair of two edge points closest to that point. Each skeleton point corresponds to a specific crack width, from which the maximum crack width is obtained.

3. Training FCN

3.1. Crack Segmentation Database

To train the FCN models, 50 photos of cracked concrete taken using a smartphone with a resolution of 4032 × 3024 × 3 and saved in JPG format are manually labeled at the pixel level using the MATLAB^R tool Image Labeler. Figure 7 depicts this labeling process, in which logical variables 0 and 1 are, respectively, assigned to background and crack pixels through pixel labels, with annotations saved in PNG-8 format. Subsequently, 110 images are cropped from these photos, each featuring either a crack or an intact background with 448 × 448 pixel resolution. These images, along with 334 web images of the same resolution, undergo data augmentation techniques including horizontal and vertical flips, resulting in a total of 1332 images. According to the fivefold cross-validation principle, the generated images are randomly divided into training, validation and test sets with 998, 110 and 224 images, respectively, in each set. Notably, a network trained on small-sized images can scan any image larger than that designed size [36]. Therefore, the randomly selected images and their annotations are resized to 224 × 224 pixels prior to being fed into the models.

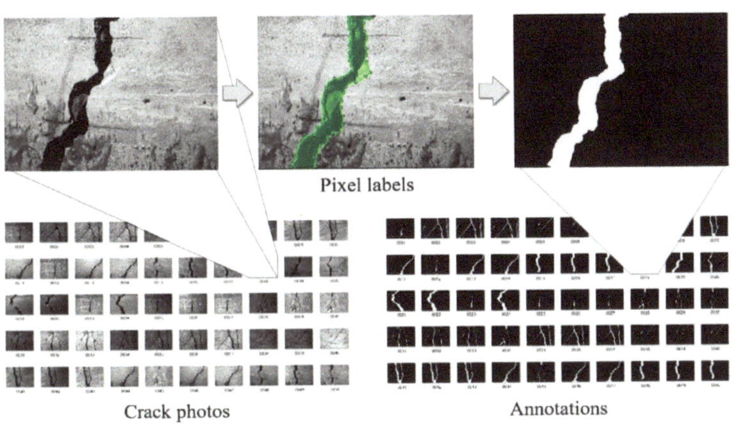

Figure 7. Pixel-level labeling process.

3.2. Implementation Parameters

The learning rate plays a pivotal role in balancing convergence speed and stability in training a CNN. In order to choose an appropriate initial value for this key hyperparameter, three sets of models are meticulously trained, each with distinct initial learning rates:

0.001, 0.0001 and 0.00001, respectively. Throughout these training sessions, exponential stepwise decay, a common technique for annealing learning rates, is employed post epochs to reduce oscillations in the loss function around the global optimum. The decay function is as follows:

$$\eta_t = \eta_0 \times r_d^{\lfloor \frac{t}{t_{\max}} \rfloor} \tag{13}$$

where the initial learning rate is denoted by η_0, r_d is the decay rate with t as the current count of iterations and $t_{\max}$ as the preset iterations for decay. $\lfloor \cdot \rfloor$ represents the floor operation, returning the largest integer not greater than the input value.

To assess the discrepancy between the prediction and the ground truth, cross entropy is utilized as the loss function on pixels. With exponential decay rates set to $\beta_1 = 0.9$ and $\beta_2 = 0.999$, the Adam optimizer is then run for training loss optimization by iteratively updating the model parameters. The FCN models are trained with 20 epochs, and the batch size is set to 2 (taking into account the limitations of GPU memory). In addition, a dropout rate of 0.5 is implemented to activate only half of the hidden nodes or feature detectors during each iteration, thereby weakening their interactions and effectively preventing overfitting [83,84].

3.3. Model Initialization and Evaluation Metrics

To expedite and optimize the learning efficiency, a model-based transfer learning strategy [85] is adopted instead of training from scratch. Following this strategy, the weights and biases of the encoder network are initialized by pre-trained VGG19. Moreover, the weights of all the deconvolutional layers in the decoding module are initialized by the truncated normal distribution with a mean of 0 and standard deviation of 0.01, and their biases are initialized as constant zero vectors.

It is widely acknowledged that pixel-level crack detection is essential to classify pixels of the input image as either a crack (positive) or the background (negative). Therefore, four cases may occur, which are outlined below:

- True Positive (crack pixels classified as crack pixels);
- False Negative (crack pixels classified as background pixels);
- False Positive (background pixels classified as crack pixels);
- True Negative (background pixels classified as background pixels).

To comprehensively evaluate the crack segmentation, three key statistical metrics are introduced: precision, recall and F1 score. These metrics are defined as follows:

$$\text{Precision} = \frac{\text{TP}}{\text{TP} + \text{FP}} \tag{14}$$

$$\text{Recall} = \frac{\text{TP}}{\text{TP} + \text{FN}} \tag{15}$$

$$\text{F1} - \text{score} = \frac{2 \times \text{Precision} \times \text{Recall}}{\text{Precision} + \text{Recall}} \tag{16}$$

where TP, FP and FN denote the number of pixels with True Positives, False Positives and False Negatives in the predicted outcomes, respectively.

3.4. Training Results and Discussion

The proposed encoder–decoder FCN is implemented on Windows 10 using Python 3.5 for programming and TensorFlow 1.4 and NumPy 1.16 for building the virtual environment. All numerical experiments are performed on a desktop computer (GPU: NVIDIA GeForce GTX 1060 GPU Ti, RAM: 8 GB, CPU: Intel® Core™ i5-8400 CPU@2.8 GHz). With the aforementioned training method and experimental configuration, the recorded training time for each model is approximately 9 h on average after 9980 iterations, and it takes about 250 ms for a trained model to process a 448 × 448-pixel image.

The training and validation losses at each learning rate are illustrated in Figure 8a. It can be intuitively seen that the loss value corresponding to Figure 8(a-2) exhibits the fastest convergence and ultimately stabilizes within 0.014, resulting in best training effect. The loss curves associated with the other two learning rates, i.e., 1×10^{-3} and 1×10^{-5}, also demonstrate satisfactory convergence results, remaining stable at around 0.021 and 0.018, respectively, which are sufficient for attaining global optimization.

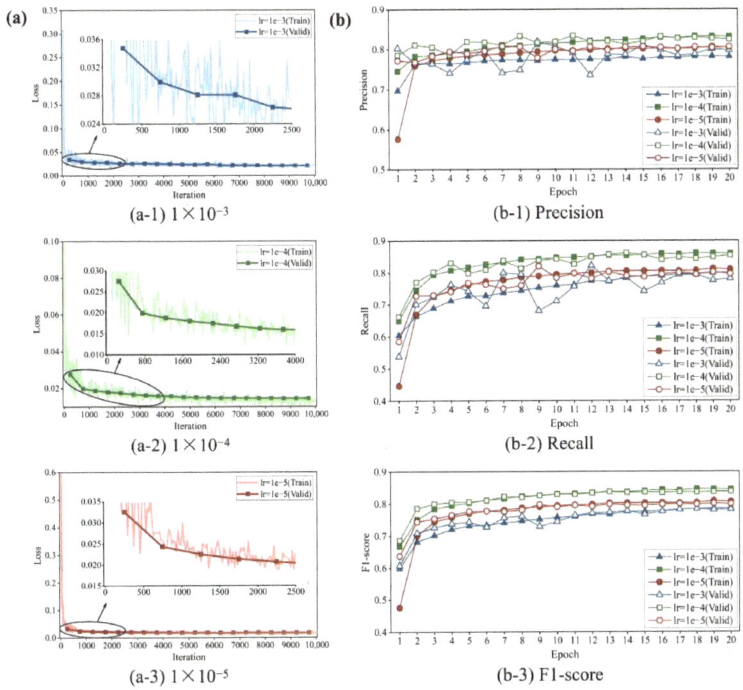

Figure 8. (**a**) Training and validation losses over iterations: (**a-1**) 1×10^{-3}, (**a-2**) 1×10^{-4} and (**a-3**) 1×10^{-5}. (**b**) Three evaluation metrics under epochs: (**b-1**) precision, (**b-2**) recall and (**b-3**) F1 score.

The average precision, recall and F1 score under epochs during training and validation processes at different learning rates are displayed in Figure 8b. These indicator curves climb rapidly in the first two epochs (nearly 1000 iterations), which, along with the observed plummet in training loss, demonstrates the efficacy of the transfer learning. Then, the convergence occurs after 16 epochs. Throughout the training process, the green curves with the square symbols consistently remain at the uppermost part of Figure 8(b-1)–(b-3), intuitively reflecting the exceptional performance of the FCN with an initial learning rate of 1×10^{-4}. The highest values (not all from the same epoch) are further selected from the training and validation averages, and these results are summarized in Table 2. As can be seen from the table, 1×10^{-4} is the optimal learning rate, and its corresponding FCN model not surprisingly achieves the highest precision, recall and F1 score at 83.85%, 85.74% and 84.14%, respectively, highlighted in bold. Therefore, this model is used for crack segmentation.

Table 2. Model performance at different learning rates.

Initial Learning Rate ($\times 10^{-4}$)	Highest Precision (%)	Highest Recall (%)	Highest F1 Score (%)
0.1	80.48	80.67	80.47
1	**83.10**	**85.74**	**84.14**
10	79.53	79.84	78.43

Note: The values highlighted in bold represent the best training results of our FCN.

To test the effectiveness of the proposed FCN in detecting cracks of various morphological types and background complexities, the crack images in the test set are pre-divided into four categories. (I) Hairline crack: the cracks are narrowly developed and susceptible to changes in illumination, often resulting in fuzzy or discontinuous patterns. (II) Block crack: the crack region exhibits a blocky pattern and occupies a significantly substantial portion of the image. (III) Intersecting crack: the interconnected cracks show an intricate morphology. (IV) Complex background crack: the cracks in backgrounds with complex textures, speckling, shadows caused by uneven lighting, or clutter are challenging to discern through traditional methods.

Figure 9 depicts the FCN predictions of the above four crack types. Figure 9a–c demonstrates the segmentation results for different types of crack morphologies. The test results indicate that the proposed model exhibits good performance in accurately segmenting hairline cracks, block cracks and intersecting cracks. The segmentation of cracks under diverse and challenging conditions, including complex backgrounds and varied lighting scenarios, is also tested and compared (Figure 9e–i). In addition, Figure 9j,k display the prediction results for intact surfaces. The results demonstrate the robustness of the proposed model in handling various noise interference. Therein, the predictions of Figure 9a,c,d,g–j exhibit a significant level of agreement with ground truth. However, there are minor inaccuracies in Figure 9b (the left sample) and Figure 9f, which might be attributed to the insufficient variation in gradient of pixel values, leading to oversight of the microcracks located at the bottom. In Figure 9k, a few pixels of the backgrounds are falsely classified as cracks, possibly due to the combined interference of overexposure and overlapping black markings.

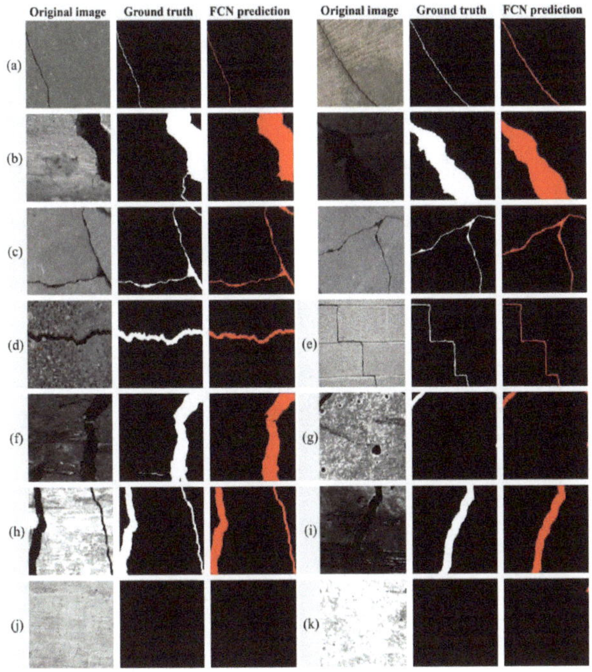

Figure 9. FCN predictions: (**a**) hairline crack; (**b**) block crack; (**c**) intersecting crack; (**d**) complex background crack (mottling); (**e**) complex background crack (interference); (**f**) complex background crack (clutter); (**g**) complex background crack (void); (**h**) different light condition (overexposure); (**i**) different light condition (uneven illumination); (**j**) intact surface (correct sample); and (**k**) intact surface (some pixels are False Positives).

Although the overall accuracy of FCN segmentation is somewhat compromised due to these omissions in detail, the extracted crack edges and skeletons still maintain an acceptable level of validity (Figure 10).

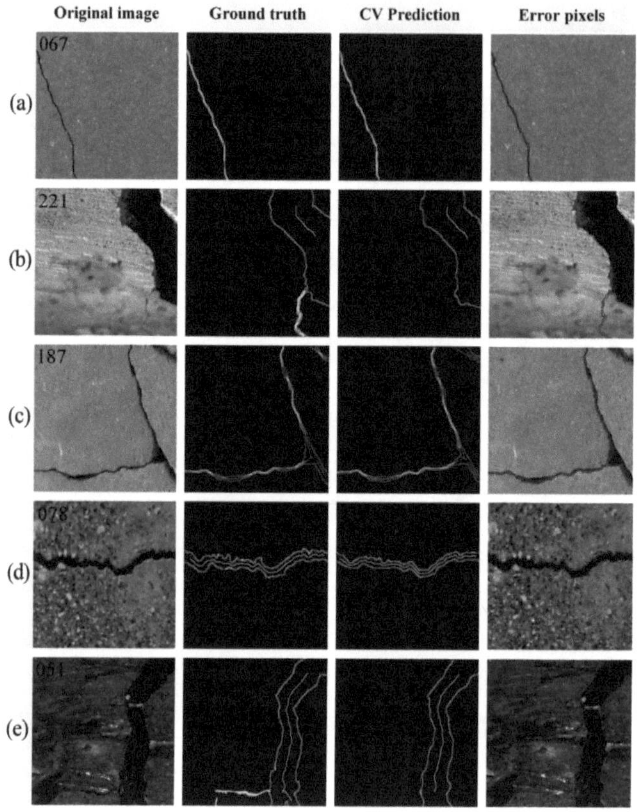

Figure 10. Extracted crack morphologies (The green lines represent the detected crack edges and the yellow lines represent the detected crack skeletons.): (**a**) hairline crack; (**b**) block crack; (**c**) intersecting crack; (**d**) complex background crack (mottling); and (**e**) complex background crack (clutter).

4. Experiment

In this section, an experiment is conducted to detect cracks in concrete specimens subjected to static load tests, with the aim of verifying the practical feasibility of the proposed method. The damaged concrete beams and slabs are neatly arranged on one side of the laboratory, and the binocular photography system is positioned approximately 0.2 m away from these cracked concretes. The aperture is adjusted accordingly to optimize exposure and capture cracks in natural indoor lighting, while simultaneously recording the manually measured values of both crack width gauges with a 0.01 mm accuracy and crack ruler as reference values for the actual crack width.

The experimental setup is illustrated in Figure 11, and a total of four cracks have been identified. Among them, three complex background cracks, designated as CrackI, CrackII and CrackIII, respectively, originating from the same beam specimen are artificially divided into multiple fragments before photographing, that is, the crack areas between black dashed lines in Figure 11a, to enhance the quantity of control groups for comparison. Additionally, as shown in Figure 11b, the fourth block crack is denoted as CrackIV_01, which is observed on a slab specimen and shot from the overhead perspective at a certain angle between the optical axis plane and the structural surface normal. The measured results are summarized in Tables 3–5, where the maximum error is 0.144 mm, corresponding to a relative error of

36.0%. This is attributed to the non-negligible prediction bias of FCN for CrackI_01. Hence, it is imperative to further optimize the performance of FCN for detecting hairline cracks.

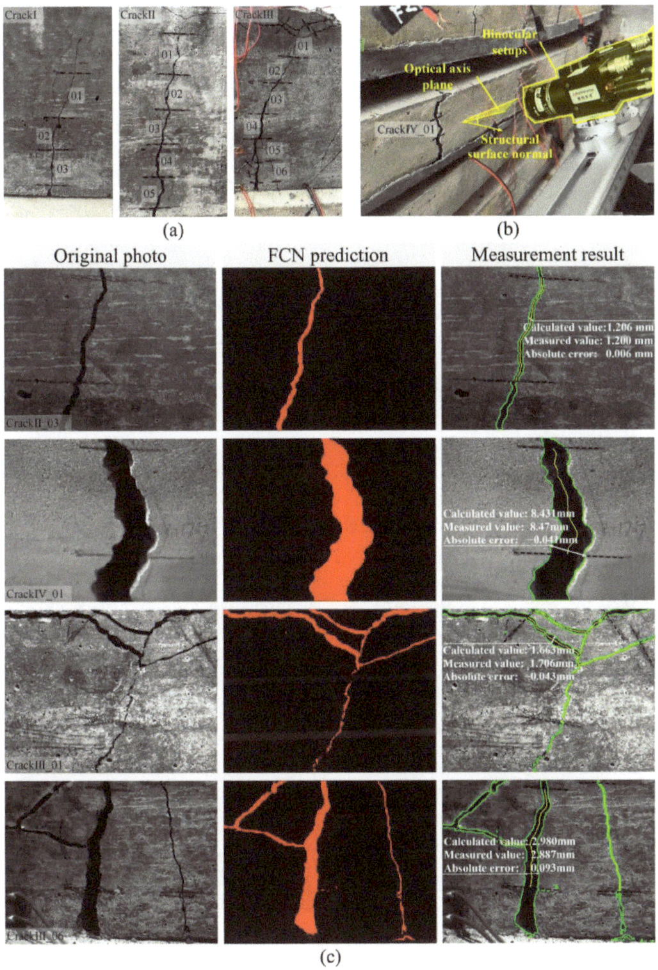

Figure 11. Concrete crack detection and measurement experiment: (**a**) divided crack fragments (the crack segment numbering corresponds to the numbering in the bottom left corner of the crack images in (**c**)); (**b**) binocular device overlooking a crack; and (**c**) visualization of the results for certain fragments.

Table 3. Results of maximum width measurement for CrackI, CrackIII_06 and CrackIV_01.

Measurement Result	CrackI_01	CrackI_02	CrackI_03	CrackIII_06	CrackIV_01
Calculated value (mm)	0.544	0.981	1.993	2.980	8.431
Reference value (mm)	0.400	1.045	2.106	2.887	8.5 *
Error (mm)	0.144	−0.064	−0.113	0.093	−0.069
Relative error	36.0%	−6.1%	−5.4%	3.2%	−0.8%

Note: * indicates that the reference value is obtained by the crack ruler.

Table 4. Results of maximum width measurement for CrackII (01–05).

Measurement Result	CrackII_01	CrackII_02	CrackII_03	CrackII_04	CrackII_05
Calculated value (mm)	0.803	1.601	1.206	1.722	2.168
Reference value (mm)	0.836	1.613	1.200	1.743	2.153
Error (mm)	−0.033	−0.012	0.006	−0.021	0.015
Relative error	−3.9%	−0.7%	0.5%	1.2%	0.7%

Table 5. Results of maximum width measurement for CrackIII (01–05).

Measurement Result	CrackIII_01	CrackIII_02	CrackIII_03	CrackIII_04	CrackIII_05
Calculated value (mm)	1.663	1.124	2.081	2.067	2.165
Reference value (mm)	1.706	1.045	2.090	2.026	2.129
Error (mm)	−0.043	0.079	−0.009	0.041	0.036
Relative error	−2.5%	7.6%	0.4%	2.0%	1.7%

Figure 11c presents the visible outcomes of certain crack fragments, among which the refined red region effectively demonstrates the generalization capability of our FCN, while the low error level further substantiates the validity of the proposed measurement method. Specifically, CrackII_03 has achieved the most accurate quantification, with an error of only 0.006 mm. As anticipated, CrackIV_01, exhibiting a calculated error of −0.069 mm, confirms the binocular vision-based approach's capability to maintain high measurement accuracy even under oblique shooting conditions, thereby highlighting its superiority over the monocular vision method in terms of shooting posture. Although the morphology of CrackIII_06 is successfully extracted despite the interference of the strain gauge wire and the shadow caused by this wire in the lower left corner, the associated error exhibits a substantial increase in comparison to CrackIII_01, reaching 0.093 mm. One possible explanation for this is that the uneven concrete surface renders the proposed method inapplicable. Apart from displaying maximum values of crack width, their specific location are also indicated through white bidirectional arrows, thereby offering a valuable reference for re-inspection.

5. Conclusions and Discussion

In this paper, a non-contact method for detecting and measuring cracks is proposed by combining a semantic segmentation network, specifically the encoder–decoder FCN, with binocular stereo vision, which achieves a balance between efficiency and accuracy. According to the research results, the following conclusions can be drawn:

1. To fit the ground truth to the fullest extent, the proposed FCN adopts the encoder–decoder structure and skip connections to enable enhanced focus on details during crack segmentation. The optimal FCN model is fine-tuned using a training dataset consisting of 1108 concrete surface images with a resolution of 448 × 448 pixels, resulting in satisfactory levels for all three evaluation metrics: precision at 83.85%, recall at 85.74% and F1 score at 84.14%. These results demonstrate that the proposed FCN can accurately detect cracks at the pixel level. Since a plate is a commonly used substructure in civil engineering, an experiment of a steel plate is carried out to validate the feasibility of the proposed methodology.
2. An integrated CV procedure is specifically designed to extract the edges and skeletons of cracks from binary graphs predicted by FCN, with the aim of preparing data for crack measurements. The performance of the CV procedure is subsequently assessed on FCN predictions of various types of cracks in the test set, demonstrating that its output is both acceptable and effective. Moreover, skeletonization results exhibit a higher level of adherence to the actual crack topology in regions that are distant from the image boundary.

3. The proposed method is applied to quantitatively evaluate the cracking of concrete specimens in real-life scenarios, with a comparison made against manual inspection results. The experimental results demonstrate that our FCN possesses remarkable generalization capability, and the binocular measurement method can also control errors at a low level, thereby ensuring both robustness in detection and accuracy in measurement. For crack width, the maximum error is 0.144 mm, while the mean relative error stands at 5.03%, thus confirming the feasibility of the proposed method.
4. The experiment also involves an overhead shot of a target crack through the binocular photography system. The calculated error of −0.041 mm, along with its corresponding relative error of −0.8%, validates the high level of accuracy achieved by the binocular vision-based measurement method even under tilted shooting conditions, emphasizing its superiority over the monocular vision method and making it more suitable for implementation on remotely operated piggyback platforms, such as UAVs or robots.

However, there are still some limitations to this research. Future studies should aim to integrate advanced algorithms like attention mechanisms and EfficientNet to further enhance the model's performance. Additionally, the incorporation of advanced feature matching algorithms like LightGlue promises to yield more precise three-dimensional reconstructions of cracks. In practical terms, the proposed binocular photography system requires an external power source of 5V or higher. It is necessary to optimize the energy management strategy for the entire detection system. This may involve reducing standby power consumption and employing dynamic programming to determine the optimal flight path of UAVs. This research, currently focused on crack segmentation and measurement, should expand to include other surface defects like delamination and spalling in future studies, broadening its scope and real-world applicability.

Author Contributions: Conceptualization, Z.Z. and H.Z.; Writing—original draft, J.L.; Writing—review & editing, Z.Z., Z.S. and H.Z.; Supervision, Z.Z., J.S. and H.Z. All authors have read and agreed to the published version of the manuscript.

Funding: The authors acknowledge the support from the National Key R&D Program of China (grant No. 2020YFA0711700), the National Natural Science Foundation of China (grant Nos. U23A20659, 52122801, 11925206, 51978609 and U22A20254) and the Foundation for Distinguished Young Scientists of Zhejiang Province (grant No. LR20E080003).

Data Availability Statement: Data are contained within the article.

Acknowledgments: The authors express their appreciation to Feilei Chen for assistance with this work.

Conflicts of Interest: The authors declare no conflict of interest.

References

1. Kayondo, M.; Combrinck, R.; Boshoff, W.P. State-of-the-art review on plastic cracking of concrete. *Constr. Build. Mater.* **2019**, *225*, 886–899. [CrossRef]
2. Wang, H.L.; Dai, J.G.; Sun, X.Y.; Zhang, X.L. Characteristics of concrete cracks and their influence on chloride penetration. *Constr. Build. Mater.* **2016**, *107*, 216–225. [CrossRef]
3. Zhang, H.; Zhou, Y.H.; Quan, L.W. Identification of a moving mass on a beam bridge using piezoelectric sensor arrays. *J. Sound. Vib.* **2021**, *491*, 115754. [CrossRef]
4. Aboudi, J. Stiffness Reduction of Cracked Solids. *Eng. Fract. Mech.* **1987**, *26*, 637–650. [CrossRef]
5. Chupanit, P.; Roesler, J.R. Fracture energy approach to characterize concrete crack surface roughness and shear stiffness. *J. Mater. Civil. Eng.* **2008**, *20*, 275–282. [CrossRef]
6. Güllü, H.; Canakci, H.; Alhashemy, A. Use of ranking measure for performance assessment of correlations for the compression index. *Eur. J. Environ. Civ. Eng.* **2018**, *22*, 578–595. [CrossRef]
7. Güllü, H.; Canakci, H.; Alhashemy, A. A Ranking Distance Analysis for Performance Assessment of UCS Versus SPT-N Correlations. *Arab. J. Sci. Eng.* **2019**, *44*, 4325–4337. [CrossRef]
8. Jahanshahi, M.R.; Kelly, J.S.; Masri, S.F.; Sukhatme, G.S. A survey and evaluation of promising approaches for automatic image-based defect detection of bridge structures. *Struct. Infrastruct. Eng.* **2009**, *5*, 455–486. [CrossRef]
9. Jiang, W.B.; Liu, M.; Peng, Y.N.; Wu, L.H.; Wang, Y.N. HDCB-Net: A Neural Network With the Hybrid Dilated Convolution for Pixel-Level Crack Detection on Concrete Bridges. *IEEE Trans. Ind. Inform.* **2021**, *17*, 5485–5494. [CrossRef]

10. Zhang, H.; Zhou, Y.H.; Huang, Z.Y.; Shen, R.H.; Wu, Y.D. Multiparameter Identification of Bridge Cables Using XGBoost Algorithm. *J. Bridge Eng.* **2023**, *28*. [CrossRef]
11. Huston, D.; Hu, J.Q.; Maser, K.; Weedon, W.; Adam, C. GIMA ground penetrating radar system for monitoring concrete bridge decks. *J. Appl. Geophys.* **2000**, *43*, 139–146. [CrossRef]
12. Chen, S.-E.; Liu, W.; Bian, H.; Smith, B. 3D LiDAR Scans for Bridge Damage Evaluations. In *Forensic Engineering 2012*; ASCE Library: New York, NY, USA, 2013; pp. 487–495. [CrossRef]
13. Valenca, J.; Puente, I.; Julio, E.; Gonzalez-Jorge, H.; Arias-Sanchez, P. Assessment of cracks on concrete bridges using image processing supported by laser scanning survey. *Constr. Build. Mater.* **2017**, *146*, 668–678. [CrossRef]
14. Zhang, B.N.; Zhou, Z.X.; Zhang, K.H.; Yan, G.; Xu, Z.Z. Sensitive skin and the relative sensing system for real-time surface monitoring of crack in civil infrastructure. *J. Intell. Mater. Syst. Struct.* **2006**, *17*, 907–917. [CrossRef]
15. Hurlebaus, S.; Gaul, L. Smart layer for damage diagnostics. *J. Intell. Mater. Syst. Struct.* **2004**, *15*, 729–736. [CrossRef]
16. Roopa, A.K.; Hunashyal, A.M.; Mysore, R.R.M. Development and Implementation of Cement-Based Nanocomposite Sensors for Structural Health Monitoring Applications: Laboratory Investigations and Way Forward. *Sustainability* **2022**, *14*, 12452. [CrossRef]
17. Dorafshan, S.; Thomas, R.J.; Maguire, M. Comparison of deep convolutional neural networks and edge detectors for image-based crack detection in concrete. *Constr. Build. Mater.* **2018**, *186*, 1031–1045. [CrossRef]
18. Zhang, H.; Zhou, Y. AI-Based Modeling and Data-Driven Identification of Moving Load on Continuous Beams. *Fundam. Res.* **2022**, *3*, 796–803. [CrossRef]
19. Yeum, C.M.; Dyke, S.J. Vision-Based Automated Crack Detection for Bridge Inspection. *Comput.-Aided Civ. Infrastruct. Eng.* **2015**, *30*, 759–770. [CrossRef]
20. Oh, J.K.; Jang, G.; Oh, S.; Lee, J.H.; Yi, B.J.; Moon, Y.S.; Lee, J.S.; Choi, Y. Bridge inspection robot system with machine vision. *Autom. Constr.* **2009**, *18*, 929–941. [CrossRef]
21. Iyer, S.; Sinha, S.K. A robust approach for automatic detection and segmentation of cracks in underground pipeline images. *Image Vis. Comput.* **2005**, *23*, 921–933. [CrossRef]
22. Fujita, Y.; Hamamoto, Y. A robust automatic crack detection method from noisy concrete surfaces. *Mach. Vis. Appl.* **2011**, *22*, 245–254. [CrossRef]
23. Lee, B.Y.; Kim, Y.Y.; Yi, S.T.; Kim, J.K. Automated image processing technique for detecting and analysing concrete surface cracks. *Struct. Infrastruct. Eng.* **2013**, *9*, 567–577. [CrossRef]
24. Zhang, H.; Shen, M.Z.; Zhang, Y.Y.; Chen, Y.S.; Lu, C.F. Identification of Static Loading Conditions Using Piezoelectric Sensor Arrays. *J. Appl. Mech.* **2018**, *85*, 011008. [CrossRef]
25. Nguyen, H.N.; Kam, T.Y.; Cheng, P.Y. An Automatic Approach for Accurate Edge Detection of Concrete Crack Utilizing 2D Geometric Features of Crack. *J. Signal Process. Syst.* **2014**, *77*, 221–240. [CrossRef]
26. Sohn, H.G.; Lim, Y.M.; Yun, K.H.; Kim, G.H. Monitoring crack changes in concrete structures. *Comput.-Aided Civ. Infrastruct. Eng.* **2005**, *20*, 52–61. [CrossRef]
27. Ni, T.Y.; Zhou, R.X.; Gu, C.P.; Yang, Y. Measurement of concrete crack feature with android smartphone APP based on digital image processing techniques. *Measurement* **2020**, *150*, 107093. [CrossRef]
28. Abdel-Qader, L.; Abudayyeh, O.; Kelly, M.E. Analysis of edge-detection techniques for crack identification in bridges. *J. Comput. Civ. Eng.* **2003**, *17*, 255–263. [CrossRef]
29. Wang, K.C.P.; Li, Q.; Gong, W.G. Wavelet-based pavement distress image edge detection with a trous algorithm. *Transp. Res. Rec.* **2007**, *2024*, 73–81. [CrossRef]
30. Xiang, T.; Huang, K.X.; Zhang, H.; Zhang, Y.Y.; Zhang, Y.N.; Zhou, Y.H. Detection of Moving Load on Pavement Using Piezoelectric Sensors. *Sensors* **2020**, *20*, 2366. [CrossRef]
31. Yamaguchi, T.; Hashimoto, S. Fast crack detection method for large-size concrete surface images using percolation-based image processing. *Mach. Vis. Appl.* **2010**, *21*, 797–809. [CrossRef]
32. Adhikari, R.S.; Moselhi, O.; Bagchi, A. Image-based retrieval of concrete crack properties for bridge inspection. *Autom. Constr.* **2014**, *39*, 180–194. [CrossRef]
33. Payab, M.; Abbasina, R.; Khanzadi, M. A Brief Review and a New Graph-Based Image Analysis for Concrete Crack Quantification. *Arch. Comput. Methods Eng.* **2019**, *26*, 347–365. [CrossRef]
34. Andrushia, A.D.; Anand, N.; Arulraj, G.P. A novel approach for thermal crack detection and quantification in structural concrete using ripplet transform. *Struct. Control Health Monit.* **2020**, *27*, e2621. [CrossRef]
35. Liu, Y.F.; Cho, S.; Spencer, B.F.; Fan, J.S. Concrete Crack Assessment Using Digital Image Processing and 3D Scene Reconstruction. *J. Comput. Civ. Eng.* **2016**, *30*, 04014124. [CrossRef]
36. Li, S.Y.; Zhao, X.F.; Zhou, G.Y. Automatic pixel-level multiple damage detection of concrete structure using fully convolutional network. *Comput.-Aided Civ. Infrastruct. Eng.* **2019**, *34*, 616–634. [CrossRef]
37. Prasanna, P.; Dana, K.J.; Gucunski, N.; Basily, B.B.; La, H.M.; Lim, R.S.; Parvardeh, H. Automated Crack Detection on Concrete Bridges. *IEEE Trans. Autom. Sci. Eng.* **2016**, *13*, 591–599. [CrossRef]
38. Peng, X.; Zhong, X.G.; Zhao, C.; Chen, A.H.; Zhang, T.Y. A UAV-based machine vision method for bridge crack recognition and width quantification through hybrid feature learning. *Constr. Build. Mater.* **2021**, *299*, 123896. [CrossRef]
39. Alipour, M.; Harris, D.K.; Miller, G.R. Robust Pixel-Level Crack Detection Using Deep Fully Convolutional Neural Networks. *J. Comput. Civ. Eng.* **2019**, *33*, 04019040. [CrossRef]

40. Zhang, H.; Shen, Z.J.; Lin, Z.H.; Quan, L.W.; Sun, L.F. Deep learning-based automatic classification of three-level surface information in bridge inspection. *Comput.-Aided Civ. Infrastruct. Eng.* **2023**. [CrossRef]
41. Zhang, L.; Yang, F.; Zhang, Y.D.; Zhu, Y.J. Road Crack Detection Using Deep Convolutional Neural Network. In Proceedings of the 2016 IEEE International Conference on Image Processing (ICIP), Phoenix, AZ, USA, 25–28 September 2016; pp. 3708–3712. [CrossRef]
42. Cha, Y.J.; Choi, W.; Buyukozturk, O. Deep Learning-Based Crack Damage Detection Using Convolutional Neural Networks. *Comput.-Aided Civ. Infrastruct. Eng.* **2017**, *32*, 361–378. [CrossRef]
43. Chen, F.C.; Jahanshahi, M.R. NB-CNN: Deep Learning-Based Crack Detection Using Convolutional Neural Network and Naive Bayes Data Fusion. *IEEE Trans. Ind. Electron.* **2018**, *65*, 4392–4400. [CrossRef]
44. Kim, B.; Cho, S. Automated Vision-Based Detection of Cracks on Concrete Surfaces Using a Deep Learning Technique. *Sensors* **2018**, *18*, 3452. [CrossRef] [PubMed]
45. Cha, Y.J.; Choi, W.; Suh, G.; Mahmoudkhani, S.; Buyukozturk, O. Autonomous Structural Visual Inspection Using Region-Based Deep Learning for Detecting Multiple Damage Types. *Comput.-Aided Civ. Infrastruct. Eng.* **2018**, *33*, 731–747. [CrossRef]
46. Deng, J.H.; Lu, Y.; Lee, V.C.S. Concrete crack detection with handwriting script interferences using faster region-based convolutional neural network. *Comput.-Aided Civ. Infrastruct. Eng.* **2020**, *35*, 373–388. [CrossRef]
47. Zhang, C.B.; Chang, C.C.; Jamshidi, M. Concrete bridge surface damage detection using a single-stage detector. *Comput.-Aided Civ. Infrastruct. Eng.* **2020**, *35*, 389–409. [CrossRef]
48. Li, Y.D.; Li, H.G.; Wang, H.R. Pixel-Wise Crack Detection Using Deep Local Pattern Predictor for Robot Application. *Sensors* **2018**, *18*, 3042. [CrossRef] [PubMed]
49. Kim, B.; Cho, S. Image-based concrete crack assessment using mask and region-based convolutional neural network. *Struct. Control. Health Monit.* **2019**, *26*, e2381. [CrossRef]
50. Zhang, A.; Wang, K.C.P.; Fei, Y.; Liu, Y.; Chen, C.; Yang, G.W.; Li, J.Q.; Yang, E.H.; Qiu, S. Automated Pixel-Level Pavement Crack Detection on 3D Asphalt Surfaces with a Recurrent Neural Network. *Comput.-Aided Civ. Infrastruct. Eng.* **2019**, *34*, 213–229. [CrossRef]
51. Yang, X.C.; Li, H.; Yu, Y.T.; Luo, X.C.; Huang, T.; Yang, X. Automatic Pixel-Level Crack Detection and Measurement Using Fully Convolutional Network. *Comput.-Aided Civ. Infrastruct. Eng.* **2018**, *33*, 1090–1109. [CrossRef]
52. Dung, C.V.; Anh, L.D. Autonomous concrete crack detection using deep fully convolutional neural network. *Autom. Constr.* **2019**, *99*, 52–58. [CrossRef]
53. Liu, Z.Q.; Cao, Y.W.; Wang, Y.Z.; Wang, W. Computer vision-based concrete crack detection using U-net fully convolutional networks. *Autom. Constr.* **2019**, *104*, 129–139. [CrossRef]
54. Liu, J.W.; Yang, X.; Lau, S.; Wang, X.; Luo, S.; Lee, V.C.S.; Ding, L. Automated pavement crack detection and segmentation based on two-step convolutional neural network. *Comput.-Aided Civ. Infrastruct. Eng.* **2020**, *35*, 1291–1305. [CrossRef]
55. Miao, Z.H.; Ji, X.D.; Okazaki, T.; Takahashi, N. Pixel-level multicategory detection of visible seismic damage of reinforced concrete components. *Comput.-Aided Civ. Infrastruct. Eng.* **2021**, *36*, 620–637. [CrossRef]
56. Guan, J.C.; Yang, X.; Ding, L.; Cheng, X.Y.; Lee, V.C.S.; Jin, C. Automated pixel-level pavement distress detection based on stereo vision and deep learning. *Autom. Constr.* **2021**, *129*, 103788. [CrossRef]
57. Zhang, X.X.; Rajan, D.; Story, B. Concrete crack detection using context-aware deep semantic segmentation network. *Comput.-Aided Civ. Infrastruct. Eng.* **2019**, *34*, 951–971. [CrossRef]
58. Chen, T.Y.; Cai, Z.H.; Zhao, X.; Chen, C.; Lianga, X.F.; Zou, T.R.; Wang, P. Pavement crack detection and recognition using the architecture of segNet. *J. Ind. Inf. Integr.* **2020**, *18*, 100144. [CrossRef]
59. Zheng, X.; Zhang, S.L.; Li, X.; Li, G.; Li, X.Y. Lightweight Bridge Crack Detection Method Based on SegNet and Bottleneck Depth-Separable Convolution With Residuals. *IEEE Access* **2021**, *9*, 161649–161668. [CrossRef]
60. Ji, A.K.; Xue, X.L.; Wang, Y.N.; Luo, X.W.; Xue, W.R. An integrated approach to automatic pixel-level crack detection and quantification of asphalt pavement. *Autom. Constr.* **2020**, *114*, 103176. [CrossRef]
61. Sun, Y.J.; Yang, Y.; Yao, G.; Wei, F.J.; Wong, M.P. Autonomous Crack and Bughole Detection for Concrete Surface Image Based on Deep Learning. *IEEE Access* **2021**, *9*, 85709–85720. [CrossRef]
62. Bang, S.; Park, S.; Kim, H.; Kim, H. Encoder-decoder network for pixel-level road crack detection in black-box images. *Comput.-Aided Civ. Infrastruct. Eng.* **2019**, *34*, 713–727. [CrossRef]
63. Li, G.; Li, X.Y.; Zhou, J.; Liu, D.Z.; Ren, W. Pixel-level bridge crack detection using a deep fusion about recurrent residual convolution and context encoder network. *Measurement* **2021**, *176*, 109171. [CrossRef]
64. Zhang, L.; Jiang, F.L.; Yang, J.; Kong, B.; Hussain, A. A real-time lane detection network using two-directional separation attention. *Comput.-Aided Civ. Infrastruct. Eng.* **2023**. [CrossRef]
65. Chen, J.; He, Y. A novel U-shaped encoder–decoder network with attention mechanism for detection and evaluation of road cracks at pixel level. *Comput.-Aided Civ. Infrastruct. Eng.* **2022**, *37*, 1721–1736. [CrossRef]
66. Chen, L.J.; Yao, H.D.; Fu, J.Y.; Ng, C.T. The classification and localization of crack using lightweight convolutional neural network with CBAM. *Eng. Struct.* **2023**, *275*, 115291. [CrossRef]
67. Du, Y.C.; Zhong, S.; Fang, H.Y.; Wang, N.N.; Liu, C.L.; Wu, D.F.; Sun, Y.; Xiang, M. Modeling automatic pavement crack object detection and pixel-level segmentation. *Autom. Constr.* **2023**, *150*, 104840. [CrossRef]

68. Yang, L.; Bai, S.L.; Liu, Y.H.; Yu, H.N. Multi-scale triple-attention network for pixelwise crack segmentation. *Autom. Constr.* **2023**, *150*, 104853. [CrossRef]
69. Zhu, G.J.; Liu, J.C.; Fan, Z.; Yuan, D.; Ma, P.L.; Wang, M.H.; Sheng, W.H.; Wang, K.C.P. A lightweight encoder-decoder network for automatic pavement crack detection. *Comput.-Aided Civ. Infrastruct. Eng.* **2023**. [CrossRef]
70. Lei, M.F.; Zhang, Y.B.; Deng, E.; Ni, Y.Q.; Xiao, Y.Z.; Zhang, Y.; Zhang, J.J. Intelligent recognition of joints and fissures in tunnel faces using an improved mask region-based convolutional neural network algorithm. *Comput.-Aided Civ. Infrastruct. Eng.* **2023**. [CrossRef]
71. Que, Y.; Dai, Y.; Ji, X.; Leung, A.K.; Chen, Z.; Jiang, Z.L.; Tang, Y.C. Automatic classification of asphalt pavement cracks using a novel integrated generative adversarial networks and improved VGG model. *Eng. Struct.* **2023**, *277*, 115406. [CrossRef]
72. Nguyen, Q.D.; Thai, H.T. Crack segmentation of imbalanced data: The role of loss functions. *Eng. Struct.* **2023**, *297*, 116988. [CrossRef]
73. Weng, X.X.; Huang, Y.C.; Li, Y.A.; Yang, H.; Yu, S.H. Unsupervised domain adaptation for crack detection. *Autom. Constr.* **2023**, *153*, 104939. [CrossRef]
74. Kang, D.; Benipal, S.S.; Gopal, D.L.; Cha, Y.J. Hybrid pixel-level concrete crack segmentation and quantification across complex backgrounds using deep learning. *Autom. Constr.* **2020**, *118*, 103291. [CrossRef]
75. Yuan, C.; Xiong, B.; Li, X.; Sang, X.; Kong, Q. A novel intelligent inspection robot with deep stereo vision for three-dimensional concrete damage detection and quantification. *Struct. Health Monit.* **2022**, *21*, 788–802. [CrossRef]
76. Kim, H.; Sim, S.-H.; Spencer, B.F. Automated concrete crack evaluation using stereo vision with two different focal lengths. *Autom. Constr.* **2022**, *135*, 104136. [CrossRef]
77. Chen, C.X.; Shen, P. Research on Crack Width Measurement Based on Binocular Vision and Improved DeeplabV3+. *Appl. Sci* **2023**, *13*, 2752. [CrossRef]
78. Pan, S.J.; Yang, Q.A. A Survey on Transfer Learning. *IEEE Trans. Knowl. Data Eng.* **2010**, *22*, 1345–1359. [CrossRef]
79. Wang, M.; Deng, W.H. Deep visual domain adaptation: A survey. *Neurocomputing* **2018**, *312*, 135–153. [CrossRef]
80. Zhang, T.Y.; Suen, C.Y. A Fast Parallel Algorithm for Thinning Digital Patterns. *Commun. ACM* **1984**, *27*, 236–239. [CrossRef]
81. Lowe, D.G. Distinctive image features from scale-invariant keypoints. *Int. J. Comput. Vis.* **2004**, *60*, 91–110. [CrossRef]
82. Shan, B.; Zheng, S.; Ou, J. A stereovision-based crack width detection approach for concrete surface assessment. *KSCE J. Civ. Eng.* **2016**, *20*, 803–812. [CrossRef]
83. Hinton, G.E.; Srivastava, N.; Krizhevsky, A.; Sutskever, I.; Salakhutdinov, R. Improving neural networks by preventing co-adaptation of feature detectors. *arXiv* **2012**, arXiv:1207.0580.
84. Srivastava, N.; Hinton, G.; Krizhevsky, A.; Sutskever, I.; Salakhutdinov, R. Dropout: A Simple Way to Prevent Neural Networks from Overfitting. *J. Mach. Learn. Res.* **2014**, *15*, 1929–1958.
85. Zhuang, F.Z.; Qi, Z.Y.; Duan, K.Y.; Xi, D.B.; Zhu, Y.C.; Zhu, H.S.; Xiong, H.; He, Q. A Comprehensive Survey on Transfer Learning. *Proc. IEEE* **2021**, *109*, 43–76. [CrossRef]

electronics

MDPI

Article

TrustDFL: A Blockchain-Based Verifiable and Trusty Decentralized Federated Learning Framework

Jinsheng Yang [1], Wenfeng Zhang [1], Zhaohui Guo [1] and Zhen Gao [2,*]

1 School of Microelectronics, Tianjin University, Tianjin 300072, China; jsyang@tju.edu.cn (J.Y.); 2020232164@tju.edu.cn (W.Z.); 2017232048@tju.edu.cn (Z.G.)

2 School of Electrical Automation and Information Engineering, Tianjin University, Tianjin 300072, China

* Correspondence: zgao@tju.edu.cn

Abstract: Federated learning is a privacy-preserving machine learning framework where multiple data owners collaborate to train a global model under the orchestra of a central server. The local training results from trainers should be submitted to the central server for model aggregation and update. Busy central server and malicious trainers can introduce the issues of a single point of failure and model poisoning attacks. To address the above issues, the trusty decentralized federated learning(called TrustDFL) framework has been proposed in this paper based on the zero-knowledge proof scheme, blockchain, and smart contracts, which provides enhanced security and higher efficiency for model aggregation. Specifically, Groth 16 is applied to generate the proof for the local model training, including the forward and backward propagation processes. The proofs are attached as the payloads to the transactions, which are broadcast into the blockchain network and executed by the miners. With the support of smart contracts, the contributions of the trainers could be verified automatically under the economic incentive, where the blockchain records all exchanged data as the trust anchor in multi-party scenarios. In addition, IPFS (InterPlanetary File System) is introduced to alleviate the storage and communication overhead brought by local and global models. The theoretical analysis and estimation results show that the TrustDFL efficiently avoids model poisoning attacks without leaking the local secrets, ensuring the global model's accuracy to be trained.

Keywords: decentralized federated learning; blockchain; verifiability; zero-knowledge proof (ZKP); zk-SNARK

1. Introduction

As a branch of artificial intelligence, machine learning (ML) is a statistical model driven by big data to tackle complex problems, and has been used widely in various fields, such as image recognition, image segmentation, and natural language processing [1–3]. With the development of the deep neural network (DNN), training a larger model requires more data and computing power, which usually is delegated to the centralized data center or cloud providers [4]. For example, the machine learning as a service (MLaaS) framework provides prediction results based on the trained models maintained on the centralized server, which needs to collect the training data set from the edge devices in advance and causes the risk of privacy leakage [5]. Federated learning (FL) is proposed for multiple participants to train a global model corroboratively without exposure of the local data [6,7]. As shown in Figure 1, the FL framework is formed by edge devices acting as trainers and the server acting as the aggregator. The global model training in FL runs multiple rounds. For each round, there are four steps, as follows [8]: (1) global model distribution: the aggregator distributes the global model to the selected trainer; (2) local training: relying on the local training data set, the selected trainers update the parameters based on the published global model; (3) update upload: the selected trainers upload the trained results (the updated weights or gradients), instated of the raw training data, to the server, which avoids privacy leakage; (4) model aggregation: the server aggregates the updates and generates the new global model for the next round. The target global model to be trained in FL could have any architectures, such as multi-layer perceptron (MLP) and convolutional neural networks (CNN). The standard implementation of FL is FederatedAveraging (FedAvg), where the updates for the global model could be aggregated through a weighted average [9].

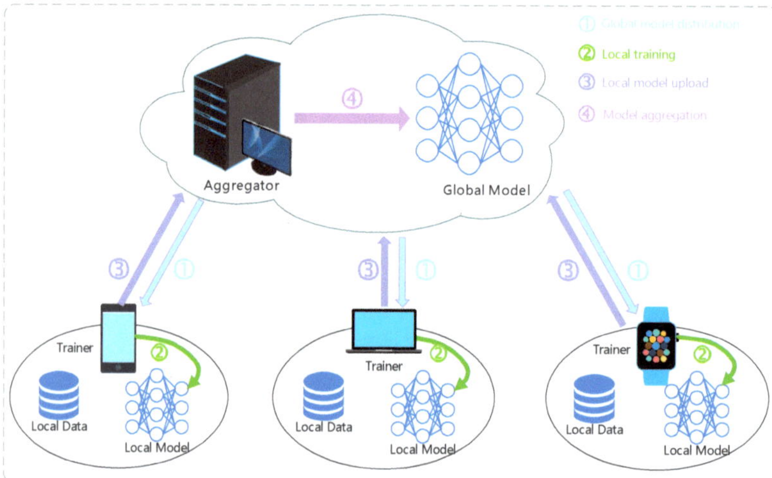

Figure 1. FL framework and global model training process.

However, there are two major issues in the FL framework. First, it is hard for the aggregators to check the validity of the updates uploaded by the trainers, which brings the risk of model poisoning attacks and reduces model availability [10–12]. The malicious trainers could launch model poisoning attacks by uploading invalid updates (for example, random float numbers generated randomly), which could interfere with the aggregation or convergence process, and reduce the global model accuracy. Since the raw training data on each trainer cannot be revealed to the aggregator for privacy-preserving reasons, there is no efficient method to check the contributions of the trainers. There are some robustness aggregation methods, such as Krum and Trim-means, which remove the updates dissimilar from others based on the statistical analysis before the aggregation process [13,14]. On one hand, these statistical methods are imprecise because trainers have different data distributions. On the other hand, by the robustness aggregation, all workloads for verification are on the aggregator sides, which causes a computation bottleneck.

Second, in the traditional FL framework (known as centralized FL, CFL), global model training is orchestrated by the centralized server, which faces a single point of failure problem and cannot meet the requirement for decentralization in the cross-silo scenarios. The CFL could be transformed into a decentralized FL (DFL) by replacing the designated server with a cluster of aggregators from different organizations [15,16]. There are two issues in the current DFL framework, as follows: (1) like the trainers, the contributions of the aggregators should also be verified efficiently. (2) All contributions should be recorded honestly to construct trust, and the incentive mechanism is needed to encourage all participants to adhere to the protocol honestly.

Various schemes were proposed to address the issues above, and most introduce blockchain for decentralized FL. Under the framework of establishing decentralized federated learning, the authors in [17] proposed a new committee consensus mechanism to complete the tasks of gradient selection and block generation. The elected committee serves as the miner's responsibility and uses K-fold cross-validation to verify and score the trainer's updates. The authors in [18] elected a committee to judge the reliability of the model parameters by verifying whether the trainer's training time is proportional to the data size. There are also some solutions that integrate blockchain, smart contracts, and the zero-knowledge proof (ZKP) system. The blockchain is adopted as the trust anchor, the ZKP is used for verifying contributions, and the smart contracts are used for the automatic execution of contribution verification and reward distribution. Different from the robustness aggregation and cross-validation, the verification methods based on the ZKP enable the trainers to convince the aggregators that the updates are valid by providing proof indicating the correct execution of the defined computation. For example, [19–21] adopted the zero-knowledge succinct non-interactive argument of knowledge (zk-SNARK) to prove the correctness of the prediction results without revealing the model parameters in the MlaaS scenarios, where only the forward prorogation (FP) process and two-party scenarios are considered. As proposed in [22,23], the schemes integrate the ZKPs and blockchain to construct the verifiable DFL framework, where the verification for updates is viewed as part of the consensus process, and miners are responsible for the verification before recording the updates into the blocks. In [24], ZKP-based verification is executed by smart contracts. However, these schemes use the blockchain as the black box or the third-party platform. The global model training of the FL is not integrated into the transaction lifecycle of the blockchain.

This paper proposes a trustworthy and verifiable decentralized federated learning framework (TrustDFL) based on blockchain and ZKP. The issues of a single point of failure problem and model poisoning attacks could be addressed efficiently by the proposed TrustDFL framework with limited computational overhead for on-chain verification. Moreover, the storage burden of the blockchain could be alleviated by dumping global and local models into the IPFS. The contributions of this paper are as follows:

1. We propose the TrustDFL framework, which integrates the global model training in the FL into the consensus process in the blockchain. The framework adopts zero-knowledge proof to establish the proof of the computation correctness of the local training process and the model aggregation process, which are implemented by smart contracts. Only the successfully verified models will be recorded on the blockchain.
2. We specifically use zk-SNARK to construct the proof system, which has the advantages of succinctness (small proof size) and efficiency (fast proof verification). We provide the construction detail of the proof for the correctness of local training (including the forward and back propagation processes). The proof for the aggregation process could be generated similarly based on the defined aggregation algorithm.
3. Considering the blockchain storage pressure, we introduce an IPFS to store the hash related to the model on the blockchain instead of the raw model parameters, which effectively avoids storage scalability issues for the blockchain.

TrustDFL can be applied to many scenarios. For example, in the current model of cloud computing, MLaaS, it is necessary to ensure the availability of the global model. In this case, the operator provides predictable service for pay using the trained global model, which is obtained by the DFL with data collected from sensors owned by users with different interests. TrustDFL can also be applied to the healthcare industry, where smart health applications utilize global models jointly trained by users with data to provide health monitoring services. Since incorrect predictions of health status can bring serious consequences, and personal health data is highly confidential, TrustDFL can ensure the reliability of the global model without leaking secrets, which is very suitable for this scenario [25].

The rest of this paper is organized as follows. Section 2 describes the basic technique. Section 3 introduces the detailed design of the TrustDFL. Theoretical analysis and simulation results for the TrustDFL are given in Sections 4 and 5, respectively. The paper discusses and concludes in Sections 6 and 7.

2. Background

This section introduces the basic techniques for constructing the TrustDFL, including the architecture of the multilayer perceptron (MLP), blockchain, and zero-knowledge proof (ZKP) system, in Sections 2.1–2.3, respectively.

2.1. Multi-Layer Perceptron (MLP)

In this paper, a multi-layer perceptron (MLP) is used as a case study to demonstrate the principle of TrustDFL. A multi-layer perceptron (MLP) is a fully connected feedforward neural network, typically including an input layer, a hidden layer, and an output layer, where all layers are formed by multiple neurons with weights and a bias vector to be trained [26]. Unlike the linear perceptron, the hidden and output layers are followed by the activation function f. In the MLP, there are two most commonly used activation functions: ReLU and sigmoid, shown as Equation (1) and Equation (2), respectively.

$$f(x) = max(0, x) \tag{1}$$

$$f(x) = \frac{1}{1 - e^{-x}} \tag{2}$$

The MLP can provide the prediction results based on the inputs known as forward propagation (FP) [27,28]. The weights and bias vectors could be trained by the backpropagation (BP). Typically, the training process in MLP includes serval epochs. For each epoch, the weights and biases are updated using the gradient descent algorithm. During the FP, the calculation of the i-th layer could be described as follows:

$$Y_i = f(X_i \bullet W_i + \mathbf{b}_i), \tag{3}$$

where X_i, Y_i, W_i, and $\mathbf{b}_i$ are the input matrix, output matrix, weight matrix, and bias vector of the i-th layer. The output of the i-th layer is the input of the $i+1$-th layer. $f(x)$ is the activation function.

In the output layer, a loss function L is used to estimate the gaps between the prediction results and the labels. Typically, for multi-category classification tasks, the cross-entropy is used as the loss function:

$$L = -\frac{1}{n}\sum_{j}\sum_{c=1}^{M} y_{jc} \bullet log(p_{jc}) \tag{4}$$

where n and M are the numbers of the samples and classes, respectively. The y_{jc} is the sign function, which is 1 when the j-th sample belongs to the c-th class ($y_{jc} = 0$, otherwise). p_{jc} is the probability that the j-th sample belongs to the c-th class. The target of the training process is minimizing the loss function by updating the weight matrices and bias vectors of all layers in the next epoch, which is realized by the gradient descent algorithm as follows:

$$\begin{cases} W_i' = W_i - \eta \nabla W_i \\ \mathbf{b}_i' = \mathbf{b}_i - \eta \nabla \mathbf{b}_i \end{cases} \tag{5}$$

where η is the learning rate. The ∇W_i and $\nabla \mathbf{b}_i$ are the gradients of the loss function regarding the weights and bias, which could be calculated based on the chain rules as:

$$\begin{cases} \nabla W_i = \frac{\partial L}{\partial W_i} = \frac{\partial L}{\partial Y_i}\frac{\partial Y_i}{\partial F_i}\frac{\partial F_i}{\partial W_i} \\ \nabla \mathbf{b}_i = \frac{\partial L}{\partial \mathbf{b}_i} = \frac{\partial L}{\partial Y_i}\frac{\partial Y_i}{\partial F_i}\frac{\partial F_i}{\partial \mathbf{b}_i} \end{cases} \tag{6}$$

where $F_i = X_i \bullet W_i + \mathbf{b}_i$. The process of updating weights and biases according to their gradients is known as backpropagation, which will be repeated for multiple epochs until the model coverages.

MLP is an early and widely used machine learning model. It is often used for text classification, audio processing, image recognition, and some traditional machine learning tasks, such as classification, regression, clustering, etc. For large-scale data sets and image processing tasks, CNN is more suitable.

2.2. Blockchain and IPFS

Blockchain is the emerging distributed database system and computation paradigm underlying cryptocurrencies, such as BTC and ETH [29,30]. As the world computer and the finite state machine, the transaction is the basic process unit to drive the state translation. All transactions are maintained by the ledger formed by the blocks chained in chronological order, where the integrity of the records is guaranteed by the cryptographic techniques (such as hash and digital signature algorithms) and the consensus processes (such as proof-of-work and proof-of-stake). With the support of peer-to-peer, all blockchain nodes share the same data access rights, and only the data passing the public verification could be appended [31].

The smart contract is the key technology for the blockchain to develop decentralization applications (Dapp), build autonomous communities, and even construct smart cities [32,33]. Taking Ethereum as an example, the smart contract is the protocol that all participants should adhere to, which is implemented as the bytecodes running in the specific virtual machine, such as the Ethereum virtual machine (EVM). The smart contract could be triggered by the preset conditions and executed automatically, which excludes the reliance on the trusted third party and the human factor. Thus, the smart contract enables behavior customization for all nodes and public verification for all data, which empowers the blockchain to act as the trust anchor in multi-party cooperative scenarios. Most existing blockchains implement smart contracts and the corresponding Turing-complete programming language, such as the smart contract written by Solidity in Ethereum and the chain code written by Golang in the Hyperledger Fabric [34,35].

In Ethereum, there are two types of transactions: the ones for token transfer and the ones for contract deployment and calling [33,34]. For the former, the field of data is defaulted to empty. For the latter, the field of data is the contract bytecodes (for deployment) or the parameters (for the calling). All transactions follow the same lifecycle from issuance to confirmation as follows: (1) the node issues a new transaction and broadcasts it to the network; (2) the nodes receiving the new transaction check its validity and cache the valid transition into the local pool; (3) the nodes pack the transaction into a new block and compete for the authority block proposal through the consensus process; (4) the nodes receiving the new block validate the consensus result and the validity of all transaction, then append the block passing all the checks into the tip of the local chain.

Currently, the blockchain faces scalability issues, especially in terms of storage scalability. Since all nodes should maintain the ledger, there will be a huge storage burden for the resource-limited nodes with increasing block height. The storage scalability issue of the blockchain could be alleviated by dumping the ledger into the off-chain, such as the interplanetary file system (IPFS). The IPFS is the file system based on the distributed hash table, where the files are indexed by their unique IPFS hash values instead of the URLs (known as content-based addressing) [36]. As proposed in the paper by [37], by integrating the IPFS with the blockchain, the original ledger could be dumped into the IPFS, with only the IPFS hash values maintained on-chain, which reduces the storage burden greatly.

2.3. Zero-Knowledge Proof (ZKP) System

The ZKP system was first proposed by Goldwasser et al. in 1985. It is an initially interactive protocol between the prover (P) and the verifier (V). With the support of the ZKP system, P could convince V that the statement is true by providing the corresponding proof without revealing the secret (also known as witness) [38,39]. Typically, the ZKP system has properties of completeness, soundness, and zero-knowledgeness [40,41]. The completeness guarantees that any correct statement can always pass the checks of the verifier following the protocol. The soundness guarantees that any fraud statement cannot cheat the verifier even though P has enough computing power. The zero-knowledgeness guarantees that V cannot learn any more information except for the correctness of the statement [42]. The typical implementation of the ZKP system is the zero-knowledge succinct non-interactive argument of knowledge (zk-SNARK), which is the integration of the multiple cryptographic primitives, such as polynomial commitment and the knowledge of exponent assumption (KEA). Compared with the early implementations of the ZKP system, zk-SNARK has the advantages of succinctness, non-interactiveness, and efficiency, which are realized by relying on the linear probabilistic checkable proof (LPCP) and the common reference string (CRS) model [43,44].

Any NP relation R could be represented as the arithmetic circuit C with the depth $|C|$, which is formed by multiple addition and multiplication gates. The core idea of the zk-SNARK is transforming the circuit satisfaction to the polynomials divisibility by reducing the constraints between gates into a set of polynomials known as the quadratic arithmetic program (QAP). The proof size and the time for proof generation are determined by the circuit depth $|C|$ [42,44]. The construction of the zk-SNARK is depicted in Figure 2, which includes four steps as follows:

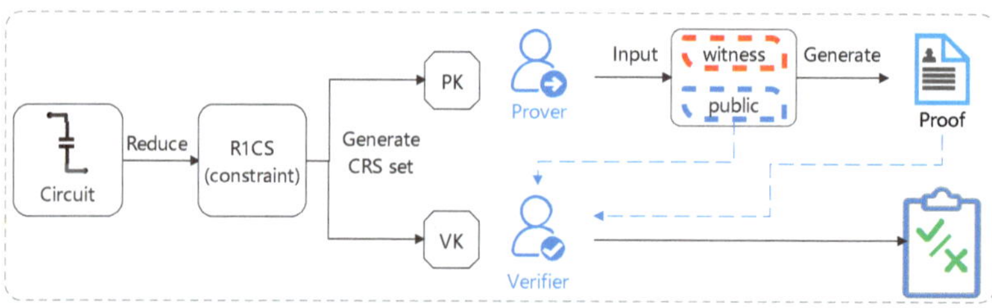

Figure 2. The construction process of the zk-SNARK.

1. Initialization. Converting the NP relation R into the arithmetic circuit C and generating the CRS. Each gate in C is formed by two input wires and one output wire. Different gates could share the same input wires, and the upper gates' output wires can be the lower gates' input wires. The CRS is a set of elements from the finite cyclic groups in the elliptic curves, which are the homophonic hidings of a random secret s. The security of the zk-SNARK is determined by s, which should be discarded completely as "toxic waste" after the CRS generation. The CRS could be accessed by P and V, where the elements for provers and verifiers are known as prover keys (PK) and verifier keys (VK), respectively . It should be noted that the elliptic curves should be pairing-friendly because the proof is validated based on the pairing operations [45].
2. Rank-1 constraints system (R1CS). The R1CS acts as the compiler to help construct the QAP from the arithmetic circuit C. Essentially, the R1CS is a set of n polynomials to represent the relationship among the inputs, outputs, and intermediated values. The construction of the R1CS follows the rule of "one operation and one line". Typically, one operation includes one multiplication gate and serval addition gates (in Groth 16 [46,47]). When the circuit satisfaction holds, the evaluation of the polynomial in each line is zero.
3. Quadratic artchimetic program (QAP). The QAP includes three polynomials, L(x), R(x), and Q(x), to represent the left input wires, right input wires, and output wires of all multiplication gates. Three polynomials satisfy that $Q(x) \bullet \mathbf{v} = (L(x) \bullet \mathbf{v}) \times (R(x) \bullet \mathbf{v})$ and $\mathbf{v} = [1, \mathbf{x}^T, \mathbf{w}^T]^T$, where $\mathbf{x}$ and $\mathbf{w}$ are the public input vector and the witness, respectively. L(x), R(x), and Q(x) represent the constraint of circuit C, which has no relationship with the $\mathbf{v}$ and could be obtained by P and V. Let $\mathbf{Q}(x) = Q(x) \bullet \mathbf{v}$, $\mathbf{L}(x) = L(x) \bullet \mathbf{v}$, and $\mathbf{R}(x) = R(x) \bullet \mathbf{v}$. When all constraints hold, P could generate a polynomial $P(x) = \mathbf{L}(x) \bullet \mathbf{R}(x) - \mathbf{Q}(x)$, and the evaluations for all lines are zeroes, which means that there is a quotient polynomial $\mathbf{H}(x)$ stratifying

that $\mathbf{H}(x) = \mathbf{P}(x)/\mathbf{Z}(x)$. $\mathbf{Z}(x) = \prod_{i=1}^{n}(x-i)$ is known as the target polynomial and could be obtained by both sides.

4. Proof generation. The QAP converts the circuit stratification into the polynomial disviability, which could be verified efficiently based on the Schwartz–Zippel lemma and the polynomial commitment schemes (such as KZG and IPA [48]). The proof is the homomorphic hiding of the H(s), which is also an element of the finite cyclic group and can be obtained by the linear combination of PK. Correspondingly, the proof could be checked through the pairing operations using VK.

It should be noted that more cryptographic primitives, such as KEA, are introduced to force the provers only to use PK and the same **w** to generate the proof as described in [44]. After construction, the main functions of the zk-SNARK could be described as follows:

1. $Setup(C, 1^{\lambda}) \rightarrow pp$, setting the public parameters pp. λ is the security parameter. $pp = (p, e, G_1, G_2, G_T, g_1, g_2, g_T)$, where G_1, G_2, and G_T are the finite cyclic groups with the generators of g_1, g_2, and g_T in the elliptic curves EC_1, EC_2, and EC_T satisfying the bilinear mapping $e : G_1(F_p) \times G_2\left(F_p^k\right) \rightarrow G_T\left(F_p^k\right)$. F_p is the finite field of prime p and F_p^k is the extension field of F_p.
2. $GenKey(pp, C) \rightarrow (PK, VK)$, generating the CRS corresponding to the circuit C, where PK and VK are elements for proof generation (in P side) and proof verification (in V side).
3. $GenProof(PK, \mathbf{x}, \mathbf{w}) \rightarrow \pi$, generating the ZKP proof in P side. The algorithm inputs the PK, the public input $mathbfx$, and witness $mathbfw$.
4. $VerProof(VK, \mathbf{x}, \pi) \rightarrow (0, 1)$, verifying the proof. When $C(\mathbf{x}, \mathbf{w}) = 0$ holds, the algorithm outputs 1, meaning the statement is correct. Otherwise, the algorithm outputs 0, meaning the statement is a fraud.

zk-SNARK is mainly used in the proving statement on private data, anonymous authorization, anonymous payments, outsourcing calculations, etc.

3. Design of the TrsutDFL Framework

This section describes the design of the TrustDFL framework in detail. The overview of the TrustDFL framework is given in Section 3.1, where the components and node roles are introduced. The workflow of the TrustDFL framework is given in Section 3.2. The proof generation for model training and aggregation are introduced in Sections 3.3 and 3.4, respectively. For clarity, symbols used in the paper and their meanings are listed in Table 1.

Table 1. Main parameters and meaning of TrustDFL.

Symbol	Meaning	Symbol	Meaning
TX_U	transaction about local training	TX_A	transaction about model aggregation
PK_T	prover key of local training	PK_A	prover key of model aggregation
VK_T	verifier key of local training	VK_A	verifier key of model aggregation
π_T	ZKP proof of local training	π_A	ZKP proof of model aggregation
M_0	original global model	$\mathrm{H}_0^{\mathrm{M}}$	IPFS hash of the original global model
∇w_i^{jk}	weight gradient of the k-th node in $i-1$-th layer connecting the j-th node in i-th layer	∇b_i^j	bias gradient of the j-th node in i-th layer
t^j	the label	ω	a vector set
$\boldsymbol{\alpha}$	a vector of intermediate variables	$\boldsymbol{\beta}$	a vector of intermediate variables
$\mathbf{o}$	a vector of output variables	$\mathbf{v_1}$	a vector to constrain outputs
$\mathbf{v_2}$	a vector to constrain left inputs	$\mathbf{v_3}$	a vector to constrain right inputs
M_{r-1}	global model of r-th round to be trained	Q_r^k	the k-th updated local model of r-th round model training

3.1. System Scenario

The TrustDFL is a ZKP-based decentralized federated learning framework assisted by the blockchain and smart contracts, where the ZKP provides zero-knowledge proofs for the validity of model training and aggregation with acceptable computational and communication overhead. The blockchain acts as the trust anchor of the multi-party cooperation scenario by recording all data from the participants distributedly in the form of transactions, and the smart contracts are applied for ZKP proof verification. As shown in Figure 3, the TrustDFL framework is an overlay network composed of the blockchain, DFL, and IPFS. The DFL is based on the blockchain, which means the blockchain is formed by the DFL's participants (trainers and aggregators) instead of the third-party infrastructure. The data exchanges in the DFL are in the form of transactions and are involved in the consensus process. Moreover, as an off-chain storage protocol, the IPFS is introduced to alleviate the storage burden brought by the global and local models. All data exchanged in the DFL could be dumped into the IPFS with the IPFS hashes maintained on-chain as the commitments and indices.

In TrustDFL, the trainers and aggregators are almost the same as in the original DFL. The only difference is that the trainers and aggregators in TrustDFL have to attach the trained or aggregated results with the corresponding ZKP proofs. The TrustDFL workers act as the blockchain miners, responsible for the consensus process and the block proposal. In each round of training, each trainer will initiate a transaction including the proof of this round of local training and the hash value of the training results, which is broadcasted and verified by all workers and recorded in the blockchain. After the number of local training transactions on the blockchain reaches a threshold, the trainer collects relevant transactions and extracts IPFS hashes, then retrieves the updated local models from IPFS and aggregates them using the aggregation algorithm. The aggregator initiates a transaction containing the ZKP proof of the aggregation process and the IPFS hash of the updated global model. The global model contained in the first valid aggregated transaction recorded on the blockchain will be used for the next round of training.

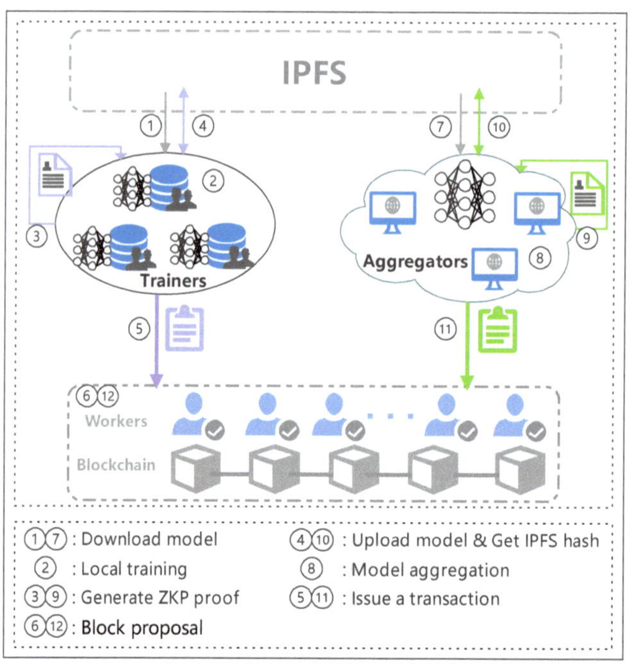

Figure 3. The roles and components of the ZKP-based TrustDFL framework assisted by blockchain.

3.2. System Workflow

Before the training rounds, the system should be initiated by defining the public parameters, including the pairing-friendly elliptic curves (EC_1, EC_2, and EC_T), their generator (g_1, g_2, and g_T), the ZKP implementations (such as Groth 16, Plonk, or STARKs), and the algorithms for the model aggregation (such as FedAvg). For the construction of the ZKP, two CRS sets (CRS_T and CRS_A) should be generated in advance, where $CRS_T = [PK_T, VK_T]$ is for the model training, and $CRS_A = [PK_A, VK_A]$ is for the model aggregation. Typically, the CRS generation and maintenance are delegated to the third party, which could also be realized by federated computation with the support of the smart contracts in TrustDFL. Depending on the selected ZKP implementation, the CRS_T and CRS_A could be the same. For example, since the ZKP implementations, such as Plonk, Sonic, and SuperSonice, could provide the universal and updatable structure reference strings (SRS) composed by the group elements independent of the specific circuits, only one CRS set with enough elements is necessary [49,50]. Moreover, since the neural network consists of multiple layers related to different operations, one CRS should be generated when Groth 16 is used as the ZKP scheme for one operation [46]. However, the CRS generation and maintenance are out of the scope of this paper. For simplicity and without the loss of generality, we assume that all necessary CRS sets are ready before the training rounds begin.

The workflow in TrustDFL could be described as follows:

1. Model publication. In TrustDFL, any node could publish the training task. The original global model M_0 is defined by the architecture of the model to be training (the number of layers, the activation functions, the size of the inputs, etc.), the initialized weights, and some super parameters (epoch, learning rate, etc.). The node first submits all the data to the IPFS and issues a special transaction TX_P with the received IPFS hash $\mathrm{H}_0^{\mathrm{M}}$. The TX_P has the same fields as the normal transactions, but the payload is filled with the IPFS hash. After the TX_P is confirmed in the blockchain, all trainers can learn the training task, extract the IPFS hash, and retrieve the model parameters from the IPFS.
2. Model training. Using the local dataset, all trainers update the model weights based on the retrieved M_0, where the BP process will be repeated according to the defined epoch (step 2). After all iterations are complete, the trainer generates the corresponding ZKP proof π_T based on the PK_T for the validity of the trained results (step 3), where the original weights act as the public inputs, and the local training dataset acts as the witness. In this way, the validity of the local training could be proved without any leakage of the local sensitive information. Then, the trainer submits all data (trained results) into the IPFS and issues a transaction TX_U with the returned IPFS hash and ZKP proof filled in the payload field (steps 4 and 5). Like normal transactions, the TX_U is broadcasted and verified by all workers, which means the contributions of all trainers are recorded in the blockchain.
3. Block proposal. Any workers receiving the TX_U should check its validity. In addition to the signature check, the worker should check the validity of the attached proof π_T by calling the smart contract SC_1 (step 6). The SC_1 implements the verification function $VerProof()$ described in Section 2.3, which should contain the VK_T from the CRS_T and be deployed in advance. Only the TX_U attached with the valid proof could be persisted in the blockchain, which is ready for the model aggregation by the aggregators.
4. Model aggregation. All trained results are accessible for the aggregators when all TX_U are confirmed in the blockchain. The aggregator could collect all the related TX_U by traversing the blockchain, then extract the IPFS hashes and retrieve the updated parameters from the IPFS (step 7). Based on the collected data, the aggregator performs the defined aggregation algorithm and generates the updated global model parameters for the training in the next rounds (step 8). Then, the aggregator generates the corresponding ZKP proof π_A based on the PK_A from the CRS_A, and submits all data into the IPFS (step 9). Finally, the aggregator issues the transaction TX_A with the payload field filled by the returned IPFS hash and ZKP proof (steps 10 and 11).

5. Block proposal. After receiving the TX_As, the worker checks their validity by checking the attached ZKP proof and queues them in an orderly manner. It should be noted that though all valid aggregated results are recorded on-chain, only the results included in the first valid TX_A will be used as the global parameters for the training in the next round.

3.3. ZKP Proof Generation for the Local Training Process

For simplicity and without the loss of generality, we use the MLP as the model to be trained, composed of an input layer, an output layer, and a hidden layer. The activation functions used in hidden and output layers are determined as ReLU and sigmoid, respectively. Without consideration of the activation functions, all operations of each layer are the linear combinations between the inputs and weights, which could be efficiently converted into the arithmetic circuit composed of only addition and multiplication gates. The float point numbers and the nonlinear operations introduced by the activation functions could be expressed as fractions and exponents, which could be mapped to their bit representations and converted to the arithmetic circuits [42]. In this section, we generate the ZKP proof for each layer and finally aggregate them with the support of the Lego SNARK [51]. The ZKP proof aggregation process for each round of local training is shown in Figure 4. Among them, the ZKP proofs at the s-th epoch can be aggregated into c_s, including the proofs of calculation at each layer of forward propagation $(b_{s1}, b_{s2}, \ldots, b_{sp})$ and proofs of calculation between each layer of backpropagation $(b_{s1}, b_{s2}, \ldots, b_{sp})$. ZKP proofs $(c_1, c_2, \ldots, c_s)$ from multiple epochs can also be aggregated into the final ZKP proof π.

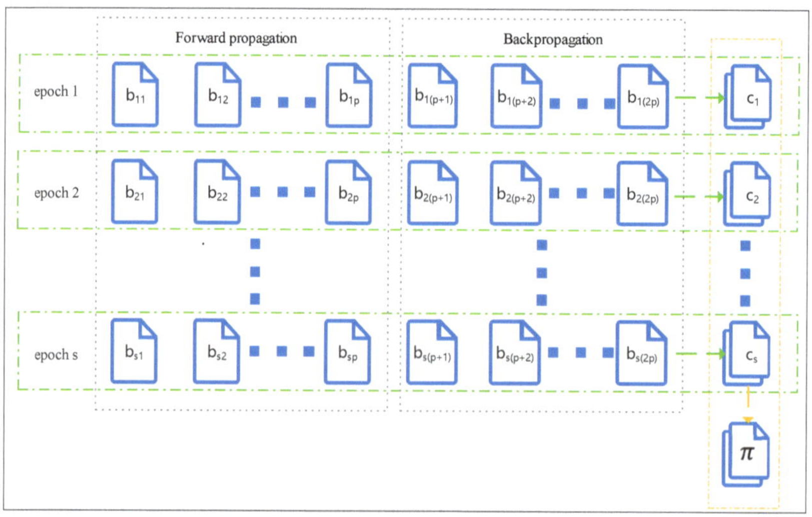

Figure 4. The aggregation process of ZKP proofs for each round of local training.

The FP process can be proved via Groth 16 as described in [19,20,24], which is not the concern of this paper. To prove the validity of the training result, the BP process must be proposed, which is the process of chain rules as shown in Section 2.1. We can get the gradient calculation formula as follows:

$$\begin{cases} \nabla w_i^{jk}(s) = \delta_i^j(s) y_{i-1}^k(s) \\ \nabla b_i^j(s) = \delta_i^j(s) \end{cases} \quad (7)$$

$$\delta_i^j(s) = \begin{cases} f'\left(u_i^j\right) \cdot \sum\limits_{h \in I_{i+1}} \delta_{i+1}^h w_{i+1}^{hj}, \textit{ hidden layer} \\ f'\left(u_i^j\right) \cdot \left(\frac{1}{n}\sum\limits_{j=1}^{n}\left(y_i^j - t^j\right) y_{i-1}^k\right), \textit{ output layer} \end{cases} \tag{8}$$

where n and s are the numbers of the samples and epochs, respectively; w_i^{jk} represents the weight of the k-th node in $i-1$-th layer connecting the j-th node in i-th layer; b_i^j refers to the bias of the j-th node in i-th layer; t^j is the label; and $u_i^j = \sum_{k=1}^{p} w_i^{jk} y_{i-1}^k + b_i^j$ (when $i = 2$, y_{i-1}^k is the k-th node in the input layer).

From Equations (7) and (8), we can know that the calculation of gradient involves linear operations and nonlinear operations. Nonlinear operations are mainly the calculation of activation functions and their derivatives. It can be seen from Groth 16 that nonlinear operations need to be converted into linear operations to generate zero-knowledge proofs. We first describe the processing of nonlinear operations and then express the operation process of generating zero-knowledge proofs from linear operations.

3.3.1. Processing Of Nonlinear Operations

The activation functions mainly considered in this article are ReLU and Sigmoid functions, which are expressed as Equation (1) and Equation (2), respectively. For Equation (1), we use the polynomial proposed in the paper by [52] for approximation. In the interval $I = (-a, a)$, our polynomial activation function is given by

$$f_{ReLU}(x) = x^2 + ax \tag{9}$$

For Equation (2), it is often used in the output layer to complete the classification task. We use the following Taylor series to express.

$$f_{sig}(x) = \frac{1}{2} + \frac{1}{4}x - \frac{1}{48}x^3 + \frac{1}{480}x^5 - \frac{17}{80640}x^7 + O\left(x^8\right) \tag{10}$$

The derivative of the Sigmoid function is shown in Equation (11), and the derivative of the ReLU function is shown in Equation (12).

$$f'_{sig}(x) = f_{sig}(x)(1 - f_{sig}(x)) \tag{11}$$

$$f'_{ReLU}(x) = \begin{cases} 1, \ x > 0 \\ 0, \ x < 0 \end{cases} \tag{12}$$

For Equation (11), there is only one nonlinear function, the Sigmoid function. Therefore, the polynomial approximation of Equation (11) can be realized with the help of Equation (10) and expressed as

$$f'_{sig}(x) = \frac{1}{4} + \frac{1}{4}x - \frac{1}{16}x^2 - \frac{1}{48}x^3 + \frac{1}{480}x^5 + \frac{1}{2304}x^6 - \frac{17}{80640}x^7 + O\left(x^8\right). \tag{13}$$

It can be seen from Equation (12) that the derivative of ReLU is not a smooth function, so we first use Equation (14) to perform a smooth approximation.

$$f'_{ReLU}(x) = \lim_{k \to \infty} \frac{1}{2}(1 + \tanh(kx)) \tag{14}$$

Then, we use Taylor series Equation (15) to perform polynomial approximation of the nonlinear function.

$$f'_{ReLU}(x) = \lim_{k \to \infty} \frac{1}{2}\left(1 + kx - \frac{1}{3}(kx)^3 + \frac{1}{5}(kx)^5 - \frac{1}{7}(kx)^7 + O\left((kx)^8\right)\right. \tag{15}$$

3.3.2. The Operation Process of Linear Calculation to Generate ZKP Proof

When i is the output layer, from Equation (7) and Equation (8) we can see that the gradient calculation formula is $\nabla w_i^{jk} = f'\left(u_i^j\right) \cdot \left(\frac{1}{n}\sum_{j=1}^{n}\left(y_i^j - t^j\right)y_{i-1}^k\right) \cdot y_{i-1}^k$ and establish R1CS for it as follows:

$$\langle \mathbf{v_1}, \omega \rangle = \langle \mathbf{v_2}, \omega \rangle \bullet \langle \mathbf{v_3}, \omega \rangle. \tag{16}$$

Among them, $\langle \cdot , \cdot \rangle$ represents the dot product of vectors. ω is the vector set of the input variable vector, output variable vector of the equation, and intermediate variable vector ($\omega = \left[1, \mathbf{t}, \mathbf{y}, \boldsymbol{\alpha}, \gamma, \boldsymbol{\beta}, \mathbf{f}, \mathbf{o}\right]^T$). $\mathbf{t}$, $\mathbf{y}$, γ, and $\mathbf{f}$ are all vectors of input variables of the equation ($\mathbf{t} = [t_1, t_2, t_3, ..], \mathbf{y} = [y_1, y_2, y_3, \ldots], \gamma = [\gamma_1, \gamma_2, \gamma_3, \ldots], \mathbf{f} = [f_1, f_2, f_3, \ldots]$). $\boldsymbol{\alpha}$ and $\boldsymbol{\beta}$ are vectors of intermediate variables, and $\mathbf{o}$ is a vector of output variables. $\mathbf{v_1}$, $\mathbf{v_2}$, and $\mathbf{v_3}$ are vectors, usually composed of 0 and 1, used to constrain the outputs, left inputs, and right inputs, respectively.

When i is the output layer, the gradient calculation of the weights involved usually contains multiple calculation equations (multiple inputs and outputs), so they can be generalized by Equation (16) and establish R1CS as follows:

$$V_1 \cdot \omega = (V_2^{\cdot} \cdot \omega) \circ (V_3 \cdot \omega) \tag{17}$$

where $\circ$ denotes the Hadamard product and V_1, V_2, and V_3 are matrices. For R1CS as shown in Equation (17), we can use Lagrangian interpolation method to convert it into QAP as follows:

$$(A(x) \cdot \omega) \circ (B(x) \cdot \omega) - C(x) \cdot \omega = 0 \tag{18}$$

where $A(x)$, $B(x)$, and $C(x)$ are matrices expressed as $A(x) = [A_1(x), A_2(x), A_3(x), \ldots]$, $B(x) = [B_1(x), B_2(x), B_3(x), \ldots]$, and $C(x) = [C_1(x),\ C_2(x),\ C_3(x), \ldots]$, respectively.

In the same way, when i is the hidden layer, R1CS and QAP can be constructed. It should be noted that the proofs for each layer are independent from each other, which could furthermore be merged to a single one by Lego SNARK. Recursively, the proofs for all epochs can also be merged by Lego SNARK. Finally, the training result could be prosed by the merged single proof, which could be checked by validators calling the function $VerProof()$.

3.4. ZKP Proof Generation for the Model Aggregation Process

During the r-th round of model training, the global model to be trained is M_{r-1} and multiple trainers engage in local training. The Federated Averaging (FedAvg) algorithm is used to aggregate the local models and obtain an updated global model denoted as M_r. The aggregators obtain K updated local model IPFS hashes from the blockchain and then access IPFS to obtain K updated local models. The k-th updated local model is Q_r^k, and the updated global model representation is shown in Equation (19).

$$M_r = \sum_{k=1}^{K} \frac{n_k}{n} Q_r^k \tag{19}$$

Based on Equation (19), it is clear that the computations involved in the aggregation algorithm are linear. Based on Equation (16), the R1CS could be constructed to obtain the QAP by homomorphism between the vector and polynomial rings.

4. System Analysis

This section provides an analysis of the TrustDFL system. The security objectives of the TrustDFL framework are given in Section 4.1. We prove that TrustDFL achieves all design goals, namely trustworthiness, resistance to attacks, data privacy, scalability, and decentralization in Sections 4.2–4.6, respectively.

4.1. Security Objectives

The specific goals of proposed system design are as follows.

- Trustworthiness: The trustworthiness guarantees the correctness of local model and updated global model, which mainly depends on the correct execution of local training and aggregation process. If the node does not perform the above process as intended, it will fail validation and be discarded. In addition, any node cannot tamper with the model written to the blockchain through effective means.
- Resistance to attacks: TrustDFL can resist to model poisoning attacks and collusion attacks, i.e., compromised nodes cannot upload invalid updates (local model or global model) to hinder the convergence of the model and cannot cooperate to manipulate verification results to attack the model.
- Data privacy: The trainer can prove to others that he has conducted the local training process honestly without leaking local data.
- Scalability: TrustDFL uses IPFS to reduce the pressure of on-chain storage and uses zk-SNARK with very small proof size and on-chain verification complexity.
- Decentralization: TrustDFL can avoid single point of failure and the communication traffic congestion caused by the central server. The overall DFL task is accomplished by nodes without any centralized server and the trusted third party.

4.2. Trustworthiness

TrustDFL uses zk-SNARK to ensure the trustworthiness of the local model and the updated global model. zk-SNARK has perfect completeness and can verify the integrity of the calculation process. This ensures reliable execution of the local training process and model aggregation process. Honest provers extract relevant matrix information that is accurate (with valid witness) and generate proofs π using the proof circuit $C(\mathbf{x}, \mathbf{w})$ and the prover key PK. When verifying the local training process, the relevant matrix information includes the initial weight values, inputs, and updated local models, etc. When verifying the model aggregation process, the relevant matrix information encompasses K local models and the updated global model, etc. Honest verifiers utilize the verifier key VK, proof π, and public input $\mathbf{x}$ to obtain a verification result of 1. For any $(\mathbf{x}, \mathbf{w})$ satisfying $C(\mathbf{x}, \mathbf{w}) = 0^l$, we have:

$$Pr\left(Verproof(VK, \mathbf{x}, \pi) = 1 \middle| \begin{matrix} Setup(1^\lambda, C) \rightarrow (EK, VK) \\ Genproof(PK, \mathbf{x}, \mathbf{w}) \rightarrow \pi \end{matrix} \right) = 1. \tag{20}$$

In addition, our solution chooses to the transaction with the payload field filled by the returned IPFS hash and ZKP proof to establish trust relationships between distributed nodes. Once transaction are verified and written into the blockchain by reaching a consensus, no one can modify or deny the data. Moreover, it is impossible to forge transactions verified by miners, and it is impossible to forge the change record of the entire transaction.

4.3. Resistance to Attacks

The TrustDFL can effectively resist model poisoning attacks. The soundness of zk-SNARK means that the probability of success for any PPT (Probabilistic Polynomial Time) malicious attacker algorithm is negligible. For every probabilistic polynomial-time adversary $\mathcal{A}$, there exists a probabilistic polynomial-time witness extractor $\mathcal{E}$ such that:

$$Pr\left(\begin{matrix} C(\mathbf{x}, \mathbf{w}) \neq 0^l \\ Verproof(VK, \mathbf{x}, \pi) = 1 \end{matrix} \middle| \begin{matrix} Setup(1^\lambda, C) \rightarrow (PK, VK) \\ \mathcal{A}(PK, VK) \rightarrow (\mathbf{x}, \pi) \\ \mathcal{E}(PK, VK) \rightarrow \mathbf{w} \end{matrix} \right) \leq negl(\lambda). \tag{21}$$

The soundness of zk-SNARK ensures that the local model and the updated global model are not forged. If the verification equation of the local proof (aggregation proof) holds, it indicates that the node correctly performed the local training task (the task of model aggregation). Therefore, it can effectively resist model attacks.

Furthermore, our scheme can effectively resist collusion attacks. zk-SNARK is non-interactive, so the proof can be attached to the blockchain transaction, and the smart contract implements on-chain verification. This effectively prevents malicious provers and malicious verifiers from colluding to forge wrong proof and verification results to attack the model. Hence, only when the number of malicious nodes exceeds 50% of the total system nodes, the probability of a successful attack becomes great. Malicious nodes would require 51% of the computational resources to attack the system, which incurs costs far exceeding the benefits.

4.4. Data Privacy

The zero-knowledge property of zk-SNARK means that for any probabilistic polynomial-time adversaries $\mathcal{A}$ and every input $\mathbf{x}$ of a given circuit C, there exists a simulator S such that the following two probabilities are approximately equal:

$$Pr\left(\mathcal{A}(PK, VK, \pi) = 1 \middle| \begin{matrix} Setup(1^{\lambda}, C) \rightarrow (PK, VK) \\ Genproof(EK, \mathbf{x}, \mathbf{w}) \rightarrow \pi \end{matrix} \right), \tag{22}$$

$$Pr\left(\mathcal{A}(\hat{PK}, \hat{VK}, \pi) = 1 \middle| \begin{matrix} S(C) \rightarrow (\hat{PK}, \hat{VK}, trap) \\ S(\hat{PK}, \mathbf{x}, trap) \rightarrow \pi \end{matrix} \right). \tag{23}$$

Therefore, the verifier can complete the verification of the training process without knowing the original data, ensuring the security of local data.

4.5. Scalability

TrustDFL introduces zero-knowledge proof and IPFS. While ensuring the safety and reliability of the model, the model is stored on IPFS and the hash is written into the blockchain, which reduces the storage pressure on the chain. Although the Groth 16 solution needs to generate different CRS for different operations, compared with Plonk, Sonic, etc., it maintains optimal performance in terms of verification workload and proof size. Its verification calculation complexity is low and has nothing to do with the circuit complexity. It only needs to verify a paring product equation. There are only three pairing calculations in this equation. The proof size is small and basically constant, and only contains three group elements. Our solution, based on Groth 16's zk-SNARK, performs a computationally intensive and complex circuit and proof generation process off-chain, while performing a simple and fast proof verification process on-chain. Compared with the solution that uses the blockchain consensus mechanism to implement model verification, e.g., [17,18], our solution has higher scalability.

4.6. Decentralization

The decentralization is achieved without any central server. In TrustDFL, the aggregation work calculated by the central server is distributed to multiple nodes. All nodes are equal and contribute to the overall functionality, thereby mitigating the risks associated with single points of failure.

5. Implementation and Evaluation

This section realizes the Proof-of-Concept implementation of the proposed TrustDFL architecture, where its performance is estimated in terms of global model accuracy and overheads introduced by the ZKP schemes in Section 5.1 and Section 5.2, respectively. All estimations run in a Linux laptop with Ubuntu 18.04 LTS operation system, RAM of 16 GB, and CPU of Intel(R) Core(TM) i7-8665@2.11 GHz. The implementation overview of the TrustDFL is as follows:

1. The decentralized federated learning architecture is deployed based on PyTorch (Version 1.6.0), where the global model to be trained is determined as MLP, and the classic datasets of MNIST, Fashion-MNIST, CIFAR10, and CIFAR100 are determined as the training sets. The stochastic gradient descent algorithm is used for local model training.

2. The specific private blockchain is implemented following the basic design of Ethereum. Only the core functions, such as consensus algorithm data persistence, are realized for simplicity. The consensus algorithm is determined as PoW, and the difficulty is set to 0×1 for efficiency. The smart contracts are simulated by C++ scripts. All participants of DFL and blockchain are simulated by the multiple processes, and the IPFS is accessed via the web entrance provided by the Protocol Lab.
3. Specifically, the samples in MNIST and Fashioin-MNIST are grayscale images of 28×28, where 60,000 and 10,000 samples are for training and testing, respectively. The samples in CIFAR10 and CIFAR100 are 32×32 colorful images, where 50,000 and 10,000 samples are for training and testing, respectively.
4. The ZKP scheme used in the TrustDFL is determined as Groth 16, which provides a smaller witness size and faster proof generation/validation. The Groth 16 is implemented based on the C++ library (libSNARK).
5. Limited to the experiment environment, the DFL and blockchain system are formed by ten nodes, where two nodes are chosen randomly as the malicious nodes to launch the model poisoning attacks by publishing the randomly generated float-point numbers as the local training results.

5.1. *Accuracy of the Global Model to Be Trained*

The target of the DFL is obtaining a converged global model to provide prediction service for external users. Thus, the accuracy of the global model is the major indicator to measure the model's effectiveness. In this section, the estimation process includes two phases based on the ten participants. In phase one, eight honest nodes execute the stochastic gradient descent based on the local datasets and publish the trained result, where the two malicious nodes only publish randomly generated float-point numbers. The designated aggregators aggregate the received trained results based on the FedAvg algorithm and update the global model for the next round of training. In phase two, all participants should publish the local training results tighter with the corresponding ZKP proofs, and all participants act as the aggregators by validating the correctness of the ZKP proofs via the smart contracts. The two phases will run based on the datasets of MINST, Fashion-MNIST, CIFAR10, and CIFAR100 in subsequence. For each phase, 10% of samples in the training sets are dispatched randomly to each node. For simplicity, the training set on each node is mutually exclusive. Two phases share the same learning ratios, batch size, and epochs.

For each phase, the training process is repeated ten times. Then, based on the testing sets, all prediction results are records, where the proportion of the results matching the preset labels to the data size is defined as the accuracy of the global model. Based on the different datasets, the accuracy of the global models obtained in two phases is depicted in Figure 5. As shown in Figure 5, the accuracy of the global model from phase two (with the TrustDFL applied) is higher than the phase one (with FedAvg algorithm for model aggregation). Therefore, the proposed TrustDFL can effectively remove the risk of poisoning attacks caused by malicious nodes and guarantee the accuracy and usability of the model. Moreover, the accuracies for datasets of CIFAR10 and CIFAR100 are lower than the accuracies for the datasets of MNIST and Fashion-MNIST. That is because, in this estimation, the MLP is determined as the global model to be trained, which is unsuitable for handling images with multiple channels. For different datasets, it could be seen that the accuracy of the global model from phase one shows volatility. That is because the aggregator cannot distinguish between the random numbers and honest training results effectively. The randomness from the malicious nodes causes accuracy volatility.

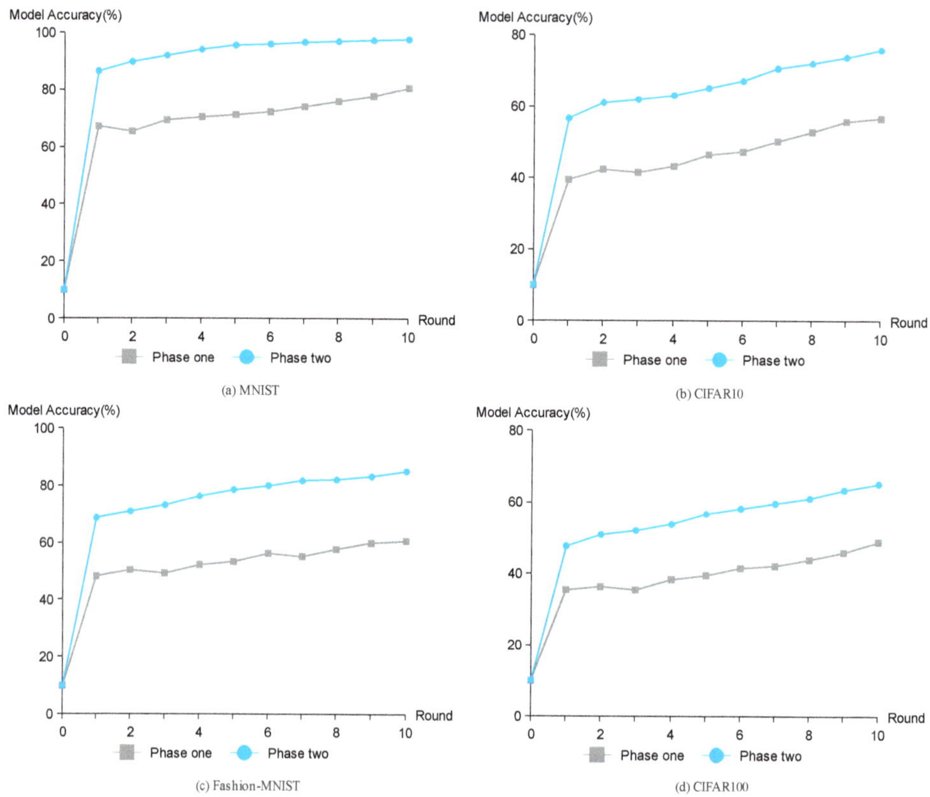

Figure 5. Comparison of the accuracy of models of different schemes on different datasets. (**a**) Comparison of model accuracy of different schemes on the MNIST dataset. (**b**) Comparison of model accuracy of different schemes on the CIFAR10 dataset. (**c**) Comparison of model accuracy of different schemes on Fashion-MNIST dataset. (**d**) Comparison of model accuracy of different schemes on the CIFAR100 dataset.

5.2. Overhead

In this section, we evaluate the storage and time consumption caused by building a zero-knowledge proof; the storage consumption is mainly the proof key size, the verification key size and the proof size; the time consumption is mainly setup time, proving time and verifying time. We build ZKP proofs for different epochs of local training based on the MLP model and adopt the Groth16 scheme in the libSNARK. At the same time, we discussed the experimental results.

As can be seen from Figure 6, the largest storage occupied by ZKP for different numbers of epoch in local training is the prover key, and it is constant, about 150.6 MB. The second one that occupies a larger amount of memory is the verifier key. As shown in Figure 6, it is also constant in different numbers of epoch, about 18.7 MB. The proof size is the smallest and much smaller than prover key and verifier key. The average proof size per epoch is 1.2 KB. For different numbers of epoch, the prover key size and verifier key size are constant. This is because the task publisher will stipulate the architecture of the model to be trained (number of layers, activation function, input size, etc.), and the specified calculation operations performed in each epoch will not change, so the circuit (multiplication gate, adding gates and wires) of each epoch is based on calculations generated will not change. According to Lego SNARK, when generating ZKP proofs for multiple epochs of training, ZKP proofs for each training epoch are constructed and stacked. Therefore, local training for different epochs only requires the same prover key and verifier key. The number of ZKP proofs will increase linearly with the linear increase in the number of epochs, so the size of the proof will also increase linearly.

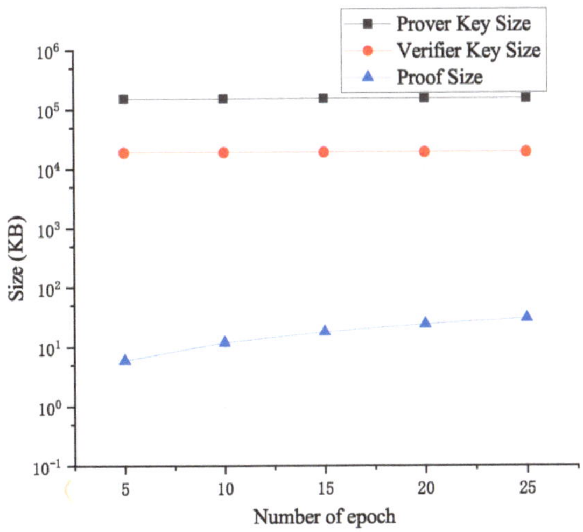

Figure 6. Storage overhead of ZKP for local training with different epoch numbers.

As can be seen from Figure 7, for the time taken to build ZKP for local training with different numbers of epochs, the setup time is the longest and is constant, about 236.7 s. Since the model structure trained in each epoch will not change, and the circuit structure built will not change, the setup time will not change either. In addition, as long as the structure of the trained model is constant, the setup time only needs to be once. The time to build and verify the proof grows with the number of epochs of local training. This is because as the number of epochs increases, the calculations required to prove increase, the number of proofs constructed increases, and the proofs that need to be verified also increase. It can also be seen from Figure 7 that it takes a long time to build the proof, but each node is synchronized off-chain. The time to verify the proof is very short, with an average of 0.17 s per epoch. Therefore, it is appropriate to choose on-chain verification proof, and the performance consumption brought to the blockchain is also considerable.

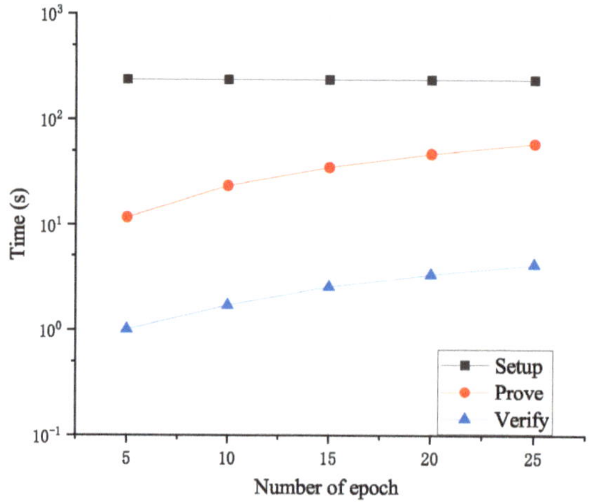

Figure 7. The time consumption of ZKP for local training with different epoch numbers.

5.3. Scheme Comparison

We conducted a functional comparative analysis between the proposed TrustDFL and BFLC, VFChain, PTDFL, and PZKP-FL schemes, as shown in Table 2. BFLC and VFChain use a combination of blockchain consensus mechanism and cross-validation to verify local training results, but this method cannot be applied to all data sets and cannot be used to verify the calculation correctness of model aggregation. When the model and data set are large, the validation time will be long. The verification time of ZKP is independent of the data set and model size. Both PTDFL and PZKP-FL use ZKP to verify the local training results, but PTDFL does not take into account the verification of model aggregation. PTDFL did not choose to store the ZKP proof in the blockchain and could not prevent malicious nodes from colluding to tamper with the verification results. Although PZKP-FL takes the above issues into consideration, its framework is CFL and does not solve the single point of failure problem. TrustDFL not only takes into account the above issues, but also considers the huge storage overhead problem caused by storing the model on the blockchain. We choose to store the IPFS hash on the blockchain and store the model in IPFS, which effectively alleviates the storage pressure on the blockchain.

Table 2. The functionality comparison with the existing schemes.

Schemes	BFLC [17]	VFChain [18]	PTDFL [22]	PZKP-FL [23]	TrustDFL
FL type	DFL	DFL	DFL	CFL	DFL
Verification method	k-fold cross validation	cross validation	ZKP	ZKP	ZKP
Local training verification	✓	✓	✓	✓	✓
Model aggregation verification	-	-	-	✓	✓
Blockchain	✓	✓	-	✓	✓
Storage scalability	-	-	-	-	✓

6. Discussion

Since this study has certain limitation, we will focus on the following three aspects in the future: (1) long proof generation time. This can be solved by improving the proof generation algorithm, specifically by considering converting QAP-based proof generation into QVP (Quadratic vector program) or QMP (Quadratic matrix program) [21]. In addition, from a hardware perspective, the proof generation time can be reduced by using hardware acceleration MSM (Matrix scalar Multiplucation)/NTT (Number Theory Transformation) components [53]. (2) Circuit-specific trusted setup. This can be solved by considering universal setup process, such as the SRS in Plonk and Sonic, or by splitting the entire computation into small and reusable steps/templates as Circom does [54]. (3) Extra communication overhead. This can be improved by more efficient proof aggregation algorithm, such as recursive ZKP, and develop more efficient elliptic curves such as bls12_377 [55].

7. Conclusions

In this paper, TrustDFL is proposed to solve the single point of failure and model attack problems in FL. TrustDFL is a trustworthy and verifiable decentralized federated learning framework based on blockchain and ZKP. It implements computational verification of the local training process and model aggregation process to resist model poisoning attacks and ensure the effectiveness of the global model. The ZKP scheme is used to prove the effectiveness of the local training without leaking sensitive information, and the blockchain acts as the trust anchor in the multi-party cooperation scenarios to record all activities for punishment and rewards. With the support of smart contracts, the verification for local training and model aggregation could be executed automatically. Moreover, the IPFS is introduced to alleviate the storage burden introduced by the models for the blockchain, where only the IPFS hashes of the models are persisted on the blockchain. According to the theoretical analysis and PoC implementation, the proposed TrustDFL framework provides enhanced security and privacy protection with limited storage overhead. Compared with the conventional FL framework using the FedAvg algorithm, the proposed TrustDFL framework distributes the workloads among all participants, providing higher efficiency for model aggregation and avoiding the single point of failure. Moreover, the TrustDFL can combine with more privacy-protected schemes, such as differential privacy secret sharing schemes, and could be applied for more complicated models, such as CNN and transformers.

Author Contributions: Conceptualization, J.Y. and W.Z.; methodology, W.Z. and Z.G. (Zhaohui Guo); software, J.Y. and W.Z.; validation, J.Y., W.Z. and Z.G. (Zhaohui Guo); formal analysis, J.Y., W.Z. and Z.G. (Zhen Gao); investigation, J.Y., W.Z. and Z.G. (Zhen Gao); writing—original draft preparation, W.Z.; writing—review and editing, Z.G. (Zhaohui Guo), W.Z. and Z.G. (Zhen Gao); project administration, J.Y. All authors have read and agreed to the published version of the manuscript.

Funding: This research was supported by the College Government Procurement Branch of Education Accounting Society of China, grant number EASCCGPB2022MS24.

Data Availability Statement: The data presented in this study are available on request from the corresponding author. The data are not publicly available due to the funding restrictions.

Conflicts of Interest: The authors declare no conflicts of interest.

Abbreviations

The following abbreviations are used in this manuscript:

ZKP	Zero-knowledge Proof
IPFS	Interplanetary file system
Dapp	Develop decentralization applications
zk-SNARK	Zero-knowledge succinct non-interactive argument of knowledge
KEA	Knowledge of exponent assumption
R1CS	Rank-1 constraints system
QAP	Quadratic artchimetic program
CRS	Common reference string
LPCP	Linear probabilistic checkable proof
MLP	Multi-layer perceptron
CNN	Convolutional neural networks
FedAvg	FederatedAveraging
ML	Machine learning
FL	Federated learning
DNN	Deep neural network
MLaaS	Machine learning as a service
CFL	Centralized FL
DFL	Decentralized FL

References

1. Jordan, M.I.; Mitchell, T.M. Machine learning: Trends, perspectives, and prospects. *Science* **2015**, *349*, 255–260. [CrossRef] [PubMed]
2. Javed, A.R.; Ahmed, W.; Pandya, S.; Maddikunta, P.K.R.; Alazab, M.; Gadekallu, T.R. A survey of explainable artificial intelligence for smart cities. *Electronics* **2023**, *12*, 1020. [CrossRef]
3. Shinde, P.P.; Shah, S. A review of machine learning and deep learning applications. In Proceedings of the 2018 Fourth International Conference on Computing Communication Control and Automation (ICCUBEA), Pune, India, 16–18 August 2018; pp. 1–6.
4. Devi, I.; Karpagam, G.; Kumar, B.V. A survey of machine learning techniques. *Int. J. Comput. Syst. Eng.* **2017**, *3*, 203–212. [CrossRef]
5. Ribeiro, M.; Grolinger, K.; Capretz, M.A. Mlaas: Machine learning as a service. In Proceedings of the 2015 IEEE 14th International Conference on Machine Learning and Applications (ICMLA), Miami, FL, USA, 9–11 December 2015; pp. 896–990.
6. AbdulRahman, S.; Tout, H.; Ould-Slimane, H.; Mourad, A.; Talhi, C.; Guizani, M. A survey on federated learning: The journey from centralized to distributed on-site learning and beyond. *IEEE Internet Things J.* **2020**, *8*, 5476–5497. [CrossRef]
7. Zhang, C.; Xie, Y.; Bai, H.; Yu, B.; Li, W.; Gao, Y. A survey on federated learning. *Knowl. Based Syst.* **2021**, *216*, 106775. [CrossRef]
8. Yang, Q.; Liu, Y.; Chen, T.; Tong, Y. Federated machine learning: Concept and applications. *ACM Trans. Intell. Syst. Technol. TIST* **2019**, *10*, 1–19. [CrossRef]
9. McMahan, B.; Moore, E.; Ramage, D.; Hampson, S.; y Arcas, B.A. Communication-efficient learning of deep networks from decentralized data. In Proceedings of the Artificial Intelligence and Statistics, PMLR, Ft. Lauderdale, FL, USA, 20–22 April 2017; pp. 1273–1282.
10. Mothukuri, V.; Parizi, R.M.; Pouriyeh, S.; Huang, Y.; Dehghantanha, A.; Srivastava, G. A survey on security and privacy of federated learning. *Future Gener. Comput. Syst.* **2021**, *115*, 619–640. [CrossRef]
11. Fu, X.; Peng, R.; Yuan, W.; Ding, T.; Zhang, Z.; Yu, P.; Kadoch, M. Federated learning-based resource management with blockchain trust assurance in smart IoT. *Electronics* **2023**, *12*, 1034. [CrossRef]
12. Nguyen, D.C.; Ding, M.; Pathirana, P.N.; Seneviratne, A.; Li, J.; Poor, H.V. Federated learning for internet of things: A comprehensive survey. *IEEE Commun. Surv. Tutorials* **2021**, *23*, 1622–1658. [CrossRef]
13. Blanchard, P.; El Mhamdi, E.M.; Guerraoui, R.; Stainer, J. Machine learning with adversaries: Byzantine tolerant gradient descent. *Adv. Neural Inf. Process. Syst.* **2017**, *30*, 104.
14. Yin, D.; Chen, Y.; Kannan, R.; Bartlett, P. Byzantine-robust distributed learning: Towards optimal statistical rates. In Proceedings of the International Conference on Machine Learning. PMLR, Stockholm Sweden, 10–15 July 2018; pp. 5650–5659.
15. Witt, L.; Heyer, M.; Toyoda, K.; Samek, W.; Li, D. Decentral and incentivized federated learning frameworks: A systematic literature review. *IEEE IoT J.* **2022**, *10*, 3642–3663. [CrossRef]
16. Wani, S.; Imthiyas, M.; Almohamedh, H.; Alhamed, K.M.; Almotairi, S.; Gulzar, Y. Distributed denial of service (DDoS) mitigation using blockchain—A comprehensive insight. *Symmetry* **2021**, *13*, 227. [CrossRef]
17. Li, Y.; Chen, C.; Liu, N.; Huang, H.; Zheng, Z.; Yan, Q. A blockchain-based decentralized federated learning framework with committee consensus. *IEEE Netw.* **2020**, *35*, 234–241. [CrossRef]
18. Peng, Z.; Xu, J.; Chu, X.; Gao, S.; Yao, Y.; Gu, R.; Tang, Y. Vfchain: Enabling verifiable and auditable federated learning via blockchain systems. *IEEE Trans. Netw. Sci. Eng.* **2021**, *9*, 173–186. [CrossRef]
19. Fan, Y.; Xu, B.; Zhang, L.; Song, J.; Zomaya, A.; Li, K.C. Validating the integrity of convolutional neural network predictions based on zero-knowledge proof. *Inf. Sci.* **2023**, *625*, 125–140. [CrossRef]

20. Zhao, L.; Wang, Q.; Wang, C.; Li, Q.; Shen, C.; Feng, B. Veriml: Enabling integrity assurances and fair payments for machine learning as a service. *IEEE Trans. Parallel Distrib. Syst.* **2021**, *32*, 2524–2540. [CrossRef]
21. Lee, S.; Ko, H.; Kim, J.; Oh, H. vcnn: Verifiable convolutional neural network based on zk-snarks. *Cryptol. ePrint* **2020**, 584, 1–16.
22. Wang, L.; Zhao, X.; Lu, Z.; Wang, L.; Zhang, S. Enhancing privacy preservation and trustworthiness for decentralized federated learning. *Inf. Sci.* **2023**, *628*, 449–468. [CrossRef]
23. Xing, Z.; Zhang, Z.; Li, M.; Liu, J.; Zhu, L.; Russello, G.; Asghar, M.R. Zero-Knowledge Proof-based Practical Federated Learning on Blockchain. *arXiv* **2023**, arXiv:2304.05590.
24. Heiss, J.; Grünewald, E.; Tai, S.; Haimerl, N.; Schulte, S. Advancing blockchain-based federated learning through verifiable off-chain computations. In Proceedings of the 2022 IEEE International Conference on Blockchain (Blockchain), Espoo, Finland, 22–25 August 2022; pp. 194–201.
25. Reegu, F.A.; Abas, H.; Gulzar, Y.; Xin, Q.; Alwan, A.A.; Jabbari, A.; Sonkamble, R.G.; Dziyauddin, R.A. Blockchain-Based Framework for Interoperable Electronic Health Records for an Improved Healthcare System. *Sustainability* **2023**, *15*, 6337. [CrossRef]
26. Bounds.; Lloyd.; Mathew.; Waddell. A multilayer perceptron network for the diagnosis of low back pain. In Proceedings of the IEEE 1988 International Conference on Neural Networks, San Diego, CA, USA, 24–27 July 1988; pp. 481–489.
27. Taud, H.; Mas, J. Multilayer perceptron (MLP). In *Geomatic Approaches for Modeling Land Change Scenarios*; Springer: Berlin, Germany, 2018; pp. 451–455.
28. Savalia, S.; Emamian, V. Cardiac arrhythmia classification by multi-layer perceptron and convolution neural networks. *Bioengineering* **2018**, *5*, 35. [CrossRef] [PubMed]
29. Underwood, S. Blockchain beyond bitcoin. *Commun. ACM* **2016**, *59*, 15–17. [CrossRef]
30. Zheng, Z.; Xie, S.; Dai, H.; Chen, X.; Wang, H. An overview of blockchain technology: Architecture, consensus, and future trends. In Proceedings of the 2017 IEEE International Congress on Big Data (BigData Congress), Boston, MA, USA, 11–14 December 2017; pp. 557–564.
31. Yli-Huumo, J.; Ko, D.; Choi, S.; Park, S.; Smolander, K. Where is current research on blockchain technology?—A systematic review. *PLoS ONE* **2016**, *11*, e0163477. [CrossRef] [PubMed]
32. Zou, W.; Lo, D.; Kochhar, P.S.; Le, X.B.D.; Xia, X.; Feng, Y.; Chen, Z.; Xu, B. Smart contract development: Challenges and opportunities. *IEEE Trans. Softw. Eng.* **2019**, *47*, 2084–2106. [CrossRef]
33. Peters, G.W.; Panayi, E. *Understanding Modern Banking Ledgers through Blockchain Technologies: Future of Transaction Processing and Smart Contracts on the Internet of Money*; Springer: Berlin, Germany, 2016.
34. Dannen, C. *Introducing Ethereum and Solidity*; Springer: Berlin, Germany, 2017; Volume 1.
35. Androulaki, E.; Barger, A.; Bortnikov, V.; Cachin, C.; Christidis, K.; De Caro, A.; Enyeart, D.; Ferris, C.; Laventman, G.; Manevich, Y.; et al. Hyperledger fabric: A distributed operating system for permissioned blockchains. In Proceedings of the Thirteenth EuroSys Conference, Porto, Portugal, 23–26 April 2018; pp. 1–15.
36. Benet, J. Ipfs-content addressed, versioned, p2p file system. *arXiv* **2014**, arXiv:1407.3561.
37. Zheng, Q.; Li, Y.; Chen, P.; Dong, X. An innovative IPFS-based storage model for blockchain. In Proceedings of the 2018 IEEE/WIC/ACM international conference on web intelligence (WI), Santiago, Chile, 3–6 December 2018; pp. 704–708.
38. Fiege, U.; Fiat, A.; Shamir, A. Zero knowledge proofs of identity. In Proceedings of the Nineteenth Annual ACM Symposium on Theory of Computing, New York, NY, USA, 1987 ; pp. 210–217.
39. Kilian, J. A note on efficient zero-knowledge proofs and arguments. In Proceedings of the Twenty-Fourth Annual ACM Symposium on Theory of Computing, Victoria, BC, Canada, 4–6 May 1992; pp. 723–732.
40. Cramer, R.; Damgård, I.; MacKenzie, P. Efficient zero-knowledge proofs of knowledge without intractability assumptions. In Proceedings of the International Workshop on Public Key Cryptography, Melbourne, Australia, 18–20 January 2000; pp. 354–372.
41. Ben-Sasson, E.; Chiesa, A.; Tromer, E.; Virza, M. Succinct {Non-Interactive} zero knowledge for a von neumann architecture. In Proceedings of the 23rd USENIX Security Symposium (USENIX Security 14), San Diego, CA, USA 20–22 August 2014; pp. 781–796.
42. Parno, B.; Howell, J.; Gentry, C.; Raykova, M. Pinocchio: Nearly practical verifiable computation. *Commun. ACM* **2016**, *59*, 103–112. [CrossRef]
43. Abe, M.; Fehr, S. Perfect NIZK with adaptive soundness. In Proceedings of the Theory of Cryptography Conference, Amsterdam, The Netherlands, 21–24 February 2007; pp. 118–136.
44. Gennaro, R.; Gentry, C.; Parno, B.; Raykova, M. Quadratic span programs and succinct NIZKs without PCPs. In Proceedings of the Advances in Cryptology–EUROCRYPT 2013: 32nd Annual International Conference on the Theory and Applications of Cryptographic Techniques, Athens, Greece, 26–30 May 2013; pp. 626–645.
45. Bowe, S.; Gabizon, A.; Miers, I. Scalable multi-party computation for zk-SNARK parameters in the random beacon model. *Cryptol. ePrint* **2017**, *2017*, 1–24.
46. Groth, J.; Maller, M. Snarky signatures: Minimal signatures of knowledge from simulation-extractable SNARKs. In Proceedings of the Annual International Cryptology Conference, Santa Barbara, CA, USA, 20–24 August 2017; pp. 581–612.
47. Groth, J. On the size of pairing-based non-interactive arguments. In Proceedings of the Advances in Cryptology–EUROCRYPT 2016: 35th Annual International Conference on the Theory and Applications of Cryptographic Techniques, Vienna, Austria, 8–12 May 2016; pp. 305–326.

48. Buterin, V. Quadratic Arithmetic Programs: From Zero to Hero. 2016. Available online: https://medium.com/VitalikButerin/quadratic-arithmetic-programs-from-zero-to-hero-f6d558cea649 (accessed on 17 March 2023).
49. Maller, M.; Bowe, S.; Kohlweiss, M.; Meiklejohn, S. Sonic: Zero-knowledge SNARKs from linear-size universal and updatable structured reference strings. In Proceedings of the 2019 ACM SIGSAC Conference on Computer and Communications Security, London, UK, 11–15 November 2019; pp. 2111–2128.
50. Gabizon, A.; Williamson, Z.J.; Ciobotaru, O. Plonk: Permutations over lagrange-bases for oecumenical noninteractive arguments of knowledge. *Cryptol. ePrint* **2019**, *953*, 1–34.
51. Campanelli, M.; Fiore, D.; Querol, A. Legosnark: Modular design and composition of succinct zero-knowledge proofs. In Proceedings of the 2019 ACM SIGSAC Conference on Computer and Communications Security, London, UK, 11–15 November 2019; pp. 2075–2092.
52. Ali, R.E.; So, J.; Avestimehr, A.S. On polynomial approximations for privacy-preserving and verifiable relu networks. *arXiv* **2020**, arXiv:2011.05530.
53. Zhang, Y.; Wang, S.; Zhang, X.; Dong, J.; Mao, X.; Long, F.; Wang, C.; Zhou, D.; Gao, M.; Sun, G. Pipezk: Accelerating zero-knowledge proof with a pipelined architecture. In Proceedings of the 2021 ACM/IEEE 48th Annual International Symposium on Computer Architecture (ISCA), Valencia, Spain, 14–18 June 2021; pp. 416–428.
54. Bellés-Muñoz, M.; Baylina, J.; Daza, V.; Muñoz-Tapia, J.L. New privacy practices for blockchain software. *IEEE Softw.* **2021**, *39*, 43–49. [CrossRef]
55. El Housni, Y.; Guillevic, A. Optimized and secure pairing-friendly elliptic curves suitable for one layer proof composition. In Proceedings of the International Conference on Cryptology and Network Security, Virtual, 14–16 December 2020; pp. 259–279.

electronics

Article

Dual-Safety Knowledge Graph Completion for Process Industry

Lingzhi Wang [1,2,†], Haotian Li [1,2,†], Wei Wang [1,2], Guodong Xin [1,2] and Yuliang Wei [1,2,*]

1 Research Institute of Cyberspace Security, Harbin Institute of Technology, Weihai 264200, China; 23s130410@stu.hit.edu.cn (L.W.); 22b903069@stu.hit.edu.cn (H.L.); wwhit@hit.edu.cn (W.W.); gdxin@hit.edu.cn (G.X.)
2 School of Computer Science and Technology, Harbin Institute of Technology, Weihai 264200, China
* Correspondence: wei.yl@hit.edu.cn
† These authors contributed equally to this work.

Abstract: With the rise of Industry 4.0, control systems have taken on increasing importance in industrial processes, and ensuring their security has become a pressing issue. While recent research has focused on cybersecurity threats, the security risks inherent to industrial processes themselves have been overlooked. Additionally, existing tools cannot simultaneously analyze both cyber vulnerabilities and processes anomaly in industrial settings. This paper aims to address these issues through two main contributions. First, we develop a knowledge graph to integrate information on security risks across cybersecurity and industrial processes, providing a foundation for comprehensively assessing threats. Second, we apply the link prediction task to the knowledge graph, introducing an embedding-based approach to unveil previously undiscovered knowledge. Our experiments demonstrate that the proposed method exhibits comparable performance on link prediction and is capable of mining valuable and diverse potential risks in industrial processes.

Keywords: industrial process security; cybersecurity; knowledge graph; link prediction

1. Introduction

Control systems play a pivotal role in the process industry by ensuring the smooth operation of complex production processes. These systems typically consist of several key components, each serving a specific purpose. Monitors, for instance, provide real-time data visualization and process status information, giving operators critical insights into the ongoing processes. Another crucial component is the controller, which oversees and regulates individual devices and process variables, ensuring that they function optimally. Additionally, network switchers are instrumental in facilitating seamless communication and data exchange among the various components of the control system. The significance of control systems extends across various application domains, including the chemical industry [1], oil and gas production [2], and the power industry [3]. Consequently, the security of control systems has gained increased and more extensive attention.

Existing work on security insurance of control systems has focused on protecting them from cyberattacks. Jarmo Alnen et al. [4] introduced a cybersecurity risk analysis method that relies on the proposed ontology. This method provides a structured data repository aimed at bolstering the assessment process and establishing traceability between the generated artifacts. Angelo Corallo et al. [5] delved into the cybersecurity challenges encountered in the realm of the Internet of Everything during the Industry 4.0 era. Certainly, the increased efforts in terms of cybersecurity to protect control systems are indeed noteworthy; however, the broad issue of security throughout the entire production process often remains inadequately addressed. For instance, in 2018, Tesla faced significant challenges with its highly automated production line for the Model 3. The company employed an extensive fleet of robots equipped with sensors to aid in component assembly. Unfortunately, a few robots experienced sensor failures or misalignment issues, leading to production delays. These disruptions stemming from sensor-related issues caused production bottlenecks, resulting in a slower output of Model 3 vehicles—a pivotal product for Tesla. Consequently, these setbacks led to delivery delays for customers and carried financial implications for the company. The reality is that sensor failures in industrial processes can result in substantial loss of both human life and property.

Therefore, a more comprehensive strategy should not only focus on strengthening control system defenses, but also prioritize the implementation of robust security measures throughout the entire process industry.

In order to aptly represent the comprehensive protection of critical infrastructure, we have coined the term 'dual-security' to encompass both the security of cyberspace and the industrial process. Since a substantial portion of prior research has concentrated on cybersecurity, the challenge of simultaneously addressing dual-security risks within the process industry has remained an uncharted territory.

Recently, knowledge graphs (KGs) have emerged as a fundamental technique for incorporating knowledge from diverse fields. It provides a structured representation of knowledge, i.e. factual triples in the form of (h, r, t), denoting the head and tail entities and the relationship between them [6]. For instance, in a KG, we may encounter entities like "Albert Einstein" and "Theory of Relativity" connected by the relationship "developed". These KGs are constructed by combining information from various sources, including structured databases, unstructured text, and linked open data [7]. The structured nature of KGs ensures semantic consistency, enabling machines to not only retrieve information, but also understand the context and perform sophisticated reasoning tasks. KGs have found applications in a wide range of domains, from natural language processing [8–10] and information retrieval to recommendation systems [11,12] and medical healthcare [13–15].

Link prediction is a fundamental subfield within knowledge representation and reasoning, dedicated to the task of inferring missing facts and relationships within the graph. Addressing the link prediction challenges has led to the development of various distinct approaches, among which the primary category is embedding-based methods. This kind of method is designed to acquire distributed low-dimension vector representations, commonly referred to as embeddings, for entities and relations within the KG. Most representative approaches within this research paradigm encompass TransE [16], DistMult [17], ConvE [18], TuckER [19], etc.

Link prediction on knowledge graphs serves as an effective and transparent approach for revealing hidden knowledge within knowledge graphs. In this paper, we study the problem of protecting the process industry in terms of dual-security. Specifically, we collect knowledge from both security aspects and build a dual-security knowledge graph. Then, we propose to discover potential security risks through inference on the KG with an embedding-based link prediction method. Empirical experiments substantiate the efficacy of our method, as it achieves good performance on the real-world dataset and mines meaningful dual-security risks. In addition to conducting comprehensive dataset experiments, we also performed an analysis of the risk prediction capability of existing models using experiments conducted on crack hydrogenation industrial processes. Our analysis showed that the employed approach is more accurate in predicting the potential risks of unitary industrial processes compared to other methods.

This paper contributes to two main aspects. Firstly, it innovatively integrates cybersecurity and industrial process security, extracting knowledge from both dimensions, into a knowledge graph. Secondly, it employs a relatively efficient method for link prediction, enabling it to mine undiscovered knowledge from the knowledge graph and effectively mitigate potential risks in industrial processes.

2. Related Work

Cybersecurity knowledge graphs (CSKG) [20] represent a specialized category of knowledge graphs tailored for the cybersecurity domain. They comprise entities and relations extracted from a myriad of attack and defense scenarios within the cybersecurity landscape [21,22]. Furthermore, abstract concepts such as 'attacker', 'attack pattern', and 'vulnerability' can be seamlessly represented as entities, while the paths of attacks and other connections are aptly captured as relations. CSKG adeptly employs knowledge graph construction techniques to extract and effectively integrate pre-existing knowledge from various security data sources, ensuring a comprehensive and unified understanding of the cybersecurity landscape. The link prediction of CSKG has three major applications in cybersecurity, namely the attack prediction [23], threat hunting [24] and intrusion detection [25–27]. However, there is an inadequate focus on process safety.

The application of knowledge graphs in the industry can be partitioned into two distinct phases, namely the construction and deduction periods, as indicated by Li et al. [28]. In the construction phase, the primary objective revolves around the integration of multiple text mining and machine learning tools to process raw data, yielding triples of the form (h, r, t) for the knowledge graph. As a result, Natural Language Processing (NLP) techniques [29–31] and associated toolkits are frequently harnessed to automatically extract entities from various unstructured knowledge resources. Notably, the utilization of knowledge graphs in industry transcends the conventional dissemination of existing knowledge items; instead, it caters to the elevated demands for synthesis and innovation within the industrial domain. Consequently, the deduction phase of industrial knowledge graphs strives to meet these requirements. Knowledge deduction can be further subcategorized into attribution prediction [32] and link prediction [33–35]. More specifically, the deduction process can be effectively reformulated into a series of matrix manipulations [36–38]. This transformation is achieved by vectorizing the entities and relations, aligning them according to their semantic and topological features within the KG. Such an approach not only enhances the computational efficiency, but also leverages the inherent structural properties of the KG for knowledge deduction. Nonetheless, the majority of the domain-specific knowledge graphs primarily focus on industrial products and services, and they often fall short in addressing the crucial aspect of industrial security.

Our work is related to the task of link prediction in dual-security KG; previous efforts in link prediction can be divided into two main aspects, embedding-based prediction and rule-based prediction. Former approaches focused on learning low-dimension representation for entities and relations. Related methods include TransE [16], TransH [39], ComplEx [40], etc., and detect facts by projecting entities and relations into a semantic space and conduct algebraic operations on the space. Specifically, TransE [16] represents the triples into d-dimensional space, $h, r, t \in \mathbb{R}^d$ and generates embeddings following the translational principle $h + r \approx t$. TransH [39] assigns a hyperplane for each relation to satisfy the N-to-N relationship scenarios. ComplEx [40] firstly uses complex vector space to generate embeddings. Through the space, we can capture both symmetric and antisymmetric relations, $h, r, t \in \mathbb{C}^d$, h can be formulated as $h = R(h) + iI(h)$ where $R(h)$ and $I(h)$ are real and imaginary parts of h, respectively. On the other hand, the latter(rule-based) approaches do not encounter these issues, including NeuralLP [41], DRUM [42]. They leverage first-order logic and rule-based approaches to infer missing relationships and facts. NeuralLP [41] combines neural networks with logical rules and utilizes a differentiable framework to learn the rules from data, allowing for the incorporation of prior domain knowledge. The key idea behind NeuralLP is to bridge the gap between symbolic reasoning and neural network learning, making it an attractive approach for rule-based knowledge graph completion. DRUM [42] integrates rules into an end-to-end differentiable neural network. This approach facilitates rule-based reasoning while benefiting from the learning capabilities of neural networks and extends the knowledge graph with rule-instantiated triples, enabling the incorporation of logical rules in the knowledge graph completion process.

3. Methodology

In this section, we will commence by presenting essential background knowledge, encompassing key concepts such as knowledge graphs, link prediction, and the fundamental steps involved in embedding-based methods for link prediction. Subsequently, we will delve into an extensive exposition of the best-performing model, namely PairRE.

3.1. Preliminaries

Knowledge graph $\mathcal{G} = \{\mathcal{E}, \mathcal{P}\}$ is a structured representation of knowledge, where $\mathcal{E}$ and $\mathcal{P}$ denote the set of entities and relations, respectively. It functions as a semantic network that organizes information into triples, denoted as (s, p, o), where $s, o \in \mathcal{E}$ and $p \in \mathcal{P}$. These triples take the form of $s \xrightarrow{p} o$, presenting a graph-like structure.

Link prediction task refers to a key task in knowledge graph completion. It involves predicting missing triples by identifying the most plausible tail entity t^* from the set of entities $\mathcal{E}$ for a given incomplete triple $(h, r, ?)$. During inference, when provided with h and r as inputs, the model computes a ranked list of all entities, with higher scores indicating higher rankings. In essence, link prediction is a supervised learning technique that aims to minimize the dissimilarity between the predicted entity and the ground truth provided in the available data.

Embedding-based methods for link prediction generally consist of three key steps. In the first step, we initialize embeddings for all entities and relations. To achieve this, we establish an index table, wherein the indexes are associated with entity and relation IDs, and the table's entries comprise stochastic vectors that represent the embeddings. The second step involves the formulation of an objective function designed to assess the score of a triple (h, r, t) within a low-dimensional space. This objective function can be realized through various methods, such as TransE, ComplEX, TuckER, and others. The final step is training and inference. During training, the model is trained, often using negative samples to ensure its robustness and accuracy. Negative samples are essentially fictitious triples constructed at random. These fictive triples share the same h and r with the factual triple, but the t is substituted with a different entity. We anticipate that the scores of true triples will be higher than those of the negative samples. Adhering to this principle, the model iteratively optimizes the embedding tables until convergence is achieved. During inference, the process typically entails taking the embeddings of h and r as inputs, followed by the computation of the score for (h, r, t') across all potential entities $t' \in \mathcal{E}$. Higher scores indicate a higher probability of a connection between entities h and t' through the relation r. This approach enables effective predictions and assessments.

3.2. The Link Prediction Model

We now advocate the utilization of PairRE [43] for augmenting the dual-security knowledge graph. This choice is motivated by two primary factors: Firstly, the data in the graph encompass diverse complex relations, including N-to-1, 1-to-N, and N-to-N relationships. Secondly, the data exhibit various relation patterns, including symmetry and antisymmetry. These intricacies pose a unique challenge that other existing methods are ill-equipped to address simultaneously. However, PairRE stands out by employing paired embedding vectors for each relation representation, allowing for adaptable margin adjustments in the loss function. This adaptability is pivotal in effectively accommodating complex relations and addressing the challenges presented by the dual-security knowledge graph.

PairRE adopts a unique approach by acquiring paired representations for each relation. For a given training triple (h, r, t), the model learns vector embeddings for entities and relations within real space. Specifically, it treats relation embedding as a pair of vectors, denoted as (r_h, r_t). These paired vectors are responsible for projecting the head entity h and the tail entity t into the Euclidean space, with the projection operation being an element-wise product between the two vectors. Furthermore, the method calculates the distance between these two embeddings to gauge the plausibility of the triple. The desired outcome is that $h \circ r_h \approx t \circ r_t$ when the triple (h, r, t) holds true. According to the calculated measure, the scoring function is defined as follows:

$$\mathcal{S}_r(h,t) = -||h \circ r_h - t \circ r_t|| \tag{1}$$

where $h, r_h, r_t, t \in \mathbb{R}^d$ with h and t subject to specific constraints, such as $||h||^2 = ||t||^2 = 1$.

To further optimize the model, we employ a contrastive learning measure, specifically adopting the self-adversarial negative sampling loss [44] as the training objective:

$$\mathcal{L} = -log(\sigma(\gamma - \mathcal{S}_r(h,t))) - \sum_{i=1}^{n} p(h'_i, r, t'_i) log(\sigma(\mathcal{S}_r(h'_i, t'_i) - \gamma)) \tag{2}$$

In this equation, γ represents a fixed margin, and σ denotes the sigmoid function. (h'_i, r, t'_i) corresponds to the ith negative triple, while $p(h'_i, r, t'_i)$ signifies the weight assigned to this negative sample. The weight, $p(h'_i, r, t'_i)$, is calculated using the softmax function:

$$p((h'_i, r, t'_i)|(h, r, t)) = \frac{exp(\mathcal{S}_r(h'_i, t'_i))}{\sum_{j=1}^{n} exp(\mathcal{S}_r(h'_j, t'_j))} \tag{3}$$

An illustration depicting the structure of this paper is presented in Figure 1. First, we extract entities and relations from the extensive documentation of industrial processes through manual annotation. Next, we employ this information to construct a dual-security knowledge graph, incorporating data from a vulnerability database. Finally, we utilize the PairRE method for the link prediction task, enabling the prediction of potential risks in the dual-security domain.

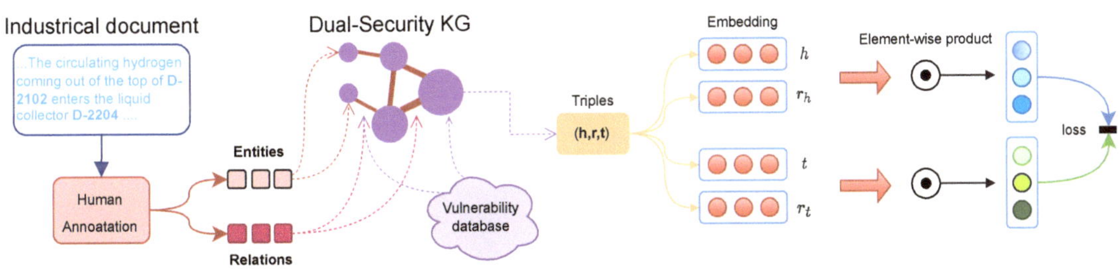

Figure 1. The construction of dual-KG and PairRE model.

4. Experiments

This section is organized into three subheadings. Firstly, we provide a detailed overview of our experimental setup, which includes a description of the dataset, evaluation metrics, and other relevant details. Secondly, we present a concise and accurate analysis of the experimental results. Finally, we draw insightful interpretations from these results, offering our conclusions based on the experiments.

4.1. Experiment Setting

Datasets, we culminated in the construction of our knowledge graph, which is founded on real-world resources and encompasses impressive 37,000 triples. We can find detailed statistics about the dataset in Table 1. Additionally, to showcase the model's proficiency in predicting potential risks within specific industrial processes, such as crack hydrogenation, we generated a focused sub-KG. This sub-KG encapsulates the knowledge related to the crack hydrogenation process alongside all relevant cybersecurity information. Refer to Table 2 for detailed statistics about this sub-KG. To optimize the data for our link prediction task, we meticulously partitioned it into three distinct subsets, ensuring a balanced division. Specifically, we allocated the data into a test set, a training set, and a validation set, maintaining a proportional split ratio of 1:8:1.

Table 1. Statistics of dual-Security knowledge graph dataset.

Entity	Relation	Triples	Train	Validation	Test
2247	45	37,000	29,600	3700	3700

Table 2. Statistics of sub-KG about crack hydrogenation and cybersecurity information.

Entity	Relation	Triples	Train	Validation	Test
104	25	10,686	8544	1068	1074

Evaluation metric, in the case of each query of the form $(h, r, ?)$ or $(?, r, t)$, we calculate a score for every entity while simultaneously assessing the rank of the correct answer. To evaluate the performance of our models, we employ two widely-used metrics, namely the Mean Reciprocal Rank (MRR) and Hit@k. MRR is computed as the mean of the reciprocal ranks for the answer entities. Hit@k, on the other hand, measures the percentage of the desired entities correctly ranked within the top k positions. Furthermore, it is important to note that these metrics exclude the scores of all known true triples in the training, validation, and testing sets. This exclusion allows for a more robust evaluation of our models, providing valuable insights into their effectiveness.

Comparison of models, in our experiments, we applied our dataset to various models, primarily focusing on comparing their performance. Given the substantial scale of our dataset, we employed embedding-based methods to conduct our experiments. These methods employ various embedding techniques to represent entities and relations in low-dimensional spaces, which include methods such as TransE [16], TransR [45], TransH [39], HAKE [46], PairRE [43]. We finally compared their performance with each other.

Hyperparameters, we maintained consistent hyperparameter settings during training. These settings entailed a dimension of embedding vectors set to 512 and a maximum of 1000 training epochs. Concurrently, other methods like TuckER [19], which decomposes a tensor into a core tensor with three-factor matrices, and ConvE [18], founded on the graph neural networks framework, we carefully selected the most suitable hyperparameters through multiple experiments. In this scenario, we set the number of graph convolutional network layers to 1, fixed the dimension of embedding vectors to 200, and established the maximum number of training epochs as 2000, among other key parameters.

4.2. Results on Link Prediction

All methods were evaluated across evaluation metrics on the dual-security dataset, and the results are presented in Table 3. Notably, these evaluations yield clear observations on the performance of these methods. It is apparent that the graph convolutional network method (ConvE) excels and outperforms some of the traditional embedding-based methods, including TransE, TransR, TransH, and TruckER. This observed superiority can be attributed to the incorporation of graph structures, allowing these models to harness a more comprehensive understanding of the training data—a critical aspect for effective link prediction tasks. However, it is worth mentioning that recently proposed methods, such as HAKE and PairRE, manage to outperform even the graph convolutional network methods. This superior performance may be attributed to the adoption of more intricate embedding techniques and measures, underscoring the benefits of sophisticated embedding approaches in addressing dual-security link prediction challenges.

While PairRE demonstrates outstanding overall performance on the dual-security dataset, achieving a commendable level of accuracy, its direct hit rate remains relatively modest at just 22.8%. This observation prompts a vital question regarding the ability of existing models to effectively absorb and adapt to the security data within the dual-security dataset. The dual nature of the KG, incorporating both cybersecurity and industrial process security knowledge, results in a semantic space that is notably vast, rendering it challenging to learn comprehensively. It is imperative to acknowledge that this dataset inherently represents the confluence of two distinct dimensions. Therefore, we deduce that the existing models' deficiencies arise from their failure to adequately learn the knowledge encompassed by both aspects in the dual-KG.

Table 3. The performance of the dominant model on the link prediction. Hit@k is in %.

	MMR	Hit@1	Hit@3	Hit@10
TransE	0.164	0.059	0.204	0.374
TransR	0.199	0.087	0.234	0.439
TransH	0.270	0.159	0.312	0.492
TuckER	0.219	0.131	0.200	0.242
ConvE	0.326	0.218	0.368	0.545
HAKE	0.331	0.215	0.381	0.569
PairRE	**0.352**	**0.228**	**0.414**	**0.593**

The bold in the table indicates the best performance of the corresponding item.

The link prediction results for the sub-KG are presented in Table 4. Notably, PairRE demonstrates efficient prediction of potential risks, with over half of the test data producing accurate predictions. In addition, 77.4% of the test data achieves a top-three ranking, and nearly all test data predictions secure a spot within the top ten. Consequently, focusing on the crack hydrogenation process, obtaining the top three predicted results for all devices would thwart three quarters of potential attacks. Moreover, achieving the top ten predicted results would provide comprehensive protection for almost all devices.

Table 4. The performance of the dominant embedding-based model on sub-KG. Hit@k is in %.

	MMR	Hit@1	Hit@3	Hit@10
TransE	0.249	0.087	0.32	0.567
TransR	0.367	0.195	0.477	0.679
TransH	0.26	0.123	0.305	0.551
TuckER	0.539	0.392	0.615	0.854
ConvE	0.649	0.505	0.744	0.934
HAKE	0.502	0.343	0.577	0.858
PairRE	**0.673**	**0.533**	**0.774**	**0.938**

The bold in the table indicates the best performance of the corresponding item.

After comparing the sub-KG performance with the whole dataset, we found that PairRE is excellent at predicting potential risks in a single scenario. This approach enables full absorption of knowledge within the sub-KG. So, we could generate suitable sub-KG for various specific industrial processes that can be safeguarded against potential security risks by applying this approach.

4.3. Mined Knowledge

By utilizing the best-performing model, PairRE, we have achieved successful predictions of various cybersecurity attacks and industrial process anomalies, as evidenced in Table 5. This approach allows us not only to identify diverse cyber virus attack patterns, but also to access various methods for mitigating cyber risks in the field of cybersecurity. Additionally, we can extend our capabilities to complete the industrial process knowledge graph, enabling us to uncover potential risks within this domain.

To demonstrate the risk prediction ability of our approach, we generate a sub-KG representing the crack hydrogenation process and all cybersecurity knowledge. Furthermore, we present the mined knowledge regarding the crack hydrogenation process and Industroyer attacks, as depicted in Figure 2 and summarized in Table 6. The approach successfully identified vulnerable devices, specifically *D*-2106, *D*-2104, and *K*-2102*A*/*B*, under Industroyer attacks. These devices are classified as power supply equipment. Given that Industroyer primarily targets power systems, our approach can precisely identify potentially vulnerable devices. In addition, Industroyer primarily employs remote system discovery to detect a list of devices. Hence, remote system discovery, which includes malicious emails, the office network system, and external O&M terminals, has been predicted for all devices. Subsequently, it identifies target devices and executes various attacks. Depending on the function of different devices, the focus of Industroyer attacks varies. Therefore, the potential risks predicted by the approach differ, as shown in Table 6. However, these identified risks strongly correlate with the devices' functionality. For example, the *K*-2102*A*/*B* serves as the backbone node of the power network. Consequently, attacks against it primarily concentrate on the control aspect, encompassing actions like blocking serial COM, stopping services, and obstructing command messages. In summary, the approach excels in not only identifying vulnerable devices in the industrial process, but also mining potential risks based on the functions of the devices and the adversaries' targets.

Table 5. The mined knowledge from the dual-security by PairRE.

Pattern	Mined Triples
$Virus \xrightarrow{\texttt{Use}} AttackPattern$	$Stuxnet \xrightarrow{\texttt{Use}} I/OImage$ $LockerGoga \xrightarrow{\texttt{Use}} LossOfView$ $Industroyer \xrightarrow{\texttt{Use}} DataDestruction$ $Nodeproperties \xrightarrow{\texttt{Use}} LateralToolTransfer$
$Action \xrightarrow{\texttt{Mitigate}} AttackPattern$	$Encrypt \xrightarrow{\texttt{Mitigate}} Network$ $NetworkSegmentation \xrightarrow{\texttt{Mitigate}} BruteForceI/O$ $NetworkAllowlists \xrightarrow{\texttt{Mitigate}}$ $BlockCommandMessage$
$Controller \xrightarrow{\texttt{control}} Device$	$1021FV0101 \xrightarrow{\texttt{control}} 1021PdIC0101$ $1031HV0051 \xrightarrow{\texttt{control}} GasifierLevel$ $Loadcontroller \xrightarrow{\texttt{control}} GasifierPressure$ $Levelcontroller \xrightarrow{\texttt{control}} GasifierWaterFlow$ $Circuit1031LIC0001 \xrightarrow{\texttt{control}} 1031LV0001A$
$Device \xrightarrow{\texttt{reaction}} ReactionType$	$Reboiler(F\text{-}2102) \xrightarrow{\texttt{reaction}} Heat$ $Pump(P\text{-}2103A/B) \xrightarrow{\texttt{reaction}} Supercharge$ $Tower(T\text{-}2568A/B) \xrightarrow{\texttt{reaction}} Desulfurization$

Table 6. Potential risks in crack hydrogenation and Industroyer attacks.

Device	Mined Risk
D-2106	$Industroyer \xrightarrow{\texttt{Use}} Remote\ System\ Discovery \xrightarrow{\texttt{Attack}} D\text{-}2106$ $Industroyer \xrightarrow{\texttt{Use}} Device\ Restart/Shutdown \xrightarrow{\texttt{Attack}} D\text{-}2106$ $Industroyer \xrightarrow{\texttt{Use}} Manipulation\ of\ Control \xrightarrow{\texttt{Attack}} D\text{-}2106$
D-2104	$Industroyer \xrightarrow{\texttt{Use}} Remote\ System\ Discovery \xrightarrow{\texttt{Attack}} D\text{-}2104$ $Industroyer \xrightarrow{\texttt{Use}} Connection\ Proxy \xrightarrow{\texttt{Attack}} D\text{-}2104$ $Industroyer \xrightarrow{\texttt{Use}} Denial\ of\ Control \xrightarrow{\texttt{Attack}} D\text{-}2104$ $Industroyer \xrightarrow{\texttt{Use}} Network\ Connection\ Enumeration \xrightarrow{\texttt{Attack}} D\text{-}2104$
K-2102*A*/*B*	$Industroyer \xrightarrow{\texttt{Use}} Remote\ System\ Discovery \xrightarrow{\texttt{Attack}} K\text{-}2102A/B$ $Industroyer \xrightarrow{\texttt{Use}} Block\ Serial\ COM \xrightarrow{\texttt{Attack}} K\text{-}2102A/B$ $Industroyer \xrightarrow{\texttt{Use}} Service\ Stop \xrightarrow{\texttt{Attack}} K\text{-}2102A/B$ $Industroyer \xrightarrow{\texttt{Use}} Block\ Command\ Message \xrightarrow{\texttt{Attack}} K\text{-}2102A/B$ $Industroyer \xrightarrow{\texttt{Use}} Loss\ of\ Protection \xrightarrow{\texttt{Attack}} K\text{-}2102A/B$

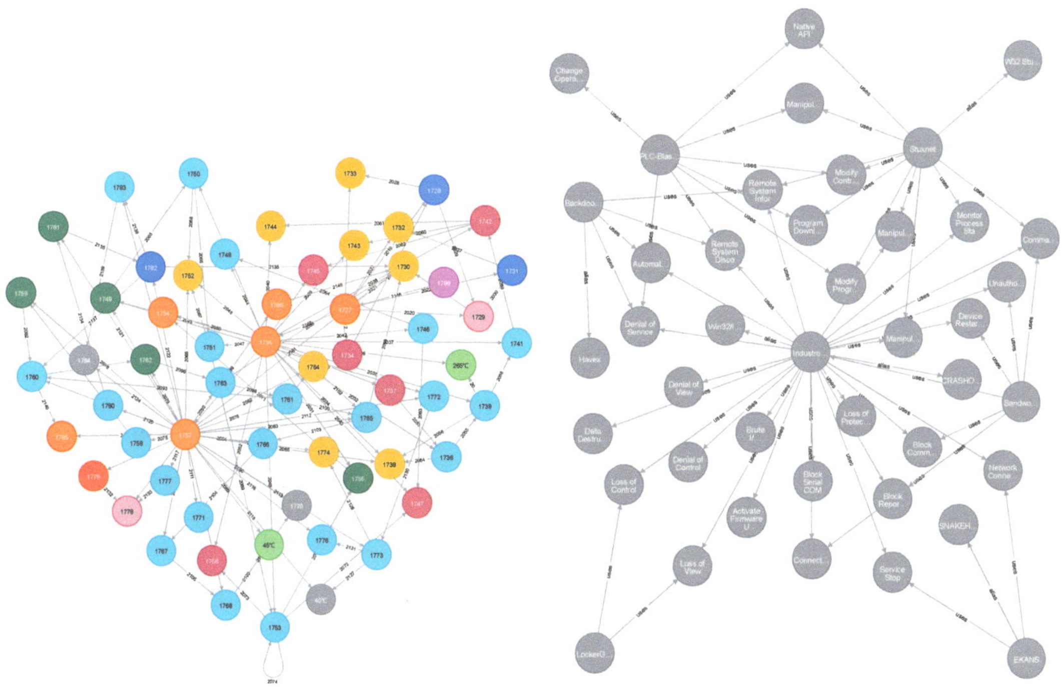

(**a**) (**b**)

Figure 2. The part of dual-security KG. (**a**) Crack hydrogenation process KG. (**b**) Industroyer attacks KG.

5. Conclusions

In this paper, we begin with constructing a dual-security knowledge graph, which innovatively integrates industrial process security information, a domain often overlooked by researchers, into the cybersecurity knowledge graph, a domain that has garnered widespread attention. Subsequently, we employ an embedding-based method (RairRE) capable of learning the underlying semantics in the knowledge graph, allowing us to predict the occurrence of novel cybersecurity attack patterns or potential security issues in the industrial process. The experimental results suggest that this method can effectively mine potential risks in some industrial process. However, the overall potential risk prediction accuracy on the dual-KG is not very satisfactory, as existing models struggle to comprehensively grasp the knowledge embedded in both aspects of the whole dataset.

In light of these insights, our proposed model currently used has the potential for improved performance in fully tapping into the rich, interrelated data contained within the knowledge graph.

Our future endeavors will be dedicated to the development of a more tailored dual-security mining model, specifically designed to extract and integrate all knowledge with greater precision from both the realms of cyber and industrial process security. The ultimate goal is to enhance the accuracy of predicting potential risks in industrial processes.

Author Contributions: Conceptualization, L.W. and H.L.; methodology, W.W.; software, L.W. and H.L; validation, L.W., G.X. and Y.W.; formal analysis, W.W.; investigation, G.X.; resources, Y.W.; data curation, Y.W.; writing—original draft preparation, L.W.; writing—review and editing, H.L.; visualization, L.W.; supervision, Y.W.; project administration, W.W.; funding acquisition, Y.W. All authors have read and agreed to the published version of the manuscript.

Funding: This research was funded by the National Natural Science Foundation of China (NSFC No.2021YFB2012400).

Data Availability Statement: Source codes and desensitized datasets are available in the GitHub repository, and the link is https://github.com/Faker-lz/Dual-safety-knowledge-graph-completion-for-process-industry.

Conflicts of Interest: The authors declare no conflicts of interest.

References

1. Lee, W.; Weekman, V.W., Jr. Advanced control practice in the chemical process industry: A view from industry. *AIChE J.* **1976**, *22*, 27–38. [CrossRef]
2. Lu, H.; Guo, L.; Azimi, M.; Huang, K. Oil and Gas 4.0 era: A systematic review and outlook. *Comput. Ind.* **2019**, *111*, 68–90. [CrossRef]
3. Schrotenboer, A.H.; Veenstra, A.A.; uit het Broek, M.A.; Ursavas, E. A Green Hydrogen Energy System: Optimal control strategies for integrated hydrogen storage and power generation with wind energy. *Renew. Sustain. Energy Rev.* **2022**, *168*, 112744. [CrossRef]
4. Alanen, J.; Linnosmaa, J.; Malm, T.; Papakonstantinou, N.; Ahonen, T.; Heikkilä, E.; Tiusanen, R. Hybrid ontology for safety, security, and dependability risk assessments and Security Threat Analysis (STA) method for industrial control systems. *Reliab. Eng. Syst. Saf.* **2022**, *220*, 108270. [CrossRef]
5. Corallo, A.; Lazoi, M.; Lezzi, M. Cybersecurity in the context of industry 4.0: A structured classification of critical assets and business impacts. *Comput. Ind.* **2020**, *114*, 103165. [CrossRef]
6. Noy, N.F.; McGuinness, D.L. *Ontology Development 101: A Guide to Creating Your First Ontology*; Stanford Knowledge Systems Laboratory Technical Report KSL-01-05; Knowledge Systems Laboratory: Stanford, CA, USA, 2001.
7. Bizer, C.; Heath, T.; Berners-Lee, T. Linked Data—The Story So Far. *Int. J. Semant. Web Inf. Syst.* **2009**, *79*, 637–638. [CrossRef]
8. Yu, D.; Zhu, C.; Yang, Y.; Zeng, M. Jaket: Joint pre-training of knowledge graph and language understanding. In Proceedings of the AAAI Conference on Artificial Intelligence, Vancouver, BC, Canada, 22 February–1 March 1 2022; Volume 36, pp. 11630–11638.
9. Lin, Q.; Mao, R.; Liu, J.; Xu, F.; Cambria, E. Fusing topology contexts and logical rules in language models for knowledge graph completion. *Inf. Fusion* **2023**, *90*, 253–264. [CrossRef]
10. Bakhshi, M.; Nematbakhsh, M.; Mohsenzadeh, M.; Rahmani, A.M. SParseQA: Sequential word reordering and parsing for answering complex natural language questions over knowledge graphs. *Knowl.-Based Syst.* **2022**, *235*, 107626. [CrossRef]
11. Gogleva, A.; Polychronopoulos, D.; Pfeifer, M.; Poroshin, V.; Ughetto, M.; Martin, M.J.; Thorpe, H.; Bornot, A.; Smith, P.D.; Sidders, B.; et al. Knowledge graph-based recommendation framework identifies drivers of resistance in EGFR mutant non-small cell lung cancer. *Nat. Commun.* **2022**, *13*, 1667. [CrossRef]
12. Yang, Y.; Huang, C.; Xia, L.; Li, C. Knowledge graph contrastive learning for recommendation. In Proceedings of the 45th International ACM SIGIR Conference on Research and Development in Information Retrieval, Madrid, Spain, 11–15 July 2022; pp. 1434–1443.
13. Wu, X.; Duan, J.; Pan, Y.; Li, M. Medical knowledge graph: Data sources, construction, reasoning, and applications. *Big Data Min. Anal.* **2023**, *6*, 201–217. [CrossRef]
14. Santos, A.; Colaço, A.R.; Nielsen, A.B.; Niu, L.; Strauss, M.; Geyer, P.E.; Coscia, F.; Albrechtsen, N.J.W.; Mundt, F.; Jensen, L.J.; et al. A knowledge graph to interpret clinical proteomics data. *Nat. Biotechnol.* **2022**, *40*, 692–702. [CrossRef]
15. Li, M.M.; Huang, K.; Zitnik, M. Graph representation learning in biomedicine and healthcare. *Nat. Biomed. Eng.* **2022**, *6*, 1353–1369. [CrossRef] [PubMed]
16. Bordes, A.; Usunier, N.; García-Durán, A.; Weston, J.; Yakhnenko, O. Translating Embeddings for Modeling Multi-relational Data. In Proceedings of the Neural Information Processing Systems, Lake Tahoe, NV, USA, 5–8 December 2013.
17. Yang, B.; tau Yih, W.; He, X.; Gao, J.; Deng, L. Embedding Entities and Relations for Learning and Inference in Knowledge Bases. *arXiv* **2014**, arXiv:1412.6575.
18. Dettmers, T.; Minervini, P.; Stenetorp, P.; Riedel, S. Convolutional 2d knowledge graph embeddings. In Proceedings of the AAAI Conference on Artificial Intelligence, New Orleans, LA, USA, 2–7 February 2018; Volume 32.
19. Balazevic, I.; Allen, C.; Hospedales, T. TuckER: Tensor Factorization for Knowledge Graph Completion. In Proceedings of the 2019 Conference on Empirical Methods in Natural Language Processing and the 9th International Joint Conference on Natural Language Processing (EMNLP-IJCNLP), Hong Kong, China, 3–7 November 2019; pp. 5185–5194. [CrossRef]
20. Liu, K.; Wang, F.; Ding, Z.; Liang, S.; Yu, Z.; Zhou, Y. Recent Progress of Using Knowledge Graph for Cybersecurity. *Electronics* **2022**, *11*, 2287. [CrossRef]
21. Rastogi, N.; Dutta, S.; Gittens, A.; Zaki, M.J.; Aggarwal, C. TINKER: A framework for Open source Cyberthreat Intelligence. In Proceedings of the 2022 IEEE International Conference on Trust, Security and Privacy in Computing and Communications (TrustCom), Wuhan, China, 9–11 December 2022; pp. 1569–1574. [CrossRef]
22. Lin, S.C.; Tseng, S.S. Constructing detection knowledge for DDoS intrusion tolerance. *Expert Syst. Appl.* **2004**, *27*, 379–390. [CrossRef]

23. Narayanan, S.N.; Ganesan, A.; Joshi, K.; Oates, T.; Joshi, A.; Finin, T. Early Detection of Cybersecurity Threats Using Collaborative Cognition. In Proceedings of the 2018 IEEE 4th International Conference on Collaboration and Internet Computing (CIC), Philadelphia, PA, USA, 18–20 October 2018; pp. 354–363. [CrossRef]
24. Gao, P.; Shao, F.; Liu, X.; Xiao, X.; Qin, Z.; Xu, F.; Mittal, P.; Kulkarni, S.R.; Song, D. Enabling efficient cyber threat hunting with cyber threat intelligence. In Proceedings of the 2021 IEEE 37th International Conference on Data Engineering (ICDE), Chania, Greece, 19–22 April 2021; IEEE: New York, NY, USA, 2021; pp. 193–204.
25. Kiesling, E.; Ekelhart, A.; Kurniawan, K.; Ekaputra, F. The SEPSES knowledge graph: An integrated resource for cybersecurity. In Proceedings of the International Semantic Web Conference, Auckland, New Zealand, 26–30 October 2019; Springer: Cham, Switzerland, 2019; pp. 198–214.
26. Garrido, J.S.; Dold, D.; Frank, J. Machine learning on knowledge graphs for context-aware security monitoring. In Proceedings of the 2021 IEEE International Conference on Cyber Security and Resilience (CSR), Virtual Conference, 26–28 July 2021; IEEE: New York, NY, USA, 2021; pp. 55–60.
27. Li, J.; Liu, Y.; Gu, L. DDoS attack detection based on neural network. In Proceedings of the 2010 2nd International Symposium on Aware Computing, Tainan, Taiwan, 1–4 November 2010; IEEE: New York, NY, USA, 2010; pp. 196–199.
28. Li, X.; Lyu, M.; Wang, Z.; Chen, C.H.; Zheng, P. Exploiting knowledge graphs in industrial products and services: A survey of key aspects, challenges, and future perspectives. *Comput. Ind.* **2021**, *129*, 103449. [CrossRef]
29. Chen, Y.; Qian, T. Relation constrained attributed network embedding. *Inf. Sci.* **2020**, *515*, 341–351. [CrossRef]
30. Gao, J.; Li, X.; Xu, Y.E.; Sisman, B.; Dong, X.L.; Yang, J. Efficient Knowledge Graph Accuracy Evaluation. *Proc. VLDB Endow.* **2019**, *12*, 1679–1691. [CrossRef]
31. Liu, W.; Liu, J.; Wu, M.; Abbas, S.; Hu, W.; Wei, B.; Zheng, Q. Representation learning over multiple knowledge graphs for knowledge graphs alignment. *Neurocomputing* **2018**, *320*, 12–24. [CrossRef]
32. Tay, Y.; Tuan, L.A.; Phan, M.C.; Hui, S.C. Multi-task neural network for non-discrete attribute prediction in knowledge graphs. In Proceedings of the 2017 ACM on Conference on Information and Knowledge Management, Singapore, 6–10 November 2017; pp. 1029–1038.
33. Long, J.; Chen, Z.; He, W.; Wu, T.; Ren, J. An integrated framework of deep learning and knowledge graph for prediction of stock price trend: An application in Chinese stock exchange market. *Appl. Soft Comput.* **2020**, *91*, 106205. [CrossRef]
34. Wang, Q.; Zou, D.; Ge, L. Multi-integrated Reform for the Course of Data Structure. In Proceedings of the 15th International Conference on Computer Science & Education (ICCSE), Delft, The Netherlands, 18–22 August 2020; pp. 9–13. [CrossRef]
35. Zhang, Y.; Xu, H.; Zhang, X.; Wu, X.; Yang, Z. TRFR: A ternary relation link prediction framework on Knowledge graphs. *Ad Hoc Netw.* **2021**, *113*, 102402. [CrossRef]
36. Wang, Q.; Hao, Y. ALSTM: An attention-based long short-term memory framework for knowledge base reasoning. *Neurocomputing* **2020**, *399*, 342–351. [CrossRef]
37. Ai, Q.; Zhang, Y.; Bi, K.; Croft, W.B. Explainable product search with a dynamic relation embedding model. *ACM Trans. Inf. Syst. (TOIS)* **2019**, *38*, 1–29. [CrossRef]
38. Wang, Q.; Ji, Y.; Hao, Y.; Cao, J. GRL: Knowledge graph completion with GAN-based reinforcement learning. *Knowl.-Based Syst.* **2020**, *209*, 106421. [CrossRef]
39. Wang, Z.; Zhang, J.; Feng, J.; Chen, Z. Knowledge graph embedding by translating on hyperplanes. In Proceedings of the AAAI Conference on Artificial Intelligence, Québec City, QC, Canada, 27–31 July 2014; Volume 28.
40. Trouillon, T.; Welbl, J.; Riedel, S.; Gaussier, É.; Bouchard, G. Complex embeddings for simple link prediction. In Proceedings of the International Conference on Machine Learning, New York, NY, USA, 20–22 June 2016; pp. 2071–2080.
41. Yang, F.; Yang, Z.; Cohen, W.W. Differentiable learning of logical rules for knowledge base reasoning. In Proceedings of the 31st Conference on Neural Information Processing Systems (NIPS 2017), Long Beach, CA, USA, 4–9 December 2017.
42. Sadeghian, A.R.; Armandpour, M.; Ding, P.; Wang, D.Z. DRUM: End-to-End Differentiable Rule Mining on Knowledge Graphs. *arXiv* **2019**, arXiv:1911.00055.
43. Chao, L.; He, J.; Wang, T.; Chu, W. Pairre: Knowledge graph embeddings via paired relation vectors. *arXiv* **2020**, arXiv:2011.03798.
44. Sun, Z.; Deng, Z.; Nie, J.Y.; Tang, J. RotatE: Knowledge Graph Embedding by Relational Rotation in Complex Space. *arXiv* **2018**, arXiv:1902.10197.
45. Lin, Y.; Liu, Z.; Sun, M.; Liu, Y.; Zhu, X. Learning entity and relation embeddings for knowledge graph completion. In Proceedings of the Twenty-Ninth AAAI Conference on Artificial Intelligence, Austin, TX, USA, 25–30 January 2015.
46. Zhang, Z.; Cai, J.; Zhang, Y.; Wang, J. Learning hierarchy-aware knowledge graph embeddings for link prediction. In Proceedings of the AAAI Conference on Artificial Intelligence, New York, NY, USA, 7–12 February 2020; Volume 34, pp. 3065–3072.

sensors

MDPI

Article

Enhancing Surveillance Systems: Integration of Object, Behavior, and Space Information in Captions for Advanced Risk Assessment

Minseong Jeon [1], Jaepil Ko [2] and Kyungjoo Cheoi [1,*]

[1] Department of Computer Science, Chungbuk National University, 1 Chungdae-ro, Seowon-gu, Cheongju, Chungbuk 28644, Republic of Korea; jeonminseong@chungbuk.ac.kr

[2] Department of Computer Engineering, Kumoh National Institute of Technology, 61 Daehak-ro, Gumi, Gyeongbuk 39177, Republic of Korea; nonezero@kumoh.ac.kr

* Correspondence: kjcheoi@chungbuk.ac.kr

Abstract: This paper presents a novel approach to risk assessment by incorporating image captioning as a fundamental component to enhance the effectiveness of surveillance systems. The proposed surveillance system utilizes image captioning to generate descriptive captions that portray the relationship between objects, actions, and space elements within the observed scene. Subsequently, it evaluates the risk level based on the content of these captions. After defining the risk levels to be detected in the surveillance system, we constructed a dataset consisting of [Image-Caption-Danger Score]. Our dataset offers caption data presented in a unique sentence format, departing from conventional caption styles. This unique format enables a comprehensive interpretation of surveillance scenes by considering various elements, such as objects, actions, and spatial context. We fine-tuned the BLIP-2 model using our dataset to generate captions, and captions were then interpreted with BERT to evaluate the risk level of each scene, categorizing them into stages ranging from 1 to 7. Multiple experiments provided empirical support for the effectiveness of the proposed system, demonstrating significant accuracy rates of 92.3%, 89.8%, and 94.3% for three distinct risk levels: safety, hazard, and danger, respectively.

Keywords: surveillance system; image captioning; descriptive captions; risk assessment; BLIP-2; BERT

1. Introduction

In various sectors of contemporary society, a plethora of closed-circuit television (CCTV) cameras have been strategically deployed to monitor and record various safety incidents and criminal activities. The growing concern for societal safety has led to an increasing demand for CCTV and security surveillance. According to The Business Research Company's "Global Security Surveillance System Market Report 2023" [1], the global security surveillance system market is forecasted to expand from USD 130.08 billion in 2022 to USD 148.29 billion in 2023, at an annual growth rate of 14.0%. With the escalating need for safety in high-risk areas, it is anticipated that the surveillance system market will reach a substantial USD 230.47 billion by the year 2027.

Surveillance systems are engineered to identify patterns within monitored scenes, encompassing risky behavior, abnormal actions, and incidents. Extensive research has been conducted on these systems, primarily focusing on the detection of abnormal behavior such as 'fighting' and 'fainting'. This involves the utilization of technologies such as object detection and recognition, tracking, pose estimation, movement detection, and anomaly detection of objects [2–9]. Jha et al. [8] proposed an N-YOLO model designed for the detection of abnormal behaviors, such as fighting. This model tracks the interrelationship of detection results in subimages, integrating them with the inference outcomes through a modified YOLO [10]. In a related study, Kim et al. [9] introduced the AT-Net model, specifically designed for abnormal situation detection. The model aims to mitigate classification ambiguities and minimize information loss by integrating object detection and human skeletal information.

These various studies [2–9] have improved performance by incorporating diverse feature information related to abnormal behavior. Nevertheless, conventional methods exhibit certain limitations. Firstly, these methods do not comprehensively consider the spatial context associated with abnormal situations. It is crucial to recognize that, even with identical objects and behavior, interpretations may vary significantly based on the spatial context, resulting in different responses to specific risk scenarios. For instance, the action of sitting on a bench, compared to sitting on a railing or cliff—both categorized as sitting actions—results in distinct risk levels. Sitting on a bench is considered safe, whereas sitting on a cliff represents an extremely dangerous situation. Secondly, within the framework of current surveillance systems designed to detect potentially hazardous scenarios, the outcomes are typically displayed as alerts on the monitoring screen or delivered as recorded footage featuring the identified risk elements. However, in numerous instances, actions exhibiting a higher degree of movement compared to the surrounding behavior might attract attention but may not necessarily be indicative of an actual accident. For instance, conventional surveillance systems may flag regular running in a park as abnormal since it involves more movement than walking or sitting. Consequently, further scrutiny becomes imperative for the observer to re-evaluate the scene and interpret the situation accurately in order to discern the precise nature of the detected abnormal behavior. Ultimately, for a more accurate and prompt interpretation of risk situations, surveillance systems must not only consider the actions of the detected objects but also take into account the surrounding environment and situation, assess the overall risk level, and provide the observer with specifically interpreted information. Thirdly, surveillance systems should have the capability to detect and interpret a broad range of risk elements and accidents, without being limited to specific locations or behaviors. For example, a system designed to detect instances of drowning in specific locations, such as swimming pools, may efficiently identify dangerous situations by employing technologies like object detection and movement-based anomaly detection, incorporating learned behavior patterns associated with drowning. However, in the case of CCTV systems installed in public spaces or on the streets, which necessitate versatile detection capabilities, the spatial context required for detection differs from that in the pool example mentioned above. Numerous categories of risks manifest in a variety of environments, involving hazardous elements such as fires and safety-related incidents like traffic accidents or altercations. Relying solely on specific technologies presents a considerable challenge in detecting a wide array of situations and types, consequently limiting the applicability of conventional methods in surveillance systems. In summary, surveillance systems must have the capability to comprehensively leverage spatial information, convey data in a format that can be rapidly and accurately grasped by the observer, and represent learnable data for potential risk factors in diverse situations. To address this challenge, this paper introduces a novel approach for a surveillance system.

Humans utilize a structured and high-level expressive tool, namely, language. Applying human-friendly natural language to the surveillance system can effectively address the aforementioned challenges. A situation to be detected by the surveillance system can be articulated in sentence form using natural language. Natural language can be employed to express a scene's danger, explain an accident in sentence form, and elucidate an accident in sentence form, encompassing various information such as the characteristics of the object, the presence or absence of a person, the distinction between adults and children, the type of accident, the place of occurrence, and nearby risk factors. When the scene is conveyed through such sentences, the observer can promptly comprehend the situation.

The rapid advancement of natural language processing (NLP) technology has paved the way for the utilization of words and sentences as data for learning, leading to the development of various models. Large language models (LLMs), in particular, which have been trained on extensive language datasets, exhibit remarkable performance in a multitude of natural language processing tasks. These large language models are proficient in diverse tasks such as sentence generation, translation, text summarization, and question answering.

Image captioning is a subfield of NLP that involves generating text describing the content of an image in natural language. By incorporating image captioning technology, enhanced with large language models boasting high generalization performance, into surveillance systems, it is possible to represent images through information-rich sentences. Notably, image captioning technology proves to be exceptionally well suited for the detection of hazardous behavior and the analysis of accident scenarios. This is attributed to its capacity to offer a detailed representation of the scene, encompassing information about individual characteristics, actions, and the spatial context. Furthermore, by enabling the system to autonomously assess risk levels for captions (sentences) generated through image captioning technology, proactive mitigation of safety incidents becomes feasible, allowing for prompt and accurate responses to any accidents that may occur.

In this paper, we introduce a novel surveillance system designed to surpass the limitations of traditional surveillance systems, which are frequently restricted to a primary emphasis on object-centric behavior analysis. Our system generates descriptive captions that encompass details regarding objects, actions, and spatial context extracted from surveillance target footage. These captions are subsequently utilized to assess the risk level of the observed scene. To generate captions for surveillance scenes, it is necessary to construct a dataset comprised of [Image-Caption-Danger Score]. The dataset should feature caption data presented in a novel sentence format, deviating from conventional caption structures, and should encompass a myriad of information encompassing objects, actions, and spatial context, facilitating comprehensive interpretation. To facilitate the interpretation of scenes by the image captioning model, we utilized BLIP-2 [11], a large language model well regarded for its efficiency in handling multimodal tasks while minimizing parameters and training costs, all while delivering state-of-the-art performance. We fine-tuned the BLIP-2 [11] model with the newly-constructed dataset to guide it in generating captions that adhere to the newly-defined sentence structure. Subsequently, we utilized BERT (bidirectional encoder representations from transformers) [12] to interpret the semantic content of the generated sentences and assess the risk level associated with each scene.

The overall structure of this paper is as follows. In Section 2, we delve into general research related to image captioning, existing studies on surveillance systems that incorporate image captioning, and the exploration of LLMs. Section 3 provides a detailed description of our newly constructed dataset and the overarching system structure. Section 4 presents a comprehensive evaluation of the proposed system's performance, utilizing both quantitative and qualitative analyses. Additionally, we delve into potential avenues for system enhancement through result analysis. Finally, in Section 5, we conclude the paper by summarizing the proposed system, discussing future development directions, and outlining potential areas for further research.

2. Related Work

Image captioning is a technology that generates descriptive captions for an input image. Notable studies in this domain are referenced as [13–16]. Vinyals, O. et al. [13] proposed a method that connects an encoder constructed with convolutional neural networks (CNN) for extracting image information with a long short-term memory (LSTM) [17] decoder for caption generation. Xu, K. et al. [14] introduced an approach that enhances the relationship between images and captions by incorporating an attention mechanism. Liu, W. et al. [15] suggested an image captioning technique that sequences images to serve as inputs for the transformer [18] model. Wang, P. et al. [16] proposed an integrated system that utilizes multimodal pre-training to represent data in a unified space. This approach allows for the expression of both image and language information in patches, facilitating simultaneous processing of image and language information within the transformer.

In the domain of surveillance systems incorporating image captioning, Dilawari et al. [19] introduced a system utilizing the VGG-16 [20] to extract specific situational information from videos and generate captions through a bidirectional-LSTM [21], with a primary focus on object-related attributes. W. Chen et al. [22] proposed a system that computes anomaly scores by combining captions generated using SwinBERT [23] and video features extracted via the ResNet-50 architecture [24]. These studies often rely on datasets like UCF Crime [25], NTU CCTV-Fights [26], ShanghaiTech [27], and XD-Violence [28], encompassing diverse types of behavior, including fighting, fainting, loitering, and abandonment. However, a noteworthy limitation arises from the prevalent lack of comprehensive captions within these datasets, which typically offer basic object descriptions but fail to capture nuanced object behavior or contextual space information. Consequently, despite the application of image captioning to enhance surveillance systems, the resulting captions primarily center on objects, neglecting the vital space context crucial for a comprehensive scene risk assessment. This limitation underscores the need for further research in integrating space information to enhance the efficacy of surveillance systems.

The advent of LLMs has brought about substantial advancements in NLP, enhancing the capacity to understand and generate human language by discerning word similarities and contextual relationships, and effectively handling sentence structure, grammar, and meaning.

Prominent models in the domain of LLMs include BERT [12], GPT [29], T5 [30], and LaMDA [31]. BERT [12] is a deep learning-based model in NLP distinguished by its capability to extract bidirectional contextual information from extensive volumes of raw text, capturing sequential relationships between sentences and representing words and their context in vectors. This comprehensive approach empowers BERT to consider both preceding and subsequent text, granting it an extensive understanding of language. Originally designed for language comprehension, BERT has significantly advanced the field of NLP. Furthermore, the LLM landscape boasts a multitude of models, such as GPT [29] and T5 [30], which have demonstrated remarkable performance in tasks such as sentence translation and summarization. Additionally, LaMDA [31], developed for interactive applications, represents another notable addition to the suite of LLMs. These models collectively signify the diverse and expanding capabilities of LLMs in the field of NLP.

In the domain of LLMs, a notable constraint is their inherent inability to comprehend image features due to the absence of image data during their training process. To overcome this limitation, extensive research is being conducted on large multimodal models (LMMs) that enrich LLMs with image information to establish a connection between images and text. These LMMs are pre-trained on a massive scale of diverse data types, including text, images, audio, and video, thereby equipping them with the capacity to perform a multitude of tasks, ranging from image captioning to vision question answering (VQA). Prominent systems, such as BLIP-2 [11], OpenAI's GPT-4 [32], Google's Gemini [33], and LAVA 1.5 [34], are all founded upon these large multimodal models, demonstrating the growing significance of this approach. Furthermore, a prevailing trend in the field involves the utilization of large web datasets for multimodal training, resulting in the development of a multitude of models. For instance, BLIP-2 [11] is trained on an extensive dataset comprising image–text pairs gathered from the web. This model incorporates frozen pre-trained models in both its encoder and decoder and effectively addresses the modality gap between the encoder and LLMs through the query transformer (Q-Former), achieving remarkable state-of–the-art (SOTA) performance in various vision–language tasks while demonstrating significant zero-shot capabilities.

In this paper, we present a surveillance system tailored to address the inherent constraints of prevailing surveillance systems. Our approach involves the creation of a novel dataset specifically tailored for surveillance applications, encompassing comprehensive captions that incorporate object information, action details, and space context. To further enhance the interpretative capacity of this surveillance system, we fine-tune BLIP-2 [11], a model trained on extensive web data, utilizing our newly curated dataset. Additionally, we leverage BERT [12] to comprehend the semantic nuances of the generated captions, enabling us to quantify the risk level based on this refined interpretation. This combination of novel dataset construction, fine-tuning, and semantic understanding represents a significant step toward a more effective and context-aware surveillance system.

3. Methodology

Figure 1 depicts a schematic representation of the proposed system's comprehensive workflow. The system initially takes video input from CCTV sources. Subsequently, this video data undergo processing through an image captioning network, generating interpretative captions to elucidate the content of the scenes. These captions encompass a wealth of details, encompassing object attributes, behaviors, and spatial context. Following caption generation, a specialized risk assessment module conducts an in-depth analysis of these generated captions. This module analyzes the information embedded within the captions, enabling a thorough evaluation of the scene's risk level.

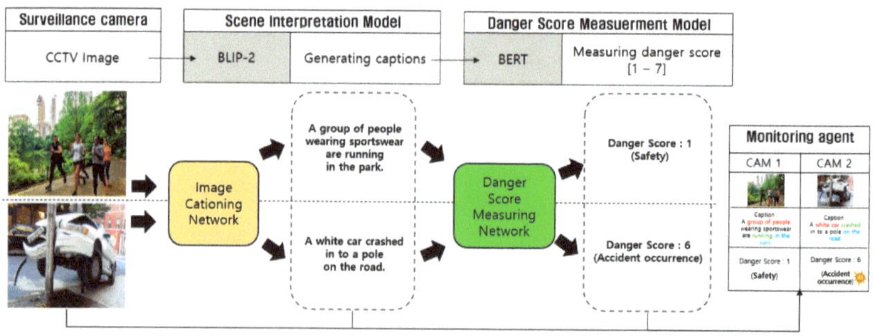

Figure 1. The comprehensive workflow of the proposed system illustrates the entire process flow, encompassing scene interpretation and danger score measurement, while showcasing the results generated at each stage.

In Figure 1, a pair of surveillance images depicting different levels of risk is presented as the input: the first depicts people running in the park, and the second depicts a car accident scenario. The deployed image captioning network effectively formulates precise captions corresponding to the content of each video, which are subsequently evaluated by the risk assessment module to determine the pertinent risk levels. In this evaluation, the first image, portraying a safe situation, is assigned a danger score of '1', while the second image, depicting an actual accident, is associated with a danger score of '6'. The danger scores defined within our proposed system are categorized into seven distinct stages, with '1' indicating a safe situation, and higher scores indicating escalating levels of risk. For an in-depth elucidation of the danger score methodology, please refer to Section 3.1. This systematic approach enables the precise evaluation of risk levels in different surveillance scenarios.

3.1. The Construction of the Dataset

The proposed system utilizes the large multimodal model BLIP-2 [11]. While BLIP-2 [11] has undergone pre-training on a diverse dataset encompassing various natural language processing tasks, its inherent design lacks task-specific optimization. Therefore, to achieve appropriate results when applying the large multimodal model to the proposed surveillance system, fine-tuning is imperative.

Conventional training datasets for surveillance systems often lack interpretive captions, and when available, the captions typically offer only rudimentary object descriptions, lacking comprehensive information about object actions or the surrounding space context, essential for practical surveillance applications. Therefore, we have constructed a dataset incorporating detailed information regarding objects, behavior, and spatial context. Additionally, our dataset encompasses diverse environmental conditions, including overcast days, nighttime shooting, and low-light environments, ensuring robustness in the face of changes due to various shooting conditions. Furthermore, a dataset is specifically created to detect and interpret individuals even when they are at a distance from other objects, focusing on people who are the subjects of the monitored safety incidents. This ensures the delivery of sufficient information for an accurate understanding and analysis of the monitored situation, surpassing the mere listing of visual elements. Additionally, this approach significantly aids in identifying and analyzing essential elements, individuals, for risk assessment within the visual field. As a result, a total of 2741 comprehensive datasets were built to enable BLIP-2 [11] to generate relevant captions for the surveillance system. Additionally, these datasets facilitate risk assessment following the processing of captions through BERT [12]. The danger score here represents a numerical value corresponding to the risk level, with specific classification criteria provided in Table 1, thereby facilitating the quantification of risk levels in the surveillance context. The construction of this comprehensive dataset is instrumental in enhancing the suitability and effectiveness of the proposed system for real-world surveillance applications.

Table 1. Risk level and danger score classification criteria.

Risk Level	Danger Score	Classification Criteria	Object	Behavior	Space
Safe	1	In situations where objects within the image are located in a secure environment.	adult	walking, sitting, running, or riding	pathway, park, or indoors
	2		child		
Hazard	3	In situations where objects within the image are located in a hazard environment.	adult		railing, cliff, road, or a construction site
	4		child		
Danger	5	In situations where objects in the image engage in perilous activities.	adult, vehicle, or fire	collapse, fight, a fire breaks out, or traffic accident	pathway, park, or indoors
	6	In situations where objects within the image engage in perilous activities while located in a hazardous environment.			railing, cliff, road, or a construction site
	7	In situations where the subjects within the image are minors and are involved in perilous activities while situated in a hazardous environment.	child, vehicle, or fire		

3.2. *Definition of Risk Level and Danger Score*

The classification of risk levels for input images involves three fundamental categories: safe situations, hazardous situations, and accident occurrences, each being assigned distinct danger scores. Table 1 illustrates that the risk levels representing safety and hazards are divided into two stages each, whereas the danger level is further segmented into three stages, contingent upon the varying levels of risk. This methodical classification system serves to delineate the entire range of risk across diverse surveillance scenarios.

The 'Safety' risk level typically involves routine activities such as walking, sitting, or running in secure areas like pathways, parks, or indoors. However, it is essential to note that even within the same risk level, situations involving children may inherently pose a relatively higher level of risk compared to those involving adults. Consequently, when adults are engaged in these activities, a danger score of '1' is assigned, while the presence of children is associated with a higher danger score of '2', signifying the increased potential for elevated risk in such scenarios. This nuanced approach enables a more accurate evaluation of potential danger in the presence of children during actions categorized as 'safe'.

The 'Hazard' risk level encompasses routine activities that typically occur in safety situations but are contingent upon the context of the surrounding space. For instance, the act of sitting is considered safe when performed on a bench, meriting a danger score of '1'; however, if the same action takes place on a railing, bridge, or roof, which inherently pose higher risks, it is reclassified as hazardous and is assigned a danger score of '3'. This systematic approach ensures that the level of risk is appropriately assessed in various space contexts, thereby enhancing the precision of the surveillance system.

The 'Danger' risk level is designated for detected actions categorized as accidents, encompassing scenarios where individuals have collapsed or incidents involving activities such as fights, fires, or traffic accidents. The degree of danger, as quantified by the danger score, escalates when these actions occur in hazardous environments like railings, cliffs, roads, or construction sites. Furthermore, a higher danger score is allocated when such actions involve children. This systematic risk assessment approach enables a more nuanced and accurate evaluation of the risk level within situations classified as dangerous, accounting for the influence of environmental factors and the age of the individuals involved.

Table 2 provides a comprehensive overview of the dataset distribution, systematically organized according to distinct risk levels. A total of 2741 images were collected, each necessitating a thorough analysis of object attributes, behavior, and spatial context. These images were meticulously selected in accordance with the classification criteria outlined in Table 1. The captions accompanying these collected images were meticulously crafted to delineate the distinctive characteristics of the object type, elucidate the actions of the object, and describe the space context in which the object is situated. Subsequently, datasets were systematically constructed by assigning an appropriate danger score to each situational scenario, ensuring the comprehensive representation of various risk levels in the surveillance system.

Table 2. The quantity of datasets by each risk level and danger score.

Risk Level	Safe		Hazard		Danger			
Danger Score	1	2	3	4	5	6	7	
Number of Images	343	215	291	259	752	613	268	
Total	558		550		1633			2741

3.3. The Sentence Structure Format of Captions

During the dataset construction phase, careful consideration was given to the design of caption structures. These structures were intentionally designed to include essential components, specifically, the object type, object attributes, object behavior, and the spatial context. To offer further clarity and insight into the employed caption structure within the dataset, Table 3 presents illustrative examples.

Table 3. The sentence structure format employed in the captions of our dataset.

Image	Caption	Risk Level (Danger Score)
	A man wearing a hat is walking in the park.	Safe (1)
	A man in a white shirt and cap is walking on the road.	Hazard (3)
	A man wearing a blue shirt is fainting in the room.	Danger (5)
	A woman wearing a black dress is lying down on the road.	Danger (6)

The captions presented in Table 3 meticulously delineate the specifics of objects within surveillance footage. Objects are discerned by red highlights, with each highlighted section indicating the object's category and pertinent details. This information encompasses the identification of whether the subject is a person, a vehicle, or a fire, and further specifies whether the person is an adult or a child, along with details about their attire. Green highlights are employed to depict the behavior of the object, while blue highlights convey information pertaining to the spatial context. For instance, the first and second examples within Table 3 both depict an adult male engaged in walking. However, the contextual differences between these examples are significant. The first example takes place in a park, characterized as a safe environment. In contrast, the second example involves walking on a road, an action considered perilous due to jaywalking. Additionally, the third and fourth examples depict a fainting incident, classified as an emergent situation within the danger (accident occurrence) category. The third example takes place within a room, devoid of additional environmental hazards, resulting in an assigned danger score of '5'. In contrast, the fourth example occurs on a road frequented by cars, posing a higher risk due to the increased likelihood of subsequent accidents and is thus assigned a danger score of '6'. This meticulous approach to caption structuring ensures the effective capture and conveyance of comprehensive information regarding the objects, their actions, and the relevant spatial context. Consequently, it enhances the dataset's applicability in the surveillance system.

3.4. Scene Descriptive Caption Generation and Risk Assessment

The architectural framework of the proposed system is illustrated in Figure 2. The system follows a two-step process: Firstly, it fine-tunes BLIP-2 [11] using the dataset specifically tailored to generate descriptive captions that interpret scenes. BLIP-2, an advanced model in image captioning, combines visual and textual data, making it adept at understanding and describing complex scenes in surveillance footage. This model's strength lies in its ability to contextualize visual elements within the framework of natural language, offering a more nuanced interpretation than traditional image recognition models. Subsequently, BERT [12] is employed to perform a semantic analysis of these captions. BERT's key feature is its bidirectional training, allowing it to understand the context of a word based on all of its surroundings in a sentence. This is a significant departure from previous models that processed text in one direction, either left-to-right or right-to-left, which could overlook the broader context of certain words or phrases. BERT's deep understanding of language nuances makes it particularly effective in assessing the risk levels in the captions generated by BLIP-2. Following the analysis, the system conducts a comprehensive assessment of the risk level associated with each scene. The risk levels are categorized and quantified on a scale ranging from 1 to 7, reflecting the severity of the risk. This multistage approach is a crucial component of the system's capacity to deliver a nuanced evaluation of scene risk levels, significantly enhancing the efficacy of the surveillance system.

In the second image of Figure 2, the BLIP-2 framework is depicted, featuring three fundamental components: the image encoder, the query transformer (Q-Former), and the large language model (LLM). The query transformer is a trainable module designed to bridge the gap between the fixed-weight image encoder and the large language model. It utilizes ViT-G [35] for the image encoder and OPT 2.7B [36] for the large language model. BLIP-2 [11] undergoes fine-tuning using the constructed dataset, a process known to be computationally intensive. To mitigate the associated costs, the LoRA (low-rank adaptation) [37] technique, as proposed by Hu, is implemented in this system.

The analysis of the semantic content within the generated captions is facilitated by the utilization of BERT [12]. As illustrated on the right side of Figure 2, the scene's risk level is subsequently measured through a classifier. BERT represents a significant milestone in the domain of natural language processing, distinguished by its exemplary performance across a spectrum of language-related tasks. Diverging from its predecessors, BERT employs a bidirectional language model that comprehensively evaluates each word within a sentence, resulting in a more profound understanding of context. The special CLS token, situated in the upper right portion of Figure 2, consistently appears at the start of BERT sequences. This token serves to encapsulate the overarching context of the input sequence and is particularly well suited for classification tasks. The proposed system is crafted to have captions generated by BLIP-2 to produce a CLS token through BERT. The vector derived from this CLS token then undergoes further processing via a linear layer, culminating in classification that determines the scene's risk level and associated danger score using a softmax function. The combination of BLIP-2 and BERT in the proposed system leverages the strengths of advanced image processing and deep language understanding. This synergy results in a sophisticated risk assessment tool capable of interpreting complex surveillance scenarios with a high degree of accuracy.

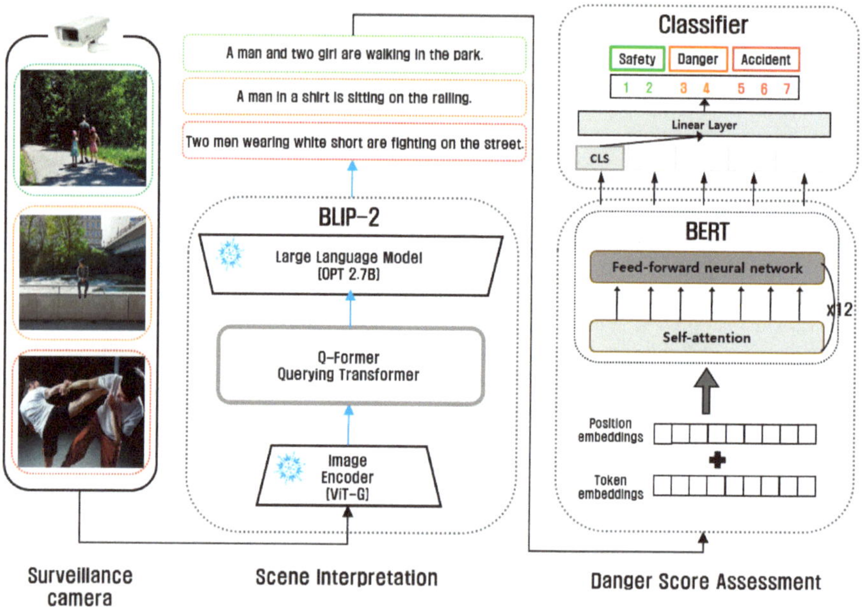

Figure 2. The architectural framework of our proposed system: the figure illustrates the procedural sequence wherein CCTV video input undergoes processing via the BLIP-2 model to produce captions, and these captions are subjected to analysis by the BERT model to facilitate risk classification.

4. Experiments and Results

4.1. Experimental Setup

The experimental setup for evaluating the proposed system is outlined as follows. BLIP-2 [11] is configured with an input image size of 224×224, utilizing a fixed batch size of 16. The optimization process is executed using the Adam [38] optimizer, and fine-tuning takes place over a span of 50 epochs. The learning rate is set at 10^{-5}. To facilitate the generation of content-rich sentences, the model's maximum output length is set to 30 characters. For BERT [12], the learning rate is established at 10^{-5}, and a batch size of 32 is employed. BERT undergoes fine-tuning over 100 epochs. The experimental dataset comprises a total of 2741 images, distributed into training, validation, and test subsets at an 8:1:1 ratio. All experiments are conducted on a single A100 GPU.

4.2. Experimental Results and Analysis

4.2.1. BLEU Score for Generated Captions

BLIP-2 [11] underwent fine-tuning to generate sentences containing object attributes, actions, and space context, as exemplified in Table 3. To assess the system's capability to produce captions adhering to the prescribed structural format, encompassing object characteristics, actions, and space information, we computed the BLEU (bilingual evaluation understudy) score [39]. The BLEU score is the most widely used metric in the field of machine translation evaluation, serving as a quantitative index to measure translation quality. It gauges the degree to which the terms within machine-generated candidate sentences align with those in reference sentences, with penalties applied to lengthier candidate sentences. The resultant BLEU score is presented as a numerical value within the range of 0 to 1, with higher scores signifying improved sentence generation quality. This multifaceted BLEU scoring approach ensures a nuanced evaluation of sentence generation performance.

The evaluation of caption quality generated by BLIP-2 was carried out using the BLEU score, with the frequency of generated captions as the underlying criterion for evaluation. The results of the BLEU scores for our dataset are outlined in Table 4. Notably, our newly constructed dataset, despite incorporating less common terminology related to risk scenarios, accidents, and space contexts, boasts an impressive BLEU-4 score of 0.4096. This accomplishment serves as a testament to the effectiveness of our dataset's captions in fulfilling their designated role. Furthermore, it reaffirms the successful fine-tuning of BLIP-2 using our dataset, ensuring the generation of pertinent captions suitable for the surveillance system.

Table 4. The results of the BLEU scores for our dataset.

BLEU-1	BLEU-2	BLEU-3	BLEU-4
0.7062	0.5827	0.4873	0.4096

4.2.2. Qualitative Evaluation of Generated Captions

Table 5 presents a comprehensive comparative analysis of caption generated results between the proposed system and existing image captioning models. The comparative models encompass state-of-the-art models developed following the emergence of transformer-based approaches, including CPTR by Xu, K [15], OFA by Liu, W [16], and BLIP-2 by Li, J. [11] (before fine-tuning). Each of these models [11,15,16] has undergone pre-training on distinct datasets. Specifically, CPTR [15] in Table 5a was pre-trained utilizing the COCO dataset [40]. OFA [16] in Table 5b was pre-trained using a diverse range of datasets, including CC12M [41], CC3M [41], SBU [42], COCO [40], and VG-Cap [43]. Additionally, BLIP-2 [11] in Table 5c was pre-trained on the LAION dataset [44]. Table 5d signifies the BLIP-2 [11] model fine-tuned using the self-constructed dataset proposed within this paper.

Table 5a illustrates a scene featuring a pedestrian jaywalking on a roadway, including the potential accident hazard. Upon examining the generated captions, it is evident that CPTR and BLIP-2(base) produce an incorrect caption, failing to accurately detect the object. Meanwhile, OFA produces a caption lacking essential human-centric information. The results obtained from our model, showcase success in generating a caption that comprehensively incorporates essential details regarding the individual, the walking action, and the spatial context of the road.

Table 5b,c illustrates a scene featuring a person who has fallen due to an accident. Upon examining the results for the image in Table 5b, all models produced captions that showed a person lying down. However, it is noteworthy that only our model was able to produce captions offering detailed information about the object and its surrounding environment. Other models misrepresented space information or failed to adequately represent the object. CPTR and OFA generate interpretations centered around a person standing nearby, deviating from a primary focus on the fallen individual. BLIP-2(base) focuses on the fallen person but is observed to lack information related to space and objects. In contrast, presenting results from our model, it accurately formulates a sentence that encompasses contextual spatial details, including the fallen person, car, and road. Notably, it surpasses anticipated outcomes by also providing a descriptive account of the person standing nearby.

Table 5. The comparison results of caption generation between our system and other models.

	 (a)	 (b)
CPTR [15]	A man walking down a street with a skateboard.	A man lying on a park bench with his legs crossed.
OFA [16]	A man is crossing the street in a city.	A man lying on the ground next to a tree.
BLIP-2 [11] (Base)	A man walking down the street with an umbrella.	A man lying on the ground in front of a house.
Ours	A man in black pants is walking on the road.	A man wearing a tan T-shirt fell on the sidewalk.
	 (c)	 (d)
CPTR [15]	A man is standing next to a parked car.	A man and woman are standing in a room.
OFA [16]	A group of people standing around a man on the street.	Two men are dancing in a room.
BLIP-2 [11] (Base)	A man lying on the ground next to a red car.	Two young men are fighting in a bedroom.
Ours	A man wearing a jacket is lying next to a red car on the road, and a man in black is standing.	A man wearing a black top is fighting with a man wearing a gray shirt in the room.

Table 5d illustrates a scene featuring two men engaged in a physical altercation. However, CPTR and OFA inaccurately capture the action-related information. BLIP-2(base) generates the correct caption, yet it is perceived as lacking in expressiveness. In contrast, presenting results from our model, generates a caption that accurately describes both the specific object information of the two men and the actions "fighting" and "in the room". These experiments have demonstrated that a model fine-tuned with a dataset specialized for surveillance systems can indeed generate captions effectively, aiding in the interpretation and judgment of scenes. Furthermore, the results underscore that existing benchmark datasets may not be well suited for specific tasks, highlighting the importance of developing task-specific datasets.

4.2.3. Risk Assessment Results

Table 6 presents the results of risk assessments conducted by our system. Specifically, Table 6a,b pertain to safe situations, exemplified by individuals walking on a path. Our proposed system has assessed both scenarios as 'safe,' but it distinguished between them by assigning a danger score of '1' to Table 6a and '2' to Table 6b. It is noteworthy that Table 6b receives a danger score of '2' primarily due to the presence of a child in the scene. The inclusion of a child is considered to increase the assessed danger score compared to Table 6a, where no child is present.

Table 6. The results of risk assessment for each individual case.

	Danger Score (Ground Truth)	Image	Generated Caption	Predicted Danger Score
(a)	1		A man and woman are walking with a dog in the park.	1
(b)	2		A woman in a red shirt is walking with two children on the street.	2
(c)	3		Two women are sitting on a railing.	3
(d)	4		A girl wearing a white shirt is sitting on a railing.	4
(e)	5		A woman wearing a gray t-shirt fell on the floor.	5
(f)	6		A woman wearing a blue shirt collapsed on the stairs.	6
(g)	7	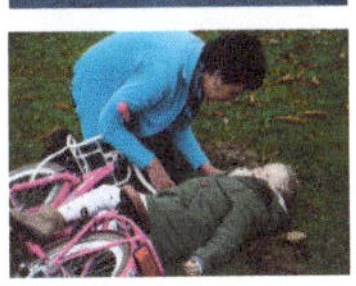	A girl in a green jacket fell next to a bicycle and a woman is helping her.	7

Table 6c,d showcase images portraying individuals engaged in potentially hazardous behavior, specifically sitting on railings. While the action of sitting can be detected in situations classified as safe, the change in space context contributes to the classification of danger level risk. Table 6c generates a successful caption, offering a descriptive account of two women seated on a railing. As a result, the system accurately assessed the danger score as '3'. Similarly, Table 6d generates a precise caption, featuring the object 'girl' and spatial context of 'railing'. Given that a child, rather than an adult, is detected in the scene, the system assigned a danger score of '4'. This exemplifies the system's proficiency in distinguishing between actions that may appear similar but carry varying levels of risk, underlining its capacity to conduct precise risk assessments in diverse situations.

Table 6e,f,g feature images with people who have collapsed. The action of collapsing is inherently associated with the classification of danger (=accident occurrence), which, in turn, triggers a danger score of '5' or higher. The precise danger score is contingent upon several factors, notably the space context and whether the affected individual is an adult or a child. Table 6e portrays a woman collapsing on the indoor floor. Since the indoor floor space does not add additional accident risk, the system assigned a danger score of '5'. Conversely, Table 6f also depicts a collapsing action similar to that in Table 6e, but it takes place on outdoor stairs rather than indoors. The system intelligently infers that the 'stair' area poses a higher level of risk compared to the 'indoor floor,' assigning a danger score of '6'. In Table 6g, a child is observed collapsing outdoors. The subject is identified as a child, and the incident occurred in an outdoor setting where there may be additional risk. This crucial distinction prompts the system to assign a danger score of '7', emphasizing our system's capacity to recognize differences in risk based on the age of the individual involved and the specific spatial context.

As shown in Table 7, the proposed system exhibited remarkable accuracies of 92.9%, 89.5%, and 94.3% for the three distinct risk levels: safety, hazard, and danger (=accident occurrence), respectively, during the evaluation involving 301 test data samples. These results affirm the valuable nature of the proposed dataset, providing easily accessible information essential for the precise evaluation of risk levels.

Table 7. Accuracy for three distinct risk levels on the test datasets.

Support / Ground Output	Safety	Hazard	Danger	Number of Images	Accuracy
Safety	53	3	1	57	92.9%
Hazard	3	60	4	67	89.5%
Danger	1	7	169	177	94.3%
Total				301	93.0%

Moreover, the consistent achievement of high accuracies across all risk categories underscores the robustness of the system. Additionally, the model trained with the proposed dataset demonstrates the system's ability to adequately interpret the relationships between types of objects and behaviors in various locations, further confirming its effectiveness and reliability in real-world applications.

To integrate the proposed scene interpretation and danger score measurement system into actual surveillance systems, it is necessary to conduct experiments to verify its proper functioning under adverse conditions, including various lighting scenarios, noise, and weather. Table 8 presents experimental results demonstrating the robustness of the proposed surveillance system in challenging shooting environments influenced by factors such as lighting, shadows, and weather. Table 8a–c utilizes an image of the same accident pattern captured under different environmental factors, including low resolution, nighttime darkness, and obscured objects due to shadows. In Table 8a, a man has collapsed on the road, captured using a low-resolution camera. Despite the small size of the individual and the low resolution of the image, the proposed system accurately interpreted the fallen man and predicted the correct level of risk. Table 8b depicts a nighttime traffic accident scene where a man is hit by a car in dark conditions. The system successfully detected and interpreted the incident involving the man and the car. Table 8c includes a scene where a man has collapsed under a tree's shadow, obscuring the view. Even in this scenario, the system effectively detected both the car and the fallen man, correctly interpreting the accident scene centered around the fallen individual. These results highlight the system's capability to operate robustly using datasets collected from various environments and emphasize its applicability as a reliable surveillance tool, even in changing shooting environments and amidst obstructive factors.

Table 8. Some experimental results that demonstrate the robustness of proposed surveillance system in unstable recording environments influenced by variations in lighting, shadows, and shooting resolution.

	 (a)	 (b)	 (c)
Influenced Factor	Low resolution	Low light condition	Overshadow
Caption	A man wearing a black top is lying on the road.	A man wearing a white top was hit by a car on the road.	A man falls down in a collision with a red car.
Danger Score	6	6	6

Table 9 presents instances of distinct types of limitations and areas for improvement encountered by our proposed system. The cases in Table 9 are examples classified under the "Danger" category.

Table 9. Examples of distinct types of limitations and areas for improvement encountered by our system in the domains of image captioning and risk assessment.

	 (a)	 (b)	 (c)
Caption	A man wearing a yellow safety vest is standing next to a bicycle on the road.	A boy wearing a helmet is sitting on the ground.	Three boys are burning a stick on the sidewalk.
Danger Score	3	2	7
Ground Truth	6	6	7

Table 9a,b were misclassified as hazard and safety, respectively. In Table 9a, the scene depicts a man who has fallen off his bicycle. The generated caption by our system primarily emphasized a safety officer standing in proximity to the fallen individual rather than the person in distress. Given the safety officer's placement on the road, the system classified the area as hazardous, assigning a danger score of '3'. The system's misinterpretation in this instance can be attributed to the camera angle, which caused the primary subject (the fallen adult male) to appear relatively smaller compared to other objects in the vicinity, notably the standing safety officer. Table 9b portrays a scene featuring a boy who has collapsed due to an accident. The system interpreted this scene as a boy sitting on the ground and accordingly assessed the danger score as '2'. It is essential to emphasize that the actual scenario involves a boy who has fallen subsequent to a bicycle accident. This example underscores the inherent limitations of image-based systems, as the concept of action depends on continuous information, which remains beyond the reach of static images. The errors discerned in Table 9a,b, where the object size was either inadequately small or entailed unrecognized words, underline the potential for system improvement through the accrual of more diverse data, encompassing a broad spectrum of object sizes and textual information. Training the system with this augmented dataset holds promise for addressing issues related to caption generation across a multitude of scenarios. Nevertheless, it is imperative to acknowledge the intrinsic limitation associated with the detection of intricate behavioral patterns at the single-frame level. A solution to this challenge lies in the extension of the system's applicability to video data, permitting a more comprehensive consideration of situational context.

In the case of Table 9c, our system accurately interpreted the presence of children lighting a fire and accordingly assessed the danger score as '7'. However, our system failed to detect the child lying on the railing to the right, which should have been identified. Table 9c exemplifies the limitations associated with situations where multiple scenarios occur simultaneously within a single scene. This table represents a scenario where distinct behaviors unfold concurrently, posing a challenge for detection. Describing such an image with a single caption is challenging. The solution to this challenge lies in integrating multi-object detection and dense captioning techniques. This integration allows for the generation of captions and sentences for multiple concurrently detected scenes, rather than attempting to fit multiple object and action descriptions into a single sentence.

5. Conclusions and Future Work

As the field of artificial intelligence advances and hardware capabilities improve, surveillance systems have evolved to handle a broad range of tasks. Our proposed surveillance system introduces a novel approach that empowers the system to autonomously interpret a wide spectrum of information, facilitating comprehensive situation analysis. This system leverages large multimodal models to generate descriptive captions for hazardous situations and employs semantic analysis to assess the associated risk levels effectively.

To create these captions, we incorporate object information, behavior details, and space context to monitor various situations, leveraging this information to measure risk. Our self-constructed dataset was designed to categorize risk levels based on factors such as the age group of individuals, types of actions, and the nature of locations. Through a series of experiments using these datasets, we demonstrate that they provide comprehensive information for risk assessment and exhibit exceptional performance in this regard. Compared to models pre-trained on existing datasets, our generated captions comprehensively encompass the requisite object attributes, behavior, and spatial context essential for the surveillance system. Furthermore, they exhibit adaptability to novel sentence structures, ensuring versatility across diverse contexts. The robustness of the dataset has also been evidenced by testing with images captured under various conditions, showing its adaptability to both indoor and outdoor environments. Consequently, monitoring personnel can make more accurate and quicker decisions by receiving combined information of the video, interpreted captions, and risk level assessment. Expanding our system to create caption data for additional situations can further enhance surveillance system performance, potentially culminating in a universally applicable system.

As part of our future research agenda, we plan to explore a system that combines multi-object detection and dense captioning technology to generate captions and seamlessly integrate sentences for multiple concurrently detected scenes. Furthermore, recognizing the constraints associated with detecting abnormal situations at the single-frame level, we aim to investigate the expansion of existing systems by incorporating video captioning technology that accounts for the context preceding and following an incident, thus enabling a more comprehensive and nuanced analysis.

Author Contributions: Conceptualization, K.C. and J.K.; software and data curation, M.J.; methodology, formal analysis, validation, and investigation, M.J. and K.C.; supervision and project administration K.C.; writing—original draft preparation, M.J.; writing—review and editing, K.C. and J.K. All authors have read and agreed to the published version of the manuscript.

Funding: This work was supported by funding for the academic research program of Chungbuk National University in 2022.

Institutional Review Board Statement: This study involved the use of human images. All images used in this research were publicly available and did not require Institutional Review Board approval, or were used with explicit consent of the individuals depicted.

Informed Consent Statement: The images used in this study involve human subjects. However, all images are publicly available and free from copyright issues and were used in a manner that does not require informed consent from the individuals depicted. These images were either obtained from publicly accessible sources or used under a license that permits academic use without the need for individual consent.

Conflicts of Interest: The authors declare no conflicts of interest.

References

1. The Business Research Company. Surveillance Technology Global Market Report. 2023. Available online: https://www.thebusinessresearchcompany.com/report/surveillance-technology-global-market-report (accessed on 9 November 2023).
2. Chang, C.W.; Chang, C.Y.; Lin, Y.Y. A Hybrid CNN and LSTM-Based Deep Learning Model for Abnormal Behavior Detection. *Multimed. Tools Appl.* **2022**, *81*, 11825–11843. [CrossRef]
3. Alairaji, R.M.; Aljazaery, I.A.; Alrikabi, H.T.S. Abnormal Behavior Detection of Students in the Examination Hall from Surveillance Videos. In *Advanced Computational Paradigms and Hybrid Intelligent Computing: Proceedings of ICACCP 2021*; Springer: Singapore, 2022; pp. 113–125.
4. Xie, S.; Zhang, X.; Cai, J. Video Crowd Detection and Abnormal Behavior Model Detection Based on Machine Learning Method. *Neural Comput. Appl.* **2019**, *31*, 175–184. [CrossRef]
5. Qiu, J.; Yan, X.; Wang, W.; Wei, W.; Fang, K. Skeleton-Based Abnormal Behavior Detection Using Secure Partitioned Convolutional Neural Network Model. *IEEE J. Biomed. Health Inform.* **2021**, *26*, 5829–5840. [CrossRef] [PubMed]
6. Sultani, W.; Chen, C.; Shah, M. Real-World Anomaly Detection in Surveillance Videos. In Proceedings of the IEEE Conference on Computer Vision and Pattern Recognition, Salt Lake City, UT, USA, 18–22 June 2018; pp. 6479–6488.
7. Perez, M.; Kot, A.C.; Rocha, A. Detection of Real-World Fights in Surveillance Videos. In *ICASSP 2019–2019 IEEE International Conference on Acoustics, Speech and Signal Processing (ICASSP)*; IEEE: Piscataway, NJ, USA, 2019; pp. 2662–2666.
8. Jha, S.; Seo, C.; Yang, E.; Joshi, G.P. Real Time Object Detection and Tracking System for Video Surveillance System. *Multimed. Tools Appl.* **2021**, *80*, 3981–3996. [CrossRef]
9. Kim, J.-H.; Choi, J.; Park, Y.-H.; Nasridinov, A. Abnormal Situation Detection on Surveillance Video Using Object Detection and Action Recognition. *J. Korea Multimed. Soc.* **2021**, *24*, 186–198.
10. Redmon, J.; Divvala, S.; Girshick, R.; Farhadi, A. You Only Look Once: Unified, Real-Time Object Detection. In Proceedings of the IEEE Conference on Computer Vision and Pattern Recognition, Las Vegas, NV, USA, 26 June–1 July 2016; pp. 779–788.
11. Li, J.; Li, D.; Savarese, S.; Hoi, S. BLIP-2: Bootstrapping Language-Image Pre-Training with Frozen Image Encoders and Large Language Models. *arXiv* **2023**, arXiv:2301.12597.
12. Devlin, J.; Chang, M.W.; Lee, K.; Toutanova, K. BERT: Pre-Training of Deep Bidirectional Transformers for Language Understanding. *arXiv* **2018**, arXiv:1810.04805.
13. Vinyals, O.; Toshev, A.; Bengio, S.; Erhan, D. Show and Tell: A Neural Image Caption Generator. In Proceedings of the IEEE Conference on Computer Vision and Pattern Recognition, Boston, MA, USA, 7–12 June 2015; pp. 3156–3164.
14. Xu, K.; Ba, J.; Kiros, R.; Cho, K.; Courville, A.; Salakhudinov, R.; Zemel, R.; Bengio, Y. Show, attend and tell: Neural image caption generation with visual attention. In Proceedings of the International Conference on Machine Learning, Lille, France, 7–9 July 2015; pp. 2048–2057.
15. Liu, W.; Chen, S.; Guo, L.; Zhu, X.; Liu, J. CPtr: Full Transformer Network for Image Captioning. *arXiv* **2021**, arXiv:2101.10804.
16. Wang, P.; Yang, A.; Men, R.; Lin, J.; Bai, S.; Li, Z.; Yang, H. OFA: Unifying Architectures, Tasks, and Modalities through a Simple Sequence-to-Sequence Learning Framework. In Proceedings of the International Conference on Machine Learning, Baltimore, MD, USA, 17–23 July 2022; pp. 23318–23340.
17. Hochreiter, S.; Schmidhuber, J. Long Short-Term Memory. *Neural Comput.* **1997**, *9*, 1735–1780. [CrossRef]
18. Vaswani, A.; Shazeer, N.; Parmar, N.; Uszkoreit, J.; Jones, L.; Gomez, A.N.; Kaiser, Ł.; Polosukhin, I. Attention is all you need. In Proceedings of the Advances in Neural Information Processing Systems, Long Beach, CA, USA, 4–9 December 2017; pp. 5998–6008.
19. Dilawari, A.; Khan, M.U.G.; Al-Otaibi, Y.D.; Rehman, Z.U.; Rahman, A.U.; Nam, Y. Natural Language Description of Videos for Smart Surveillance. *Appl. Sci.* **2021**, *11*, 3730. [CrossRef]
20. Simonyan, K.; Zisserman, A. Very Deep Convolutional Networks for Large-Scale Image Recognition. *arXiv* **2014**, arXiv:1409.1556.
21. Graves, A.; Fernández, S.; Schmidhuber, J. Bidirectional LSTM Networks for Improved Phoneme Classification and Recognition. In Proceedings of the International Conference on Artificial Neural Networks, Warsaw, Poland, 11–15 September 2005; Springer: Berlin/Heidelberg, Germany, 2005; pp. 799–804.
22. Chen, W.; Ma, K.T.; Yew, Z.J.; Hur, M.; Khoo, D.A.A. TEVAD: Improved Video Anomaly Detection with Captions. In Proceedings of the IEEE/CVF Conference on Computer Vision and Pattern Recognition, Vancouver, BC, Canada, 18–22 June 2023; pp. 5548–5558.
23. Lin, K.; Li, L.; Lin, C.C.; Ahmed, F.; Gan, Z.; Liu, Z.; Wang, L. Swinbert: End-to-End Transformers with Sparse Attention for Video Captioning. In Proceedings of the IEEE/CVF Conference on Computer Vision and Pattern Recognition, New Orleans, LA, USA, 18–24 June 2022; pp. 17949–17958.

24. He, K.; Zhang, X.; Ren, S.; Sun, J. Deep Residual Learning for Image Recognition. In Proceedings of the IEEE Conference on Computer Vision and Pattern Recognition, Las Vegas, NV, USA, 26 June–1 July 2016; pp. 770–778.
25. University of Central Florida. UCF-Crime Dataset. Available online: https://www.v7labs.com/open-datasets/ucf-crime-dataset (accessed on 9 November 2023).
26. Perez, M.; Kot, A.C.; Rocha, A. NTU CCTV-Fight Dataset. Available online: https://rose1.ntu.edu.sg/dataset/cctvFights/ (accessed on 9 November 2023).
27. Ranjan, V.; Le, H.; Hoai, M. Iterative Crowd Counting. In Proceedings of the European Conference on Computer Vision (ECCV), Munich, Germany, 8–14 September 2018; pp. 270–285.
28. Wu, P.; Liu, J.; Shi, Y.; Sun, Y.; Shao, F.; Wu, Z.; Yang, Z. Not only look, but also listen: Learning multimodal violence detection under weak supervision. In *Proceedings of the European Conference on Computer Vision*; Springer: Berlin/Heidelberg, Germany, 2020; pp. 322–339.
29. Brown, T.; Mann, B.; Ryder, N.; Subbiah, M.; Kaplan, J.D.; Dhariwal, P.; Amodei, D. Language Models Are Few-Shot Learners. *Adv. Neural Inf. Process. Syst.* **2020**, *33*, 1877–1901.
30. Raffel, C.; Shazeer, N.; Roberts, A.; Lee, K.; Narang, S.; Matena, M.; Liu, P.J. Exploring the Limits of Transfer Learning with a Unified Text-to-Text Transformer. *J. Mach. Learn. Res.* **2020**, *21*, 5485–5551.
31. Thoppilan, R.; De Freitas, D.; Hall, J.; Shazeer, N.; Kulshreshtha, A.; Cheng, H.T.; Le, Q. Lamda: Language Models for Dialog Applications. *arXiv* **2022**, arXiv:2201.08239.
32. OpenAI. GPT-4 Technical Report. *arXiv* **2023**, arXiv:2303.08774.
33. Zhang, M.; Zhang, J.; Wang, R.; Govindan, R.; Mogul, J.C.; Vahdat, A. Gemini: Practical Reconfigurable Datacenter Networks with Topology and Traffic Engineering. *arXiv* **2021**, arXiv:2110.08374.
34. Liu, H.; Li, C.; Li, Y.; Lee, Y.J. Improved Baselines with Visual Instruction Tuning. *arXiv* **2023**, arXiv:2310.03744.
35. Dosovitskiy, A.; Beyer, L.; Kolesnikov, A.; Weissenborn, D.; Zhai, X.; Unterthiner, T. Transformers for Image Recognition at Scale. *arXiv* **2020**, arXiv:2010.11929.
36. Zhang, S.; Roller, S.; Goyal, N.; Artetxe, M.; Chen, M.; Chen, S.; Zettlemoyer, L. OPT: Open Pre-Trained Transformer Language Models. *arXiv* **2022**, arXiv:2205.01068. Available online: https://arxiv.org/abs/2205.01068 (accessed on 9 November 2023).
37. Hu, E.J.; Shen, Y.; Wallis, P.; Allen-Zhu, Z.; Li, Y.; Wang, S.; Chen, W. LoRA: Low-Rank Adaptation of Large Language Models. *arXiv* **2021**, arXiv:2106.09685.
38. Kingma, D.P.; Ba, J. Adam: A Method for Stochastic Optimization. *arXiv* **2014**, arXiv:1412.6980.
39. Papineni, K.; Roukos, S.; Ward, T.; Zhu, W. BLEU: A Method for Automatic Evaluation of Machine Translation. In Proceedings of the 40th Annual Meeting on Association for Computational Linguistics (ACL), Philadelphia, PA, USA, 6–12 July 2002; pp. 311–318.
40. Lin, T.Y.; Maire, M.; Belongie, S.; Bourdev, L.; Girshick, R.; Hays, J.; Perona, P.; Ramanan, D.; Zitnick, C.L.; Dollár, P. Microsoft COCO: Common Objects in Context. *arXiv* **2015**, arXiv:1405.0312.
41. Changpinyo, S.; Sharma, P.; Ding, N.; Soricut, R. Conceptual 12M: Pushing Web-Scale Image-Text Pre-Training to Recognize Long-Tail Visual Concepts. In Proceedings of the IEEE/CVF Conference on Computer Vision and Pattern Recognition, Nashville, TN, USA, 20–25 June 2021; pp. 3558–3568.
42. Ordonez, V.; Kulkarni, G.; Berg, T. Im2Text: Describing Images Using 1 Million Captioned Photographs. *Adv. Neural Inf. Process. Syst.* **2011**, *24*, 1143–1151.
43. Krishna, R.; Zhu, Y.; Groth, O.; Johnson, J.; Hata, K.; Kravitz, J.; Li, F.-F. Visual Genome: Connecting Language and Vision Using Crowdsourced Dense Image Annotations. *Int. J. Comput. Vis.* **2017**, *123*, 32–73. [CrossRef]
44. Schuhmann, C.; Beaumont, R.; Vencu, R.; Gordon, C.; Wightman, R.; Cherti, M.; Jitsev, J. LAION-5B: An Open Large-Scale Dataset for Training Next Generation Image-Text Models. *Adv. Neural Inf. Process. Syst.* **2022**, *35*, 25278–25294.

MDPI AG
Grosspeteranlage 5
4052 Basel
Switzerland
Tel.: +41 61 683 77 34

MDPI Books Editorial Office
E-mail: books@mdpi.com
www.mdpi.com/books

www.ingramcontent.com/pod-product-compliance
Lightning Source LLC
Chambersburg PA
CBHW041712270926
51850CB00027B/3422
* 9 7 8 3 7 2 5 8 4 1 6 5 3 *